ELECTRICAL MACHINES

SECOND EDITION

Smarajit Ghosh
Thapar University, Patiala

PEARSON
Delhi • Chennai • Chandigarh

Associate Editor–Acquisitions: Sameer Gupta
Associate Editor–Production: Barun Kumar Sarkar

ISBN: 978-81-317-6090-1

First Impression

Published by Dorling Kindersley (India) Pvt. Ltd, licensees of Pearson Education in South Asia.

Head Office: 7th Floor, Knowledge Boulevard, A-8(A), Sector - 62, Noida 201309, India
Registered Office: 11 Community Centre, Panchsheel Park, New Delhi 110017, India

Compositor: HT Media Limited

Printer: Rahul Print O Pack

To the memory of my father,
the late Saileswar Ghosh,
to whom I am indebted
for what I am

CONTENTS

Preface xxiii

INTRODUCTION 1

Electromagnetism 1
Direction of Current in a Conductor 1
Direction of Magnetic Flux in a Conductor 1
Flux Distribution of an Isolated Current-carrying Conductor 2
Force Between Two Current-carrying Conductors 2
Force on a Conductor in a Magnetic Field 2
Generation of Induced emf and Current 3
Faraday's Laws 4
Lenz's Law 5
Induced emf 5
Dynamically Induced emf 5
Statically Induced emf 6
Magnetic Circuits 7
Magnetomotive Force 7
Magnetic Field Intensity 8
Magnetic Flux 8
Single-phase Circuits 9
Series R-L Circuit 9
Series R-C Circuit 10
Series L-C-R Circuit 11
Power Triangle 12
Complex Power 12
Three-phase Circuits 13
Advantages of Three-phase System 15
Phase Sequence 15
Interconnection of Three Phases 15
Star and Delta Connections 16
Voltages, Currents and Power in Star Connections 16
Voltages, Currents and Power in Delta Connections 18
For Star Connections 20
Measurement of Three-phase Power 21

Principle of Energy Conversion 22
Energy in the Coupling Field 23
Energy in the Field 24
Co-energy 26
Electrical Energy Input to the system 27
Estimation of Mechanical Forces in an Electromagnetic System 28
Case 1: Motion of the Moving Part is Very Slow 29
Case 2: Motion of the Moving Part Instantaneously 29
Estimation of Magnetic Force in Linear Systems 30
Doubly Excited Systems 31
Cylindrical Rotating Machine 32
Case 1: Synchronous Motor/Machine 33
Case 2: $\omega_m = \omega_s - \omega_r$ *34*

Chapter 1 TRANSFORMERS 38

1.1 Definition 38
1.2 Basic Principle 38
1.3 Types of Transformers 39
1.4 Construction of Single-phase Transformer 39
1.4.1 Core Type 41
1.4.2 Shell Type 43
1.4.3 Spiral Core Type 44
1.5 Transformer Windings 44
1.5.1 Concentric Windings 45
1.5.2 Sandwich Windings 47
1.6 Terminals and Leads 48
1.7 Bushings 48
1.8 Tapping 48
1.9 Cooling of Transformer 49
1.10 Transformer Oil 51
1.11 Conservator and Breather 51
1.12 Buchholz Relay 52
1.13 Transformer Tank 52
1.14 Theory of Transformer 53
1.15 EMF Equation of a Transformer 54
1.16 Step-up and Step-down Transformer 56
1.17 Transformer on No Load 57
1.18 Transformer on Load 59
1.19 Equivalent Resistance 60

1.20 Magnetic Leakage 61
1.21 Equivalent Reactance 61
1.22 Transformer with Resistance and Leakage Reactance 63
1.23 Equivalent Circuit 64
1.24 Open Circuit Test or No-load Test 66
1.25 Short Circuit or Impedance Test 67
1.26 Separation of Core (or Iron) Losses in a Transformer 70
1.27 Total Approximate Voltage Drop of a Transformer 72
1.28 Exact Voltage Drop 73
1.29 Per Unit Resistance, Leakage Reactance and Impedance Voltage Drop 77
1.30 Voltage Regulation of Transformer 78
1.30.1 Inherent Voltage Regulation 78
1.30.2 Voltage Regulation Down 78
1.30.3 Voltage Regulation Up 78
1.31 Calculation for Voltage Regulation 78
1.31.1 Zero Voltage Regulation 79
1.31.2 Condition for Maximum Voltage Regulation 79
1.31.3 Kapp's Regulation 84
1.32 Losses in a Transformer 84
1.32.1 Core or Iron Loss 85
1.32.2 Copper Loss 85
1.33 Efficiency of a Transformer 85
1.34 Condition for Maximum Efficiency 86
1.34.1 Load Current at Maximum Efficiency 87
1.34.2 kVA Supplied at Maximum Efficiency 87
1.35 All-day Efficiency 89
1.36 Polarity Test of a Single-phase Transformer 90
1.37 Sumpner's Test 91
1.38 Parallel Operation of Single-phase Transformer 92
1.39 Load Sharing by Two Transformers 92
1.39.1 Equal Voltage Ratios 92
1.39.2 Unequal Voltage Ratios 95
1.40 Autotransformers 97
1.40.1 Construction 97
1.40.2 Copper Saving in Autotransformer 99
1.40.3 Conversion of Two-winding Transformer into Single-phase Transformer 100
1.40.4 Advantages of Autotransformers 100

1.40.5 Disadvantages of Autotransformers 101
1.40.6 Applications of Autotransformers 101
1.41 Pulse Transformer 102
1.41.1 Pulse Response Characteristics 103
1.41.2 Usage of Pulse Transformer 104
1.42 Welding Transformers 104
1.42.1 Reactors Used with Welding Transformers 105
1.43 Current Transformer 107
1.44 Potential Transformer 107
1.45 Tap Changing Transformers 108
1.46 Off-load Tap-changing Transformers 108
1.47 On-load Tap-changing Transformers 109
1.48 On-load Tap Changer with Single Primary Winding 109
1.49 Preventive Autotransformer 110
1.50 Booster Transformer 111
1.51 Inrush Phenomenon 112
Additional Solved Problems 113
Significant Points 139
Short Questions and Answers 141
Supplementary Problems 144
Multiple-choice Questions and Answers 147

Chapter 2 THREE-PHASE TRANSFORMERS 154
2.1 Advantages of Three-phase Transformers 154
2.2 Principle of Operation 154
2.3 Construction of Three-phase Transformers 155
2.3.1 Core-type Construction 156
2.3.2 Shell-type Construction 156
2.4 Three-phase Transformer Connection 156
2.4.1 Star–Star (γ/γ) Connection 156
2.4.2 Delta–Delta (Δ/Δ) Connection 158
2.4.3 Star–Delta (γ/Δ) Connection 159
2.4.4 Delta–Star (Δ/γ) Connection 160
2.4.5 Delta–Zig-zag Star Connection 161
2.5 Open-delta or *V–V* Connection 163
2.6 Scott Connection or *T–T* Connection 165
2.7 Three-phase to Two-phase Conversion 167
2.8 Parallel Operations of Transformers 170

2.9 Three-phase to Six-phase Conversion 171

2.9.1 Double-star Connection 171

2.9.2 Double-delta Connection 171

2.9.3 Six-phase Star Connection 171

2.9.4 Diametrical Connection 171

2.10 Three-winding Transformer 173

2.11 Three-phase Transformer Connections 173

2.12 Rating of Transformers 176

Additional Solved Problems 176

Significant Points 179

Short Questions and Answers 180

Supplementary Problems 181

Multiple-choice Questions and Answers 182

Chapter 3 BASIC CONCEPTS OF ROTATING MACHINES 184

3.1 Electromagnetic Torque 184

3.2 Reluctance Torque 185

3.3 Constructional Features of Rotating Electrical Machines 185

3.4 Construction of DC Machines 186

3.4.1 Magnetic Frame or Yoke 187

3.4.2 Pole Cores and Pole Shoes 187

3.4.3 Pole Coils 188

3.4.4 Armature Core 188

3.4.5 Armature Windings 189

3.4.6 Commutator 189

3.4.7 Brushes and Bearings 190

3.5 Ring Windings 190

3.6 Drum Windings 192

3.6.1 Number of Coil Sides Per Layer 194

3.6.2 Coil Span 194

3.6.3 Winding Pitch 195

3.6.4 Commutator Pitch 195

3.6.5 Numbering of Armature Conductors 195

3.6.6 Difference Between Coil Span and Winding Pitch 195

3.7 Types of DC Windings 196

3.7.1 Simple Lap Winding 196

3.7.2 Wave Winding 200

3.8 Equalizing Connections for Lap Winding 205

3.9 Uses of Lap and Wave Windings 207

3.10 Dummy Coils 207
3.11 Principle of DC Generator 207
3.12 Operation of a Simple DC Generator with a Two-segment Commutator 209
3.13 Principle of DC Motor 213
3.14 Construction of Synchronous Machines 214
3.14.1 Stator 215
3.14.2 Rotor 216
3.14.3 Classification of Synchronous Machines Based on the Prime Mover 218
3.14.4 Excitation System 218
3.14.5 Damper Windings 219
3.14.6 Frequency and Synchronous Speed 219
3.14.7 Armature Windings 219
3.15 Polyphase Induction Machines 228
3.15.1 Squirrel-cage Rotor 228
3.15.2 Wound Rotor 229
3.16 Air Gap 230
3.17 Principle of Operation of Three-phase Induction Motor 230
3.18 Synchronous Speed and Slip in Induction Motor 232
3.18.1 Synchronous Speed 232
3.18.2 Slip in Induction Motor 232
3.19 Frequency of Rotor Currents 233
3.20 Speed of the Rotor mmf 234
3.21 Electrical and Mechanical Degrees 234
3.22 Pitch Factor 236
3.23 Distribution Factor 238
3.24 Winding Factor 240
3.25 Flux Per Pole 240
3.26 Generated emf in Full-pitched Coil 241
3.27 EMF Generated in AC Machines 242
3.27.1 Synchronous Machines 243
3.27.2 Induction Machines 243
3.27.3 A General Expression for the emf of Synchronous Generator 244
3.28 EMF Generated in DC Generator 246
3.29 Concept of Rotating Magnetic Field 246
3.29.1 Case 1: $\omega t = \theta = 0°$ *247*
3.29.2 Case 2: $\omega t = \theta = 60°$ *247*
3.29.3 Case 3: $\omega t = \theta = 120°$ *248*
3.29.4 Case 4: $\omega t = \theta = 180°$ *249*

Additional Solved Problems 250
Significant Points 255
Short Questions and Answers 256
Supplementary Problems 257
Multiple-choice Questions and Answers 258

Chapter 4 DC GENERATORS **261**

4.1 Types of DC Machines 261
4.2 DC Generator 261
4.3 Brush Drop 262
4.4 EMF Equation 262
4.4.1 Shunt Generator 262
4.4.2 Series Generator 263
4.4.3 Long-shunt Compound Generator 263
4.4.4 Short-shunt Compound Generator 263
4.5 Derivation for E_g 264
4.6 Losses in DC Generator 266
4.7 Stray Losses 266
4.8 Constant or Standing Losses 267
4.9 Power Stages 267
4.10 Efficiency 267
4.11 Condition for Maximum Efficiency 268
4.12 Armature Reaction in DC Machines 270
4.13 Demagnetizing and Cross-magnetizing Conductors 274
4.14 Demagnetizing Ampere-turns Per Pole 274
4.15 Cross-magnetizing Ampere-turns Per Pole 275
4.16 Compensating Windings 277
4.17 Number of Compensating Windings 277
4.18 Commutation 277
4.18.1 Linear Commutation 279
4.18.2 Retarded Commutation 281
4.18.3 Accelerated Commutation 282
4.18.4 Sinusoidal Commutation 284
4.19 Value of Reactance Voltage 284
4.20 Methods of Improving Commutation 285
4.20.1 Resistance Commutation 285
4.20.2 EMF Commutation 286
4.21 Equalizer Rings 287

4.22 Characteristics of DC Generators 287
4.23 Separately Excited Generators 289
4.23.1 No-load Saturation Characteristic 289
4.23.2 Internal and External Characteristic (or Load Characteristic) 290
4.24 No-load Curve for Self-excited Generators 290
4.25 Advantages and Disadvantages of Separately Excited Generators 294
4.26 Voltage Build-up of Shunt Generator 294
4.27 Conditions for Build-up of Shunt Generator 295
4.28 Reasons for Failure to Build-up of Shunt Generators 295
4.29 External Characteristic of Shunt Generator 295
4.30 Voltage Regulation 297
4.31 Internal or Total Characteristic 297
4.32 External Characteristic and Internal Characteristic from OCC 298
4.32.1 Neglecting Armature Reaction 298
4.32.2 With Armature Reaction 299
4.33 Effect of Brush Shift on the Terminal Voltage 300
4.34 Series Generator 301
4.35 Compound Generator 301
4.36 Parallel Operations of DC Generators 303
4.37 Requirements for Paralleling DC Generators 303
4.38 Parallel Operation of Shunt Generators 304
4.39 Parallel Operation of Series Generators 307
4.40 Parallel Operation of Compound Generators 308
4.41 Uses of DC Generators 310
4.42 Indications of an Overloaded Generators 310
4.43 Causes of Overloading 311
4.44 Causes of Sparking at Brushes of a DC Machine 311
4.45 Causes of Excessive Heating of Generator on Running 311
4.46 Causes of Heating of Armature 312
4.47 Causes for Abnormal Sound in DC Generator 312
4.48 Reasons for Rapid Brush Wear in a DC Machine 312

Additional Solved Problems 312
Significant Points 321
Short Questions and Answers 323
Supplementary Problems 330
Multiple-choice Questions and Answers 331

Chapter 5 DC MOTORS **337**

5.1 Voltage Equation 337
5.2 Back emf 338
5.3 Condition for Maximum Mechanical Power 339
5.4 Armature Torque of a Motor 339
5.5 Rotational Losses of DC Machines 341
5.6 Compound Motor 343
5.7 Relation of Speed (N) with Back emf (E_b) and Flux (Φ) 344
5.8 Characteristics of Shunt or Separately Excited DC Motor 345
5.9 Characteristics of DC Series Motor 345
5.10 Characteristics of Compound Motor 346
5.11 Speed Regulation 347
5.12 Torque and Speed of DC Series Motor 347
5.13 Speed Control of DC Motors 347
5.13.1 Armature Resistance Control 347
5.13.2 Field Resistance Control 350
5.14 Ward-Leonard Control (Voltage Control) 355
5.15 Necessity of a Starter for DC Motors 357
5.16 Manual Starter 358
5.16.1 Three-point Starter 358
5.16.2 Four-point Starter 359
5.17 Automatic Starters 360
5.17.1 Time Element Starter 361
5.17.2 Back emf Starter 361
5.17.3 Shunt Current-limit Starter 362
5.17.4 Series Current-limit Starter 363
5.18 Starters for DC Series Motors 364
5.19 DC Shunt Motor Starter Design 364
5.20 Electric Braking 366
5.21 Electric Braking of Shunt Motors 367
5.21.1 Rheostatic or Dynamic Braking 367
5.21.2 Plugging or Reverse Breaking 367
5.21.3 Regenerative Braking 368
5.22 Electric Braking of Series Motor 368
5.22.1 Rheostatic Braking 368
5.22.2 Plugging or Reverse Current Braking 368
5.22.3 Regenerative Braking 369
5.23 Testing of DC Machines 372

5.24 Brake Test 372
5.25 Swinburne's Test 373
5.25.1 Motoring Mode 375
5.25.2 Generating Mode 375
5.26 Hopkinson's Test (Back-to-Back Test) 377
5.27 Separation of Losses in a DC Machine 380
5.28 Retardation or Running Test 383
5.28.1 To Find $\frac{d\omega}{dt}$ 383
5.28.2 To Find J 384
5.29 Field's Test 386
5.30 Uses of DC Motors 388
5.31 Special DC Machines 389
5.31.1 Three-wire Generator 389
5.31.2 Dynamometer 390
5.31.3 Cross-field Machines 390
5.32 Characteristics of Cross-field Generators 391
5.33 Brushless DC Motor 394
5.33.1 Unipolar or Half-wave Brushless DC Motor 394
5.33.2 Bipolar or Full-wave Brushless DC Motor 396
5.34 Features of Brushless DC Motor 396

Additional Solved Problems 397
Significant Points 418
Short Questions and Answers 418
Supplementary Problems 424
Multiple-choice Questions and Answers 425

Chapter 6 SYNCHRONOUS GENERATORS 430

6.1 Parameters of Armature Winding 430
6.1.1 Armature Resistance 430
6.1.2 Armature Leakage Reactance (X_L) 431
6.2 Armature Reaction 431
6.2.1 Unity Power Factor Load 431
6.2.2 Zero Lagging Power Factor Load 432
6.2.3 Zero Leading Power Factor Load 432
6.3 Concept of Synchronous Reactance and Impedance 433
6.4 Equivalent Circuit of an Alternator 433
6.5 Voltage Equation of Alternator 434

6.6 Phasor Diagram of Alternator 434
6.6.1 Lagging Power Factor 434
6.6.2 Unity Power Factor 435
6.6.3 Leading Power Factor 435
6.7 Voltage Regulation 435
6.8 Determination of Voltage Regulation 437
6.8.1 Direct Loading 438
6.8.2 The emf or Synchronous Impedance Method 438
6.8.3 Ampere-turn Method 443
6.8.4 Zero Power Factor Method or Potier Method 447
6.8.5 ASA Method 453
6.9 Load Characteristics of Alternators 454
6.10 Output Power Equation of an Alternator 456
6.10.1 Real Power Output per Phase of Alternator (P_{og}) 456
6.10.2 Reactive Power Output per Phase of Alternator (Q_{og}) 457
6.10.3 Maximum Power Output per Phase of Alternator 457
6.10.4 Power Output Equation of Alternator Neglecting Armature Resistance 457
6.11 Input Power Equation of Alternator 458
6.11.1 Real Power Input per Phase to Alternator 459
6.11.2 Reactive Power Input per Phase to Alternator 459
6.12 Two-reaction Theory 460
6.13 Two-reaction Theory of Salient-pole Alternator 460
6.14 Torque-angle Characteristic of Salient-pole Alternator 464
6.15 Maximum Reactive Power for Salient-pole Alternator 468
6.16 Losses and Efficiency 468
6.16.1 Determination of Losses 469
6.17 Determination of X_d and X_q 470
6.18 Capability Curves 471
6.19 Excitation Circle of an Alternator 476
6.20 Prime Mover Characteristic 477
6.20.1 Sharing of Powers by Two Alternators 478
6.21 Infinite Bus 482
6.22 Need for Parallel Operation of Alternators 483
6.22.1 Reasons for Paralleling of Alternators 483
6.22.2 Conditions for Paralleling of Alternators 483
6.22.3 Identical Voltage Requirement 484
6.22.4 Identical Phase Sequence 484
6.22.5 In-phase Requirement 484

6.22.6 Identical Frequency Requirement 484
6.22.7 Relation of Prime Mover Torque Speed 484
6.23 Synchronizing Procedures of Alternator 485
6.23.1 Voltage Matching 485
6.23.2 Phase Sequence Matching 485
6.23.3 Determination of In-phase 486
6.23.4 Frequency Synchronization 488
6.23.5 Synchroscope Synchronization 488
6.24 Distribution of Load 488
6.24.1 Effect of Change in Excitation 488
6.24.2 Effect of Change of Steam Supply 489
6.25 Synchronizing Power and Synchronizing Torque Coefficient 489
6.26 Units of Synchronizing Power Coefficient 491
6.27 Significance of Synchronizing Power Coefficient 492
6.28 Hunting 492
6.28.1 Causes of Hunting 492
6.28.2 Effects of Hunting 492
6.28.3 Reduction of Hunting 493
6.29 Oscillations of Synchronous Machines 493
6.30 Sudden Short Circuit of Synchronous Generator 494
6.31 Short-circuit Ratio 495
6.32 Protection of Generators 497
Additional Solved Problems 499
Significant Points 516
Short Questions and Answers 518
Supplementary Problems 521
Multiple-choice Questions and Answers 523

Chapter 7 SYNCHRONOUS MOTORS 526
7.1 Principles of Operation 526
7.2 Armature Reaction in Synchronous Motors 527
7.3 Phasor Diagram of Synchronous Motor 528
7.3.1 Effect of Loading Synchronous Motor 528
7.3.2 Lagging Power Factor Load 528
7.3.3 Leading Power Factor Load 529
7.3.4 Unity Power Factor Load 529
7.4 Operation at a Constant Load with Variable Excitation 532
7.5 *V* Curves and Inverted *V* Curves 535
7.6 Complex Power Input of Synchronous Motor 535

7.7 Complex Power Output of a Synchronous Motor 538
7.8 Maximum Output Power 538
7.9 Power Output when Armature Resistance Is Negligible 540
7.10 Input Reactive Power when Armature Resistance Is Negligible 540
7.11 Motor Characteristics, Performance and Circle Diagram of a Synchronous Motor 541
7.11.1 Current Locus for Constant Power Input 541
7.11.2 Current Locus for Constant Power Developed (P_{om}) 541
7.11.3 Current Locus for Constant Excitation (E_f) 543
7.12 Torque of a Synchronous Motor 544
7.13 Salient-pole Synchronous Motor—Two-reaction Model 544
7.13.1 Lagging Power Factor 545
7.13.2 Leading Power Factor 546
7.13.3 Unity Power Factor 546
7.14 Power Developed by a Salient-pole Synchronous Motor 547
7.14.1 Stability and Maximum Load Angle 548
7.15 Damper Windings 549
7.16 Damping Effect 550
7.17 Hunting/Surging of Synchronous Motors 550
7.18 Periodicity of Hunting 551
7.19 Methods of Starting of Synchronous Motors 553
7.20 Applications of Synchronous Motor 553
7.21 Synchronous Condensers 554

Additional Solved Problems 555
Significant Points 570
Short Questions and Answers 572
Supplementary Problems 573
Multiple-choice Questions and Answers 574

Chapter 8 POLYPHASE INDUCTION MOTORS 577

8.1 Rotor Current 577
8.2 Rotor Power 579
8.3 Expression for Torque 581
8.4 Starting Torque 582
8.4.1 Starting Torque of a Squirrel-cage Motor 582
8.4.2 Starting Torque of Slip-ring Motor 582
8.5 Effect of Change in Supply Voltage on Starting Torque 582
8.6 Effect of Change in Supply Voltage on Torque and Slip 583
8.7 Condition for Maximum Torque 583

8.8 Torque-slip and Torque-speed Characteristics 584
8.9 Relation Between Full-load Torque and Maximum Torque 586
8.10 Relation Between Starting Load Torque and Maximum Torque 587
8.11 Condition for the Starting Torque to Be Equal to the Maximum Torque 588
8.12 Effect of Variation of Rotor Resistance and Rotor Reactance on Maximum Torque, Efficiency and Power Factor of an Induction Motor 588
8.12.1 Effect of Rotor Resistance on Maximum Torque 589
8.12.2 Effect of R_2 and X_2 on the Power Factor of an Induction Motor 589
8.12.3 Effect of R_2 and X_2 on the Efficiency of an Induction Motor 590
8.13 Effect of Change in Supply Voltage on Torque and Speed 591
8.14 Effect of Change in Supply Frequency on Torque and Speed 591
8.15 Losses 591
8.16 Power Stages 592
8.17 Efficiency 592
8.18 Synchronous Watt 594
8.19 Measurement of Slip 594
8.19.1 Electromechanical Counter 594
8.19.2 Mechanical Differential Counter 594
8.19.3 Stroboscopic Method 595
8.20 Equivalent Circuit 596
8.20.1 Stator Equivalent Circuit 596
8.20.2 Rotor Equivalent Circuit 596
8.21 Thevenin's Equivalent Circuit of an Induction Motor 600
8.22 Starting of Induction Motors 604
8.23 Starting of Squirrel-cage Motors 605
8.23.1 Direct-on-line Starting of Induction Motors 605
8.23.2 Stator Resistor (or Reactor) Starting 606
8.23.3 Autotransformer Starting 607
8.23.4 Star-delta Starter 608
8.24 Starting of Slip-ring Induction Motors 610
8.25 No-load Test or Open-circuit Test 613
8.26 Blocked-rotor or Short-circuit Test 614
8.27 Direct Testing of Induction Motors 616
8.28 Circle Diagram 617
8.29 Speed Control of Induction Motor 623
8.29.1 Speed Control from Rotor Side 623
8.29.2 Speed Control of Induction Motors from Stator Side 627

8.30 Comparison Between Wound-rotor and Cage-rotor Induction Motors 629
8.31 Crawling 630
8.32 Magnetic Locking (Cogging) 630
8.33 Deep-cage Rotors 631
8.34 Double-cage Rotors 631
8.35 Applications 633
8.36 Comparison Between Synchronous and Induction Motors 633
8.37 Factors Governing the performance of Induction Motors 634
8.38 Effects of Operating Conditions 634
8.39 Ratings of Induction Motor 634
8.40 Common Faults in Three-phase Induction Motors 635
8.41 Most Probabale Reasons for which Three-phase Induction Motors Fail to Start 635
8.42 Most Probabale Reasons for Which Three-phase Induction Motors Fail to Carry Load 636
8.43 Schrage Motor 636
8.44 Power Factor Compensation 639
8.45 Linear Induction Motor 639
8.46 Induction Generator 641
8.47 Electrical Braking of Polyphase Induction Motors 643
8.47.1 Regenerative Braking 643
8.47.2 Plugging or Counter-current Braking 643
8.47.3 Dynamic Braking 644
8.48 Synchronous-induction Motor 645

Additional Solved Problems 647
Significant Points 674
Short Questions and Answers 676
Supplementary Problems 680
Multiple-choice Questions and Answers 681

Chapter 9 SINGLE-PHASE MOTORS AND SPECIAL MACHINES 686

9.1 Classification of Single-phase Induction Motors 686
9.2 Production of Rotating Field 686
9.3 Working Principle of Single-phase Induction Motor 687
9.4 Double Revolving Field Theory 687
9.5 Rotor Slip with Respect to Two Rotating Fields 689

9.6 Equivalent Circuit of Single-phase, Single-winding Induction Motor 690

9.7 Power Developed and Losses of Single-phase, Single-winding Induction Motor 693

9.8 Determination of Equivalent Circuit Parameters 695

9.8.1 Blocked Rotor Test 695

9.8.2 No-load Test 696

9.9 Split-phase Induction Motors 697

9.10 Capacitor Motors 698

9.10.1 Capacitor-start Motors 699

9.10.2 Capacitor-start Capacitor-run Motors 699

9.11 Permanent Split Capacitor Motors 701

9.12 Shaded Pole Motor 701

9.13 Single-phase Synchronous Motors 702

9.13.1 Reluctance Motors 702

9.13.2 Hysteresis Motor 703

9.14 Series Motor or Universal Motor 704

9.14.1 Circuit Model and Phasor Diagram 705

9.14.2 Torque 705

9.14.3 Phasor Diagram and Performance Characteristics 705

9.15 Stepper Motor 707

9.15.1 Variable Reluctance Stepper Motor 708

9.15.2 Permanent Magnet Stepper Motor 711

9.15.3 Hybrid Stepper Motor 711

9.16 Characteristics of Stepper Motors 712

9.17 DC Servomotors 713

9.17.1 Excitation of DC Servomotors 714

9.17.2 Armature-controlled DC Servomotor 714

9.17.3 Field-controlled DC Servomotors 717

9.18 AC Servomotors 718

9.19 Servomechanism 721

9.19.1 Positional Servomechanism 721

9.19.2 Rate Servomechanism 724

9.19.3 Damping by Derivative Feedback 726

Significant Points 729

Short Questions and Answers 730

Supplementary Problems 731

Multiple-choice Questions and Answers 731

Appendix A 735
Appendix B 749
Appendix C 752
Appendix D 756
Appendix E 757
Appendix F 759
Appendix G 761
Appendix H 763
Multiple Choice Questions 777
References 815
Index 817

PREFACE

Electrical Machines, Second Edition, has been written for undergraduate students of electrical engineering, electrical and electronics engineering and for other engineering programmes in India and abroad. The book covers a whole range of machines that include static machines (transformers), rotating machines (DC machines, AC machines). The introduction at the beginning discusses the elementary topics to be read before moving on to other topics. Solved numerical examples interspersed in the chapters make understanding of the theory simpler. Exhaustive solved examples are given for further practice. Significant Points, Short Questions and Answers, Supplementary Problems and Multiple Choice Questions (MCQs) with Answers have also been incorporated in each chapter.

This book contains *nine chapters*: Chapter 1 explains single-phase transformers. Chapter 2 deals with three-phase transformers. Basic concepts of rotating machines are presented in Chapter 3. Chapter 4 introduces DC generators. DC motors are dealt with Chapter 5. Chapter 6 presents synchronous generators/alternators. Synchronous motors are discussed in Chapter 7. Chapter 8 deals with induction motors. Chapter 9 details single-phase motors and special machines. Besides the chapters, the book has a number of *appendices* to cover the other sub-parts, followed by 300 *MCQs*.

The contents of the book have been systematically organized so that they are useful both for the novice as well as advance-level students. They will be of much help to students of AMIE (the certification Associate Member of the Institution of Engineers) and anyone preparing for any competitive examination, such as those for the IES (Indian Engineering Services) or other UPSC (Union Public Service Commission) examinations. I will be happy to receive feedback and suggestions for the improvement of this book.

Acknowledgements

My heartiest thanks go to the editorial team of Pearson Education for editing this manuscript and giving it good shape. I am sincerely thankful to my friends in this area for their comments and encouragement. Finally, I wish to express my sincere gratitude to my mother Beena Ghosh and my wife, Dr Kuntal Ghosh, for their patience and constant encouragement during the preparation of the manuscript.

SMARAJIT GHOSH

Introduction

In the study of electrical machines the fundamentals of electromagnetism and single- and three-phase circuits play the most important role. The main aim of this introductory chapter is to introduce the basics, which are the cornerstones of electrical machines.

ELECTROMAGNETISM

When an electric current passes through a conductor, a magnetic field immediately builds up due to the motion of electrons. When a magnetic field encompasses a conductor and moves relative to the conductor, the electrons come in motion. This is the converse of the previous phenomenon. Electromagnetism is the study of electromagnetic force (emf) induced in the conductor when it cuts the magnetic flux or is cut by the magnetic flux.

DIRECTION OF CURRENT IN A CONDUCTOR

Figure 1(a) shows that there is no current through the conductor. Figure 1(b) shows that the conductor carries current away from the observer, that is, in a downward direction, while Figure 1(c) indicates that the conductor carries current towards the observer, that is, in an upward direction.

DIRECTION OF MAGNETIC FLUX IN A CONDUCTOR

The direction of magnetic flux can be easily found out by using either the right-hand rule or the corkscrew rule.

- **Right-hand rule:** If the conductor is graphed by the right hand in such a way that the thumb points in the direction of the current, then the closed fingers give the direction of flux, as shown in Figure 2.
- **Corkscrew rule:** Imagine a corkscrew is pointed in the direction of a current. The direction of magnetic field is in the direction in which the screw is to be turned to move it in a forward direction. The application of the corkscrew rule is shown in Figure 3.

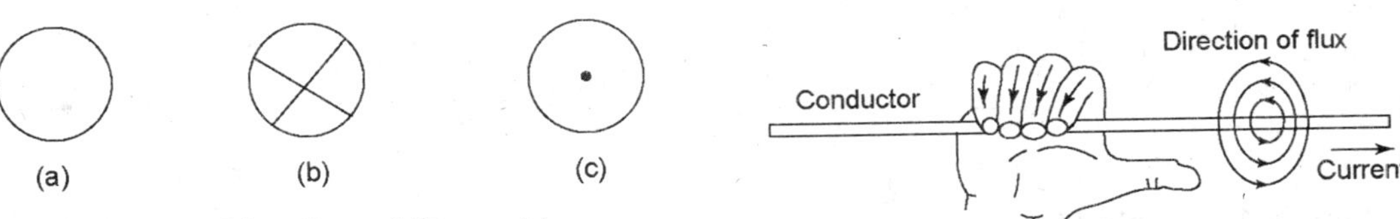

Figure 1 Direction of Current in a Conductor

Figure 2 Right-hand Rule

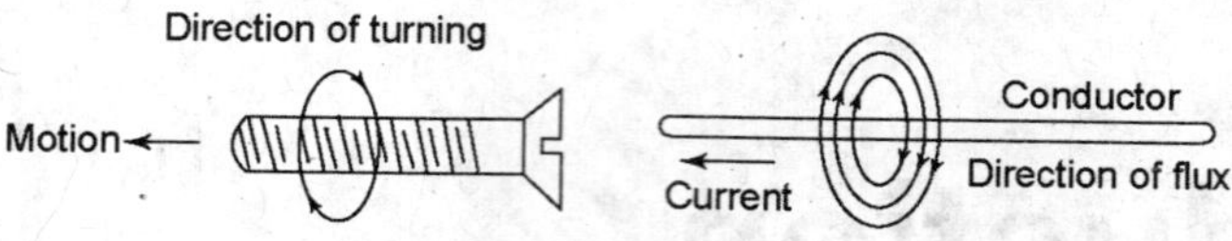

Figure 3 Corkscrew Rule

FLUX DISTRIBUTION OF AN ISOLATED CURRENT-CARRYING CONDUCTOR

The distribution of flux when a conductor carries current downwards, that is, away from the observer, is shown in Figure 4(a). The flux distribution when a conductor carries current upwards, that is, towards the observer, is shown in Figure 4(b).

FORCE BETWEEN TWO CURRENT-CARRYING CONDUCTORS

Figures 5(a) and 5(b) show the two conductors *A* and *B*, carrying current in the same and the opposite directions, respectively. The flux distribution when they carry current in the same direction is shown in Figure 5(a). The magnetic fields set up by the two conductors are in opposite directions. Hence, they attract each other. The flux distribution when they carry current in opposite directions is shown in Figure 5(b). The magnetic fields set up by the two conductors are in the same direction. Hence, they repel each other.

FORCE ON A CONDUCTOR IN A MAGNETIC FIELD

The force experienced by a current-carrying conductor placed in a magnetic field, shown in Figure 6, is expressed as

$$F = BIL$$

where F is the force on the conductor (N), B is the magnetic flux density (Wb/m^2), I is the current in the conductor (A) and L is the effective length of the conductor to the field (m).
If $F = 1$ N, $I = 1$ A and $L = 1$ m, $B = 1$ T.

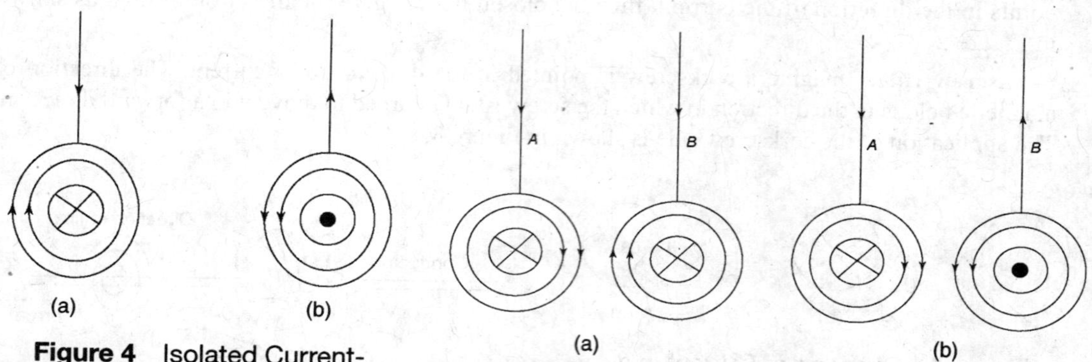

Figure 4 Isolated Current-carrying Conductors

Figure 5 Two Current-carrying Conductors

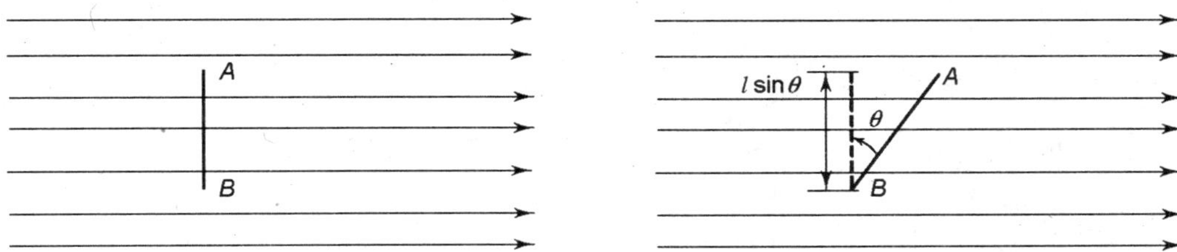

Figure 6 Conductor at an Angle of 90° to a Magnetic Field

Figure 7 Conductor at an Angle θ to a Magnetic Field

If the conductor is placed at an angle θ to the field, as shown in Figure 7, the effective length will be $L \sin \theta$ and the force on the conductor will be $F = BIL \sin \theta$.

The effective length is the length of the conductor lying within the magnetic field.

If $\theta = 0$, that is, the conductor is placed parallel to the field, $F = 0$.

The basic principle of an electric motor is that the armature conductor carrying the current is placed in a magnetic field. So, it is also called the motor principle. The direction of rotation of the armature is defined by Fleming's left-hand rule, as shown in Figure 8.

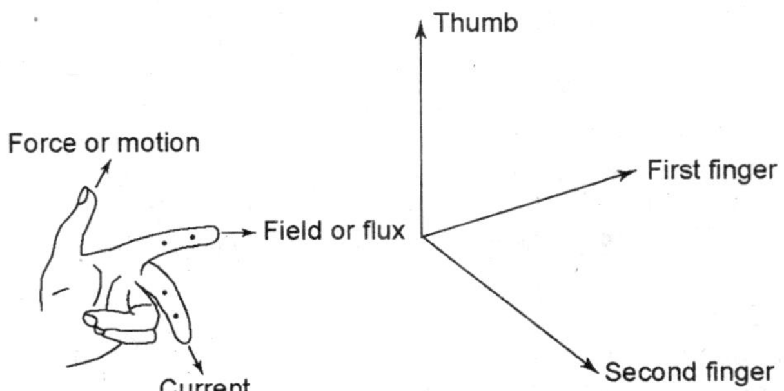

Figure 8 Fleming's Left-hand Rule

Spread the thumb, forefinger and second finger of the left hand so that they are mutually perpendicular to each other. If the forefinger indicates the direction of flux and the second finger indicates the direction of current, the thumb will indicate the direction of rotation of the armature. Please see Appendix A for basic terminologies and losses.

GENERATION OF INDUCED EMF AND CURRENT

Figure 9 shows an insulated coil whose terminals are connected to a sensitive galvanometer (G). There is a bar magnet, AB, close to the coil. When the bar magnet is suddenly moved towards the coil to position $A'B'$, there is a deflection in the galvanometer. This deflection in the galvanometer lasts so long as there are relative motions of the magnet with respect to the coil; that is, the flux linking with the coil changes.

Figure 10 shows that when the magnet is withdrawn, the deflection of the galvanometer is in the direction opposite to the above case. This deflection exists so long as the bar magnet is in relative motion to the coil; that is, the flux linking with the coil changes.

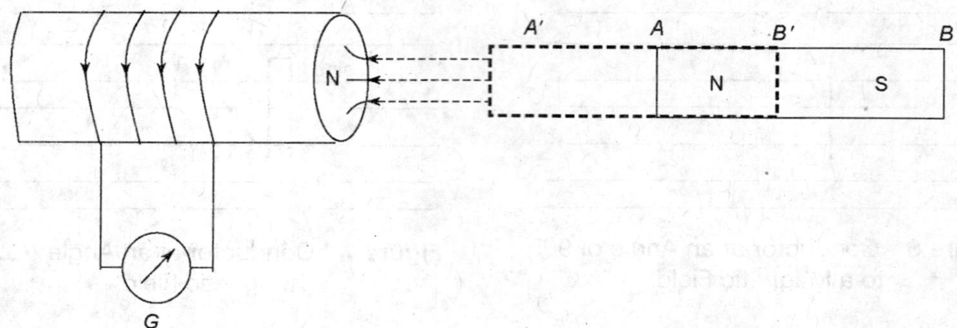

Figure 9 Generation of Induced emf

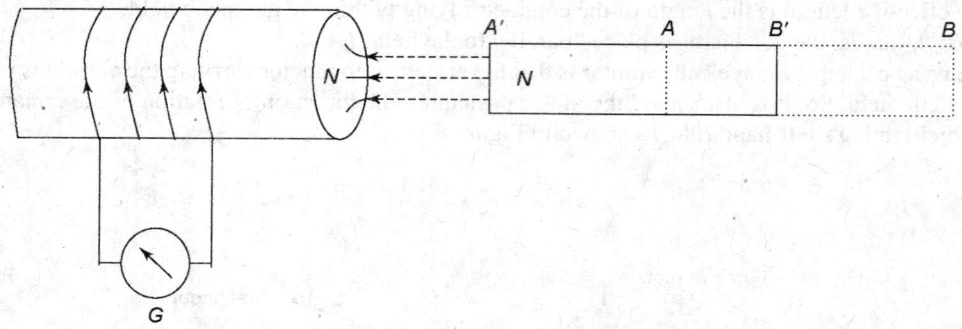

Figure 10 Magnet Withdrawn

In both the cases, the deflection is reduced to zero when the bar magnet becomes stationary. The flux linked with the coil increases as the bar magnet approaches the coil in the first case, while in the second case, the flux linked with the coil decreases when the bar magnet is withdrawn.

It is clear from above that deflections in the two cases are in different directions. The deflection in the galvanometer indicates that there is an induced current produced in the coil. In the first case, the induced current flows through the coil in an anticlockwise direction as seen from the bar magnet. This indicates that the face of the coil is the N pole. To move the bar magnet towards the coil, a force from outside must be given to the bar magnet. When the magnet is withdrawn, the current flows in a clockwise direction, as seen from the bar magnet. This indicates that the face of the coil is the S pole. So once again force must be supplied to the bar magnet from outside to take it away from the coil. Here, the principle of conservation of energy is fully satisfied.

From the above results, Faraday proposed two laws known as Faraday's first and second laws.

FARADAY'S LAWS

- *First law:* Whenever there is variation of magnetic flux linked with a coil, an emf is induced in it. Or an emf is induced in a conductor whenever it cuts the magnetic flux.
- *Second law:* The magnitude of this induced emf is the rate of change of flux linkage.

Let the initial flux be ϕ_1 and the final flux be ϕ_2 in time t. Let the turns of the coil be N. Therefore, the initial flux linkage is $N\phi_1$ and the final flux linkage is $N\phi_2$.

$$\text{Induced emf } (e) = \frac{N\phi_2 - N\phi_1}{t} = N\left(\frac{\phi_2 - \phi_1}{t}\right) \tag{1}$$

In differential form, we can put Equation (1) as $e = N\dfrac{d\phi}{dt}$

LENZ'S LAW

The law states that the induced current will flow in such a direction that it will oppose the cause that produces it. It is explained as follows:

The current through the coil 1 is varied by the rheostat, as shown in Figure 11. An emf is induced in coil 2 due to variation of flux in coil 1. The induced emf between P and Q will be such that the current will flow from P to Q via the resistance R so that it will oppose the flux that is linked with it. In this case P will have a positive polarity and Q will have a negative polarity. The induced emf is given by $e = -\dfrac{d\phi}{dt}$.

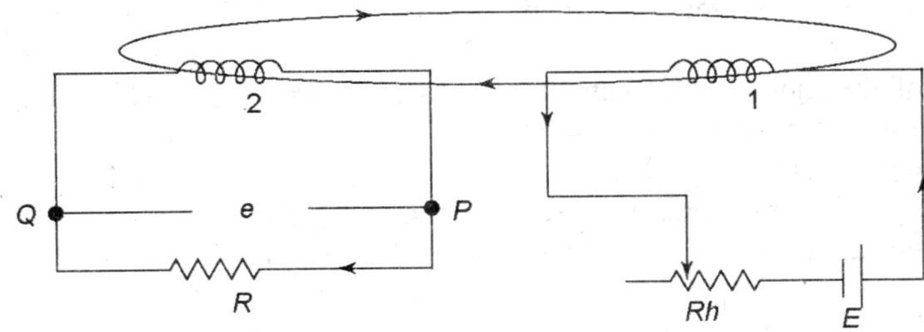

Figure 11 Example of Lenz's Law

INDUCED EMF

An emf is induced in the conductor either (i) dynamically or (ii) statically. In the first case, the conductor will move in the magnetic field and the flux will be non-varying. In the second case, the conductor will be stationary and the flux will be varying.

Dynamically Induced emf

In Figure 12(a), conductor A is shown by the cross-section within the uniform magnetic field. It moves at the right angle to the magnetic field of flux density B Wb/m^2. If the conductor moves a length dx in time dt, the area swept by it is equal to ldx, where l is the length of the conductor lying within the magnetic field. Here, change in flux is equal to $Bldx$ in time dt. According to Faraday's law, the induced emf (also known as dynamically induced emf) is expressed as

$$e = \frac{Bldx}{dt} = Bl\frac{dx}{dt} = Blu \text{ V} \tag{2}$$

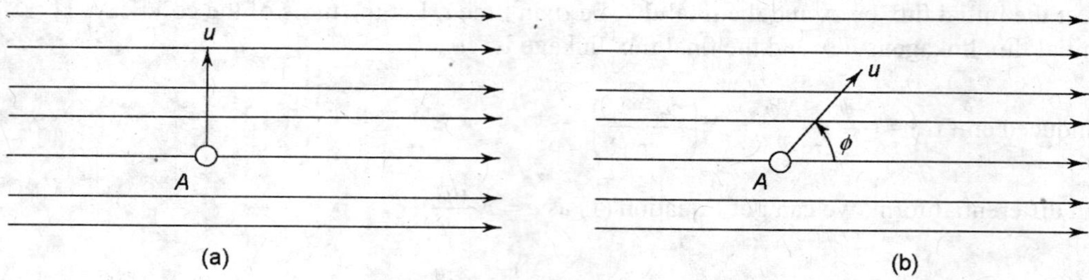

Figure 12 Dynamically Induced emf

where $u = \frac{dx}{dt} =$ the velocity of the conductor if the conductor moves at an angle ϕ with the magnetic field shown in Figure 12(b). The component of its velocity in the direction perpendicular to the field is $u \sin \phi$. The induced emf (e) becomes

$$e = Blu \sin \phi \text{ V} \tag{3}$$

If the conductor moves parallel to the field, the value of ϕ will be 0°. The induced emf becomes zero; i.e., $e = 0$.

The direction of the induced emf can be found out by Fleming's right-hand rule as shown in Figure 13. This rule states that if anyone spreads the thumb, forefinger and second finger of the right hand so that they are mutually perpendicular to each other, the thumb indicates the direction of rotation of the conductor in the magnetic field, the forefinger shows the direction of the magnetic field and the second finger indicates the direction of the flow of the induced current in the conductor.

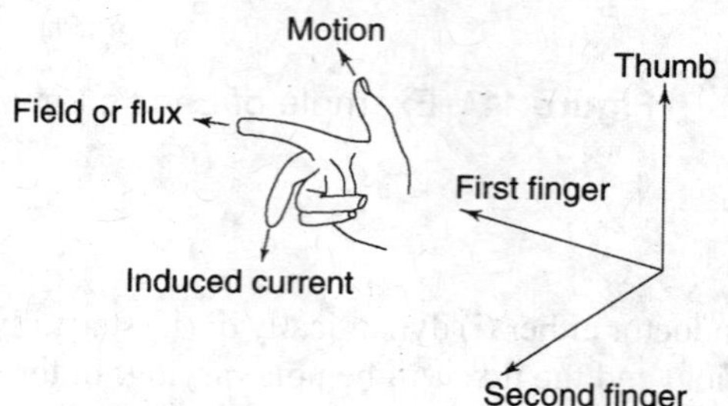

Figure 13 Fleming's Right-hand Rule

DC generators work on the principle of production of dynamically induced emf in the conductors, which are housed in a revolving armature lying within the magnetic field.

Statically Induced emf

If the flux in coil A varies by varying current with the help of the rheostat, the flux linked with coil B also changes and emf is induced in coil B, and its direction will be obtained by Lenz's law.

Coil A is connected to a battery (E) via a rheostat (Rh) as shown in Figure 14. Coil B is connected to a voltmeter (V). When the current in coil A is established, the flux will be produced in coil A, and it

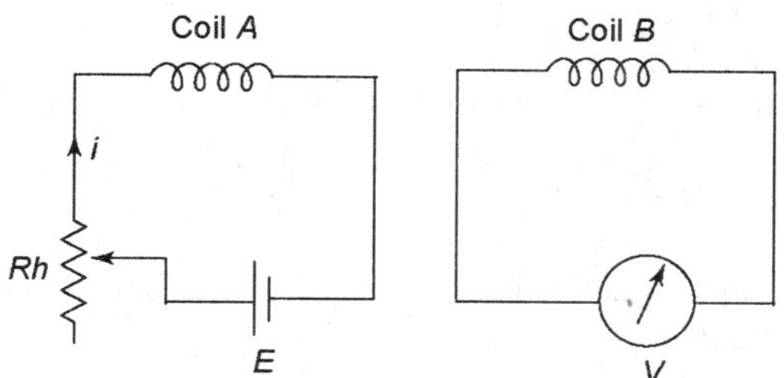

Figure 14 Statically Induced emf in Coil B due to Variation of Flux in Coil A

varies by varying the current in the coil with the help of the rheostat. This varying flux links with coil B and emf is induced in the coil. This emf is also called mutually induced emf. In this case there is no movement of coil B. Similarly, if the connection is reversed, that is, a voltmeter is connected to coil A and the battery and the rheostat to coil B, the emf will be produced in coil A.

Another type of statically induced emf in a coil is the self-induced emf in a coil, which is produced due to the changes of flux in the coil itself by changing the current through it. In Figure 15, the current through the coil is changed by the rheostat (Rh). The flux linked with the coil also changes producing self-induced emf in the coil. The direction of the induced emf in the coil is obtained by Lenz's law that opposes any change of flux in it. This emf is also known as opposing emf or counter emf.

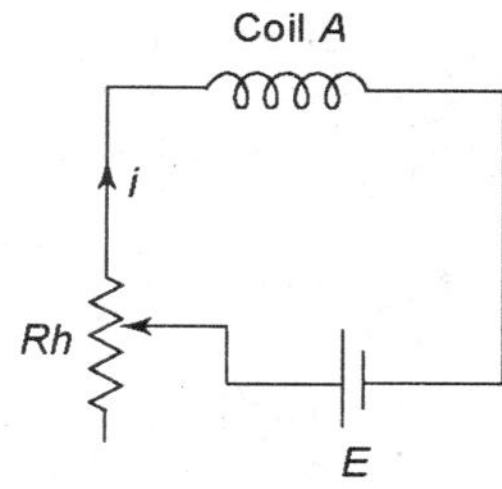

Figure 15 Statically Induced emf in a Single Coil

MAGNETIC CIRCUITS

Magnetic flux lines form closed loops. A magnetic circuit is defined as the closed path followed by flux lines. An electric circuit provides a path for electric current, whereas a magnetic circuit provides a path for magnetic flux. The knowledge of magnetic circuits is of utmost importance in the design, analysis and applications of electromagnetic devices such as transformers, rotating electrical machines and electromagnetic relays. The basics of magnetic circuits are introduced in the following sections.

MAGNETOMOTIVE FORCE

A current-carrying conductor produces a magnetic field around it. The coil should have the correct number of turns in order to produce the required flux density. Magnetomotive force (mmf) is defined as the product of current and number of turns.

Magnetomotive force = Current × Turns

Let I be the current through the coil (A) and N be the number of turns in the coil.

$$\therefore \quad F = IN \tag{4}$$

The unit of mmf is ampere-turns (AT). Since N is dimensionless, its unit is ampere (A) also.

MAGNETIC FIELD INTENSITY

Magnetomotive force per unit length of magnetic flux path is called magnetic field intensity (*H*).

$$H = \frac{F}{l} = \frac{IN}{l} \text{ A/m} \tag{5}$$

where *l* is the mean length of the magnetic circuit in metres. Magnetic field intensity is also called magnetic field strength or magnetizing force.

From Equation (5),

$$F = Hl \tag{6}$$

MAGNETIC FLUX

A current-carrying conductor produces a magnetic field around it. The magnetic field is measured in terms of flux lines (or simply flux) ϕ and flux density *B*, the number of flux lines per unit area.

$$B = \frac{\phi}{A} \tag{7}$$

The density of flux is greatest near the conductor and tapers off inversely with distance.

$$B = \frac{\mu_0 I}{2\pi r} \tag{8}$$

The unit of flux ϕ is webers (Wb) and the unit of magnetic intensity (*B*) is webers per metre square (Wb/m^2) or tesla (T).

1 weber = 10^8
1 gauss = 1 Wb/cm^2 = 10^{-4} Wb/m^2
1 tesla = 10^4 gausses

These units are in rationalized MKS system.

By Ampere's law, the line integral of the field intensity ($\vec{H}$) around a closed path is equal to the current enclosed by that path.

$$\oint \vec{H} \cdot d\vec{l} = I \tag{9}$$

The line integral of the closed path *dl* is the circumference of the circle $2\pi r$.

From Equation (8),

$$I = \frac{B(2\pi r)}{\mu_0} \tag{10}$$

From Equations (9) and (10),

$$\frac{B(2\pi r)}{\mu_0} = H(2\pi r)$$

i.e.,

$$B = \mu_0 H \tag{11}$$

For medium other than free space,

$$\mu = \mu_0 \mu_r$$

where μ_r is the relative permeability.
Flux is produced by current

$$\phi \propto I.$$

If there are N turns instead of single conductor,

$$\phi \propto NI \tag{12}$$

where NI, called ampere-turns (AT), is the flux-producing ability of the current-carrying circuit.

We can visualize NI or AT as the force that produces flux and hence is called the magnetomotive force, mmf (F).

$$F = NIA \tag{13}$$

Thus, we have arrived at a magnetic circuit similar to an electric circuit.

The medium surrounding the conductor causes resistance to the creation of flux. In the magnetic circuit, this resistance is called reluctance (S).

$$\phi = \frac{F}{S} \text{ Wb} \tag{14}$$

where S is the reluctance.

As in the case of an electric circuit, reluctance is directly proportional to length, l, and inversely to area, A, and the property of the material called permeability, μ.

$$S = \frac{l}{\mu A} \text{ ampere-turns/weber}$$

So, higher the permeability, lesser the reluctance and greater the flux.

We have seen that permeability of free space is very low, $4\pi \times 10^{-7}$ H/m. Ferromagnetic materials, that is, alloys of iron with cobalt, tungsten, nickel and aluminum, have very high permeability (μ_r), ranging from 2,000 to 80,000. High-permeability materials have the property of confining the flux within their cross-sectional area instead of spreading out as in space.

SINGLE-PHASE CIRCUITS

In AC circuits, voltage and current vary sinusoidally. The passive parameters of AC circuits are resistance (R), inductance (L) and capacitance (C).

Series R-L Circuit

Figure 16(a) shows an R–L series circuit.

Let

$$i = I_m \sin \omega t \tag{15}$$

$$e = E_m \sin\left(\omega t + \tan^{-1}\frac{\omega L}{R}\right) \tag{16}$$

where

$$E_m = \sqrt{R^2 + \omega^2 L^2}\, I_m \tag{17}$$

The current lags behind the voltage by an angle $\beta = \tan^{-1}\left(\frac{\omega L}{R}\right)$ shown in Figure 16(b).

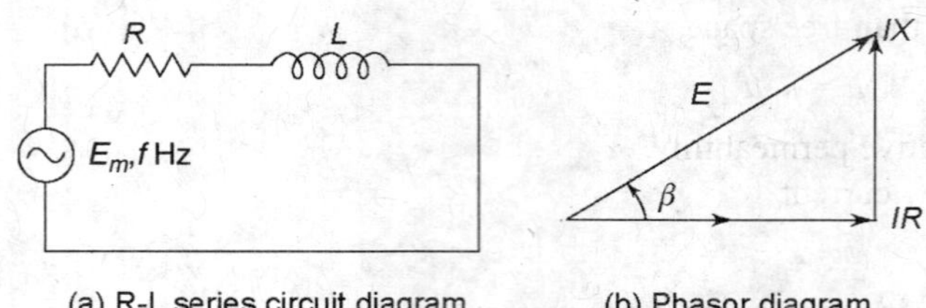

(a) R-L series circuit diagram (b) Phasor diagram

Figure 16 R-L Series Circuit

Here, $\beta = \tan^{-1}\dfrac{\omega L}{R} = \cos^{-1}\dfrac{R}{Z} = \sin^{-1}\dfrac{\omega L}{Z}$

$\therefore$ Power factor $= \cos\beta = \dfrac{R}{Z} = \dfrac{R}{\sqrt{R^2+X^2}}$ (18)

where $X = \omega L$.

Series R-C Circuit

Figure 17(a) shows an R–C series circuit.

Let $i = I_m \sin\omega t$ (19)

where I_m is the maximum value of current.

$$e = E_m \sin(\omega t - \beta) \quad (20)$$

where $E_m = \sqrt{R^2 + \dfrac{1}{\omega^2 C^2}}\, I_m$ (21)

$\therefore$ $Z = \dfrac{E_r}{I_r} = \sqrt{R^2 + \dfrac{1}{\omega^2 C^2}}$ (22)

where Z is called the impedance of the circuit.

The phasor diagram is shown in Figure 17(b). From Figure 17(b), the power factor of the circuit is given by $\cos\beta = \dfrac{R}{Z}$ (leading).

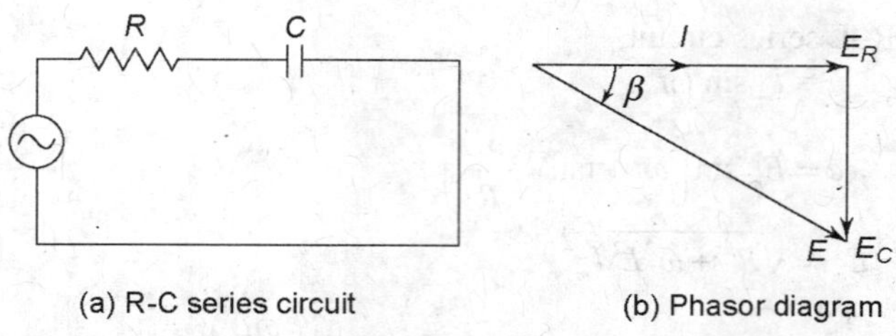

(a) R-C series circuit (b) Phasor diagram

Figure 17 R-C Series Circuit and Its Phasor Diagram

Series L-C-R Circuit

Figure 18 shows a series R–L–C circuit.

Let $$i = I_m \sin\omega t \tag{23}$$

$$e = \sqrt{R^2 + \left(\omega L - \frac{1}{\omega C}\right)^2}\, I_m \sin\,(\omega t + \beta)$$

∴ $$e = E_m \sin(\omega t + \beta) \tag{24}$$

where $$\beta = \tan^{-1}\left(\frac{\omega L - \frac{1}{\omega C}}{R}\right) \tag{25}$$

From Equation (25), the following conclusions can be drawn:

(i) If $\omega L > \frac{1}{\omega C}$, β is positive. In this case the voltage leads the current by an angle β. The phasor diagram is shown in Figure 19(a).

(ii) If $\omega L < \frac{1}{\omega C}$, β is negative. In this case the current leads the voltage by an angle β. The phasor diagram is shown in Figure 19(b).

(iii) If $\omega L = \frac{1}{\omega C}$, $\beta = 0°$. In this case the voltage and the current are in phase. The phasor diagram is shown in Figure 19(c).

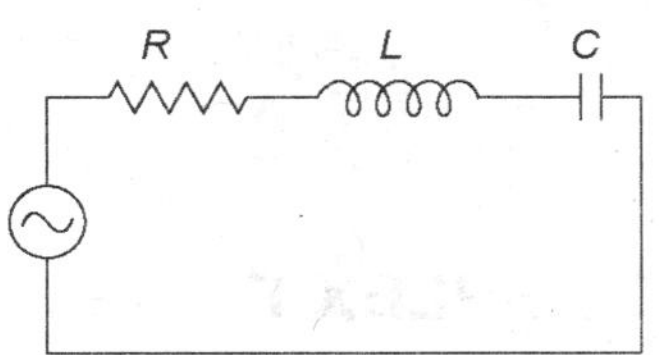

Figure 18 Schematic Diagram of Series R-L-C Circuit

When $\beta = 0$, $Z = R$. The impedance is purely resistive and minimum. Whereas the current is maximum. This is the condition of resonance for a series resonance circuit.

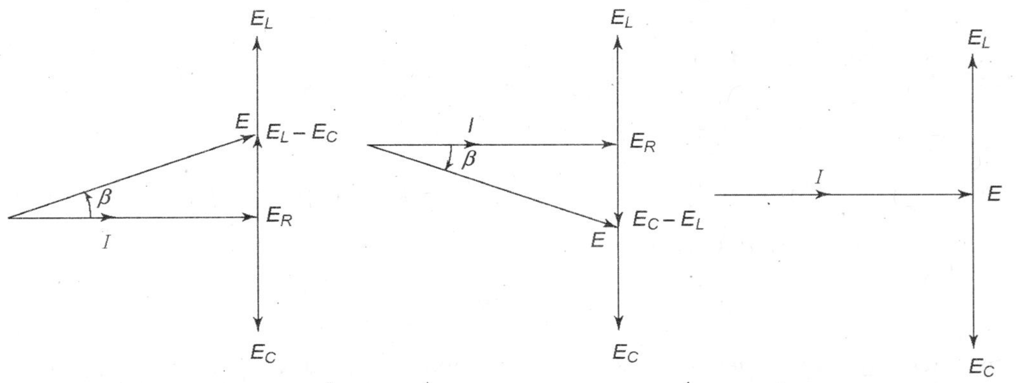

(a) Phasor diagram when $\omega L > \frac{1}{\omega C}$ (b) Phasor diagram when $\omega L < \frac{1}{\omega C}$ (c) Phasor diagram when $\omega L = \frac{1}{\omega C}$

Figure 20 Phasor Diagrams of Series R-L-C Circuit for Three Different Cases

POWER TRIANGLE

The geometrical representation of apparent power, average power (true power or real power) and reactive power is called the power triangle. Figures 20(a) and 20(b) show power triangles for an inductive load and a capacitive load, respectively.

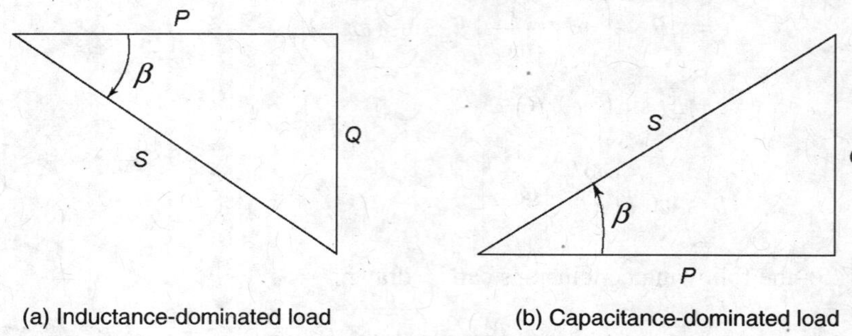

Figure 20 Power Triangles

COMPLEX POWER

S, P and Q of a power triangle can be obtained from the product VI^*. It is called complex power (S).

For a capacitive load, let

$$V = V_r e^{j\theta} \text{ and } I = I_r e^{+j(\theta+\beta)}$$

i.e.,

$$S = VI^* = V_r e^{j\theta} \cdot I_r e^{-j(\theta+\beta)} = V_r I_r e^{-j\beta} = V_r I_r (\cos\beta - j\sin\beta)$$

$$= V_r I_r \cos\beta - j\, V_r I_r \sin\beta$$

i.e.,

$$S = P - jQ$$

Similarly, for an inductive load, let $V = V_r e^{j\theta}$ and $I = I_r e^{j(\theta-\beta)}$

$$S = VI^*$$

i.e.,

$$S = VI^* = V_r e^{j\theta} \cdot I_r e^{-j(\theta-\beta)} = V_r I_r e^{j\beta} = V_r I_r (\cos\beta + j\sin\beta)$$

$$= V_r I_r \cos\beta + jV_r I_r \sin\beta$$

$$= P + jQ$$

By convention, $+Q$ indicates lagging reactive power while $-Q$ indicates leading reactive power. The inductive load absorbs lagging reactive power and capacitive load supplies lagging reactive power. Table 1 shows the P, Q and S for different types of circuits.

Table 1 P, Q and S for Different Types of Circuits

Circuit	pf	Real Power $P = E_r I_r \cos\beta$	Reactive Power Q $= E_r I_r \sin\beta$	Apparent Power $(S) = \sqrt{P^2+R^2}$
R	1	$E_r I_r$	0	$E_r I_r$
L	0	0	$E_r I_r$ (lag)	$E_r I_r$
C	0	0	$E_r I_r$ (lead)	$E_r I_r$
R-L	$\cos\beta$ (lag)	$E_r I_r \cos\beta$	$E_r I_r \sin\beta$ (lag)	$E_r I_r$
R-C	$\cos\beta$ (lead)	$E_r I_r \cos\beta$	$E_r I_r \sin\beta$ (lead)	$E_r I_r$
R-L-C	$\cos\beta$ (lag)	$E_r I_r \cos\beta$	$E_r I_r \sin\beta$ (lag)	$E_r I_r$
	$\cos\beta$ (lead)	$E_r I_r \cos\beta$	$E_r I_r \sin\beta$ (lead)	$E_r I_r$
	1 in phase	$E_r I_r$	0	$E_r I_r$

THREE-PHASE CIRCUITS

The generation of electrical power takes place in three phases in practice; these three phases are represented by *a*, *b* and *c* in Figure 21. The field rotates with synchronous speed. The three phases *a*, *b* and *c* are the armature conductors displaced from each other by 120° in a clockwise direction. With the position of the rotor shown in Figure 21 and the clockwise rotation of the pole, the emf induced in coil *aa′* is maximum. The emf in conductor *b* of coil *bb′* will be maximum when the pole turns through an angle 120° clockwise direction. Therefore, conductor *b* attains its maximum emf 120° later than the maximum emf in conductor *a*. Similarly, conductor *c* attains its maximum emf 120° later than the maximum emf in conductor *b*; in other words, 240° later than the maximum value in conductor *a*.

In this case the emfs induced in conductors *a*, *b* and *c* are equal in magnitude but 120° apart. The waveform and phasor diagrams are shown in Figures 22 and 23, respectively.

From Figure 23, we have

$$V_a = V_m \angle 0^\circ \text{ V} \tag{26}$$

$$V_b = V_m \angle -120^\circ \text{ V} \tag{27}$$

$$V_c = V_m \angle -240^\circ \text{ V} \tag{28}$$

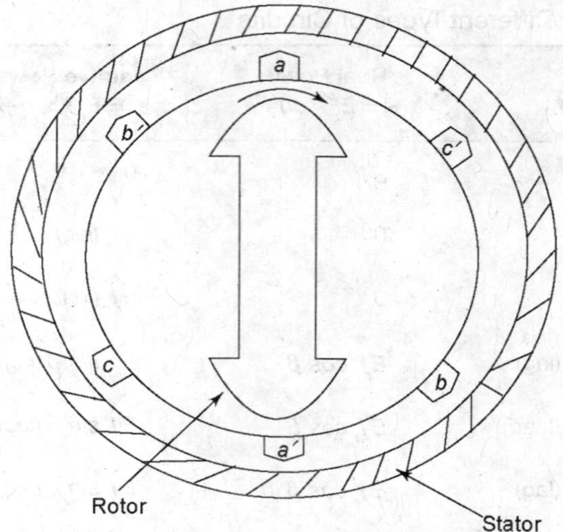

Figure 21 Generation of 3-phase Voltage

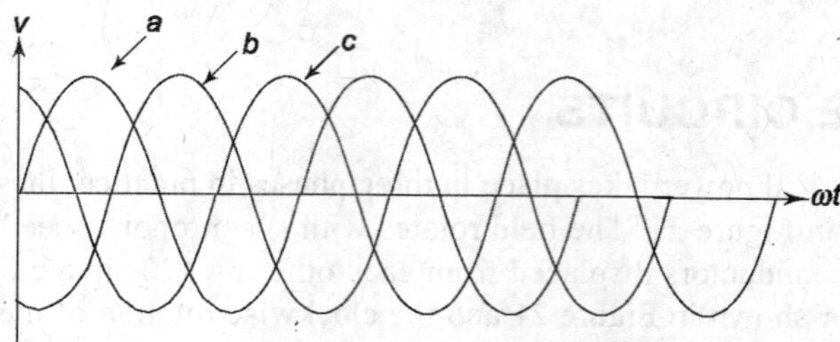

Figure 22 Waveform Diagram of Figure 21

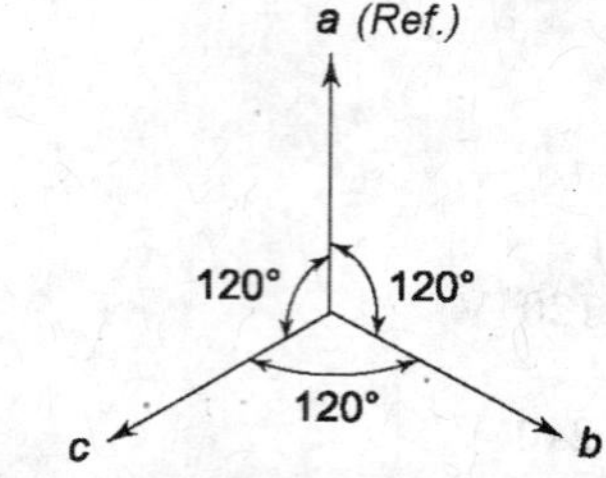

Figure 23 Phasor Diagram of Figure 21

The sum of Equations (26), (27) and (28) becomes

$$\begin{aligned}V_a + V_b + V_c &= V_m \angle 0^\circ + V_m \angle -120^\circ + V_m \angle -240^\circ \\ &= V_m e^{j0^\circ} + V_m e^{-j120^\circ} + V_m e^{-j240^\circ} \\ &= V_m + V_m(\cos 120^\circ - j\sin 120^\circ) \\ &\quad + V_m(\cos 240^\circ - j\sin 240^\circ) \\ &= V_m + V_m\left(-\frac{1}{2} - j\frac{\sqrt{3}}{2}\right) + V_m\left(-\frac{1}{2} + j\frac{\sqrt{3}}{2}\right) \\ &= V_m\left(1 - \frac{1}{2} - j\frac{\sqrt{3}}{2} - \frac{1}{2} + j\frac{\sqrt{3}}{2}\right) \\ &= 0\end{aligned}$$

ADVANTAGES OF THREE-PHASE SYSTEM

The three-phase system has the following advantages compared to a single-phase system:

(i) It is more economical as it requires less conductor material compared to a single-phase system.
(ii) A three-phase machine gives more output compared to a single-phase machine of the same size.
(iii) Three-phase motors have uniform torque, whereas most of the single-phase motors have pulsating torque.
(iv) Domestic power and industrial or commercial power can be supplied from the same source.
(v) A three-phase motor produces more torque as compared to a single-phase motor.
(vi) Three-phase induction motors are self-starting, whereas single-phase motors are not self-starting.
(vii) Voltage regulation is better in a three-phase system.

PHASE SEQUENCE

The three phases may be denoted by *a*, *b* and *c* or *R*, *Y* and *B* where *R*, *Y* and *B* indicate red, yellow and blue, respectively. *RYB* has commercial usage. *abc* or *RYB* is taken as a positive sequence, whereas *acb* or *RBY* is taken as a negative sequence.

INTERCONNECTION OF THREE PHASES

The three separate armature coils of the three-phase alternator are shown in Figure 24.

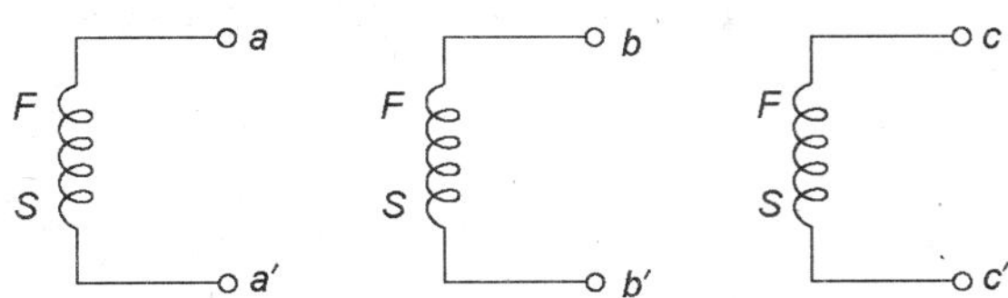

Figure 24 Three separate armature coils

STAR AND DELTA CONNECTIONS

The *three armature conductors* in a three-phase generator are not generally interconnected. They may be connected in the star (Y) or delta (Δ) form. The star connection of the three-armature conductors is shown in Figure 25 where all the similar ends a', b' and c' are at the starting point. The terminals a', b' and c' are joined to form the neutral terminal (N). In a star connection, similar ends are connected (the '*s*' in star indicates similar ends). The neutral conductor may or may not be present. If it is present, it is called a three-phase four-wire system. Otherwise, it is a three-phase three-wire system. The delta connection of the three-armature conductor is shown in Figure 26. In a delta connection (the *'d'* in delta indicates dissimilar ends), dissimilar ends are interconnected.

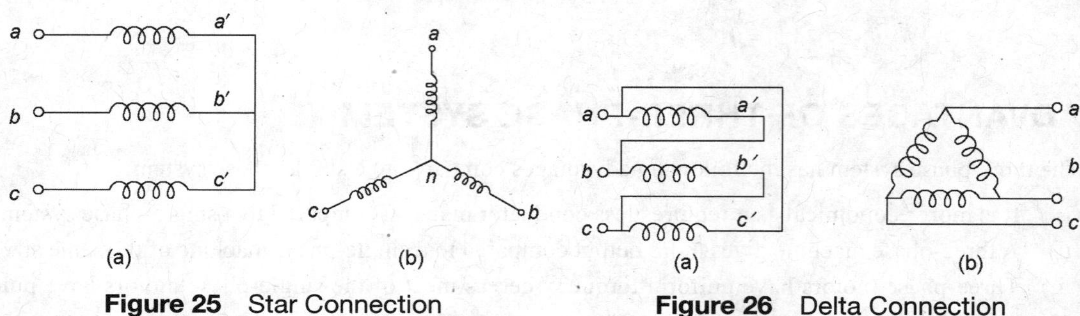

Figure 25 Star Connection **Figure 26** Delta Connection

VOLTAGES, CURRENTS AND POWER IN STAR CONNECTIONS

Figure 27 shows the three-phase three-wire system. The voltages V_{an}, V_{bn} and V_{cn} are called phase voltages, whereas the voltages V_{ab}, V_{bc} and V_{ca} are called line voltages, and the phase current (I_P) in each phase is equal to the line current in each line because each line is in series with its individual phase winding.

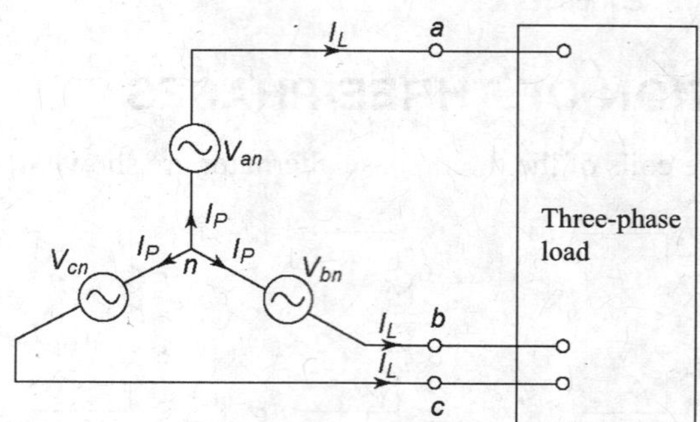

Figure 27 Star Connection of the Circuit

Let

$$V_{an} = V_P \angle 0^\circ$$
$$V_{bn} = V_P \angle -120^\circ$$
$$V_{cn} = V_P \angle -240^\circ$$

V_P indicates the magnitude of phase voltage.

Now,

$$\begin{aligned} V_{ab} &= V_{an} + V_{nb} = V_{an} - V_{bn} \\ &= V_P \angle 0^\circ - V_P \angle -120^\circ \\ &= V_P (\cos 0^\circ + j \sin 0^\circ) - V_P [\cos(-120^\circ) + j \sin(-120^\circ)] \\ &= V_P - V_P (\cos 120^\circ - j \sin 120^\circ) \\ &= V_P - V_P \left(-\frac{1}{2} - j\frac{\sqrt{3}}{2} \right) = V_P \left(1 + \frac{1}{2} + j\frac{\sqrt{3}}{2} \right) \\ &= V_P \left(\frac{3}{2} + j\frac{\sqrt{3}}{2} \right) = \sqrt{3}\, V_P \left(\frac{\sqrt{3}}{2} + j\frac{1}{2} \right) \\ &= \sqrt{3}\, V_P (\cos 30^\circ + j \sin 30^\circ) = \sqrt{3}\, V_P \angle 30^\circ \end{aligned}$$

Similarly,

$$\begin{aligned} V_{bc} &= V_{bn} + V_{nc} = V_{bn} - V_{cn} \\ &= V_P \angle -120^\circ - V_P \angle -240^\circ = V_P \angle -90^\circ \end{aligned}$$

and

$$\begin{aligned} V_{ca} &= V_{cn} + V_{na} = V_{cn} - V_{an} = V_P \angle -240^\circ - V_P \angle 0^\circ \\ &= \sqrt{3}\, V_P \angle -210^\circ \end{aligned}$$

Figure 28 shows the phasor diagram.

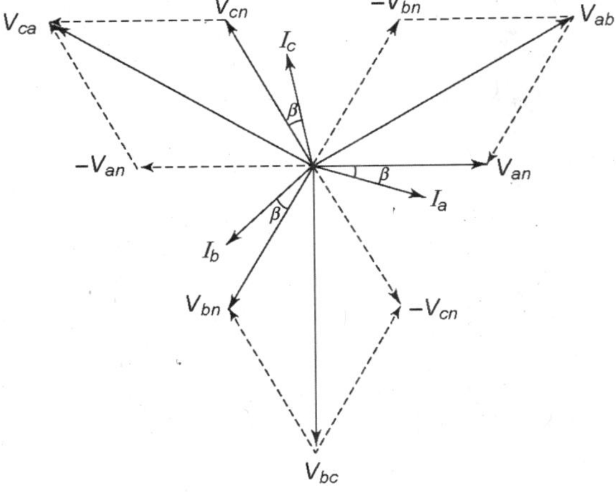

Figure 28 Phasor Diagram

So the magnitude of line voltage is $\sqrt{3}$ times the magnitude of phase voltage. In this case, $I_P = I_L$. The active or true power in each phase of the circuit is given by $V_P I_P \cos\beta$, where β is the angle between each phase voltage and its corresponding phase current. The total three-phase power (P) is expressed as

$$P = 3V_P I_P \cos\beta = 3\left(\frac{V_L}{\sqrt{3}}\right) I_L \cos\beta$$

i.e.,

$$P = \sqrt{3}\, V_L I_L \cos\beta \tag{29}$$

Similarly, total reactive power (Q) is expressed as

$$Q = 3V_P I_P \sin\beta = 3\left(\frac{V_L}{\sqrt{3}}\right) I_L \sin\beta$$

i.e.,

$$Q = \sqrt{3}\, V_L I_L \sin\beta \tag{30}$$

Q is positive when the load is of the inductive type, and it is negative when the load is of the capacitive type.

The total apparent power of the three phases $(S) = \sqrt{3}\, V_L I_L$.

VOLTAGES, CURRENTS AND POWER IN DELTA CONNECTIONS

Figure 29 shows three voltage sources connected in delta. In Figure 29, the phase voltages and line voltages are identical, but the line currents and phase currents are not identical.

Here, taking I_{ab} as reference, we have

$$\begin{aligned}
I_{ab} &= I_P \angle 0^\circ \\
I_{bc} &= I_P \angle -120^\circ \\
I_{ca} &= I_P \angle -240^\circ \\
I_a &= I_{ca} - I_{ab} = I_P \angle -240^\circ - I_P \angle 0^\circ \\
&= I_P(\cos 240^\circ - j\sin 240^\circ) - I_P(\cos 0^\circ + j\sin 0^\circ) \\
&= I_P\left(-\frac{1}{2} + j\frac{\sqrt{3}}{2}\right) - I_P = I_P\left(-\frac{3}{2} + j\frac{\sqrt{3}}{2}\right) \\
&= \sqrt{3}\, I_P\left(-\frac{\sqrt{3}}{2} + j\frac{1}{2}\right) = \sqrt{3}\, I_P(\cos 150^\circ + j\sin 150^\circ) \\
&= \sqrt{3}\, I_P \angle 150^\circ
\end{aligned}$$

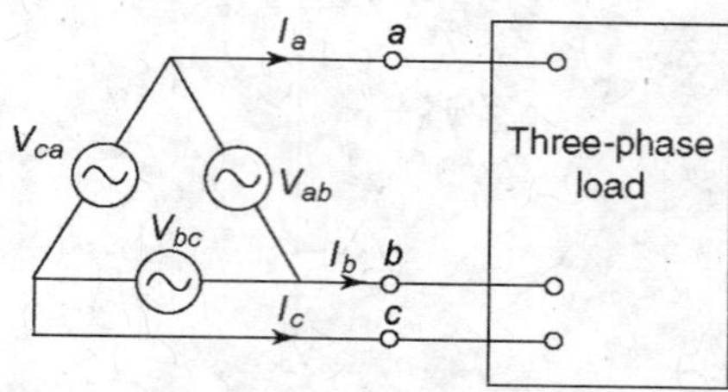

Figure 29 Delta Connection of the Circuit

Similarly, $I_b = I_{ab} - I_{bc}$

$$= I_P \angle -0° - I_P \angle -120°$$

$$= I_P - I_P(\cos 120° - j\sin 120°) = I_P - I_P\left(-\frac{1}{2} - j\frac{\sqrt{3}}{2}\right)$$

$$= I_P\left(\frac{3}{2} + j\frac{\sqrt{3}}{2}\right) = \sqrt{3}\, I_P\left(\frac{\sqrt{3}}{2} + j\frac{1}{2}\right)$$

$$= \sqrt{3}\, I_P \angle 30°$$

and $I_c = I_{bc} - I_{ca}$

$$= I_P \angle -120° - I_P \angle -240°$$

$$= I_P(\cos 120° - j\sin 120°) - I_P(\cos 240° - j\sin 240°)$$

$$= I_P\left(-\frac{1}{2} - j\frac{\sqrt{3}}{2}\right) - I_P\left(-\frac{1}{2} + j\frac{\sqrt{3}}{2}\right)$$

$$= -j\sqrt{3}\, I_P = \sqrt{3}\, I_P \angle -90°$$

The phasor diagram is shown in Figure 30.

From Figures 29 and 30, it can be seen that phase voltages are equal to line voltages. If I_P is the magnitude of phase current and I_L is that of line current,

$$I_L = \sqrt{3}\, I_P,$$

then the total three-phase true or real power (P) is expressed as follows:

$$P = 3V_P I_P \cos\phi = 3V_L\left(\frac{I_L}{\sqrt{3}}\right)\cos\beta = \sqrt{3}\, V_L\, I_L \cos\beta$$

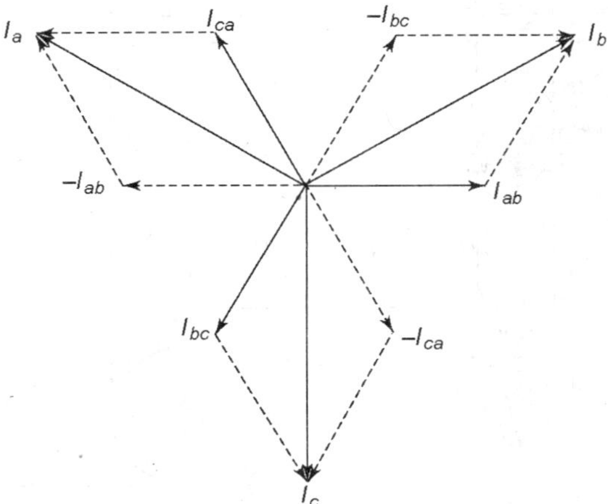

Figure 30 Phasor Diagram

The total three-phase reactive power (Q) becomes

$$Q = 3V_P I_P \sin\beta = 3V_L\left(\frac{I_L}{\sqrt{3}}\right)\sin\phi = \sqrt{3}\,V_L I_L \sin\beta$$

Apparent power $(S) = 3V_P I_P = \sqrt{3}\,V_L I_L$.

For Star Connections

(i) Phase voltage $=\left(\frac{1}{\sqrt{3}}\right)$ line voltage; delta connection phase voltage = line voltage. Thus, for a given phase voltage, *Y*, the connected alternator gives more line voltage and the insulation of the armature winding will not be costly.

(ii) As the star point is available in a star-connected alternator, the star point is connected to the ground. From the point of view of protection, it is advantageous.

A practical connection is shown in Figure 31. The load is normally distributed among the phases in order to keep the system as balanced as possible.

Here, suppose the line voltage is 400 V. The phase voltage to each domestic consumer will be $\frac{400}{\sqrt{3}}\text{V} = 230\,\text{V}$. Figure 31 shows that the domestic consumer (single-phase) as well as the industrial consumer (three-phase) can be supplied.

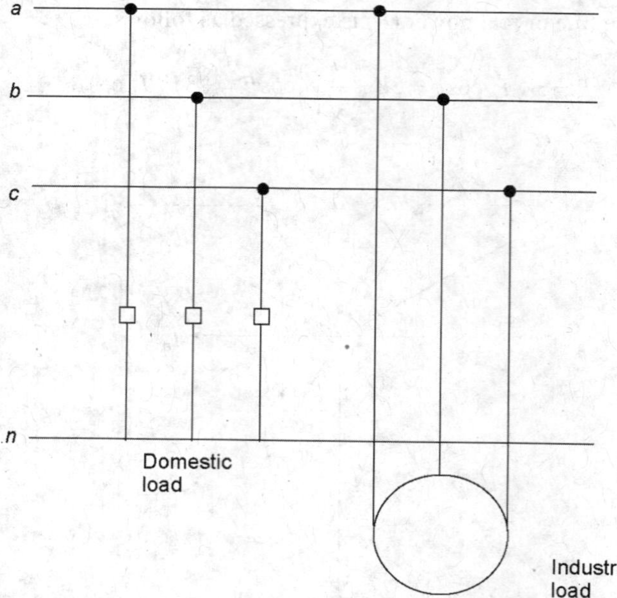

Figure 31 Connection of Domestic and Industrial Load

MEASUREMENT OF THREE-PHASE POWER

In this method, wattmeter W_1 is inserted in phase a and wattmeter W_2 is inserted in phase c. Their pressure coils are connected in phase b shown in Figure 32. Figure 33 shows the phasor diagram.

The sum of instantaneous power measured by W_1 and W_2 is expressed as

$$
\begin{aligned}
W &= W_1 + W_2 \\
&= i_a (V_{an} - V_{bn}) + i_c (V_{cn} - V_{bn}) \\
&= i_a V_{an} + i_c V_{cn} - (i_a + i_c) V_{bn} \\
&= i_a V_{an} + i_b V_{bn} + i_c V_{cn} \quad [\because \quad i_a + i_b + i_c = 0]
\end{aligned}
$$

Therefore, W is the total instantaneous power of a three-phase circuit.

Figure 33 depicts a balanced load, and the currents in each phase lag behind the corresponding phase voltages by an angle θ from the phasor diagram; the current passing through the current coil of W_1 lags behind the line voltage V_{ab} by an angle 30° + θ. Similarly, the current passing through the pressure coil of W_2 lags behind the line voltage V_{cb} by an angle 30° − θ

Now, the magnitude of line voltage (V_L) is $\sqrt{3}$ times the phase voltage V_P; that is, $V_L = \sqrt{3}\, V_P$. Let the value of current in each phase be I_P. So, the line current $(I_L) = I_P$. The line voltage magnitude V_L and line current I_L are the rms values.

Now, $$W_1 = V_L I_L \cos(30° + \theta)$$

and $$W_2 = V_L I_L \cos(30° - \theta)$$

$\therefore$
$$
\begin{aligned}
W_1 + W_2 &= V_L I_L \cos(30° + \theta) + V_L I_L \cos(30° - \theta) \\
&= V_L\, I_L\, [\cos(30° + \theta) + \cos(30° - \theta)] \\
&= V_L I_L (2 \cos 30° \cos\theta)
\end{aligned}
$$

i.e., $$W_1 + W_2 = 3 V_P I_P \cos\theta \tag{31}$$

Similarly, $W_1 - W_2 = V_L I_L [\cos(30° + \theta) - \cos(30° - \theta)]$

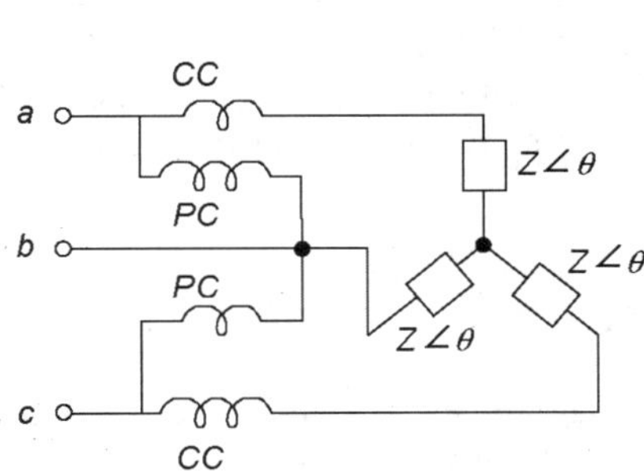

Figure 32 Two-wattmeter Connection

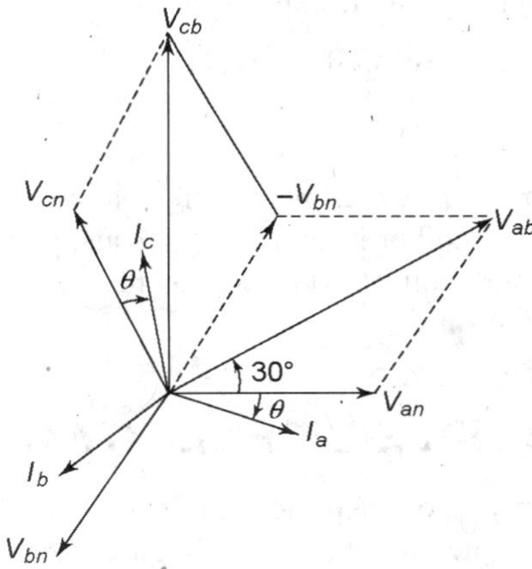

Figure 33 Phasor Diagram

i.e.,
$$W_1 - W_2 = V_L I_L[-2\sin 30°\sin\theta]$$
$$= V_L I_L \sin\theta$$

i.e.,
$$W_2 - W_1 = \sqrt{3}\, V_P I_P \sin\theta \qquad (32)$$

Dividing Equation (32) by Equation (31), we have

$$\frac{W_2 - W_1}{W_1 + W_2} = \frac{\sqrt{3}\, V_P\, I_P \sin\theta}{\sqrt{3}\, V_P\, I_P \cos\theta} = \frac{\tan\theta}{\sqrt{3}}$$

i.e.,
$$\tan\theta = \sqrt{3}\left(\frac{W_2 - W_1}{W_2 + W_1}\right) \qquad (33)$$

Since $W_1 = V_L I_L \cos(30° + \theta)$ and $W_2 = V_L I_L \cos(30° - \theta)$ the following cases are discussed:

(i) When $\theta = 0°$, that is, the power factor is unity (resistive load), it can be written as
$W_1 = W_2 = V_L I_L \cos 30°$
The readings of both wattmeters are equal in magnitude.

(ii) When $\theta = 60°$, that is, the power factor = 0.5 lagging,
$W_1 = V_L I_L \cos(30° + 60°) = V_L I_L \cos 90° = 0$ and $W_2 = V_L I_L \cos(30° - 60°) = V_L I_L \cos 30°$.

(iii) When $90° > \theta > 60°$, the reading of W_2 is positive and that of W_1 is negative.

(iv) When $\theta = 90°$ (pure inductive load),
$W_2 = V_L I_L \cos(30° - 90°) = V_L I_L \cos(-60°) = V_L I_L \cos 60° = V_L I_L \sin 30°$
$W_1 = V_L I_L \cos(30° + 90°) = -V_L I_L \sin 30°$

$\therefore \quad W_1 + W_2 = 0$

The above facts are for a lagging power factor load. For a leading power factor load, θ will be replaced by $-\theta$ in the expression of W_1 and W_2.

$W_1 = V_L I_L \cos(30° - \theta)$

and $W_2 = V_L I_L \cos(30° + \theta)$

During the experiment if any wattmeter shows a negative reading, it is necessary to reverse either its pressure coil or its current coil, usually the former. The reading of the wattmeter after the reversal of the pressure coil is to be taken as negative. In the above case, the *abc* phase sequence has been taken as positive.

PRINCIPLE OF ENERGY CONVERSION

The main aim of the principle of energy conversion is to estimate forces. The energy conversion can be neither created nor destroyed, but it can be converted from one form to the other. A typical electromechanical system is shown in Figure 34.

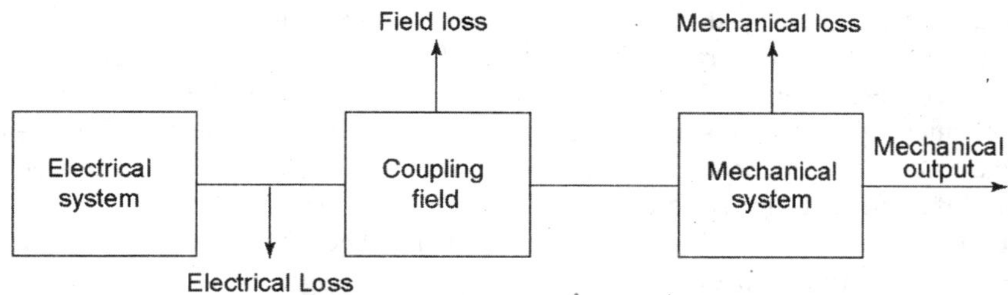

Figure 34 Electromechanical Energy Conversion System

The main parts of the system are as follows:

(i) an electrical system,
(ii) a coupling field system and
(iii) a mechanical system.

The energy transfer equation is expressed as

$$\begin{matrix}\text{Electrical Energy}\\ \text{Input from Source}\end{matrix} = \begin{matrix}\text{Total Energy}\\ \text{Losses}\end{matrix} + \begin{matrix}\text{Increase in}\\ \text{Stored Energy}\\ \text{in the Coupling Field}\end{matrix} + \begin{matrix}\text{Mechanical}\\ \text{Energy}\\ \text{Output}\end{matrix} \tag{34}$$

The total energy losses in the system include I^2R losses in the resistance of the winding of the electrical system, core loss in the field due to having of magnetic field in core as well as the mechanical loss in the form of friction and windage loss due to motion of mechanical components in the mechanical system. The total losses are converted into heat. Equation (34) becomes

$$\begin{matrix}\text{Electrical Energy}\\ \text{Input from Source–}\\ \text{Ohmic Losses}\end{matrix} = \begin{matrix}\text{Friction and}\\ \text{Windage}\\ \text{Loss}\end{matrix} + \begin{matrix}\text{Increase in}\\ \text{Stored Energy}\\ \text{in the Coupling}\\ \text{Field + Core Loss}\end{matrix} + \begin{matrix}\text{Mechanical}\\ \text{Energy}\\ \text{Output}\end{matrix} \tag{35}$$

In differential form, Equation (35) becomes

$$dW_e = dW_f + dW_m \tag{36}$$

where dW_f represents change in energy stored in the field neglecting the core loss which is generally small and dW_m represents mechanical energy output when mechanical losses (friction and windage losses) are neglected.

ENERGY IN THE COUPLING FIELD

In rotating machines, the coupling field is the link between the stationary and movable members. In order to rotate the rotating member, an air gap must exist between the stator and the rotor. The energy stored in the coupling field must produce action on the electrical and mechanical system to ensure conversion of energy from electrical to mechanical (motoring mode) or from mechanical to electrical (generating mode).

In DC machines, the field system is a stationary member whereas the armature is a rotating member. For mechanical output (motoring mode), the coupling field must react with the electrical system to absorb the electrical energy from it. For electrical output (generating mode), the coupling field must react with the mechanical system to absorb mechanical energy from it. The reaction between the coupling fields with the electrical or mechanical system depends on the operating mode of the device. Since the torque (*T*) and the induced emf (*e*) are associated with the coupling field, it is termed as electromechanical coupling. Due to the inertia of the electromechanical coupling devices, these are slow-moving devices. Since the variation of the coupling field with time during energy conversion is slow, the electromagnetic radiation losses are neglected.

ENERGY IN THE FIELD

Figure 35 shows a simple electromechanical energy conversion system, which is a singly excited system because only one winding receives electrical energy.

A spring control is used to hold the movable part of the system in a static equilibrium. The movable part is held at an air gap '*g*'. The current in the winding is increased from 0 to *i* and hence the flux is established. Since the movable part is stationary, dW_m is zero and we have

$$dW_e = dW_f \tag{37}$$

Equation (37) indicates that all the electrical energy input is stored in the energy field because the core loss is neglected. For the flux linkage 'λ', the back emf induced in the winding becomes

$$e = \frac{d\lambda}{dt} \tag{38}$$

The electrical energy input to the system is expressed as

$$dW_e = ei\,dt \tag{39}$$

From Equations (38) and (39), we have

$$dW_e = \frac{d\lambda}{dt} i\,dt = i\,d\lambda$$

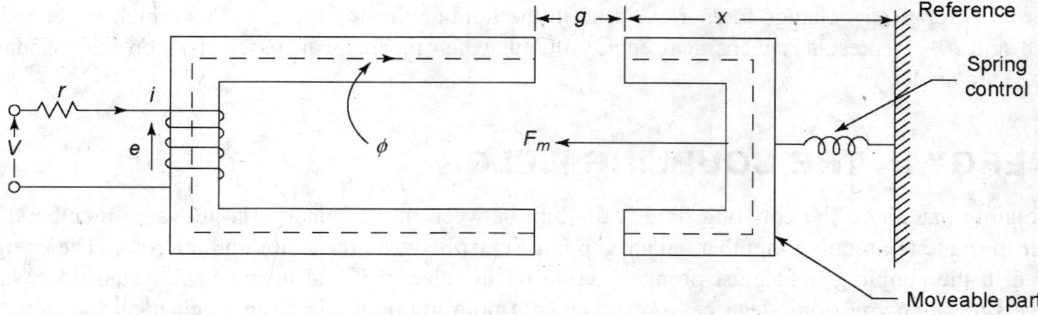

Figure 35 Electromechanical Energy Conversion System

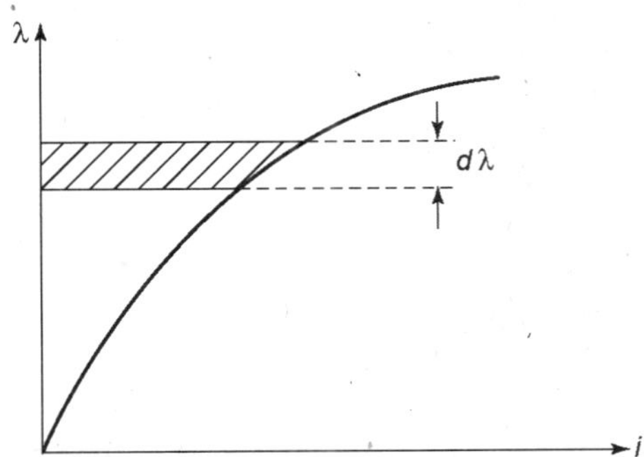

Figure 36 λ–i Characteristic for the System

i.e., $dW_f = id\lambda$ [using Equation (36)] (40)

i.e., W_f = energy stored in the magnetic field = $\int_0^{\lambda} id\lambda$ (41)

where λ is increased from 0 to λ.

Figure 36 shows the relation between flux linkages and the current in the coil for Figure 35. The hatched portion of Figure 36 shows the incremental energy stored in the field. The total energy stored in the field is the space between λ-axis and the curve ($\lambda - i$).

Let

N be the total turns of the system,
i be the input current,
H_c be the magnetic field intensity in the core,
H_g be the magnetic field intensity in the air gap,
l_c be the length of the magnetic core material and
l_g be the length of the air gap.

We have

$$Ni = H_g l_g + H_c l_c \tag{42}$$

and $\lambda = N\phi$ = total flux linkages with the winding (43)

If the cross-section area of the core material and uniform magnetic flux density are A and B, respectively, we have

$$\lambda = NAB \tag{44}$$

i.e.,

$$d\lambda = NA\,dB \tag{45}$$

Substituting dλ in Equation (41), we have

$$W_f = \int_0^B iNA\,dB$$

i.e.,

$$W_f = \int_0^B (H_g l_g + H_c l_c)\,A\,dB \tag{46}$$

i.e., $$W_f = \int_0^B \left(\frac{B}{\mu_0} l_g + H_c l_c) \right) A\, dB$$

i.e., $$W_f = \int_0^B \left(\frac{B}{\mu_0} l_g + H_c l_c \right) A\, dB$$

i.e., $$W_f = l_c A \int_0^B H_c\, dB + \frac{A l_g}{\mu_0} \int_0^B B\, dB$$

i.e., $$W_f = \left[\int_0^B H_c\, dB \right] \times \text{volume of core material} + \frac{B^2}{2\mu_0} \times \text{ volume of the air gap} \tag{47}$$

Let

W_{fc} be the energy density in the core $= \int_0^B H_c dB$,

W_{fg} be the energy density in the air gap $= \frac{B^2}{2\mu_0}$,

V_c be the volume of the core material and

V_g be the volume of air gap.

Equation (47) becomes

W_f = energy stored in the magnetic field $= W_{fc} V_c + W_{fg} V_g$ (48)

The supplied energy to the system is partly stored in the magnetic field and partly in the air gap. The energy stored in the air gap is much larger than the energy stored in the field. The energy stored in the field is expressed as

$$W_f = \int_0^\lambda i d\lambda = W_{fc} V_c + W_{fg} V_g \tag{49}$$

CO-ENERGY

The magnetic material and the air gap length decide the flux linkage—the current characteristic of the electromagnetic system. The $\lambda - i$ characteristic of the electromechanical system shown in Figure 35 is illustrated in Figure 37(a) for different air gap length. The characteristic becomes more linear when the air gap length is increased as shown in Figure 37(a).

From Figure 37(b), the energy stored in the magnetic field 'W_f' is the hatched portion between the λ-axis and the $\lambda - i$ characteristic. The hatched portion between the $\lambda - i$ characteristic and the i-axis represents the co-energy (W'_f) of the system, which is used to estimate force or torque in an electromagnetic system as explained below. The co-energy (W'_f) of the system is expressed as

$$W'_f = \int_0^\lambda \lambda\, di \tag{50}$$

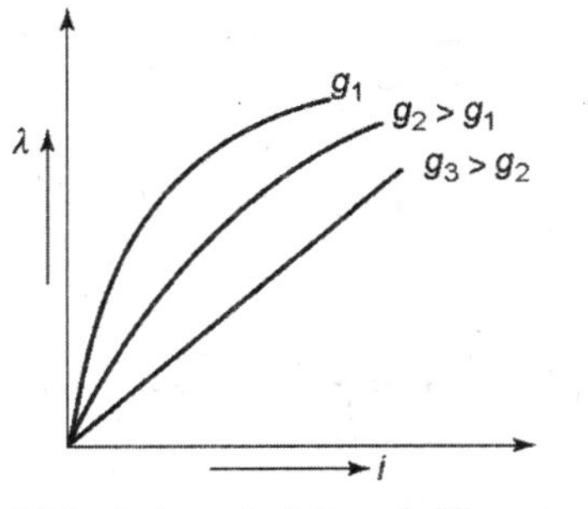

(a) λ – i characteristics of different gaps

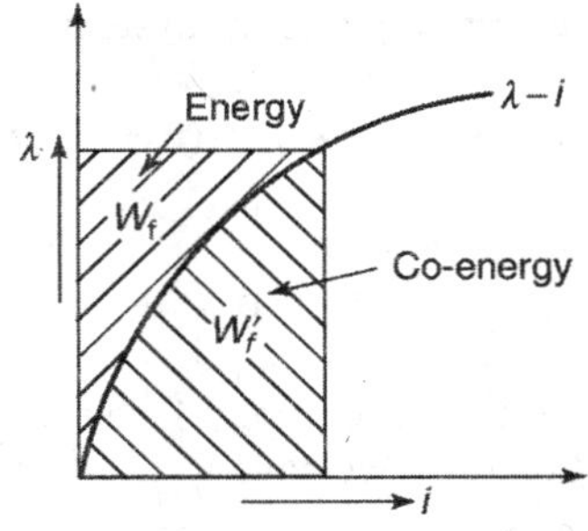

(b) Energy and co-energy for a given gap

Figure 37 Co-energy

The total energy of the system is expressed as

$$W_f + W_f' = \lambda i \tag{51}$$

Due to the non-linearity of the electromechanical system, co-energy is more than the energy stored in the magnetic field.

For linear systems,

$$W_f = W'_f \tag{52}$$

ELECTRICAL ENERGY INPUT TO THE SYSTEM

A simple electromagnetic system with a single excited coil shown in Figure 35 is considered here. At any instant the terminal equation is represented by

$$v = ir + e \tag{53}$$

In Equation (53), v is the instantaneous input voltage, i is the instantaneous input current and e is the back emf.

If λ is the instantaneous flux linkage, we have

$$e = \frac{d\lambda}{dt} \tag{54}$$

Substituting 'e' in Equation (53), we have

$$v = ir + \frac{d\lambda}{dt} \tag{55}$$

From Equation (55), we have

$$iv\,dt = i^2 r\,dt + i\,d\lambda$$

i.e.,

$$i(v - ir)dt = i\,d\lambda$$

The elementary energy input to the system

$$= dW_e = (v - ir)idt = id\lambda \tag{56}$$

We know that

$$\lambda = N\phi \tag{57}$$

where N is the total number of turns of the exciting coil and ϕ is the instantaneous flux linking with the coil.

From Equation (56), we have

$$dW_e = iN\,d\phi \tag{58}$$

i.e.,

$$W_e = \int iN\,d\phi \tag{59}$$

Most of the flux is confined to the core of the excited coil made of ferromagnetic material and the core extracts energy from the supply system for the case of change of flux linkage.

This change of flux causes the back emf that is responsible for extracting energy from the electrical system.

ESTIMATION OF MECHANICAL FORCES IN AN ELECTROMAGNETIC SYSTEM

Figure 38 shows the λ–i characteristic of the system when the moving part of the electromechanical system shown in Figure 35 is shifted from $x = x_1$ to $x = x_2$. The change of the λ–i characteristic shown in Figure 38(a) is for a system whose moving part moves very slowly and the excitation during this period of motion is assumed to be constant, whereas the λ–i characteristic shown in Figure 38(b) is for a system whose moving part is assumed to change its position very quickly. Let us consider the following cases.

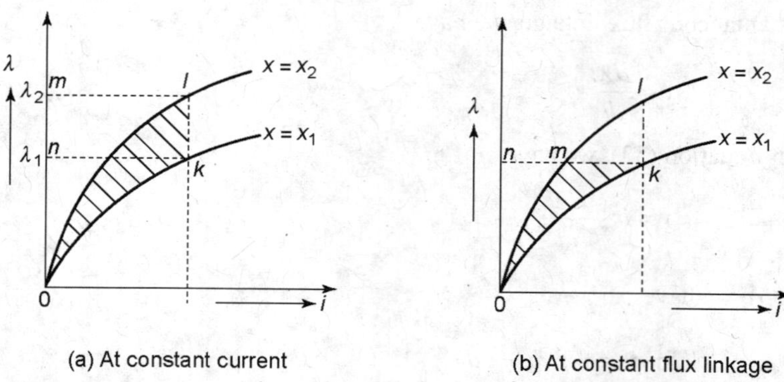

Figure 38 Locus of Operating Point on λ–i Characteristic

Case 1: Motion of the Moving Part Is Very Slow

For this case the excitation current is assumed to be constant during this period. When the moving part is shifted from $x = x_1$ to $x = x_2$, the λ–i characteristic moves upwards due to the decrease of the air gap length. The operating power in this case moves from k to l and the change of electrical energy during this case is given by

$$dW_e = \int ei\,dt = \int_{\lambda_1}^{\lambda_2} i\,d\lambda = \text{area } klmn \tag{60}$$

Energy stored in the magnetic field is expressed as

$$dW_f = \text{area } olm - \text{area } okn \tag{61}$$

The mechanical work done by the moving part is expressed as

$$\begin{aligned} dW_m &= dW_e - dW_f \\ &= \text{area } klmn - (\text{area } olm - \text{area } okn) \\ &= \text{area } klmn + \text{area } okn - \text{area } olm \\ &= \text{area } okl \\ &= \text{hatched portion of Figure 38(a)} \end{aligned}$$

The shaded portion of Figure 38(a) represents the mechanical work done when the motion of the moving part is under constant current conditions which is also increase of co-energy and we have

$$dW_m = dW'_f \tag{62}$$

If the mechanical force F_m, acts on the moving part to cause a displacement of dx, we have

$$\begin{aligned} F_m\,dx &= \text{elementary work done} \\ &= dW_m \\ &= dW'_f \end{aligned}$$

i.e.,

$$F_m = \left[\frac{\delta W'_f(i,x)}{\delta x}\right]_{i=\text{constant}} \tag{63}$$

The force represented by Equation (63) acts on the moving part in such a direction so that it increases the stored energy of the system at constant current.

Case 2: Motion of the Moving Part Instantaneously

Figure 38(b) shows that the flux linkage with the exciting winding remains constant during the instantaneous motion of the moving part. The mechanical work done is represented by the hatched area *okm* and the field energy decreases. We have

$$F_m dx = dW_m = -dW_f$$

i.e.,

$$F_m = -\left[\frac{\delta W_f(\lambda,x)}{\delta x}\right]_{\lambda=\text{constant}} \tag{64}$$

The negative sign in Equation (64) is due to decrease of the stored energy. During the instantaneous motion of the moving part, electrical energy input (dW_e) becomes zero due to the constant flux linkages. The requirement of mechanical energy output is met from the stored energy of the field, which causes the decrease of field energy. For small displacement, the area *klm* can be neglected.

ESTIMATION OF MAGNETIC FORCE IN LINEAR SYSTEMS

The reluctance of the magnetic path of the electromechanical system shown in Figure 35 is negligible compared to the air gap and hence the λ–*i* characteristic of the system becomes linear, causing the inductance to depend on the air gap.

$$L(x) = \text{inductance of the system} = \frac{\lambda}{i} \tag{65}$$

$$\text{Field energy} = W_f = \int i dx \tag{66}$$

i.e.,

$$W_f = \int \frac{\lambda}{L(x)} d\lambda = \frac{\lambda^2}{2L(x)} = \frac{1}{2} L(x) i^2 \tag{67}$$

From Equation (64), the magnetic force (F_m) is expressed as

$$F_m = -\frac{\delta W_f}{\delta x} = -\frac{\delta}{\delta x}\left[\frac{1}{2} L(x) i^2\right]_{\lambda=\text{constant}} = -\frac{i^2}{2}\frac{dL(x)}{dx} \tag{68}$$

For a linear system, energy is equal to co-energy and $W_f = W'_f = \frac{1}{2} L(x) i^2$.
From Equation (63), we have

$$F_{m2} = \frac{\delta}{\delta x}\left[W'_f(i,x)\right]_{i=\text{constant}} = \frac{\delta}{\delta x}\left[\frac{1}{2} L(x) i^2\right] = \frac{1}{2} i^2 \frac{dL(x)}{dx} \tag{69}$$

Equations (68) and (69) show that the force obtained is the same. The entire mmf drop will be in the air path if the reluctance of the electromechanical system shown in Figure 35 is neglected. We have

$$Ni = (H_g l_g) \times 2 = 2H_g l_g \tag{70}$$

The energy in the field is

$$W_f = \text{Energy density} \times \text{Volume of the air gap}$$

$$= \frac{B_g^2}{2\mu_0} \times A l_g \times 2 = \frac{B_g^2}{2\mu_0}(2Al_g) \tag{71}$$

From Equation (63), the magnetic force becomes

$$F_m = \frac{\delta}{\delta l_g}\left[\frac{B_g^2}{2\mu_0}(2Al_g)\right] = \frac{AB_g^2}{\mu_0} \tag{72}$$

$$\text{The force per unit area} = \frac{F_m}{A} = \frac{B_g^2}{\mu_o} \text{ N/m}^2$$

DOUBLY EXCITED SYSTEMS

Figure 39 shows a rotatory electromagnetic system having the moving part known as the rotor. The fixed part of this system is known as the stator. Since the stator and rotor carry separate windings, this system is known as doubly excited system. The rotor is usually mounted on the shaft, which moves freely between the poles of the stator.

The rotor current I_r is fed through slip rings. I_s is known as stator current. Let e_s and e_r be the induced emfs in the stator and rotor windings. The energy stored in the magnetic field of the system for no mechanical output is expressed as

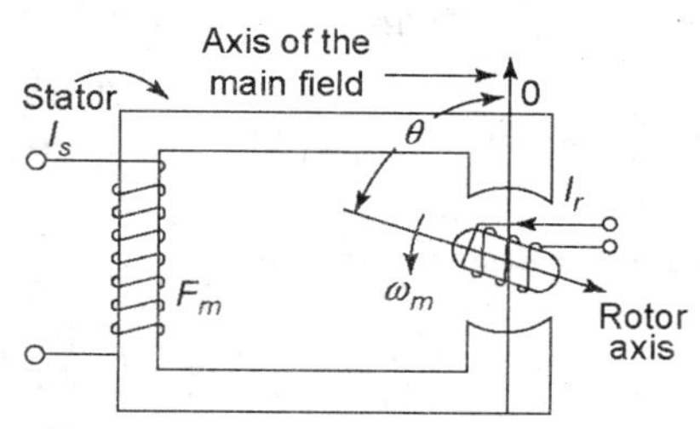

Figure 39 Rotatory Electromagnetic System Having the Moving Part

$$dW_e = dW_f = e_s i_s\, dt + e_r i_r\, dt \tag{73}$$

$$= i_s d\lambda_s + i_r d\lambda_r \tag{74}$$

Equations (73) and (74) can be expressed in terms of inductances for linear systems. The inductances also depend on the position of angle of rotor (θ). We have the following expressions:

$$\lambda_s = L_{ss} i_s + L_{sr} i_r \tag{75}$$

and

$$\lambda_r = L_{rs} i_s + L_{rr} i_r \tag{76}$$

where L_{ss} is the self-inductance of the stator winding, L_{rr} is the self-inductance of the rotor winding and L_{sr} or L_{rs} is the mutual inductance between the stator and rotor windings. For linear systems, $L_{sr} = L_{rs}$, Equations (75) and (76) can be expressed as

$$\begin{bmatrix} \lambda_s \\ \lambda_r \end{bmatrix} = \begin{bmatrix} L_{ss} & L_{sr} \\ L_{rs} & L_{rr} \end{bmatrix} \begin{bmatrix} i_s \\ i_r \end{bmatrix} \tag{77}$$

Substituting λ_s and λ_r in Equation (74), we have

$$\begin{aligned} dW_f &= i_s d(L_{ss} i_s + L_{sr} i_r) + i_r d(L_{sr} i_s + L_{rr} i_r) \\ &= L_{ss} i_s di_s + L_{sr} i_s di_r + L_{sr} i_r di_s + L_{rr} i_r di_r \\ &= L_{ss} i_s di_s + L_{rr} i_r di_r + L_{sr} d(i_s i_r) \end{aligned} \tag{78}$$

The total field energy is expressed as

$$W_f = L_{ss} \int_0^{i_s} i_s di_s + L_{rr} \int_0^{i_r} i_r di_r + L_{sr} \int_0^{i_s i_r} d(i_s i_r)$$

i.e.,

$$W_f = \frac{1}{2} L_{ss} i_s^2 + \frac{1}{2} L_{rr} i_r^2 + L_{sr} i_s i_r \tag{79}$$

The torque developed in the rotational system can be expressed as

$$T = \left[\frac{dW'_f(i,\theta)}{\delta\theta}\right]_{i=\text{constant}}$$

In a linear system, since $W_f = W'_f$,

$$T = \frac{1}{2} i_s^2 \frac{\delta(L_{ss})}{\delta\theta} + \frac{1}{2} i_r^2 \frac{\delta(L_{rr})}{\delta\theta} + i_s i_r \frac{\delta L_{sr}}{\delta\theta} \tag{80}$$

The first two terms in the right-hand side of Equation (80) are the torque due to the variation of self-inductance of the stator and rotor and are known as reluctance torque, whereas the third term is due to variation of mutual inductance between the stator and rotor windings. Please see Appendix B for Reluctance motor.

CYLINDRICAL ROTATING MACHINE

A cylindrical two-pole rotating machine consisting mainly of the rotor and stator is shown in Figure 40. They are provided with the distributed winding and assumed to be one slot/pole/phase for the sake of simplicity. If the effect of slots is neglected, the reluctance of magnetic path between the stator and rotor will be invariant. Let the inductance of the stator and the rotor windings be L_{ss} and L_{rr}, respectively.

Since L_{ss} and L_{rr} remain constant if the effects of the slots are neglected, no reluctance torque is produced. But due to variation in mutual inductance (L_{rs}) between stator and rotor windings, the torque is given by

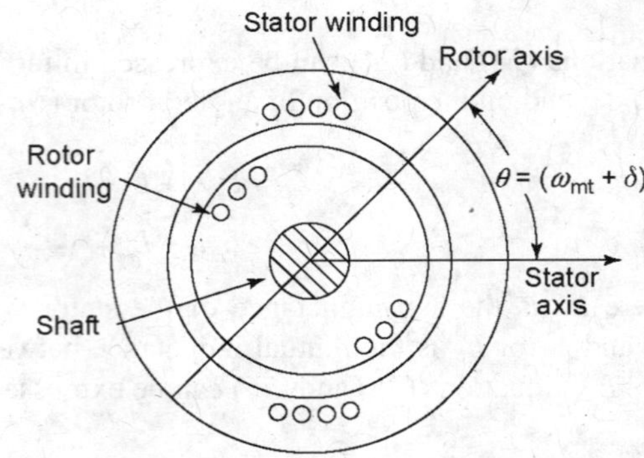

Figure 40 Cylindrical Rotating Machine

$$T = i_s i_r \frac{d(L_{sr})}{d\theta} \tag{81}$$

Also, $L_{sr} = M\cos\theta$, where $\theta = \omega_m t + \delta$, M is the maximum value of mutual inductance between the stator and the rotor, ω_m is the angular velocity of the rotor, δ is the initial position of the rotor axis with respect to the stator axis at $t = 0$ and θ is the angle between magnetic axes of the stator and the rotor.

Let the instantaneous current in stator and rotor windings be

$$i_s = I_{sm}\cos\omega_s t \tag{82}$$

and

$$i_r = I_{rm}\cos(\omega_s t + \alpha) \tag{83}$$

where $\omega_s = 2\pi f_s$ and $\omega_r = 2\pi f_r$.

Here, f_s and f_r are the supply frequency corresponding to the stator and rotor, and α is the phase difference between i_s and i_r. Substituting for i_s and i_r in Equation (81), we have

$$
\begin{aligned}
T_t &= -I_{sm}\cos(\omega_s t)MI_{rm}\cos(\omega_r t+\alpha)\sin(\omega_m t+\delta) \\
&= -\frac{I_{sm}I_{rm}M\cos\omega_s t}{2}[\sin\{(\omega_m+\omega_r)t+\alpha+\delta\}+\sin\{(\omega_m-\omega_r)t+\delta-\alpha\}] \\
&= -\frac{I_{sm}I_{rm}M}{4}[\sin\{(\omega_m+\omega_r+\omega_s)t+(\alpha+\delta)\}+\sin\{(\omega_m-\omega_r-\omega_s)t+(\alpha+\delta)\} \\
&\quad + \sin\{(\omega_m+\omega_r-\omega_s)t+(\delta-\alpha)\}+\sin\{(\omega_m-\omega_r-\omega_s)t+(\delta-\alpha)\}
\end{aligned} \tag{84}
$$

Equation (84) shows that instantaneous torque T_t, which varies sinusoidally with time. The average torque will be non-zero when the coefficient of t is 0.
Therefore,

$$
\omega_m = \pm(\omega_s \pm \omega_r) \tag{85}
$$

The above condition shows that the torque is developed in the machine when it runs at a speed equal to the sum or difference of synchronous speeds corresponding to the supply frequencies of rotor and stator currents.

Case 1: Synchronous Motor/Machine

In this stator, supply is three-phase balanced currents and rotor supply is DC current. Therefore, $\omega_r = 0$, $\alpha = 0$, $\omega_m = \omega_s$ and $I_r = I_{dr}$.

The condition $\omega_m = \omega_s$ indicates that the machine is rotating at synchronous speed corresponding to stator supply frequency. Substituting the above values in Equation (84), the expression is reduced as

$$
\begin{aligned}
T_t &= -\frac{2I_{sm}I_{rm}Ma}{4}[\sin(2\omega_s t+\delta)+\sin\delta] \\
&\quad -\frac{I_{sm}I_{rm}M}{2}[\sin(2\omega_s t+\delta)+\sin\delta]
\end{aligned} \tag{86}
$$

This instantaneous torque expression shows that the torque is pulsating in nature with non-zero average value. The average torque is

$$
T_{av} = \frac{-I_{sm}I_{rm}}{2}M\sin\delta \tag{87}
$$

Equation (87) shows that when the rotor runs at a speed ($\omega_m = \omega_s$), the machine develops an average non-zero torque. The condition derived from Equation (84) also confirms that the synchronous motor is not self-starting when $\omega_m = 0$. If the rotor is wound with single winding, it is called a single-phase synchronous motor. But due to pulsating field instead of rotating field, even though single-phase machine runs at synchronous speed, it produces noise and vibrations. This is eliminated by having three-phase winding on the stator.

Case 2: $\omega_m = \omega_s - \omega_r$

When the stator and the rotor of the machine shown in Figure 40 carry AC currents at different frequencies, the motor runs at asynchronous speed $\omega_m = \omega_s - \omega_r$. The instantaneous torque developed is obtained by substituting $\omega_m = \omega_s - \omega_r$ in Equation (84) as

$$T_t = -\frac{I'_{sm} I_{rm} M}{4}[\sin(2\omega_s + \alpha + \delta) + \sin(-2\omega_r t - \alpha + \delta) + \sin(2\omega_s t - 2\omega_r t - \alpha + \delta) + \sin(\alpha + \delta)] \quad (88)$$

The average value of this pulsating torque is expressed as

$$T_{av} = \frac{-I_{sm} I_{rm} M}{4}\sin(\alpha + \delta) \quad (89)$$

This is the torque developed in the inductor motor. In the three-phase induction motor, both stator and rotor currents are at different frequencies but the machine runs at synchronous speed. It has the self-starting torque.

When $\omega_m = 0$, the average torque given in Equation (89) becomes zero, which is the same as that of the single-phase induction motor. To remove the nature of the pulsating torque, as given in Equation (88), the stator should carry three-phase winding as in the polyphase induction motor. Therefore, it can be concluded that the torque can be produced in rotating machines under two conditions: by variation in the reluctance of the magnetic path by rotating the rotor and by variation in mutual inductance between the stator and the rotor. But the torque that is produced is very small as in the case of the reluctance motor.

Example 1 For an electromagnetic system, the λ–i relation is given by (g = air gap length). Determine the mechanical force on the moving part of the system $\lambda = \frac{i^{1/2}}{g}$ using

(i) energy of the system and

(ii) co-energy of the system.

Use i = 4 A and g = 5 cm.

Solution

i = 4 A, g = 5 cm and $\lambda = \frac{i^{1/2}}{g}$; i.e., $i = (\lambda g)^2$

(a) Mechanical force by using energy of the system

$$W_f = \text{energy of the system} = \int_0^\lambda i\,d\lambda = \int_0^\lambda (\lambda g)^2 d\lambda = \frac{g^2\lambda^3}{3}$$

$$\text{Force on the system} = -\frac{\delta(W_f)}{dg} = -\frac{\lambda^3}{3}(2g)$$

$$= -\frac{i^{3/2}}{g^3}\times\frac{(2g)}{3} = -\frac{4^{3/2}}{g^2}\times\frac{2}{3} = \frac{4^{3/2}}{25\times10^{-4}}\times\frac{2}{3} = -2133.3\text{ N/m}$$

(b) Co-energy of the system

$$W'_f = \text{co-energy of the system} = \int \lambda di = \int \frac{i^{1/2}}{g} di = \frac{i^{3/2}}{\frac{3g}{2}} = \frac{2}{3} \times \frac{i^{3/2}}{g}$$

$$\text{Force on the system} = +\left[\frac{\delta W'_f}{\delta g}\right] = -\frac{2}{3} \times \frac{i^{3/2}}{g^2} = -\frac{2}{3} \times \frac{4^{3/2}}{25 \times 10^{-4}} = -2133.3\,\text{N/m}$$

The negative sign in both the cases indicates that force acts in such a direction so as to decrease the air gap length.

Example 2 The electromagnet shown in Figure E1 is wound with a coil which has 200 turns and a resistance of 2Ω. The air gap length is each of 5 mm and the magnetic core has a square cross-section of 2 cm × 2 cm. An average force of 300 N is required to lift the sheet of steel. For DC supply (1) source voltage and (2) energy stored in the magnetic field for 50 Hz AC supply. Determine the AC source voltage. Assume that the reluctance of core is negligible.

Solution

Given: $F = 300$ N, $r = 2\ \Omega$, $g = 0.5$ mm, $A = 2 \times 2$ cm² and $N = 200$.

$$\text{Energy in the air gap} = \frac{Bg^2}{2\mu_0} \times 2A = 300\,\text{N}$$

$$\text{i.e., } B_g^2 = \frac{300 \times \mu_0}{A} = \frac{300 \times 4\pi \times 10^{-7}}{4 \times 10^{-7}} = 942 \times 10^{-3} = 0.942$$

i.e., $B_g = 0.9705$ T

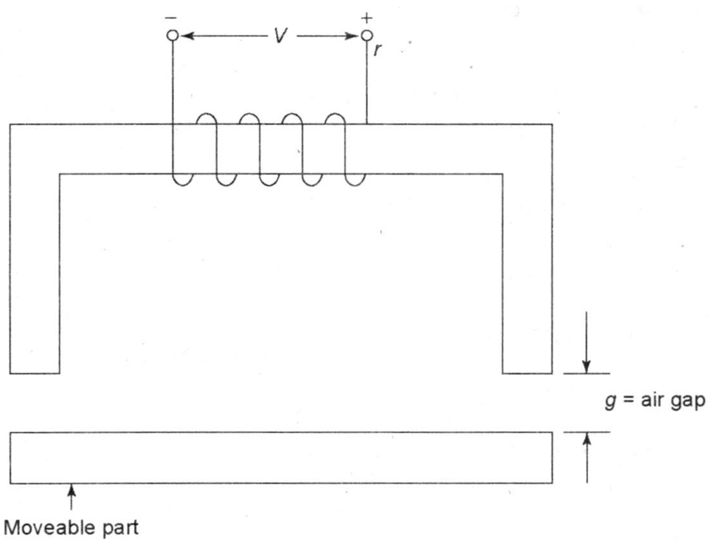

Figure E1

The mmf of the exciting coil $= H_c l_c + H_g l_g$

$$NI = 2 \times (H_g l_g) \text{ since the core has zero reluctance}$$

$$= \frac{B_g}{\mu_0} \times 2 \times 10^{-3} = \frac{0.9705}{\pi \times 0^{-7}} \times 2 \times 10^{-3} = 1545.3\,\text{AT}$$

i.e., $$I = \frac{1545.3}{N} = \frac{1545.3}{200} = 7.72\,\text{A}$$

DC voltage source required = 7.72 × 2 = 15.44 V

$$\text{Energy stored in the air gap} = \frac{B_g^2}{2\mu_0} \times \text{volume of the air gap}$$

$$= \frac{(0.9705)^2}{2 \times 4\pi \times 10^{-7}} \times 2 \times 2.5 \times 2.5 \times 1 \times 10^{-7} = 0.4686\,\text{J}$$

(b) AC excitation:

Impedance of the coil $= R + j\omega L$

$$\text{Inductance of the coil} = \frac{\mu_0 N^2 A}{l} = \frac{4\pi \times 10^{-7} \times 200^2 \times 2 \times 2 \times 10^{-4}}{2 \times 0.5 \times 10^{-3}} = 0.02\,\text{H}$$

i.e., $X_L = 2\pi f L = 2 \times 3.14 \times 0.02 = 6.28\,\Omega$

∴ $Z = 2 + j\,6.28$

i.e., $|Z| = \sqrt{2^2 + 6.28^2} = 6.590\,\Omega$

The rms value of the current in the coil = 7.72 A

AC voltage required = 7.72 × 6.59 = 50.87 V

Example 3 A magnetic actuator is shown in Figure E2. The cross-sectional area of the plunger = 3 cm × 3 cm. The coil has 1000 turns and a resistance of 5 Ω. A voltage of 12 V DC is applied to the coil terminals.

(i) Assuming flux density in the air gap is 1.5 T. Determine the air gap *g* (in mm) and stored energy in the air gap.

(ii) Obtain an expression for the force on the plunger as a function of air gap length. Assume that the reluctance of magnetic material of the core is negligible.

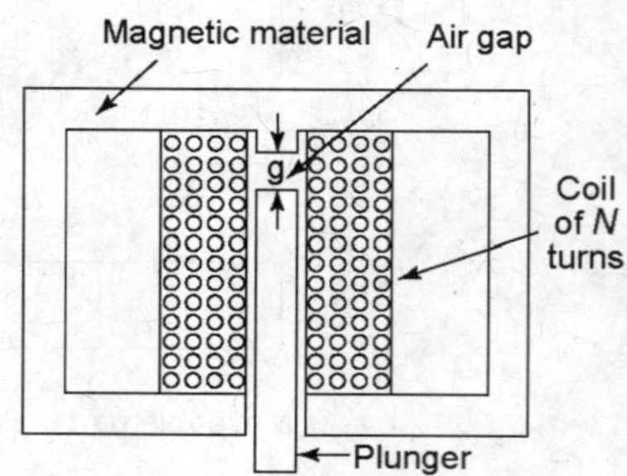

Figure E2

Solution

Let x be the air gap at any time t where the plunger moves upwards in the direction of mmf and i be the current in the coil.

$$H_g = \text{mmf in the air gap} = \frac{Ni}{x}$$

$$B_g = \mu_0 H_g = \frac{\mu_0 N i}{x}\,\text{Wb/m}^2$$

Inductance = flux linkage/current

Let A be the cross-sectional area of the plunger.

ϕ_g = flux in the air gap = $B_g \times A$

Ψ_g = flux linkages with the coil = $NB_g A$

$$L(x) = \text{inductance of the coil} = \frac{\psi_g}{i} = \frac{NB_g A}{i} = \frac{\mu_0 N^2 A}{x}$$

$$\text{Energy stored in magnetic field} = \frac{1}{2}L(x)i^2 = \frac{B_g^2}{2\mu_0} \times \text{volume of air gap}$$

$$W_m = \frac{1}{2}i^2\left(\frac{\mu_0 N^2 A}{x}\right)$$

$$\text{Force acting on the plunger} = F_m = \frac{\delta W_f}{\delta x} = -\frac{1}{2}\times i^2 \times \frac{\mu_0 N^2 A}{x^2}$$

$$= -\frac{1}{2}A\left(\frac{iN}{x}\right)^2 \mu_0$$

Mechanical work done by the plunger during motion:

$$W_m = \int_{x=g_1}^{g_2} F_m dx = -\frac{1}{2}(A)(iN)^2 \mu_0 \int_{x=g_1}^{g_2} \frac{dx}{x^2} = \frac{\mu_0 A(iN)}{2}\left[\frac{1}{g_1} - \frac{1}{g_2}\right] \text{ J}$$

Given: $B_g = 1.5\ T$, $N = 1000$, $A = 9$ cm², $V_t = 1.2\ V$ and $R = 5\ \Omega$.

$$\text{Current through the coil} = \text{voltage/resistance} = \frac{V_t}{R} = 12/5 = 2.4 \text{ A}$$

$$B_g = \frac{\mu_0 Ni}{x}$$

i.e.,
$$x = \frac{\mu_0 Ni}{B_g} = \frac{4\pi \times\ 0^{-7} \times 1000 \times 2.4}{1.5} = 2.009\,\text{mm}$$

$$\text{Stored energy in the air gap} = \frac{B_g^2}{2\mu_0} \times \text{volume of the gap}$$

$$= \frac{(1.5)^2}{2 \times 4\pi \times 10^{-7}} \times 9 \times 10^{-4} \times 0.2 \times 10^{-2} = 1.612\,\text{J}$$

$$L(x) = \text{inductance of the coil} = \frac{\mu_0 N^2 A}{x} = \frac{4\pi \times 10^{-7} \times 1500^2 \times 9 \times 10^{-4}}{0.2 \times 10^{-2}} = 1.2717 \text{ H}$$

$$\text{Force acting on the plunger} = \frac{1}{2}A\left(\frac{iN}{x}\right)^2 = \frac{9 \times 10^{-4}}{2} \times \left(\frac{2.4 \times 1000}{0.2 \times 10^{-2}}\right)^2 = 6.48 \times 10^8 \text{ N/m}^2$$

1 Transformers

Nowadays, electrical power is generated for industrial and utility purposes by thermal power stations and large hydroelectric plants in the three-phase form at a frequency of 50 Hz (in India). The generated frequency in some other countries is 60 Hz. The generated voltage at the generating station is 6.6 kV, 11 kV or higher. For transmission purposes, it is required to step it up to a voltage of 132 kV or higher. Again, in urban and rural areas it is required to step it down to 3.3 kV and 6.6 kV, respectively, and 11 kV at the substation. For domestic purposes, it is required to step it down to 400 V or 230 V. Three-phase transformers are used to step up the generated voltage before transmission of electrical power and also to step down the high voltage before distribution, that is, at the substation. Before the study of three-phase transformers, knowledge of single-phase transformers is essential. The aim of this chapter is to discuss single-phase transformers only.

1.1 DEFINITION

A transformer is a static or stationary electromagnetic device, consisting of two coils, by means of which electrical power in one circuit is transformed into electrical power of the same frequency in another circuit.

1.2 BASIC PRINCIPLE

Figure 1.1 shows a basic single-phase transformer having two windings wound on a common magnetic core. From the principle of mutual induction, when two coils are inductively coupled and the current

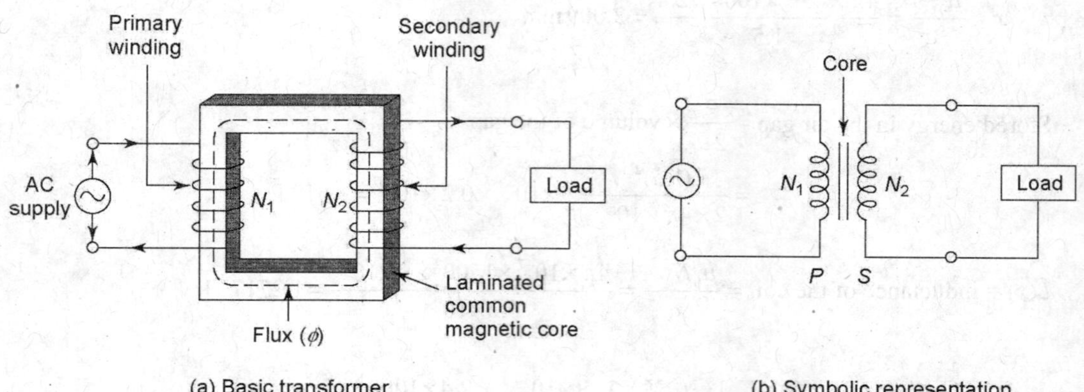

Figure 1.1 Transformer

in one coil is changed uniformly, an emf (electromagnetic force) is induced in the other coil. If a closed path is provided at the secondary circuit, this induced emf at the secondary drives a current. As shown in Figure 1.1, the transformer has two coils, which are electrically separated and magnetically linked through a common magnetic path. The basic principle of the transformer is the same as the principle of mutual induction. The coils of the transformer have high mutual inductance.

In brief, we can say the following:

- The transformer is a static device.
- It transfers electrical power from one circuit to another.
- During transfer of power, there is no change of frequency.
- It uses electromagnetic induction to transfer electrical power.
- The two electrical circuits are in mutual inductive influence of each other.

1.3 TYPES OF TRANSFORMERS

A transformer may be step up or step down. If the output voltage is greater than the input voltage, the transformer is said to be a *step-up transformer*. A transformer is said to be a *step-down transformer* if the output voltage is less than the input voltage. This classification is carried out on the basis of the ratio of input and output voltages. From application point of view, the following transformers are most important:

- *Power and distribution transformer:* These transformers are used for transmission and distribution of power.
- *Autotransformer:* These transformers are used to change the voltage within relatively small limits and are used for starting AC motors, and so on.
- *Transformers for feed installations with static converters:* These are used for converting AC to DC and also DC to AC. The first one is used for rectification purposes and the second one for inversion purposes.
- *Testing transformers:* These are used to conduct tests at high and ultra-high voltages.
- *Power transformers for special applications:* These are used in furnaces, welding and so on.
- *Radio transformers:* These are used in radio engineering and similar purposes.

From frequency range point of view, transformers can be divided as (50–400 Hz) audio transformer, wide band and narrow band transformers and pulse transformers. Transformers can also be divided depending on the number of windings such as two winding (conventional) and single winding known as autotransformer.

1.4 CONSTRUCTION OF SINGLE-PHASE TRANSFORMER

Magnetic core and windings (or coils) are the two basic parts of any transformer. The core is made of silicon or sheet steel with 4 per cent silicon and laminated to reduce eddy current loss. It may be in either square or rectangular shape. It has two parts. The vertical portion on which the coil is wound is called the *limb of the core*, whereas the top and bottom horizontal portions are called the *yoke*. Figure 1.2(a) shows the limb and yoke of the core. The permeability of the material used for core must have high value ($\mu_r > 1{,}000$) to reduce reluctance of the magnetic path.

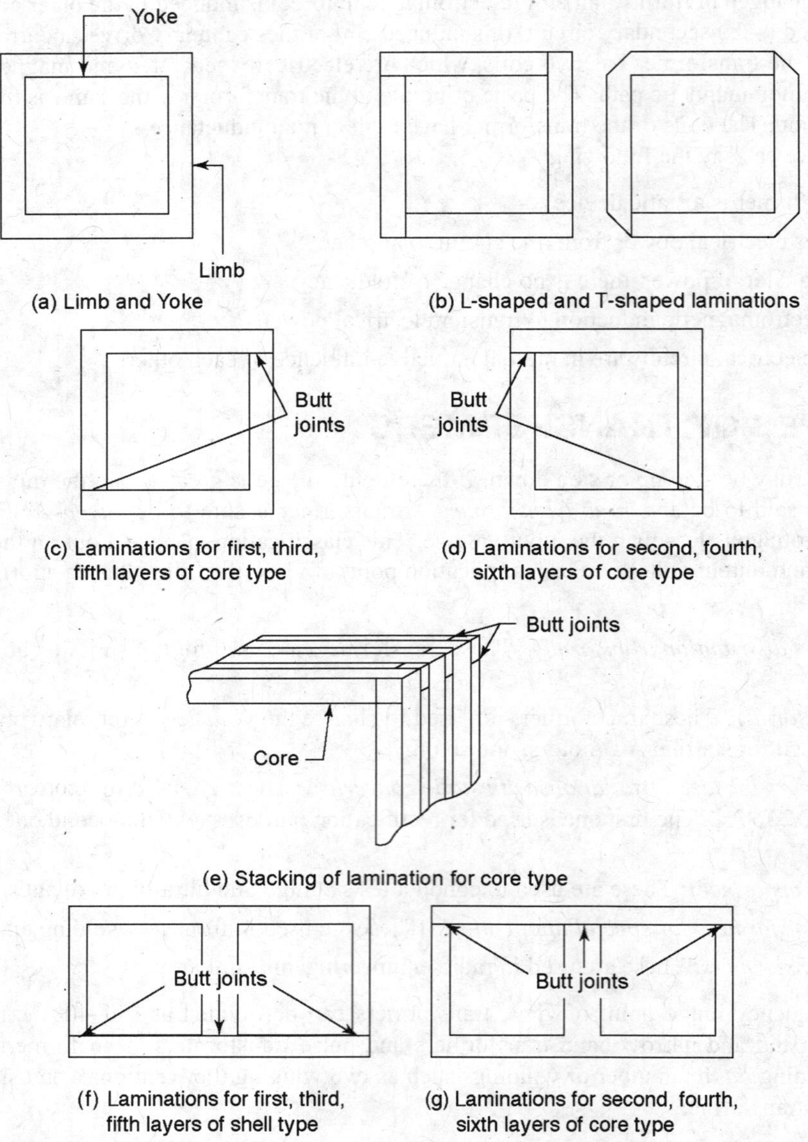

Figure 1.2 Amalgamation of Laminations

The laminations are insulated from each other by a light coat of core plate varnish or by an oxide layer on the surface. The thickness of lamination is 0.35 mm for a frequency of 50 Hz and 0.5 mm for a frequency of 25 Hz. Figure 1.2 (b) shows that the joints of the laminations are staggered to avoid the pressure of narrow gaps right through the cross section of the core. Such staggered joints are said to be *imbricated*. The cross section of the limb depends on the size of the coil. The thickness of the laminations should be small to reduce the eddy current loss. There is a practical limit of the thickness of lamination, which is 0.3 mm. Any further reducing the value of thickness of laminations leads to make them mechanically weak. The range of acceptable thickness of laminations is 0.33–0.50 mm. The laminations are made by grade steel containing 3–5 per cent of silicon, which increases the resistivity of the core leading to reduction of eddy current core loss. This high-content silicon steel is a soft iron material, which has narrow hysteresis loop and a high permeability. This causes reduction of hysteresis loss and magnetizing current. The steel used for transformer core are either hot rolled or cold rolled. The maximum value of flux density permitted to hot-rolled steel is 1.45 Wb/mm^2. The maximum value of flux density permitted to cold-rolled steel is 1.8 Wb/m^2, and it is 25–35 per cent more expensive than the hot-rolled steel. The cold-rolled grain-oriented (CRGO) steel sheet with a silicon content of 3 per cent is also used for magnetic circuits of transformer. In a magnetic circuit, we know that flux = mmf/reluctance = primary ampere turns/reluctance, which suggests that the requirement of primary ampere-turns are less for a given value of flux when the reluctance is low. To avoid the joints to come in same line, the laminations of the core are arranged in the form of stack. To achieve this, the laminations shown in Figures 1.2(c) and 1.2(d) are put in alternate layers of Figure 1.2(e) for core-type laminations. Figures 1.2(f) and 1.2(g) shows the laminations for obtaining the shell-type transformers.

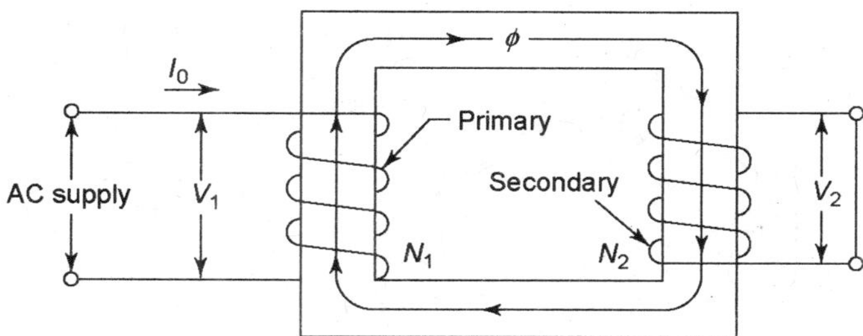

Figure 1.3 Two-winding Transformer

The two windings are wound on two limbs, that is, one on the primary and the other on the secondary, as shown in Figure 1.3. Leakage flux increases in this connection. This leakage flux has a negative effect on the transformer performance. To get high mutual inductance, the two coils must be close to each other. To obtain this, the two windings are divided into a number of coils and wound adjacent to each other on the same limb, as shown in Figure 1.4.

In Figure 1.4, the common arrangement is of cylindrical coils, which are used in the core-type transformers. These coils are mechanically strong and are wound in the helical layers where different layers are insulated from each other by paper, cloth or mica. The low-voltage winding is generally placed heat to the core because it is easier to insulate the low-voltage winding. The high-voltage winding is placed after it. Due to laminations and insulation, the net effective core area is reduced and this reduction is generally in the order of 10 per cent. The other type of winding is generally used for the shell-type transformer known as sandwich coils (shown in Figure 1.5). Since each high-voltage portion lies between two low-voltage portions, the two low-voltage portions sandwich the high-voltage portion. The leakage flux is reduced due to subdivision of windings into small portions. The top and bottom portions are low-voltage coils and all portions are insulated from each other by paper.

Transformers are classified into the following three categories based on the relative arrangement or disposition of the core and the winding: (i) Core type, (ii) shell type and (iii) spiral core type.

1.4.1 Core Type

Figure 1.6(a) shows the complete magnetic circuit of a core-type transformer in the shape of a hollow rectangle having two limbs. It has a single magnetic circuit. In Figure 1.6, I_0 is the no-load current and ϕ is the flux produced by it. Number of turns of the primary and secondary are N_1 and N_2, respectively.

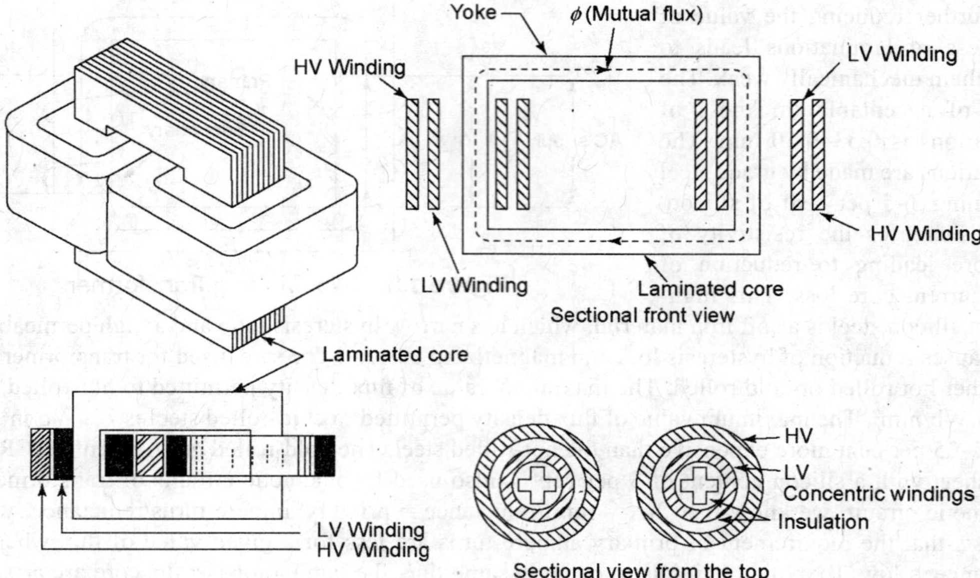

Figure 1.4 Schematic and Sectional View of a Core-type Transformer

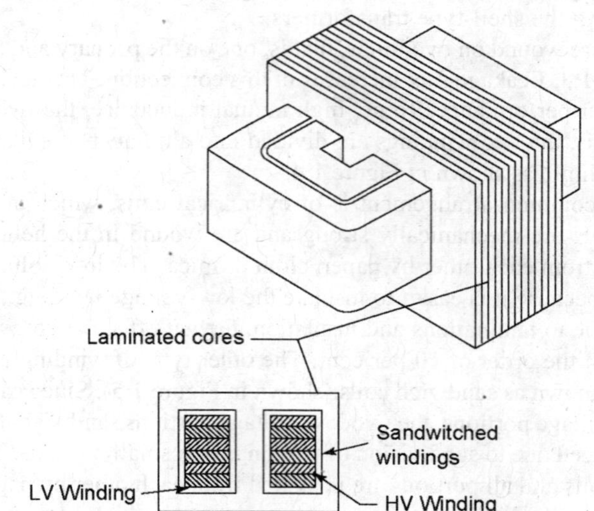

Figure 1.5 Schematic View of a Shell-type Transformer

The windings surround the core. The coils used are wound and are of cylindrical type having the general form circular, oval or rectangular.

Core-type transformer has a longer mean length of core and a shorter mean length of coil turn. Core has a small cross section of iron; more number of turns is required because the high flux may not reach the core. Core type is used for high-voltage service, since it has sufficient room for insulation.

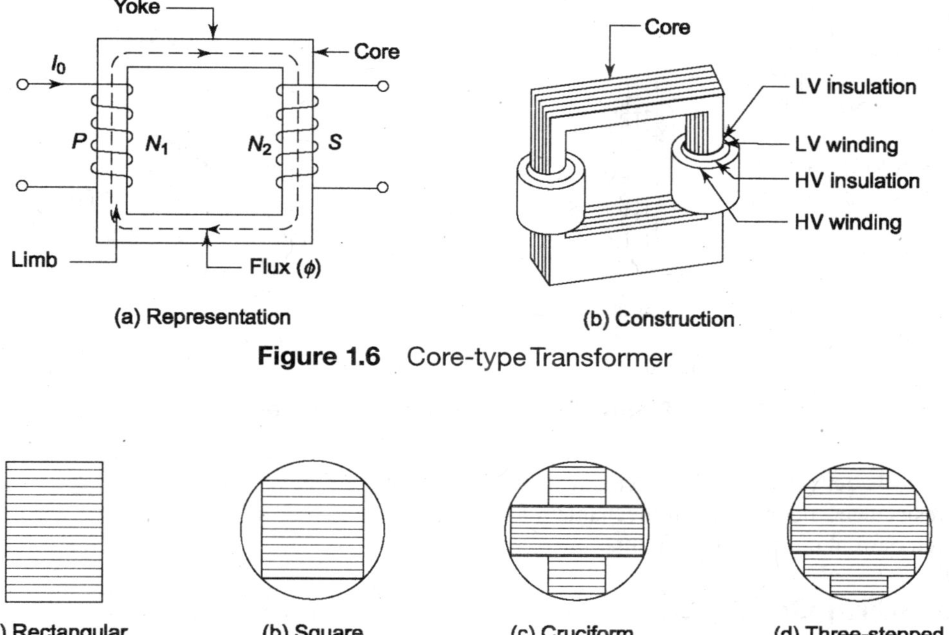

Figure 1.6 Core-type Transformer

Figure 1.7 Different Cross Sections of Core-type Transformer

Figure 1.6(b) shows the actual view of a core-type transformer. The different cross sections used in core-type transformer are shown in Figure 1.7.

In small core-type transformers, rectangular cores with rectangular cylindrical coils are used as shown in Figure 1.7(a), whereas circular cylindrical coils are used for large transformers; hence, square cores are preferred as shown in Figure 1.7(b). If rectangular cores are used for large transformers, it becomes wasteful. Figure 1.7(c) shows the cruciform core, which is an improvement of square core. Figure 1.7(d) shows further core stepping (three-stepped cores) for large transformers resulting in reduced length of mean turn and copper (Cu) loss.

1.4.2 Shell Type

Shell-type transformer has double magnetic circuit and three limbs. Both windings are placed on the central limb. The coils occupy the entire space of windows. The coils are usually multi-layer disc type or sandwich coils. The low-voltage coils are placed nearest to the iron core to reduce the amount of high-voltage insulation. Core is laminated. Special care is taken to arrange the laminations of the core. All the points at alternate layers are staggered properly to avoid narrow air gap at the joint, right through the cross section of the core. The joints are known as *overlapped or imbricated* joints. The shell-type construction is preferred for a few high-voltage transformers. Since the windings are surrounded by core, natural cooling does not exist. To remove any winding during maintenance, removal of a large number of laminations is required. Figure 1.8 shows a shell-type transformer.

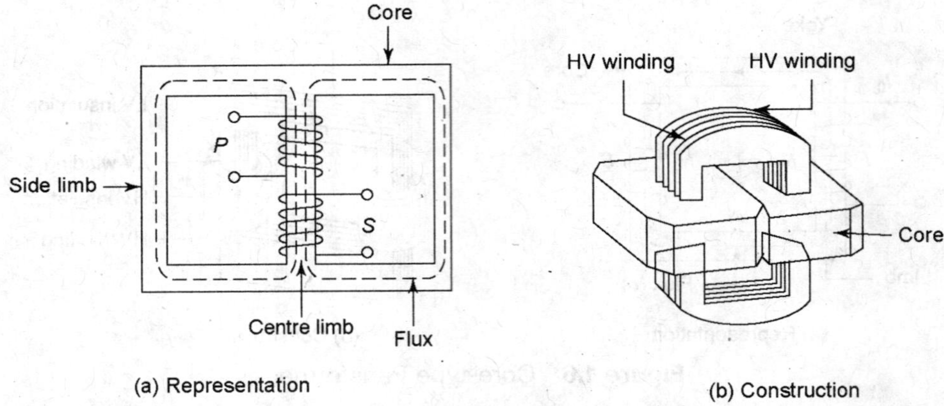

Figure 1.8 Shell-type Transformer

Due to better provision for mechanical support and bracing of coils in the shell-type transformer, better resistance to combat high mechanical force is obtained. High mechanical forces are developed for a high current during short circuit.

1.4.3 Spiral Core Type

Figure 1.9 shows spiral core-type transformer where the core is assembled either of a continuous strip of the transformer steel wound in the form of a circular or elliptical cylinder or of a group of short strips assembled to produce the same elliptical shape. In this construction, the core flux always follows along the grain of iron. Cold-rolled steel of high silicon content allows the designer to use higher operating flux densities with lower loss per kilogram. The main advantage of using the higher flux density is that weight/kVA is reduced.

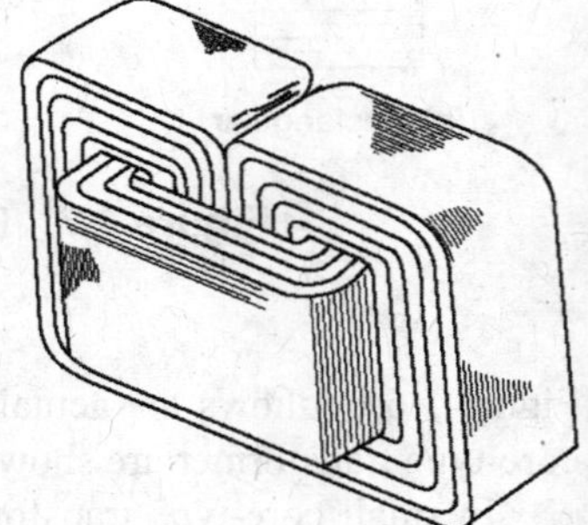

Figure 1.9 Spiral Core-type Transformer

1.5 TRANSFORMER WINDINGS

There are two types of windings usually used in a transformer. The winding receiving electrical energy is known as primary winding, whereas the winding delivering electrical energy is called secondary winding. Windings are usually made of high grade of copper. Standard conductors are used for carrying higher current. To avoid the each turn to come in contact with each other, the windings are provided with insulation. In addition to inter-turn insulator, bare copper wires are provided with enamel coating. Usually single- or double-layer cotton is used. Sometimes press board or cotton insulation is also used to support the windings. Usually additional insulation is provided for line end turns for their protection from lightning and switching over voltages. During transient disturbances, the distribution of voltage is not uniform along the windings and 80 per cent of voltage at that time appears across the first 10 per cent of turns from line end. Heat generation occurs due to energy loss, which is proportional to the volume of the material in which the losses occurs. The heat dissipation is proportional to the surface area of the same material and the tank. The ratio of heat generated to heat dissipated is approximately proportional to the ratio of volume of the material for conductors and the core to the surface area of the material for conductors, the core and the tank, which must approach to unity to limit the temperature rise. This can be achieved by corrugating the surface area of the tank. To get effective cooling, the radiators are used shown in Figure 1.18 later on.

The following are the most important requirements of transformer windings:

- The windings must be economical.
- The heating conditions of the windings should satisfy standard requirements.
- The windings must have good mechanical strength to combat the force that originates due to short circuit.
- The windings must have the necessary electrical strength during over-voltage. The following are the two different types of windings: (i) Concentric windings and (ii) sandwich windings.

1.5.1 Concentric Windings

Figure 1.10 shows concentric windings, which are used for core-type transformers.

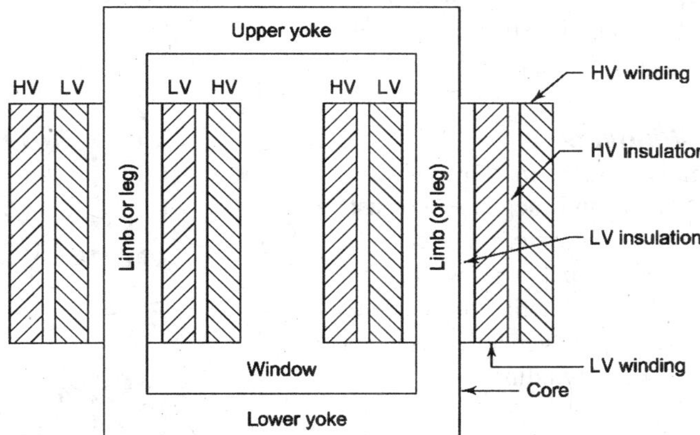

Figure 1.10 Concentric Windings

Concentric windings are classified into four following groups:

(i) Spiral windings.
(ii) Helical windings.
(iii) Cross-over windings.
(iv) Continuous disc windings.

These windings are discussed as follows.

1.5.1.1 Spiral Windings

These coils are suitable for windings to carry high currents, which are generally used or currents greater than 100 A. They are almost used for LV windings. Figure 1.11 shows double-layer spiral coils, which are wound on solid insulating former, and hence are mechanically strong.

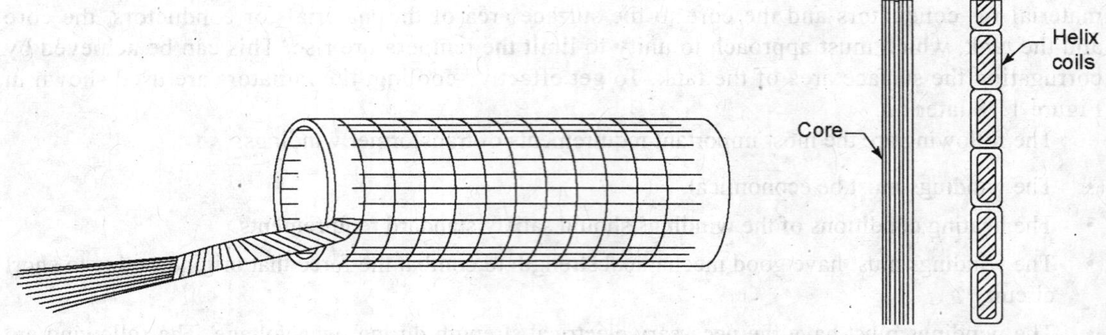

Figure 1.11 Double-layer Spiral Coil

Figure 1.12 Helical Coils

1.5.1.2 Helical Windings

These coils are wound in the form of helix, which are generally used for low voltages 11 kV to 33 kV for large transformers. Figure 1.12 shows the cross-sectional view of helical coils where each conductor consists of a number of rectangular strips wound in parallel radially.

1.5.1.3 Cross-over Windings

These coils are generally wound on formers. Each coil consists of several layers and each layer consists of several turns. Figure 1.13 shows cross-over coils. The conductors may be of round wire with paper or cotton insulation and not suitable for currents exceeding 20 A. These coils are generally used for small transformers and for HV windings.

1.5.1.4 Continuous Disc Windings

Figure 1.14 shows disc coils. These windings consist of a number of disc and each disc consists of number of turns wound radially over one another from inside outwards and outside inwards alternately. Conductor consists of single number of rectangular strips and passes continuously from disc to disc for multiple strip of conductors. The transposition of conductors is done to ensure uniform current distribution. These windings are mechanically strong and hence can withstand stresses during short circuit conditions. These are used for HV windings of large power transformers.

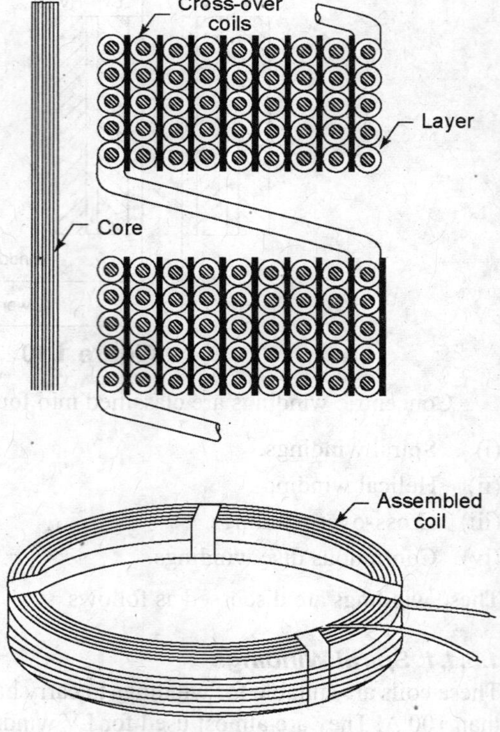

Figure 1.13 Cross-over Coils

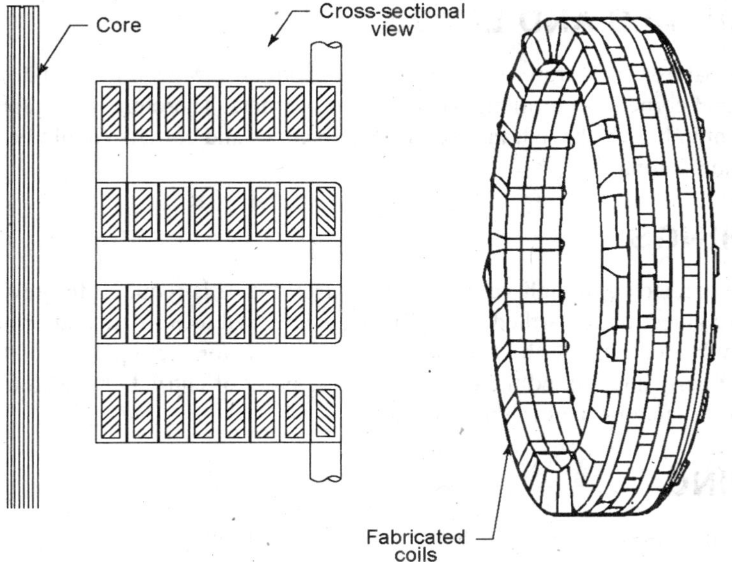

Figure 1.14 Disc Coils

1.5.2 Sandwich Windings

Figure 1.15 shows sandwich windings used in shell-type transformers. The high-voltage and low-voltage windings are split into a number of sections where each high-voltage section lies between two low-voltage sections. In sandwich coils, easily leakage can be controlled. Desired value of leakage reactance can be obtained by proper division of windings.

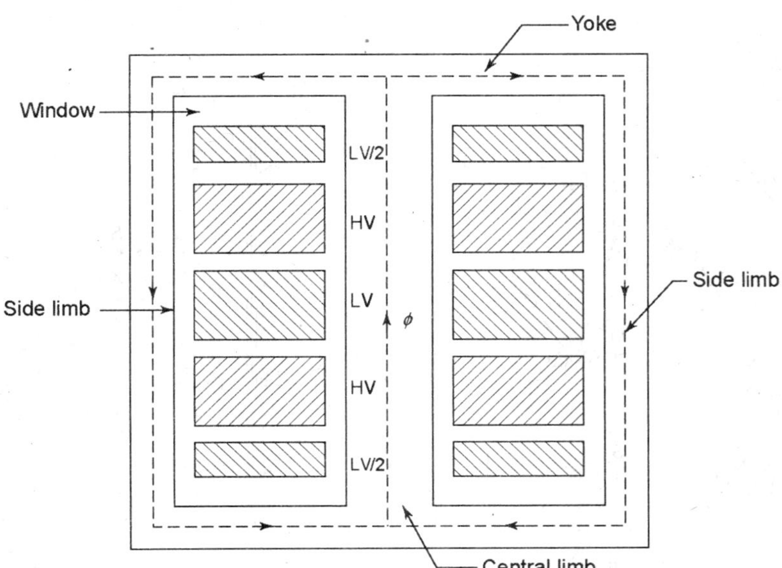

Figure 1.15 Sandwich Windings

1.6 TERMINALS AND LEADS

Insulated copper bars or rods are used for the connections to the windings. High electric stress and corona at bends and corners occur in high-voltage transformers. Therefore, the shape and size of leads are most important in high-voltage transformers. Sharp edges and corners should be avoided to reduce dielectric stress and corona.

1.7 BUSHINGS

Transformers are connected to high-voltage lines. Extreme care is required to prevent flashover from the high-voltage connection to earthed bank. Bushings are used to insulate and bring out terminals of the winding from the container to the external circuit. For transformers upto 33 kV, this is achieved by using bushings of porceling around the conductor at the point of entry. For transformers above 33 kV, either oil-filled or capacitor-type bushings are used.

1.8 TAPPING

By changing the turns ratio of the transformer, we can easily control the voltage supplied to power networks by the transformer. To affect a change in the ratio of transformation, we provide tapping at different places in the windings of the transformer. Therefore, it is possible to get different turns ratio and thus different voltages at different tappings. Figure 1.16 shows the tapping used in a transformer.

For a three-phase, 11,000/400 V distribution transformer, there is always tapping on the high-voltage winding. If 11,000 V supply is given on the high-voltage side, 400 V is obtained on the low-voltage side. This tapping is known as principal tapping. The tapping at which the number of turns included is more than or less than the number of turns included at principal tapping is known as positive and negative tapping, respectively.

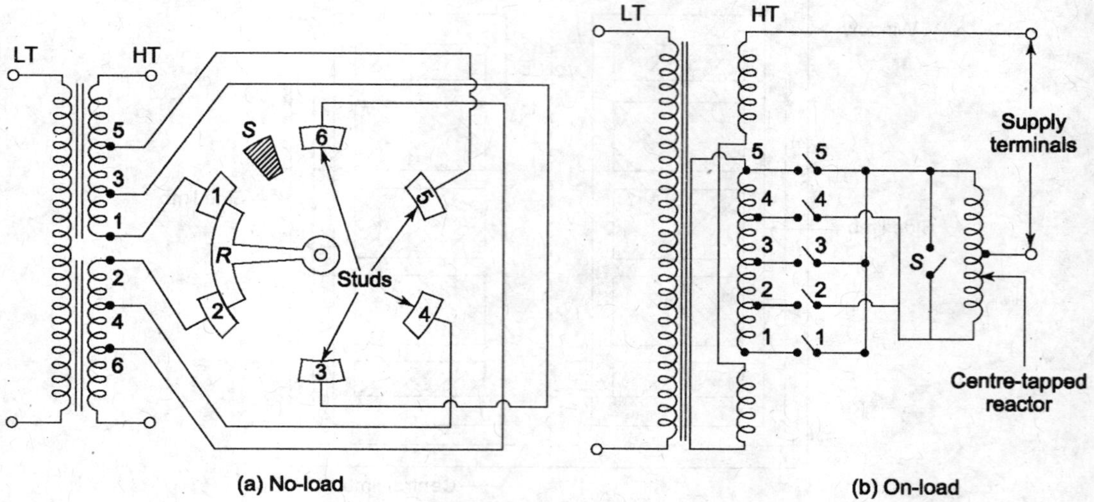

Figure 1.16 Tapping

1.9 COOLING OF TRANSFORMER

Transformer is a static device that converts one voltage level to another voltage level. Due to occurrence of iron and copper losses, the transformer gets heated. In order to avoid deterioration of insulation, dissipation of heat is required to keep the temperature of the winding within a limited value. Since transformer is a static device, its cooling is more difficult than that of a rotating machine. Rotating machine creates a turbulent air flow, which helps in removing the heat generated due to losses. The losses in a transformer are comparatively small. The coolants used in transformer are: (i) air and (ii) oil.

The dry-type transformer uses air as coolant, whereas oil-immersed transformer uses oil as coolant. The heat produced in dry-type transformers is conducted across the core and windings. Finally, the heat dissipates from the outer surfaces of the windings to the surrounding air through convection. The heat generated inside the core and windings of an oil-immersed transformer is conducted across them to their surfaces. This heat produced is transferred by the oil to the walls of the tank through *convection.*

The cooling methods used in transformers up to 25 kVA size and of dry type are discussed below:

- *Natural air:* Ambient air is used as the cooling medium in this method, and the natural circulation of surrounding air is used to carry away the generated heat by natural convection.
- *Air blast:* Natural circulation of air used in cooling becomes insufficient for dissipation of heat from large transformers. Therefore, circulation of air (air blast) is used to keep the temperature rise within prescribed limits. A continuous blast of cool air is forced through the cores and the windings to cool the transformer in this method. External fans are used to produce air blast shown in Figure 1.17(a).

Generally most of the transformers are of oil-immersed types because oil provides better insulation than air due to its better conduction heat. Mineral oil is used for this purpose. The following cooling methods are used for oil-immersed transformers.

- *Oil-immersed self-cooled transformers:* The transformer is immersed in oil. The heat generated in cores and windings is passed to the oil by conduction. Therefore, oil in contact with the heated parts rises and cool oil takes its place. The heat is transferred to the tank walls by natural oils.

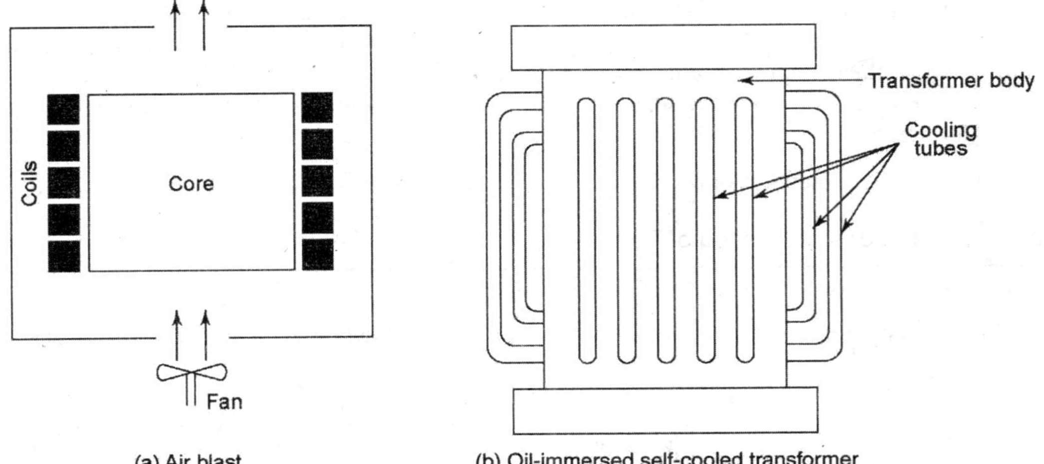

(a) Air blast (b) Oil-immersed self-cooled transformer

Figure 1.17 Air Blast and Oil-immersed Self-cooled Transformer

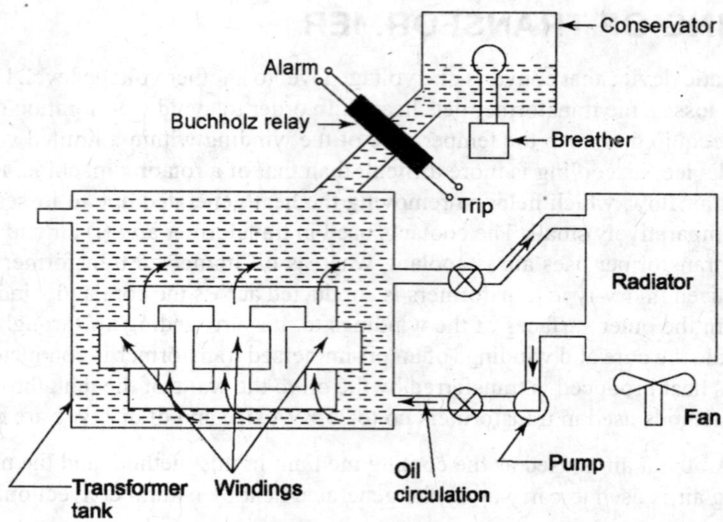

Figure 1.18 Air Blast Cooling for High-capacity Transformers

Finally, ambient air takes this heat. To increase the heat dissipating capacity, corrugations, fins, tubes (shown in Figure 1.18) and radiators are to be provided in Figure 1.17(b). In oil natural cooling, there is no chance to clog the ducts and hence windings are free from the effects of moisture.

- *Oil-immersed forced air-cooled transformers:* In this method, air is directed over the outer surfaces of the tank of the transformer immersed in oil.
- *Oil-immersed water-cooled transformers:* In this type of cooling, water is pumped through a metallic coil immersed in the oil just below the top of the tank to extract heat from the oil. The heated water is cooled in a spray pond or a cooling tower.
- *Oil-immersed forced oil-cooled transformers:* To extract heat from the oil, oil itself is pumped upwards through the winding. Then it is sent back by way of external radiators. These radiators are cooled by fans. The extra cost of pumping equipment should be justified economically. The main advantage is the reduction in temperature difference between the top and the bottom of the enclosing tank. Figure 1.18 shows the schematic arrangement of air blast cooling used for high-capacity transformers.

Table 1.1 summarizes methods of cooling of transformers.

Table 1.1 Methods of Cooling of Transformers

Method	Use
(1) Natural radiations	Used for transformers having low voltage and output ratings
(2) Oil-filled and self-cooled	Used for large-sized transformers with ratings up to 132 kV and 100 MVA
(3) Forced cooling with air blast	Used for machines with ratings higher than 33 kV and 100 MVA
(4) Oil-forced water cooled	It is used for medium-sized transformers. This cooling is similar to natural radiations except water cooling

1.10 TRANSFORMER OIL

Transformer oil is a mineral (clean hydrocarbon) oil, which is obtained by refining crude petroleum. It has the following purposes:

- It provides additional insulation.
- It carries away the heat generated in the core and the coils.
- It protects the paper from dirt and moisture.

Transformer oil has the following properties:

- High dielectric strength.
- Low viscosity to provide heat transfer.
- Good resistance to emulsion.
- It must be free from inorganic acid, alkali and corrosive sulphur.
- It must have high flash point.
- It must be free from sludging under normal operating conditions.

The most important factors are:

- Operating temperature.
- Atmospheric conditions, particularly inside stations.
- Electric strength.
- Moisture and other contamination.
- Sludge formation.

1.11 CONSERVATOR AND BREATHER

Satisfactory operation of a transformer solely depends on the condition of the oil. It is required to keep the oil clean and dry. With the rise of temperature, the oil level increases. Rise of temperature depends on the load of the transformer. If load increases, oil expands. If load decreases, contraction of oil occurs. Since smaller transformers are not totally filled by oils, some space is left between tank walls and oil, which is occupied by air. The tank is connected to the atmosphere through a vent pipe. Air is expelled out if oil expands. Air is drawn from the atmosphere if oil contracts. When air enters the transformer, it is required to extract moisture from the air. An apparatus known as breather is used to extract moisture from the air. It is a small container connected to the vent pipe containing a dehydrating material such as silica gel crystal impregnated with cobalt chloride. The colour of the material is blue when it is dry. It becomes whitish pink when it is damp. In front of the container, a glass window is provided to observe the colour of the material.

If the transformer becomes overloaded, it overheats the oil and sludge formation occurs in the presence of air. If transformer suffers short circuit, temperature rise becomes very high and this causes vaporization of a part of the oil, which forms an explosive mixture with air. This explosive mixture can ignite and cause considerable damage. To prevent the air coming in contact with air as well as moisture, conservators are used. It takes the responsibility of expansion and contraction of oil without allowing it to come into contact with air. The conservator is an air-tight cylindrical drum mounted on or near the cover of the transformer. It is connected to the transformer through a small pipe. In Figure 1.18, breather and conservator are shown.

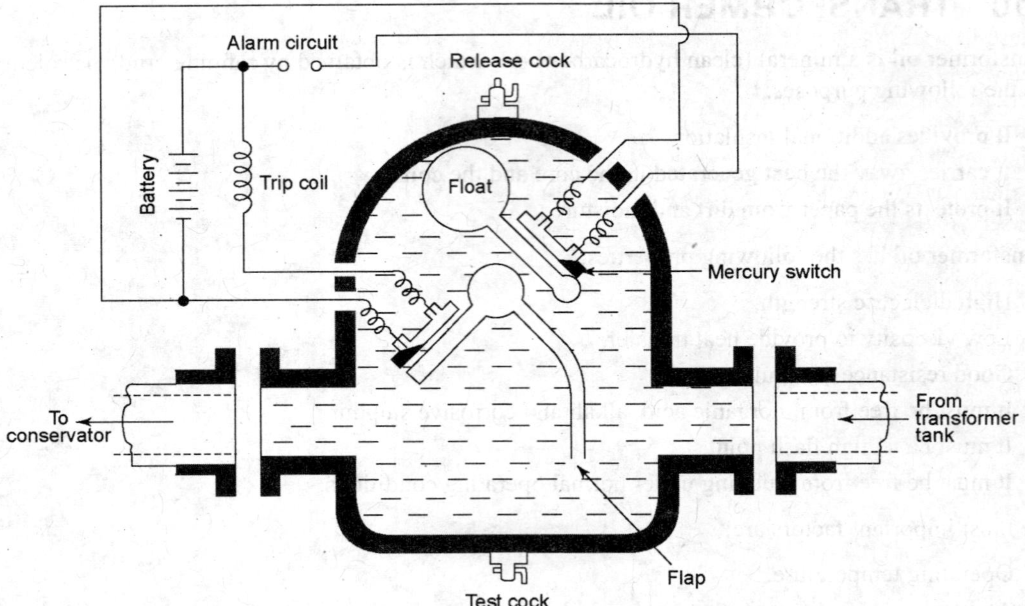

Figure 1.19 Buchholz Relay

1.12 BUCHHOLZ RELAY

Figure 1.19 shows Buchholz relay having two elements mounted in a small chamber. It is located in the pipe connection between the conservator and the transformer tank. Heat is produced due to leakage current for any minor fault and some of the oil in the transformer tank evaporates. Some vapour comes to the top of the chamber while passing through to the conservator. The oil level falls due to accumulation of vapour, and the mercury tape attached to the float is tilted closing the alarm circuit to ring the bell. A release cock is attached to the top of the chamber to release the pressure of the chamber after operation and gas is emitted. It allows refilling of oil in the chamber once again. During severe fault, a large volume of gas is evolved, which tilts the lower element containing a mercury switch mounted on a hinged-type flap and the trip coil is energized. A test cock is provided at the bottom of the chamber, which allows air to be pumped into the chamber for test purposes.

1.13 TRANSFORMER TANK

The factors such as weight, stray load losses and minimum cost are kept in mind while selecting the material for transformer tank. The material must be capable to withstand stresses due to jacking and lifting and must have capacity to house cores, windings and internal connections giving adequate clearance between the windings and the walls. Rolled steel plates are used for tank bodies of most of the transformers. Small tanks are generally welded from steel plates. The larger tanks are assembled from boiler plates. Usage of aluminium for transformer tank reduces the weight and the stray magnetic losses. However, it increases the cost and also needs special attention for lifting to present stressing. The aluminium tanks are usually made of cast aluminium tank parts, which are mounted on a shallow mild steel tray and are arranged to carry the main lifting.

1.14 THEORY OF TRANSFORMER

Figure 1.20 shows the elementary diagram of an ideal transformer with secondary side open circuited. It has no ohmic resistance and leakage reactance. There is no loss in an ideal transformer. In Figure 1.16, alternating voltage (V_1) is applied at the primary and hence alternating current flows in the primary. The primary draws the magnetizing current I_μ only because it is purely inductive in nature. I_μ is small in magnitude and lags behind V_1 by an angle 90°. The function of I_μ is to magnetize the core, and it produces an alternating flux (ϕ), which is proportional to I_μ. The alternating flux (ϕ) is linked with both primary and secondary windings and causes self-induced emf (E_1) in the primary. This self-induced emf (E_1) is equal and opposite to V_1 at every instant. This induced emf is known as *back emf* or *counter emf.* Due to mutual induction, an emf E_2 is produced in the secondary. This emf is known as *mutually induced emf.* It is anti-phase with V_1 and its magnitude is proportional to the rate of flux as well as the number of turns of the secondary windings.

Figure 1.21(a) shows the instantaneous values of applied voltage, induced emfs, flux and magnetizing current by sinusoidal waves, while Figure 1.21(b) shows the vectorial representation of the effective values of the above quantities.

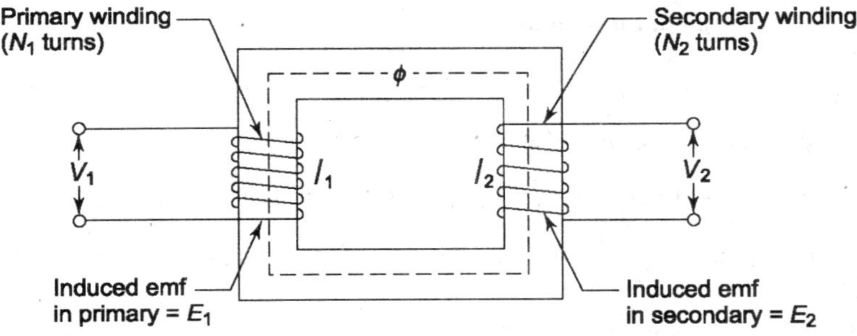

Figure 1.20 Elementary Diagram of an Ideal Transformer

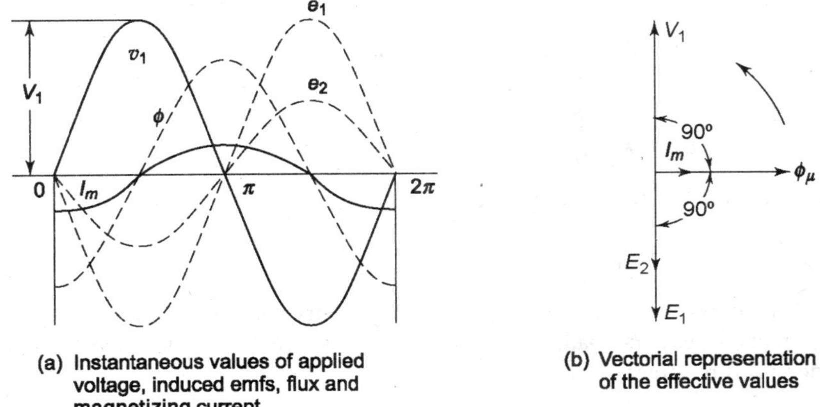

(a) Instantaneous values of applied voltage, induced emfs, flux and magnetizing current

(b) Vectorial representation of the effective values

Figure 1.21 Instantaneous Values and Vectorial Representation

1.15 EMF EQUATION OF A TRANSFORMER

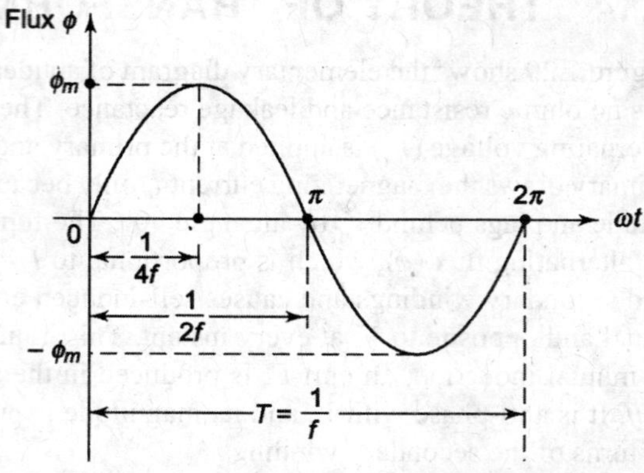

Figure 1.22 Representation of Alternating Flux

Figure 1.22 shows the representation of alternating flux, varying sinusoidally, which increases from its zero value to maximum value (ϕ_m) in one-quarter of the cycle, that is in one-fourth of a second where f is the frequency of AC input in hertz.

The average rate of change of flux is given by $\dfrac{\phi_m}{\frac{1}{4f}}$, that is $4f\phi_m$ Wb/s or V.

This rate of change of flux per turn is the induced emf in V.

Therefore, average emf/turn = $4f\phi_m$ V.

Let N_1 and N_2 be the number of turns in primary and secondary.

$$\text{Again, form factor} = \frac{\text{RMS value}}{\text{Average value}} = 1.11$$

$$\therefore \quad \text{RMS value/turn} = 1.11 \times 4f\phi_m = 4.44 f\phi_m \text{ V}$$

The rms value of induced emf in primary winding is given by

$$E_1 = (4.44 f \phi_m) \times N_1 = 4.44 f \phi_m N_1 = 4.44 f B_m A_r N_1 \tag{1.1}$$

where $B_m = \dfrac{\phi_m}{A_r}$ is the maximum value of flux density having unit Tesla (T) and A_r is the area of cross-section.

Similarly, RMS value of induced emf in secondary winding is

$$E_2 = (4.44 f \phi_m) \times N_2 = 4.44 f \phi_m N_2 = 4.44 f B_m A_r N_2 \tag{1.2}$$

From Equations (1.1) and (1.2), we have

$$\frac{E_1}{N_1} = \frac{E_2}{N_2} = 4.44 f \phi_m = \text{constant} \tag{1.3}$$

i.e.,

$$\frac{E_1}{E_2} = \frac{N_1}{N_2} = a \tag{1.4}$$

where 'a' is the turns ratio of the transformer,

i.e.,

$$a = \frac{N_1}{N_2} \tag{1.5}$$

Equation (1.3) shows that emf induced per turn in primary and secondary windings are equal.

In an ideal transformer at no load, $V_1 = E_1$ and $V_2 = E_2$, where V_2 is the terminal voltage of the transformer. Equation (1.3) becomes

$$\frac{V_1}{V_2} = \frac{N_1}{N_2} \tag{1.6}$$

Example 1.1 The voltage ratio of a single-phase, 50 Hz transformer is 5,000/500 V at no load. Calculate the number of turns in each winding if the maximum value of the flux in the core is 7.82 mWb.

Solution

Here

$$E_1 = V_1 = 5{,}000 \text{ V}$$
$$E_2 = V_2 = 500 \text{ V}$$
$$\phi_{max} = 7.82 \text{ mWb} = 7.82 \times 10^{-3} \text{ Wb}, \; f = 50 \text{ Hz}$$

Let N_1 and N_2 be the number of turns of the primary and secondary windings, respectively.

Since $$E_1 = 4.44 f \phi_m N_1$$

$$N_1 = \frac{E_1}{4.44 f \phi_m}$$

i.e., $$N_1 = \frac{5{,}000}{4.44 \times 50 \times 7.82 \times 10^{-3}} = 2880 \quad \text{(since it is an integer number)}$$ ■

Again, $$\frac{E_2}{E_1} = \frac{N_2}{N_1}$$

∴ $$N_2 = N_1 \times \frac{E_2}{E_1} = 2880 \times \frac{500}{5{,}000} \simeq 288$$

Example 1.2 A single-phase transformer is connected to a 800 V supply. The voltage/turn of the transformer is 8 V. The secondary voltage of the transformer is found to be 400 V. Determine the following:

(i) Primary and secondary turns.

(ii) Cross-section of the core if the maximum flux density is 1.4 T.

Solution

Here Voltage/turn = 8 V, E_1 = 660 V, E_2 = 440 V, B_{max} = 1.4 T

(i) $$N_1 = \frac{E_1}{8} = \frac{800}{8} = 100 \quad \text{and} \quad N_2 = \frac{E_2}{8} = \frac{400}{8} = 50$$

Therefore, the number of primary and secondary turns is 100 and 50, respectively.

(ii) Again $E_1 = 4.44 f N_1 B_{max} A_r$

i.e., $$A_r = \frac{E_1}{4.44 \, f \, N_1 \, B_{max}} = \frac{800}{4.44 \times 50 \times 100 \times 1.4} = 0.02574 \text{ m}^2 = 257.4 \text{ cm}^2$$ ■

Example 1.3 The cross-section of the core of a 4 kVA, 5,000/500 V, 50 Hz, single-phase transformer having maximum flux-density 1.5 T is 500 cm². Determine the following:

(i) The number of turns of primary and secondary windings.

(ii) The emf per turn.

The stacking factor of the core is 0.85.

Solution

Here $A_r = 500\ \text{cm}^2 = 500 \times 10^{-4}\ \text{m}^2$, $B_{max} = 1.5\ T$, $f = 50$ Hz, $E_1 = 5{,}000$ V, $E_2 = 500$ V, stacking factor = 0.85.

(i) $$N_1 = \frac{E_1}{4.44\ f\ B_m\ A_r \times \text{stacking factor}} = \frac{5{,}000}{4.44 \times 50 \times 1.5 \times 500 \times 10^{-4} \times 0.85} = 353$$

(ii) $$\text{emf/turn} = \frac{E_1}{N_1} = \frac{5{,}000}{353} = 14.16$$ ■

1.16 STEP-UP AND STEP-DOWN TRANSFORMER

From Equation (1.6), we have

$$\frac{V_1}{V_2} = \frac{N_1}{N_2}$$

If $N_1 > N_2$, $V_1 > V_2$, that is the output voltage is less than the primary voltage. The transformer is said to be a *step-down transformer*.

If $N_1 < N_2$, $V_1 < V_2$, that is the output voltage is greater than the primary voltage. The transformer is said to be a *step-up transformer*.

For an ideal transformer, input VA = output VA

∴ $$V_1 I_1 = V_2 I_2$$

i.e., $$\frac{V_1}{V_2} = \frac{I_2}{I_1} = a$$

i.e., $$\frac{I_1}{I_2} = \frac{1}{a} = \frac{N_1}{N_2} \quad (1.7)$$

Comparing Equations (1.6) and (1.7), we can conclude that currents are in inverse ratio of turns ratio.

Example 1.4 A 50 kVA, single-phase transformer has 500 turns on the primary and 100 turns on the secondary. The primary is connected to 2,500 V, 50 Hz supply. Calculate the following:

(i) The secondary voltage on open circuit.
(ii) The current flowing through the windings on full load.
(iii) The maximum value of flux.

Solution

(i) $$\frac{E_2}{E_1} = \frac{N_2}{N_1}$$

∴ $$E_2 = E_1 \times \frac{N_2}{N_1} = 2{,}500 \times \frac{100}{500} = 500\ \text{V}$$

∴ The secondary voltage on open circuit is $V_2 = E_2 = 500$ V.

(ii) kVA rating the transformer = 50 kVA

$$\text{Primary full-load current } (I_1) = \frac{\text{kVA} \times 1{,}000}{V_1} = \frac{50 \times 1{,}000}{2{,}500} = 20 \text{ A}$$

$$\text{Secondary full-load current } (I_2) = \frac{\text{kVA} \times 1{,}000}{500} = \frac{50 \times 1{,}000}{500} = 100 \text{ A}$$

(iii) Again,

$$E_1 = 4.44 f \phi_{max} N_1$$

i.e.,

$$\phi_{\max} = \frac{E_1}{4.44 f N_1} = \frac{2{,}500}{4.44 \times 50 \times 500} \text{ Wb} = 0.0225 \text{ Wb}$$

■

1.17 TRANSFORMER ON NO LOAD

For an ideal transformer, we have assumed that there are no core losses and copper losses. For practical transformers, these two losses cannot be neglected. At no-load condition, the primary current is not fully reactive and it supplies (i) iron loss in the core, that is, hysteresis loss and eddy current loss and (ii) very small amount of copper loss in the primary. There is copper loss in the secondary because it is an open circuit. The no-load current lags behind V_1 by an angle θ_0, which is less than 90° (around 80°–85°). The no-load input power is given by

$$W_0 = V_1 I_0 \cos\theta_0 \tag{1.8}$$

where $\cos\theta_0$ is the no-load power factor. Figure 1.23 shows the no-load phasor diagram of a practical transformer. From Figure 1.23, the no-load primary current (I_0) has the following two components:

- One component of I_0, that is $I_w = I_0 \cos\theta_0$ is in phase with V_1. Since I_w supplies the iron loss and primary copper loss at no load, it is known as active or working or iron loss component.
- The other component of I_0 that is, $I_\mu = I_0 \sin\theta_0$ is in quadrature with V_1. It is known as magnetizing component. Its function is to sustain the alternating flux in the core and it is wattless.

  From Figure 1.23, we have

$$I_0 = \sqrt{I_w^2 + I_\mu^2} \tag{1.9}$$

and

$$\theta_0 = \tan^{-1}\left(\frac{I_\mu}{I_w}\right) \tag{1.10}$$

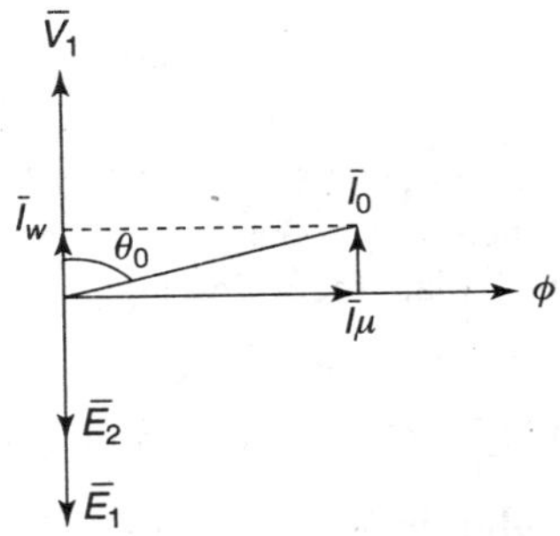

Figure 1.23 Phasor Diagram at No-Load

The following points are most important:

- The no-load primary current is 1–5 per cent of full-load current.
- Since the permeability of the core varies with the instantaneous value of exciting or magnetizing current, the waveform of exciting or magnetizing current is not truly sinusoidal.
- Since I_0 is very small, the no-load copper loss is negligible. Hence, no-load input is practically equal to the iron loss in the transformer.
- Since core loss is solely responsible for shifting the current vector I_0, the angle θ_0 is known as *hysteresis angle of advance*.

Figure 1.24 shows the equivalent circuit of the transformer at no load.

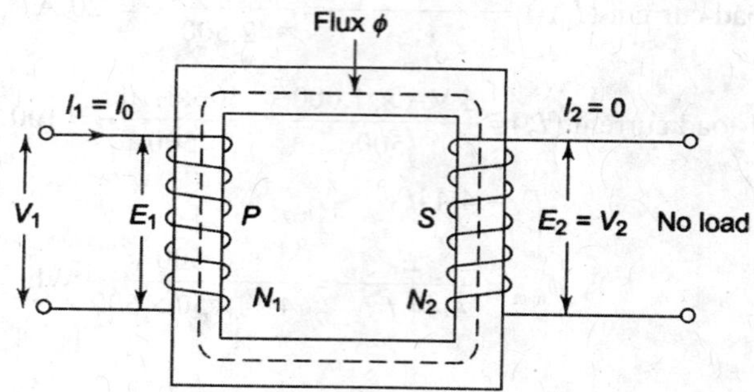

Figure 1.24 Equivalent Circuit of Transformer at No-load

Example 1.5 The no-load current of a 4,400/440 V, sinlge-phase, 50 Hz transformer is 0.04. It consumes power 80 W at no load when supply is given to LV side and HV side is kept open. Calculate the following:

(i) Power factor of no-load current.

(ii) Iron loss component of current.

(iii) Magnetizing component of current.

Solution

$$W_0 = 80 \text{ W}, I_0 = 0.04 \text{ A}, V_1 = 4{,}400 \text{ V}$$

(i) Since $W_0 = V_1 I_0 \cos\theta_0$

$$\therefore \quad \cos\theta_0 = \frac{W_0}{V_1 I_0} = \frac{80}{4{,}400 \times 0.04} = 0.454$$

The no-load power factor is 0.454 (lagging).

(ii) $I_W = I_0 \cos\theta_0 = 0.04 \times 0.454 = 0.0187$ A

(iii) $\sin\theta_0 = \sqrt{1 - \cos^2\theta_0} = 0.891$

$\therefore \quad I_\mu = I_0 \sin\theta_0 = 0.04 \times 0.891 = 0.0356$ A

■

Example 1.6 A 2,200/220 V, 20 kVA, single-phase transformer takes a no-load current of 1.3 A when high-voltage winding is kept open. The iron loss component of no-load current is equal to 0.5A. Calculate the following:

(i) No-load input power

(ii) Magnetizing component and power factor of no-load current.

Solution

Given

$$I_0 = 1.3 \text{ A}, I_W = I_0 \cos\theta_0 = 0.5 \text{ A}$$

$$V_1 = E_1 = 2{,}200 \text{ V and } E_2 = V_2 = 220 \text{ V}$$

(i) No-load input power $(W_0) = V_2 I_0 \cos\theta_0 = V_2 I_W = 220 \times 0.5 = 110$ W.

(ii) Now $\cos\theta_0 = \frac{I_W}{I_0} = \frac{0.5}{1.3} = 0.385$ (lagging)

$$\therefore \quad \sin\theta_0 = \sqrt{1-\cos^2\theta_0} = 0.923$$

$$\therefore \quad I_\mu = I_0 \sin\theta_0 = 1.3 \times 0.923 = 1.199 \text{ A.}$$

■

1.18 TRANSFORMER ON LOAD

Figure 1.25(a) shows the transformer during no-load condition. The flux ϕ is set up in the core. When the secondary is loaded shown in Figure 1.25(b), the secondary current will set up its own flux (ϕ_2), which opposes ϕ. The resultant flux becomes $\phi - \phi_2$. The value of e_1 will decrease because magnitude of ϕ decreases. Hence, v_1 becomes greater than e_1 and the primary winding draws more current from the source. Let the additional current drawn by the primary be I_2'. This current I_2' will set up its own flux (ϕ_2') in the same direction of ϕ and it will oppose ϕ_2 shown in Figure 1.25(c). The resultant flux will be ϕ at any load condition if and only if

$$\phi_2' = \phi_2$$

i.e.,
$$N_1 I_2' = N_2 I_2$$

i.e.,
$$I_2' = \left(\frac{N_2}{N_1}\right) I_2 = \frac{I_2}{a} \qquad (1.11)$$

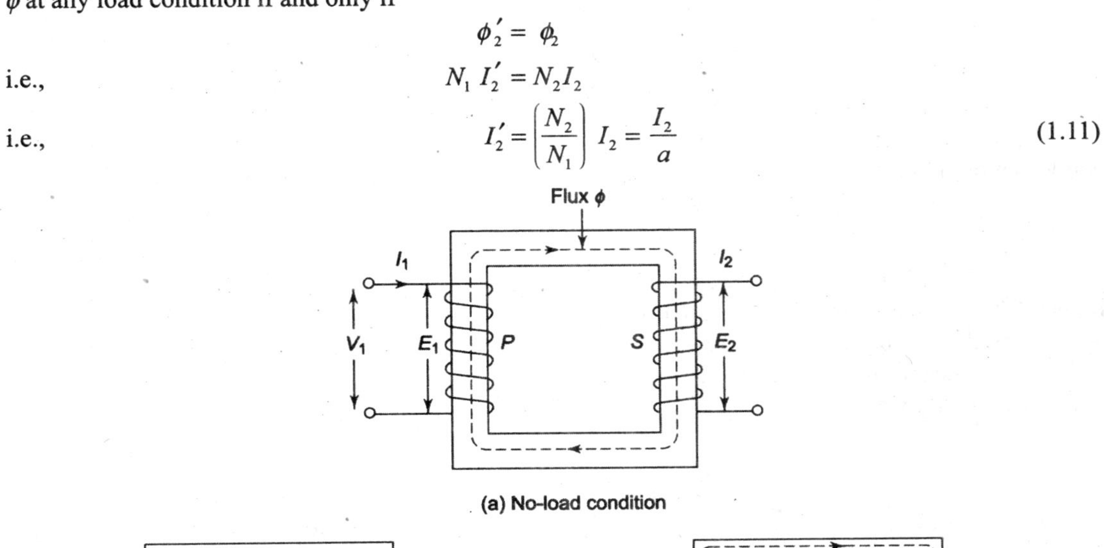

(a) No-load condition

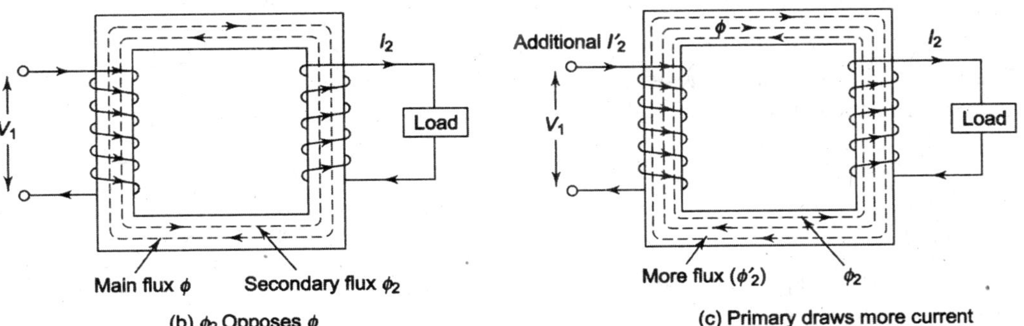

(b) ϕ_2 Opposes ϕ

(c) Primary draws more current

Figure 1.25 On-load Transfer

1.19 EQUIVALENT RESISTANCE

Figure 1.26 shows a transformer having primary resistance R_1 and secondary resistance R_2, where resistances have been shown external to the windings. In Figure 1.26, it is assumed that there is no fringing, i.e. no leakage of flux. It is possible to transfer resistance from one winding to another to simplify the calculation. Let N_1 and N_2 be the number of turns of primary and secondary winding respectively. Let the turns ratio be 'a'. Let I_1 and I_2 be the currents in primary and secondary winding, respectively. Neglecting I_0, $\frac{I_2}{I_1} = a$. Let the referred value of R_2 be R_2' when it is transferred to primary. The copper loss of secondary is $I_2^2 R_2$ when R_2 is in secondary. The copper loss across R_2' is $I_1^2 R_2'$ when R_2 has been transferred to primary. These two losses must be equal.

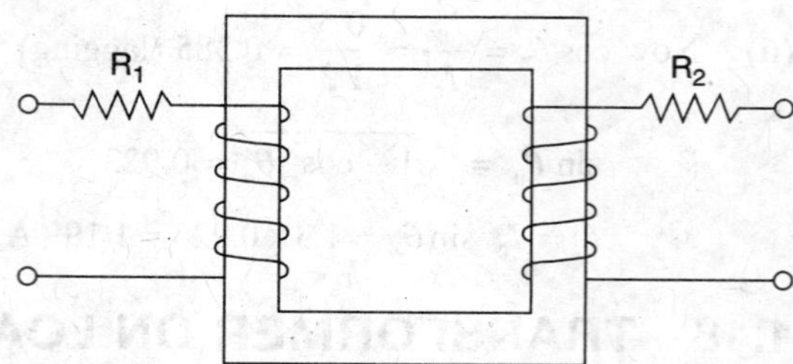

Figure 1.26 Individual Resistances

$\therefore$
$$I_2^2 R_2 = I_1^2 R_2'$$

i.e.,
$$R_2' = \left(\frac{I_2^2}{I_1^2}\right) R_2 = a^2 R_2 \tag{1.12}$$

The total resistance referred to as primary becomes $R_1 + R_2'$ or $R_1 + a^2 R_2$. This is also known as equivalent or effective resistance of the transformer referred to as primary and is denoted by R_{01}.

$$R_{01} = R_1 + a^2 R_2 \tag{1.13}$$

Figure 1.27 is the equivalent of Figure 1.26 when the secondary resistance is transferred to the primary.

If R_1 is transferred to secondary, having referred value R'_1, we have

$$I_1^2 R_1 = I_2^2 R_1'$$

i.e.,
$$R_1' = \left(\frac{I_1^2}{I_2^2}\right) R_1 = \frac{R_1}{a^2} \tag{1.14}$$

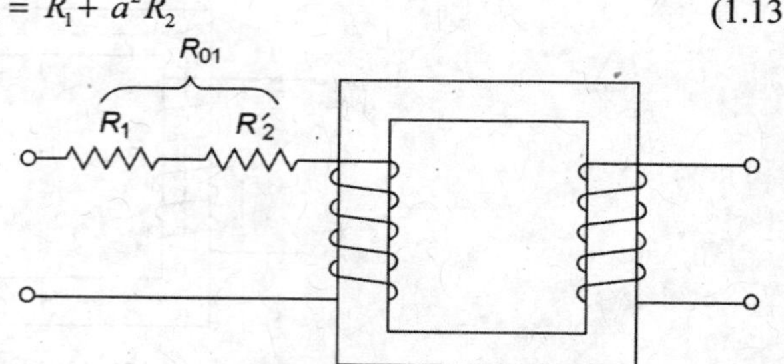

Figure 1.27 Resistance Referred to as Primary

Therefore, the equivalent resistance of the transformer referred to as secondary becomes

$$R_{02} = R_2 + \frac{R_1}{a^2} \tag{1.15}$$

Figure 1.28 is the equivalent of Figure 1.26 when the primary resistance is transferred to the secondary.

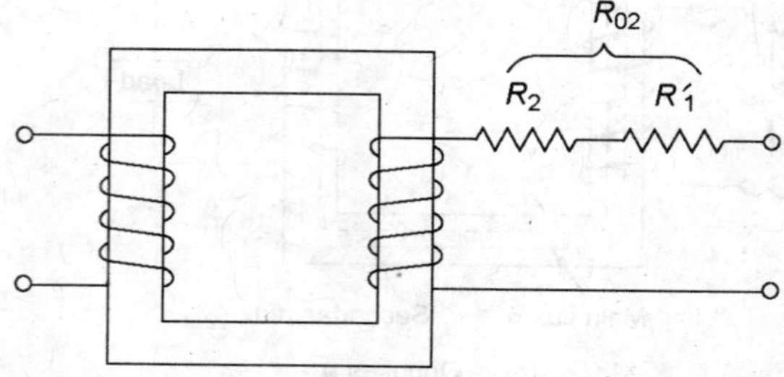

Figure 1.28 Resistance Referred to as Secondary

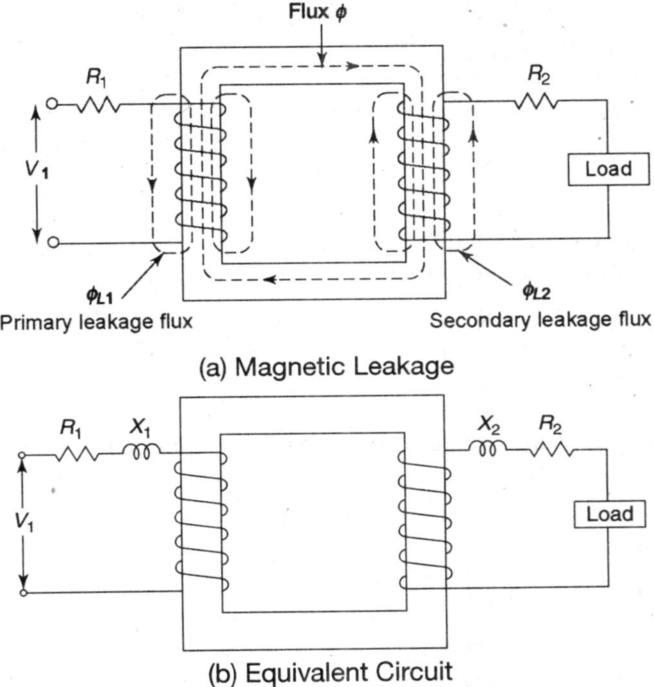

Figure 1.29 Magnetic Leakage and Equivalent Circuit

1.20 MAGNETIC LEAKAGE

Till now we have assumed that all the flux linked with the primary also links with the secondary. But in practice, the permeability of the core of the transformer is finite. All the flux linked with the primary do not link with the secondary. As shown in Figure 1.29(a), ϕ_{L1} and ϕ_{L2} induce emf e_{L1} and e_{L2} in primary and secondary windings respectively. Therefore, in effect, we can consider it as an equivalent inductive coil in phase with the winding shown in Figure 1.29(b).

1.21 EQUIVALENT REACTANCE

If X_2 be transferred to primary, let its referred value be X_2'. We have

$$I_2^2 X_2 = I_1^2 X_2'$$

i.e.,

$$X_2' = \left(\frac{I_2^2}{I_1^2}\right) X_2 = a^2 X_2 \tag{1.16}$$

If X_1 be transferred to secondary, let the referred value be X_1'. We have

$$I_1^2 X_1 = I_2^2 X_1'$$

i.e.,

$$X_1' = \left(\frac{I_1^2}{I_2^2}\right) X_1 = \frac{X_1}{a^2} \tag{1.17}$$

The total reactance referred to as primary (X_{01}) is $X_1 + a^2 X_2$ and that of referred to as secondary (X_{02}) is $X_2 + \frac{X_1}{a^2}$. The total reactance is known as *equivalent reactance*. It is denoted by

$$X_{01} = X_1 + a^2 X_2 \tag{1.18}$$

$$X_{02} = X_2 + \frac{X_1}{a^2} \tag{1.19}$$

Example 1.7 A 30 kVA, 3,000/300 V, 50 Hz, single-phase transformer has the following winding resistances (R's) and leakage reactances (X's):

$$R_1 = 2.5\ \Omega \qquad R_2 = 0.018\ \Omega$$
$$X_1 = 3.8\ \Omega \qquad X_2 = 0.052\ \Omega$$

Calculate the following:

(i) Equivalent resistance, leakage reactance and impedance referred to as high-voltage side.
(ii) Equivalent resistance, leakage reactance and impedance referred to as low-voltage side.
(iii) Total Cu loss of the transformer at full load condition.

Solution

Here $R_1 = 2.5\ \Omega$, $R_2 = 0.018\ \Omega$, $X_1 = 3.8\ \Omega$, $X_2 = 0.052\ \Omega$.

$$\text{Turns ratio } (a) = \frac{3,000}{300} = 10$$

(i) Equivalent resistance referred to as HV side:

$$R_{01} = R_1 + a^2 R_2 = 2.5 + (10)^2 \times 0.018 = 4.3\ \Omega$$

Equivalent leakage reactance referred to as HV side:

$$X_{01} = X_1 + a^2 X_2 = 3.8 + (10)^2 \times 0.052 = 9\ \Omega$$

Equivalent impedance referred to as HV side:

$$Z_{01} = \sqrt{R_{01}^2 + X_{01}^2} = \sqrt{(4.3)^2 + 9^2} = 9.974\ \Omega$$

(ii) Equivalent resistance referred to as LV side:

$$R_{02} = \frac{R_1}{a^2} + R_2 = \frac{2.5}{10^2} + 0.018 = 0.043\ \Omega$$

Equivalent reactance referred to as LV side:

$$X_{02} = \frac{X_1}{a^2} + X_2 = \frac{3.8}{10^2} + 0.052 = 0.09\ \Omega$$

Equivalent impedance referred to as LV side:

$$Z_{02} = \sqrt{R_{02}^2 + X_{02}^2} = \sqrt{(0.043)^2 + (0.09)^2} = 0.09974\ \Omega$$

(iii) Total Cu loss at full load:

$$\text{Primary current } (I_1) = \frac{30 \times 10^3}{3{,}000} = 10\,\text{A}$$

$$\text{Secondary current } (I_2) = \frac{30 \times 10^3}{300} = 100\,\text{A}$$

$$\text{Total Cu loss} = I_1^2 R_{01} = (10)^2 \times 4.3 = 430\text{ W}$$

Also, $\text{Total Cu loss} = I_2^2 R_{02} = (100)^2 \times 0.043 = 430\text{ W}$

Also, $\text{Total Cu loss} = I_1^2 R_1 + I_2^2 R_2 = 10^2 \times 2.5 + 100^2 \times 0.018 = 430\text{ W.}$ ■

1.22 TRANSFORMER WITH RESISTANCE AND LEAKAGE REACTANCE

Figure 1.30 shows a transformer having resistances R_1 and R_2 of primary and secondary windings respectively, and leakage reactances X_1 and X_2 of primary and secondary windings respectively. The primary impedance is given by

$$Z_1 = \sqrt{R_1^2 + X_1^2} \tag{1.20}$$

and the secondary impedance is given by

$$Z_2 = \sqrt{R_2^2 + X_2^2} \tag{1.21}$$

The applied voltage V_1 on primary side is given by

$$\bar{V}_1 = -\bar{E}_1 + \bar{I}_1(R_1 + jX_1) = -\bar{E}_1 + \bar{I}_1\bar{Z}_1 \tag{1.22}$$

If V_2 be the secondary terminal voltage during load, the secondary induced emf (E_2) is given by

$$\bar{E}_2 = \bar{V}_2 + \bar{I}_2(R_2 + jX_2) = \bar{V}_2 + \bar{I}_2\bar{Z}_2 \tag{1.23}$$

Figure 1.31 shows the phasor diagram of the transformer for unity power factor, lagging power factor and leading power factor respectively.

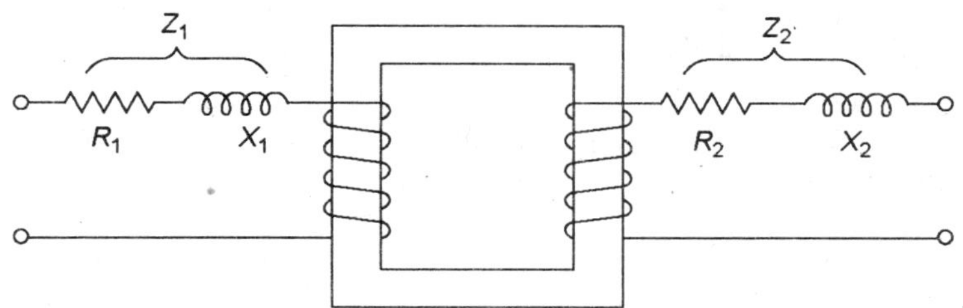

Figure 1.30 Transformer with Resistance and Leakage Reactance

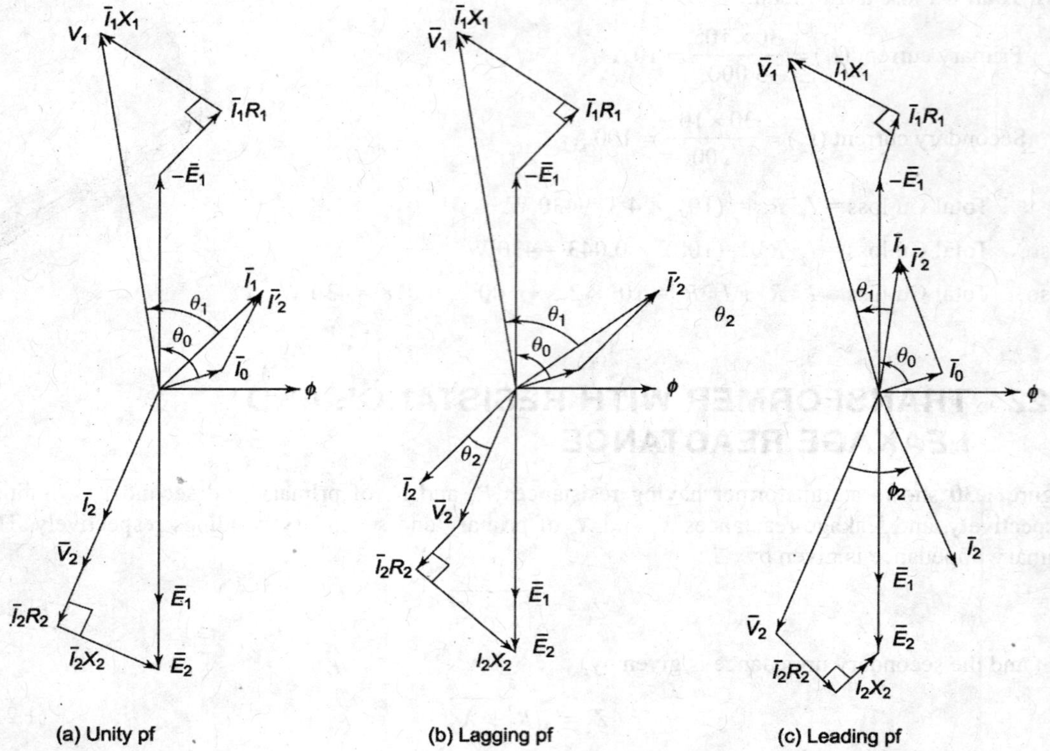

(a) Unity pf (b) Lagging pf (c) Leading pf

Figure 1.31 Phasor Diagram of a Transformer at Load

1.23 EQUIVALENT CIRCUIT

Figure 1.32 shows the equivalent circuit of a single-phase transformer having load impedance Z_L. Figure 1.33 is the equivalent circuit of Figure 1.32. The secondary winding resistance (R_2), leakage reactance (X_2) and load impedance (Z_L) connected to secondary terminal is transferred to primary side.

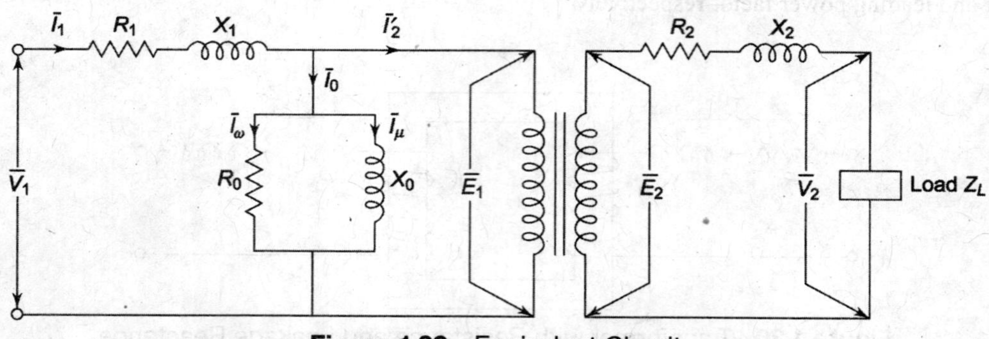

Figure 1.32 Equivalent Circuit

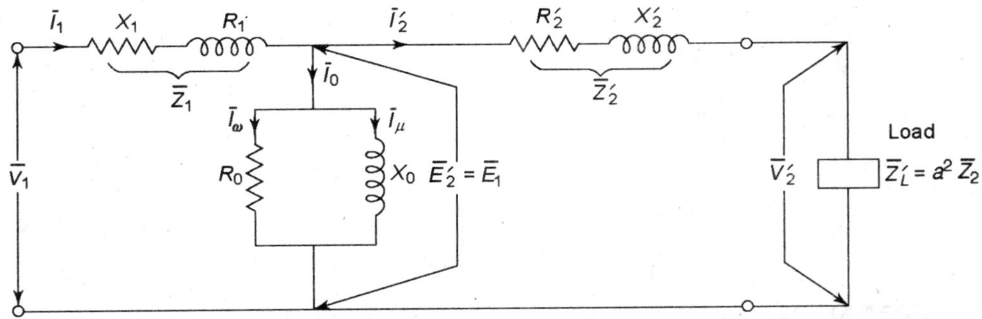

Figure 1.33 Exact Equivalent Circuit

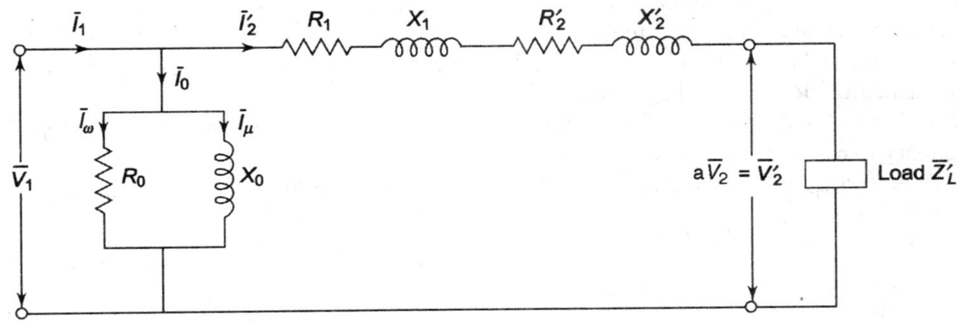

Figure 1.34 Approximate Equivalent Circuit

Since I_0 is small compared to full-load current, we can shift the excitation circuit towards the terminal voltage side shown in Figure 1.34. Figure 1.34 is the approximate equivalent circuit, whereas Figure 1.33 is the exact equivalent circuit of Figure 1.32.

From Figure 1.33, the total input impedance between input terminals becomes

$$\bar{Z} = \bar{Z}_1 + \bar{Z}_m \parallel \left(\bar{Z}_2' + \bar{Z}_L'\right) = \bar{Z}_1 + \frac{\bar{Z}_m\left(\bar{Z}_2' + \bar{Z}_L'\right)}{\bar{Z}_m + \bar{Z}_2' + \bar{Z}_L'} \tag{1.24}$$

where $\bar{Z}_L' = R_L' + jX_L'$

and $\bar{Z}_m$ is the impedance of the exciting circuit.

$$\bar{V}_1 = \bar{I}_1\bar{Z} = \bar{I}_1\left[\bar{Z}_1 + \frac{\bar{Z}_m\left(\bar{Z}_2' + \bar{Z}_L'\right)}{\bar{Z}_m + \bar{Z}_2 + \bar{Z}_L'}\right] \tag{1.25}$$

R_0, X_0, R_{01} and X_{01} are the four important parameters of the transformer. From open circuit and short circuit test of a transformer, R_0, X_0 and R_{01}, X_{01} can be determined respectively.

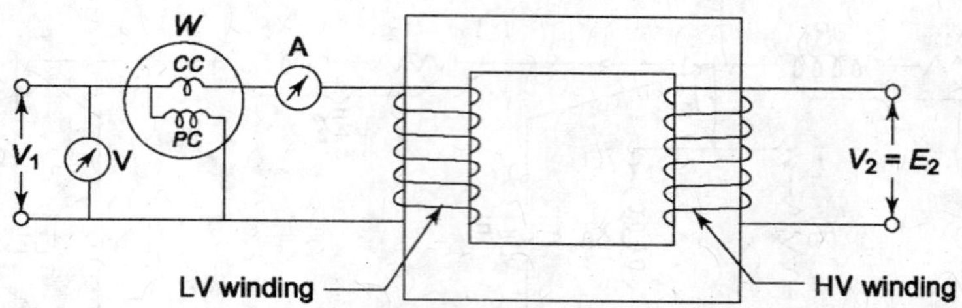

Figure 1.35 Open Circuit Test

1.24 OPEN CIRCUIT TEST OR NO-LOAD TEST

From this test, we can determine core loss and no-load current (I_0) of the transformer. Figure 1.35 shows the schematic diagram of the transformer. The high-voltage side is generally kept open because the current in high-voltage winding is less compared to that on low-voltage winding. In low-voltage side, a voltmeter, an ammeter and a wattmeter are connected to measure the input voltage, no-load current and the core loss of the transformer. Since no-load current is generally small, the copper loss at no-load condition is negligible. The wattmeter reading practically gives the iron loss of the transformer.

To measure the induced emf in secondary winding, a high-resistance voltmeter is connected across the secondary to calculate the turns ratio (a).

Let I_0 be the reading of the ammeter, V_1 be the reading of the voltmeter and W be the reading of the wattmeter.

We have

$$W = V_1 I_0 \cos\theta_0$$

i.e.,

$$\cos\theta_0 = \frac{W}{V_1 I_0} \tag{1.26}$$

Therefore,

$$I_w = I_0 \cos\theta_0 \tag{1.27}$$

and

$$I_\mu = I_0 \sin\theta_0 \tag{1.28}$$

During no-load condition, the voltage drop across the primary impedance is small. Therefore, we have

$$R_0 = \frac{V_1}{I_w} \tag{1.29}$$

and

$$X_0 = \frac{V_1}{I_\mu} \tag{1.30}$$

The total iron loss depends on the frequency as well as the maximum flux density. Hysteresis loss and eddy current loss are the two parts of the total iron loss, which are described below.

(i) Hysteresis loss: $P_h = k_1 B_{\max}^{1.6} f$ (1.31)

(ii) Eddy current loss: $P_e = k_2 B_{\max}^2 f^2$ (1.32)

where k_1 and k_2 are the constants in the above two equations, which can be obtained from the experiment. The hysteresis and eddy current losses can be calculated by knowing k_1 and k_2. Figure 1.36 shows the variation of iron loss with the applied voltage.

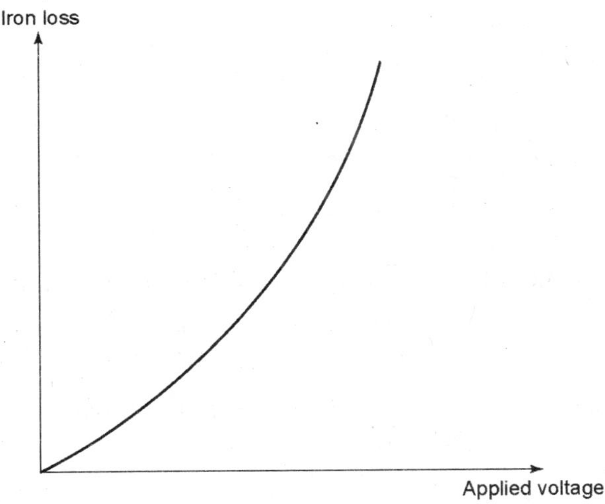

Figure 1.36 Variation of Iron Loss with Applied Voltage

1.25 SHORT CIRCUIT OR IMPEDANCE TEST

The main purpose of this test is to compute the following:

- Equivalent impedance of the transformer referred to primary and secondary.
- Cu loss of the transformer at any desired load.
- Total voltage drop of the transformer referred to primary and secondary.

Figure 1.37 shows the schematic diagram for the short circuit test of a transformer in which the low-voltage winding is short-circuited. During the short circuit test, we apply 5–10 per cent of the rated voltage to high-voltage side so that the full-load current flow both in primary and secondary. Voltage is slowly increased from zero to a value to get full-load current to flow. Since 5–10 per cent of rated voltage at the primary is easier to achieve smoothly and read by a voltmeter, instruments are always placed on the high-voltage side. A voltmeter, an ammeter and a wattmeter are placed on the high-voltage side. The low-voltage side is directly short-circuited by a thick conductor or by an ammeter.

Let P_{SC} be the reading of the wattmeter, I_{SC} be the reading of the ammeter and V_{SC} be the reading of the voltmeter.

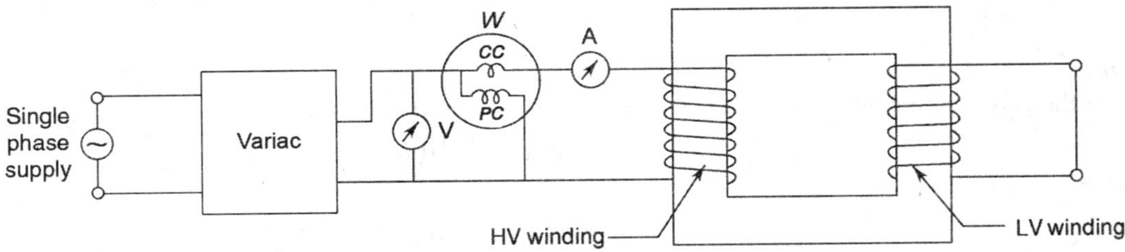

Figure 1.37 Short Circuit Test

We have

$$R_{01} = \frac{P_{SC}}{I_{SC}^2} \tag{1.33}$$

and

$$Z_{01} = \frac{V_{SC}}{I_{SC}} \tag{1.34}$$

∴

$$X_{01} = \sqrt{Z_{01}^2 - R_{01}^2} \tag{1.35}$$

Example 1.8 A 4 kVA, 400/200 V, 50 Hz, single-phase transformer has the following test data

OC test: 200 V, 2 A, 90 W
SC test: 20 V, 10 A, 100 W

The instruments during SC test are placed on the side opposite to that of OC test. Find equivalent circuit referred to as primary side (high-voltage side).

Solution

OC test:

Since rated voltage is applied during OC test, instruments are placed on the low-voltage side.

$$\text{Turns ratio } (a) = \frac{400}{200} = 2$$

$$\text{No-load current } (I_0) = 2A = (I_0)_{lv}$$

∴
$$(I_0)_{hv} = (I_0)_{lv} \times \frac{1}{a} = \frac{2}{2} = 1\text{ A}$$

Applied voltage at no load = 400 V [for OC Test]

$$V_0 I_0 \cos\theta_0 = 90$$

i.e.,
$$400 \times 1 \times \cos\theta_0 = 90$$

i.e.,
$$\cos\theta_0 = \frac{90}{400} = 0.225$$

∴
$$\sin\theta_0 = \sqrt{1 - cos^2\theta_0} = 0.974$$

∴
$$I_W = I_0 \cos\theta_0 = 1 \times 0.225 = 0.225\text{ A}$$

and
$$I_\mu = I_0 \sin\theta_0 = 1 \times 0.974 = 0.974\text{ A}$$

∴
$$R_0 = \frac{V_0}{I_w} = \frac{400}{0.225} = 1777.78\ \Omega \quad \text{and} \quad X_0 = \frac{V_0}{I_\mu} = \frac{400}{0.974} = 410.68\ \Omega$$

SC test:

As per the given statement, the instruments are placed on high-voltage side.

$$V_{SC} = 20\text{ V}, I_{SC} = 10\text{ A}, P_{SC} = 100\text{ W}$$

∴
$$Z_{01} = \frac{V_{sc}}{I_{sc}} = \frac{20}{10} = 2\ \Omega,\ R_{01} = \frac{P_{SC}}{I_{SC}^2} = \frac{100}{100} = 1\Omega \text{ and}$$

$$X_{01} = \sqrt{Z_{01}^2 - R_{01}^2} = \sqrt{4-1} = \sqrt{3} = 1.732\Omega$$

■

Figure E1.1 shows the equivalent circuit of the transformer referred to as HV side.

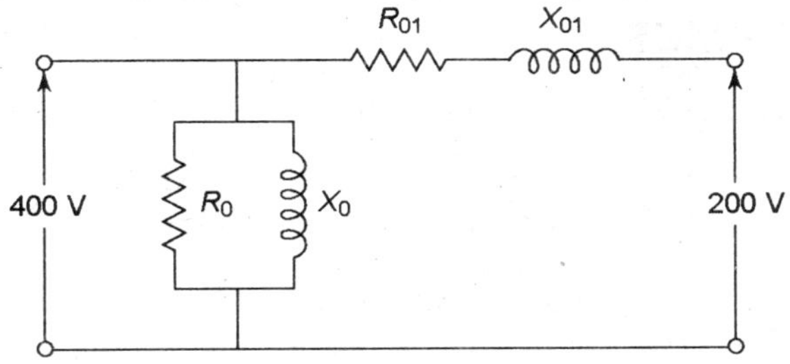

Figure E1.1

Example 1.9 A sigle-phase 4 kVA, 460/230 V, 50 Hz transformer has the following constants:

Resistance primary: 0.4 Ω; secondary 0.10 Ω

Leakage reactance primary: 0.5 Ω; secondary 0.12 Ω

Resistance and reactance of excitation circuit referred to as primary are $R_0 = 650\ \Omega$ and $X_0 = 250\ \Omega$. Calculate readings of the instruments when the transformer is connected for:

(i) OC test.

(ii) SC test.

In both tests, supply is given to high-voltage side.

Solution

$$R_1 = 0.4\ \Omega,\ X_1 = 0.5\ \Omega,\ R_2 = 0.1\ \Omega,\ X_2 = 0.12\ \Omega,\ R_0 = 650\ \Omega,\ X_0 = 250\ \Omega$$

(i) ***OC test:***

$V_1 = 460$ V (rated voltage is applied to high-voltage side)

$$I_W = \frac{V}{R_0} = \frac{460}{650} = 0.707\text{ A} \quad \text{and} \quad I_\mu = \frac{V}{X_0} = \frac{460}{250} = 1.84\text{ A}$$

$$\therefore \quad W_0 = V_0 I_0 \cos\theta_0 = V I_w\ [\because I_W = I_0 \cos\theta_0]$$

$$= 460 \times 0.707 = 325.22\text{ W.}$$

$$\therefore \quad I_0 = \sqrt{I_w^2 + I_\mu^2} = \sqrt{0.707^2 + 1.84^2} = 1.97\text{ A}$$

The instrument readings are: **460 V, 1.97 A, 325.22 W.**

(ii) ***SC test:***

$$R_{01} = R_1 + a^2 R_2 = 0.4 + 2^2 \times 0.1 = 0.8\ \Omega$$

$$X_{01} = X_1 + a^2 X_2 = 0.5 + 2^2 \times 0.12 = 0.98\ \Omega$$

$$\therefore \quad Z_{01} = \sqrt{R_{01}^2 + X_{01}^2} = \sqrt{(0.8)^2 + (0.98)^2} = 1.265\ \Omega$$

Full-load current in high-voltage winding $(I_1) = \dfrac{4\times 10^3}{460} = 8.7$ A

$$V_{SC} = I_{SC} Z_{01} = 8.7 \times 1.265 = 11\text{ V}$$

$$P_{SC} = I_{SC}^2 R_{01} = (8.7)^2 \times 0.8 = 60.55\text{ W}$$

Instrument readings are: 11 V, 8.7 A, 60.55 W. ■

1.26 SEPARATION OF CORE (OR IRON) LOSSES IN A TRANSFORMER

Hysteresis loss and eddy current loss are the components of the iron losses. For the applied flux density B_{max} to the core, we have

$$\text{Hysteresis loss} = Af$$

and

$$\text{Eddy current loss} = Bf^2$$

The no load loss can be expressed as

$$W_c = Af + Bf^2 \tag{1.36}$$

where A and B are constants.

Therefore,

$$\frac{W_c}{f} = A + Bf \tag{1.37}$$

Figure 1.38 shows the graph, which is a straight line when $\frac{W_c}{f}$ and f are plotted along the y-axis and x-axis, respectively. The intercept on the y-axis gives the value of A, whereas the slope of the line gives the value of B. Now the hysteresis and eddy current loss can be determined at any desired frequency.

The experimental circuit arrangement for determining $\frac{W_c}{f}$ and f is shown in Figure 1.39.

In Figure 1.39, a variable frequency alternator supplies to the transformer under the test, which is driven by DC shunt motor whose speed can be varied over a wide range. The switches S_1 and S_2 are opened and the alternator is started with the help of the DC shunt motor. The speed is adjusted to the

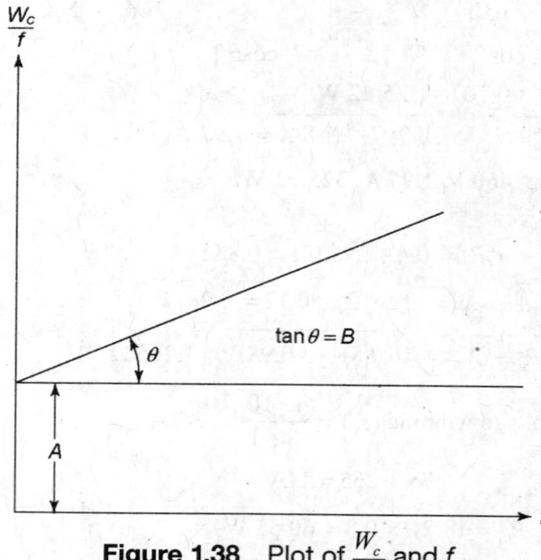

Figure 1.38 Plot of $\frac{W_c}{f}$ and f

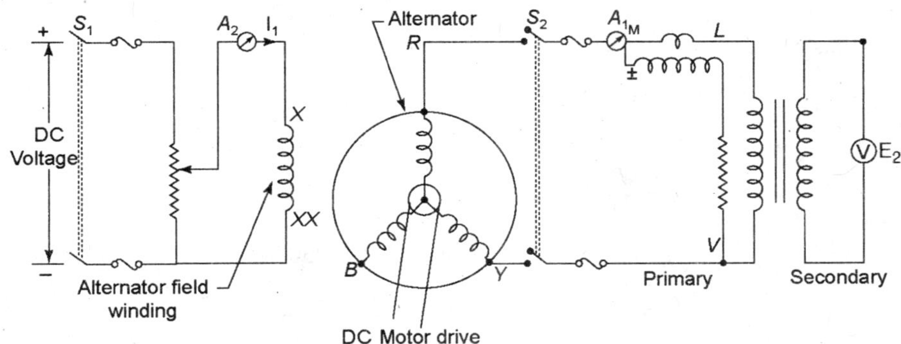

Figure 1.39 Experimental Circuit for Determining $\frac{W_c}{f}$ and f

value of the required frequency. The excitation of the field coil (X–XX) is varied until the voltmeter on the secondary side of the transformer achieves the rated value. If E_2 is the transformer emf on the secondary, we have

$$E_2 = 4.44\,\phi_m fN$$

i.e.

$$\frac{E_2}{f} = 4.44\,\phi_m N \tag{1.38}$$

For constant $\frac{E_2}{f}$, the flux density in the transformer remains constant. To achieve this, the frequency of the alternator emf is varied so that $\frac{E_2}{f}$ remains constant. The necessary f can be adjusted to vary E_2 so that $\frac{E_2}{f}$ is kept constant. For different values of frequencies above and below the rated value, the reading of wattmeter (W) is noted. The graph $\frac{W_c}{f}$ and f is drawn to get the constants A and B. After getting the value of A and B, the hysteresis loss and eddy current loss is obtained.

Example 1.10 The hysteresis and eddy current loss of a transformer are 300 W and 250 W, respectively, if the supply voltage is 500 V at 50 Hz. Find the hysteresis loss and eddy current loss if the supply voltage is 1,000 V, 100 Hz.

Solution

Hysteresis loss at 500 V, 50 Hz = 300 W

Eddy current loss at 500 V, 50 Hz = 250 W

Induced emf $(E) = 4.44 f\phi_m N = 4.44 fB_m A_r N$

$$E \propto B_m f$$

$$\therefore \quad B_m \propto \frac{E}{f}$$

$$\therefore \quad B_m = \alpha\,\frac{E}{f}\ [\alpha \text{ is the constant}]$$

For the first case: $B_{m1} = 10\ \alpha$
For the second case: $B_{m2} = 10\ \alpha$

$\therefore \quad B_{m1} = B_{m2}$

Hence, $\quad P_h = Cf$ and $P_e = Df^2$

$\therefore \quad C = \frac{P_h}{f} \quad$ and $\quad D = \frac{P_e}{f^2}$

When $\quad f = 50$ Hz, $P_h = 300$ W and $P_e = 250$ W

$\therefore \quad C = \frac{300}{50} = 6 \quad$ and $\quad D = \frac{250}{50^2} = \frac{1}{10}$

When $\quad f = 100$ Hz

$$P_h = Cf = 6 \times 100 = 600 \text{ W}$$

and $\quad P_e = Df^2 = \frac{1}{10} \times 100^2 = 1{,}000$ W ■

1.27 TOTAL APPROXIMATE VOLTAGE DROP OF A TRANSFORMER

During no-load condition, induced voltages at the primary and secondary windings are equal to the applied voltage and secondary terminal voltage respectively. If ${}_0V_2$ be the secondary terminal voltage at no load, we can write $E_2 = {}_0V_2$. Let V_2 be the secondary voltage on load. Figure 1.40 shows the phasor diagram of a transformer referred to as secondary.

In Figure 1.40, R_{02} and X_{02} are the equivalent resistance and reactance of the transformer, respectively, referred to as secondary side. With O as centre, an arc is drawn in Figure 1.40, which intersects the extended OA at H. From C, a perpendicular is drawn on OH, which intersects it at G. Now $AC = AH$ represents the actual drop and AG represents the approximate voltage drop. BF is drawn perpendicular to OH. BE is drawn parallel to AG, which is equal to FG.

The approximate voltage drop

$$= AG = AF + FG = AF + BE$$

$$= I_2 R_{02} \cos\theta + I_2 X_{02} \sin\theta \tag{1.39}$$

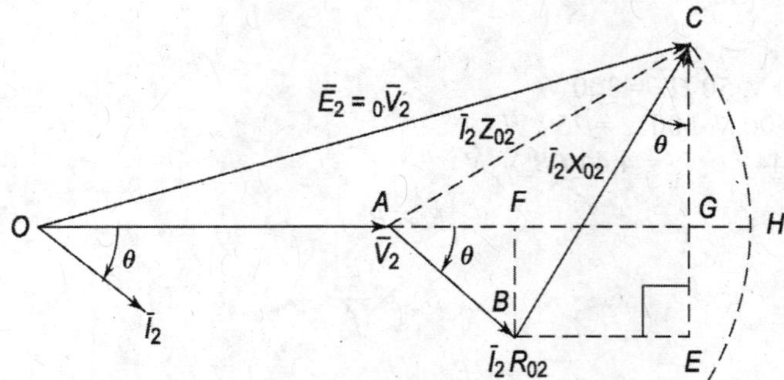

Figure 1.40 Phasor Diagram of a Transformer Referred to as Secondary

This approximate voltage drop shown in Equation (1.39) is for lagging power factor only.
For leading power factor, the approximate voltage drop will be

$$= I_2R_{02}\cos\theta - I_2X_{02}\sin\theta \tag{1.40}$$

where '+' sign is for lagging power factor and '–' sign is for leading power factor.
The above calculation is referred to as secondary. It may be noted that voltage drop referred to as primary is

$$I_1R_{01}\cos\theta \pm I_1X_{01}\sin\theta \tag{1.41}$$

$$\therefore \text{ \% voltage drop in secondary is } = \frac{I_2R_{02}\cos\theta \pm I_2X_{02}\sin\theta}{{}_0V_2}\times 100$$

$$= \frac{I_2R_{02}\cos\theta}{{}_0V_2}\times 100 \pm \frac{I_2X_{02}\sin\theta}{{}_0V_2}\times 100$$

$$= v_r\cos\theta \pm v_x\sin\theta \tag{1.42}$$

where $v_r = \dfrac{I_2R_{02}}{{}_0V_2}\times 100 = \text{\% resistive drop} = \dfrac{I_1R_{01}}{V_1}\times 100$

and $v_x = \dfrac{I_2X_{02}}{{}_0V_2}\times 100 = \text{\% reactive drop} = \dfrac{I_1X_{01}}{V_1}\times 100$

1.28 EXACT VOLTAGE DROP

In Figure 1.40, the exact voltage drop is *AH* instead of *AG*. During calculation of approximate voltage drop, *AG* has already been calculated. If *GH* is being added to *AG*, the exact voltage drop can be obtained.

Consider the right-angled triangle *OCG*. We have

$$OC^2 = OG^2 + GC^2$$

i.e. $OC^2 - OG^2 = GC^2$

i.e. $(OC - OG)(OC + OG) = GC^2$

i.e. $(OH - OG)(OC + OG) = GC^2$

i.e. $GH.2.OC = GC^2$ [Taking $OC \approx OG$]

i.e. $GH = \dfrac{GC^2}{2(OC)} = \dfrac{(CE - GE)^2}{2(OC)} = \dfrac{(CE - FB)^2}{2(OC)} = \dfrac{(I_2X_{02}\cos\theta - I_2R_{02}\sin\theta)^2}{2\,{}_0V_2}$

For lagging power factor, the exact voltage drop is

$$= AG + GH = (I_2R_{02}\cos\theta + I_2X_{02}\sin\theta) + \frac{(I_2X_{02}\cos\theta - I_2R_{02}\sin\theta)^2}{2\,{}_0V_2}$$

For leading power factor, the exact voltage drop is given by

$$= (I_2R_{02}\cos\theta - I_2X_{02}\sin\theta) + \frac{(I_2X_{02}\cos\theta + I_2R_{02}\sin\theta)^2}{2\,{}_0V_2}$$

In general, the voltage drop is

$$= (I_2R_{02}\cos\theta \pm I_2X_{02}\sin\theta) + \frac{(I_2X_{02}\cos\theta \mp I_2R_{02}\sin\theta)^2}{2\,_0V_2} \quad (1.43)$$

and percentage voltage drop is

$$= \frac{(I_2R_{02}\cos\theta \pm I_2X_{02}\sin\theta)}{_0V_2}\times 100 + \frac{(I_2X_{02}\cos\theta \mp I_2R_{02}\sin\theta)^2}{2\,_0V_2^2}\times 100 \quad (1.44)$$

$$= (v_r\cos\theta \pm v_x\sin\theta) + \frac{1}{200}(v_x\cos\theta \mp v_r\sin\theta)^2 \quad (1.45)$$

It may be noted that the upper sign is to be used for lagging power factor and the lower sign for leading power factor.

Example 1.11 The OC and SC tests on a 300/600 V, 50 Hz, single-phase transformer gave the following results:

OC test (LV side): 300 V, 0.8 A, 70 W

SC test (HV side): 20 V, 12 A, 90 W

where LV and HV is low voltage and high voltage, respectively.

Find the equivalent circuit of the transformer referred to as LV side and also calculate the secondary voltage when delivering 6 kW at 0.8 p.f. (power factor) lagging.

Solution

OC test (LV side):

Instruments are placed on LV side and HV side is kept open.

$$V_1 = 300\ V,\ I_0 = 0.8\text{ A},\ W_0 = 70\text{ W}$$

Now, $\cos\theta_0 = \dfrac{W_0}{V_1I_0} = \dfrac{70}{300\times 0.8} = 0.292$ and $\sin\theta_0 = \sqrt{1-\cos^2\theta_0} = 0.956$

∴ $I_W = I_0\cos\theta_0 = 0.8\times 0.292 = 0.2336\text{ A}$

and $I_\mu = I_0\sin\theta_0 = 0.8\times 0.956 = 0.7648\text{ A}$

∴ $R_0 = \dfrac{V_1}{I_W} = \dfrac{300}{0.2336} = 1284.2\ \Omega$ and $X_0 = \dfrac{V_1}{I_\mu} = \dfrac{300}{0.7648} = 392.25\ \Omega$

SC test:

Instruments are placed on HV side and LV side is kept open.

$$\text{Turns ratio } (a) = \frac{300}{600} = \frac{1}{2}$$

$$V_{SC} = 20\text{ V},\ I_{SC} = 12\text{ A},\ P_{SC} = 90\text{ W}$$

∴ $Z_{02} = \dfrac{V_{SC}}{I_{SC}} = \dfrac{20}{12} = 1.67\ \Omega$

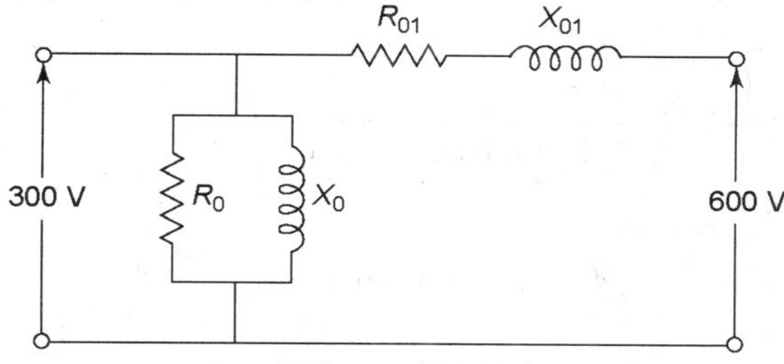

Figure E1.2

$$\therefore \qquad R_{02} = \frac{P_{SC}}{I_{SC}^2} = \frac{90}{12^2} = 0.625\ \Omega$$

$$\therefore \qquad X_{02} = \sqrt{Z_{02}^2 - R_{02}^2} = \sqrt{1.67^2 - 0.625^2} = 1.548\ \Omega$$

$$\therefore \qquad R_{01} = a^2 R_{02} = \left(\frac{1}{2}\right)^2 \times 0.625 = 0.156\ \Omega$$

$$\therefore \qquad X_{01} = a^2 X_{02} = \left(\frac{1}{2}\right)^2 \times 1.548 = 0.387\ \Omega$$

The equivalent circuit referred to as LV side is shown in Figure E1.2.

$$\therefore \qquad I_2 = \frac{6\times 10^3}{600\times 0.8} = 12.5\ \text{A}$$

Approximate voltage drop referred to as secondary for 0.8 power factor lagging is

$$= I_2(R_{02}\cos\theta + X_{02}\sin\theta)$$
$$= 12.5\times(0.625\times 0.8 + 1.548\times 0.6) = 17.86\ \text{V}$$

$\therefore$ Secondary terminal voltage = 600 − 17.86 = 582.14 V (approx.) ■

Example 1.12 A single-phase 200/400 V, 6 kVA, 50 Hz transformer gives the following test results:

OC test (LV side): 200, 0.8 A, 80 W

SC test (HV side): 25 V, 10 A, 90 W

where LV and HV are low voltage and high voltage, respectively.

Determine the following: (i) The circuit constants referred to as LV side and (ii) the applied voltage and efficiency if the output current is 12 A at the terminal voltage 400 V at 0.8 power factor lagging.

Solution

(i) Turns ratio $(a) = \frac{200}{400} = \frac{1}{2}$

OC test:

Instrument placed on LV side

$$V_1 = 200\ \text{V},\quad I_0 = 0.8\ \text{A},\quad W_0 = 80\ \text{W}$$

Since $$W_0 = V_1 I_0 \cos\theta_0$$

$\therefore$ $$\cos\theta_0 = \frac{W_0}{V_1 I_1} = \frac{80}{200\times 0.80} = 0.5$$

$$\sin\theta_0 = 0.866$$

$\therefore$ $$I_w = I_0 \cos\theta_0 = 0.8\times 0.5 = 0.4 \text{ A}$$

and $$I_\mu = I_0 \sin\theta_0 = 0.8\times 0.866 = 0.6928 \text{ A}$$

$\therefore$ $$R_0 = \frac{V_1}{I_w} = \frac{200}{0.4} = 500\ \Omega \quad \text{and} \quad X_0 = \frac{V_1}{I_\mu} = \frac{200}{0.6928} = 288.681\ \Omega$$

SC test:
Instruments placed on HV side

$$V_{SC} = 25 \text{ V},\ I_{SC} = 10 \text{ A},\ P_{SC} = 90 \text{ W}$$

$$Z_{02} = \frac{25}{10} = 2.5\ \Omega$$

$$R_{02} = \frac{P_{sc}}{I_{SC}^2} = \frac{90}{10^2} = 0.9\ \Omega$$

$$X_{02} = \sqrt{Z_{02}^2 - R_{02}^2} = \sqrt{2.5^2 - 0.9^2} = 2.33\ \Omega$$

$$R_{01} = a^2 R_{02} = \frac{1}{4}\times 0.9 = 0.225\ \Omega$$

$$X_{01} = a^2 X_{02} = \frac{1}{4}\times 2.33 = 0.5825\ \Omega$$

$\therefore$ The circuit parameters referred to as LV side are

$$R_0 = 500\Omega \text{ and } X_0 = 288.68\Omega,\ R_{01} = 0.255\Omega,\ X_{01} = 0.5825\Omega.$$

To maintain the output voltage at 400 V, the applied voltage can be calculated from Figure E1.3.

$$I_2 = 12\ A,\ I_1 = \frac{I_2}{a} = \frac{12}{\frac{1}{2}} = 24 \text{ A (neglecting } I_0)$$

$$(V_1')^2 = (V_1 \cos\theta_1 + I_1 R_{01})^2 + (V_1 \sin\theta_1 + I_1 X_{01})^2$$

$$= (200\times 0.8 + 24\times 0.255)^2 + (200\times 0.6 + 24\times 0.5825)^2$$

$$= 45{,}546.50$$

i.e. $$V_1' = 213.4 \text{ V}$$

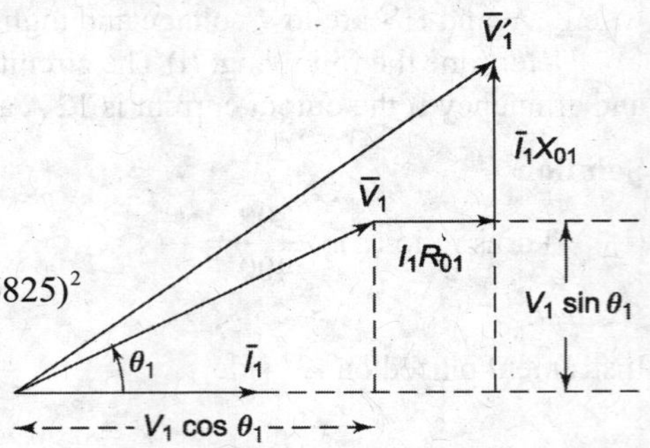

Figure E1.3

To maintain the output voltage at 400 V for the given load, the applied voltage becomes 213.4 V.
From OC test: iron loss = 80 W
From SC test: copper loss = 90 W for 10 A current on the HV side.

$$\text{Full-load copper loss} = \left(\frac{12}{10}\right)^2 \times 90 = 129.6\text{ W}$$

$$\text{Total loss of the transformer} = 129.6 + 80 = 209.6\text{ W}$$

$$\therefore \quad \text{Efficiency of the transformer} = \frac{6\times 10^3 \times 0.8}{6\times 10^3 \times 0.8 + 209.6} = 0.9580\text{ p.u.} = 95.8\ \%$$ ■

1.29 PER UNIT RESISTANCE, LEAKAGE REACTANCE AND IMPEDANCE VOLTAGE DROP

Full-load voltage of a transformer can be expressed as a fraction of the full-load terminal voltage.

Let I_{1fl} be the full-load primary current, I_{2fl} be the full-load secondary current, V_1 be the rated primary voltage and V_2 be the rated secondary voltage.

Per unit resistance drop of a transformer

$$= \frac{(\text{Full-load primary current})\times(\text{Equivalent resistance referred to as primary})}{\text{Rated primary voltage}}$$

$$= \frac{(\text{Full-load secondary current})\times(\text{Equivalent resistance referrred to as secondary})}{\text{Rated secondary voltage}}$$

$$= \frac{I_{1fl}R_{01}}{V_1} = \frac{I_{2fl}R_{02}}{V_2} \tag{1.46}$$

$$R_{epu} = \frac{I_{1fl}^2 R_{01}}{V_1 I_{1fl}} = \frac{P_{Cufl}}{S_{1fl}}$$

$$\therefore \quad = \frac{I_{2fl}^2 R_{02}}{V_2 I_{2fl}} = \frac{P_{Cufl}}{S_{2fl}}$$

$$= \frac{\text{Total copper loss at full load}}{\text{Rated volt-amperes}} \tag{1.47}$$

Per unit reactance drop of a transformer

$$= \frac{I_{1fl}X_{01}}{V_1} = \frac{I_{2fl}X_{02}}{V_2} \tag{1.48}$$

Per unit reactance drop of a transformer is called per unit reactance and it is given by

$$X_{epu} = \frac{I_{1fl}X_{01}}{V_1} = \frac{I_{2fl}X_{02}}{V_2} \tag{1.49}$$

Per unit impedance drop of a transformer is called per unit impedance and it is given by

$$Z_{epu} = \frac{I_{1fl}Z_{01}}{V_1} = \frac{I_{2fl}X_{02}}{V_2} \tag{1.50}$$

1.30 VOLTAGE REGULATION OF TRANSFORMER

With constant voltage applied in primary, the secondary terminal voltage will decrease due to voltage drop across its internal resistance and leakage reactance.

Let ${}_0V_2$ and V_2 be the secondary terminal voltages at no load and on load respectively. There are three kinds of voltage regulation, which are discussed below.

1.30.1 Inherent Voltage Regulation

The difference ${}_0V_2 - V_2$ is known as *inherent voltage regulation* of the transformer.

1.30.2 Voltage Regulation Down

If inherent voltage drop is divided by ${}_0V_2$, it is known as *voltage regulation down.* Mathematically, we can write

$$\mu = \frac{{}_0V_2 - V_2}{{}_0V_2} \text{ p.u.} \tag{1.51}$$

and also

$$\%\,\mu = \frac{{}_0V_2 - V_2}{{}_0V_2} \times 100 \tag{1.52}$$

1.30.3 Voltage Regulation Up

If inherent voltage drop is divided by V_2, it is known as *voltage regulation up.* Mathematically, we can write it as

$$\mu = \frac{{}_0V_2 - V_2}{V_2} \text{ p.u.} \tag{1.53}$$

and also

$$\%\mu = \frac{{}_0V_2 - V_2}{V_2} \times 100 \tag{1.54}$$

The secondary terminal voltage not only depends on load current but also on the power factor of the load. The regulation is said to be at full load provided V_2 is determined for full load and at specified power factor condition. V_2 drops more and more with increasing load current. For lagging power factor load, $V_2 < E_2$, the voltage regulation is positive. For leading power factor load, $V_2 > E_2$, the voltage regulation is negative.

To maintain constant secondary terminal voltage on load, the primary terminal voltage is adjusted. It is expected that voltage drop would be as small as possible. Therefore, the lesser the value of regulation, the better is the performance of a transformer.

1.31 CALCULATION FOR VOLTAGE REGULATION

The voltage regulation up is expressed mathematically by

$$\%\mu = \frac{{}_0V_2 - V_2}{{}_0V_2} \times 100 = \frac{I_2\left(R_{02}\cos\theta \pm X_{02}\sin\theta\right)}{{}_0V_2} \times 100 \tag{1.55}$$

Positive sign is for lagging power factor and negative sign is for leading power factor.

1.31.1 Zero Voltage Regulation

For lagging power factor and unity power factor, ${}_0V_2 > V_2$. Therefore, we get positive voltage regulation. For leading power factor, V_2 starts increasing. At a certain leading power factor, ${}_0V_2 = V_2$ and hence regulation becomes zero. If the load power factor is further increased, ${}_0V_2$ becomes less than V_2 and hence regulation becomes negative. For zero voltage regulation, we have

$$ {}_0V_2 - V_2 = 0 $$

i.e.,
$$ I_2\left(R_{02}\cos\theta - X_{02}\sin\theta\right) = 0 $$

i.e.,
$$ \tan\theta = \frac{R_{02}}{X_{02}} $$

∴
$$ \cos\theta = \cos\left\{\tan^{-1}\left(\frac{R_{02}}{X_{02}}\right)\right\} \tag{1.56} $$

Equation (1.56) shows the leading power factor at which voltage regulation becomes zero.

1.31.2 Condition for Maximum Voltage Regulation

Maximum voltage regulation can be obtained for lagging power factor. For maximum voltage regulation, we have

$$ \frac{d}{d\theta}\left[\frac{I_2\left(R_{02}\cos\theta + X_{02}\sin\theta\right)}{{}_0V_2}\right] = 0 $$

i.e.,
$$ -R_{02}\sin\theta + X_{02}\cos\theta = 0 $$

i.e.,
$$ \tan\theta = \frac{X_{02}}{R_{02}} \tag{1.57} $$

This is satisfied only when the power factor of the load is lagging. The regulation is maximum when the load power factor angle is equal to the impedance angle of the transformer.

i.e.,
$$ \cos\theta = \cos\left\{\tan^{-1}\left(\frac{X_{02}}{R_{02}}\right)\right\} \tag{1.58} $$

Equation (1.58) shows the power factor of the load at which voltage regulation is maximum.

Figure 1.41 shows the variation of regulation with the power factor of the load. Figure 1.41 also shows that regulation becomes maximum when the power factor of the load is lagging and it becomes zero when the power factor is leading.

Example 1.13 A 6 kVA, 500/250 V, 50 Hz, single-phase transformer has the following test results:

OC test (LV side): 250 V, 1.5 A, 80 W
SC test (HV side): 22 V, 10 A, 90 W

Determine the following:

(i) The approximate equivalent circuit referred to as HV side.
(ii) Voltage regulation and efficiency at full load and 0.8 power factor lagging load.
(iii) The efficiency of the transformer at half of full load and 0.8 power factor lagging load.

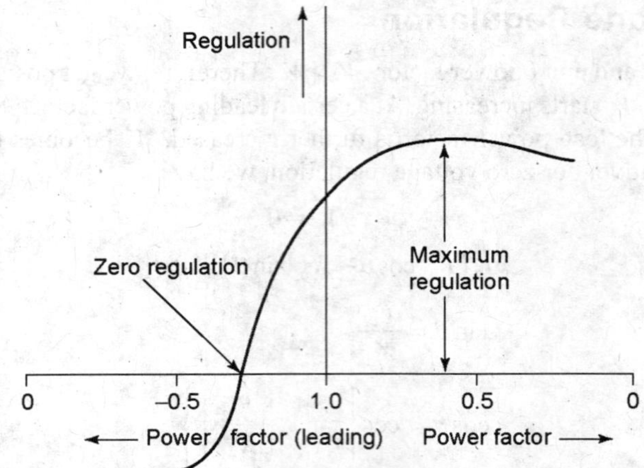

Figure 1.41 Effect of Load Power Factor on Regulation

Solution

(i) Turns ratio $(a) = \dfrac{500}{250} = 2$

OC test:

The HV side is kept open

$$(I_0)_{lv} = 1.5 \text{ A}$$

i.e., $$\frac{(I_0)_{hv}}{(I_0)_{lv}} = \frac{1}{a}$$

i.e., $$(I_0)_{hv} = \frac{(I_0)_{lv}}{a} = \frac{1.5}{2} = 0.74 \text{ A}$$

Instruments readings when they are placed on HV side

$$V_1 = 500,\ I_0 = 0.75 \text{ A and } W_0 = 80 \text{ W}$$

$\therefore$ $$\cos\theta_0 = \frac{W_0}{V_1 I_0} = \frac{80}{500 \times 0.75} = 0.214$$

$\therefore$ $$I_W = I_0 \cos\theta_0 = 0.75 \times 0.214 = 0.1605 \text{ A}$$

and $$I_\mu = I_0 \sin\theta_0 = 0.75 \times \sqrt{1 - \cos^2\theta_0} = 0.75 \times 0.976 = 0.732 \text{ A}$$

$\therefore$ $$R_0 = \frac{V_1}{I_W} = \frac{500}{0.1605} = 3{,}115.26\ \Omega \quad \text{and} \quad X_0 = \frac{V_1}{I_\mu} = \frac{500}{0.732} = 683.06\ \Omega$$

referred to as HV side.

SC test:
Instruments placed on HV side.

$$V_{SC} = 22 \text{ V}, I_{SC} = 10 \text{ A}, P_{SC} = 90 \text{ W}$$

∴ $$Z_{01} = \frac{V_{SC}}{I_{SC}} = \frac{22}{10} = 2.2\,\Omega,\ R_{01} = \frac{P_{SC}}{I_{SC}^2} = \frac{90}{10^2} = 0.9\,\Omega$$

and $$X_{01} = \sqrt{Z_{01}^2 - R_{01}^2} = 2.0\,\Omega$$

The equivalent circuit is shown in Figure E1.4.

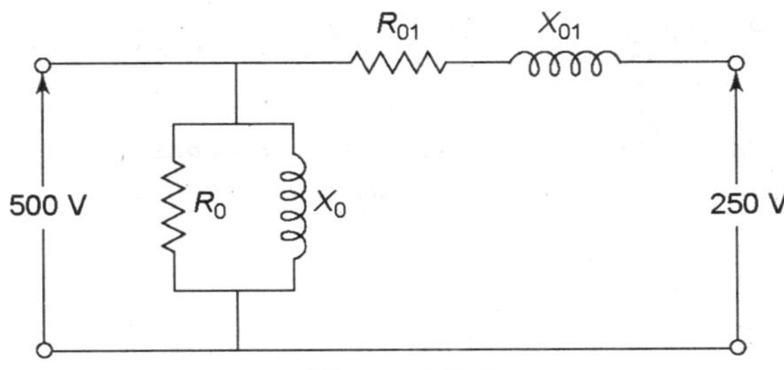

Figure E1.4

(ii) Full-load secondary current of the transformer

$$I_2 = \frac{6 \times 10^3}{250} = 24 \text{ A}$$

∴ Approximate voltage drop of the transformer referred to as secondary

$$\Delta V = I_2 (R_{02} \cos\theta + X_{02} \sin\theta)$$

Now $$R_{02} = \frac{R_{01}}{a^2} = \frac{0.9}{4} = 0.225\,\Omega \quad \text{and} \quad X_{02} = \frac{X_{01}}{a^2} = \frac{2}{4} = 0.5\,\Omega$$

Here $\cos\theta = 0.8$ (lagging)

∴ $\sin\theta = 0.6$

∴ $\Delta V = 24 \times (0.225 \times 0.8 + 0.5 \times 0.6) = 11.52 \text{ V}$

∴ Regulation of the transformer $= \dfrac{11.52}{250} = 0.046$ p.u. $= 4.6\%$

From OC test: iron loss of the transformer = 80 W
From SC test: copper loss of the transformer = 90 W when primary current is 10 A, that is, secondary current is 20 A.

$$\text{Full-load copper loss of the transformer} = \left(\frac{24}{20}\right)^2 \times 90 = 129.6 \text{ W}$$

∴ Total losses of the transformer at full load = 80 + 129.6 = 209.6 W
∴ Output of the transformer at full load and 0.8 p.f. lagging = $V_2 I_2 \cos\theta_2$

Secondary terminal voltage (V_2) = 250 − Approximate voltage drop = 250 − 11.52 = 238.48 V

$\therefore$ Output power = 238.48 × 24 × 0.8 = 4,578.816 W

$\therefore$ Efficiency of the transformer

$$= \frac{\text{Output}}{\text{Output} + \text{Total loss}} = \frac{4{,}578.816}{4{,}578.816 + 209.6} = 0.9562 \text{ p.u.} = 95.62\%$$ ■

Another method to calculate the efficiency

$$\text{Output power} = 6 \times 10^3 \times 0.8 = 4{,}800 \text{ W}$$

$\therefore$ Efficiency of the transformer

$$= \frac{\text{Output}}{\text{Output} + \text{Total loss}} = \frac{4{,}800}{4{,}800 + 209.6} = 0.9581 \text{ p.u.} = 95.81\%$$

This method gives an approximate answer. If equivalent resistance and reactance are available, it is suggested that the previous method should be used. Otherwise, the approximate method may be used.

(iii) Copper loss of the transformer at half of full load $= \left(\frac{1}{2}\right)^2 \times 129.6 \text{ W} = 32.4 \text{ W}$

$$\text{Voltage drop at half-load} = \frac{\Delta V}{2} = 5.76 \quad \therefore \; V_2' = 250 - 5.76 = 244.24 \text{ V}$$

$$\therefore \quad \text{Output} = \frac{1}{2} \times 244.24 \times 24 \times 0.8 = 2{,}344.704 \text{ W}$$

$\therefore$ Efficiency of the transformer

$$= \frac{\text{Output}}{\text{Output} + \text{Total loss}} = \frac{2{,}344.704}{2{,}344.704 + 80 + 32.4} = 0.9542 \text{ pu} = 95.42\%$$ ■

Example 1.14 A 12 kVA, 1,000/100 V, 50 Hz, single-phase transformer has the following test results:

OC test (LV side): 100 V, 0.7 A, 85 W
SC test (HV side): 60 V, 10 A, 95 W

Determine the following:

(i) Core loss of the transformer.
(ii) Equivalent resistance and leakage reactance referred to as HV side.
(iii) Equivalent resistance and leakage reactance referred to as LV side.
(iv) Regulation of the transformer at full load and half load at 0.8 power factor lagging.
(v) Transformer terminal voltage at full load at 0.8 power factor lagging.
(vi) Efficiency of the transformer at full load and half load at 0.8 power factor lagging.

Solution

$$\text{Turns ratio } (a) = \frac{1{,}000}{100} = 10$$

(i) Core loss or iron loss of the transformer = 85 W

(ii) From SC test

$$V_{SC} = 60\text{ V}, I_{SC} = 10\text{ A}, P_{SC} = 95\text{ W}$$

$$\therefore \quad Z_{01} = \frac{V_{SC}}{I_{SC}} = \frac{60}{10} = 6\,\Omega,\ R_{01} = \frac{P_{SC}}{I_{SC}^2} = \frac{95}{10^2} = 0.95\,\Omega \text{ and}$$

$$\therefore \quad X_{01} = \sqrt{Z_{01}^2 - R_{01}^2} = \sqrt{6^2 - 0.95^2} = 5.924\,\Omega$$

$\therefore$ Resistance and leakage reactance referred to as HV side are 0.095 Ω and 5.924 Ω, respectively.

(iii) $$R_{02} = \frac{R_{01}}{a^2} = 0.0095\,\Omega \text{ and } X_{02} = \frac{X_{01}}{a^2} = 0.05924\,\Omega$$

(iv) $$I_{2fl} = \frac{12\times 10^3}{100} = 120\text{ A}$$

$$\therefore \quad \text{Approximate voltage drop} = I_2(R_{02}\cos\theta + X_{02}\sin\theta)$$

$$= 120\times(0.0095\times 0.8 + 0.05924\times 0.6) = 5.177\text{ V}$$

$$\text{Regulation of the transformer} = \frac{5.177}{100} = 0.05177\text{ p.u.} = 5.177\,\%$$

(v) At full-load transformer terminal voltage = 100 − 5.177 = 94.823 V

(vi) Output = 94.823 × 120 × 0.8 = 9,103 W

$$\therefore \quad \text{Copper loss at full load} = \left(\frac{12}{10}\right)^2 \times 95 = 136.8\text{ W}$$

$\therefore$ Total loss of the transformer = 85 + 136.8 = 221.8 W

$$= \frac{\text{Output}}{\text{Output} + \text{Losses}} = \frac{9103}{9103 + 221.8} = 0.9763\text{ p.u.} = 97.63\%$$

$$\text{At half load, } I_{2(\frac{1}{2}fl)} = \frac{I_{2fl}}{2} = \frac{120}{2} = 60\text{A}$$

At half load, approximate voltage drop of the transformer

$$= I_{2(\frac{1}{2}fl)}(R_{02}\cos\theta + X_{02}\sin\theta)$$

$$= 60 \times (0.0095 \times 0.8 + 0.05924 \times 0.6) = 2.588\text{ V}$$

$$\therefore \quad V_1 = (100 - 2.588)\text{ V} = 97.412\text{ V}$$

Output power = 97.412 × 60 × 0.8 = 4,675.776 W

$$\text{Copper loss at half-load} = \left(\frac{1}{2}\right)^2 \times 136.8 = 34.2\text{ W}$$

$\therefore$ Total losses = 85 + 34.2 = 119.2 W

$$\therefore \quad \text{Efficiency of the transformer at half-load} = \frac{4{,}675.776}{4{,}675.776 + 119.2} = 0.9751\text{ p.u.} = 97.51\%$$

■

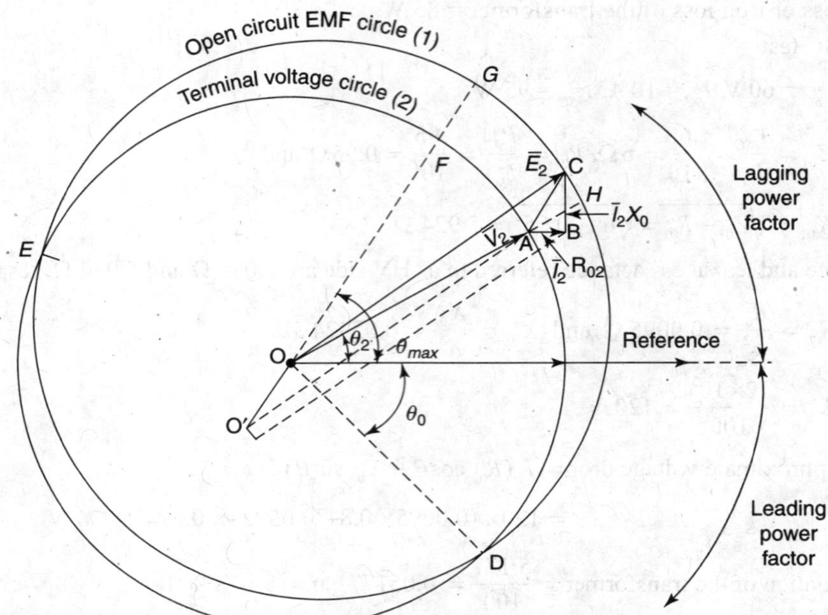

Figure 1.42 Kapp's Regulation Diagram

1.31.3 Kapp's Regulation

Kapp had designed a diagram shown in Figure 1.42 to determine the regulation at any power factor. The description of the construction of the diagram is shown below.

Load current (I_2) is taken as a reference phasor. OA representing V_2 is drawn at angle θ_2 with I_2. AB represents I_2R_{02} drawn parallel to I_2, whereas BC represents I_2X_{02} drawn perpendicular to AB, i.e., I_2. Here OC represents secondary emf (${}_0V_2 = E_2$) at no-load. The circle 1 known as often circuit EMF circle is drawn with O as centre and OC as radius. The line OO′ is drawn parallel to AC representing I_2Z_{02}. With O′ as centre and OA as radius, the circle 2 known as terminal voltage circle is drawn, which intersects with circle 1 at the points D and E. The region above and below the reference line represents the lagging and leading power factors region, respectively. The point D is the point corresponding to zero regulation. The intercept FG gives the maximum regulation, which is drawn through O and drawn parallel to AC. The regulation at any power factor angle θ is obtained by extending OA to meet the outer circle at H. The regulation at the required power factor cosθ is represented by AH, which is the intercept between the two circles.

1.32 LOSSES IN A TRANSFORMER

Two types of losses occur in a transformer:

- Core loss or iron loss occurs in a transformer because it is subjected to an alternating flux.
- The windings carry current due to loading and hence copper losses occur.

1.32.1 Core or Iron Loss

The separation of core losses has already been introduced. The alternating flux gets set up in the core and it undergoes a cycle of magnetization and demagnetization. Therefore, loss of energy occurs in this process due to hysteresis. This loss is called *hysteresis loss* (P_h), which is expressed by

$$P_h = K_h B_m^{1.6} f V \quad \text{W} \tag{1.59}$$

where K_h is the hysteresis constant depending on the material, B_m is the maximum flux density, f is the frequency and V is the volume of the core.

The induced emf in the core sets up eddy current in the core, and hence eddy current loss (P_e) occurs, which is given by

$$P_e = K_e B_m^2 f^2 t^2 \quad \text{W per model} \tag{1.60}$$

where K_e is the eddy current constant and t is the thickness of the core.

Since the supply voltage V_1 at rated frequency f is always constant, the flux in the core is almost constant. Therefore, flux density in the core remains constant. Hence, hysteresis and eddy current losses are constant at all loads. Thus, the core loss or iron loss is also known as constant loss. The iron loss is denoted by P_i.

Iron loss is reduced using high-grade core material such as silicon steel having very low hysteresis loop for reducing hysteresis loss and laminated core for reducing the eddy current loss.

1.32.2 Copper Loss

The loss of power in the form I^2R due to the resistances of the primary and secondary windings is known as *copper losses.* The copper loss also depends on the magnitude of currents flowing through the windings. The total Cu loss is given by

$$P_{\text{Cu}} = I_1^2 R_1 + I_2^2 R_2 = I_1^2 R_{01} = I_2^2 R_{02} \tag{1.61}$$

Copper losses are determined on the basis of R_{01} or R_{02} which is determined from short circuit test. Since the standard operating temperature of electrical machine is taken as 75 °C, it is then corrected to 75 °C.

The copper loss due to full-load current is known as full-load Cu loss. If the load on the transformer is half, the Cu loss is known as half-load Cu loss, which is less than the full-load Cu loss. The Cu loss is also known as *variable loss.*

There are two other losses known as stray loss and dielectric loss. Since leakage field is present in a transformer, eddy currents are induced in the conductors, tanks walls and bolts etc. Stray losses occur due to this eddy currents. Dielectric loss occurs in insulating materials coil and solid insulation. These two losses are small and hence neglected.

Therefore, the total loss of the transformer = Iron loss + Cu loss = $P_i + P_{Cu}$

1.33 EFFICIENCY OF A TRANSFORMER

Due to the losses in a transformer, its output power is less than the input power.

∴ Power output = Power input − Total losses

∴ Power input = Power output + Total losses = Power output + $P_i + P_{Cu}$

The ratio of power output to power input of any device is called its *efficiency* (η).

$$\therefore \quad \eta = \frac{\text{Power ouput}}{\text{Power input}} = \frac{\text{Power ouput}}{\text{Power ouput} + P_i + P_{Cu}} \tag{1.62}$$

Output power of a transformer at full-load = $V_2 I_{2fl} \cos\theta$, where $\cos\theta$ is the power factor of the load, I_{2fl} is the secondary current at full load and V_2 is the rated secondary voltage of the transformer.

Full-load copper loss of the transformer = $I_{2fl}^2 R_{02}$.

$\therefore$ Efficiency of the transformer at full load is given by

$$\eta_{fl} = \frac{V_2 I_{2fl} \cos\theta}{V_2 I_{2fl} \cos\theta + P_i + I_{2fl}^2 R_{02}} \tag{1.63}$$

Now $V_2 I_{2fl}$ = *VA* rating of the transformer.

$$\therefore \quad \eta = \frac{(VA \text{ rating of the transformer}) \times \cos\theta}{(VA \text{ rating of the transformer}) \times \cos\theta + P_i + I_{2fl}^2 R_{02}} \text{ in p.u.}$$

$$\text{i.e.,} \quad \eta = \frac{(VA \text{ rating of the transformer}) \times \cos\theta}{(VA \text{ rating of the transformer}) \times \cos\theta + P_i + I_{2fl}^2 R_{02}} \times 100 \tag{1.64}$$

Efficiency of the transformer at any load *m* is given by

$$\eta = \frac{m \times (VA \text{ rating of the transformer}) \times \cos\theta}{m \times (VA \text{ rating of the transformer}) \times \cos\theta + P_i + m^2 P_{Cu\,fl}} \text{ in p.u.} \tag{1.65}$$

where $m = \dfrac{\text{Actual load}}{\text{Full-load}}$ and $P_{Cu\,fl}$ is the Cu loss of the transformer at full load.

1.34 CONDITION FOR MAXIMUM EFFICIENCY

During working of a transformer at constant voltage and frequency, its efficiency varies with the load. Its efficiency increases as the load increases. At a certain load, its efficiency becomes maximum. If the transformer is further loaded, its efficiency starts decreasing. Figure 1.43 shows the plot of efficiency versus load current.

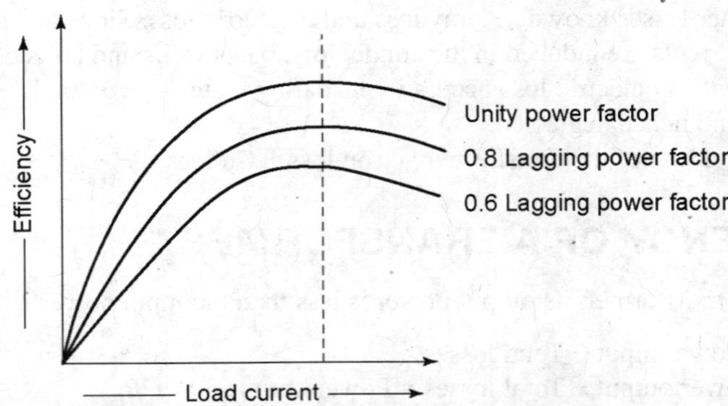

Figure 1.43 Comparison Efficiency and Current

To determine the condition of maximum efficiency, let us assume that the power factor of the load remains constant and the secondary terminal voltage (V_2) is constant. Therefore, efficiency becomes only a function of load current (I_2).

For maximum efficiency

$$\frac{d\eta}{dI_2} = 0$$

Now,

$$\eta = \frac{V_2 I_2 \cos\theta}{V_2 I_2 \cos\theta + P_i + I_2^2 R_{02}}$$

$\therefore$

$$\frac{d\eta}{dI_2} = \frac{d}{dI_2}\left(\frac{V_2 I_2 \cos\theta}{V_2 I_2 \cos\theta + P_i + I_2^2 R_{02}}\right) = 0$$

i.e.,

$$\frac{V_2 \cos\theta\left(V_2 I_2 \cos\theta + P_i + I_2^2 R_{02}\right) - V_2 I_2 \cos\theta\left(V_2 \cos\theta + 2 I_2 R_{02}\right)}{\left(V_2 I_2 \cos\theta + P_i + I_2^2 R_{02}\right)^2} = 0$$

i.e.,

$$V_2 I_2 \cos\theta + P_i + I_2^2 R_{02} - V_2 I_2 \cos\theta - 2 I_2^2 R_{02} = 0$$

i.e.,

$$P_i = I_2^2 R_{02} \tag{1.66}$$

To achieve maximum efficiency, *Iron loss* = *Cu loss*

i.e., *Constant loss* = *Variable loss*

1.34.1 Load Current at Maximum Efficiency

Let I_{2M} be the load current at maximum efficiency.

$\therefore$

$$I_{2M}^2 R_{02} = P_i$$

i.e.,

$$I_{2M}^2 = \frac{P_i}{R_{02}}$$

Let I_{2fl} be the full-load current.

$\therefore$

$$I_{2M}^2 = \frac{P_i I_{2fl}^2}{I_{2fl}^2 R_{02}} = I_{2fl}^2 \times \frac{P_i}{P_{\text{Cu}\,fl}} \tag{1.67}$$

i.e.,

$$I_{2M} = I_{2fl}\sqrt{\frac{P_i}{P_{\text{Cu}\,fl}}}$$

Equation (1.67) shows the load current in terms of full-load current at maximum efficiency.

1.34.2 kVA Supplied at Maximum Efficiency

For constant V_2, the kVA supplied is the function of load current only.

$\therefore$

$$\text{kVA at } \eta_{\max} = I_{2M} V_2 \times 10^{-3} = V_2 I_{2fl}\sqrt{\frac{P_i}{P_{\text{Cu}\,fl}}} \times 10^{-3}$$

$$= (\text{kVA rating}) \times \sqrt{\frac{P_i}{P_{\text{Cu}\,fl}}}$$

In general,

$$\eta_{max} = \frac{V_2 I_{2M} \cos\theta}{V_2 I_{2M} \cos\theta + 2P_i} \times 100 = \frac{m V_2 I_{2fl} \cos\theta}{m V_2 I_{2fl} \cos\theta + P_i + I_{2fl}^2 R_{02}} \times 100 \quad (1.68)$$

where

$$m = \sqrt{\frac{P_i}{P_{Cu\,fl}}} \quad (1.69)$$

Example 1.15 A single-phase 10 kVA, 2,000/200 V, 50 Hz transformer has the following test results:

OC test (LV side): 200 V, 0.8 A, 60 W
SC test (HV side): 40 V, 4 A, 70 W

Calculate the following:

(i) The efficiency of the transformer at half load and 0.8 power factor lagging.
(ii) The load kVA at which maximum efficiency occurs and also the maximum efficiency at 0.8 power factor lagging.
(iii) The voltage regulation at 0.8 power factor leading on full-load condition.

Solution

From *OC test*, iron loss of the transformer = 60 W

Full load primary current (I_1) $= \frac{10 \times 10^3}{2,000} = 5\text{ A}$

From SC test

$$Z_{01} = \frac{40}{4} = 10\ \Omega,\ R_{01} = \frac{70}{4^2} = \frac{70}{16} = 4.375\ \Omega$$

and

$$X_{01} = \sqrt{Z_{01}^2 - R_{01}^2} = \sqrt{10^2 - 4.375^2} = 8.99\ \Omega$$

Again, $R_{02} = \frac{R_{01}}{a^2} = \frac{4.375}{10^2} = 0.04375\ \Omega$ and $X_{02} = \frac{X_{01}}{a^2} = \frac{8.99}{10^2} = 0.089\ \Omega$

∴ Approximate voltage drop at secondary

$$\Delta V = I_2 (R_{02} \cos\theta + X_{02} \sin\theta)$$

∴ Secondary side current (I_2) at half-load $= \frac{5 \times 10}{2} = 25\text{ A}$

∴ $\Delta V = 25 \times (0.04375 \times 0.8 + 0.0899 \times 0.6) = 2.2235\text{ V}$

∴ $V_2' = 200 - 2.2235 = 197.7765\text{ V}$

∴ Output power $= 197.7765 \times 25 \times 0.8 = 3,955.53$ W at full-load.

∴ Copper loss at half-load $= \left(\frac{\frac{5}{2}}{4}\right)^2 \times 70 = \left(\frac{2.5}{4}\right)^2 \times 70 = 27.34\text{ W}$

∴ Efficiency of the transformer at half-load $= \dfrac{\frac{1}{2} \times 3,955.536}{\frac{1}{2} \times 3,955.53 + 60 + 27.34}$

$= 0.9577\text{ p.u.} = 95.77\%$

(ii) At maximum efficiency, Cu loss = Iron loss = 60 W

$\therefore I_1^2 R_{01} = 60$, where I_1 is the primary current at maximum efficiency

$$I_1^2 = \frac{60}{R_{01}} = \frac{60}{4.375} = 13.714$$

$$\therefore \quad I_1 = 3.703 \text{ A}$$

Here V_2 is not constant.

$\therefore$ Approximate voltage drop at secondary

$$\Delta V = I_2(R_{02}\cos\theta + X_{02}\sin\theta)$$

$$\therefore \; \Delta V = 3.703 \times (4.375 \times 0.8 + 8.99 \times 0.6) = 32.934 \text{ V}$$

$$\therefore V_1' = V_1 - \Delta V = 2{,}000 - 32.934 = 1{,}967.066 \text{ V}$$

$\therefore$ Load kVA at maximum efficiency = 1,967.066 × 3.703 VA = 7.284 kVA

$\therefore$ Output = $7.284 \times 10^3 \times 0.8$ = 5,827.2 W

$$\therefore \eta_{\max} = \frac{\text{Output}}{\text{Output} + 2P_i} = \frac{5{,}827.2}{5{,}827.2 + 2 \times 60} = 0.9798 \text{ p.u.} = 97.98\%$$

(iii) Voltage regulation at 0.8 p.f. leading on full-load condition:

$$I_2 = 50 \text{ A at full load.}$$

$$\therefore \quad \Delta V = I_2 (R_{02}\cos\theta - X_{02}\sin\theta)$$

$$= 50 \times (0.04375 \times 0.8 - 0.0899 \times 0.6) = -0.947 \text{ V.}$$

$$\therefore \text{Regulation of the transformer} = \frac{-0.947}{200} = -4.735 \times 10^{-3} \text{ p.u.} = -0.4735\%$$ ■

1.35 ALL-DAY EFFICIENCY

The ratio of output in watts to input in watts is called *commercial efficiency* of a transformer. Distribution transformers are used for supplying lighting and general networks. Distribution transformers are energized throughout the day. Their secondaries are at no load most of the time in a day except during the hours of lighting period. Core loss occurs throughout the day. Copper loss occurs only when they are loaded and hence is less important. To judge their performance, *all-day efficiency* or *operational efficiency* is calculated. The all-day efficiency is defined by

$$\eta_{\text{all-day}} = \frac{\text{Output in kWh}}{\text{Input in kWh}} \quad \text{(for 24 hours)} \tag{1.70}$$

The all-day efficiency is less than the commercial efficiency of a transformer.

Example 1.16 A 200 kVA single-phase transformer is in circuit throughout 24 hours. For 8 hours in a day, the load is 150 kW at 0.8 power factor lagging and for 7 hours, the load is 90 kW at 0.9 power factor. Remaining time or the rest period, it is at no-load condition. Full-load Cu loss is 4 kW and the iron loss is 1.8 kW. Calculate the all-day efficiency of the transformer.

Solution

Full-load output = 200 kVA, Full-load Cu loss = 4 kW, Iron loss = 1.8 kW.

$$\text{Copper loss for 24 hours} = \left(\frac{\frac{150}{0.9}}{200}\right)^2 \times 4 \times 8 + \left(\frac{\frac{90}{1}}{200}\right)^2 \times 4 \times 7 = 27.89 \text{ kWh}$$

$$\text{Iron loss for 24 hours} = 1.8 \times 24 = 43.2 \text{ kWh}$$

$$\text{All-day output} = (150 \times 8) + (90 \times 7) = 1{,}830 \text{ kWh}$$

$$\text{All-day input} = 1{,}830 + 27.89 + 43.2 = 1{,}901.09 \text{ kWh}$$

$$\text{All-day efficiency } (\eta_{\text{all-day}}) = \frac{\text{All-day output}}{\text{All-day input}}$$

$$= \frac{1{,}830}{1{,}901.09} = 0.9626 \text{ p.u.} = 96.26\%$$ ■

1.36 POLARITY TEST OF A SINGLE-PHASE TRANSFORMER

Polarity testing of transformers is vital before connecting them in parallel. Otherwise, with incorrect polarity, it is not possible to connect them in parallel. The rated voltage is applied to the primary and its two terminals are marked as A_1 and A_2, respectively, as shown in Figures 1.44(a) and 1.45(b), respectively. The secondary winding terminals are also marked as a_1 and a_2, shown in Figures 1.44(a) and 1.45(b), respectively. Now a voltmeter is connected across A_2 and a_2. If it measures the difference of E_1 and E_2, A_2 and a_2 are of the same polarity. If it measures the addition of E_1 and E_2, A_2 and a_2 are of opposite polarity.

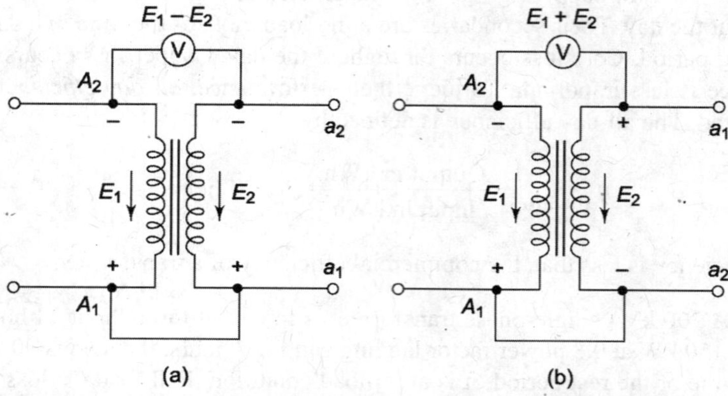

Figure 1.44 Polarity Test of a single-phase Two Winding Transformer

1.37 SUMPNER'S TEST

To determine the rise of maximum temperature of a transformer, its load test is of utmost importance. Using suitable load impedance, small transformers can be put on full load. The full-load test of large transformers is not possible because considerable wastage of energy occurs and it is difficult to get a suitable load for absorbing full-load power. Sumpner's test is used to put large transformer on full load. This test can also be used to determine the efficiency of a transformer. Figure 1.45 shows the schematic diagram of Sumpner's test. This test is also known as *back to back test* or *load test.*

This test requires two identical transformers. The two primaries are connected in parallel and are energized at rated voltage and rated frequency. The wattmeter W_1 records the reading of core loss of both the transformers. Next the two secondaries are connected in series in such a way that their polarities are in phase opposition and the reading of the voltmeter V_2 becomes zero. With the help of voltage regulator fed from source, a voltage is injected to the secondary of the transformers, which is adjusted until the rated secondary current flows. The voltmeter reads a voltage, which is the leakage impedance drop of the two transformers. The reading of W_1 remains unaltered. The wattmeter W_2 reads the total copper (Cu) loss of the two transformers. Although the transformers are not supplying any load current, this test measures the full iron loss as well as copper loss of the transformers. The net input during this test is $W_1 + W_2$. To measure the temperature rise, the two transformers are kept under rated loss conditions for several hours.

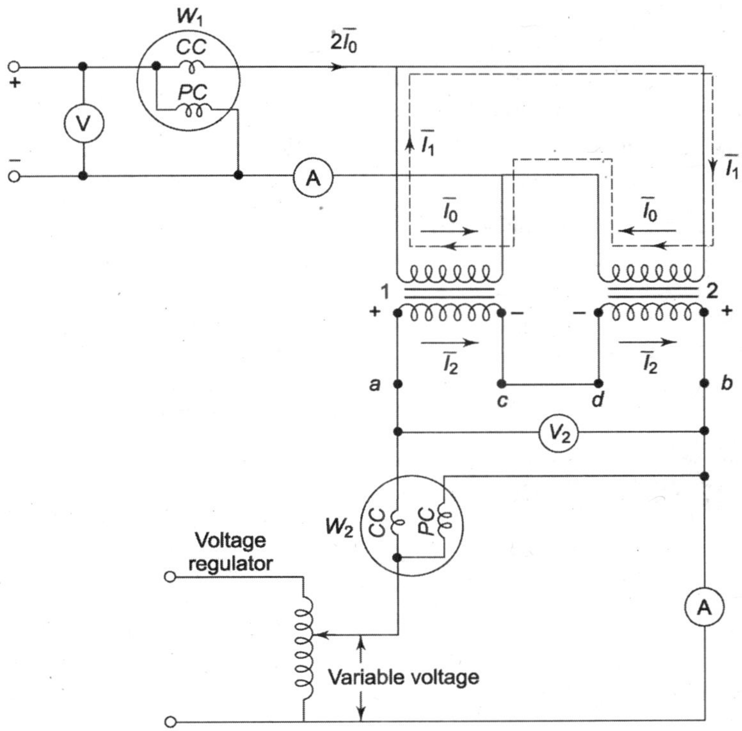

Figure 1.45 Sumpner's Test

1.38 PARALLEL OPERATION OF SINGLE-PHASE TRANSFORMER

It is required to connect a second transformer in parallel with the first transformer if the load exceeds the rating of the transformer shown in Figure 1.46. The primary windings are connected to the supply bus bars while the secondary windings are connected to the load bus bars. During paralleling of the transformer, similar polarities of the transformers should be connected to the same bus bars shown in Figure 1.46. Otherwise, the two emfs induced in the secondary windings with incorrect polarities will produce the equivalent of a dead short circuit shown in Figure 1.47.

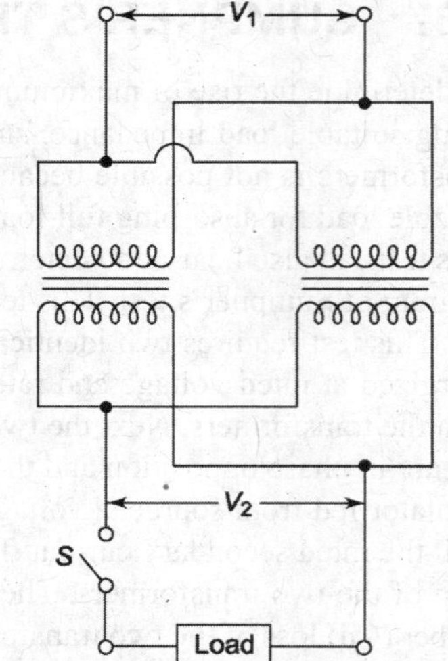

Figure 1.46 Parallel Operation of Transformers

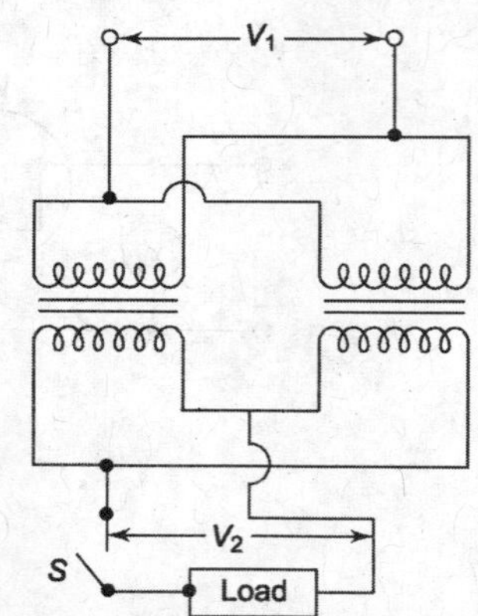

Figure 1.47 Parallel Operation of Transformers with Incorrect Polarities

The following conditions are important for parallel operation of transformers:

- The voltage ratings of both the primary and the secondary of the transformers should be identical. Small differences are permissible if the resultant circulating currents can be tolerated.
- The connections of the transformers should be proper with respect to their polarities.
- The percentage impedances should be equal in magnitude and the *X/R* ratio must be the same to avoid circulating current and operation at different power factors.
- The equivalent impedances must be inversely proportional to the respective kVA ratings.

The above conditions must be satisfied by paralleling transformers of identical ratings of the same make/model. With different kVA ratings of even the same make/model, the effects in steps 1, 2 and 3 may appear in undesirable amounts. Step 2 must be carried out satisfactorily even if steps 1, 3 and 4 are slightly modified.

1.39 LOAD SHARING BY TWO TRANSFORMERS

Let us consider the following two cases:

- Equal voltage ratios.
- Unequal voltage ratios.

1.39.1 Equal Voltage Ratios

Assume no-load voltages E_A and E_B are identical and in phase. Under these conditions if the primary and secondary are connected in parallel, there will be no circulating current between them on no load.

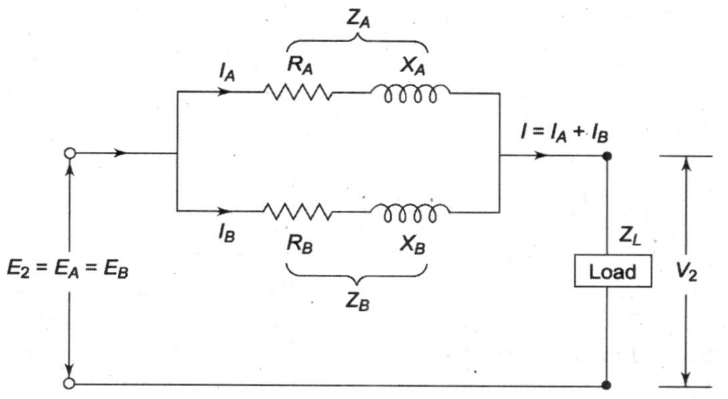

Figure 1.48 Equal Voltage Ratios

Figure 1.48 shows two impedances in parallel. Let R_A, X_A and Z_A be the total equivalent resistance, reactance and impedance of transformer *A* and R_B, X_B and Z_B be the total equivalent resistance, reactance and impedance of transformer *B*.

From Figure 1.48, we have

$$E_A = V_2 + I_A Z_A \tag{1.71}$$

and

$$E_B = V_2 + I_B Z_B \tag{1.72}$$

∴

$$I_A Z_A = I_B Z_B$$

∴

$$\frac{I_A}{I_B} = \frac{Z_B}{Z_A} = \frac{R_B + jX_B}{R_A + jX_A} \tag{1.73}$$

Equation (1.73) suggests that if two transformers with different kVA ratings are connected in parallel, the total load will be divided in proportion to their kVA ratings if their equivalent impedances are inversely proportional to their respective ratings.

Since

$$\frac{I_A}{I_B} = \frac{Z_B}{Z_A}$$

i.e.,

$$\frac{I_A}{I_A + I_B} = \frac{Z_B}{Z_A + Z_B}$$

i.e.,

$$I_A = \frac{Z_B (I_A + I_B)}{Z_A + Z_B} = I \times \left(\frac{Z_B}{Z_A + Z_B} \right) \tag{1.74}$$

Similarly,

$$I_B = I \times \left(\frac{Z_A}{Z_A + Z_B} \right) \tag{1.75}$$

Similarly, load shared by transformer A,

$$S_A = V_2 I_A \times 10^{-3}\ \text{kVA} = V_2 I \left(\frac{Z_B}{Z_A + Z_B} \right) \times 10^{-3}\ \text{kVA} \tag{1.76}$$

Similarly,
$$S_B = V_2 I_B \times 10^{-3} \text{ kVA} = V_2 I \left(\frac{Z_A}{Z_A + Z_B}\right) \times 10^{-3} \text{ kVA} \tag{1.77}$$

Total
$$S = S_A + S_B = V_2 I \times 10^{-3} \text{ kVA.}$$

$$\therefore \quad S_A = S\left(\frac{Z_B}{Z_A + Z_B}\right) \quad \text{and} \quad S_B = S\left(\frac{Z_A}{Z_A + Z_B}\right)$$

Example 1.17 Two single-phase transformers *A* and *B* of equal voltage ratio are running in parallel and supply a load of 800 A at 0.8 power factor lagging having equivalent impedances (1.5 + *j*3) Ω and (2 + *j* 4.5) Ω respectively. Find the current supplied by each transformer and the ratio of the kW output of the two transformers.

Solution

$$Z_A = (1.5 + j3)\ \Omega \quad \text{and} \quad Z_B = (2 + j4.5)\ \Omega$$

$$\frac{I_A}{I_B} = \frac{Z_B}{Z_A} = \frac{2 + j4.5}{1.5 + j3} = 1.47 + j0.067$$

$$\therefore \quad I_A = (1.47 + j0.067)\, I_B$$

Let us take the secondary terminal voltage as reference.

$$I = 800(0.8 - j0.6) = (640 - j480)\ A$$

Again,
$$I = I_A + I_B = (1.47 + 0.067)\, I_B + I_B = (2.47 + j0.067)\, I_B$$

$$I_B = \frac{I}{2.47 + j0.067} = \frac{640 - j480}{2.47 + j0.067} = 253.65 - j201.2 = 323.76\angle{-38.42°}$$

$$I_A = (1.47 + j0.067)\, I_B = (1.47 + j0.067) \times (253.65 - j201.2) = 386.34 - j287.77$$
$$= 481.74\angle{-36.68°}\text{A}$$

The ratio of kW outputs is equal to the ratio of in-phase components of the two currents.

$$\frac{\text{Ouput of A}}{\text{Output of B}} = \frac{386.34}{253.65} = 1.523$$ ■

Example 1.18 Two transformers *A* and *B* are connected in parallel and supply 400 V. Both transformers have no-load ratio of 800/400 V. Transformer *A* is rated at 100 kVA, its total resistance and reactance being 2 per cent and 3 per cent, respectively. The rating of transformer *B* is 150 kVA, its total resistance and reactance being 1 per cent and 4 per cent, respectively. Determine the load on each transformer and secondary voltage for a load of 250 kVA at 0.8 power factor lagging.

Solution

$$\%Z_A = (2 + j3)\ ,\ \%Z_B = \frac{100}{150}(1 + j4) = 0.67 + j2.67$$

$$S = 250(0.8 - j0.6) = 200 - j150$$

$$S_A = S\left(\frac{Z_B}{Z_A + Z_B}\right) = \left(250\angle -36.87^\circ\right) \times \left(\frac{0.67 + j2.67}{2.67 + j5.67}\right)$$

$$= (250\angle -36.87^\circ)\times(0.439\angle 11.12^\circ) = 109.75\angle -25.75^\circ$$

$$S_B = S\left(\frac{Z_A}{Z_A + Z_B}\right) = (250\angle -36.87^\circ) \times \left(\frac{2 + j3}{2.67 + j5.67}\right)$$

$$= (250\angle -36.87^\circ)\times (0.575\angle -8.47^\circ) = 143.75\angle -45.35^\circ$$

Since Z_A and Z_B are in parallel

$$Z_{AB} = \frac{Z_A Z_B}{Z_A + Z_B} = \frac{(2 + j3)(0.67 + j2.67)}{(2 + j3) + (0.67 + j2.67)} = 0.60 + j1.46$$

Percentage drop over $Z_{AB} = \dfrac{0.60 \times 200}{100} + \dfrac{1.46 \times 150}{100} = 3.39\%$

$$V_2 = 400 - \frac{400 \times 3.39}{100} = 386.44\,\text{V}$$

■

1.39.2 Unequal Voltage Ratios

For unequal voltage turns ratio, if the primary is connected to the supply, a circulating current will flow in the primary even at no load. The circulating current will be superimposed on the currents drawn by the load when the transformers share a load.

Let V_1 be the primary supply voltage, a_1 be the turns ratio of transformer A, a_2 be the turns ratio of transformer B, Z_A be the equivalent impedance of transformer A ($= R_A + jX_A$) referred to as secondary, Z_B be the equivalent impedance of transformer B ($= R_B + jX_B$) referred to as secondary, I_A be the output current of transformer A and I_B be the output current of transformer B.

The induced emf in the secondary of transformer A is

$$E_A = \frac{V_1}{a_1} = V_2 + I_A Z_A \tag{1.78}$$

The induced emf in the secondary of transformer B is

$$E_B = \frac{V_1}{a_2} = V_2 + I_B Z_B \tag{1.79}$$

Again, $V_2 = IZ_L$ where Z_L is the impedance of the load

$$\therefore \quad V_2 = (I_A + I_B) Z_L \tag{1.80}$$

From Equations (1.78), (1.79) and (1.80), we have

$$E_A = I_A Z_A + (I_A + I_B) Z_L \tag{1.81}$$

and
$$E_B = I_B Z_B + (I_A + I_B) Z_L \tag{1.82}$$

∴
$$E_A - E_B = I_A Z_A - I_B Z_B$$

i.e.
$$I_A = \frac{(E_A - E_B) + I_B Z_B}{Z_A} \tag{1.83}$$

Substituting I_A from Equation (1.83) in Equation (1.82), we have

$$E_B = I_B Z_B + \frac{(E_A - E_B) + I_B Z_B}{Z_A} \times Z_L + I_B Z_L$$

i.e.,
$$I_B \left(Z_B + Z_L + \frac{Z_B}{Z_A} \times Z_L \right) = \frac{E_B Z_A - (E_A - E_B) Z_L}{Z_A}$$

i.e.,
$$I_B = \frac{E_B Z_A - (E_A - E_B) Z_L}{Z_A Z_B + Z_L (Z_A + Z_B)} \tag{1.84}$$

Similarly,
$$I_A = \frac{E_A Z_B + (E_A - E_B) Z_L}{Z_A Z_B + Z_L (Z_A + Z_B)} \tag{1.85}$$

Example 1.19 Two transformers *A* and *B* are connected in parallel and supply a common load. Open circuit emf of *A* and *B* are 5,500 V and 5,400 V, respectively. Equivalent impedance in terms of secondary of *A* and *B* are (0.4 + *j*4) Ω and (0.1 + *j* 1.5) Ω respectively. The load impedance is (10 + *j* 6) Ω. Find the current supplied by each transformer.

Solution

$$I_A = 5{,}500 \text{ V},\ E_B = 5{,}400 \text{ V},\ Z_L = (10 + j6)\Omega,\ Z_A = (0.4 + j4)\Omega,\ Z_B = (0.1 + j1.5)\Omega$$

$$I_A = \frac{E_A Z_B + (E_A - E_B) Z_L}{Z_A Z_B + Z_L (Z_A + Z_B)} = \frac{5{,}500(0.1 + j1.5) + (5{,}500 - 5{,}400) \times (10 + j6)}{(0.4 + j4) \times (0.1 + 1.5) + (10 + j6) \times (0.4 + j4 + 0.1 + j1.5)}$$

$$= \frac{1{,}550 + j8{,}850}{-33.96 + j59} = (101.31 - j84.58) \text{ A}$$

Similarly,

$$I_B = \frac{E_B Z_A - (E_A - E_B) Z_L}{Z_A Z_B + Z_L (Z_A + Z_B)}$$

$$= \frac{5{,}400(0.4 + j4) - (5{,}500 - 5{,}400) \times (10 + j6)}{(0.4 + j4) \times (0.1 + 1.5) + (10 + j6) \times (0.4 + j4 + 0.1 + j1.5)}$$

$$= \frac{1{,}160 + j21{,}000}{-33.96 + j59} = (258.85 - j168.66) \text{ A}$$

■

1.40 AUTOTRANSFORMERS

Two-winding transformers have already been discussed. In such transformers, the two windings are electrically isolated and emf is induced in the secondary winding due to mutual induction. There also exists other types of transformers in which a part of the winding is common to both the primary and secondary circuits. These transformers are termed as *autotransformers.* In an autotransformer, the two windings are electrically connected and it works on the principle of induction and conduction.

1.40.1 Construction

Figure 1.49 depicts an autotransformer in which only one winding is wound on a laminated magnetic core. It also shows that a single winding is used as primary and secondary and a part of the winding is common to both the primary and secondary. The autotransformers are also classified as step-up and step-down transformers because voltage can be stepped up and stepped down using these transformers.

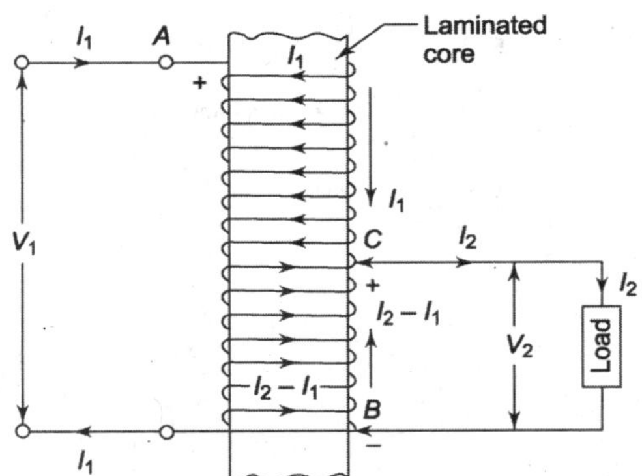

Figure 1.49 Step-down Autotransformer

Figure 1.50(a) shows a two-winding transformer. Figure 1.50(b) shows an autotransformer, which steps down the voltage. In a step-up autotransformer, the entire winding is used as a primary winding and the part of the winding is used as a secondary winding. Figure 1.50(c) shows an autotransformer, which steps up the voltage. The entire winding is used as a secondary winding.

From Figures 1.50(a) and 1.51(b), it seems that an autotransformer is similar to a resistance potential divider. An autotransformer can step up and step down the voltage. An autotransformer has less loss whereas more loss occurs in a potential divider. Therefore, the efficiency of an autotransformer is higher than that of a potential divider. In a potential divider, input current is more than the output current. In an autotransformer, if output voltage is less than the input voltage, the output current will be higher than the input current.

Let I_1 be the input current, I_2 be the output current, V_1 be the input voltage, V_2 be the output voltage, $\cos\theta_1$ be the input power factor and $\cos\theta_2$ be the output power factor.

If we neglect the losses

$$V_1 I_1 \cos\theta_1 = V_2 I_2 \cos\theta_2 \tag{1.86}$$

If we neglect the internal impedance drops and losses, Equation (1.86) can be written as follows

$$V_1 I_1 = V_2 I_2$$

$$\therefore \quad \frac{V_1}{V_2} = \frac{I_2}{I_1} = \frac{N_1}{N_2} = a \tag{1.87}$$

where N_1 is the total number of turns in the primary and N_2 the total number of turns in the secondary.

Figure 1.51 shows a step-down autotransformer. Let the point X be positive with respect to Z. At no load the exciting current flows from X to Z and it produces a working mmf vectorially downwards, i.e., from X to Z. During the presence of load at the secondary, the current flows from Z to Y and it

(a) Two-winding transformer

(b) Step-down autotransformer

(c) Step-up autotransformer

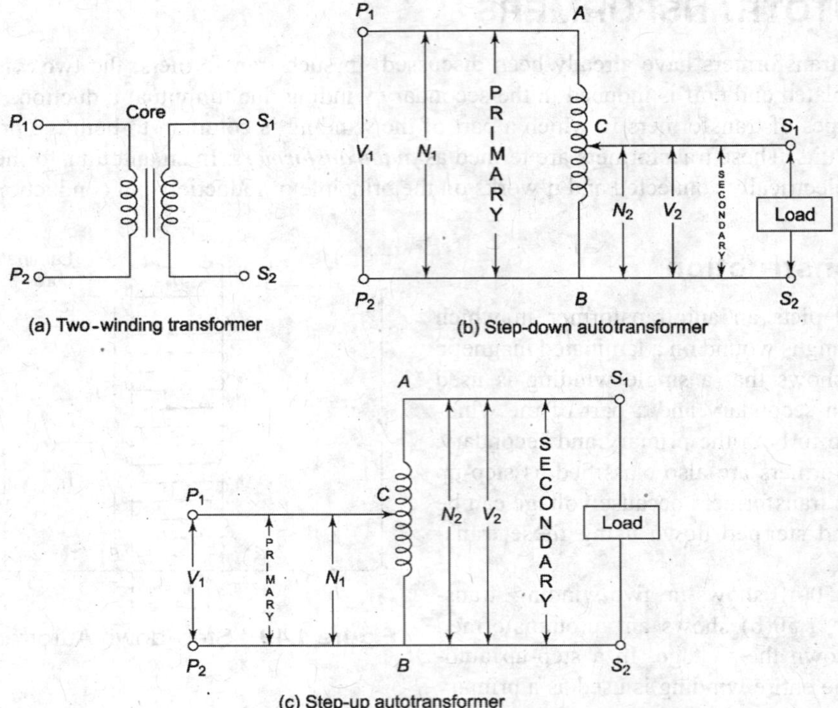

Figure 1.50 Two-winding and Autotransformers

Figure 1.51 Step-down Autotransformer

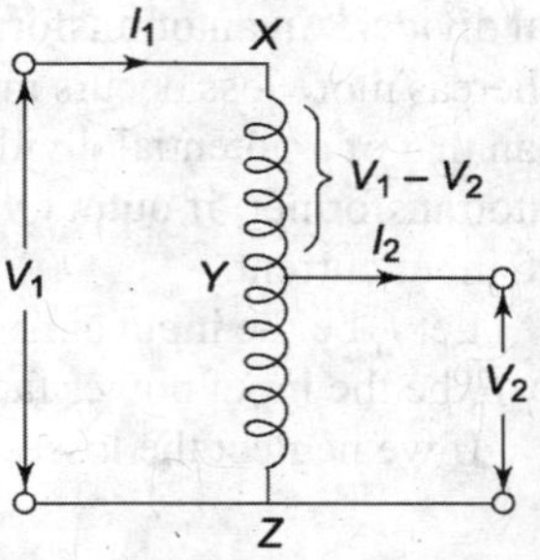

weakens the produced working magnetomotive force (mmf), and the transformer draws extra current from primary. Ultimately, it will maintain the same working mmf. In winding *XZ*, the current is I_1 (from *X* to *Y*) whereas in winding *YZ* the current is I_2 (from *Z* to *Y*). $V_2 < V_1$, $I_2 > I_1$ and the net current through *YZ* is $I_2 - I_1$ (from *Z* to *Y*).

∴ mmf of winding

$$XY = I_1(N_1 - N_2) = I_1N_1 - I_1N_2 = I_2N_2 - I_1N_2 \text{ [since } I_1N_1 = I_2N_2]$$
$$= (I_2 - I_1)N_2 = \text{mmf of winding } ZY.$$

∴ Voltamperes across winding *XY* are transferred by transformer action to load connected across winding *YZ*.

$$\therefore \quad \text{Transformed } VA = V_{XY}I_{XY} = (V_1 - V_2)I_1$$

$$\therefore \quad \frac{\text{Transformed } VA}{\text{Input } VA} = \frac{(V_1 - V_2)\,I_1}{V_1I_1} = \frac{V_1 - V_2}{V_1} = 1 - \frac{V_2}{V_1} = 1 - \frac{1}{a} \tag{1.88}$$

Total input *VA* is V_1I_1 and $(V_1 - V_2)\,I_1$ is transformed to the load by transformer action and the remaining *VA* is conducted directly.

$\therefore$ Conducted VA = Total Input VA − Transformed $VA = V_1I_1 - (V_1 - V_2)I_1 = V_2I_1$

$$\therefore \quad \frac{\text{Conducted } VA}{\text{Input } VA} = \frac{V_2I_1}{V_1I_1} = \frac{V_2}{V_1} = \frac{1}{a} \tag{1.89}$$

1.40.2 Copper Saving in Autotransformer

We know that for any winding, the total length of winding is proportional to the number of turns (N), while its cross-section is proportional to the current (I). Therefore, its copper weight is proportional to the product of N and I.

$\therefore$ Weight of copper $\propto NI$ where I is the current in winding and N is the number of turns of the winding.

Figure 1.52(a) shows a two-winding transformer, whereas Figure 1.52(b) shows a step-down autotransformer.

Let W_{TW} and W_{AT} be the total weight of copper in the two-winding transformer and in the autotransformer. In Figure 1.52(a), let N_1 and N_2 be the number of turns of primary and secondary windings, respectively. In Figure 1.52(b), let N_1 and N_2 be the number of turns of primary winding and secondary winding, respectively.

In two-winding transformer, weight of copper of primary $\propto N_1I_1$ and that of secondary $\propto N_2I_2$.

$$\therefore W_{TW} \propto N_1I_1 + N_2I_2 \tag{1.90}$$

In autotransformer, weight of copper of section $XY \propto (N_1 - N_2)I_1$ and that of section $YZ \propto (I_2 - I_1)N_2$.

$$W_{AT} \propto (N_1 - N_2)I_1 + N_2(I_2 - I_1) \tag{1.91}$$

From Equations (1.90) and (1.91), we have

$$\frac{W_{TW}}{W_{AT}} = \frac{N_1I_1 + N_2I_2}{(N_1 - N_2)I_1 + N_2 \times (I_2 - I_1)} = \frac{N_1I_1 + N_2I_2}{N_1I_1 - N_2I_1 + N_1I_1 - N_2I_1}$$

$$= \frac{N_1I_1 + N_2I_2}{N_1I_1 + N_2I_2 - 2N_2I_1}$$

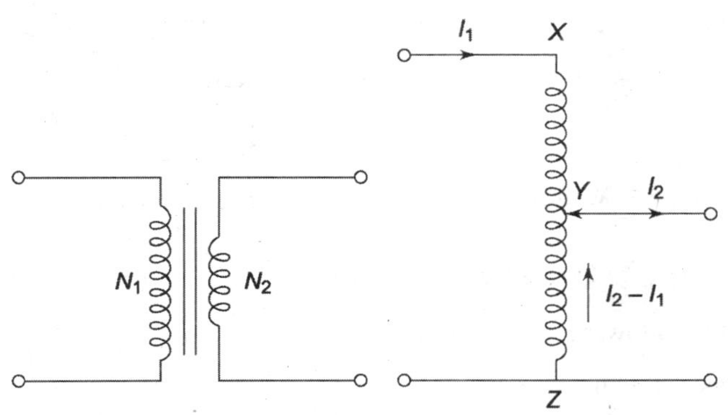

(a) Two-winding transformer (b) Step-down autotransformer

Figure 1.52 Two-winding and Step-down Autotransformer

Again, $N_1 I_1 = N_2 I_2$

$$\therefore \quad \frac{W_{TW}}{W_{AT}} = \frac{2N_1 I_1}{2N_1 I_1 - 2N_2 I_1} = \frac{1}{1 - \dfrac{N_2}{N_1}} = \frac{1}{1 - \dfrac{1}{a}}$$

i.e.,
$$W_{TW} = W_{TW}\left(1 - \frac{1}{a}\right) \tag{1.92}$$

Therefore, saving of copper $W_{TW} - W_{AT} = W_{TW} - W_{TW}\left(1 - \dfrac{1}{a}\right) = \dfrac{W_{TW}}{a}$

i.e.,
$$\text{Saving of copper} = \frac{1}{\text{a}} W_{TW} \tag{1.93}$$

Thus, saving of copper is dependent on the turns ratio '*a*'.

1.40.3 Conversion of Two-winding Transformer into Single-phase Transformer

Figure 1.53 shows a two-winding transformer having primary and secondary voltages of 1,000 V and 400 V respectively. There are two ways to convert it into an autotransformer.

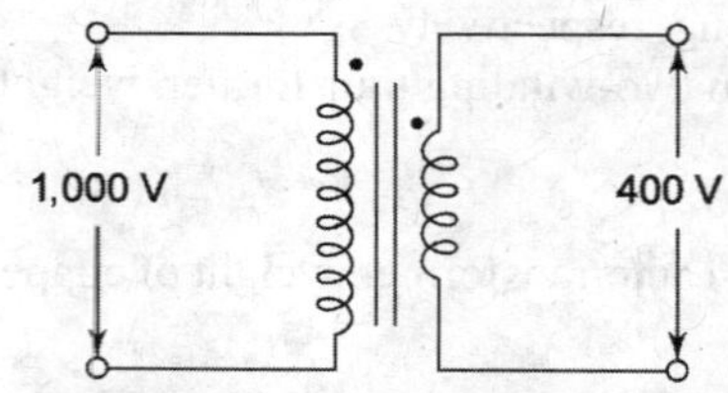

Figure 1.53 Two-winding Transformer

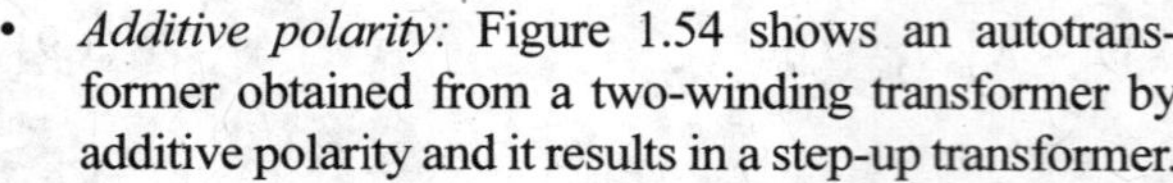

- *Additive polarity:* Figure 1.54 shows an autotransformer obtained from a two-winding transformer by additive polarity and it results in a step-up transformer.
- *Subtractive polarity:* Figure 1.55 shows an autotransformer obtained from a two-winding transformer by subtractive polarity and it results in a step-down transformer.

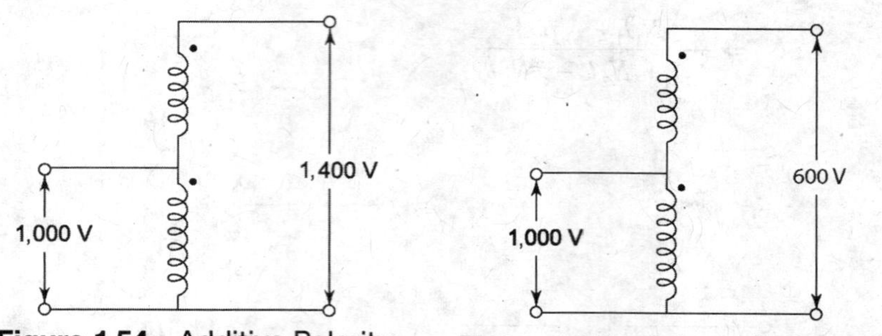

Figure 1.54 Additive Polarity **Figure 1.55** Subtractive Polarity

1.40.4 Advantages of Autotransformers

Autotransformers have the following advantages:

- Less amount of copper is required.
- Due to smaller size, cost is less compared to two-winding transformer.
- The resistance and reactance are less compared to a two-winding transformer and hence it has superior voltage regulation.

- Copper (Cu) loss is less.
- Volt-ampere rating is more compared to a two-winding transformer.
- Since loss is less, efficiency is more.
- It is possible to get smooth and continuous variation of voltage.

1.40.5 Disadvantages of Autotransformers

In spite of various advantages of autotransformers, the following are the disadvantages of autotransformer:

- There is possibility of high short circuit currents for short circuits on the secondary side due to low impedance.
- The full primary current will appear across the secondary causing higher voltage on secondary resulting in danger of accidents if the common winding is open circuited.
- Since there is no electrical isolation between primary and secondary, risk factor appears at high voltage levels.
- It is economical only if the voltage ratio is less than 2.

1.40.6 Applications of Autotransformers

Autotransformers have the following applications:

- They are used as a starter for safely starting machines such as induction motors and synchronous motors.
- They are used as boosters to give a small boost to a distribution cable for compensating the voltage drop.
- They can be used as furnace transformers to supply power to the furnaces at the required supply voltage.
- They can be used as variac.

Example 1.20 A single-phase 2,400/240 V, 120 kVA two-winding transformer is connected as an autotransformer with additive and subtractive polarity. Determine the kVA rating of the autotransformer in both cases and also the power transformed and the power conducted. Show the current distribution in both cases. If the efficiency of the two-winding transformer is 98 per cent on full load at unity power factor, find the efficiency of the transformer on full load at unity power factor.

Solution

For the two-winding transformer

$$I_1 = \frac{120\times10^3}{2,400} = 50\text{ A} \quad \text{and} \quad I_2 = \frac{120\times10^3}{240} = 500\text{ A}$$

(i) *Additive polarity:* The current distribution is shown in Figure E1.4(a).

$$\text{kVA rating} = \frac{2,400\times550}{1,000} = 1,320\text{ kVA}$$

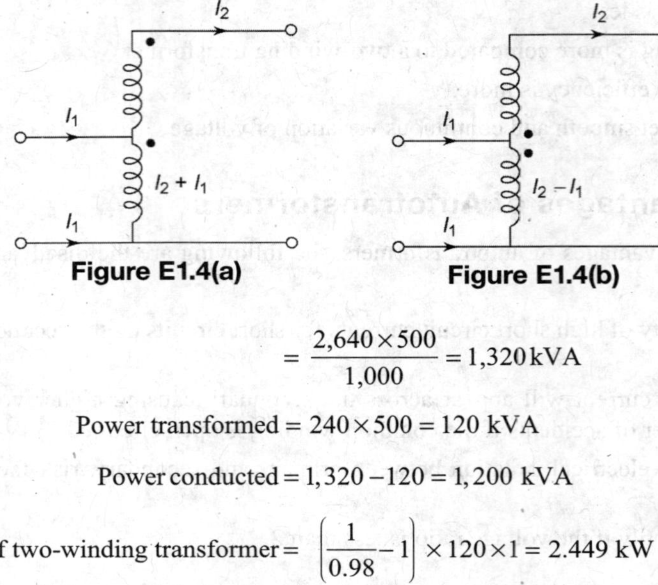

Figure E1.4(a) Figure E1.4(b)

$$= \frac{2{,}640 \times 500}{1{,}000} = 1{,}320\,\text{kVA}$$

$$\text{Power transformed} = 240 \times 500 = 120 \text{ kVA}$$

$$\text{Power conducted} = 1{,}320 - 120 = 1{,}200 \text{ kVA}$$

$$\text{Losses of two-winding transformer} = \left(\frac{1}{0.98} - 1\right) \times 120 \times 1 = 2.449 \text{ kW}$$

Since autotransformer is operated at rated voltage and current, losses will remain constant.

$$\therefore \quad \text{Efficiency} = 1 - \frac{2.449}{1{,}320 \times 1} = 0.9981 \text{ p.u.} = 99.81\%$$

(ii) *Subtractive polarity:* The current distribution is shown in Figure E1.4(b).

$$\text{kVA rating} = \frac{2{,}400 \times 450}{1{,}000} = 1{,}080\,\text{kVA}$$

$$= \frac{2{,}160 \times 500}{1{,}000} = 1{,}080\,\text{kVA}$$

$$\text{Power transformed} = 240 \times 500 = 120 \text{ kVA}$$

$$\text{Power conducted} = 1{,}080 - 120 = 960 \text{ kVA}$$

Since autotransformer is operated at rated voltage and current, losses will remain constant.

$$\therefore \quad \text{Efficiency} = 1 - \frac{2.449}{1{,}080 \times 1} = 0.9977 \text{ p.u.} = 99.77\%$$

■

1.41 PULSE TRANSFORMER

Figure 1.56 shows ideal rectangular pulse input to the pulse transformer, which couples a source of pulses of electrical energy to the load. Its shape and other properties are maintained properly. The size of these transformers is small and they have few turns. The inter-winding capacitance of these transformers is low. The leakage reactance of the windings is small. Ferrites or wound strips of high

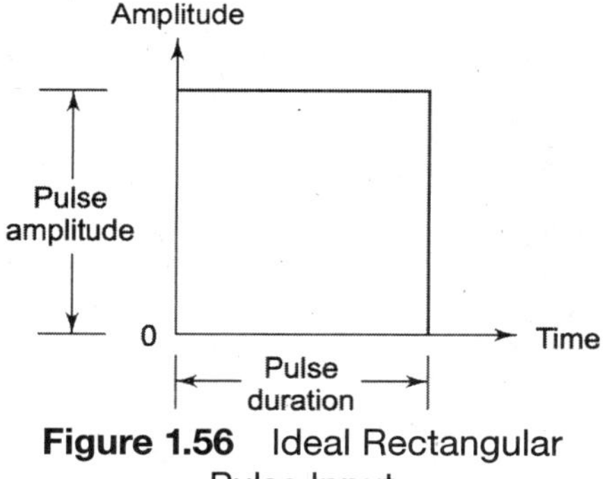

Figure 1.56 Ideal Rectangular Pulse Input

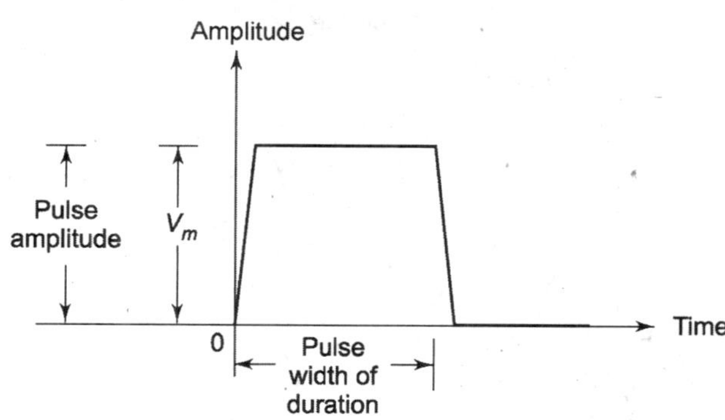

Figure 1.57 Practical Rectangular Input Pulse

permeability alloys such as permalloy are used to construct the cores. Therefore, these transformers have high magnetizing inductance. A high-voltage insulation is used between windings and to ground. Figure 1.57 shows practical rectangular input pulses to such transformers. Small pulse transformers are used in computers, pulse generators and so on, whereas large pulse transformers are used in radar systems and so on.

1.41.1 Pulse Response Characteristics

Figure 1.57 shows the practical input signal. Since dI/dt cannot be infinite, the ideal rectangular pulse cannot be obtained. The output response is usually damped oscillation type due to the stray capacitance. Generally, leakage reactance of the pulse transformer is kept small to obtain the shortest rise time. Overshoot and back swing are present in the response, which extends below zero amplitude. This part of the response is known as *trailing edge response.* The response in between trailing edge and leading edge is of flat type and it is called *top response.* Figure 1.58 shows the overall voltage pulse shape of the pulse transformer output.

The following parameters are important to define the limits of permissible phase distortion:

- *Pulse amplitude:* The maximum peak value of the pulse excluding the unwanted spikes is known as pulse amplitude.
- *Rise time (T_r):* Time taken by the output pulse to rise from 10 per cent of peak pulse amplitude to 90 per cent of peak pulse amplitude during its first attempt is known as rise time (T_r).
- *Overshoot:* The amount by which the output pulse exceeds its peak amplitude is termed as overshoot.
- *Pulse width:* The time interval between the first and last instant at which the instantaneous amplitude achieves 50 per cent of the peak amplitude is known as pulse width or pulse duration.
- *Droop:* The displacements of the pulse amplitude during its flat response are known as droop. It is also termed as tilt.
- *Fall time (T_f):* The time taken by the output pulse to decrease from 90 per cent of its peak amplitude to 10 per cent of its peak amplitude during trailing edge response is known as decay time.
- *Back swing:* The portion of the trailing edge which extends below the zero amplitude level is known as back swing.

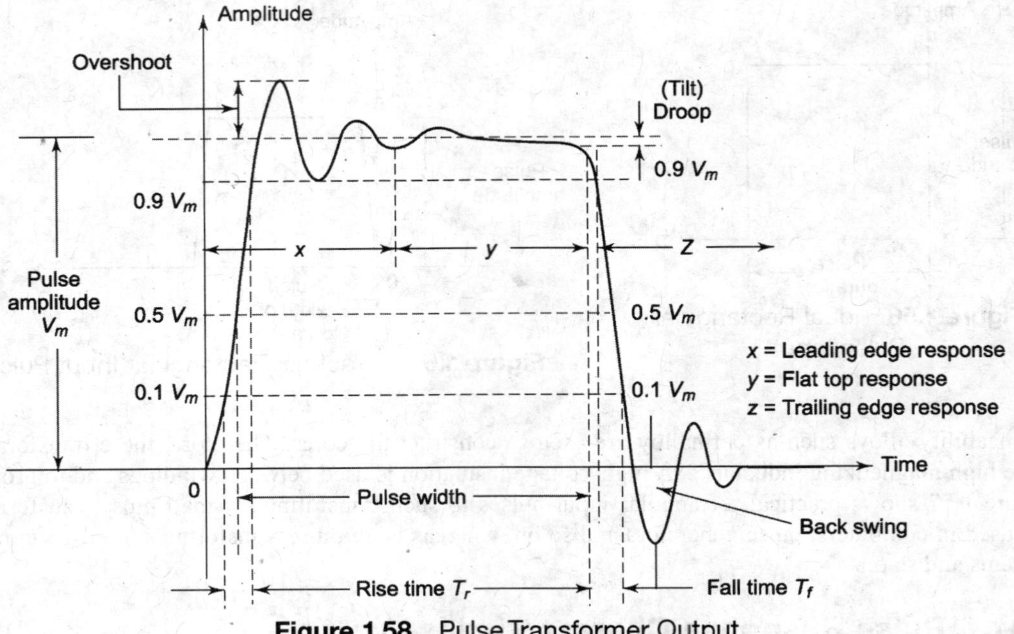

Figure 1.58 Pulse Transformer Output

1.41.2 Usage of Pulse Transformer

Pulse transformers have the following usages:

- SCR and switching transistors.
- Radar systems.
- Cathode ray tubes.
- Microwave tube circuits.
- Data-handling circuits.
- Analogue switching applications.
- Transmission line pulse transformers are useful in fast pulse signal transmission applications.
- Pulse transformers are also useful in digital signal processing.

1.42 WELDING TRANSFORMERS

Figure 1.59 shows a schematic diagram of a welding transformer having thin primary windings with a large number of turns. On the other hand, the secondary has more area of cross-section and less number of turns ensuring less voltage and very high current in the secondary. One end of the secondary is connected to the welding electrode, whereas the other end is connected to the pieces to be welded. If any high current flows, heat is produced due to the contact resistance between the electrode and the pieces to be welded. The generated heat melts a tip of the electrode and the gap between the two pieces is filled.

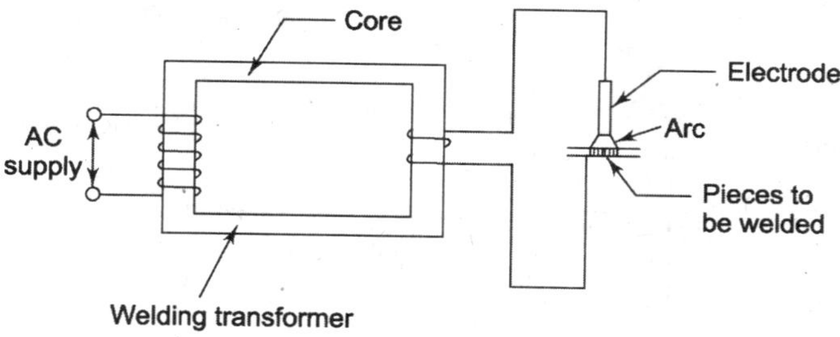

Figure 1.59 Welding Transformer

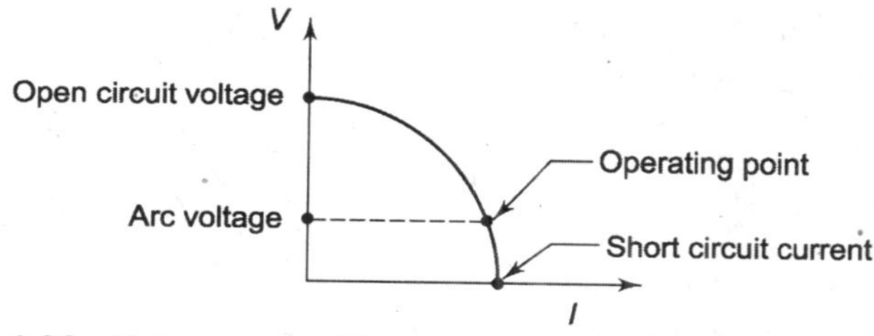

Figure 1.60 Volt-ampere Characteristic of a Welding Transformer

The winding used for the welding transformer is highly reactive. Otherwise, a separate reactor may be added in series with the secondary winding.

Figure 1.60 shows the volt-ampere characteristic of a welding transformer.

1.42.1 Reactors Used with Welding Transformers

To control the arc, various reactors are used with welding transformers. Some methods to control the arc are given below:

(i) *Tapped reactor:* With the help of taps on the reactor, the output current is regulated. This has limited number of current settings shown in Figure 1.61.

(ii) *Moving coil reactor:* Figure 1.62 shows a moving coil reactor in which the reactive distance between primary and secondary is adjusted. The current becomes less if the distance between the coils is large.

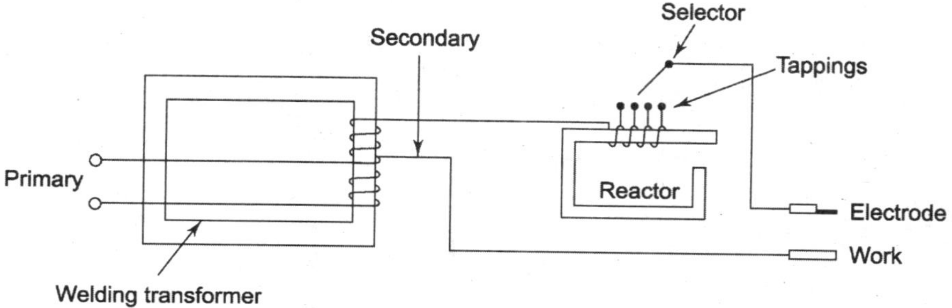

Figure 1.61 Tapped Reactor

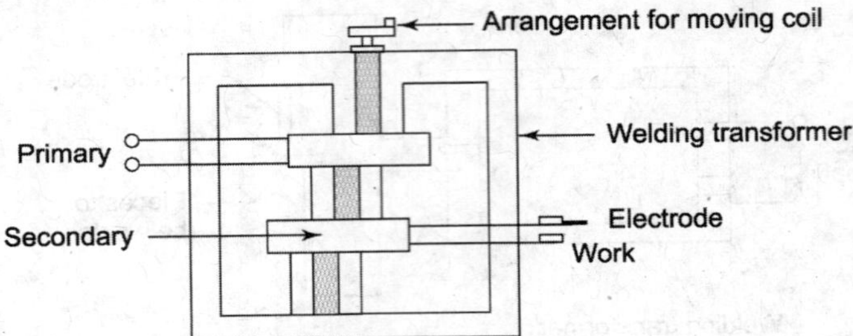

Figure 1.62 Moving Coil Reactor

(iii) *Moving shunt reactor:* Figure 1.63 shows a moving shunt reactor in which the position of the central magnetic shunt can be adjusted. Change of the output current is obtained due to the adjustment of the shunted flux.

(iv) *Continuously variable reactor:* Figure 1.64 shows a continuously variable reactor in which the height of the reactor is continuously varied. Greater reactance is obtained due to greater core insertion and hence the output current is less.

(v) *Saturable reactor:* Figure 1.65 shows a saturable reactor. To adjust the reactance of the reactor, the required DC excitation is obtained from a DC controlled transducer. Reactor approaches saturation if the DC excitation current is more. Therefore, changes of current are obtained due to the change of reactance.

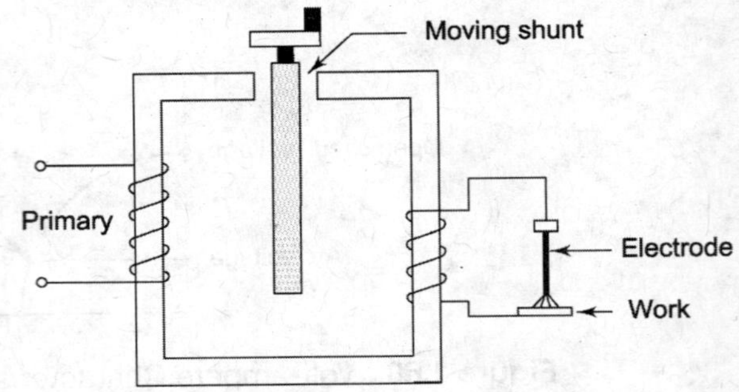

Figure 1.63 Moving Shunt Reactor

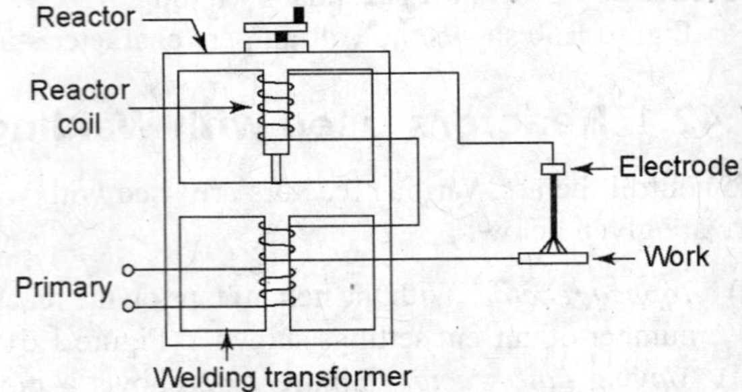

Figure 1.64 Continuously Variable Reactor

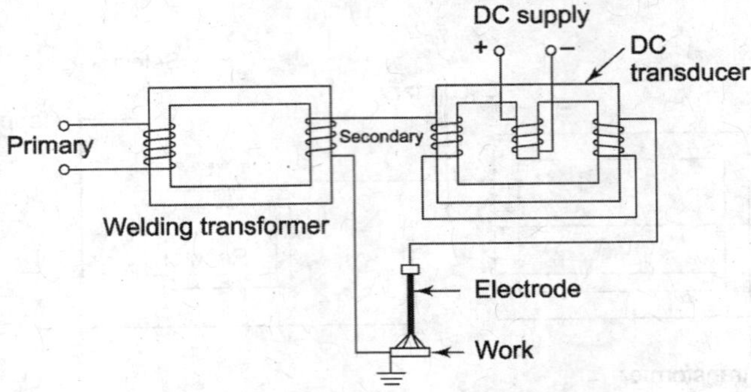

Figure 1.65 Saturable Reactor

1.43 CURRENT TRANSFORMER

Since the high alternating current should not be allowed to pass through normal ammeters and current coils of wattmeters, the current is reduced with the help of a current transformer. Current transformer is a device having two windings called primary and secondary. It transfers energy from one side to another with suitable change in level of current or voltage. It basically has a primary coil with a few turns having large cross-sectional area. This side is connected in series with the line carrying high current shown in Figure 1.66. On the other hand, the secondary is made up of many turns of fine wire having rating of 5 A current. This is connected to the coil of a low-range meter. Current transformers are basically step-up transformers, that is, stepping up the voltage and hence stepping down the current. If I_1 and I_2 be the currents in primary and secondary sides and N_1 and N_2 be the number of turns of primary and secondary windings, we can write

$$\frac{I_2}{I_1} = \frac{N_1}{N_2}$$

If the current ratio of *CT* and meter reading are known, the actual line current value can be determined. The secondary of *CT* should not be kept open. If it is kept open, secondary current will be zero and ampere turns produced by the secondary will be zero. Since the secondary ampere turns oppose the primary ampere turns, the counter mmf will be zero. Hence, unopposed primary produces heavy flux in the core, which causes excessive core losses. Hence, large emf will be induced in primary and secondary. This will damage the insulation, which is dangerous from the operating point of view.

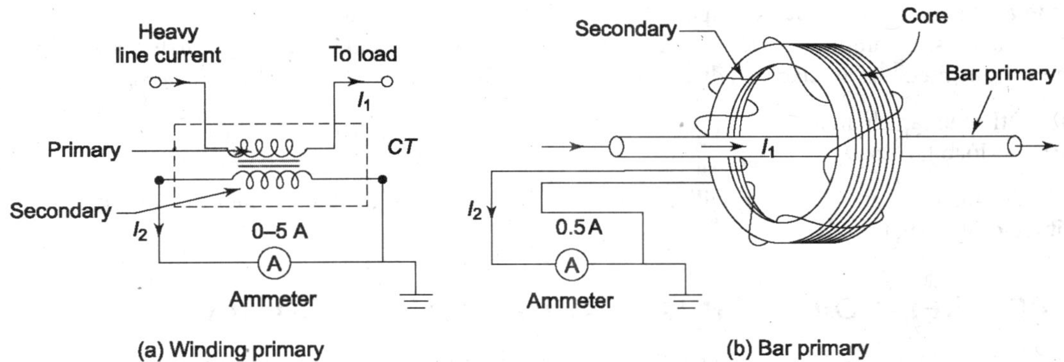

Figure 1.66 Current Transformer

1.44 POTENTIAL TRANSFORMER

These transformers reduce the voltage and have the same basic principle as current transformers. Figure 1.67 shows a potential transformer. The primary winding of *PT* consists of a large number of turns and the secondary winding consists of a small number of turns. The secondary is always grounded for safety purposes.

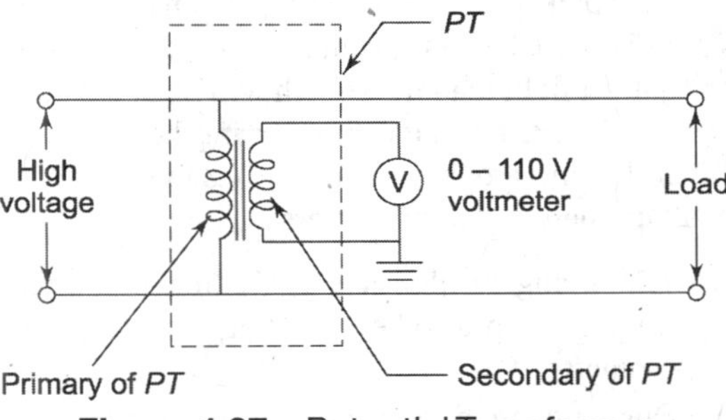

Figure 1.67 Potential Transformer

1.45 TAP CHANGING TRANSFORMERS

Electrical energy at constant voltage without interruption must be ensured by the electric supply company. The voltage at the consumer terminal should not increase the specified limit as per the specification of the country. The undesirable phenomena will happen to the equipment connected to the system when there is change in terminal voltage. This will cause the poor performance of the apparatus and life of the equipment will be reduced. The following are the examples on the performance of the equipments due to reduction in the voltage:

(i) Fluorescent lamp will not glow at low voltage.
(ii) The 75 per cent of load on power system consists of induction motors. Since the output torque of induction motor is proportional to the square of the voltage, the output torque of the motor will be reduced due to the reduction of the voltage.
(iii) Due to reduction of the supply voltage, the picture of the television set becomes smaller and smaller and rolls off.
(iv) The illumination of power will be reduced by 75 per cent due to 6 per cent under voltage, whereas the life of the incandescent lamps will be reduced by 50 per cent due to 6 per cent over voltage.
(v) Due to much variation in voltage, mal-operation of detection of equipment connected to the system may occur.

The voltage at the receiving end is reduced due to the drop at the line, which is dependent on the magnitude of load current and the power factor of the load. The lagging volt-amperes drawn by the consumer reduces the voltage at the consumer terminals. The imbalance of reactive voltage ampere generated and reactive voltage ampere drawn governs the voltage variations.

The most common method used in distribution system is the tap changing transformer in spite of the various methods available in distribution system. There are two types of tap-changing:

(i) Off-load tap changing.
(ii) On-load tap changing.

The tap changing gear is usually provided on the HV side of the transformer to minimize the magnitude of interruption.

1.46 OFF-LOAD TAP-CHANGING TRANSFORMERS

Figure 1.68 shows the tappings on the HV side of a distribution transformer to control the voltage at the consumer terminals.

The load voltage is minimum when the switch is at tap one, whereas it is maximum when it is at tap four. At light load the switch will be at tap one, whereas it will be at tap four during heavy load. The following are the main disadvantages of off-load tap-changing transformers:

(i) The arcing results due to shifting of switch from one tap to other resulting in burning of contacts.
(ii) During change over of taps, the momentary interruption of supply voltage occurs.

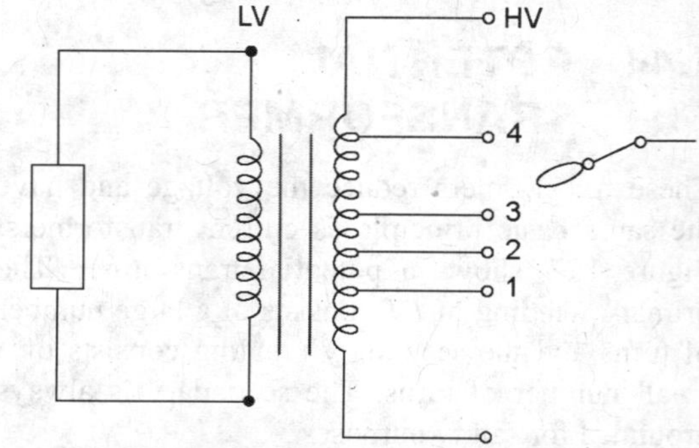

Figure 1.68 Off-load Tap Changing

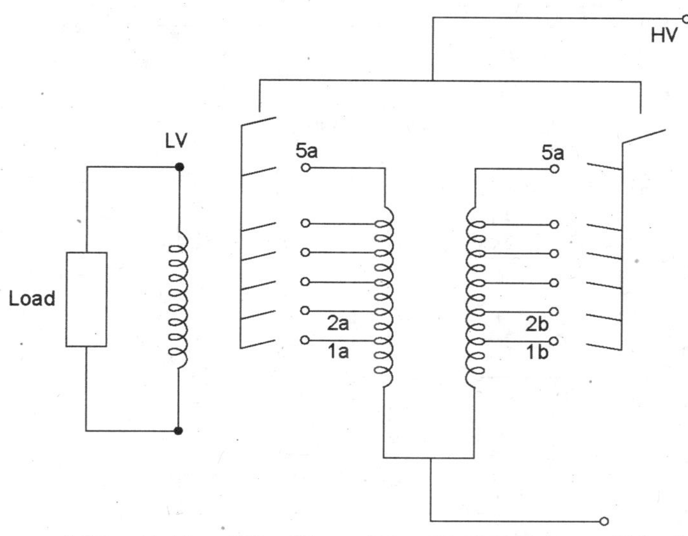

Figure 1.69 On-load Tap Changing with Two Primary Windings

1.47 ON-LOAD TAP-CHANGING TRANSFORMERS

Figure 1.69 shows the on-load tap-changing transformers, which are commonly used in modern days to ensure the reliable power supply.

The voltage at the secondary side is minimum when the switches *A* and *B* are closed as well as the switches 1a and 1b are closed. The following steps are followed to increase the voltage:

(i) The switch *A* is opened at first followed by the switch 1a. The entire load is shared by winding *B* only. The switch 2a is closed at first followed by switch *A*.
(ii) The switch *B* is opened at first followed by the switch 1b. The entire load is shared by winding *A* only. The switch 2b is closed at first followed by switch *B*.

The main function of the additional switches *A* and *B* is to avoid the arcing on tap-changing switches. The following are the main disadvantages of on-load tap-changing transformers:

(i) Number of tapping is equal to two times the number of step voltages.
(ii) Since the change in impedance of the transformer occurs during switching operation, it causes the voltage surge.
(iii) It is very difficult to design a transformer with two windings.

1.48 ON-LOAD TAP CHANGER WITH SINGLE PRIMARY WINDING

Figure 1.70 shows the on-load tap changer with single primary winding, which overcomes the disadvantages of on-load tap-changing transformers. An additional inductor with mid-point tapping is used in this method. The maximum voltage is obtained on the LV side when the switches *A* and *B* are closed and the switches S_1 and S_2 are in position.

1. The following sequences performed to change the sequence of voltage:
   (i) Open *A*, Open S_1, Close S_1 on 2 and Close *A*.
   (ii) Open *B*, Open S_2, Close S_2 on 2 and Close *B*.

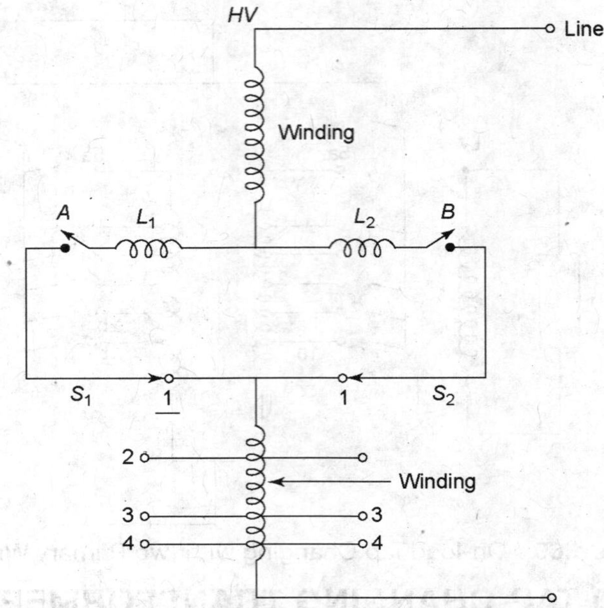

Figure 1.70 On-load Tap Changer with Single Primary Winding

When S_1 is changed from 1 to 2, very few turns between the tapping 1 and 2 will be momentarily short-circuited. L_1 and L_2 are provided to reduce the heavy short circuited currents during the change over taps. Current flows in the L_1 and L_2 are in opposite direction during normal operations and hence impedance drop can be neglected.

1.49 PREVENTIVE AUTOTRANSFORMER

Figure 1.71 shows on-load tap changing devices with preventive autotransformer. The short circuiting switch (S) is closed during the normal operation and the tap 1 is also closed, which gives the highest voltage on LV side. The following steps are followed to raise the voltage:

(i) The switch (S) is opened.
(ii) Tap-2 is closed.
(iii) Tap-1 is opened.
(iv) S is closed.

The taps 1 and 2 are closed during the changeover of switches, and at that time part of the main winding is short-circuited by the autotransformer. The reactance of autotransformer is made large to minimize the short circuit currents. The flow of currents in the two halves of autotransformer is in the opposite directions, and hence the voltage drop in the autotransformer is negligible. The sliding contacts are used instead of switches. Since the autotransformer avoids the short circuit during switching operation, hence the name is preventive autotransformer.

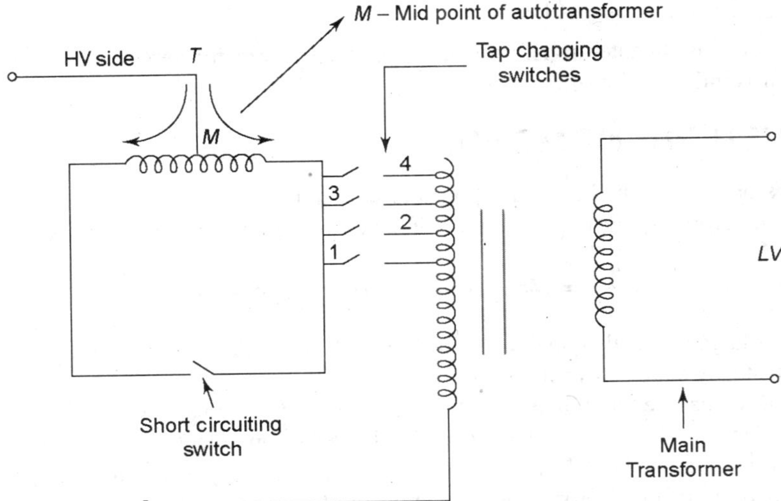

Figure 1.71 Tap Changing with Preventing Autotransformer

1.50 BOOSTER TRANSFORMER

Figure 1.72 shows the regulating transformer along with booster transformer. Booster transformers are useful in controlling the voltage when the tap-changing gear is not convenient to be accommodated with the main transformer. Booster transformers are also used to increase the voltage at the middle of the line. The main transformer maintains the constant voltage at the terminals S-S. The voltage at the middle of the line and load end can be controlled by the booster transformer.

The following are the advantages of the booster transformer:

- The operation of the regulation transformer does not depend on the main transformer. If the main transformer fails, it will throw the latter.
- It is cheaper.

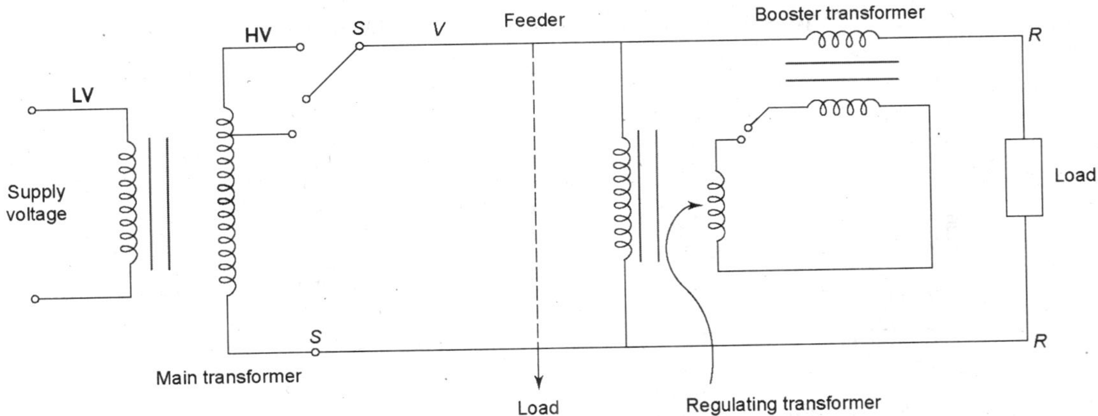

Figure 1.72 Booster Transformer

- It needs less space and it has less losses.
- It is economical to use booster transformer along with large transformers because the cost of the tap changing gear is independent of the size.

1.51 INRUSH PHENOMENON

An unloaded transformer switched on to a supply *s* is shown in Figure 1.73.

If the flux in transformer be $\phi = \phi_m \sin\omega t$, the induced voltage can be expressed by

$$e = N\frac{d\phi}{dt} = N\phi_m \omega \cos \omega t = N\phi_m \omega \sin(\omega t + 90°)$$

Since the applied voltage is exactly equal to the induced voltage, the flux lags behind the applied voltage by an angle 90° in steady state shown in Figure 1.74.

Figure 1.75 shows that the flux is at its negative maxima when the voltage is passing through the zero. The flux changes from $-\phi_m$ to ϕ_m in a time equal to *T*/2. Therefore, in time *T*/2, the change in flux is $2\phi_m$.

Let the transformer be switched on at positive zero crossing of the voltage waveform shown in Figure 1.75 and also the residual flux be zero. Therefore, the initial value of flux is zero and the flux

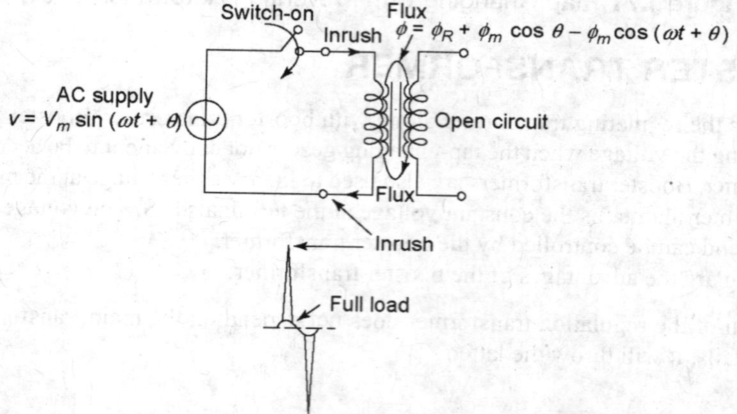

Figure 1.73 Switch on an Unloaded transformer

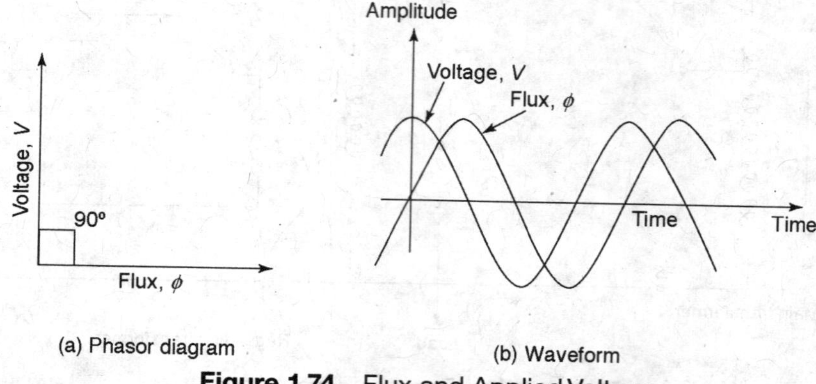

(a) Phasor diagram (b) Waveform

Figure 1.74 Flux and Applied Voltage

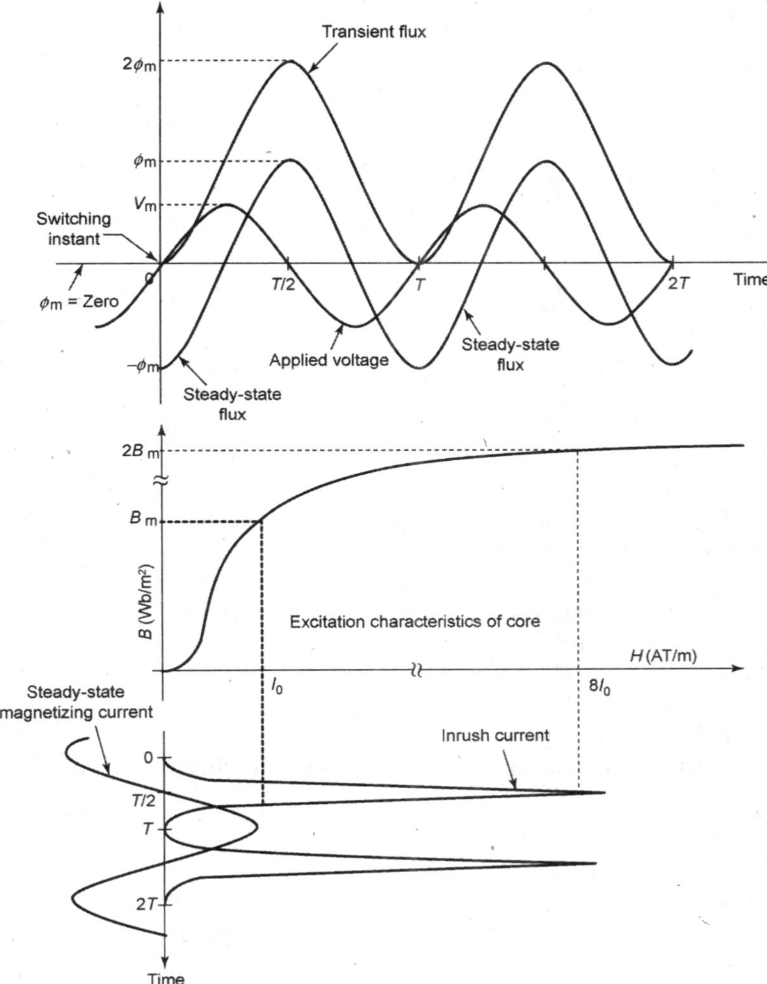

Figure 1.75 Phenomenon of Inrush

subsequently must have the same rate of change of flux. The waveform will be the same as in the steady state and hence the flux will reach the peak value of $+2\phi_m$ in half a cycle. The power transformers operate near the knee of the saturation curve. The flux demand of $+2\phi_m$ causes the transformer core deep into the saturation. This causes to draw large magnetizing current with a peaky non-sinusoidal waveform, which is of the order of 8–30 times the full load current known as inrush current having time constant of this transient component of current of the order of a few seconds at worst.

ADDITIONAL SOLVED PROBLEMS

Example 1.21 The net cross-sectional area of a single-phase, 50 Hz transformer is 500 cm^2 having 400 and 100 turns in the primary and secondary winding, respectively. If it is connected to a 230 V, 50 Hz supply, calculate the following:

(i) Emf induced in the secondary.
(ii) Maximum value of the flux density in the core.

Solution

$$N_1 = 400,\ N_2 = 100$$

$$E_1 = V_1 = 230 \text{ V}$$

$$A_r = 500 \text{ cm}^2 = 500 \times 10^{-4} \text{ m}^2$$

(i)
$$E_2 = E_1 \times \frac{N_2}{N_1} = 230 \times \frac{100}{400} = 57.5 \text{ V}$$ ■

(ii)
$$E_1 = 4.44 f \phi_m N_1 = 4.44 f B_m A_r N_1$$

$$B_m = \frac{E_1}{4.44 f A_r N_1} = \frac{230}{4.44 \times 50 \times 500 \times 10^{-4} \times 400} = 0.0518 \text{ T}$$ ■

Example 1.22 The details of a single-phase transformer are shown below.
Primary turns = 600, secondary turns = 60
Primary applied voltage = 3000 V, Frequency = 50 Hz
Mean length of the circuit = 152 cm with the joints equivalent to an air gap of 0.12 mm
Maximum flux density = 1.2 Wb/m^2
Iron loss at this flux density = 4.1 W/kg of iron
Density of iron = 7.7 g/cm^3
Ampere turns / cm at this flux density = 6
Calculate the no-load primary current and no-load power factor of this transformer.

Solution
We know that

$$E_1 = 4.44 f N_1 B_m A_r$$

i.e.
$$3{,}000 = 4.44 \times 50 \times 600 \times 1.2 \times A_r$$

i.e.
$$A_r = \frac{3{,}000}{4.44 \times 50 \times 60 \times 1.2} = 0.01877 \text{ m}^2 = 18.77 \text{ cm}^2$$

$$\text{Secondary voltage } (E_2) = E_1 \times \frac{N_2}{N_1} = 3{,}000 \times \frac{60}{600} = 300 \text{ V}$$

Ampere-turns/cm length = 6
Therefore, ampere-turns for the iron core = 6 × 152 = 912

$$\text{Ampere-turns for the air-gap} = \frac{B_0}{\mu_0} \times 0.12 \times 10^{-3} = \frac{1.2}{4\pi \times 10^{-7}} \times 0.12 \times 10^{-3} = 114.65$$

Total ampere-turns at the given B_{max} is = 114.65 + 912 = 1,026.65

$$\text{Maximum value of magnetizing current} = \frac{1{,}026.65}{600} = 1.711 \text{ A}$$

RMS value of magnetizing current $(I_\mu) = \dfrac{1.711}{\sqrt{2}} = 1.21$ A

Length of iron path = 152 cm and its area of cross-section = 18.77 cm^2

$\therefore$ Volume of iron core = 152 × 18.77 = 2853.04 cm^2

Here, density of iron = 7.7 g / cm^3

$$\text{Mass of iron} = \frac{7.7 \times 2853.04}{1,000}\ \text{kg} = 21.97\ \text{kg}$$

Total iron loss = 21.97 × 14.4 = 316.97 W

Iron loss component of the no-load primary voltage current $(I_c) = \dfrac{316.97}{3,000} = 0.105$ A

No-load primary current $(I_0) = \sqrt{I_c^2 + I_\mu^2} = \sqrt{0.105^2 + 1.21^2} = 1.475$ A

No-load power factor $(\cos\theta_0) = \dfrac{I_c}{I_\mu} = \dfrac{0.105}{1.21} = 0.0867$ ■

Example 1.23 The ratio of the number of turns of primary and secondary windings is 5. The load current on the secondary is 120 A at 0.8 lagging power factor. The primary current is 27A at 0.72 power factor lagging. Find the no-load current of the transformer and its phase with respect to the voltage. Neglect the angle between $-\overline{E}_1$ and $\overline{V}_1$.

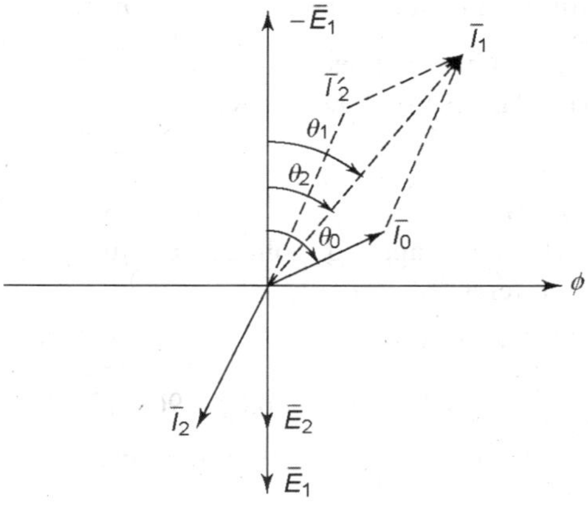

Figure E1.5

Solution

Turns ratio $(a) = 5$
Secondary current = 120 A
Power factor $(\cos\theta_2) = 0.8$, i.e., $\theta_2 = 36.87°$
Primary current = 27 A
Power factor $(\cos\theta_1) = 0.72$, i.e., $\theta_1 = 43.94°$
After resolving $\bar{I}_0$, $\bar{I}'_2$ and $\bar{I}_1$ into their x and y components, we have

$$\bar{I}_1 \cos\theta_1 = \bar{I}'_2 \cos\theta_2 + \bar{I}_0 \cos\theta_0$$

$\therefore$ $$\bar{I}_0 \cos\theta_0 = \bar{I}_1 \cos\theta_1 - \bar{I}'_2 \cos\theta_2 = 27 \times 0.72 - 24 \times 0.8 = 19.44 - 19.2$$

i.e., $$I_0 \cos\theta_0 = 0.240 \qquad (1)$$

and $$I_1 \sin\theta_1 = I'_2 \sin\theta_2 + I_0 \sin\theta_0$$

i.e., $$I_0 \sin\theta_0 = I_1 \sin\theta_1 - I'_2 \sin\theta_2 = 27 \times 0.694 - 24 \times 0.6$$

i.e., $$I_0 \sin\theta_0 = 4.338 \qquad (2)$$

Dividing Equation (2) by Equation (1), we have

$$\tan\theta_0 = \frac{4.338}{0.240} = 18.075$$

i.e.,
$$\theta_0 = 86.84°$$

From Equation (1), we have

$$I_0 = \frac{0.240}{\cos\theta_0} = \frac{0.240}{\cos 86.84°} = 4.354\ \text{A}.$$

∴ No-load current is 4.354 A, and the angle between no-load current and voltage is 86.84°. ■

Example 1.24 The data of a single-phase transformer are shown below:
Turns ratio (a) = 10, R_1 = 4.95 Ω, X_1 = 7.86 Ω, R_2 = 0.18 Ω, X_1 = 0.38 Ω, no-load current leads the flux by an angle 8°. The value of no-load current is 1.4 A. A load of power factor 0.8 lagging is connected to the secondary so that the load current is 120 A at a terminal voltage of 350 V. Calculate the following with the help of phasor diagram:

(i) Primary applied voltage.
(ii) The primary power factor.
(iii) The efficiency of the transformer.

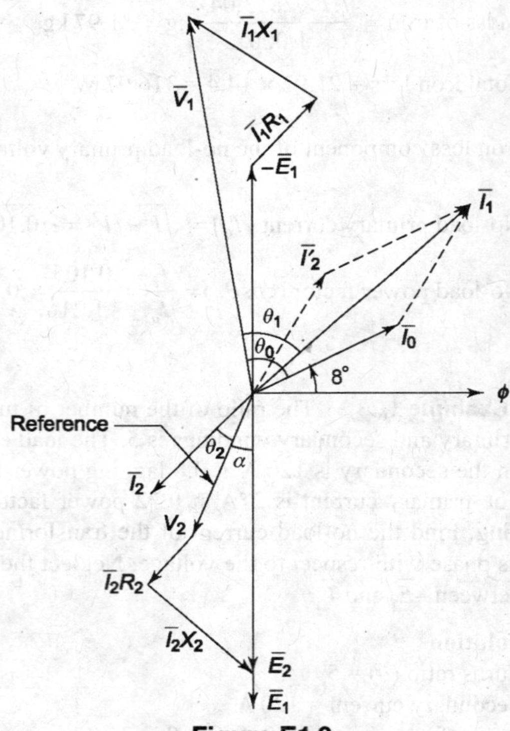

Figure E1.6

Solution

(i) Primary applied voltage: Let us take V_2 as reference.

$$\therefore \quad \overline{V}_2 = 350\angle 0° = 350 + j0\ \text{V}$$

$$\overline{I}_2 = 120(0.8 - j\,0.6) = 96 - j72\ \text{A}$$

$$\overline{Z}_2 = R_2 + jX_2 = 0.18 + j0.38\ \Omega$$

$$\therefore \quad \overline{E}_2 = \overline{V}_2 + \overline{I}_2\overline{Z}_2 = (350 + j0) + (96 - j72)(0.18 + j0.38) = 394.64 + j\,23.52\ \text{V}$$

$$= 395.34\angle 3.41°\ \text{V}$$

$$\therefore \quad \alpha = 3.41° \text{ (angle between } E_2 \text{ and } V_2)$$

$$\therefore \quad \overline{E}_1 = a\,E_2 = 10\,(394.64 + j23.52) = 3{,}946.4 + j235.2\ \text{V}$$

$$\therefore \quad -\overline{E}_1 = -3{,}946.4 - j235.2 = 3{,}953.4\angle 183.41°\ \text{V}$$

$$\therefore \quad \overline{I}'_2 = -\frac{\overline{I}_2}{a} = -\frac{96 - j72}{10} = -9.6 + j7.2$$

The angle between I_0 and V_2 is 3.41° + 90° + 8° = 101.41°

$$\therefore \quad \bar{I}_0 = 1.4\angle 101.41° = -0.277 + j1.372$$

∴ Primary current,

$$\bar{I}_1 = \bar{I}_2' + \bar{I}_0 = -9.6 + j7.2 - 0.277 + j1.372 = -9.877 + j8.572 = 13.08\angle 139°$$

$$\therefore \quad \bar{V}_1 = -\bar{E}_1 + \bar{I}_1\bar{Z}_1$$

$$= -3{,}946.4 - j\,235.2 + (-9.877 + j\,8.572)(4.95 + j\,7.86)$$

$$= -4{,}062.66 - j\,270.40 = 4{,}071.65\angle 183.80°$$

(i) The primary applied voltage $= |V_1| = 4{,}071.65$ V

(ii) The primary power factor $= \cos\theta_1 = \cos(183.80° - 139°) = 0.7095$ (lagging)

(iii) Here $\theta_0 = 183.80° - 183.41° + 90° - 8° = 82.39°$

No-load primary input $= W_0 = V_1 I_0 \cos\theta_0 = 4{,}071.65 \times 1.4 \times \cos 82.39° = 754.88$ W

∴ $W_0 = 754.88$ W is the iron loss of the transformer

Total copper losses $= I_1^2 R_1 + I_2^2 R_2 = 13.08^2 \times 4.95 + 120^2 \times 0.18 = 3{,}438.87$ W

∴ Total losses of the transformer = 754.88 + 3,438.87 = 4,193.75 W

∴ Output $= V_2 I_2 \cos\theta_2 = 350 \times 120 \times 0.8 = 3{,}3600$ W

$$\therefore \quad \text{Efficiency} = \frac{\text{Output}}{\text{Output} + \text{Total loss of the transformer}}$$

$$= \frac{33{,}600}{33{,}600 + 4{,}193.75} = 0.8890 \text{ p.u.} = 88.90\%$$

■

Example 1.25 A single-phase transformer has the following data:

Turns ratio (a) = 10:1, $Z_1 = 1.6 + j\,4.3\ \Omega$, $Z_2 = 0.019 + j\,0.048\ \Omega$

The input voltage of the transformer is 5,000 V and the load current at the secondary is 250 A at 0.8 lagging power factor. Neglect no-load current. Calculate secondary terminal voltage and output power.

Solution

$$\bar{Z}_1 = 1.6 + j\,4.3\ \Omega,\ \bar{Z}_2 = 0.019 + j\,0.048\ \Omega$$

$$a = 10,\ |V_1| = 5{,}000\text{ V},\ \bar{I}_2 = 250\angle -36.87°\text{ A}$$

Let the secondary terminal voltage be $V_2\angle 0°$ which is the reference phasor.

$$E_2 = V_2\angle 0° + \bar{I}_2\bar{Z}_2 = V_2 + 250\angle -36\ 87° \times (0.019 + j0.048)$$

$$= V_2 + 250\,(0.8 - j\,0.6)(0.019 + j\,0.048) = (V_2 + 11) + j\,6.75$$

Since $$\frac{\bar{E}_1}{\bar{E}_2} = 10$$

∴ $$\bar{E}_1 = 10\bar{E}_2$$

i.e., $$\bar{E}_1 = 10\,V_2 + 110 + j\,67.5$$

Again, $\overline{V}_1 = -\overline{E}_1 + \overline{I}_1\overline{Z}_1$

$$= -10\ V_2 - 110 - j\,67.5 + \overline{I}_1\,(1.6 + j\,4.3)$$

Since $\overline{I}_0$ is neglected,

$$\overline{I}_1 = \overline{I}_2' = \frac{-\overline{I}_2}{a} = \frac{-250\angle -36.87^\circ}{10} = -25\times(0.8 - j0.6)$$

$\therefore$ $\overline{V}_1 = -10\ V_2 - 110 - j\,67.5 - 25\,(0.8 - j\,0.6)\,(1.6 + j\,4.3)$

$$= -(10\ V_2 + 110 + j\,67.5 + 96.5 + j\,62)$$

$$= -(10\ V_2 + 206.5 + j\,129.5) = -(10\ V_2 + 206.5) - j\,129.5$$

$\therefore$ $|\overline{V}_1|^2 = (10\,V_2 + 206.5)^2 + (129.5)^2$

Since, $|\overline{V}_1| = 5{,}000$

$\therefore$ $5{,}000^2 = (10\ V_2 + 206.5)^2 + (129.5)^2$

i.e., $(10\ V_2 + 206.5)^2 = 5{,}000^2 - 129.5^2$

i.e., $10\ V_2 + 206.5 = 4{,}998.32$

i.e., $10\ V_2 = 4{,}998.32 - 206.5$

i.e., $V_2 = 479.18$ V

$\therefore$ Secondary terminal voltage is 479.18 V and

Output power = 479.18 × 250 × 0.8 W= 95.836 kW ■

Example 1.26 The maximum flux density of a 600 kVA, 2,000/400 V, 50 Hz, single-phase transformer is 1.2 Wb/m² having a net cross-sectional area of 0.05 m².

(i) Find (a) the number of primary and secondary turns, (b) the magnetizing current and (c) the efficiency of the transformer at full-load and 0.8 power factor lagging.
The mean length of the magnetic circuit is 1.1 m and relative permeability is 1,200. The net respective power input during short circuit test and open circuit test are 720 W and 360 W.

(ii) If this transformer is connected to the supply having similar voltage but frequency of 100 Hz, comment on the effect on efficiency.

Solution

(i) (a) $E_1 = 4.44\,fB_m A_r N_1$

i.e., $2{,}000 = 4.44 \times 50 \times 1.2 \times 0.05 \times N_1$

i.e., $N_1 = 150$

$\therefore$ $N_2 = N_1 \times \dfrac{E_2}{E_1} = 150 \times \dfrac{400}{2{,}000} = 30$

Therefore, the number of primary and secondary turns are 150 and 30, respectively.

(b) $$AT = Hl = \frac{Bl}{\mu_0 \mu_r} = \frac{1.2\times 1.1}{1{,}200\times 4\pi\times 10^{-7}} = 875.8\ \text{A}$$

Magnetizing current $(I_\mu) = \dfrac{875.8}{150} = 5.84$ A ■

(c) Here Cu loss = 720 W and iron loss = 360 W.

Total losses = 720 + 360 = 1,080 W

Output at 0.8 power factor = 60,000 × 0.8 = 48,000 W

$$\therefore \quad \text{Efficiency at full-load} = \frac{48,000}{48,000 + 1,080} = 0.9779 \text{ p.u.} = 97.79\%$$

(ii) Hysteresis loss (P_h) $\propto$ f and eddy current loss (P_e) $\propto f^2$. Therefore, P_h becomes double and P_e becomes four times.

Hence, overall losses will increase and efficiency of the transformer will decrease. ■

Example 1.27 A 8 kVA, 500/250 V, 50 Hz, single-phase transformer gave the following test data:
OC test (LV side): 250 V, 1.2 A, 80 W
SC test (HV side): 25 V, 12.5 A, 90 W, where LV refers to low voltage and HV refers to high voltage.
Determine the following:

(i) Equivalent circuit referred to as LV side.
(ii) Secondary load voltage at 0.8 power factor lagging with full-load current.

Solution

OC test:

Instruments placed on LV side and HV side is kept open.

$$V_2 = 250 \text{ V}, I_0 = 1.2 \text{ A}, W_0 = 80 \text{ W}$$

$$\cos\theta_0 = \frac{W_0}{V_2 I_0} = \frac{80}{250 \times 1.2} = 0.267$$

$$\sin\theta_0 = \sqrt{1 - \cos^2\theta_0} = 0.964$$

$$I_W = I_0 \cos\theta_0 = 1.2 \times 0.267 = 0.3204 \text{ A}$$

$$I_\mu = I_0 \sin\theta_0 = 1.2 \times 0.964 = 1.1568 \text{ A}$$

$$\therefore \quad R_0 = \frac{V_0}{I_W} = \frac{250}{0.3204} = 780.27\ \Omega \quad \text{and} \quad X_0 = \frac{V_0}{I_\mu} = \frac{250}{1.1568} = 216.11\ \Omega$$

SC test:

Instruments placed on HV side and LV side is short-circuited:

$$V_{SC} = 25 \text{ V}, I_{SC} = 12.5 \text{ A}, P_{SC} = 90 \text{ W}$$

$$\therefore \quad Z_{01} = \frac{V_{SC}}{I_{SC}} = \frac{25}{12.5} = 2\ \Omega$$

$$\therefore \quad R_{01} = \frac{P_{SC}}{I_{SC}^2} = \frac{90}{(12.5)^2} = 0.576\ \Omega$$

$$\therefore \quad X_{01} = \sqrt{Z_{01}^2 - R_{01}^2} = \sqrt{2^2 - 0.576^2} = 1.915\ \Omega$$

$$\text{Turns ratio } (a) = \frac{500}{250} = 2$$

$$\therefore \quad R_{02} = \frac{R_{01}}{a^2} = \frac{0.576}{4} = 0.144\ \Omega$$

$$\therefore \quad X_{02} = \frac{X_{01}}{a^2} = \frac{1.915}{4} = 0.4787\ \Omega$$

$$(I_2)_{\text{Full load}} = \frac{8 \times 10^3}{250} = 32\text{ A}$$

$\therefore$ Total voltage drop referred to as secondary

$$= I_2 (R_{02} \cos\theta + X_{02} \sin\theta)$$

$$= 32 \times (0.144 \times 0.8 + 0.4787 \times 0.6)$$

$$= 12.88\text{ V}$$

$\therefore$ Secondary terminal voltage = 250 − 12.88 = 237.12 V (Approx.) ■

Example 1.28 The turn ratio of a single-phase transformer is 4. The resistance and leakage reactance of HV winding are 1.4 Ω and 6 Ω respectively. The resistance and leakage reactance of LV winding are 0.06 Ω and 0.04 Ω respectively. If 200 V, 50 Hz is applied to HV winding and LV winding is short-circuited, find the following: (i) current in LV winding, (ii) copper loss in the transformer and (iii) power factor.

Solution

Given $R_1 = 1.4\ \Omega$, $X_1 = 6\ \Omega$, $R_2 = 0.06\ \Omega$, $X_2 = 0.04\ \Omega$
Turns ratio (a) = 4.

$$\therefore \quad R_{01} = R_1 + a^2 R_2 = 1.4 + 4^2 \times 0.06 = 2.36\ \Omega$$

$$\therefore \quad X_{01} = X_1 + a^2 X_2 = 6 + 4^2 \times 0.04 = 6.64\ \Omega$$

$$\therefore \quad Z_{01} = \sqrt{R_{01}^2 + X_{01}^2} = \sqrt{2.36^2 + 6.64^2} = 7.047\ \Omega$$

$$I_{SC} = \frac{200}{7.047} = 28.38\text{ A}$$

(i) Neglecting I_0, $I_1 = I_2' = 28.38$ A

$$\therefore \quad I_2' = I_2 \times a = 28.38 \times 4 = 113.52\text{ A}$$

(ii) Total Cu loss = $I_1^2 R_{01} = (28.38)^2 \times 2.36 = 1{,}900.8$ W

(iii) Short circuit power factor:

$$\cos\theta_{SC} = \frac{P_{SC}}{V_{SC} I_{SC}} = \frac{1{,}900.8}{200 \times 28.38} = 0.335(\text{lagging})$$ ■

Example 1.29 A single-phase, 4 kVA, 250/125 V, 50 Hz transformer has the following parameters:
Resistance: primary 0.9 Ω, secondary 0.94 Ω
Leakage reactance: primary 0.4 Ω, secondary 0.11 Ω
Resistance and reactance of equivalent exciting circuit referred to primary are $R_0 = 700\ \Omega$ and $X_0 = 250\ \Omega$ respectively. Find the instruments reading for OC and SC tests when supply in given to HV side. During SC test, full-load current flows.

Solution

Here

$$R_1 = 0.9\ \Omega,\ X_1 = 0.94\ \Omega$$

$$R_2 = 0.4\ \Omega,\ X_2 = 0.11\ \Omega$$

$$R_0 = 700\ \Omega \text{ and } X_0 = 250\ \Omega \text{ referred to as primary}$$

OC test:

$$V_1 = 250\ \text{V}$$

$$I_W = \frac{V_1}{R_0} = \frac{250}{700} = 0.357\ \text{A}$$

and
$$I_\mu = \frac{V_1}{X_0} = \frac{250}{250} = 1\ \text{A}$$

∴
$$I_0 = \sqrt{I_\mu^2 + I_W^2} = \sqrt{1^2 + 0.357^2} = 1.062\ \text{A}$$

∴
$$W_0 = V_1 I_0 \cos\theta_0 = V_1 I_W = 250 \times 0.357 = 89.25\ \text{W}$$

∴ Reading of the instruments are: 250 V, 1.062 A, 89.25 W

SC test:

$$\text{Turn ratio } (a) = \frac{250}{125} = 2$$

∴
$$R_{01} = R_1 + a^2 R_2 = 0.9 + 2^2 \times 0.4 = 2.5\ \Omega$$

and
$$X_{01} = X_1 + a^2 X_2 = 0.94 + 2^2 \times 0.11 = 1.38\ \Omega$$

∴
$$Z_{01} = \sqrt{R_{01}^2 + X_{01}^2} = \sqrt{2.5^2 + (1.38)^2} = 2.856\ \Omega$$

∴
$$I_1 = \frac{2 \times 10^3}{250} = 8\ \text{A} = I_{SC}$$

∴
$$V_{SC} = I_{SC} Z_{01} = 8 \times 2.856 = 22.85\ \text{V}$$

∴
$$P_{SC} = I_{SC}^2 R_{01} = (8)^2 \times 2.5 = 160\ \text{W}$$

The instrument's readings are 22.85 V, 8A, 160 W ■

Example 1.30 A 10 kVA, 500/250 V, 50 Hz, single-phase transformer has the following test results:

OC test: 500 V, 1.2 A, 80 W

SC test: 50 V, 15 A, 90 W (LV side is short-circuited)

Determine the regulation and efficiency of the transformer at full load and at 0.8 power factor lagging.

Solution

From *OC test* of the transformer, iron loss of the transformer = 80 W

SC test:

$$V_{SC} = 30\ \text{V},\ I_{SC} = 15\ \text{A},\ P_{SC} = 90\ \text{W}$$

∴
$$Z_{01} = \frac{V_{SC}}{I_{SC}} = \frac{30}{15} = 2\Omega,\ R_{01} = \frac{P_{SC}}{I_{SC}^2} = \frac{90}{15^2} = 0.4\ \Omega, \text{ and}$$

$$X_{01} = \sqrt{Z_{01}^2 - R_{01}^2} = \sqrt{2^2 - 0.4^2} = 1.96\ \Omega$$

$$\text{Turns ratio } (a) = \frac{500}{250} = 2$$

$$R_{02} = \frac{R_{01}}{a^2} = \frac{0.4}{4} = 0.1\ \Omega \quad \text{and} \quad X_{02} = \frac{X_{01}}{a^2} = \frac{1.96}{4} = 0.49\ \Omega$$

Full-load secondary current of the transformer is

$$I_2 = \frac{10 \times 10^3}{250} = 40\text{ A}$$

Approximate voltage drop of the transformer

$$= I_2 (R_{02} \cos\theta + X_{02} \sin\theta) = 40 \times (0.1 \times 0.8 + 0.49 \times 0.6) = 14.96\text{ V}$$

Regulation of the transformer $= \dfrac{14.96}{250} = 0.0598$ p.u. = 5.98%

Cu loss of the transformer at full load

$$= \left(\frac{\text{Secondary FL current}}{\text{Secondary current during SC test}}\right)^2 \times 90 = \left(\frac{40}{30}\right)^2 \times 90 = 160\text{ W}$$

Total loss of the transformer = 80 + 160 = 240 W
Secondary terminal voltage at full load = 250 − 14.96 = 235.04 V
Output power = 235.04 × 40 × 0.8 = 7,521.28 W

$\therefore$ Efficiency of the transformer $= \dfrac{7{,}521.28}{7{,}521.28 + 240} = 0.9690$ p.u. = 96.9% ■

Example 1.31 A 4 kVA, 200/400 V, 50 Hz, single-phase transformer gave the following test results:
OC test (LV side): 200 V, 0.8 A, 75 W
SC test (HV side): 20 V, 10 A, 90 W

Calculate the parameters of the transformer referred to as LV side. Find the regulation and efficiency of the transformer for full load and half load both at 0.8 power factor lagging.

Solution

OC test:

$$V_1 = 200\text{ V}, I_0 = 0.8\text{ A}, W_0 = 75\text{ W}$$

$$\cos\theta_0 = \frac{W_0}{V_1 I_0} = \frac{75}{200 \times 0.8} = 0.468$$

$$\therefore \quad \sin\theta_0 = \sqrt{1 - \cos^2\theta_0} = 0.884$$

$I_w = I_0 \cos\theta_0 = 0.8 \times 0.468 = 0.3744$ A and $I_\mu = I_0 \sin\theta_0 = 0.8 \times 0.884 = 0.7072$ A

$R_0 = \dfrac{V_1}{I_w} = \dfrac{200}{0.3744} = 534.18\ \Omega$ and $X_0 = \dfrac{V_1}{I_\mu} = \dfrac{200}{0.7072} = 282.80\ \Omega$ referred to as LV side.

SC test:

$$V_{SC} = 20\text{ V}, I_{SC} = 10\text{ A}, P_{SC} = 90\text{ W}$$

$$\therefore \quad Z_{02} = \frac{V_{SC}}{I_{SC}} = \frac{20}{10} = 2\ \Omega, \quad R_{02} = \frac{P_{SC}}{I_{SC}^2} = \frac{90}{10^2} = 0.9\ \Omega$$

and

$$X_{02} = \sqrt{Z_{02}^2 - R_{02}^2} = \sqrt{2^2 - 0.9^2} = 1.786\ \Omega$$

$$\therefore \quad R_{01} = a^2 R_{02} = \left(\frac{1}{2}\right)^2 \times 0.9 = 0.225\ \Omega$$

$$\therefore \quad X_{02} = a^2 X_{02} = \left(\frac{1}{2}\right)^2 \times 1.786 = 0.4465\ \Omega$$

Full-load voltage drop $= I_2 (R_{02} \cos\theta + X_{02} \sin\theta)$

$$= 10 \times (0.9 \times 0.8 + 1.786 \times 0.6) = 17.916\ \text{V}$$

$$\therefore \quad \text{Regulation of the transformer} = \frac{17.196}{400} = 0.0448\ \text{p.u.} = 4.48\%$$

During F.L., the output voltage = 400 − 17.916 = 382.084 V

$$\therefore \quad \text{Output power} = 382.084 \times 10 \times 0.8 = 3{,}056.672\ \text{W}$$

$$\therefore \quad \text{Full-load efficiency} = \frac{3{,}056.672}{3{,}056.672 + 75 + 90} = 0.9488\ \text{p.u.} = 94.88\%$$

■

Example 1.32 A 6 kVA, 250/500 V, 50 Hz, single-phase transformer has the following test results:

OC test (LV side): 250 V, 1.2 A, 80 W

SC test (HV side): 25 V, 10 A, 95 W

Calculate the circuit parameters referred to as LV side. Also calculate the regulation and efficiency of the transformer at full load and half load at 0.8 power factor lagging.

Solution

Here, turn ratio $(a) = \frac{1}{2}$

OC test:

$$V_1 = 250\ \text{V},\ I_0 = 1.2\ \text{A},\ W_0 = 80\ \text{W}$$

$$\therefore \quad \cos\theta_0 = \frac{W_0}{V_1 I_0} = \frac{80}{250 \times 1.2} = 0.267$$

$$\therefore \quad \sin\theta_0 = \sqrt{1 - \cos^2\theta_0} = 0.964$$

$$\therefore \quad R_0 = \frac{V_1}{I_w} = \frac{250}{0.267} = 963.33\ \Omega \quad \text{and} \quad X_0 = \frac{V_1}{I_\mu} = \frac{250}{0.964} = 259.34\ \Omega$$

SC test:

$$V_{SC} = 25\ \text{V},\ I_{SC} = 10\ \text{A},\ P_{SC} = 95\ \text{W}$$

$$\therefore \quad Z_{02} = \frac{V_{SC}}{I_{SC}} = 2.5\ \Omega,\ R_{02} = \frac{P_{SC}}{I_{SC}^2} = \frac{95}{100} = 0.95\ \Omega \text{ and}$$

$$X_{02} = \sqrt{Z_{02}^2 - R_{02}^2} = 2.312\ \Omega$$

$$\therefore \quad R_{01} = a^2 R_{02} = \frac{0.95}{4} = 0.2375\ \Omega \quad \text{and} \quad X_{01} = a^2 X_{02} = \frac{2.312}{4} = 0.578\ \Omega$$

The voltage drop due to lagging p.f. is expressed by

$$\Delta V = I_2 (R_{02} \cos\theta + X_{02} \sin\theta)$$

Full load:

$$(I_2)_{fl} = \frac{6\times 10^3}{500} = 12 \text{ A}$$

$\therefore$ $$\Delta V = 12 \times (0.95 \times 0.8 + 2.312 \times 0.6) = 25.7664 \text{ V}$$

$\therefore$ $$\text{Regulation} = \frac{25.7664}{500} = 0.0515 \text{ p.u.} = 5.15\%$$

$\therefore$ $$V_2' = 500 - 25.7664 = 474.2336 \text{ V}$$

$\therefore$ $$\text{Output power} = 474.2336 \times 12 \times 0.8 = 4{,}552.64 \text{ W}$$

$\therefore$ $$\text{Iron loss} = 80 \text{ W}$$

$\therefore$ $$\text{Full-load copper loss} = \left(\frac{12}{10}\right)^2 \times 95 = 136.8 \text{ W}$$

$\therefore$ $$\text{FL efficiency} = \frac{4{,}552.64}{4{,}552.64 + 80 + 136.8} = 0.9545 \text{ p.u.} = 95.45\%$$

Half load:

Secondary current at half load $(I_2)_{\frac{1}{2}fl} = \dfrac{12}{2} = 6$ A

Voltage drop at half load = $6 \times (0.95 \times 0.8 + 2.312 \times 0.6) = 12.8832$ V

$\therefore$ $$\text{Regulation} = \frac{12.8832}{500} = 0.0257 \text{ p.u.} = 2.57\%$$

$\therefore$ $$V_2' = 500 - 12.8832 = 487.1168 \text{ V}$$

$\therefore$ $$\text{Output power} = V_2' \times (I_2)_{\frac{1}{2}fl} \times \cos\theta = 487.1168 \times 6 \times 0.8 = 2{,}338.16 \text{ W}$$

$$\text{Half-load copper loss} = \left(\frac{1}{2}\right)^2 \times 136.8 = 34.2 \text{ W}$$

$\therefore$ $$\text{HL efficiency} = \frac{2{,}338.16}{2{,}338.16 + 80 + 34.2} = 0.9534 \text{ p.u.} = 95.34\%$$ ■

Example 1.33 A 40 kVA single-phase step-down transformer has a full-load secondary current of 200 A, and the total resistance referred to as secondary is 0.008 Ω. Find the efficiency of the transformer at full-load and three-fourth load when the load is at (i) unity power factor and (ii) 0.8 power factor. The iron loss of the transformer is 190 W.

Solution

Full-load secondary current = 200 A

$\therefore$ $$\text{Full-load Cu loss} = I_2^2 R_{02} = (200)^2 \times 0.008 = 320 \text{ W}$$

$\therefore$ $$\text{Cu loss at } \frac{3}{4}\text{th full load} = \left(\frac{3}{4}\right)^2 \times 320 = 180 \text{ W}$$

At full load:

$$\text{Total losses} = (190 + 320) \text{ W} = 510 \text{ W}$$

(i) Output at unity power factor = $40 \times 10^3 \times 1 = 40{,}000$ W

$\therefore$ $$\text{Efficiency} = \frac{40{,}000}{40{,}000 + 510} = 0.9874 \text{ p.u.} = 98.74\%$$

(ii) Output at 0.8 power factor lagging = $40 \times 10^3 \times 0.8 = 32{,}000$ W

$\therefore$ $$\text{Efficiency} = \frac{32{,}000}{32{,}000 + 510} = 0.9843 \text{ p.u.} = 98.43\%$$

At 3/4th load:

$$\text{Total losses} = 190 + 180 = 370 \text{ W}$$

(i) Output at unity power factor = $\frac{3}{4} \times 40 \times 10^3 \times 1 = 30{,}000$ W

$\therefore$ $$\text{Efficiency} = \frac{30{,}000}{30{,}000 + 370} = 0.9878 \text{ p.u.} = 98.78\%$$

(ii) Output at 0.8 p.f. lagging = $\frac{3}{4} \times 40 \times 10^3 \times 0.8 = 24{,}000$ W

$\therefore$ $$\text{Efficiency} = \frac{24{,}000}{24{,}000 + 370} = 0.9848 \text{ p.u.} = 98.48\%$$ ■

Example 1.34 A 20 kVA single-phase transformer has iron loss of 250 W and full-load Cu loss of 500 W.

(i) Determine the load at which maximum efficiency occurs and the value of maximum efficiency at unity power factor.
(ii) If the maximum efficiency occurs at 85 per cent of full load, calculate the new core loss and full-load Cu loss assuming that total loss at full load remains constant.

Solution

Rating of transformer = 20 kVA

Iron loss (P_i) = 250 W

Copper loss (P_{Cu}) = 500 W

(i) Let the maximum efficiency occurs at x times of full load. We have

$$x^2 P_{\text{Cu fl}} = P_i$$

i.e., $$x = \sqrt{\frac{P_i}{P_{\text{Cu fl}}}} = \sqrt{\frac{250}{500}} = \sqrt{\frac{1}{2}} = 0.707$$

Maximum efficiency occurs at 70.7 per cent of full load.

Output at unity power factor = $0.707 \times 20 \times 1 = 14.14$ kW

Total losses = $2 \times 250 = 500$ W = 0.5 kW

$$\therefore \qquad \text{Maximum efficiency} = \frac{14.14}{14.14+0.5} = 0.9658 \text{ p.u.} = 96.58\%$$

(ii) Maximum efficiency occurs at 85 per cent of full load.

$$\text{Now} \qquad P_i' + P_{Cu}' = P_i + P_{Cu}$$

where P_i' and P_{Cu}' are new core loss and full-load Cu loss of the transformer

Since the total loss remains constant

$$\therefore \qquad P_i' + P_{Cu}' = P_i + P_{Cu} = 250 + 500 = 750 \text{ W}$$

$$\text{Again,} \qquad \sqrt{\frac{P_i'}{P_{Cu}'}} = 0.85$$

$$\text{i.e.,} \qquad \frac{P_i'}{P_{Cu}'} = 0.85^2 = 0.7225 \qquad (1)$$

$$\text{i.e.,} \qquad \frac{P_i' + P_{Cu}'}{P_{Cu}'} = 1.7225$$

$$\text{i.e.,} \qquad \frac{750}{P_{Cu}'} = 1.7225$$

$$\text{i.e.,} \qquad P_{Cu}' = \frac{750}{1.7225} = 435.41 \text{ W}$$

From Equation (1), $\qquad P_i' = 0.7225\, P_{Cu}' = 314.58$ W

∴ New iron and Cu loss of the transformer are 314.58 W and 435.41 W respectively ■

Example 1.35 A single-phase, 1,000/200 V transformer has primary and secondary resistances of 0.25 Ω and 0.018 Ω respectively. If iron loss of the transformer is 240 W, calculate the secondary current at which maximum efficiency occurs. Also calculate the maximum efficiency at 0.8 power factor lagging. Assume that secondary terminal voltage remains constant.

Solution

$$\text{Turns ratio } (a) = \frac{1,000}{200} = 5$$

Here $R_1 = 0.25\ \Omega$ and $R_2 = 0.018\ \Omega$

$$\therefore \qquad R_{02} = R_2 + \frac{R_1}{a^2} = 0.018 + \frac{0.25}{5^2} = 0.028\ \Omega$$

At maximum efficiency

$$\text{Copper loss} = \text{iron loss}$$

$$\text{i.e.,} \qquad I_2^2 R_{02} = 240$$

i.e., $$I_2 = \sqrt{\frac{240}{R_{02}}} = \sqrt{\frac{240}{0.028}} = 92.58 \text{ A}$$

∴ The secondary current at maximum efficiency is 92.58 A.
Output at maximum efficiency at 0.8 power factor lagging

$$= V_2 I_2 \cos\theta = 200 \times 92.58 \times 0.8 \text{ W} = 14.812 \text{ kW}$$

$$\text{Total loss} = 2 \times \text{iron loss} = 2 \times 240 = 480 \text{ W} = 0.48 \text{ kW}$$

$$\text{Maximum efficiency} = \frac{\text{Output}}{\text{Output} + \text{Losses}} = \frac{14.812}{14.812 + 0.48} = 0.9686 \text{ p.u.} = 96.86\%$$ ■

Example 1.36 A single-phase, 150 kVA transformer has an efficiency 96 per cent at full load on 0.8 power factor and 96.5 per cent at half load on 0.8 power factor lagging. Find the following:

(i) Iron loss.
(ii) Copper loss at full load.
(iii) The load kVA at which maximum efficiency occurs.
(iv) The maximum efficiency of the transformer at 0.8 power factor lagging.

Solution

Let P_i and P_{Cu} be the iron loss and full-load copper loss of the transformer, respectively. Full-load efficiency is 96 per cent at 0.8 p.f. lagging

∴ $$\frac{96}{100} = \frac{150 \times 10^3 \times 0.8}{150 \times 10^3 \times 0.8 + P_i + P_{Cu}}$$

i.e., $$P_i + P_{Cu} = 150 \times 10^3 \times 0.8\left(\frac{1}{0.96} - 1\right)$$

i.e., $$P_i + P_{Cu} = 4{,}999.9 \text{ W} \tag{1}$$

Half-load efficiency is 96 per cent at 0.8 p.f. lagging

$$\frac{96.5}{100} = \frac{\frac{1}{2} \times 150 \times 10^3 \times 0.8}{\frac{1}{2} \times 150 \times 10^3 \times 0.8 + P_i + \frac{1}{4} P_{Cu}}$$

i.e., $$P_i + \frac{1}{4} P_{Cu} = 75 \times 10^3 \times 0.8\left(\frac{1}{0.965} - 1\right)$$

i.e., $$P_i + \frac{1}{4} P_{Cu} = 2{,}176.16 \text{ W} \tag{2}$$

Solving Equations (1) and (2), we have

$$P_{Cu} = 3{,}764.98 \text{ W}$$

and $$P_i = 1{,}234.92 \text{ W}$$

(i) Iron loss of the transformer = 1,234.98 W
(ii) Full-load copper loss of the transformer = 3,764.98 W
(iii) Let x be the fraction of full load to get the maximum efficiency of the transformer.

$$\therefore \qquad x = \sqrt{\frac{P_i}{P_{Cu}}} = \sqrt{\frac{1,234.92}{3,764.98}} = 0.5727$$

∴ Maximum efficiency occurs at 57.27 per cent of full load.

(iv) For maximum efficiency

$$\text{Copper loss} = \text{Iron loss} = 1,234.92 \text{ W}$$

$$\therefore \qquad \eta_{max} = \frac{0.5727\times150\times10^3\times0.8}{0.5727\times150\times10^3\times0.8+2\times1,234.92} = 0.9653 \text{ p.u.} = 96.53\%$$ ■

Example 1.37 A 12 kVA, 1,200/400 V, single-phase, 50 Hz transformer has the following test results:

OC test (HV side): 1,200 V, 1.2 A, 70 W

SC test (LV side): 30 V, 22 A, 80 W

Find the following:

(i) Core loss of the transformer.

(ii) Equivalent resistance and leakage reactance of the transformer referred to as LV side.

(iii) Equivalent resistance and leakage reactance of the transformer referred to as HV side.

(iv) Calculate the secondary winding voltage to keep the secondary terminal voltage 400 V at full load and 0.8 power factor lagging. Hence calculate the primary supply voltage.

(v) Regulation of the transformer at full load and 0.8 power factor lagging.

(vi) Regulation of the transformer at full load and 0.8 power factor leading.

(vii) Power factor of the load so that regulation of the transformer is maximum at full-load current. Also find the regulation.

(viii) Power factor of the load so that regulation of the transformer is zero.

(ix) Efficiency of the transformer for: (a) full load, (b) half load, (c) $\frac{3}{4}$th load and (d) $\frac{1}{4}$th load at 0.8 power factor lagging. In all cases, terminal voltage remains 400 V.

(x) Find the load at kVA at which efficiency of the transformer becomes maximum. Find the maximum efficiency at 0.8 power factor lagging.

Solution

$$\text{Here kVA rating} = 12 \text{ kVA}$$

$$\text{Turns ratio } (a) = \frac{1,200}{400} = 3$$

(i) Core loss of the transformer = 70 W (from **OC Test**)

(ii) **SC test:** Instruments placed on LV side (side 2).

$$V_{SC} = 30 \text{ V}, I_{SC} = 22 \text{ A}, P_{SC} = 80 \text{ W}$$

$$\therefore \qquad Z_{02} = \frac{V_{SC}}{I_{SC}} = \frac{30}{22} = 1.364\,\Omega, R_{02} = \frac{P_{SC}}{I_{SC}^2} = \frac{80}{484} = 0.1653\,\Omega$$

and

$$X_{02} = \sqrt{Z_{02}^2 - R_{02}^2} = \sqrt{1.364^2 - 0.1653^2} = 1.354\,\Omega$$

$$\therefore \qquad R_{02} = 0.1653\,\Omega \text{ and } X_{02} = 1.354\,\Omega \text{ referred to as LV side}$$

(iii) ∴ $R_{01} = a^2 R_{02} = 3^2 \times 0.1653 = 1.4877\,\Omega$

∴ $X_{01} = a^2 X_{02} = 3^3 \times 1.354 = 12.186\,\Omega$

(iv)

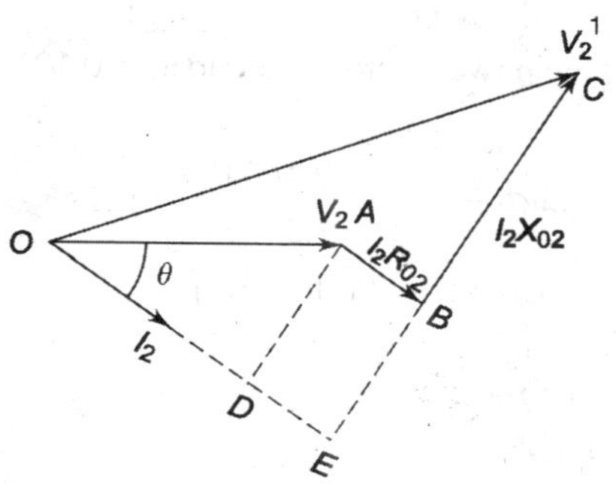

Figure E1.7

$$AB = I_2 R_{02},\ BC = I_2 X_{02},\ OC = {}_0V_2,\ OA = V_2$$

$$\therefore \quad OD = OA\cos\theta = V_2\cos\theta \text{ and } BE = AD = V_2\sin\theta$$

$$\therefore \quad I_2 = \frac{12\times10^3}{400} = 30\text{ A}.$$

From ΔOEC

$$OC^2 = OE^2 + EC^2$$
$$= (OD + DE)^2 + (EB + BC)^2$$
$$= (V_2\cos\theta + I_2 R_{02})^2 + (V_2\sin\theta + I_2 X_{02})^2$$
$$= (400 \times 0.8 + 30 \times 0.1653)^2 + (400 \times 0.6 + 30 \times 1.354)^2$$
$$V_2' = 429.35\text{ V}$$

$$\therefore \quad V_1 = \frac{N_1}{N_2}\times V_2' = 3\times 429.35 = 1{,}288.05\text{ V}$$

The secondary winding voltage must be 429.35 V and the applied voltage becomes 1,288.05 V.

(v) $I_{2fl} = 30$ A

$$\therefore \quad \Delta V = I_{2fl}(R_{02}\cos\theta + X_{02}\sin\theta) \text{ [for lagging power factor]}$$
$$= 30 \times (0.1653 \times 0.8 + 1.354 \times 0.6) = 28.34\text{ V}$$

$$\therefore \quad \text{Regulation} = \frac{28.34}{400} = 0.0708\text{ p.u.} = 7.08\%$$

(vi)

$$\Delta V = I_2(R_{02}\cos\theta - X_{02}\sin\theta) \text{ [for leading power factor]}$$
$$= 30 \times (0.1653 \times 0.8 - 1.354 \times 0.6) = -20.40\text{ V}$$

$$\therefore \quad \text{Regulation} = \frac{-20.40}{400} = -0.051 = -5.10\%$$

(vii) For maximum regulation:

$$\tan\theta = \frac{X_{02}}{R_{02}} = \frac{1.354}{0.1653} = 8.191$$

$$\therefore \qquad \cos\theta = 0.1211 \text{ (lagging)}$$

For maximum regulation, the power factor of the load is 0.1211 (lagging)

(viii) For Zero regulation:

$$\tan\theta = \frac{R_{02}}{X_{02}} = \frac{0.1654}{1.3453} = 0.1229$$

$$\therefore \qquad \cos\theta = 0.9925 \text{ (leading).}$$

(ix) (a) *At full load:*

From SC test copper loss at 22 A is 80 W.

$$\therefore \qquad I_{2\text{fl}} = 30 \text{ A}$$

$$\therefore \qquad \text{Full-load Cu loss} = \left(\frac{30}{22}\right)^2 \times 80 = 148.76 \text{ W}$$

$$\therefore \qquad \eta_{\text{fl}} = \frac{12\times10^3\times0.8}{12\times10^3\times0.8+70+148.76} = 0.9777 \text{ p.u.} = 97.77\%$$

(b) *At half load:*

$$\therefore \qquad \eta_{\frac{1}{2}\text{fl}} = \frac{\frac{1}{2}\times12\times10^3\times0.8}{\frac{1}{2}\times12\times10^3\times0.8+70+\left(\frac{1}{2}\right)^2\times148.76} = 0.9781 \text{p.u.} = 97.81\%$$

(c) *At 3/4th load:*

$$\therefore \qquad \eta_{\frac{3}{4}\text{fl}} = \frac{\frac{3}{4}\times12\times10^3\times0.8}{\frac{3}{4}\times12\times10^3\times0.8+70+\left(\frac{3}{4}\right)^2\times148.76} = 0.9791 \text{p.u.} = 97.91\%$$

(d) *At 1/4th load:*

$$\therefore \qquad \eta_{\frac{1}{4}\text{fl}} = \frac{\frac{1}{4}\times12\times10^3\times0.8}{\frac{1}{4}\times12\times10^3\times0.8+70+\left(\frac{1}{4}\right)^2\times148.76} = 0.9680 \text{ p.u.} = 96.80\%$$

(x) Let the maximum efficiency occurs at the fraction x of the full load.

At maximum efficiency

$$x^2 P_{\text{Cu fl}} = W_i$$

$$\therefore \qquad x = \sqrt{\frac{W_i}{P_{\text{Cu fl}}}} = \sqrt{\frac{70}{148.76}} = 0.686$$

The maximum efficiency of the transformer occurs at 68.6 per cent of full load.

$$\therefore \qquad \eta_{\text{max}} = \frac{0.686\times12\times10^3\times0.8}{0.686\times12\times10^3\times0.8+70\times2} = 0.9792 \text{ p.u.} = 97.92\%$$

■

Example 1.38 The short circuit test data referred to as HV side of two single-phase transformers 1,000/200 V are

Transformer A: 50 V, 25 A, 375 W
Transformer B: 37.5V, 25 A, 500 W

If two transformers are connected in parallel and fed a total load of 18.176 kW at 0.8 power factor lagging, calculate the secondary load current, power factor and power output of each transformer.

Solution

For transformer A:

$$Z_A = \frac{50}{25} = 2\,\Omega$$

$$R_A = \frac{375}{25^2} = 0.6\,\Omega$$

$$X_A = \sqrt{Z_A^2 - R_A^2} = \sqrt{2^2 - 0.6^2} = 1.908\,\Omega$$

For transformer A:

$$Z_B = \frac{37.5}{25} = 1.5\,\Omega$$

$$R_B = \frac{500}{25^2} = 0.8\,\Omega$$

$$X_B = \sqrt{Z_B^2 - R_B^2} = \sqrt{1.5^2 - 0.8^2} = 1.269\,\Omega$$

$$\therefore \quad \frac{\bar{I}_A}{\bar{I}_B} = \frac{\bar{Z}_B}{\bar{Z}_A} = \frac{1.5\angle\tan^{-1}\left(\frac{1.269}{0.8}\right)}{2\angle\tan^{-1}\left(\frac{1.908}{0.6}\right)} = \frac{1.5\angle 57.8^\circ}{2\angle 72.54^\circ} = 0.75\angle -14.74^\circ$$

$$\bar{I}_A = (0.725 - j0.191)\bar{I}_B$$

$$\text{Secondary load current} = \frac{18{,}176}{200 \times 0.8} = 113.6\text{ A}$$

$$\bar{I} = 113.6(0.8 - j0.6) = 90.88 - j68.16$$

Again,

$$\bar{I} = \bar{I}_A + \bar{I}_B$$

$$90.88 - j68.16 = (0.725 - j0.191)\bar{I}_B + \bar{I}_B = \bar{I}_B\,(1.725 - j0.191)$$

$$\bar{I}_B = \frac{90.88 - j68.16}{1.725 - j0.191} = 65.45\angle -30.55^\circ\text{ A}$$

$$\bar{I}_A = (65.45\angle -30.55^\circ) \times (0.725 - j\,0.191)\ (65.45\angle -30.55^\circ)\ (0.749\angle -40.7^\circ) = 49.02\angle -45.31^\circ\text{ A}$$

Power output of transformer $A = 200 \times 49.02 \times \cos(-45.31^\circ)$ W $= 6.895$ kW
Power output of transformer $B = 200 \times 65.45 \times \cos(-30.55^\circ)$ W $= 11.273$ kW ■

Example 1.39 Find the all-day efficiency of a transformer having maximum efficiency of 98.5 per cent at 20 kVA at unity power factor and loaded as follows:

11 hours: 5 kW at 0.7 power factor lagging
6 hours: 8 kW at 0.8 power factor lagging
7 hours: no load.

The maximum efficiency of the transformer occurs at 80 per cent of full load.

Solution

At maximum efficiency:

$$\text{Output} = 20 \times 1.0 = 20 \text{ kW and } \eta_{max} = 0.985$$

$$\text{Input} = \frac{\text{Output}}{\eta_{max}} = \frac{20}{0.985} = 20.304 \text{ kW}$$

$$\text{Total losses} = 20.304 - 20 = 0.304 \text{ kW}$$

$$\text{Iron loss} = 0.304/2 = 0.152 \text{ kW}$$

$$\text{Full-load copper loss} = \left(\frac{100}{80}\right)^2 \times 0.152 = 0.2375 \text{ kW}$$

$$\text{All-day output} = 5 \times 11 + 8 \times 6 + 0 \times 7 = 103 \text{ kWh}$$

$$\text{Iron loss for 24 hours} = 0.152 \times 24 = 3.648 \text{ kWh}$$

$$\text{Total copper loss} = \left[\left(\frac{\frac{5}{0.7}}{20}\right)^2 \times 11 + \left(\frac{\frac{8}{0.8}}{20}\right)^2 \times 6\right] \times 0.2375 = 0.68947 \text{ kWh}$$

Therefore all-day efficiency,

$$= \frac{\text{Output}}{\text{Output} + \text{Losses}} = \frac{103}{103 + (0.68947 + 3.648)} = 0.9595 \text{ p.u.} = 95.95\%$$ ■

Example 1.40 Two single-phase transformers A and B are connected in parallel on both primary and secondary. Transformer A has full-load resistance drop of 1.2 per cent and reactance drop of 6.5 per cent, while the corresponding values of transformer B are 1 per cent and 5.5 per cent, respectively. Find the sharing of each of the load for the following cases:

(i) 150 kW at 0.8 power factor lagging.
(ii) 160 kW at 0.9 power factor lagging.

Solution

Percentage impedance of transformer A ($\overline{Z}_A$) = 1.2 + j 6.5
Percentage impedance of transformer B ($\overline{Z}_B$) = 1.0 + j 5.5

$$\therefore \quad \overline{Z}_A + \overline{Z}_B = (1.2 + j\,6.5) + (1.0 + j\,5.5) = 2.2 + j\,12.0$$

(i) 150 kW at 0.8 power factor lagging

$$\text{Total load in kVA} = \frac{150}{0.8} = 187.5$$

Load to be shared ($\bar{S}$) = 187.5∠−36.90 = 149.94 − j112.5
Load shared by transformer A

$$\bar{S}_A = \bar{S} \times \frac{\bar{Z}_B}{\bar{Z}_A + \bar{Z}_B} = (149.94 - j112.5) \times \left(\frac{1 + j5.5}{2.2 + j12.0} \right) = 68.78 - j51.45$$

Load shared by transformer B

$$\bar{S}_B = \bar{S} \times \frac{\bar{Z}_A}{\bar{Z}_A + \bar{Z}_B} == (149.94 - j112.5) \times \left(\frac{1.2 + j6.5}{2.2 + j12.0} \right) = 81.16 - j61.05$$

(ii) 160 kW at 0.9 power factor lagging

Total load in kVA $= \dfrac{160}{0.9} = 177.78$

Load to be shared ($\bar{S}$) = 177.78∠ − 25.84° = 160.0 − j77.48
Load shared by transformer A

$$\bar{S}_A = \bar{S} \times \frac{\bar{Z}_B}{\bar{Z}_A + \bar{Z}_B} = (160.0 - j77.48) \times \left(\frac{1 + j5.5}{2.2 + j12.0} \right) = 73.36 - j35.4$$

Load shared by transformer B

$$\bar{S}_B = \bar{S} \times \frac{\bar{Z}_A}{\bar{Z}_A + \bar{Z}_B} = (160.0 - j77.48) \times \left(\frac{1.2 + j6.5}{2.2 + j12.0} \right) = 86.63 - j42.08$$

■

Example 1.41 A 1,000 kVA single-phase transformer has a percentage impedance of (1.2 + *j* 4.8), while a 400 kVA single-phase transformer has a percentage impedance of (2.2 +*j* 6.5). These two transformers are connected in parallel. The no-load voltage of each transformer is 600 V. If the total load is 800 kVA at 0.8 power factor lagging, find the load shared by each of the transformers.

Solution

Let 1,000 kVA and 400 kVA transformers be denoted by A and B, respectively.
Percentage impedance of transformer A = 1.2 + *j* 4.8 = $\bar{Z}_A$
Percentage impedance of transformer B = 2.2 + *j* 6.5 = $\bar{Z}_B$
Percentage impedance of transformer B referred to as 1,000 kVA

$$= \frac{1{,}000}{400}(2.2 + j6.5) = 5.5 + j16.25 = \bar{Z}_B$$

$$\therefore \quad \bar{Z}_A + \bar{Z}_B = (2.2 + j6.5) + (5.5 + j16.25) = 7.7 + j22.75$$

Total load to be shared $= \dfrac{800}{0.8}\angle -36.9^\circ$ kVA $= 1{,}000\angle -36.9^\circ$ kVA

Load shared by transformer A

$$\bar{S}_A = \bar{S} \times \frac{\bar{Z}_B}{\bar{Z}_A + \bar{Z}_B} = (800.0 - j600.0) \times \left(\frac{5.5 + j16.25}{7.7 + j22.75} \right) = 571.43 - j428.57$$

■

Load shared by transformer B

$$\overline{S}_B = \overline{S}\times\frac{\overline{Z}_A}{\overline{Z}_A+\overline{Z}_B} = (800.0 - j600.0)\times\left(\frac{2.5+j6.5}{7.7+j22.75}\right) = 224.67 - j183.3$$

Example 1.42 Two single-phase transformers A and B have the following parameters:

Transformer A: 10 kVA, 400/200 V, percentage resistance and reactance are 4 per cent and 5 per cent, respectively.

Transformer B: 5 kVA, 400/200 V, percentage resistance and reactance are 5 per cent and 4 per cent, respectively.

When these two transformers are connected in parallel, they share a common load of 12 kW at 0.8 power factor lagging. Calculate the active and reactive power delivered by transformer A and transformer B.

Solution

Transformer A:

$$10 \text{ kVA}, 400/200\text{V}, \%R_A = 4, \%X_A = 5, \%\overline{Z}_A = 4 + j\,5$$

Transformer B:

$$5 \text{ kVA}, 400/200\text{V}, \%R_B = 5, \%X_B = 4, \%\overline{Z}_B = 5 + j\,4$$

Load = 12 kW, 0.8 power factor lagging

$$\text{Load}(\overline{S}) = \frac{12}{0.8}\angle\cos^{-1}0.8 = 15\angle -36.9°$$

$\%\overline{Z}_A$ and $\%\overline{Z}_B$ on common base 10 kVA are

$$\%\overline{Z}_A = 4 + j5 \qquad \text{and} \qquad \%\overline{Z}_B = \frac{10}{5}(5 + j4) = 10 + j8$$

Load shared by transformer A

$$\overline{S}_A = \overline{S}\times\frac{\overline{Z}_B}{\overline{Z}_A+\overline{Z}_B} = (12 - j9)\times\left(\frac{10+j\,8}{14+j\,13}\right) = 7.378 - j6.608$$

Load shared by transformer B

$$\overline{S}_B = \overline{S}\times\frac{\overline{Z}_A}{\overline{Z}_A+\overline{Z}_B} = (12 - j\,9)\times\left(\frac{4+j\,5}{14+j\,13}\right) = 4.422 - j2.392$$ ■

Example 1.43 Two single-phase transformers with voltage ratio 1,000/200 V have the following short circuit data:

Transformer A: Primary voltage = 55 V, primary current = 25 A, power input to primary = 500 W
Transformer B: Primary voltage = 40V, primary current = 25 A, power input to primary = 610 W

The two transformers are connected in parallel and share a common load of 24 kW at 0.8 power factor lagging. Calculate the secondary load current, power factor and real power output of each transformer.

Solution

Transformer A:

$$Z_A = \frac{V_{sc}}{I_{sc}} = \frac{55}{25} = 2.2\,\Omega$$

$$R_A = \frac{500}{25^2} = 0.8\,\Omega$$

$$X_A = \sqrt{Z_A^2 - R_A^2} = \sqrt{2.2^2 - 0.8^2} = 2.05\,\Omega$$

$$\overline{Z}_A = R_A + jX_A = (0.8 + j2.05)\,\Omega$$

Transformer B:

$$Z_B = \frac{V_{sc}}{I_{sc}} = \frac{40}{25} = 1.6\,\Omega$$

$$R_B = \frac{600}{25^2} = 0.976\,\Omega$$

$$X_B = \sqrt{Z_A^2 - R_A^2} = \sqrt{1.6^2 - 0.976^2} = 1.268\,\Omega$$

$$\overline{Z}_B = R_A + jX_A = (0.976 + j1.268)\,\Omega$$

Let the currents delivered by transformer A and transformer B be I_A and I_B, respectively.

$$\therefore \quad \frac{\overline{I}_A}{\overline{I}_B} = \frac{\overline{Z}_A}{\overline{Z}_B} = \frac{0.976 + j1.268}{0.8 + j2.05}$$

i.e.,
$$\overline{I}_A = (0.698 - j0.204)\overline{I}_B$$

Therefore, secondary load current $(I) = \dfrac{24 \times 1,000}{200 \times 0.8} = 150$ A

Since $\overline{I} = \overline{I}_A + \overline{I}_B$, we have

$$150 = (0.698 - j0.204)\overline{I}_B + \overline{I}_B$$

i.e.,
$$150 = (1.698 - j0.204)\overline{I}_B$$

i.e.,
$$\overline{I}_B = \frac{150}{1.698 - j0.204} = 87\ 08 + j\ 10\ 46 = 87.70 \text{ A, } 0.993 \text{ lagging}$$

$\because$
$$\overline{I}_A = (0.698 - j0.204)\overline{I}_B, \text{ we have}$$

$$\overline{I}_A = (0.698 - j0.204)\ (87.08 + j\ 10.46) = 62.915 - j\ 10.46 = 63.78 \text{ A, } 0.986 \text{ lagging}$$

$\therefore$ Power output of transformer A = 200 × 63.78 × 0.986 = 12.577 kW

$\therefore$ Power output of transformer B = 200 × 87.70 × 0.993 = 17.417 kW ■

Example 1.44 Two single-phase transformers A and B are connected in parallel for supplying a load, which has resistance and reactance of 12 Ω and 6 Ω, respectively. The effective resistance and reactance of transformer A are 0.12 Ω and 1.3 Ω respectively, and that of transformer B are 0.17 Ω and 1.6 Ω respectively. The no-load emf of transformer A is 2,400 V and that of transformer B is 2,300 V. Calculate the following:

(i) Circulating current at no load.
(ii) Current output of each transformer and its power factor.
(iii) Power delivered by each transformer.

Solution

No-load emf of transformer A (E_A) = 2,400 V

No-load emf of transformer B (E_B) = 2,300 V

R_A = 0.12 Ω, X_A = 1.3 Ω, R_B = 0.17 Ω, X_B = 1.6 Ω

$$\overline{Z}_A = (0.12 + j1.3)\,\Omega \text{ and } \overline{Z}_B = (0.17 + j1.6)\,\Omega$$

∴ $$\overline{Z} = (\overline{Z}_A + \overline{Z}_B) = (0.12 + j1.3) + (0.17 + j1.6) = (0.29 + j2.9)\Omega$$

Here, $$\overline{Z}_L = (12 + j\,6)$$

(i) $$\overline{I}_C = \frac{\overline{E}_A - \overline{E}_B}{\overline{Z}_A + \overline{Z}_B} = \frac{2,400 - 2,300}{0.29 + j\,2.9} = 3.414 - j\,34.14 = 34.31\,A\text{ , 0.099 pf lagging}$$

(ii) $$\overline{I}_A = \frac{\overline{E}_A\overline{Z}_B + (\overline{E}_A - \overline{E}_B)\overline{Z}_L}{\overline{Z}_A\overline{Z}_B + (\overline{Z}_A + \overline{Z}_B)\overline{Z}_L}$$

$$= \frac{2,400(0.17 + j\,1.6) + (2,400 - 2,300)(12 + j\,6)}{(0.12 + j\,1.3)(0.17 + j\,1.6) + (0.12 + j\,1.3 + 0.17 + j\,1.6)(12 + j\,6)}$$

$$= \frac{(408 + j\,3,840) + (1,200 + j\,600)}{-15.9796 + j36.953} = \frac{(1,608 + j4,440)}{-15.9796 + j36.953} = 85.37 - j\,80.432$$

= 117.3 A, 0.7278 power factor lagging

$$\overline{I}_B = \frac{\overline{E}_B\overline{Z}_A - (\overline{E}_A - \overline{E}_B)\overline{Z}_L}{\overline{Z}_A\overline{Z}_B + (\overline{Z}_A + \overline{Z}_B)\overline{Z}_L}$$

$$= \frac{2,300(0.12 + j\,1.3) - (2,400 - 2,300)(12 + j\,6)}{(0.12 + j\,1.3)(0.17 + j\,1.6) + (0.12 + j\,1.3 + 0.17 + j\,1.6)(12 + j\,6)}$$

$$= \frac{(276 + j\,2,990) - (1,200 + j\,600)}{-15.9796 + j\,36.953} = \frac{(-924 + j\,2390)}{-15.9796 + j\,36.953} = 63.597 - j\,2.4966$$

= 63.646 A, 0.999 power factor lagging

(iii) Power output of transformer A = 2,400 × 117.3 × 0.7278 = 204.89 kW

Power output of transformer B = 2,300 × 63.646 × 0.999 = 146.239 kW ■

Example 1.45 Two single-phase transformers A and B having ratings 600 kVA and 300 kVA share a load of 800 kVA at unity power factor. The open circuit voltage of transformer A and transformer B are 610 V and 600 V, respectively. Percentage resistance and reactance of transformer A are 1.2 and 3.3 respectively, and that of transformer B are 1.7 and 5.7 respectively. Calculate the circulating current in the secondary at no load, current output of each transformer and its power factor and sharing of load by transformer A and transformer B.

Solution

Transformer A: Rating is 600 kVA

Open circuit voltage of transformer A (E_A) = 610 V

$\%R_A = 1.2$, $\%X_A = 3.3$

Transformer B: Rating is 600 kVA

Open circuit voltage of transformer B (E_B) = 600 V

$\%R_B = 1.7$, $\%X_B = 5.7$

Full-load current of Transformer $A = \dfrac{6,00,000}{610}$ A

$$\therefore \quad \overline{Z}_A \text{ in ohms} = (0.012 + j0.033) \times \frac{610^2}{6,00,000} = 0.007442 + j0.0204$$

Full-load current of Transformer $B = \dfrac{3,00,000}{600}$ A

$$\therefore \quad \overline{Z}_B \text{ in ohms} = (0.017 + j0.057) \times \frac{600^2}{3,00,000} = 0.0204 + j0.0684$$

∴ Circulating current ($\overline{I}_c$)

$$= \frac{\overline{E}_A - \overline{E}_B}{\overline{Z}_A + \overline{Z}_B} = \frac{610 - 600}{(0.007442 + j0.0204) + (0.0204 + j0.0684)} = 32.15 - j102.53$$

= 107.45 A, 0.2992 power factor lagging.

No-load secondary terminal voltage

$$\overline{V}_{20} = \overline{E}_A - \overline{I}_c \overline{Z}_A = 610 - (32.15 - j102.53)\ (0.007442 + j0.0204)$$

$$= 607.67 - j0.1072$$

$$\therefore \quad |V_{20}| = 607.67 \text{ V}$$

$$\therefore \text{ Load impedance} \quad (\overline{Z}_L) = \frac{|V_{20}|^2}{(800\angle 0°) \times 1,000} = \frac{607.67^2}{800 \times 1,000} \angle 0° = 0.4616\angle 0°$$

$$\overline{I}_A = \frac{\overline{E}_A \overline{Z}_B + (\overline{E}_A - \overline{E}_B)\overline{Z}_L}{\overline{Z}_A \overline{Z}_B + (\overline{Z}_A + \overline{Z}_B)\overline{Z}_L}$$

$$= \frac{610(0.0204 + j\ 0.0684) + (610 - 600)(0.4616 + j\ 0)}{(0.007442 + j0.0204)(0.0204 + j0.0684) + (0.007442 + j0.0204 + 0.0204 + j0.0684)(0.4616 + j\ 0)}$$

$$= \frac{17.06 + j41.724}{0.011608 + j0.041915}$$

$$= 1029.2243 - j121.98$$

= 1036.43 A, 0.993 power factor lagging.

$$\overline{I}_B = \frac{\overline{E}_B \overline{Z}_A - (\overline{E}_A - \overline{E}_B)\overline{Z}_L}{\overline{Z}_A \overline{Z}_B + (\overline{Z}_A + \overline{Z}_B)\overline{Z}_L}$$

$$= \frac{600(0.007442 + j0.0204) - (610 - 600)(0.4616 + j\ 0)}{(0.007442 + j0.0204)(0.0204 + j0.0684) + (0.007442 + j0.0204 + 0.0204 + j0.0684)(0.4615 + j\ 0)}$$

$$= \frac{-0.1508 + j12.24}{0.011608 + j0.41915}$$

$$= 270.293 + j78.453$$

= 281.45 A, 0.960 power factor leading.

Load shared by transformer A = 607.67 × 1036.43 ∠−6.76° VA = 629.807 ∠−6.76° kVA
Load shared by transformer B = 607.67 × 281.45 ∠16.26° VA = 171.028 ∠16.26° kVA ■

Example 1.46 A 100 kVA, 2,000/400V, single-phase two-winding transformer is to be used as an autotransformer for stepping up the voltage from 2,000 V to 2,400 V. At rated load, the two-winding transformer has 2.4 per cent loss, 3.4 per cent voltage regulation and 4.2 per cent impedance. Calculate the following for an autotransformer:

(i) Voltage and current rating.
(ii) kVA rating.
(iii) Efficiency.
(iv) Percentage impedance.
(v) Regulation.
(vi) Short circuit current on each side.

Solution

(i) For autotransformer, the rated voltage on LV side is 2,000 V and that on HV side is given by (2,000 + 400) V or 2,400 V. The rated current on HV side of the autotransformer is given by

$$I_2 = \text{rated current on 400 V side of two-winding transformer} = \frac{100\times 10^3}{400} = 250 \text{ A}$$

∴ The rated current on LV side of the autotransformer $I_1 = I_2$ + (rated current on HV side of two-winding transformer).

$$\therefore \qquad I_1 = 250 + \frac{100\times 10^3}{2{,}000} = 300 \text{ A}$$

(ii) kVA rating as an autotransformer

$$= \frac{I_1 V_1}{1{,}000} = \frac{350\times 2{,}000}{1{,}000} = 600 \text{ kVA}$$

(iii) Transformer ratio (a) of autotransformer

$$= \frac{\text{LV side voltage}}{\text{HV side voltage}} = \frac{2{,}000}{2{,}400} = \frac{10}{12}$$

∴ Per cent full-load losses as an autotransformer
= $(1 - a) \times$ (per cent full-load losses as two-winding transformer)

$$= \left(1 - \frac{10}{12}\right) \times 2.4 = 0.4\%$$

∴ Efficiency as an autotransformer = 100 − 0.4 = 99.6%

(iv) Percentage impedance as an autotransformer
= $(1 - a) \times$ (percentage impedance as two-winding transformer)

$$= \left(1 - \frac{10}{12}\right) \times 4.2 = 0.7\%$$

(v) Regulation as an autotransformer at a certain power factor
= (1− *a*) × (regulation as a two-winding transformer at same power factor)

$$= \left(1 - \frac{10}{12}\right) \times 3.4 = 0.56\%$$

(vi) Short circuit current as an autotransformer

$$= \left(\frac{1}{1-a}\right) \times \text{(short circuit current as two-winding transformer)}$$

$$= \frac{1}{\left(1 - \frac{10}{12}\right)} \times \frac{1}{\text{(p.u. impedance as two-winding transformer)}}$$

$$= \frac{12}{2} \times \frac{1}{0.042} = 142.85 \text{ p.u.}$$

SIGNIFICANT POINTS

EMF Equation

$$\frac{E_1}{N_1} = \frac{E_2}{N_2} = 4.44 f \phi_m = \text{constant}$$

$$\frac{E_1}{E_2} = \frac{N_1}{N_2} = a$$

where $a = \frac{N_1}{N_2}$ is the turns ratio of the transformer.

During no load

$$I_0 = \sqrt{I_w^2 + I_\mu^2}$$

and

$$\phi_0 = \tan^{-1}\left(\frac{I_\mu}{I_w}\right)$$

Equivalent Resistance

$$R_2' = a^2 R_2$$

$$R_{01} = R_1 + a^2 R_2$$

$$R_1' = \frac{R_1}{a^2}$$

$$R_{02} = R_2 + \frac{R_1}{a^2}$$

Equivalent Reactance

$$X_2' = a^2 X_2$$

$$X_{01} = X_1 + a^2 X_2$$

$$X_1' = \frac{X_1}{a^2}$$

$$X_{02} = X_2 + \frac{X_1}{a^2}$$

Open Circuit Test

$$W = V_1 I_0 \cos\theta_0$$

∴ $$\cos\theta_0 = \frac{W}{V_1 I_0}$$

∴ $$I_w = I_0 \cos\theta_0$$

and $$I_\mu = I_0 \sin\theta_0$$

$$R_0 = \frac{V_1}{I_w}$$

and $$X_0 = \frac{V_1}{I_\mu}$$

Short Circuit or Impedance Test

$$R_{eq} = \frac{P_{SC}}{I_{SC}^2}$$

and $$Z_{eq} = \frac{V_{SC}}{I_{SC}}$$

∴ $$X_{eq} = \sqrt{Z_{eq}^2 - R_{eq}^2}$$

Voltage Regulation of a Transformer
Voltage Regulation Down

$$\mu = \frac{{}_0V_2 - V_2}{{}_0V_2} \text{ p.u.} = \frac{{}_0V_2 - V_2}{{}_0V_2} \times 100$$

Voltage Regulation Up
If inherent voltage drop is divided by V_2, it is known as voltage regulation up. Mathematically, we can write it as

$$\mu = \frac{{}_0V_2 - V_2}{V_2} \text{ p.u.} = \frac{{}_0V_2 - V_2}{V_2} \times 100$$

The voltage regulation up is expressed mathematically by

$$\%\mu = \frac{{}_0V_2 - V_2}{{}_0V_2} \times 100 = \frac{I_2 \left(R_{02}\cos\theta \pm X_{02}\sin\theta\right)}{{}_0V_2} \times 100$$

Positive sign is for lagging power factor and negative sign for leading power factor.

Zero Voltage Regulation

$$\cos\theta = \cos\left\{\tan^{-1}\left(\frac{R_{02}}{X_{02}}\right)\right\}$$

Condition for Maximum Voltage Regulation

$$\cos\theta = \cos\left\{\tan^{-1}\left(\frac{X_{02}}{R_{02}}\right)\right\}$$

Maximum Efficiency

Iron loss = Cu loss

i.e., Constant loss = Variable loss.

Load Current at Maximum Efficiency

$$I_{2M} = I_{2fl}\sqrt{\frac{P_i}{P_{\text{Cu}\,fl}}}$$

kVA Supplied at Maximum Efficiency (V_2 Constant)

$$\eta_{\max} = \frac{V_2 I_{2M}\cos\theta}{V_2 I_{2M}\cos\theta + 2P_i}\times 100 = \frac{mV_2 I_{2fl}\cos\theta}{mV_2 I_{2fl}\cos\theta + P_i + m^2 I_{2fl}^2 R_{02}}\times 100$$

where $m = \sqrt{\dfrac{P_i}{P_{\text{Cu}fl}}}$

SHORT QUESTIONS AND ANSWERS

Q.1 What is the power factor of a transformer?

Ans. At no load, the power factor of a transformer is very low and lagging, whereas the power factor on load is nearly equal to the power factor of the load, which it is carrying.

Q.2 What is the normal phase difference between the voltage and the no-load current in a transformer?

Ans. The no-load current in a transformer normally lags behind the voltage by about 80° – 85°.

Q.3 What are the essential parts of a transformer?

Ans. The essential parts of a transformer are as follows:

(i) Magnetic circuit consisting of laminated iron core and clamping structure

(ii) Primary winding

(iii) Tank filled with insulating oil

(iv) HT terminals with bushings

(v) LT terminals with bushings

(vi) Conservator tank

(vii) Breather

(viii) Vent pipe

(ix) Thermometer

Q.4 What is the name of the winding to which supply is given?
Ans. Primary winding.

Q.5 What is the name of the winding from which the supply is taken for load connections?
Ans. Secondary winding.

Q.6 Which material is used for the core of a transformer and why?
Ans. Laminations of specially alloyed silicon steel (silicon content 4–5 per cent) are used due to its high electrical resistance, high permeability, non-ageing characteristics and minimum iron loss.

Q.7 What is the use of iron core in a transformer?
Ans. The iron core is used in a transformer to provide continuous easy magnetic path of low reluctance.

Q.8 How is magnetic leakage reduced?
Ans. Magnetic leakage is reduced to a minimum by sectionalizing and interleaving the primary and secondary windings.

Q.9 Why are iron cores in transformers made laminated?
Ans. Iron cores are made laminated to reduce eddy current loss.

Q.10 What determines the thickness of the lamination or stamping?
Ans. Frequency determines the thickness of the lamination or stamping.

Q.11 Why are the laminations insulated from each other?
Ans. The laminations are insulated from each other by insulating varnish or thin paper to break the path of eddy currents and thus reduce eddy current loss.

Q.12 What is stacking factor? What is its approximate value?
Ans. Stacking factor is the ratio of iron content in the laminated varnished core by volume. Its value is about 90 per cent, that is, 10 per cent volume is occupied by varnish and air.

Q.13 What is called grain-oriented laminations?
Ans. Grain-oriented laminations are cold rolled laminations specially annealed to orient the iron crystals, that is, the grains in a uniform way in the direction of rolling to get very high permeability and low hysteresis loss.

Q.14 What is the permissible maximum flux density in transformer core?
Ans. 1.6–1.8 Wb/m^2.

Q.15 What is the phase relationship between the primary and secondary voltages of a transformer?
Ans. The primary and secondary voltages of a transformer are 180° out of phase.

Q.16 What is turn ratio of a transformer?
Ans. The ratio of the number of turns in the primary to the number of turns in the secondary windings is called the turn ratio or the ratio of transformation of the transformer, which is indicated by a constant.

That is, $\dfrac{N_1}{N_2} = \dfrac{E_1}{E_2} = a$

Q.17 What is voltage ratio of a transformer?
Ans. Voltage ratio is the ratio of the voltage between the line terminals of one winding to that between the line terminals of another winding at no load.

Q.18 What current flows in the transformer primary when its secondary is open?
Ans. When the secondary is open, there is no current in the secondary of the transformer. The primary takes a small current (I_0) from the source called no-load current which has a magnetizing component ($I_0 \sin\theta_0$) producing the magnetic flux and a working component ($I_0 \cos\theta_0$) supplying real power for iron losses.

Q.19 What is the formula for calculating no-load current?
Ans. No-load current $(I_0) = \sqrt{(I_w)^2 + (I_\mu)^2} = \sqrt{(I_0 \cos\theta_0)^2 + (I_0 \sin\theta_0)^2}$

Q.20 Why is the frequency not changed during transformation of electrical energy in a transformer?
Ans. As the same flux having a definite frequency is responsible for the production of emf on both the primary and secondary windings, there is no question of change in the frequency.

Q.21 What is the emf equation of a transformer?
Ans. $E_{rms} = 4.44\ \phi_m fN$ V
where ϕ_m = maximum value of the flux linkage with both the windings, f = frequency of the supply, N = number of turns.

Q.22 What are the two basic types of transformers?
Ans. The two basic types of transformers are:

(i) The isolation type in which the two windings are physically isolated and electrically insulated from each other.

(ii) The autotransformer type in which one coil is used for both the windings.

Q.23 What are the types of transformers according to the arrangement of iron cores?
Ans. There are three types: (i) core type, (ii) shell type and (iii) distributed core or Berry type.

Q.24 What magnetic circuit is formed in Berry-type constructions and why?
Ans. Distributed magnetic circuits are formed in Berry-type construction because of its distributed cores.

Q.25 What is called limb of a transformer?
Ans. The vertical portion of the iron core where the windings are placed is called limb of a transformer.

Q.26 Why are LT windings placed near the core?
Ans. LT windings are placed near the core to reduce the total dielectric strength of the insulating materials provided on the winding.

Q.27 What are the types of windings according to the construction?
Ans. Sandwich type and cylindrical type.

Q.28 What is the difference between cylindrical-type and sandwich-type winding?
Ans. In cylindrical-type winding, the length of the coils is equal to the length of the core limb. The primary and secondary windings are placed one over the other, placing low-voltage winding nearer the core, whereas in sandwich-type winding HT and LT windings are placed lengthwise one above and other like a sandwich.

Q.29 What are the types of transformers according to its location?
Ans. (i) Indoor-type transformer (ii) Step-down transformer.

Q.30 What do you mean by step-up and step-down transformers?
Ans. When a transformer converts low voltage to high voltage it is called a step-up transformer, and when the transformer converts high voltage to low voltage it is called a step-down transformer.

Q.31 What are the advantages of using a transformer in an AC supply?

Ans. The transformer has the following advantages:

(i) Voltage can be transformed from low voltage to high voltage for transmission and high voltage to low voltage for distribution without changing the frequency.

(ii) As the transformer has no rotating parts, there is no factional loss and wear and tear is minimum.

(iii) A high voltage or an extra high voltage can be transformed easily by providing good insulation.

(iv) Its maintenance cost is low.

(v) Practically it requires very little attention for its operation.

Q.32 What is an ideal transformer?

Ans. A transformer having an overall efficiency of 100 per cent is called an ideal transformer.

Q.33 What do you mean by power transformer?

Ans. Transformers that are used on transmission lines for the transmission and distribution of relatively large quantities of energy are called power transformers.

Q.34 What do you mean by distribution transformers?

Ans. When transformers are used for distributing the energy from transmission lines as well as networks for local consumption and the secondaries are directly connected to the consumer's load, they are called distribution transformers.

Q.35 What do you mean by lighting transformer?

Ans. A transformer used to supply a distribution circuit having no motors connected to it is called lighting transformer.

Q.36 How does a transformer contribute towards the widespread popularity of AC system over DC?

Ans. (i) High voltage of AC system can be obtained using a transformer for transmission of electrical power.

(ii) Using a transformer, AC voltage can be increased or decreased without any power loss.

SUPPLEMENTARY PROBLEMS

1. A single-phase transformer has 400 primary and 1,000 secondary turns. The net cross-sectional area of the core is 60 cm^2. If the primary winding be connected to a 50 Hz supply at 500 V, calculate (i) the peak value of the flux density in the core and (ii) the voltage induced in the secondary winding.

   [*Ans.* 0.9384 T; 1,250 V]

2. The emf per turn of a single-phase 10 kVA, 2,200/220 V, 50 Hz transformer is 10 V. Calculate (i) the number of primary and secondary turns and (ii) the net cross-sectional area of the core for a maximum flux density of 1.5 T.

   [*Ans.* (i) 220, 22 (ii) 0.03 m^2]

3. The no-load current of a transformer is 5 A at 0.25 power factor when supplied at 235 V, 50 Hz. The number of turns on the primary winding is 200. Calculate (i) the maximum value of flux in the core, (ii) the core loss and (iii) the magnetizing component.

   [*Ans.* (i) 5.293 mWb, (ii) 293.75 W (iii) 4.84 A]

4. A 50 kVA, 4,400/220 V, single-phase transformer has $R_1 = 3.45\ \Omega$, $R_2 = 0.009\ \Omega$. The values of reactances are $X_1 = 5.2\ \Omega$ and $X_2 = 0.015\ \Omega$. Calculate the following:
   (i) The equivalent resistance referred to primary.
   (ii) The equivalent resistance referred to secondary.
   (iii) The equivalent reactance referred to both primary and secondary.
   (iv) Equivalent impedance referred to both primary and secondary.
   (v) Total Cu loss.

   [*Ans*. (i) 7.05 Ω, (ii) 0.0176 Ω, (iii) 11.2 Ω, 0.028 Ω, (iv) 13.23 Ω, 0.0331 Ω, (v) 909 W]

5. A 5 kVA, 400/200 V, single-phase, 50 Hz transformer gave the following test results:
   Open circuit: 400 V, 1A, 50 W
   Short circuit: 12 V, 10 A, 40 W
   The instruments are placed on the same side during these two tests. Determine the following:
   (i) The component of the no-load current.
   (ii) The efficiency and regulations at full load and 0.8 power factor lagging.

   [*Ans*. (i) $I_w = 0.125$ A, $I_\mu = 0.922$ A (ii) Efficiency = 97.72% Regulation = 3.13 per cent]

6. The primary and secondary windings of a 500 kVA, single-phase transformer have resistances of 0.4 Ω and 0.0015 Ω respectively. The primary and secondary voltages are 6,000 V and 400 V, respectively. The iron loss is 3.2 kW. Calculate the efficiency on (i) full load and (ii) half load, assuming power factor of load to be 0.8.

   [*Ans*. (i) 97.962%, (ii) 97.81%]

7. A 4 kVA, 200/400 V, 50 Hz, single-phase transformer gave the following test results:
   Open circuit: 200 V, 0.7, 60 W (instruments placed on LV side)
   Short circuit test: 9 V, 6 A, 21.6 W (instruments placed on HV side)
   Determine (a) the magnetizing current and the component corresponding to iron loss at normal voltage and frequency, (b) the efficiency on full load at unity power factor and (c) the secondary terminal voltage on full load at power factors of unity, 0.8 lagging and 0.8 leading.

   [*Ans*. (a) 0.63 A, 0.3 A; (b) 97.1%; (c) 394 V, 387 V, 403.4 V]

8. A 100 kVA, 50 Hz, 440/11,000 V, single-phase transformer has an efficiency of 98.5 per cent when supplying full-load current at 0.8 power factor and an efficiency of 99 per cent when supplying half of full-load current at unity power factor. Find the iron and copper losses corresponding to full-load current. At what value of load current will maximum efficiency be attained?

   [*Ans*. Iron loss = 267.3 W Cu loss = 951 W, load current = 120 A]

9. The maximum efficiency of a 500 kVA, 3,300/500 V, 50 Hz, single-phase transformer is 97 per cent and occurs at $\frac{3}{4}$th of full load, unity power factor. If impedance is 10 per cent, calculate the regulation at full load, power factor 0.8 lagging.

   [*Ans*. 6.88%]

10. A 5 kVA, 500/250 V, 50 Hz, single-phase transformer has the following test data:
    OC test (LV side): 250 V, 2.5A, 80 W
    SC test (HV side): 20 V, 14 A, 95 W
    Find the following: (i) equivalent circuit referred to as primary side and (ii) secondary load voltage on full load at 0.8 power factor lagging.

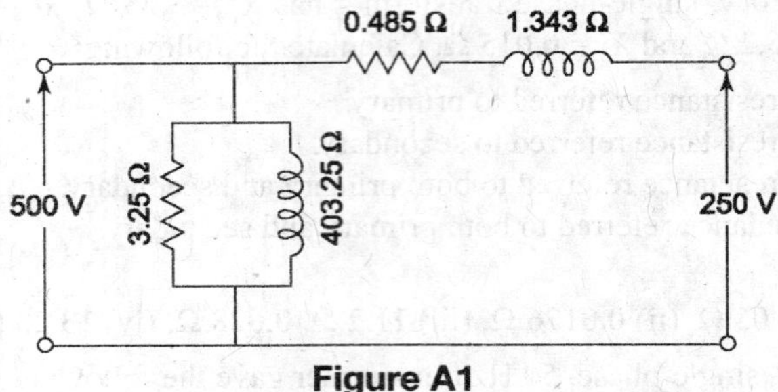

Figure A1

[*Ans.* (i) Figure A1(ii) 242.59 V]

11. The following test results were obtained in 300/600 V, single-phase, 50 Hz transformer:
OC test (LV side): 300 V, 1.5 A, 90 W
SC test (HV side): 22 V, 13 A, 120 W
Determine (i) the circuit constants referred to as LV side, (ii) the applied voltage and efficiency when the output is 10 A at 600 V at 0.8 lagging power factor.

[*Ans.* (i) R_0 = 1,000 Ω, X_0 = 204.3 Ω, R_{01} =0.177 Ω, X_{01} = 0.384 Ω
(ii) Applied voltage = 307.46 Efficiency = 96.75%]

12. A 200/400 V, 20 kVA, 50 Hz single-phase transformer is connected as an autotransformer to work off 600/200 V supplies. Calculate the kVA rating of the autotransformer. With load of 20 kVA, 0.8 power factor lagging connected to the 200 V terminals, calculate the currents in the load and in the two-winding sections.

[*Ans.* 30 kVA, 33.33A (600 V section), 66.67 A (200 V section)]

13. A transformer has its maximum efficiency of 0.98 at 15 kVA at unity power factor. During the day it is loaded as follows:
12 hours – 2 kW at power factor 0.5
6 hours – 12 kW at power factor 0.8
6 hours – 18 kW at power factor 0.9
Find the all-day efficiency.

[*Ans.* 0.97]

14. Two transformers *A* and *B* are joined in parallel to the same load. Determine the current delivered by each transformer. Given: open circuit voltage 6,600 V for *A* and 6,400 V for *B*. Equivalent leakage impedance in terms of secondary = 0.3 + *j* 3 for *A* and 0.2 + *j* 1 for *B*. The load impedance is 8 + *j* 6.

[*Ans.* 195 A, 422 A]

15. Two 100 kW single-phase transformers are connected in parallel both on primary and on secondary. One transformer has an ohmic drop of 1/2 per cent at full load and an inductive drop of 8 per cent at full load. The other has an inductive drop of 4 per cent and an ohmic drop of 3/4 per cent. Show how they will share the following loads:
(a) 180 kW at 0.9 power factor.
(b) 120 kW at 0.6 power factor.

[*Ans.* (a) 58.06 kW and 121.81 kW (b) 36 kW and 84 kW]

MULTIPLE-CHOICE QUESTIONS AND ANSWERS

1. In a transformer
   (a) All turns are equally insulated
   (b) The end turns are more strongly insulated
   (c) The end turns are closely wound
   (d) The end turns are widely separated
2. Laminated insulations coated with varnish are normally used in the transformer
   (a) To reduce reluctance of magnetic path
   (b) To reduce the effect of eddy current
   (c) To increase the reluctance of magnetic path
   (d) To reduce the hysteresis effect
3. The required thickness of lamination in a transformer decreases when
   (a) The applied frequency increases
   (b) The applied frequency decreases
   (c) The applied voltage increases
   (d) The applied voltage decreases
4. Oil in transformer is used to
   (a) Transfer electrical energy
   (b) Insulate the windings
   (c) Cool the windings
   (d) None of the above
5. The following arrangement will reduce the reluctance of magnetic path of the transformer.

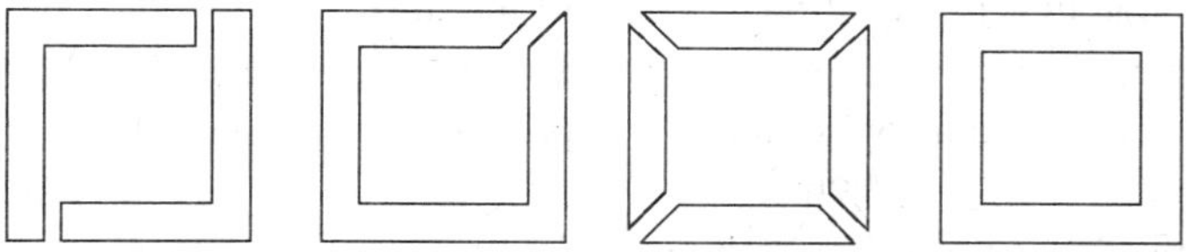

(a) (b) (c) (d)

6. The windings of a transformer are divided into several coils because
   (a) It is difficult to wind as one coil
   (b) It reduces voltage per coil
   (c) It requires less insulation
   (d) None of the above
7. The size and construction of bushings in a transformer depend upon the
   (a) Size of winding
   (b) Size of tank
   (c) Current flowing
   (d) Voltage supplied
8. Transformer humming sound is reduced by the
   (a) Proper bracing of transformer's assemblies
   (b) Proper insulation
   (c) Proper design
   (d) Proper design of winding
9. Sludge in transformer oil is due to
   (a) Decomposition of oil
   (b) Decomposition of insulation
   (c) Moisture content in oil
   (d) None of the above
10. A transformer used only for electrical isolation between two circuits has turns ratio which is
    (a) More than unity (b) Less than unity (c) Equal to unity (d) More than 0.5
11. If 90 per cent of normal voltage and 90 per cent of normal frequency are applied to a transformer, the per cent charge in hysteresis losses will be
    (a) 20% (b) 4.7% (c) 19% (d) 21%

12. If 110 per cent of normal voltage and 110 per cent of normal frequency is applied to a transformer, the percentage change of eddy current losses will be
(a) 10% (b) 20% (c) 25% (d) 21%.
13. A transformer has two 2,400 V primary coils and two 240 V coils. By proper connection of the windings, the transformation ratio that can be obtained is
(a) 10 (b) 5 (c) 20 (d) 9
14. A single-phase, 2,200/200 V transformer takes 1 A at the *HT* side or no load at a power factor of 0.385 lagging. The iron losses are
(a) 167 W (b) 77 W (c) 88 W (d) 98 W
15. Neglecting resistance, at constant flux density, the power required per kilogram to magnetize the iron core of a transformer is 0.8 W at 25 Hz and 2.04 W at 60 Hz. The power required per kilogram for 100 Hz is
(a) 3.8 W (b) 3.63 W (c) 3.4 W (d) 5.2 W
16. Select the correct statement:
(a) emf per turn of both the windings are equal
(b) emf per turn in HV winding is more than the emf per turn in LV winding
(c) emf per turn in HV winding is less than the emf per turn in LV winding
17. The flux involved in the emf equation of a transformer has
(a) rms value (b) Average value (c) Total value (d) Maximum value
18. The no-load current in a transformer lags the applied voltage by
(a) 90° – 95° (b) About 80° – 85° (c) 0° – 15° (d) About 110°
19. High leakage impedance transformers are used for applications such as
(a) Power distribution (b) Electrical toys
(c) Arc welding (d) Fluorescent lamps
20. A transformer 2,000 kVA, 250 Hz is operated at 50 Hz. Its kVA rating should be revised to
(a) 400 kVA (b) 10,000 kVA
(c) 2,000 kVA (d) Cannot be revised
21. The transformer which will have the largest size is
(a) 100 kVA, 25 Hz (b) 100 kVA, 100 Hz
(c) 100 kVA, 50 Hz (d) 100 kVA, 60 Hz
22. The magnetic flux of a transformer follows
(a) High reluctance path (b) Low reluctance path
(c) Low conductivity path (d) High conductivity path
23. The efficiency of a transformer is usually in the range of
(a) 50–60% (b) 65–75%
(c) 70–90% (d) 90–98%
24. Transformer is laminated to reduce
(a) Hystersis loss (b) Eddy current loss
(c) Partly (a) and partly (b) (d) Increases exciting current
25. Transformer changes the value of
(a) Power (b) Frequency
(c) Voltage (d) Current

26. Transfer of electrical power from primary to secondary in a transformer takes place
    (a) Electrically (b) Electromagnetically
    (c) Magnetically (d) None of these
27. The thickness of a 50 Hz transformer lamination is
    (a) 0.35 cm (b) 0.35 mm
    (c) 0.33 m (d) 0.30 cm
28. A power transformer is a constant
    (a) Current device (b) Voltage device
    (c) Power device (d) Main flux device
29. The value of flux in the emf equation of a transformer is
    (a) rms (b) Average
    (c) Maximum (d) Integral wave cycle
30. The leakage flux in a transformer depends upon the value of
    (a) Load current (b) Supply frequency
    (c) Mutual flux (d) None of these
31. The voltage regulation of a transformer at full-load 0.8 power factor lag is 6 per cent. Its voltage regulation at full-load 0.8 power factor lead will be
    (a) Negative (b) 54% (c) Positive (d) Zero
32. The full-load efficiency of a transformer at 0.85 p.f. lag is 97 per cent. Its efficiency at full load 0.85 power factor lead will be
    (a) 99% (b) 96% (c) 97% (d) 98%
33. If a load on secondary side of a transformer increases, the current on the primary side
    (a) Remains constant (b) Increases (c) Decreases (d) None of these
34. If a sinusoidal exciting current is applied to a transformer, the mutual flux produced is
    (a) Negative (b) Sinusoidal (c) Zero (d) Flat top
35. To get sinusoidal flux in a transformer the mutual flux produced is
    (a) Non-sinusoidal (b) Sinusoidal (c) Direct current (d) Zero
36. A 220V, 150V bulb is connected in series with the primary of a 220/1,100 V, 50 Hz transformer. If the load on the secondary side is disconnected, the brightness of the bulb will
    (a) Decrease (b) Increase
    (c) Be unaffected (d) Decrease to a very low value
37. The open circuit test of a transformer gives information about
    (a) Core losses of the transformer (b) Cu losses of the transformer
    (c) Exciting current (d) None of these
38. The short circuit test of a transformer gives the information of
    (a) Cu losses of the transformer (b) Core losses of the transformer
    (c) Winding circuit impedance (d) None of these
39. During testing of a transformer
    (a) Both SC and OC tests are performed at rated current
    (b) Both SC and OC tests are performed at rated voltage
    (c) OC test is performed at rated voltage
    (d) SC test is performed at rated current

40. All-day efficiency of a transformer is
(a) Equal to its power efficiency
(b) Less than its power efficiency
(c) More than its power efficiency
(d) None of these

41. Routine efficiency of a transformer depends upon the value of
(a) Load current
(b) Supply frequency
(c) Power factor of load
(d) Both (a) and (b)

42. A ferrite core has less eddy current loss than an iron core because ferrites have
(a) Lower permeability
(b) High hysteresis
(c) Alnico as the main constituent
(d) High resistance

43. The maximum regulation of a transformer occurs at a power factor of
(a) Unity
(b) $\frac{X_{02}}{R_{02}}$ lag
(c) $\frac{X_{02}}{R_{02}}$ lead
(d) $\frac{R_{02}}{X_{02}}$ lead

44. The zero of regulation of a transformer occurs at a power factor of
(a) Zero
(b) $\frac{R_{02}}{X_{02}}$ lead
(c) $\frac{R_{02}}{X_{02}}$ lag
(d) $\frac{X_{02}}{R_{02}}$ lead

45. Circular coil sections are used because they
(a) Reduce iron material
(b) Reduce copper material
(c) Have the strongest mechanical shape
(d) All of these

46. A transformer is connected to a constant voltage source. If the supply frequency decreases, the magnetic flux in the core will
(a) Increase towards saturation
(b) Decrease
(c) Remain unchanged
(d) None of these

47. The power in autotransformer is transferred through
(a) Inductive process
(b) Convection process
(c) Conduction process
(d) All of these

48. The cross-sectional area of the common portion of an autotransformer is
(a) Kept constant
(b) Proportionally decreased
(c) Proportionally increased
(d) None of these

49. The efficiency of an autotransformer for the same output compared to a two-winding transformer is
(a) Greater
(b) Lesser
(c) Poor
(d) None of these

50. The condition for successful parallel operation of transformer is
(a) Correct polarity
(b) Per unit impedance based on their rating should be equal
(c) Identical voltage and frequency rating
(d) Equal ratio of equivalent resistance to reactance
(e) All of these

51. During parallel operation of transformers incorrect polarity will result in
(a) Open circuit
(b) Dead short circuit
(c) Regeneration of power
(d) Power factor of transformer will be different from that of the connected load

52. In an autotransformer of voltage ratio $\frac{V_1}{V_2}$ and $V_1 > V_2$, the fraction of power transferred inductively is

(a) $\frac{V_1}{V_1 + V_2}$ (b) $\frac{V_2}{V_1}$ (c) $\frac{V_1 - V_2}{V_1 + V_2}$ (d) $\frac{V_1 - V_2}{V_1 V_2}$

53. A 10 kVA, 400/200 V sinlge-phase transformer with 10 per cent impedance draws a steady short circuit line current of

(a) 50 A (b) 150 A (c) 250 A (d) 350 A

54. A 400/100 V, 10 kVA two-winding transformer is reconnected as an autotransformer across a suitable voltage source. The maximum kVA is

(a) 50 kVA (b) 15 kVA (c) 12.5 kVA (d) 8.75 kVA

55. An autotransformer having a transformation ratio of 0.8 supplies a load of 10 kW. The power transferred inductively from the primary to the secondary is

(a) 10 kW (b) 8 kW (c) 2 kW (d) Zero

56. A single-phase induction regulator is a constant input transformer to obtain smooth variation of the output voltage by varying the

(a) Ratio of turns between primary and secondary winding
(b) Frequency
(c) Flux-density in the core
(d) Angle between the magnetic forces of the primary and secondary windings

57. In operating a 400 Hz transformer at 50 Hz

(a) Only voltage is reduced in the same proportion as the frequency
(b) Only kVA is reduced in the same proportion as the frequency
(c) Both voltage and kVA rating are reduced in the same proportion as the frequency
(d) None of the above

58. In performing the short circuit test of a transformer

(a) High-voltage side is usually short-circuited
(b) Low-voltage side is usually short-circuited
(c) Any side is short-circuited with preference
(d) None of these

59. In performing open circuit test of a transformer

(a) High voltage is usually kept open (b) Low voltage is usually kept open
(c) Any side can be kept open (d) None of the above

60. Transformer core usually uses grain-oriented laminated sheets. The grain orientation reduces

(a) Copper loss (b) Eddy current loss
(c) Hysteresis loss (d) None of the above

61. In the circuit model of a transformer, the core loss is represented as

(a) Series resistance (b) Series inductance
(c) Shunt resistance (d) Shunt inductance

62. While performing short circuit test on a transformer, the impressed voltage magnitude is kept constant but the frequency is increased. The short circuit current will

(a) Increase (b) Decrease
(c) Remain the same (d) None of the above

63. A transformer has a hysteresis loss of 30 W at 240 V, 60 Hz. The hysteresis loss at 200 V, 50 Hz will be
(a) 20.8 W (b) 25 W (c) 30 W (d) 36 W

64. In a transformer, the exciting current will be in phase quadrature with the impressed voltage provided
(a) Only the leakage impedance drop is ignored
(b) Only the core loss is ignored
(c) Both the leakage and impedance drop and the core loss are ignored
(d) Only no-load copper loss is ignored

65. Magnetizing impedance of a transformer is determined by
(a) Short circuit test
(b) Open circuit test
(c) Both open circuit and short circuit tests
(d) None of these

66. A transformer is to be tested at full-load conditions consuming only losses from the mains. We do
(a) Load test (b) Open circuit and short circuit tests
(c) Back-to-back test (d) None of these

67. In a transformer operating at constant voltage if the input frequency increases the core loss will
(a) Increase (b) Decrease
(c) Remain constant (d) Increase as the square of the frequency

68. Grain-oriented laminated sheet steel in a transformer reduces
(a) Copper loss (b) Eddy current loss
(c) Hysteresis loss (d) None of above

69. R_1 is the resistance of the primary winding of a transformer. The turns ratio in terms of primary to secondary is *a*. The equivalent resistance of the primary referred to as secondary is
(a) R_1/a (b) $a^2 R_1$ (c) R_1/a^2 (d) None of these

70. Choose the correct statement:
(a) emf per turn in high-voltage winding is more than the emf per turn in low-voltage winding
(b) emf per turn in both the finding are equal
(c) emf per turn in both the windings are not equal
(d) None of these

71. If E_2 and V_2 be the induced emf in secondary winding and V_2 be the terminal voltage during load, the regulation of the transformer is given by
(a) $\frac{E_2 - V_2}{V_2}$ (b) $\frac{E_2 - V_2}{E_2}$ (c) $\frac{V_2 - E_2}{E_2}$ (d) $\frac{V_2 - E_2}{V_2}$

72. One transformer has leakage impedance of $1 + j\,4\ \Omega$ and $3 + j\,4\ \Omega$ for its primary and secondary windings, respectively. The transformer has
(a) Low voltage primary (b) High voltage primary
(c) Medium voltage primary (d) Insufficient data

73. A single-phase transformer has its maximum efficiency at 60 per cent of full load. At full load copper loss will be
(a) Equal to core loss (b) Less than core loss
(c) More than core loss (d) Zero

74. The flux in a magnetic core is alternating sinusoidally at a frequency of 600 Hz. The maximum flux density is 2 T and eddy current loss is 15 W. If the frequency is raised to 800 Hz and maximum flux density reduced to 1.5 T, the eddy current loss will be
   (a) Same (b) Reduced to half
   (c) Doubled (d) Increased by 50%

75. The core flux in a transformer depends mainly on
   (a) Supply voltage
   (b) Supply voltage and frequency
   (c) Supply voltage, frequency and load
   (d) Supply voltage and load but independent of frequency

Answers

1 (b),(d)	2 (b)	3 (c)	4 (c)	5 (c)	6 (b)
7 (b),(d)	8 (a)	9 (a)	10 (c)	11 (b)	12 (d)
13 (a),(b),(c)	14 (b)	15 (b)	16 (a)	17 (d)	18 (b)
19 (c)	20 (a)	21 (b)	22 (b)	23 (d)	24 (b)
25 (c),(d)	26 (b)	27 (b)	28 (d)	29 (c)	30 (a)
31 (a)	32 (c)	33 (b)	34 (d)	35 (a)	36 (d)
37 (a),(c)	38 (a),(c)	39 (c), (d)	40 (b)	41 (a),(c)	42 (d)
43 (b)	44 (b)	45 (c)	46 (a)	47 (a),(c)	48 (b)
49 (a)	50 (e)	51 (b)	52 (d)	53 (c)	54 (a)
55 (c)	56 (d)	57 (c)	58 (b)	59 (a)	60 (b)
61 (c)	62 (b)	63 (b)	64 (b)	65 (b)	66 (c)
67 (a)	68 (c)	69 (c)	70 (b)	71 (a)	72 (a)
73 (c)	74 (a)	75 (b)			

2 Three-phase Transformers

The generation of electric power is three-phase in nature and the generated voltage is 13.2 kV, 22 kV or higher. Transmission of power is carried out at high voltages like 132 kV or 400 kV. Before transmission, it is required to step-up the voltage and for this a three-phase step-up transformer is required. Similarly, at the distribution substation, the voltage must be stepped down and it is necessary to reduce the voltage level up to 6,600 V, 400 V, 230 V and so on. Here, a three-phase step-down transformer is required. Therefore, it is economical to use three-phase transformers for transmission and utilization purposes. Earlier it was common practice to use three suitable single-phase transformers rather than a single three-phase transformer bank. Nowadays, a single three-phase transformer bank is popular due to improvement in its design and manufacture.

2.1 ADVANTAGES OF THREE-PHASE TRANSFORMERS

A single three-phase transformer bank has the following advantages:

- It occupies less space for same rating, compared to a bank of three single-phase transformers.
- It weighs less.
- The cost is also less.
- Since only one unit is required to be handled, it is easy for the operator.
- It can be transported very easily.
- The core is of smaller size and hence less material is required.

In spite of the above advantages of a single three-phase transformer bank, a bank of three single-phase transformers is used in underground work, such as in mines, for easy transport of these units. Again, open-delta operation with reduced rating is possible in a bank of three single-phase transformers when one unit of three single-phase transformers is out of order. It is common practice to use a single three-phase transformer unit because it has reduced cost.

2.2 PRINCIPLE OF OPERATION

Figure 2.1 shows the basic principle of a three single-phase transformer. The three cores are arranged at 120° to each other. In Figure 2.1, only the primary windings have been shown for simplicity. The primaries are connected to a three-phase supply and carry currents I_R, I_Y and I_B, producing fluxes (ϕ_R, ϕ_Y and ϕ_B) in the individual cores. The centre leg carries the sum of all the three fluxes. In a three-phase system, $I_R + I_Y + I_B = 0$ at any instant and hence the sum of the fluxes is zero at that instant. Therefore, the centre leg does not carry any flux. If centre leg is removed, it hardly makes any difference in the other conditions of the transformers. If the centre leg is removed, any two legs provide the return path for the current, which is the current of the third leg. This principle is used in the design of the three-phase core type transformer.

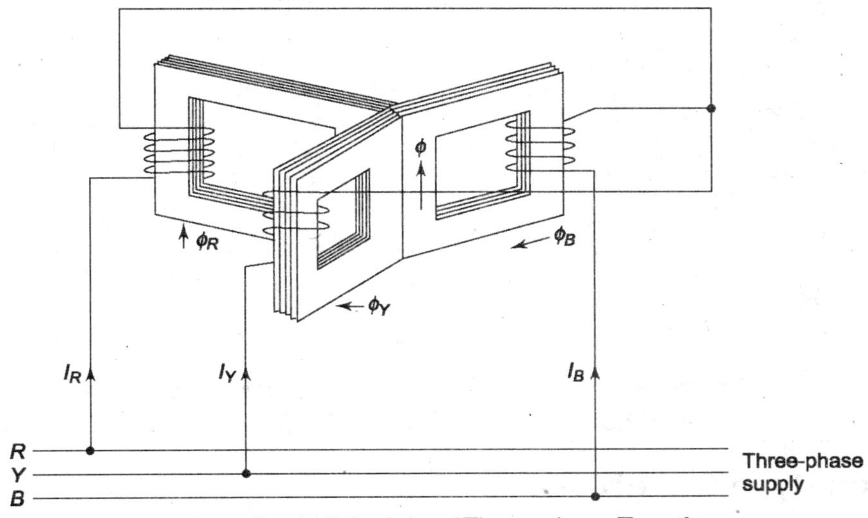

Figure 2.1 Basic Principle of Three-phase Transformers

2.3 CONSTRUCTION OF THREE-PHASE TRANSFORMERS

Three-phase transformers are of the core type and shell type. The three-phase core-type transformer can be made by combining three single-phase core-type transformers. The three-phase shell-type transformer can be made by combining three single-phase shell-type transformers.

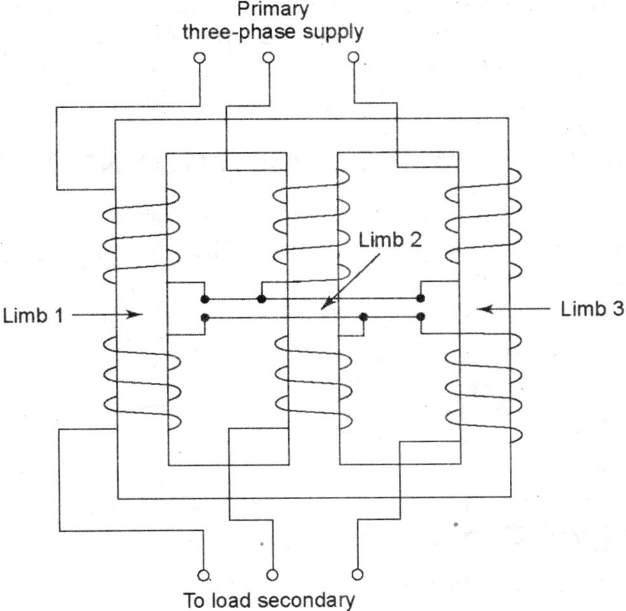

Figure 2.2 Core-type Three-phase Transformer

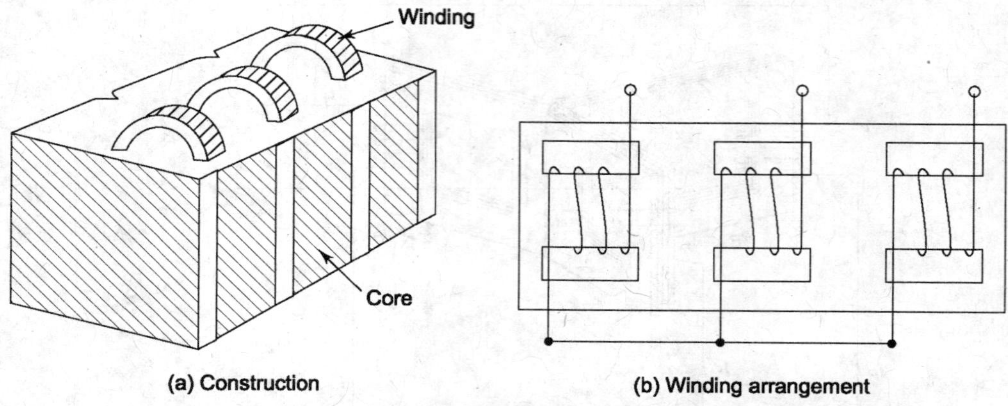

Figure 2.3 Shell-type Three-phase Transformer

2.3.1 Core-type Construction

Figure 2.2 shows a three-phase core-type transformer having a three-limb construction. The core in Figure 2.2 consists of three legs and the magnetic circuit is completed through two yokes, where one is at the top and two are at the bottom. Each limb has primary and secondary windings, which are arranged concentrically. Core-type transformers are usually wound with circular cylindrical coils.

2.3.2 Shell-type Construction

Figure 2.3 shows a three-phase shell-type transformer. In this type, the three phases are more independent than they are in the core-type three-phase transformers because each phase has an individual independent magnetic circuit. The phase magnetic circuits are in parallel. Therefore, these are independent if saturation effects occurring in the common magnetic paths are neglected. The construction and winding arrangements have been shown in Figures 2.3(a) and 2.3(b), respectively. This type of transformers is rarely used.

2.4 THREE-PHASE TRANSFORMER CONNECTION

The primary and secondary windings of three-phase transformers can he connected in several ways such as star and delta, and the voltage can be raised or lowered. The following three-phase transformers are the most commonly used in practice:

- star–star,
- delta–delta,
- star–delta,
- delta–star and
- delta–zig-zag star connection.

2.4.1 Star–Star (γ/γ) Connection

Figure 2.4 shows a star–star connection, which is economical for small, high-voltage transformers because the phase voltage is $\frac{1}{\sqrt{3}}$ times the line voltage. Hence, the number of turns per phase and quantity of insulation required is minimum. There is a phase shift of 30° between the phase voltages and line voltages on both primary and secondary, whereas the line voltages on both primary and secondary are in phase with each other, as shown in Figure 2.5.

The ratio of voltages on the primary and secondary sides is equal to the transformation ratio of each transformer. The star–star connection works well for balanced load. If the load is unbalanced, the neutral shifts. To prevent this, that star point of the primary is required to be connected to the star point of the generator.

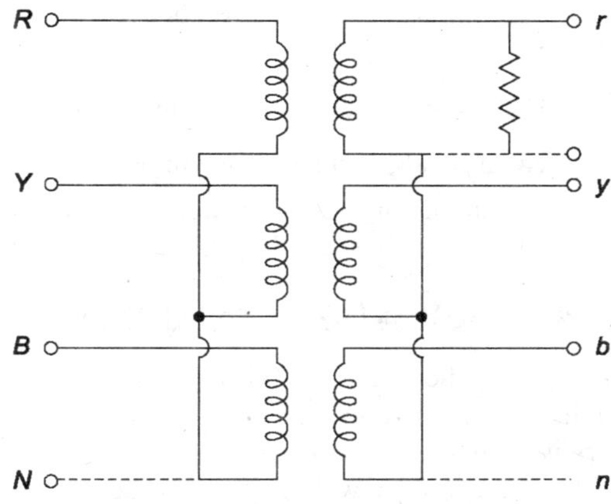

Figure 2.4 Star–Star Connection

Let V_{L1} be the line voltage on the primary side. The phase voltage on the primary (V_{Ph1}) is given by the following:

$$V_{Ph1} = \frac{V_L}{\sqrt{3}} \tag{2.1}$$

If a is the turns ratio, the phase voltage on the secondary is given the following:

$$V_{Ph2} = \frac{1}{a}\,\frac{V_{L1}}{\sqrt{3}}$$

where $a = \dfrac{V_{Ph1}}{V_{Ph2}}$.

The suffix '1' indicates the primary, whereas suffix '2' indicates the secondary.

A star–star connection has the following advantages:

- Less number of turns and less quantity of insulation is required because $V_P = \dfrac{V_L}{\sqrt{3}}$.
- Since $I_P = I_L$, the current through the windings is high. The windings must have a large cross-section and must be mechanically strong so that they can bear heavy load and short-circuit.
- There is no phase shift between the primary and secondary voltages.

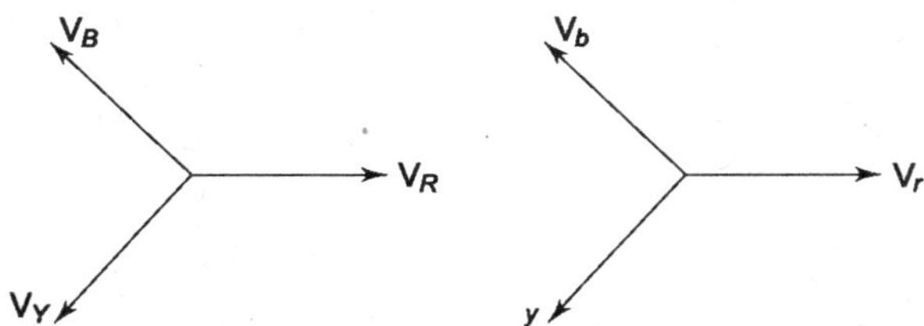

Figure 2.5 Phasor Diagram of Star–Star Connection

- It is suitable for the three-phase and four-wire system because of the presence of the neutral point.

A star–star connection has the following disadvantages:

- The neutral point shifts due to unbalanced load and performance is not satisfactory.
- In spite of connecting the neutral point to earth, the third harmonic present in the alternator voltage may appear in secondary and cause distortion of the secondary voltage.

2.4.2 Delta–Delta (Δ/Δ) Connection

Figure 2.6 shows a three-phase transformer having delta connections on both the primary and secondary windings.

Figure 2.7 shows the phasor diagram of both the primary and secondary sides. From Figure 2.7, it can be seen that there is no phase difference between the primary and secondary voltages. This connection is economical for large, low-voltage transformers because it increases the number of turns per phase.

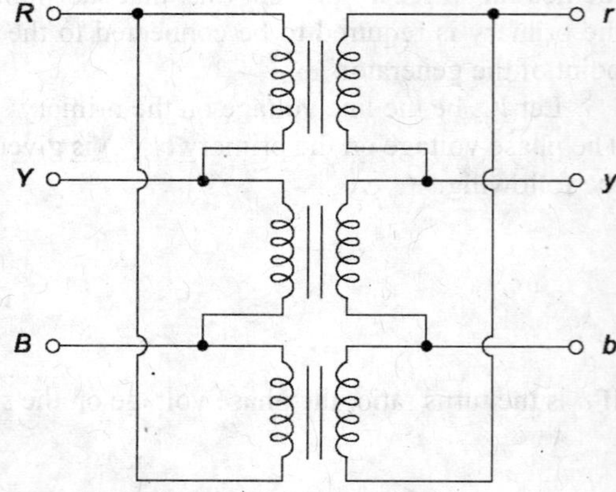

Figure 2.6 Delta–Delta Connection

Let V_{L1} and V_{Ph1} be the line voltage and phase voltage, respectively, on the primary side, V_{L2} and V_{Ph2} be the line voltage and phase voltage, respectively, on the secondary side and a be the turns ratio.

For delta connection, $V_{L1} = V_{Ph1}$ and $V_{L2} = V_{Ph2}$.

Again,

$$a = \frac{V_{L1}}{V_{L2}} = \frac{V_{Ph1}}{V_{Ph2}}$$

A delta–delta connection has the following advantages:

- The delta connection provides a closed path for circulation of the third harmonic component of current and hence the flux remains sinusoidal since magnetizing component of current must contain the third harmonic to get the secondary voltage sinusoidal.
- This connection also permits unbalanced loading.
- In the delta–delta connection, if one transformer is inoperative, V–V operation is still possible with reduced rating. *(Refer to Section 2.5)*
- No distortion in the secondary voltage occurs.

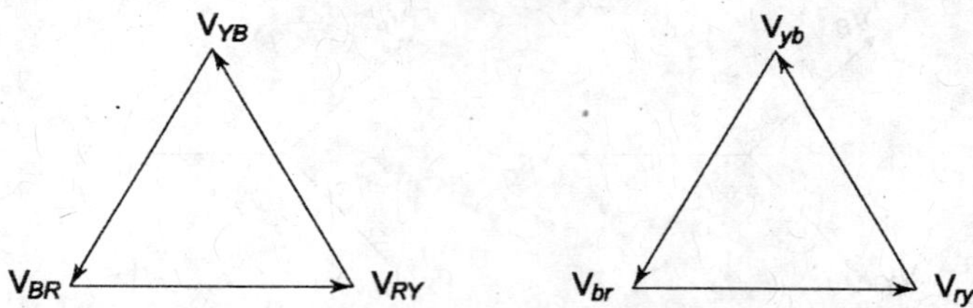

Figure 2.7 Phasor Diagram of Delta–Delta Connection

- For the delta connection, $I_P = \frac{I_L}{\sqrt{3}}$ and cross-section of the winding is low, which makes the connection economical for low-voltage transformers.

A delta–delta connection has the following disadvantages:

- It is not suitable for the three-phase four-wire system because the neutral point is absent.
- This connection is generally used for low-voltage transformers.

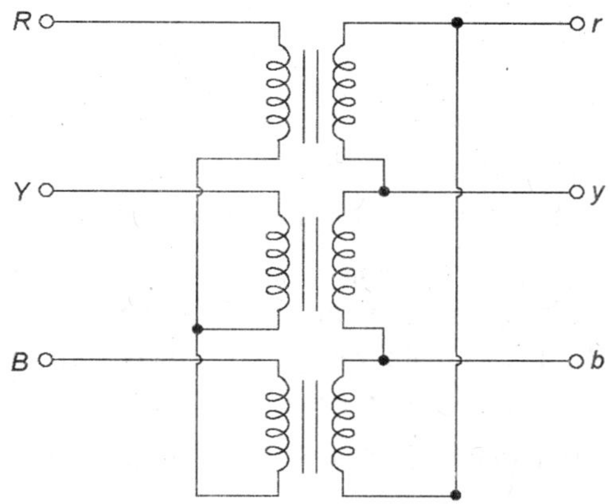

Figure 2.8 Star–Delta Connection

2.4.3 Star–Delta (γ/Δ) Connection

Figure 2.8 shows a three-phase (3-ϕ) transformer having star and delta connections on the primary and secondary windings respectively.

Figure 2.9 shows the phasor diagram of both the primary and secondary sides. From Figure 2.8, it can be seen that there is a 30° phase difference between the primary and secondary line voltages. The neutral is available on the primary side, and the secondary allows the flow of the third harmonic. The main use of this transformer is to step down voltages and hence it is used at the distribution side, that is, at the receiving end after the transmission.

Let V_{L1} and V_{Ph1} be the line voltage and phase voltage, respectively, on the primary side, V_{L2} and V_{Ph2} be the line voltage and phase voltage, respectively, on the secondary side and a be the turns ratio.

For star connection,

$$V_{Ph1} = \frac{V_{L1}}{\sqrt{3}}$$

For delta connection,

$$V_{L2} = V_{Ph2}$$

$$V_{Ph2} = \frac{1}{a} V_{Ph1} = \frac{V_{L1}}{a\sqrt{3}}$$

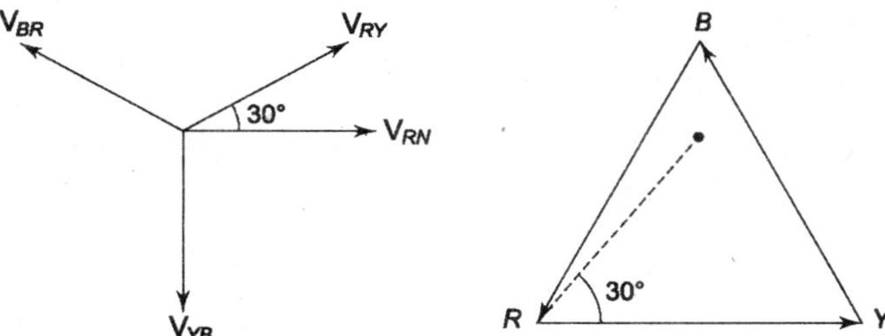

Figure 2.9 Phasor Diagram of Star–Delta Connection

Again,

$$a = \frac{V_{Ph1}}{V_{Ph2}}$$

Since the secondary is connected in delta, this type of transformer is not affected by unbalanced load. A star–delta transformer has the following advantages:

- Since the primary is star connected, fewer numbers of turns are required in the primary, which makes it economical for high-voltage, step-down power transformers.
- The available neutral point on the primary side can be earthed to avoid distortion.
- It is possible to handle large, unbalanced load.

A star–delta transformer has the following disadvantage:

- Since the secondary voltage is not in phase with the primary, it is not possible to make it parallel with star–star and delta–delta transformers.

2.4.4 Delta–Star (Δ/γ) Connection

Figure 2.10 shows a three-phase transformer having delta and star connections on the primary and secondary windings, respectively.

Figure 2.11 shows the phasor diagram of both the primary and secondary sides. From Figure 2.10, it can be seen that there is a 30° phase difference between the primary and secondary line voltages. The neutral point is available on the secondary side. Single-phase and three-phase loads can be supplied with this type of transformers. The main use of this transformer is to step up the voltages and hence it is used at the beginning of transmission or at the sending end.

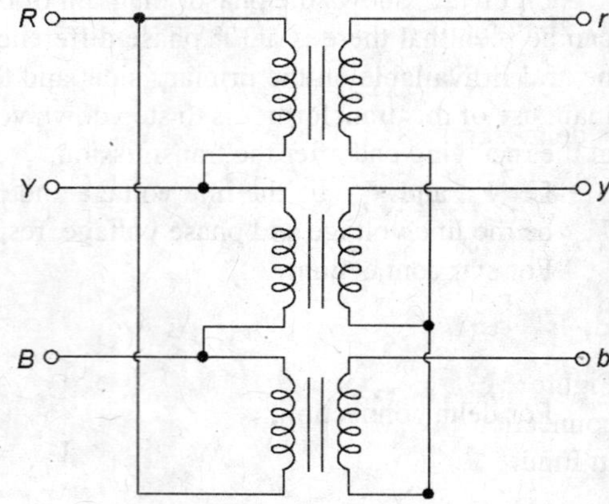

Figure 2.10 Delta–Star Connection

Let V_{L1} and V_{Ph1} be the line voltage and phase voltage, respectively, on the primary side, V_{L2} and V_{Ph2} be the line voltage and phase voltage, respectively, on the secondary side and a be the turns ratio.

For a star connection,

$$V_{Ph2} = \frac{V_{L2}}{\sqrt{3}}$$

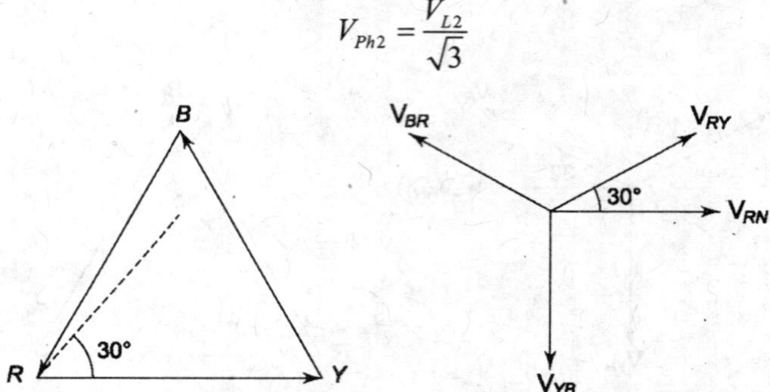

Figure 2.11 Phasor Diagram of Delta–Star Connection

For delta connection, $V_{L1} = V_{Ph1}$

$$V_{Ph2} = \frac{1}{a} V_{Ph1} = \frac{V_{L1}}{a}$$

Again,

$$a = \frac{V_{Ph1}}{V_{Ph2}}$$

$$\therefore\ V_{L2} = \sqrt{3} V_{Ph2} = \sqrt{3}\left(\frac{V_{L1}}{\sqrt{3}}\right)$$

Since the secondary is connected in delta, this type of transformers is not affected by unbalanced load.

A delta–star connection has the following advantages:

- Since the primary is delta connected, the winding cross-section is small.
- Since the neutral is available on the secondary side, three-phase and four-wire supply can be carried out.
- No distortion occurs due to the third harmonic component.
- Saving in the cost of insulation is possible due to availability of star connection on one side.
- No difficulty occurs due to large unbalanced load.

A delta–star connection has the following disadvantage:

- Since the secondary voltage is not in phase with the primary, it is not possible to make it parallel with star–star and delta–delta transformers.

2.4.5 Delta–Zig-zag Star Connection

Figure 2.12 shows the two equal halves of each winding on each phase on the star side. To form the star connection each, its phase is formed by using half of the windings from two different phases, as shown in Figure 2.13. Figure 2.14 shows the phasor diagram of this connection. This connection is +30° and its phase–to-phase transformation ratio $\frac{V}{Va} = \frac{1}{a}$ and line-to-line transformation ratio $\frac{\sqrt{3}}{2}\left(\frac{Va}{V}\right) = \frac{a\sqrt{3}}{2}$.

This connection is specifically useful for the rotary converters for supplying the three-wire direct

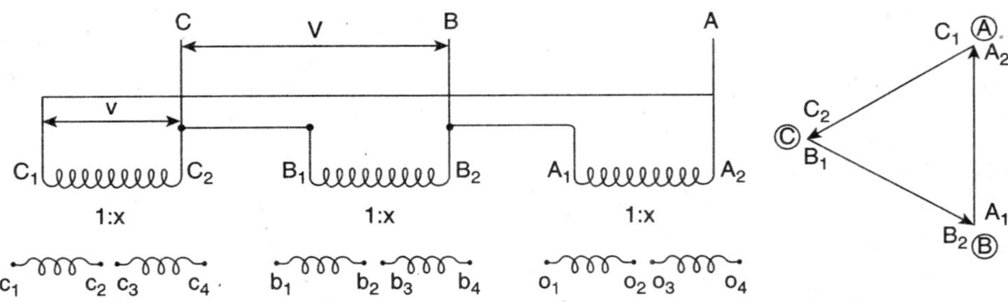

Figure 2.12 Primary Voltage Phasor

current, where the unbalanced DC in the neutral wire will divide in such a way that half of each winding will carry equal direct currents in opposite directions, and hence there will be no resultant unidirectional magnetization of the core due to this unbalanced direct current.

Table 2.1 compares voltage and current relationships for different types of connections.

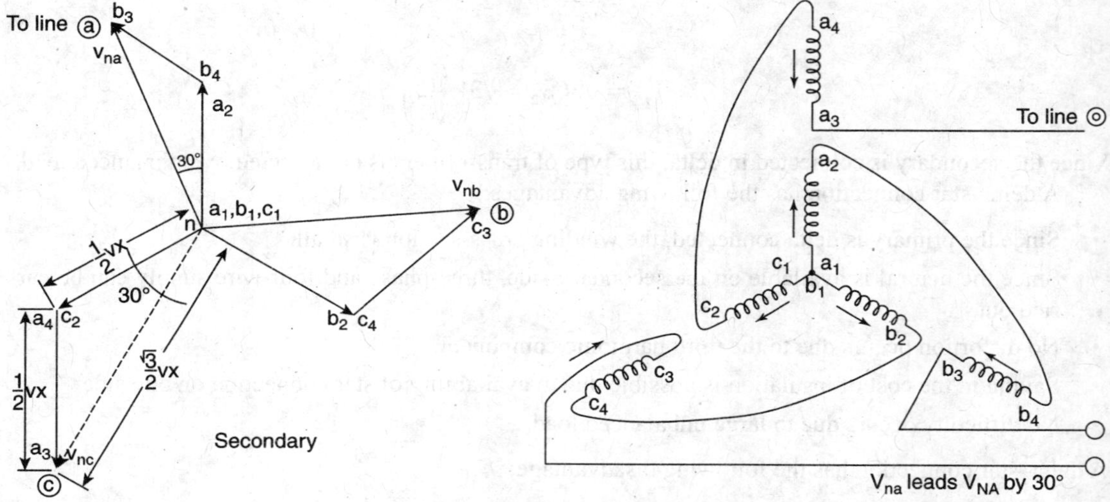

Figure 2.13 Secondary Side Phasor Diagram

Figure 2.14 Delta–Zig-zag Connection of Transformer with +30° Phase Shift

Table 2.1 Comparison of Voltage and Current Relationships for Different Types of Connections

Type of Connection	Primary Side			Secondary Side			
	Line Voltage	Phase Voltage	Phase Current	Phase Voltage	Phase Current	Line Voltage	Line Current
Star–Star	V_L	$\frac{V_L}{\sqrt{3}}$	I_L	$\frac{V_L}{a\sqrt{3}}$	aI_L	$\frac{V_L}{a}$	aI_L
Delta–Delta	V_L	V_L	$\frac{I_L}{\sqrt{3}}$	$\frac{V_L}{a}$	$a\left(\frac{I_L}{\sqrt{3}}\right)$	$\frac{V_L}{a}$	aI_L
Star–Delta	V_L	$\frac{V_L}{\sqrt{3}}$	I_L	$\frac{V_L}{a\sqrt{3}}$	aI_L	$\frac{V_L}{a\sqrt{3}}$	$\sqrt{3}aI_L$
Delta–Star	V_L	V_L	$\frac{I_L}{\sqrt{3}}$	$\frac{V_L}{a}$	$\frac{aI_L}{\sqrt{3}}$	$\frac{\sqrt{3}V_L}{a}$	$\frac{aI_L}{\sqrt{3}}$

Example 2.1 A three-phase 100 kVA, 5,000/500 V, *Y–Y*, 50 Hz transformer has an iron loss of 1,500 W. The maximum efficiency occurs at 80 per cent of full load. Calculate

(i) the efficiency of the transformer at full load and 0.85 power factor lagging,
(ii) the efficiency of the transformer at $\frac{3}{4}$ th of full load at unity power factor, and
(iii) the maximum efficiency at unity power factor.

Solution

Iron loss = 1,500 W

Since the maximum efficiency occurs at 80 per cent of full load, $x = 0.80$.

∴ Cu loss at 80 per cent of full load = 1,500 W

Let Cu loss at full load be P_{Cufl} W.

$$\therefore \quad x^2 P_{Cufl} = 1,500$$

$$\text{i.e.,} \quad P_{Cufl} = \frac{1,500}{x^2} = \frac{1,500}{0.8^2} = 2,343.75\ \text{W}$$

(i) Full load:

$$\text{Total loss} = 2,343.75 + 1,500 = 3,843.75\ \text{W}$$

$$\text{Output on full load at 0.85 power factor lagging} = 100 \times 0.85 \times 10^3 = 85,000\ \text{W}$$

$$\text{Efficiency} = \frac{85,000}{85,000 + 3,843.75} = 0.9567\ \text{p.u.} = 95.67\%$$

(ii) $\text{Cu loss at three-fourths of full load} = 0.75^2 \times 2,343.75 = 1,318.36\ \text{W}$

$$\text{Total loss} = 1,318.36 + 1,500 = 2,818.36\ \text{W}$$

Output at three-fourths of full load at unity power factor = (3/4) × 100 × 10^3 = 75,000 W

$$\therefore \text{Efficiency} = \frac{75,000}{75,000 + 2,818.36} = 0.9638\ \text{p.u.} = 96.38\%$$

(iii) During the maximum efficiency, total loss = 2 × 1,500 = 3,000 W

Output = 100 × 0.8 × 1 × 10^3 = 80,000 W

$$\therefore \text{Efficiency} = \frac{80,000}{80,000 + 3,000} = 0.9638\ \text{p.u.} = 96.38\%\ \textbf{Ans.}$$

■

2.5 OPEN-DELTA OR *V–V* CONNECTION

If one of the transformers of delta–delta connection is inoperative while a three-phase supply is connected to primary, three equal supplies will be available at the secondary terminals on no load and supply is still possible with reduced capacity. This method is known as open-delta or *V–V* connection. It has the following applications:

- If one of the transformers in delta–delta bank is inoperative, it is possible to continue service with reduced capacity.
- If the three-phase load is small, it is preferable to use a *V–V* connection.
- If the load increases, in future, the open delta can be closed to increase the rating.

Figure 2.15 shows the open-delta or *V–V* bank. The total load carried by the *V–V* bank is not two-thirds of the full load. It is only 57.7 per cent of full load and thus there is reduction of 15.5 per cent. Let there be delta–delta bank of three 15 kVA transformers. If it operates in *V–V*, the capacity of the *V–V* bank is (15 + 15) × 0.866, that is, 25.98 kVA instead of (15 + 15) kVA or 30 kVA. We can prove that the ratio of *V–V* capacity to delta–delta capacity is 57.7 per cent instead of 66.66 per cent.

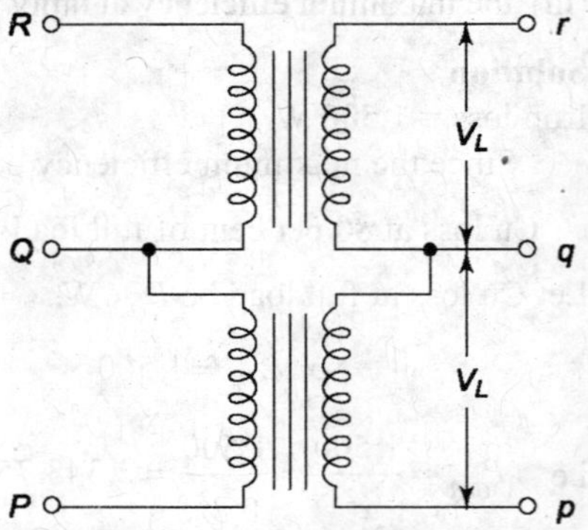

We can write,

$$\Delta\text{–}\Delta\,\text{capacity} = \sqrt{3}V_L I_L = \sqrt{3}V_L \times \sqrt{3}I_P = 3V_L I_P$$

Figure 2.15 *V-V* Connection

To get *V–V* bank, one transformer of Δ–Δ bank must be open; that is, secondary line current = secondary phase current or, $I_L = I_P$.

$$\therefore \quad V\text{–}V\ \text{capacity} = \sqrt{3}V_L I_L = \sqrt{3}V_L I_L$$

$$\therefore \quad \frac{V\text{–}V\ \text{capacity}}{\Delta\text{–}\Delta\ \text{capacity}} = \frac{\sqrt{3}V_L I_P}{3V_L I_P} = \frac{1}{\sqrt{3}} = 0.577$$

i.e., $\quad V\text{–}V$ capacity = 57.7% of Δ–Δ capacity

∴ *V–V* bank carries 57.7 per cent of original load. Now, we can conclude the following during *V–V* operation:

- The bank capacity becomes 25.98 kVA instead of 30 kVA.
- Since 30 × 0.866 = 25.98, we can say that 86.6 per cent of rated capacity of two working transformers is available. This factor is also known as utility factor.
- Each transformer supplies 57.7 per cent of load instead of 50 per cent during *V–V* operation.If the rated load is supplied by Δ–Δ bank and one transformer is removed, the overload on each of the two transformers becomes 73.2 per cent. This is because

$$\frac{\text{Total load in } V\text{–}V}{\text{VA/transformer}} = \frac{\sqrt{3}V_L I_S}{V_L I_S} = \sqrt{3} = 1.732$$

In the *V–V* bank the average power factor is less than that of the load and the secondary terminal voltages become unbalanced with increase of load. Some provision should be made to reduce this overload to avoid overheating and consequent breakdown of two transformers. This overload may be carried temporarily.

If the V–V bank supplies power to a balanced three-phase load having power factor cos β, one transformer will operate at power factor cos (30°–β) and the other at cos (30° +β). Hence, the two transformers will have different voltage regulation.

The power supplied by the two transformers are the following:

$$P_1 = \text{kVA}\cos(30^\circ - \beta) \quad \text{and} \quad P_2 = \text{kVA}\cos(30^\circ + \beta)$$

The following conclusions can be drawn:

- $\beta = 0°$ The power factor of each transformer is 0.866.
- $\beta = 30°$ In this case the power factor of one transformer is cos (30° – 30°), that is, 1, and the power factor of the other transformer is cos (30° + 30°), that is, 0.866.
- $\beta = 60°$ In this case the power factor of one transformer is cos (30° – 60°), that is, 0.866, and the power factor of the other transformer is cos (30° + 60°), that is, 0. Therefore, one transformer will not supply any load and the entire load is carried out by the other transformer.

Example 2.2 Two transformers are connected in open-delta supply with a 300 kVA balanced load operating at 0.866 power factor lagging. If the load voltage is 440 V, calculate the following:

(i) kVA supplied by each transformer and

(ii) kW supplied by each transformer.

Solution

(i) kVA supplied by each transformer $= \dfrac{\frac{300}{2}}{0.866} = 173.21\text{ kVA}$

(ii) $\cos\beta = 0.866\,(\text{lagging}), \beta = \cos^{-1}(0.866) = 30^\circ$

$$\therefore P_1 = 173.21 \times \cos(30^\circ - \beta) = 173.21 \times \cos(30^\circ - 30^\circ) = 173.21\text{ kW}$$

$$\therefore P_2 = 173.21 \times \cos(30^\circ + \beta) = 173.21 \times \cos(30^\circ + 30^\circ) = 86.6\text{ kW}$$

Again, $P_1 + P_2 = 173.21 + 86.6 = 259.81\text{ kW}$

and also $P_1 + P_2 = 300 \times 0.866 = 259.8\text{ kW}$ ■

2.6 SCOTT CONNECTION OR *T–T* CONNECTION

Figure 2.16 shows a Scott connection or T–T connection, which was originally proposed by Charles F. Scott. This connection needs two transformers on each side instead of three transformers and accomplishes three-phase to three-phase transformations. The transformer, which is a horizontal member of the connection having centre taps both on the primary and the secondary, is known as the main transformer. The other transformer of the primary and secondary whose one end is connected to the main transformer has a 0.866 tap and it is called the teaser transformer. Three-phase supply is given to the other end of the teaser and the two ends of the main transformer.

Figure 2.17 shows the voltage diagram of the primary and secondary where the three-phase supply voltage is 100 V. The turns ratio (a) is taken as unity.

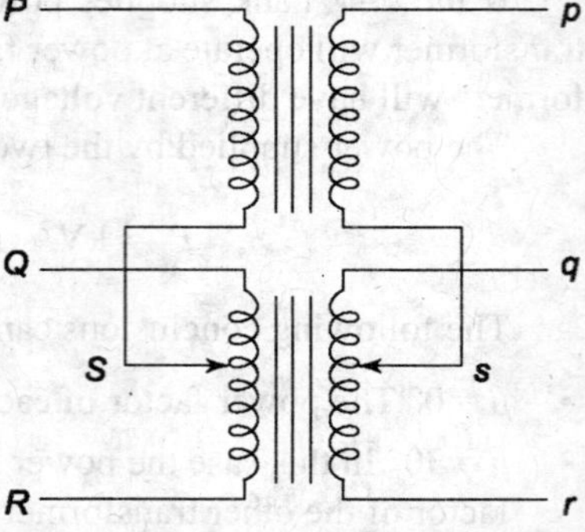

Figure 2.16 Scott Connection

From Figure 2.17, we have $E_{SR} = E_{SQ} = 50$ V, having a phase difference of 180° as *SR* and *SQ* are on the same coils. The value of $E_{SP} = \frac{\sqrt{3}}{2} \times 100 = 86.6$ V, which lags behind the voltage across the main by an angle of 90°. Similar relations also occurs for the secondary.

From Figure 2.17(b), for a load power factor of unity I_{SQ} lags behind E_{SQ} by an angle 30° whereas I_{SR} leads E_{SR} by an angle 30°. Therefore, we can conclude that each half of the main transformer will operate at different power factors. The teaser transformer operates at 0.866 of its rated voltage, and the main transformer operates at 0.866 power factor. Therefore, the coils of the main transformer work at 86.6 per cent of its kVA rating. The rating ratio of Scott connection is the same as that of *V–V* connection if and only if two identical coils are used while heating in the two cases is different.

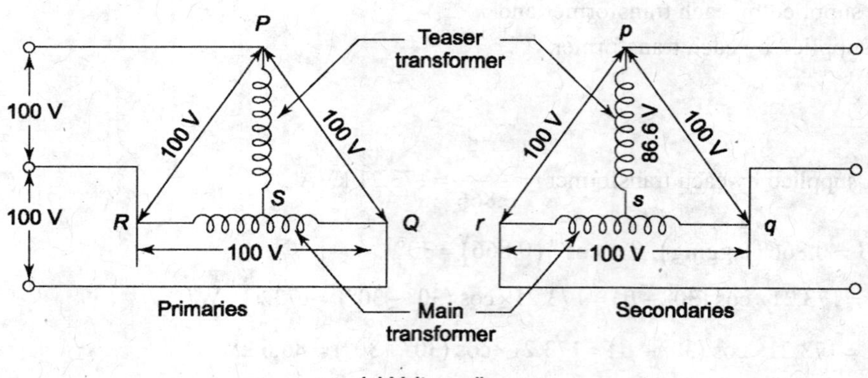

(a) Voltage diagram

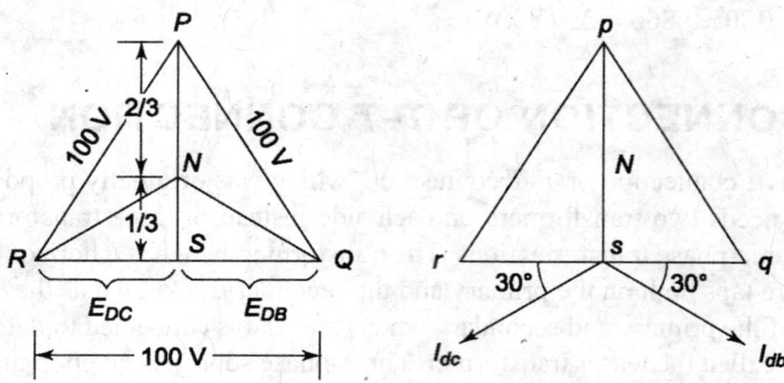

(b) Phasor diagram

Figure 2.17 Voltage and Phasor Diagram of *T–T* Connection

In this case, the available VA capacity is = $V_L I_L + 0.866\ V_L I_L = 1.866\ V_L I_L$. I_L is the primary line current. Actually utilized VA due to three-phase supply = $1.732\ V_L I_L$.

$\therefore \quad \dfrac{\text{kVA actually utilized}}{\text{kVA available}} = \dfrac{1.732 V_L\ I_L}{1.866 V_L\ I_L} = 0.928$. Hence, this connection is more economical compared to V–V connection. For teaser transformer with load of unity power factor at the secondary, we can conclude the following from Figure 2.18:

(i) The neutral is one-third way up from S.

(ii) The current in the teaser transformer is in phase with its voltage.

(iii) The current in one half of the main transformer leads the voltage, whereas the current in the other half lags the voltage.

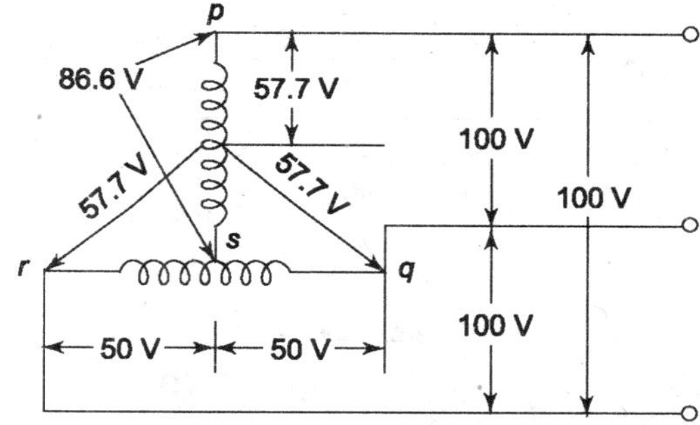

Figure 2.18 Phasor Diagram at Unity Power Load at Secondary

Another important point is that for a balanced load of power factor cosβ, the current in one half leads the voltage by an angle cos (30° – β), whereas the current in other half lags the voltage by an angle cos (30° + β), which is similar to the V–V connection. The Scott connection can also be used for conversion of three phases to two phases.

2.7 THREE-PHASE TO TWO-PHASE CONVERSION

To convert three phases to two phases or vice versa, the connection shown in Figure 2.19(a) is used, which is nothing but a Scott connection having two transformers of different ratings. If both the transformers are identical, they should have suitable tappings.

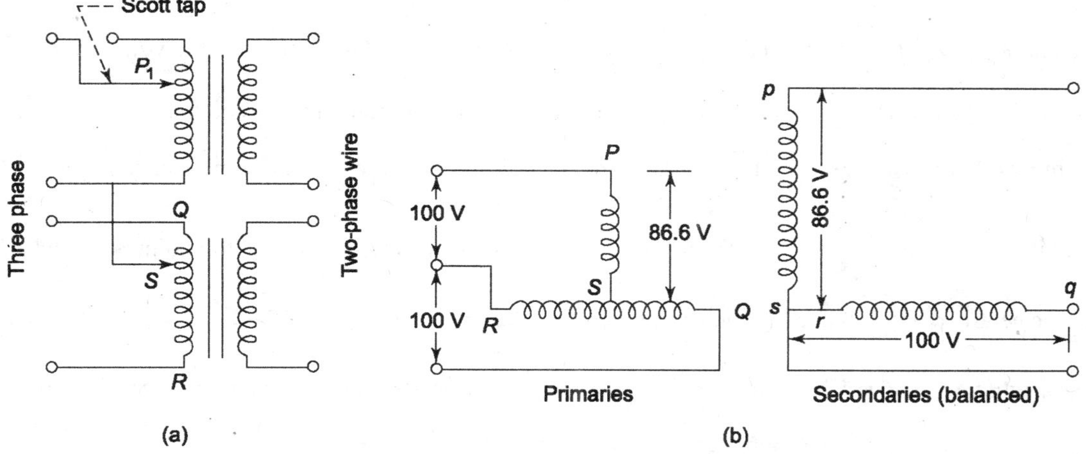

Figure 2.19 Three-phase to Two-phase Conversion

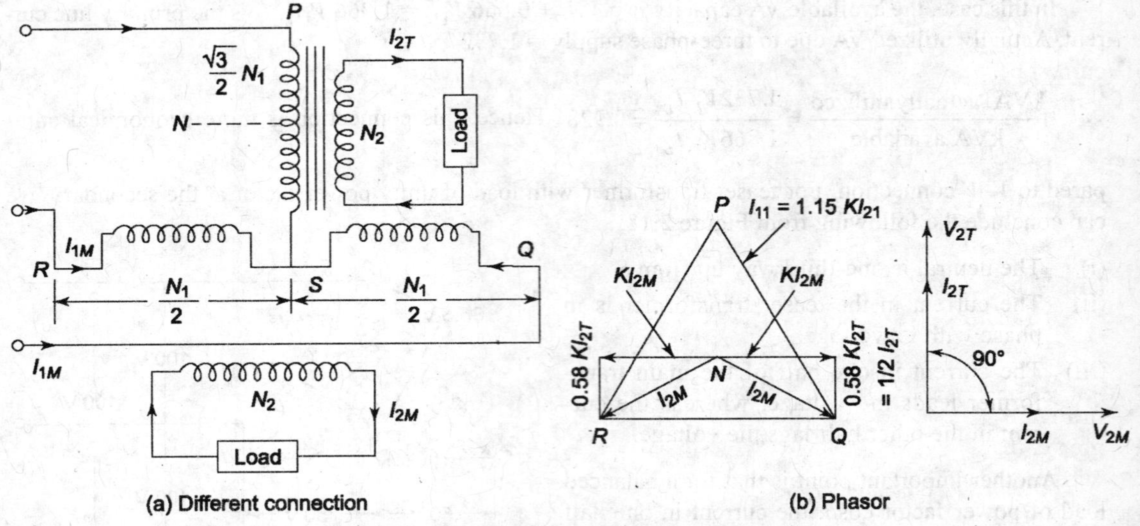

Figure 2.20 A Different Connection and a Phasor

The points s and r of the secondary are connected as shown in Figure 2.19(b), which gives the two-phase three-wire system. Here, E_{sp} = 86.6 V whereas E_{sq} = 100 V, producing two unequal voltages. To get the same volt/turn both in the primary and the secondary, one line of three-phase supply is connected to point P_1, where SP_1 is 86.6 per cent of the teaser primary turns. The secondary voltages will be equal in magnitude if and only if the secondary transformers have the same number of turns, which results in a symmetrical two-phase three-wire system instead of an unsymmetrical two-phase three-wire system.

Figure 2.20(a) shows a different connection where the primary has N_1 number of turns and is connected between Q and R of the main transformer. Now, $V_{PQ} = V_{QR} = V_{RS} = V_1$ if the supply line voltage is V_1 and $E_{PS} = \frac{\sqrt{3}}{2}V_1$. The number of turns between P and S is $\frac{\sqrt{3}}{2}N_1$ to give the same volt/turn in both primaries. The secondary terminal voltages will be equal in magnitude if the secondaries have an equal number of turns. But the secondary terminal voltages will be quadrature to each other. Since $E_{PS} = \frac{\sqrt{3}}{2}V_1$ and it is not equal to $\frac{V_1}{\sqrt{3}}$, S is not the neutral point. To get the position of the neutral point, let us take N as the neutral point. N will be the neutral point only if $E_{PN} = \frac{V}{\sqrt{3}}$. Now $E_{NS} = \frac{\sqrt{3}}{2}V_1 - \frac{V_1}{\sqrt{3}} = 0.288V_1$. This means that N is above S by a number of turns equal to 28.8 per cent of N_1. N divides the teaser winding in the ratio of 2:1 because $\frac{0.866}{3} = 0.288$.

Let the current supplied by the teaser secondary be I_2T at unity power factor. The teaser primary current becomes $I_{1T} = I_{2T} \times \dfrac{N_2}{\dfrac{\sqrt{3}N_1}{2}} = 1.15\left(\dfrac{N_2}{N_1}\right) I_{2T} = 1.15\left(\dfrac{I_{1T}}{a}\right)$, where $a = \dfrac{N_1}{N_2}$ is the turns ratio.

The current I_{Ml} of each half of the main transformer of the primary has the following two components:

(i) One part is required to balance the secondary current (I_{2M}), which is $I_{2M} \times \dfrac{N_2}{N_1} = \dfrac{I_{2M}}{a}$.

(ii) The other part is equal to one half of the teaser primary current or (1/2) I_lT because the main transformer primary serves as the return path for teaser primary current and it divides itself into two halves at the mid point *S*.

The condition due to a balanced two-phase load at a lagging power factor of 0.866 has been shown in Figure 2.21(a). The three-phase side is balanced here. The main transformer rating is 15 per cent greater than the rating of the teaser transformer. This is because its voltage is 15 per cent greater. But its current remains the same.

Figure 2.21(b) shows the condition due to an unbalanced two-phase load having different currents and power factors.

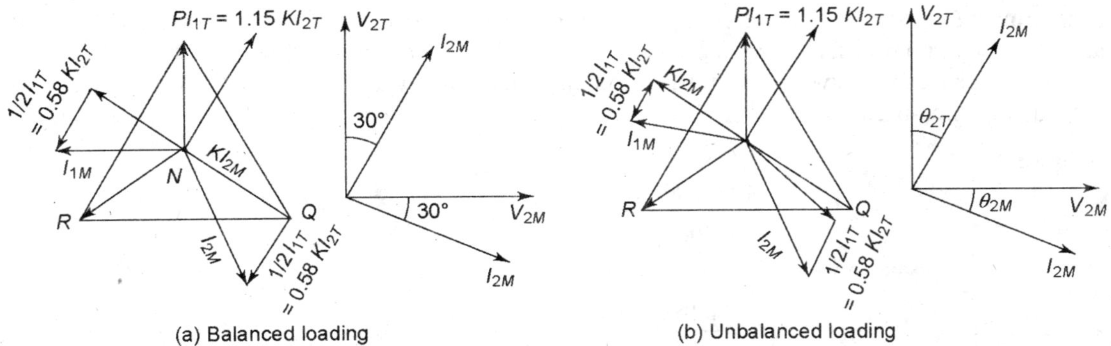

Figure 2.21 Vector Diagram for Balanced and Unbalanced Loading

Example 2.3 Two *T*-connected transformers are used to supply a 440 V, 4.4 kVA balanced load to form a balanced three-phase supply of 4.4 kV. Determine (i) voltage and current rating of each coil and (ii) kVA rating of the main and teaser transformers.

Solution

(i) The voltage across main primary = 4.4 kV = 4,400 V and that across teaser primary = 0.866 × 4,400 = 3,810.4 V.The current is the same in the teaser and main primary and it is equal to the line current. $I_{LP} = \dfrac{44 \times 10^3}{\sqrt{3} \times 4,400} = 5.77\text{A}$

The secondary teaser voltage = 440 × 0.866 = 381 V

The secondary main voltage = 440 V

The secondary line current (I_{LS}) = (I_{LP} × 4,400)/440 = 5.77 × 10 = 57.7 A

(ii) Main kVA = 4,400 × 5.77 × 10^{-3} = 25.388 kVA

Teaser kVA = 0.866 × main kVA = 0.866 × 25.388 = 21.986 kVA ■

Example 2.4 Two single-phase transformers working at 200 V are supplied from 2,200 V, three-phase mains through a Scott connection. Find the current in each line of the three-phase mains when the power taken by each furnace is 450 kW at a power factor of 0.85 lagging. The losses in the transformer are negligible.

Solution

Here,

$$a = 2{,}200/200 = 11 \quad \text{and} \quad I_2 = \frac{450\times10^3}{0.85\times200} = 2{,}647.05 \text{ A}$$

$$I_{2T} = I_{2M} = I_2 = 2{,}647.05 \text{ A}$$

Since the two-phase load is balanced, the three-phase side is also balanced.
Primary currents are = $(1.15)I_2/a$ = (1.15 × 2,647.05)/11 = 276.74 A
Again, for star connection, phase current and line current are equal.

∴ line current = 276.74 A ■

Example 2.5 In a Scott connection, the loads on the two-phase side are 400 kW and 500 kW, both at 200 V and 0.8 power factor lagging. The three-phase line voltage is 2,200 V. The 400 kW load is on the leading phase on the two-phase side. Neglecting transformer losses, calculate the value of line currents on the three-phase side.

Solution

$$a = 2{,}200/200 = 11$$

$$I_{2T} = \frac{400\times10^3}{200\times0.8} = 2{,}500 \text{ A}$$

$$I_{1T} = \frac{1.15 I_{2T}}{a} = \frac{1.15\times2{,}500}{11} = 261.36 \text{ A}$$

Now, $\frac{1}{2}I_{1T} = \frac{261.36}{2} = 130.68$ A and $\frac{I_{2M}}{a} = \frac{1}{11}\times\frac{500\times10^3}{200\times0.8} = 284.1$ A

Main primary current = $\sqrt{284.1^2 + 130.68^2}$ = 312.71 A

The phase line currents are 130.68 A and 312.71 A. ■

2.8 PARALLEL OPERATIONS OF TRANSFORMERS

For parallel operation of three-phase transformers, the conditions for paralleling single-phase transformers are required as well as the following additional conditions:

- The ratio of voltage should refer to terminal voltage of the primary and secondary.
- The phase sequence must be identical.
- The phase between the primary and secondary of all transformers to be paralleled must be identical.

2.9 THREE-PHASE TO SIX-PHASE CONVERSION

The following are commonly used schemes for three-phase to six-phase conversion:

(i) double-star connection,
(ii) double-delta connection,
(iii) six-phase star connection and
(iv) diametrical connection

2.9.1 Double-star Connection

Figure 2.22 shows a double-star connection of transformers for three-phase to six-phase conversion, where three identical single-phase transformers are used. The three primaries are connected in delta, whereas each transformer in the secondary unit splits into two equal connections. Each set of secondaries is connected in star.

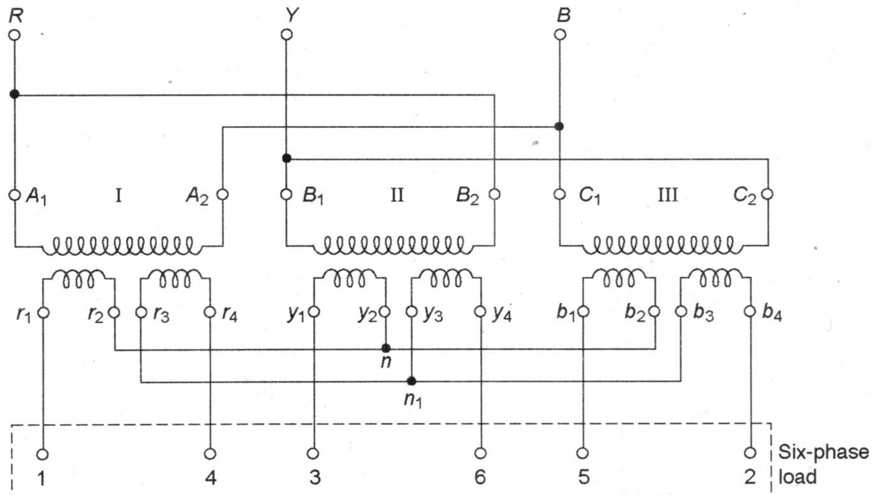

Figure 2.22 Double-Star Connection

2.9.2 Double-delta Connection

Figure 2.23 shows the double-delta connection of transformers for three-phase to six-phase conversion, where three identical single-phase transformers are used. The three primaries are connected in delta, whereas each transformer in the secondary unit splits into two equal connections. Each set of secondaries is connected in delta.

2.9.3 Six-phase Star Connection

Figure 2.24 shows the six-phase star connection of transformers for three-phase to six-phase conversion, where three identical single-phase transformers are used. The three primaries are connected in delta and the centre tap of each secondary transformer is connected to neutral.

2.9.4 Diametrical Connection

Figure 2.25 shows a diametrical connection. No centre tapping is used here. Neutral connections are not required.

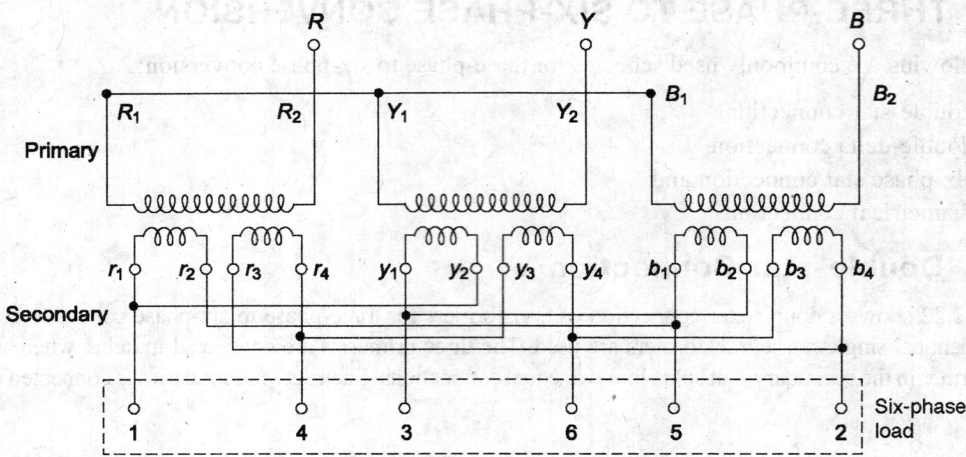

Figure 2.23 Double-delta Connection

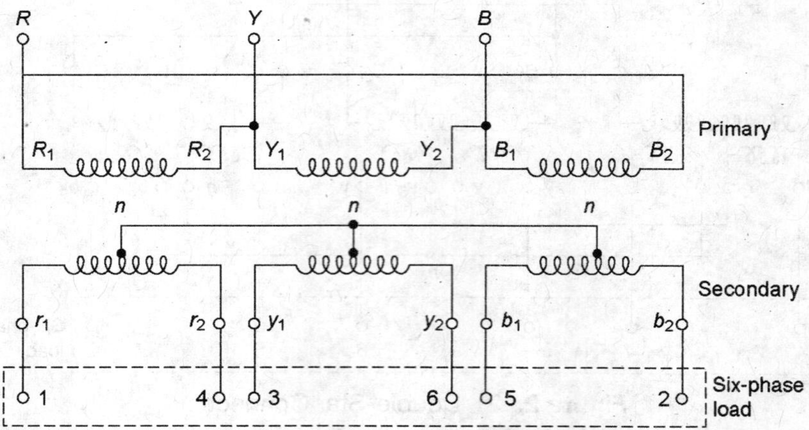

Figure 2.24 Six-phase Star Connection

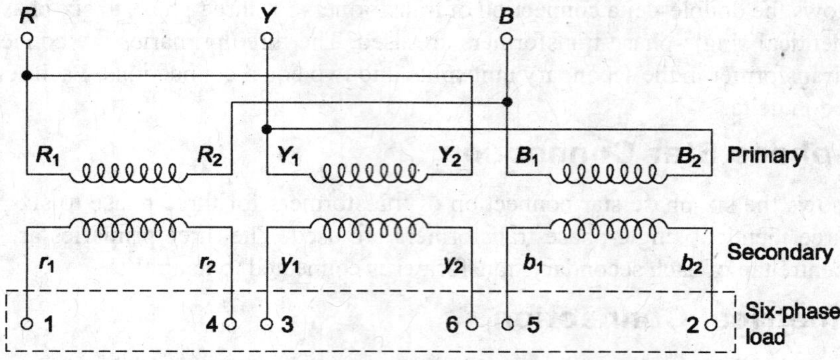

Figure 2.25 Diametrical Connection

2.10 THREE-WINDING TRANSFORMER

In addition to the primary and secondary windings in a transformer, there may be a third winding known as tertiary winding, which has the low-voltage rating whereas the primary has the highest voltage rating. The kVA ratings in three-phase transformers are unequal, whereas it is equal in two-winding transformers. The chief advantage of tertiary winding is that it reduces imbalance of the primary and secondary and hence the secondary load imbalance is distributed more evenly among the primary phases. This is because when the primaries and secondaries are star connected and the load is unbalanced, this reflects in unbalanced primary currents and the increased circulating current is reduced in tertiary windings. Figure 2.26 shows schematic diagram of a three-winding transformer having the primary (N_1), secondary (N_2) and tertiary windings (N_3).

From Figure 2.26, we can have for an ideal transformer

$$\frac{V_2}{V_1} = \frac{N_2}{N_1}, \frac{V_3}{V_1} = \frac{N_3}{N_1} \quad \text{and} \quad I_1N_1 = I_2N_2 + I_3N_3$$

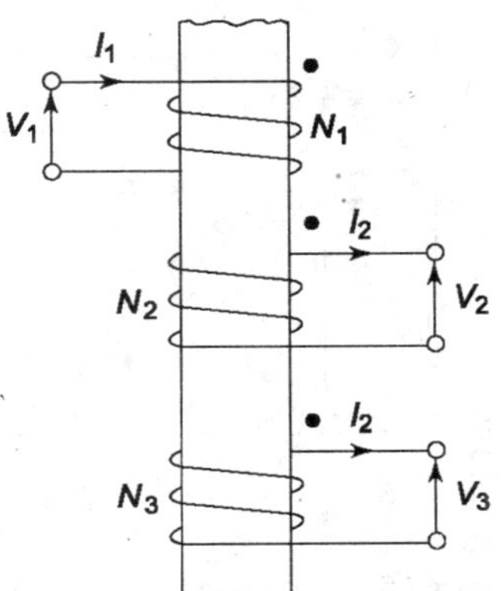

Figure 2.26 Three-winding Transformer

Tertiary windings have the following advantages:

- It is possible to supply substation auxiliaries at different voltage with respect to the primary and secondary with the help of tertiary windings.
- To inject the reactive power into the system, synchronous capacitors are connected across the delta-connected output of the tertiary windings.
- It is possible to interconnect three-phase supply systems with the help of tertiary windings.
- The impedance offered to the zero-sequence current by the delta-connected tertiary is reduced and hence the sufficient earth-fault current can flow, which helps the proper operation of the protective devices.
- Since continuity of supply is more important, it is possible to give the supply of single load from two sources.
- It possible to measure voltage for high-voltage testing transformers with the help of tertiary windings.

Currently, tertiary winding transformers are manufactured having tertiary winding voltage ratings up to 35 per cent of the total transformer VA rating. If both sets of the windings are connected in delta, the imbalance and the third harmonic do not come into picture.

2.11 THREE-PHASE TRANSFORMER CONNECTIONS

The twelve standard methods have been illustrated in Table 2.2. The four main groups are given as follows:

Group 1: Zero phase displacement (*Yy*0, *Dd*0, *Dz*0)

Group 2: 180° phase displacement (*Yy*6, *Dd*6, *Dz*6)

Group 3: 30° lag phase displacement (*Dy*1, *Yd*1, *Yz*1)

Group 4: 30° lead phase displacement (*Dy*11, *Yd*11, *Yz*11)

Table 2.2 Different Types of Three-phase Transformer Connections

Main Group No.	Phase Displace-ment	Phasor Group	Symbol	Phasor Diagrams of Induced Voltages: hv	Phasor Diagrams of Induced Voltages: lv	Winding Connections and Relative Position of Terminals: hv	Winding Connections and Relative Position of Terminals: lv
1	0°	1_1	Yy0	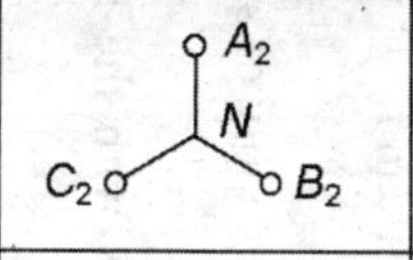	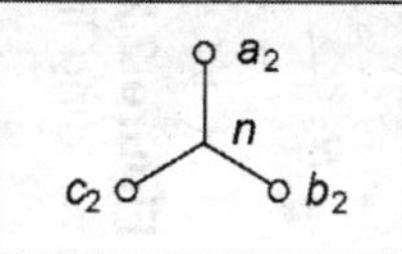	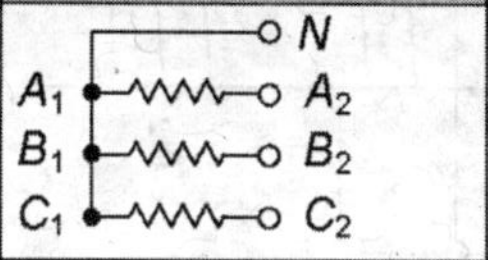	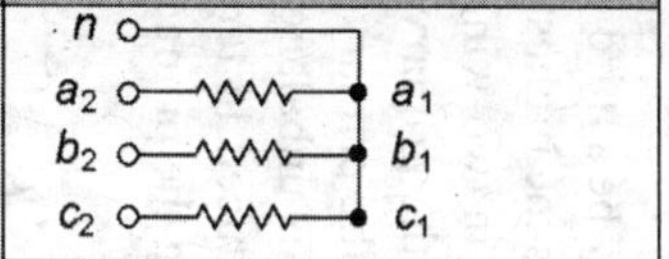
		1_2	Dd0	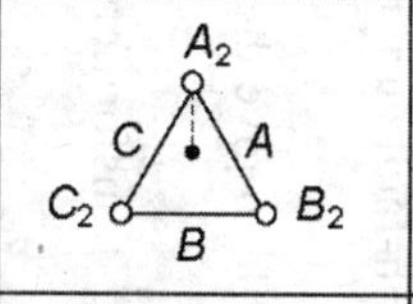	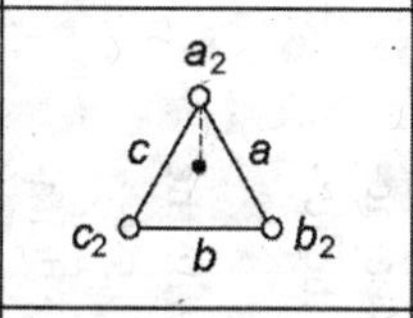	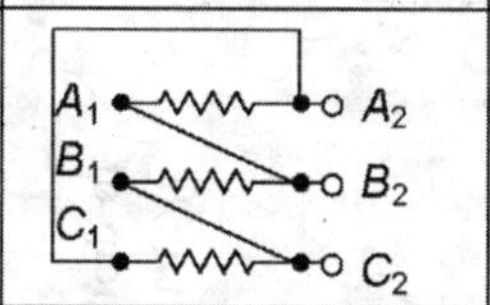	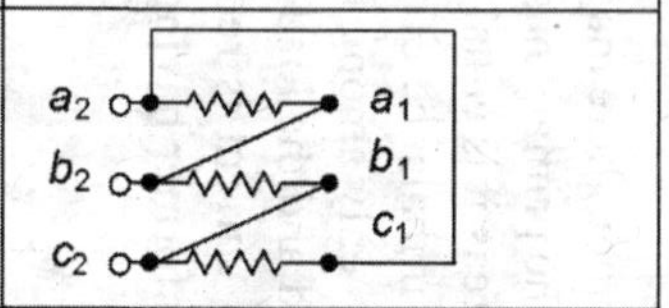
		1_3	Dz0	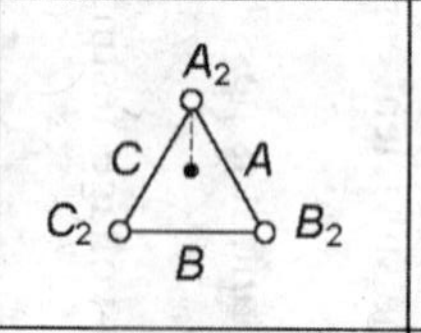	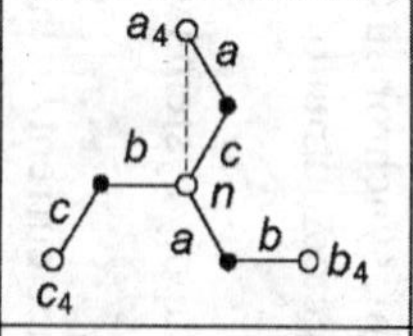	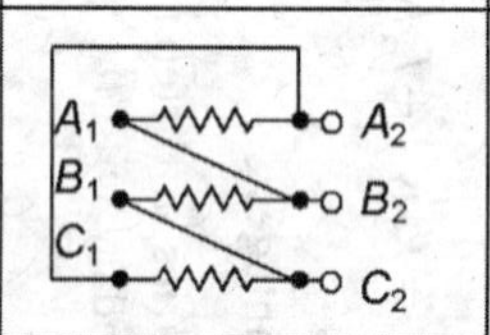	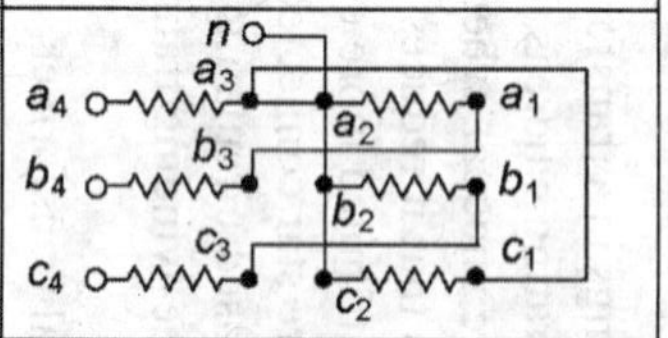
2	180°	2_1	Yy6	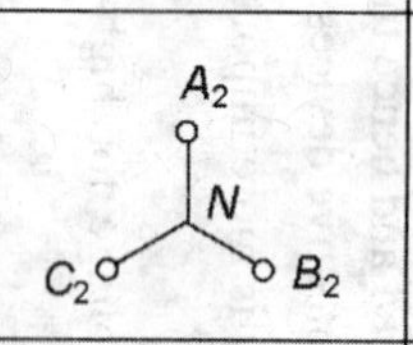	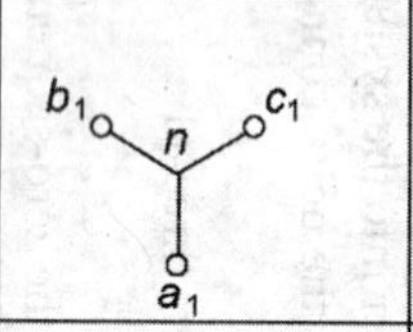	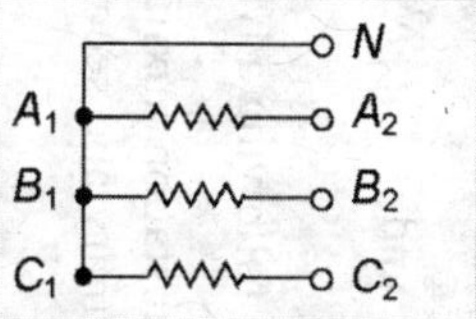	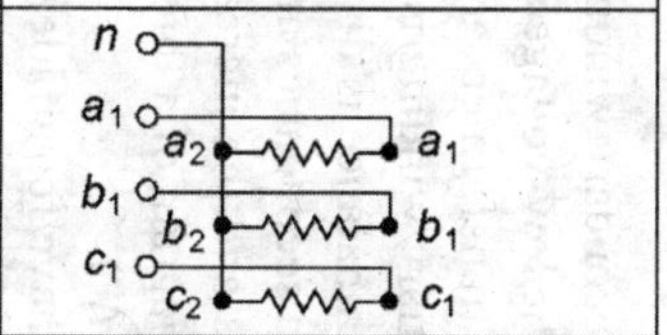
		2_2	Dd6	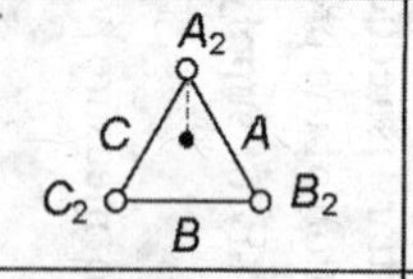	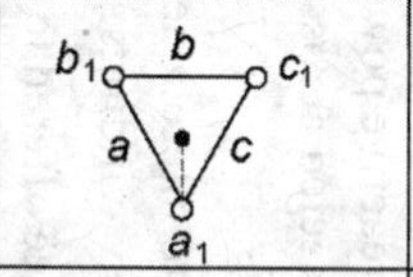	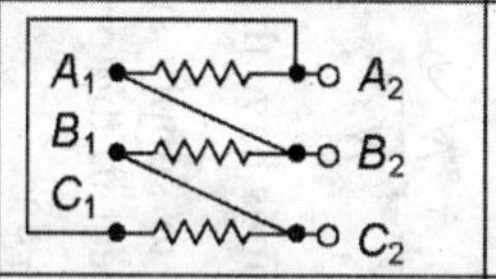	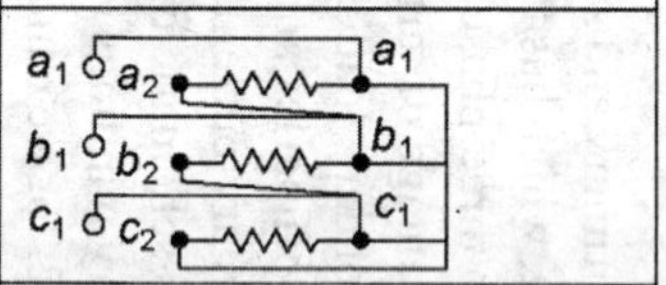
		2_3	Dz6	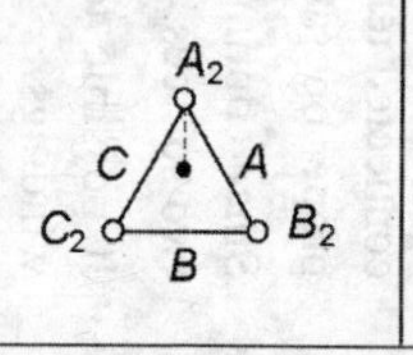	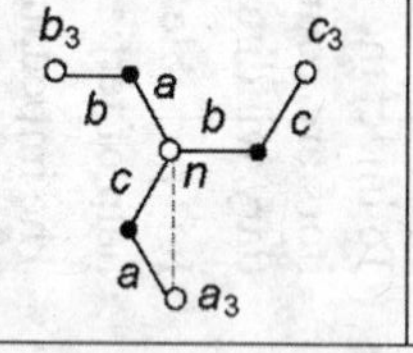	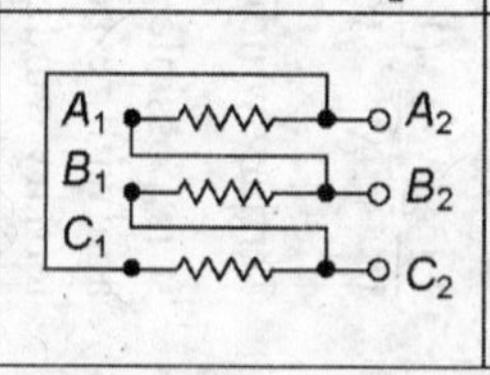	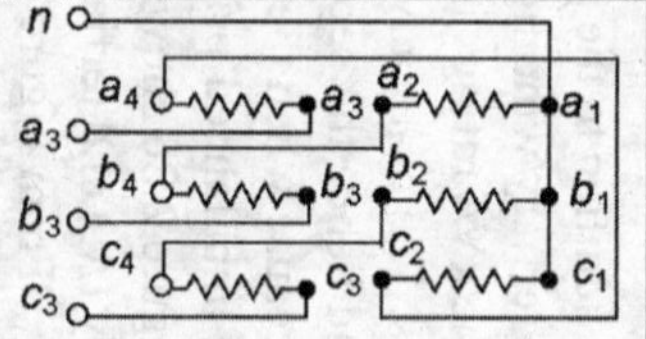

Table 2.2 *(Continued)*

Main Group No.	Phase Displacement	Phasor Group	Symbol	Phasor Diagrams of Induced Voltages		Winding Connections and Relative Position of Terminals	
				hv	lv	hv	lv
3	–30°	3_1	*Dy*1	A_2, B_2, C_2, A, B, C	a_2, b_2, c_2, n	A_1, B_1, C_1, A_2, B_2, C_2	n, a_2, b_2, c_2, a_1, b_1, c_1
		3_2	*Yd*1	A_2, B_2, C_2, N	a_2, b_2, c_2, a, b, c	N, A_1, B_1, C_1, A_2, B_2, C_2	a_2, b_2, c_2, a_1, b_1, c_1
		3_3	*Yz*1	A_2, B_2, C_2, N	a_4, b_4, c_4, a, b, c, n	N, A_1, B_1, C_1, A_2, B_2, C_2	n, a_4, b_4, c_4, a_3, b_3, c_3, a_2, b_2, c_2, a_1, b_1, c_1
4	+ 30°	4_1	*Dy*11	A_2, B_2, C_2, A, B, C	a_2, b_2, c_2, n	A_1, B_1, C_1, A_2, B_2, C_2	n, a_2, b_2, c_2, a_1, b_1, c_1
		4_2	*Yd*11	A_2, B_2, C_2, N	a_2, b_2, c_2, a, b, c	N, A_1, B_1, C_1, A_2, B_2, C_2	a_2, b_2, c_2, a_1, b_1, c_1
		4_3	*Dz*11	A_2, B_2, C_2, N	a_4, b_4, c_4, a, b, c, n	N, A_1, B_1, C_1, A_2, B_2, C_2	n, a_4, b_4, c_4, a_3, b_3, c_3, a_2, b_2, c_2, a_1, b_1, c_1

The delta–star connection (4, *dy*11) is the most popular method of connecting transformers. Here, the lettering for phase is taken as *A*, *B* and *C* for primary, that is, the hv side, and *a*, *b* and *c* for the *lv* side or secondary.

2.12 RATING OF TRANSFORMERS

The broad specifications of transformers are given as follows:

- kVA rating.
- Rated voltage.
- Number of phases (i.e. one phase or three phases).
- Rated frequency.
- Type of connections for three phases.
- Tappings if it exists.
- Type of core.
- Power or distribution type.
- Ambient temperature (it is generally 40°C).
- Nature of cooling.
- Rise of temperature above ambient temperature.
- Voltage regulation.
- No-load current.
- Efficiency.

The standard ratings of distribution transformers are 16, 25, 40, 50, 63, 80 and 100 kVA, whereas the standard ratings of power transformers are 25, 40, 63, 100, 125, 160, 200, 250, 315, 400, 500, 630, 800, 1,000, 1,250, 1,600, 2,000, 2,500, 3,125, 4,000, 6,300, 8,000, 10,000, 12,500, 160,000, 20,000, 25,000, 31,500, 40,000, 50,000, 63,000 and 80,000 kVA.

ADDITIONAL SOLVED PROBLEMS

Example 2.6 A three-phase step-down transformer having turns ratio per phase of 10 takes 10 A when connected to 3.3 kV supply mains. Determine the secondary line voltage, line current and output when the transformer windings are connected in (i) star/delta and (ii) delta/star.

Solution

Turns ratio/phase (a) = 10

(i) Star/delta:

$$\text{Primary voltage per phase } (V_1) = \frac{3.3\times10^3}{\sqrt{3}} = 1{,}905.25\,\text{V}$$

$$\text{Secondary voltage per phase } (V_2) = \frac{V_1}{a} = \frac{1{,}905.25}{10} = 190.52\,\text{V}$$

Secondary line voltage = secondary phase voltage = 190.52 V

Primary phase current = 10 A

Secondary phase current = $a \times 10 = 10 \times 10 = 100$ A

Secondary line current = $\sqrt{3} \times 100 = 173.2\,\text{A}$

$$\text{Output in kVA} = \sqrt{3} \times 173.2 \times 190.52 \times 10^{-3} = 57.154 \text{ kVA}$$

(ii) Delta/star:

Primary phase voltage = primary line voltage = 3,300 V

$$\text{Secondary phase voltage} = \frac{3{,}300}{a} = \frac{3{,}300}{10} = 330\,\text{V}$$

$$\text{Secondary line voltage} = \sqrt{3} \times 330 = 571.57\,\text{V}$$

Primary line current = 10 A

$$\text{Primary phase current} = \frac{10}{\sqrt{3}} = 5.77\,\text{A}$$

Secondary line current = $a \times 5.77 = 10 \times 5.77 = 57.7$ A

Secondary phase current = secondary line current = 57.7 A

$$\text{Output in kVA} = \sqrt{3} \times 57.7 \times 571.57 \times 10^{-3} = 57.122 \text{ kV}$$

■

Example 2.7 A three-phase, 1,200 kVA, 6.6/1.1 kV transformer has delta-connected primary and star-connected secondary. The per phase values of the primary resistance and secondary reactance are 2 Ω and 0.03 Ω respectively. Calculate the efficiency on full load at 0.9 power factor lagging if iron loss is 20 kW.

Solution

Primary resistance/phase (R_1) = 2 Ω

Secondary resistance/phase (R_2) = 0.03 Ω

Iron loss = 20 kW

Primary phase voltage = primary line voltage = 6,600 V

Secondary line voltage = 1,100 V

$$\text{Secondary phase voltage} = \frac{1{,}100}{\sqrt{3}} = 635.08\ \Omega$$

$$\text{Turns ratio } (a) = \frac{6{,}600}{635.08} \approx 10.4$$

$$\therefore R_{02} = R_2 + \frac{R_1}{a^2} = 0.03 + \frac{2}{10.4^2} = 0.0485\ \Omega$$

$$\text{Secondary phase current } (I_{Ph2}) = \text{secondary line current} = \frac{1{,}200 \times 10^3}{\sqrt{3} \times 1{,}100} = 629.84\ \text{A}$$

$$\text{Total full-load Cu loss } (P_{\text{Cu}}) = 3I_{Ph2}^2 R_{02} = 3 \times 629.84^2 \times 0.0485 = 57.72\,\text{kW}$$

$$\text{Iron loss } (P_i) = 20 \text{ kW}$$

$$\text{Total loss} = P_{Cu} + P_i = 57.72 + 20 = 77.72 \text{ kW}$$

Output on full load at 0.9 power factor lagging $= 1{,}200 \times 0.9 = 1{,}080$ kW

$$\therefore \text{ Efficiency} = \frac{\text{output}}{\text{output + losses}} = \frac{1080}{1080 + 77.72} = 0.9328 \text{ p.u.} = 93.28\% \textbf{ Ans.}$$ ■

Example 2.8 A 1,500 kVA, 6.6/1.1 kV, three-phase delta–star-connected transformer has the following test result: SC test (instruments placed on the hv side) = 300 V, 131.21 A and 30 kW. Calculate its percentage resistance, percentage reactance drops, percentage regulation and percentage efficiency on full load at 0.8 power factor lagging. The iron loss during OC test is 25 kW.

Solution

From SC test data:

Primary voltage per phase = primary line voltage = 300 V

$$\text{Primary current/phase} = \frac{131.21}{\sqrt{3}} = 75.75 \text{ A}$$

$$\therefore Z_{01} = \frac{300}{75.75} = 3.96\ \Omega$$

$$\text{Now, } I_1^2 R_{01} = \frac{30 \times 10^3}{3} = 10{,}000 \text{ W}$$

$$\text{i.e., } R_{01} = \frac{10{,}000}{75.75^2} = 1.74\ \Omega$$

$$\therefore X_{01} = \sqrt{Z_{01}^2 - R_{01}^2} = \sqrt{3.96^2 - 1.74^2} = 3.56\ \Omega$$

$$\%R = \frac{I_1 R_{01}}{V_1} \times 100 = \frac{75.75 \times 1.74}{6{,}600} \times 100 = 1.99$$

$$\%X = \frac{I_1 X_{01}}{V_1} \times 100 = \frac{75.75 \times 3.56}{6{,}600} \times 100 = 4.08$$

The percentage voltage regulation on full load at 0.8 power factor lagging $= v_r \cos\beta + v_x \sin\beta = 1.99 \times 0.8 + 4.08 \times 0.6 = 4.04$

$$\text{Primary full-load current} = \frac{1{,}500 \times 10^3}{\sqrt{3} \times 6.6 \times 10^3} = 131.21 \text{ A}$$

Therefore, short circuit test has been carried out on full-load conditions.

Total losses = 30 + 25 = 55 kW

Full-load output = 1,500 × 0.8 = 1,200 kW

$$\therefore \text{Efficiency} = \frac{1,200}{1,200+55} = 0.9561\,\text{p.u.} = 95.61\%$$ ■

Example 2.9 A three-phase Δ–Δ bank consists of three 25 kVA, 3,300/300 V transformers and supplies a load of 50 kVA. After removing one transformer, determine the following for *V–V* connection:

(i) kVA load carried by each transformer.
(ii) per cent of rated load carried by each transformer.
(iii) total kVA rating.
(iv) ratio of *V–V* bank to Δ–Δ bank transformer ratings.
(v) percentage increase in load on each transformer when the bank is converted into *V–V* connection.

Solution

(i) kVA load supplied by each of the two transformers

$$\frac{\text{total kVA load in } V-V \text{ bank}}{\sqrt{3}} = \frac{50}{\sqrt{3}} = 28.86\,\text{kVA}$$

(ii) Per cent of rated load $= \dfrac{\text{kVA load/transformer}}{\text{kVA rating/transformer}} = \dfrac{28.86}{25} = 1.1544$, i.e. 115.44% carried by each transformer

(iii) kVA rating of the *V–V* bank = 2 × 25 × 0.866 = 43.3 kVA

(iv) $\dfrac{V-V \text{ rating}}{\Delta-\Delta \text{ rating}} = \dfrac{43.3}{75} = 0.577$ or 57.7%

∴ The rating is reduced to 57.7% of the original rating.

(v) Load supplied by each of Δ–Δ bank $= \dfrac{50}{3} = 16.67\,\text{kVA}$

∴ Percentage increase in load supplied by each transformer

$$= \frac{\text{kVA load/transfer in } V-V \text{ bank}}{\text{kVA load/transfer in } \Delta V - \Delta V \text{ bank}} = \frac{28.86}{16.67} = 1.7313, \text{ i.e., } 173.13\%$$ ■

SIGNIFICANT POINTS

Advantages of Three-phase Transformers

A single three-phase transformer bank has the following advantages:

- It occupies less space for the same rating, compared to a bank of three single-phase transformers.
- It weighs less.
- The cost is also less.
- Since only one unit is required to be handled, it is easy to operate.
- It can be transported very easily.

Comparison of Voltage and Current Relationships for Different Types of Connections

Type of Connection	Primary Side			Secondary Side			
	Line Voltage	Phase Voltage	Phase Current	Phase Voltage	Phase Current	Line Voltage	Line Current
Star–Star	V_L	$\frac{V_L}{\sqrt{3}}$	I_L	$\frac{V_L}{a\sqrt{3}}$	aI_L	$\frac{V_L}{a}$	aI_L
Delta–Delta	V_L	V_L	$\frac{I_L}{\sqrt{3}}$	$\frac{V_L}{a}$	$\frac{aI_L}{\sqrt{3}}$	$\frac{V_L}{a}$	aI_L
Star–Delta	V_L	$\frac{V_L}{\sqrt{3}}$	I_L	$\frac{V_L}{a\sqrt{3}}$	aI_L	$\frac{V_L}{a\sqrt{3}}$	$\sqrt{3}aI_L$
Delta–Star	V_L	V_L	$\frac{I_L}{\sqrt{3}}$	$\frac{V_L}{a}$	$\frac{aI_L}{\sqrt{3}}$	$\frac{\sqrt{3}V_L}{a}$	$\frac{aI_L}{\sqrt{3}}$

- The core is of a smaller size and hence less material is required.

$$\frac{V\text{–}V\text{ capacity}}{\Delta\text{–}\Delta\text{capacity}} = \frac{\sqrt{3}V_L I_P}{3V_L I_P} = \frac{1}{\sqrt{3}} = 0.577$$

The power supplied by the two transformers is as follows:

$$P_1 = \text{kVA}\cos(30^\circ - \beta) \quad \text{and} \quad P_2 = \text{kVA}\cos(30^\circ + \beta)$$

- $\beta = 0^\circ$
  The power factor of each transformer is 0.866.
- $\beta = 30^\circ$
  In this case the power factor of one transformer is cos (30° – 30°), i.e., 1, and the power factor of the other transformer is cos (30° + 30°), i.e., 0.5.
- $\beta = 60^\circ$
  In this case the power factor of one transformer is cos (30° – 60°), i.e., 0.866, and the power factor of the other transformer is cos (30° + 60°), i.e., 0.
  ∴ One transformer will not supply any load and the entire load is carried out by the other transformer.

For Scott connection:

$$\frac{\text{kVA actually utilized}}{\text{kVA available}} = \frac{1.732\,V_L I_L}{1.866\,V_L I_L} = 0.928$$

SHORT QUESTIONS AND ANSWERS

Q.1 State the different forms of connections used in three-phase transformers.

Ans. The following are the possible connections:
(i) star–star, (ii) star–delta,
(iii) delta–delta, (iv) delta–star and
(v) open-delta

Q.2 Where is the delta–delta connection applied?

Ans. It is applied in large low-voltage transformers where insulation problem is less urgent.

Q.3 Where is the star–star connection applied?

Ans. It is best suited for small high-voltage transformers with balanced load.

Q.4 Where is the star–delta connection applied?

Ans. It is applied at the sending end of transmission lines to step up the voltage.

Q.5 Where is the delta–star connection applied?

Ans. It is used as a step-down transformer to provide three-phase four-wire supply.

Q.6 Where is the open-delta connection applied?

Ans. When two three-phase loads are too small or one transformer in Δ–Δ bank is disabled, then the open-delta connection is applied.

Q.7 Where is the Scott connection applied?

Ans. It is applied to supply three-phase voltage from a two-phase source or vice versa.

Q.8 What do you mean by phasing out of a three-phase transformer?

Ans. Phasing out detects the same phase terminals of the primary and secondary.

SUPPLEMENTARY PROBLEMS

1. A three-phase transformer, 33/6.6 kV, $\Delta - Y$, 2 MVA, has a primary resistance of 8 Ω per phase and a secondary resistance of 0.08 Ω per phase. The percentage impedance is 7 per cent. Find the secondary voltage with rated primary voltage and hence the regulation for full load at 0.75 power factor lagging conditions.

   [*Ans.* 5.25%]

2. A three-phase transformer rated at 1,000 kVA, 11/3.3 kV, has its primary windings star-connected and its secondary windings delta-connected. The actual resistances per phase of these windings are as follows: $R_1 = 0.375\ \Omega$, $R_2 = 0.095\ \Omega$, $X_1 = 9.5\ \Omega$ and $X_2 = 2\ \Omega$ per phase. Calculate the voltage at normal frequency which must be applied to primary terminals in order to obtain full-load current in the windings when the secondary terminals are short circuited. Also, calculate the power input under these conditions.

   [*Ans.* 1.54 kV, 6 kW]

3. The maximum efficiency of a three-phase, 11,000/400 V, 500 kVA transformer is 98.8 per cent and occurs at 80 per cent of full load at unity power factor. Its percentage impedance is 4.5 per cent. Load power factor is now varied, while the load current and the supply voltage are held constant at their rated values. Determine the load power factor at which the secondary terminal voltage is minimum.

   [*Ans.* 0.1687 (lagging)]

4. A 2000 kVA, 6,600/400 V, three-phase is delta connected on the hv side and store connected on the lv side. Determine its percentage reactance and percentage reactance drops, percentage efficiency and percentage regulation at full-load 0.8 power factor leading.The following data are given:

SC test: 400 V, 175 A, 17 kW

OC test: 400 V, 150 A, 15 kW

[*Ans.* 0.85%, 6%, 98.04%, –2.92%]

5. Two transformers connected in open delta supply a 400 kVA balanced load operating at 0.866 power factor lagging. The load voltage is 440 V. What is the (i) kVA supplied and (ii) kW supplied by each transformer?

[*Ans.* (i) 231 kVA, (ii) 231 kW, 115.5 kW]

6. The primary and secondary windings of two transformers, each rated 250 kVA, 11/2 kV and 50 Hz, are connected in open delta. Find (i) the kVA load that can be supplied from this connection and (ii) currents on the hv side if a delta-connected three-phase load of 250 kVA, 0.8 power factor lagging, 2 kV, is connected to the *lv* side of the connection.

[*Ans.* (i) 433 kVA, (ii) 13 A]

7. Two *T*-connected transformers are used to supply a balanced load of 100 kVA at 400 V from a balanced 11 kV three-phase supply. Determine (i) current and voltage rating of each transformer coil and (ii) kVA rating of the main and teaser transformers.

[*Ans.* (i) 11,000 V, 9,526 V, 5.25 A, 400 V, 346.4 V, 144.34 A (ii) 57.75 kVA, 50 kVA]

8. A balanced three-phase 100 kW load at 400 V and 0.8 power factor lagging is to be obtained from a balanced two-phase 1,100 V lines. Determine the kVA rating of each unit of the Scott-connected transformer.

[*Ans.* 72.172 kVA, 62.5 kVA]

9. It is desired to transform 2,400 V, 500 kVA three-phase power to two-phase power at 600 V by Scott-connected transformers. Determine the voltage and current ratings of both primary and secondary of each transformer. Neglect the transformer's no-load currents.

[*Ans.* Teaser: 2,080 V, 1,200 A (primary); 600 V, 4,167 A (secondary);
Main: 2,400 V, 1,200 A (primary); 600 V, 4,167 A (secondary)]

MULTIPLE-CHOICE QUESTIONS AND ANSWERS

1. A three-phase transformer over a bank of three single-phase transformers of equal rating has the advantages of

(a) low cost (b) less weight

(c) occupation of less space (d) all of the above

2. Two three-limb, three-phase delta–star-connected transformers are supplied from the same source. One of the transformers is *Dy*1 and the other is *Dy*11 connection. The phase difference of the corresponding phase voltage of the secondaries would be

(a) 0° (b) 30°

(c) 60° (d) 120°

3. Three single-phase transformers are connected to form a three-phase transformer bank. The transformers are connected as shown in Figure 1. The nomenclature of the above connection for phase sequence ABC is

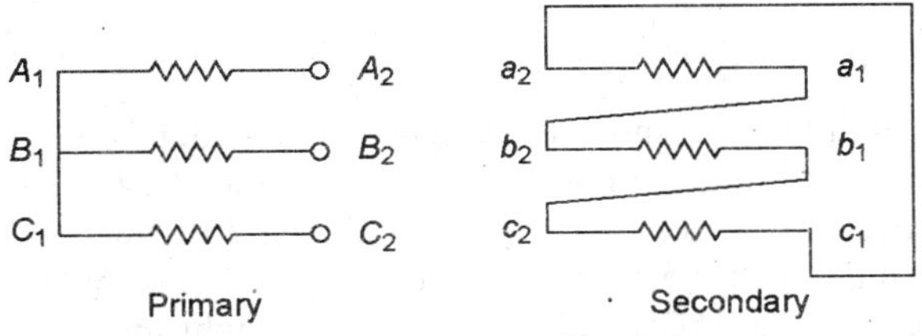

Figure 1

(a) $Yd0$ (b) $Yd6$

(c) $Yd11$ (d) $Yd1$

4. Which of the following connections of the transformer will give the highest secondary voltage?

(a) delta primary, delta secondary (b) delta primary, star secondary

(c) star primary, star secondary (d) star primary, delta secondary

5. A three-phase transformer has its primary connected in delta and secondary in star. Secondary to primary turn ratio per phase is 5. For a primary voltage of 400 V, the secondary voltage would be

(a) 2,000 V (b) 80 V

(c) 3,464 V (d) $80\sqrt{3}$ V

6. In a properly connected $Y - \Delta$ transformer, a voltmeter connected is in the secondary (opening a node of delta); the voltmeter will read

(a) zero voltage (b) line voltage of supply

(c) phase voltage in the secondary (d) none of the above

7. In a Δ–Δ connection, if one transformer becomes disabled, the capacity will reduce to

(a) 66.67% (b) 50%

(c) 40% (d) 57.74%

Answers

1 (d) 2 (c) 3 (d) 4 (d) 5 (c) 6 (a) 7 (d)

3 Basic Concepts of Rotating Machines

Transformers have already been introduced. The transformer is a static device having primary and secondary windings. In rotating machines, there are two parts: the stator and the rotor. Rotating electrical machines are also of two types: DC and AC machines. Electrical machines are widely used. In DC machines the stator is used as a field and the rotor is used as an armature, while reverse is the case for AC machines, that is, synchronous generators and synchronous motors. The induction motor is another kind of AC machine, which is singly excited; that is, AC supply voltage is only given to the stator and no supply is given to the rotor. In DC machines and synchronous machines, the field is always excited by DC supply. DC machines and AC machines have their own advantages and disadvantages. The classification of rotating machines is shown in Figure 3.1.

The aim of this chapter is to introduce the basic concepts of rotating machines.

3.1 ELECTROMAGNETIC TORQUE

Figure 3.2(a) shows a two-pole cylindrical rotor with one conductor. Only the stator magnetic flux path due to excitation of field coils has been shown in Figure 3.2(a), and there is no current in the rotor conductor. Figure 3.2(b) shows the distribution of flux due to current only in the rotor conductor, and there is no excitation in stator field coils. Figure 3.2(c) shows the resultant flux distribution when both field coils and rotor conductors are excited. The rotor conductor experiences a force in the upward direction and torque is produced in the clockwise direction due to interaction between stator and rotor magnetic fields. This torque is known as *electromagnetic torque.*

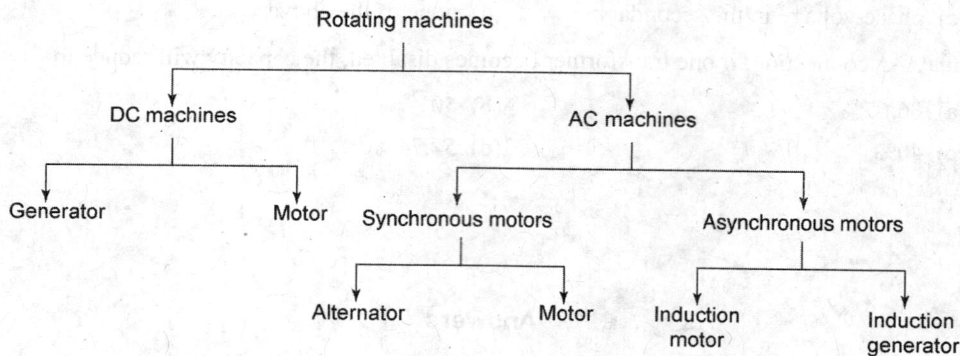

Figure 3.1 Classification of Rotating Machines

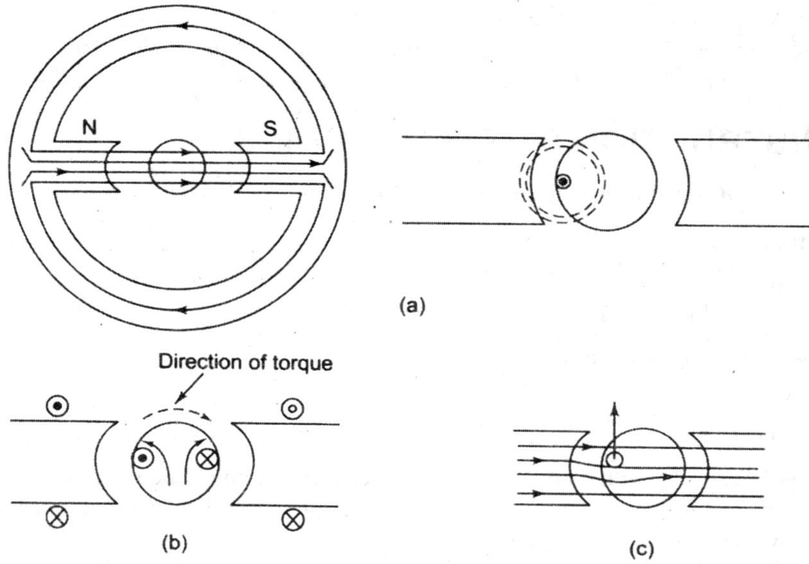

Figure 3.2 Electromagnetic Torque

3.2 RELUCTANCE TORQUE

Figure 3.2(a) shows the magnetic flux path due to current in stator coils. Figure 3.3 shows the stator-produced flux when a ferromagnetic material is placed in this magnetic field. The rotor will immediately experience a counterclockwise torque because the magnetic flux has a tendency to follow a minimum reluctance path. This torque is called *reluctance torque.* The reluctance torque will be zero when the rotor axis coincides with the stator axis.

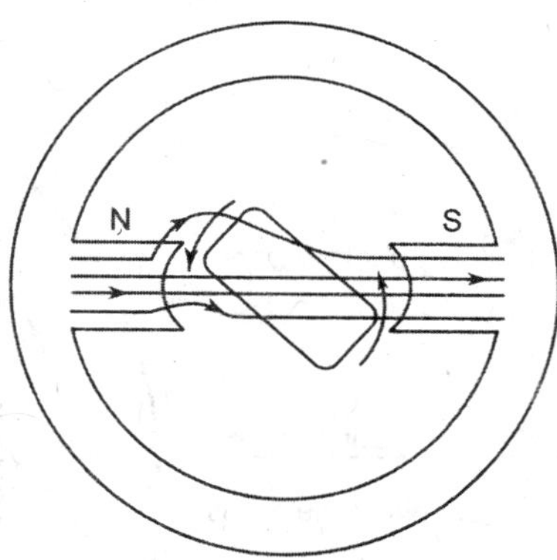

Figure 3.3 Reluctance Torque

3.3 CONSTRUCTIONAL FEATURES OF ROTATING ELECTRICAL MACHINES

Any rotating electrical machine has (i) stator, (ii) rotor, (iii) air gap separating the stator and rotor and (iv) shaft, bearing, foundation and so on, including excitation or field winding. The current in the excitation or field winding does not vary with the load on the machine. This current is also known as *magnetizing current*, *exciting current* or *field current.* The exciting or field winding produces the working flux. The current in the armature winding varies with load. In armature winding emf is induced by the working flux. Since the armature winding handles all the power, the rating of the armature winding is equal to the power rating of the machine. The power rating of the field winding is 0.5–2 per cent of the rated power of the machine. After establishing the field current, the power input to DC field winding is dissipated as I^2R loss. The field winding for AC and DC machines must be energized from DC source.

In both DC and AC machines, the armature windings deal only with alternating current. Therefore, all rotating electrical machines have laminated armature structure to reduce eddy current loss. To reduce the hysteresis loss, the core material is selected in such a way that it has small hysteresis loop.

3.4 CONSTRUCTION OF DC MACHINES

The construction of DC machines is discussed here. An actual generator consists of the following parts:

- magnetic frame or yoke,
- pole cores and pole shoes,
- armature core,
- armature windings or conductors,
- commutator and
- brushes and bearings.

The yoke, pole cores, armature core and air gaps between the poles and the armature core form the magnetic field. The rest form the electrical circuit. Figure 3.4 shows the construction of a DC machine in which the above parts have been depicted.

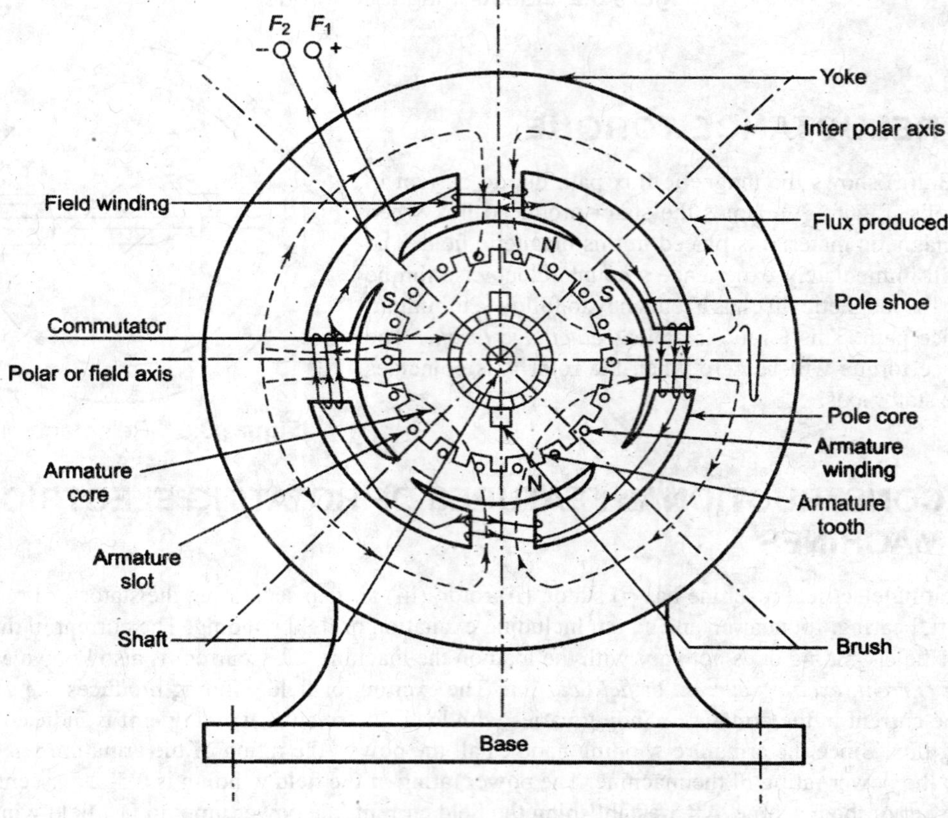

Figure 3.4 Construction of a DC Machine

3.4.1 Magnetic Frame or Yoke

The magnetic frame or yoke gives mechanical support for poles as well as protects the whole machine as a protecting cover. It also carries the magnetic flux produced by the poles. In small generators yokes are made of cast iron, whereas for large machines cast steel is used. The yoke carries 50 per cent of total flux per pole.

3.4.2 Pole Cores and Pole Shoes

The field magnet has two parts: pole cores and pole shoes. The pole shoes spread out the flux in the air gap and reduce the reluctance of the magnetic path due to its large cross-section. The pole shoes support the exciting coils. The pole shoes are made curved to ensure uniform air gap around the armature core. The following are the functions of pole shoes:

(i) to support the field coils,
(ii) to reduce the reluctance of the magnetic path and
(iii) to achieve uniform flux distribution around the air gap.

The pole core may be a solid piece and is made out of either cast iron or cast steel, whereas the pole shoe is laminated and fastened to the pole face by countersunk screws, shown in Figure 3.5.

The field coil is rigidly put in position between the frame attached to the yoke and the pole shoe. In modern design, the pole cores and pole shoes are built of thin laminations of annealed steel, where the thickness of lamination varies from 1 mm to 0.25 mm. To secure the laminated poles to yoke, any of the following ways can be used:

(i) The holding screws may be bolted into the steel bar, and it passes through the pole across the plane of laminations.
(ii) The pole is secured to the yoke by means of screws bolted through the yoke and into the pole body.

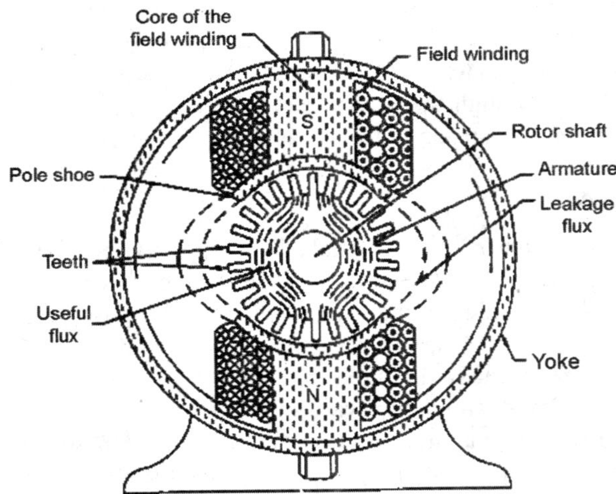

Figure 3.5 Pole Cores and Pole Shoes and Distribution of Magnetic Flux for a Two-pole DC Machine

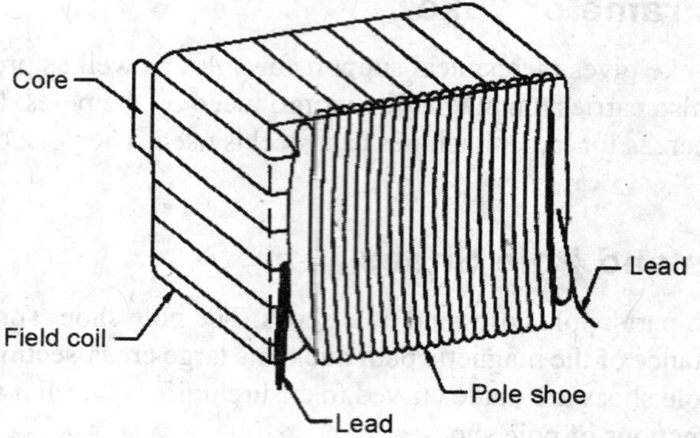

Figure 3.6 Field Coils of a DC Machine

If the air gap is increased, the reluctance will be increased. This causes the machine to draw the more current in case of motoring mode and to give less output power in case of generating mode. The actual flux distribution in the air gap depends on the following factors:

(i) size of air gap,
(ii) shape of pole shoes and
(iii) clearance between the tips of the adjacent pole shoe.

3.4.3 Pole Coils

The field coils or pole coils consist of high-grade copper or aluminium enamelled wire or strip. These are former wound for the correct dimension. After getting the correct dimension, the former is removed. Now, the wound coil is placed over the core. Figure 3.6 shows the field coils of a DC machine. From Figures 3.5 and 3.6, it can be seen that the field coils are wound in such a way that alternate north and south poles are produced to ensure the correct polarity of the induced emf. To obtain the necessary flux, the current is passed through these coils and the field windings form an electromagnet. The field windings receive current either from an external DC source or from the armature when they connected across it. Due to the flow of current through the field windings, a magnetic flux is established in the yoke, through the pole pieces, in the air gap and in the armature gap, shown in Figure 3.5.

3.4.4 Armature Core

Armature conductors or coils consist of super-enamelled copper wire or strip placed on the armature core, which is cylindrical or drum shaped. Figure 3.7(a) shows the constructional details of an armature, which is mounted on the shaft rotating in the magnetic field. During rotation it cuts the magnetic flux of the field magnets. It also provides a path of very low reluctance to the flux from the north pole to the south pole. The armature core is usually built of circular sheet steel discs or laminations to reduce the eddy current loss. The thickness of laminations varies from 0.3 mm to 1.0 mm. Figure 3.7(b) shows the laminated sheet steel used for the armature core. Perforations exist in these laminations to provide axial flow of air through the armature for cooling purposes and also to reduce the weight. For small machines, these laminations are directly fixed to the shaft. On the other hand, these laminations are assembled on a spider that is fixed to the shaft. The laminations are punched with the slots that house the armature

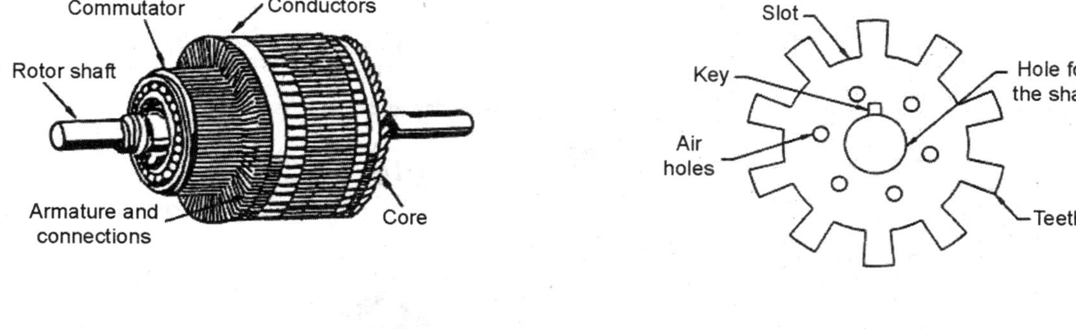

(a) Armature

(b) Lamination of armature

Figure 3.7 Armature of a DC Machine

conductors. The armature core is keyed to the shaft. The sheets of insulations are provided in the slots between the core and the conductors.

3.4.5 Armature Windings

The armature windings, which are the first wound in the form of flat rectangular coils, are usually former wound. Finally, the armature coils are pulled into their appropriate shape in a coil puller. The various conductors of the coils are insulated from each other. The armature conductors are placed in armature slots. The armature conductors are held in the slots by wedges made of fibre, driven into the tops of the slots to avoid the flying of conductors under the centrifugal forces. Wood is not used to make the armature core because of its high reluctance and non-capability of bearing high rotational torque.

3.4.6 Commutator

A commutator is a mechanical device. It facilitates the collection of current from the armature conductors. Figure 3.8 shows the commutator made of a number of wedge-shaped segments or bars made of copper, which are assembled to form a cylinder. These segments are insulated from each other by thin layers of mica, and each commutator segment is connected to one end of coil. This insulation is designed to withstand voltage not less than 15 V. The number of commutator segments is equal to the number of coils. The commutator along with the brush gear helps to convert the alternating current induced in armature conductors into unidirectional current across the brush.

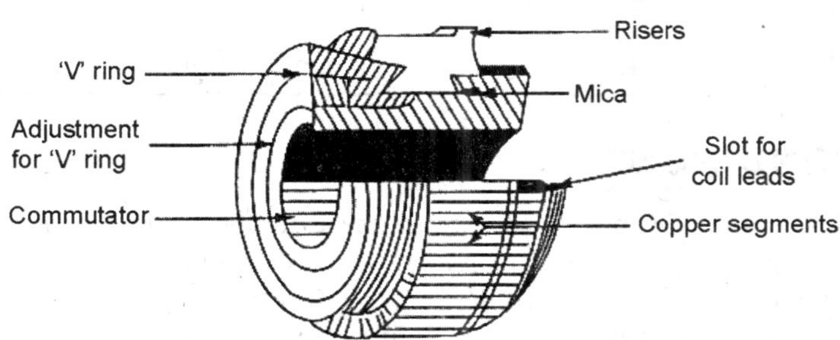

Figure 3.8 Commutator of a DC Machine

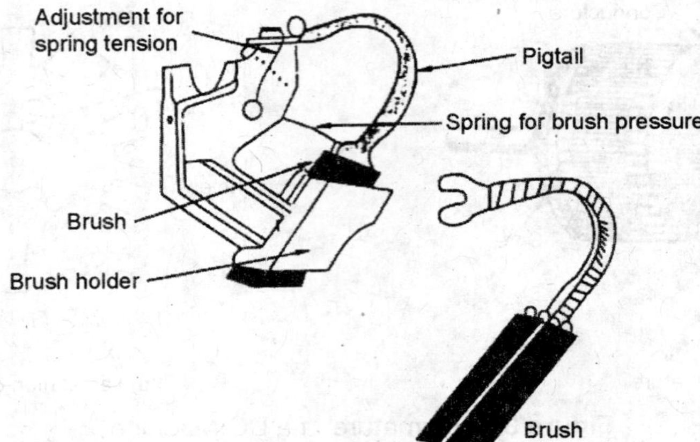

Figure 3.9 · Pigtail Brush

3.4.7 Brushes and Bearings

The function of the brushes, which are housed in a brush holder, is to collect current from the commutator and to supply the same to the external circuit. The brushes are placed in such a way that they can rest on the commutator by placing them in the brush holder against the action of a spring. The tension of the spring can be adjusted. Figure 3.9 shows the pigtail brush with its holder.

The brushes are usually made of carbon or graphite and are in the shape of a rectangular block. The ball bearings are frequently employed for their reliability. Roller bearings are preferable for heavy duties. Table 3.1 summarizes the essential parts of a DC machine.

Table 3.1 Essential Parts of a DC Machine

Part	Material Used
Yoke	Cast iron (small machine), cast steel (small large)
Field system	High-grade copper or aluminium enamelled wire or strip
Armature, core	Laminated sheet steel (0.4–0.6 mm)
Armature conductor	Super-enamelled copper wire or strip
Commutator	High-grade copper
Brush	Copper/carbon

3.5 RING WINDINGS

These are not used in modern DC machines because these are the earliest forms of armature windings and these windings are very useful to understand the action of the commutator windings. Figure 3.10 shows the ring winding of a primitive DC generator where the core is usually made of steel laminations and insulated from each other. Figure 3.10 shows that the core has eight coils and each coil consists of two turns. S_i and F_i represent the start and end of the coil 'i'. The finish of the

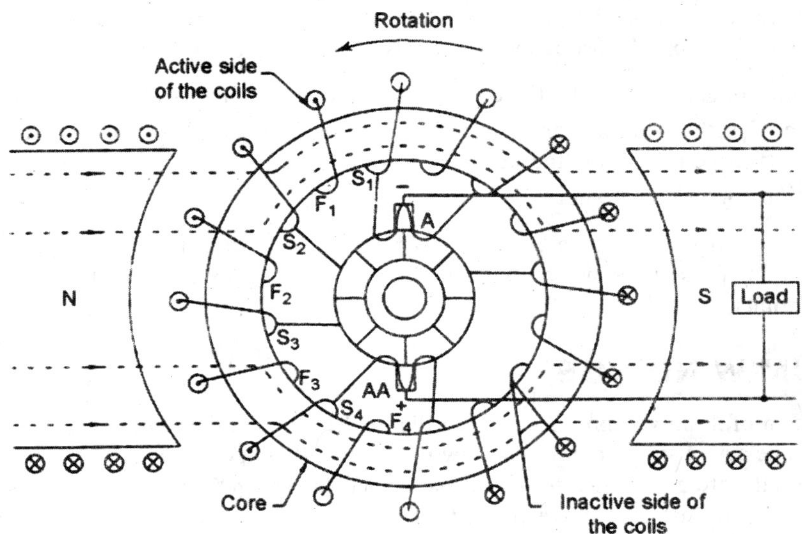

Figure 3.10 Ring Winding

coil 'i' is connected to the start of the coil 'i + 1' and so on until the finish of the last coil is connected to the start of the first coil. The commutator segment is connected to the junction between the neighbouring coils, that is, to the finish of one coil and the start of the another coil. The dotted lines in Figure 3.10 are the flux lines. An electromagnetic field is set up by the electromagnet. For the anticlockwise direction of rotation of the armature, the emf induced in the coils under the north poles will be directed away from the papers and that of in the coils under the south poles will be directed into the papers. Hence, the emfs of the coils under the north pole and the south pole are indicated by crosses and dots.

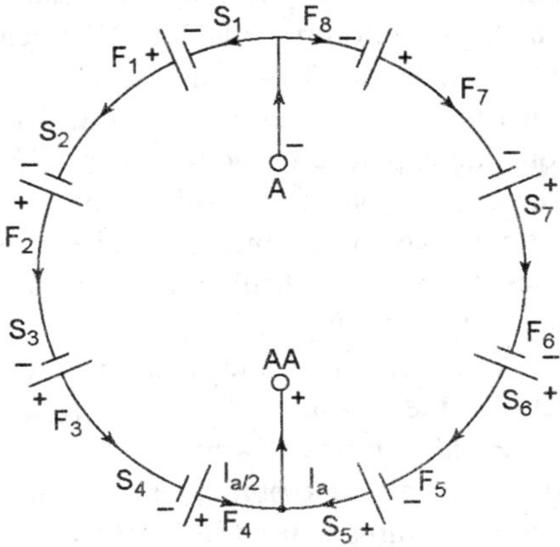

Figure 3.11 Equivalent Circuit of Figure 3.10

The equivalent circuit of Figure 3.10 is shown in Figure 3.11. The total emf induced around the closed loop will be zero.

But for the brush position shown in Figure 3.11, there will be two circuits. These two circuits are parallel between the brushes where each circuit contains equal number of coils generating emf acting from A to AA. Therefore, the current in each circuit will be half of the total current. The emf between the brushes A and AA depends on the emf/conductor as well as the number of conductors in each series circuit. If the brushes are placed such that one half of the coil comes under the north pole and the remaining half comes under the south pole, the maximum emf will be across the brushes. To achieve this, the brushes are to be placed one pole pitch apart, that is, 180 electrical degrees. The number of conductors generating emf should not vary due to the rotation of the armature, and in this case the voltage between the brushes A and AA will remain constant. The number of parallel paths in ring windings is equal to the number of poles, and hence two parallel paths exist in Figure 3.11. For six poles, the number of parallel paths will be six. The following are the disadvantages of ring windings:

(i) Each turn has to be taken around the core by the hand, and hence it is very expensive.
(ii) Less than half of the windings are useful for generating the emf because a small portion of the winding is effective for cutting the flux.
(iii) Commutation conditions are not good.

These windings are obsolete today, but they are very useful for better understanding of the working of the commutation. These windings are inferior to drum windings in respect of both mechanical and constructional points of view.

3.6 DRUM WINDINGS

The simplest form of drum winding is shown in Figure 3.12, where the armature carries single coil of many turns and the active sides are housed in two diametrically opposite slots.

For the normal slotted armature, the coil ends are placed in such a way to embrace the teeth, shown in Figure 3.13. The coil has two active coil sides. The active coil sides only cut the flux, whereas the inactive coil sides do not cut the flux. The coils in drum windings are placed in the slots of the armature on outer periphery.

To facilitate the satisfactory end connections, double-layer windings are used in DC windings. Multiturn coils of a DC machine are shown in Figure 3.14, where each single turn consists of two sides of the coil. The upper coil side is placed in the top layer of one slot, whereas the other coil side is placed in the bottom layer of another slot. These two coil sides are usually displaced by about a pole pitch, shown in Figure 3.15.

The coil span indicates the separation between the two sides of the same coil, and it is expressed in terms of armature slots. The coil span is usually one pole pitch. In practice, the number of turns is used to form a coil without

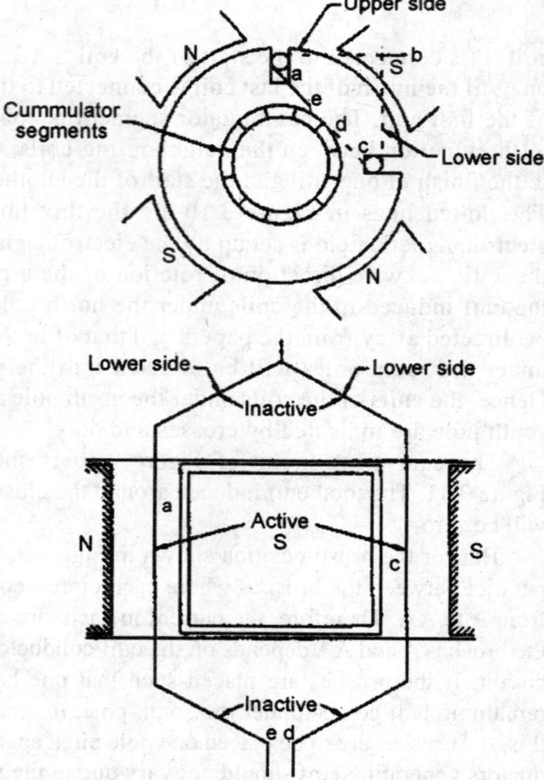

Figure 3.13 Drum-type Armature Coil

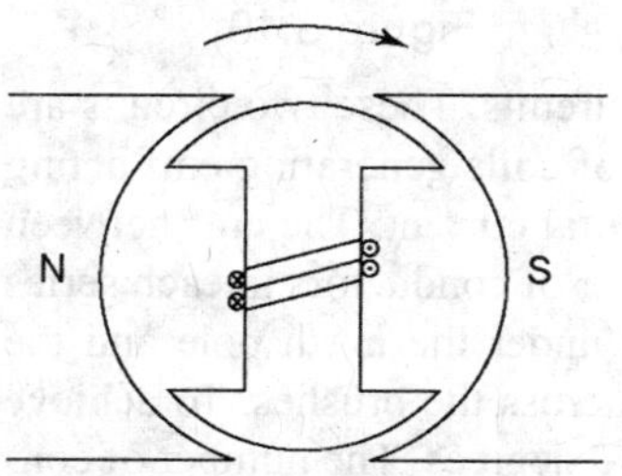

Figure 3.12 Slotted Armature with Coils

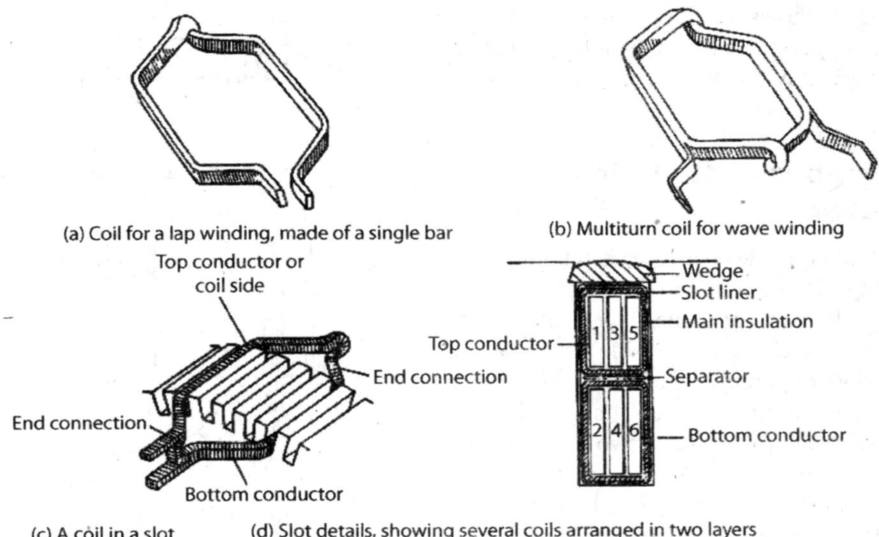

Figure 3.14 Armature Coils

joints. The change from the upper coil side to the lower coil side is done by means of bend, that is, kink, which is introduced at the back end of the coil shown in Figure 3.16.

Figure 3.17 shows that in double-layer windings, each slot contains the upper side/sides of one or more coils along with the lower side/sides of a corresponding number of other coil side/sides of a coil separated by a coil. Figure 3.18 shows two-turn coil used in the drum winding.

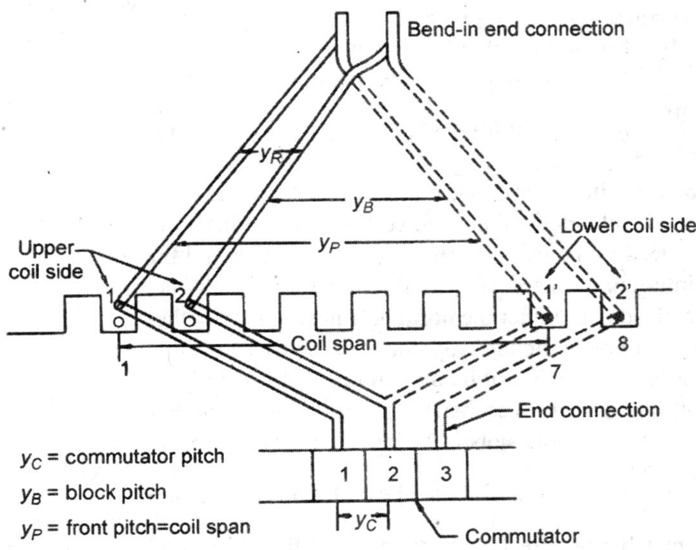

Figure 3.15 Two-layer Windings

Figure 3.17 shows the upper coil sides (a, b) of a coil in the upper portion of the slot. The bottom layer is occupied by the lower coil sides (8′, 9′) of some other coils (8, 9).

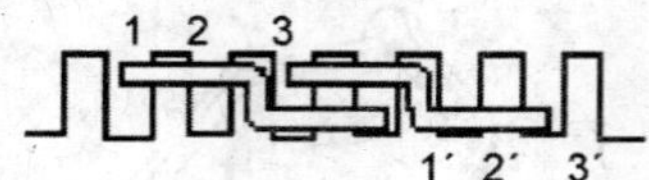

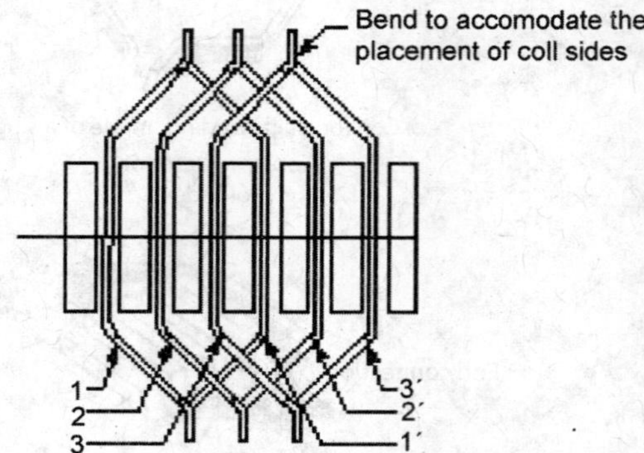

Figure 3.16 End Connections

3.6.1 Number of Coil Sides per Layer

The number of coil sides per layer is totally dependent on the number of coils and the number of slots. In each slot, there are an upper layer and a lower layer. If each slot contains one side per layer, the number of slots required is equal to the number of coils. Due to the limited number of available slots, it is not always possible to have two coil sides per slot. Hence, as many as four coil sides per layer become necessary. The arrangement of two coil sides per layer has been shown in Figure 3.16, where the upper coil sides of coils 1, 2 and 3 are denoted by 1, 2 and 3, respectively, and the lower coil sides by 1′, 2′ and 3′, respectively.

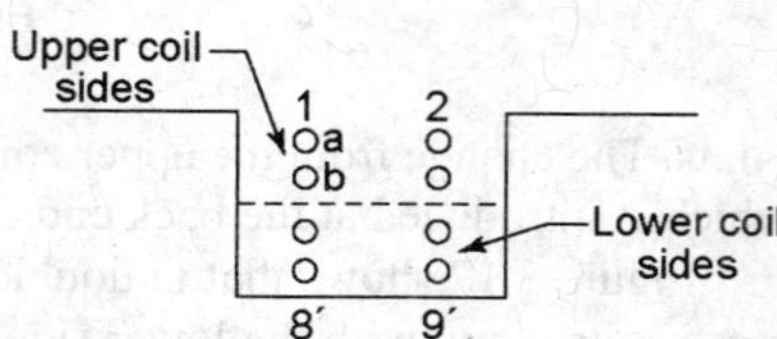

Figure 3.17 Arrangement of Conductor in a Slot with Two-coil Sides per Layer of a Two-turn Coil

3.6.2 Coil Span

The arrangement of coil sides in the slots has been shown in Figures 3.15 and 3.16. The separation of the upper coil side and the lower coil side of the same coil is expressed in terms of coil span, which is equal to the pole pitch for a full-pitched coil. Pole pitch is defined by the number of slots per pole. If the number of slots is 20 for a four-pole machine, the number of slots per pole $= \frac{20}{4} = 5$. For a full-pitched coil, the coil span in this case becomes 5. For a full-pitched coil, one coil side lies under the north pole, whereas the other coil side of the same coil lies under the south pole, and the emf induced in the two coil sides will be additive and the emf induced is maximum in this case. If the coil span becomes less than one pole pitch, one coil side lies under the north pole (say) and the other coil side of the same coil lies in the midway between the two coils and the emf induced between them will not be maximum. To provide good commutation, the coil span is made less than the pole pitch by a slot. For example, if the number of available slots is 23 for a four-pole machine, the pole pitch becomes $\frac{23}{4} = 5.75$. In this case, the coil span either 5 or 6 may be employed, which has no effect on the type of winding. If the coil span is not an integer, the integer value is considered.

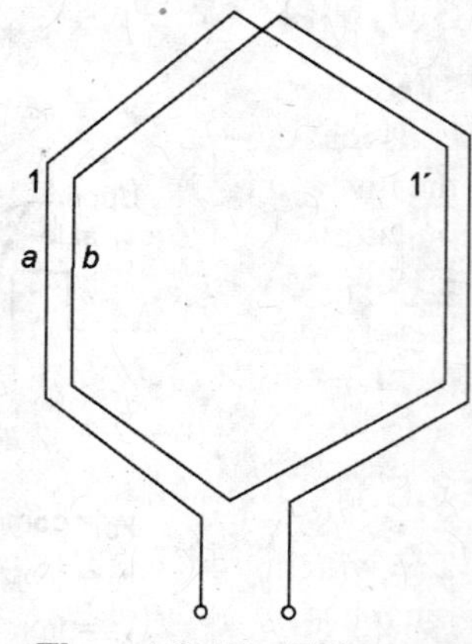

Figure 3.18 Two-turn Coil Winding

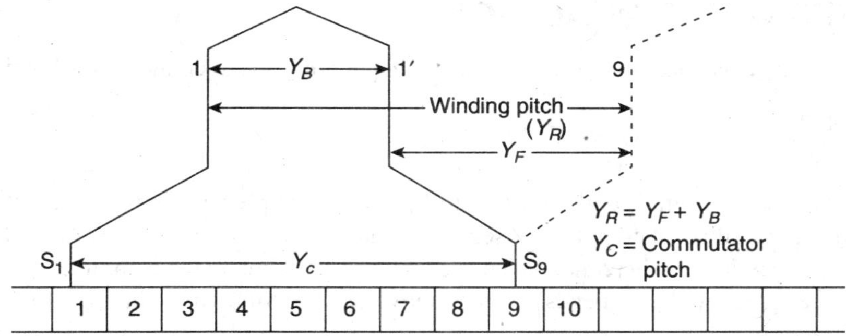

Figure 3.19 Winding Pitch, Commutator Pitch and Coil Span of a Coil Winding

3.6.3 Winding Pitch

In Figure 3.19, the finish of one coil is connected to the start of the next coil which is used for large machines. For small machines, Figure 3.19 shows that the finish of the first coil is connected to the start of the ninth coil, that is, two pole pitches apart. The displacement of the two consecutive coils is connected in series termed as pole pitch. In Figure 3.19, the winding pitch is equal to 8.

3.6.4 Commutator Pitch

The coil ends of a coil are connected to the commutator segments, and the displacement between the two commutator segments to which the two ends of a coil are joined is known as commutator pitch (Y_c). The coil ends shown in Figure 3.19 are connected to the commutator segments 1 and 9. Therefore, the commutator pitch is $9 - 1 = 8$. For the simple ring and lap winding, the commutator pitch is 1. The number of insulated pieces between the two ends of a coil is known as commutator pitch because consecutive commutator segments are insulated from each other. For lap and ring windings, the commutator pitch is unity.

3.6.5 Numbering of Armature Conductors

There are various methods to number the armature conductors. The easiest method is to number the coils instead to number the coil sides. From Figure 3.19, the upper coil side of coil 1 is denoted as 1 and the lower coil side of coil 1 is denoted as 1′. The junction of two different coil sides is joined to the commutator segments. The number of commutator segments is equal to the number of coils. From Figure 3.19, it is clear that the number of commutator pitch and winding pitch is equal. Therefore, in practice, it is enough to deal with the commutator pitch. The numbering of multiturn coil is shown in Figure 3.18, but the individual turn has not been shown.

3.6.6 Difference Between Coil Span and Winding Pitch

The winding pitch or commutator pitch has a vital role in deciding the type of winding, but the coil pitch has no influence on the type of winding. There will not be any alteration on the performance of the machine for any change in coil span, whereas a slight change in the winding pitch can produce a considerable effect on the performance of the machine.

3.7 TYPES OF DC WINDINGS

The following are the classifications of DC windings depending on the connection of their coil ends:

(i) lap winding and
(ii) wave winding.

Figure 3.20(a) shows a simple coil. Figure 3.20(b) shows a simple lap winding where the two ends of the coil are taken to the adjacent commutator segments and the coil ends are bent inwards. Figure 3.20(c) shows a simple wave winding where the two ends of the coil are taken to the commutator segments separated by a distance apart and the coil ends are bent outwards. The following definitions are very useful:

- *Pole pitch:* It is defined as the number of armature conductors per pole. If there are 60 conductors and 6 poles, the pole pitch is 60/6 = 10.
- *Conductor:* It is defined as the length of a wire lying in the magnetic field, shown in Figure 3.20(a).
- *Coil:* Two conductors with their end connections, shown in Figure 3.20(a), form one coil. The coil may be single-turn coil or multiturn coil. The end connections are called overhang.
- *Coil pitch (Y_s):* It is defined as the distance between two sides of a coil in terms of armature slots. In other words, it is defined as the periphery of the armature between two sides of a coil. If the coil pitch is equal to the pole pitch, the winding is said to be full pitched. Otherwise, it is a fractional pitch.
- *Front pitch:* It is defined as the distance between the second conductor of one coil and the first conductor of the next coil, shown in Figure 3.20(b).
- *Back pitch:* It is defined as the distance measured in terms of armature conductors that a coil advances on the back of the armature, shown in Fig. 3.20(b).
- *Resultant pitch (Y_R):* It is defined as the distance between the beginning of one coil and the beginning of the next coil. For the lap winding, $Y_R = Y_B - Y_F$, and for wave winding, $Y_R = Y_F + Y_B$.
- *Coil pitch (Y_c):* It is defined as the distance between the beginning of one coil and the beginning of next coil, shown in Figures 3.20(b) and 3.20(c) for lap and wave windings, respectively.

3.7.1 Simple Lap Winding

The lap winding as well as the wave winding is most commonly used for drum-type armatures. To get these types windings, the following rules are commonly used:

- The windings must be full pitched; that is, the front and back pitch must be approximately equal to the pole pitch.
- To place the coils properly on the armatures, the front pitch as well as the back pitch must be odd.
- The commutator segments are the images of the coil; that is, the number of commutator segments is equal to the number of the coils.
- The windings must close upon itself.

  The lap winding has been shown in Figure 3.20(a). The following points are important for the simple lap winding:
- The back pitch (Y_B) and the front pitch (Y_F) are odd and opposite signs. They differ by two or some multiples thereof.
- Y_B and Y_F must be nearly equal to the pole pitch.

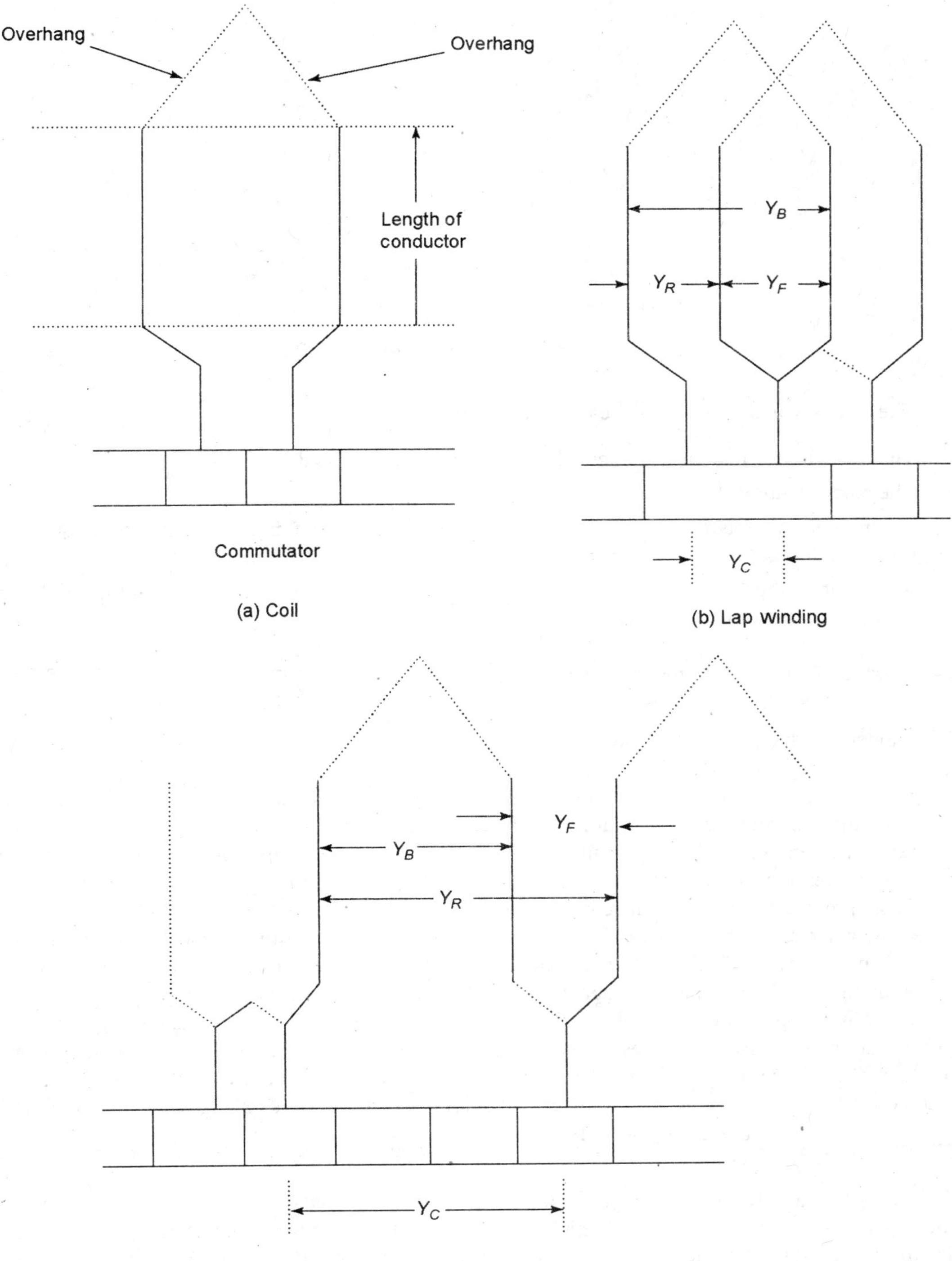

Figure 3.20 Winding

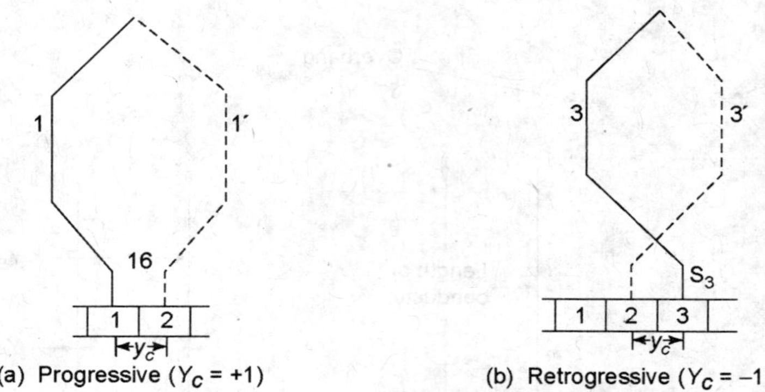

(a) Progressive (Y_C = +1)
(b) Retrogressive (Y_C = −1)

Figure 3.21 Simple Lap Winding

- The average pitch $\frac{Y_F + Y_B}{2}$ is equal to the pole pitch.
- The resultant pitch is the arithmetical difference between Y_B and Y_F.
- The commutator pitch $Y_C = \pm 1$.
- The number of slots for a two-layer winding as well as the commutator segments is equal to the number of coils.
- The number of parallel paths in the armature = mP, where m is the multiplicity of the winding and P is the number of parallel paths.

For the progressive or right-handed winding, $Y_F = \frac{Z}{P} - 1$ and $Y_B = \frac{Z}{P} + 1$; and for the retrogressive or left- handed winding, $Y_F = \frac{Z}{P} + 1$ and $Y_B = \frac{Z}{P} - 1$.

Similarity between the lap winding and the ring winding are there except the way they are placed in the armature. In ring winding, the finish of the coil is connected to the start of the next coil and so on, and hence the winding pitch or commutator pitch is always unity. A simple lap winding can be progressive or retrogressive, shown in Figures 3.21(a) and 3.21(b), respectively.

In the progressive winding, the coil ends are connected to the commutator segments in the ascending order, whereas in the retrogressive winding, these are connected to the commutator segments in the descending order. Therefore, for the simple lap winding, $Y_c = \pm 1$, where '+' is for the progressive winding and '−' is for the retrogressive winding. Figure 3.22 shows the developed view of the simple lap winding having four poles and 16 slots, with two coil sides per slot using single-turn coils only. Since there are two coil sides per slot, the number of coils required is 16. The winding is drawn in a progressive way (Y_c = +1), and the coil assumed here is full pitched. The upper coil sides are denoted by 1, 2, 3, etc., whereas the lower coil sides are denoted by 1′, 2′, 3′, etc. The pole pitch for this arrangement = $\frac{\text{number of slot}}{\text{number of poles}} = \frac{16}{4} = 4$. The coil span taken here is four slots/pole, that is, full pitched. The solid line shows the upper coil sides, whereas the dotted lines show the lower coil sides. The upper coil side of coil 1, that is, 1 (thick line), is connected to commutator segment 1. The lower coil side of coil 1, that is, 1′ (dotted line), will be placed in 1 + 4, that is, 5th slot. The front portion of coil side

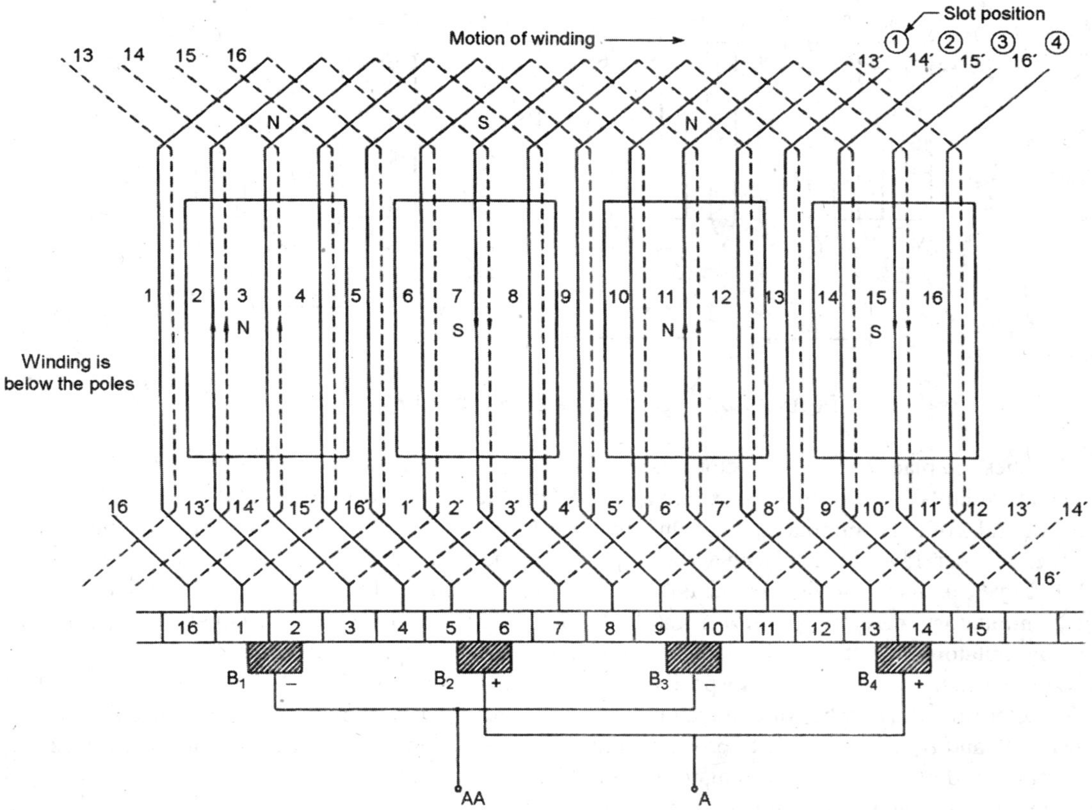

Figure 3.22 Developed View of Simple Lap Winding

1′ (shown by dotted line) is connected to commutator segment 2. The upper coil side 2, that is, 2 (shown by thick line), is connected to commutator 2, and the lower coil side of coil 2, that is, 2′, will be placed in 2 + 4, that is, 6th slot. The front portion of coil side 2′ (shown by dotted line) is connected to commutator segment 3. In the similar way, the winding diagram is completed. The upper coil side of coil 13, that is, 13 (shown by thick line), is connected to commutator segment 13, and the lower coil side of coil 13, that is, 13′, will be placed in 13 + 4, that is, 17th slot. The available of number of slots is 16. Therefore, 17th slot = 17 − 16, that is, 1st slot. The remaining diagram can be completed in a similar way.

To represent the poles in Figure 3.22, they must be placed equally spaced. The width of the pole is taken as $\frac{3}{4}$ times the pole pitch, that is, $\frac{3}{4} \times 4 = 3$ slots.

In Figure 3.22, it is assumed that the winding is progressive, that is, right handed, and the winding is behind the poles. The right hand rule is used to mark the emf on the windings. The emf induced in the conductors will be away from the commutator. The number of parallel paths in the lap winding is equal to the number of poles. The number of conductors per series (Z_c) will be the total number of conductors (Z)/number of poles (P). Since two conductors make a turn, the total number of turns = $Z/2$.

The number of brushes for the lap winding is equal to the number of poles. The brushes must be equally spaced around the commutator, and there will be no emf induced to the coil ends across which

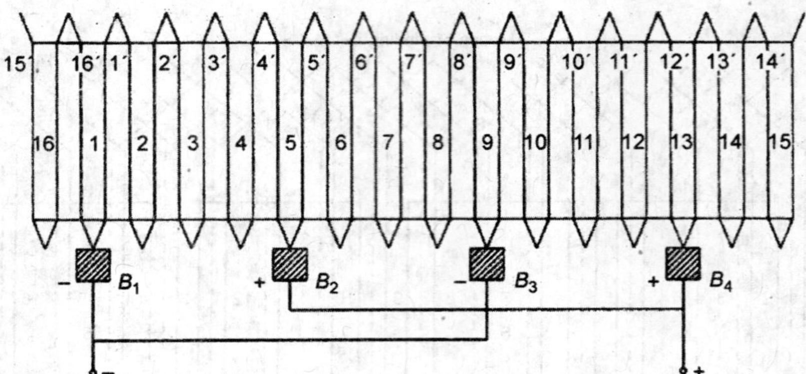

Figure 3.23 Sequence Diagram of Figure 3.22

the brushes are placed. The conductor 1 is moving in the interpolar space and the emf is induced in it. The ends of coils are connected to segments 1 and 2 of the commutator. Brush B_1 is placed across segments 1 and 2 of the commutator. The remaining brushes B_2, B_3 and B_4 are displaced one pole pitch apart from each other. Brush B_2 is placed across segments 5 and 6. Brush B_3 is placed across segments 9 and 10. Brush B_4 is placed across segments 13 and 14. The direction of emf in conductor 2 is away from the commutator. Therefore, brush B_1 is marked as negative. The emf induced in conductor 6 is towards the commutator and hence the brush B_2 is marked as positive. Similarly, the polarity of brushes B_3 and B_4 will be negative and positive, respectively. The number of parallel paths between A and AA will be 4. To determine the number of conductors per parallel path, let us start from terminal A connected to brushes B_2 and B_4. It is possible to proceed either from B_2 or from B_4. If starting is carried out from B_2, the tracing is done using the commutator segment 5 or 6. Segment 6 is connected to the coil ends 6 and 5′. The tracing using segment 6 is shown as follows:

$$B_2\begin{Bmatrix} 6-6'-7-7'-8-8' \\ 4'-4-3'-3-2'-2 \end{Bmatrix}B_1$$

Terminal A　　　　　　　　　　　　　　　　　　　　Terminal AA

$$B_4\begin{Bmatrix} 14-14'-15-15'-16-16' \\ 12'-12-11'-11-10'-10 \end{Bmatrix}B_3$$

Positive brushes　　　　　　　　　　　　Negative brushes

Figure 3.23 shows the sequence diagram, which indicates the sequence of connection of coils.

The coils 1–1′, 5–5′, 9–9′ and 13–13′ do not appear in the parallel path because they are momentarily short circuited by brushes B_1, B_2, B_3 and B_4, respectively.

3.7.2 Wave Winding

Figure 3.20(c) shows the wave winding. For the wave winding the following points are important:

- Y_B and Y_F are odd and the same sign.
- Y_B and Y_F are near equal to the pole pitch and differ by 2.

- The resultant pitch (Y_R) is the sum of Y_B and Y_F.
- The commutator pitch $(Y_C) = \dfrac{\text{mumber of commutator bars} - 1}{\text{number of pairs of poles}}$
- The average pitch (Y_A) is an integer and given by $Y_A = \dfrac{Z \pm 2}{P} = \dfrac{\frac{Z}{2} \pm 1}{\frac{P}{2}} = \dfrac{\text{number of commutator bars} - 1}{\text{numbere of pairs of poles}}$
- The number of coils (N_C) can be obtained from $N_C = \dfrac{PY_A - 2}{2}$.
- The number of parallel path = $2m$, where m is the multiplicity of the winding.

For the lap winding all the pole groups of coils generating emf in the same direction at any instant of time are connected in parallel by the brushes. In case of the wave winding, all the coils carrying current in the same direction are connected in series. Therefore, the coils carrying current in a particular direction are connected in one series circuit, whereas the coils carrying current in the other particular direction are connected in another series circuit. Hence, there are two parallel circuits in wave windings.

Figure 3.24 shows the principle of wave windings, where coils 1, 2 and 3 are situated two pole pitches apart. The dotted lines of the front connections are for the lap winding in which each coil is connected across a pair of adjacent commutator segments. These coils are placed symmetrically with respect to the poles. Therefore, the emf generated in the three coils will be identical in direction. These three coils are connected in series as a group. The ends of the groups are joined to the adjacent commutator segments. The coil ends are bent outwards and this has been shown by the arrow. The coil ends are connected to the commutator segments which are two pole pitches apart. The finish of coil 1 (F_1) is connected to the start of coil 2 (S_2). The finish of coil 2 (F_2) is connected to the start of coil 3 (S_3). The finish of coil 3 (F_3) is connected to the commutator segment which is adjacent to the commutator segment at which the start of coil 1 (S_1) is connected. From junction points of the coils, the connection of the commutator is taken. The commutator pitch (Y_c) is the number of commutator segments between S_1 and F_1. In Figure 3.24, there are three pairs of poles, and if C be the total number of commutator segments, then we have the following:

$$3Y_c = C + 1 \tag{3.1}$$

The number of coils joined in series is equal to the number of pairs of poles. The relation between the commutator segments or the number of coils and the commutator pitch for a machine having 'p' pairs of poles is expressed as follows:

$$pY_c = C \pm 1 \tag{3.2}$$

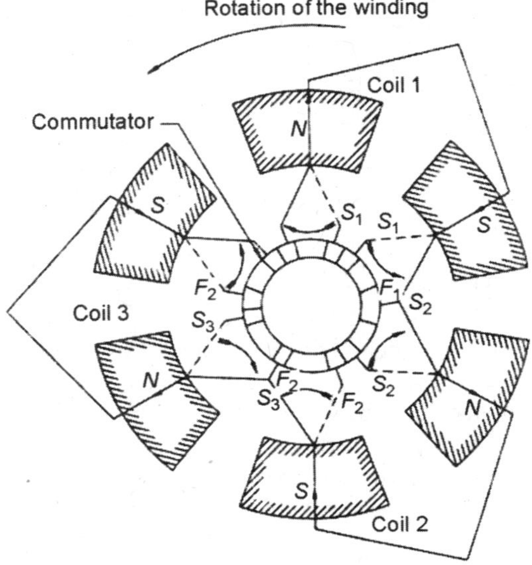

Figure 3.24 Principle of Wave Winding

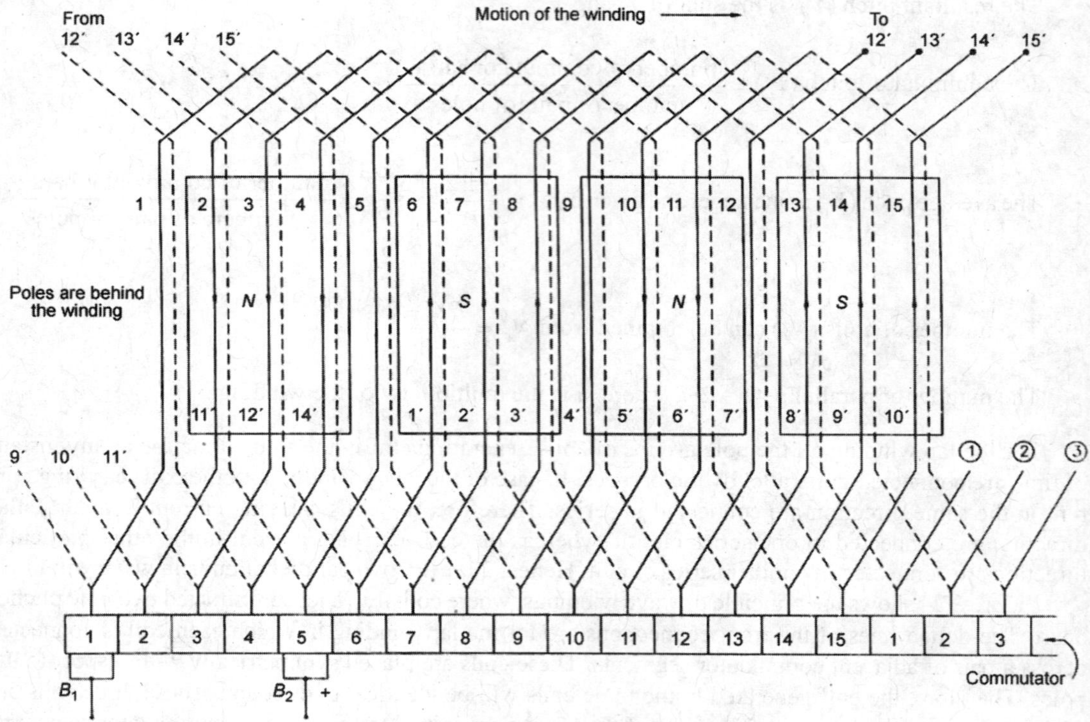

Figure 3.25 Simple Wave Winding with Four Poles

In Equation (3.2), '+' sign is for the progressive winding whereas the '−' sign is for the retrogressive winding.

Figure 3.25 shows the simple wave winding with four poles and an armature having 15 slots and single-turns coils wound with two conductors per slot. Here, C = number of coils = number of commutator segments = 15.

For the retrogressive winding, the commutator pitch is given by $Y_c = \dfrac{15-1}{2} = 7$ and the pole pitch $= \dfrac{15}{4} = 3.75$. Therefore, the coil span is taken as 3 or 4. The coil span of 4 near to the pole pitch is taken. The numbering of conductor similar to the lap winding is taken. The upper coil sides are marked as 1, 2, 3, etc., whereas the lower coil side is marked as 1′, 2′, 3′, etc. Let the upper coil side of coil 1 (shown by thick line) be placed in slot 1. The lower coil side of coil 1, that is, 1′, is placed in the slot 1 + 4, that is, 5th slot. The other end of coil 1′ is connected to the start of 1 + 7, that is, 8th coil that is separated from coil side 1 by a commutator pitch $Y_c = 7$. The junction of 1′ and 8 is connected to segment 8 of the commutator. The lower coil side of coil 8, that is, 8′, will be placed in 8 + 4, that is, 12 slots. The lower coil side of coil 8, that is, 8′, will be connected to the start of coil side 8 + 7, that is, 15. This procedure is repeated to obtain Figure 3.25. In Figure 3.25, it is assumed that the poles are behind the coils and the poles are equally spaced. For the left to the right motion of the winding, the direction of emf induced in the conductors

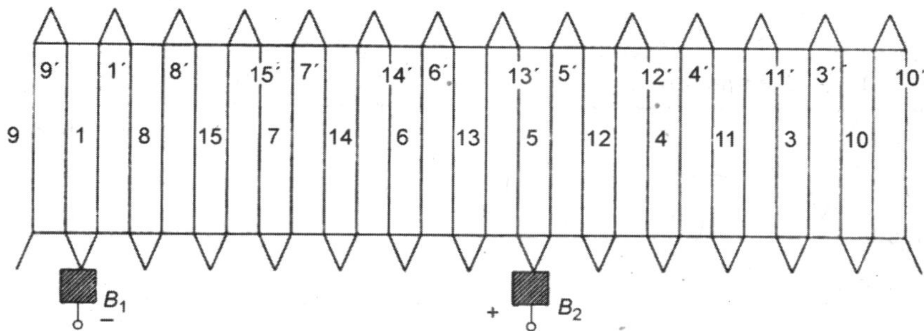

Figure 3.26 Sequence Diagram

situated above the north pole will be towards the commutator whereas the direction of emf induced in the conductors situated below the south pole will be away from the commutator.

Here, the conductor is taken in which the emf is induced and brush B_1 is placed below the commutator segment 1. The other brush is placed with a separation of the coil span, that is, 1 + 4 = 5, that is, under segment 5. The polarity of brush B_1 is taken as positive because the emf induced in conductor 15 is towards the commutator, whereas the polarity of brush B_2 is taken as negative because the emf induced in conductor 6 is away from the commutator. One parallel path is from brush B_2 through 5–5′–12–12′... up to brush B_1 and the other parallel path starting from B_2 through 13′–13–6′–6... up to brush B_1. These two parallel circuits can be represented as follows:

$$\text{Positive brush } B_1 \begin{Bmatrix} 5-5'-12-12'-4-4'-11-11'-3-3'-10-10'-2-2'-9-9' \\ 13'-13-6'-6-14'-14-7'-7-15'-15-8'-8-1'-1 \end{Bmatrix} \text{Negative brush } B_2$$

Figure 3.26 shows the sequence diagram. Only two brushes are essential in the wave winding.

Example 3.1 Draw the developed view of a simple lap winding of four-pole DC generators having 12 slots with two coil sides per slot.

Solution

$$\text{Pole pitch} = \frac{\text{number of slots}}{\text{number of poles}} = \frac{12}{4} = 3$$

The coil span taken here is equal to the full pole pitch of four slops/pole. It is assumed here that the winding is moving from left to right and the winding is behind the poles. Using a full-pitched coil span, the winding is drawing in the progressive way ($Y_c = +1$). The upper sides of the coil are denoted as 1, 2, 3, etc., and the lower sides of the same coil are denoted as 1′, 2′, 3′, etc. Table E3.1 shows the sequence of back connection and front connection.

Table E3.1 Sequence of back connection and front connection of Example 3.1

Back Connection	Front Connection
1 is connected to 1′ (4)	1′ is connected to 2 (2)
2 is connected to 2′ (5)	2′ is connected to 3 (3)
3 is connected to 3′ (6)	3′ is connected to 4 (4)
4 is connected to 4′ (7)	4′ is connected to 5 (5)

(Continued)

(Continued)

Back Connection	Front Connection
5 is connected to 5′ (8)	5′ is connected to 6 (6)
6 is connected to 6′ (9)	6′ is connected to 7 (7)
7 is connected to 7′ (10)	7′ is connected to 8 (8)
8 is connected to 8′ (11)	8′ is connected to 9 (9)
9 is connected to 9′ (12)	9′ is connected to 10 (10)
10 is connected to 10′ (13, i.e. 13 – 12 = 1)	10′ is connected to 11 (11)
11 is connected to 11′ (14, i.e. 14 – 12 = 2)	11′ is connected to 12 (12)
12 is connected to 12′ (15, i.e. 15 – 12 = 3)	12′ is connected to 1 (1)

The numbers in brackets indicates the slots.
Figure E3.1 shows the sequence diagram.

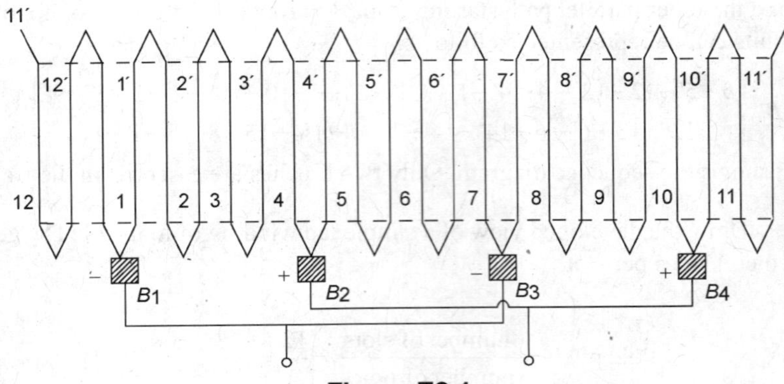

Figure E3.1

Example 3.2 Draw a developed view for a two-layer, wave-winding, four-pole DC generator having 13 slops with two coil sides per slot.

Solution

$$\text{Pole pitch} = \frac{\text{number of slots}}{\text{number of poles}} = \frac{13}{4} = 3.25$$

Therefore, the span taken here is 4. The commutator pitch Y_c can be computed by using the progressive winding as follows:

$$Y_c = \frac{13+1}{2} = 7$$

Similar assumptions in Example 3.1 are also done here. Table E3.2 shows the sequence of back connection and front connection.

Table E3.2 Sequence of back connection and front connection of Example 3.2

Back Connection	Front Connection
1 is connected to 1′	1′ is connected to 1 + 7 (i.e. 8)
8 is connected to 8′	8′ is connected to 8 + 7 (i.e. 15 − 13 = 2)
2 is connected to 2′	2′ is connected to 2 + 7 = 9
9 is connected to 9′	9′ is connected to 9 + 7 = 16 (i.e. 16 − 13 = 3)
3 is connected to 3′	3′ is connected to 3 + 7 = 10
10 is connected to 10′	10′ is connected to 10 + 7 = 17 (i.e. 17 − 13 = 4)
4 is connected to 4′	4′ is connected to 4 + 7 = 11
11 is connected to 11′	11′ is connected to 11 + 7 = 18 (i.e. 18 − 13 = 5)
5 is connected to 5′	5′ is connected to 5 + 7 = 12
12 is connected to 12′	12′ is connected to 12 + 7 = 19 (i.e. 19 − 13 = 6)
6 is connected to 6′	6′ is connected to 6 + 7 = 13
13 is connected to 13′	13′ is connected to 13 + 7 = 20 (i.e. 20 − 13 = 7)
7 is connected to 7′	7′ is connected to 7 + 7 = 14 (i.e. 14 − 13 = 1)

Figure E3.2 shows the sequence diagram.

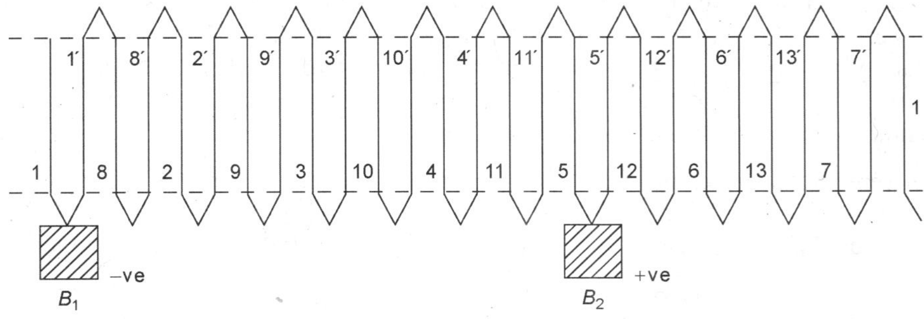

Figure E3.2

3.8 EQUALIZING CONNECTIONS FOR LAP WINDING

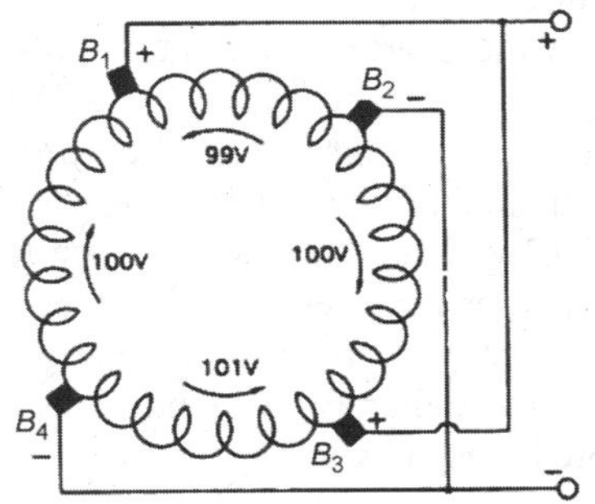

Figure 3.27 Lap-wound Armature

In a simple lap winding, shown in Figure 3.27, the emfs generated by all the pole groups of coils are in the same direction at any instant and they are connected in parallel by the brushes. If the emfs generated in each parallel circuit are not equal, this results in circulating currents among the parallel paths even at no-load condition. This causes unnecessary power loss and overheating of the windings. To divert these circulating currents, equalizing connections

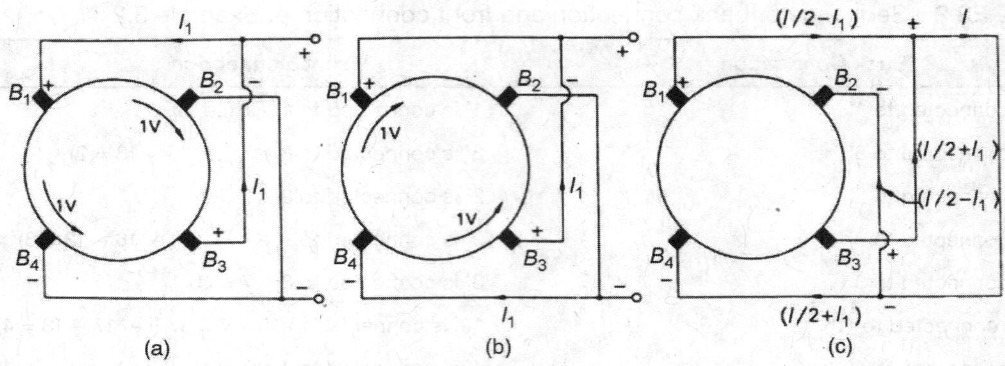

Figure 3.28 Circulating Currents in Brush Connections

are used. Using a total number of slots divisible by the number of poles, the magnitude of the circulating currents in lap windings can be minimized and hence inequalities in the emfs in the parallel paths can be minimized. Using equalizing connection, the above can be satisfied. In Figure 3.27, the numbers of parallel paths and the conductors in each parallel path are distributed under adjoining poles. Any difference in the reluctance of the magnetic circuit due to unequal air gaps etc. results inequality in the flux per pole. Hence, the emf generated in each circuit will be different.

The generated emfs and their direction are assumed and shown in Figure 3.27. In the parallel circuits $B_1B_2B_3$ and $B_1B_4B_3$, the resultant emf is 1 V, shown in Figure 3.28(a). The circulating current in the conductor bridging brushes B_1 and B_3 is I_1, shown in Figure 3.28(a). Similarly, the resulting voltage acting in the parallel circuits $B_4B_1B_2$ and $B_4B_3B_2$ is also 1 V, shown in Figure 3.28(b). The circulating current I_1 flows through the conductor joining brushes B_2 and B_4, shown in Figure 3.28(b).

If the emfs generated in each parallel path are equal, the current per branch becomes 1/2 for the load current I delivered by the generator. The resultant current at brushes B_1 and B_2 will be $\frac{I}{2} - I_1$ due to the presence of circulating current I_1 and that of at brushes B_3 and B_4 will be $\frac{I}{2} + I_1$. Therefore, brushes B_3 and B_4 will be overloaded and it results in excessive sparking.

Figure 3.29 shows the equalizing connections between $R-R'$, $S-S'$ and $T-T'$ to minimize or even avoid the sparking. For non-uniform potential of the above points, the current I_1 will flow through $R-R'$ instead of through the brushes. In practice, the number of equalizing connections employed varies between 6 and 12 which are arranged in the form of a ring to get the proper arrangement of connections at all positions of the armature to divert the currents. For the number of poles more than two, the numbers of parallel paths are more in the lap winding compared to the wave winding. Table 3.2 shows the comparison between lap winding and wave winding.

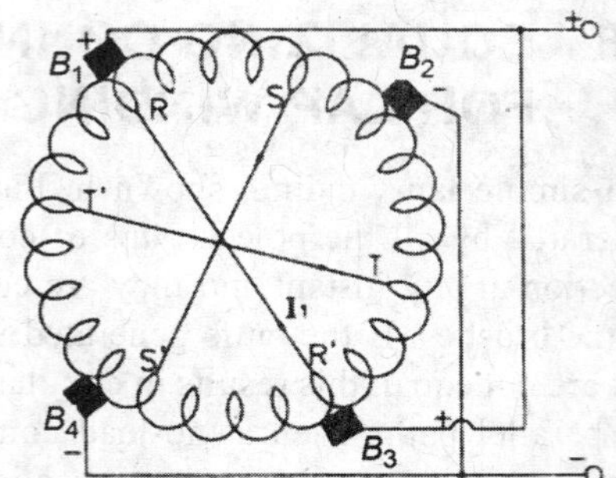

Figure 3.29 Equalizing Connections

Table 3.2 Comparison Between Lap Winding and Wave Winding

Lap Winding	Wave Winding
• Number of parallel paths = number of poles	• Number of parallel paths = 2
• Number of brush sets is equal to the number of poles	• Number of brush sets = 2
• This type of winding is used for high-current and low-voltage capacity generators	• This type of winding is used for low-current and high-voltage capacity generators
• Used for generators of a capacity of >500 A	• Used for generators of a capacity of <500 A

3.9 USES OF LAP AND WAVE WINDINGS

Wave windings give more emf for a given number of pole and armature conductors, whereas lap windings give the same emf for more number of conductors. Hence, the winding cost increases for the lap winding. Equalizing connections are not required in the wave winding, but this is mandatory for the lap winding. In the lap winding, there are P parallel paths when the numbers of poles are P, whereas in the wave winding, the numbers of parallel paths are two. Therefore, any inequality of pole fluxes in the wave winding cannot produce unequal voltages because this inequality of fluxes affects the two paths equally. In the lap winding, the inequality of fluxes produces unequal voltages and causes sparking at the brushes.

3.10 DUMMY COILS

These coils are used in the wave winding when the requirement of windings is not satisfied by armature punching. Since these coils are not connected to the commutator, they do not affect the electrical characteristic of the windings.

3.11 PRINCIPLE OF DC GENERATOR

Figure 3.30 shows a single-turn generator. The coil is rotated in the anticlockwise direction with the help of the prime mover. During its rotation, the conductors *ab* and *cd* cut the lines of flux. According to Faraday's law of electromagnetic induction, emf is induced in the conductors, which drives the current through the conductor. The conductors are connected to brushes through the slip rings. The magnitude of induced emf depends on the position of the coil in the magnetic field. Figure 3.31 shows the different instants of the induced emf due to different positions of the coil. The magnetic field is produced either by a permanent magnet or by an electromagnet energized by the DC supply.

When the angle $\theta = 0°$, the coil is perpendicular to the magnetic field and the instantaneous component of velocity of conductors *ab* and *cd* is parallel to the magnetic field. The flux linked with the coil is maximum, but the rate of change of flux is minimum. Hence, emf cannot be generated. Therefore, no current flows through the conductors.

When $0° < \theta < 90°$ due to rotation of the coil in the anticlockwise direction, the velocity of the conductors has two components, that is, $v \sin\theta$ and $v \cos\theta$. Therefore, the conductors will cut the flux due to $v \sin\theta$. Hence, emf is induced in the conductors and the current flows in the conductors.

When the angle $\theta = 90°$, the coil is parallel to the magnetic field and the instantaneous component of velocity of conductors *ab* and *cd* is perpendicular to the magnetic field. The flux linked with the coil

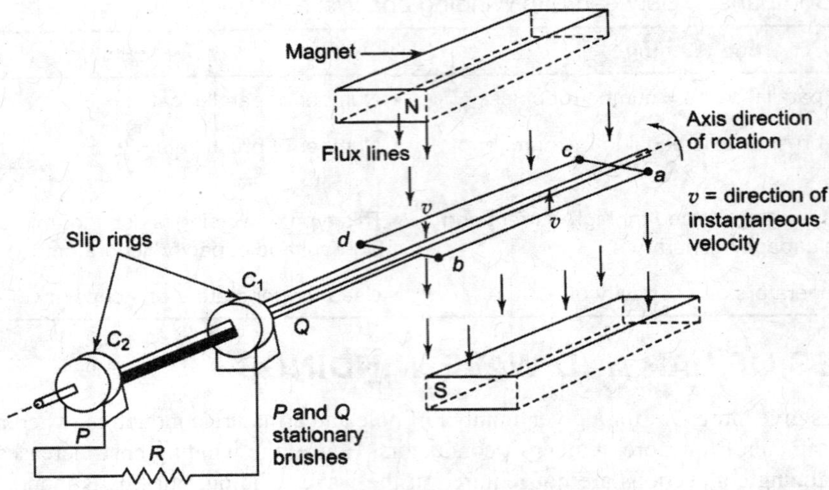

Figure 3.30 Single-turn Generator

is minimum, but the rate of change of flux is maximum. Hence, emf generated is maximum. Therefore, the maximum current flows through the conductors.

When the coil rotates from $\theta = 90°$ to $\theta = 180°$, the component of velocity, that is, $v \sin\theta$ of conductors perpendicular to the field, decreases and hence emf is induced in the conductors and the current flow in the conductors decreases.

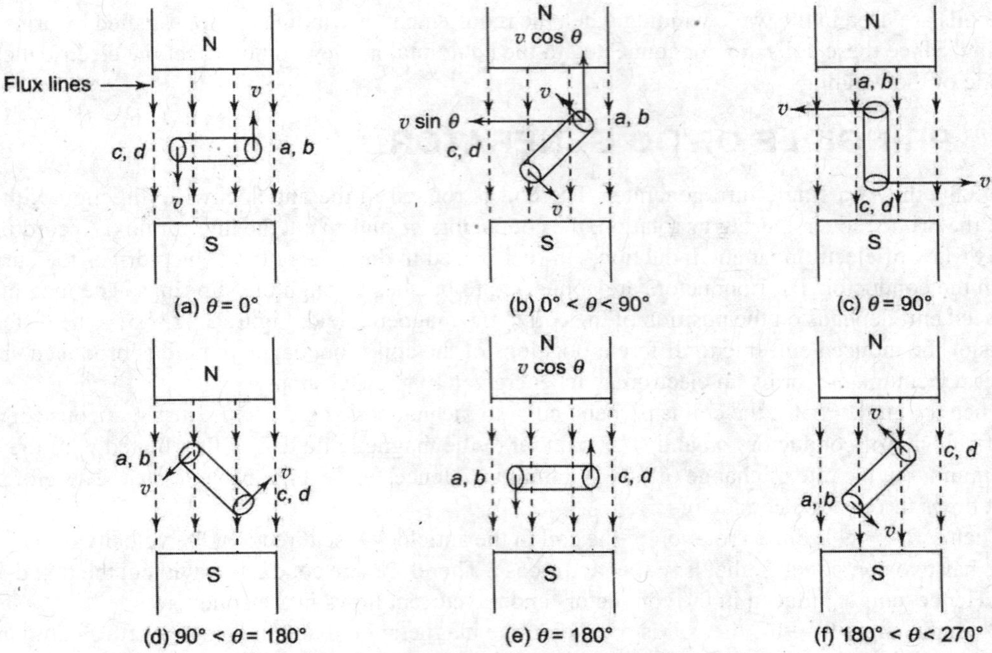

Figure 3.31 Induced emf for Different Instants

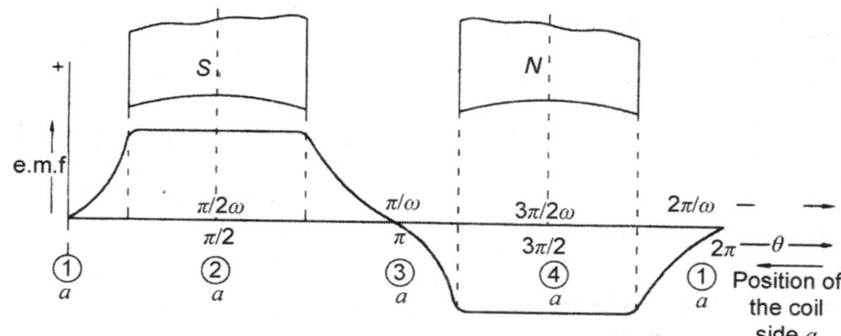

Figure 3.32 Graphical Representation of Generated Current

When the angle $\theta = 180°$, the coil is perpendicular to the magnetic field and the instantaneous component of velocity of conductors *ab* and *cd* is parallel to the magnetic field. The flux linked with the coil is maximum, but the rate of change of flux is minimum. Hence, emf cannot be generated. Therefore, no current flows through the conductors.

When $180° < \theta < 270°$ due to rotation of the coil in the anticlockwise direction, the velocity of the conductors has two components, that is, $v \sin\theta$ and $v \cos\theta$ in the direction opposite to the case of $0° < \theta < 90°$. Therefore, the conductors cut the flux due to $v \sin\theta$. Hence, emf is induced in the conductors in the direction opposite to the case of $0° < \theta < 90°$ and current flows in the conductors in the direction opposite to the case of $0° < \theta < 90°$.

When the angle $\theta = 270°$, the coil is parallel to the magnetic field and the instantaneous component of velocity of conductors *ab* and *cd* is perpendicular to the magnetic field. The flux linked with the coil is minimum, but the rate of change of flux is maximum. Hence, emf generated is maximum in the direction opposite to the case of $\theta = 90°$. Therefore, the maximum current flows through the conductors in the direction opposite to the case of $\theta = 90°$.

When the coil rotates from $\theta = 270°$ to $\theta = 360°$, the component of the velocity, that is, $v \sin\theta$ of conductors perpendicular to the field, decreases and hence emf is induced in the conductors and the current flow in the conductors decreases.

When the angle $\theta = 360°$, the coil is perpendicular to the magnetic field and the instantaneous component of velocity of conductors *ab* and *cd* is parallel to the magnetic field. The flux linked with the coil is maximum, but the rate of change of flux is minimum. Hence, emf cannot be generated. Therefore, no current flows through the conductors.

The direction of flow of current through the conductors can be found by *Fleming's right-hand rule.* Figure 3.32 shows the graphical representation of generated current.

3.12 OPERATION OF A SIMPLE DC GENERATOR WITH A TWO-SEGMENT COMMUTATOR

Figure 3.33 shows a two-pole DC generator with a single coil to explain working of a simple DC generator with a two-segment commutator. The two ends of the coil (conductor *a* and conductor *b*) are brought out to split rings C_1 and C_2, respectively, which are made of copper. This split ring is termed as the commutator, and segments of the split ring are insulated from each other. This arrangement is different from the arrangement shown in Figure 3.30.

The split rings are rigidly fixed to the axis of rotation so that the rotation of the coil ensures the rotation of the segments C_1 and C_2 of the commutator. Brushes B_1 and B_2 remain stationary and will come into contact alternately with C_1 and C_2 when the coil completes one revolution. The angle between the plane of the coil and the vertical axis is represented by θ. For the vertical position of the coil, that is, $\theta = 0$, the soil side 'a' is at the position '1' and the emf induced in the coil will be zero. Figure 3.34(a) shows this position. Brushes B_1 and B_2 bridge the commutator segments C_1 and C_2, and the coil becomes short circuited. The emf induced in coil sides '*a*' and '*b*' will be zero and the emf induced in brush B_1 will be zero, as shown in Figure 3.35(a). The corresponding position of commutator segments has been shown in Figure 3.35(b).

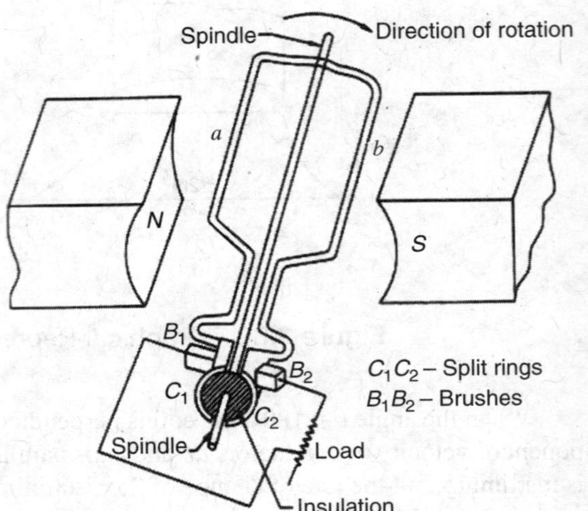

Figure 3.33 DC Generator with Two-segment Commutator

For rotation of the coil through an angle $\theta = \frac{\pi}{2}$, coil side '*a*' occupies the position '2' and it is under the south pole, as shown in Figure 3.34(b). The segment C_1 comes into contact brush B_1 in this case. The magnitude of induced emf acts away from the paper, and it is represented by the dot which is taken as positive. Therefore, the polarity of the voltage of brush B_1 will also be positive because coil side '*a*' is connected to B_1 through the commutator segment C_1. Figure 3.35(a) shows the magnitude of the voltage, while Figure 3.35(b) shows the corresponding positions of the commutator segments with respect to the brushes. Coil side '*a*' occupies position '3' for further rotation of an angle $\frac{\pi}{2}$, that is, $\theta = \pi$. Figure 3.34(c) shows the positions of the commutator segments C_1 and C_2. The position of the commutator segments C_1 and C_2 has been interchanged in this case. The emf induced in coil sides '*a*' and '*b*' will be zero and the emf induced in brush B_1 will be zero, as shown in Figure 3.35(a), and the corresponding position of commutator segments has been shown in Figure 3.35(b). For another rotation of the coil by an angle $\frac{\pi}{2}$, that is, $\theta = \frac{3\pi}{2}$, coil side '*a*' comes to the position '4', which is under the influence of the north pole as shown in Figure 3.34(d). The magnitude of the induced emf will be maximum and is in the reversed direction shown by the cross. It acts on the paper. The segment C_2 in this case is in contact with brush B_1, which is connected to coil side '*b*'. Coil side '*b*' is under the influence of the south pole and its induced emf is represented by dots. Therefore, the polarity of the voltage brush B_1 in this case will remain positive, shown in Figure 3.35(a), and the corresponding position of commutator segments has been shown in Figure 3.35(b).

The polarity of brush B_1 in this case becomes positive because it is always under the influence of the south pole in the case shown in Figure 3.35(b). For positions '2' and '3' of Figure 3.35(b), brush B_1 is in contact with C_1 and C_2, which are connected to coil sides '*a*' and '*b*', respectively. In these two positions, coil sides '*a*' and '*b*' are under the influence of the south pole and the emf induced in them

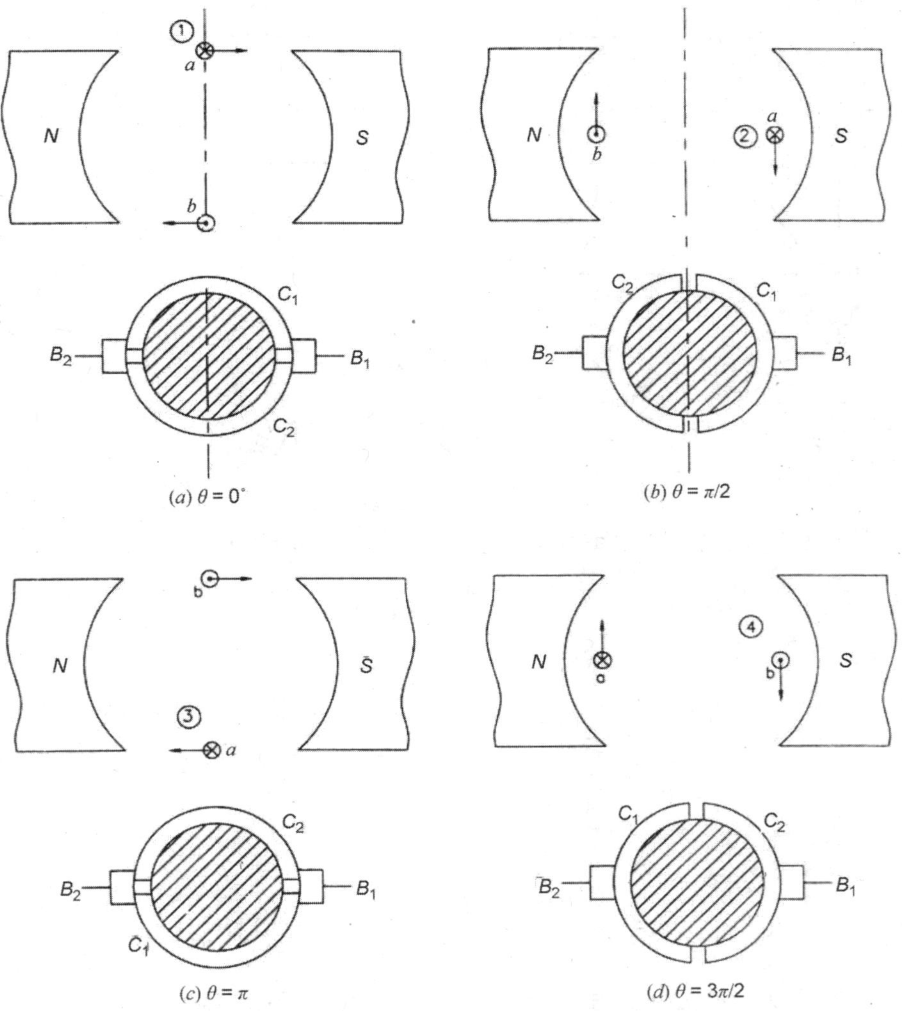

Figure 3.34 Relative Position of the Commutator Segments with Respect to the Brushes

is away from the paper and hence positive. Therefore, the emf induced in brush B_1 remains positive in these cases.

Figure 3.36 shows a DC generator having a single-layer winding with more number of coils. Figure 3.37 shows the waveform of the voltage of brush B_1, and the voltage waveform of each of the coils is displaced by a slot pitch angle (γ), shown in Figure 3.37. For the clockwise direction of rotation of the armature, the voltage induced in the coil 'a_2' leads the voltage induced in the coil 'a_1' by an angle γ. If the number of slots and the number of armature conductors are increased, the shape of the waveform can be improved.

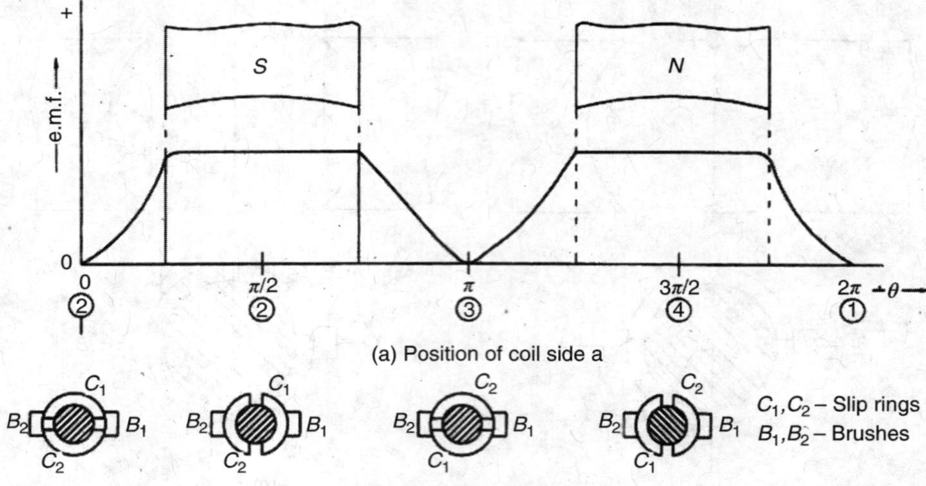

Figure 3.35 Waveform of the emf Generated at Brush B_1

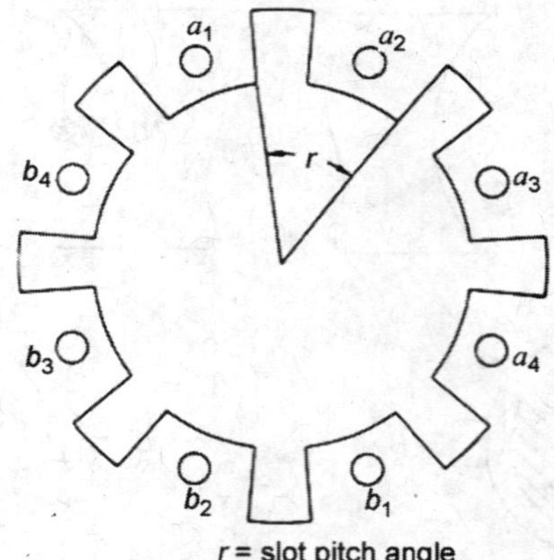

Figure 3.36 Slotted Armature with Four Coils

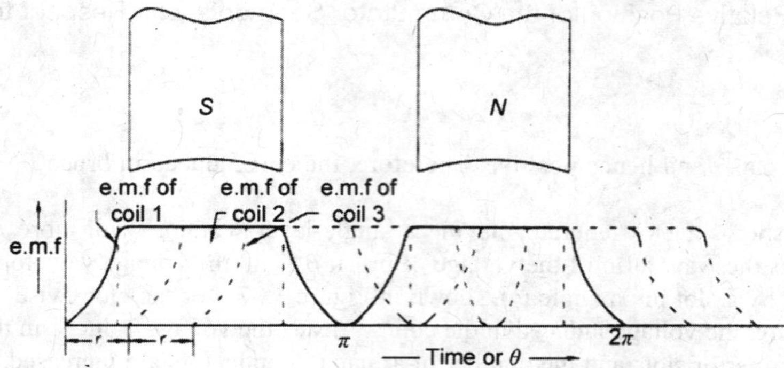

Figure 3.37 Waveforms of the Generated emf at Brush B_1 (Armature Carries Four Coils)

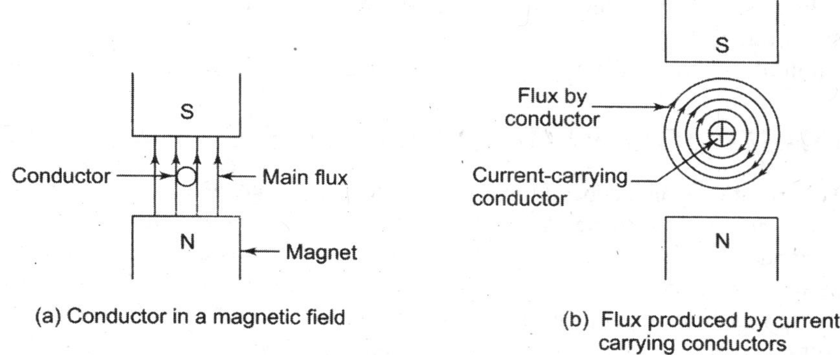

(a) Conductor in a magnetic field

(b) Flux produced by current-carrying conductors

Figure 3.38 Flux due to Field and Conductor

3.13 PRINCIPLE OF DC MOTOR

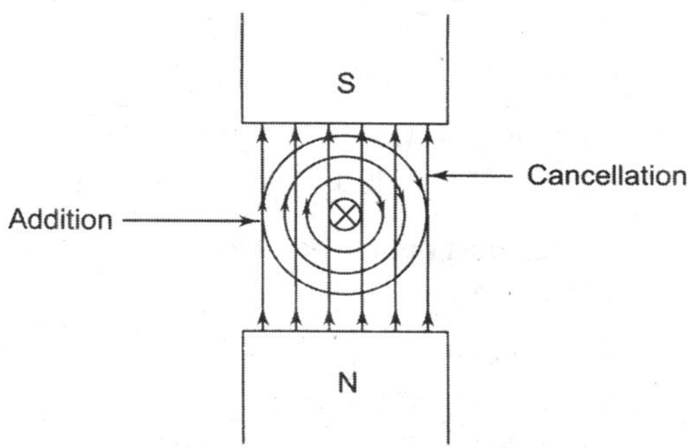

Figure 3.39 Interaction of Two Fluxes

The basic principle of operation of a DC motor is that a current-carrying conductor, if placed in a magnetic field, experiences a mechanical force. Figure 3.38(a) shows a conductor placed in a magnetic field produced by the field windings of a DC motor carrying current. The flux produced by the field windings is known as the main flux. Figure 3.38(b) shows the flux produced by the conductor alone. There are two fluxes present in the magnetic field. One is produced by the permanent magnet called the main flux and the other is produced by the current-carrying conductor. On the left side of the conductor, the two fluxes are additive in nature, whereas on the right side of the conductor, the fluxes are subtractive in nature. Therefore, the density of flux lines is stronger on the left side whereas it is weaker on the right side. This is shown in Figure 3.39. The flux distribution around the

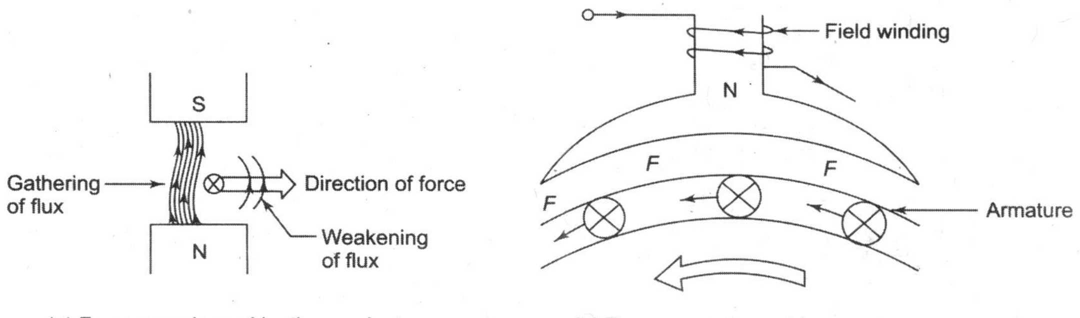

(a) Force experienced by the conductor

(b) Torque experienced by armature

Figure 3.40 Force and Torque Experienced by Armature

conductor acts like a stretched rubber band under tension and exerts a mechanical force on the conductor, which acts from left to right, shown in Figure 3.40.

The direction of rotation of the conductor can be found by *Fleming's left-hand rule.*

3.14 CONSTRUCTION OF SYNCHRONOUS MACHINES

The principle of operation which depends on the principle of electromagnetic induction of a synchronous generator and DC motor is the same. According to the principle of electromagnetic induction, emf is induced in the conductors due to the rate of change of flux linkage, which is produced due to the relative motion between the conductors and the magnetic field. The magnetic field is on the stator, and the armature which carries the conductors is on the rotor in DC generators. On the other hand, the field winding is on the rotor and armature conductors is on the stator in synchronous generators. This is the main difference between DC generators and synchronous generators. But in both cases, the generated emf is alternating in nature. In DC generators, the induced emf depends on the total flux per pole and it is independent on the nature of flux distribution in space. On the other hand, the flux distribution in an alternator must be as nearly a sine wave as possible so that the maximum sinusoidal emf from a given value of the flux/pole is obtained. In DC generators, the commutator plays the role of production of unidirectional emf across the brush axis. In synchronous generators there is no need to use the commutator.

In case of the alternator, the following possibilities exist:

(i) rotating armature and stationary field or
(ii) rotating field and stationary armature.

In practice, most of the alternators have rotating field-type construction with the stationary armature due to the following reasons:

- The generation level of AC voltage may be in the range of 11 kV to 33 kV, which gets induced in the armature. It is possible to provide large space to accommodate many conductors and insulation for the stationary armature. The ventilation arrangement for high-voltage side can be improved.
- Keeping high-voltage armature stationary, it is possible to avoid interaction of mechanical stress.
- It is very easy to collect high currents at high voltages from the stationary member.

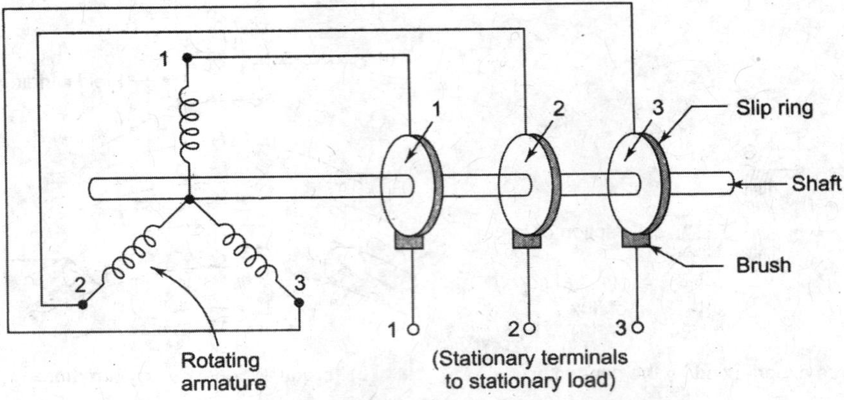

Figure 3.41 Collection of Induced emf from Armature

- Since field coil needs very low voltage, it can easily be supplied with the help of the slip ring and brush assembly. At low voltage, it is possible to avoid sparking at the slip rings.
- Since the low-voltage level is on the field side, insulation cost is reduced. Field has low inertia. It is always preferable to keep a low inertia system as the rotating member because it is easy to rotate the low inertia system. It can also be rotated at high speed.
- Overall construction is simplified with rotating field and hence high output can be obtained.
- Only two slip rings are required to excite the field by DC supply.
- Cooling arrangement becomes very easy.

Synchronous generators export AC power to a bus bar, whereas synchronous motors import AC power from the source. The frequency of generated emf in a synchronous generator depends on the speed of the rotor. If the frequency of the generated emf of the synchronous generator is equal to the frequency of the bus bar, then it is possible to connect it to the bus bar provided other conditions are satisfied. (For detailed procedure, refer to Chapter 6.)

The main parts of synchronous machines are (i) the stator (armature) and (ii) the rotor (field).

3.14.1 Stator

Figure 3.42 shows the stator and the rotor of an alternator. The stator shown in Figure 3.43 consists of a core and slots to hold the armature winding like the armature of a DC generator, which is a stationary

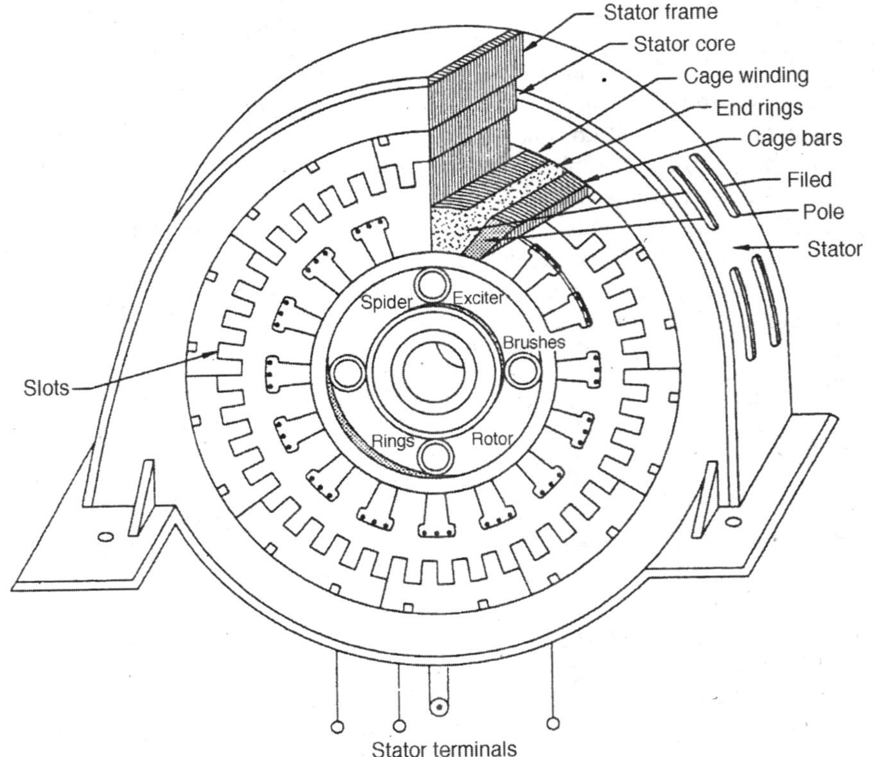

Figure 3.42 Stator and Rotor of an Alternator

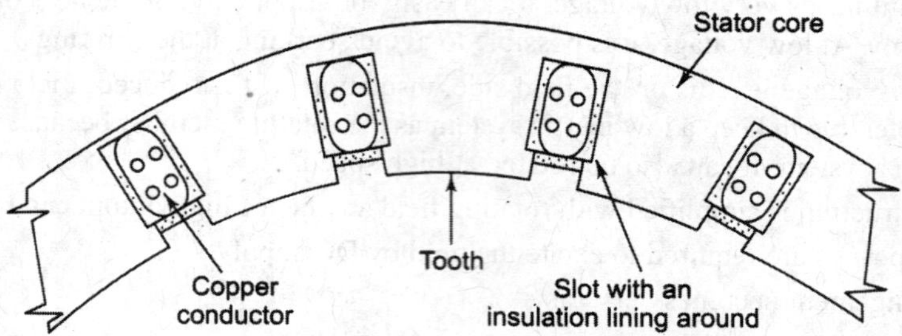

Figure 3.43 Stator

member of the alternator. The stator core is laminated to reduce eddy current loss and is made of steel stampings insulated from each other with varnish or paper, and ducts are provided through them for ventilating purposes. The ventilating ducts have been shown in Figure 3.44. To reduce hysteresis loss, the selected material for the core, which is a special steel alloy, must have small hysteresis loop. In small machines, the laminations are generally stamped out in complete rings. The entire core is fabricated in a frame made of cast iron to house the armature conductors. The frame only supports the core and does not carry any flux. In modern machines, the fabrication of stator frames is done from mild steel, which are welded together. There are two kinds of ventilations: natural ventilation and closed circuit ventilation. In natural ventilation, a fan is attached to both ends of the machine and atmospheric air is the ventilating medium. Since atmospheric air may contain injurious elements, such as dust, moisture and acidic fumes, which are harmful for the insulation of the winding, the closed circuit system is preferred for ventilation and hydrogen is used here as a ventilating medium.

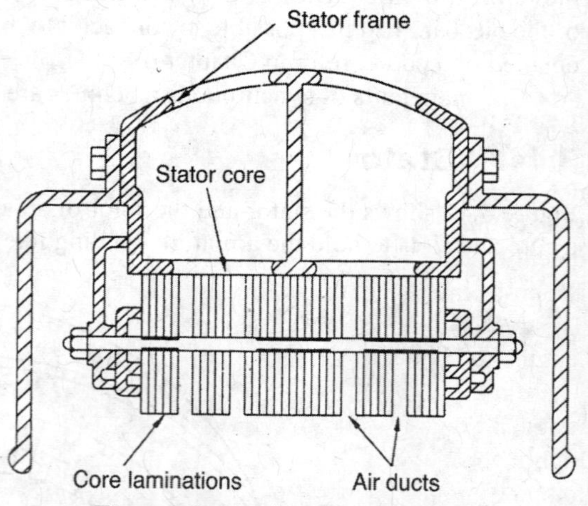

Figure 3.44 Ventilating Ducts

3.14.2 Rotor

Synchronous machine has different geometric structure of the rotor and hence it is classified as

(i) salient pole-type ($P > 4$, low speed) and
(ii) smooth cylindrical pole-type ($P < 4$, high speeds).

3.14.2.1 Salient Pole-type Rotor

Figure 3.45 shows a salient pole-type rotor. All poles are projected out from the surface of the rotor. It is also called *projected pole-type.* The poles are built of thick steel laminations and bolted to the rotor. The salient pole-type rotor is used for low-speed alternators ranging from 125 rpm to 500 rpm because its mechanical strength is low. The prime movers used to drive this type of rotor are usually water turbines and IC engines.

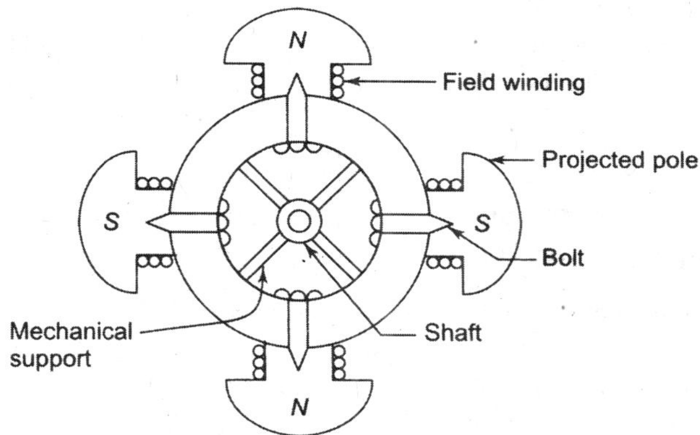

Figure 3.45 Salient Pole-type Rotor

3.14.2.2 Smooth Cylindrical Type Rotor

Figure 3.46 shows a cylindrical type rotor, which is non-salient-type. The rotor consists of a smooth solid steel cylinder having a number of slots to accommodate the field coil. In this type of rotor, the poles are not projecting out and uniform air gap is maintained. This type of rotor has small diameter and large axial lengths to keep the peripheral speeds within limits. These rotors are preferred for high-speed alternators ranging from 1,500 rpm to 3,000 rpm due to their high mechanical strength. Such high-speed alternators are known as *turbo alternators.* Prime movers such as steam turbines and electric motors are used to drive such types of rotors.

The comparison between salient pole type and smooth cylinder type rotors is shown in Table 3.3.

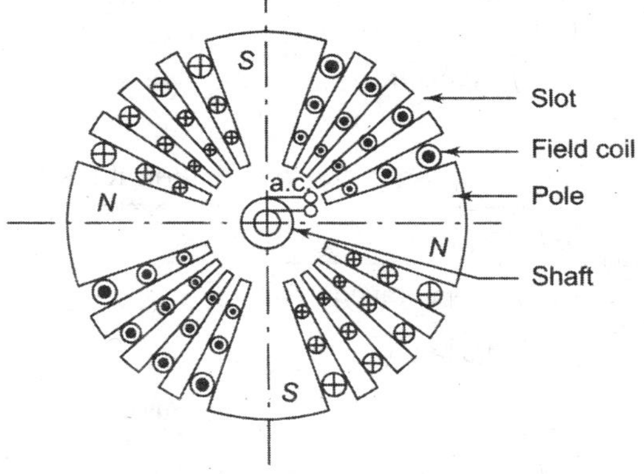

Figure 3.46 Smooth Cylindrical Type Rotor

Table 3.3 Comparison Between Salient Pole-type and Smooth Cylinder Type Rotors

Salient Pole-type	Smooth Cylindrical type
• Air gap is non-uniform	• Air gap is uniform
• Poles are projecting out of surface	• Poles are non-projecting-type
• Diameter is high and axial length is small	• Diameter is small and axial length is large
• Mechanically feeble	• Mechanically strong
• Used in low-speed alternators	• Used in high-speed alternators
• Prime movers used are water turbines and IC engines	• Prime movers used are steam turbines and electric motors
• Separate damper winding is provided	• Separate damper winding is not required
• Rating is smaller for same size compared to a cylindrical type	• Rating is higher for same size compared to a salient pole-type

The principle of DC generators has already been introduced. The basic principle of a synchronous generator or an alternator is the same as that of the DC generator. The commutator part is not required in a synchronous generator. (The working principle of synchronous motors has been discussed in Chapter 7, Section 7.1.)

3.14.3 Classification of Synchronous Machines Based on the Prime Mover

The following are the classifications of the synchronous machines based on the type of the prime mover:

- *Turbogenerators:* The efficiency of these machines is found to be very high at high speeds, and hence they run at very high speeds having peripheral speed as high as 160 m/s. They are driven by steam turbines having the rotor cylindrical types made of solid steel forging of chromium–nickel–molybdenum steel to withstand centrifugal forces at high speeds. The capacity of these machines is very high (>1,000 MW).
- *Hydrogenerators:* The salient pole rotors are used because they are run at low speeds. They are driven by water turbines. The capacity of these machines ranges from 5 MW to 150 MW.

  Table 3.4 shows the comparison between turbogenerators and hydrogenerators.

Table 3.4 Comparison Between Turbogenerators and Hydrogenerators

S. No.	Items	Turbogenerators	Hydrogenerators
1	Power ratings	60 MW to above 1,000 MW	5 MW to 150 MW
2	Voltage	11.5–33 kV	6.6 kV–11.5 kV
3	DC excitation voltage nominal value	100–800 V 425 V (DC)	100–400 V 100V (DC)
4	Cooling system	Forced air hydrogen cooling	Forced air cooling
5	Rotor	Cylindrical, solid-type	Salient pole laminated
6	Speed	3,000 rpm (high speeds)	Low speeds
7	Poles	2 or 4	More than 4
8	Damper windings	Not required	Used
9	Response time (excitation system)	0.5 ms	High (>0.5 ms)
10	Short circuit ratio	0.6–0.9	0.9–1.5

3.14.4 Excitation System

DC excitation is required for field coils both for alternators and for motors. The field coils are connected to a 125–600 V DC source to provide excitation current to field coils. In many cases, exciting current is obtained from DC generators, which are mounted on the shaft of the alternator. Figure 3.47 shows the brushless system for a high-rating alternator where field current up to 4,000 A is required.

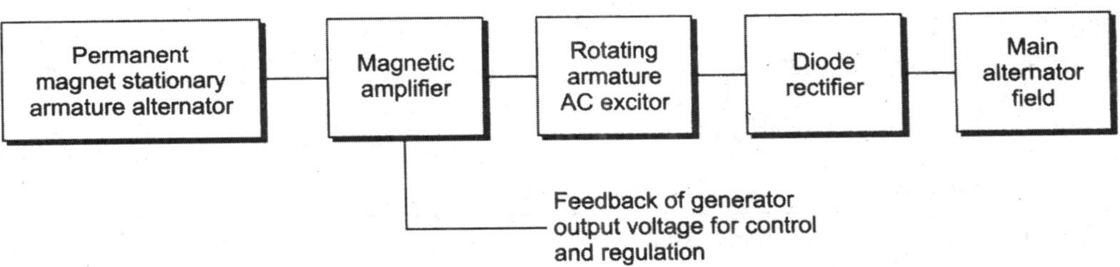

Figure 3.47 Excitation System

3.14.5 Damper Windings

In salient types the poles are slotted to provide the copper bars, which are short circuited at the ends, known as damper windings. These windings reduce momentary fluctuations in speed due to load shedding, and this phenomenon is known as hunting. Damper windings are also useful for the starting of synchronous motors. Turboalternators are usually not provided with damper windings due to the presence of solid field poles, which act as efficient dampers.

3.14.6 Frequency and Synchronous Speed

The emf induced in the conductors of an alternator is alternating and hence has the sinusoidal waveform. The frequency of the induced emf is expressed as follows:

$$f = \frac{PN}{120} \tag{3.3}$$

where P is the total number of poles on the alternator, f is the frequency of the emf generated in Hz and N is the speed of the rotor of the alternator in rpm.

From Equation (3.3), the speed of an alternator can be expressed as follows:

$$N = \frac{120f}{P} \tag{3.4}$$

The rotor of the alternator rotates at a speed so that the frequency of the induced emf becomes $f = \frac{PN}{120}$, and this speed is known as synchronous speed. Table 3.5 shows the synchronous speed in rpm for different number of poles for frequency $f = 50$ Hz.

From Table 3.5, the synchronous speed of a machine decreases with increase in the number of poles.

Table 3.5 Synchronous Speed in rpm for Different Number of Poles for Frequency $f = 50$ Hz

Number of poles	2	4	6	8	12	24
Synchronous speed in rpm	3,000	1,500	1,000	750	500	250

3.14.7 Armature Windings

The windings of AC machines are open-type while that of DC machines are closed-type. Hence, the AC armature windings are different from DC windings. No closed path exists for the armature currents in AC windings itself. All the ends of the windings of AC machines are brought out on one side and the others on the other side are connected together to form the neutral point for the star connection.

Table 3.6 shows the characteristics of three-phase AC windings which are commonly used.

Table 3.6 Characteristics of Three-phase AC Windings

Type of coil	Concentric, lap, wave
Overhang	Diamond, multiplane, involute and mush
Layers	Single, double
Slots	Open, closed, semi-closed
Connection	Star, delta
Phase spread	60°, 120°
Slotting	Integral, fractional
Coil span	Full pitched, chorded
Circuits	Series or parallel
Coils	Single-turn, multiturn

The basic condition that should be satisfied in the AC winding is that it should produce in general a symmetrical n-phase balanced system of emfs of identical magnitude, wave shape and frequency, but should be displaced in time phase by $\theta = \frac{2\pi}{n}$, except for two phases for which $\theta = \frac{\pi}{2}$ electrical radians. This is achieved by using identical windings for all the phase groups and arranging the groups in such a way so that they have effective space displacement of $\frac{2\pi}{n}$ electrical radians. Usually, phase windings are arranged so that spread of each phase per pole $\sigma = \frac{\pi}{n}$. The sequence of phase bands thus formed is AC′BA′CB′, as shown in Figure 3.48(d).

Ungrouped conductors arranged in 12 slots are shown in Figure 3.48(a). The emf induced in the conductors will be as shown in Figure 3.48(á), with each being displaced by slot pitch angle α assuming that the flux density in the gap is sinusoidal.

Let us consider that phase A and phase B are formed by conductors 1, 2, 3, 4 and conductors 5, 6, 7, 8, respectively, and remaining conductors from phase C, as shown in Figure 3.48(b). The four conductors forming any one phase will have a spread of 120° since the slot pitch is 30°. Hence, the phase difference between the three phases is generated.

The three-phase emfs formed due to the summation of emfs in each phase band (1, 2, 3, 4) will be displaced from each 120° in time phase. The resultant emf of the each phase will be less than the arithmetic sum of the component emfs, that is, $e_a < (e_1 + e_2 + e_3 + e_4)$. This is due to the spreading of the windings. The number of slots used per pole per phase is 4; that is, SPP = 4.

In Figure 3.48(c), SPP is 2 and there is six-phase grouping, each being 60° apart. In arrangement of Figure 3.48(c), single-layer winding cannot be used because of absence of return conductors left out for every one pole pitch separation.

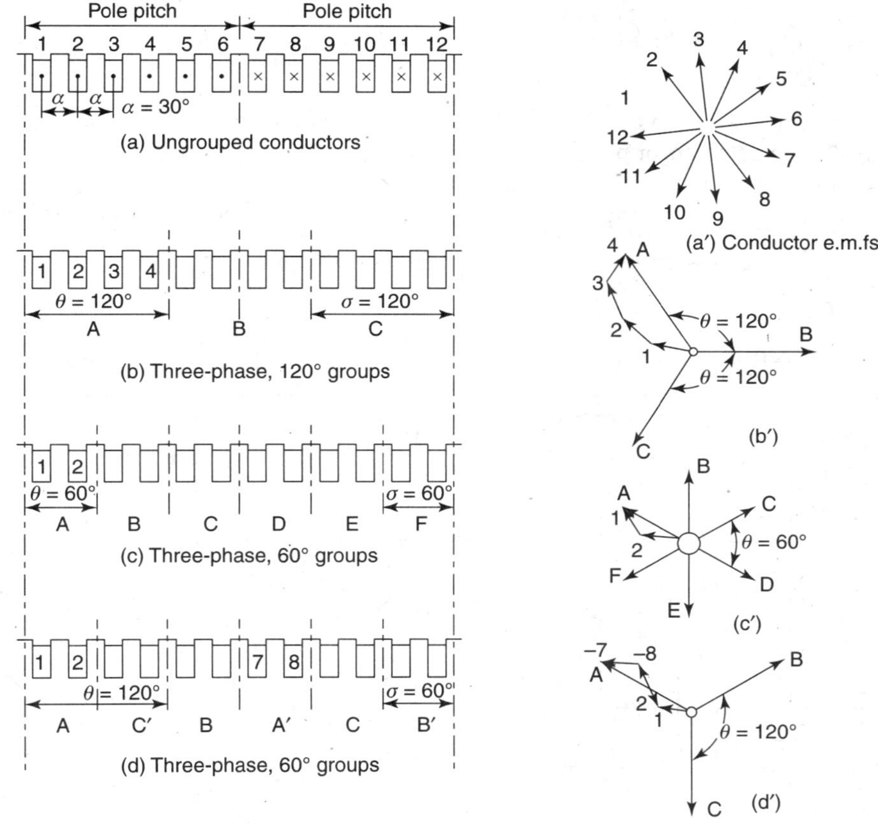

Figure 3.48 Phase Groupings

3.14.7.1 Single-layer Winding

Single-layer windings can be divided into three types:

(i) spiral winding or concentric winding,

(ii) lap winding and

(iii) wave wining.

All the windings shown are full-pitched windings in the sense that the span from centre to centre of adjacent belts of the conductor is equal to the pole pitch. If some of the individual coils have higher or lower pitch than the average pitch, it does not alter the conclusion as to the pitch of the winding as a whole. Single-layer windings are generally used for small machines of a few kW ratings.

They are of two types: (1) concentric-type and (2) mush-type. The characteristics of three-phase single-layer windings are shown as follows:

(i) Concentric single-layer windings are always full-pitched windings.

(ii) In concentric winding, even if coil span of individual coils in coil group is different from pole pitch, the effective coil span is full pitched.

(iii) In mush winding, coil span is constant and hence shape and size of all coils is the same.

(iv) Fractional slot windings cannot be employed for single-layer winding.

(v) In mush winding, coil span is always odd. For example, in two-pole, 14-slot machine, coil span is 14/2 = 7 (full pitched). But in two-pole, 12-slot machine, coil span is 12/2 = 6 since coil span is to be odd. One can use 5 or 7 slots per pole.

From Figure 3.49, we can note that an overhang design is required so that end connections are accommodated in separate tiers or planes.

Overhang is arranged in two or three planes as shown in Figure 3.49(b). For Figure 3.49(a), we have 24 slots with four poles having two-turn coil. So in this arrangement, the number of slots/pole/phase $\frac{24}{3\times 4} = 2$. The pole pitch = total number of slots/number of poles = $\frac{24}{4} = 6$.

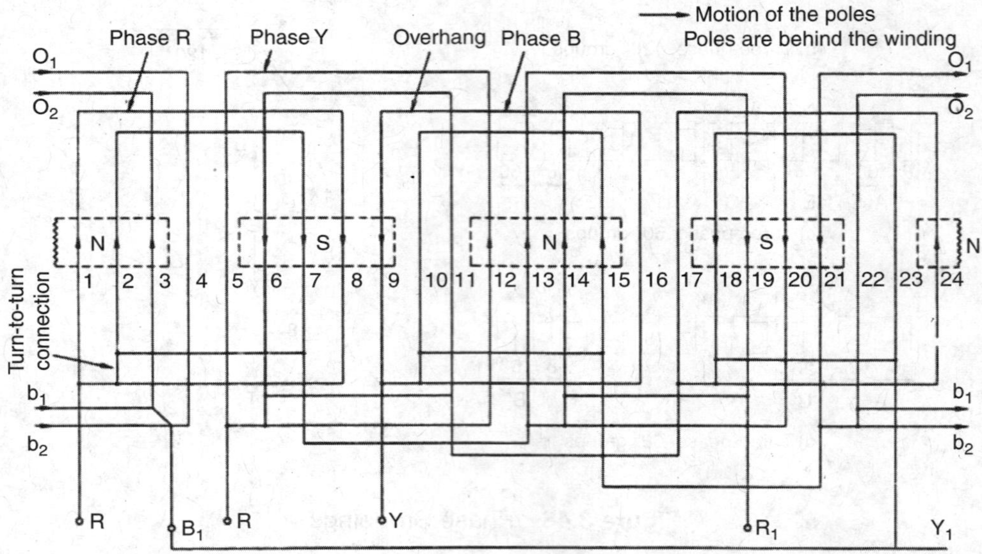

(a) Three-phase single-layer winding

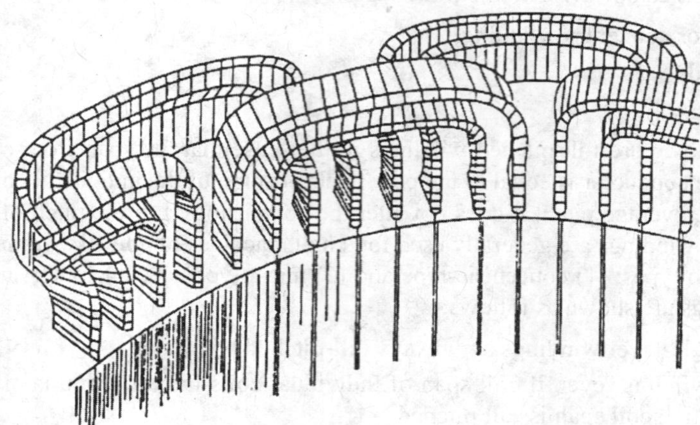

(b) End connections of a three-phase single-layer winding

Figure 3.49 Single-layer Winding

Slot pitch angle = pole pitch/number of slots per pole $\frac{\pi}{6} = 30°$

Phase spread = slot pitch angle × slot/pole/phase = 30° × 2 = 60°

In order to obtain the maximum emf in a coil, the two sides should be one pole pitch apart, that is, if one side of the coil is under the centre of north pole. Then other coil side must be under the centre of south pole (approximately 180 electrical degrees apart). This is for achieving the maximum emf per coil.

By keeping in view, the winding of the above condition is designed for three phases of Figure 3.49(a) which is explained as follows:

- *For phase R:* Phase R starts at slot 1 and passes through slots 8, 2 and 7 before it is taken through slots 13, 20, 14 and 19, which are situated two pole pitches apart.
- *For phase Y:* Now Y phase must start at 120° apart from R phase. Now the Y phase must start $\frac{120}{30} = 4$ slots away from slot 1 since the slot pitch angle is 30°. Therefore, the starting slot of Y phase is 5th. Then further Y phases are completed going through slots 12, 6, 11, 17, 24, 18 and 23.
- *For phase B:* Now B phase must start at 120° apart from Y phase. Now B phase must start from 9th slot and is further completed by going through slots 16, 10, 15, 21, 4, 22 and 3.

So R-Y-B are the starting points of the phases and R′Y′B′ are the end points of the phases. R′Y′B′ are joined together at a point to from the neutral point of the star-connected system. For winding to be delta connected, R′ is connected to Y, Y′ is end connected to B, and B′ is connected to R.

For the winding shown in Figure 6.49(a), the number of coil sides is similar to the number of slots, and also it is to be noted that polar groups of each phase (1, 2 and 13, 14) are two pole pitches apart. In this type of winding, the following points should be noted before designing:

(i) To avoid fouling of end connections, it is necessary to use two different shapes of coils.
(ii) To make it possible that the emfs add to each other, polar groups of each phase must be connected in the same sense.
(iii) Short pitch coils should not be used as they reduce emf.

3.14.7.2 Double-layer Winding

Two-layer windings are of two types: lap wound and wave wound. Here, we discuss the modifications that are required to give three-phase emfs. Two-layer windings are mostly applied in alternators and induction motors of a few kW ratings.

Double-layer windings are usually classified in two types:

(i) integral slot windings and
(ii) fractional slot windings.

3.14.7.3 Integral Slot Winding

If the number of slots per phase (SPP) is an integer, then the winding is called an integral slot winding. If a winding is there with 24 slots and 4 poles, then the number of slots/pole/phase (q) = $\frac{24}{4\times3} = 2$ (an integer). So the arrangement is integral slot winding.

If the number of slots is divisible by the number of phases then we obtain the balanced winding.

Let us consider the three-phase winding with 24 slots and 4 poles. The full pitch of the winding is $\frac{24}{4} = 6$.

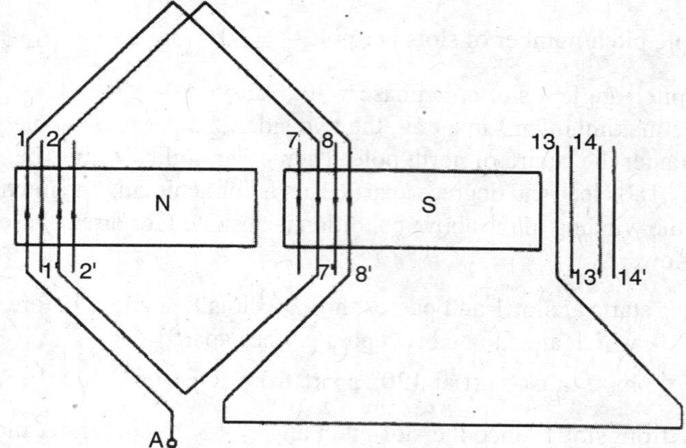

(a) Double-layer, full-pitched integral slot winding

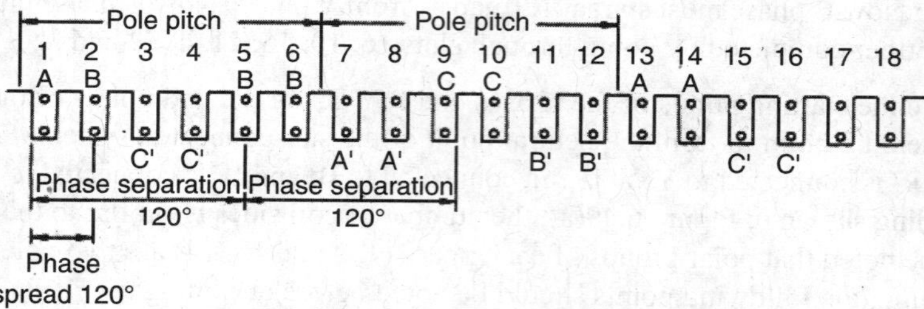

(b) Arrangement of the conductors of different phases

Figure 3.50 Three-phase Integral Lap Winding with 24 Slots and 4 Poles

Also, since they are six slots per pole, the slot pitch angle $(\alpha) = \dfrac{180}{6} = 30$. Now, if it is assumed that the coil span is full pitched, the upper coil side of phase A situated in slot 1 must be connected to the bottom coil side placed in slot 7 (6 + 1), as shown in Figure 3.50(a), since SPP = 2; the conductor of phase A will occupy two slots/pole as shown in Figure 3.50(b). Also, the phase spread is $= 30° \times 2 = 60°$.

Based on the above logic, the winding diagram of A can be completed further from Figure 3.50(b). It can be interpreted that slots 1, 2, 7, 8, 13, 14, 19 and 20 contain coils belonging to phase A. So it can be generalized for full-pitched integral slot belonging to the same phase. Phase B is started 120° from phase A. Since the slot pitch angle is 30°, phase B is started from 5th slot (4 slots away from start of phase A), and similarly, phase C must start from 9th (5 + 4) slot.

3.14.7.4 Fractional Slot Winding

If the number of slots per pole per phase is not an integer, it is termed as fractional slot winding. But for symmetry of winding, the number of slots must be divisible by the number of phases. The various advantages of this type of winding are as follows:

(i) Since the number of poles is not a multiple of number of slots, the existing slotted armature can be used eliminating the saving of punching of new armature.

(ii) High-frequency harmonics in emf and mmf waveforms are reduced.

(iii) This allows higher flexibility in the choice of the coil span. In a machine with 44 slots and 10 poles, the pole pitch is $\frac{44}{10} = 4.4$. Now the coil span can be taken as 4.

The limitations of fractional slot windings are as follows:

(i) They are only useful for double-layer windings.

(ii) Numbers of parallel paths are limited.

Distribution of Coils

The ratio $\frac{\text{number of slots/phase}}{\text{number of poles}} = \frac{14}{10} = \frac{2\times 7}{2\times 5} = \frac{7K_{S_k}}{5K_{P_k}}$ is called as characteristic ratio of fractional slot winding. For a 42-slot and 10-pole winding, the characteristic ratio is $\frac{7}{5}$. The repeatable section of the winding is indicated by the highest common factor (k) between number of slots/phase and number of poles. Here, $k = 2$ which means winding repeats itself for 5 poles. The numerator or term $S_k = 7$ represents the coils per phase distributed among S poles. A tabular method is proposed for knowing the distribution of $S_k(7)$ coils under $P_k(5)$ poles.

Construction of Tables

Total number of columns = $3 \times S_k = 21$

Total number of rows = $p_k = 5$

The sequence of the phase belt is calculated by dividing $3S_k$ columns into three equal sections for a phase spread of 60°. Start from the extreme left and top square with a cross. Now, move to the right in the first row and put a square separated by $P_k(5)$. After the first row, go to the second row and mark a cross at an interval of five squares as shown in Table 3.7.

From Table 3.7, it is interpreted that number of cells of phase A under pole 1 is 2, under pole 2 is 1 and so on, shown in Table 3.8.

Consider a three-phase, four-pole alternator having 24 slots and 2 conductors per slot. In double-layer winding, the number of slots is equal to the number of coils when there are 2 conductors per slot. Pole pitch (assuming coil span of full pitch) $= \frac{\text{number of slots}}{\text{number of poles}} = \frac{24}{4} = 6$. The number of slots/pole/phase $= \frac{6}{3} = 2$. Phase spread $= \frac{180}{6} \times 2 = 60°$. The distribution of conductors in the slot of phase R is shown in the tabular form below.

Table 3.7 Distribution of Coils of Different Phases Under Various Poles

Pole	Phase A						Phase B						Phase C					
1	×					×						×	×					×
2					×						×						×	
3				×						×						×		
4			×						×						×			
5		×					×							×				

Table 3.8 Distribution of Coils of Phases Under Poles

Pole No.	1	2	3	4	5
Phase A	2	–1	1	–1	2
Phase C	–1	1	–2	2	–1
Phase B	+2	–2	1	–1	+1

Table 3.9 Conductors Under Phase R

Poles →	N	S	N	S
Conductors →	3, 4	9, 10	15, 16	21, 22

There are 16 conductors in phase R and these are displaced one pole pitch apart with 4 conductors per pole. The conductors under phase R are shown in Table 3.9.

The conductors in the upper layer are numbered as 1, 2, 3, etc., and are indicated by thick lines. The conductors in bottom layers are numbered by superscripts as 2′, 3′, etc., and are indicated by the bottom lines. The motion of the poles is from left to right and the direction of induced emf in the conductor under each pole is shown in Figure 3.51. Phase R is obtained with the following sequence of connections of conductors:

3–9′–15–21′–4–10′–16–22′–4′–22–16′–10–3′–21–15′–9

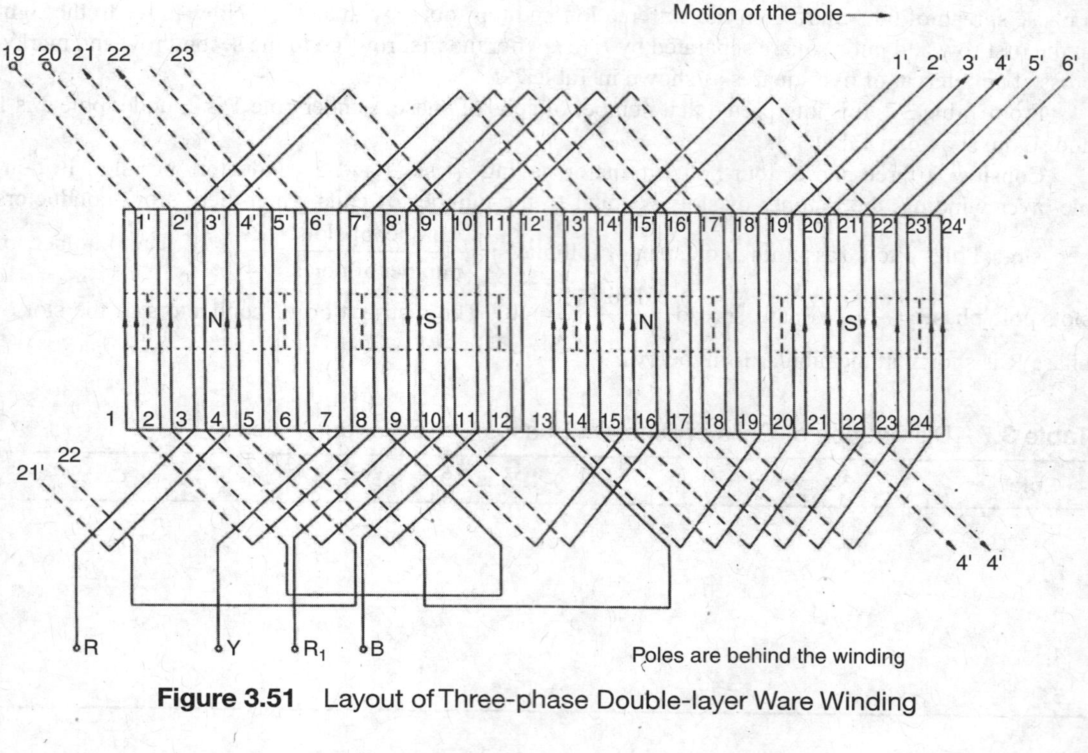

Figure 3.51 Layout of Three-phase Double-layer Ware Winding

3–9′ are connected at the coil span of 6 slots. Now, since conductors 3 and 4 come under the phase spread of R, 21′ is connected to 4. Conductors 4–10′–16′–22 are connected at a coil span of 6 slots. To facilitate end connection strap ST is used to connect 22′ and 4′. The connection of conductors 4′–22–16′–10 is in reverse direction with the coil span of –6. If we add –6 to 4′ we get –2. Now, conductors with number (–2) do not exist; total number of conductors will be –2 added to 24 = 22. So the conductor 4 is connected to conductor 22. The set 3′–21–15′–9 is similar to the set 4′–22–16′–10. The number of slots per pole is taken as 6 and the slot pitch angle is defined as follows:

$$\frac{\text{Pole pitch}}{\text{Number of slots per pole}} = \frac{180}{6} = 30^\circ$$

Now, the phase difference between the *R* and *Y* phases is 120° and the phase difference between emfs induced in conductors of adjacent slots is 30 electrical degrees. So the separation between *R* and *Y* phases should be 4 slot pitches (i.e. 4 × 30° = 120°); for the concept discussed above, the phase *Y* starts from 7th conductor in 7th slot. So *Y* phase conductor sequence combination is as follows:

1–13′–19–1′–8–14′–20–2′–8′–2–20′–14–7′–1–19′–13

Similarly, 120° away from phase B, phase A is started. Thus, the conductors forming the phase B are as follows:

11–17′–23–5′–12–18′–24–6′–12′–6–24′–18–23′–17

Figure 3.52 shows a three-phase, double-layer, two-pole lap winding with slots per pole per phase equal to 2.

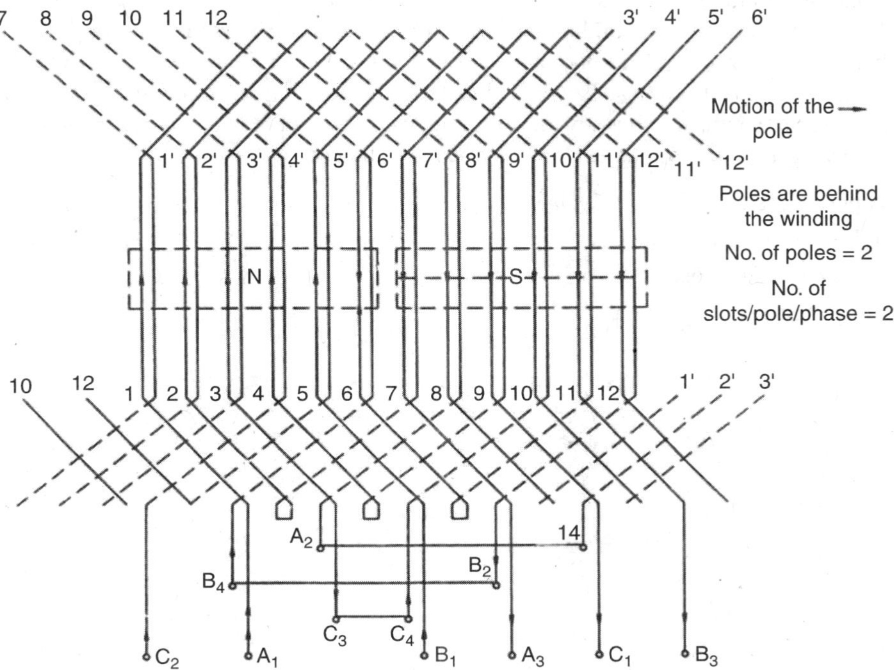

Figure 3.52 Layout of Three-phase, Double-layer Lap Winding

3.15 POLYPHASE INDUCTION MACHINES

Polyphase induction machines are generally used as motors. An induction machine can also operate as a generator. But its performance is not satisfactory for generator operation.

The stator of an induction motor consists of stator frame, stator core, polyphase distributed windings, two-end covers, bearing and so on. The stator frame consists of a symmetrical and substantial casting. The stator core is assembled in the frame under hydraulic pressure. The stator core consists of high-grade, low-loss electrical sheet stampings. The stator carries three-phase windings and is fed from a three-phase supply. Actually, it is wound for a definite number of poles. The number of poles is determined by the requirements of speed. The number of poles produced in a rotating field is given by $P = 2q$, where q is the number of stator slots/pole/phase. Therefore, greater the number of poles, lesser is the speed and vice versa as $N_S = \dfrac{120\,f}{P}$ rpm.

The rotors of induction motors are of two types: (i) squirrel-cage rotor and (ii) wound rotor/slip ring rotor.

3.15.1 Squirrel-cage Rotor

In practice, nearly 90 per cent of induction motors are squirrel-cage-type, which are the simplest and of the most rugged construction, as shown in Figure 3.53. The rotor core is made of laminated silicon steel with slots suitably punched to house the rotor bars. In small machines the punched laminations are stacked and fitted directly into a shaft. In case of large machines, a stack of annular punching of suitable cross-section area is fitted into the spider web arrangement on the shaft. The slots house insulated bars made of copper or aluminium. The solid bars are short circuited at both ends by end rings of the same material. There is no possibility to add any external resistance in series with the rotor circuit for starting purposes because the rotor bars are permanently short circuited on themselves. The rotor bars are not exactly parallel to the shaft. They are given a slight skew which helps to make the motor run quietly by reducing the magnetic hum. Another advantage is that this skew reduces the locking tendency of the rotor.

The following are the advantages of squirrel-cage induction motor:

(i) Cheaper in cost.
(ii) Simple and rugged in construction.
(iii) Low maintenance cost.
(iv) Explosion proof.

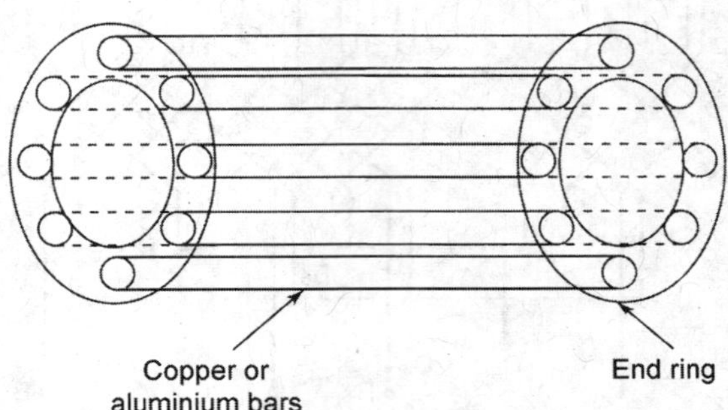

Figure 3.53 Squirrel-cage Rotor of Induction Motor

(v) Due to base end rings, these machines can be cooled better.
(vi) More pull-out torque.
(vii) Maximum power output.
(viii) Starting arrangement is simple.
(ix) Speed is nearly constant.
(x) Overload capacity is high.
(xi) Power factor is high.

The following are the disadvantages of squirrel-cage induction motor:

(i) Poor starting torque due to low rotor resistance.
(ii) Draws a large current at the time of switching.
(iii) Very sensitive to the supply voltage.
(iv) Power factor is low at light load.
(v) Speed regulation is not possible.

3.15.2 Wound Rotor

This type of rotor also has slotted stampings, which are former wound. The rotor slots in the wound rotor accommodate an insulated copper winding very much like the stator winding. The rotor windings are distributed uniformly and usually connected in star. The three ends from the star connection are brought out at the three slip rings mounted on but insulated from the shaft. This type of rotor is also called *slip ring rotor* as the rotor is wound with polyphase windings and carries slip rings. Figure 3.54 shows a wound rotor induction motor.

The stator of a polyphase induction motor receives electrical energy from an alternating voltage source. Therefore, it is called *singly excited machine.* After supply is given to the stator, a rotating flux is produced in the air gap rotating at synchronous speed. The concept of rotating magnetic field (RMF) has been discussed in **Section 3.29**. This flux sweeps past the rotor surface cutting the stationary rotor

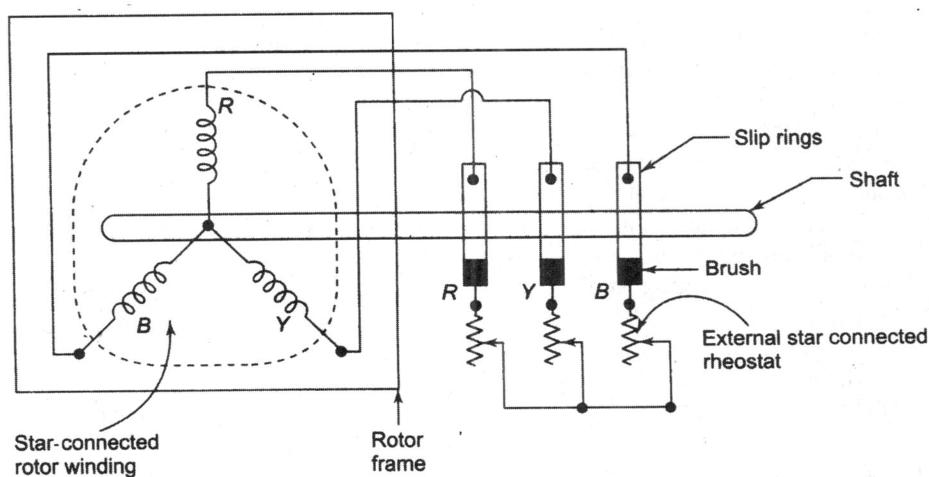

Figure 3.54 Wound Rotor Induction Motor

conductors. According to Faraday's law of electromagnetic induction, an emf is induced in the rotor conductors due to the relative speed between the rotor and the rotating flux. Due to the relative velocity between the rotor and the RMF, the rotor will try to oppose the very cause of its induced emf as per Lenz's law. It will turn to follow the RMF. Stator winding connected to the supply is called the primary winding similar to the transformer where the rotor winding is called the secondary winding as it receives energy from the stator by mutual flux. Thus, induction motor may be taken as a generalized transformer.

The following are the advantages of the wound rotor or slip ring induction motors:

(i) High starting torque.
(ii) High overload capacity.
(iii) Nearly constant speed.
(iv) Low starting current.

The following are the disadvantages of the wound rotor induction motor:

(i) Low power factor at light loads.
(ii) Lower efficiency.
(iii) Sensitive to fluctuation in supply voltage.
(iv) Maintenance cost is higher.
(v) Speed regulation is poor during operation with external resistances in the rotor circuit.

3.16 AIR GAP

The main source of low power factor at which induction motor operates is the air gap between the stator and the rotor. This air gap increases the reluctance between the stator and the rotor, which enhances the magnetizing current for production of the given mutual flux between the stator and the rotor for a given supply voltage. Therefore, the no-load current of an induction motor becomes 30 to 40 per cent of full-load current, which is remarkably larger than the no-load current of a transformer for a given kVA rating. The air gap in an induction motor should be made small so that the induction motor gives better performance. The small air gap may result mechanical problems in addition to the noise and losses at the slot tooth faces.

3.17 PRINCIPLE OF OPERATION OF THREE-PHASE INDUCTION MOTOR

An induction motor with one slot/pole/phase with the stator and rotor wound for three phases and two poles is shown in Figure 3.55. Figure 3.56 shows that the rotor of the induction motor is open circuited initially and the three-phase supply is given to the stator. This produces an rotating magnetic field (RMF) in the air gap between the stator and the rotor, which rotates at synchronous speed. This RMF induces an emf in the stator winding, which balances the supply voltage. This RMF also induces an emf in the rotor winding. No rotor current flows in spite of emf induced in it because it is open circuited. Due to transformer

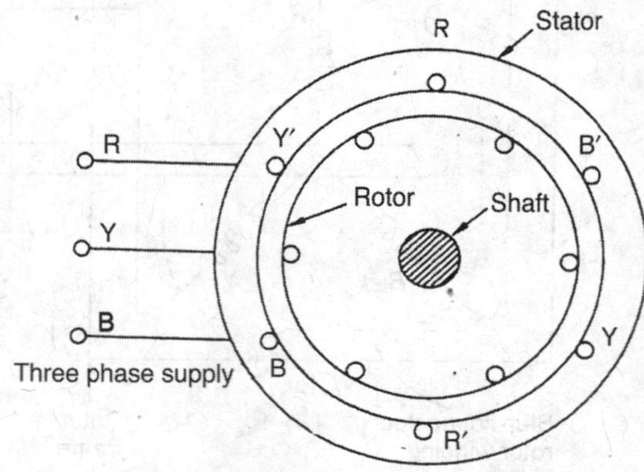

Figure 3.55 Induction Motor

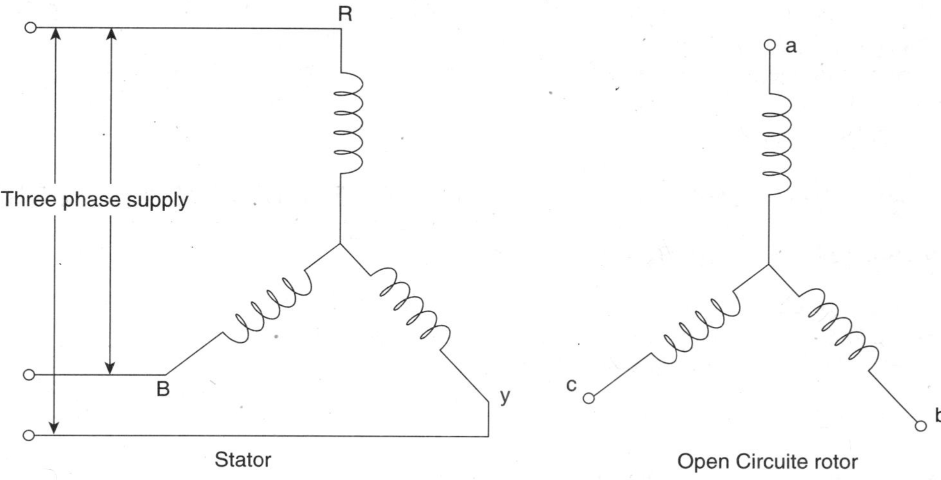

Figure 3.56 Rotor Winding Open Circuited

action between the stator and the rotor, the frequency of the induced emf between the stator and the rotor remains same and the rotor remains stationary in the RMF. Therefore, the machine behaves like a transformer, where the stator resembles as a primary and the rotor resembles as a secondary and the induced emfs in both the stator and the rotor have the same frequency.

Figure 3.57 shows that the rotor of the induction motor is short circuited and there will be induction of balanced three-phase currents in the rotor due to the induced emf in the rotor, which produce RMF of the same speed as that of the stator in space. The torque is produced due to interaction between stator and rotor mmfs. It can be said that the rotor behaves like a current-carrying conductor, which is placed in the RMF of the stator. Hence, it will experience a torque whose direction will be as per Lenz's law. The direction of this torque is such that it will rotate the rotor in the direction of the RMF and the rotor starts to rotate in the same direction of the RMF. It will try to catch up the speed of RMF. But the rotor will not be able to attain

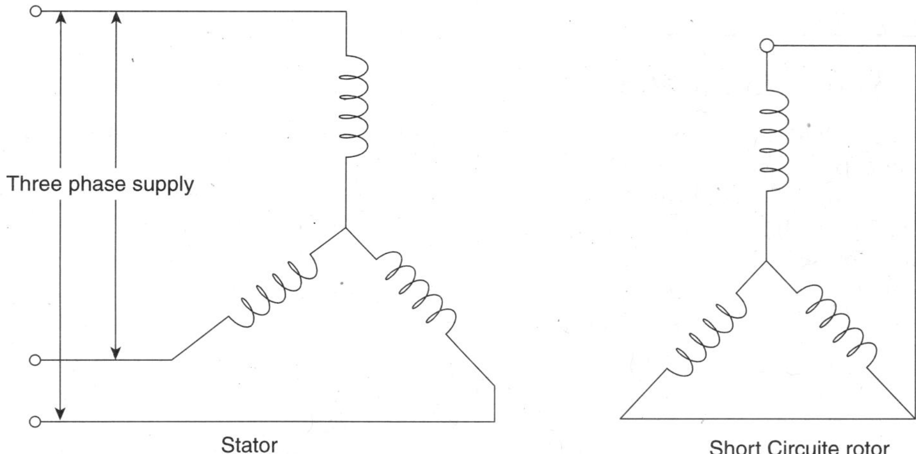

Figure 3.57 Rotor Winding Short Circuited

the synchronous speed. If it attains the synchronous speed, there will be no relative motion between the rotor conductors and RMF. Hence, the induced emf and the corresponding induced current in the rotor become zero, resulting no torque developed on the rotor. The speed of the rotor decreases. When the rotor speed falls below the speed of RMF, a torque will be experienced by the rotor due to induced current in it. Therefore, an average speed (N_s) attained by the rotor is less than the speed (N_s) of RMF.

3.18 SYNCHRONOUS SPEED AND SLIP IN INDUCTION MOTOR

3.18.1 Synchronous Speed

If a balanced three-phase supply is fed to the stator of an induction motor, an RMF is produced in the air gap between the stator and the rotor, which revolves in space at a speed known as synchronous speed (N_s). It is related to the stator supply frequency (f) and can be expressed as follows:

$$f = \frac{PN_s}{120} \tag{3.5}$$

where N_s is the synchronous speed of the RMF, P is the number of poles on the stator and f is the frequency of the stator supply voltage.

Equation (3.5) can also be expressed as follows:

$$N_s = \frac{120f}{P} \tag{3.6}$$

Table 3.10 shows the synchronous speed (N_s) for different number of poles (P) along with the probable speed attained by the rotor on full load for a given supply frequency (f= 50 Hz).

Table 3.10 Values of N_s and N_r for Different Number of Stator Poles (P)

Number of Poles (P)	Synchronous Speed N_s (rpm)	Full-load Speed N_r (rpm)
2	3,000	2,950
4	1,500	1,440
6	1,000	950

3.18.2 Slip in Induction Motor

The speed (N_r) of the rotor is always less than the synchronous speed (N_s). The slip in an induction motor is the difference in speed of N_s and N_r. Figure 3.58 shows the variation of speed and slip with torque of an induction motor. The speed of an induction motor is AC when the torque is OA and the corresponding slip is AB. There is linear increase of slip of an induction motor with torque at moderate loads. The slip in per unit is expressed as follows:

$$\text{Slip (p.u.)} = \frac{\text{slip in rpm}}{N_s} = \frac{AD}{AB}$$

i.e.,

$$s = \frac{N_s - N}{N_s} = 1 - \frac{N}{N_s} \tag{3.7}$$

i.e.,

$$\%\ \text{slip} = \text{per unit slip} \times 100 \tag{3.8}$$

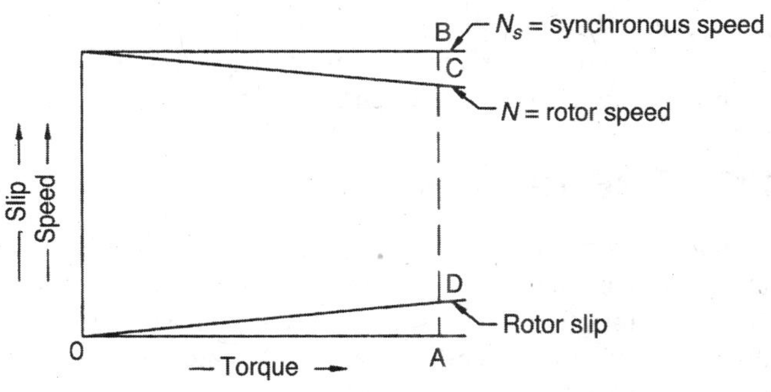

Figure 3.58

The slip for large size motors at full load varies from 2 per cent to 4 per cent, while for small motors its value can be as high as 8 per cent. Practically, induction motor can be taken as constant speed motors in spite of slip. The speed control of induction motor is more complicated and expensive than the methods used for the speed control of DC shunt motors although both of them have similar speed–torque characteristic. From Equation (3.8), we have $s = 0$ for $N = N_s$ on no-load and $s = 1$ for $N = 0$ at the time of starting. Therefore, the slope of an induction motor varies from 0 (on no-load) to 1 (at the time of starting).

Example 3.3 A 16-pole, three-phase alternator is coupled to an engine running at 375 rpm. It supplies to an induction motor that has a full-load speed of 1,450 rpm. Find the slip and the number of poles of the motor.

Solution

Synchronous speed (N_s) = 375 rpm

Number of poles = 16

$$\therefore \quad f = \frac{PN_s}{120} = \frac{16 \times 375}{120} = 50\,\text{Hz}$$

The speed of the induction motor is 1,450 rpm. Since an induction motor runs at a speed less than its synchronous speed under normal operating conditions, the synchronous speed is 1,500 rpm for a frequency of 50 Hz.

$$\therefore \quad N_s = 1{,}500\ \text{rpm}$$

$$\therefore \quad \text{Slip } (s) = \frac{N_s - N_r}{N_s} = \frac{1{,}500 - 1{,}450}{1{,}500} = 0.033 = 3.3\%$$

3.19 FREQUENCY OF ROTOR CURRENTS

The rotor of an induction motor runs at a speed (N_r), which is less than the synchronous speed (N_s) of rotating magnetic field. The frequency of the induced currents in the rotor of the induction motor depends on the difference between the speeds N_s and N. If the frequency of the induced currents in the rotor is f_2, it can be expressed as follows:

$$f_2 = \frac{(N_s - N)P}{120} = \frac{N_s - N}{N_s} \times \frac{PN_s}{120}$$

i.e.,

$$f_2 = sf \qquad (3.9)$$

Equation (3.9) shows the relation between the frequency of the rotor current and the supply frequency. If the normal full-load slip of an induction motor is of the order of 1 to 4 per cent, the frequency (f_2) of the induced currents in the rotor will be 1–4 Hz when the supply frequency (f) is 50 Hz.

3.20 SPEED OF THE ROTOR MMF

It is known that a revolving magnetic field is produced when a balanced three-phase supply is applied to stator winding of the three-phase induction motor. The axis of this revolving field rotates at synchronous speed relative to stator conductors. In a similar manner, the three-phase induced current in the rotor produces a resultant emf. The axis of this emf rotates in the direction of the rotation of the rotor at a speed $(N_s - N)$ with respect to rotor conductors. The rotor rotates in the air gap at a speed N with respect to the stator core, which is less than the synchronous speed. Therefore, the speed of the resultant rotor emf in space $= (N_s - N) + N = N_s$. Therefore, the resultant rotor mmf rotates at a speed, which is same as the resultant stator mmf in space. Hence, the mmfs produced by the stator and the rotor are stationary relative to each other in space. They differ only in phase where the phase angle between the two mmfs increases with the increase of load. The phase difference between the two mmfs becomes $\frac{\pi}{2} + \theta_2$, where θ_2 is the internal power factor of the rotor.

3.21 ELECTRICAL AND MECHANICAL DEGREES

The conductor rotates at a uniform angular velocity, shown in Figure 3.59(a). The emf induced in the conductor is zero at '*a*' because the corresponding flux is zero at '*a*'. The emf induced in the conductor is maximum at '*b*' because the corresponding flux is maximum at '*b*'. The emf induced in conductor is zero at '*c*' because the corresponding flux is zero at '*c*'. The emf induced in conductor is maximum at '*d*' because the corresponding flux is maximum at '*d*'. The flux entering the stator or leaving the rotor is assumed to be positive. Figures 3.59(b) and 3.59(c) show the corresponding flux and emf variations, respectively. For one complete rotation of the rotor, one cycle of emf is obtained for two poles. Figure 3.59(d) shows four poles. Figures 3.59(e) and 3.59(f) show the corresponding flux and emf variations, respectively. For one complete rotation of the rotor, two cycles of emf will be generated.

One revolution indicates 360 mechanical degrees and one cycle of emf indicates 360 electrical degrees. So two cycles of emf is equivalent to 720 electrical degrees.

$$720 \text{ electrical degrees} = \frac{4}{2} \times (360 \text{ mechanical degrees})$$

$$\phi_e = \left(\frac{P}{2}\right)\theta_m$$

$$\frac{d\theta_e}{dt} = \frac{P}{2}\left(\frac{d\theta_m}{dt}\right)$$

i.e.,

$$\omega_e = \frac{P}{2}\omega_m \tag{3.10}$$

where ω_e is the angular velocity in electrical radians per second and ω_m is the angular velocity in mechanical radians per second.

For a four-pole machine, two cycles are generated for one revolution of the rotor. Therefore, for a *P*-pole machine, *P*/2 cycles per second are generated in one revolution per second. Therefore, we can

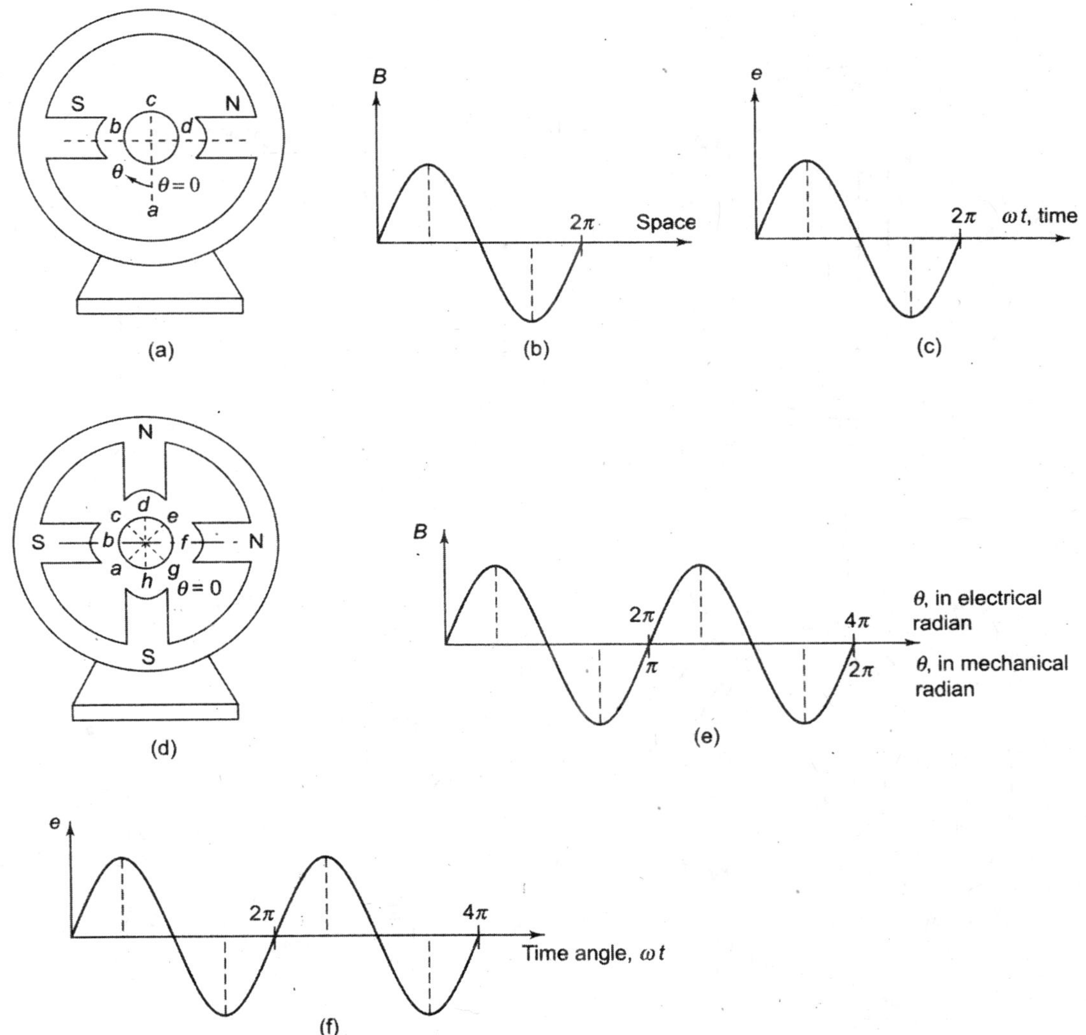

Figure 3.59 Electrical and Mechanical Degrees

conclude that for a P-pole machine, in 'n' revolution/second, $\frac{Pn}{2}$ cycles/second are generated. But cycles/second is termed as frequency (f) of the generated emf wave.

$$f = \frac{Pn}{2} \text{ Hz} \tag{3.11}$$

If the speed n is in rpm,

$$f = \frac{Pn}{120} \text{ Hz} \tag{3.12}$$

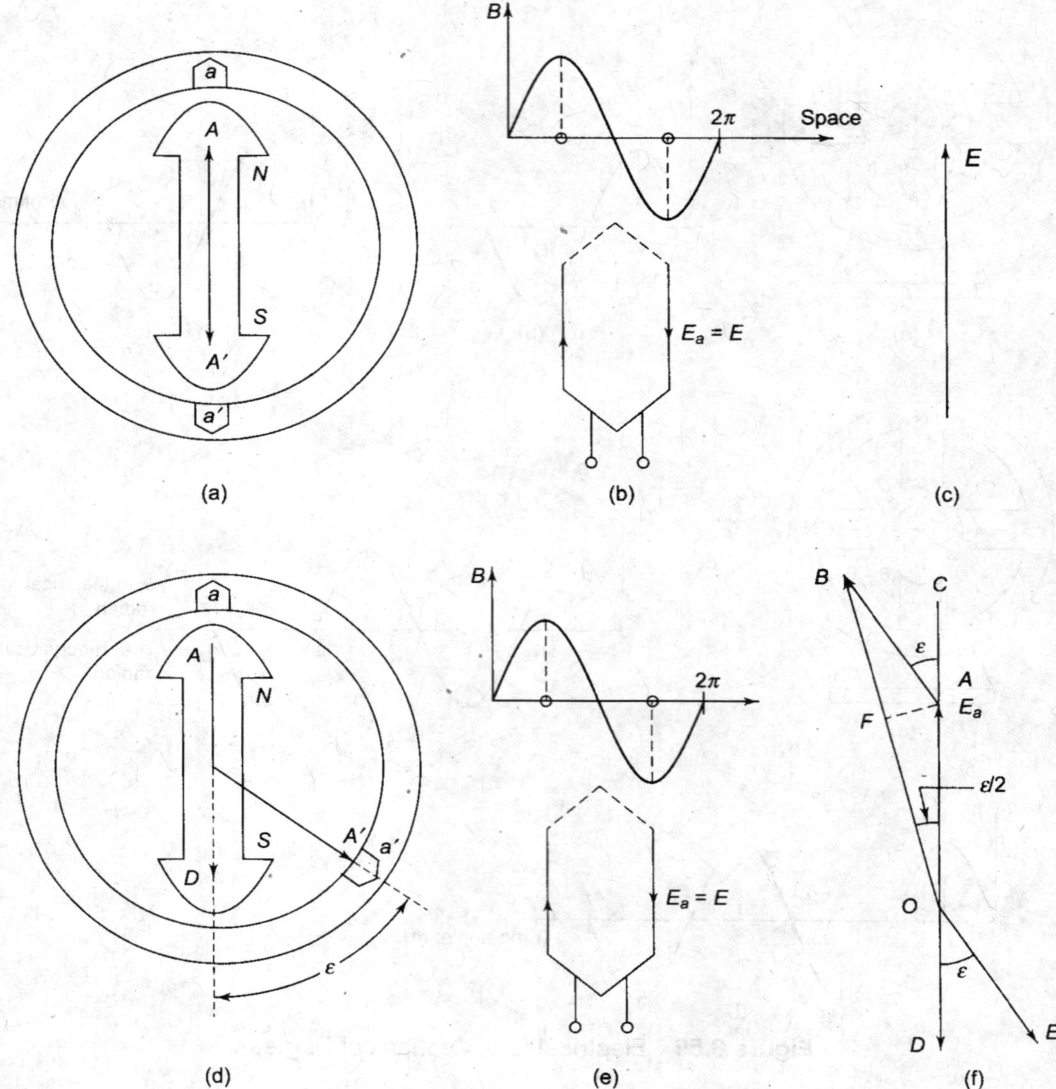

Figure 3.60 Full-pitched and Fractional-pitched Winding

3.22 PITCH FACTOR

Figures 3.60(a) and 3.60(d) show full-pitched and fractional pitched coils, respectively. In full-pitched winding, the resultant emf is $2E$ as shown in Figures 3.60(b) and 3.60(c). It is not equal to $2E$ in fractional pitched winding. It is less compared to the resultant emf in full-pitched winding, shown in Figures 3.60(e) and 3.60(f).

In the fractional pitched winding, emfs induced in the two coils are E (= OA) and $E \cos \frac{\varepsilon}{2}$ (= OD).

From the phasor diagram, Figure 3.60(f), the pitch factor for fractional pitched winding can be calculated as follows:

$$\angle DOE = \angle BAC = \varepsilon$$

Now $$\angle BAO = 180° - \varepsilon$$

∴ $$\angle FAO = \frac{1}{2}\angle BAO = 90° - \frac{\varepsilon}{2}$$

$$\angle FOA = 90° - \angle FAO = \frac{\varepsilon}{2}$$

∴ $$E_R = OB = OF + FB = E\cos\frac{\varepsilon}{2} + E\cos\frac{\varepsilon}{2} = 2E\cos\frac{\varepsilon}{2}$$

The ratio of resultant emf of a fractional-pitched coil to the resultant emf for the same coil when it is full-pitched,

$$K_p = \frac{\text{resultant emf of a fractional pitched coil}}{\text{resultant emf of a full-pitched coil}} = \frac{2E\cos\frac{\varepsilon}{2}}{2E}$$

i.e., $$K_p = \cos\frac{\varepsilon}{2} \tag{3.13}$$

For nth harmonic,

∴ $$K_{np} = \cos\frac{n\varepsilon}{2} \tag{3.13}$$

Fractional pitched windings have the following advantages:

(i) The cost of the machine is less due to reduction of copper in overhang.
(ii) There is reduction of harmonics in the generated emf wave. So, the output emf wave becomes almost a sine wave.

Example 3.4 Find the pitch factors of the following windings:

(i) 32 stator slots, 4 poles, coil span is 1–7.
(ii) 72 stator slots, 8 poles, coil span is 1–8.

Solution

(i) 32 stator slots, 4 poles, coil span is 1–7.

$$\beta = \frac{2}{8}\times 180° = 45°$$

$$K_p = \cos\frac{\beta}{2} = \cos\frac{45°}{2} = 0.924$$

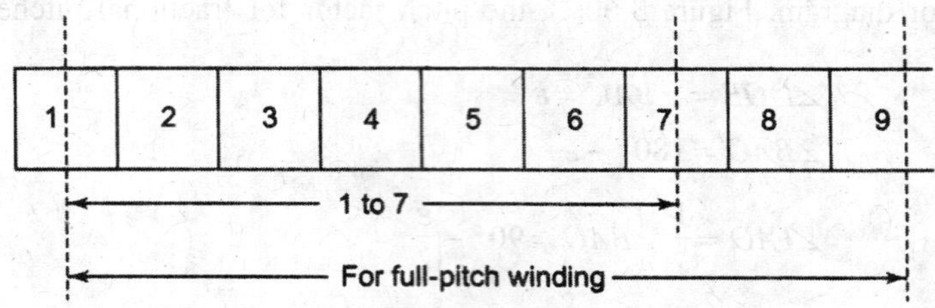

Figure E3.3

(ii) 72 stator slots, 8 poles, coil span is 1–8.

$$\beta = \frac{2}{9} \times 180^\circ = 40^\circ$$

$$K_p = \cos\frac{\beta}{2} = \cos\frac{40^\circ}{2} = 0.940$$ ■

3.23 DISTRIBUTION FACTOR

Armature winding is not usually concentrated in a single slot for rotating electrical machines. The armature winding is distributed uniformly in the slots of the armatures. The main advantages of distribution of the windings in slots are as follows:

(i) The output emf becomes almost a sine wave as there is reduction of harmonics in the generated emf wave.

(ii) It increases rigidity and mechanical strength of the winding.

(iii) There is full utilization of the armature iron and copper.

The ratio of emf induced in a distributed winding to the emf induced if the winding is concentrated is called *distribution factor/breadth factor.*

$$K_d = \frac{\text{emf induced in the distributed winding}}{\text{emf induced if the winding is concentrated}}$$

$$= \frac{\text{vector sum or phasor sum of coil emf}}{\text{arithmetic sum of coil emfs}}$$

The value of K_d is always less than unity.

Suppose n = number of slots/pole, q = number of slots/pole/phase, γ = angular displacement between the slots = $\frac{180°}{n}$, E = induced emf in each coil side.

The emf induced in different coils of one phase under one pole is shown in Figure 3.61 by *AB*, *BC*, *CD*, *DE*, etc., that are equal in magnitude but differ in phase (say α) from each other. Let the bisectors of *AB*, *BC*, *CD*, *DE*, etc., meet at a common point *O*. The point *O* is called the circumcentre of the circle having *AB*, *BC*, *CD*, *DE*, etc., as the chords.

Here, emf induced in each coil side is given by

$$E = AB = 2OA \sin\frac{\gamma}{2}$$

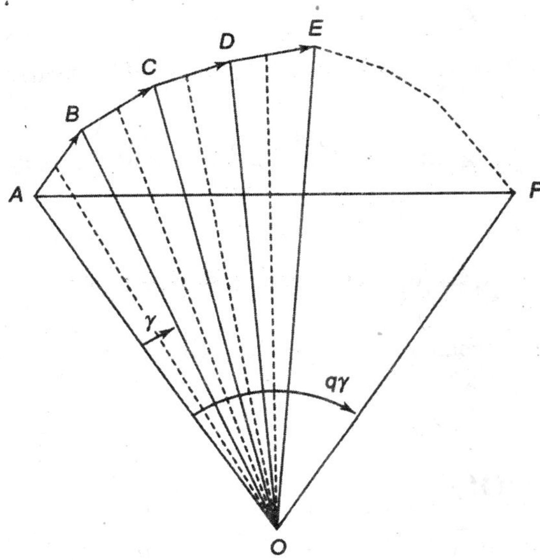

Figure 3.61 Calculation for Distribution Factor

∴ Arithmetic sum of emf $= q \times 2 \times OA \times \sin\frac{\gamma}{2}$

and the resultant emf $= AF = 2 \times OA \sin\frac{q\gamma}{2}$

$$\text{Distribution factor } (K_d) = \frac{\text{vector sum or emf}}{\text{arithmetic sum of emf}} = \frac{2 \times OA \times \sin\frac{q\gamma}{2}}{2q \times OA \times \sin\frac{\gamma}{2}}$$

∴
$$K_d = \frac{\sin\frac{q\gamma}{2}}{q\sin\frac{\gamma}{2}} \tag{3.15}$$

The term $q\gamma$ is also known as the *phase spread* and is expressed in electrical radians.

For the nth harmonic,
$$K_{dn} = \frac{\sin\frac{nq\gamma}{2}}{q\sin\frac{n\gamma}{2}} \tag{3.16}$$

Example 3.5 Calculate the distribution factor for a 36-slot, four-pole, single-layer, three-phase winding.

Solution

Slots/pole $(n) = 36/4 = 9$

$\gamma = 180°/9 = 20°$

$$\text{Slots/pole/phase } (q) = \frac{36}{4\times 3} = 3$$

$$\therefore \qquad \text{Distribution factor } (K_d) = \frac{\sin\left(\frac{q\gamma}{2}\right)}{q\sin\frac{\gamma}{2}} = \frac{\sin\frac{3\times 20^\circ}{2}}{3\times\sin\frac{20^\circ}{2}} = 0.9597 = 0.96$$

■

3.24 WINDING FACTOR

It is defined as the product of distribution factor (K_d) and pitch factor (K_p). It is denoted by the symbol K_w.

$$\therefore \qquad K_w = K_d \times K_p \tag{3.17}$$

3.25 FLUX PER POLE

Figures 3.62(a) and 3.62(b) show the field windings on the rotor and the stator, respectively. In both cases cylindrical rotors have been considered. It does not make any difference if salient poles are considered. The starting of space angle is taken from field winding axis. When $\theta = 0°$, flux density B is maximum. For $\theta = 90°$, flux density is minimum. When $\theta = 180°$, B is again maximum but in the opposite direction. For $\theta = 270°$, flux density is minimum. B is again maximum for $\theta = 360°$ and so on. The flux density wave can be written as follows:

$$\therefore \qquad B = B_{max}\cos\theta \tag{3.18}$$

An elemental angle $d\theta$ at an angle θ from the field winding axis is considered to calculate the flux per pole. If l be the axial length of armature core, the curved area of the elemental area is $lrd\theta$,

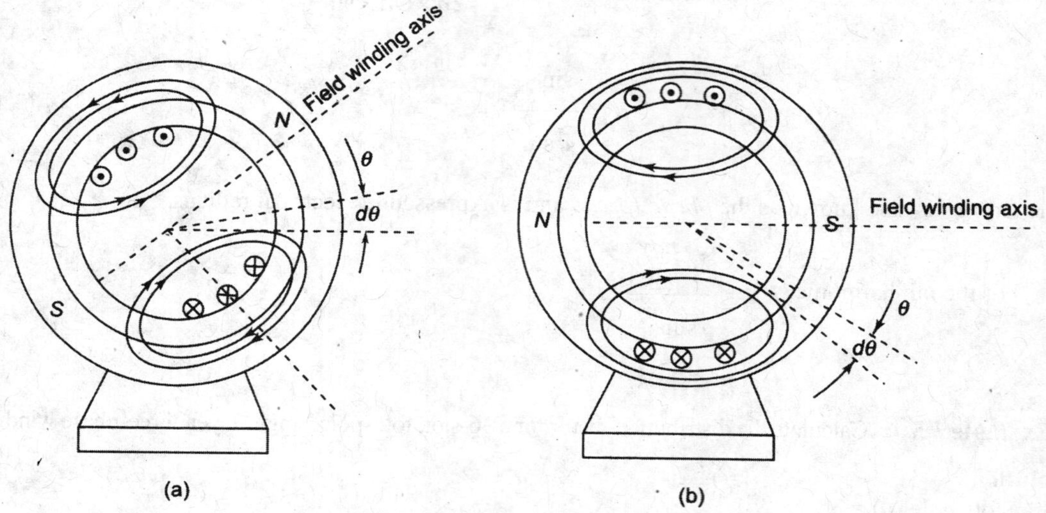

Figure 3.62 Calculation of Flux per Pole

where r is the radius of the rotor. The portion of flux passing through the elemental surface area $lrd\theta$ is given by

$$d\phi = Blrd\,\theta$$

i.e.,
$$d\phi = B_{max}\cos\theta\, lrd\theta \tag{3.19}$$

Total flux can be obtained by integrating Equation (3.19) from $\theta = -\pi/2$ to $+\pi/2$ as follows:

$$\phi = \int_{-\pi/2}^{\pi/2} B_{max} lr\cos\theta d\theta = 2B_{max} lr$$

i.e.,
$$\phi = 2B_{max}\, lr \tag{3.20}$$

The value of the flux per pole is obtained in Equation (3.20) for a two-pole machine.

Per pole area for a two-pole machine $= \dfrac{2\pi rl}{2} = \pi rl$

Therefore, per pole area for a P-pole machine $= \dfrac{2\pi rl}{P} = \dfrac{2}{P}(\pi lr)$

Now for a two-pole machine, flux per pole (with pole area πlr) $= 2B_{max}rl$

Therefore, for a P-pole machine, flux per pole $\left(\text{with pole area } \dfrac{2\pi rl}{P}\right)$

$$= (2B_{max}lr)\times\left(\frac{1}{\pi rl}\right)\times\frac{2}{P}(\pi rl) = \frac{4}{P}B_{max}lr \tag{3.21}$$

3.26 GENERATED EMF IN FULL-PITCHED COIL

Figures 3.63(a) and 3.63(b) show a two-pole machine. In Figures 3.62(a) and3.62(b), the pole is on the stator and the rotor, respectively. For simplicity two-pole machines have been considered. It is also assumed that there is an N-turn full-pitched armature coil. From Figure 3.63, it is seen that when the field winding axis coincides with the stator coil axis, flux linking with the coil is maximum; that is, at

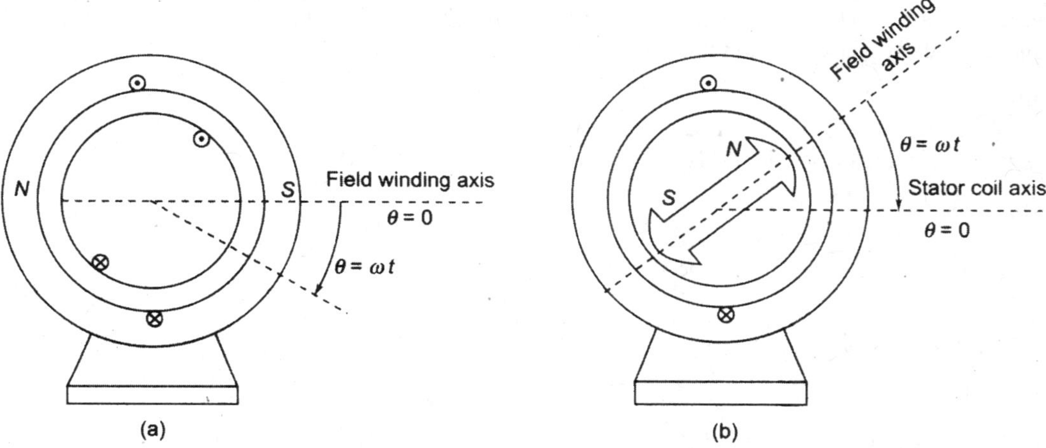

Figure 3.63 Illustration of Generated emf in a Single-turn Full-pitched Coil

$t = 0, \theta = \omega_r t = 0$ and the flux passing through the coil is maximum, that is, $\phi \cos 0°$ or ϕ. At time angle $\theta = \omega_r t = 90°, \phi$ linking the coil is minimum. Again at $\theta = \omega_r t = 180°$, ϕ linking the coil is maximum but in the opposite direction. So the flux linkage Ψ with the full-pitched armature coil having N turns per phase at any instant t is given by

$$\Psi = N\phi \cos \omega_r t$$

$$e = -\frac{d\Psi}{dt} = -N\frac{d\phi}{dt} = -N\frac{d\phi}{dt}\cos\omega_r t + N\phi\,\omega_r \sin\omega_r t \quad (3.22)$$

Since flux ϕ is time invariant, the term $N\frac{d\phi}{dt}\cos\omega_r t$ called *transformer-voltage term* is zero. The term $N\omega_r\phi \sin \omega_r t$ called *speed-voltage term* will exist. Equation (3.22) can be expressed as follows:

$$e = N\omega_r\phi \sin \omega_r t \quad (3.23)$$

From Equation (3.23), we conclude that the induced emf lags behind the flux by an angle 90°.

The emf generated in a short-pitched distributed winding is given by

$$e = N\omega_r\phi K_w \sin \omega_r t = N\omega_r\phi K_w \cos(\omega_r t - 90°) \quad (3.24)$$

Equation (3.24) suggests that the induced emf lags behind the flux by an angle 90°.

3.27 EMF GENERATED IN AC MACHINES

In rotating electrical machines, the armature turns are not concentrated in slots. The armature turns are usually distributed in slots to utilize the armature periphery completely. Distribution of the coils in different slots helps in reduction of harmonics. Besides the distribution, the windings are made short pitch.

The emf generated in a short pitch distributed winding is given by

$$E = 4.44 f N_{ph}\phi K_p K_d \quad (3.25)$$

where N_{ph} is the number of turns per phase.

Example 3.6 A three-phase, 16-pole alternator has a star-connected winding with 144 slots and 10 conductors per slot. The flux per pole is 0.03 Wb sinusoidally distributed and speed is 375 rpm. Find the frequency and the phase emf and line emf. Assume coil is full pitched.

Solution

Here, $N = 375$ rpm, $P = 16$, $K_P = 1$, $\phi_m = 0.03$ Wb

Slots = 144, conductor per slots = 10

$\therefore$ Slots/pole $(n) = \frac{144}{16} = 9$

$\therefore$ Angular displacement between the slots $(\gamma) = \frac{180°}{9} = 20°$

$\therefore$ Slots/pole/phase $(q) = \frac{144}{16 \times 3} = 3$

$\therefore$ Distribution factor $(K_d) = \frac{\sin\frac{q\gamma}{2}}{q\sin\frac{\gamma}{2}} = \frac{\sin\frac{3\times 20°}{2}}{3\sin\frac{20°}{2}} = 0.96$

Since the coil is full pitched, $K_p = 1$

$\therefore$ Number of conductors/phase $(Z) = \dfrac{144\times10}{3} = 480$

$\therefore$ Number of turns/phase $(N) = \dfrac{Z}{2} = \dfrac{480}{2} = 240$

$\therefore$ Induced emf/phase (E_{ph}) is given by

$$E_{ph} = 4.44\,K_p\,K_d\,f\,\phi_m N$$

$$E_{ph} = 4.44\times1\times0.96\times50\times0.03\times240 = 1{,}534.46\,\text{V}$$

Line voltage $(E_L) = \sqrt{3}\,E_{ph} = \sqrt{3}\times1534.46 = 2657.77\text{V}$ ■

3.27.1 Synchronous Machines

3.27.1.1 Armature Winding

The field winding is excited with DC so that the flux does not vary with time. Let N_{ph} be the total number of series turns per phase of the polyphase armature winding. The emf generated in any one phase is given by

$$E = 4.44\,f\,N_{ph}\,\phi\,K_w \tag{3.26}$$

Let the time origin be taken from the instant when flux linkages with phase are maximum. The emf induced for phase a can be written as follows:

$$e_a = \sqrt{2}\,E\sin\omega t \tag{3.27}$$

For phases b and c,

$$e_b = \sqrt{2}\,E\sin(\omega t - 120°) \tag{3.28}$$

$$e_c = \sqrt{2}\,E\sin(\omega t + 120°) \tag{3.29}$$

The relative velocity between the flux density wave and armature winding must be synchronous because the synchronous machine runs at synchronous speed (N_s). In a synchronous machine the rotational frequency is $f = \dfrac{PN_s}{120}$ Hz for which the machine has been designed. The emf generated per phase of a synchronous machine is given by

$$E = 4.44\,f\,N_{ph}\,\phi\,K_w \tag{3.30}$$

3.27.1.2 Field Winding

Let V_f be direct voltage applied to the field winding. We can write

$$V_f = R_f\,i_f + L_f\frac{di_f}{dt}$$

3.27.2 Induction Machines

3.27.2.1 Stator

If a balanced polyphase voltage is applied to the stator winding, it produces an RMF of constant amplitude in the air gap. This speed is called synchronous speed. There is relative velocity between the constant amplitude rotating field and the stationary stator winding. So an emf is induced in the stator winding. The magnitude of this induced emf in any one phase is given by

$$E_{ph} = 4.44 f K_w N_{ph} \phi \tag{3.31}$$

where f is the frequency of the applied voltage and N_{Ph} is the stator series turns per phase.

3.27.2.2 Rotor

In an induction motor, the RMF in the air gap rotates at a synchronous speed $N_s = \frac{120f}{P}$ rpm where P is the number of stator poles. Since the rotor rotates at a speed N_r rpm less than the synchronous speed in the same direction of the RMF, the relative speed between the RMF and the rotor is $(N_s - N_r)$ rpm. $N_s - N_r$ rpm is called the *slip speed.*

$$\therefore \quad \text{Slip } (s) = \frac{N_s - N_r}{N_s} = \frac{\text{synchronous speed} - \text{sotor speed}}{\text{synchronous speed}}$$

$$\therefore \quad N_r = (1 - s) N_s \tag{3.32}$$

The frequency of induced or generated emf in rotor coil is $f_r = sf$.
So the emf induced or emf generated in one rotor coil is

$$e = 4.44 f_r N_{ph} K_w \phi \tag{3.33}$$

i.e.,

$$e = 4.44 s f N_{ph} K_w \phi \tag{3.34}$$

where N_{ph} is the rotor series turns/phase.

The frequency of the rotor emf is sf, which is called the *slip frequency.* This is denoted by f_r.

If the rotor is being rotated at a speed (N_r rpm) in the opposite direction of the rotating field (N_s rpm), the slip is given by

$$s = \frac{N_s + N_r}{N_s} \tag{3.35}$$

3.27.3 A General Expression for the emf of Synchronous Generator

The emf generated per phase in a full-pitched alternator having concentrated winding is expressed by

$$E_{ph} = 4.44 N_{ph} f \phi \tag{3.36}$$

If k_w be the winding factor, the expression for the emf induced per phase is expressed by

$$E_{ph} = 4.44 N_{ph} K_w f \phi \tag{3.37}$$

For non-sinusoidal flux density, harmonics will be present in the generated emf. The harmonics can be estimated using Fourier analysis of the flux density waveform. Figure 3.64 shows a general

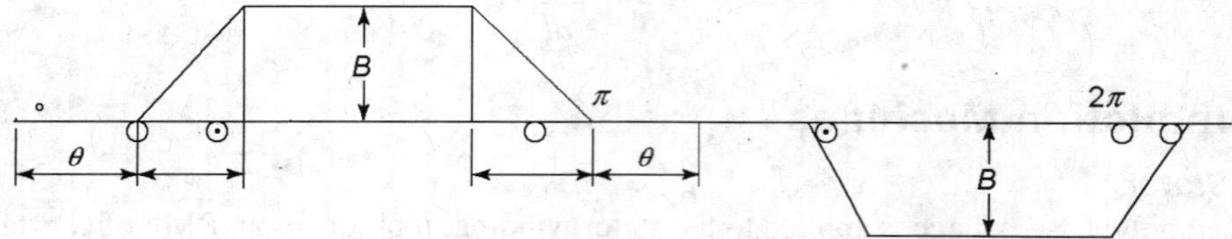

Figure 3.64 Fractional-pitched Trapezoidal Waveform

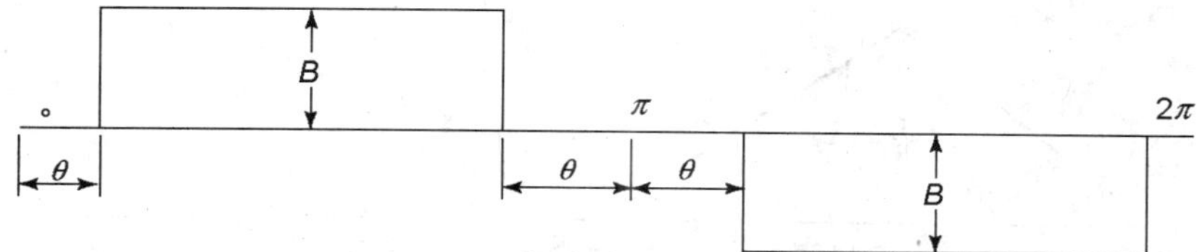

Figure 3.65 Fractional-pitched Rectangular Distribution When $\gamma = 0$

trapezoidal waveform of flux density. Using the suitable value of θ and γ, all other waveforms can be deduced from it.

Figure 3.64 shows fractional pitched trapezoidal waveform.

B_n = magnitude of the nth harmonic

$$= \frac{8}{\pi} \times \frac{B}{\pi^2 \gamma} \sin\left(\frac{n\gamma}{2}\right) \cos n\left(\theta + \frac{\gamma}{2}\right)$$

B_1 = magnitude of the fundamental

$$= \frac{8}{\pi} \times \frac{B}{\pi^2 \gamma} \sin\left(\frac{\gamma}{2}\right) \cos\left(\theta + \frac{\gamma}{2}\right)$$

Figure 3.65 shows fractional pitched rectangular waveform.

B_n = magnitude of the nth harmonic $= \frac{4}{\pi} \times \frac{B}{n} \cos n\theta$

B_1 = magnitude of the fundamental $= \frac{4}{\pi} \times B \cos\theta$

Figure 3.66 shows full-pitched rectangular waveform.

B_n = magnitude of the nth harmonic $= \frac{4}{H} \times \frac{B}{n}$

B_1 = magnitude of the fundamental $= \frac{4}{H} \times B$

Figure 3.67 shows full-pitched triangular distribution.

B_n = magnitude of the nth harmonic $= (-1)^{(n-1)/2} \frac{8}{\pi^2} \times \frac{B}{n^2}$

B_1 = magnitude of the fundamental $= \frac{8}{\pi^2} \times B$

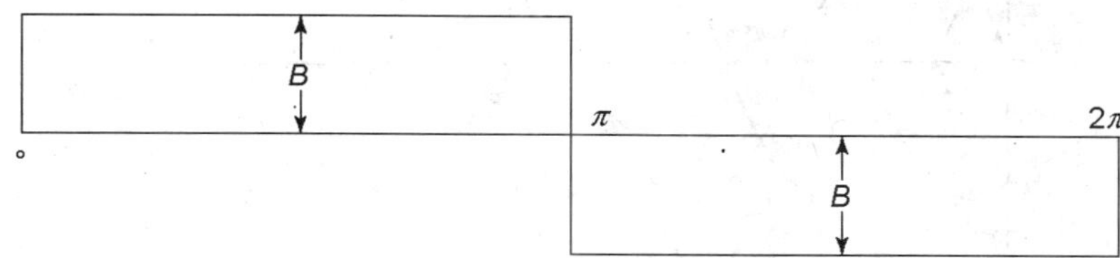

Figure 3.66 Full-pitched Rectangular Waveform

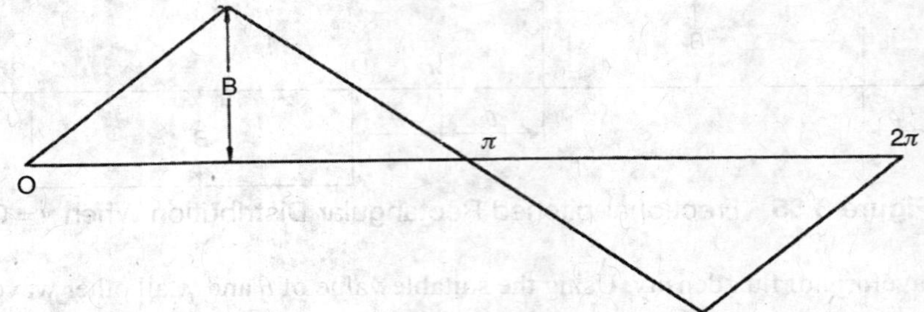

Figure 3.67 Full-pitched Triangular Distribution

Therefore, the resultant emf of the waveform containing harmonics is expressed by

$$E_r = \sqrt{E_1^2 + E_2^2 + E_3^2 + \cdots + E_n^2} \tag{3.38}$$

where E_1, E_2, .., E_n are the fundamental, second, ..., and nth harmonics.

3.28 EMF GENERATED IN DC GENERATOR

The emf generated in a DC generator is given by

$$E = \frac{\phi NZ}{60}\left(\frac{P}{A}\right) \tag{3.39}$$

where ϕ = flux/pole, P = number of generator poles, Z = total number of armature conductors, A = number of parallel paths, N = speed of armature in rpm.

For derivation of Equation (3.39), see Figure 4.5, Chapter 4.

3.29 CONCEPT OF ROTATING MAGNETIC FIELD

The stator of synchronous and induction motor may be star or delta connected. The three-phase windings are displaced from each other by an angle 120°. Figure 3.68 shows that the three-phase windings are supplied by a balanced three-phase supply having phase sequence R–Y–B. The current through the

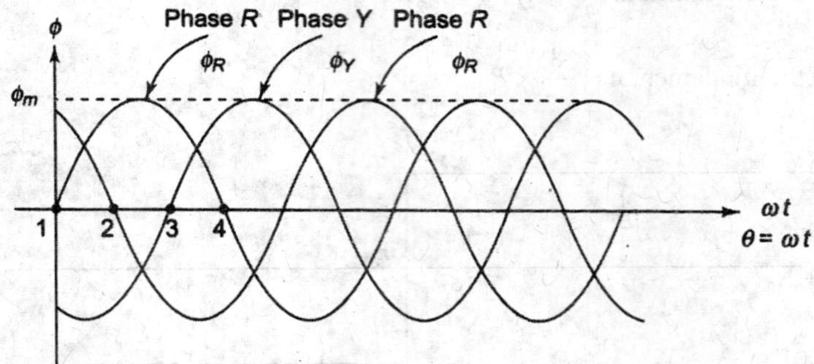

Figure 3.68 Waveform of Fluxes

three windings is displaced from each other by an angle of 120° electrical. The mathematical expressions for instantaneous values of the three fluxes are given by

$$\phi_R = \phi_m \sin \omega t \tag{3.40}$$

$$\phi_Y = \phi_m \sin(\omega t - 120°) \tag{3.41}$$

$$\phi_B = \phi_m \sin(\omega t - 240°) \tag{3.42}$$

Figure 3.68 shows the waveform of fluxes. The resultant flux is $(\bar{\phi}_T)$.

$$\bar{\phi}_T = \bar{\phi}_R + \bar{\phi}_Y + \bar{\phi}_B \tag{3.43}$$

Let us consider the following cases.

3.29.1 Case 1: $\omega t = \theta = 0°$

Substituting $\theta = 0°$ in Equations (3.40)–(3.42), we have

$$\phi_R = \phi_m \sin \omega t = 0$$

$$\phi_Y = \phi_m \sin(-120°) = -0.866\,\phi_m$$

$$\phi_B = \phi_m \sin(-240°) = +0.866\,\phi_m$$

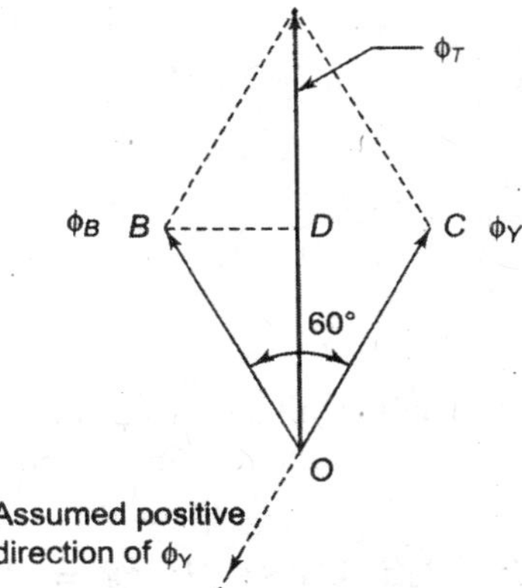

Figure 3.69 Phasor Addition of Fluxes: Case 1

Figure 3.69 shows the phasor addition of the above three fluxes. From Figure 3.69, we have

$$OD = DA = \phi_T/2$$

Again, $$\angle BOD = 30°$$

Therefore, $$\cos 30° = \frac{OD}{OB} = \frac{\frac{\phi_T}{2}}{0.866\phi_m}$$

i.e., $$\phi_T = 2 \times 0.866\,\phi_m \times \cos 30° = 1.5\,\phi_m.$$

∴ The magnitude of ϕ_T is 1.5 ϕ_m and its position is vertically upwards.

3.29.2 Case 2: $\omega t = \theta = 60°$

Substituting $\theta = 60°$ in Equations (3.40)–(3.42), we have

$$\phi_R = \phi_m \sin 60° = 0.866\phi_m$$
$$\phi_Y = \phi_m \sin(60° - 120°) = -0.866\,\phi_m$$
$$\phi_B = \phi_m \sin(60° - 240°) = 0$$

Figure 3.70 shows the phasor addition of the above three fluxes. From Fig. 3.70, we have

$$OD = DA = \phi_T/2$$

Again, $$\angle BOD = 30°$$

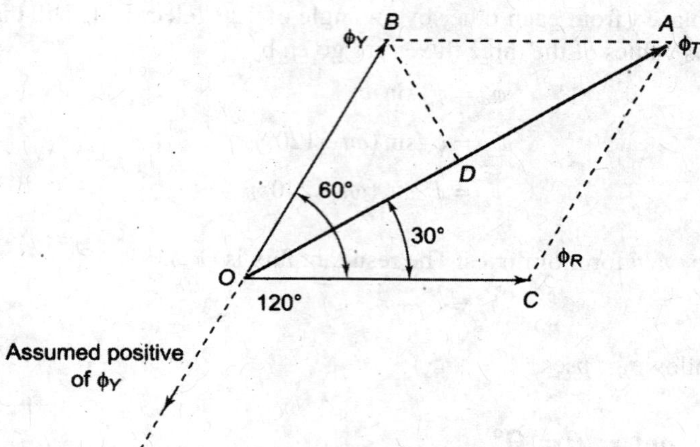

Figure 3.70 Phasor Addition of Fluxes: Case 2

$$\therefore \qquad \cos 30° = \frac{OD}{OB} = \frac{\frac{\phi_T}{2}}{0.866\phi_m}$$

i.e.,

$$\phi_T = 2\times 0.866\,\phi_m \times \cos 30° = 1.5\ \phi_m.$$

∴ The magnitude of ϕ_T is 1.5 ϕ_m and it is rotated through 60° in space in the clockwise direction compared to its previous position in Case 1.

3.29.3 Case 3: $\omega t = \theta = 120°$

Substituting $\theta = 120°$ in Equations (3.40)–(3.42), we have

$$\phi_R = \phi_m \sin 120° = 0.866\,\phi_m$$

$$\phi_Y = \phi_m \sin(120° - 120°) = 0$$

$$\phi_B = \phi_m \sin(120° - 240°) = -0.866\,\phi_m$$

Figure 3.71 shows the phasor addition of the above three fluxes. From Figure 3.71, we have

$$OD = DA = \phi_T/2$$

Again,

$$\angle BOD = 30°$$

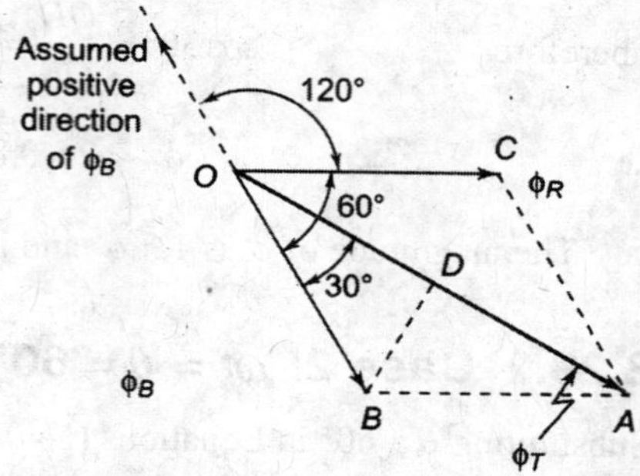

Figure 3.71 Phasor Addition of Fluxes: Case 3

$$\therefore \qquad \cos 30° = \frac{OD}{OB} = \frac{\frac{\phi_T}{2}}{0.866\phi_m}$$

i.e.,

$$\phi_T = 2\times 0.866\,\phi_m \times \cos 30° = 1.5\ \phi_m.$$

∴ The magnitude of ϕ_T is 1.5 ϕ_m, and it as rotated through 120° in space in the clockwise direction compared to its previous position in Case 1.

3.29.4 Case 4: $\omega t = \theta = 180°$

Substituting $\theta = 180°$ in Equations (3.40)–(3.42), we have

$$\phi_R = \phi_m \sin 180° = 0.$$
$$\phi_Y = \phi_m \sin(180° - 120°) = 0.866\phi_m$$
$$\phi_B = \phi_m \sin(180° - 240°) = -0.866\phi_m$$

Figure 3.72 shows the phasor addition of the above three fluxes. From Figure 3.72, we have

$$OD = DA = \phi_T/2$$

Again, $\angle BOD = 30°$

Therefore, $$\cos 30° = \frac{OD}{OB} = \frac{\frac{\phi_T}{2}}{0.866\phi_m}$$

i.e., $$\phi_T = 2 \times 0.866\phi_m \times \cos 30° = 1.5\ \phi_m.$$

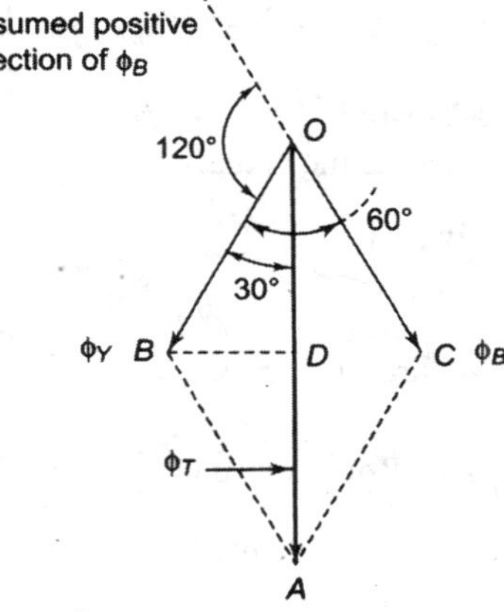

Figure 3.72 Phasor Addition of Fluxes: Case 4

Therefore, the magnitude of ϕ_T is 1.5 ϕ_m, and it as rotated through 180° in space in the clockwise direction compared to its previous position in Case 1.

Therefore, rotating magnetic flux is produced when three stationary windings are excited by balanced three-phase AC supply. The produced resulting field is called *rotating magnetic field* which rotates at a synchronous speed $\frac{120f}{P}$, where f is the frequency of the supply in Hz and P is the number of poles for which the winding is wound.

Figure 3.73 shows the directions of rotation of the RMF. The RMF rotates in the clockwise direction when phase sequence is R–Y–B, while it rotates in the anticlockwise direction for phase sequence R–B–Y.

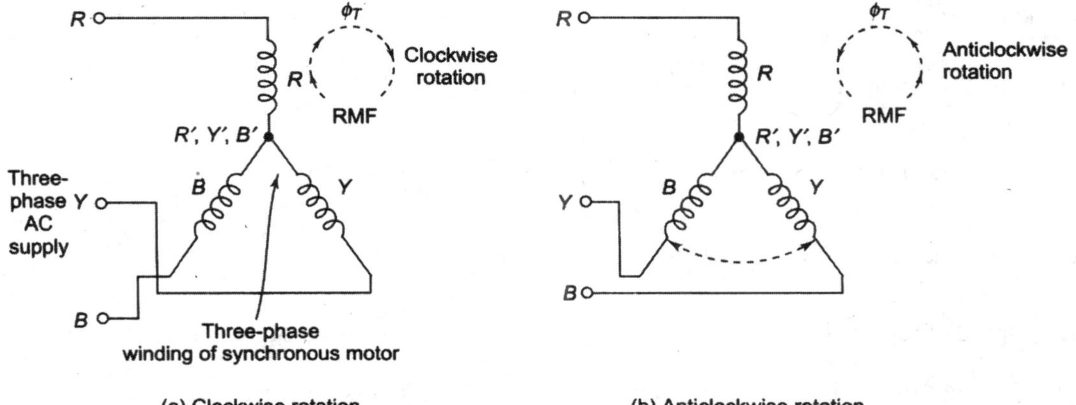

Figure 3.73 Direction of Rotation of RMF

Please see Appendix C for mmf of distributed windings and Appendix D for torque in AC and DC machines.

ADDITIONAL SOLVED PROBLEMS

Example 3.7 A three-phase induction motor has two poles and is connected to 400 V, 50 Hz supply. Calculate the actual rotor speed and rotor frequency when the slip is 4 per cent.

Solution

$$P = 2,\ s = 4\% = 0.04,\ f = 50\,\text{Hz}$$

Synchronous speed $(N_s) = \dfrac{120f}{P} = \dfrac{120\times 50}{2} = 3{,}000$ rpm

Rotor speed $(N_r) = N_s(1-s) = 3{,}000 \times (1-0.04) = 2{,}880$ rpm

Rotor frequency $(f_r) = sf = 0.04\times 50 = 2\,\text{Hz}$ ■

Example 3.8 A three-phase induction motor runs at almost 1,000 rpm at no-load and 950 rpm at full load when supplied with power from a 50 Hz three-phase line. Calculate the following:

(i) How many poles does the motor have?
(ii) What is the percentage slip at full load?
(iii) What is the corresponding frequency of rotor voltage?
(iv) What is the corresponding speed of the rotor field with respect to the rotor?
(v) What is the corresponding speed of the rotor with respect to the stator?
(vi) What is the corresponding speed of the rotor field with respect to the stator field?
(vii) What is the rotor frequency at the slip of 10 per cent?

Solution

$$N_0 = 100\,\text{rpm},\ N_r = 950\,\text{rpm},\ f = 50\,\text{Hz}$$

Here, no-load speed of the motor is almost 1,000 rpm. Therefore, synchronous speed near to 1,000 rpm is 1,000 rpm.

(i) Poles on motor, $P = \dfrac{120f}{N_s} = \dfrac{120\times 50}{1{,}000} = 6$

(ii) Percentage slip at full load, $s_f = \dfrac{N_s - N_r}{N_s}\times 100 = \dfrac{1{,}000-950}{1{,}000}\times 100 = 5\%$

(iii) Frequency of rotor voltage, $f_r = s\times f = \dfrac{5\times 50}{100} = 2.5\,\text{Hz}$

(iv) Speed of rotor field with respect to rotor $= \dfrac{120\times f_r}{P} = \dfrac{120\times 2.5}{6} = 50\,\text{rpm}$

(v) Speed of rotor with respect to stator = 950 rpm (since stator is stationary)

(vi) Rotor field and stator field are revolving at the same speed of 1,000 rpm. Therefore, speed of rotor field with respect to stator field is zero

(vii) Rotor frequency at the slip of 10%, $f_r = s'\times f = \dfrac{10}{100}\times 50 = 5\,\text{Hz}$

Example 3.9 A three-phase, four-pole, 50 Hz induction motor has a slip of 2 per cent at no-load and 4 per cent at full load. Calculate (i) synchronous speed, (ii) no-load speed and (iii) full-load speed.

Solution

$$P = 4, f = 50 \text{ Hz}, s_0 = 2\% \text{ and } s_{fl} = 4\%$$

(i) $N_s = \dfrac{120f}{P} = \dfrac{120 \times 50}{4} = 1{,}500 \text{ rpm}$

(ii) $N_0 = (1 - s_0) N_s = (1 - 0.02) \times 1{,}500 = 1{,}470 \text{ rpm}$

(iii) $N_{fl} = (1 - s_{fl}) N_s = (1 - 0.04) \times 1{,}500 = 1{,}440 \text{ rpm}$

Example 3.10 The stator of a three-phase, 16-pole alternator has 144 slots, and there are four conductors per slot connected in two layers. The conductors of each phase are connected in series. If the speed of the alternator is 375 rpm, calculate the emf induced per phase. Resultant flux in the air gap is 5 × 10^{-2} Wb per pole sinusoidal distributed. Assume the coil span as 150° electrical.

Solution

$$N_s = 375 \text{ rpm}$$

$$\varepsilon = 180° - 150° = 30° \text{ rpm}$$

$$K_p = \cos\frac{\varepsilon}{2} = \cos\frac{30°}{2} = 0.966$$

∴ Number of slots/pole $(n) = \dfrac{144}{16} = 9$

∴ $$\gamma = \frac{180°}{n} = \frac{180°}{9} = 20°$$

∴ $$K_d = \frac{\sin\frac{q\gamma}{2}}{q\sin\frac{\gamma}{2}} = \frac{\sin\frac{3 \times 20°}{2}}{3\sin\frac{20°}{2}} = 0.96$$

∴ $$f = \frac{N_s P}{120} = \frac{375 \times 16}{120} = 50 \text{ Hz}$$

Number of sots/phase $= \dfrac{144}{3} = 48$

Here, number of conductors per slot = 4

∴ Number of conductors in the series/phase $(Z) = 48 \times 4 = 192$

∴ Turns/phase $(N) = \dfrac{Z}{2} = \dfrac{192}{2} = 96$

∴ $$E_{ph} = 4.44 K_p K_d f \phi_m N$$

$$= 4.44 \times 0.966 \times 0.96 \times 50 \times 5 \times 10^{-2} \times 96$$

$$= 988.19 \text{ V}$$

■

Example 3.11 A four-pole, three-phase, 50 Hz star-connected alternator has 60 slots with four conductors per slot. Coils are short pitched by three slots. If the phase spread is 60°, find the line voltage induced for a flux per pole of 0.943 Wb distributed sinusoidally in space. All the turns per phase are in series.

Solution

$$q = \frac{60}{4\times3} = 5$$

Hence, phase spread $(q\gamma) = 60°$

$$\gamma = \frac{60°}{q} = \frac{60°}{5} = 12°$$

$$\therefore \quad K_d = \frac{\sin\frac{q\gamma}{2}}{q\sin\frac{\gamma}{2}} = \frac{\sin\frac{5\times12°}{2}}{5\sin\frac{12°}{2}} = 0.957$$

The number of slots per phase $(n) = \frac{60}{4} = 15$

Since the coil is short pitched by three slots,

$$\varepsilon = \frac{3}{15}\times180° = 36°$$

$$\therefore \quad K_p = \cos\frac{\varepsilon}{2} = \cos18° = 0.951$$

$\therefore$ Number of conductors/phase $(Z) = \frac{60\times4}{3} = 80$

$\therefore$ Number of turns/phase $(N) = Z/2 = 40$

$\therefore$ emf induced per phase $(E_{ph}) = 4.44 K_p K_d f\phi N$

$$= 4.44\times0.951\times0.957\times50\times0.943\times40 = 7621.09\text{ V}$$

$\therefore$ Line voltage $(E_L) = \sqrt{3}\,E_{ph} = \sqrt{3}\times7621.09 = 13200.11$ V ■

Example 3.12 Calculate the rms value of the induced emf per phase of a 10-pole, three-phase, 50 Hz alternator with two slots per pole per phase and four conductors per slot in two layers. The coil span is 150°. The flux per pole has a fundamental component of 0.12 Wb and a 20 per cent of third harmonic component.

Solution

(i) Fundamental emf

Here, $\varepsilon = 180° - 150° = 30°$

$$K_p = \cos\frac{\varepsilon}{2} = \cos\frac{30°}{2} = 0.966$$

$$q = 2$$

∴ Number of slots/pole = 2 × 3 = 6

$$K_d = \frac{\sin\frac{q\gamma}{2}}{q\sin\frac{\gamma}{2}} = \frac{\sin\frac{2\times 30^\circ}{2}}{2\sin\frac{30^\circ}{2}} = 0.966$$

Number of conductors per phase (Z) = 10 × 2 × 4 = 80

Number of turns per phase (N) $= \frac{Z}{2} = \frac{80}{2} = 40$

$$\therefore \quad E_{ph1} = 4.44\,K_p\,K_d\,f\phi N$$

$$= 4.44 \times 0.966 \times 0.966 \times 50 \times 0.12 \times 40 = 994.37\,\text{V}$$

(ii) Third harmonic emf

$$K_{3p} = \cos\frac{3\varepsilon}{2} = \cos\frac{3\times 30^\circ}{2} = 0.707$$

$$K_{3d} = \frac{\sin\frac{3q\gamma}{2}}{q\sin\frac{3\gamma}{2}} = \frac{\sin\frac{3\times 2\times 30^\circ}{2}}{2\sin\frac{3\times 30^\circ}{2}} = 0.707$$

$$f_3 = 3f = 50\times 3 = 150\,\text{Hz}$$

$$\phi_3 = \frac{1}{3}\times 20\%\,\text{of fundamental flux} = 0.008\ \text{Wb}$$

$$E_{ph3} = 4.44\,K_{3p}\,K_{3d}\,f_3\phi_3 N$$

$$= 4.44 \times 0.707 \times 707 \times 150 \times 0.008 \times 40 = 106.53\,\text{V}$$

$$\therefore \quad E/\text{phase} = \sqrt{E_{ph1}^2 + E_{ph3}^2} = \sqrt{994.37^2 + 106.53^2} = 1{,}000.06\,\text{V}$$ ■

Example 3.13 A three-phase, four-pole, 50 Hz star-connected alternator has 60 slots with two conductors per slot and has armature winding of the double-layer type. Coils are short-pitched; that is, if one coil side lies in slot number 1, the other coil side lies in slot number 13. Find the useful flux per pole required to induce a line voltage of 6.6 kV.

Solution

Number of poles (P) = 4

Conductors/slot = 2, E_L = 6.6 kV

Number of slots = 60

Let the useful flux per pole be ϕ

$$\text{Number of slots/pole} = \frac{60}{4} = 15$$

$$\text{Number of slots/pole/phase} = \frac{60}{4\times 3} = 5$$

$$\text{Number of conductors connected in series/phase } Z = \frac{60\times 2}{3} = 40$$

$$\text{Number of series turns/phase } (N) = \frac{Z}{2} = \frac{40}{2} = 20$$

$$\text{Angular displacement between the slots } (\gamma) = \frac{180°}{15} = 12°$$

$$K_d = \frac{\sin\frac{q\gamma}{2}}{q\sin\frac{\gamma}{2}} = \frac{\sin\frac{5\times12°}{2}}{5\sin\frac{12°}{2}} = 0.957$$

$$\text{Coil span} = \frac{\text{coil span in terms of slots}}{\text{number of slots/pole}} = \left(\frac{13-1}{15}\right)\times 180° = 144°$$

Short pitch angle (ε) = 180° − 144° = 36°

$$\therefore \quad K_p = \cos\frac{\varepsilon}{2} = \cos\frac{36°}{2} = 0.951$$

Since $\quad E_L = \sqrt{3}\times 4.44 K_p K_d\, f\phi N$

$$\therefore \quad \phi = \frac{E_L}{\sqrt{3}\times 4.44 K_p K_d\, f} = \frac{6.6\times10^3}{\sqrt{3}\times 4.44\times0.951\times0.957\times50\times20} = 0.943 \text{ Wb}$$

∴ Useful flux/pole = 0.943 Wb ■

Example 3.14 A three-phase, 20-pole, star-connected alternator has the following data: number of slots = 240; conductors/slot = 10 (conductors of each phase are connected in series); coil span 150 electrical degrees; speed of the alternator = 300 rpm, flux/pole = 56 mWb.

Calculate the following:

(i) frequency of induced emf,
(ii) pitch factor,
(iii) distribution factor,
(iv) winding factor,
(v) number of turns per phase,
(vi) phase voltage and
(vii) line voltage.

Solution

Given:

Number of poles (P) = 20
Number of slots = 240
Conductors/slots = 10
Coil span = 50
Synchronous speed (N_s) = 300 rpm
Flux/pole (ϕ) = 56 mWb

(i) $f = \frac{NP}{120} = \frac{300 \times 20}{120} = 50\,\text{Hz}$

(ii) $\varepsilon = 180° - 150° = 30°$

$\therefore \quad K_p = \cos\frac{\varepsilon}{2} = \cos 15° = 0.966$

(iii) $n(\text{number of slots/pole}) = \frac{240}{20} = 12$

$\therefore \quad \gamma = \frac{180°}{n} = \frac{180°}{12} = 15°$

$\therefore \quad q(\text{number of slots/pole/phase}) = \frac{240}{20 \times 3} = 4$

$\therefore \quad K_d = \frac{\sin\frac{q\gamma}{2}}{q\sin\frac{\gamma}{2}} = \frac{\sin\frac{4 \times 15°}{2}}{4\sin\frac{15°}{2}} = 0.957$

(iv) $K_w = K_p \text{K}_d = 0.966 \times 0.957 = 0.924$

(v) Number of conductors/phase $= \frac{240}{3} \times 10 = 800$

$\therefore$ Number of turns/phase $(N) = \frac{800}{2} = 400$

(vi) $E_{ph} = 4.44 f\phi N K_w = 4.44 \times 50 \times 56 \times 10^{-3} \times 400 \times 0.924 = 4{,}594.86\ \text{V}$

(vii) $E_L = \sqrt{3}\, E_{ph} = 7958.54\,\text{V}$

SIGNIFICANT POINTS

If the speed N is in rpm, $f = \frac{PN}{120}$ Hz

Pitch Factor:

$$\therefore \quad K_p = \frac{\text{resultant emf of a fractional pitched coil}}{\text{resultant emf of a full-pitched coil}} = \frac{2E\cos\frac{\varepsilon}{2}}{2E} = \cos\frac{\varepsilon}{2}$$

For nth harmonic,

$$\therefore \quad K_{np} = \cos\frac{n\varepsilon}{2}$$

The fractional pitched windings have the following advantages:

(i) Cost of the machine is less due to reduction of copper in overhang.

(ii) There is reduction of harmonics in the generated emf wave. So, the output emf wave becomes almost a sine wave.

Distribution Factor:

$$\text{Distribution factor } (K_d) = \frac{\text{vector sum of emf}}{\text{arithmetic sum of emf}} = \frac{\sin\frac{q\gamma}{2}}{q\sin\frac{\gamma}{2}}$$

The term $q\gamma$ is also known as the phase spread and is expressed in electrical radians.

$$\text{For the } n\text{th harmonic, } K_{dn} = \frac{\sin\frac{nq\gamma}{2}}{q\sin\frac{n\gamma}{2}}$$

Winding Factor:
It is defined as the product of distribution factor (K_d) and pitch factor (K_p). It is denoted by the symbol K_w.

$$\therefore \quad K_w = K_d \times K_p$$

The emf generated in a short pitch distributed winding is given by

$$E = 4.44 f N_{ph}\phi K_p K_d$$

Therefore, a P-pole machine, flux per pole is given by $= \frac{4}{P} B_{max} lr$

SHORT QUESTIONS AND ANSWERS

Q.1 State the types of armature slots.

Ans. Open-type and main-type are the two types of armature slots.

Q.2 What types of materials are used generally in armature winding?

Ans. Conducting materials are used in the form of wires, and insulating materials are used in the form of sleeves, varnish and so on.

Q.3 What is called armature conductor?

Ans. The part of the wire lying in the armature slot and cutting the magnetic lines of force when the armature rotates is called the armature conductor.

Q.4 What do you mean by coil span?

Ans. The distance measured in terms of armature slots or armature conductors between the centre of the two sides of the coil is termed as coil span.

Q.5 What do you mean by pole pitch, conductor, coil, coil pitch and front pitch and back pitch?

Ans. *Pole pitch:* It is defined as the number of armature conductors per pole. If there are 80 conductors and 8 poles, pole pitch is 80/8 = 10.

Conductor: It is defined as the length of a wire lying in the magnetic field.

Coil: Two conductors with their end connections form one coil. The coil may be single-turn coil or multiturn coil. The end connections are called overhang.

Coil pitch (Y_S): It is defined as the distance between two sides of a coil in terms of armature slots. In other words, it is defined as the periphery of the armature between two sides of a coil. If the coil pitch is equal to the pole pitch, the winding is said to be full pitched. Otherwise, it is called fractional pitch.

Front pitch: It is defined as the distance between the second conductor of one coil and the first conductor of the next coil.

Back pitch: It is defined as the distance measured in terms of armature conductors that a coil advances on the back of the armature. The distance in slots between the centres of the two adjacent poles is known as pole pitch.

Q.6 What do you mean by full-pitched and fractional pitched windings?

Ans. If the pitch factor is one, the winding isw termed as full pitch, and if it is less than unity, the winding is termed as fractional pitch.

Q.7 State the advantages of fractional pitched windings.

Ans. Fractional pitched windings have the following advantages:

(i) Cost of the machine is less due to reduction of copper in overhang.

(ii) There is reduction of harmonics in the generated emf wave. So, the output emf wave becomes almost a sine wave.

Q.8 What do you mean by distribution factor?

Ans. The ratio of emf induced in a distributed winding to the emf induced if the winding is concentrated is called distribution factor or breadth factor.

Q.9 State the advantages of distributed windings.

Ans. The main advantages of distribution of the windings in slots are as follows:

(i) The output emf becomes almost a sine wave as there is reduction of the harmonics in the generated emf wave.

(ii) There is increase in rigidity and mechanical strength of the winding.

(iii) There is full utilization of the armature iron and copper.

Q.10 State the difference between single-layer and double-layer windings.

Ans. One conductor or one coil side is placed in each armature slot in a single-layer winding, while two conductors or two coil sides are placed in each armature slot in a double-layer winding.

SUPPLEMENTARY PROBLEMS

1. A four-pole, three-phase induction motor is supplied from 50 Hz supply. Find its synchronous speed. Determine its full-load slip if it has a speed of 1,420 rpm on full load.

   [*Ans.* 1,500 rpm, 5.34%]

2. A four-pole, three-phase, 50 Hz star-connected induction motor has a full-load slip of 6 per cent. Determine full-load speed of the motor.

   [*Ans.* 1,410 rpm]

3. A four-pole, three-phase, 50 Hz induction motor runs at a speed of 1,460 rpm. Determine the frequency of the induced emf in the rotor under this condition.

   [*Ans.* 1.34 Hz]

4. A four-pole, three-phase induction motor is supplied from 50 Hz AC supply. The frequency of induced emf in the rotor is 2.5 Hz on full load. Calculate the full-load slip and the corresponding speed.

[*Ans.* 5%, 1,425 rpm]

5. A four-pole, three-phase alternator has 36 armature slots. It is using an armature winding, which is short pitched by one slot. Find its pitch factor.

[*Ans.* 0.9848]

6. A three-phase, eight-pole alternator has 120 slots. Determine its distribution factor.

[*Ans.* 0.957]

7. A six-pole, three-phase, 50 Hz alternator has 12 slots per pole and four conductors per slot. The winding is 5/6 pole pitch. There is 0.025 Wb entering the armature from each north pole and this flux is sinusoidally distributed along the air gap. The armature coils are connected in series. The winding is star connected. Determine the open circuit emf of the alternator. Find the breadth factors for third and fifth harmonics.

[*Ans.* 246.67 V, 0.654, 0.355]

8. An alternator runs at 250 rpm and generates an emf at 50 Hz. There are 216 slots, each connecting five conductors. The winding is distributed and full pitched. All the conductors of each phase are in series and flux per pole is 30 mWb which is sinusoidally distributed. If the winding is star connected, calculate the value of the induced emf available across the terminals.

[*Ans.* 1,992.70 V]

9. A three-phase, 16-pole, star-connected alternator has 144 slots on the armature periphery. Each slot contains 10 conductors. It is driven at 375 rpm. Determine the frequency of the induced emf and flux per pole if the line value of induced emf across the terminal is 2,657 V.

[*Ans.* 50 Hz, 30 mWb]

10. A three-phase, eight-pole, star-connected alternator has 120 stator slots and each slot has 10 conductors. The coil span is 12 slots. If the flux/pole is 0.12 Wb, calculate the emf induced between the lines.

[*Ans.* 8,389.7 V]

MULTIPLE-CHOICE QUESTIONS AND ANSWERS

1. The purpose of a brush in a DC machine is to
   (a) collect current from the commutator
   (b) collect voltage from the commutator
   (c) provide interaction between segments of the commutator
   (d) clean the commutator
2. If P be the number of poles, f be the frequency in Hz, N be the speed in rpm, the relationship among the three for a synchronous alternator is given by

   (a) $f = \dfrac{PN}{120}$ (b) $P = \dfrac{fN}{120}$ (c) $f = \dfrac{NP}{60}$ (d) $N = \dfrac{fP}{120}$

3. For a P-pole machine, the relation between electrical and mechanical degrees is given by

(a) $\theta_{elec} = \frac{2}{P}\theta_{mech}$ (b) $\theta_{mech} = \frac{P}{2}\theta_{elec}$ (c) $\theta_{mech} = \frac{2}{P}\theta_{elec}$ (d) $\theta_{elec} = \frac{P}{2}\theta_{mech}$

4. A coil consists of
 (a) two conductors (b) two coil sides (c) two turns (d) four turns

5. The main advantage of distributing the winding in slots is to
 (a) add mechanical strength to the winding (b) reduce the amount of copper required
 (c) reduce the harmonics in the generated emf (d) reduce the size of the machine

6. The distribution factor is defined as the ratio of
 (a) arithmetic sum of coil emf to phase sum of coil emf
 (b) phasor sum of the emf per coil to the arithmetic sum of emf per coil
 (c) phasor sum of coil emfs to the arithmetic sum of coil emf
 (d) phasor sum of coil emf to the per phase voltage

7. The pitch factor in a rotating electrical machine is defined as the ratio of resultant emf of a
 (a) full-pitched coil to that of a chorded coil (b) full-pitched coil to the phase emf
 (c) chorded coil to that of a full-pitched coil (d) chorded coil to the phase emf

8. The main advantage of using fractional pitched winding is to reduce the
 (a) harmonics in generated emf (b) amount of copper in the winding
 (c) size of the machine (d) cost of the machine

9. In an alternator, the frequency per revolution is equal to the
 (a) number of poles (b) number of armatures
 (c) number of pairs of poles (d) fitting

10. A 90-slot stator has a three-phase, six-pole, whole coiled winding, each coil of which spans from slots 1 to slot 14. The pitch factor is
 (a) 0.978 (b) 0.84 (c) 0.960 (d) 0.859

11. The maximum speed at which the field of an alternator can be operated to develop 60 Hz is
 (a) 2,600 rpm (b) 1,600 rpm (c) 3,600 rpm (d) 2,800 rpm

12. For a full-pitched winding, the generated voltage in both coil sides is
 (a) exactly 180° out of phase (b) exactly in phase
 (c) approximately in phase (d) approximately 180° out of phase

13. Distribution of winding instead of windings bunched into one slot has the effect of
 (a) improving the shape of voltage as a sine wave
 (b) increasing the speed of machine
 (c) adding rigidity and mechanical strength to the winding
 (d) reducing armature reaction

14. The full-load slip of 60 Hz, 12-pole, squirrel-cage induction motor is 5%. Its full-load speed is
 (a) 600 rpm (b) 570 rpm (c) 500 rpm (d) 475 rpm

15. The relationship between rotor frequency f_2, slip s and the stator frequency f_1 is given by
 (a) $f_1 = sf_2$ (b) $f_2 = sf_1$ (c) $f_2 = f_1(1-s)$ (d) $f_2 = \sqrt{sf_1}$

16. An eight-pole wound rotor induction motor operating on 60 Hz supply is driven at 1,800 rpm by a prime mover in the opposite direction to the revolving magnetic field. The rotor current frequency is
(a) 60 Hz (b) 120 Hz (c) 180 Hz (d) 120 Hz

17. A three-phase induction motor is running at slip s. If its two supply leads are interchanged, its slip at that instant is equal to
(a) 2 s (b) 1 – s (c) $2 - s$ (d) zero

18. The stator of a three-phase, four-pole shipping induction motor is fed from a 50 Hz source and other from a 30 Hz source. Then the other will run at
(a) 1,500 rpm (b) 900 rpm (c) 0 rpm (d) 2,400 rpm

19. An eight-pole, 50 Hz, single-phase induction motor is running at 690 rpm. Its slip with respect to forward field is (in p.u.)
(a) +0.08 (b) +1.92 (c) –0.08 (d) 1.00

20. The frequency of rotor emf of an eight-pole induction motor is 2 Hz. If supply frequency is 50 Hz, then the motor speed is
(a) 1,500 rpm (b) 750 rpm (c) 375 rpm (d) 720 rpm

Answers

1 (a)	2 (a)	3 (d)	4 (b)	5 (c)
6 (c)	7 (c)	8 (a)	9 (c)	10 (a)
11 (c)	12 (b)	13 (a), (c)	14 (b)	15 (b)
16 (c)	17 (c)	18 (d)	19 (a)	20 (d)

DC Generators 4

Direct current (DC) generators are machines that convert mechanical energy to electrical energy. This conversion of energy is based on the principle of the production of dynamically induced electromotive force (emf). In a DC generator the field is on the stator, whereas the armature is on the rotor. The field coil is excited by a DC supply. The rotor is rotated with the help of a prime mover, thereby cutting the flux produced by the field coils. The armature cuts the rotor flux and an alternating emf is induced in it, which is converted into a unidirectional emf with the help of the commutator. The basics of DC machines have been introduced in Chapter 3. This chapter introduces other aspects of the DC generators.

4.1 TYPES OF DC MACHINES

The classification of the DC machines based on the field configuration is as follows:

(i) Homopolar machines.

(ii) Heteropolar machines.

Homopolar machines: These types of machines are used where low voltage and high currents are required. Faraday's disc dynamo is an example of this type of machine.
Heteropolar machines: The DC machines that are commonly used fall under this category.

4.2 DC GENERATOR

The classification of the DC generators, depending upon their field excitation, is as follows:

- *Separately excited generator:* In this type of generator, the field coils are energized from an independent external source, as shown in Figure 4.1(a).
- *Self-excited generator:* In this type of generator, the field coils are energized by the current produced by the generator itself. These types of generators are classified into the following three categories, depending upon their field connections:
    (i) *Shunt-wound generator:* The field winding is connected in parallel to the armature. So, the full terminal voltage of the generator is applied across the field winding, as shown in Figure 4.1(b). The resistance of the field coil is very high.
    (ii) *Series-wound generator:* The field winding is connected in series with the armature. Since the field carries the full-load current, it consists of relatively few turns of thick wires, as shown in Figure 4.1(c). The resistance of the field coil is very low.
    (iii) *Compound-wound generator:* This type of generator is actually a combination of a few series-wound and a few shunt-wound generators. It can be either long shunt, as shown in Figure 4.1(d) or short shunt, as shown in Figure 4.1(e). If the series field aids the shunt field, the generator is said to be *cumulatively compounded.* If the series field opposes the shunt field, the generator is said to be *differentially compounded.*

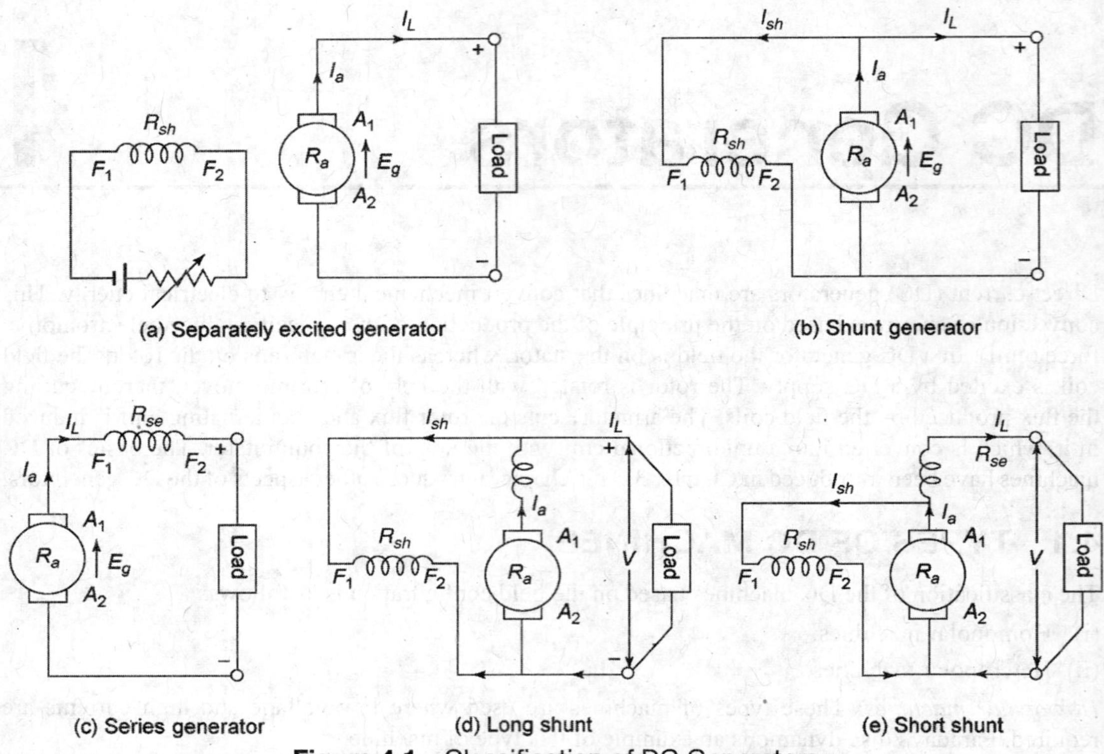

Figure 4.1 Classification of DC Generators

4.3 BRUSH DROP

The brush at first collects current from the commutator. Therefore, a contact drop occurs across the brush. For the metal graphite brushes, this drop is 0.5 V and for the carbon brushes it is 2.0 V.

4.4 EMF EQUATION

Let V_t be the terminal voltage,
I_a be the armature current,
R_a be the armature resistance,
E_g be the generated voltage in armature and
I_L be the load current.

4.4.1 Shunt Generator

Let R_{sh} be the resistance of the field coil and I_{sh} be the field current. The following equations are valid:

$$I_a = I_L + I_{sh} \tag{4.1}$$

$$E_g = V_t + I_a R_a + V_{\text{brush}} \tag{4.2}$$

where V_{brush} is the total voltage drop across the brush.

4.4.2 Series Generator

Let R_{se} be the series field resistance and I_{se} be the field current. The following expressions are valid:

$$I_a = I_{se} = I_L \tag{4.3}$$

$$E_g = V_t + I_a\left(R_a + R_{se}\right) + V_{\text{brush}} \tag{4.4}$$

4.4.3 Long-shunt Compound Generator

The shunt field winding is connected across the series combination of the armature and series field winding.

Armature current $$I_a = I_{se} \tag{4.5}$$

and $$I_a = I_{sh} + I_L \tag{4.6}$$

where I_{se} and I_{sh} are the currents through the series and shunt field windings.
The voltage across the shunt field winding is V_t. The shunt field current (I_{sh}) is expressed by

$$I_{sh} = \frac{V_t}{R_{sh}} \tag{4.7}$$

where R_{sh} is the resistance of shunt field winding.
The voltage equation becomes

$$E_g = V_t + I_a R_a + I_a R_{se} + V_{\text{brush}} \tag{4.8}$$

where R_{se} is the resistance of series field winding.

4.4.4 Short-shunt Compound Generator

The shunt field winding is connected across the armature only.

Armature current $$I_a = I_{se} + I_{sh} \tag{4.9}$$

and $$I_{se} = I_L \tag{4.10}$$

$$I_a = I_L + I_{sh} \tag{4.11}$$

The drop across the shunt field winding is the drop across the armature only and not the total V_t. Therefore, the drop across the shunt field winding is $E_g - I_a R_a$.

$$\therefore \quad I_{sh} = \frac{E_g - I_a R_a}{R_{sh}} \tag{4.12}$$

The voltage equation is expressed by

$$E_g = V_t + I_a R_a + I_{se} R_{se} + V_{\text{brush}} \tag{4.13}$$

Here, $$I_{se} = I_L \tag{4.14}$$

Equation (4.13) in terms of I_L becomes

$$\therefore \quad E_g = V_t + I_a R_a + I_L R_{se} + V_{\text{brush}} \tag{4.15}$$

i.e., $$E_g - I_a R_a = V_t + I_L R_{se} + V_{\text{brush}} \tag{4.16}$$

Substituting $E_g - I_a R_a$ in Equation (4.12), we have

$$I_{sh} = \frac{V_t + I_L R_{se} + V_{brush}}{R_{sh}} \tag{4.17}$$

From the emf equation, we can conclude that E_g is always greater than V_t. The expression for E_g is derived in Section 4.5.

4.5 DERIVATION FOR E_g

E_g depends on the following factors:

(i) The magnetic flux ϕ (Wb).
(ii) The speed of the machine (rps or rpm).
(iii) A constant that depends on the construction of the machine.

The emf (E_g) in a conductor of active length l moving at a velocity u in a magnetic field having flux density B is given by the following equation:

$$E_g = Blu \text{ V} \tag{4.18}$$

Let ϕ be the flux/pole (Wb),
P be the number of generator poles,
Z be the total number of armature conductors,
A be the number of parallel paths and
N be the speed of armature in rpm.

The number of conductors in series = Z/A. The emf generated is expressed by the following equation:

$$E_g = \left(\frac{Z}{A}\right) Blu \tag{4.19}$$

where B is the magnetic flux density, l is the length of the conductor and u is the velocity.
The flux density can be expressed as

$$B = \frac{\text{(Flux/pole) (Number of poles)}}{\text{Cylindrical area of armature surface}}$$

i.e.,
$$B = \frac{\phi P}{2\pi rl} \tag{4.20}$$

where r is the radius of the armature

The velocity
$$u = \frac{2\pi rN}{60} \tag{4.21}$$

Using Equations (4.20) and (4.21), Equation (4.19) can be expressed as

$$E_g = \frac{Z}{A}\left[\frac{\phi P}{2\pi rl}\right](l)\left[\frac{2\pi rN}{60}\right] \tag{4.22}$$

∴
$$E_g = \frac{\phi NZ}{60}\left[\frac{P}{A}\right]$$

For lap winding, $A = P$ and for wave winding, $A = 2$.

Example 4.1 A four-pole generator, having wave-wound armature winding has 55 slots, each slot containing 19 conductors. What will be the voltage generated in the machine when driven at 1,500 rpm assuming that the flux per pole is 3.0 mWb?

Solution

It is known that $$E_g = \frac{\phi ZN}{60} \frac{P}{A} \text{ V}$$

Here, $$\phi = 3\times10^{-3} \text{ Wb}, Z = 55\times19 = 1{,}045, A = 4, P = 2, N = 1{,}500 \text{ rpm}$$

$\therefore$ $$E_g = \frac{3\times10^{-3}\times1{,}045\times1{,}500}{60}\times\left[\frac{4}{2}\right] = 156.75 \text{ V}$$ ■

Example 4.2 The terminal voltage of an eight-pole DC shunt generator with 780 wave-connected armature conductors and running at 500 rpm at terminal voltage is 240 V. The armature resistance is 0.24 Ω and the field resistance is 240 Ω. Find the armature current, the induced emf and the flux per pole if the load resistance is 12 Ω.

Solution

Here, $Z = 780$

Number of parallel paths (A) = 2 (due to wave winding)

Load current = V/R = 240/12 = 20 A

Shunt current = 240/240 = 1 A

Armature current = 20 + 1 = 21 A

Induced emf = 240 + (21 × 0.24) = 245.04 V

$$E_g = \frac{\phi ZN}{60}\left[\frac{P}{A}\right]$$

i.e., $$245.04 = \frac{\phi\times780\times500}{60}\left[\frac{8}{2}\right]$$

i.e., $$\phi = 9.4246 \text{ mWb}$$ ■

Example 4.3 A four-pole, long-shunt lap-wound DC generator supplies 20 kW at a terminal voltage of 400 V. The armature resistance is 0.04 Ω, series field resistance is 0.05 Ω and shunt field resistance is 200 Ω. The voltage drop per brush may be taken as 1.0 V. Determine the emf generated.

Also, calculate the number of conductors if the speed is 1,000 rpm and flux per pole is 0.03 Wb. Neglect the armature reaction.

Solution

$$I = 20{,}000/400 = 50 \text{ A}$$

$$I_{sh} = 400/200 = 2$$

$$I_a = I + I_{sh} = 50 + 2 = 52 \text{ A}$$

Series field drop = 52 × 0.05 = 2.6 V

Armature drop = 52 × 0.04 = 2.08 V

Brush drop = 2 × 1 = 2 V

General emf E_g = 400 + 2.6 + 2.08 + 2 = 406.68 V

Now, $$E_g = \frac{\phi ZN}{60}\left(\frac{P}{A}\right)$$

i.e., $$406.68 = \frac{0.03 \times Z \times 1{,}000}{60}\left(\frac{4}{4}\right)$$

i.e., $$Z = 814$$

4.6 LOSSES IN DC GENERATOR

The following losses occur in a DC generator:

- *Copper loss:*
  (i) Armature Cu-loss = $I_a^2 R_a$, where R_a is the resistance of the armature and I_a is the armature current.
  (ii) For the shunt generators the field Cu-loss is equal to $I_{sh}^2 R_{sh}$, where R_{sh} is the resistance of the shunt field winding and I_{sh} is the current through it. This loss for shunt generator is practically constant. For the series generator it is equal to $I_{se}^2 R_{se}$, where R_{se} is the resistance of the series field winding and I_{se} is the current through it. Field Cu-loss is also constant for the compound generators. This loss is about 20–30 per cent of full-load losses.
  (iii) The loss due to brush contact is generally included in armature Cu-loss.
- *Magnetic loss or iron loss or core loss:*
  (i) *Hysteresis loss:* This loss occurs due to reversal of the armature core. Each portion of the rotating core passes under the north (N) and south (S) pole alternately and attains S and N polarity, respectively. When the core passes under one pair of poles, the core is said to complete one cycle of frequency reversal. Hysteresis loss $\propto B_{max}^n f$ (n is generally 1.6). The frequency of magnetic reversal $(f) = \frac{PN}{120}$ Hz, where P is the number of poles and N is the speed of the armature in rpm. To reduce this loss, material having small hysteresis loop should be used.
  (ii) *Eddy current loss:* The armature core cuts the magnetic flux during its rotation and emf is induced in the body of the core according to the laws of electromagnetic induction. This emf is very small, but it sets up a large current in the body of the core. This current is called the eddy current and the loss is known as the eddy current loss. Eddy current loss $\propto B_{max}^2 f^2$. These losses are reduced by using laminated core.

Iron losses are constant for shunt and compound generators, as their field currents are approximately constant. This loss is about 20–30 per cent of full-load loss.

- *Mechanical losses:*
  (i) Air friction or windage loss of rotating armature.
  (ii) Friction loss at bearings and commutator, which are about 10–20 per cent of full-load losses. Mechanical losses together are known as *stray losses.*

4.7 STRAY LOSSES

The magnetic and mechanical losses are collectively known as stray losses, which are also known as *rotational losses.*

4.8 CONSTANT OR STANDING LOSSES

The shunt and compound generators have constant field Cu-loss. Therefore, stray losses and shunt Cu-losses are constant in these types of generators. The sum of these two losses is known as *standing or constant losses* denoted by W_c.

4.9 POWER STAGES

Figure 4.2 shows the various power stages of the DC generators.

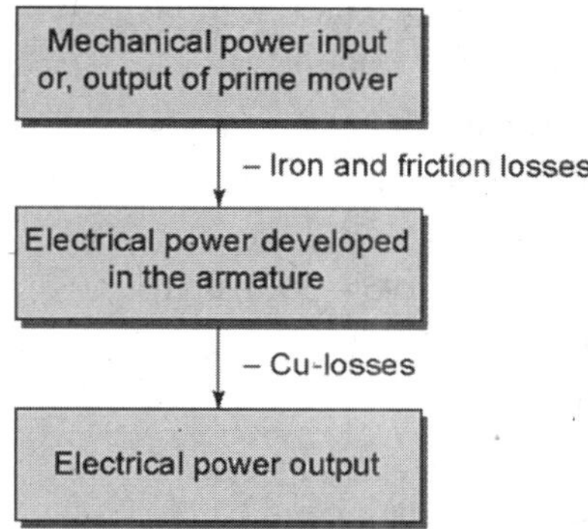

Figure 4.2 Various Power Stages of DC Generators

4.10 EFFICIENCY

DC generators have the following efficiencies due to the different power stages:

- *Mechanical efficiency:*

$$\eta_m = \frac{\text{Electrical power developed in armature}}{\text{Mechanical power input}}$$

- *Electrical efficiency:*

$$\eta_e = \frac{\text{Electrical power output}}{\text{Electrical power developed in armature}}$$

- *Commercial or overall efficiency:*

$$\eta_c = \frac{\text{Electrical power output}}{\text{Mechanical power input}}$$

$\therefore\ \eta_c = \eta_m \times \eta_e.$

Commercial efficiency is always to be taken if it is not specified. Since the commercial efficiency is the product of the mechanical and electrical efficiencies, η_c is always less than η_e and η_m.

Example 4.4 A DC shunt generator delivers 195 A at terminal voltage of 200 V. The armature resistance and the shunt field resistance are 0.03 Ω and 40 Ω, respectively. The iron and friction losses are equal to 900 W. Find the following: (i) emf generated, (ii) Cu-losses, (iii) output of the prime mover and (iv) commercial, mechanical and electrical efficiencies.

Solution

(i) $I_{sh} = 200/40 = 5$ A; $I_a = 195 + 5 = 200$ A

Armature voltage drop $= I_a R_a = 200 \times 0.03 = 6$ V

∴ Generated emf $(E_g) = 200 + 6 = 206$ V

(ii) Armature Cu-loss $= I_a^2 R_a = 200^2 \times 0.03 = 1{,}200$ W

Shunt Cu-loss $= VI_{sh} = 200 \times 5 = 1{,}000$ W

∴ Total Cu-loss = 1,200 + 1,000 = 2,200 W

(iii) Stray losses = 900 W

Total losses = 2,200 + 900 = 3,100 W

Electrical power output = 200 × 195 = 39,000 W

Output of prime mover = 39,000 + 3,100 = 42,100 W = 42.1 kW

(iv) Generator input = 42,100 W

Here, stray losses = 900 W

Electrical power produced in armature = 42,100 − 900 = 41,200 W

$$\eta_m = \frac{41{,}200}{42{,}100} = 0.9786 \text{ p.u.} = 97.86\%$$

Total Cu-losses = 2,200 W

$$\eta_e = \frac{39{,}000}{39{,}000 + 2{,}200} = 0.9466 \text{ p.u.} = 94.66\%$$

$$\eta_c = \frac{39{,}000}{42{,}100} = 0.92636 \text{ p.u.} = 92.636\%$$ ■

Also, $\eta_c = \eta_m \eta_e = 0.92636$ p.u. = 92.636%.

4.11 CONDITION FOR MAXIMUM EFFICIENCY

Figure 4.3 shows the variation of efficiency (η) of a DC generator with the load current. The value of η increases with an increase in the load current (I). For a particular value of I, η is maximum. With further increase of I, η decreases. Therefore, η is a function of I. Let us find the condition of maximum efficiency.

Let V_t be the output voltage of the generator,
I_L be the load current,
W_c be the iron loss,

R_a be the resistance of the armature,
I_{sh} be the current through the shunt winding,
I_a be the armature current and
$I_a = I_L + I_{sh}$ for the shunt generator.

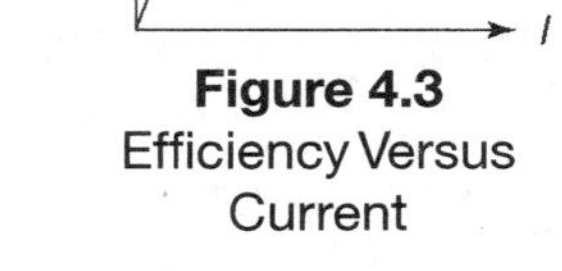

Figure 4.3 Efficiency Versus Current

Since I_{sh} is negligible order compared to I_L, $I_L = I_a = I$ *(say)*

Generator output $= V_t I$
Generator input $=$ Output + Losses
$= V_t I + I^2 R_a + W_c$

$$\therefore \quad \text{Efficiency } (\eta) = \frac{\text{Output}}{\text{Input}} = \frac{V_t I}{V_t I + I^2 R_a + W_c} \tag{4.23}$$

i.e.,

$$\eta = \frac{1}{1 + \frac{IR_a}{V_t} + \frac{W_c}{V_t I}}$$

Now, η will be maximum when denominator of Equation (4.23) becomes minimum.

$$\therefore \quad \frac{d}{dI}\left(1 + \frac{IR_a}{V_t} + \frac{W_c}{V_t I}\right) = 0$$

i.e.,

$$\frac{R_a}{V_t} - \frac{W_c}{V_t I^2} = 0$$

i.e.,

$$W_c = I^2 R_a \tag{4.24}$$

Hence, generator efficiency is maximum when

Constant loss = Cu-loss or variable loss

The load current at which maximum frequency occurs is given by

$$I = \sqrt{\frac{W_c}{R_a}} \tag{4.25}$$

Example 4.5 The terminal voltage and the load current of a DC shunt generator are 250 V and 190 A. The field coil resistance and the stray load losses are 125 Ω and 800 W. Find the armature resistance if it has a full-load efficiency of 90 per cent. Calculate the load current at maximum efficiency.

Solution

Here, $I = 190$ A, $V_t = 250$ V, stray load losses = 800 W, $\eta_{fl} = 90\%$
Output power = 190 × 250 = 47,500 W

$$I_{sh} = \frac{250}{125} = 2 \text{ A}, \text{ field Cu-loss } = V_t I_{sh} = 250 \times 2 = 500 \text{ W}$$

$$\therefore \quad I_a = I + I_{sh} = 190 + 2 = 192 \text{ A}$$

Let the armature resistance be R_a Ω.

$$\therefore \qquad \eta = \frac{47{,}500}{47{,}500 + I_a^2 R_a + \text{Stray load losses} + \text{Field Cu-loss}}$$

i.e.,
$$0.90 = \frac{47{,}500}{47{,}500 + I_a^2 R_a + 800 + 500}$$

i.e.,
$$I_a^2 R_a = \left(\frac{1}{0.90} - 1\right) \times 47{,}500 - 1{,}300 = 3{,}977.78 \text{ W}$$

$$R_a = \frac{3{,}977.78}{I_a^2} = \frac{3{,}977.78}{192^2} = 0.108\ \Omega$$

At maximum efficiency,

$I_a^2 R_a$ = Constant losses, where I_a is the armature current at its maximum efficiency

W_C = Stray load losses + Shunt field losses = 800 + 250 × 2 = 1,300 W

$$\therefore \qquad I_a = \sqrt{\frac{\text{Constant losses}}{R_a}} = \sqrt{\frac{1{,}300}{0.108}} = 109.71 \text{ A}$$

■

4.12 ARMATURE REACTION IN DC MACHINES

The armature magnetomotive force (mmf) has an effect on the magnitude and distribution of the field in the air gap and also in the armature core. This effect is known as the armature reaction. The following are the effects of armature magnetomotive force on the main field flux:

(i) It partly weakens or demagnetizes the main field flux.

(ii) It cross magnetizes or distorts the main field flux.

Figure 4.4(a) shows a two-pole machine having six slots and two conductors per slot. Figure 4.4(b) shows the distribution of the main field flux when the armature conductor carries no current, and at this stage the position of brushes is at the geometrical neutral axis (GNA), which coincides with the magnetic neutral axis (MNA). The bottom conductors have not been shown for simplicity. The developed armature along with the main field flux is shown in Figure 4.4(b), where the waveform of flux is plotted with an assumption that the flux entering the armature through the air gap of the field system is taken as positive, and hence the waveform under the north pole is taken as positive, whereas the one under the south pole is taken as negative.

Figure 4.5(a) shows an armature with two slots and one conductor per slot. The distribution of flux is from F to E for the given distribution of armature currents. The armature becomes an electromagnet, and its poles have been indicated. The flux due to the armature coils (a,b) that enters the air gap field is taken as positive. Figure 4.5(b) shows the waveform of the mmf of the coil shown in Figure 4.5(a).

Figure 4.6 shows the distribution of the armature mmf when armature carries current as a generator and it rotates in the anticlockwise direction.

Figure 4.7(a) shows the developed view of the armature of Figure 4.6 to obtain the mmf waveform of armature currents. Any two conductors forming one turn are situated a pole pitch apart. The end connectors

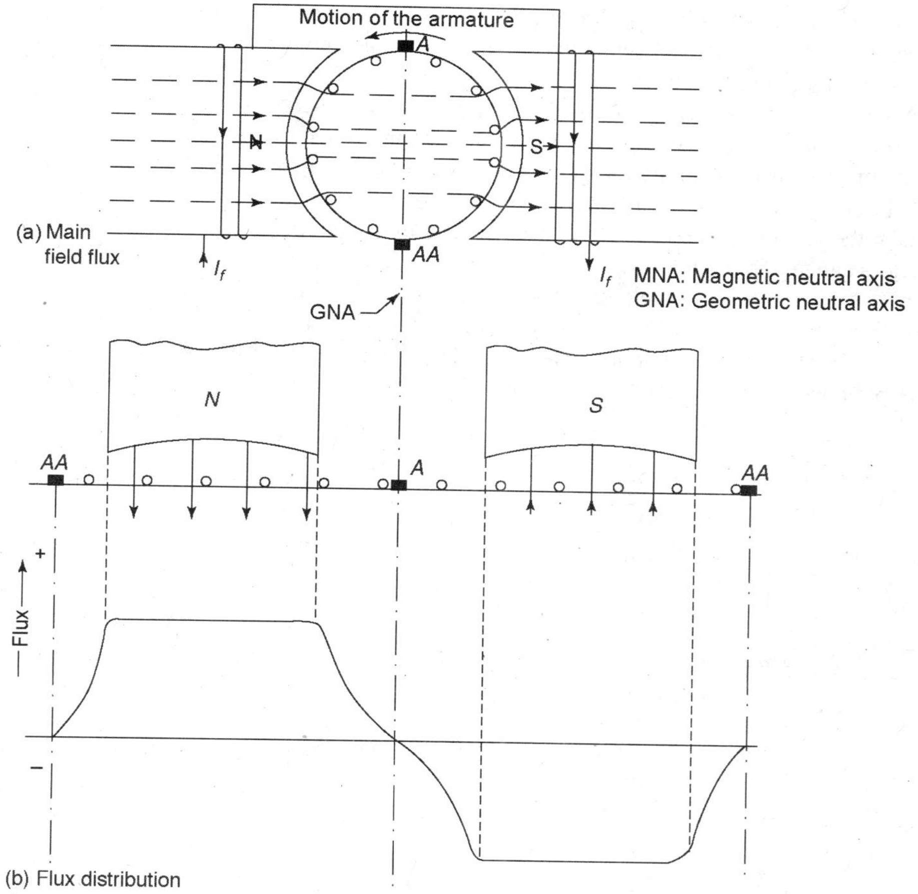

Motion of the armature
A
N
S
(a) Main field flux
I_f
AA
I_f
MNA: Magnetic neutral axis
GNA: Geometric neutral axis
GNA
N
S
AA
A
AA
+
Flux
−
(b) Flux distribution

Figure 4.4 Main Field Flux and Flux Distribution when the Armature Is on No Load

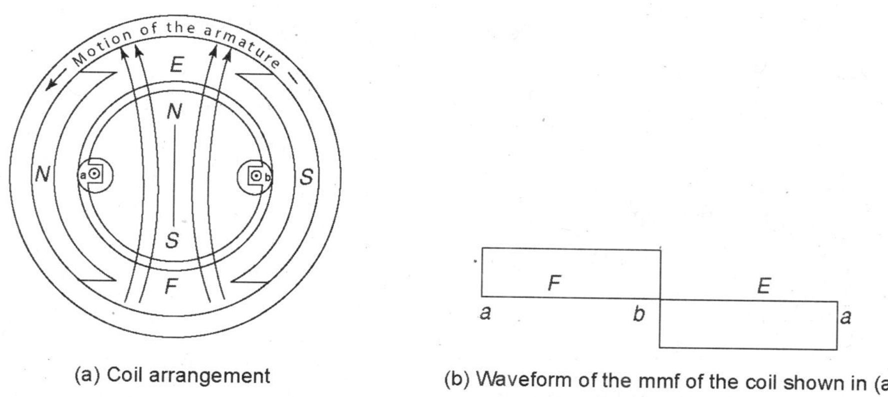

Motion of the armature
E
N
N
a
b
S
S
F
(a) Coil arrangement
F
E
a
b
a
(b) Waveform of the mmf of the coil shown in (a)

Figure 4.5 Coil Arrangement and Waveform of mmf

are shown by dotted lines in Figure 4.7(a). Each coil, in the allotted time, develops mmf, which is represented by small rectangular steps spreading across the breadth of the coil, as shown in Figure 4.7(b). The conductors lying in between the vertical axes C_1C_2 and D_1D_2 act as if they form the concentric coils producing a stepped mmf waveform and its maximum value along the geometric neutral axis A_2A_1. The stepped mmf waveform is taken as negative because the mmf is directed from the armature core to the field system. The mmf due to the conductors situated in the left of the axis C_1C_2 and in the right of the axis D_1D_2 forms a stepped wave directed along B_2B_1. The stepped mmf waveform transforms into a triangular waveform, as shown in Figure 4.7(c), when the armature conductors are distributed uniformly around the periphery of the armature. The peak value is calculated below.

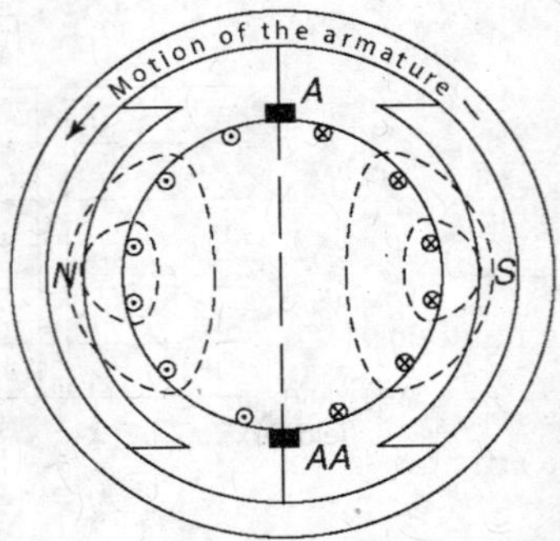

Figure 4.6 Flux Distribution Due to the Armature Current

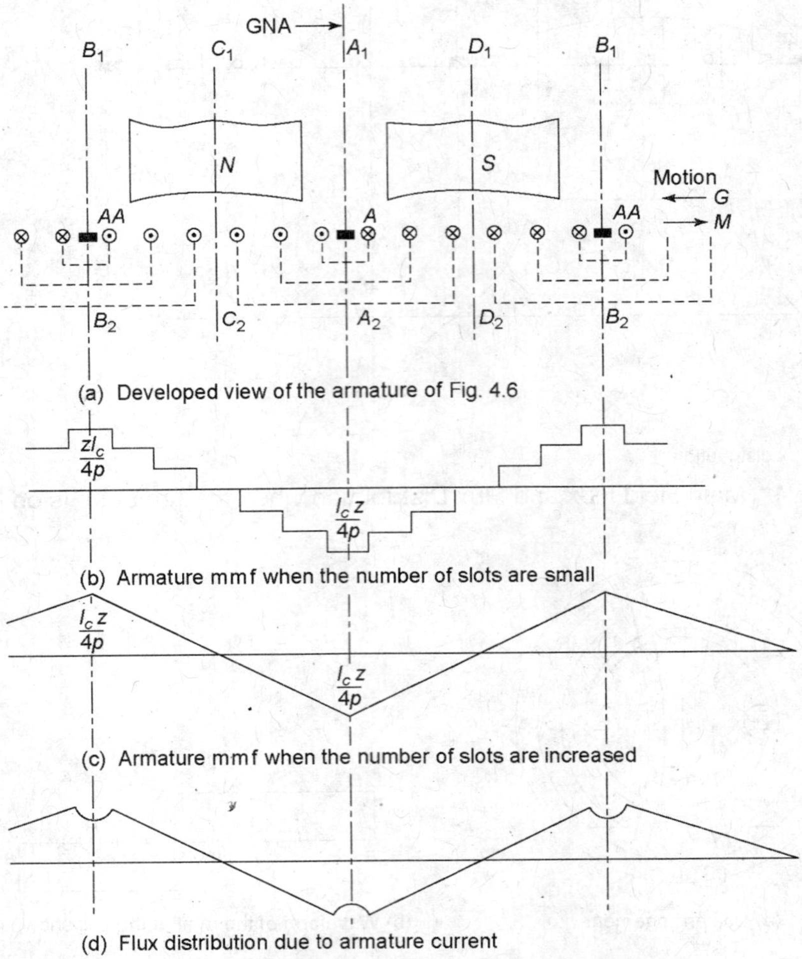

(a) Developed view of the armature of Fig. 4.6

(b) Armature mmf when the number of slots are small

(c) Armature mmf when the number of slots are increased

(d) Flux distribution due to armature current

Figure 4.7 Distribution of the Armature mmf

Let I_a be the total current in the armature and $2a$ be the number of parallel paths.

The current per conductor $(I_c) = \frac{I_a}{2a}$. The number of ampere conductors on the armature $= I_c Z$ and the number of ampere conductors per pole $= \frac{I_c Z}{2p}$, where p is the number of pair of poles.

The number of ampere turns per pole $= \frac{1}{2}\left[\frac{I_c Z}{2p}\right] = \frac{I_c Z}{4p} = \frac{I_a Z}{8ap}$

$$\text{The armature mmf per pole} = \frac{I_a Z}{8ap} \tag{4.26}$$

The mmf represented by Equation (4.26) acts along the brush axis. At the positions of the uniform reluctance of the air gap, the waveforms of mmf and the flux distribution due to armature are identical. Since the reluctance of the air gap at the interpolar gap between the poles is more as compared to the value under the poles, a kink is introduced in the flux waveform along with the GNA and symmetrical with respect to it, as shown in Figure 4.7(d). The zeros of the main field flux pass through the GNA shown in Figure 4.5(b) when the armature carries no current, which means that the MNA coincides with the GNA.

Figure 4.8(a) shows the developed view of the armature, and Figure 4.8(b) shows the distribution of the field and the armature fluxes for the motion of the armature from right to left when acting as a generator. The resultant flux due to the main field flux and the armature flux is shown in Figure 4.8(c) in which distortion of the main field flux has been clearly depicted. From Figure 4.8(c), it is obvious that the total flux decreases at the leading pole tip, whereas it increases at the trailing pole tip. The axis *CD* is the MNA that is normal to the resultant flux in the core, which is shifted from the axis *A–AA* by an angle θ in the direction of rotation of the machine when it acts as a generator and in the opposite direction when it acts as a motor. The resultant waveform, shown in Figure 4.8(c), is obtained by the superposition of the two flux waveforms. The peak value *RS* of the resultant flux at the trailing pole tip is obtained by adding fluxes *LM* and *L'K* of the main field flux and the armature flux at the trailing pole tip, respectively. The armature flux and the main field flux reinforce each other at the trailing pole tip, whereas they oppose each other at the leading pole tip. The zero of the resultant waveform is shifted by an angle θ in the direction of rotation for the case of generator and in the opposite direction for the case of motor. This means that the MNA shifts in the forward direction for the generator and in the backward direction for the motor. If the direction of rotation of the machine for motoring mode is assumed from left to right, the waveform remains unaltered. For good commutation, the brush axis must coincide with the MNA. Hence, the brush axis must be placed along *CD* when the armature is under load instead of placing it along the GNA during load. Therefore, the effect of armature flux not only distorts the main field flux waveform but also partly weakens the main field flux at the leading pole tip, which is called the armature reaction. If the saturation is present in the teeth of the armature, the peak of the resultant waveform will be rounded off. If the working flux density is high, the effect of the armature reaction can be minimized.

The iron losses in the teeth and the pole shoes of the armature depend on the maximum value of the working flux density in the regions. Due to the armature reaction, the maximum flux density increases under the trailing pole tip, whereas it decreases at the leading pole tip. Hence, the iron loss on the teeth is greater on load than on no load. Due to very high degree of magnetic saturation in the teeth, some of the flux gets its way into the core end plates that are fairly massive and induces eddy current, which

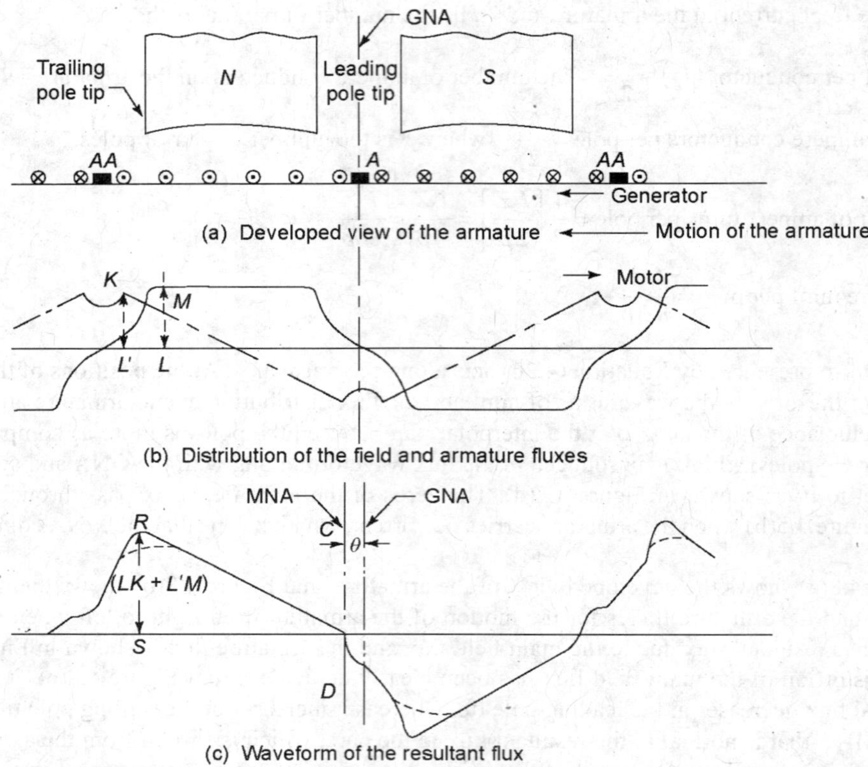

Figure 4.8 Resultant Flux of Loaded and Unsaturated DC Machines

gives rise to appreciable losses. During load, the extra iron loss of about 50 per cent of the no-load iron losses occur in the machine.

4.13 DEMAGNETIZING AND CROSS-MAGNETIZING CONDUCTORS

Figures 4.9(a) and 4.9(b) show the armature conductors that produce the demagnetizing and cross-magnetizing effects when brush axis lies in the new position of the MNA, having a forward lead of θ. The direction of the currents through the conductors lying within angles *AOC* and *BOD*, shown in Figure 4.9, is such that the flux through the armature is from left to right, and hence produces demagnetizing effects, whereas the direction of the currents through the conductors lying within the angles *AOD* and *COB* is such that the combined flux is at right angles to the main flux, and hence causes cross-magnetizing effects, as shown in Figure 4.9 (b).

4.14 DEMAGNETIZING AMPERE-TURNS PER POLE

To neutralize the demagnetizing effects, extra ampere-turns (AT) are added to the main field winding and calculation of the extra AT is required.

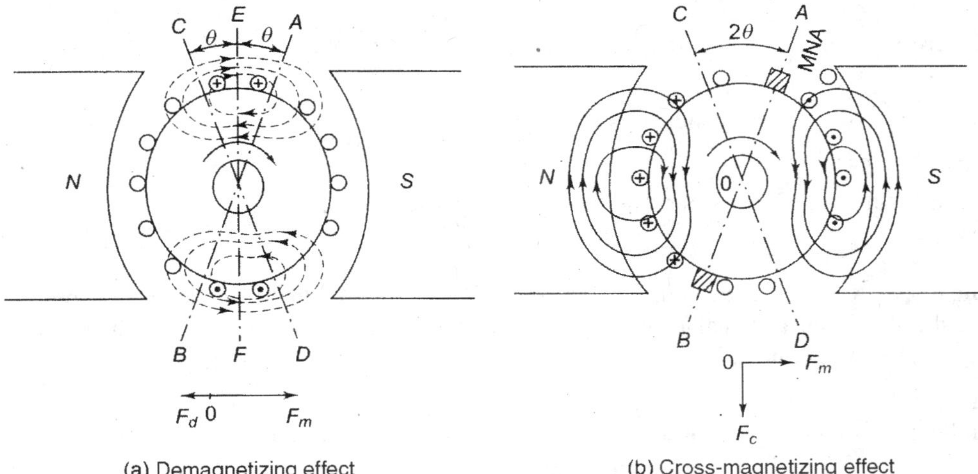

(a) Demagnetizing effect (b) Cross-magnetizing effect

Figure 4.9 Demagnetizing and Cross-magnetizing Effects

Let Z = number of armature conductors,

θ_m = forward lead to mechanical or geometrical or angular degrees,

I = current in each armature conductor,

= $I_a/2$ for simplex wave winding,

= I_a/P for simplex lap winding.

The total number of armature conductors within the angles AOC and $BOD = \dfrac{4\theta_m}{360} \times Z$

Therefore, the total number of turns in angles AOC and $BOD = \dfrac{2\theta_m}{360} \times Z$

The total number of AT in angles AOC and $BOD = \dfrac{2\theta_m}{360} \times ZI$

Therefore, the demagnetizing AT per pair of poles $= \dfrac{2\theta_m}{360} \times ZI$

Hence, the demagnetizing AT per pole $= \dfrac{\theta_m}{360} \times ZI$ (4.27)

4.15 CROSS-MAGNETIZING AMPERE-TURNS PER POLE

The cross-magnetizing conductors lie within the angles AOD and BOC, as shown in Figure 4.7(b). The total number of conductors including cross-magnetizing and demagnetizing is given by $\dfrac{Z}{P}$.

As derived earlier, demagnetizing conductors per pole $= \dfrac{2\theta_m}{360} \times Z$

Therefore, cross-magnetizing conductors per pole $= \frac{Z}{P} - \frac{2\theta_m}{360} \times Z = Z\left(\frac{1}{P} - \frac{2\theta_m}{360}\right)$

The cross-magnetizing ampere-conductors per pole $= ZI\left(\frac{1}{P} - \frac{2\theta_m}{360}\right)$

Hence, the cross-magnetizing AT per pole $= ZI\left(\frac{1}{2P} - \frac{\theta_m}{360}\right)$ (4.28)

Example 4.6 A wave-wound four-pole DC generator with 780 armature conductors supplies 100 A. The brushes are given an actual lead of 10° (mechanical). Calculate the armature demagnetizing and cross-magnetizing AT per pole.

Solution

$P = 4$, parallel path $(A) = 2$, $Z = 780$, armature current $(I_a) = 100$ A (assuming armature current = load current).

Therefore, current per parallel path $(I) = \frac{I_a}{A} = \frac{100}{2} = 50$ A

Brush shift $(\theta_m) = 10°$ (mechanical)

Demagnetizing AT/pole $= ZI \times \frac{\theta_m}{360°} = 780 \times 50 \times \frac{10}{360} = 1{,}084$

Cross-magnetizing AT/pole $= ZI\left(\frac{1}{2P} - \frac{\theta_m}{360}\right) = 780 \times 50\left(\frac{1}{2\times 4} - \frac{10}{360}\right) = 3{,}792$ ■

Example 4.7 A wave-connected, eight-pole, 60 kW, 300 V DC generator has 540 conductors and delivers full-load current. If the brush shift is 4° (mechanical), calculate demagnetizing and cross-magnetizing AT/pole.

Solution

$$\text{Load current} = \frac{60{,}000}{300} = 200 \text{ A}$$

Armature current $(I_a) = 200$ A (assuming armature current = load current)

Therefore, current per parallel path $(I) = \frac{I_a}{A} = \frac{200}{2} = 100$ A

Angle shift $(\theta_m) = 4°$ (mechanical)

Demagnetizing AT/pole $= ZI \times \frac{\theta_m}{360} = 540 \times 100 \times \frac{4}{360} = 600$

Cross-magnetizing AT/pole $= ZI\left(\frac{1}{2P} - \frac{\theta_m}{360}\right) = 540 \times 100 \times \left(\frac{1}{2\times 4} - \frac{4}{360}\right) = 6{,}150$ ■

4.16 COMPENSATING WINDINGS

Compensating windings are used for large DC machines to neutralize the cross-magnetizing effect of armature reaction. Large DC machines are subjected to major fluctuation of loads. In the absence of compensating windings the flux shifts backwards and forwards due to change of load, which induces statically induced emf in armature conductors. The change of load determines the magnitude of this induced emf. The magnitude of this induced emf may be very high, resulting in a flash-over around the whole commutator circuit. Therefore, the whole armature circuit gets short circuited. These windings that are embedded in the pole shoes and connected in series with the armature are shown in Figure 4.10. Connections are made in such a way that the direction of the current in these conductors is opposite to the direction of the current in the armature conductors. The mmfs of the compensating windings should be sufficient to counterbalance the armature mmfs.

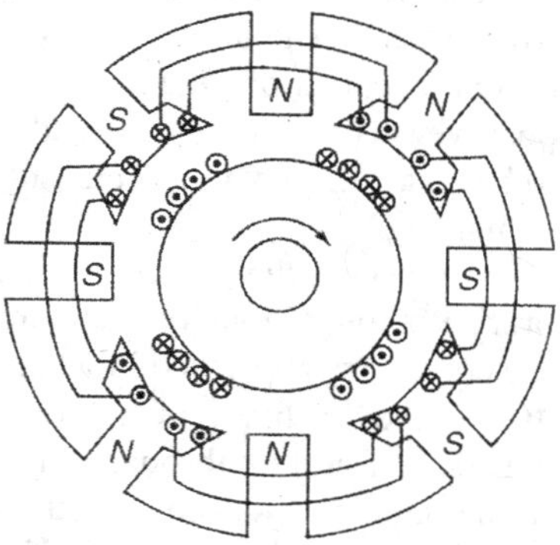

Figure 4.10 Compensating Windings

Let Z_a be the number of active armature conductors/pole,
I_a be the current/armature conductors,
A be the number of parallel paths and
Z_c be the number of compensating conductors/pole face.

$$\therefore \qquad Z_c I_a = Z_a\left(\frac{I_a}{A}\right)$$

i.e.,

$$Z_c = \frac{Z_a}{A} \tag{4.29}$$

Compensating windings are generally used in large DC machines.

4.17 NUMBER OF COMPENSATING WINDINGS

The number of armature conductors/pole $= \dfrac{Z}{P}$

Therefore, the numbers of armature turns per pole $= \dfrac{Z}{2P}$

The number of armature-turns immediately under one pole $= \dfrac{Z}{2P} \times \dfrac{\text{Pole arc}}{\text{Pole pitch}}$

Approximately, the number of armature-turns immediately under one pole $= 0.7\left(\dfrac{Z}{2P}\right)$

Therefore, the number of AT/pole for compensating windings $= 0.7 \times \dfrac{ZI}{2P} = 0.7 \times$ armature amp-turns/pole

4.18 COMMUTATION

In DC generators, the brush width is greater than the width of a commutator segment and the mica insulation. The armature coils connected to the commutator segments become short circuited during the spanning of the brush. Commutation is the process of collection of current by the brush or the changes that takes place in a coil during the period of short circuit by a brush. The induced current in the armature conductors of the DC generators is alternating in nature. Commutator are used to make their flow unidirectional. The armature conductors that are under the influence of the north poles carry current in one direction and that are under south poles carry current in the opposite direction. Therefore, if the conductors under the influence of the north pole come under the influence of the south pole, the direction of current through the conductors changes. Current in theses conductors changes from $+I$ to 0 and then from 0 to $-I$. Similar change occurs when conductors under the influence of the south pole come under the influence of the north pole. Current in these conductors changes from $-I$ to 0 and then from 0 to $+I$. The direction of current through the armature conductors also changes. This reversal of current occurs along the MNA. If this changing of current from positive to zero and then from zero to positive is completed by the end of the short circuit or the commutation period, the commutation is said to be ideal. There will be sparking at the brush and commutator segments. Otherwise, this sparking will damage both.

To discuss the process of commutation, let us consider Figure 4.11. Figure 4.11(a) shows that the brush is about to touch the commutator segment *a*, and hence coil *B* is about to be short circuited. If all the coils carry currents of 40 A, the current in the brush at this stage is 80 A because every coil supplies 40 A to the brush. Figure 4.11(b) shows that coil *B* has entered its short circuit, which is one-third of its period. The current through coil *B* is 20 A, as shown by the arrow. Figure 4.11(c) shows that the coil is in the middle of its short circuit period; therefore, the current through it is zero. Figure 4.11(d) shows that the brush contact area with commutator *b* has decreased rapidly, whereas its contact area with the commutator *a* has increased. Now, the coil *B* carries a current of 20 A in the direction opposite to

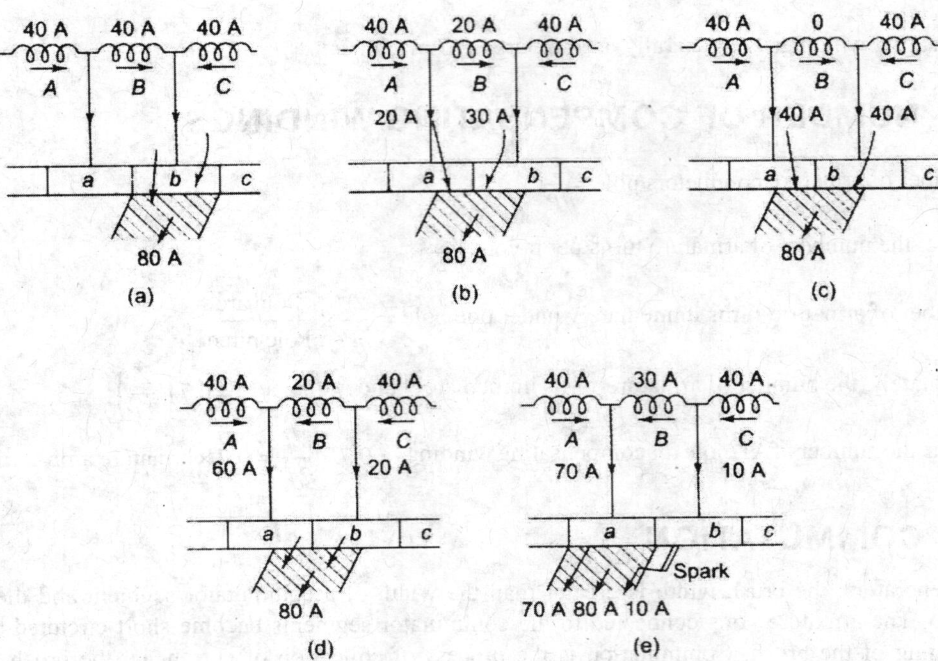

Figure 4.11 Process of Commutation

that in Figure 4.11(b). Figure 4.11(e) shows almost the end of the commutation period. In this case, the current through coil *B* is 30 A instead of 40 A. The 10A current flows from the commutator segment *b* to the brush via air, which results in sparking. Since the coils are embedded in the armature, armature is made up of materials of high permeability. Therefore, the coils posses appreciable amount of self-inductance. Hence, it is very difficult to get ideal commutation. Figure 4.12 shows the changes of the current through coil *B*.

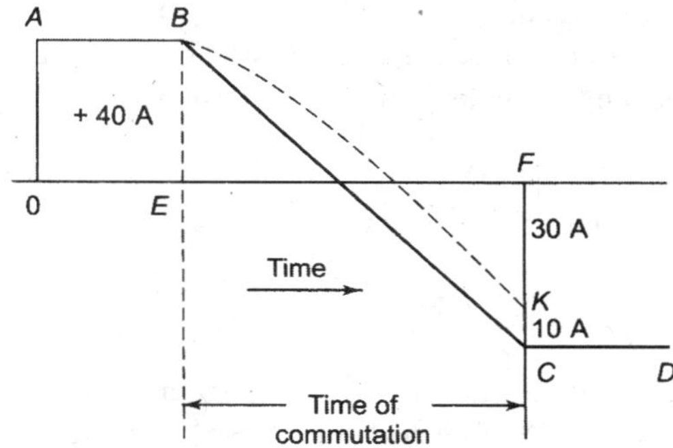

Figure 4.12 Changes of Current Through Coil *B*

4.18.1 Linear Commutation

Figure 4.13(a) shows the position of the brush on the commutator having length, width and thickness of *l*, *b* and *d*, respectively. The thickness of the brush is assumed to be equal to the width of each commutator segment, neglecting the thickness of the mica separation between the segment to simplify the analysis. The current in the coil undergoing commutation is not zero due to the use of short-pitched coils and also due to self-inductance of the coil. Figure 4.13(b) shows the coil undergoing commutation.

Let i be the current in coil 1 that is undergoing commutation, x be the breadth of the contact between segment 2 and the entering edge of the brush, I_1 be the current delivered to the brush by segment 1, I_2 be the current delivered to the brush by segment 2, $\frac{I_a}{2}$ be the current per circuit in the armature.

The total current at the brush $= I_1 + I_2 = I_a$. The current density under the brush is assumed to be uniform. We have

$$I_1 = \frac{I_a(b-x)d}{b \times d} = I_a\left(\frac{b-x}{b}\right) \tag{4.30}$$

and

$$I_2 = \frac{I_a xd}{b \times d} = I_a\left(\frac{x}{b}\right) \tag{4.31}$$

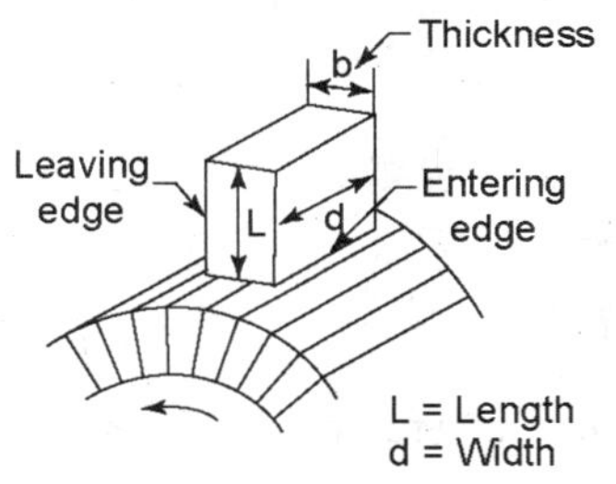

(a) Commutator-brush arrangement

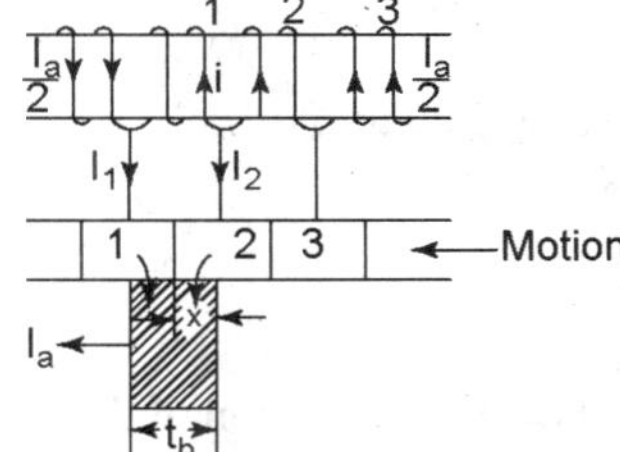

(b) Ring winding

Figure 4.13 Coil Undergoing Commutation

The commutation in coil 2 starts when $x = 0$ and ends when $x = b$. The time is measured from the instant of commencement of commutation. Let V_c be the linear velocity of the commutator in m/s and t be the time taken to cover the distance x, then

$$x = V_c t \tag{4.32}$$

and time of commutation

$$T_c = \frac{b}{V_c} \tag{4.33}$$

Let the current in the coil undergoing commutation at any instant t be i and direction of the current in the coil before commutation be positive. We have

$$I_1 = \frac{I_a}{2} + i$$

i.e.,

$$i = \frac{I_a}{2} - I_1 \tag{4.34}$$

and

$$I_2 = \frac{I_a}{2} - i$$

i.e.,

$$i = \frac{I_a}{2} + I_2 \tag{4.35}$$

From Equations (4.34) and (4.35), we have

$$I_a = I_1 + I_2 \tag{4.36}$$

and

$$i = \frac{I_1 - I_2}{2} \tag{4.37}$$

Substituting I_2, using Equation (4.31) in Equation (4.35), we have

$$i = \frac{I_a}{2}\left(1 - \frac{2x}{b}\right) \tag{4.38}$$

From Equations (4.32) and (4.33), we have

$$\frac{x}{b} = \frac{t}{T_c} \tag{4.39}$$

Substituting $\frac{x}{b}$, using Equation (4.39) in Equation (4.38), we have

$$i = \frac{I_a}{2}\left(1 - \frac{2t}{T_c}\right) \tag{4.40}$$

Equation (4.40) shows the variation of the current with time, and this plot has been shown in Figure 4.14, where i varies linearly from $+\frac{I_a}{2}$ to $-\frac{I_a}{2}$ in time T_c. The current density under the brush remains constant, which is denoted by J_m during the period of commutation.

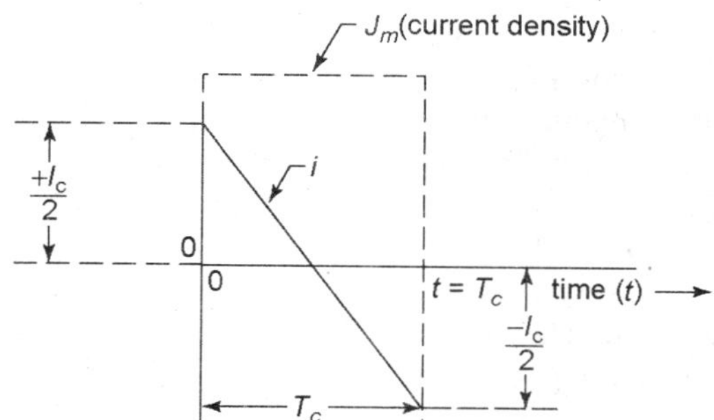

Figure 4.14 Linear Commutation

4.18.2 Retarded Commutation

The variation of the current in the coil undergoing commutation is not linear in practice. Due to the presence of self-inductance, the commutation is delayed. Therefore, the current density at the leaving edge of the brush will be greater and that at the entering edge will be less than the value with straight-line commutation.

Figure 4.15 shows the graph of the short-circuited current as a function of time, where i represents the current in the short-circuited coil. The current entering at the leaving edge of the brush is expressed by $I_1 = \frac{I_a}{2} + i$, which is also shown in Figure 4.15, adding $\frac{I_a}{2}$ to i. The contact area of the brush on the segment at the leaving edge varies linearly from $b \times d$ at $t = 0$ to zero at the end of commutation. The variation of the contact area with respect to the time is represented by a straight line. At any instant t, the current carried by the segment at the leaving edge = PN, having the contact area QN. The current density at the segment = $\frac{PN}{QN} = NS = J_1$.

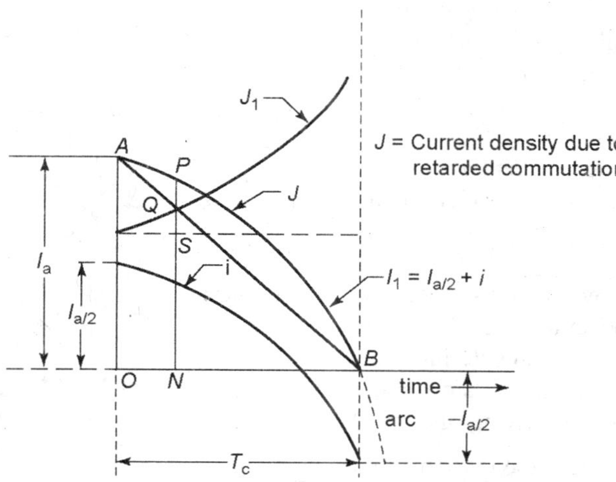

Figure 4.15 Variation of Current Density at the Leaving Edge of the Brush

The current density J_1 at the leaving edge increases with time because the contact area at the segment decreases more rapidly than the current. The current density may reach the value 2.5 times its initial value at the end of breaking of contact with the brush at time $t = T_c$. Figure 4.16 shows the plot of current density with time at the entering edge of the brush. The current through the segment at the entering edge of the brush is expressed by $I_2 = \frac{I_a}{2} - i$, which is shown in Figure 4.16, adding $\frac{I_a}{2}$ to $-i$. I_2 varies from zero at $t = 0$ at I_a to $t = T_c$, which means that the current density attains its normal value at the end of commutation. It is less than the normal value J_m for the major part of the commutation period.

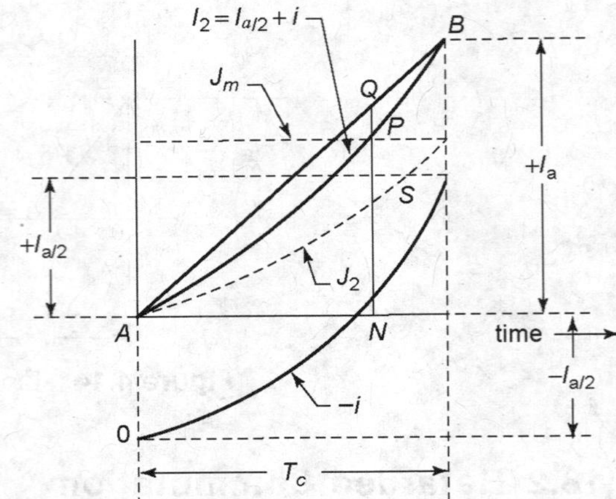

Figure 4.16 Variation of the Current Density at the Entering Edge of the Brush

For retarded commutation, the current at the segment at the leaving edge attains zero instead of $\frac{-I_a}{2}$ at the end of commutation, as shown in Figure 4.15, and the arcing results at the end as the current suddenly changes from 0 to $-\frac{I_a}{2}$.

4.18.3 Accelerated Commutation

Figure 4.17 shows that the brush axis is in forward direction, and the reversal of the current in the coil undergoing commutation can be aided. Coil 1 will immediately cut the fringing flux of the main pole immediately in the front of the coil if the brush axis is shifted by a small angle θ from the axis A–AA in the direction of the motion. At this time, the emf generated in coil 1 will assist in forcing the reversal of current in the coil.

Figure 4.18 shows the current i in coil 1 with accelerated commutation. In each armature path, the current becomes $\frac{I_a}{2} = OA$, and the total current collected by the brush becomes $I_a = AC$. Figure 4.18 shows the variation of the current density at the leaving and entering edges with accelerated commutation. At any time t, the current in the segment leaving the brush becomes $I_1 = \frac{I_a}{2} + i = \text{PN}$, and the current in the segment entering the brush $I_2 = \frac{I_a}{2} - i = PR$.

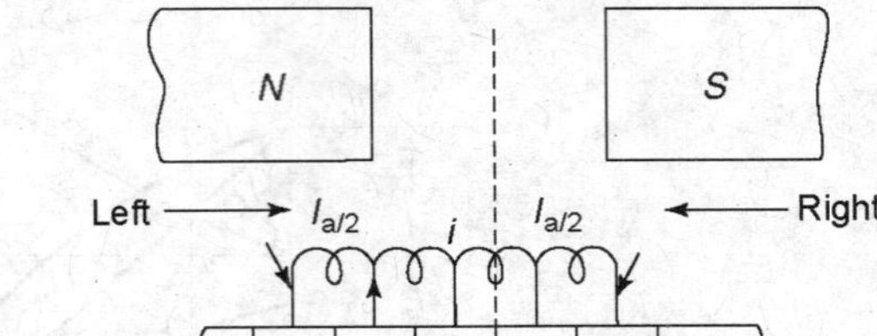

Figure 4.17 Accelerated Commutation with the Forward Shift in the Position of Brush Axis

The following are the corresponding values of the current for linear commutation.

$$I_1 = \frac{I_a}{2} + i = QN$$

and

$$I_2 = \frac{I_a}{2} - i = QR$$

The ratio of the currents at the entering edge with accelerated commutation to that with linear commutation $= \frac{PN}{QN}$. The current density at the leading edge is

$$J_1 = \left(\frac{PN}{QN}\right) J_m < J_m \tag{4.41}$$

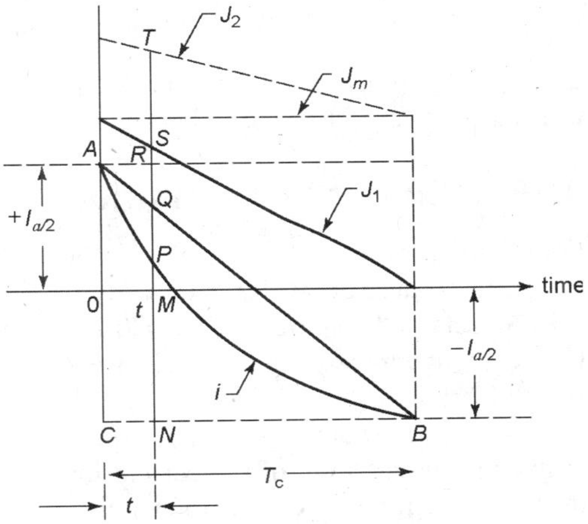

Figure 4.18 Variation of the Current Density at the Leaving and Entering Edges

The ratio of the currents at the entering edge with accelerated commutation to that with linear commutation $= \frac{RP}{RQ}$. The current density at the entering edge

$$J_2 = \left(\frac{RP}{RQ}\right) J_m > J_m \tag{4.42}$$

The current density at the leaving edge of the brush at the end of commutation is

$$J_1 = J_m \left(\frac{\text{Rate of change of current with accelerated commutation}}{\text{Rate of change of current with linear commutation}}\right)$$

$$= J_m \left[\frac{\left(\frac{di}{dt}\right)}{\left(\frac{di_1}{dt}\right)}\right] = J_m \left[\frac{\left(\frac{di}{dt}\right)}{\left(\frac{I_a}{T_c}\right)}\right] \tag{4.43}$$

because $\frac{di_1}{dt} = \frac{I_a}{T_c}$.

If $\frac{di}{dt} = 0$ at $T = T_c$, then $J_1 = 0$. This causes breaking of sparing at the contact of the segment with the brush at the leaving edge. Similarly, the current density at the segment entering the brush will be

$$J_1 = J_m \left[\frac{\left(\frac{di}{dt}\right)}{\left(\frac{di_1}{dt}\right)}\right] = J_m \left[\frac{\left(\frac{di}{dt}\right)}{\left(\frac{I_a}{T_c}\right)}\right] \tag{4.44}$$

If $\frac{di}{dt} = 0$ at $T = T_c$, then $J_2 = 0$.

From Figure 4.18, $\frac{di}{dt} \neq 0$ at $t = 0$ and the current density is more than J_m. This causes sparking at the entering edge of the brush. The current density in the segment must be zero at the time of entering of the brush as well as at the time of leaving of the brush to ensure commutation without sparks.

4.18.4 Sinusoidal Commutation

For retarded commutation, $J_2 = 0$ at $t = 0$ and $J_1 \geq J_m$ at $t = T_c$, and sparking occurs at the leaving edge of the brush because $J_1 \neq 0$ at the time of breaking of the contact of the segment with the brush.

For retarded commutation, $J_2 \neq 0$ at $t = 0$ and $J_1 = 0$ at $t = T_c$, and sparking occurs at the time of contact segment approaching the brush, since $J_2 \neq 0$ at the time when the segment is about to make contact with the brush.

Therefore, to obtain sparkless commutation, the variation of the current in the coil undergoing commutation must be retarded in the first half of the commutation and must be accelerated in the next half of the commutation, as shown in Figure 4.19.

Figure 4.19 shows the current densities J_1 and J_2. At $t \neq 0$, $J_2 = 0$ at the entering segment and $J_1 = 0$ at $t = T_c$ at the leaving segment. This results in sparkles commutation at the making and breaking contacts of the segment.

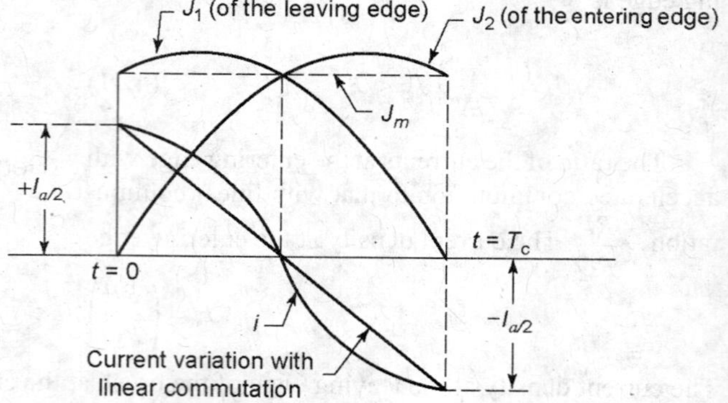

Figure 4.19 Sinusoidal Commutation

4.19 VALUE OF REACTANCE VOLTAGE

The self-induced emf is generated in the coils that makes the proper current reversal of the conductors impossible at the time of commutation. This self-induced emf is called *reactance voltage.*

Therefore, reactance voltage = co-efficient of self-inductance × rate of change of current.

The time of short circuit or commutation is the time required by the commutator to move a distance equal to the circumferential thickness of the brush minus the thickness of one insulating plate or strip of mica.

Let W_b be the width of brush in cm,

W_m be the width of mica insulation in cm and

v_c be the peripheral velocity of the commutator segments in cm/sec.

$$\text{The time of commutation or short circuit } (T_c) = \frac{W_b - W_m}{V_c} \text{ sec} \tag{4.45}$$

If I is the current through the conductor, the total change of current during commutation

$$= I - (-1) = 2I$$

Reactance voltage or self-induced voltage

$$= L \times \frac{2I}{T_c} \text{ if commutation is linear}$$

$$= 1.11L \times \frac{2I}{T_c} \text{ if commutation is sinusoidal.} \quad (4.46)$$

Example 4.8 Calculate the reactance voltage for a machine having the following parameters: Number of commutator segments = 55, revolutions per minute = 900, brush width in commutator segments = 1.76, co-efficient of self-induction = 155 × 10^{-3} H and current per coil = 30 A.

Solution

Current per coil (I) = 30 A.

Self-inductance of coil (L) = 155 × 10^{-3} H

$$\text{Peripheral velocity of commutator } (v_c) = \text{Number of commutator segments} \times \frac{N}{60}$$

$$= 55 \times \frac{900}{60} = 825 \text{ segments/sec.}$$

$$\text{Commutation period } (T_c) = \frac{W_b}{v_c} = \frac{1.76}{825} \text{ sec}$$

Assuming linear commutation, we have

$$\text{Reactance voltage} = L \times \frac{2I}{T_c} = 155 \times 10^{-6} \times 2 \times 30 \times \frac{825}{1.76} = 4.36 \text{ V}$$

■

4.20 METHODS OF IMPROVING COMMUTATION

There are two methods available for improving commutation:

(i) Resistance commutation.

(ii) emf commutation.

4.20.1 Resistance Commutation

In this method, low-resistance copper brushes are replaced by comparatively high-resistance carbon brushes.

Figure 4.20 shows the current I from C reaching commutator segment b. There are two paths for the current to reach the commutator segments. The first path is through bar b to the brush and the other is via the short-circuited coil B to bar a and then to the brush. If copper brushes are used, current will not follow the second path. If carbon brushes are used, current will follow the second path instead of the first path for the following reasons:

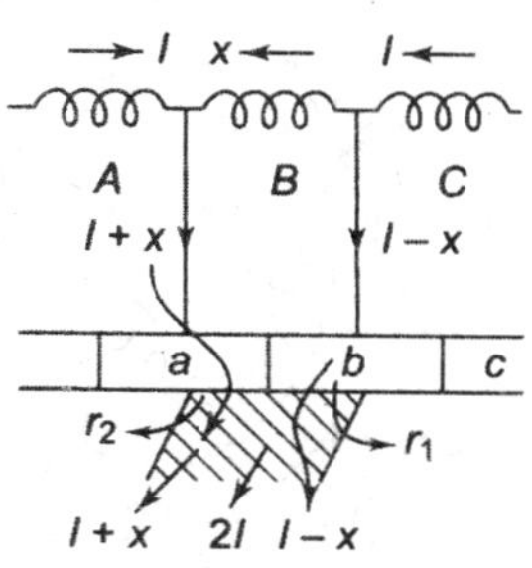

Figure 4.20 Resistance Commutation

(i) The resistance of the first path increases because the area of contact of bar b diminishes.

(ii) The resistance of the second path decreases because the area of contact of bar a increases.

Carbon brushes have the following advantages:

(i) Their high contact resistance is useful for good commutation.
(ii) They lubricate and polish the commutator during rotation.
(iii) In spite of sparking, carbon brushes cause less damage to the commutator as compared to copper brushes.

Carbon brushes have the following disadvantages:

(i) They are more suitable for small machines because a contact drop of 2 V occurs due to their high contact resistance.
(ii) The commutator needs to be made larger than the copper brushes due to the loss for contact drop.
(iii) Large brush holders are required due to their lower current density.

4.20.2 EMF Commutation

An arrangement is made in this method to neutralize the reactance voltage by producing a reversing emf in the short-circuited coil under commutation. This reversing emf can be produced by the following two methods:

(i) By giving the brush a forward lead, sufficient to bring the short-circuited coil under the influence of the next pole of opposite polarity.
(ii) By using interpoles or compoles.

Figure 4.21 shows a generator such that the induced voltage due to motion drives the current as the coil enters the commutative zone while the reactance voltage tends to maintain the current. If the brushes are set forward in the direction of rotation, it is possible to delay the current reversal until the coil is under the fringe field of the next pole in sequence. This results in a generated voltage that opposes the reactance voltage. This arrangement is possible for small machines. The compensation voltage is determined primarily by the field excitation that does not have any functional relation with the armature current. The average reactance voltage is directly proportional to the value of I before commutation.

Figure 4.22 shows the compensation of reactance voltage by interpoles or compoles that makes use of the auxiliary poles mounted in space quadrature with the main poles. These auxiliary poles that are

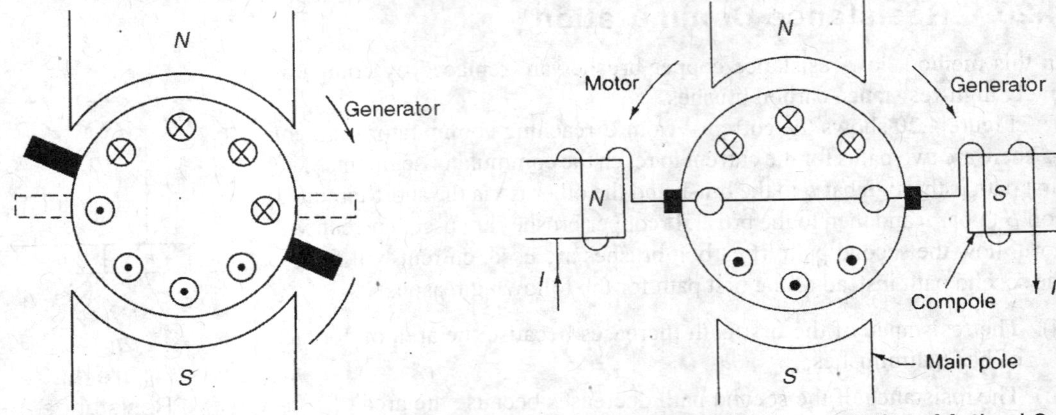

Figure 4.21 EMF Commutation: Method 1 **Figure 4.22** EMF Commutation: Method 2

known as *interpoles* or more appropriately *commutating poles* or *compoles* are excited from the load current. If saturation is neglected, the compole flux, the voltage induced and the reactance voltage all are directly proportional to the load current. Therefore, the compensation is automatically adjusted with changes of load. The polarities of compoles are decided from the brush shift. For generators, the polarity must be the same as the next pole in the direction of rotation. For motors, with the same current and main field polarities, the rotation is reversed and the induced emfs bring in the opposite sense to the current. If the machine is functioning in different modes i.e., generator or motor, the compole polarities are still correct. Interpoles have the following functions:

(i) They make commutation sparkle and ensure automatic neutralization of reactance voltage.

(ii) They neutralize the cross-magnetizing effect of the armature reaction.

4.21 EQUALIZER RINGS

The asymmetrical field or pole system or eccentric armature bearings causes asymmetry between the pole fluxes of the machine. Since in wave winding the winding elements of each parallel path are distributed under all poles of the machine, this effect of flux asymmetry has no effect on the magnitude of emfs induced in the individual circuit of a simple wave winding. But this flux asymmetry has a different effect in lap winding. In this case, the winding elements belong to an armature path, that is, only two adjacent poles. Due to this flux asymmetry, the emfs of the individual flux paths are no longer equal. The points of the winding, which were in-phase points under conditions of symmetry, are no longer in phase. The circulating current flows through these brushes because some of these points are connected by the brushes of the same polarity. In this case sparking can occur if the individual brushes are overloaded. Some in-phase points of the winding having no voltage difference under complete symmetry are tied together by the equalizer connections of negligible resistance. The flow of the alternating current through these windings results in the improvement of commutation and also gives additional heating losses. Equalizer connections are applied to the multipole lap windings and to the wave winding such as duplex, triplex wave windings and so on, having more than two parallel paths. Figure 4.23(a) shows the diagram of a simplex lap-winding for a six-pole machine, whereas Figure 4.23(b) shows equalizer rings at the commutator end.

4.22 CHARACTERISTICS OF DC GENERATORS

Important characteristics of a DC generator are as follows:

(i) *No-load saturation characteristic* (E_0/I_f): It is also called *open-circuit characteristic or magnetic characteristic*. It gives the relation between the no-load generated emf in armature (E_0) and the field or exciting current (I_f). This characteristic is same for the separately excited and the self-excited generators.

(ii) *Internal characteristic* (E_g/I_a): It practically gives the relation between the emf (E_g) actually induced in the armature conductors (after considering the demagnetizing effect of armature reaction) and armature current I_a. It is also known as *total characteristic.*

(iii) *External characteristic* (V/I_a): It shows the relation between the terminal voltage V_t and the load current I. It is also called *performance characteristic* or sometimes *voltage regulating curve*.

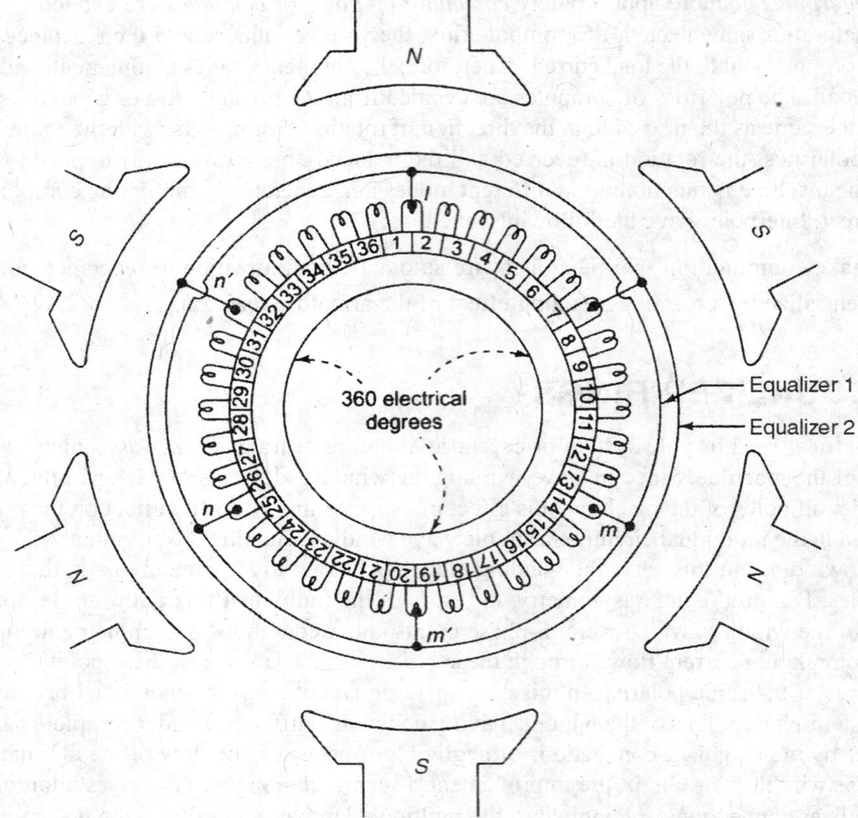

(a) Simplex lap winding for a six-pole machine

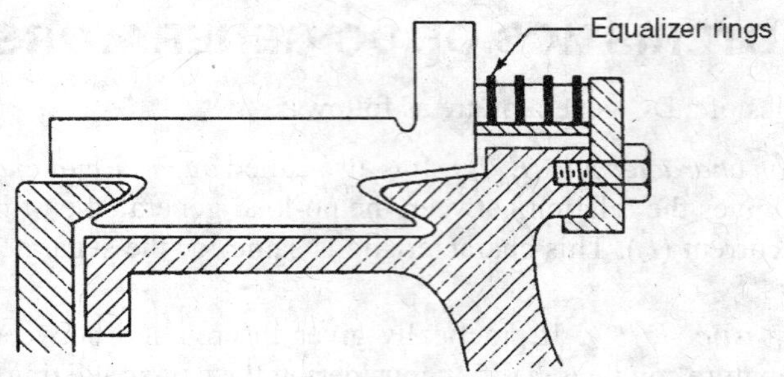

(b) Equalizer rings at commutator ends

Figure 4.23 Equalizer Rings

4.23 SEPARATELY EXCITED GENERATORS

Figure 4.24 shows a separately excited generator. The field coil is being excited separately.

In Figure 4.24, the field circuit is provided with variable excitation. The armature terminals are connected to the load through a two-pole main switch. A voltmeter is connected across the armature and an ammeter is connected in series with the field coil.

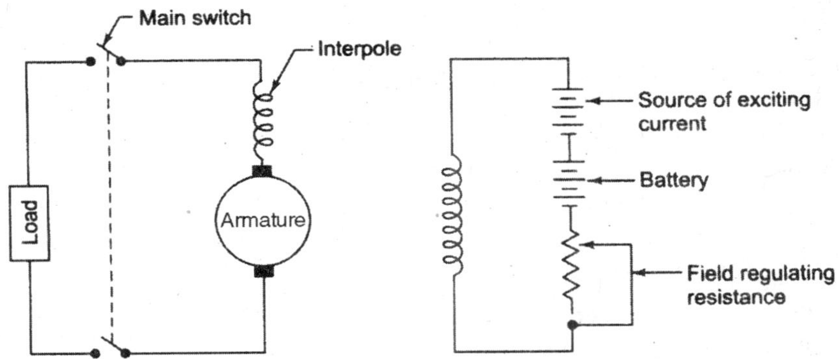

Figure 4.24 Separately Excited Generator

4.23.1 No-load Saturation Characteristic

- The generator is run at a constant speed with the two-pole switch kept open.
- The excitation of the armature is varied and the readings of the ammeter and voltmeter are noted.
- The plot of voltmeter reading and ammeter reading gives the open-circuit characteristic of a separately excited generator shown in Figure 4.25. This is also called magnetization curve.
- During reading of ammeter and voltmeter, it is observed that voltmeter gives a reading at the zero value of field excitation current. This is due to a small amount of permanent magnetism in field poles. This is called residual magnetism, which is usually sufficient to produce 2 to 3 per cent of normal terminal voltage. In some special cases it is purposely increased to 10 per cent or more.
- The first part of the curve is approximately straight and in this zone the flux produced is proportional to the current. But after a certain point, the saturation of iron becomes perceptible and the curve departs from a straight line.

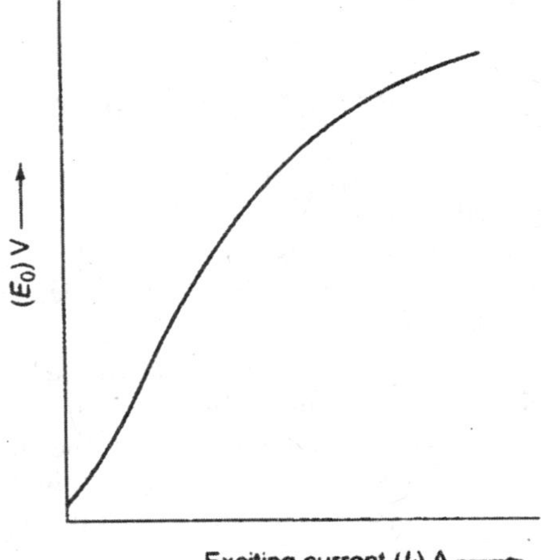

Figure 4.25 Open-circuit Characteristic

4.23.2 Internal and External Characteristic (or Load Characteristic)

Figure 4.26 shows the external characteristic or load characteristics for a separately excited generator. External characteristic or total characteristic is most important because it gives the way in which the terminal voltage varies with the variation of load current from zero to full-load value. The speed of rotation and the excitation current are kept constant. Voltage drop between the no-load characteristic and external characteristic at a particular load current is due to the following voltage drops:

(i) Armature reaction, which has demagnetizing effect on the field.

(ii) Resistance drop.

To obtain the internal characteristic, the resistance drop for a few values of current is added to the voltage shown by the external characteristic. The vertical distance between the internal characteristic and no-load voltage represents the effect of armature reaction. The voltage drop across the load resistance is IR, where R is the load resistance and I is the current. If V is plotted for different values of I, it gives a straight line (OL). The point of intersection of OL and the external characteristic gives the load current and terminal voltage.

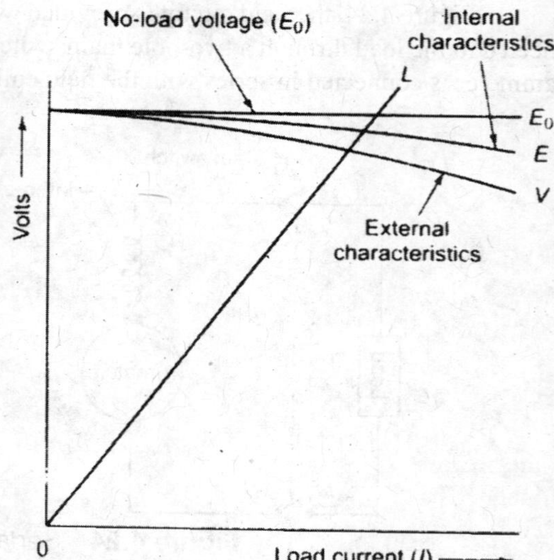

Figure 4.26 Load Characteristic of a Separately Excited Generator

4.24 NO-LOAD CURVE FOR SELF-EXCITED GENERATORS

The field winding of the generator is at first disconnected. It is excited from external DC source, and the excitation current is varied through the rheostat. The armature is rotated at a constant speed (say N rpm). Figure 4.27 shows the circuit connection. The readings of the voltmeter (V) and the ammeter (A) are noted and are plotted, as shown in Figure 4.28. At zero value of excitation current, the initial value of emf is OD. Slight curvature at the lower end is due to magnetic inertia. The first part of the curve is practically straight because reluctance of the iron path is in negligible order at low flux density. So, the flux and generated emf are proportional to the exciting current. Permeability (μ) is small and reluctance of iron path is large at high flux densities and the straight-line relation between E and I_f is not satisfied.

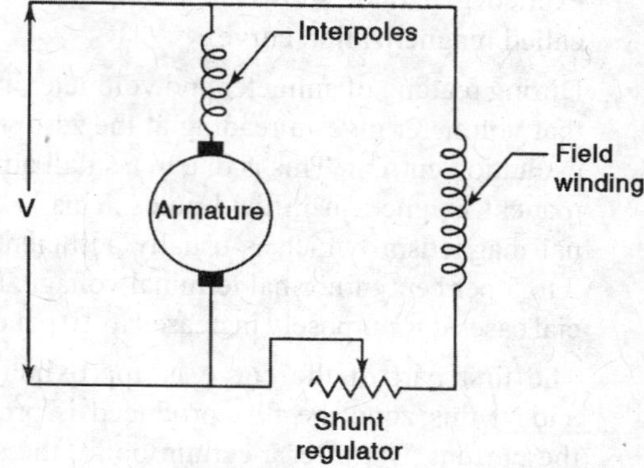

Figure 4.27 Circuit Connection

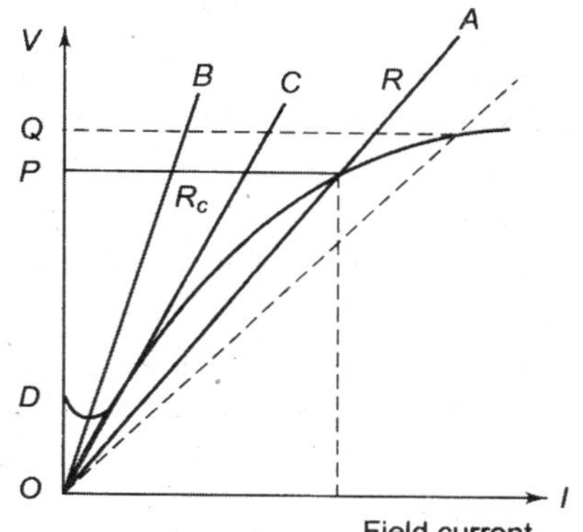

Figure 4.28 No-load Characteristic of Self-excited Generator

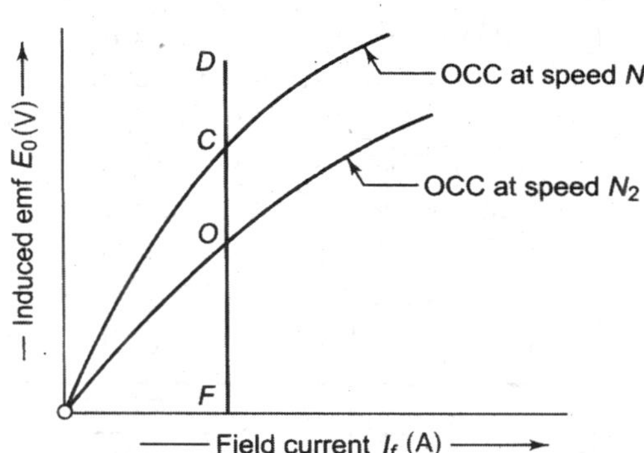

Figure 4.29 OCC for Different Speeds

To get the critical resistance for a shunt generator, the field winding is connected in parallel to the armature and the machine is run as a shunt generator. Since some initial emf exists due to residual magnetism in the poles, current will be generated. This current, during their flow through the field coils, strengthens the magnetism of the poles which increases the pole flux. Hence, the generated emf is further increased. The mutual reinforcement of emf and flux is continued till equilibrium is obtained at some point *P* (say) in Figure 4.28.

The voltage for which the machine is built up is *OP* for field resistance *R*. For a resistance smaller than *R* (*OA*), the voltage *OQ* is slightly greater than *OP*. For higher value of *R*, the machine will build up a voltage less than *OP*. For *OC* the resistance represented by the tangent of the curve is called the critical resistance of the machine. For *OB*, the line does not cut the curve, the machine obviously fails to build up or to excite. To calculate the critical resistance, the data are plotted and the open circuit characteristic (OCC) is obtained. A tangent is drawn at the initial position. The slope of the tangent gives the critical resistance. To draw the OCC at different speeds, the OCC for a given speed is obtained first for a speed say, N_1.

Since $$E \propto N \quad \therefore \quad \frac{E_1}{E_2} = \frac{N_1}{N_2} \quad \text{i.e.,} \quad E_2 = E_1 \times \frac{N_2}{N_1}$$

From Figure 4.29, if $I_f = OF$ and $E_1 = FC$, the new value of voltage for the same I_f at speed N_2 is given by $E_2 = FC \times \frac{N_2}{N_1} = FO$. In this way one point is located. Similarly, other points of the no-load curve can be located.

Example 4.9 The magnetization curve of a DC shunt generator at 1,500 rpm is as follows:

I_f (A)	0	0.4	0.8	1.2	1.6	2.0	2.4	2.8	3.0
E_0 (V)	6	60.0	120.0	172.5	202.5	221.0	231.0	237.0	240.03

For this generator, find (i) the no-load emf for a total shunt field resistance of 100 Ω, (ii) the critical field resistance at 1,500 rpm and magnetization curve at 1,200 rpm and (iii) the open-circuit voltage for a field resistance of 100 Ω.

Solution

(i) From the given data the OCC is drawn. This OCC is for a speed of 1,500 rpm. The shunt field resistance line ($R = 100\ \Omega$) is drawn joining the origin (0, 0) and the point (1 A, 100 V). The point of intersection of the OCC and the resistance line is F, and the voltage corresponding to the point F is 227.5 V. Therefore, no-load emf is 227.5 V.

(ii) The tangent OE to the OCC at the origin is drawn. The critical resistance = slope of line OE = 225/1.5 = 150 Ω.

(iii) The induced emf for field currents for 1,200 rpm is (1,200/1,500) or 0.8 times the values given at 1,500 rpm. The data of magnetization curve corresponding to 1,200 rpm are shown below:

I_f (A)	0	0.4	0.8	1.2	1.6	2.0	2.4	2.8	3.0
E_0 (V)	4.8	48.0	96.0	138.0	162.0	176.8	184.8	189.6	192.0

The OCC is shown in Figure E4.1.

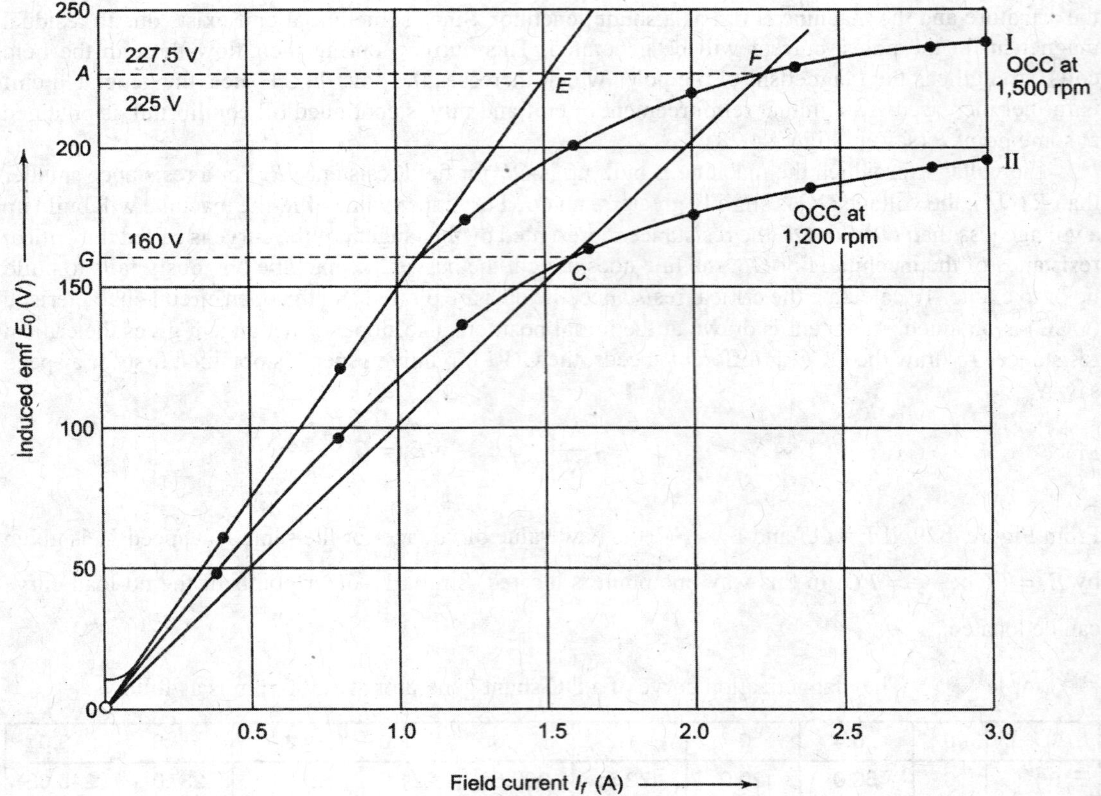

Figure E4.1

Example 4.10 The open-circuit characteristic of a separately excited DC generator driven at 1,000 rpm is as follows:

I_f (A)	0.2	0.4	0.6	0.8	1.0	1.2	1.4	1.8
E_0 (V)	30.0	55.0	75.0	90.0	100.0	110.0	115.0	120.0

If the machine is connected as a shunt generator and driven at 1,000 rpm and has a field resistance of 100 Ω, find (i) the open-circuit voltage and exciting current, (ii) the critical resistance and (iii) the resistance to induce 115 V on open circuit.

Solution

(i) From the given data the OCC is drawn and shown in Figure E4.2. This OCC is for a speed of 1,500 rpm. The shunt field resistance line (R = 100 Ω) is drawn joining the origin (0, 0) and the point (1 A, 100 V). The point of intersection of the OCC and the resistance line is F and the voltage corresponding to the point F is 100 V. Therefore, no-load emf is 100 V.

(ii) The tangent OE to the OCC at the origin is drawn. The critical resistance = slope of line OE = 12/0.8 = 150 Ω.

(iii) From Figure E4.2, the exciting field current corresponding to open-circuit voltage 115 V is 1.4 A.

$$\therefore \text{ The required field resistance} = \frac{115}{1.4} = 82.14\ \Omega.$$

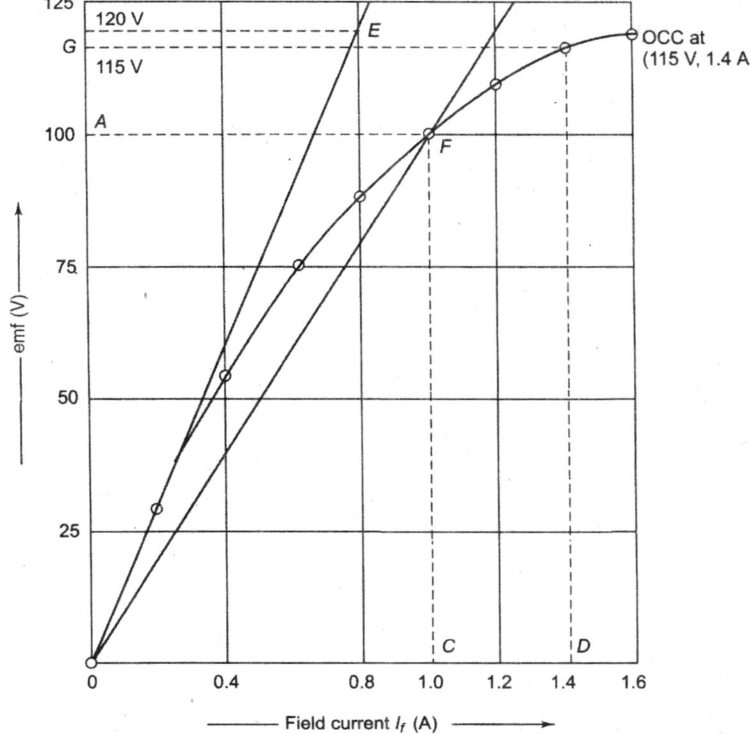

Figure E4.2

4.25 ADVANTAGES AND DISADVANTAGES OF SEPARATELY EXCITED GENERATORS

In separately excited DC generators, the reduction in terminal voltage is due to the $I_a R_a$ drop, which is 2 to 5 per cent, and hence it is very small. The terminal voltage remains more or less constant with load current compared to the other generators. Separately excited generators are very widely used at the distribution point where constant voltage is required.

4.26 VOLTAGE BUILD-UP OF SHUNT GENERATOR

The shunt generator is allowed to build up at first before loading it. Due to the presence of residual magnetism in the poles, a small emf is generated, which circulates a small current in the field circuit. If the field is properly connected to the armature, this circulation of current further increases the pole flux, which in turn increases the generated emf and this continues until the full open-circuit emf is generated. Any improper connection of field coil to armature may destroy the residual magnetism. The generated emf in the armature supplies the ohmic drop and overcomes the self-induced emf in the field coil. For the current *OE* shown in Figure 4.30, *EF* goes to supply the ohmic drop and the rest of *FG* overcomes the self-induced emf. For the field current *OX*, the generated emf is fully utilized to supply the ohmic drop and cannot overcome the self-induced emf. There is no energy stored in field poles and no further increased interpole flux and generated emf. The maximum voltage to which the machine builds up for the given shunt field resistance (*R*) represented by the line *OA* is *XB*. For smaller resistance of field coil, the machine will build up to a somewhat higher value. If the shunt field is greater than the critical resistance value, the machine will not be able to build up. The critical resistance is represented by the line *OQ*.

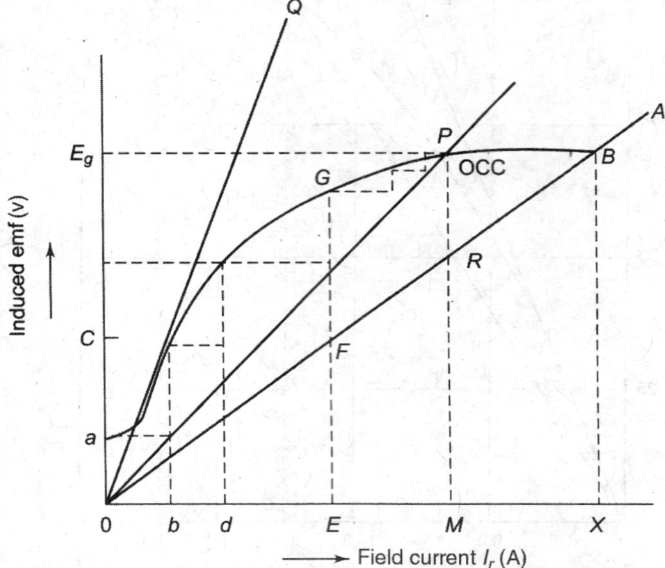

Figure 4.30 Voltage Build-up of a Shunt Generator

4.27 CONDITIONS FOR BUILD-UP OF SHUNT GENERATOR

The following conditions must be met to build-up of a shunt generator:

- Generator poles must have residual magnetism.
- The shunt field coils must be properly connected to the armature for a given direction of rotation of armature.
- The resistance of shunt field coil must be less than the critical resistance if excited on open circuit.
- The shunt field resistance must be greater than some minimum value if excited on load.

4.28 REASONS FOR FAILURE TO BUILD-UP OF SHUNT GENERATORS

Some probable causes for the failure of shunt generators to build up voltage are as follows:

- No residual magnetism.
- Field connection is reversed.
- Field circuit resistance is too high.
- Open field circuit connection.
- Dirty commutator.

4.29 EXTERNAL CHARACTERISTIC OF SHUNT GENERATOR

In a shunt generator, the field circuit is directly connected across the armature. The terminal voltage decreases due to an increase in the generator current. The following three factors are responsible for the decrease in terminal voltage:

(i) *Armature circuit resistance:* The generator has armature resistance, and hence voltage drop occurs. This resistance includes the resistance of (i) the copper conductors of the armature windings, (ii) contact resistance between the brushes and the commutator and (iii) the brushes themselves.

(ii) *Armature reaction:* Due to the flow of currents through the armature conductors, a flux known as armature flux surrounds these conductors. The direction of this flux is such that it reduces the strength of the main flux, and hence reduces the generated voltage as well as the terminal voltage.

(iii) *Reduction of field current:* The decrease of the terminal voltage decreases the field current because the field coil is connected across the armature. This decrease in the field current decreases the field flux, which in turn further decreases the terminal voltage.

Figure 4.31 shows the circuit connection of a shunt generator. The shunt generator is at first excited on no-load and its full open-circuit voltage is obtained. Gradually the load is applied and its terminal voltage decreases due to I_aR_a drop. The readings of armature terminal voltage and the load current are recorded. Using a rheostat, the field current is kept unaltered. The readings of terminal voltage and load current are plotted and external characteristic is obtained. Usually the external load-voltage characteristic decreases with application of load only to a small extent up to its rated load value or current value.

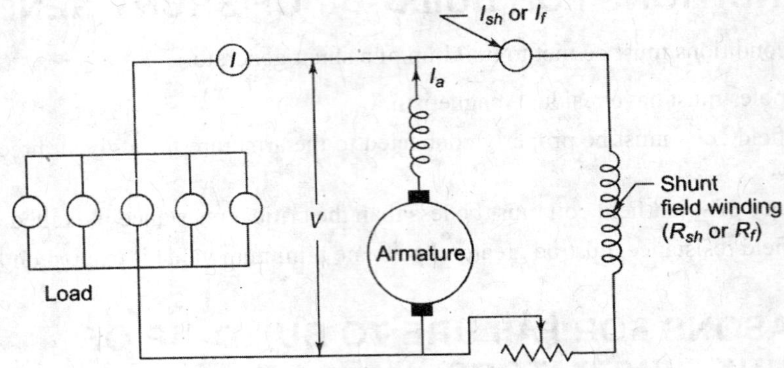

Figure 4.31 Circuit Connection of a Shunt Generator

The portion *ij* of Figure 4.32 is called the working part of this curve, and any decrease in load resistance increases the load current resulting in a comparatively small additional voltage drop over this working part, which holds good until the breakdown point *j* is reached. Figure 4.32 shows that if load current is increased by decreasing the load resistance beyond the breakdown point *j*, it results in a rapid decrease in terminal voltage. This happens because any further decrease in the load resistance at the point *i* causes the current to become greater than *OX* (rated current) of Figure 4.32. This causes severe armature reaction and increases the armature voltage drop and decreases the terminal voltage. Therefore, over the part *jlk* any decrease in load resistance results in a less load current and Ohm's law seems to be inapplicable. The curve *jlk* turns back and cuts the load current axis and the terminal voltage becomes zero.

Figure 4.33 shows the external characteristic due to varying excitation of field coil.

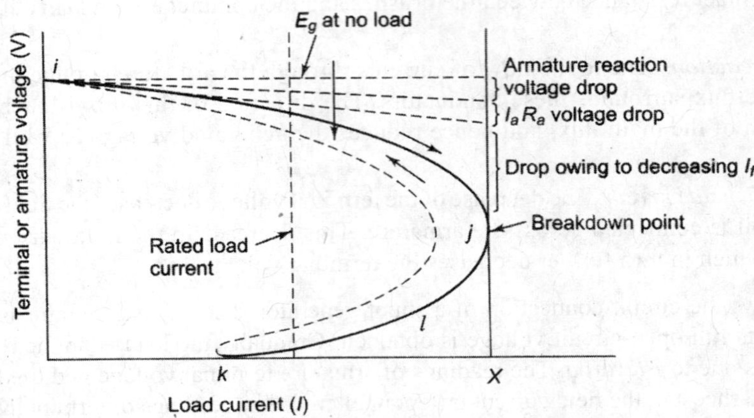

Figure 4.32 External Characteristic of a Shunt Generator

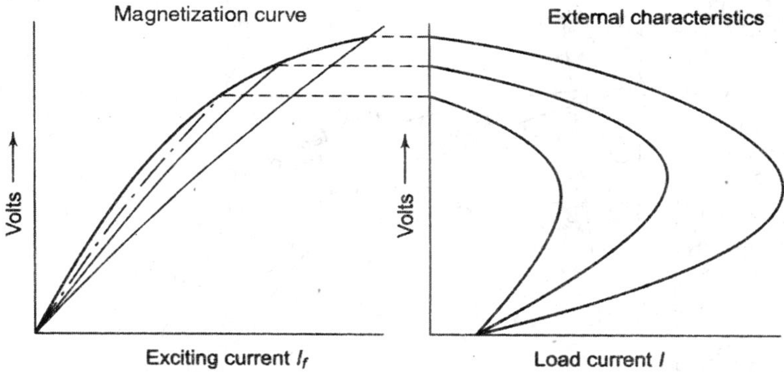

Figure 4.33 Effect of Varying Excitation

4.30 VOLTAGE REGULATION

The degree of change in armature terminal voltage due to application of load is known as voltage regulation. If there is little change in voltage from no load to full load, the generator is said to have good voltage regulation. If there is appreciable change in voltage from no load to full load, the generator is said to have poor voltage regulation.

The change in load voltage from no load to full load expressed as a percentage of the rated terminal voltage is known as per cent voltage regulation.

Per cent voltage regulation $= \frac{V_{nl} - V_{fl}}{V_{fl}} \times 100$, where V_{nl} is the no-load voltage and V_{fl} is the full-load voltage.

4.31 INTERNAL OR TOTAL CHARACTERISTIC

The following procedure is followed to determine the internal characteristic from the external characteristic shown in Figure 4.34.

- External characteristic is drawn from the available data.
- The shunt field resistance line (*OL*) and armature resistance line (*OM*) are drawn.
- Any point *F* is taken on the external characteristic.
- From *F*, a vertical line and horizontal lines intersecting *x*-and *y*-axes, respectively are drawn. These lines are *FC* and *FA*, respectively.
- A point *D* is taken on *x*-axis, where $CD = AB$ represents the shunt field current (I_{sh}).
- A vertical line *DE* is drawn from *D* and it is extended so that it intersects extended *AF* at *H*.
- A point *G* is taken on the line *DH* extended so that $HG = DE = I_a R_a$.
- A number of points are taken on the external characteristic, and the corresponding points lying on internal characteristic are obtained.
- A curve is drawn passing through these points and this represents the external characteristic.

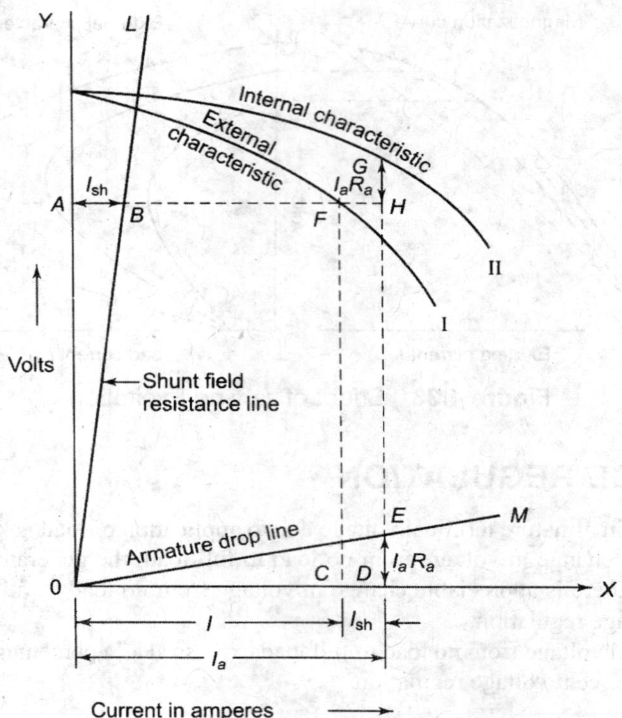

Figure 4.34 Determination of Internal Characteristic from External Characteristic

4.32 EXTERNAL CHARACTERISTIC AND INTERNAL CHARACTERISTIC FROM OCC

The external characteristic of a shunt generator can be obtained directly from its no-load saturation curve. The following two cases are taken into account:

(i) Neglecting armature reaction.
(ii) With armature reaction.

4.32.1 Neglecting Armature Reaction

- From the given data the OCC is plotted, as shown in Figure 4.35.
- A shunt field resistance line (*OS*) is drawn that meets OCC at the point *A*.
- A horizontal line intersecting the *y*-axis at *B* is drawn from the point *A*. *OB* is the maximum no-load or open-circuit voltage.
- A point *L* is taken on OCC, and an ordinate *LMN* is drawn, which intersects the field resistance line and *x*-axis at *M* and *N*, respectively. *LN*, *MN* and *LM* represent the generated emf, the terminal voltage and voltage drop in armature, respectively.
- From points *L* and *M*, horizontal lines are drawn that cut the vertical axis at points *D* and *E*, respectively.
- An armature resistance line *OC* is drawn.

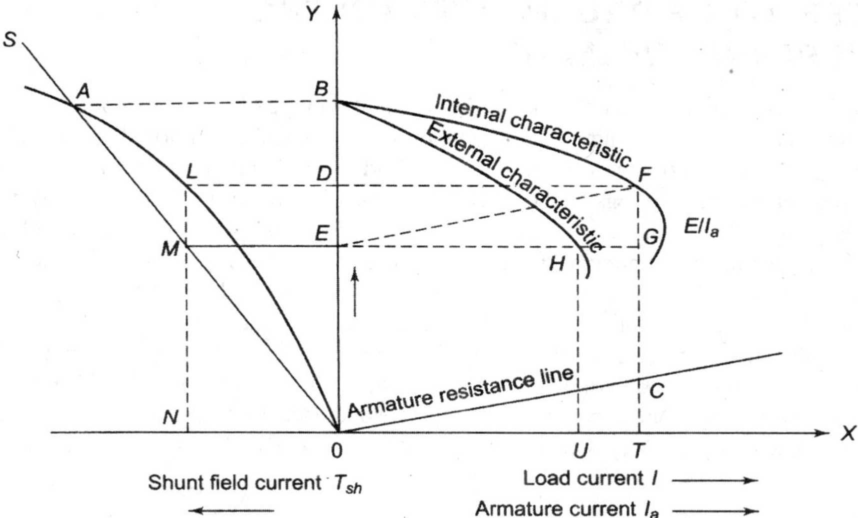

Figure 4.35 Determination of External and Internal Characteristic from OCC

- A line *EF* parallel to line *OC*, cutting line *LD* extended at *F* is drawn from point *E*. The point *F* is lying on the internal characteristic.
- Other points can be obtained and internal characteristics be drawn through these points.
- A vertical line is drawn from the point *F*, which intersects the extended line *ME* at the point *G* and *x*-axis at a point *T*. The point *G* lies on the curve representing the relation between armature current and terminal voltage because $FG = LM = CT$.
- *TU* is the shunt field current and it is equal to *ON* (different scale). *OU* represents the load current corresponding to the armature current represented by *OT* and terminal voltage *OE*.
- A vertical line intersecting line *EG* at *H* is drawn from *U*. The point *H* lies on the external characteristic.
- The other points can be obtained similarly. The curve through these points gives external characteristic.

4.32.2 With Armature Reaction

In this case, the voltage drop due to armature reaction, in addition to voltage drop due to armature resistance, is considered.

A right-angled triangle *lmn* is drawn such that *ln* and *mn* represent the voltage drop in armature and shunt field current, respectively. The triangle *lmn* is known as the drop reaction triangle. All the processes in Section 4.32.1 are repeated with the following modifications to draw the internal and external characteristics.

- A point *L* is taken on the OCC, and a line *LM* parallel to *lm* is drawn and the triangle *LMN* is completed. The vertical lines are drawn from the points *L* and *M*, which cut *x*-axis at points *N'* and *M'*, respectively. *LN'*, *MM'*, *LN* and *ON* represent the generated emf, terminal voltage, the voltage drop in armature due to armature resistance and the shunt field current to induce an emf represented by *LN'*, respectively. *N'M'* represents the increase in shunt field current to counteract the demagnetizing effect.
- *TU* is the shunt field current that is equal to *ON'* (on different scale), and *OU* represents the load current corresponding to the armature current represented by *OT* and terminal voltage *MM'*.

4.33 EFFECT OF BRUSH SHIFT ON THE TERMINAL VOLTAGE

When the brush axis coincides with the GNA, the terminal voltage of the generator gradually decreases with the increase in load. The armature mmf acts on the GNA and it is normal to the field mmf (M_f). Hence, there is no demagnetizing effect in the main field mmf. If the brush axis is shifted by an angle θ from the geometrical neutral axis in the direction of rotation shown in Figure 4.36(a), the generator terminal voltage decreases more rapidly shown by curve 2 in Figure 4.36(c), which is solely due to the armature reaction having a component $M_a \sin\theta$ opposing M_f shown in Figure 4.36(a). The resultant mmf in the air gap decreases. This reduces the resulting induction emf, and hence the terminal voltage also decreases, which is more pronounced with an increase in the load current.

Figure 4.36(b) shows the case when the brush is shifted backwards having the armature mmf component $M_a \sin\theta$, acting in the direction of the main field mmf (M_f). This reinforces M_f and hence the flux in the air gap increases, resulting in the generated emf on the load. Hence, the terminal voltage also increases, as shown in curve 3 of Figure 4.36(c). This is not preferred because it leads to bad commutation.

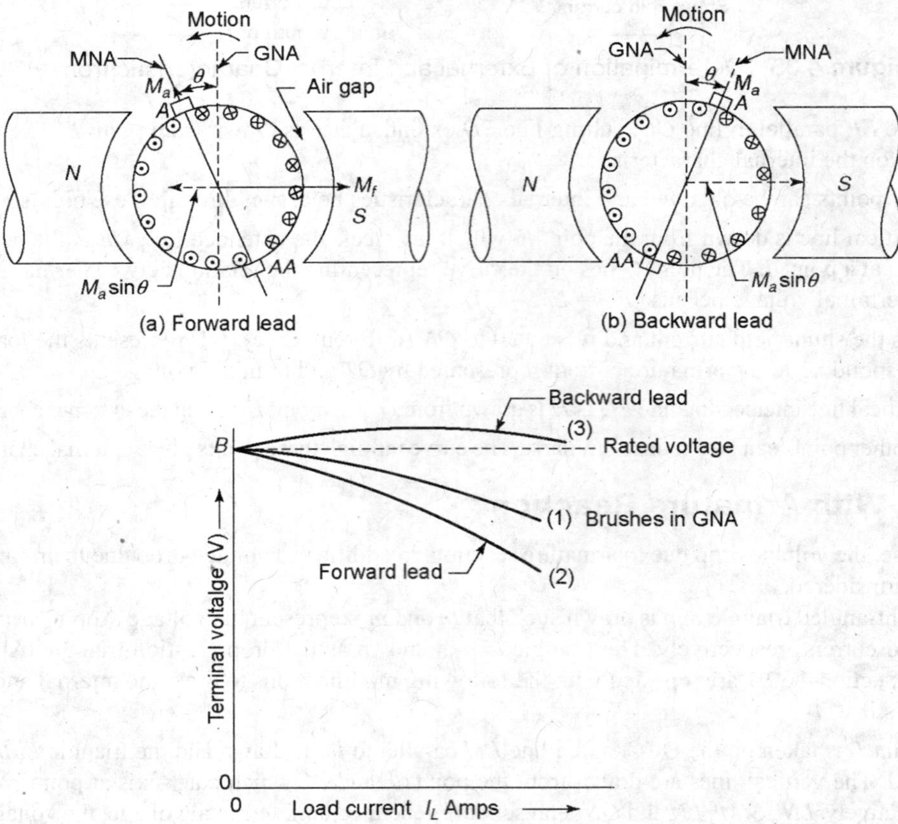

Figure 4.36 Effect of Brush Shift on the Terminal Voltage

4.34 SERIES GENERATOR

The OCC of a series generator is drawn in a manner similar to that of the shunt generator. Figure 4.37 shows the circuit diagram for series generator to obtain the OCC. Curve 1 is the OCC for series generator, as shown in Figure 4.38.

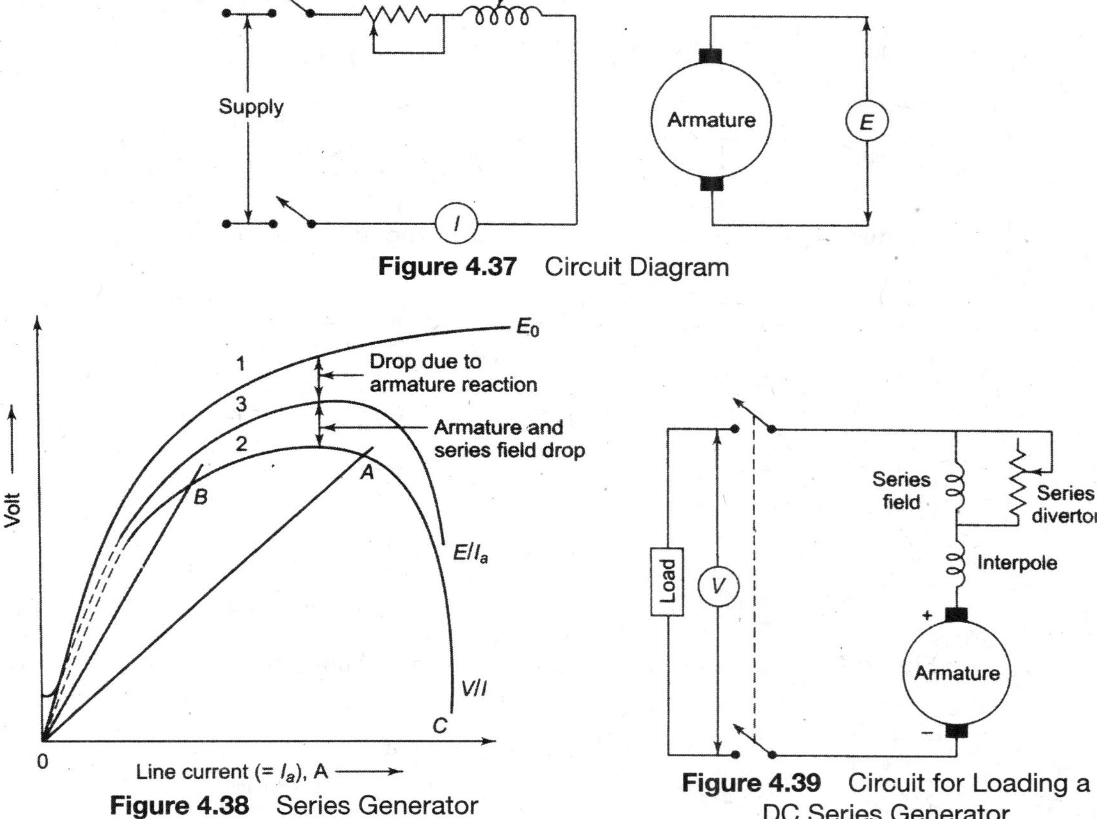

Figure 4.37 Circuit Diagram

Figure 4.38 Series Generator

Figure 4.39 Circuit for Loading a DC Series Generator

Figure 4.39 shows the circuit for loading a series generator. Curve 2 of Figure 4.38 is the external characteristic of DC series generator.

To get the internal characteristic (curve 3 of Figure 4.38), the voltage drop $I_a(R_a + R_{se})$ is added to the external characteristic curve, where R_a and R_{se} are the armature resistance and the series field resistance, respectively.

4.35 COMPOUND GENERATOR

Figure 4.40 shows the connection diagram for a compound generator.

Figure 4.41 shows the current directions in the series and shunt generators.

The external load voltage characteristics of cumulative and differential compound generators are shown in Figure 4.42.

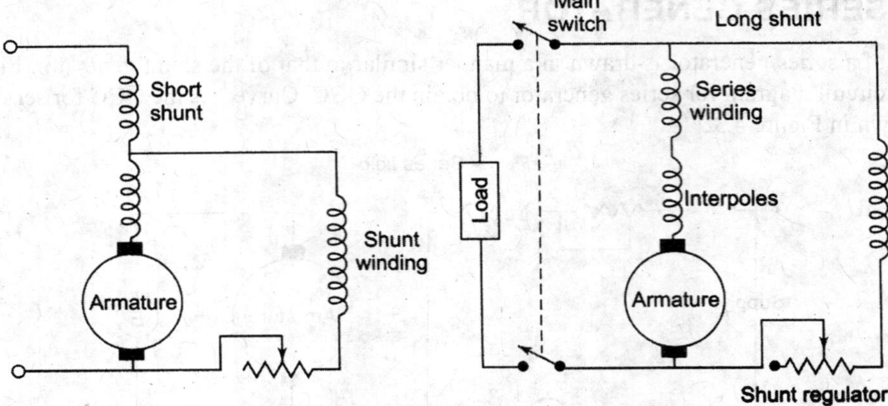

Figure 4.40 Connection Diagram for Compound Generator

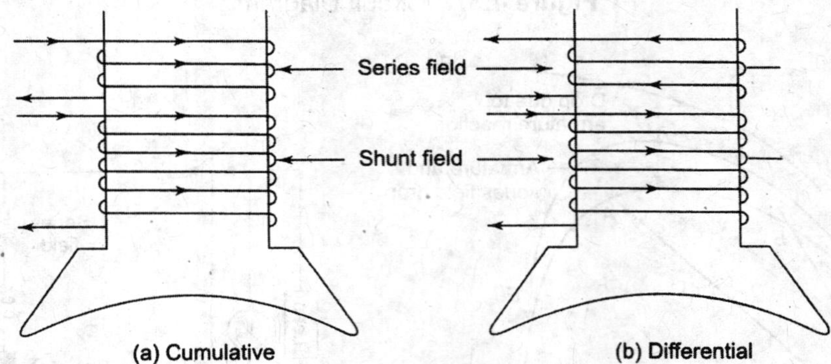

Figure 4.41 Current Directions in Series and Shunt Generators

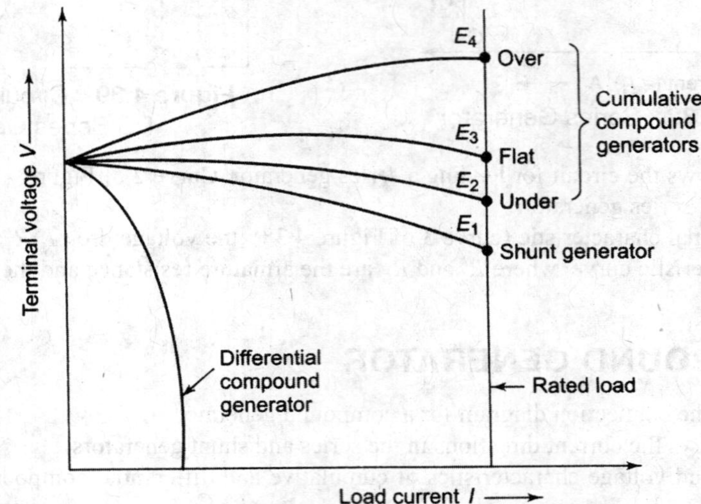

Figure 4.42 External Load Characteristics of Cumulative and Differential Compound Generators

The following definitions are useful:

- *Over-compound generator:* In over-compound generator, the terminal voltage rises with the application of load so that its full-load voltage exceeds its no-load voltage. The voltage regulation is negative.
- *Flat-compound generator:* In flat-compound generator, the full-load voltage and no-load voltage are equal. The voltage regulation is zero.
- *Under-compound generator:* In under-compound generator, the full-load voltage is less than no-load voltage. The voltage regulation is positive.

4.36 PARALLEL OPERATIONS OF DC GENERATORS

Parallel operation of DC generators is required to meet the extra load demand. It is very difficult to meet the extra load demand by a single generator, or it is not possible to give supply when one generator is out of order. The following reasons are given for paralleling of DC generators:

- *Reliability:* The generators that are sources of power are primary safety items, and hence paralleled for reliability.
- *Continuity of service:* Continuity of power supply is most important to satisfy the customers. If a single generator is used, it is difficult to give supply during its breakdown.
- *Efficiency:* The load is not uniform. It is maximum during the day and minimum during the night. The generators must operate most efficiently during full-load condition. It is preferred to use a single generator during light load.
- *Maintenance and repair:* Inspection of generators is carried out frequently to check any possibility of future breakdown. This is possible only when the generator is at rest and other generators take care of the load. During breakdown stage of a generator, it is repaired and other generators take over and meet the load requirements.

4.37 REQUIREMENTS FOR PARALLELING DC GENERATORS

Paralleling DC generators is required mainly for the following two types of situations:

- Paralleling of shunt generators for the same or varying sizes.
- Paralleling of compound generators of the same and varying sizes.

For successful paralleling of DC generators, the voltage of each parallel unit must be equal. During paralleling of DC generators the following conditions must be satisfied to make an agreement with Kirchhoff's current law if the generated voltage of each generator is not the same:

- If the developed internal generated voltage is greater than the voltage at the paralleling point, generating action will take place and the unit will supply current to the load.
- If the developed internal generated voltage is equal to the voltage at the paralleling point, the generator is said to be 'floating' on the line.
- If the developed internal generated voltage is less than the voltage at the paralleling point, it will operate in motoring mode, drawing current from the paralleling point.

The following conditions must be satisfied for the paralleling of DC generators:

- The polarities of the generators must be identical.
- The terminal voltages must be the same.

- The change of voltage with the change of load must be of the same character.
- The prime movers driving the armatures of the generators must have similar and stable rotational characteristics.

A positive regulation generator should not be combined with a negative regulation generator because in this case circulating current will dominate. An exact match characteristic in this case is very difficult to achieve.

4.38 PARALLEL OPERATION OF SHUNT GENERATORS

Figure 4.43 shows parallel operation of DC generators. Generator 1 is connected to the bus bars and supplies the load. Generator 2 is required to be connected in parallel with generator 1 to supply a part of the load.

The following procedures are followed during paralleling:

- Generator 2 is speeded up to the rated voltage by means of a prime mover. The switch S_2 is closed and a voltmeter is placed across the open switch S_1.
- The excitation of the generator is changed until the reading of the voltmeter connected across the open switch S_1 is zero. At this point, the voltage of generator 2 is the same as the voltage of generator 1.
- The switch S_1 is closed and generator 2 is in parallel with generator 1. Generator 1 is supplies the entire load, whereas generator 2 is in floating stage.
- It is required to shift load from generator 1 to generator 2. The field rheostat resistance of generator 1 is cutting in, whereas the field rheostat resistance of generator 2 is cutting out. In this way any degree of load can be shifted from generator 1 to generator 2. If the entire load is shifted from generator 1 to generator 2, the switch S is opened to remove generator 1 from the line.
- During shifting the load from one generator to another, the reading of A_1 and A_2 should be observed very carefully.

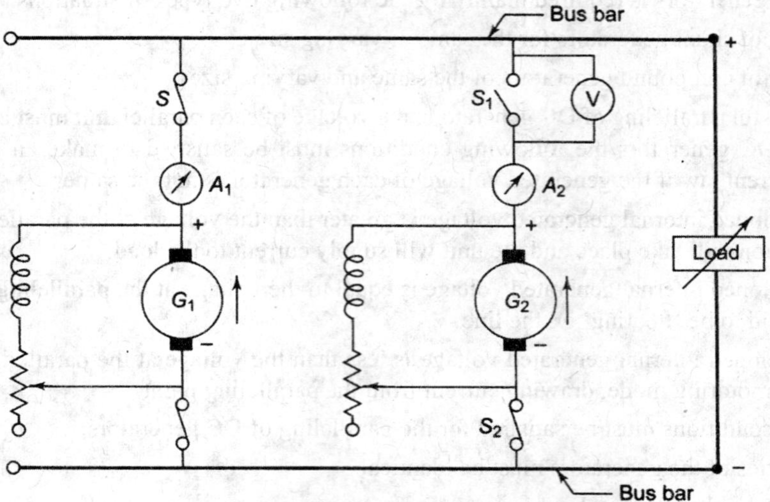

Figure 4.43 Two-shunt Generators in Parallel

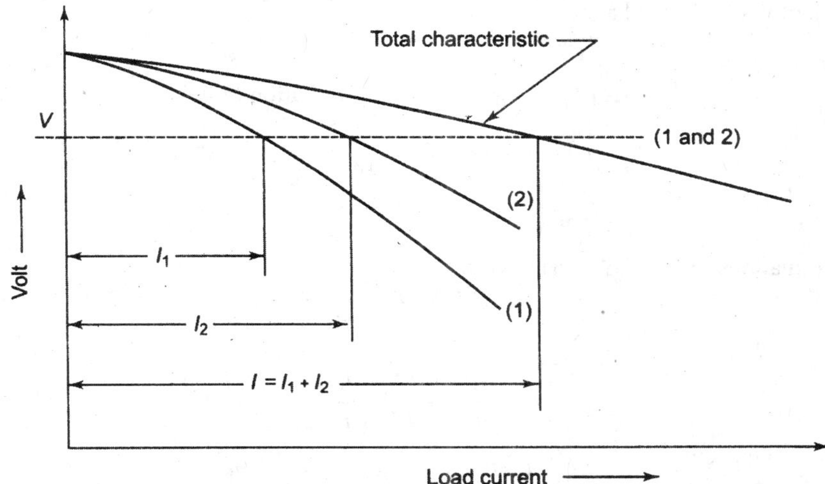

Figure 4.44 Characteristics of Shunt Generators in Parallel

Due to the drooping characteristic of shunt generators, these generators are well suited for parallel operation. Figure 4.44 shows the characteristics of two shunt generators 1 and 2. Generator 1 has more drooping characteristic than generator 2. Neglecting very small voltage drop across the connecting leads, the terminal voltages of the two generators connected in parallel must be identical. For a common terminal voltage, generator 1 supplies the load I_1 and generator 2 supplies the load I_2. Therefore, the generator having more drooping characteristic carries the smaller load.

A condition may arise that temporarily causes generator 1 to take more than its share of load. This condition might arise for a temporary increase in the speed of its prime mover, or due to momentary change of load on the system. After restoring the normal condition, generator 1 would tend to operate at some point a on its characteristics. This results in a drop in its terminal voltage, which tends to make it take less load. Moreover, at the time generator 1 takes more load, generator 2 must take less load. The total load remains constant and generator 2 tends to operate at some point *b* on its characteristics. This raises its terminal voltage and causes it to take more load. Therefore, any tendency of one generator to take more than its share of load results in changes of voltage in the system which opposes this tendency. Hence, the shunt generators in parallel may be said to be in stable equilibrium. The reaction of the system is such as to hold the generators in parallel.

When the machines have unequal emfs on no load, the distribution of currents between them can be determined analytically as follows:

Let R be the resistance of the combined load,
E_1 be the emf of generator 1,
E_2 be the emf of generator 2,
V_t be the terminal voltage of the two machines,
R_1 be the armature resistance of generator 1 and
R_2 be the armature resistance of generator 2.

We have

$$V_t = E_1 - I_1 R_1 \quad (4.47)$$

and

$$V_t = E_2 - I_2 R_2 \quad (4.48)$$

$\therefore$

$$V_t = (I_1 + I_2)\,\mathrm{R} \quad (4.49)$$

From Equation (4.47), we have

$$E_1 = V_t + I_1 R_1$$

$$= (I_1 + I_2)R + I_1 R_1 \text{[Using Equation (4.49)]} \quad (4.50)$$

i.e.,

$$E_1 = I_1 (R_1 + R) + I_2 R$$

Similarly,

$$E_2 = I_1 R + I_2 (R_2 + R) \quad (4.51)$$

Solving Equations (4.50) and (4.51), we have

$$I_1 = \frac{E_1 R_2 + (E_1 - E_2) R}{R_1 R_2 + (R_1 + R_2) R} \quad (4.52)$$

and

$$I_2 = \frac{E_2 R_1 - (E_1 - E_2) R}{R_1 R_2 + (R_1 + R_2) R} \quad (4.53)$$

The total load current

$$I = I_1 + I_2 = \frac{E_1 R_2 + E_2 R_1}{R_1 R_2 + (R_1 + R_2) R} \quad (4.54)$$

The terminal voltage

$$V_t = IR = \frac{(E_1 R_2 + E_2 R_1) R}{R_1 R_2 + (R_1 + R_2) R}$$

i.e.,

$$V_t = \frac{E_1 R_2 + E_2 R_1}{\frac{R_1 R_2}{R} + R_1 + R_2} \quad (4.55)$$

From Equation (4.55) the following cases can be drawn.

Case I: When the load is open circuited, $R = \infty$ and $V_t = V_0$ = the no-load terminal voltage.

$$\therefore \quad V_t = V_0 = \frac{E_1 R_2 + E_2 R_1}{R_1 + R_2} \quad (4.56)$$

From Equations (4.52) and (4.53), we have $I = I_c$ (Circulating current)

$$= I_2 - I_1 = \frac{2(E_2 - E_1)}{R_1 + R_2} \quad (4.57)$$

Case II: When load is short circuited, $R = 0$.

From Equation (4.54), we have

$$I = I_s = \text{Short circuit current}$$

$$= \frac{E_1 R_2 + E_2 R_1}{R_1 R_2}$$

$$= \frac{E_1}{R_1} + \frac{E_2}{R_2} \quad (4.58)$$

DC shunt generators having drooping external characteristic perform well when they are connected in parallel. If any fault occurs in any prime mover and the speed gets reduced, there will be a reduction in the emf of this machine. The second machine will be overloaded without showing any tendency for unstable operation.

4.39 PARALLEL OPERATION OF SERIES GENERATORS

These generators are not usually operated in parallel. They are put in parallel for electric traction purposes to provide electric breaking of motors. Figure 4.45 shows two series generators put in parallel operations under normal working conditions, that is, on the rising portion of external characteristics. If the load current shared by machine 1 is increased, the induced emf will also increase, which will lead machine 1 to share the greater load current and it relieves the other machine. This causes the emf of machine 1 to increase and the emf of machine 2 to decrease continuously. When the emf of the machine 2 is less than the terminal voltage, it receives current and its polarity gets reversed, which makes the emfs of the two machines to act in series around the load circuit of the generators having a very low resistance virtually acting as a short circuit on the two machines.

If the field coils are connected in parallel using equalizing bus bars having negligible resistance shown in Figure 4.45(a), this will make the currents in the field windings of two machines equal. The increase in the current of generator 1 leads to an increase in the current of generator 2; hence, the emf of generator 2 also increases preventing machine 1 to share more load. There is no tendency to relieve machine 2 from sharing its load as in the previous case.

If there is any fault developed by the prime mover of machine 2 and is not driving the machine 2, machine 2 is slowed down only until its emf reduces to a value less than its terminal voltage. Therefore, machine 2 acts as a separately excited motor due to the equal currents in the field windings because they are connected in parallel and will continue to run on light load.

Figure 4.45(b) shows that the field coils are cross-connected when the series motors are connected in parallel. In case of any increase in the current of the armature of generator 1, the increased current flows through the field coil of generator 2. This increases the emf of machine 2, which opposes the change in load sharing trying to stabilize the operation of the two machines at the original operating point itself or near to it. This gives more positive and better operation than in the case shown in Figure 4.45(a).

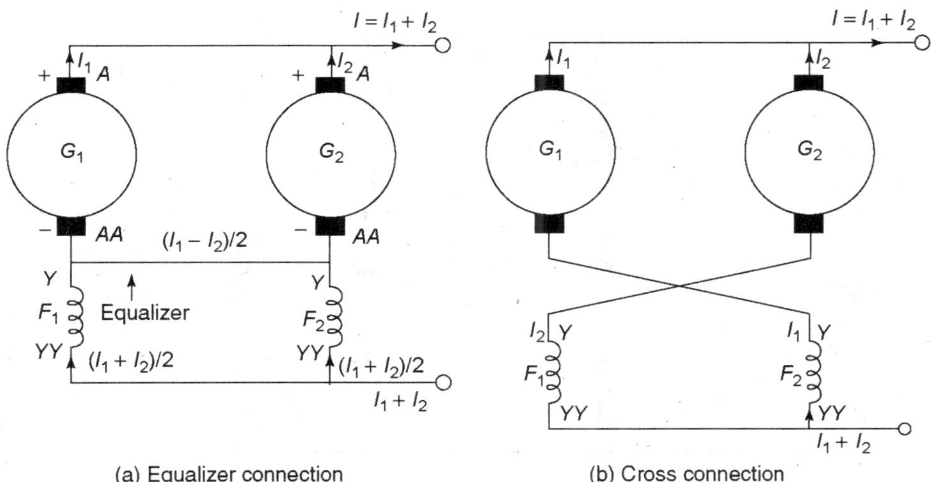

Figure 4.45 Series Generators in Parallel

4.40 PARALLEL OPERATION OF COMPOUND GENERATORS

The under-compound generators have drooping characteristic, and hence it is possible to operate them in parallel without any difficulty. The level- and over-compound generators cannot be operated in parallel satisfactorily.

If they are operated in parallel, inherently they are unstable. Any increase in the load current in one generator strengthens the field flux, and hence increases the generated voltage. It causes the machine to take up still greater load. For a fixed load, the other generator sheds part of its load, which results in the weakening of its field and it sheds more load. If this process is continued cumulatively, one machine can run the entire load and it can drive the other as a differentially compounded motor at an excessively high speed.

To get stable operation of compounded generators in parallel, the series field windings are connected in parallel using equalizers that are low-resistance connections. These are usually provided in the form of another bus bar on the switch board, as shown in Figure 4.46.

Example 4.11 Two DC shunt generators operating in parallel deliver a total current of 250 A. One of the generators is rated 50 kW and the other 100 kW. The voltage rating of both machines is 500 V and have regulations of 6 per cent (smaller one) and 4 per cent, respectively. Assume linear characteristic. Determine (i) the current delivered by each machine and (ii) the terminal voltage.

Solution

For a 50 kW generator:

$$\text{Full-load current } (I_1) = \frac{50 \times 1{,}000}{500} = 100 \text{ A}$$

Full-load voltage drop = 0.06 × 500 = 30 V.

Voltage drop per ampere of current supply = 30/100 = 0.3 V/A.

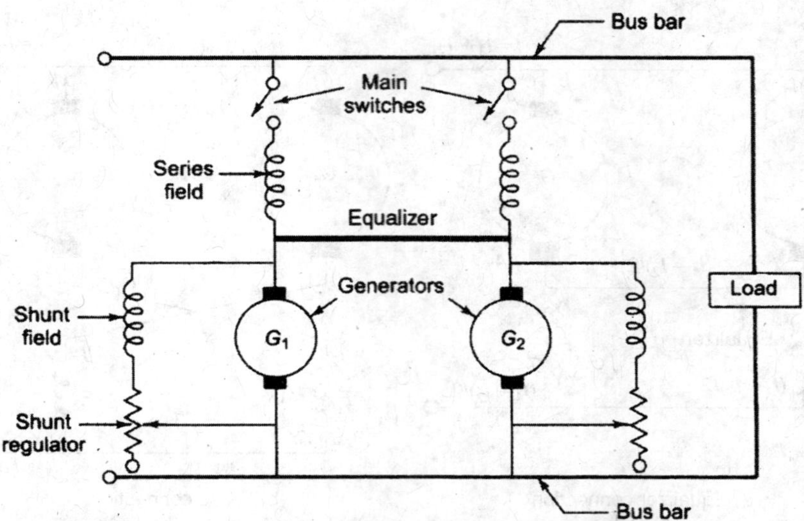

Figure 4.46 Two-compounded Generators in Parallel

For 100 kW generator:

$$\text{Full-load current } (I_2) = \frac{100 \times 1{,}000}{500} = 200 \text{ A}$$

Full-load voltage drop = 0.04 × 500 = 20 V

Voltage drop per ampere of current supply = 20/200 = 0.1 V/A.

Let 50 kW and 100 kW generators supply current I_1 *A* and I_2 *A*, respectively at the terminal voltage of V_t V.

Therefore, we can write the following equations:

$$V_t = 500 - 0.3\, I_1 \quad (1)$$

$$V_t = 500 - 0.1\, I_2 \quad (2)$$

and

$$I_1 + I_2 = 250 \quad (3)$$

Solving Equations (1) and (2), we have

$$I_1 = 62.5 \text{ A and } I_2 = 187.5 \text{ A.}$$

Substituting $I_1 = 62.5$ A in Equation (1), we have

$$V_t = 500 - 0.3 \times 62.5 = 481.25 \text{ V}$$

Example 4.12 Two DC shunt generators with emfs 120 V and 115 V, armature resistances of 0.05 Ω and 0.04 Ω and field resistances of 20 Ω and 25 Ω, respectively are in parallel, supplying a total load of 25 kW. How do they share the load?

Solution

Let the terminal voltage of the generators be V_t.

For

$$G_1 : I_{a1} = I_1 + \frac{V_t}{R_{sh1}} = I_1 + \frac{V_t}{20} = I_1 + 0.05 V_t$$

For

$$G_2 : I_{a2} = I_2 + \frac{V_t}{R_{sh2}} = I_2 + \frac{V_t}{25} = I_2 + 0.04 V_t$$

For G_1:

The generated emf is

$$E_{g1} = V_t + I_{a1} R_{a1}$$

i.e.,

$$120 = V_t + (I_1 + 0.05 V_t) \times 0.05 = 1.0025 V_t + 0.05 I_1 \quad (1)$$

i.e.,

$$2{,}400 = 20.05 V_t + I_1$$

For G_2:

The generated emf is

$$E_{g2} = V_t + I_{a2} R_{a1}$$

i.e.,

$$115 = V_t + (I_2 + 0.04 V_t) \times 0.04 = 1.0016 V_t + 0.04 I_2 \quad (2)$$

i.e.,

$$2{,}875 = 25.04 V_t + I_2$$

Also,

$$I_1 + I_2 = \frac{25{,}000}{V_t}$$

∴

$$I_1 = \frac{25{,}000}{V_t} - I_2 \quad (3)$$

Substituting I_1 in Equation (1), we have

$$120 = 1.0025V_t + 0.05\left(\frac{25,000}{V_t} - I_2\right)$$

i.e.,
$$2,400 = 20.05V_t + \frac{25,000}{V_t} - I_2 \quad (4)$$

Adding Equations (2) and (4), we have

$$5,275 = 45.09V_t + \frac{25,000}{V_t}$$

i.e.,
$$45.09\,V_t^2 - 5,275V_t + 25,000 = 0$$

i.e.,
$$V_t = \frac{5,275 \pm \sqrt{(-5,275)^2 - 4 \times 45.09 \times 25,000}}{2 \times 45.09} = \frac{5,275 \pm 4,828.73}{2 \times 45.09}$$

Taking '+' sign, we have $V_t = 112$ V.

From Equation (1) we have
$$I_1 = \frac{120 - 1.0025 \times 112}{0.05} = 154.4 \text{ A}$$

From Equation (2) we have
$$I_2 = \frac{25,000}{V_t} - I_1 = \frac{25,000}{112} - 154.4 = 68.81 \text{ A}$$

Power supplied by G_1:
$$P_1 = \frac{V_1 I_1}{1,000} = \frac{112 \times 154.4}{1,000} = 17.293 \text{ kW}$$

Power supplied by G_2:
$$P_2 = \frac{V_2 I_2}{1,000} = \frac{112 \times 68.81}{1,000} = 7.707 \text{ kW}$$

4.41 USES OF DC GENERATORS

- *Separately excited generators:* This type of generators have special applications such as electroplating, electrorefining of materials because separate supply is required for field excitation.
- *Shunt generators:* This type of generators are used in battery charging and ordinary lighting purposes.
- *Series generators:* This type of generators is used commonly as boosters on DC feeders, as current generators for welding and arc lamp.
- *Cumulatively compounded generators:* This type of generators is used for domestic lighting purposes and to transmit energy over long distances.
- *Differential compound generators:* This type of generators is rarely used. They are used for special applications such as electric arc welding.

4.42 INDICATIONS OF AN OVERLOADED GENERATORS

The following are the indications of overloading:

(i) Excessive sparking at brushes.

(ii) Overheating of the armature and other parts of the generator.

4.43 CAUSES OF OVERLOADING

The following are the causes of overloading:

(i) Excessive voltage due to over excitation of field magnets or too high speed of the prime mover.
(ii) Excessive current due to bad feeding of load or short circuit/ground in the generator itself or in the external load circuit.
(iii) Reversal of polarity.

4.44 CAUSES OF SPARKING AT BRUSHES OF A DC MACHINE

The following are the most probable causes:

(i) Inadequate pressure on brushes.
(ii) Incorrect position of the brushes.
(iii) Unequal brush spacing.
(iv) Carbon brushes short circuit more segments at time.
(v) Brushes not properly bedded.
(vi) Worn out brushes.
(vii) Wrong grade of brushes.
(viii) Commutator is not fully round.
(ix) Rough commutator surface.
(x) Projection of intersegment micas.
(xi) Weak mica between segments.
(xii) Segments short circuited by carbon filling.
(xiii) Oil and dirt on commutator surface.
(xiv) Play in the armature.
(xv) Short circuit in the armature coils.
(xvi) Earth fault in armature windings.
(xvii) Overloading of armature.
(xviii) Reversed interpole polarity.

4.45 CAUSES OF EXCESSIVE HEATING OF GENERATOR ON RUNNING

The following are the probable causes of excessive heating of generator on running:

(i) Sparking at the brushes.
(ii) Weak insulation in the lamination.
(iii) Short circuit in field and armature winding.
(iv) Defective bearing.

4.46 CAUSES OF HEATING OF ARMATURE

The following are the causes of heating of armature:

(i) Eddy current.
(ii) Moisture absorption in insulation of armature winding.
(iii) Unequal strength of field poles.
(iv) Operating at above rated voltage and below normal speed.

4.47 CAUSES FOR ABNORMAL SOUND IN DC GENERATOR

The following are the most probable causes for abnormal sound in DC generators:

(i) Unequal resistances of armature or field coils.
(ii) Loose fitting of field cores with yoke or frame.
(iii) Improper fitting of end covers.
(iv) Loose fitting of foundation.

4.48 REASONS FOR RAPID BRUSH WEAR IN A DC MACHINE

The following are the causes for the rapid brush wear in DC machine:

(i) Rough surface of the commutator.
(ii) Imperfect contact with the commutator.
(iii) Excessive pressure of the brushes.
(iv) Severe sparking.

ADDITIONAL SOLVED PROBLEMS

Example 4.13 A DC shunt generator having terminal voltage 250 V, delivers load current 195 A. The armature and shunt field resistances are 0.05 Ω. and 50 Ω, respectively. Calculate (i) the shunt field current, (ii) the armature current, (iii) the generated emf in armature and (iv) the load resistance.

Solution

Here, $V_t = 250$ V, $I = 195$ A, $R_{sh} = 50\ \Omega$ and $R_a = 0.05\ \Omega$.

(i) $$I_{sh} = \frac{V_t}{R_{sh}} = \frac{250}{50} = 5 \text{ A}$$

(ii) $$I_a = I + I_{sh} = 195 + 5 = 200\text{A}$$

(iii) $$E_g = V_t + I_a R_a = 250 + 200 \times 0.05 = 260 \text{ V}$$

(iv) $$R_L = \frac{V_t}{I} = \frac{250}{195} = 1.28$$ ■

Example 4.14 A long-shunt compound DC generator having terminal voltage 300 V delivers a load current of 60 A. The armature, series field and shunt field resistances of the generator are 0.05 Ω,

0.04 Ω. and 200 Ω, respectively. Calculate the generated voltage and the armature current if the contact voltage drop per brush is 1 V.

Solution

Here, $V_t = 300$ V, I = 60 A, $R_{sh} = 200\ \Omega$, $R_a = 0.05\ \Omega$ and $R_{se} = 0.04\ \Omega$.

$$I_{sh} = \frac{300}{200} = 1.5\text{A}$$

$$I_a = I + I_{sh} = 60 + 1.5 = 61.5\text{ A}$$

$$E_g = V_t + I_a\left(R_a + R_{se}\right) + 2\times 1 = 300 + 61.5\times\left(0.05 + 0.04\right) + 2 = 307.535\text{ V}$$

The armature current and the generated emf are 61.5 A and 307.535 V, respectively. ■

Example 4.15 A short-shunt compound generator having terminal voltage 250 V delivers a load current of 40 A. The armature, series field and shunt field resistances are 0.05 Ω, 0.04 Ω and 100 Ω, respectively. Calculate the generated emf and the armature current if there is voltage drop of 1 V per brush.

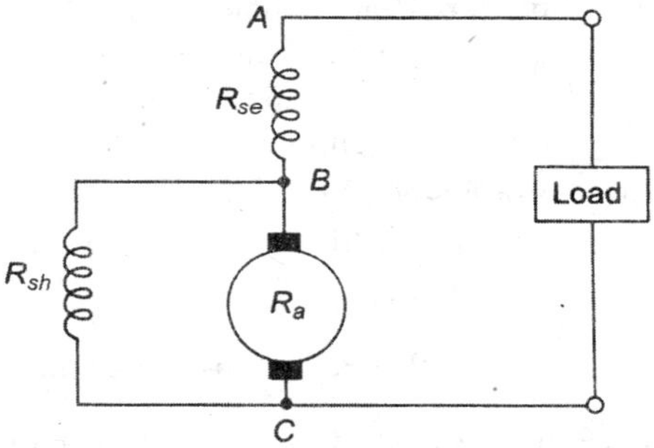

Figure E4.3

Solution

Here, $V_t = 250\text{ V} = V_{AC}$, $I = 40$ A, $R_{se} = 0.04\ \Omega$, $R_a = 0.05\ \Omega$ and $R_{sh} = 100\ \Omega$.

$$\therefore \quad V_{BC} = V_t + IR_{se} = 250 + 40\times 0.04 = 251.6\text{ V}$$

$$I_{sh} = \frac{V_{BC}}{R_{sh}} = \frac{251.6}{100} = 2.516\text{ A}$$

$$I_a = I + I_{sh} = 40 + 2.516 = 42.516\text{ A}$$

$$E_g = V_{BC} + I_a R_a + 2\times 1 = 251.6 + 42.516\times 0.05 + 2 = 255.73\text{ V}$$

Therefore, the armature current and the generated voltage are 42.516 A and 255.73 V ■

Example 4.16 An eight-pole lap-wound DC generator having 760 conductors, and a useful flux of 35 mWb/pole is driven at 500 rpm. Calculate the generated emf.

Solution

Here, $P = 8$, $A = 8$, $\phi = 35\times 10^{-3}$ Wb, $N = 500$ rpm and $Z = 760$.

$$\text{Generated emf }\left(E_g\right) = \frac{\phi NZ}{60}\times\frac{P}{A} = \frac{35\times 10^{-3}\times 500\times 760}{60}\times\frac{8}{8} = 221.67\text{ V}$$

Example 4.17 The useful flux of an eight-pole lap connected DC generator is 40 mWb. Determine the number of conductors on the armature periphery if it has no-load voltage of 250 V and rotates at a speed of 500 rpm.

Solution

Here, $P = 8$, A = 8, $\phi = 40\times 10^{-3}$ Wb, $N = 500$ rpm and $E_g = 250$ V.

$$E_g = \frac{\phi NZ}{60}\times\frac{P}{A}$$

i.e.,
$$Z = \frac{E_g \times 60}{\phi N} \times \frac{A}{P} = \frac{250 \times 60}{40 \times 10^{-3} \times 500} \times \frac{8}{8} = 750$$ ■

Example 4.18 An eight-pole lap-wound DC generator has 70 slots in its armature with 22 conductors per slot. The ratio of pole arc to pole pitch is 0.64. The diameter of the bore of the pole shoe is 0.48 m. The length of the pole shoe is 0.28 m. The air gap flux density is 0.32 Wb/m² and the generated emf in the armature is 400 V. Find the speed of the generator.

Solution

Here, $P = 8$, number of slots = 70, conductors per slot = 22.
Therefore, the total number of conductors = 70 × 22 = 1,540.
The ratio of pole arc to pole pitch = 0.64.
Diameter of the bore of the pole shoe = 0.48 m and the length of the pole shoe = 0.28 m

Air gap flux density = 0.32 Wb/m².

emf generated (E_g) = 400 V

Speed of rotation (N):

$$\frac{\text{Pole arc}}{\text{Pole pitch}} = 0.64$$

∴ $$\text{Pole arc} = 0.64 \times \text{pole pitch} = 0.64 \times \frac{\pi D}{P} = 0.64 \times \frac{3.14 \times 0.48}{8} = 0.120576 \text{ m}$$

Area of pole shoe (A) = Pole arc × length of pole shoe = 0.120576 × 0.28 = 0.03376 m².

Now, $$\phi = 0.03376 \times 0.32 = 0.010803 \text{ Wb.}$$

Again, $$E_g = \frac{\phi N Z}{60} \times \frac{P}{A}$$

∴ $$400 = \frac{0.010803 \times N \times 1540}{60} \times \frac{8}{8}$$

∴ $$N = 1{,}442.60 \text{ rpm.}$$ ■

Example 4.19 A four-pole DC shunt generator having 676 wave-connected armature conductors runs at a speed of 600 rpm and supplies a load resistance of 10 Ω at a terminal voltage of 200 V. The armature and field resistances are 0.34 Ω. and 100 Ω, respectively. Calculate the following:

(i) The armature current.
(ii) The generated emf.
(iii) The flux/pole.

Solution

Here, V_t = 200 V, R_L = 10 Ω, Z = 676 wave connected, A = 2, P = 4, N = 600 rpm, R_a = 0.34 Ω, R_{sh} = 100 Ω.

$$I_{sh} = \frac{V_t}{R_{sh}} = \frac{200}{100} = 2 \text{ A}$$

$$I_L = \frac{V_t}{R_L} = \frac{200}{10} = 20 \text{ A}$$

(i) $I_a = I_L + I_{sh} = 20 + 2 = 22$ A

(ii) $E_g = V_t + I_a R_a = 200 + 22 \times 0.34 = 207.48$ V

(iii) $E_g = \dfrac{\phi NZ}{60} \times \dfrac{P}{A}$

i.e., $$207.48 = \frac{\phi \times 600 \times 676}{60} \times \frac{4}{2}$$

i.e., $$\phi = 15.3 \text{ mWb}.$$ ■

Example 4.20 A six-pole DC shunt generator supplies full-load current at a terminal voltage of 250 V. The armature and field resistances are 0.04 Ω and 100 Ω, respectively. It runs at a speed of 1,000 rpm and has 700 lap-connected conductors. The voltage across armature resistance is 7.2 V. Find the following: (i) the load current, (ii) the emf generated and (iii) the flux/pole. Neglect brush contact drop.

Solution

Here, $V_t = 250$ V, $R_a = 0.04\ \Omega$, $R_{sh} = 100\ \Omega$, $I_a R_a = 7.2$ V, N = 1,000 rpm, $Z = 700$, $A = P$ (for lap winding)

$$I_{sh} = \frac{V_t}{R_{sh}} = \frac{250}{100} = 2.5\,\text{A}$$

Since $$I_a R_a = 7.2, \quad \therefore I_a = \frac{7.2}{R_a} = \frac{7.2}{0.04} = 180\,\text{A}$$

(i) $I = I_a - I_{sh} = 180 - 2.5 = 177.5$ A

(ii) $E_g = V_t + I_a R_a = 250 + 7.2 = 257.2$ V

(iii) $E_g = \dfrac{\phi NZ}{60} \times \dfrac{P}{A}$

i.e., $$257.2 = \frac{\phi \times 1{,}000 \times 700}{60} \times \frac{P}{P}$$

i.e., $$\phi = 22.04 \text{ mWb}.$$ ■

Example 4.21 A six-pole, long-shunt, wave-connected DC generator supplies 22 kW at terminal voltage of 220 V. The armature, series field and shunt field resistances are 0.05 Ω, 0.04 Ω and 110 Ω, respectively. Find the armature current and the generated emf. Assume 1 V per brush for contact drop.

If the speed of the armature is 1,000 rpm and the flux per pole is 7.8 mWb, calculate the number of conductors of the armature.

Solution

Here, $P = 6$, power output = 22 kW, $V_t = 220$ V, $R_a = 0.05\ \Omega$, $R_{se} = 0.04\ \Omega$, $R_{sh} = 110\ \Omega$, $N = 1{,}000$ rpm, $\phi = 7.8 \times 10^{-3}$ Wb, contact drop per brush = 1 V, parallel path (A) = 2 due to wave connection of conductors.

$$I_L = \frac{22 \times 10^3}{220} = 100\,\text{A}, \quad I_{sh} = \frac{220}{110} = 2\,\text{A}$$

$$I_a = I_L + I_{sh} = 100 + 2 = 102 \text{ A}.$$

$$E_g = V_t + I_a (R_a + R_{se}) + 2 = 220 + 102 \times (0.05 + 0.04) + 2 = 231.18 \text{ V}$$

Again, $$E_g = \frac{\phi NZ}{60} \times \frac{P}{A}$$

$$Z = \frac{E_g \times 60}{\phi N} \times \frac{A}{P} = \frac{231.18 \times 60 \times 2}{7.8 \times 10^{-3} \times 1{,}000 \times 4} = 890$$

■

Example 4.22 A six-pole DC shunt generator has the following data:
Field resistance = 120 Ω, armature resistance = 0.8 Ω, number of conductors = 350 (wave connected), flux/pole = 0.02 Wb.

Load resistance across the terminals = 12 Ω, armature rotates at 1,000 rpm. Calculate the power absorbed by the load.

Solution

Here, $R_{sh} = 120\ \Omega$, $R_a = 0.8\ \Omega$, $R_L = 12\ \Omega$, $Z = 350$, $A = 2$, $\phi = 0.02$ Wb, $N = 1{,}000$, $P = 6$

Let the terminal voltage be V_t volt.

∴ $$I_{sh} = \frac{V_t}{120}\text{ A and } I_L = \frac{V_t}{12}\text{ A}$$

$$I_a = I_{sh} + I_L = \frac{V_t}{120} + \frac{V_t}{12} = \frac{11V_t}{120}$$

Now, $$E_g = \frac{\phi NZ}{60} \times \frac{P}{A} = \frac{0.02 \times 1{,}000 \times 350}{60} \times \frac{6}{2} = 350\text{ V}$$

Since, $$E_g = V_t + I_a R_a \quad \text{i.e., } 350 = V_t + \frac{11V_t}{120} \times 0.8 = \frac{128.8V_t}{120}$$

$$V_t = \frac{350 \times 120}{128.8} = 326.08\text{ V}$$

Therefore, load current $$(I_L) = \frac{V_t}{R_L} = \frac{326.08}{12} = 27.17\text{ A}$$

and power absorbed by the load = $V_t I_L = 326.08 \times 27.17 = 8{,}859.6$ W ■

Example 4.23 A separately excited DC generator has terminal voltage 250 V with constant field excitation. If the load changes from 200 kW to 125 kW, find the percentage reduction in the speed. The armature resistance is 0.015 Ω and the total contact drop at brushes is 2 V. Neglect the armature reaction. The flux and the total number of armature conductors remain constant.

Solution

Terminal voltage (V_t) = 250 V, $R_a = 0.015\ \Omega$, total contact drop at brushes = 2 V. Since the flux and total number of armature conductors remain constant, we have

$$E_g = \frac{\phi NZ}{60} \times \frac{P}{A}$$

∴ $$E_g \propto N$$

For 200 kW load:

$$I_{a1} = \frac{200 \times 10^3}{250} = 800\text{ A}$$

$$E_{g1} = V_t + I_{a1} R_a + 2 = 250 + 800 \times 0.015 + 2 = 264\text{ V}$$

For 125 kW load:

$$I_{a2} = \frac{125 \times 10^3}{250} = 500 \text{ A}$$

$$E_{g2} = V_t + I_{a2}R_a + 2 = 250 + 500 \times 0.015 + 2 = 259.5 \text{ V}$$

Since, $E_g \propto N$,

$$\frac{E_{g1}}{E_{g2}} = \frac{N_1}{N_2}$$

i.e.,

$$\frac{N_1}{N_2} = \frac{264}{259.5}$$

$$\frac{N_1 - N_2}{N_2} = \frac{264 - 259.5}{264} = 0.01734 \text{ p.u.} = 1.734\%$$

Therefore, reduction in speed is 1.734 per cent. ■

Example 4.24 A separately excited DC generator having constant excitation supplies power to a constant resistance load. It delivers 150 A at 400 V at a speed of 1,000 rpm. If current is reduced to 100 A, determine the speed. Total contact drop of brush is 2 V. Neglect the armature reaction. Armature resistance of the generator is 0.12 Ω.

Solution

$I_{a1} = 150$ A, $N_1 = 1{,}000$ rpm, $V_t = 400$ V, $R_a = 0.12\ \Omega$, $I_{a2} = 100$ A, total contact drop for brush = 2 V.

The value of load resistance $(R) = \frac{400}{150} = 2.67$ W

$$E_{g1} = V_t + I_{a1}R_a + 2 = 400 + 150 \times 0.12 + 2 = 420 \text{ V}$$

For load current 100 A, the terminal voltage becomes

$$V_t' = 2.67 \times 100 = 267 \text{ V}$$

$$E_{g2} = V_t' + I_{a2}R_a + 2 = 267 + 100 \times 0.12 + 2 = 281 \text{V}$$

Again, $E_g \propto N$ because ϕ, Z and P remain constant.

$$\therefore \quad \frac{E_{g1}}{E_{g2}} = \frac{N_1}{N_2}$$

i.e.,

$$N_2 = N_1 \times \frac{E_{g2}}{E_{g1}} = 1{,}000 \times \frac{281}{420} = 690.05 \text{ rpm}$$ ■

The speed at load of 100 A is 690.05 rpm.

Example 4.25 A long-shunt DC dynamo having terminal voltage of 250 V at 1,000 rpm supplies 25 kW. The resistances of armature, shunt field and series field are 0.04 Ω, 50 Ω and 0.05 Ω, respectively. Calculate (i) Cu-losses, (ii) iron and friction losses and (iii) the torque exerted by the prime mover if it has efficiency 89 per cent of the above load.

Solution

$$I_{sh} = \frac{250}{50} = 5\text{ A}, \quad I = \frac{25\times10^3}{250} = 100\text{ A}$$

$$\therefore \quad I_a = I + I_{sh} = 100 + 5 = 105\text{ A}$$

(i) Armature Cu-loss $= I_a^2 R_a = 105^2 \times 0.04 = 441$ W

Series field Cu-loss $= I_a^2 R_{se} = 105^2 \times 0.05 = 551.25$ W

Shunt field Cu-loss $= I_{sh}^2 R_{sh} = 5^2 \times 50 = 1{,}250$ W

∴ Total Cu-loss = 441 + 551.25 + 1,250 = 2,242.25 W

(ii) The output power = 25,000 W

$$\text{Input power} = \frac{\text{Output power}}{\eta} = \frac{25{,}000}{0.89} = 28{,}089.89\text{ W}$$

Total losses = 28,089.89 − 25,000 = 3,089.89 W
∴ Iron and friction losses = 3,089.89 − 2,242.25 = 847.64 W

(iii) Let T be the torque exerted by the prime mover.

$$T \times \frac{2\pi N}{60} = 28{,}089.89$$

$$\therefore T = \frac{28{,}089.89 \times 60}{2 \times 3.14 \times 1{,}000} = 268.37\text{ N-m}$$

■

Example 4.26 A 10 kW, six-pole DC generator develops an emf of 200 V at 1,500 rpm. The armature has a lap-connected winding. The average flux density over a pole pitch is 0.9 T. The length and diameter of the armature are 0.25 m and 0.2 m, respectively. Calculate (i) the flux per pole, (ii) the total number of active conductors in the armature and (iii) the torque developed by the machine when the armature supplies a current of 50 A.

Solution

$E_g = 200$ V, $N = 1{,}500$ rpm, $P = 6$
$A = 6$ (because generator is lap wound)
$L = 0.25$ m and $d = 0.2$ m, $B = 0.9$ T

(i) If Z = number of conductors, pole pitch $= \frac{Z}{P} = \frac{Z}{6}$

$$\text{Angle subtended by pole shoe} = \frac{360°}{Z} \times \frac{Z}{6} = 60°$$

$$\text{Arc length of the pole shoe} = \frac{\pi \times 0.25 \times 60}{360} = 0.130834\text{ m}$$

Area of the pole shoe = 0.130834 × 0.2 = 0.026168 m^2
∴ Flux per pole = Flux density × Area of pole shoe
= 0.9 × 0.026168 = 0.0235512 Wb

(ii) $$E_g = \frac{\phi NZ}{60} \times \frac{P}{A}$$

$\therefore$ $$200 = \frac{0\ 02355012 \times 1500 \times Z}{60} \times \left(\frac{6}{6}\right)$$

$\therefore$ $$Z = \frac{200 \times 60}{0.02356 \times 1500} = 340$$

($\because$ number of conductors cannot be fractional number)

$\therefore$ The total of active conductors = 340

(iii) Armature current = 50 A

$\therefore I_a = 50$ A

$\therefore$ $$\text{Torque developed} = 9.55 \times \frac{E_g I_a}{N} = 9.55 \times \left(\frac{200 \times 50}{1{,}500}\right) = 63.67 \text{ N-m}$$ ■

Example 4.27 A 40 kW, 400 V, four-pole DC generator has a two-layer simplex lap winding in 30 slots with 12 conductors in each layer. If the brushes are given an actual lead of 10°, calculate the following:

(i) Demagnetizing AT/pole.

(ii) Cross-magnetizing AT/pole.

(iii) Number of turns per pole on the compensating winding if the pole arc to pitch is 0.8 and brushes are placed on geometric neutral plane.

Solution

$Z = 2 \times$ number of slots $\times$ conductors/slot $= 2 \times 30 \times 12 = 720$

Number of parallel paths (A) = 4.

Brush lead (θ_m) = 10°

$$\text{Armature current } (I_a) = \frac{40 \times 10^3}{400} = 100 \text{ A}$$

$$\text{Current in each conductor conductor} = \frac{I_a}{A} = \frac{100}{4} = 25 \text{ A}$$

(i) Demagnetizing AT/pole:

$$AT_d/\text{pole} = ZI \times \frac{\theta_m}{360} = 720 \times 25 \times \frac{10}{360} = 500 \text{ AT/pole}$$

(ii) Cross-magnetizing AT/pole:

$$AT_c/\text{pole} = ZI\left(\frac{1}{2P} - \frac{\theta_m}{360}\right) = 720 \times 25 \times \left(\frac{1}{2 \times 4} - \frac{10}{360}\right) = 1750 \text{ AT/pole}$$

(iii) Number of turns per pole for compensating winding:

$$\text{AT/pole for compensating winding} = \frac{ZI}{2P} \times \frac{\text{Pole arc}}{\text{Pole pitch}} = \frac{720 \times 25}{2 \times 4} \times 0.8 = 1{,}800$$

$$\text{Therefore, the number of turns per pole for compensating winding} = \frac{1{,}800}{25} = 72$$ ■

Example 4.28 Two 240V DC generators operate in parallel. The first generator has a terminal voltage of 280 V on no load and 240 V when supplying 40 A. The second generator has a voltage of 300 V when on no load and 240 V when supplying 50 A. Calculate the following:

(i) The current supplied by each generator.
(ii) The output voltage.
(iii) The output kW of each machine.

The total load current is 60 A.

Solution

(i) *Generator 1:*

Voltage drop for 40 A = 280 − 240 = 40 V

Therefore, voltage drop per ampere $= \frac{40}{40} = 1$ V/A

Generator 2:

Voltage drop for 50 A = 300 − 240 = 60 V

Therefore, voltage drop per ampere $= \frac{60}{50} = 1.2$ V/A

Let I_1 and I_2 be the output currents of generator 1 and generator 2, respectively. Let V_t be the common terminal voltage.

$$V_t = 280 - I_1$$

$$V_t = 300 - 1.2I_2$$

$$\therefore \quad 280 - I_1 = 300 - 1.2I_2 \qquad (1)$$

$$1.2I_2 - I_1 = 20 \qquad (2)$$

Again,

$$I_1 + I_2 = 60$$

From Equations (1) and (2), we have

$$I_1 = 23.64 \text{ A and } I_2 = 36.36 \text{ A}$$

(ii) Output of each generator.

$$V_t = 280 - I_1 = 280 - 23.64 = 256.36 \text{ V}$$
$$V_t = 300 - 1.2I_2 = 300 - 1.2 \times 36.36 = 256.36 \text{ V}$$

(iii) Output of generator 1,

$$P_1 = \frac{V_t I_1}{1{,}000} = \frac{256.36 \times 23.64}{1{,}000} = 6.06 \text{kW}$$

Output of generator 2,

$$P_2 = \frac{V_t I_2}{1{,}000} = \frac{256.36 \times 36.36}{1{,}000} = 9.32 \text{ kW}$$

■

Example 4.29 Two DC shunt generators having armature and field resistances of 0.05 Ω and 100 Ω, respectively are operating in parallel. The emf induced in one machine is 250 V, whereas that in the other is 260 V, supplying a common load of 1,500 A. Calculate the following:

(i) The terminal voltage.
(ii) The output of each machine.

Solution

$E_{g1} = 250$ V and $E_{g2} = 260$ V, $R_a = R_{a1} = R_{a2} = 0.05\ \Omega$ and $R_{sh} = 200\ \Omega$, $I = 1{,}500$ A.
Let V_t the common terminal voltage.

(i) $$I_{sh} = \frac{V_t}{100}\ \text{A}$$

$\therefore$ $$V_t = 250 - \left(I_1 + \frac{V_t}{100}\right) \times 0.05 \qquad (1)$$

and $$V_t = 260 - \left(I_2 + \frac{V_t}{100}\right) \times 0.05 \qquad (2)$$

where I_1 and I_2 are the currents supplied by the generator 1 and generator 2, respectively.

$\therefore$ $$250 - \left(I_1 + \frac{V_t}{100}\right) \times 0.05 = 260 - \left(I_2 + \frac{V_t}{100}\right) \times 0.05$$

i.e., $$(I_2 - I_1) \times 0.05 = 10$$

i.e., $$I_2 - I_1 = 200 \qquad (3)$$

Again, $$I_2 + I_1 = 1{,}500 \qquad (4)$$

From Equations (3) and (4), we have

$$I_1 = 650\ \text{A and } I_2 = 850\ \text{A}.$$

Substituting the value of I_1 in Equation (1), we have

$$V_t = 250 - \left(650 + \frac{V_t}{100}\right) \times 0.05 = 250 - 32.5 - \frac{0.05V_t}{100}$$

i.e., $$\frac{100.05V_t}{100} = 217.40$$

i.e., $$V_t = 217.4\ \text{V}$$

Therefore, the terminal voltage is 217.4 V.

(ii) Output of generator 1,

$$P_1 = \frac{V_t I_1}{1{,}000} = \frac{217.4 \times 650}{1{,}000} = 141.31\ \text{kW}$$

Output of generator 2,

$$P_2 = \frac{V_t I_1}{1{,}000} = \frac{217.4 \times 850}{1{,}000} = 184.79\ \text{kW}$$

SIGNIFICANT POINTS

- *Shunt Generator:*

$$I_a = I_L + I_{sh}$$
$$E_g = V_t + I_a R_a$$

- *Series Generator:*

$$I_a = I_{se} = I_L$$
$$E_g = V_t + I_a(R_a + R_{se})$$

- *Induced emf:*

$$E_g = \frac{\phi NZ}{60} \times \frac{P}{A}$$

- *Efficiency:*

(i) *Mechanical efficiency:*

$$\eta_m = \frac{\text{Electrical power developed in armature}}{\text{Mechanical power input}}$$

(ii) *Electrical efficiency:*

$$\eta_e = \frac{\text{Electrical power output}}{\text{Electrical power developed in armature}}$$

(iii) *Commercial or overall efficiency:*

$$\eta_c = \frac{\text{Electrical power output}}{\text{Mechanical power input}}$$

- *Maximum efficiency:*

Generator efficiency is maximum when

Constant loss = Cu-loss or variable loss.

The load current at which maximum frequency occurs is given by

$$I = \sqrt{\frac{W_c}{R_a}}$$

- *Demagnetizing and cross-magnetizing ampere-turns:*

$$\text{Demagnetizing AT per pole} = \frac{\theta_m}{360} \times ZI$$

$$\text{Cross-magnetizing AT per pole} = ZI\left(\frac{1}{2P} - \frac{\theta_m}{360}\right)$$

- *Number of compensating winding:*

The number of armature-turns immediately under one pole

$$= \frac{Z}{2P} \times \frac{\text{Pole arc}}{\text{Pole pitch}}$$

Approximately, the number of armature-turns immediately under one pole is given by

$$= 0.7\left(\frac{Z}{2P}\right)$$

Therefore, the number of AT/pole for compensating winding is given by

$$= 0.7 \times \frac{ZI}{2P} = 0.7 \times \text{armature amp-turns/pole}$$

- *Value of reactance voltage:*

  Reactance voltage or self-induced voltage

$$= L \times \frac{2I}{T_c} \quad \text{if commutation is linear}$$

$$= 1.11 L \times \frac{2I}{T_c} \quad \text{if commutation is sinusoidal.}$$

SHORT QUESTIONS AND ANSWERS

Q.1 What is a DC generator?

Ans. A DC generator is a rotating machine that converts mechanical energy into electrical energy and delivers DC.

Q.2 On what principle does the DC generator work?

Ans. The DC generator woks on the principle of Faraday's laws of electromagnetic induction.

Q.3 What are the differences between a DC generator and an alternator?

Ans. The differences between a DC generator and an alternator are shown below:

(i) In a DC generator field is static and armature is rotating, whereas in an alternator either field or armature can be made rotating, usually field.

(ii) In DC generator generated alternating emf is rectified to DC by the arrangements of commutator. This kind of arrangement is not required in alternator.

(iii) In DC generator a closed type of winding is arranged on the armature, but alternator coils are not arranged in closed fashion.

(iv) A closed type of winding arrangement on armature is required in DC generator, whereas in the alternator coils are not arranged in closed fashion.

Q.4 What are the main parts of a DC generator?

Ans. The mains parts of DC generators are shown below:

Stationary parts: (i) Yoke or body, (ii) Eye bolts, (iii) Pole shoes, (iv) Field winding, (v) Rocker ring, (vi) Carbon brushes, (vii) Brush holder and (viii) Side covers.

Rotating parts: (i) Shaft, (ii) Armature core, (iii) Armature windings, (iv) Commutator, (v) Ventilating fan and (vi) Bearings.

Q.5 Mention the parts which form the magnetic circuit.

Ans. (i) Yoke, (ii) Pole cores, (iii) Armature core and (iv) Air gaps (between the poles and armature core).

Q.6 What percentage of flux per pole does the yoke carry?

Ans. The yoke carries about 50 per cent of total flux per pole.

Q.7 What do you mean by magnetic leakage and leakage co-efficient of a DC machine?

Ans. In a DC machine a portion of the total flux developed in the poles completes its path through the air without entering the armature core, and this is known as magnetic leakage.

The ratio of the total flux developed in the poles to the total useful flux entering the armature core is called the leakage co-efficient of a DC machine. The value of leakage co-efficient lies between 1.1 and 1.2.

Q.8 How many types of pole shoes are used in DC generators?

Ans. (i) Solid core made out of either cast iron or cast steel.

(ii) Laminated core built of thin laminations of annealed steel revetted together under hydraulic pressure.

Q.9 Why is the pole shoe curved?

Ans. The pole shoes are curved to give uniform air gap around the armature core.

Q.10 What are the functions of pole shoes in a DC machine?

Ans. Pole shoes have the following functions:

(i) These support the field coils.

(ii) These reduce the reluctance of the magnetic path.

(iii) These achieve uniform flux distribution in the air gap by spreading out the flux.

Q.11 What will be the effect of increasing the air gap?

Ans. Reluctance will be increased. Therefore, the machine will draw more current in case of a motor and the machine will give less output in case of a generator.

Q.12 What are the factors on which the actual flux distribution depends?

Ans. The actual flux distribution depends upon (i) size of air gap, (ii) shape of pole shoes and (iii) clearance between the tips of the adjacent pole shoes.

Q.13 What are the functions of the armature core?

Ans. The armature core houses the armature windings and provides a low reluctance path for the magnetic lines of force.

Q.14 Is the armature core made of solid homogeneous material?

Ans. No, it is made of laminations of silicon steel.

Q.15 Why is the armature core laminated?

Ans. Armature core is laminated to reduce eddy current losses.

Q.16 Why are holes provided in the armature core?

Ans. Holes are provided in the armature core to reduce the weight of the core as well as to provide air circulation for cooling the core.

Q.17 Why is the armature core not made of wood?

Ans. Armature core is not made of wood because of its high reluctance and incapability of bearing high rotational torque.

Q.18 State the uses of lap and wave windings?

Ans. Lap winding is used for low-voltage and high-current DC machines, whereas wave winding is used for low-current and high-voltage DC machines.

Q.19 What is commutator?

Ans. Commutator is a cylindrical structure attached to the armature and mounted on the same shaft, segments of which are separated from one another by mica insulation and connected to armature coils through risers.

Q.20 What is dummy coil and why is it used?

Ans. Dummy coils are inactive filler coils. These are placed in the excess slots not required by the armature winding. These are not connected to the rest of the winding and commutator segments, but are taped.Dummy coils are used in the armature to get mechanical balance of the armature.

Q.21 What are the different types of DC generators according to their field excitation?

Ans. There are two types:

(i) Separately excited generators
(ii) Self-excited generators

Q.22 What do you mean by self-excited generator?

Ans. A generator in which the field coils are excited by the generator's armature current itself, is known as self-excited generator.

Q.23 How does a self-excited generator produce emf at starting?

Ans. At starting, some residual magnetism left with the pole induces some emf in the armature, which causes some current in the armature, a part or whole of which, passing through the field, further strengthens the field flux and gradually the voltage is built up.

Q.24 What are the types of self-excited generators?

Ans. (i) Series generator, (ii) Shunt generator and (iii) Compound generator.

Q.25 What do you mean by series generator?

Ans. A generator in which the field windings, consisting of a few turns,, is connected in series with the armature, that is, same current will flow through the armature, the field and the load, is said to be a series generator.

Q.26 What will be the effect if the resistance of series field winding is high?

Ans. The voltage at the external load point will be less due to consumption of a major part of the generated voltage in the series field winding having high resistance.

Q.27 How is the shunt field winding wound?

Ans. Shunt field winding is wound with a large number of turns of thin wire having sufficient high resistance to bear the generated voltage as it is connected across the armature.

Q.28 What is the usual value of shunt field current?

Ans. The usual value of the shunt field current is about 2 to 5 per cent of the total full-load current.

Q.29 Give the expression of generated emf of a DC generator.

Ans. Generated emf, $E_s = \frac{\phi NZ}{60} \times \frac{P}{A}$ volts

where ϕ = flux/pole in Wb
Z = total number of conductors
N = armature revolution in rpm
P = number of poles
A = number of parallel paths in armature
E_g = emf induced in any parallel in armature

Q.30 What are the factors on which the emf of a DC generator depends?

Ans. The induced emf depends upon the following factors:

(i) The intensity of the magnetic field.
(ii) The length of the armature conductors and the number of turns.
(iii) The rate at which the armature conductors cut off the magnetic field or the speed of rotation in rpm.

Q.31 What are the losses in a DC machine?

Ans. There are three types of losses:

(i) Cu-losses $\rightarrow$ (Armature copper loss + Field copper loss)

(ii) Iron losses → (Hysteresis loss + Eddy current loss)
(iii) Mechanical losses → (Frictional + Windage)

Q.32 What is hysteresis loss?

Ans. The loss of energy in the form of heat, to overcome the molecular magnetic friction during reversal of magnetization of the armature core is called hysteresis loss. It depends upon the volume and grade of iron, maximum value of flux density and frequency of magnetic reversal.

Q.33 What is eddy current loss?

Ans. The loss of energy in the form of heat, which is caused by the flow of eddy current induced in the armature, pole and magnetic cores by the changing emf is called eddy current loss. The eddy current loss depends upon the range of the variation of flux density, the frequency of the variation of flux thickness of the iron lamination and the volume of the magnetic material. The loss varies directly as the square of the thickness of lamination.

Q.34 How is the eddy current in an armature core minimized?

Ans. Eddy current is minimized by laminating the armature core.

Q.35 What should be the place of lamination?

Ans. The armature core should be laminated at right angles to its axis.

Q.36 Why are eddy currents induced in the armature core of a DC machine?

Ans. Eddy currents are induced due to relative rotation between the field and the armature.

Q.37 What is the usual formula for calculating eddy current losses?

Ans. The eddy current loss $W_e = K_e B_m^2 f^2 V_l t^2$

where K_e = Constant depending upon the electrical resistance of the magnetic material and the units employed in the expression,

B_m = Maximum flux density,

f = Frequency of magnetic cycles per second,

t = Thickness of the magnetic material and

V_l = Volume of the magnetic material.

Q.38 Does a change in speed affect iron losses and how?

Ans. Yes. hysterisis loss directly varies with the speed and eddy current loss varies with the square of the speed of the machine.

Q.39 What are stray losses?

Ans. Magnetic and mechanical losses are collectively known as stray losses. They are also known as rotational losses.

Q.40 What is the cause of mechanical losses in the armature?

Ans. The cause of mechanical losses in an armature is the speed of the armature.

Q.41 What is the relation between mechanical losses and speed of the armature?

Ans. Windage loss is proportional to the square of the speed of the armature, but frictional loss is proportional to the speed of the armature.

Q.42 Why is the output power of a DC generator always less than the input power?

Ans. Output power of a DC generator is always less than its input power because of the copper losses, iron and frictional losses and windage losses, which occur during running of the generator.

Q.43 What are the power stages in a DC generator?

Ans. Mechanical power input − Iron and frictional losses = Electrical power developed in armature (EI_a), Electrical power developed in armature − Copper losses = Electrical power output ($V_t I_L$)

Q.44 What is the condition for maximum efficiency in a DC generator?

Ans. The efficiency is maximum when variable losses = constant losses.

Q.45 What is an equalizer ring?

Ans. An equalizer ring is a balancing ring made of copper conductors fitted at the driving end of the armature. It may also be a separate winding behind the commutator of a DC machine connected to two or more points of the armature windings to provide a low-resistance path at the same potential.

Q.46 Where are equalizer rings used in a DC generator and why?

Ans. Equalizer rings are used in lap winding and duplex, triplex wave winding of a DC machine to avoid unequal distribution of current at the brushes as well as to get sparkless commutation.

Q.47 What is the relation between the numbers of commutation segments and the number of armature coils in a DC machine?

Ans. In DC machines the number of commutator segments is equal to the number of armature coils.

Q.48 What is the function of a commutator?

Ans. Function of a commutator is to facilitate the collection of current from the armature conductors and to convert the alternating emf induced in the armature conductors into uni-directional emf in the external load circuit.

Q.49 What is the main purpose of using carbon brushes in a DC machine?

Ans. Carbon brushes are used to improve the commutation process.

Q.50 Where are copper brushes used in a DC machine instead of carbon brushes?

Ans. Copper brushes are used in a DC machine where high currents and low voltage are involved.

Q.51 What is the magnetic neutral axis (MNA)?

Ans. The MNA is defined as the axis which is at right angles to the main pole flux passing through the armature along which no emf is produced in the armature conductors. It is also known as the axis of commutation.

Q.52 What is geometrical neutral axis (GNA)?

Ans. The GNA is the axis, which bisects the angle between two adjacent poles.

Q.53 What is commutating plane?

Ans. It is the magnetic neutral plane (MNP).

Q.54 What is the position of collecting maximum emf?

Ans. Maximum emf is collected at magnetic neutral plane.

Q.55 Why are the brushes shifted from the geometrical neutral axis in a DC generator?

Ans. The brushes are shifted from the geometrical neutral axis in a DC generator to avoid sparking on the commutator.

Q.56 What is armature flux?

Ans. When the generator is loaded, the armature carries its load current that produces a magnetic flux around the armature, which is called the armature flux.

Q.57 What do you mean by back torque of a DC generator and what are its effects?

Ans. Back torque of a DC generator means the torque developed in the armature due to load current in opposite direction to the direction of rotating of prime time mover.

Back torque determines the power requirement of the prime mover due to which power of the prime mover is kept more than the maximum generator output.

Q.58 What is armature reaction?

Ans. The effect of magnetic field set up by armature current on the distribution of magnetic lines of force produced by the main field poles of a generator is called armature reaction.

Q.59 What is the function of compensating windings?

Ans. The function of compensating windings is to neutralize the cross-magnetizing effect of armature reaction, particularly in the zones outside the influence of the interpoles, and thus maintain a uniform flux distribution under the faces of the main poles.

Q.60 What are the factors that create difficulties in the process of commutation of a non-interpolar DC machine?

Ans. The cross magnetizing effect of armature reaction and the self-induced emf in the coil undergoing commutation known as the reactance voltage create difficulties in the process of commutation of a non-interpolar DC machine.

Q.61 What are inperpoles and how are they connected?

Ans. Interpoles or commutating poles are small narrow poles situated midway between the main poles and are connected in series with the armature.

Q.62 What is the polarity of interpoles relative to that of main poles in a DC machine?

Ans. The polarity of interpoles in case of generators is the same as that of the main poles further ahead in the direction of rotation, whereas in case of motors their polarity is the same as that of the main poles further behind in the direction of rotation of the machine.

Q.63 Why is the air gap of interpoles made larger than that of the main poles?

Ans. Since both the armature reaction and the emf of self-induction in the coils undergoing commutation are proportional to the armature current, the flux density necessary to compensate for them, produced by the interpoles, may also be proportional to the armature current to keep the interpole flux nearly proportional to the armature current at all loads, so the air gap between the interpoles and the armature is made larger.

Q.64 What is the difference between compensating winding and interpoles?

Ans. Although they are connected in series and their emf are such as to neutralize armature reaction, the interpoles additionally supply mmf to counteract the reactance voltage induced in the coil undergoing commutation.

Q.65 What are the important characteristics of a DC generator?

Ans. The three important characteristics are:

(i) No-load saturation characteristic (E_0/I_f)

(ii) Internal or total characteristic, (E/I_a)

(iii) External characteristic (V/I)

Q.66 What do you mean by no-load saturation characteristic?

Ans. It is the magnetic characteristic of open-circuit characteristic that gives the relation between the no-load generated emf in armature (E_0) and the field or exciting current (I_1) at a given fixed speed.

Q.67 What do you understand by external characteristic?

Ans. It is the performance characteristic that gives the relation between the terminal voltage (V_t) and the load current (I_L).

Q.68 What are the conditions for building up of a self-excited shunt generator?

Ans. The conditions are as follows:

(i) There must be some residual magnetism in the poles.

(ii) The shunt field windings should be correctly connected to the armature for the given direction of rotation so that the induced current can reinforce the emf produced initially due to residual magnetism.

(iii) If the generator is excited in open circuit, its shunt field resistance should be less than the critical resistance.

(iv) If excited on load, its shunt field resistance should be more than a certain minimum value of resistance, which is given by internal characteristic.

Q.69 Can a shunt generator build up its voltage without residual magnetism?

Ans. A shunt generator cannot build up its voltage without residual magnetism.

Q.70 Can a DC generator be made to build up with reversed direction of rotation?

Ans. Yes, a DC generator can be made to build up in reverse direction of rotation.

Q.71 Why does the voltage of a shunt generator drop as load is applied to its terminals?

Ans. (i) The terminal voltage is less than the induced emf by the resistance drop in the armature.

(ii) Armature reaction weakens the field and so reduces the induced emf.

(iii) The drop in terminal voltage due to resistance drop and armature reaction results in a decreased field current, which in turn reduces the induced emf.

Q.72 What types of generators would you like to install in (i) hotels and office and (ii) in power houses?

Ans. (i) Hotels and office buildings – For short distances such as in hotels and office buildings level-compound generators are used because the voltage drop over small lengths of the feeder is negligible.

(ii) Power houses – To supply power from the power houses, it is required to maintain constant voltage at the end of a long feeder. Hence, over-compound generators are best suited for this purpose.

Q.73 What will be the effect on the terminal voltage when the series field is short circuited at full load in case of (i) cumulative compound and (b) differential compound generator?

(i) Due to reduction in the resultant flux in the air gap the terminal voltage will be reduced.

(ii) Due to the rise in the resultant flux in the air gap the terminal voltage will rise.

Q.74 What do you mean by parallel operation of DC generators?

Ans. Parallel operation of DC generators means running two or more DC generators in parallel on the same bus.

Q.75 What are the conditions to be fulfilled for parallel operation of two or more DC shunt generators?

Ans. The following three essential conditions should be ensured before paralleling DC shunt generators:

(i) The terminal voltage of the generators must be equal.

(ii) The polarities must be same, that is, positive should be connected to positive and negative to negative.

(iii) The generator characteristics must be similar in form, and each must have the same voltage drop from no load to full load.

Q.76 What are the conditions to be fulfilled before paralleling two DC compound generators?

Ans. The conditions are as follows:

(i) The terminal voltages must be same.

(ii) The polarities must be same.

(iii) The generator characteristics must be similar.

To maintain proper division of load from no load to full load, the following conditions must be satisfied in addition to the above conditions:

(iv) The regulation of each armature must be the same.
(v) The series field resistance must be inversely proportional to the generator ratings.

Q.77 What are the causes of overloading?

Ans. Overloading is caused by the following:

(i) Excessive voltage due to over excitation in field magnets on high speed of the prime mover.
(ii) Excessive current due to bad feeding of load or short circuit in the generator itself or in the external load circuit.
(iii) Reversal polarity.

SUPPLEMENTARY PROBLEMS

1. A 20 kW DC compound generator works on full load with a terminal voltage of 250 V. The armature, series and shunt windings have resistances of 0.05 Ω, 0.025 Ω and 100 Ω, respectively. Calculate the total emf generated in the armature when the machine is connected as short shunt.

   [*Ans.* 256.126 V]

2. A DC shunt generator has an induced voltage on open circuit of 127 V. When the machine is on load, the terminal voltage is 120 V. Find the load current if the field circuit resistance is 15 Ω and armature resistance is 0.02 Ω. Ignore armature reaction.

   [*Ans.* 342 A]

3. A six-pole lap-wound DC generator has 70 slots with 20 conductors per slot. The ratio of pole arc to pole pitch is 0.68. The diameter of the pole shoe is 0.46 m. The length of the pole shoe is 0.3 m. If the air gap flux density is 0.3 Wb/m^2 and the emf induced in the armature is 400 V, find the speed at which it runs.

   [*Ans.* 1,163 rpm]

4. A 110 V DC shunt generator has a full-load current of 100 A, shunt field resistance of 55 Ω and constant losses of 500 W. If full-load efficiency is 88 per cent, find the armature resistance. Find the load current corresponding to maximum efficiency.

   [*Ans.* 0.078 Ω, 96.1 A]

5. A DC shunt generator has a full-load current of 196 A at 220 V. The stray losses are 720 W and the shunt field resistance is 55 Ω. It has full-load efficiency of 88 per cent. Find the armature resistance and the load current corresponding to maximum efficiency.

   [*Ans.* 0.107 Ω, 122.3 A]

6. A 250 V, 500 kW, eight-pole DC generator has a lap-wound armature with 480 conductors. Calculate the demagnetizing and cross-magnetizing AT per pole at full load if brushes are given a forward lead of 7.5° (mechanical).

   [*Ans.* 2,500 AT, 5,000 AT]

7. A four-pole wave-wound DC shunt generator runs at 1,000 rpm when supplying 330 lamps each rated at 60 W, 110 V. It has a shunt field current of 2 A. There are 90 commutator segments and brush width is equal to 1.1 commutator segments. The self-inductance of each coil is 0.025 mH. Determine the reactance voltage if the commutation is (i) linear and (ii) sinusoidal.

   [*Ans.* 6.2 V, 6.89 V]

8. A separately excited DC generator gave the following data for open-circuit characteristic at 1,000 rpm.

I_f (A)	0	0.2	0.4	0.6	0.8	1.0	1.2	1.4
E_0 (V)	5	50	100	140	170	190	200	205

The armature resistance including brushes is 0.5 Ω. If the generator is now shunt connected and is driven at 1,100 rpm, then for a total shunt field resistance of 180 Ω, calculate (i) no-load emf and (ii) the output current and shunt field current for a terminal voltage of 190 V (neglect armature reaction).

[*Ans.* (i) 221 V (ii) 43.94 A, 1.06 A]

9. Two DC shunt generators run in parallel together to supply 2,500 A. The machines have armature resistances of 0.04 Ω and 0.025 Ω, field resistances of 25 Ω and 20 Ω and induced emfs of 440 V and 420 V, respectively. Find the bus bar voltage and output of each machine.

[*Ans.* 388.7 V, 492.54 kW, 479.21 kW]

10. Two DC shunt generators and a battery are working in parallel. The open-circuit voltage, armature and field resistances of generators are 250 V, 0.02 Ω, 120 Ω and 248 V, 0.10 Ω and 120 Ω, respectively. If the generators supply the same current when the load on the bus basis is 50 A, calculate the emf of the battery if its internal resistance is 0.16 Ω.

[*Ans.* 248.2 V]

MULTIPLE-CHOICE QUESTIONS AND ANSWERS

1. The rotating part of a DC machine is
   (a) armature (b) field (c) both (a) and (b) (d) none of these
2. The actual generated voltage in a DC generator is
   (a) direct (b) alternating (c) partly (a) and (b) (d) none of these
3. DC machines are characterized by
   (a) magnetic poles (b) armature (c) electric brushes (d) commutator
4. The function of a commutator in a DC machine is
   (a) to provide easy speed control
   (b) to improve commutation
   (c) to convert alternating voltage to direct voltage
   (d) to convert ac current to DC current
5. The laminated parts of a DC machine are armature and
   (a) pole shoe (b) core (c) yoke (d) pole
6. Residual magnetism is necessary in a DC
   (a) shunt motor
   (b) series motor
   (c) separately excited generator
   (d) shunt generator
7. If a shunt generator delivers 100 A at 200 V and the resistance of the shunt field and armature are 100 Ω and 0.01 Ω, respectively, the generated emf will be
   (a) 205 V (b) 212 V (c) 201.02 V (d) 208 V
8. The condition for maximum efficiency of a DC generator is
   (a) constant loss = $\sqrt{\text{variable loss}}$
   (b) variable loss = $\sqrt{\text{cosntant loss}}$
   (c) constant loss = variable loss
   (d) variable loss $\frac{1}{\text{variable loss}}$

9. The main cause that delays or retards the quick reversal of current is
   (a) the reactance voltage
   (b) resistance of the brush being large
   (c) the air gap between the brushes and the commutator segments
   (d) none of these
10. If the armature current is increased to double its previous value and the time of commutation is halved, the reactance voltage will
   (a) remain the same (b) be doubled (c) be halved (d) become four times
11. DC shunt generator has
   (a) slightly drooping characteristic (b) constant voltage characteristic
   (c) appreciably rising characteristic (d) appreciably drooping characteristic
12. The function of interpoles of a DC generator is to neutralize
   (a) reactance voltage (b) current unbalance at the pole tip
   (c) cross-magnetizing effect of armature reaction (d) none of these
13. For charging batteries, we use
   (a) shunt generator (b) series generator
   (c) compound generator (d) any of the above
14. In a DC generator, sparking at brushes results due to
   (a) armature reaction (b) reactance voltage in coil undergoing commutation
   (c) winding distribution (d) high constant resistance at the brushes
15. The copper losses of a DC shunt generator amount to the following percentage of full-load losses:
   (a) 10 to 20% (b) 20 to 25% (c) 30 to 40% (d) none of these
16. Stray losses in a DC generator are the same as
   (a) magnetic losses (b) mechanical losses
   (c) either (a) or (b) (d) sum of (a) and (b)
17. Standing or constant losses of a DC shunt generator are
   (a) stray losses (b) field losses of shunt generator
   (c) armature losses of a compound (d) both (a) and (b)
18. The voltage of a DC generator may conveniently be increased by
   (a) decreasing the air gap flux density (b) increasing the speed of rotation
   (c) decreasing the speed of rotation (d) both (a) and (b)
19. If external resistance is added to the field of a DC shunt generator, it results in
   (a) decrease in the voltage of the generator (b) increase in the power delivered
   (c) increase in the voltage of the generator (d) increase in the speed of the generator
20. Hysteresis loss in a DC generator may be reduced by
   (a) using material having small hysteresis loop (b) laminating the core material
   (c) both (a) and (b) (d) none of these
21. During maximum efficiency, the load current of a DC generator is given by
   (a) $I = \sqrt{\dfrac{W_a + 2}{R_a}}$ (b) $I = \sqrt{\dfrac{R_a}{W_a}}$ (c) $I = \sqrt{\dfrac{W_a}{R_a}}$ (d) $I = \dfrac{W_a}{R_a}$
22. Overall efficiency of a DC generator is usually in the order of
   (a) 70 to 75% (b) 75 to 85% (c) 80 to 90% (d) 85 to 95%

23. The armature current of a DC shunt generator is
(a) $\frac{V_t}{R_a}$ (b) $\frac{V_t - E_g}{R_a}$ (c) $E_g - \frac{V_t}{R_a}$ (d) $\frac{E_g - V_t}{R_a}$

24. For supplying power to short distances such as in office buildings, generators used are
(a) series generators (b) under-compounded generators
(c) over-compounded generators (d) flat-compounded generators

25. In DC generators, the effect of armature emf on the main pole flux distribution is to
(a) produce no distribution
(b) distort it
(c) shift the zero crossing of resultant flux density wave in the direction of rotation
(d) shift the zero crossing of resultant flux density wave against the direction of rotation

26. A four-pole DC generator is running at 1,500 rpm. The frequency of current in the armature winding is
(a) 25 Hz (b) 35 Hz (c) 50 Hz (d) 60 Hz

27. The internal characteristic of a generator is plotted between the
(a) field current and voltage generated at no load
(b) armature current and voltage generated after armature reaction
(c) field current and voltage generated on load
(d) armature current and voltage actually generated at the output terminals

28. Level-compounded DC generator is used for the following application.
(a) Electric arc welding (b) Battery charging
(c) Electric traction (d) Lighting

29. Hysteresis loss in electric machines is reduced by (1) and eddy current loss by (2)
(a) (1) less flux, (2) less current
(b) (1) special steel (2) thin insulated lamination
(c) (1) more poles (2) increased speed
(d) (1) and (2) by increased winding in rotor and stator

30. Hysteresis and eddy current losses in electric machines depend on
(a) flux density (b) speed (c) speed and flux density (d) voltage only

31. Among the three losses – electrical, magnetic and mechanical –
(a) mechanical loss is minimum
(b) all the three losses are equal
(c) magnetic losses are highest
(d) mechanical loss is the minimum at about 15 per cent of total losses

32. The characteristics of a compound DC generator are shown in Figure 1. Identify the curves.
(a) (I) Series (II) Dropping (III) Parabolic
(b) (I) Under compounding (II) Over compounding (III) Level compounding
(c) (I) Over compounding (II) Level compounding (III) Under compounding
(d) (I) Shunt (II) Series (III) Series-parallel

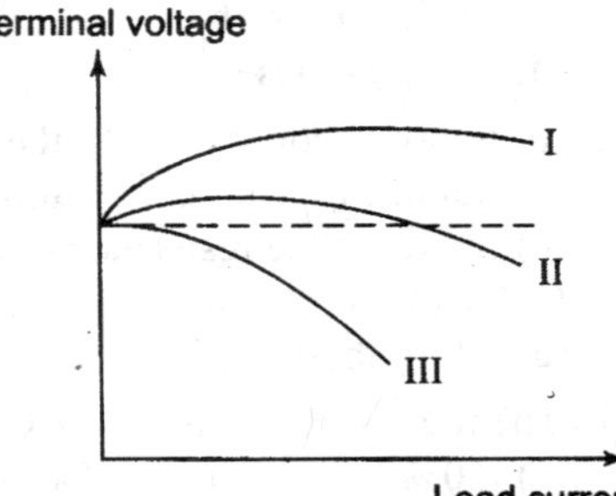

Figure 1

33. In a series generator the field resistance is 650 Ω and the load resistance is 800 Ω.
    (a) The machine will not excite
    (b) The machine will deliver a low voltage
    (c) The machine will deliver a high voltage
    (d) None of these
34. The chief difference in the external characteristics of a series and shunt DC generator is that
    (a) it becomes unstable beyond a certain load current
    (b) it has dropping characteristics
    (c) it has two critical resistances
    (d) there is no difference
35. In the series field generator, the resistance should be (i) while for the shunt field generator it should be (ii)
    (a) (i) high (ii) low (b) (i) low (ii) high (c) (i) and (ii) same (d) immaterial
36. In the armature conductors of a DC generator
    (a) maximum emf is produced (b) maximum current is produced
    (c) minimum current is produced (d) minimum emf is produced
37. If P be the number of poles of a DC generator, cross magnetizing AT per pole is equal to
    (a) $ZI\left(\frac{1}{P}-\frac{\theta_m}{P}\right)$ (b) $ZI\left(\frac{1}{2P}-\frac{\theta_m}{2P}\right)$ (c) $ZI\left(\frac{2}{P}-\frac{\theta_m}{2P}\right)$ (d) $ZI\left(\frac{1}{P}-\frac{2\theta_m}{P}\right)$
38. The number of AT per pole for compensating winding of a DC generator is given by
    (a) $\frac{ZI}{2\pi P}$ (b) $\frac{ZI}{\frac{1}{2}\pi P}$ (c) $\frac{2ZI}{\pi P}$ (d) $\frac{ZI}{2P}$
39. To get a constant voltage output at all loads, we use
    (a) series generator (b) shunt generator
    (c) level-compound generator (d) short-shunt compound generator
40. In the commutation process
    (a) current reverses (b) voltage reverses
    (c) both (a) and (b) (d) partly (a) and (b)
41. The main reason for delay in reversal of current in a DC generator is
    (a) the air gap between the brushes and the commutator surface
    (b) reactance voltage
    (c) both (a) and (b)
    (d) none of these
42. To improve commutation, the following point (s) is (are) valid
    (a) reduce the number of turns in armature and segments of commutator
    (b) increase the resistance of brushes
    (c) neutralize the reactance voltage by producing a reverse emf in the coil undergoing commutation
    (d) all of these
43. The regulation of an ideal DC generator is
    (a) 20% (b) 15% (c) 25% (d) none of these

44. Choose the correct statement.
    (a) DC series generator is more popular than the shunt generator
    (b) DC shunt generator is more popular than the series generator
    (c) DC compound generator is the most popular
    (d) All of these
45. The terminal voltage of DC shunt generator drops due to
    (a) armature resistance (b) armature reactance
    (c) weakening of field due to (b) (d) all of these
46. The critical resistance of a DC generator can be increased by
    (a) increasing its armature resistance (b) increasing its speed
    (c) increasing its field current (d) all of these
47. The function of an equalizer ring in a DC generator is
    (a) to avoid unequal distribution of current at the brushes
    (b) to get sparkless commutation
    (c) partly (a) and (b)
    (d) both (a) and (b)
48. The function of interpole or compole of a DC generator is
    (a) to neutralize the demagnetizing effect of armature reaction
    (b) to neutralize the reactance voltage and help in commutation process
    (c) to neutralize the cross-magnetizing effect of armature reaction
    (d) all of these
49. Equalizer ring is used in
    (a) wave-wound armatures only (b) lap-wound armatures only (c) both (a) and (b)
50. The main reason for voltage not building up in a DC series generator is
    (a) reversed terminal of field winding (b) short-circuited terminal
    (c) disconnected load (d) reversed terminal of armature winding
51. For shifting load from one shunt generator to the other running in parallel, the simple method is
    (a) adding resistance in their armature circuit (b) adjusting their field rheostat
    (c) using equalizer connection (d) none of these
52. The maximum generated voltage of a DC generator is 650 V; the limitation is due to
    (a) starter (b) commutator (c) field winding (d) armature winding
53. The function of equalizing connections of a lap-wound DC generator is
    (a) to avoid unequal distribution of currents at brushes
    (b) to avoid short-circuit current
    (c) to neutralize the armature reaction
    (d) all of these
54. For a specified open-circuit voltage of a DC generator, the short-circuit current will be maximum when it is
    (a) separately excited (b) shunt connected
    (c) commutatively over compounded (d) differentially compounded
55. In DC generators the polarities of the interpole is
    (a) same as that of the main poles behind (b) same as that of the main pole ahead
    (c) opposite to that of the main pole ahead (d) none of these

56. For building up voltage, residual magnetism is essential in field systems of all types of DC generator except
(a) series (b) shunt (c) compound (d) separately excited
57. The terminal voltage of a DC series generator running at rated speed and at no load is equal to
(a) rated voltage (b) half of its rated voltage (c) zero (d) very little voltage
58. For parallel operation, the polarities of two generators
(a) must be the same
(b) must oppose each other
(c) may or may not be the same
(d) none of these
59. Generators are quite often run is parallel because of
(a) greater reliability
(b) greater efficiency
(c) capability to meet greater demand
(d) all of these
60. Find the type of generator that most suits parallel operation.
(a) Shunt generator
(b) Series generator
(c) Compound generator
(d) All of these
61. The critical resistance of a DC generator is the resistance of the field winding of the generator
(a) at which it can supply maximum power
(b) beyond which it just does not develop any voltage
(c) at which it develops maximum voltage
(d) all of these
62. The number of conductors for a wave-wound armature of a four-pole dynamo, having a flux of 10^6 lines and generating a voltage of 240 V when driven at 800 rpm is
(a) 900 (b) 400 (c) 200 (d) 300

Answers

1 (a)	2 (b)	3 (d)	4 (c)	5 (a)	6 (d)
7 (c)	8 (c)	9 (a)	10 (d)	11 (a)	12 (a),(c)
13 (a)	14 (b)	15 (c)	16 (d)	17 (d)	18 (b)
19 (a)	20 (a)	21 (c)	22 (d)	23 (d)	24 (d)
25 (c)	26 (c)	27 (b)	28 (b),(d)	29 (b)	30 (c)
31 (a)	32 (c)	33 (a)	34 (c)	35 (b)	36 (d)
37 (b)	38 (a)	39 (c)	40 (a)	41 (b)	42 (d)
43 (d)	44 (b)	45 (d)	46 (b)	47 (d)	48 (c)
49 (b)	50 (b)	51 (b)	52 (b)	53 (a)	54 (a)
55 (b)	56 (d)	57 (c)	58 (a)	59 (d)	60 (a)
61 (a)	62 (a)				

DC Motors

5

DC motors convert electrical energy into mechanical energy and have construction very similar to that of a DC generator. Generators are operated in protected locations. Therefore, their construction is usually of the open type. Since DC motors do not operate in protected locations, they face dust, moisture, fumes and tend to get mechanical damage. Hence, their construction is of closed type. The operation of motors is based on the principle – when a current-carrying conductor is placed in a magnetic field, it experiences a mechanical force. The magnitude of this force is $F = BIL$, and its direction is determined by Fleming's left-hand rule. The aim of this chapter is to discuss DC motors in detail.

5.1 VOLTAGE EQUATION

DC motors are of shunt type, series type and compound type. They are similar to their DC generator counterparts. The difference is that the prime mover is absent and DC supply is given from supply mains to the terminals of the motor. The shunt motor and series motor are shown in Figures 5.1(a) and 5.1(b), respectively.

In Figure 5.1, let

V_t be the supply voltage,
R_a be the armature resistance,
E_b be the back emf,
I be the current drawn by the motor from supply mains,
I_a be the armature current,
I_{sh} be the shunt field current – Figure 5.1(a),
I_{se} be the series field current – Figure 5.1(b),
R_{sh} be the shunt field resistance – Figure 5.1(a) and
R_{se} be the series field resistance – Figure 5.1(b).

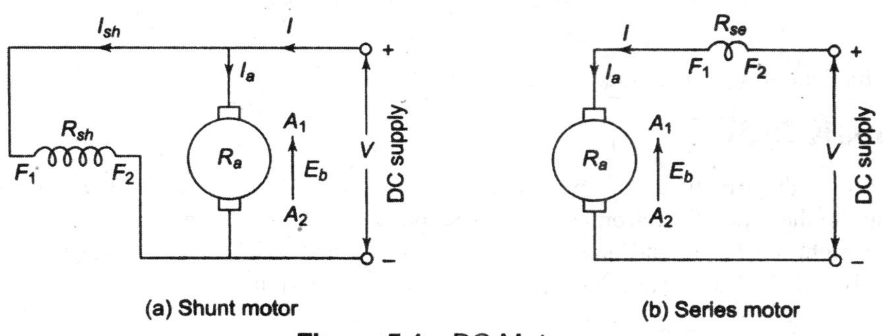

Figure 5.1 DC Motors

The supply voltage (V_t) of the motor overcomes the following:

(i) the back emf (E_b) and

(ii) the armature ohmic drop (or, I_aR_a drop).

Shunt motor:

$$I = I_a + I_{sh} \tag{5.1}$$

$$V_t = E_b + I_aR_a \tag{5.2}$$

Series motor:

$$I = I_a = I_{se} \tag{5.3}$$

$$V_t = E_b + I_a\left(R_a + R_{se}\right) \tag{5.4}$$

From Equations (5.2) and (5.4), we can conclude that $V_t > E_b$ for motoring mode.

Multiplying both sides of Equation (5.2) by I_a, we have

$$V_tI_a = E_bI_a + I_a^2R_a \tag{5.5}$$

In Equation (5.5),

V_tI_a is the electrical power input,

E_bI_a is the gross mechanical power developed in the armature denoted by P_m and

$I_a^2R_a$ is the armature Cu loss.

Example 5.1 The field and armature resistances of a 220 V DC shunt machine are 88 Ω and 0.05 Ω, respectively. Calculate the total armature power developed when working

(i) as a generator delivering power of 22 kW and

(ii) as a motor taking 22 kW input.

Solution

(i) *Generator:*

Here, $I = \dfrac{22\times10^3}{220} = 100 \text{ A}$ and $I_{sh} = \dfrac{220}{88} = 2.5 \text{ A}$

$$I_a = I + I_{sh} = 100 + 2.5 = 102.5 \text{ A}$$

$$E_g = V_t + I_aR_a = 220 + 0.05\times102.5 = 225.125 \text{ V}$$

∴ Power developed in armature = E_gI_a = 225.125 × 102.5 = 23.075 kW

(ii) *Motor:*

Here, $I = \dfrac{22\times10^3}{220} = 100 \text{ A}$ and $I_{sh} = \dfrac{220}{88} = 2.5 \text{ A}$

$$I_a = I - I_{sh} = 100 - 2.5 = 97.5 \text{ A}$$

∴ $$E_b = V_t - I_aR_a = 220 - 0.05\times97.5 = 215.125 \text{ V}$$

∴ Mechanical power developed = E_bI_a = 215.125 × 97.5 = 20.974 kW ■

5.2 BACK EMF

During rotation of the armature, its conductors cut the magnetic flux and emf is induced in the conductors according to the laws of electromagnetic induction. The direction of this induced emf can be found by Fleming's right-hand rule and have the direction opposite to the applied voltage. Therefore, this induced emf is called *back emf* (E_b) having the following expression:

$$E_b = \frac{\phi NZ}{60}\left(\frac{P}{A}\right) \tag{5.6}$$

5.3 CONDITION FOR MAXIMUM MECHANICAL POWER

From Equation (5.2),

$$P_m = E_b I_a = V_t I_a - I_a^2 R_a \tag{5.7}$$

Differentiating Equation (5.7) with respect to I_a,

$$\frac{dP_m}{dI_a} = V_t - 2I_a R_a$$

For maximum value of P_m, $\dfrac{dP_m}{dI_a} = 0$

i.e.,

$$V_t - 2I_a R_a = 0$$

i.e.,

$$I_a R_a = \frac{V_t}{2} \tag{5.8}$$

Using Equation (5.8), we have from Equation (5.2)

$$E_b = V_t - I_a R_a = V_t - \frac{V_t}{2} = \frac{V_t}{2} \tag{5.9}$$

The back emf is equal to half of the applied voltage for maximum gross mechanical power. It is impossible to achieve this condition practically. Since half of the power input is wasted as heat in the armature, efficiency will be less than 50 per cent taking account of other mechanical losses also.

5.4 ARMATURE TORQUE OF A MOTOR

If T_a be the gross torque developed by the armature rotating at a speed n rps, the mechanical power (P_m) developed is expressed as follows:

$$P_m = T_a \times 2\pi n \tag{5.10}$$

where

$$P_m = E_b I_a \tag{5.11}$$

From Equations (5.10) and (5.11), we have

$$T_a \times 2\pi n = E_b I_a \tag{5.12}$$

i.e.,

$$T_a = \frac{E_b I_a}{2\pi n} = \frac{1}{2\pi n} \phi n Z\,(P/A) I_a$$

i.e.,

$$T_a = \frac{1}{2\pi} Z\left(\frac{P}{A}\right)\phi I_a = 0.159 Z\left(\frac{P}{A}\right)\phi I_a$$

i.e.,

$$T_a = K\phi I_a \tag{5.13}$$

where

$$K = \frac{1}{2\pi} Z(P/A) = \text{constant for a particular DC motor}$$

$\therefore$

$$T_a \propto \phi I_a \tag{5.14}$$

For a DC shunt motor, ϕ is near about constant and Equation (5.14) becomes

$$T_a \propto I_a \tag{5.15}$$

For DC series motors, ϕ is proportional to the armature current and Equation (5.14) becomes

$$T_a \propto I_a^2 \tag{5.16}$$

If speed (N) is in rpm, the torque is expressed by

$$T_a = 9.55\left(\frac{E_b I_a}{N}\right) \tag{5.17}$$

Example 5.2 A DC shunt motor having armature resistance of 0.24 Ω takes an armature current of 80 A at 300 V. The machine has eight poles and 800 lap-connected conductors. The flux per pole is 0.042 Wb. Calculate the speed and gross torque developed by the armature.

Solution

Here, $I_a = 80$ A, $V_t = 300$ V, $R_a = 0.24\ \Omega$, $Z = 800$, $P = 8$, $A = P = 8$ (due to lap connection), $\phi = 0.042$ Wb

$$E_b = V_t - I_a R_a = 300 - 80 \times 0.24 = 280.8 \text{ V}$$

Again, $$E_b = \frac{\phi N Z}{60} \times \frac{P}{A}$$

i.e., $$280.8 = \frac{0.042 \times N \times 800}{60} \times \frac{8}{8}$$

i.e., $$N = \frac{280.8 \times 60}{0.042 \times 800} = 501.43 \text{ rpm}$$

Now, $$T_a = 0.159 Z\left(\frac{P}{A}\right)\phi I_a = 0.159 \times 0.042 \times 800 \times 80 \times \frac{8}{8} = 427.4 \text{ N-m}$$ ■

Example 5.3 A four-pole, 250 V, wave-connected DC series motor with 750 armature conductors supplies a load of 7.8 kW and takes a current of 40 A from the supply mains. The flux per pole is 22 mWb and the sum of armature and field resistance is 0.48 Ω. Calculate the developed torque and shaft torque.

Solution

Here, $I_a = 40$ A, $V_b = 250$ V, $R'_a = R_a + R_{se} = 0.48\ \Omega$, $Z = 750$, $P = 4$, $A = 2$ (wave connected), $\phi = 22$ mWb
Developed torque and gross torque are nothing but armature torque.

$$T_a = 0.159 Z\left(\frac{P}{A}\right)\phi I_a = 0.159 \times 22 \times 10^{-3} \times 750 \times 40 \times \frac{4}{2} = 209.88 \text{ N-m}$$

$$E_b = V_t - I_a R'_a = 250 - 40 \times 0.48 = 230.8 \text{ V}$$

Since, $$E_b = \frac{\phi N Z}{60} \times \frac{P}{A}$$

i.e., $$230.8 = \frac{22 \times 10^{-3} \times N \times 750}{60} \times \frac{4}{2}$$

i.e., $$N = \frac{230.8 \times 60}{22 \times 10^{-3} \times 750} = 419.64 \text{ rpm}$$

The output power is 7.8 kW = 7.8 × 10³ W

Again, $$T_{sh} \times 2\pi\left(\frac{N}{60}\right) = 7.8 \times 10^3$$

i.e., $$T_{sh} = \frac{7.8 \times 10^3 \times 60}{2 \times 3.14 \times 419.64} = 177.58 \text{ N-m}$$ ■

5.5 ROTATIONAL LOSSES OF DC MACHINES

DC motors have the following rotational losses:

(i) loss due to friction of bearings,

(ii) the windage loss due to consumption of power by the circulation of air or other cooling gas in the machine and

(iii) losses in the magnetic core of the machine known as core losses or iron losses. Core losses are divided into hysteresis loss and eddy current loss.

Let P_r be the rotational loss, P_b be bearing friction loss, P_w be windage loss, P_h be hysteresis loss, P_e be eddy current loss and P_i be iron or core loss $= P_h + P_e$.

$$P_r = P_b + P_w + P_i = P_b + P_w + P_h + P_e \tag{5.18}$$

Let T_a be average torque or internal torque at which the conversion of electromechanical power takes place, T_{sh} be useful torque or shaft torque developed at the shaft of an electric motor and T_P be prime mover torque applied at the shaft of a generator.

For a DC generator, we have

$$E_g I_a = V_t I_a + I_a^2 R_a \tag{5.19}$$

and

$$\omega T_a = \omega T_P - P_r \tag{5.20}$$

For a DC motor, we have

$$E_b I_a = V_t I_a - I_a^2 R_a \tag{5.21}$$

and

$$\omega T_a = \omega T_{sh} + P_r \tag{5.22}$$

Equations (5.21) and (5.22) are known as the power equation for a DC generator and a DC motor, respectively. T_a is called the *armature* or *gross torque* and T_{sh} is called the *useful torque*, *shaft torque* or *net torque*.

From Equation (5.22), we have

$$\omega T_{sh} = E_b I_a - P_r$$

i.e.,

$$T_{sh} = \frac{E_b I_a - P_r}{\omega}$$

i.e.,

$$= \frac{E_b I_a - P_r}{2\pi n} \quad \text{(where } n \text{ is in rps)} \tag{5.23}$$

i.e.,

$$T_{sh} = \frac{E_b I_a - P_r}{2\pi\left(\frac{N}{60}\right)} \quad \text{(where } N \text{ is in rpm)}$$

$$= 9.55\left(\frac{E_b I_a - P_r}{2\pi N}\right) \tag{5.24}$$

Example 5.4 A DC series motor rotates at 1,000 rpm and draws 50 A at 250 V. Its armature and field resistances are 0.22 Ω and 0.12 Ω, respectively. The iron and friction losses are 0.54 kW. Find the developed torque and output power of the motor.

Solution

Here, $V_t = 250$ V, $I = 50$ A, $R_a = 0.22\ \Omega$, $R_{se} = 0.12\ \Omega$, $N = 1000$ rpm and iron and friction losses = 0.54 kW = 540 W

$$E_b = V_t - I_a R_a = 250 - 50 \times (0.22 + 0.12) = 233 \text{ V}$$

$$P_m = E_b I_a = 233 \times 50 = 11{,}650 \text{ W}$$

$$T_a = 9.55 \times \frac{P_m}{N} = \frac{9.55 \times 11{,}650}{1{,}000} = 111.257 \text{ N-m}$$

$P_0 = P_m$ – iron and friction losses = (11,650 – 540) W = 11.11 kW ■

Example 5.5 A four-pole, 220 V, lap-connected DC shunt motor delivers 12 kW. It runs at a speed of 1,000 rpm and draws armature and field current of 60 A and 2 A, respectively. The total number of armature conductors is 500 and armature resistance is 0.12 Ω. Calculate (i) total torque, (ii) useful torque, (iii) useful flux/pole, (iv) rotational losses and (v) efficiency. Assume 1 V per brush for contact drop.

Solution

$$E_b = V_t - I_a R_a - 1 \times 2 = 220 - 60 \times 0.12 - 2 = 210.8 \text{ V}$$

(i) Armature torque $(T_a) = 9.55\left(\frac{E_b I_a}{N}\right) = 9.55 \times \frac{210.8 \times 60}{1{,}000} = 120.79 \text{ N-m}$

(ii) Useful torque $(T_{sh}) = 9.55 \times \frac{\text{output power}}{N} = 9.55 \times \frac{12 \times 10^3}{1{,}000} = 114.6 \text{ N-m}$

(iii) $E_b = \frac{\phi NZ}{60} \times \frac{P}{A}$

i.e., $210.8 = \frac{\phi \times 1{,}000 \times 500}{60} \times \frac{4}{4}$

i.e., $\phi = \frac{210.8 \times 60}{1{,}000 \times 500} = 0.0253 \text{ Wb}$

(iv) Armature power developed = $E_b I_a$ = 210.8 × 60 = 12,648 W

Rotational losses = 12,648 – 12,000 = 648 W

(v) Power input to the motor = VI = 220 × (60 + 2) = 13,640 W

$$\eta = \frac{\text{ouput power}}{\text{input power}} = \frac{12{,}000}{13{,}640} = 0.8797 \text{ p.u.} = 87.97\%$$ ■

Example 5.6 A four-pole, 240 V, wave-connected DC shunt motor gives 9 kW when running at 1,000 rpm and drawing armature and field currents of 40 A and 1.0 A, respectively. It has 500 conductors and armature resistance of 0.1 Ω. Assuming contact drop of 1 V per brush, find (i) total torque, (ii) useful torque, (iii) useful flux/pole, (iv) rotational losses and (v) efficiency.

Solution

$$E_b = V_t - I_a R_a - \text{brush drop} = 240 - (40 \times 0.1) - 2 = 234 \text{ V}$$

(i) Armature torque $T_a = 9.55\left(\frac{E_b I_a}{N}\right)$ N-m

$$= 9.55 \times \left(\frac{234 \times 40}{1{,}000}\right) = 89.39 \text{ N-m}$$

(ii) $T_{sh} = 9.55 \times \left(\frac{\text{output}}{N}\right)$ N-m

$$= 9.55 \times \left(\frac{9{,}000}{1{,}000}\right) = 85.95 \text{ N-m}$$

(iii) $$E_b = \frac{\phi ZN}{60} \times \left(\frac{P}{A}\right)$$

i.e., $$234 = \frac{\phi \times 500 \times 1{,}000}{60} \times \left(\frac{4}{2}\right)$$

i.e., $$\phi = 14.04 \text{ mWb}$$

(iv) Armature input = $V_t I_a = 240 \times 40 = 9{,}600$ W

Armature Cu loss = $I_a^2 R_a = 40^2 \times 0.1 = 160$ W

Brush contact loss = $40 \times 2 = 80$ W

Power developed = $9{,}600 - (160 + 80) = 9{,}360$ W

Output = 9 kW = 9,000 W

Rotational losses = $9{,}360 - 9{,}000 = 360$ W

(v) Total motor input = $VI = 240 \times 41 = 9{,}840$ W

Motor output = 9,000 W

$$\text{Efficiency} = \left(\frac{9{,}000}{9{,}840}\right) = 0.9146 \text{ p.u.} = 91.46\%$$ ■

Example 5.7 A DC series motor running at 2,000 rpm has the hysteresis and eddy current losses 500 W and 200 W, respectively. If flux remains constant, calculate the speed at which total iron losses are halved.

Solution

Hysteresis loss = $C \times 2{,}000 = 500$ W

Eddy current loss = $D \times 2{,}000^2 = 200$ W

Therefore, $$C = 0.25 \text{ and } D = \frac{200}{2{,}000^2} = 5 \times 10^{-5}$$

Let at speed N, $$P_i = \frac{200 + 500}{2} = 350 \text{ W}$$

Therefore, $$0.25\,N + 5 \times 10^{-5} N^2 = 350$$

i.e., $$N^2 + 0.5 \times 10^4 N - 700 \times 10^4 = 0$$

i.e., $$N = \frac{-5{,}000 \pm \sqrt{5{,}000^2 + 28 \times 10^6}}{2} = \frac{-5{,}000 \pm 7{,}280}{2}$$

Since, speed cannot be negative, the feasible solution is

$$N = \frac{7{,}280 - 5{,}000}{2} = 1{,}140 \text{ rpm}$$ ■

5.6 COMPOUND MOTOR

Compound motors have both series and shunt windings, and they may be short shunt type or long shunt type depending upon the connection to the field windings. Depending on whether the series field aids or opposes the shunt field, this machine is said to be a cumulatively compounded or differentially compounded DC machine.

5.7 RELATION OF SPEED (*N*) WITH BACK EMF (E_b) AND FLUX (ϕ)

From Equation (5.6), we have

$$E_b = \frac{\phi NZ}{60}(P/A)$$

i.e.,

$$N = \frac{E_b}{\phi}\left(\frac{60 \times A}{Z \times P}\right)$$

i.e.,

$$N = K\left(\frac{E_b}{\phi}\right) \tag{5.25}$$

where $K\left(=\frac{60A}{ZP}\right)$ is a constant for a particular DC machine.

$$\therefore \quad N \propto \frac{E_b}{\phi} \tag{5.26}$$

From Equation (5.26), we conclude that speed (*N*) is proportional to the back emf (E_b) and it is inversely proportional to the flux (ϕ).

If N_1 (in rpm) and N_2 (in rpm) be the two different speeds of a DC machine corresponding to the flux ϕ_1 and ϕ_2, respectively, we have

$$N_1 = \frac{E_1}{K\phi_1} \tag{5.27}$$

$$N_2 = \frac{E_2}{K\phi_2} \tag{5.28}$$

From Equations (5.27) and (5.28), we have

$$\frac{N_1}{N_2} = \frac{E_1}{E_2} \times \frac{\phi_2}{\phi_1}$$

Example 5.8 The no-load current and speed of a four-pole, 250 V, DC shunt motor are 6 A and 500 rpm, respectively. It has a shunt field current of 1 A. If its full-load current is 80 A, find its full-load speed. The value of armature resistance is 0.12 Ω. Consider 1 V per brush for contact drop. Armature reaction weakens the flux by 2.5 per cent.

Solution

Let ϕ_0 be the flux at no-load.

$$\therefore \text{ Flux}(\phi_1) \text{ at full load} = \frac{(100-2.5)\phi_0}{100} = 0.975\ \phi_0$$

Let *N* be the full-load speed in rpm.

At no-load,

$$I_{a0} = I_0 - I_{sh} = 6 - 1 = 5 \text{ A}$$

$$\therefore \quad E_{b0} = V_t - I_{a0}R_a - \text{contact drop at brushes} = 250 - 5 \times 0.12 - 2 = 247.4 \text{ V}$$

At full-load,

$$I_a = I - I_{sh} = 80 - 1 = 79 \text{ A}$$

$$\therefore \quad E_{b1} = 250 - 79 \times 0.12 - 2 = 238.52 \text{ V}$$

Since $E_b \propto \phi N$, we have

$$\frac{E_{b0}}{E_{b1}} = \frac{\phi_0 N_0}{\phi_1 N_1}$$

i.e.,

$$N_1 = \frac{E_{b1}}{E_{b0}} \times \frac{\phi_0}{\phi_1} \times N_0 = \frac{238.52 \times \phi_0 \times 500}{247.40 \times 0.975\phi_0} = 494.41 \text{ rpm}$$

∴ The full-load speed of the motor is 494.41 rpm. ■

5.8 CHARACTERISTICS OF SHUNT OR SEPARATELY EXCITED DC MOTOR

For shunt and separately excited DC motors, the field winding is supplied from a constant voltage. So the field flux remains constant. Thus, the shunt and separately excited DC machine have the same characteristic. Using Equation (5.2), Equation (5.26) can be written as follows:

$$N \propto \frac{V_t - I_a R_a}{\phi} \tag{5.29}$$

Since ϕ is constant for a DC shunt motor, we have

$$N \propto (V_t - I_a R_a) \tag{5.30}$$

Since R_a is very small, $I_a R_a$ is also small.

The torque of any DC shunt motor is expressed by Equation (5.15).

The speed versus armature current and torque versus armature current are shown in Figures 5.2(a) and 5.2(b), respectively.

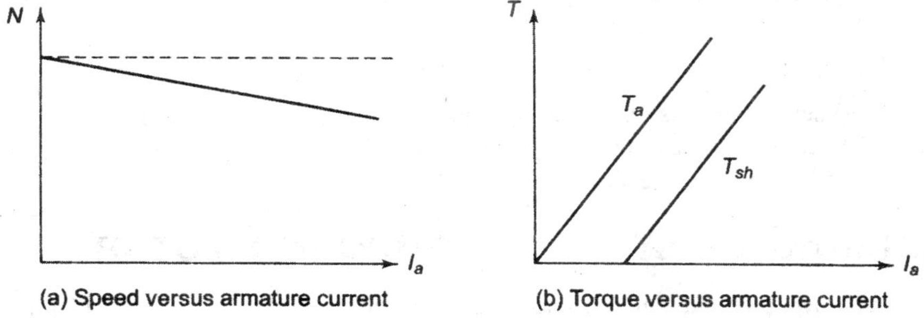

(a) Speed versus armature current (b) Torque versus armature current

Figure 5.2 Characteristics of a Shunt or Separately Excited DC Motor

Let T_g be the gross torque and T_{sh} be the useful torque or shaft torque.

∴ $$T_{sh} = T_g - (T_f - T_w)$$

where T_f is frictional torque and T_w is windage torque.

5.9 CHARACTERISTICS OF DC SERIES MOTOR

Using Equation (5.4), Equation (5.26) can be expressed as follows:

$$N \propto \frac{V_t - I_a(R_a + R_{se})}{\phi} \tag{5.31}$$

Equation (5.31) shows the expression for the speed of a DC series motor.

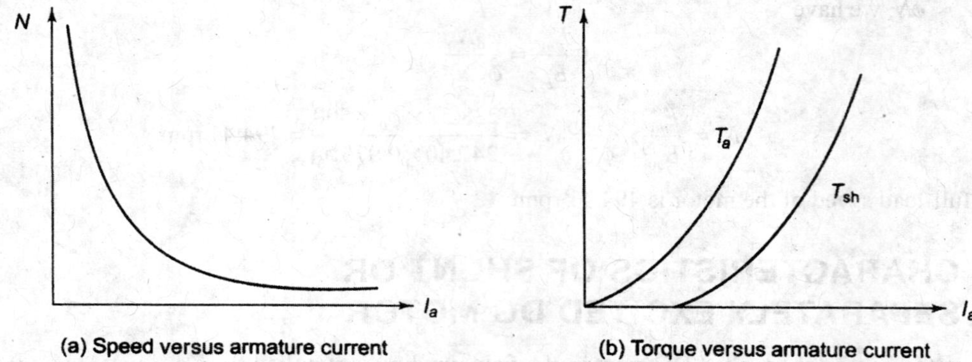

Figure 5.3 Characteristics of a DC Series Motor

At low value of I_a, $I_a(R_a + R_{se})$ can be neglected compared to V_t. Equation (5.31) reduces to

$$N \propto \frac{V_t}{\phi} \tag{5.32}$$

Since supply voltage V_t is constant,

$$N \propto \frac{1}{\phi} \tag{5.33}$$

In a series motor, flux (ϕ) is proportional to the armature current (I_a). Therefore, we have

$$\phi \propto I_a \tag{5.34}$$

Using Equation (5.34), Equation (5.33) can be expressed as follows:

$$N \propto \frac{1}{I_a} \tag{5.35}$$

The torque of a DC series motor is expressed by Equation (5.16).

The speed (N) versus armature current (I_a) and torque (T) versus armature current (I_a) are shown in Figures 5.3(a) and 5.3(b), respectively.

5.10 CHARACTERISTICS OF COMPOUND MOTOR

Speed (N) versus armature current (I_a) and torque (T) versus armature current (I_a) are shown in Figures 5.4(a) and 5.4(b), respectively.

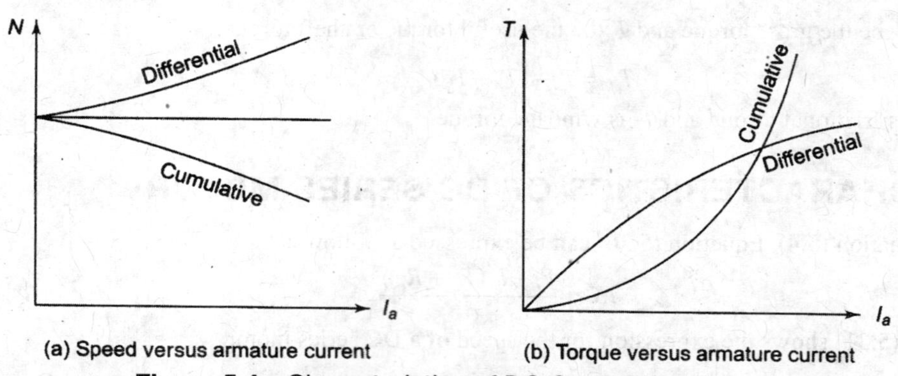

Figure 5.4 Characteristics of DC Compound Motor

5.11 SPEED REGULATION

The change in speed of a motor, with change in applied torque while other condition remains unaltered, is known as speed regulation. The change in speed occurs due to inherent properties of the motor itself and not by manipulation of rheostats or other speed controlling devices. The change in speed when the load on the motor is reduced to zero from full-load is known as speed regulation, and it is expressed by the percentage of the rated load speed.

$$\therefore \quad \text{Speed regulation} = \frac{\text{no-load speed} - \text{full-load speed}}{\text{full-load speed}}$$

$$= \frac{N + dN - N}{N} \times 100 = \frac{dN}{N} \times 100$$

5.12 TORQUE AND SPEED OF DC SERIES MOTOR

Using Equation (5.20), Equation (5.25) can be expressed as follows:

$$N = K\left(\frac{E_b}{\phi}\right) = K\left(\frac{V_t - I_a R_a}{\phi}\right) \tag{5.36}$$

From Equation (5.14), the expression for the torque is rewritten:

$$T_a \propto \phi I_a \tag{5.37}$$

From Equations (5.36) and (5.37), we can observe that increase in the value of ϕ will decrease N and increase T, which are contradictory to each other. The following steps will put the light into the apparent inconsistency between two equations:

Step 1: Since $N = K\left(\frac{E_b}{\phi}\right)$, it drops instantly due to decrease in flux ϕ.

Step 2: Since $I_a = \frac{V_t - E_b}{R_a}$, I_a will increase because V_t and R_a are constants.

Step 3: Since $T_a \propto \phi I_a$, a large increase in I_a counterbalances a small decrease in ϕ. This results a net increase in T.

Step 4: The increased T increases the speed (N).

5.13 SPEED CONTROL OF DC MOTORS

From Equation (5.36), the speed control of DC motors can be carried in the following ways:

(i) Speed can be controlled by varying resistance in series with the armature circuit, which is called *armature resistance control*.

(ii) Speed can be controlled by varying the field flux (ϕ). This method is known as *field resistance control*.

(iii) Speed can also be controlled by varying the applied voltage, which is called *voltage control*.

5.13.1 Armature Resistance Control

If a speed below the no-load speed is required, armature or rheostatic method is used. The supply voltage is constant and the rheostat connected in series with the armature is varied. By increasing the controller resistance, it is possible to decrease the potential drop of the armature and hence also the

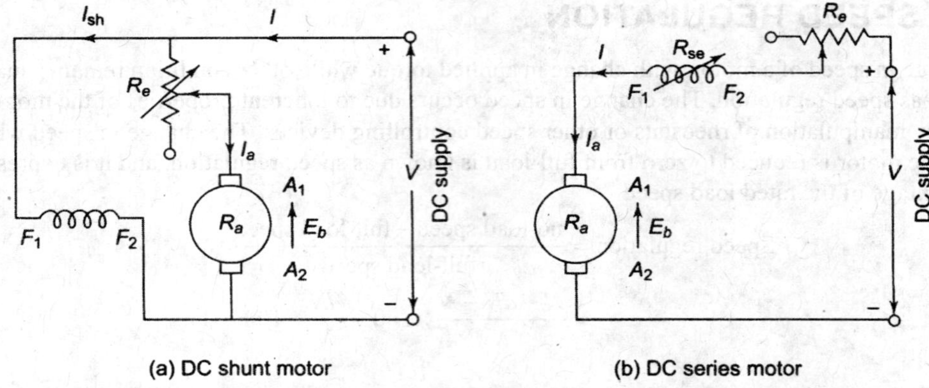

(a) DC shunt motor (b) DC series motor

Figure 5.5 Armature Resistance Control of DC Motors

armature speed. Figures 5.5(a) and 5.5(b) show the method of connection to external resistance (R_e) in the armature circuit of a DC shunt and series machine, respectively. In a shunt machine, flux (ϕ) will not be affected. But in DC series machine, flux (ϕ) will be affected. In both cases the motor runs at a lower speed if the value of R_e is increased.

The speed (N) versus armature current (I_a) for DC shunt and series machines is shown in Figures 5.6(a) and 5.6(b), respectively.

Let us calculate the stalling current for the DC shunt motor.

If the load torque is constant, the speed is proportional to the potential difference across the armature. Figure 5.7 shows the circuit diagram of the armature control method, while Figure 5.8 shows the speed versus armature current characteristic.

Let I_{a0} be the armature current at no-load and the armature resistance be R_a. Let I_{a1} be the armature current when the controller resistance R is added to the armature resistance. The total armature resistance becomes $R_t = R + R_a$. Let N_0 be the no-load speed and N_1 be the speed of armature current I_{a1}.

$\therefore$ $$N_0 \propto (V_t - I_{a0}R_a); \text{ i.e., } N_0 \propto E_{b0}$$

and $$N_1 \propto (V_t - I_{a1}R_t); \text{ i.e., } N_1 \propto E_{b1}$$

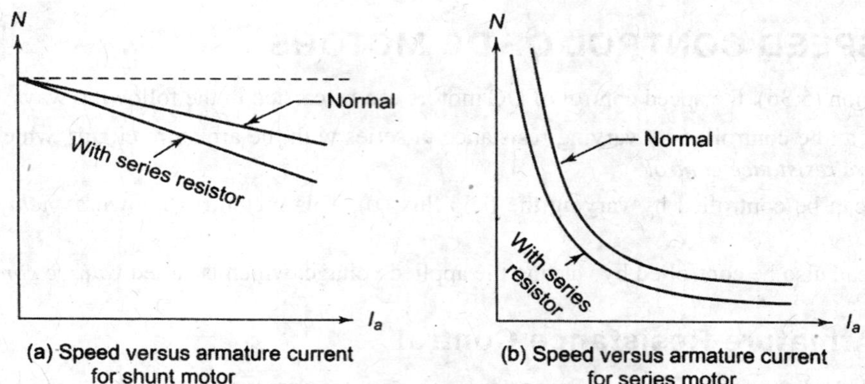

(a) Speed versus armature current for shunt motor (b) Speed versus armature current for series motor

Figure 5.6 Speed Versus Armature Current for DC Shunt and Series Motors

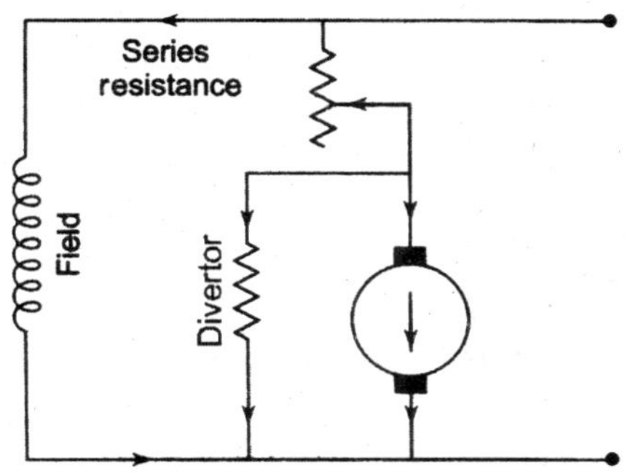

Figure 5.7 Circuit Connection

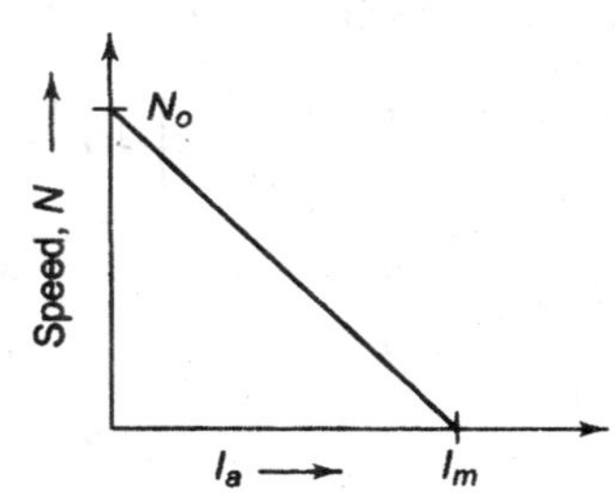

Figure 5.8 Speed Versus Armature Current Characteristic

Since the flux is constant for the shunt motor, we have

$$\frac{N_0}{N_1} = \frac{V_t - I_{a0}R_a}{V_t - I_{a1}R_t}$$

i.e., $$N_1 = N_0\left(\frac{V_t - I_{a1}R_t}{V_t - I_{a0}R_a}\right) = N_0\left(1 - \frac{I_{a1}R_t}{V_t}\right) \quad [\text{neglecting } I_{a0}] \tag{5.38}$$

Equation (5.38) shows the linear relation between R_t and speed.

The load current I_{a1} for which the speed N_1 becomes zero is obtained by putting $N_1 = 0$ in Equation (5.38).

$$0 = N_0\left(1 - \frac{I_{a1}R_t}{V_t}\right)$$

i.e., $$I_{a1} = \frac{V_t}{R_t} \tag{5.39}$$

Equation (5.39) shows the required maximum current for which speed becomes zero. This current is known as *stalling current*.

The drawbacks of this method are as follows:

(i) A large amount of power is lost in the external resistance R_e.

(ii) This method gives the speed below normal value.

(iii) For a specified value R_e, speed reduction is not constant. It varies with the motor load.

Example 5.9 The torque of the load driven by a 400 V DC shunt motor varies as the cube of the speed. The current taken by the motor is 40 A at a certain speed. Calculate the additional resistance required to be connected in series with the armature circuit to reduce the speed to 60 per cent of the original speed. The resistance is 0.35 Ω.

Solution

$$V_t = 400\,\text{V}, I_{a1} = 40\,\text{A}, R_a = 0.35\,\Omega.$$

Let the resistance R be inserted in series with the armature to reduce the speed to 60 per cent of its original value.

$\therefore$ $$T \propto \phi I_a \propto I_a \text{ since } \phi \text{ is constant}$$

Again, $$I_a \propto N^3$$

We have the following relation [$\therefore N_2 = 0.6\,N_1$]

$$\frac{I_{a2}}{I_{a1}} = \left(\frac{N_2}{N_1}\right)^3 = \left(\frac{0.6N_1}{N_1}\right)^3 = 0.6^3$$

i.e.,
$$I_{a2} = 0.6^3 \times I_{a1} = 0.6^3 \times 40 = 8.64 \text{ A}$$

$\therefore$
$$E_{b1} = V_t - I_{a1}R_a = 400 - 40 \times 0.35 = 386 \text{ V}$$

And
$$E_{b2} = V_t - I_{a2}(R_a + R_{se}) = 400 - (0.35 + R) \times 8.64 = 396.976 - 8.64R$$

Again,
$$\frac{E_{b2}}{E_{b1}} = \frac{N_2}{N_1}$$

i.e.,
$$\frac{396.976 - 8.64R}{386} = 0.6$$

$$R = 19.14\ \Omega.$$

Therefore, the required additional resistance is 19.14 Ω to reduce the speed to 60 per cent of the original speed. ■

Example 5.10 A 220 V DC series motor has an armature resistance of 0.3 Ω and a field resistance of 0.2 Ω. It runs at a speed of 700 rpm taking a current of 15 A. Calculate the resistance to be inserted in series with the armature to reduce the speed to 600 rpm. The input current remains constant. Assume that the magnetization characteristic is a straight line.

Solution

$$V_t = 220 \text{ V},\ R_a = 0.3\ \Omega,\ R_{se} = 0.2\ \Omega,\ I_1 = I_{a1} = I_{a2} = 15 \text{ A},\ N_1 = 700 \text{ rpm},\ N_2 = 600 \text{ rpm}$$

$$E_{b1} = V_t - I_{a1}(R_a + R_{se}) = 220 - 15 \times (0.2 + 0.3) = 212.5 \text{ V}$$

Let the resistance R Ω be connected in series with the field to reduce the speed from 700 rpm to 600 rpm.

$$E_{b2} = V_t - I_{a2}(R_a + R_{se} + R) = 220 - 15 \times (0.2 + 0.3 + R) = (212.5 - 15R) \text{ V}$$

Since,
$$I_{a1} = I_{a2},\ \phi_1 = \phi_2$$

$\therefore$
$$\frac{N_2}{N_1} = \frac{E_{b2}}{E_{b1}} \times \frac{\phi_1}{\phi_2} = \frac{E_{b2}}{E_{b1}}$$

i.e.,
$$\frac{600}{700} = \frac{212.5 - 15R}{212.5}$$

i.e.,
$$R = 212.5 \times \frac{1}{7} \times \frac{1}{15} = 2.02\ \Omega$$

Therefore, the required additional resistance is 2.02 Ω. ■

5.13.2 Field Resistance Control

In this method, current through the field winding is controlled by using a variable resistor in series with the field coil in the DC shunt motor and DC series motor, shown in Figures 5.9 and 5.10, respectively.

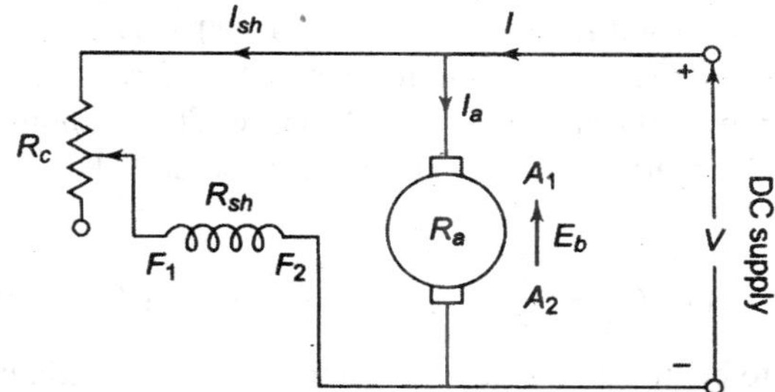

Figure 5.9 Field Resistance Control of DC Shunt Motor

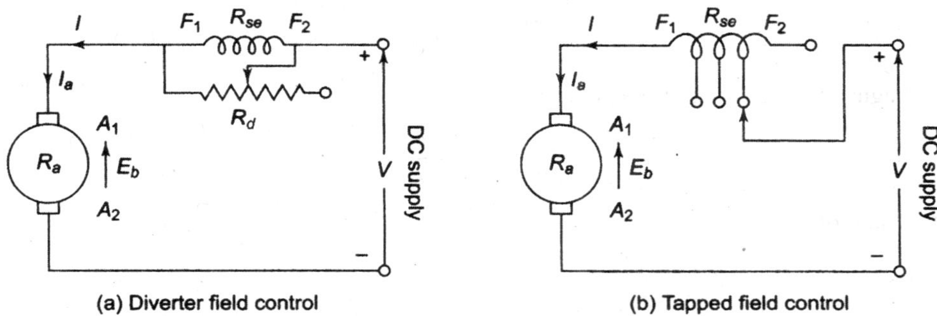

(a) Diverter field control

(b) Tapped field control

Figure 5.10 Field Control of DC Series Motor

In a series motor the field flux can also be changed by the following two methods:

(i) A variable resistance (R_d) is connected in parallel with the field coil as shown in Figure 5.10(a). This resistance R_d is called the *diverter resistance.*

(ii) The tapped field control can also be used as shown in Figure 5.10(b). In this case the ampere turns are varied by varying the number of field turns. The arrangement is used in electric traction.

Figures 5.11(a) and 5.11(b) show the speed–torque curves of DC shunt and series motors, respectively. The speed of the DC shunt and series machines is controlled by the variation of the field flux.

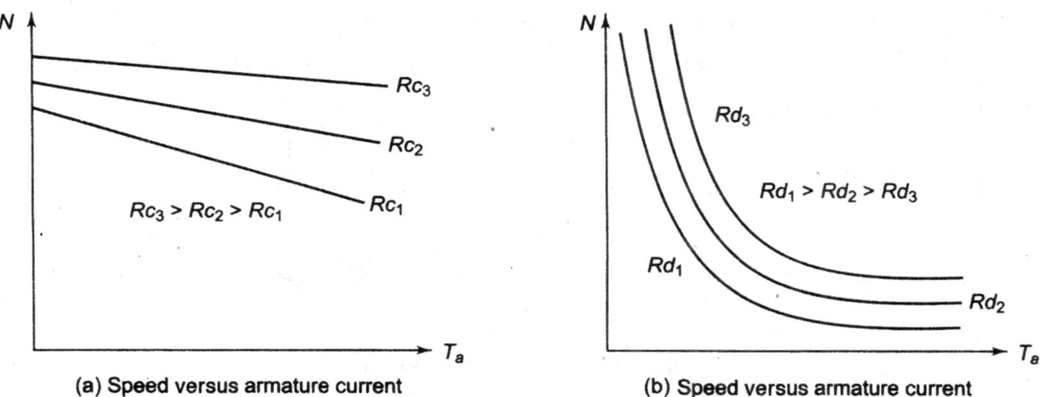

(a) Speed versus armature current

(b) Speed versus armature current

Figure 5.11 Speed Torque Characteristic of DC Shunt and Series Motors

Example 5.11 The armature and field resistances of a 200 V DC shunt motor are 0.4 Ω and 200 Ω, respectively. When driving a load of constant torque at 500 rpm, the armature current is 20 A. It is desired to raise the speed from 500 rpm to 700 rpm. What resistance should be inserted in series with the shunt field circuit? Assume that the magnetic circuit is unsaturated.

Solution

$$V_t = 200\text{ V}, R_a = 0.4\ \Omega, R_{sh} = 200\ \Omega, N_1 = 500\text{ rpm and } N_2 = 700\text{ rpm}$$

Let the required resistance to be inserted in series with the shunt field circuit be R Ω.

We know that

$$\frac{N_2}{N_1} = \frac{E_{b2}}{E_{b1}} \times \frac{\phi_1}{\phi_2}$$

Since the magnetic circuit is unsaturated, $\phi \propto I_{sh}$.

$$\therefore \quad \frac{N_2}{N_1} = \frac{E_{b2}}{E_{b1}} \times \frac{I_{sh1}}{I_{sh2}} \tag{1}$$

Due to constant torque,

$$\phi_1 I_{a1} = \phi_2 I_{a2}$$

i.e.,

$$I_{a2} = I_{a1} \times \frac{\phi_1}{\phi_2} = I_{a1} \times \frac{I_{sh1}}{I_{sh2}} \quad [\because \phi \propto I_{sh}] \tag{2}$$

Here,

$$I_{sh1} = \frac{200}{200} = 1\text{ A} \quad \text{and} \quad I_{sh2} = \frac{200}{R_t}$$

From Equation (2), we have

$$I_{a2} = I_{a1} \times \frac{I_{sh1}}{I_{sh2}} = 20 \times \frac{1}{\frac{200}{R_t}} = 0.1R_t$$

Again,

$$E_{b1} = V_t - I_{a1}R_a = 200 - 20 \times 0.4 = 192\text{ V}$$

And

$$E_{b2} = V_t - I_{a2}R_a = 200 - 0.1R_t \times 0.4 = 200 - 0.04R_t$$

Substituting I_{sh1}, I_{sh2}, E_{b1} and E_{b2} in Equation (1), we have

$$\frac{700}{500} = \frac{200 - 0.04R_t}{192} \times \frac{1}{\frac{200}{R_t}}$$

i.e.,

$$\frac{7}{5} = \frac{200 - 0.04R_t}{192} \times \frac{R_t}{200}$$

i.e.,

$$(200 - 0.04\,R_t)\,R_t = \frac{7}{5} \times 192 \times 200$$

i.e.,

$$0.04R_t^2 - 200R_t + 53{,}760 = 0$$

i.e., $$R_t = \frac{200 \pm \sqrt{200^2 - 4\times 0.04 \times 53,760}}{2\times 0.04} = \frac{200\pm 177.2}{0.08}$$

i.e., $$R_t = \frac{22.8}{0.08} = 285\ \Omega \quad \text{(feasible)}$$

Therefore, the required additional resistance $(R) = R_t - 200 = 85\ \Omega$. ■

Example 5.12 A 240 V DC shunt motor has armature resistance of 0.25 Ω and draws a current of 40 A from the supply mains on half of full-load. The speed is to be increased twice the half-load speed. Determine the percentage change in flux if the torque of the motor remains constant.

Solution

$$V_t = 200\text{ V}, R_a = 0.25\ \Omega, I_{a1} = 40\text{ A} \quad \text{[neglecting field current]}$$

$$\frac{\text{Speed}(N_1)}{\text{Speed}(N_2)} = \frac{1}{2}$$

$$E_{b1} = V_t - I_{a1}R_a = 240 - 40\times 0.25 = 230\text{ V}$$

and $$E_{b2} = V_t - I_{a2}R_a = 240 - 0.25I_{a2}$$

We know that $$\frac{N_2}{N_1} = \frac{E_{b2}}{E_{b1}} \times \frac{\phi_1}{\phi_2}$$

i.e. $$\frac{2}{1} = \frac{240 - 0.25I_{a2}}{230} \times \frac{\phi_1}{\phi_2}$$

i.e., $$460 \times \left(\frac{\phi_2}{\phi_1}\right) = 240 - 0.25I_{a2} \tag{1}$$

Again, the load torque remains constant, and we have

$$\phi_1 I_{a1} = \phi_2 I_{a2}$$

i.e., $$I_{a2} = I_{a1}\left(\frac{\phi_1}{\phi_2}\right) = 40\left(\frac{\phi_1}{\phi_2}\right) \tag{2}$$

Using Equation (2), from Equation (1) we have

$$460 \times \left(\frac{\phi_2}{\phi_1}\right) = 240 - 10\left(\frac{\phi_1}{\phi_2}\right)$$

i.e., $$460x = 240 - \frac{10}{x} \qquad \left[\text{where } x = \frac{\phi_2}{\phi_1}\right]$$

i.e., $$460x^2 - 240x + 10 = 0$$

i.e., $$x = \frac{240 \pm \sqrt{240^2 - 4\times 460 \times 10}}{2\times 460} = \frac{240 \pm 197.99}{920}$$

i.e., $$x = 0.0456 \text{ (feasible)}$$

Since $$\frac{\phi_2}{\phi_1} = 0.0456$$

We have

$$1 - \frac{\phi_2}{\phi_1} = 1 - 0.0456 = 0.9544$$

i.e., $$\frac{\phi_1 - \phi_2}{\phi_1} = 0.9544 \text{ p.u.} = 95.44\%$$

Therefore, percentage change in flux is 95.44. ■

Example 5.13 The input current of a 400 V DC series motor is 200 A when running at a speed of 800 rpm. When it runs at a speed of 1,000 rpm, it draws a current of 220 A. Calculate the value of resistance connected across the field circuit. The armature and field resistances are 0.15 Ω and 0.08 Ω, respectively.

Solution

$$E_{b1} = V_t - I_{a1}(R_a + R_{se}) = 400 - 200 \times (0.15 + 0.08) = 354 \text{ V}$$

and $$E_{b2} = V_t - I_{a2}R_{a2} - I_{se}R_{se} = 400 - 220 \times 0.15 - I_{se} \times 0.08 = 367 - 0.08I_{se}$$

We know that $$\frac{E_{b2}}{E_{b1}} = \frac{N_2\phi_2}{N_1\phi_1}$$

i.e., $$\frac{367 - 0.08I_{se}}{354} = \frac{1{,}000 \times I_{se}}{800 \times 200}$$

i.e., $$367 - 0.08I_{se} = \left(\frac{1{,}000 \times 354}{800 \times 200}\right) \times I_{se} = 2.2125I_{se}$$

i.e., $$I_{se} = 160.08 \text{ A}$$

Again, $$I_{se} = \left(\frac{R_{div}}{R_{div} + R_{se}}\right) \times I_{a2}$$

where R_{div} is the diverter resistance.

∴ $$\left(\frac{R_{div}}{R_{div} + R_{se}}\right) \times I_{a2} = 160.08$$

i.e., $$\left(\frac{R_{div}}{R_{div} + R_{se}}\right) \times 220 = 160.08$$

i.e., $$\frac{R_{div} + R_{se}}{R_{div}} = \frac{220}{160.08} = 1.3743$$

i.e., $$1 + \frac{R_{se}}{R_{div}} = 1.3743$$

i.e., $$\frac{R_{se}}{R_{div}} = 0.3743$$

i.e., $$R_{div} = \frac{R_{se}}{0.3743} = \frac{0.08}{0.3743} = 0.2137$$

Therefore, 0.2137 Ω must be inserted across the resistance field. ■

5.14 WARD-LEONARD CONTROL (VOLTAGE CONTROL)

The drawbacks of the earlier methods can be overcome by this method. When large motors are to be controlled, a separate motor-driven generator of suitable rating as shown in Figure 5.12 is used. The main motor field is supplied with full voltage independent of the armature, and thus it is a separately excited motor. The generator field is fed from a potential divider so that the field current and hence the terminal voltage of the generator can be varied from +V to −V, thus making it possible to control the speed of the main motor in either direction of rotation.

This method has the following advantages:

(i) Starting can be done without the necessity of series resistance starters.
(ii) Stepless control from standstill to maximum speed can be done in either direction.
(iii) No excessive high armature currents are required for rapid and instant reversal.
(iv) A wide range of speed from standstill to high speeds in either direction is possible.

This method has the following disadvantages:

(i) It is costly since it requires two extra machines of ratings comparable with the main motor.
(ii) The system has low overall efficiency because three machines are involved.

Example 5.14 A 220 V DC series motor takes 20 A while running at a constant speed. Its armature and field resistances are 0.3 Ω and 0.2 Ω, respectively. Calculate the resistance to be connected in series with the armature to reduce the speed by 40 per cent. Assume that the torque varies with the cube of the speed and flux is proportional to the current.

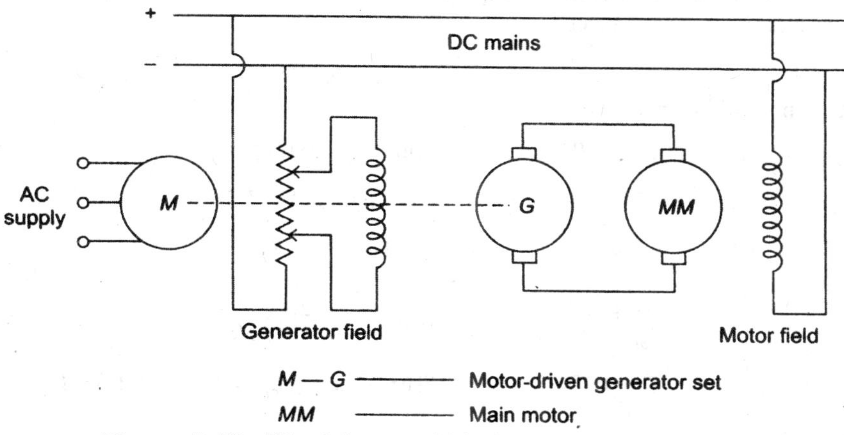

Figure 5.12 Ward–Leonard Method of Speed Control

Solution

$$V_t = 220\text{ V}, I_{a1} = 20\text{ A}, R_a = 0.3\,\Omega, R_{se} = 0.2\,\Omega, N_2 = N_1(1-0.4) = 0.6N_1$$

Let the resistance R Ω be connected in series with the armature to get speed N_2. For series motor,

$$\phi \propto I_a$$

and $$T \propto \phi I_a \quad \text{i.e.,} \quad T \propto I_a^2$$

Again, $$T \propto N^3$$

∴ $$I_a^2 \propto N^3$$

∴ $$\left(\frac{I_{a2}}{I_{a1}}\right)^2 = \left(\frac{N_2}{N_1}\right)^3 \quad \text{i.e.,} \quad I_{a2} = I_{a1} \times \left(\frac{N_2}{N_1}\right)^{\frac{3}{2}} = 20 \times (0.6)^{\frac{3}{2}} = 9.3\text{ A}$$

$$E_{b1} = V_t - I_{a1}(R_a + R_{se}) = 220 - 20 \times 0.5 = 210\text{ V}$$

and $$E_{b2} = V_t - I_{a2}(R_a + R_{se} + R) = (215.35 - 9.3R)\text{ V}$$

Again, $$\frac{N_2}{N_1} = \frac{E_{b2}}{E_{b1}} \times \frac{\phi_1}{\phi_2} = \frac{E_{b2}}{E_{b1}} \times \frac{I_{a1}}{I_{a2}}$$

i.e., $$\frac{0.6N_1}{N_1} = \left(\frac{215.35 - 9.3R}{210}\right) \times \frac{20}{9.3}$$

$$R = \frac{1}{9.3}\left(215.35 - \frac{0.6 \times 9.3 \times 210}{20}\right) = 16.85\,\Omega$$

The required resistance is 16.85 Ω. ■

Example 5.15 A 220 V DC series motor takes 40 A and runs at 500 rpm while driving a fan load. The load varies as the square of the speed. The resistance between terminals is 1.3 Ω. It is required to raise the speed to 600 rpm by increasing the supply voltage. Find the required supply voltage. Assume that the flux varies directly as the current.

Solution

Resistance between terminals = 1.3 Ω

$$I_{a1} = 40\text{ A}, N_1 = 500\text{ rpm}, N_2 = 600\text{ rpm}$$

$$T \propto \phi I_a \quad \text{i.e.,} \quad T \propto I_a^2 \quad [\text{since } \phi \propto I_a]$$

Again, $$T \propto N^2 \quad \therefore \; N^2 \propto I_a^2 \quad \text{i.e.,} \quad N \propto I_a$$

∴ $$\frac{I_{a2}}{I_{a1}} = \frac{N_2}{N_1} \quad \text{i.e.,} \quad I_{a2} = I_{a1} \times \frac{N_2}{N_1} = 40 \times \frac{600}{500} = 48\text{ A}$$

$$E_{b1} = 220 - 40 \times 1.3 = 168\text{ V} \quad \text{and} \quad E_{b2} = V_t - 48 \times 1.3 = V_t - 62.4$$

where V_t is the new supply voltage.

Again,
$$\frac{N_2}{N_1} = \frac{E_{b2}}{E_{b2}} \times \frac{\phi_1}{\phi_2} = \frac{E_{b2}}{E_{b1}} \times \frac{I_{a1}}{I_{a2}}$$

i.e.,
$$\frac{600}{500} = \left(\frac{V_t - 62.4}{168}\right) \times \frac{40}{48}$$

i.e.,
$$V_t = \frac{6}{5} \times 168 \times \frac{48}{40} + 62.4 = 304.32 \text{ V}$$

Therefore, the new supply voltage is 304.32 V. ■

5.15 NECESSITY OF A STARTER FOR DC MOTORS

During rest and at the time of starting, the back emf (E_b) of the DC machine is zero. The general expression for the armature current (I_a) of the DC machine is $I_a = \frac{V_t - E_b}{R_a}$, where V_t is the supply voltage, E_b is the back emf and R_a is the armature resistance.

For $E_b = 0$, the expression for the armature current reduces to $I_a = \frac{V}{R_a}$. For a 250 V, 5 HP, 25 A, DC motor, $V_t = 250$ V, $R_a = 0.5\,\Omega$ and $I_a = \frac{250}{0.5} = 500$ A, which is 20 times the rated full-load current. For the full voltage applied to the armature, the armature current becomes several times of its rated value. If this large current is allowed to flow through the armature, it will severely damage the motor. Hence, this high value of armature current should be prohibited for which a high value of external resistance must be connected to the armature in series only at the time of starting for 5 to 10 seconds. This limits the starting current to a safe value. This series resistance is known as the starting resistance, which is cut out gradually so that the motor achieves speed with simultaneous increase in the back emf. Once the back emf (E_b) reaches to the normal value, it regulates the motor armature current. Therefore, a variable resistance (R_s) must be included in series with the armature, shown in Figure 5.13, which is cut out manually or automatically, as the motor comes up to normal speed and full voltage is applied across the armature. This arrangement is known as starter. For the machine considered, if the current is limited to 25 A, the total resistance of the starter and armature should be 250/25 = 10 Ω. Therefore, the resistance of the starter alone must be 10 − 0.5 = 9.5 Ω.

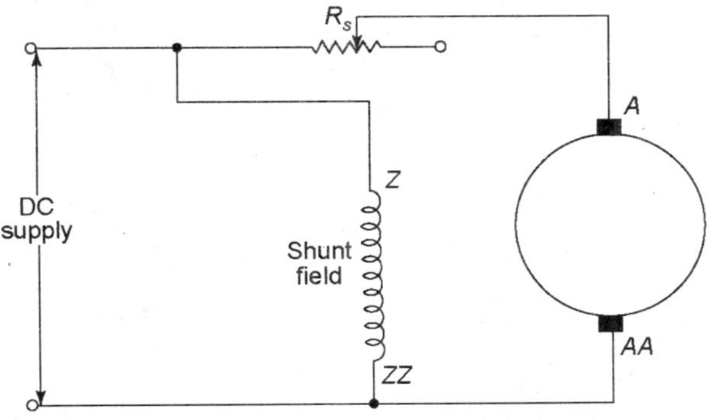

Figure 5.13 Principle of a Starter

5.16 MANUAL STARTER

In manual starters, the starting resistance is cut out manually until the motor speed increases, which is the face plate box starters and is used for starting shunt and compound motors of normal industrial capacity. These starters are classified as follows:

(a) three-point starter and

(b) four-point starter.

5.16.1 Three-point Starter

Figure 5.14 shows the internal wiring diagram of a three-point starter having three terminals, which are to be connected to the motors. Figure 5.14 is basically the same as that shown in Figure 5.13. The starter terminals *L*, *A* and *F* are connected to supply, armature and field, respectively. In Figure 5.14, *SR* is the starting arm, OLR is the overload or overcurrent release and NVR is the no-volt coil that holds on to the starting arm in the full running position during the normal operation of the motor and releases it

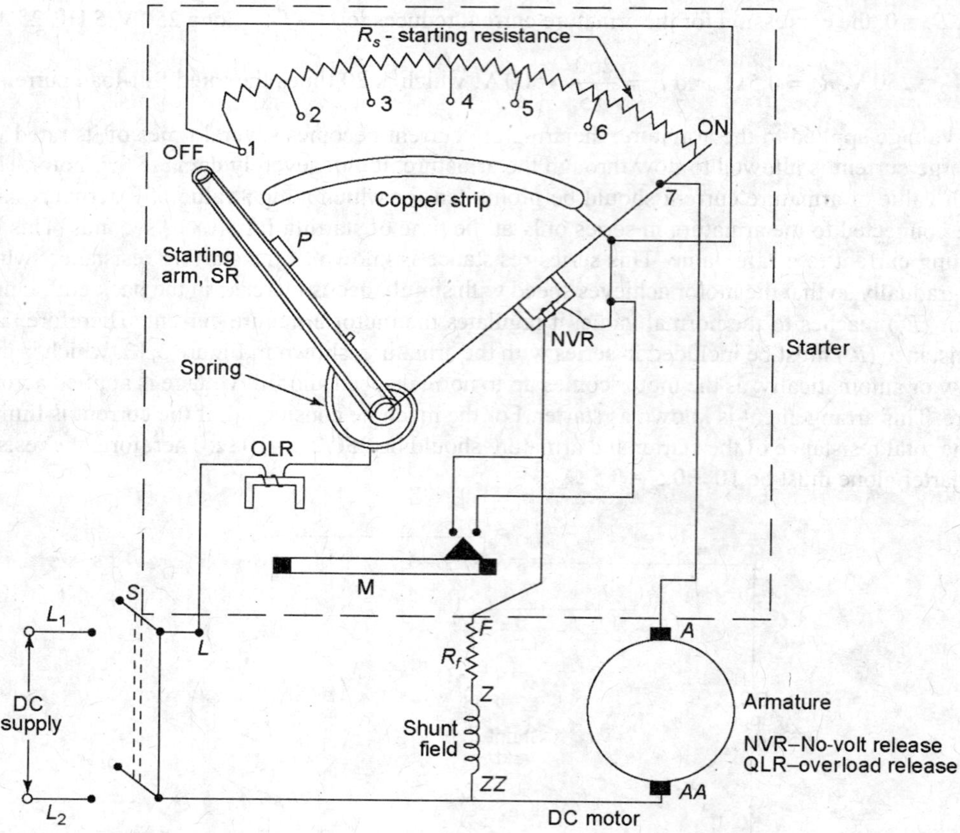

Figure 5.14 Three-point Starter

when the supply is off. The main switch S is closed at first to start the motor and the starting arm (SR) is moved to the right slowly to make contact with the series starting resistance R_s. When SR touches the stud number 1, the field circuit is directly connected to the mains supply and the full value of starting resistance R_s comes in series with the armature. This reduces the starting current. As the arm SR further moves to the right, the resistance R_s is gradually cut out. The motor attains the rated speed when the arm SR is in the ON position and the starting resistance R_s is completely cut out and the armature current totally depends on the armature resistance now. The spring is provided in the starter to restore the arm SR to OFF position. When the supply fails, it keeps the arm SR at the OFF position. An electromagnet NVR energized by the field current is used to keep the arm SR in the ON position against the force of the spring. This electromagnet NVR is also known as the hold-on coil or no-voltage release. During normal operating condition, this hold on coil will hold the arm SR so that the motor runs at the normal speed. When the failure of power or disconnection of supply voltage or break in field circuit occurs, this electromagnet gets energized and releases the arm. This arm will be pulled back to the OFF position by the force of the spring, which prevents the stationary armature being connected to the supply voltage even after restoration of the supply voltage. If the field circuit becomes open, the arm SR immediately comes back to the OFF position and this prevents the motor from running away, which is great advantage of connecting the arm SR in series with field coil. During the ON position of the arm SR, the whole of the starting resistance R_s comes in series with the field winding resistance. Although the field winding resistance is very high, usually 200–300 Ω, the inclusion of the starting resistance decreases the field current by 2 to 4 per cent. To avoid this even small amount of current, a copper strip SR stud number 1 and the field current will now pass through this arc only and R_s will be bypassed. The overload release (OLR) consists an electromagnet connected to the supply line. The rod M gets lifted when the armature current exceeds a certain preset value and it short circuits the NVR. Hence, the arm SR will be released and it comes back to its original OFF position since the more accurate and reliable devices such as circuit breakers or magnetic conductors are used now instead of this OLR.

5.16.2 Four-point Starter

Figure 5.15 shows a four-point starter in which the resistance R_f is connected in series. If the value of R_f is increased, the field flux will decrease and hence the speed will increase $\left(N \propto \frac{1}{\phi}\right)$. If this resistance is very large, the field current will be very small. The hold on coil will not be able to hold the arm SR in ON position against the pull of the spring and hence the motor will be switched off from the supply, which is the main disadvantage of the three-point starter. Hence, the three-point starter cannot be used with the variable speed motors. For such applications, four-point starters are used.

A four-point starter is similar to a three-point starter. The only difference is that the hold-on coil is taken out of the field circuit and connected directly across the mains supply voltage with the help of a protective resistance R_p. The function of a four-point starter is similar to a three-point starter. When the arm SR is in contact number 1, the line current will have the following three paths:

(i) through the starting resistance R_s,

(ii) through the shunt field winding and rheostat winding and

(iii) through the hold-on coil and protective resistance R_p.

The field current in this case will not flow through the hold-on coil unlike the three-point starter. The variation of resistance R_f to control the speed will not affect the electromagnetic pull of the hold-on coil in this case.

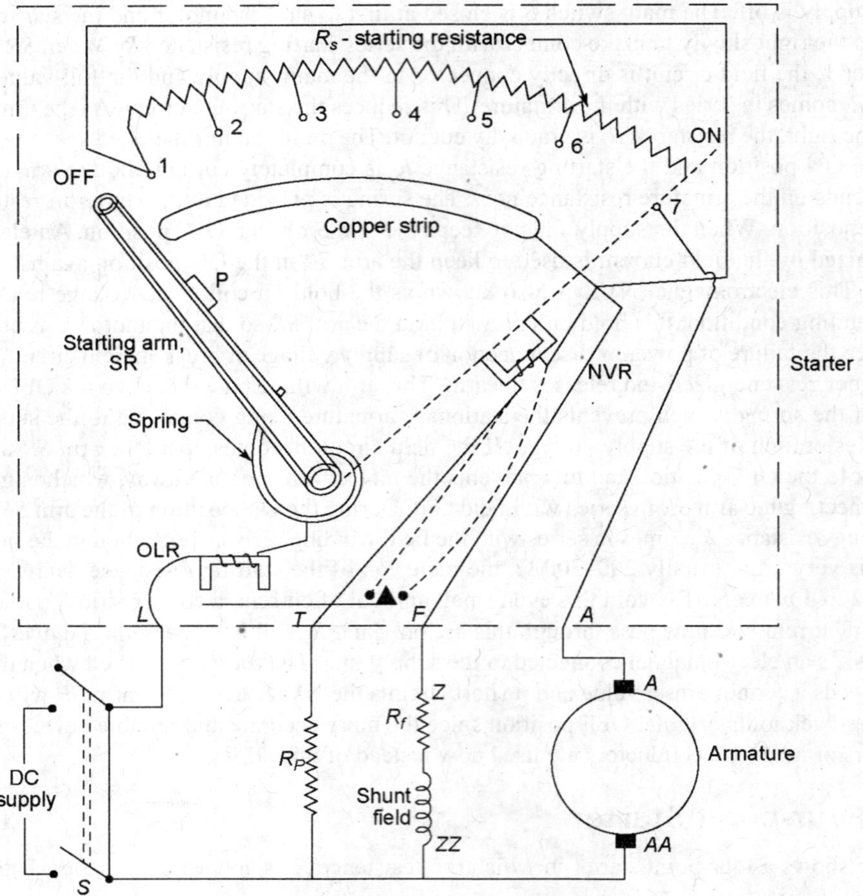

Figure 5.15 Four-point Starter

5.17 AUTOMATIC STARTERS

To start a motor at remote location or to start a large motor, it is desired that the starter to take it to its normal speed. The small motors which are not in remote locations are started by manual starters. Difficulties are encountered to start large motors. The skill with which the operator operates in manually operated starters depends on his/her experience. An overcautious operator may take too much times or a careless operator may exceed the maximum permissible currents. It is always preferable to use automatic starters to avoid these cases, and it will interrupt the conditions of the load and function accordingly. There are the following types of automatic starters:

(i) the time element starter,

(ii) the back emf type,

(iii) the shunt current-limit type and

(iv) the series current-limit type.

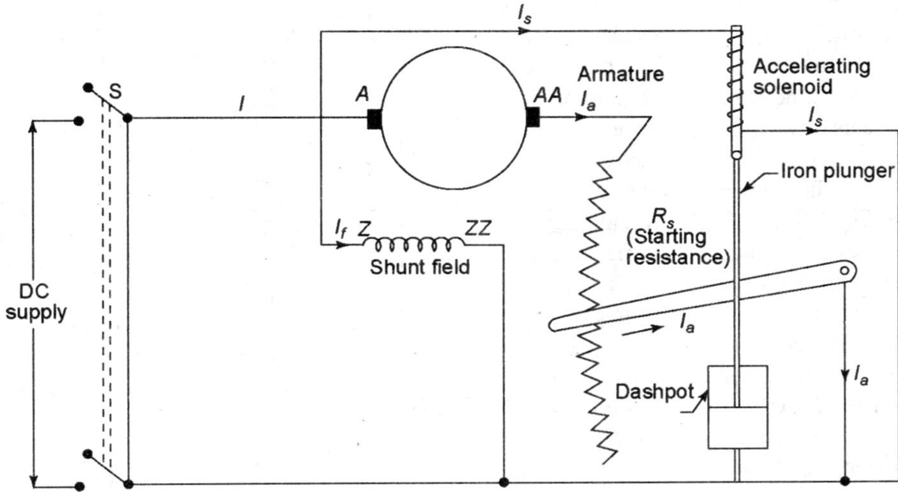

Figure 5.16 Three-element Automatic Starter

5.17.1 Time Element Starter

Figure 5.16 shows a three-element automatic starter in which the resistance in series with the armature is reduced by a certain amount during each succeeding interval regardless of load of the motor. The full value of starting current is connected with the armature initially to limit the current to a safe value. The full voltage appears across both the field winding and the accelerating solenoid. Due to the flow of current through the solenoid, it immediately begins to draw the iron core upwards and R_s is gradually cut out from the armature circuit. The dashpot has a hole through which the oil can flow known as orifice. More time will be consumed to cut out R_s if the size of orifice is smaller. To build up the speed and back emf and to maintain the starting current at the maximum safe value during the motor acceleration period, the proper size of the orifice is designed. When the motor achieves its rated speed, the starting resistance is completely cut out.

The following are the advantages of this starter:

(i) It is cheaper.

(ii) The wiring is simple.

(iii) The starting is generally guranteed.

The main disadvantage is that the load cannot be sensed; that is, the starting resistance is cut out whether the load is low or high.

5.17.2 Back emf Starter

Figure 5.17 shows a back emf starter and its operation depends on the back emf developed across the motor armature. A solenoid is connected directly across the armature and hence it functions as per the voltage of armature that is slightly higher than the back emf. The variation of back emf and the speed of the motor are shown in Figure 5.18.

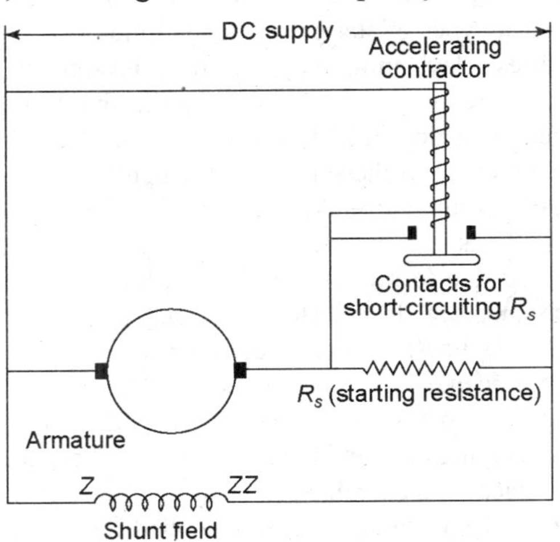

Figure 5.17 Back emf Starter

At the time of starting, the armature current that flows through the starting resistance reduces the voltage to be applied to the armature. Therefore, the accelerating conductor will not immediately close its contacts due to low voltage across its operating coil. As the motor picks up speed, the armature current and the voltage drop across R_s reduce due to the rising back emf. The voltage across the armature increases at the same time when voltage increases to a given value, say at time t_1. The accelerating contactor closes at this time and R_s gets short circuited. Hence, full voltage is applied across the armature, therefore resulting increase in starting current. This causes the motor speed to increase continuously till it reaches to its rated speed. The armature current again reduces to its normal value. In Figure 5.17, only one resistor and a accelerating contactor have been shown, but in actual practice the starters for large motors have several contactors and large numbers of steps in starting resistance.

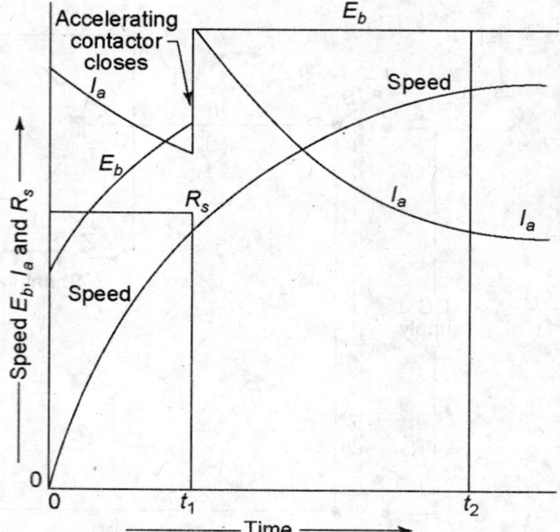

Figure 5.18 Variation of Back emf and Speed of the Motor with Time

Figure 5.18 shows the variation of back emf and speed of the motor with time. The motor speed increases slowly for heavy load, and the rise in voltage on the accelerating contactor operating coil is delayed so that R_s is not cut out until the rise in speed and back emf permits the contactor to loose.

It is possible to anticipate the load and the starting resistance can be cut out accordingly. This type of starters has low cost and simple wiring. The speed of the motor becomes erratic if the line voltage varies.

5.17.3 Shunt Current-limit Starter

Figure 5.19 shows the wiring diagram of the shunt current-limit starter, which uses accelerating contactors with shunt-type operating coils. The full voltage is applied across the shunt-type operating coils. These types of starters can overcome the disadvantages of the back emf starter. There are interlocking series relays and the operation of these relays can be overcome by the starting current. As soon as the series relay closes the contacts, the shunt operating coils of the accelerating coils are energized. This cuts a portion of R_s from the armature circuit. Since the contactors are in shunt with the armature and the closure of the series relay is determined by the magnitude of the armature current, these relays are called shunt current-limit starter.

This starter operates in conjunction with a push button starter. The no-voltage release coil is energized due to pressing the ON button momentarily and the holding relay contacts as well as the line contacts are closed. The series relay contact is released mechanically and hence its contacts would close. Due to the flow of armature current through the series coil, its core will be held up and its contacts are kept open.

The back emf increases due to the increase in the speed and hence the armature current reduces weakening the flux of the series relay. Therefore, the relay contacts get closed. The coil of the accelerating contacts gets energized and hence its contacts get closed. This short circuits R_s and series relay. Therefore, the full-line voltage is safely applied across the armature. As long as the motor is connected to the supply line, the accelerating contractor contacts remain closed. To interrupt the current through the no-voltage

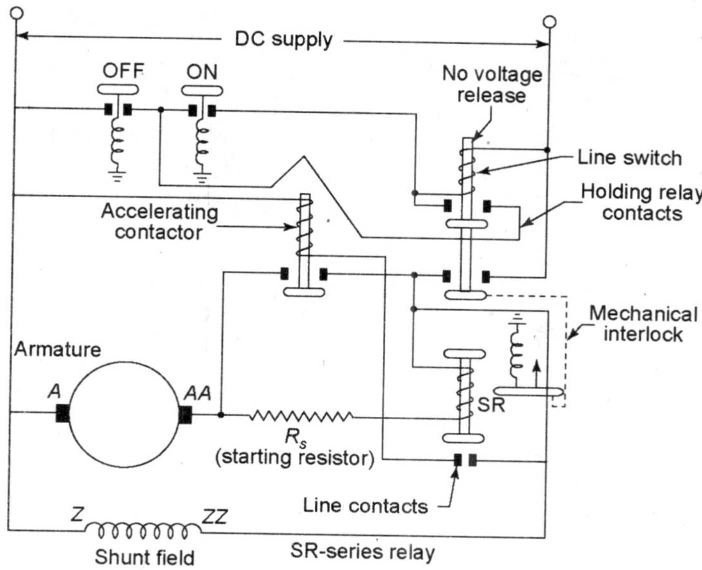

Figure 5.19 Shunt Current-limit Starter

release coil, the OFF button is to be pushed. This also opens the holding relay contacts, the line supply contacts as well as the series relay contacts. Finally, the motor will be disconnected from the supply mains. Although one starting resistor is shown in Figure 5.19 to simplify the drawing, but in practice two or more resistors are used. Using additional starting resistor and interlocking series relay, the starting resistance is cut out in steps as the motor builds up its speed. In spite of the fluctuations of the supply voltage, the armature current is reduced to a safe value because the starting resistance of the armature circuit is cut out only after the speed had built up. For a heavy load the motor cannot accelerate immediately. The back emf also cannot build up sufficiently to reduce the current through the series relay to make its contacts get closed and hence R_s remains open in the circuit. This protects the motor against the overload. The main disadvantage is that the accelerating unit is very complicated and expensive.

5.17.4 Series Current-limit Starter

Figure 5.20 shows the series current-limit starter that uses the less complicated equipment and performs the same job as done by shunt current-limit starter without using interlocking series relay. The accelerating contacts are designed in such a way that they open immediately after the starting current starts rising. The contactor closes as soon as the current falls to a predetermined value and the starting resistance is cut out. From Figure 5.20, the relay coil that operates the accelerating contractor is connected in series with the armature.

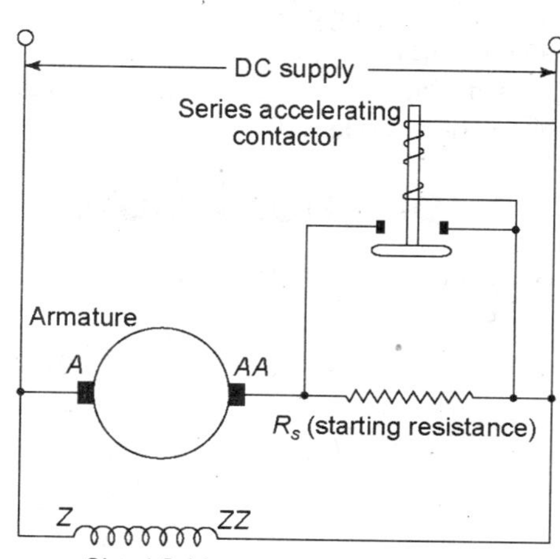

Figure 5.20 Series Current-limit Starter

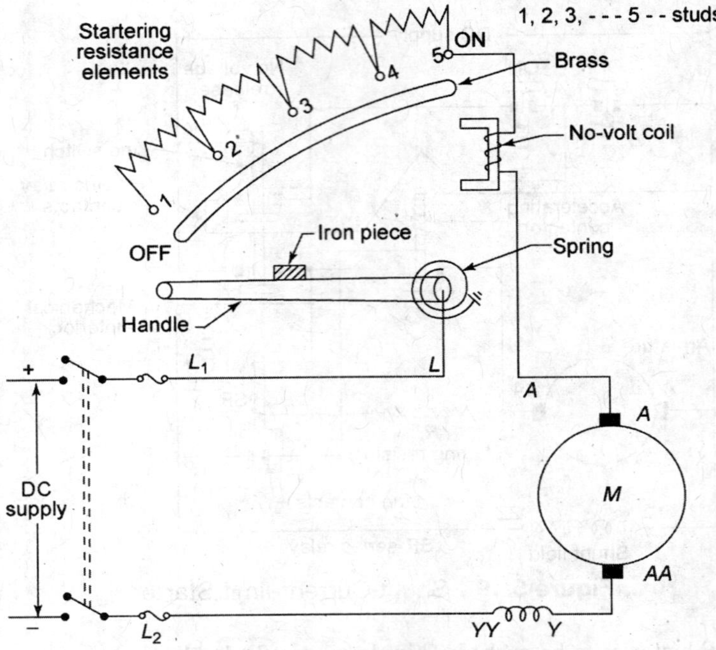

Figure 5.21 No-load Release-type Series Motor Starter

5.18 STARTERS FOR DC SERIES MOTORS

Figure 5.21 shows no-load release type of starter for the series motor in which series resistance is inserted in the armature at the time of starting. The starting resistance is gradually cut out from the circuit. The no-load type of starter is preferable because the speed increases dangerously to high value due to a low value of load drops. The strength of the no-voltage coil decreases due to decrease in the load current. Hence, the handle flies back to the OFF position due to the spring force. Due to failure of supply during the operation of motor, the no-volt coil gets demagnetized and hence the handle from the ON position is released.

5.19 DC SHUNT MOTOR STARTER DESIGN

A shunt motor with n resistance elements, $(n + 1)$ steps or $(n + 1)$ studs has been shown in Figure 5.22.

Let

$$R_1 = r_1 + r_2 + r_3 + r_4 + \cdots + r_n + R_a$$
$$R_2 = r_2 + r_3 + r_4 + r_5 + \cdots + r_n + R_a$$
$$\cdots$$
$$\cdots$$
$$\cdots$$
$$R_{n+1} = R_a$$

where R_a is the armature resistance including brushes.

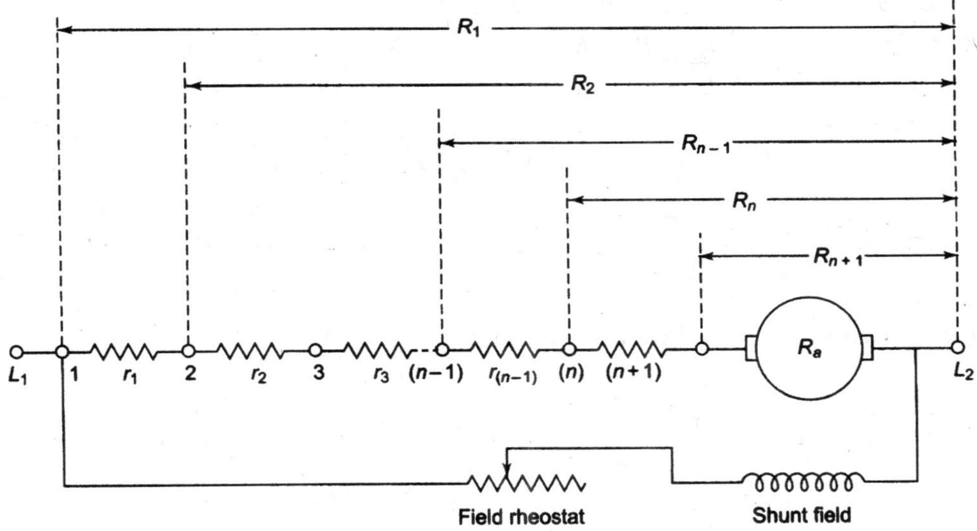

Figure 5.22 Starter Resistance for DC Shunt Motor

At the instant when the motor is switched on, the total armature circuit resistance, R_1, should be equal to

$$R_1 = \frac{\text{terminal voltage}}{\text{maximum permissible armature current}}$$

$$= \frac{V_1}{I_{a1}} = r_1 + r_2 + r_3 + \cdots + r_n + R_a$$

With the handle on stud 1, the motor accelerates, counter emf develops and, as a result, the armature current starts decreasing from I_{a1}. When the current has dropped to minimum current I_{a2}, then the counter emf with the handle on stud 1 is given by

$$E_{a1} = V_t - I_{a2} R_1$$

i.e.,

$$R_1 = \frac{V_t - E_{a1}}{I_{a2}} \tag{5.40}$$

At stud 1, as soon as the current drops to I_{a2}, the resistance r_1 is cut out by moving the handle to stud 2. During the notching up process from stud 1 to stud 2, the speed and, therefore, E_{a1} do not change. At stud 2, just after r_1 is cut out, the current shoots up again to I_{a1}.

∴

$$R_2 = \frac{V_1 - E_{a1}}{I_{a1}} = r_2 + r_3 + r_4 + \cdots + r_n + R_a \tag{5.41}$$

From Equations (5.40) and (5.41),

$$\frac{R_2}{R_1} = \frac{I_{a2}}{I_{a1}} = \frac{\text{minimum armature current}}{\text{maximum armature current}}$$

Following the above procedure for other study, we have

$$\frac{I_{a2}}{I_{a1}}=\frac{R_2}{R_1}=\frac{R_3}{R_2}=\cdots=\frac{R_n}{R_{n-1}}=\frac{R_{n+1}}{R_n}=\frac{R_a}{R_n}=\alpha \tag{5.42}$$

From Equation (5.42), we have

$$\alpha^n=\frac{R_2}{R_1}\times\frac{R_3}{R_2}\times\frac{R_4}{R_3}\times\cdots\times\frac{R_n}{R_{1n-1}}\times\frac{R_{n+1}}{R_n}=\frac{R_{n+1}}{R_1}=\frac{R_a}{R_1}$$

i.e.,

$$\alpha=\frac{I_{a2}}{I_{a1}}=\left(\frac{R_a}{R_1}\right)^{\frac{1}{n}} \tag{5.43}$$

Continuing the analysis further, it can be proved that

$$\alpha=\left(\frac{I_{a1}R_a}{V_t}\right)^{\frac{1}{n}} \tag{5.44}$$

And

$$\alpha=\left(\frac{I_{a2}R_a}{V_t}\right)^{\frac{1}{n+1}} \tag{5.45}$$

Example 5.16 A 220 V, 1,500 rpm, DC motor has a full-load armature current of 35 A. It is proposed to design a starter, which restricts the maximum armature current during start to 70 A. For design purposes, the minimum current during starting is to be restricted to 35 A, and the series resistance in the armature circuit is cut out in steps when the current falls to 35 A. Calculate the maximum series resistance used in the starter and the amount of resistance cut out during each of the first two steps. The armature resistance of the DC motor is 0.45 Ω.

Solution

$$I_{a1}=70\text{ A}, I_{a2}=35\text{ A}, V_t=220\text{ V and } R_a=0.45\ \Omega.$$

Total armature resistance $(R_1)=\frac{V_t}{I_{a1}}=\frac{220}{70}=3.143\Omega$

The external resistance to be used at start = 3.143 − 0.45 = 2.693 Ω

$$\alpha=\frac{I_{a2}}{I_{a1}}=\frac{35}{70}=0.5$$

$$R_2=\alpha R_1=0.5\times3.143=1.5715\ \Omega$$

$$R_3=\alpha R_2=0.5\times1.5715=0.78575\ \Omega$$

The resistance cut out in the first step = R_1-R_2 = 3.143 − 1.5715 = 1.5715 Ω

The resistance cut out in the second step = R_2-R_3 = 1.5715 − 0.78575 = 0.78575 Ω ■

5.20 ELECTRIC BRAKING

Using friction or electric braking, it is possible to bring a motor and its load to rest quickly. In friction braking, it is difficult to get smooth stop. In electric braking it is possible to stop the motor smoothly. There are three kinds of electric braking: (i) rheostatic or dynamic, (ii) plugging and (iii) regenerative braking. Friction braking is required for holding the motor even after it has been brought to rest.

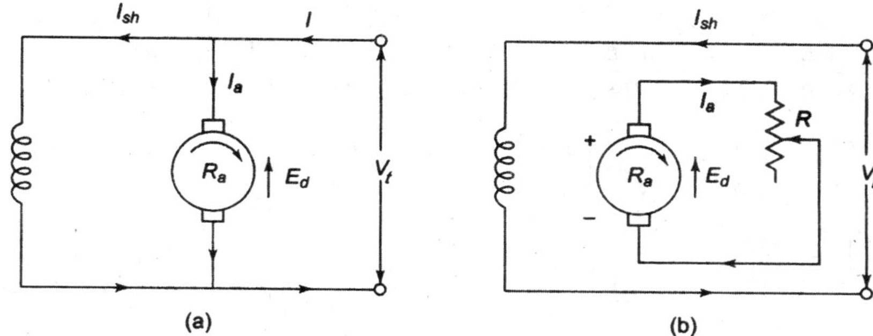

Figure 5.23 Rheostatic or Dynamic Braking

5.21 ELECTRIC BRAKING OF SHUNT MOTORS

Shunt motors are brought to rest using electric braking. The three types of braking – rheostatic or dynamic, plugging and regenerative – are discussed below.

5.21.1 Rheostatic or Dynamic Braking

Figure 5.23(a) shows the rheostatic braking in which the armature of the shunt motor is disconnected from the supply and it is connected across a variable resistance R. The field winding is kept undisturbed, and this braking is controlled by varying the series resistance R. This method uses the generator action in a motor to bring it to rest.

From Figure 5.23(b), the armature current is given by

$$I_a = \frac{E_b}{R+R_a} = \frac{\phi n Z\left(\frac{P}{A}\right)}{R+R_a} = \frac{k\phi\, n}{R+R_a}$$

The braking torque is given by

$$T_B = \frac{1}{2\pi}\phi Z I_a\left(\frac{P}{A}\right) = \frac{1}{2\pi}\phi Z\left(\frac{P}{A}\right)\times\frac{\phi n Z\left(\frac{P}{A}\right)}{R+R_a} = \frac{1}{2\pi}\left(\frac{ZP}{A}\right)^2\left(\frac{\phi^2 n}{R+R_a}\right) = k'\phi^2 n$$

$$T_B \propto n \tag{5.46}$$

Therefore, T_B decreases and the motor slows down. When the motor comes to stop, T_B becomes zero.

5.21.2 Plugging or Reverse Breaking

Figure 5.24 shows plugging or reverse braking. In this method, the armature terminals are reversed to rotate the motor in the reverse direction, and the applied voltage V_t and the back emf E_b start acting in the same direction. To limit the armature current, a resistance is inserted in series with the armature during reversing the armature.

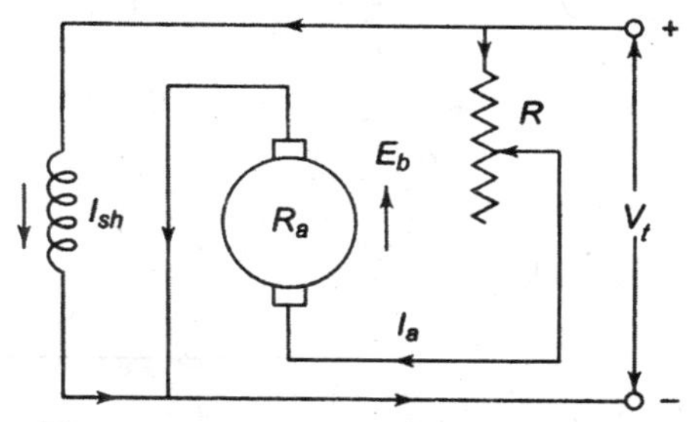

Figure 5.24 Plugging or Reverse Braking

$$I_a = \frac{V_t + E_b}{R + R_a} = \frac{V_t}{R + R_a} + \frac{E_b}{R + R_a} = \frac{V_t}{R + R_a} + \frac{\phi n Z\left(\frac{P}{A}\right)}{R + R_a} = \frac{V_t}{R + R_a} + \frac{k\phi\, n}{R + R_a}$$

$$T_B = \frac{1}{2\pi}\phi Z I_a\left(\frac{P}{A}\right) = \frac{1}{2\pi}\phi Z\left(\frac{P}{A}\right)\times\left(\frac{V_t}{R + R_a} + \frac{kn\phi}{R + R_a}\right), \text{ where } k = Z\left(\frac{P}{A}\right)$$

$$= \frac{1}{2\pi}\times\left(\frac{ZP}{A}\right)\times\frac{V_t}{(R + R_a)}\times\phi + \frac{1}{2\pi}\times\left(\frac{ZP}{A}\right)^2\times\frac{\phi^2 n}{(R + R_a)} = k_1\phi + k_2\phi^2 n = k_4 + k_5 n \qquad (5.47)$$

Compared to rheostatic braking, this method gives greater braking torque. In this braking, power is drawn from the supply and it is dissipated by R in the form of heat. In spite of zero motor speed, there is still braking torque.

5.21.3 Regenerative Braking

In regenerative braking, E_b is greater than V_t. Figure 5.25 shows regenerative braking. The direction of I_a and the armature torque T_B is reversed. Regenerative braking is used for downgrade motion of an electric train.

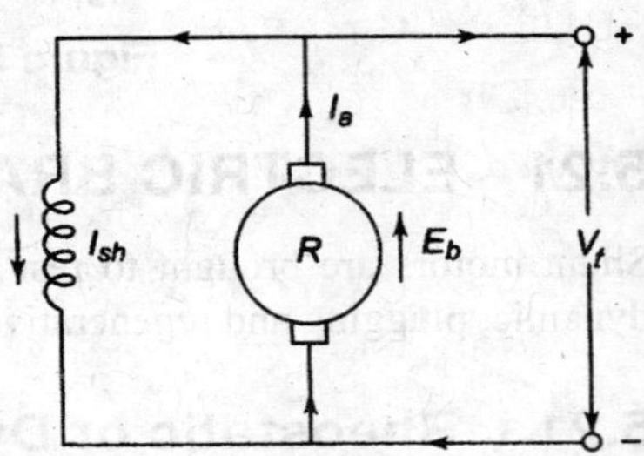

Figure 5.25 Regenerative Braking

5.22 ELECTRIC BRAKING OF SERIES MOTOR

5.22.1 Rheostatic Braking

Figure 5.26 shows the rheostatic braking of a DC series motor. In this method, the motor is disconnected from supply. The field connection is reversed and the motor is connected through a variable resistance R.

5.22.2 Plugging or Reverse Current Braking

Figure 5.27 shows plugging or reverse current braking and it is similar to that of the shunt motor.

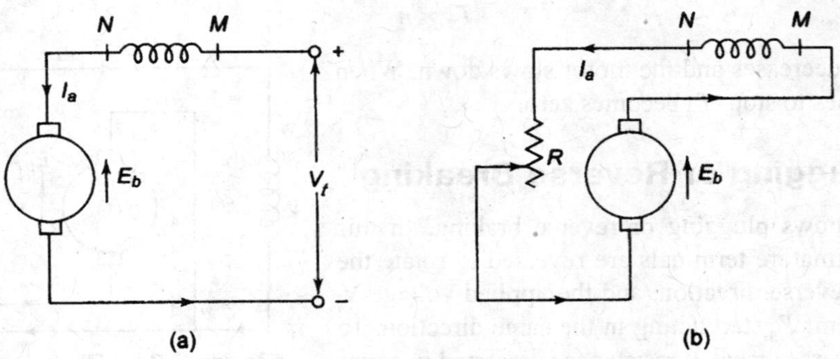

Figure 5.26 Rheostatic Braking

5.22.3 Regenerative Braking

This type of braking is impossible in series motors. To achieve it, modifications are required because reversal of current would reverse the field current and hence E_b.

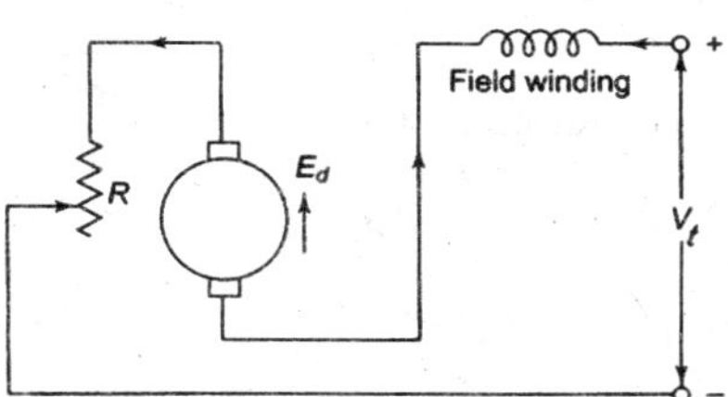

Figure 5.27 Plugging or Reverse Current Braking

Example 5.17 If A is the number of parallel paths, P is the number of poles, J is the moment of inertia of the moving parts (kg m^2), ϕ is the flux in Wb, Z is the number of conductors and I_a is the accelerating current which is assumed to be constant of a DC motor, prove that the time taken by the motor to accelerate to a speed of ω rad/s is $\dfrac{2\pi J\omega A}{\phi ZPI_a}$.

Solution

Energy in moving parts $(E) = \dfrac{J\omega^2}{2}$ joules

$$\text{Power }(P) = \frac{dE}{dt} = J\omega\frac{d\omega}{dt} = \frac{J\times 2\pi\times N}{60}\times\frac{2\pi}{60}\times\frac{dN}{dt} = J\left(\frac{2\pi}{60}\right)^2 N\frac{dN}{dt}$$

$$\omega = \frac{2\pi N}{60}$$

Angular acceleration $(\alpha) = \dfrac{d\omega}{dt} = \dfrac{E_b I_a}{J\omega}$

t = time taken to allow 'ω' radians $= \dfrac{\omega}{\alpha} = \dfrac{\omega}{\dfrac{E_b I_a}{J\omega}} = \dfrac{J\omega^2}{E_b I_a}$

Back emf $(E_b) = \dfrac{\phi NZ}{60}\times\dfrac{P}{A}$

$$\therefore \quad t = \frac{J\omega^2}{E_b I_a} = \frac{J\omega^2}{\left(\dfrac{\phi NZ}{60}\times\dfrac{P}{A}\right)I_a} = \frac{2\pi J\omega A}{\phi ZPI_a}$$

Example 5.18 A 500 V, 52 HP (metric), shunt motor is running at full-load at 550 rpm and takes 82 A. Calculate the time to attain full-load speed when started from rest against full-load torque. The starting current fluctuates between 1.1 and 1.3 times the normal rated current. The efficiency constant at full-load value during starting period is assumed to be constant. Take the moment of inertia of the system 22 kg m^2.

Solution

Full-load current = 82 A

Starting current $= \left(\dfrac{1.1+1.3}{2}\right)\times$ full-load current $= 1.2I_{fl}$

The current causing acceleration $(I_a) = 0.2 \times 82 = 16.4$ A

$$E_bI_a = 500 \times 16.4 = 8{,}200 \text{ W}$$

Accelerating power $(P_a) = J\omega\dfrac{d\omega}{dt}$

i.e.,
$$\frac{d\omega}{dt} = \frac{P_a}{J\omega}$$

$$\omega = \frac{2\pi N}{60} = \frac{2\pi \times 550}{60} = 57.56 \text{ rad/s}$$

∴ $$\text{Angular acceleration} = \frac{d\omega}{dt} = \frac{P_a}{J\omega} = \frac{8{,}200}{22 \times 57.56} = 6.475 \text{ rad/s}^2$$

∴ $$t = \text{time taken to attain the final speed} = \frac{\omega}{\alpha} = \frac{57.56}{6.475} = 8.89 \text{ s}$$ ■

Example 5.19 Calculate the required time for a shunt motor to fall in speed from 700 rpm to 250 rpm, keeping the excitation constant. The armature is connected across a resistance of 12 Ω. The induced emf at 750 rpm is 210 V. The moment of inertia of the armature is 24.2 kg-m². Assume that (i) the motor has negligible loss and (ii) the frictional torque is 12.52 N-m.

Solution

Back emf (E_b) at 700 rpm = 210 V

Current in the load resistance $= \dfrac{210}{12} = 17.5$ A

Power wasted during running down = 210 × 17.5 = 3,675 W.

Case I: *Negligible motor loss*

Energy stored in rotating p arts $(E) = \dfrac{1}{2}J\omega^2$

∴ $$\text{Power } (P) = \frac{dE}{dt} = J\omega\frac{d\omega}{dt} = E_bI_a$$

∴ $$dt = \frac{J\omega d\omega}{E_bI_a}$$

Since $\omega = \dfrac{2\pi N}{60}$, we have

$$\frac{d\omega}{dt} = \left(\frac{2\pi}{60}\right)\frac{dN}{dt}$$

where N is the speed of the motor at any time.

E_b = back emf at speed $N = \left(\dfrac{210}{700}\right)N$

∴ $$I_a = \frac{210N}{700 \times 12}$$

$$\text{Power} = E_b I_a = \frac{210N}{700} \times \frac{210N}{700 \times 12} = \left(\frac{210N}{700}\right)^2 \times \frac{1}{12}$$

$$\therefore \qquad dt = \frac{J\omega d\omega}{E_b I_a} = \frac{J \times \frac{2\pi N}{60} \times \frac{2\pi}{60} dN}{\left(\frac{210N}{700}\right)^2 \times \frac{1}{12}}$$

Since $E_b I_a$ is negative, we have

$$dt = -\frac{J \times \frac{2\pi N}{60} \times \frac{2\pi}{60} dN}{\left(\frac{210N}{700}\right)^2 \times \frac{1}{12}} = -\frac{24.2 \times \frac{2\pi N}{60} \times \frac{2\pi}{60} dN}{\left(\frac{210N}{700}\right)^2 \times \frac{1}{12}} = -34.35\frac{dN}{N}$$

i.e.,
$$\int_{t_2}^{t_1} dt = -34.35 \int_{700}^{250} \frac{dN}{N}$$

i.e.,
$$t_1 - t_2 = -34.35\left[ln\ N\right]_{700}^{250} = -34.35\ ln\frac{250}{700} = 35.36\text{ s}$$

Case II: *Frictional torque = 12.52 N-m*

Wastage of power due to friction at any speed (N) is given by

$$P_f = \frac{2\pi N}{60} \times T_f$$

where T_f is the torque due to friction.

Wastage of power in retardation = power wasted in load + power wasted in friction

$$= E_b I_a + P_f$$

$$\therefore \qquad -J\omega\frac{d\omega}{dt} = \left(\frac{210}{700}N\right)^2 \times \frac{1}{12} + \frac{2\pi N}{60} \times 12.52$$

i.e.,
$$-J\left(\frac{2\pi}{60}\right)^2 N\frac{dN}{dt} = \left(\frac{210}{700}N\right)^2 \times \frac{1}{12} + \frac{2\pi N}{60} \times 12.52$$

i.e.,
$$-J\left(\frac{2\pi}{60}\right)^2 \frac{dN}{dt} = \left(\frac{210}{700}\right)^2 \times \frac{N}{12} + \frac{2\pi}{60} \times 12.52$$

i.e.,
$$-x\frac{dN}{dt} = yN + z$$

where
$$x = J\left(\frac{2\pi}{60}\right)^2 = 24.2 \times \left(\frac{2\pi}{60}\right)^2,\ y = \left(\frac{210}{700}\right)^2 \times \frac{1}{12} \text{ and } z = \frac{2\pi}{60} \times 12.52$$

$$\therefore \qquad \int_{t_2}^{t_1} dt = -\int_{700}^{210} \frac{x dN}{yN + z}$$

i.e., $$t_1 - t_2 = \frac{x}{y} ln\left[\frac{700y+z}{250y+z}\right]$$

i.e., $$t_1 - t_2 = \frac{24.2\times\left(\frac{2\pi}{60}\right)^2}{\left(\frac{210}{700}\right)^2\times\frac{1}{12}} ln\left[\frac{700\times\left(\frac{210}{700}\right)^2\times\frac{1}{12}+\frac{2\pi}{60}\times 12.52}{250\times\left(\frac{210}{700}\right)^2\times\frac{1}{12}+\frac{2\pi}{60}\times 12.52}\right] = 25.54 \text{ s}$$ ■

5.23 TESTING OF DC MACHINES

The tests that are performed on DC machines are as follows:

- open-circuit characteristic,
- load characteristics,
- determination of efficiency curve and
- temperature rise test

The first two have already been discussed. There are three methods to determine the efficiency of DC motors:

- *Direct method:* In this method, the generator or motor is put at full-load and the total power developed by it is wasted. An example of direct test is break test. Direct test is generally used for small machines. Although this method is very simple, it is very difficult to measure the mechanical power input in case of the generator and output in case of the motor. The brake must be tight before starting a series motor if brake is applied in the series motor. If this is not done, the armature may get damaged and fly wheel broken to pieces.
- *Indirect method:* Measurement of losses and calculation of efficiency are carried out in this method. Swinburne test is the simplest example of the indirect method. No loading is required for calculation of losses. There is no difficulty in applying this method to large machines. The disadvantage of this method is that the machine is run light during the test and so it does not give any indication of temperature rise at load or of commutating qualities of the machine.
- *Regenerative method:* In this method, two identical machines are required. One of them will act as a motor to drive the other as a generator, and these are mechanically coupled. The machine working as a generator feeds back power to the motor. The total power drawn from the supply is only used for supplying the internal losses of the two machines. It is possible to test very large machines as the power required is very small. Hopkinson test is a regenerative test for determining efficiency of a DC machine.

5.24 BRAKE TEST

Figure 5.28 shows the brake test of a DC motor. This test is used to determine the efficiency of comparatively small motors. The motor is loaded directly by means of a mechanical brake or by means of an eddy current or a calibrated air fan.

Figure 5.21 shows a common type of mechanical brake used in testing – the rope brake. One end of the rope is fixed to a spring balance which is further attached to a fixed hook, whereas the other end is attached to a hook or pan to which load is added.

The load on the motor is adjusted to carry the full-load current after its starting.

Let W be the suspended weight (kg-wt), S be the spring balance reading (kg-wt), R be the radius of pulley (m) and N be the speed (rpm).

Therefore, torque $(T) = (W - S) \times 9.81$ N-m

$$\text{Motor output} = \frac{2\pi NT}{60}\ \text{W} = \frac{2\pi NT}{60 \times 746}\ \text{HP (British), or}\ \frac{2\pi \text{NT}}{60 \times 735.5}\ \text{HP (metric)}$$

$$\text{Motor input} = VI \quad \text{W}$$

Therefore, $$\text{efficiency} = \frac{\text{motor ouput}}{\text{motor input}}$$

This method has the following disadvantages:

- It is not possible to measure the output power directly.
- It is not possible to use this method for determining internal losses and efficiency of large motors.

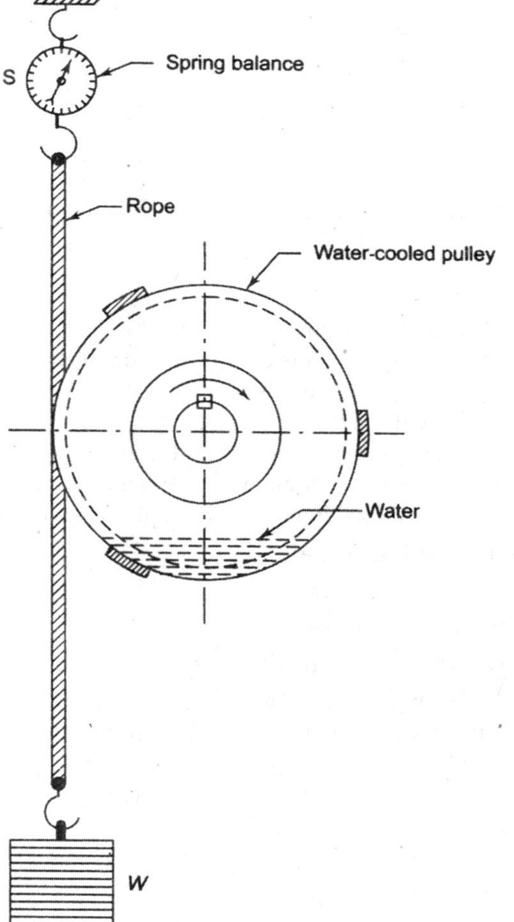

Figure 5.28 Brake Test

Example 5.20 A 400 V DC shunt motor gave the following readings during brake test:
supply voltage = 400 V, armature current = 55 A, shunt field current = 4 A, length of the brake arm = 100 cm, reading of spring balance at the end of brake arm = 30 kg-wt and speed = 700 rpm. Determine (i) the torque developed, (ii) British Horse Power (BHP) and (iii) efficiency.

Solution

(i) Torque developed (T) = spring balance reading × length of arm = 30 × 9.81 × 1 = 294.3 N-m

(ii) $$\text{BHP} = \frac{2\pi NT}{60 \times 746} = \frac{2 \times 3.14 \times 700 \times 294.3}{60 \times 746} = 28.90$$

(iii) $I = I_a + I_{sh} = 55 + 4 = 59$ A

Input = VI = 400 × 59 = 23,600 W

Output = BHP × 746 = 28.90× 746 = 21,559.4 W

$$\therefore \quad \text{Efficiency} = \frac{\text{output}}{\text{input}} = \frac{21{,}559.4}{23{,}600}$$

$$= 0.9135 = 91.35\%$$ ■

5.25 SWINBURNE'S TEST

Losses are measured separately in this method, and efficiency at any desired load is predetermined. The iron and friction losses can be determined by measuring the input to the machine at no-load. The machine is run as a motor at normal voltage and speed. The copper losses are calculated from measured

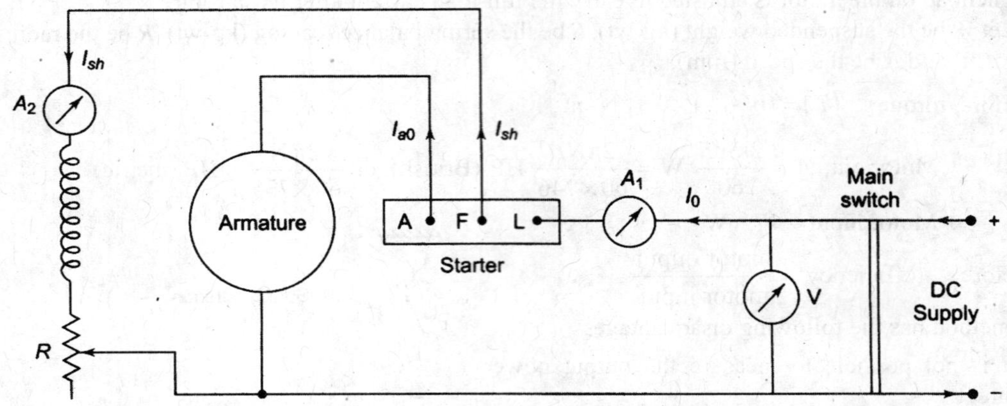

Figure 5.29 Swinburne's Test

values of the various resistances. This method is only applicable to level compounded and shunt motors. Figure 5.29 shows the circuit diagram of Swinburne's test.

Let V_t be the supply voltage, I_0 be the input armature current at no-load (measured by ammeter A_1) and I_{sh} be the shunt field current (measured by ammeter A_2) for Figure 5.29.

Therefore, no-load armature current $(I_{a0}) = I_0 - I_{sh}$, no-load input $= V_t I_0$ W, power input to the armature $= V_t (I_0 - I_{sh})$ W and power input to shunt $= V_t I_{sh}$ W.

The no-load input power supplies the following:

(i) iron loss in core,
(ii) friction losses,
(iii) windage losses and
(iv) armature copper losses.

During calculation of copper losses, the hot resistance of the armature must be used. For this, stationary measurement of armature circuit resistance at room temperature of, say, 15°C, is made by passing current through the armature from a low-voltage DC supply.

The hot resistance after allowing a temperature rise of 50°C is given by

$$R_{15} = R_0\left(1+15\alpha_0\right) \quad \text{and} \quad R_{65} = R_0\left(1+65\alpha_0\right)$$

$$\therefore \qquad \frac{R_{65}}{R_{15}} = \frac{1+65\alpha_0}{1+15\alpha_0}$$

i.e.,

$$R_{65} = R_{15} \times \left(\frac{1+65\alpha_0}{1+15\alpha_0}\right) \tag{4.49}$$

where α_0 is the temperature coefficient of resistance at 0°C.

Constant losses = no-load input − no-load armature copper loss

$$= V_t I_0 - \left(I_0 - I_{sh}\right)^2 R_{sh} - V_t I_{sh} \quad \text{for motor}$$

$$= E_g I_0 - \left(I_0 + I_{sh}\right)^2 R_{sh} - V_t I_{sh} \quad \text{for generator}$$

After getting the constant losses of a machine, its efficiency at any other load can be calculated as follows: $I_a = I - I_{sh}$, where I is the input current at which the efficiency is required for motoring mode; and $I_a = I + I_{sh}$, where I is the load current at which the efficiency is required for generating mode.

5.25.1 Motoring Mode

Motor input $= V_t I$, armature Cu loss $= I_a^2 R_a = \left(I - I_{sh}\right)^2 R_a$ and shunt Cu loss $= V_t I_{sh}$, where I is the input current and V_t is the supply voltage. Constant losses have been calculated above.

$$\text{Total losses} = \text{constant losses} + \left(I - I_{sh}\right)^2 R_a + V_t I_{sh}$$

$\therefore$

$$\text{Efficiency of the motor } (\eta_m) = \frac{\text{input} - \text{losses}}{\text{input}}$$

$$= \frac{V_t I - \left(I - I_{sh}\right)^2 R_{sh} - (\text{constant losses}) - V_t I_{sh}}{V_t I}$$

5.25.2 Generating Mode

Output $= V_t I$, armature Cu loss $= I_a^2 R_a = \left(I + I_{sh}\right)^2 R_a$ and shunt field Cu loss $= V_t I_{sh}$, where V_t is the output voltage and I is the load current.

$\therefore$

$$\text{Efficiency of the generator } (\eta_g) = \frac{\text{output}}{\text{output} + \text{losses}}$$

$$= \frac{V_t I}{V_t I + (I + I_{sh})^2 R_a + V_t I_{sh} + \text{constant losses}}$$

This method has the following advantages:

- This method is convenient and economical for testing of DC machines because power required is small.
- Since constant losses are known, efficiency can be predetermined at any load.

In spite of the above advantages, this method has the following disadvantages:

- The change in iron losses from no-load to full-load is not taken into account. At full-load, the flux is distorted due to armature reaction, which increases iron losses in some cases by 50 per cent.
- It is difficult to know whether commutation is satisfactory or not at full-load because the test is conducted at no-load.

Since series motors cannot be operated at no-load due to their high speed, this test is not applicable to series motors.

Example 5.21 A 400 V DC shunt machine takes 7 A at no-load when operates as a motor. The armature current and field resistances are 0.25 Ω and 200 Ω, respectively, when measured at room temperature. Calculate the efficiency of the machine:

(i) when runs as a motor taking a line current of 70 A at 400 V and

(ii) when runs as a generator delivering a line current of 70 A at 400 V.

The stray load loss is 1.1 per cent of the output. Neglect the increase in resistance due to temperature rise.

Solution

$$V_t = 400\text{ V},\ I_0 = 7\text{ A},\ R_a = 0.25\,\Omega,\ R_{sh} = 200\,\Omega$$

$$I_{sh} = \frac{V_t}{R_{sh}} = \frac{400}{200} = 2\text{ A}$$

$$I_{ao} = I_a - I_{sh} = 7 - 2 = 5\text{ A}$$

$$\text{No-load Cu loss} = 5^2 \times 0.25 + 400 \times 2 = 806.25\text{ W}$$

$$\text{No-load input} = 400 \times 7 = 2{,}800\text{ W}$$

∴ Mechanical and iron losses (i.e. constant losses) = 2,800 − 806.25 = 1,993.75 W

(i) *Efficiency when working as a motor*

$$\text{Input} = V_t I = 400 \times 70 = 28{,}000\text{ W}$$

$$\text{Shunt field Cu loss} = 400 \times 2 = 800\text{ W}$$

$$I_a = I - I_{sh} = 68\text{ A}$$

$$\text{Armature Cu loss} = 68^2 \times 0.25 = 1{,}156\text{ W}$$

$$\text{Total losses} = 800 + 1{,}156 + 1{,}993.75 = 3{,}949.75\text{ W}$$

$$\text{Output} = \text{input} - \text{total losses} = 24{,}050.25\text{ W}$$

$$\text{Stray load losses} = \text{output} \times \frac{1.1}{100} = 264.55\text{ W}$$

Therefore, net output = output − stray load losses = 24,050.25 − 264.55 = 23,785.7 W

$$\text{Efficiency} = \frac{\text{net output}}{\text{input}} = \frac{23{,}785.7}{28{,}000} = 0.8495\text{ p.u.} = 84.95\%$$

(ii) *Efficiency when working as a generator*

$$\text{Output} = 400 \times 70 = 28{,}000\text{ W}$$

$$I_a = I + I_{sh} = 70 + 2 = 72\text{ A}$$

$$\text{Total Cu losses} = 72^2 \times 0.25 + 400 \times 2 = 2{,}096\text{ W}$$

$$\text{Stray load losses} = \frac{1.1}{100} \times 28{,}000 = 308\text{ W}$$

$$\text{Input} = 28{,}000 + 2{,}096 + 1{,}993.75 + 308 = 32{,}397.75\text{ W}$$

$$\text{Efficiency} = \frac{\text{output}}{\text{input}} = \frac{28{,}000}{32{,}397.75} = 0.8642\text{ p.u.} = 86.42\%$$

■

5.26 HOPKINSON'S TEST (BACK-TO-BACK TEST)

In this test, full-load test can be carried out on two identical shunt machines without wasting their outputs. Figure 5.30 shows the connection diagram of this test on a pair of shunt machines.

Figure 5.30 shows two machines, which are mechanically coupled. Their fields are adjusted in such a way that one runs as a generator and the other as a motor. The power produced by the generator is utilized by the motor driving the generator. The power taken from the supply is used to supply losses only. Two identical machines of any size can be tested under full-load conditions. This method is very useful to determine the efficiency as well as a heat run test for the temperature rise. The following procedures are followed:

- The machine *M* in Figure 5.23 is started through a starter. Its field excitation is adjusted to take it to normal speed. The machine *M* drives the machine *G*. The switch *S* is kept open.
- The excitation of machine *G* is gradually increased so that the reading of the voltmeter (V_1) becomes zero. The switch *S* is closed and the machine *G* is at the floating stage. The shunt field regulators are adjusted to put any desired load.

The input power is used to meet losses only.

Let V_t be the supply voltage, I_1 be the output current of the generator, I_2 be the current taken from the supply, I_3 be the exciting current of the generator, I_4 be the exciting current of the motor and R_a be the armature resistance of each machine.

$$\text{Motor input} = V_t(I_1 + I_2)$$
$$\text{Generator output} = V_t I_1$$

If the efficiency of both the machines is η, we have

$$\text{Output of the motor} = \eta \times \text{input} = \eta\, V_t(I_1 + I_2) = \text{generator input}$$
$$\text{Output of the generator} = \eta \times \text{input} = \eta \times \eta\, V_t(I_1 + I_2) = \eta^2\, V_t(I_1 + I_2)$$

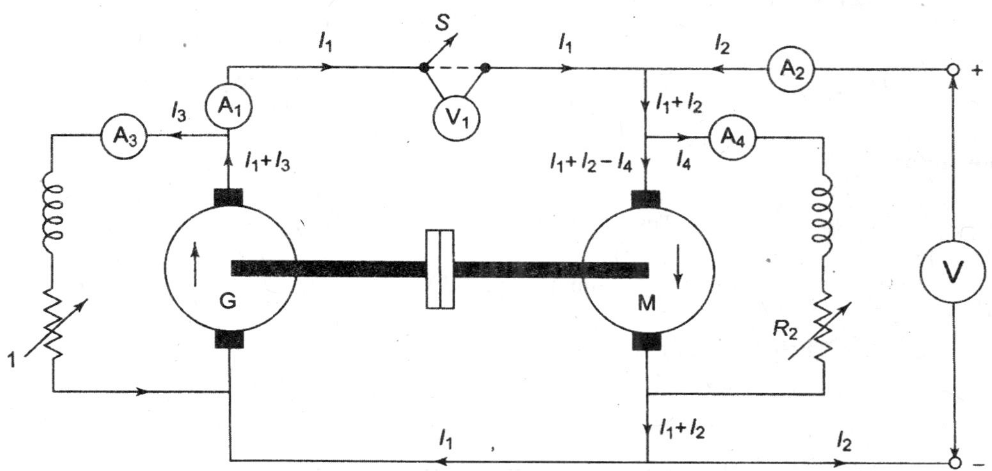

Figure 5.30 Hopkinson's Test

$$\eta^2 V_t (I_1 + I_2) = V_t I_1$$

$$\therefore \qquad \eta = \sqrt{\frac{I_1}{I_1 + I_2}}$$

In practice, it is very difficult to get two machines having identical efficiency. Hence, the following analysis is done taking different efficiency of the machines.

Armature Cu loss in generator $= (I_1 + I_3)^2 R_a$

Shunt Cu loss in generator $= V_t I_3$

Armature Cu loss in motor $= (I_1 + I_2 - I_4)^2 R_a$

Shunt Cu loss in motor $= V_t I_4$

The overall losses of the motor and generator are equal to the power supplied by the mains.

$$\text{Power taken from supply} = V_t I_2$$

Therefore, stray losses for the set

$$= V_t I_2 - \left[(I_1 + I_3)^2 R_a + (I_1 + I_2 - I_4)^2 R_a + V_t I_3 + V_t I_4 \right] = P_{\text{stray}}$$

If the stray losses are equally divided between the two machines, stray loss per machine $= \dfrac{P_{\text{stray}}}{2}$.

For the generator:

$$\text{Total losses} = (I_1 + I_3)^2 R_a + V_t I_3 + \frac{P_{\text{stray}}}{2}$$

$$\text{Output} = V_t I_1$$

$$\therefore \qquad \text{Efficiency of generator}\left(\eta_g\right) = \frac{\text{output}}{\text{output} + \text{losses}}$$

$$= \frac{V_t I_1}{V_t I_1 + (I_1 + I_2)^2 R_a + V_t I_3 + \dfrac{P_{\text{stray}}}{2}} \qquad (5.48)$$

For the motor:

$$\text{Total losses} = (I_1 + I_2 - I_4)^2 R_a + V_t I_4 + \frac{P_{\text{stray}}}{2}$$

$$\text{Input} = V_t(I_1 + I_2)$$

$$\text{Efficiency of motor}\left(\eta_m\right) = \frac{\text{input} - \text{losses}}{\text{input}}$$

$$= \frac{V_t (I_1 + I_2) - \left[(I_1 + I_2 - I_4)^2 R_a + V_t I_4 + \dfrac{P_{\text{stray}}}{2} \right]}{V_t (I_1 + I_2)} \qquad (5.49)$$

This test has the following advantages:

- This test is very economical since power required is small.
- Commutation and temperature rise of the machine can be studied properly.
- Since the test is conducted at full-load, any change in iron losses due to flux distortion at full-load is taken into account.

The main disadvantage of this test is that two identical machines cannot be obtained very easily.

Example 5.22 Hopkinson's test on two identical shunt machines gave the following readings: supply voltage = 240 V, field currents = 6 A (generator) and 5A (motor), line current = 40 A, armature current of motor = 240 A, armature resistance of each machine = 0.014 Ω, voltage drop/brush = 1 V.

Find the efficiency of each machine.

Solution

V_t = 240 V, I_{sh1} (field current of generator) = 6 A, I_{sh2} (field current of motor) = 5 A, supply current = 40 A, armature current of motor = 240 A, armature resistance of each machine = 0.014 Ω, voltage drop per brush = 1 V

From Figure E5.1, we have the following:

For motor:

$$\text{Motor input current} = 245 \text{ A}$$
$$\text{Armature Cu loss} = 240^2 \times 0.014 = 806.4 \text{ W}$$
$$\text{Shunt field Cu loss} = 240 \times 5 = 1{,}200\text{W}$$
$$\text{Brush contact loss} = 240 \times 1 \times 2 = 480 \text{ W}$$
$$\text{Total Cu loss} = 806.4 + 1{,}200 + 480 = 2{,}486.4 \text{ W}$$

For generator:

$$\text{Generator output current} = 205 \text{ A}$$
$$\text{Armature current of generator} = 211 \text{ A}$$
$$\text{Armature Cu loss} = 211^2 \times 0.014 = 623.3 \text{ W}$$
$$\text{Shunt field Cu loss} = 240 \times 6 = 1{,}440 \text{ W}$$
$$\text{Brush contact loss} = 211 \times 1 \times 2 = 422 \text{ W}$$
$$\text{Total Cu loss} = 623.3 + 1{,}440 + 422 = 2{,}485.3 \text{ W}$$

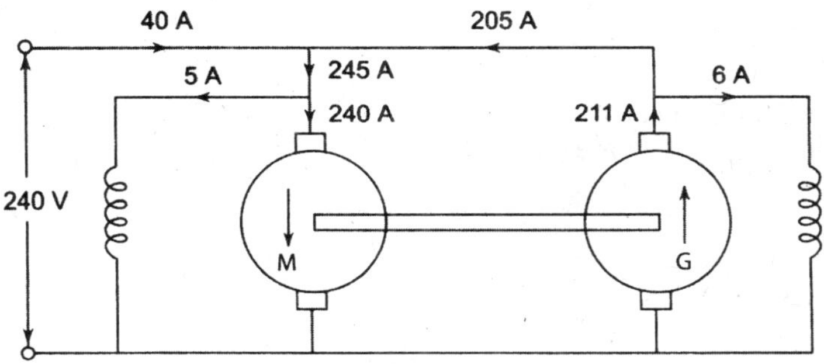

Figure E5.1

Total power input from supply = total losses in the set

i.e., $240 \times 40 = 2{,}485.3 + 2{,}486.4 +$ stray load losses

i.e., Stray load losses = 4,628.3 W

Therefore, stray losses per machine = 2,314.15 W

Motor efficiency:

$$\text{Total losses} = 2{,}486.4 + 2{,}314.15 = 4{,}800.55 \text{ W}$$

$$\text{Motor input} = 240 \times 245 = 58{,}800 \text{ W}$$

$$\text{Motor output} = 58{,}800 - 4{,}800.55 = 53{,}999.45 \text{ W}$$

$$\text{Therefore, motor efficiency } (\eta_M) = \frac{\text{output}}{\text{input}} = \frac{53{,}999.45}{58{,}800} = 0.9183 \text{ p.u.} = 91.83\ \%$$

Generator efficiency:

$$\text{Total losses} = 2{,}485.3 + 2{,}314.15 = 4{,}799.45 \text{ W}$$

$$\text{Generator output} = 240 \times 205 = 49{,}200 \text{ W}$$

$$\text{Generator input} = 49{,}200 + 4{,}799.49 = 53{,}999.45 \text{ W}$$

$$\text{Therefore, generator efficiency } (\eta_g) = \frac{\text{output}}{\text{input}}$$

$$= \frac{49{,}200}{53{,}999.45} = 0.9111 \text{ p.u.} = 91.11\%$$

■

5.27 SEPARATION OF LOSSES IN A DC MACHINE

The main purpose of this test is to segregate various losses in a DC machine into their component parts. The DC machine is run at no-load by varying the speed, and its excitation is kept constant. The frictional loss and the windage loss of the shunt motor are proportional to N and N^2, respectively, where N is the speed at any instant. The mechanical losses (W_m) can be expressed by

$$W_m = AN + BN^2. \tag{5.50}$$

where A and B are constants.

The iron loss consists of hysteresis loss and eddy current loss, which are proportional to

$$W_h \propto B_{\max}^{1.6} f \tag{5.51}$$

and

$$W_h \propto B_{\max}^2 f^2 t^2 \tag{5.52}$$

where t is the thickness of laminations.

The frequency of eddy current is expressed by

$$f = \frac{PN}{120} \tag{5.53}$$

If $B_{\max}$ is kept constant, iron loss can be expressed as

$$W_i = CN + DN^2 \tag{5.54}$$

Therefore, the total losses become

$$W = W_m + W_i = AN + BN^2 + CN + DN^2$$
$$= (A+C)N + (B+D)N^2 \quad (5.55)$$

Dividing both sides by N, we have

$$\frac{W}{N} = A + C + (B+D)N \quad (5.56)$$

Figure 5.31 shows the graph where y-axis and x-axis represent $\frac{W}{N}$ and N, respectively, and PQ shows the straight-line relationship.

The intercept on the y-axis is $OP = A + C$ (5.57)

and the slope of the line is $PQ = \tan\theta_1 = B + D$ (5.58)

For constant excitation, B_{max} remains constant and the constants A and B will remain unchanged. The constants C and D change to C' and D', respectively, when the motor runs at different excitations. Equation (5.56) becomes

$$\frac{W}{N} = A + C' + (B + D')N \quad (5.59)$$

This relationship is plotted in Figure 5.31 where it represents another straight-line SR such that

$$OR = A + C' \quad (5.60)$$

and

$$\tan\theta_2 = B + D' \quad (5.61)$$

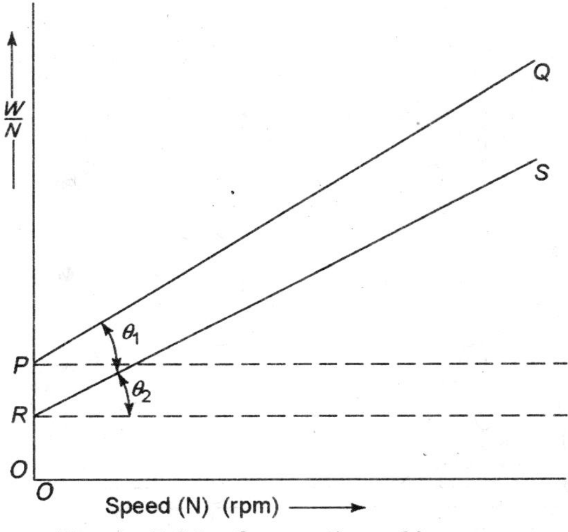

Figure 5.31 Separation of Losses

From Equations (5.57) and (5.60), we have

$$C - C' = OP - OR \tag{5.62}$$

From Equations (5.58) and (5.61), we have

$$D - D' = \tan\theta_1 - \tan\theta_2 \tag{5.63}$$

If B_{1max} and B_{2max} are the flux densities at two different excitations, we have

$$\frac{C}{C'} = \left(\frac{B_{1\max}}{B_{2\max}}\right)^{1.6} = \left(\frac{E_{b1}}{E_{b2}}\right)^{1.6} \tag{5.64}$$

where E_{b1} and E_{b2} are the back emf developed at two different excitations at any given speed. Similarly,

$$\frac{D}{D'} = \left(\frac{B_{1\max}}{B_{2\max}}\right)^{2} = \left(\frac{E_{b1}}{E_{b2}}\right)^{2} \tag{5.65}$$

The value of $\frac{E_{b1}}{E_{b2}}$ can be found out from the test described below and hence the ratio $\frac{C}{C'}$ and $\frac{D}{D'}$ can be determined. Therefore, the values of C, C', D and D' can be known. From Equations (5.57) and (5.58), A and B can be found out if and only if the curves PQ and RS shown in Figure 5.31 are known at two different excitations.

Figure 5.32 shows the circuit diagram to determine the separation of losses of a DC shunt motor by exciting the machine separately. The motor is run at no-load. The switch 'S' is closed. The series resistance (R_s) is adjusted to have a high value in the armature circuit and the flow of maximum field current in the field. The resistance R_s is fully cut off so that the rated voltage is applied across the armature. By changing the excitation, the speed of the motor is adjusted to the rated value. When the motor runs at rated speed and the rated voltage is applied, the motor is said to be normally excited. The speed, the voltage across the armature and the armature current are noted. The excitation is now kept constant, and the speed of the motor is varied by varying the armature voltage with the help of R_s. Restoring more

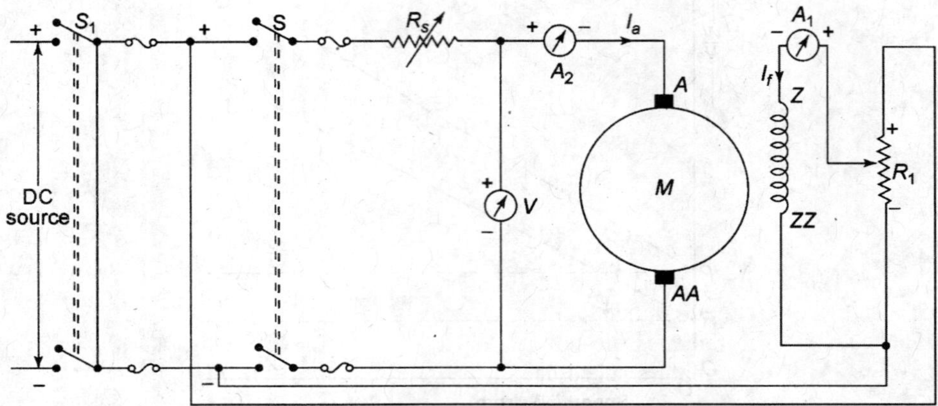

Figure 5.32 Schematic Diagram of the Separation of Losses of a DC Shunt Motor

resistance in armature circuit, speeds differing by 100 rpm are obtained and the measurement at each speed is noted down.

The no-load loss at any given speed is expressed as follows:

$$W = VI_a - I_a^2 R_a \tag{5.66}$$

If W is known, $\frac{W}{N}$ can be computed for each value of the speed. The plot of $\frac{W}{N}$ versus N is plotted to obtain the curve PQ. Let the line RS be plotted with $\frac{W}{N}$ in the y-axis and N along the x-axis. Let line RS represent the graph at the reduced 0.75 × excitation for the rated speed. The two straight lines will show some divergence between them and the constants A, B, C and D can be computed. Using Equations (5.51)–(5.52), the different loss components can be obtained at any desired speed at the normal excitation. The advantages of this method are as follows:

(i) The various losses can be divided into their component parts.

(ii) The method is simple to carry out with very low power loss.

The main disadvantage of this method is that the experiment must be carried out very carefully to avoid the erroneous results.

5.28 RETARDATION OR RUNNING TEST

This test is applied to shunt motors and generators to obtain stray losses. After determining the armature and shunt field Cu losses, the efficiency of the motor can be calculated. During testing, the machine is speeded up slightly beyond its normal speed and the supply is removed. But field excitation is continued. Therefore, the armature slows down and its kinetic energy is used to meet the rotational losses, that is, friction, windage and iron losses.

Kinetic energy of the armature is given by

$$KE = \frac{1}{2} J\omega^2$$

where J is the moment of inertia of the armature and ω is its kinetic energy.

Since, rotational losses = rate of loss of KE

i.e.,
$$W = \frac{d}{dt}\left(\frac{1}{2} J\omega^2\right) = J\omega \frac{d\omega}{dt} \tag{5.67}$$

Therefore, the value of J and $\frac{d\omega}{dt}$ or $\frac{dN}{dt}$ ($\because \omega \propto N$) is required.

5.28.1 To Find $\frac{d\omega}{dt}$

Figure 5.33 shows the circuit connection. A voltmeter V is connected across the armature. Since the speed indicator suitably graduates the voltage across the armature, this voltmeter is required. After removing the supply, the armature speed decreases and reading of the voltmeter also falls. The reading of different voltage falls in different periods is noted down. Figure 5.34 shows the plot of time and speed.

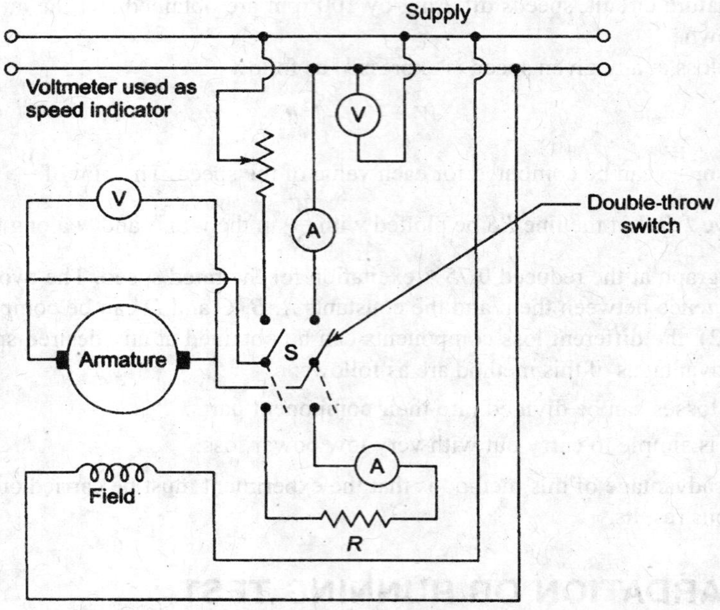

Figure 5.33 Retardation or Running Test

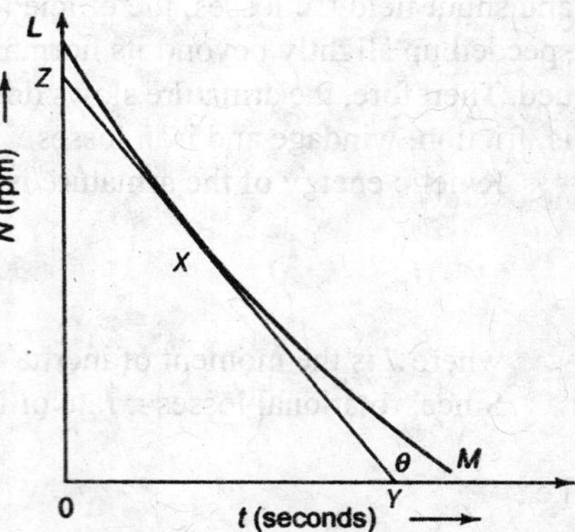

Figure 5.34 Speed–Time Curve in Retardation Test

At the point X, tangent YZ is drawn. From Figure 5.34, we have

$$\frac{dN}{dt} = \frac{OZ\left(\text{in rpm}\right)}{OY\left(\text{in sec}\right)}$$

Again, $\omega = \frac{2\pi N}{60}$ rpm

From Equation (5.67), we have

$$W = J\omega\frac{d\omega}{dt} = J \times \frac{2\pi N}{60} \times \frac{d}{dt}\left(\frac{2\pi N}{60}\right) = 0.011N\frac{dN}{dt}$$

$$= 0.011N\frac{dN}{dt} \tag{5.68}$$

5.28.2 To Find *J*

5.28.2.1 First Method

In this method, J is calculated. After obtaining the speed–time curve, a flywheel of known moment of inertia (J') is keyed to the shaft. Again, speed–time curve is obtained. $\frac{dN}{dt}$ (without adding the flywheel) and $\frac{dN'}{dt}$ (after adding the flywheel) are calculated for any given speed.

Since losses are the same in both cases,

$$0.0011J\frac{dN}{dt} = 0.0011(J+J')\left(\frac{dN'}{dt}\right)$$

i.e.,

$$\frac{J+J'}{J} = \left(\frac{dN}{dt}\right)\Big/\left(\frac{dN'}{dt}\right)$$

i.e.,

$$J = J'\left(\frac{dN'}{dt}\right)\Big/\left(\frac{dN}{dt} - \frac{dN'}{dt}\right) \tag{5.69}$$

5.28.2.2 Second Method

In this method, J is eliminated. This method is more popular and more convenient. The following procedures are generally adopted:

- The time required for decrease in speed, say 5 per cent, is noted down.
- A known retardation torque is applied and the time for the same drop is noted down.

Using the circuit, shown in Figure 5.33, the retarding torque is applied electrically. The armature is connected to a double-throwing switch.

At first, the machine is brought to the rated speed. The time required to decrease the voltage from 250 V to 230 V is noted down. The switch is thrown to the supply side and the motor is brought to the rated speed. The switch is again thrown to the other side so that the non-inductive resistance (R) across the armature is connected. The flow of current (I) results in loss $I^2(R_a + R)$. The means of initial and final value of δN are taken to find this extra power loss because the value of I will change with change in speed. Let this power be P'.

Let P be the iron and mechanical losses of motor at mean speed, δN be the change in speed, t_1 be the time of slowing down without extra loss, and t_2 be the time of slowing down with extra loss.

$\therefore$

$$P = 0.0011JN\frac{\delta N}{t_1} \quad \text{and} \quad P + P' = 0.0011JN\frac{\delta N}{t_2}$$

$\therefore$

$$\frac{P+P'}{P} = \frac{t_1}{t_2}$$

i.e.,

$$P = P'\left(\frac{t_2}{t_1 - t_2}\right) \tag{5.70}$$

Example 5.23 A DC machine running at 700 rpm is put to a retardation test. The time taken at the speed to fall from 730 rpm to 670 rpm is (a) 27 seconds with no excitation, (b) 10 seconds with full excitation and (c) 6 seconds with full excitation and armature supplying an extra load of 7 A at 200 V. Calculate (i) moment of inertia of the armature and (ii) mechanical losses and iron losses at mean speed of 700 rpm.

Solution

(i)

$$P_m = 0.011JN\frac{dN}{dt} \quad \text{[where } J \text{ is the moment of inertia of the armature]}$$

$$dN = 730 - 670 = 60 \text{ rpm}$$

$$dt = 27 \text{ s}$$
$$N = 700 \text{ rpm}$$
$$P_m = 0.011JN \times \left(\frac{60}{27}\right) \quad (1)$$

Similarly, stray load losses $P_s = 0.011JN \times \left(\frac{60}{10}\right)$ (2)

Also, $$P_s = P'\left(\frac{t_2}{t_1 - t_2}\right) = 200 \times 7 \times \left(\frac{6}{10-6}\right) = 2{,}100 \text{ W} \quad (3)$$

From Equations (2) and (3), we have

$$2{,}100 = 0.011 \times J \times 700 \times \frac{60}{10}$$

i.e., $$J = \frac{2{,}100 \times 10}{0.011 \times 700 \times 60} = 45.45 \text{ N-m}$$

(ii) Dividing Equation (1) by Equation (2), we have

$$\frac{P_m}{P_s} = \frac{10}{27}$$

That is, $P_m = \frac{10}{27} \times P_s = \frac{10}{27} \times 2{,}100 = 777.78 \text{ W}$

$\therefore$ Iron losses $= P_s - P_m = 2{,}100 - 777.78 = 1{,}322.22$ W ■

5.29 FIELD'S TEST

Small series motors can be tested by the brake test, while large series motors cannot be tested by this method. Field's test is applicable to two similar series machines, which are used for traction purposes. Figure 5.35 shows field's test on series machines.

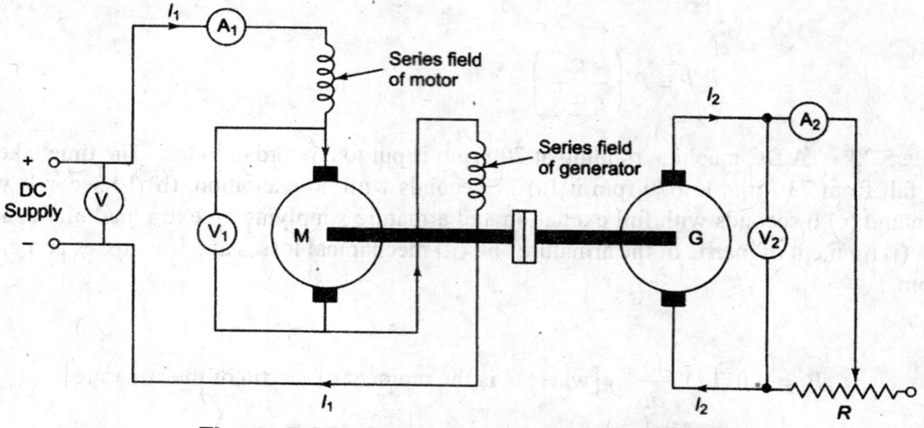

Figure 5.35 Field's Test on Series Machines

Figure 5.35 shows two machines, which are mechanically coupled together and their series fields are connected in series to make the iron losses of both the machines equal. One machine runs as a motor and drives the generator whose output is wasted in variable load R. R is varied until the ammeter A_1 shows the full-load current of the motor. The readings of the following instruments are taken:

- voltmeter V_t (supply voltage),
- voltmeter V_1 (terminal voltage of motor),
- ammeter A_1 (motor input current),
- voltmeter V_2 (terminal voltage of generator) and
- ammeter A_2 (load current of generator).

Let R_a and R_{se} be the armature resistance and field resistance of each machine.

Input to the whole set $= V_tI_1$, output $= V_2I_2$, total losses of the set $(P_t) = V_tI_1 - V_2I_2$, armature and field Cu loss of the motor $= I_1^2(R_a + R_{se})$, armature and field Cu-loss of the generator $= I_2^2R_a + I_1^2R_{se}$, total Cu loss of the set $(P_c) = I_1^2(R_a + R_{se}) + I_2^2R_a + I_1^2R_{se} = I_1^2(R_a + 2R_{se}) + I_2^2R_a$, stray load losses $= P_t - P_c$.

Therefore, stray load losses per machine $(P_s) = \dfrac{P_t - P_c}{2}$.

Motor efficiency

$$\text{Motor input} = V_1I_1$$

$$\text{Motor losses} = I_1^2(R_a + R_{se}) + P_s$$

$$\text{Motor output} = V_1I_1 - \left[I_1^2(R_a + R_{se}) + P_s\right]$$

$$\therefore \quad \text{Motor efficiency } (\eta_m) = \frac{V_1I - \left[I_1^2(R_a + R_{se}) + P_s\right]}{V_1I_1} \tag{5.71}$$

Generator efficiency

$$\text{Generator output} = V_2I_2$$

$$\text{Generator losses} = I_2^2R_a + I_1^2R_{se} + P_s$$

$$\text{Therefore, generator efficiency } (\eta_g) = \frac{V_2I_2}{V_2I_2 + I_2^2R_a + I_1^2R_{se} + P_s} \tag{5.72}$$

Example 5.24 During the field test on two mechanically coupled similar DC series machines (with their fields connected in series and with one machine as a motor and the other as a generator), the following data are obtained:

Motor:

Armature current = 40 A
Armature voltage = 220 V
Voltage drop across its field windings = 10 V

Generator:

Armature current = 30 A
Armature voltage = 190 V
Voltage drop across its field windings = 10 V

Calculate the efficiency of each machine. Take armature resistance of each machine as 0.4 Ω.

Solution

$$\text{Supply voltage} = (220 + 10 + 10)\text{ V} = 240\text{V}$$

$$\text{Total input} = 240 \times 40 = 9{,}600 \text{ W}$$

$$\text{Output} = 190 \times 30 = 5{,}700 \text{ W}$$

$$\text{Total losses in two machines} = \text{input} - \text{output} = 9{,}600 - 5{,}700 = 3{,}900 \text{ W}$$

$$\text{Series field resistance } (R_{se}) = \frac{10}{40} = 0.25\ \Omega$$

$$\text{Total Cu losses} = (0.4 + 2 \times 0.25) \times 40^2 + 30^2 \times 0.4 = 1{,}800 \text{ W}$$

$$\therefore \quad \text{Stray load losses of the set} = 3{,}900 - 1{,}800 = 2{,}100 \text{ W}$$

$$\therefore \text{Stray load losses of each machine} = \frac{2{,}100}{2} = 1{,}050 \text{ W}$$

Motor efficiency

$$\text{Motor input} = \text{armature voltage} \times \text{motor current} = 220 \times 40 = 8{,}800 \text{ W}$$

$$\text{Total Cu losses} = (0.4 + 0.25) \times 40^2 = 1{,}040 \text{ W}$$

$$\text{Stray losses} = 1{,}050 \text{ W}$$

$$\text{Total losses} = 1{,}040 + 1{,}050 = 2{,}090 \text{ W}$$

$$\text{Output} = 8{,}800 - 2{,}090 = 6{,}710 \text{ W}$$

$$\therefore \quad \text{Motor efficiency } (\eta_m) = \frac{6{,}710}{8{,}800} = 0.7625\text{p.u.} = 76.25\%$$

Generator efficiency

$$\text{Total Cu loss} = 0.4 \times 30^2 + 0.25 \times 40^2 = 760 \text{ W}$$

$$\text{Stray losses} = 1{,}050 \text{ W}$$

$$\text{Total losses} = 760 + 1{,}050 = 1{,}810 \text{ W}$$

$$\text{Output} = 30 \times 190 = 5{,}700 \text{ W}$$

$$\text{Input} = \text{output} + \text{total losses} = 5{,}700 + 1{,}810 = 7{,}510 \text{ W}$$

$$\therefore \quad \text{Generator efficiency } (\eta_m) = \frac{5{,}700}{7{,}510} = 0.7589\text{ p.u.} = 75.89\%$$

■

5.30 USES OF DC MOTORS

The shunt motor has fairly constant speed as well as medium starting torque. It has the following applications:

- blowers and fans,
- lathe machines,
- centrifugal and reciprocating pumps,
- machine tools,
- milling machines and
- drilling machines.

The series motor has variable speed and high starting torque. No-load speed is dangerous. It has the following applications:

- cranes,
- trolleys,
- hoists and elevators,
- conveyers and
- electric locomotives.

Cumulative compound motor has high starting torque, and no-load condition is allowed. It has the following applications:

- punches,
- rolling mills,
- shears,
- elevators and
- heavy planers.

The speed of differential motor increases with load and hence these motors are not suitable for practical applications.

5.31 SPECIAL DC MACHINES

Special DC machines such as three-wire generators, dynamometers and some cross-field machines are discussed.

5.31.1 Three-wire Generator

A three-wire generator is used to get low and high voltages. Figure 5.36 shows a three-wire DC generator having two poles and ring winding. It has two outer wires and one neutral wire. An induction coil D is wound on the iron core, which is connected between two coils having one-pole pitch apart. The frequency of the induced AC voltage in the coil is of $\frac{PN}{120}$. The coil has high reactance and hence the current drawn by the coil is very low. The middle of the coil is connected to the neutral so that the three-wire DC system is formed. If the currents in the outer wires are equal, and the neutral will carry no current. The voltage between outer wires is twice that of the value between the outer and the neutral wires.

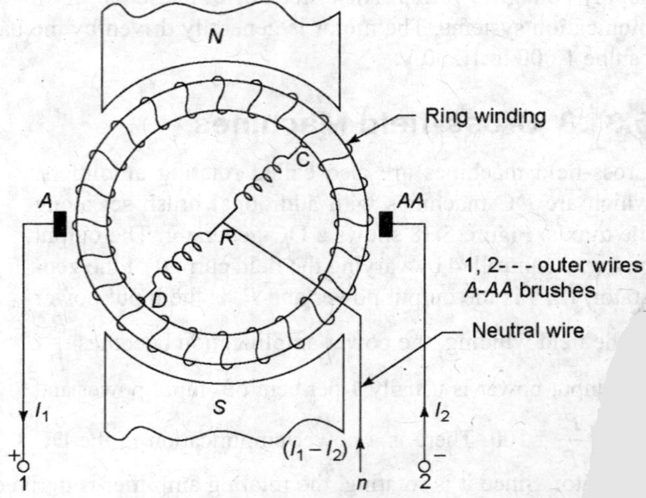

Figure 5.36 A Three-wire DC Generator

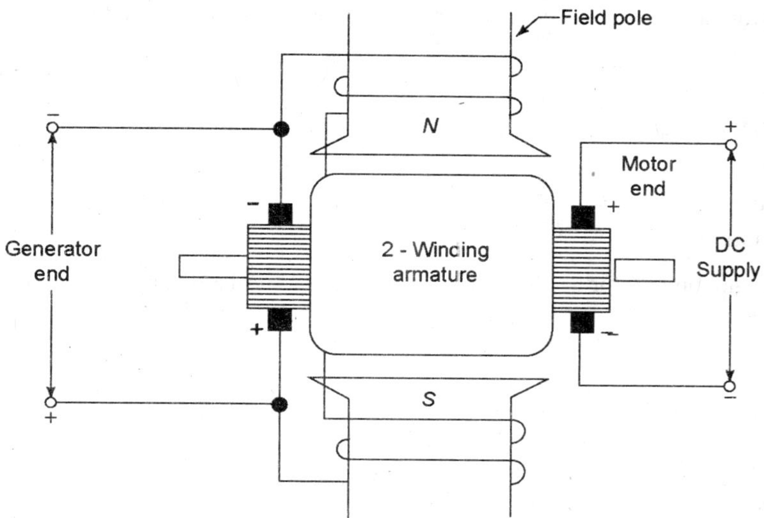

Figure 5.37 Dynamometer

5.31.2 Dynamometer

A dynamometer is a motor and generator unit assembled in the same unit where the armature core and pole pitch structure are common. Figure 5.37 shows a dynamometer. It has separate armature and commutator with brush gear for the generator motor. When machine is impressed, it runs as a motor. Other armature gets emf and acts as a generator due to common field flux. The emf of the generator can be controlled by varying the number of conductors in the armature. To control the generator voltage, the supply voltage is varied. The efficiency of the set is very high and is commonly used in radio and communication systems. The motor is generally driven by the battery. The output voltage of the generator can be 1,000 to 1,500 V.

5.31.3 Cross-field Machines

Cross-field machines are also called rotating amplifiers, which are DC machines with additional brush set along the α-axis. Figure 5.38 shows a DC generator. The output power is controlled by varying the field current of the generator. If P_2 is the output power and P_1 is the input power the field winding, the power amplification becomes $\frac{P_2}{P_1}$. input power is usually 1 per cent of output power and $\frac{P_2}{P_1} \cong 100$. There is a power amplification in the DC ator. Since it is rotating, the rotating amplifier is derived. Depending on the degree of compensa- cross-field generators can be classified as (i) metadyne and (ii) amplidyne.

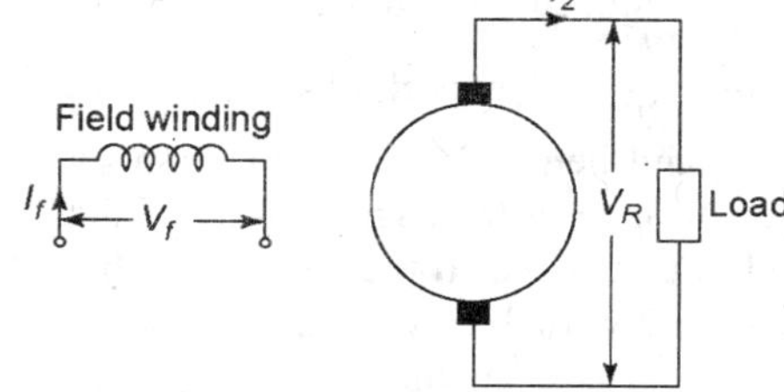

Figure 5.38 DC Generator

5.31.3.1 *Metadyne*

Figure 5.39 shows a metadyne generator, which is obtained from an ordinary DC generator by providing an additional pair of brushes in the *d*-axis whereas the brushes in the *q*-axis are short circuited. The current (I_q) in the *q*-axis produces an mmf M_q along the *q*-axis, which is the first stage of amplification. The voltage induced along the *q*-axis due to controlled mmf M_f circulates I_q when the brushes are short circuited. In the second stage, due to rotation of armature in the mmf M_q, a voltage is developed along the *d*-axis across which the load is connected. Due to the current I_d, the armature mmf in the *d*-axis opposes the control mmf M_f according to Lenz's law. I_q and I_d attain a steady-state value for a given value of I_f and load resistance. If there is any increase in the value of I_d, it will increase M_d that neutralizes M_f. Hence, the induced emf along the *d*-axis is reduced that prevents further increase in current keeping the output current of a metadyne substantially constant over a wide range of load variation. Therefore, the metadyne is also called a constant current generator.

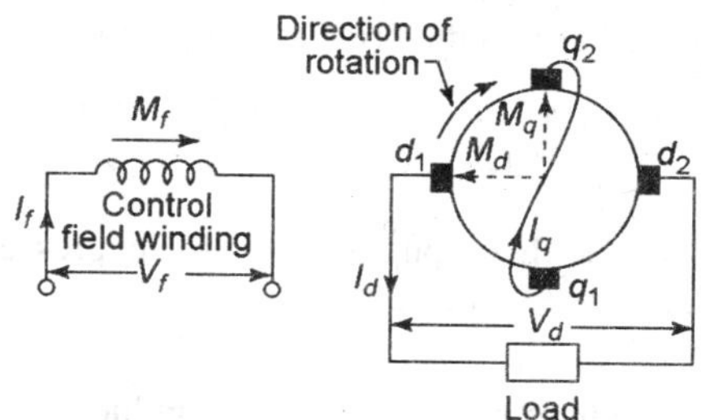

Figure 5.39 Metadyne Generator

5.31.3.2 *Amplidyne*

Figure 5.40 shows an amplidyne, which is a metadyne having a compensation for a *d*-axis mmf M_d. In Figure 5.40, a compensating winding is provided on the stator to produce an mmf M_c opposing M_d. Due to the flow of current through the compensating winding, the compensating mmf is produced and hence the negative feedback effect due to load current I_d is minimized.

The flux along the *d*-axis depends solely on M_f due to the presence of the compensating winding. Since there are differences in space distribution between M_d and M_q, it is difficult to achieve complete neutralization of M_d. The ratio of effective turns of compensating winding to effective turns of the armature winding is called the compensating factor.

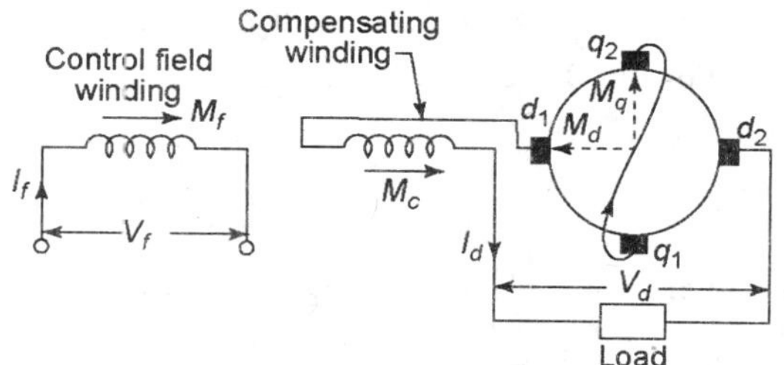

Figure 5.40 Amplidyne

5.32 CHARACTERISTICS OF CROSS-FIELD GENERATORS

To derive analytically the characteristics of the cross field-generator, let T_f be the effective turns of the control field winding, T_d be the effective turns of the control armature along the *d*-axis, *N* be the speed of the generator in rpm and *C* be the compensation factor.

The net ampere turns on the *d*-axis is

$$T_f I_f - I_d T_d (1-C) \tag{5.73}$$

The net ampere flux along the *d*-axis assuming no saturation is expressed by

$$\phi_d = k_1[T_f I_f - I_d T_d (1-C)] \tag{5.74}$$

The induced voltage along the *q*-axis

$$E_q = k_2 \phi_d N = k_1 k_2 [T_f I_f - I_d T_d (1-C)] \tag{5.75}$$

If R_q is the total resistance of the armature along the q-axis including the external resistance connected across the brushes q_1q_2, then

$$I_q = \frac{E_q}{R_q} = \frac{k_1k_2N[T_fI_f - I_dT_d(1-C)]}{R_q} \tag{5.76}$$

The flux along the q-axis is expressed by

$$\phi_q = k_4T_dI_q \tag{5.77}$$

where k_4 is constant. The induced emf along the d-axis is expressed by

$$E_d = k_4\,\phi_qN = k_4Nk_3T_dI_q$$

$$= \frac{k_1k_2k_3k_4N^2T_d}{R_q}\left[T_fI_f - I_dT_d(1-C)\right] \tag{5.78}$$

i.e.,

$$E_d = \frac{k_fN^2I_f}{R_q} - \frac{N^2k(1-C)}{R_q}I_d \tag{5.79}$$

where $k_f = k_1k_2k_3k_4T_dT_f$ and $k = k_1k_2k_3k_4T_d^2$.

For $I_d = 0$, we have

$$E_d = \frac{k_fN^2I_f}{R_q} = AV_f \tag{5.80}$$

where $A = \dfrac{k_fN^2}{R_qR_f}$ is voltage amplification factor.

In general, E_d is expressed as

$$E_d = AV_f - B(1-C)I_d \tag{5.81}$$

where $B = \dfrac{kN^2}{R_q}$.

The terminal voltage of a cross-field generator is

$$V_d = E_d - I_dR_d \tag{5.82}$$

where $R_d = R_a + R_c$ is the resistance of the armature along the d-axis including compensating winding.

$$V_d = AV_f - B(1-C)I_d - I_dR_d \tag{5.83}$$

i.e.,

$$V_d = AV_f - [B(1-C) + R_d]I_d \tag{5.84}$$

If R_L is the load resistance, then

$$I_dR_L = AV_f - [B(1-C) + R_d]I_d$$

i.e.,

$$I_d = \frac{AV_f}{R_L + R_d + B(1-C)}$$

i.e.,
$$I_d = \frac{\left(\frac{k_f N^2}{R_q R_f}\right) V_f}{R_L + R_d + \frac{kN^2}{R_q}(1-C)}$$

i.e.,
$$I_d = \frac{\left(\frac{k_f}{R_q R_f}\right) V_f}{\frac{R_L + R_d}{N^2} + \frac{k}{R_q}(1-C)}$$

i.e.,
$$I_d = \frac{C_1 V_f}{\frac{R_L + R_d}{N^2} + C_2(1-C)} \tag{5.85}$$

where
$$C_1 = \frac{k_f}{R_q R_f} \text{ and } C_2 = \frac{k}{R_q}.$$

From Equation (5.85), I_d increases with increase in speed, shown in Figure 5.41, which is the plot of load current versus speed. For an uncompensated cross-field generator, $C = 0$. In this case, the load current reaches a constant value for higher speeds. The polarity of the load current does not depend upon the direction of rotation of the generator. This is an important characteristic of the cross-field generator. Rosenberg generator used for battery charging and train lighting is an uncompensated cross-field generator whose output voltage is independent of the direction of motion of the train. Figure 5.42 shows the external characteristics of cross-field generators for different degrees of compensation. The terminal voltage is almost constant with load current for $C = 1$ (i.e. 100 per cent compensation).

These cross-field generators are known as amplidynes. The output voltage remains constant and is more due to large magnification. For $C = 0$ (i.e., zero compensation), the terminal voltage varies rapidly but the current remains constant as in metadyne generators. In addition to control and shunt field winding, series and shunt field windings are some times used to enhance the amplification. The interpoles for

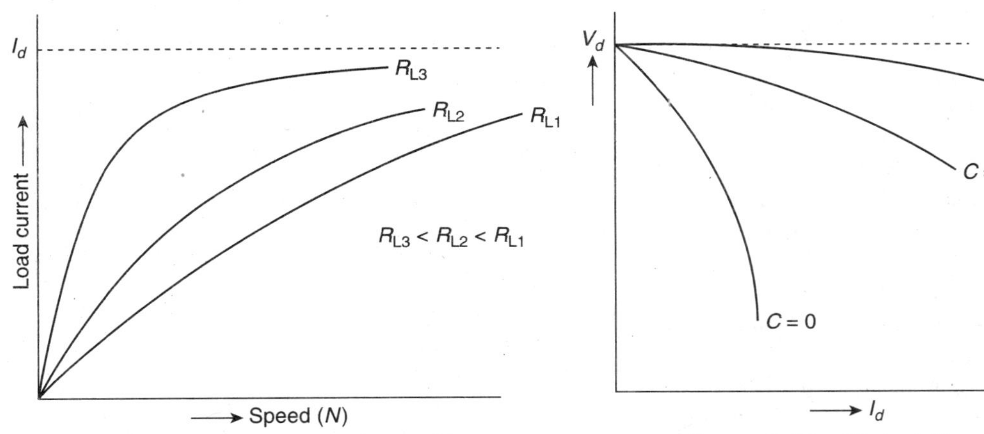

Figure 5.41 Plot of I_d vs N

Figure 5.42 Plot of V_d vs N

cross-field generators are placed along the q-axis and d-axis unlike DC generators. The required number of interpoles is twice that of main poles.

5.33 BRUSHLESS DC MOTOR

The main difference between the brushless DC motor and the ordinary DC motor is the type of commutator used. Mechanical commutation is done in the DC motor, whereas electrical commutation is done in the brushless DC motor. The construction of the brushless DC motor is similar to that of a permanent magnet synchronous motor. The rotor consists of the permanent magnet, whereas the polyphase winding is placed on the stator. The cross-sectional view of a brushless DC motor is shown in Figure 5.43.

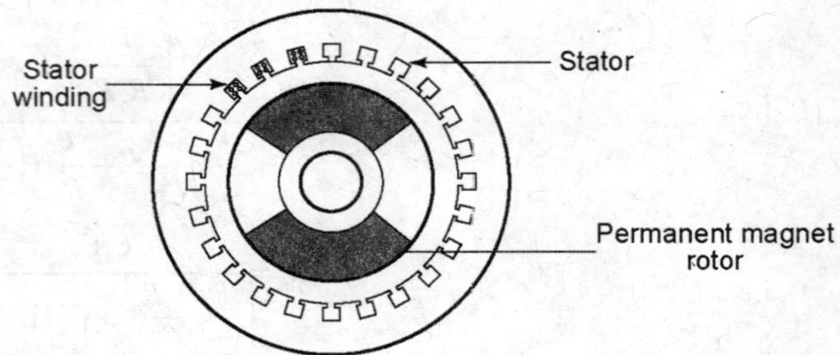

Figure 5.43 Cross-sectional View of a Brushless DC Motor

Due to the combination of drive, its electronic drive circuit and rotor position sensor, the word 'brushless DC motor' is used. The electronic drive circuit in the brushless DC motor is an inverter consists of the transistor, and it feeds stator winding, where the transistors are controlled by pulses generated by rotor position sensors ensuring that the rotor revolves at an angular speed. This angular speed is equal to the average speed of the field produced by the stator.

The driver circuit is fed by DC supply like DC motors where the fields of stator and rotor remain stationary with respect to each other at all the speeds. The torque speed characteristic is similar to the DC motor. The speed can be controlled by the DC supply voltage. Due to these similarities and without brush, this type of machine is known as brushless DC machine. The two types of brushless DC motors available in practice are (i) the unipolar or half-wave brushless DC motor and (ii) the bipolar or full-wave brushless DC motor.

5.33.1 Unipolar or Half-wave Brushless DC Motor

In this type, optical sensors are embedded on the rotor to excite these sensors. Three phototransistors – P_1, P_2 and P_3 – are mounted on the end plate of the motor separately by an angle of 120°. Figure 5.44 shows the arrangement of the optical sensor.

The armature consists of a polyphase winding (three-phase) mounted on the stator, and the field consists of a two-pole permanent magnet. The driving circuit consists of simple transistors Q_1, Q_2 and Q_3, which are used to excite the stator winding. Figure 5.45 shows the schematic arrangement of the unipolar brushless DC motor.

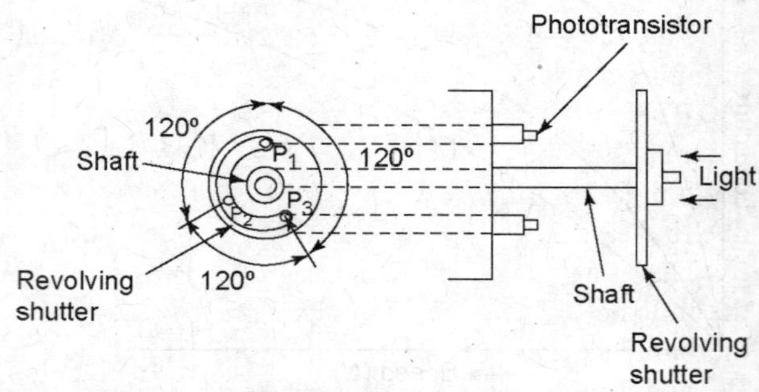

Figure 5.44 Arrangement of the Optical Sensor

The phototransistors get exposed to light due to the revolution of the shutter. For each revolution, the phototransistors

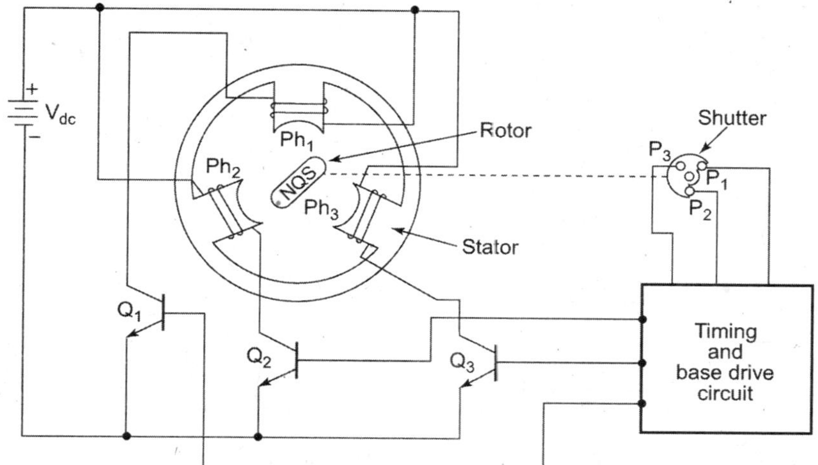

Figure 5.45 Schematic Arrangement of Unipolar Brushless DC Motor

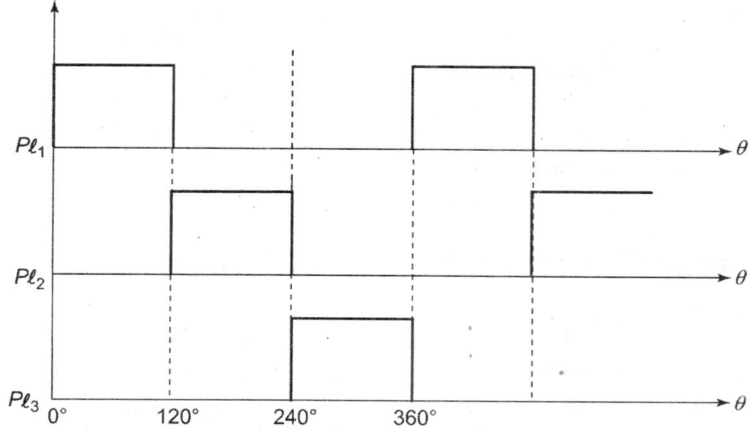

Figure 5.46 Pulses Generated by Phototransistors

generate the pulses PI_1, PI_2 and PI_3. The duration and phase displacement of PI_1, PI_2 and PI_3 are shown in Figure 5.46.

The phototransistor (P_1) generates a pulse when it receives light, and it generates pulse PI_1. Hence, the transistor (Q_1) gets turned on. Therefore, current flows through the phase winding Ph_1 producing the north pole face of Ph_1. The rotor revolves in the anticlockwise direction. The light stops falling on P_1 due to the revolution of the rotor and P_2 gets the light generating pulse Pl_2. This turns on the transistor Q_2. Therefore, current flows through Ph_2 and it produces north pole face of Ph_2. Rotor further rotates in the anticlockwise direction and it reaches the axis of the pole face Ph_2 and light stops falling on P_2. Now, the light starts to fall of P_3 which in turn turns on the transistor Q_3. Therefore, current flows through Ph_3 and it produces the north pole face of Ph_3. Switching sequence is repeated. Hence, the continuous rotation of the rotor is obtained.

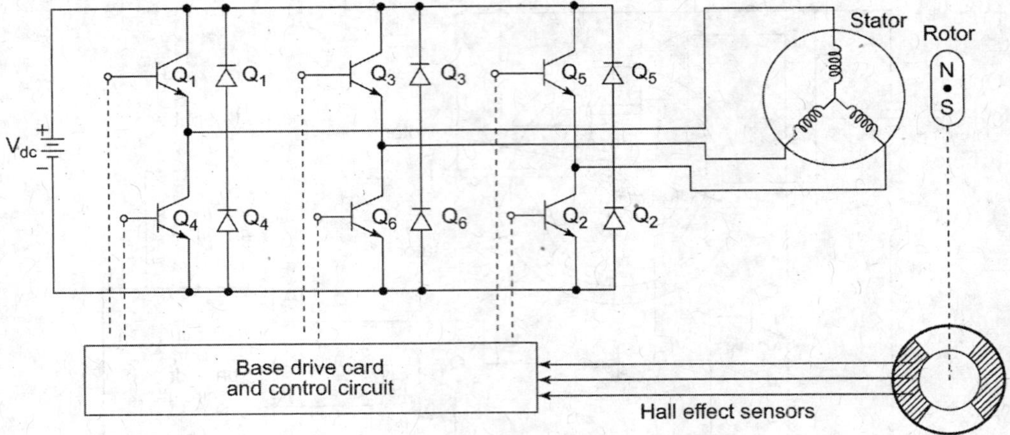

Figure 5.47 . Bipolar Brushless DC Motor

To obtain the torque reversal, the operation of timing and base circuit drive circuit is carried out in such a way that the transistors Q_1, Q_2 and Q_3 are carried out for the duration of pulses Pl_2, Pl_3 and Pl_1 respectively. The base drive signals of the transistors are shifted by 180° to obtain the torque reversal. No feedback or free wheeling diode is required for small motors because the inductive energy in the phase windings is very small, which cannot cause any destruction to the transistors. The cost of the motor is very low and the drive circuit is very simple. The prime limitation of this motor is that it cannot be used above 100 W. The feedback diode is required when the stored energy in the phase winding inductance is large. The bipolar brushless DC motor is used in such cases.

5.33.2 Bipolar or Full-wave Brushless DC Motor

Figure 5.47 shows a bipolar brushless DC motor, which uses three-phase inverter with feedback diodes where the feedback diodes return the inductive energy to the supply. Hall effect sensors are used by rotors to sense the position where the hall effect sensors and magnet ring form the rotor position sensors. The magnet ring is mounted on the rotor shaft and the hall sensors are 120° electrical apart from each other.

The response rotor position sensor causes the switching of each transistor. The transistors are triggered in sequence Q_1, Q_2, Q_3, Q_4, Q_5 and Q_6 where each transistor conducts for 120°. At any time, two transistors conduct and each pair conducts for 60°. The phase shifting of the transistor base drive signals is done 180° to achieve the torque reversal. Since each phase conducts twice in a cycle and hence better utilization of stator takes place compared to the unipolar drive. The bipolar type is suitable for high-performance servo drives. These are also suited for the ratings higher than 100 W.

5.34 FEATURES OF BRUSHLESS DC MOTOR

The following are the features of the brushless DC motor compared to the conventional DC motor:

- These motors do not need maintenance practically due to the absence of commutator and brushes.
- Life is long.

- Highly reliable operation.
- Low inertia and friction.
- Operation is noiseless due to low radiofrequency interference and noise.
- Cooling is better because the armature winding is on the stator.
- Efficiency is as high as 75 to 80 per cent due to the feedback diodes.

ADDITIONAL SOLVED PROBLEMS

Example 5.25 A four-pole, 220 V, lap-connected DC series motor has 200 slots with six conductors per slot. The current is 40 A and flux/pole is 16 mWb. The armature and field resistances are 0.46 Ω and 0.26 Ω, respectively. The iron and friction losses are 700 W. Calculate the useful torque.

Solution

$$E_b = V_t - I_a(R_a + R_{se}) = 220 - 40\times(0.46+0.26) = 191.2 \text{ V}$$

Again,

$$E_b = \frac{\phi NZ}{60}\times\frac{P}{A}$$

We have

$$191.2 = \frac{16\times10^{-3}\times N\times 1{,}200}{60}\times\frac{4}{4}$$

i.e.,

$$N = \frac{191.2\times60}{16\times10^{-3}\times1{,}200} = 597.5 \text{ rpm}$$

$$\text{Total input} = V_t I = 220\times40 = 8{,}800 \text{ W}$$

$$\text{Total Cu loss} = I_a^2(R_a + R_{se}) = 40^2\times(0.46+0.26) = 1{,}152 \text{ W}$$

Iron and friction losses = 700 W.

Total losses = 1,152 + 700 = 1,852 W

Therefore, output power = 8,800 − 1,852 = 6,948 W

∴

$$T_{sh} = \frac{9.55\times6{,}948}{597.5} = 111.05 \text{ N-m}$$

■

Example 5.26 The no-load and shunt field currents of a 240 V, four-pole, lap-connected DC shunt motor are 7 A and 2 A, respectively. Its full-load current and armature resistance are 90 A and 0.15 Ω, respectively. At no-load, the flux per pole is 0.06 Wb. The motor has 700 conductors and contact drop per brush is 1 V. Determine (i) no-load speed, (ii) full-load speed and (iii) speed regulation. Field flux is reduced by 2 per cent due to armature reaction.

Solution

(i) At no-load

$$I_{a0} = I_0 - I_{sh} = 7 - 2 = 5 \text{ A}$$

∴

$$E_{b0} = V - I_{a0}R_a - \text{contact drop at brushes} = 240 - 5\times0.15 - 2 = 237.25 \text{ V}$$

We know that $$E_{b0} = \frac{\phi N_0 Z}{60} \times \frac{P}{A}$$

$$\therefore \quad 237.25 = \frac{0.06 \times N_0 \times 700}{60} \times \frac{4}{4}$$

i.e., $$N_0 = \frac{237.25 \times 60}{0.06 \times 700} = 338.93 \text{ rpm}$$ ■

(ii) $$I_a = I - I_{sh} = 90 - 2 = 88 \text{ A}$$

$$\therefore \quad E_{b1} = 240 - 88 \times 0.15 - 2 = 224.8 \text{ V}$$

$$\phi = \frac{100-2}{100} \times \phi_0 = 0.98\phi_0$$

For a particular machine, $E_b \propto \phi N$

$$\frac{E_{b0}}{E_{b1}} = \frac{\phi_0 N_0}{\phi_1 N_1}$$

i.e., $$N_1 = \frac{E_{b1}}{E_{b0}} \times \frac{\phi_0}{\phi_1} \times N_0 = \frac{224.8 \times \phi_0 \times 338.93}{237.25 \times 0.98\phi_0} = 327.7 \text{ rpm}$$

Therefore, full-load speed of the motor is 327.7 rpm. ■

(iii) Speed regulation $$= \frac{N_0 - N_1}{N_1} = \frac{338.93 - 327.7}{327.7} = 0.0343 \text{ p.u.} = 3.43\%$$ ■

Example 5.27 The line current and speed of a 220 V DC shunt motor are 55 A and 700 rpm, respectively. The armature and field resistances are 0.14 Ω and 110 Ω, respectively. Considering 1 V/brush as contact drop, calculate (i) the no-load speed if no-load current is 7 A and (ii) the percentage reduction in the flux per pole so that the speed becomes 900 rpm at armature current of 35 A. Neglect the effect of armature reaction.

Solution

(i) At no-load

$$I_{a0} = I_0 - I_{sh} = 7 - 2 = 5 \text{ A}$$

$$\therefore \quad E_{b0} = V_t - I_{a0} R_a - \text{contact drop at brushes} = 220 - 5 \times 0.14 - 2 = 217.3 \text{ V}$$

At load $$I_1 = 55 \text{ A}$$

$$I_{a1} = I_1 - I_{sh} = 55 - 2 = 53 \text{ A}$$

$$\therefore \quad E_{b1} = V_t - I_{a1} R_a - \text{total brush drop} = 220 - 53 \times 0.14 - 2 = 210.58 \text{ V}$$

$$\therefore \quad \frac{E_{b0}}{E_{b1}} = \frac{\phi_0 N_0}{\phi_1 N_1}$$

i.e., $$N_0 = \frac{E_{b0}}{E_{b1}} \times N_1 = \frac{217.3}{210.58} \times 700 = 722.33 \text{ rpm}$$

Therefore, no-load speed of the motor is 722.33 rpm. ■

(ii) Let the flux be ϕ_2 at a speed of 900 rpm at armature current of 35 A.

$$E_{b2} = V_t - I_{a2} R_a - \text{total brush drop} = 220 - 35 \times 0.14 - 2 = 213.1 \text{ V}$$

Again, $$\frac{E_{b2}}{E_{b1}} = \frac{N_2\phi_2}{N_1\phi_1}$$

i.e., $$\frac{\phi_2}{\phi_1} = \frac{E_{b2}}{E_{b1}} \times \frac{N_1}{N_2} = \frac{213.1}{210.58} \times \frac{700}{900} = 0.787$$

i.e., $$\frac{\phi_1 - \phi_2}{\phi_1} = 1 - 0.787 = 0.213 \text{ p.u.} = 21.3\%$$

Therefore, the required percentage reduction in flux is 21.3 per cent. ■

Example 5.28 A 240 V DC shunt motor takes a full-load current of 60 A. Its armature and shunt field resistances are 0.25 Ω and 120 Ω, respectively. The motor draws a current of 6 A at no-load having a speed of 1,000 rpm. Determine (i) full-load speed, (ii) speed regulation, (iii) HP (British) rating of the machine and (iv) full-load efficiency.

Solution

$$V_t = 240 \text{ V}, R_a = 0.25\ \Omega, R_{sh} = 120\ \Omega, N_0 = 1{,}000 \text{ rpm}, I_1 = 60 \text{ A}$$

$$I_{sh} = \frac{V_t}{R_{sh}} = \frac{240}{120} = 2 \text{ A}$$

∴ $$I_{a1} = I_1 - I_{sh} = 60 - 2 = 58 \text{ A}$$

$$E_{b1} = V_t - I_{a1}R_a = 240 - 58 \times 0.25 = 225.5 \text{ V}$$

At no-load, $$I_0 = 6 \text{ A}, I_{sh} = 2 \text{ A}$$

∴ $$I_{a0} = I_0 - I_{sh} = 6 - 2 = 4 \text{ A}$$

$$E_{b0} = V_t - I_{a0}R_a = 240 - 4 \times 0.25 = 239 \text{ V}$$

(i) $$\frac{N_1}{N_0} = \frac{E_{b1}}{E_{b0}} \quad \text{[since } \phi \text{ remains constant]}$$

i.e., $$N_1 = N_0 \times \frac{E_{b1}}{E_{b0}} = 1{,}000 \times \frac{225.5}{239} = 943.5 \text{ rpm}$$ ■

(ii) $$\text{Speed regulation} = \frac{N_0 - N_1}{N_0} \times 100 = \frac{1{,}000 - 943.5}{1{,}000} \times 100 = 5.65\%$$ ■

(iii) Full-load power output = full-load power developed − stray losses

$= E_{b1}I_{a1}$ − stray losses

Stray losses = mechanical power developed at no-load = $E_{b0}I_{a0}$

This is because the entire power developed is used to overcome stray losses at no-load.

∴ Full-load shaft output = $E_{b1}I_{a1} - E_{b0}I_{a0} = 225.5 \times 58 - 239 \times 4 = 12{,}123$ W

$$\text{HP rating} = \frac{\text{Full-load shaft output}}{746} = \frac{12{,}123}{746} = 16.25 \text{ HP (British)}$$ ■

(iv) Power input = $V_t I_1 = 240 \times 60 = 14{,}400$ W

∴ $$\text{Efficiency} = \frac{\text{power output}}{\text{power input}} = \frac{12{,}123}{14{,}400} = 0.8418 \text{ p.u.} = 84.18\%$$ ■

Example 5.29 A four-pole, 240 V, wave-connected DC shunt motor has 1,000 conductors, and useful flux/pole is 8 mWb. Armature and field resistances are 0.4 Ω and 240 Ω, respectively. If the motor draws a current of 25 A from the supply mains, find the speed and torque developed by the motor. If magnetic and mechanical losses are 800 W, find (i) useful flux and (ii) efficiency at this load.

Solution

∴ $$I_{sh} = \frac{V_t}{R_{sh}} = \frac{240}{240} = 1\,\text{A}$$

$$I_a = I_1 - I_{sh} = 25 - 1 = 24\text{ A}$$

$$E_b = V_t - I_a R_a = 240 - 24 \times 0.4 = 230.4\text{ V}$$

Again, $$E_b = \frac{\phi NZ}{60} \times \frac{P}{A}$$

∴ $$230.4 = \frac{8 \times 10^{-3} \times 1{,}000 \times N}{60} \times \frac{4}{2}$$

i.e., $$N = \frac{230.4 \times 60}{8 \times 2} = 864\text{ rpm}$$

$$P_m = E_b I_a = 230.4 \times 24 = 5{,}529.6\text{ W}$$

$$T_g = 9.55 \times \frac{P_m}{N} = 9.55 \times \frac{5{,}529.6}{864} = 61.12\text{ N-m}$$

$$P_{sh} = P_m - \text{mechanical losses} = 5{,}529.6 - 800 = 4{,}729.6\text{ W}$$

$$P_{in} = V_t I_1 = 240 \times 25 = 6{,}000\text{ W}$$

∴ $$\eta = \frac{P_{sh}}{P_{in}} = \frac{4{,}729.6}{6{,}000} = 0.7882\,\text{p.u.} = 78.82\%$$ ■

Example 5.30 A 240 V, 7,355 W, DC shunt motor has armature and field resistances of 0.5 Ω and 120 Ω, respectively. The motor runs at a speed of 1,000 rpm on full-load having full-load efficiency of 88 per cent. If the speed is to be raised to 1,200 rpm keeping the load constant, find the extra resistances to be added in the field circuit.

Solution

$$V_t = 240\text{ V}, R_a = 0.5\ \Omega, R_{sh} = 120\ \Omega, \eta_{fl} = 88\%, N_1 = 1{,}000\text{ rpm}, N_2 = 1{,}200\text{ rpm}$$

$$I_{sh} = \frac{V_t}{R_{sh}} = \frac{240}{120} = 2\,\text{A}$$

$$\text{Full-load output} = 7{,}355\text{ W}$$

∴ $$\text{Input} = \frac{\text{output}}{\eta_{fl}} = \frac{7{,}355}{0.88} = 8{,}357.95\text{ W}$$

$$\text{Input} = V_t \times I_1 = 8{,}357.95\text{ W}$$

$$I_1 = \frac{8{,}357.95}{V_t} = \frac{8{,}357.95}{240} = 34.82\text{ A}$$

$$I_{a1} = I_1 - I_{sh} = 34.82 - 2 = 32.82 \text{ A}$$

Again, $T \propto \phi I_a$

$$\frac{T_1}{T_2} = \frac{\phi_1 I_{a1}}{\phi_2 I_{a2}} = \frac{I_{a1} I_{sh1}}{I_{a2} I_{sh2}}$$

Since load torque remains constant, we have

$$I_{a1} I_{sh1} = I_{a2} I_{sh2}$$

i.e.,
$$I_{a2} = \frac{I_{a1} I_{sh1}}{I_{sh2}} = \frac{32.82 \times 2}{I_{sh2}} = \frac{65.64}{I_{sh2}}$$

$$E_{b1} = V_t - I_{a1} R_a = 240 - 0.5 \times 32.82 = 223.59 \text{ V}$$

and
$$E_{b2} = V_t - I_{a2} R_a = 240 - 0.5 I_{a2} = 240 - \frac{32.82}{I_{a2}}$$

Again,
$$\frac{N_2}{N_1} = \frac{E_{b2}}{E_{b1}} \times \frac{\phi_1}{\phi_2} = \frac{E_{b2}}{E_{b1}} \times \frac{I_{sh1}}{I_{sh2}}$$

i.e.,
$$\frac{1,200}{1,000} = \frac{240 - \dfrac{32.82}{I_{sh2}}}{223.59} \times \frac{2}{I_{sh2}}$$

i.e.,
$$\frac{1,200}{1,000} \times \frac{223.59}{2} = \frac{240 I_{sh2} - 32.82}{I_{sh2}^2}$$

i.e.,
$$134.15 I_{sh2}^2 - 240 I_{sh2} + 32.82 = 0$$

So,
$$I_{sh2} = \frac{240 \pm \sqrt{240^2 - 4 \times 32.82 \times 134.15}}{2 \times 134.15} = \frac{240 \pm 199.97}{268.3} = \text{either } 1.64 \text{ or } 0.149$$

$\therefore$ $I_{sh2} = 1.64$ A (feasible)

Let the resistance R be connected in series with the field circuit.

$$I_{sh2} = \frac{240}{R_{sh} + R}$$

i.e.,
$$1.64 = \frac{240}{120 + R}$$

i.e.,
$$R = \frac{240}{1.64} - 120 = 26.34 \ \Omega$$

Therefore, the required resistance is 26.34 Ω. ■

Example 5.31 A 240 V DC shunt motor develops an output of 20 kW when it takes 23 kW. The armature and field resistances are 0.2 Ω and 125 Ω, respectively. Find the efficiency and power input when output is 12 kW. Assume stray and friction losses remain constant.

Solution

$V_t = 250$ V, $R_a = 0.2\ \Omega$, $R_{sh} = 125\ \Omega$, $(P_{out})_1 = 20$ kW and $(P_{in})_1 = 23$ kW

$$\text{Input current } (I_1) = \frac{23 \times 10^3}{250} = 92 \text{ A}$$

$$I_{sh} = \frac{V_t}{R_{sh}} = \frac{250}{125} = 2\,\text{A}$$

$$\therefore \qquad I_a = I_a - I_{sh} = 92 - 2 = 90\,\text{A}.$$

$$\text{Armature Cu loss} = 90^2 \times 0.2 = 1{,}620\,\text{W}$$

$$\text{Field Cu loss} = V_t I_{sh} = 250 \times 2 = 500\,\text{W}$$

$$\text{Total Cu losses} = 1{,}620 + 500 = 2{,}120\,\text{W}$$

$$\therefore \qquad \text{Stray and friction losses} = [(P_{in})_1 - (P_{out})_1] - 2{,}120 = 3{,}000 - 2{,}120 = 880\,\text{W}$$

In second case $P_0 = 12$ kW

$$P_m = P_0 + \text{stray and friction losses} = 12{,}000 + 880 = 12{,}880\,\text{W}$$

Again,

$$P_m = E_{b2} I_{a2}$$

$$E_{b2} = V_t - I_{a2} R_a = 250 - 0.2 I_{a2}$$

i.e.,

$$E_{b2} I_{a2} = (250 - 0.2 I_{a2}) I_{a2}$$

i.e.,

$$12{,}880 = (250 - 0.2 I_{a2}) I_{a2}$$

i.e.,

$$0.2 I_{a2}^2 - 250 I_{a2} + 12{,}880 = 0$$

i.e.,

$$I_{a2} = \frac{250 \pm \sqrt{250^2 - 4 \times 0.2 \times 12{,}880}}{2 \times 0.2} = \frac{250 \pm 228.46}{0.4}$$

i.e.,

$$I_{a2} = 53.85\,\text{A (feasible)}$$

$$\therefore \qquad \text{Total Cu losses} = I_{a2}^2 R_a + V_{sh} I_{sh} = 53.85^2 \times 0.2 + 250 \times 2 = 1{,}079.96\,\text{W}$$

$$\therefore \qquad (P_{in})_2 = P_m + 1{,}079.96 = 12{,}880 + 1{,}079.96 = 13{,}959.96\,\text{W} = 13.959\,\text{kW}$$

$$\text{Efficiency} = \frac{\text{output}}{\text{input}} = \frac{12}{13.959} = 0.8596\,\text{p.u.} = 85.96\%$$ ■

Example 5.32 A 240 V DC shunt motor takes a current of 4 A from 240 V supply mains while running at no-load. The armature and field resistances are 0.2 Ω and 240 Ω, respectively. Determine (i) the efficiency of the motor when it draws 20 A from the supply mains, (ii) the armature current during maximum efficiency and (iii) the value of maximum efficiency.

Solution

$$V_t = 240\,\text{V},\ R_a = 0.6\,\Omega,\ R_{sh} = 240\,\Omega,\ I_0 = 4\,\text{A and } I_1 = 20\,\text{A}$$

(i)

$$I_{sh} = \frac{V_t}{R_{sh}} = \frac{240}{240} = 1\,\text{A}$$

$$I_{a0} = I_0 - I_{sh} = 4 - 1 = 3\,\text{A and } I_{a1} = I_1 - I_{sh} = 20 - 1 = 19\,\text{A}$$

$$E_{b0} = V_t - I_{a0} R_a = 240 - 3 \times 0.2 = 239.4\,\text{V}$$

and

$$E_{b1} = V_t - I_{a1} R_a = 240 - 19 \times 0.2 = 236.2\,\text{V}$$

∴ Power developed at no-load = iron and friction losses = $E_{b0} I_{a0} = 239.4 \times 3 = 718.2$ W

Mechanical power developed $(P_m) = E_{b1}I_{a1} = 236.2 \times 19 = 4{,}487.8$ W

$$\therefore \quad \text{Output power} = P_m - E_{b0}I_{a0} = 4{,}487.8 - 718.2 = 3{,}769.6 \text{ W}$$

$$\text{Input power } (P_{in}) = V_t I_1 = 240 \times 20 = 4{,}800 \text{ W}$$

$$\text{Efficiency} = \frac{\text{output}}{\text{input}} = \frac{3{,}769.6}{4{,}800} = 0.7853 \text{ p.u.} = 78.53\%$$ ■

(ii) For maximum efficiency,

$$\text{Constant losses} = \text{Cu losses}$$

$$\therefore \quad \text{Iron and friction losses} + \text{field Cu loss} = I_a^2 R_a$$

i.e.,
$$718.2 + 240 \times 1 = 0.2 I_a^2$$

i.e.,
$$I_a = 69.22 \text{ A}$$

The armature current at maximum efficiency is 69.22 A. ■

(iii)
$$E_b = V_t - I_a R_a = 240 - 69.22 \times 0.2 = 226.15 \text{ V}$$

$$\therefore \quad P_m = 226.15 \times 69.22 = 15{,}654.10 \text{ W}$$

$$P_{out} = P_m - (\text{iron and friction losses}) = 15{,}654.1 - 718.2 = 14{,}935.9 \text{ W}$$

$$\therefore \quad P_{in} = V_t I_a = 240 \times (69.22 + 1) = 16{,}852.8 \text{ W}$$

$$\text{Maximum efficiency} = \frac{\text{output}}{\text{input}} = \frac{14{,}935.9}{16{,}852.8} = 0.8862 \text{ p.u.} = 88.62\%$$ ■

Example 5.33 A 230 V DC shunt motor takes an armature current of 3.40 A at rated voltage and at no-load speed of 1,000 rpm. The armature and field circuit resistances are 0.4 Ω and 170 Ω, respectively. The line current at full-load and rated voltage is 41 A. Calculate (i) the speed at full-load and the developed torque in case the armature reaction weakens the no-load flux by 4 per cent, (ii) the shaft power if rotational losses at no-load and full-load are the same and (iii) the efficiency of the motor for case (ii).

Solution

(i) The back emf at no-load is given by

$$E_{b0} = V - I_{a0}R_a = 230 - 3.40 \times 0.4 = 228.64 \text{ V}$$

$$\therefore \quad \text{Field current } (I_f) = \frac{230}{170} = 1.353 \text{ A}$$

The full load current $(I_L) = 41$ A

During full-load, the armature current is given by

$$I_{a1} = I_L - I_f = 41 - 1.353 = 39.647 \text{ A}$$

∴ Back emf at full load is given by

$$E_{b2} = V_t - I_{a1} R_a = 230 - 39.647 \times 0.4 = 214.14 \text{ V}$$

At full load, the field flux is given by

$$\phi_1 = 0.96\, \phi_0$$

where ϕ_1 is the flux at no-load.

Since counter emf $E_b = K\phi N$,

$$\therefore \quad \frac{E_{b0}}{E_{b1}} = \frac{K\phi_0 N_0}{K\phi_1 N_1}$$

i.e.,
$$\frac{228.64}{214.14} = \frac{\phi_1 \times 1{,}000}{0.96\phi_0 \times N_1} \quad [\because N_0 = 1{,}000 \text{ rpm}]$$

i.e.,
$$N_1 = \frac{1{,}000}{0.96} \times \frac{214.14}{228.64} = 975.60 \text{ rpm}$$

$$E_{b1} = 214.14\text{V}, I_{a1} = 39.647 \text{ A and } N_1 = 975.60 \text{ rpm}$$

$$\therefore \quad T_a = 9.55 \times \frac{214.14 \times 39.647}{975.60} = 83.107 \text{ N-m}$$ ■

(ii) Rotational losses $(P_r) = E_{b0} I_{a0} = 228.64 \times 3.40 = 777.376$ W

∴ Shaft power (P_{sh}) = electromagnetic power – rotational losses

Since electromagnetic power or mechanical power (P_m)

$$= E_{b1} I_{a1} = 214.14 \times 39.647 = 8{,}490 \text{ W}$$

$$P_{sh} = P_m - P_r = 8{,}490 - 777.376 = 7{,}712.624 \text{ W}$$

$$\text{Shaft torque } (T_{sh}) = 9.55 \times \frac{7{,}712.624}{975.60} = 75.50 \text{ N-m}$$ ■

(iii) Input power = 230 × 41 = 9,430 W

Output power = shaft power = 7,712.624 W

$$\text{Efficiency } (\eta) = \frac{\text{output power}}{\text{input power}} = \frac{7{,}712.624}{9{,}430} = 0.8179 \text{ pu} = 81.79\%$$ ■

Example 5.34 **A 240 V DC shunt motor takes 26 A at rated voltage and runs at 1,000 rpm. Its field resistance is 120 Ω and armature circuit resistance (including brushes) is 0.25 Ω. Compute the value of additional resistance required in the armature circuit to reduce the speed to 900 rpm, when (i) the load torque is independent of speed, (ii) the load torque is proportional to speed and (iii) the load torque varies as square of the speed.**

Solution

Supply voltage (V_t) = 240 V

Supply current (I) = 26 A

Shunt field resistance (R_{sh}) = 120 Ω

Armature circuit resistance (R_a) = 0.25 Ω

$$\text{Field current } (I_f) = \frac{240}{120} = 2 \text{ A}$$

$$\text{Armature current } (I_a) = I - I_f = 26 - 2 = 24 \text{ A}$$

Here, the speed is to be reduced from 1,000 rpm to 900 rpm without disturbing the shunt field current. So the field flux (ϕ) remains constant.

(i) The load torque is independent of speed. Therefore, the electromagnetic torque remains constant at both speeds.

i.e.,
$$T_e = K\phi_1 I_{a1} = K\phi_2 I_{a2}$$
$$I_{a1} = I_{a2} = 24 \text{ A}$$

At 1,000 rpm, the back emf

$$E_{b1} = V_t - I_{a1}R_a = 240 - 24 \times 0.25 = 234 \text{ V}$$

At 900 rpm, the back emf

$$E_{b2} = V_t - I_{a2}(R_a + R) = 240 - 24 \times (0.25 + R) = (234 - 24R) \text{ V}$$

where R is the additional resistance connected in series with armature.

Since
$$E_b = K\phi N$$
$$\frac{E_{b1}}{E_{b2}} = \frac{K\phi_1 N_1}{K\phi_2 N_2} = \frac{N_1}{N_2} \quad \text{[since } \phi \text{ remains constant]}$$

i.e.,
$$\frac{234}{234 - 24R} = \frac{1,000}{900} = \frac{10}{9}$$

i.e.,
$$2,340 - 240R = 2,106$$

i.e.,
$$R = 0.975\ \Omega$$

The required additional resistance is 0.975 Ω. ■

(ii) In this case, $T_L \propto N$

The electromagnetic torque is given by $T_e = K\phi I_a$

$$K\phi_1 I_{a1} = N_1 \text{ and } K\phi_1 I_{a2} = N_2$$

∴
$$\frac{K\phi_1 I_{a1}}{K\phi_2 I_{a2}} = \frac{N_1}{N_2}$$

i.e.,
$$\frac{I_{a1}}{I_{a2}} = \frac{N_1}{N_2} \quad \text{[since } \phi \text{ remains constant]}$$

∴
$$I_{a2} = I_{a1}\left(\frac{N_2}{N_1}\right) = 24 \times \left(\frac{900}{1,000}\right) = 21.6 \text{ A}$$

∴
$$E_{b2} = 240 - 21.6 \times (0.25 + R) = (234.6 - 21.6R) \text{ V}$$

and
$$E_{b1} = 234 \text{ V}$$

∴
$$\frac{E_{b2}}{E_{b1}} = \frac{234.6 - 21.6R}{234} = \frac{900}{1,000} = \frac{9}{10}$$

i.e.,
$$2,346 - 216R = 2,106$$

∴
$$R = 1.11\ \Omega$$

The required additional resistance is 1.11 Ω. ■

(iii) In this case,

$$T_L \propto (N)^2$$

Similarly from part (ii), we have

$$\frac{I_{a1}}{I_{a2}} = \left(\frac{N_1}{N_2}\right)^2$$

i.e., $$I_{a2} = \left(\frac{N_2}{N_1}\right)^2 I_{a1} = 24 \times \left(\frac{900}{1{,}000}\right)^2 = 19.44 \text{ A}$$

$\therefore$ $$E_{b2} = 240 - 19.44 \times (0.25 + R)$$

$$= 235.14 - 19.44\,R$$

$$\frac{E_{b2}}{E_{b1}} = \frac{235.14 - 19.44R}{234} = \frac{900}{1{,}000} = \frac{9}{10}$$

i.e., $$2{,}351.4 - 194.4R = 2{,}106$$

i.e., $$R = 1.26\ \Omega$$

The required additional resistance is 1.26 Ω. ■

Example 5.35 A DC series motor running a fan at 1,000 rpm takes 50 A from 240 V mains. The armature plus field resistance is 0.4 Ω. If an additional resistance of 4.6 Ω is inserted in series with the armature circuit, determine the motor speed in case the field flux is proportional to the armature current. Assume the load torque for a fan is proportional to the square of the speed.

Solution

The load torque (T_L) is given by

$$T_L \propto N^2$$

The mechanical torque (T_m) is given by

$$T_m = K\phi I_a$$

For series motor, $$\phi \propto I_a$$

For series motor, $$T_m \propto I_a^2$$

$\therefore$ $$I^2_{a1} \propto N_1^2 \text{ and } I^2_{a2} \propto N_2^2$$

where I_{a1} is the armature current when $N_1 = 1{,}000$ rpm.

$\therefore$ $$\frac{I_{a1}^2}{I_{a2}^2} = \frac{N_1^2}{N_2^2}$$

$\therefore$ $$\frac{I_{a1}}{I_{a2}} = \frac{N_1}{N_2}$$

i.e., $$I_{a2} = \left(\frac{N_2}{N_1}\right) I_{a1} = 50 \times \frac{N_2}{1{,}000} = \frac{N_2}{20} \text{ A}$$

$\therefore$ Back emf

$$E_{b1} = 240 - 50 \times 0.4 = 220 \text{ V}$$

And $$E_{b2} = 240 - \frac{N_2}{20} \times (0.4 + 4.6) = 240 - \frac{N_2}{20} \times 5 = \left(240 - \frac{N_2}{4}\right) \text{V}$$

∴ $$\frac{E_{b2}}{E_{b1}} = \frac{K\phi_2 N_2}{K\phi_1 N_1}$$

i.e., $$\frac{E_{b2}}{E_{b1}} = \frac{I_{a2} N_2}{I_{a1} N_1} [\because \phi \propto I_a]$$

i.e., $$\frac{240 - \frac{N_2}{4}}{220} = \frac{N_2 \times \frac{N_2}{20}}{(1{,}000) \times (50)}$$

i.e., $$240 - \frac{N_2}{4} = \frac{N_2^2}{4{,}545.45}$$

i.e., $$N_2^2 + 1{,}136.36\, N_2 - 10{,}90{,}908 = 0$$

i.e., $$N_2 = \frac{-1{,}136.36 \pm 2{,}378.01}{2} = 620.82 \text{ rpm [feasible]}$$ ■

Example 5.36 A 10 kW, 240 V, DC shunt motor draws a line current of 5.2 A while running at no-load speed of 1,200 rpm from a 240 V DC supply. It has an armature resistance of 0.25 Ω and a field resistance of 160 Ω. Determine the efficiency of the motor when it delivers the rated load.

Solution

$$\text{Shunt field current} = \frac{240}{160} = 1.5 \text{ A}$$

∴ No-load armature current $(I_{a0}) = 5.2 - 1.5 = 3.7$ A

Constant no-load rotation loss (W) is given by

$$W = VI_{a0} - I_{a0}^2 R_a = 240 \times 3.7 - (3.7)^2 \times 0.25 = 884.57 \text{ W}$$

∴ Mechanical power = $E_b I_a$ = shaft power + (no-load rotational loss)

∴ $(240 - I_a \times 0.25) I_a = 10{,}000 + 884.57$

i.e., $0.25\, I_a^2 - 240\, I_a + 10{,}884.57 = 0$

i.e., $I_a^2 - 960\, I_a + 43{,}538.28 = 0$

i.e., $$I_a = \frac{960 \pm 864.55}{2}$$

∴ $I_a = 47.72$ A (feasible)

∴ Motor input = $E_b I_a$ + full-load armature Cu loss + shunt field Cu loss

$= 10{,}884.6 + (47.72)^2 \times 0.25 + 1.5 \times 240$

$= 11{,}813.89$ W

∴ $$\text{Motor efficiency} = \frac{10{,}000}{11{,}813.89} = 0.8464 \text{ p.u.} = 84.64\%$$ ■

Example 5.37 A 440 V DC shunt motor has a no-load speed of 2,000 rpm. It is running at 1,000 rpm at full-load torque. Its load torque is reduced to 50 per cent of rated value with armature voltage and field voltage held constant at previous values; the speed increases to 1,050 rpm. Find the armature voltage drop at full-load. Neglect the effect of armature reaction.

Solution

Let the armature current at full-load torque be I_{a1}.

$\therefore$ Back emf (E_{b1}) at full-load torque is given by

$$E_{b1} = (440 - I_a R_a) \propto N_1\phi_1$$

$\therefore$ Motor torque, $T_{m1} = K\phi_1 I_{a1}$ and $T_{m2} = K\phi_2 I_{a2}$

Since the field voltage is held constant, we have $\phi_2 = \phi_1$

$$\frac{T_{m1}}{T_{m2}} = \frac{K\phi_1 I_{a1}}{K\phi_2 I_{a2}}$$

i.e.,
$$\frac{T_{m1}}{T_{m2}} = \frac{I_{a1}}{I_{a2}}$$

i.e.,
$$2 = \frac{I_{a1}}{I_{a2}}$$

i.e.,
$$I_{a2} = \frac{I_{a1}}{2}$$

$\therefore$ Counter emf (E_{b2}) at half the rated torque is given by

$$E_{b2} = (440 - \frac{1}{2}I_{a1}R_a) \propto N_2\phi_2$$

$\therefore$
$$\frac{E_{b1}}{E_{b2}} = \frac{K\phi_1 N_1}{K\phi_2 N_2}$$

i.e.,
$$\frac{440 - I_{a1}R_a}{440 - \frac{1}{2}I_{a1}R_a} = \frac{1{,}000 \times \phi_1}{1{,}050 \times \phi_2} = \frac{20}{21} \quad [\text{since } \phi_1 = \phi_2]$$

i.e.,
$$440 \times 21 - 21I_{a1}R_a = 440 \times 20 - 10I_{a1}R_a$$

i.e.,
$$I_{a1}R_a = 40 \text{ V}$$

Armature voltage drop at full-load $= I_{a1} R_a = 40$ V. ■

Example 5.38 A 220 V, 1.5 kW, 859 rpm, separately excited DC motor has armature resistance of 2.5 Ω and it draws a current of 8 A at rated load condition. If the field current and the armature current are fixed at the value of rated speed at rated load, what will be the no-load speed of the motor? Assume losses remain constant between no-load and full-load operations.

Solution

Let E_{b1} and E_{b0} be the back emf at rated load and no load, respectively. During no-load, $I_{a0} = 0$, and at rated load, $I_{a1} = 8$ A (given)

Armature resistance (R_a) = 2.5 Ω

Supply voltage (V) = 220 V

Rated speed (N) = 859 rpm = N_1 (say)

∴ $$E_{b1} = V - I_a \times R_a = 220 - 8 \times 2.5 = 200 \text{ V}$$

And $$E_{b0} = V - R_a I_{a0} = 220 - 0 \times 2.5 = 220 \text{ V}$$

Since $$E_{b1} = \frac{\phi N_1 Z}{60}\left(\frac{P}{A}\right) \quad \text{and} \quad E_{b0} = \frac{\phi N_0 Z}{60}\left(\frac{P}{A}\right)$$

∴ $$\frac{E_{b1}}{E_{b0}} = \frac{N_1}{N_0}$$

i.e., $$N_0 = N_1 \times \left(\frac{E_{b0}}{E_{b1}}\right) = 859 \times \left(\frac{220}{200}\right) = 944.9 \text{ rpm}$$ ■

Example 5.39 A 220 V DC shunt motor develops a total torque of 90 N-m and draws a current of 27 A from supply mains and runs at 700 rpm. The armature and shunt field resistances are 0.2 Ω and 110 Ω, respectively. Determine the percentage reduction of the field flux and the additional resistance to be inserted in series with the field circuit if the speed is to be increased to 900 rpm. The gross torque at 900 rpm is 75 N-m. Assume that magnetization characteristic is a straight line and neglect the armature reaction.

Solution

V_t = 220 V, R_a = 0.2 Ω , R_{sh} = 110 Ω, N_1 = 700 rpm, N_2 = 900 rpm, T_{a1} = 90 N-m and T_{a2} = 70 N-m, I_1 = 27 A

Let ϕ_1 and ϕ_2 be the fluxes when the speeds are 700 rpm and 900 rpm, respectively.

$$I_{sh1} = \frac{V_t}{R_{sh}} = \frac{220}{110} = 2 \ \Omega$$

$$I_{a1} = I_a - I_{sh1} = 27 - 2 = 25 \text{ A}$$

$$E_{b1} = V_t - I_{a1}R_a = 220 - 25 \times 0.2 = 215 \text{ V}$$

$$T \propto \phi I_a$$

∴ $$\frac{T_1}{T_2} = \frac{\phi_1 I_{a1}}{\phi_2 I_{a2}} = \frac{1}{x} \times \frac{I_{a1}}{I_{a2}} \qquad \text{where } x = \frac{\phi_2}{\phi_1}$$

i.e., $$I_{a2} = \frac{T_2}{T_1} \times \frac{I_{a1}}{x} = \frac{75}{90} \times \frac{25}{x} = \frac{20.84}{x}$$

Again, $$\frac{E_{b2}}{E_{b1}} = \frac{N_2 \phi_2}{N_1 \phi_1}$$

i.e., $$E_{b2} = E_{b1} \times \frac{N_2}{N_1} \times \frac{\phi_2}{\phi_1} = \left(215 \times \frac{900}{700}\right) x = 276.43\, x$$

Again, $$E_{b2} = V_t - I_{a2} R_a$$

i.e., $$276.43x = 220 - \left(\frac{20.84}{x}\right) \times 0.2$$

i.e., $$276.43x^2 - 220x + 4.168 = 0$$

i.e., $$x = \frac{220 \pm \sqrt{220^2 - 4 \times 276.43 \times 4.168}}{2 \times 276.43} = \frac{220 \pm 209.26}{552.86} = 0.776 \text{ [feasible]}$$

i.e., $$\frac{\phi_2}{\phi_1} = 0.776$$

i.e., $$\frac{\phi_1 - \phi_2}{\phi_1} = 0.224$$

Therefore, percentage reduction of flux is 22.4 per cent.

Since the magnetization characteristic is linear, we have

$$\frac{\phi_2}{\phi_1} = \frac{I_{sh2}}{I_{sh1}}$$

i.e., $$\frac{I_{sh2}}{I_{sh1}} = 0.776$$

i.e., $$I_{sh2} = 0.776\, I_{sh1} = 0.776 \times 2 = 1.552 \text{ A}$$

If R is the additional resistance to be inserted in series with the field coil, we have

$$I_{sh2} = \frac{V_t}{R_{sh} + R}$$

i.e., $$1.552 = \frac{220}{110 + R}$$

i.e., $$R = \frac{220}{1.552} - 110 = 31.75\ \Omega$$

Therefore, a resistance of 31.75 Ω is required to be inserted in series with the field coil. ■

Example 5.40 A 400 V DC shunt motor drives a load and takes 25 A from supply mains running at a constant speed. The load torque varies with the square of the speed. If a resistance of 15 Ω is inserted in series with the armature, determine the new speed as a percentage of the original speed. Shunt field resistance of the motor is 200 Ω. Neglect armature resistance.

Solution

$$V_t = 400 \text{ V}, R_{sh} = 200\ \Omega, I_1 = 25 \text{ A}, R = 15\ \Omega$$

Let N_1 be the original speed and N_2 be the new speed.

$$I_{sh1} = \frac{V_t}{R_{sh}} = \frac{400}{200} = 2 \text{ A}$$

$$I_{a1} = I_1 - I_{sh} = 25 - 2 = 23 \text{ A}$$

Here, $$T \propto \phi I_a \propto I_a \text{ [since } \phi \text{ is constant]}$$

∴ $$\frac{T_1}{T_2} = \frac{I_{a1}}{I_{a2}} \quad (1)$$

Again, $$T \propto N^2$$

$\therefore$ $$\frac{T_1}{T_2} = \left(\frac{N_1}{N_2}\right)^2 \quad (2)$$

From Equations (1) and (2), we have

$$\left(\frac{N_1}{N_2}\right)^2 = \frac{I_{a1}}{I_{a2}}$$

i.e., $$I_{a2} = \left(\frac{N_2}{N_1}\right)^2 I_{a1} = x^2 I_{a1} = 23x^2 \quad (3)$$

where $$x = \frac{N_2}{N_1}.$$

Again, $E_{b1} = 400$ V [since R_a is negligible]

and $E_{b2} = V_t - I_{a2}R = 400 - 15I_{a2}$

Since, $\frac{E_{b2}}{E_{b1}} = \frac{N_2}{N_1}$, we have

$$\frac{400 - 15I_{a2}}{400} = x$$

i.e., $$\frac{400 - 15 \times 23x^2}{400} = x$$

i.e., $$345x^2 + 400x - 400 = 0$$

i.e., $$x = \frac{-400 \pm \sqrt{400^2 + 4 \times 345 \times 400}}{2 \times 345} = \frac{-400 \pm 843.8}{690} = 0.643 \text{ [feasible]}$$

i.e., $$\frac{N_2}{N_1} = 0.643$$

i.e., $$N_2 = 0.643 \times N_1$$

Hence, new speed is 64.3 per cent of original speed. ■

Example 5.41 A 240 V DC shunt motor draws a current of 50 A from supply mains at full-load while running at a speed of 1,000 rpm. It delivers 10.5 kW at full-load. The armature and shunt field resistances are 0.2 Ω and 240 Ω, respectively. If a resistor is to be inserted in series with the armature to reduce the speed to 0.85 times of the full-load speed, determine the resistance of the resistor and efficiency at this speed. The constant loss varies as $(\text{speed})^{1.3}$. Assume that field and armature currents remain constant.

Solution

$V_t = 240$ V, $P_{out} = 10.5$ kW, $I_1 = 50$ A, $N_1 = 1{,}000$ rpm, $R_a = 0.2\ \Omega$, $R_{sh} = 240\ \Omega$

Power input $(P_{in}) = V_t I_1 = 240 \times 50 = 12{,}000$ W

Total losses on full-load $= P_{in} - P_{out} = 1.5$ kW

$$I_{sh} = \frac{240}{240} = 1 \text{ A}$$

$$I_{a1} = 50 - 1 = 49 \text{ A}$$

$$E_{b1} = V_t - I_{a1}R_a = 240 - 49 \times 0.2 = 230.2 \text{ V}$$

and $$E_{b2} = V_t - I_{a2}(R_a + R) = 240 - 49 \times (0.2 + R) \text{ [since } I_a \text{ remains constant]}$$

$$= 230.2 - 49R$$

$$N_2 = 0.85\, N_1$$

We have that $$\frac{N_2}{N_1} = \frac{E_{b2}}{E_{b1}} \times \frac{\phi_1}{\phi_2} = \frac{E_{b2}}{E_{b1}} \quad \text{[since } \phi \text{ is constant]}$$

i.e., $$\frac{0.85N_1}{N_1} = \frac{230.2 - 49R}{230.2}$$

i.e., $$R = \frac{230.2 \times (1 - 0.85)}{49} = 0.705\ \Omega$$

$$\text{Full-load armature Cu loss} = I_a^2 R_a = 49^2 \times 0.2 = 480.2\ \Omega$$

$$\text{Shunt field Cu loss} = V_t I_{sh} = 240 \times 1 = 240 \text{ W}$$

$$\text{Total full-load Cu loss} = 480.2 + 240 = 720.2 \text{ W}$$

$$\text{Therefore, constant losses at full-load} = 1{,}500 - 720.2 = 779.8 \text{ W}$$

$$\therefore \quad \text{Constant losses at 0.85 times of full-load} = (0.85)^{1.3} \times 779.8 = 631.28 \text{ W}$$

Therefore, Cu losses in the armature circuit at 0.85 times of full-load speed $= I_a^2 (R_a + R)$

$$= 49^2 \times (0.2 + 0.705) = 2172.905 \text{ W}$$

$$\text{Therefore, total losses} = 2{,}172.905 + 240 + 631.28 = 3{,}044.185 \text{ W}$$

$$\text{Output} = 12{,}000 - 3{,}044.185 = 8{,}955.815 \text{ W}$$

$$\therefore \quad \text{Efficiency} = \frac{\text{output}}{\text{input}} = \frac{8{,}955.815}{12{,}000} = 0.7463 \text{ p.u.} = 74.63\%$$ ■

Example 5.42 The full-load rating of a 250 V DC shunt motor running at 1,000 rpm is 10 kW. The armature and shunt field resistances are 0.5 Ω and 250 Ω, respectively. The full-load efficiency is 90 per cent. Calculate the following:

(i) no-load current,
(ii) speed when the motor takes 25 A and
(iii) the armature current when the torque produced is 90 N-m and speed 900 rpm.

Solution

$$V_t = 250 \text{ V},\ N_1 = 1{,}000 \text{ rpm},\ P_{out} = 10 \text{ kW},\ R_a = 0.5\ \Omega,\ R_{sh} = 250\ \Omega,\ \eta = 90\%$$

$$\text{Input to motor} = \frac{10 \times 10^3}{0.9} = 11{,}111.12 \text{ W}$$

$$\text{Line current (I)} = \frac{11{,}111.12}{250} = 44.44\text{ A}$$

$$I_{sh} = \frac{V_t}{R_{sh}} = \frac{250}{250} = 1\,\text{A}$$

$$I_a = I - I_{sh} = 44.44 - 1 = 43.44\text{ A}$$

$$E_b = V_t - I_a R_a = 250 - 43.44 \times 0.3 = 236.97\text{ V}$$

$$\text{Armature Cu loss} = I_a^2 R_a = 43.44^2 \times 0.3 = 566.11\,\text{V}$$

Therefore, constant losses = total losses − armature Cu losses = (input − output) − 566.11

= (11,111.12 − 10,000) − 566.11 = 545 W

Neglecting armature Cu loss at no-load, the no-load input = 545 W.

(i) No-load line current $= \frac{545}{250} = 2.18\,\text{A}$

(ii) Back emf when line current is 25 A is given by

$$E_{b1} = 250 - (25 - 1) \times 0.3 = 242.8\text{ V}$$

Assuming flux to be constant,

$$\frac{N_1}{N} = \frac{E_{b1}}{E_b} \quad \text{i.e.,} \quad N_1 = N \times \frac{E_{b1}}{E_b} = 1{,}000 \times \frac{242.8}{236.97} = 1024.6\text{ rpm}$$

(iii) Let the armature current be I_{a2} for $T_{a2} = 90$ N-m and speed $N_2 = 900$ rpm

$$T = \frac{E_{b2} I_{a2}}{N} \times 9.55$$

i.e.,
$$90 = \frac{9.55(V_t - I_{a2} R_a) I_{a2}}{900}$$

i.e.,
$$(250 - 0.5 I_{a2}) I_{a2} = \frac{90 \times 900}{9.55}$$

i.e.,
$$0.5 I_a^2 - 250 I_a + 8{,}481.67 = 0$$

i.e.,
$$I_a = \frac{250 \pm 213.4}{2 \times 0.5}$$

i.e.,
$$I_a = 36.6\,[\text{feasible}]$$

Example 5.43 A 240 V DC shunt motor has armature resistance of 0.40 Ω. The maximum current limit is to be 65 A and the minimum is three-fourths of this value. Find the number of sections of the starter and calculate the resistance of each element.

Solution

Here,
$$\frac{I_2}{I_1} = \frac{3}{4} = \alpha$$

We know that $$\alpha = \left[\frac{I_1 R_a}{V_t}\right]^{\frac{1}{n}}$$

i.e., $$\frac{3}{4} = \left[\frac{65 \times 0.4}{240}\right]^{\frac{1}{n}}$$

i.e, $$n = \frac{ln\left[\frac{65 \times 0.4}{240}\right]}{ln(0.75)} = 7.726$$

Let us take $n = 8$.

Therefore, the number of starter element is 8.

$$\frac{R_a}{R_1} = \alpha^n = \left(\frac{3}{4}\right)^8 \text{ i.e., } R_1 = \left(\frac{4}{3}\right)^8 \times R_a = \left(\frac{4}{3}\right)^8 \times 0.40 = 3.995\ \Omega$$

$$\frac{R_2}{R_1} = \alpha \text{ i.e., } R_2 = \frac{3}{4} \times 3.995 = 2.996\ \Omega$$

$$\frac{R_3}{R_2} = \alpha \text{ i.e., } R_3 = \frac{3}{4} \times 2.996 = 2.247\ \Omega$$

$$\frac{R_4}{R_3} = \alpha \text{ i.e., } R_4 = \frac{3}{4} \times 2.247 = 1.685\ \Omega$$

$$\frac{R_5}{R_4} = \alpha \text{ i.e., } R_5 = \frac{3}{4} \times 1.685 = 1.264\ \Omega$$

$$\frac{R_6}{R_5} = \alpha \text{ i.e., } R_6 = \frac{3}{4} \times 1.264 = 0.948\ \Omega$$

$$\frac{R_7}{R_6} = \alpha \text{ i.e., } R_7 = \frac{3}{4} \times 0.948 = 0.711\ \Omega$$

$$\frac{R_8}{R_7} = \alpha \text{ i.e., } R_8 = \frac{3}{4} \times 0.711 = 0.533\ \Omega$$

$\therefore$ $$r_1 = R_1 - R_2 = 3.995 - 2.996 = 0.999\ \Omega$$

$\therefore$ $$r_2 = R_2 - R_3 = 2.996 - 2.247 = 0.749\ \Omega$$

$\therefore$ $$r_3 = R_3 - R_4 = 2.247 - 1.685 = 0.526\ \Omega$$

$\therefore$ $$r_4 = R_4 - R_5 = 1.685 - 1.264 = 0.421\ \Omega$$

$\therefore$ $$r_5 = R_5 - R_6 = 1.264 - 0.948 = 0.316\ \Omega$$

$\therefore$ $$r_6 = R_6 - R_7 = 0.948 - 0.711 = 0.237\ \Omega$$

$\therefore$ $$r_7 = R_7 - R_8 = 0.711 - 0.533 = 0.178\ \Omega$$

$\therefore$ $$r_8 = R_8 - R_a = 0.533 - 0.400 = 0.133\ \Omega$$

Example 5.44 Calculate the values of the five steps in a six-stud starter for a 5 HP (metric), 220 V, shunt motor having full-load efficiency of 89 per cent. The current is not to exceed twice the full-load current. Armature Cu loss is half the total loss. Field current is 0.62 A.

Solution

$$\text{Losses} = \frac{1-\eta}{\eta} \times \text{output} = \frac{1-0.89}{0.89} \times 5 \times 735 = 454.22 \text{ W}$$

$$\text{Input current} = \frac{\text{output}}{\eta \times V_t} = \frac{5 \times 735}{0.89 \times 220} = 18.77 \text{ A}$$

$$\text{Armature current } (I_a) = 18.77 - 0.62 = 18.15 \text{ A}$$

$$\text{Armature Cu loss} = \frac{1}{2} \times \text{total losses} = \frac{1}{2} \times 454.22 = 227.11 \text{ W}$$

$$I_a^2 R_a = 227.11$$

i.e.,

$$R_a = \frac{227.11}{I_a^2} = \frac{227.11}{18.15^2} = 0.689 \ \Omega$$

$$\text{Maximum current} = I_1 = 2 \times I_{fl} = 2 \times 18.15 = 36.3 \text{ A}$$

$$\alpha = \left[\frac{I_1 R_a}{V_t}\right]^{\frac{1}{n}} = \left[\frac{36.3 \times 0.689}{220}\right]^{\frac{1}{5}} = 0.647$$

$$\frac{R_a}{R_1} = \alpha^n = (0.647)^5 \quad \text{i.e., } R_1 = \frac{R_a}{(0.647)^5} = \frac{0.689}{(0.647)^5} = 6.077 \ \Omega$$

$$\frac{R_2}{R_1} = \alpha \quad \text{i.e., } R_2 = 0.647 \times 6.077 = 3.932 \ \Omega$$

$$\frac{R_3}{R_2} = \alpha \quad \text{i.e., } R_3 = 0.647 \times 3.932 = 2.544 \ \Omega$$

$$\frac{R_4}{R_3} = \alpha \quad \text{i.e., } R_4 = 0.647 \times 2.544 = 1.646 \ \Omega$$

$$\frac{R_5}{R_4} = \alpha \quad \text{i.e., } R_5 = 0.647 \times 1.646 = 1.065 \ \Omega$$

$\therefore$ $$r_1 = R_1 - R_2 = 6.077 - 3.932 = 2.145 \ \Omega$$

$\therefore$ $$r_2 = R_2 - R_3 = 3.932 - 2.544 = 1.388 \ \Omega$$

$\therefore$ $$r_3 = R_3 - R_4 = 2.544 - 1.646 = 0.898 \ \Omega$$

$\therefore$ $$r_4 = R_4 - R_5 = 1.646 - 1.065 = 0.581 \ \Omega$$

$\therefore$ $$r_5 = R_5 - R_6 = R_5 - R_a = 1.065 - 0.689 = 0.376 \ \Omega$$

Example 5.45 A 20 HP, 500 V, DC shunt motor having efficiency of 88 per cent at rated load is required to reach 1,000 rpm in 30 seconds from rest against full-load torque. The armature and field resistances of the motor are 0.33 Ω and 250 Ω, respectively. The moment of inertia of the moving part is 13.85 kg-m². A starter is required to design having capability to limit the maximum line current to 52 A. Assume that the average current during the staring period is equal to the mean of the minimum and maximum current.

Solution

Let J be the moment of inertia of the system in kg²m and ω be the angular velocity in rad/second.

$$\therefore \quad E = \text{energy stored in the moving parts} = \frac{1}{2}J\omega^2$$

$$\therefore \quad P = \frac{dE}{dt} = J\omega\frac{d\omega}{dt} = J\left(\frac{2\pi}{60}\right)^2 N\frac{dN}{dt}$$

where $$\omega = \frac{2\pi N}{60}.$$

Let I_a be the current in excess for acceleration.

$$\therefore \quad I_a = \frac{I_{max} + I_{min}}{2} - I_{fl} = \frac{I_1 + I_2}{2} - I_{fl}$$

where I_{fl} is the full-load current.

Here, $V_t = 500$ V, $J = 13.85$ kg-m², $N = 1{,}000$ rpm and $\frac{dN}{dt} = \frac{1{,}000}{30}$.

$$\text{Power} = V_t I_a = 13.85 \times 0.0109 \times 1{,}000 \times \frac{1{,}000}{30}$$

Therefore, $$I_a = \frac{13.85 \times 0.0109 \times 1{,}000 \times 1{,}000}{30 \times V_t} = 10.06 \text{ A}$$

$$\text{Maximum armature current} = I_1 = 52 - \frac{500}{250} = 50 \text{ A}$$

$$\text{Full-load current} = I_{fl} = \frac{20 \times 735}{500 \times 0.88} = 33.41 \text{ A}$$

$$\therefore \quad \frac{I_1 + I_2}{2} = I_a + I_{fl} = 10.06 + 33.41 = 43.47 \text{ A}$$

i.e., $$I_1 + I_2 = 86.94 \text{ A}$$

Minimum armature current $(I_2) = 86.94 - I_1 = 86.94 - 50 = 36.94$ A

$$\therefore \quad \alpha = \frac{I_2}{I_1} = \frac{36.94}{50} = 0.7388$$

$$\therefore \quad \alpha'' = \frac{I_1 R_a}{V_t} = \frac{50 \times 0.33}{500} = 0.033$$

$$\therefore \qquad n = \frac{ln(0.033)}{ln(0.7388)} = 11.26$$

Let the number of starter element be 12.

We know that $\frac{R_a}{R_1} = \alpha^n = 0.033.$

Therefore, $R_1 = \frac{R_a}{\alpha^n} = \frac{0.33}{0.033} = 10\ \Omega$

$$\therefore \qquad R_2 = \alpha \times R_1 = 0.7388 \times 10 = 7.388\ \Omega$$

$$\therefore \qquad R_3 = \alpha \times R_2 = 0.7388 \times 7.388 = 5.458\ \Omega$$

$$\therefore \qquad R_4 = \alpha \times R_3 = 0.7388 \times 5.458 = 4.010\ \Omega$$

$$\therefore \qquad R_5 = \alpha \times R_4 = 0.7388 \times 4.010 = 2.963\ \Omega$$

$$\therefore \qquad R_6 = \alpha \times R_5 = 0.7388 \times 2.963 = 2.189\ \Omega$$

$$\therefore \qquad R_7 = \alpha \times R_6 = 0.7388 \times 2.189 = 1.617\ \Omega$$

$$\therefore \qquad R_8 = \alpha \times R_7 = 0.7388 \times 1.617 = 1.195\ \Omega$$

$$\therefore \qquad R_9 = \alpha \times R_8 = 0.7388 \times 1.195 = 0.8828\ \Omega$$

$$\therefore \qquad R_{10} = \alpha \times R_9 = 0.7388 \times 0.8828 = 0.6522\ \Omega$$

$$\therefore \qquad R_{11} = \alpha \times R_{10} = 0.7388 \times 0.6522 = 0.4818\ \Omega$$

$$\therefore \qquad R_{12} = \alpha \times R_{11} = 0.7388 \times 0.4818 = 0.3559\ \Omega$$

$$\therefore \qquad r_1 = R_1 - R_2 = 10 - 7.388 = 2.612\ \Omega$$

$$\therefore \qquad r_2 = R_2 - R_3 = 7.388 - 5.458 = 1.930\ \Omega$$

$$\therefore \qquad r_3 = R_3 - R_4 = 5.458 - 4.010 = 1.448\ \Omega$$

$$\therefore \qquad r_4 = R_4 - R_5 = 4.010 - 2.963 = 1.047\ \Omega$$

$$\therefore \qquad r_5 = R_5 - R_6 = 2.963 - 2.189 = 0.774\ \Omega$$

$$\therefore \qquad r_6 = R_6 - R_7 = 2.189 - 1.617 = 0.572\ \Omega$$

$$\therefore \qquad r_7 = R_7 - R_8 = 1.617 - 1.195 = 0.422\ \Omega$$

$$\therefore \qquad r_8 = R_8 - R_7 = 1.195 - 0.8828 = 0.3122\ \Omega$$

$$\therefore \qquad r_9 = R_9 - R_8 = 0.8828 - 0.6522 = 0.2306\ \Omega$$

$$\therefore \qquad r_{10} = R_{10} - R_9 = 0.6522 - 0.4818 = 0.1704\ \Omega$$

$$\therefore \qquad r_{11} = R_{11} - R_{10} = 0.4818 - 0.3559 = 0.1259\ \Omega$$

$$\therefore \qquad r_{12} = R_{12} - R_a = 0.3559 - 0.33 = 0.0259\ \Omega$$

SIGNIFICANT POINTS

Shunt Motor:

$$I = I_a + I_{sh}$$
$$V_t = E_b + I_a R_a$$

Series Motor:

$$I = I_a = I_{se}$$
$$V_t = E_b + I_a (R_a + R_{se})$$
$$V_t > E_b \text{ for motoring mode}$$

The electrical power input = $V_t I_a$

The gross mechanical power developed in the armature $(P_m) = E_b I_a$

Back emf:

$$E_b = \frac{\phi NZ}{60}\left(\frac{P}{A}\right)$$

Armature Torque:

$$T_a = \frac{1}{2\pi} Z \left(\frac{P}{A}\right)\phi I_a = 0.159Z\left(\frac{P}{A}\right)\phi I_a$$
$$T_a = 9.55\left(\frac{E_b I_a}{N}\right), \text{ where } N \text{ is in rpm}$$

SHORT QUESTIONS AND ANSWERS

Q.1 What do you mean by a DC motor?

Ans. It is a DC machine, which converts electrical energy into mechanical energy.

Q.2 What is the working principle of a DC motor?

Ans. If a current-carrying conductor is placed in a magnetic field, a force is exerted on the conductor. The direction of force acting on the conductor is determined by Fleming's left-hand rule. This is the working principle of a DC motor.

Q.3 What are the types of DC motors?

Ans. (i) Series wound, (ii) shunt wound and (iii) compound wound.

Q.4 What will be the armature current in a DC motor?

Ans. Let V_t be the applied voltage and E_b be the back emf. Then the effective emf acting around the armature circuits is $V_t - E_b$. Therefore, the armature current $I_a = \frac{V_t - E_b}{R_a}$, where R_a is the armature resistance.

Q.5 What is mechanical power in a DC motor?

Ans. Mechanical power = $E_b I_a = V_t I_a - I_a^2 R_a$, where $V_t I_a$ represents power in watt supplied to the armature and $I_a^2 R_a$ is the armature copper loss in watt.

Q.6 What is the condition for maximum power?

Ans. The condition for maximum power is $E_b = \frac{V_t}{2}$.

Q.7 Where is torque maximum?

Ans. Torque is maximum when it is at right angles to the magnetic flux.

Q.8 What is the torque relation in a DC motor?

Ans. The toque in DC motor is proportional to the flux per pole and the armature current, that is, $T \propto \phi I_a$.

Q.9 What are the losses in a DC motor?

Ans. The losses in a DC motor are as follows

(a) Copper losses
   (i) Armature copper losses ($I_a^2 R_a$)
   (ii) Field copper losses ($I_{sh}^2 R_{sh}$)
   (iii) Loss due to brush contact resistance ($E_b I_a$, where E_b is the drop due to brush resistance)

(b) Magnetic losses occur in the core of the motor
   (i) Hysteresis loss
   (ii) Eddy current loss

(c) Mechanical losses
   (i) Windage loss due to rotating armature
   (ii) Frictional loss at the bearings
   (iii) Frictional loss due to brush contacts

Mechanical losses and magnetic losses are termed as stray losses (W_1). Stray losses and shunt field copper losses are known as constant losses.

Q.10 What do you mean by the commercial efficiency of a DC motor?

Ans. The commercial or overall efficiency of a DC motor,

$$\eta_e = \frac{\text{output in watt}}{\text{motor input in watt}} \times 100\%$$

Q.11 What is the condition for maximum efficiency in a DC motor?

Ans. The condition for maximum efficiency in a DC motor is that copper losses are equal to constant losses.

Q.12 What is the speed relation in a DC motor?

Ans. The speed of a DC motor is directly proportional to the back emf and inversely proportional to the flux pole:

$$N \propto \frac{E_b}{\phi} \text{ i.e., } N \propto \frac{V - I_a R_a}{\phi}$$

Q.13 What will be the result if the field circuit of a DC motor is opened?

Ans. Due to weakening of field flux, the armature will race up to abnormally high speeds, and due to the severe centrifugal forces, the armature windings may get permanently damaged.

Q.14 What is meant by speed regulation of a DC motor?

Ans. Speed regulation of a DC motor is defined as the change in speed when the load on the motor is reduced from rated load to zero and is expressed in percentage of the rated load speed.

$$\therefore \qquad \%\text{ speed regulation} = \frac{\text{no-load speed} - \text{full-load speed}}{\text{full-load speed}} \times 100$$

Q.15 What are the characteristics of a DC motor?

Ans. There are two main characteristics: electrical and mechanical characteristics.

(a) Electrical characteristics are

(i) speed (N) versus armature current (I_a) and

(ii) torque (T) versus armature current (I_a).

(b) Mechanical characteristic is torque (T) versus speed (N).

Q.16 Why does a shunt motor run at a constant speed?

Ans. We know that $N \propto \frac{E_b}{\phi}$. As the shunt field resistance is constant, the shunt field current is constant. Accordingly, the flux per pole is practically constant. The back emf is also practically constant. Hence, by speed relation, it is seen that the speed of a DC shunt motor is constant.

Q.17 Why is the speed of a DC shunt motor dependent on back emf only?

Ans. The speed of a DC shunt motor is dependent on back emf because the flux is practically constant in a DC shunt motor.

Q.18 What are the characteristics of a DC shunt motor?

Ans. (i) The starting torque is poor.

(ii) The speed of the shunt motor is constant.

Q.19 What are the uses of a DC shunt motor?

Ans. Due to constant speed operation, DC shunt motors are used for driving line shaft, machine tools, lathe, wood working machines, milling machine, weaving machine, small printing press and so on.

Q.20 What do you mean by a DC series motor?

Ans. A DC motor in which the field winding is connected in series with the armature winding is known as a DC series motor.

Q.21 What are the characteristics of a DC series motor?

Ans. (i) Starting torque is very high.

(ii) Speed decreases when the load increases.

(iii) Without load it runs at dangerously high speed.

Q.22 What will be the speed of a DC series motor at no-load?

Ans. The speed of a DC series motor at no-load is infinity.

Q.23 Why does a DC series motor run at a dangerous speed without load?

Ans. In a DC series motor, $\phi \propto I_a$. When load is removed, the armature current as well as the field flux becomes very small and at the same time the back emf increases. According to the speed relation, the motor overruns and attains a dangerously high speed.

Q.24 Which type of DC motor has high starting torque but the poorest speed regulation for the same horsepower rating and full-load speed?

Ans. A DC series motor.

Q.25 What do you mean by a DC compound motor?

Ans. The DC motor in which a part of the field winding is connected in series with the armature winding and the other part of the field winding is connected in parallel with the armature winding is known as a DC compound motor.

Q.26 What are the advantages of a DC compound motor?

Ans. The DC compound motor can be operated as a series motor or shunt motor and the speed can be increased or decreased to a greater extent from the normal speed.

Q.27 What are the characteristics of a cumulative compound motor?

Ans. The characteristics of cumulative compound motor lie between those of shunt and series motors. The series field provides a high starting torque and the shunt field prevents overrunning in no-load condition.

Q.28 Where are cumulative compound motors used?

Ans. These motors are used for drives where high starting torque is required with the probability of the load being totally removed such as punch, press, shears, planning machine, conveyors, crushers, bulldozers, lift haulage gears, mine hoist, power fans, rolling mills, stamping press and large printing press.

Q.29 What do you mean by a differential compound motor?

Ans. The compound motor in which the series field winding and the shunt field windings oppose each other, acting in the reverse direction for the production of magnetism, is known as a differential compound motor.

Q.30 If a cumulative compound long-shunt generator is operated as a motor, what type of motor would it be?

Ans. It would be a differential compound long-shunt motor.

Q.31 What is brake test and where is the test used?

Ans. Brake test is a direct method of testing a DC machine in which the machine is actually loaded by applying some brakes to a water-cooled pulley mounted on the motor shaft. The brake band is fixed with the help of wooden blocks gripping the pulley. One end of the band is fixed to the floor via a spring balance and the other end is connected to a suspended weight.

This method of testing can be used for small motors only because in a large machine it is difficult to dissipate the amount of heat developed at the brake.

Q.32 What is Swinburne's test and where is this test used?

Ans. Swinburne's test is the simplest indirect method of testing a DC machine without actually loading it, in which only losses are measured separately. From this knowledge, efficiency at any load can be predetermined only by calculation. This test can only be applied to shunt and compound machines where the field flux is practically constant by running at the machine at light load as a motor at its rated voltage.

Q.33 Is Swinburne's test applicable to series motors? Give reason for your answer.

Ans. No, because a series motor cannot be run at no-load.

Q.34 How can you determine windage and friction losses of a DC shunt motor?

Ans. Running at no-load with normal voltage, the terminal voltage is reduced gradually till the motor stops. Multiplication of the voltage and the current obtained in this condition will be the windage and friction loss in watts.

Q.35 What is Hopkinson's test and where is it applied?

Ans. Hopkinson's test, also known as back-to-back test, is a regenerative method of testing in which two identical DC machines are mechanically coupled and electrically adjusted so that one of them acts as a motor and the other as a generator. The electrical power developed by the generator is utilized in driving the motor, while the mechanical power of the motor is supplied to drive the generator. Under this test, the electrical power will be required from the supply mains just to overcome the internal losses of the two machines. In this method, two identical machines of any size can be tested under full-load condition.

This method is applied for testing large machines on full-load condition for longer duration.

Q.36 What are the factors which control the speed of a DC motor?

Ans. Speed control depends upon the following factors:

(i) voltage control,

(ii) field flux control and

(iii) armature current control.

Q.37 What is field control?

Ans. Field control is a method of controlling speed of a DC shunt motor above normal by regulating the field flux by means of a field regulating variable resistor connected in series with the field winding.

Q.38 What types of resistances are employed in armature circuit and field circuit for speed control?

Ans. A resistance of low ohmic value and high current-carrying capacity is employed in armature circuit, whereas a resistance of high ohmic value and low current-carrying capacity is employed in field circuit.

Q.39 What is meant by Ward–Leonard method of speed control?

Ans. Ward–Leonard method is the basic adjustable armature voltage control method of controlling speed of a DC shunt motor over the whole range from zero to normal speed in both directions, accomplished by means of an adjustable voltage generator.

Q.40 Where is Ward-Leonard method employed?

Ans. Ward-Leonard method of speed control is employed where fine speed adjustment and smooth acceleration are required, as in driving rolling mills, ship propulsion, colliery winders and elevators in both directions.

Q.41 What are the methods of speed control of a DC series motor?

Ans. The speed of a DC series motor can be controlled by the following methods:

(a) Flux control method

  (i) By field diverters,

  (ii) armature diverters

  (iii) tapped field control and

  (iv) paralleling field coils

(b) Variable resistance in series with the motor

Q.42 What do you understand by the field diverter?

Ans. It is a variable resistance of low ohmic value and high current-carrying capacity connected in parallel to the field winding.

Q.43 What is the difference between resistances used for field control method in shunt and series motors?

Ans. The resistance used for the shunt motor is of high ohmic value and low current-carrying capacity, whereas the resistance used for the series motor is of low ohmic value and high current-carrying capacity.

Q.44 What are the functions of a starter?

Ans. The functions of a starter are:

(a) to limit the starting current,

(b) to protect the motor from taking overload current,

(c) to protect the motor against supply failure and

(d) to facilitate the starting and stopping of the motor.

Q.45 What is the difference between a starter and a regulator?

Ans. Both the starter and regulator consist of resistances, the value of which is comparatively greater in the regulator. The starter resistance is connected in series with the armature resistance to limit the starting current, while the regulator is connected in series with the field winding to control the speed of the motor. The regulator consists of only resistance, but the starter consists of one no-volt release and one overload release in addition to starting resistance. A regulator has two terminals, whereas a starter has three or four terminals.

Q.46 Why are small motors connected directly to supply lines without starters?

Ans. The reasons are as follows:

(i) Small motors have relatively higher armature resistance than large motors. So their starting current is not very high.

(ii) They speed up quickly due to low moment of inertia.

(iii) The momentary large starting current does not disturb the voltage regulation of supply lines.

Q.47 What are the types of starters used in DC shunt and compound motors?

Ans. The two types are:

(i) three-point face plate box-type starter and

(ii) four-point face plate box-type starter.

Q.48 Why is the no-volt release coil of a three-point starter connected in series with the field and not with the armature of the motor?

Ans. The no-volt release coil of a three-point shunt motor stator is connected in series with the field and not with the armature of the motor due to the following reasons:

(i) If the shunt field circuit becomes open circuited accidentally, the hold on the magnet of the no-volt release coil connected in series with the field gets demagnetized and the starter arm is returned to its OFF position by the pull of the spring. This prevents the motor from running away. This protection can be possible if the non-volt release coil is connected in series with the armature.

(ii) The no-volt release coil requires a steady current of small value. The armature current of a shunt motor varies with the load, the range of which is much higher than that required by a no-volt release coil. So a no-volt release coil will burn out if it is connected in series with the low-resistance high armature current.

Q.49 Why cannot the three-point face plate-type DC starter be used as a speed regulator of the motor?

Ans. In a three-point shunt motor starter, the hold on the magnet is connected in series with the shunt field winding. If the motor is required to be operated at a speed higher than the normal speed by field flux control, the field current has to be decreased by the operating speed regulating resistance. At a certain value of reduced shunt field current, the pull on the hold on the magnet may become less than the spring force. In that case, the starting arm returns to the OFF position and the motor stops. Thus, a three-point starter is not suitable for an application where speed regulation by field flux control is required.

Q.50 What is the advantage of the four-point starter over the three-point starter?

Ans. Speed control is possible by varying the field current which does not affect the operation of the hold-on coil because the field circuit and no-volt coil circuit are made independent of each other.

Q.51 What are the types of starters used in series motors?

Ans. Regulator types:

(i) one having a no-volt release and

(ii) the other having a no-load release.

Q.52 If supply voltage is less than the rated voltage, what will be the effect on the performance of a DC motor?

Ans. The speed as well as the output of the motor will be reduced.

Q.53 What are the reasons if fuses are blown out at the time of starting?

Ans. Fuses may be blown out at the time of starting due to any one of the following reasons:

(a) Lower capacity fuses are provided.

(b) The motor is much overhead.

(c) Starting resistance is not in action due to short circuit developed in it.

(d) Short circuit is developed in armature winding.

(e) Earth fault is developed in the field or armature winding.

SUPPLEMENTARY PROBLEMS

1. A shunt generator delivers 50 kW at 250 V and 400 rpm. The armature and field resistances are 0.02 Ω and 50 Ω, respectively. Find the speed of the machine running as a shunt motor and taking 50 kW input at 250 V. Allow 1 V per brush for contact drop.

   [*Ans.* 328 rpm]

2. A 100 kW belt-driven shunt generator running at 300 rpm on 220 V bus bars continues to run as a motor when it breaks and it takes 10 kW. What will be its speed? The armature resistance is 0.025 Ω, field resistance is 60 Ω and contact drop under each brush is 1 V. Ignore armature reaction.

   [*Ans.* 279 rpm]

3. A 250 V, four-pole, DC shunt motor has two-circuit armature winding with 500 conductors. The armature circuit resistance is 0.25 Ω, field resistance is 125 Ω and the flux/pole is 0.02 Wb. If the motor draws 14 A from supply mains, calculate the following:

   (i) speed and internal (total or gross) torque developed and

   (ii) the shaft power, shaft torque and efficiency with rotational losses equal to 300 W.

   Neglect armature reaction.

   [*Ans.* (i) 741 rpm, 38.197 N-m; (ii) 2,664 W, 34.33 N-m, 76.1%]

4. A 200 V DC shunt motor takes 22 A at rated voltage and runs at 1,000 rpm. Its field resistance is 100 Ω and armature circuit resistance (including brushes) is 0.1 Ω. Compute the value of resistance required to be added to the armature circuit to reduce the speed to 800 rpm, when
   (i) the load torque is proportional to the speed,
   (ii) the load torque varies with the square of the speed and
   (iii) the load torque varies with the cube of the speed.

   [*Ans.* (i) 2.5 Ω, (ii) 3.15 Ω and (iii) 3.96 Ω]
5. A 240 V unsaturated shunt motor has an armature resistance (including brushes and interpoles) of 0.04 Ω and a field resistance of 100 Ω.
   (i) Find the field resistance to be added in the field circuit to increase the speed from 1,200 rpm to 1,500 rpm when the supply current is 200 A.
   (ii) With the field as in (i), find the speed when the supply current is 100 A.
   (iii) Find the field current at 1,200 rpm if the machine is run as a generator to give 200 A at 240 V.
   (iv) Find the speed when the field current is 2 A if the machine is run as a generator to give 200 A at 240 V.

   [*Ans.* (i) 25 Ω, (ii) 1,526 rpm, (iii) 2.56 A and (iv) 1,440 rpm]
6. A 240 V DC series motor takes 40 A when giving its rated output at 1,500 rpm. Its resistance is 0.3 Ω. Find the resistance that must be added to obtain the rated torque
   (i) at starting and
   (ii) at 1,000 rpm.

   [*Ans.* (i) 5.7 Ω and (ii) 1.9 Ω]
7. A 250 V, 7.355 kW, DC shunt motor, with $R_a = 0.4\ \Omega$ and $R_{sh} = 100\ \Omega$, runs at 1,000 rpm on full-load. Efficiency on full-load in 85 per cent. If the speed is raised to 1,400 rpm keeping the load constant, determine the extra resistance to be added in the field circuit.

   [*Ans.* 43.36 Ω]
8. A DC series motor runs at 1,000 rpm when taking a current of 10 A at 200 V, with both field coils in series. Its armature resistance is 0.01 Ω and the resistance of each field coil is 0.02 Ω. Calculate the speed of the motor if the field coils are connected in parallel with each other, the load torque being constant. Assume magnetic field to be unsaturated.

   [*Ans.* 1,415.76 rpm]
9. The shaft torque, speed and efficiency of a 500 V DC series motor on full load are 1,800 N-m, 800 rpm and 90 per cent, respectively. When the speed is increased to 1,200 rpm by connecting a resistance across the series field, it develops a shaft torque of 900 N-m and has an efficiency of 80 per cent. Find the value of resistance connected across the series field.

   [*Ans.* 0.2112 Ω]

MULTIPLE-CHOICE QUESTIONS AND ANSWERS

1. The back emf of a motor at the time of starting is
   (a) zero (b) maximum (c) minimum (d) optimum
2. In a series motor, the field flux is
   (a) proportional to the armature current
   (b) is nearly constant

(c) inversely proportional to the armature current
(d) proportional to the square of the armature current

3. Series motors have a relatively
(a) high starting torque
(b) medium starting torque
(c) low starting torque
(d) a zero starting torque

4. If the field current of a shunt motor is changed, then
(a) its torque remains constant but the horsepower changes
(b) horsepower remains constant but the torque changes
(c) both torque and horsepower change
(d) both torque and horsepower remain constant

5. The torque speed characteristic of a series motor is
(a) linearly rising
(b) a rectangular parabola
(c) a parabola
(d) linearly decreasing

6. A 220 V DC machine has an armature resistance of 1 Ω. If the full-load current is 20 A, the difference in induced voltage, when the machine is running as a motor and as a generator, is
(a) 20 V
(b) 0 V
(c) 40 V
(d) 60 V

7. In a DC series motor, the torque developed is 20 N-m at 10 A load current. If the load current is doubled, the new torque will be
(a) 60 N-m
(b) 80 N-m
(c) 40 N-m
(d) 120 N-m

8. In a DC shunt motor, the torque developed is 15 N-m at 10 A load current. If the load current is doubled, the new torque will be
(a) 60 N-m
(b) 3 N-m
(c) 45 N-m
(d) 80 N-m

9. For electric traction, the motor most commonly used is
(a) a synchronous motor
(b) an induction motor
(c) a DC series motor
(d) a DC shunt motor

10. The DC motor starter used with a constant speed shunt motor is a
(a) two-point starter
(b) three-point starter
(c) four-point starter
(d) five-point starter

11. The field control of a DC shunt motor gives speeds which are
(a) below normal speed
(b) above normal speed
(c) none of these

12. The direction of rotation of a DC shunt motor can be reversed by interchanging
(a) the supply terminals
(b) the field terminals only
(c) the armature terminals only
(d) either field or armature terminal only

13. The speed current characteristics of a DC motor are shown in the figure below. Identity the curves.
(a) They are not speed current curves of the DC motor
(b) I – series, II – shunt, III – compound
(c) I – shunt, II – compound, III – series
(d) I – shunt, II – series, III – compound

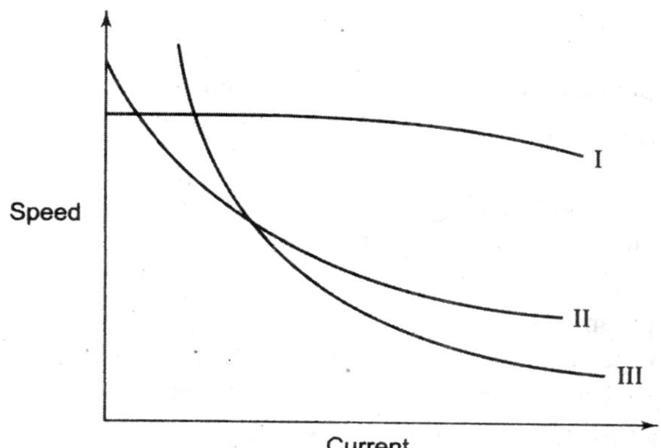

14. It is desired to have a reverse DC motor which is possible if
    (a) the field coil is split
    (b) supply to the motor is reversed by a switch
    (c) polarity of supply is changed by a relay
    (d) all of the above
15. Match the following
    (a) DC series motor (1) Factory shafting, machine tools
    (b) DC shunt motor (2) Electric traction, cranes, lift
    (c) DC compound motor (3) Conveyers, punch processes
16. Speed of a DC shunt motor is proportional to
    (a) $\frac{I}{I_a}$
    (b) I_a
    (c) I_a^2
    (d) none of these
17. Back emf of a DC shunt motor depends on
    (a) number of armature conductors
    (b) field flux
    (c) armature speed (N)
    (d) none of these
18. The armature of a DC motor is laminated
    (a) to reduce hysteresis loss
    (b) to reduce eddy current loss
    (c) to reduce copper loss
    (d) none of these
19. Maximum mechanical power is developed when back emf equals to
    (a) half of the applied voltage
    (b) the applied voltage
    (c) double of the applied voltage
    (d) one-third of the applied voltage
20. A DC series motor should not be started
    (a) at full-load
    (b) at no-load
    (c) with normal load
21. The torque speed characteristic of a DC series motor is
    (a) rectangular hyperbola
    (b) parabola
    (c) linearly rising
    (d) linearly declining
22. When an electric train moves down a hill, the DC motor behaves as a
    (a) DC shunt generator
    (b) DC series generator
    (c) DC shunt motor
    (d) none of these

23. Find the motor which is used for rolling mills:
(a) DC shunt mot
(b) DC series motor
(c) DC differentially compound motor
(d) DC cumulatively compound motor

24. The main drawback of field control method of speed control is
(a) its low frequency
(b) that commutation becomes unsatisfactory
(c) that only speed above normal can be achieved
(d) all of these

25. Stalling current is the maximum value of
(a) load current for which speed is zero
(b) load current for which speed is maximum
(c) field current for which speed is maximum
(d) field current for which speed is zero

26. If the field winding of a DC shunt motor is opened, the speed of the motor will be
(a) increased to an unstable value
(b) reduced
(c) reversed
(d) zero

27. If the load of a DC shunt motor is increased, its speed will
(a) increase proportionately
(b) decrease slightly
(c) increase slightly
(d) remain constant

28. The developed torque of a DC shunt motor is proportional to
(a) I_a
(b) I_a^2
(c) $I_a^{1/3}$
(d) $I_a^{\frac{1}{2}}$

29. The developed torque of a DC series motor is proportional to
(a) I_a
(b) $I_a^{\frac{1}{2}}$
(c) $\frac{I}{I_a}$
(d) I_a^2

30. Compound motors are used when
(a) load increases with increase in speed
(b) load is alternating sinusoidally
(c) there are large load variations and load equalization is done by flywheel
(d) all of the above

31. In Swinburn's method of testing, the DC machine is run as
(a) motor at no-load
(b) motor with break load applied at rated load
(c) generator connected to rated load
(d) all of the above

32. In Hopkinson's method of testing DC machines, the power is used to supply the losses in
(a) generator only
(b) motor only
(c) motor at rated load only
(d) both (a) and (b)

33. Ward-Leonard method of speed control
(a) has low efficiency at light loads
(b) has high maintenance and initial costs
(c) can provide speeds below and above normal in both directions
(d) minimum three machines are used
(e) is a voltage control method
(f) all of the above

34. A 200 V DC shunt motor develops a torque of 54 N-m at an armature current of 10 A. The torque produced, when the armature current is 20 A, is
    (a) 54 N-m (b) 81 N-m (c) 108 N-m (d) none of these
35. A shunt motor works on AC mains
    (a) unsatisfactorily (b) satisfactorily (c) not at all (d) none of the above
36. A DC shunt motor is operating at no-load. If the armature is shunted by a resistance, its speed will
    (a) increase (b) decrease (c) remain the same (d) be unpredictable
37. A DC shunt motor drives a load at rated speed and supply voltage. If both the load and voltage are halved, the speed will be
    (a) almost double (b) almost half
    (c) the rated speed (d) slightly less than the rated speed
38. A DC shunt motor does not operate on AC due to
    (a) low resistance of armature circuit (b) high resistance of field circuit
    (c) high inductance of field circuit (d) reversal of polarity
39. The function of brush and commutation in a DC motor is
    (a) to convert DC into AC
    (b) to produce unidirectional torque
    (c) to produce unidirectional current in the armature
    (d) to reduce demagnetizing effect of armature reaction
40. The field windings of a DC compound motor are connected in short-shunt configuration. If the series and shunt field windings are interchanged and the motor is connected to a rated voltage, then
    (a) series winding gets overloaded (b) shunt winding gets overloaded
    (c) neither of the windings gets overloaded (d) both windings get overloaded

Answers

1 (a)	2 (a)	3 (a)	4 (b)	5 (b)	6 (c)
7 (b)	8 (b)	9 (c)	10 (b)	11 (b)	12 (d)
13 (c)	14 (a)	15 (a)–(2), (b)–(1), (c)–(3)	16 (d)	17 (a),(b),(c)	18 (b)
19 (a)	20 (b)	21 (a)	22 (b)	23 (d)	24 (b)
25 (a)	26 (a)	27 (b)	28 (a)	29 (d)	30 (c)
31 (a)	32 (d)	33 (f)	34 (c)	35 (a)	36 (c)
37 (c)	38 (d)	39 (b)	40 (c)		

6 Synchronous Generators

In Chapter 3, basic principles of synchronous generators and motors have already been introduced. In synchronous machine, the stator is used as the armature and the rotor is used as the field. The field is always excited by DC voltage. The difference between the synchronous generator (or alternàtor) and the synchronous motor is that the synchronous generator exports electrical power to the bus bar whereas the synchronous motor imports power from the bus bar. The aim of this chapter is to introduce the synchronous generator or alternator.

6.1 PARAMETERS OF ARMATURE WINDING

An alternator has the following three important parameters:

(i) armature resistance (R_a),

(ii) armature leakage reactance (X_L) and

(iii) reactance (X_{ar}) due to armature reaction.

6.1.1 Armature Resistance

Every armature has its armature resistance (R_a). When load is put on the alternator, the voltage drop caused by armature resistance per phase is I_aR_a, which is in phase with I_a. Since R_a is very small, this drop is negligible. The effective armaturê resistance is greater than its resistance measured by direct current. The increment of resistance for alternating current is due to the additional energy, which expands inside and outside the conductor. Generally, the armature resistance is calculated by the ratio of applied direct voltage to the measured value of current through it. Figure 6.1 shows a star-connected alternator with DC field winding open circuited. The DC resistance between each pair terminal is to be measured either by using ammeter–voltmeter method or by using Wheatstone's bridge. The average of three sets of value is taken and it is divided by two to obtain DC resistance or ohmic resistance per phase. During this measurement, the alternator is kept stationary. To get the effective resistance, its DC value is multiplied by a factor ranging from 1.25 to 1.75 or more. Eddy currents,

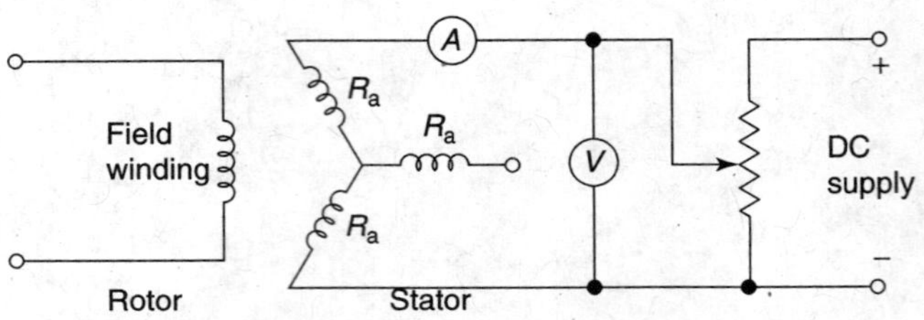

Figure 6.1 Measurement of Resistance

losses in the surrounding material and unequal distribution of current in the conductor are the main causes of additional energy.

6.1.2 Armature Leakage Reactance (X_L)

Due to the flow of load current through the armature, it creates its own flux. Some part of the flux passes through the air around the conductor. This flux is known as leakage flux, which makes the armature winding inductive in nature. In addition to resistance, the winding possesses reactance. This reactance is termed 'leakage reactance' (X_L), given by $X_L = 2\pi fL$, where L is leakage inductance in Henry. Compared to the value of armature leakage reactance, armature resistance is very small. Figure 6.2 shows the armature leakage flux.

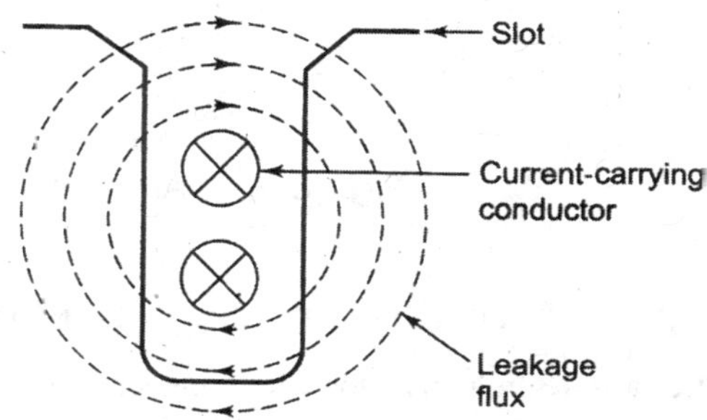

Figure 6.2 Armature Leakage Flux

6.2 ARMATURE REACTION

Due to connection of load to alternator, current flows through the armature winding, which establishes its own flux. In the air gap there are two fluxes—one due to the current through the field winding known as *main flux* and the other due to the current through the armature known as *armature flux*. The effect of armature flux on main flux is known as *armature reaction*. The effect of the nature of load power factor on armature reaction is discussed below.

6.2.1 Unity Power Factor Load

Let a resistive load be connected I to the alternator. The induced emf per phase ($\bar{E}_{ph}$) drives a current per phase ($\bar{I}_a$) through the load. Both $\bar{E}_{ph}$ and $\bar{I}_a$ are in phase. If the main flux ($\bar{\phi}_f$) generates $\bar{E}_{ph}$, $\bar{E}_{ph}$ lags $\bar{\phi}_f$ by a 90° angle. The armature current ($\bar{I}_a$) establishes the armature flux ($\bar{\phi}_a$), where $\bar{\phi}_a$ and $\bar{I}_a$ are in the same direction.

From Figure 6.3, it is clear that $\bar{\phi}_a$ can oppose as well as assist. Therefore, average flux in the air gap remains constant. In spite of maintaining a constant air-gap flux, its distribution gets distorted. This distorting effect of armature reaction under unity power factor condition of the load is known as cross-magnetizing effect of armature reaction.

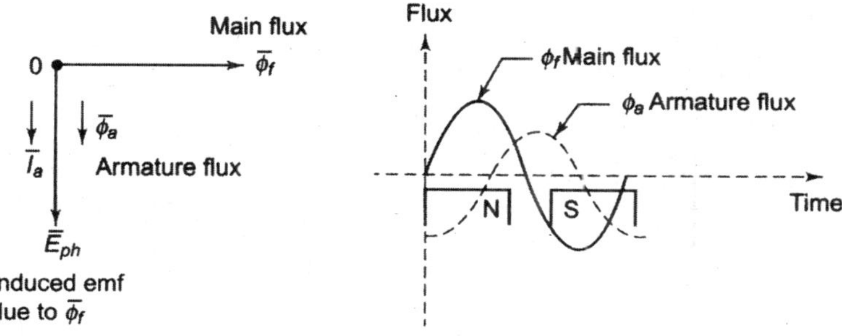

Figure 6.3 Armature Reaction for Unity Power Factor Load

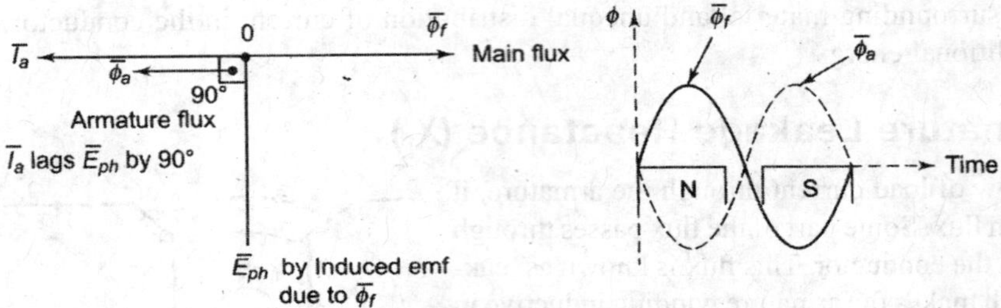

Figure 6.4 Armature Reaction for Zero Lagging pf Load

6.2.2 Zero Lagging Power Factor Load

Let a pure inductive load be connected to the alternator. The induced emf per phase ($\overline{E}_{ph}$) drives a current per phase ($\overline{I}_a$) through the load, which lags behind $\overline{E}_{ph}$ by a 90° angle. Since the field flux ($\overline{\phi}_f$) produces $\overline{E}_{ph}$, $\overline{E}_{ph}$ lags behind $\overline{\phi}_f$ by a 90° angle. Again armature current ($\overline{I}_a$) produces armature flux ($\overline{\phi}_a$), which is in phase with $\overline{I}_a$. Figure 6.4 shows the phasor diagram.

From Figure 6.4, it is clear that ϕ_a and ϕ_f are exactly in opposite directions to each other. This effect of armature reaction is known as demagnetizing effect of armature reaction.

6.2.3 Zero Leading Power Factor Load

Let a pure capacitor load be connected to the alternator. The induced emf per phase ($\overline{E}_{ph}$) drives a current per phase ($\overline{I}_a$) through the load, which leads $\overline{E}_{ph}$ by a 90° angle. Since the field flux ($\overline{\phi}_f$) produces $\overline{E}_{ph}$, $\overline{E}_{ph}$ lags behind $\overline{\phi}_f$ by a 90° angle. Again armature current ($\overline{I}_a$) produces armature flux ($\overline{\phi}_a$), where ϕ_a is in phase with $\overline{I}_a$. Figure 6.5 shows the phasor diagram.

From Figure 6.5, it is clear that $\overline{\phi}_a$ and $\overline{\phi}_f$ are exactly in the same direction. This effect is known as magnetizing effect of armature reaction.

The change in terminal voltage is due to the armature reaction for different types of loads. The practical loads are inductive in nature. The reduction of terminal voltage is due to demagnetizing effect of armature reaction. In other words, this drop is due to the interaction of armature flux and main flux and it is not across any physical element. To include this drop, it is assumed that armature winding has a *fictitious reactance,* which is known as *armature reaction reactance* (X_{ar}).

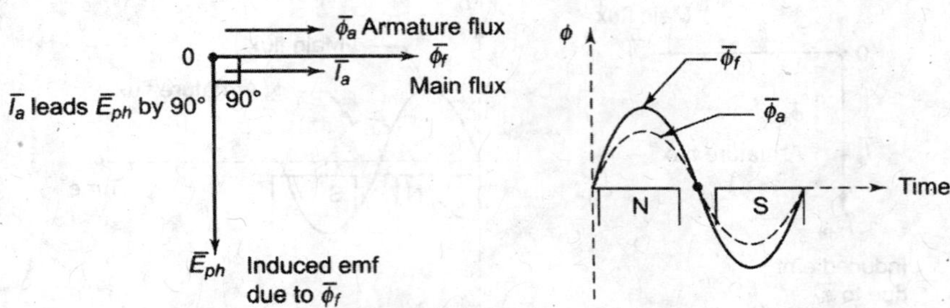

Figure 6.5 Armature Reaction for Zero Leading pf Load

6.3 CONCEPT OF SYNCHRONOUS REACTANCE AND IMPEDANCE

The overall reactance of the armature winding is the sum of its leakage reactance plus fictitious reactance, which is known as synchronous reactance (X_s).

$$\therefore \quad X_s = X_L + X_{ar} \tag{6.1}$$

where X_L and X_{ar} are in Ω/phase. Therefore, X_s is in Ω/phase.

The impedance of armature winding is obtained by combining its resistance and its synchronous reactance.

$$\therefore \quad Z_s = \sqrt{R_a^2 + X_s^2}\ \Omega/\text{phase} \tag{6.2}$$

where Z_s is known as synchronous impedance.

If an alternator is driven at a constant speed (or synchronous speed), the standard frequency is obtained. The synchronous impedance (X_s) depends upon the load and its power factor condition. Therefore, synchronous impedance also depends on the load and its power factor.

6.4 EQUIVALENT CIRCUIT OF AN ALTERNATOR

Figure 6.5 shows the single-line diagram of the equivalent circuit of an alternator. R_a, X_L and X_{ar} represent the armature resistance, armature leakage reactance and fictitious reactance due to armature reaction. All are in Ω/phase.

Since the combination of X_L and X_{ar} gives the synchronous reactance (X_s), Figure 6.6 is redrawn and shown in Figure 6.7.

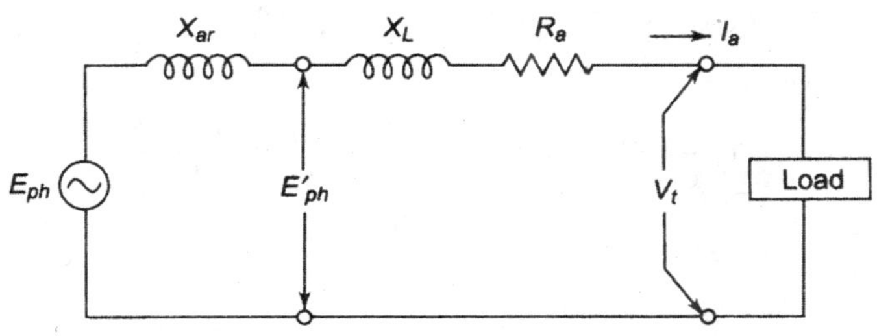

Figure 6.6 Equivalent Circuit

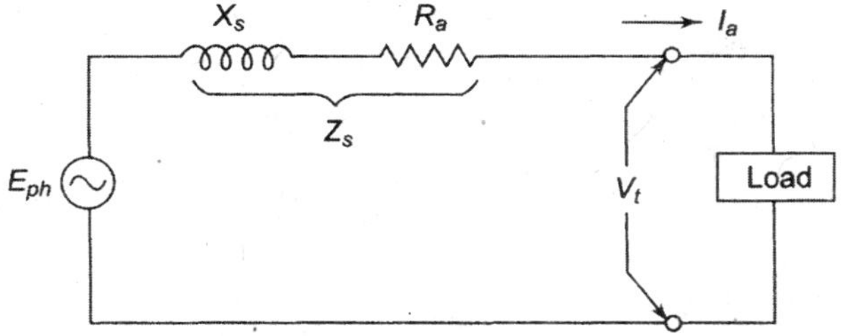

Figure 6.7 Equivalent Circuit of Figure 6.6

6.5 VOLTAGE EQUATION OF ALTERNATOR

In Figure 6.6 or in Figure 6.7, $\overline{V}_t$ is the terminal voltage and $\overline{E}_f$ is the generated emf in armature per phase. At no-load, $\overline{I}_a = 0$ and $\overline{V}_t = \overline{E}_f$. During the presence of load, $\overline{E}_f$ will overcome the voltage drop across R_a, X_L and X_{ar}. The voltage equation of the alternator is expressed by

$$\overline{E}_f = \overline{V}_t + \overline{I}_a\left(R_a + jX_s\right) \quad (6.3)$$

$$= \overline{V}_t + \overline{I}_a \overline{Z}_s \quad (6.4)$$

The cosine of the angle between $\overline{I}_a$ and $\overline{V}_t$ is called the power factor. Depending on the nature of power factor of the load, a different phasor diagram of the alternator can be drawn.

6.6 PHASOR DIAGRAM OF ALTERNATOR

The phasor diagrams of an alternator for lagging, unity and leading power factor, respectively, are discussed below.

6.6.1 Lagging Power Factor

Figure 6.8 shows the phasor diagram of an alternator for lagging power factor load. In Figure 6.8, $\overrightarrow{OC} = \overline{E}_f$ = no-load emf, $\overrightarrow{OA} = \overline{V}_t$ = terminal voltage, $\overline{I}_a$ = armature current, R_a = effective armature resistance, X_s = synchronous reactance, $\cos\beta$ = power factor and δ = power angle or torque angle.

From Figure 6.8, we have

$$OD = V_t \cos\beta,\, AB = I_a R_a = DE,\; BC = I_a X_s \text{ and } AD = BE = V_t \sin\beta$$

$$OC^2 = OE^2 + EC^2 = (OD + DE)^2 + (EB + BC)^2$$

i.e.,

$$E_f^2 = (V_t \cos\beta + I_a R_a)^2 + (V_t \sin\beta + I_a X_s)^2$$

i.e.,

$$E_f = [(V_t \cos\beta + I_a R_a)^2 + (V_t \sin\beta + I_a X_s)^2]^{1/2} \quad (6.5)$$

If $\gamma = \beta + \delta$, from OEC we have

$$\tan\gamma = \tan(\beta + \delta) = \frac{EC}{OE} = \frac{EB + BC}{OD + DE} = \frac{EB + BC}{OD + AB}$$

i.e.,

$$\tan\gamma = \tan(\beta + \delta) = \frac{V_t \sin\beta + I_a X_s}{V_t \cos\beta + I_a R_a} \quad (6.6)$$

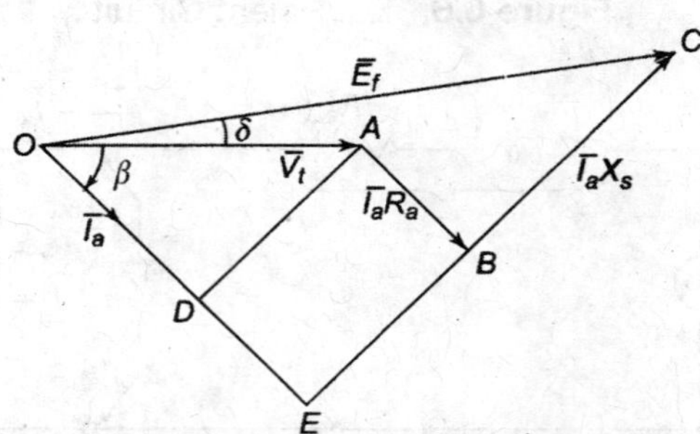

Figure 6.8 Lagging Power Factor Load

6.6.2 Unity Power Factor

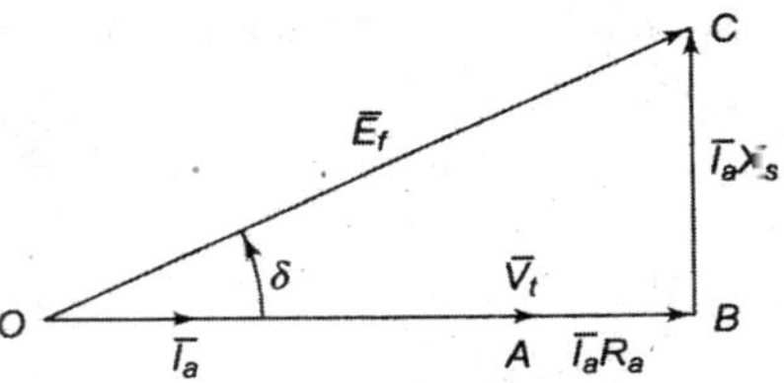

Figure 6.9 Unity Power Factor Load

Figure 6.9 shows the phasor diagram of an alternator for unity power factor.

From Figure 6.9, we have

$$OC^2 = OB^2 + BC^2 = (OA + AB)^2 + BC^2$$

i.e.,
$$E_f^2 = \left(V_t + I_a R_a\right)^2 + \left(I_a X_s\right)^2$$

i.e.,
$$E_f = \left[\left(V_t + I_a R_a\right)^2 + \left(I_a X_s\right)^2\right]^{\frac{1}{2}} \qquad (6.7)$$

If $\gamma = \delta$, from $\Delta\, OBC$ we have

$$\tan\gamma = \tan\delta = \frac{BC}{OB} = \frac{BC}{OA + AB}$$

i.e.,
$$\tan\gamma = \tan\delta = \frac{I_a X_s}{V_t + I_a R_a} \qquad (6.8)$$

6.6.3 Leading Power Factor

Figure 6.10 shows the phasor diagram of an alternator for leading power factor.

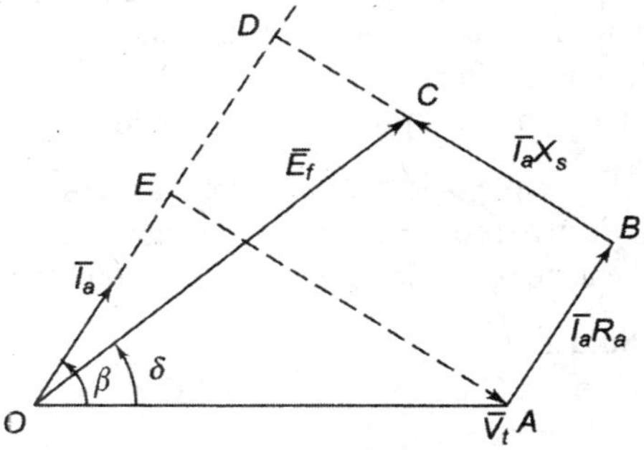

From Figure 6.10, we have

$OE = V_t \cos\beta,\ AB = ED = I_a R_a,\ AE = DB = V_t \sin\beta,\ BC = I_a X_s$

Also,

$$OC^2 = OD^2 + DC^2$$
$$= (OE + ED)^2 + (DB - CB)^2$$

Figure 6.10 Leading Power Factor Load

i.e.,
$$E_f^2 = (V_t \cos\beta + I_a R_a)^2 + (V_t \sin\beta - I_a X_s)^2$$

i.e.,
$$E_f = [(V_t \cos\beta + I_a R_a)^2 + (V_t \sin\beta - I_a X_s)^2]^{1/2} \qquad (6.9)$$

If $\gamma = \beta - \delta$, from $\Delta\, OCD$ we have

$$\tan\gamma = \tan(\beta - \delta) = \frac{CD}{OD} = \frac{DB - CB}{OE + ED}$$

i.e.,
$$\tan\gamma = \tan(\beta - \delta) = \frac{V_t \sin\beta - I_a X_s}{V_t \cos\beta + I_a R_a} \qquad (6.10)$$

The angle between $\bar{E}_f$ and $\bar{V}_t$ is called the *power angle*, *load angle* or *torque angle* of the alternator.

6.7 VOLTAGE REGULATION

When the alternator is loaded, its terminal voltage may not be equal to the induced emf. It may be less than or greater than the induced emf. At no-load, I_a is zero and hence $V_t = E_f$ because there is no voltage drop and no armature flux exists to cause armature reaction. If speed and field current remain constant after removal of load at a given power factor, voltage regulation is defined thus:

$$\text{Voltage regulation} = \frac{(\text{terminal voltage at no-load}) - (\text{terminal voltage at full load})}{\text{terminal voltage at full load}}$$

$$= \frac{E_f - V_t}{V_t} \text{ in p.u.} = \frac{E_f - V_t}{V_t} \times 100 \text{ in percentage}$$

Here, E_f is the no-load excitation voltage and V_t is the full-load terminal voltage, keeping speed and field excitation unaltered. For lagging power factor load, E_f always increases and voltage regulation is positive. But E_f always decreases for leading power factor load and voltage regulation may also be negative. The following points are most important:

- The voltage rise must be known after throwing off the load because winding insulation must be able to withstand this increased voltage.
- Voltage regulation affects the steady-state short-circuit conditions and stability.
- Automatic voltage-control equipment is determined by voltage regulation.
- Voltage regulation affects the parallel operation of alternators.

Due to the above reasons, voltage regulation of synchronous machines must be known.

The larger values of reactances are used in modern alternators so that they can be protected in the event of short-circuit near the terminals. Hence, their voltage regulation becomes inherently poor. To overcome this problem, automatic voltage regulator is used.

Example 6.1 A three-phase star-connected alternator is rated at 1,500 kVA, 11 kV. The armature effective resistance and synchronous reactance are 1.2 Ω and 24 Ω respectively, per phase. Calculate the percentage regulation for a load of 1,200 kW at power factors of (i) 0.8 lagging, (ii) unity and (iii) 0.8 leading.

Solution

$R_a = 1.2$ Ω/phase and $X_s = 24$ Ω/phase

(i)
$$I_L = \frac{1,200\times10^3}{\sqrt{3}\times11\times10^3\times0.8} = 78.73\text{A} = I_a$$

$$V_t = \frac{11\times10^3}{\sqrt{3}} = 6,350.8\text{ V}$$

$$\cos\beta = 0.8 \text{ (lagging)}, \ \sin\beta = 0.6$$

$$E_f = \sqrt{(V_t\cos\beta + I_aR_a)^2 + (V_t\sin\beta + I_aX_s)^2}$$

$$= \sqrt{(6,350.8\times0.8+78.73\times1.2)^2+(6,350.8\times0.6+78.73\times24)^2}$$

$$= 7,698.82\text{ V}$$

$$\therefore \quad \text{Percentage regulation} = \frac{E_f - V_t}{V_t}\times100 = \frac{7,698.82-6,350.8}{6,350.8}\times100 = 21.22\%$$

(ii)
$$I_L = \frac{1,200\times10^3}{\sqrt{3}\times11\times10^3\times1} = 62.98\text{A} = I_a$$

$$V_t = \frac{11\times 10^3}{\sqrt{3}} = 6{,}350.8\,\text{V}$$

$$\cos\beta = 1,\ \ \sin\beta = 0$$

$$E_f = \sqrt{(V_t\cos\beta + I_aR_a)^2 + (V_t\sin\beta + I_aX_s)^2}$$

$$= \sqrt{(6{,}350.8\times 1 + 62.98\times 1.2)^2 + (6{,}350.8\times 0 + 62.98\times 24)^2} = 6{,}601.74\,\text{V}$$

$$\therefore \text{ Percentage regulation} = \frac{E_f - V_t}{V_t}\times 100 = \frac{6{,}601.74 - 6{,}350.8}{6{,}350.8}\times 100 = 3.95\%$$

(iii)

$$I_L = \frac{1{,}200\times 10^3}{\sqrt{3}\times 11\times 10^3\times 0.8} = 78.73\,\text{A} = I_a$$

$$V_t = \frac{11\times 10^3}{\sqrt{3}} = 6{,}350.8\,\text{V}$$

$$\cos\beta = 0.8 \text{ (leading)},\ \ \sin\beta = 0.6$$

$$E_f = \sqrt{(V_t\cos\beta + I_aR_a)^2 + (V_t\sin\beta - I_aX_s)^2}$$

$$= \sqrt{(6{,}350.8\times 0.8 + 78.73\times 1.2)^2 + (6{,}350.8\times 0.6 - 78.73\times 24)^2}$$

$$= 5{,}520.14\,\text{V}$$

$$\therefore \quad \text{Percentage regulation} = \frac{E_f - V_t}{V_t}\times 100 = \frac{5{,}520.14 - 6{,}350.8}{6{,}350.8}\times 100 = -13.08\%$$

■

6.8 DETERMINATION OF VOLTAGE REGULATION

There are various methods to determine the regulation of alternators. By using direct loading, the regulation of small-capacity motors can be determined. There are other methods by which voltage regulation of large capacity alternators can be determined. These methods have their advantages and disadvantages. Voltage regulation can be determined by the following indirect methods:

- emf or synchronous impedance method,
- mmf or ampere-turn method,
- zero power factor or Potier method and
- ASA method.

All the above methods have the following requirements:

(i) armature resistance (R_a),

(ii) Open-circuit characteristic and

(iii) Short-circuit characteristic (zero power factor lagging characteristic is required for Potier method).

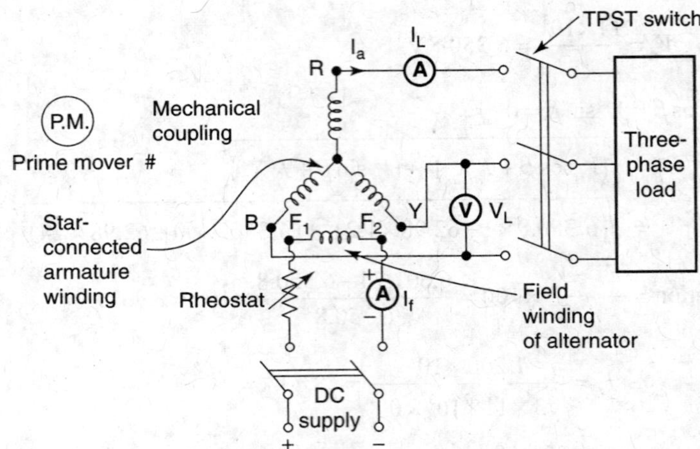

Figure 6.11 Circuit Diagram for Direct Load Testing

6.8.1 Direct Loading

The circuit diagram shown in Figure 6.11 is used to conduct the direct load test on the three-phase alternator. A triple-pole single-throw (TPST) switch has been used for connection of the star-connected armature to a three-phase load. A separate DC supply is used for excitation of the field winding. A rheostat is inserted in series with the field winding to control the current through field winding. Hence, flux can be controlled. The alternator is driven at its synchronous speed with the help of a prime mover, shown in Figure 6.11.

Since $E_{ph} \propto \phi$, the field current is adjusted to adjust the flux so that rated voltage is observed across the terminals, which can be observed on the voltmeter connected across the lines. The load is connected with the help of TPST and load is then increased to obtain the rated value of current, which is read by the ammeter. This gives the full-load condition of the alternator. The voltage is adjusted to its rated value with the help of field excitation by the rheostat.

The entire load is thrown off by opening the TPST switch. The speed and field excitation are kept unchanged. The voltmeter reading is observed, which gives the internally induced emf known as no-load terminal voltage because there is no armature current and its associated drops. Both these reading are converted into phase values. If V_t and E_f are the phase values of rated voltage on full-load and no-load, respectively, we have

$$\text{Regulation} = \frac{E_f - V_t}{V_t}\ \text{pu} = \frac{E_f - V_t}{V_t} \times 100$$

The voltage regulation of the alternator can be determined. The value of regulation obtained by this method is accurate because a particular load at a particular power factor is actually connected to the alternator for noting down the readings. It is difficult to simulate the required full-load for high-capacity alternators, which can be directly connected to the alternator. This method is restricted for small-capacity alternators.

6.8.2 The emf or Synchronous Impedance Method

The effect of winding impedance and armature reaction is represented by equivalent impedance drop in this method. Therefore, this method is called emf method. Figure 6.12 shows the vector diagram of an alternator on load.

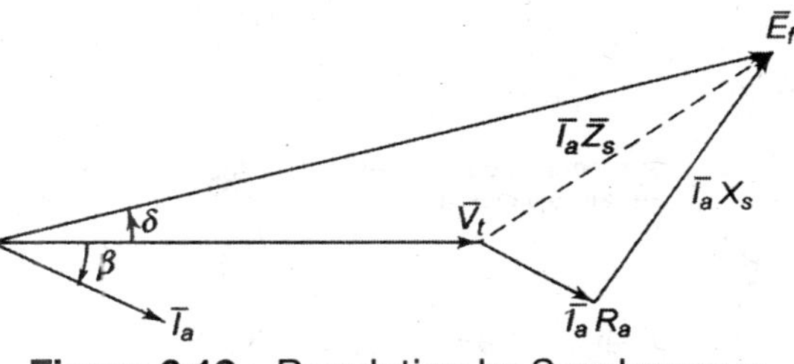

Figure 6.12 Regulation by Synchronous Impedance Method

The notations in the above vector diagram are stated as follows:

$\bar{V}_t$ is the terminal voltage per phase.
$\bar{E}_f$ is the induced emf per phase.
$\bar{I}_a$ is the full-load current per phase.
β is the power factor angle per phase.
R_a is the effective armature resistance per phase.
X_s is the synchronous reactance per phase.
δ is the load angle.

From Figure 6.12, we have,

$$E_f^2 = (V_t \cos\beta + I_a R_a)^2 + (V_t \sin\beta + I_a X_s)^2$$

$$= V_t^2 + 2I_a V_t (R_a \cos\beta + X_s \sin\beta) + I_a^2 (R_a^2 + X_s^2)$$

$$= V_t^2 \left[1 + \frac{2I_a (R_a \cos\beta + X_s \sin\beta)}{V_t} + \frac{I_a^2 (R_a^2 + X_s^2)}{V_t^2}\right]$$

i.e.,

$$E_f \approx V_t \left[1 + \frac{2I_a (R_a \cos\beta + X_s \sin\beta)}{V_t}\right]^{\frac{1}{2}}$$

i.e.,

$$E_f = V_t + I_a (R_a \cos\beta + X_s \sin\beta)$$

$$\text{Percentage regulation} = \frac{E_f - V_t}{E_f} \times 100 = \frac{I_a (R_a \cos\beta + X_s \sin\beta)}{V_t} \times 100$$

This gives the regulation for lagging power factor. For leading power factor,

$$\text{Percentage regulation} = \frac{E_f - V_t}{E_f} \times 100 = \frac{I_a (R_a \cos\beta - X_s \sin\beta)}{V_t} \times 100$$

6.8.2.1 Open-circuit Test

The alternator terminals are open circuited. The alternator is run at synchronous speed with the help of prime mover. The field current of the alternator is varied in steps. The reading of the induced emf corresponding to the field current is noted. This test is conducted until complete saturation point is obtained. Figure 6.13(a) shows the circuit connection for open-circuit test and Figure 6.13(b) shows the open-circuit characteristic of an alternator.

6.8.2.2 Short-circuit Test

The alternator terminals are shorted and it is run at synchronous speed with the help of a prime mover. The field current is gradually increased from zero in steps and the corresponding armature current is measured. The field current is increased to an amount so that the armature current becomes 125 to 150 per cent of its rated value. The field current is increased and the corresponding armature currents are noted. The plot of armature current and field current shows the short-circuit characteristic of an alternator, which is normally a straight line passing through the origin. No magnetic saturation is present due to small net

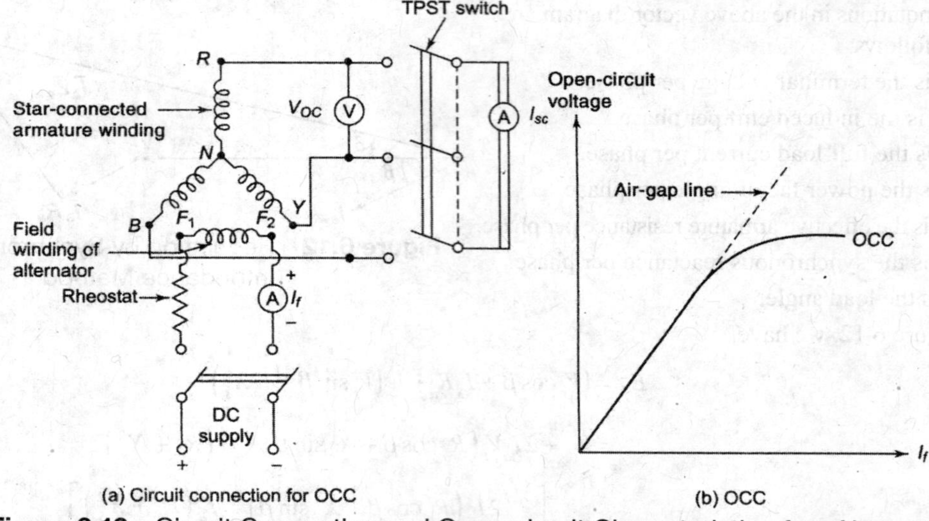

(a) Circuit connection for OCC (b) OCC

Figure 6.13 Circuit Connection and Open-circuit Characteristic of an Alternator

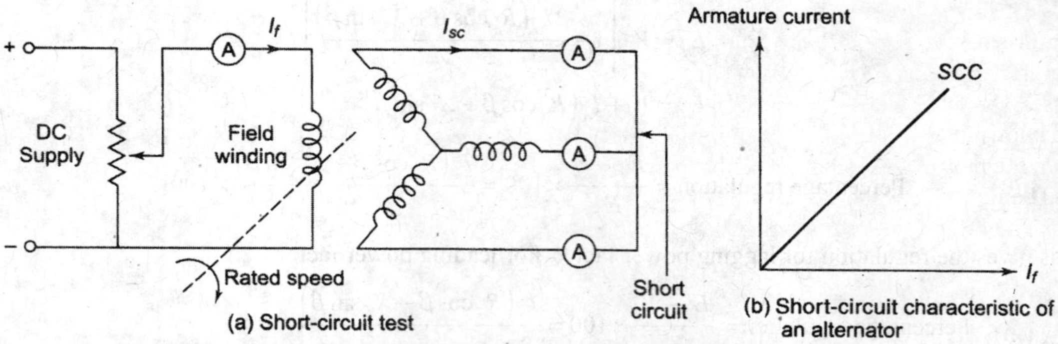

(a) Short-circuit test (b) Short-circuit characteristic of an alternator

Figure 6.14 Circuit Connection and *SCC* of an Alternator

excitation. If there is slight deviation in speed from the rated, the short-circuit characteristic remains unaffected. Figure 6.14(a) shows the circuit for short-circuit test and Figure 6.14(b) shows the short-circuit characteristic of an alternator.

6.8.2.3 Armature Resistance Measurement

By applying low-voltage DC across one phase of the armature winding, the circulating current is measured. The effective AC resistance is taken as 1.25 to 1.75 times the DC resistance.

From Figure 6.15, the synchronous impedance is as follows:

$$Z_s = \left.\frac{\left(E_f\right)_{\text{rated}}}{I_{sc}}\right|_{\text{at the field current } (I_f)_{SC}} \tag{6.11}$$

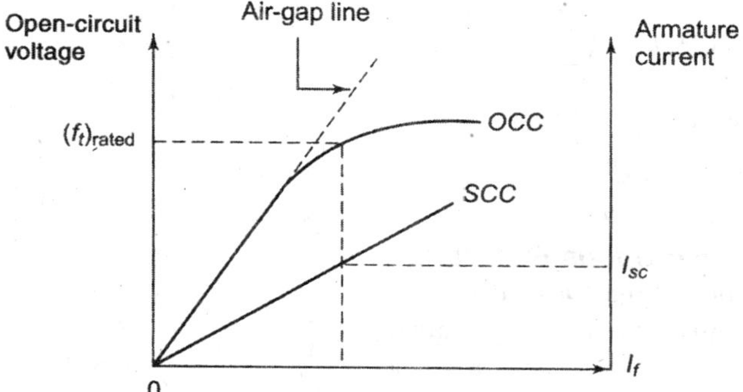

Figure 6.15 Armature Resistance Measurement

After calculation of effective armature resistance per phase (R_a), the synchronous reactance is given by

$$X_s = \sqrt{Z_S^2 - R_a^2} \tag{6.12}$$

After calculating the values of R_a and X_s, the voltage regulation of the alternator at the power factor $\cos\beta$ (lag or lead) can be calculated.

The open-circuit test is conducted at rated voltage and it gives the constant losses of the alternator, comprising no-load core loss and mechanical loss due to windage and friction. The necessary power required is drawn from the prime mover running the alternator. To carry out the separation of these two losses, the machine is run at synchronous speed in underexcited state. The friction and windage losses are known at this stage. The machine is highly underexcited at the short-circuit condition. Losses drawn from the prime mover consist of mechanical loss and copper loss in the resistance of the armature, iron loss being negligible.

The following assumptions are made in the synchronous impedance method:

(i) The synchronous impedance remains constant.

(ii) The flux under test condition and that under load condition is same.

(iii) The effect of armature reaction flux is replaced by a voltage drop, which is proportional to the armature current. The armature reaction voltage drop is added to the armature reactance voltage drop.

(iv) The magnetic reluctance due to the armature flux remains constant and it is independent on power factor.

The value of Z_s remains constant at low excitation because the *OCC* coincides with the air-gap line. This value of Z_s is termed as linear or unsaturated synchronous reactance. Due to increase in excitation, the separation effect decreases Z_s and the value beyond the linear part of the *OCC* is termed as the saturated values of synchronous impedance. These values vary with the excitation, that is, with voltage, current and power factor of the machine on load. Hence, these values are not constant. In a given situation, the value used is known as the effective synchronous impedance.

6.8.2.4 Unsaturated Synchronous Reactance

Figure 6.16 shows the plot of air-gap line, linear magnetization curve, *OCC* and *SCC*. The unsaturated synchronous reactance ($X_s(un)$) is obtained from the air-gap line and short-circuit current of the machine for a certain value of field current. From Figure 6.16, we have

$$Z_{s(un)} = \frac{ps}{pq} = R_a + jX_{s(un)} \quad (6.13)$$

For negligible R_a, we have

$$X_{s(un)} = \frac{ps}{pq}$$

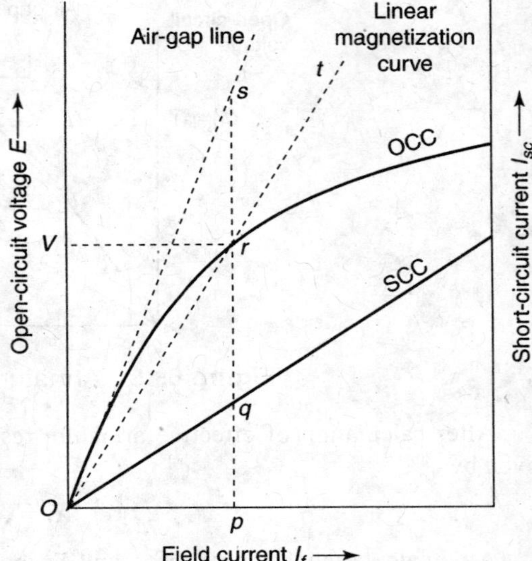

Figure 6.16 Plot of Airgap Line, Linear Magnetization Curve, OCC and SCC

6.8.2.5 Saturated Synchronous Reactance

The degree of saturation of the *OCC* decides the variation of the approximate value of synchronous reactance. The calculation of synchronous reactance, which is to be used in a particular problem, is to be calculated at the approximate load on the machine. The terminal voltage of the machine remains same at that value when it is connected to the infinite bus. If the excitation is changed due to change of field current, it will change along the dotted line *Ort*. The line *Ort* is called modified air-gap line, which represents the same magnetic saturation level as that corresponding to the operating point *r*. This has been shown in Figure 6.16.

The expressed reactance at rated voltage is expressed by

$$\bar{Z}_{s(sat)} = \frac{E_{pr}}{I_{pq}} = R_a + jX_{s(sat)} \quad (6.14)$$

For negligible R_a, we have

$$X_{s(sat)} = \frac{E_{pr}}{I_{pq}} \quad (6.15)$$

6.8.2.6 Advantages and Limitations of Synchronous Impedance Method

The value of synchronous impedance (Z_s) at any load condition can be determined. This leads to calculation of voltage regulation of the alternator at any load condition and power factor. Since the actual load need not to be connected to the alternator, this method can be used for large capacity alternators.

The main disadvantage is that this method gives large values of synchronous impedance that leads to high values of percentage regulation than actual results. That is why this method is also known as pessimistic method.

Example 6.2 A three-phase, 11 kV, star-connected alternator delivers a current of 80 A at (i) 0.8 power factor leading, (ii) unity power factor and (iii) 0.75 power factor lagging. Full-load current of 80 A is produced on short circuit by a field excitation of 2.8 A. An emf of 400 V per phase is produced on open circuit by the same excitation. The armature resistance is 0.7 Ω per phase. Determine the voltage regulation in each case.

Solution

$$Z_s = \frac{\text{OC voltage}}{\text{SC current}} = \frac{400}{80} = 5\text{ A} \quad \text{and} \quad R_a \text{ per phase} = 0.7\ \Omega$$

$$\therefore \quad X_s = \sqrt{Z_s^2 - R_a^2} = \sqrt{5^2 - 0.7^2} = 4.95\Omega$$

$$I_a \text{ per phase} = 80\text{ A.}$$

(i) $\cos\beta = 0.8$ (leading), $\sin\beta = 0.6$

$$V_t\text{(per phase)} = \frac{11{,}000}{\sqrt{3}} = 6{,}350.8\text{ V}$$

$$\therefore \quad E_f = \sqrt{(V_t\cos\beta + I_aR_a)^2 + (V_t\sin\beta - I_aX_s)^2}$$

$$= \sqrt{(6{,}350.8\times 0.8 + 80\times 0.7)^2 + (6{,}350.8\times 0.6 - 80\times 4.95)^2} = 6{,}167.96\text{ V}$$

$$\therefore \quad \text{Percentage regulation} = \frac{6{,}167.96 - 6{,}350.8}{6{,}350.8}\times 100 = -2.88$$

(ii) $\cos\beta = 1,\ \sin\beta = 0$

$$\therefore \quad E_f = \sqrt{(V_t\cos\beta + I_aR_a)^2 + (V_t\sin\beta + I_aX_s)^2}$$

$$= \sqrt{(6{,}350.8\times 1 + 80\times 0.7)^2 + (6{,}350.8\times 0 + 80\times 4.95)^2} = 6{,}419.02\text{ V}$$

$$\text{Percentage regulation} = \frac{6{,}419.02 - 6{,}350.8}{6{,}350.8}\times 100 = 1.07$$

(iii) $\cos\beta = 0.75$ (lagging), $\sin\beta = 0.661$

$$\therefore \quad E_f = \sqrt{(V_t\cos\beta + I_aR_a)^2 + (V_t\sin\beta + I_aX_s)^2}$$

$$= \sqrt{(6{,}350.8\times 0.75 + 80\times 0.7)^2 + (6{,}350.8\times 0.661 + 80\times 4.95)^2}$$

$$= 6{,}657.88\text{ V}$$

$$\text{Percentage regulation} = \frac{6{,}657.88 - 6{,}350.8}{6{,}350.8}\times 100 = 4.835$$

■

6.8.3 Ampere-turn Method

This method is the converse of the emf method. In the synchronous impedance method, the effect of armature reaction is replaced by a fictitious reactance. In this method, the effect of armature leakage reactance is replaced by an equivalent armature reaction mmf, which may be combined with the armature reaction mmf (F_{ar}). In this method, the effect of winding impedance and armature reaction is equivalent to ampere-turns, and hence this is called the mmf method. The required data for calculation of regulation are obtained from the open- and short-circuit test of the alternator. Figure 6.17 shows the vector diagram from which voltage regulation by the mmf method is obtained.

In Figure 6.17,

- $\bar{V}_t$ is the terminal voltage per phase (reference phasor).
- $\bar{I}_a$ is the load current per phase lagging $\bar{V}_t$ by an angle β.
- $\bar{V}_1$ is the sum of $\bar{V}_t$ and $\bar{I}_a R_a$ added vectorially.
- $\bar{I}_{f1}$ is the field current required to generate the rated voltage $\bar{V}_1$ (taken from *OCC* test, Figure 6.18) leading $\bar{V}_1$ by a 90° angle.

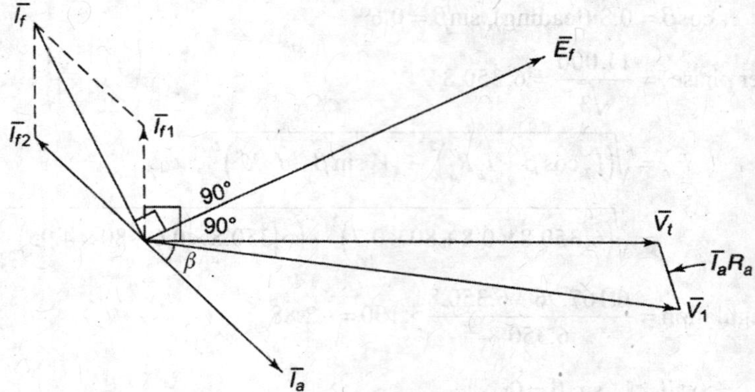

Figure 6.17 Vector Diagram

$\bar{I}_{f2}$ is the field current required to get the rated armature current in the alternator winding on short circuit (taken from SC test, Figure 6.18) marked in phase opposition to $\bar{I}_a$.

$\bar{I}_f$ is the vectorial addition of current $\bar{I}_{f1}$ and $\bar{I}_{f2}$.

$\bar{E}_f$ is the emf induced for a field current of $\bar{I}_f$ marked 90° lagging the current $\bar{I}_f$.

The field current $\bar{I}_{f2}$ not only balances the impedance drop but also overcomes armature reaction. The regulation is given by reaction on full-load. The effective armature resistance is very small. X_L is also very small for low voltage on short circuit. Hence, impedance drop can be neglected. This leads to power factor on short circuit in zero lagging. Hence, fixed ampere-turns are entirely used to overcome the armature reaction. $\bar{I}_{f2}$ solely gives the demagnetizing ampere-turns at full-load.

$$\text{Percentage regulation} = \frac{E_f - V_t}{V_t} \times 100$$

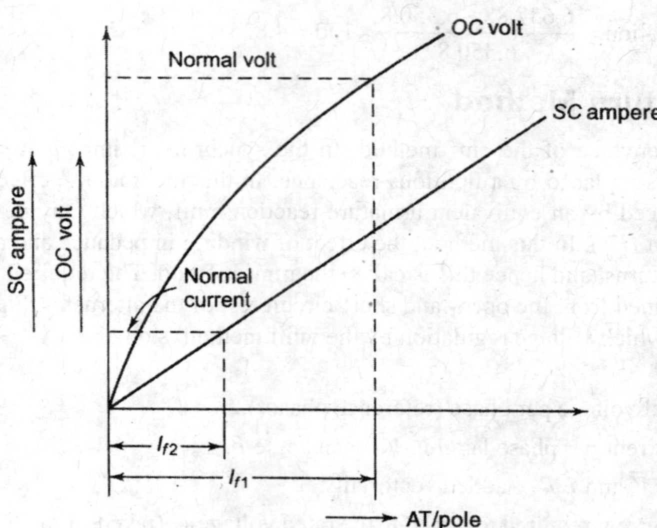

Figure 6.18 Ampere-turn Method

6.8.3.1 *Ampere-turn Method with R_a Neglected*

Figure 6.19 shows the phasor diagram at lagging power factor ($\cos\beta$) when R_a is neglected.

From Figure 6.19, we have

$$\bar{I}_{f2} = I_{f2}\angle(180° - \beta)$$

$$\bar{I}_{f1} = I_{f1}\angle 90°$$

and

$$\bar{I}_f = \bar{I}_{f1} + \bar{I}_{f2} \tag{6.16}$$

Also from Figure 6.19, we have

$$I_f^2 = I_{f1}^2 + I_{f2}^2 + 2I_{f1}I_{f2}\cos(90° - \beta)$$

i.e.,

$$I_f^2 = I_{f1}^2 + I_{f2}^2 + 2I_{f1}I_{f2}\sin\beta \tag{6.17}$$

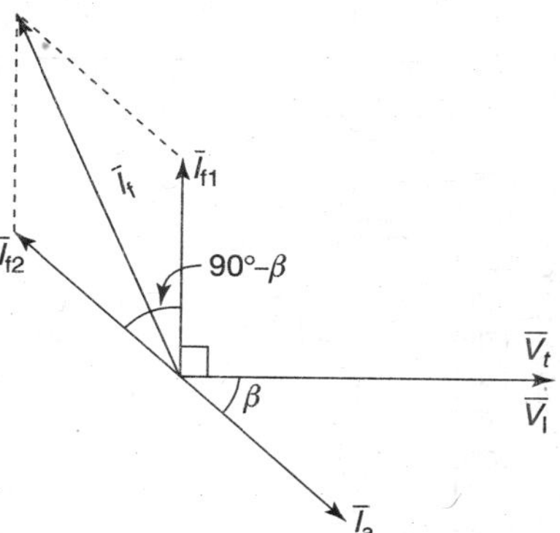

Figure 6.19 Phasor Diagram with Negligible R_a

Example 6.3 A three-phase, 1,200 kVA, 2.2 kV, 50 Hz, star-connected alternator having effective resistance per phase of 0.22 Ω gave the following results:

Field current (A)	8	10	15	20	25	30	40	50	58
OC voltage (V)	873.2	1,057.8	1,527.11	1,821.25	2,214.54	2,243.3	2,365.9	2,409.0	2,443.6
SC armature current (A)	–	–	128	240	340				

Find the voltage regulation at (i) 0.8 pf lagging and (ii) 0.8 power factor leading.

Solution

$$V_t = \frac{2.2\times10^3}{\sqrt{3}} = 1,270.17\,\text{V} \quad \text{and} \quad I_L = \frac{1,200\times10^3}{\sqrt{3}\times2.2\times10^3} = 314.92\,\text{A}$$

The phase values of OC voltages are 504.14, 610.74, 881.68, 1,051.5, 1,220.83, 1,295.17, 1,365.96, 1,390.84 and 1,410.84.

Figures E6.1(a) and E6.1(b) show the DCC and SCC plots.

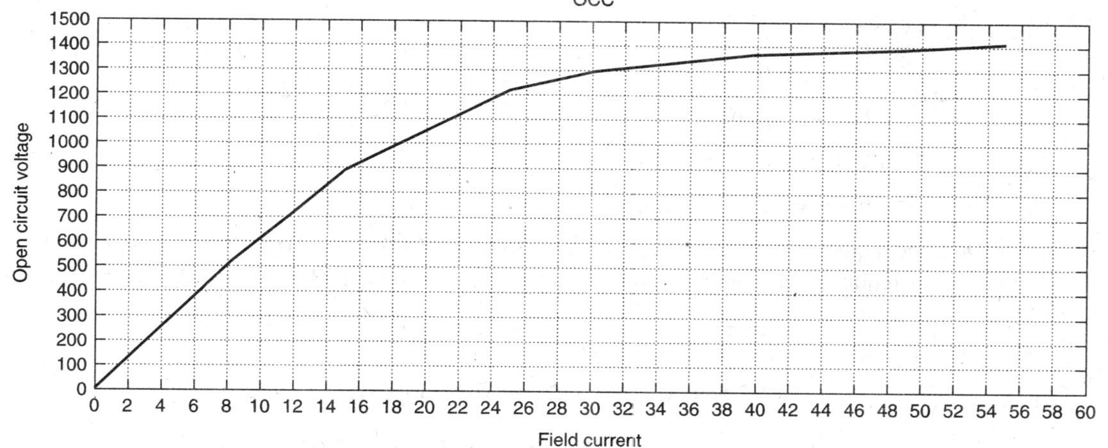

(a) OCC plot

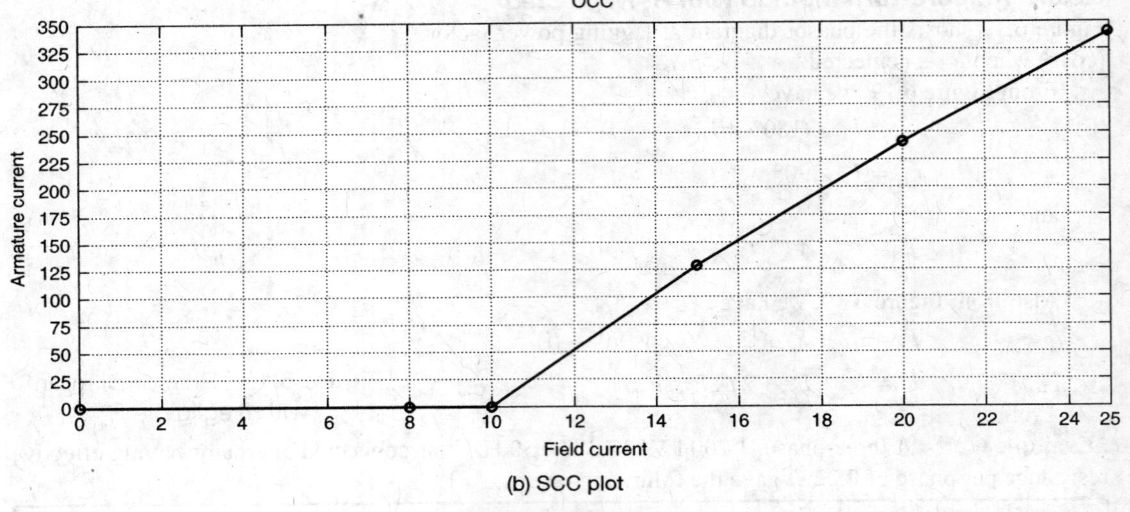

(b) SCC plot

Figure E6.1

$$I_a \text{ per phase} = I_L = 314.92\text{ A}, R_a \text{ per phase} = 0.22$$

(i) $\cos\beta = 0.8(\text{lagging})$ and $\sin\beta = 0.6$

$$\beta = \cos^{-1}(0.8) = 36.87°$$

$$E' = \sqrt{(V_t\cos\beta + I_aR_a)^2 + (V_t\sin\beta + I_aX_s)^2}$$

$$= \sqrt{(1{,}270.17\times 0.8 + 314.92\times 0.22)^2 + (1{,}270.17\times 0.6)^2}$$

$$= 1{,}326.25\text{ V}$$

The angle between E' and V_t is

$$\tan\delta'' = \frac{V_t\sin\beta}{V_t\cos\beta + I_aR_a} = \frac{1{,}270.17\times 0.8}{1{,}270.17\times 0.6 + 314.92\times 0.22} = 0.702 = \tan 35.07°$$

i.e., $\delta'' = 35.07°$

$\therefore$ $\delta = \beta - \delta'' = 1.8°$ [lagging]

From *OCC* the required field current to produce the voltage of 1,326.25 V is 37.7 A. Hence, $I_{f1} = 34.1$ A
From *SCC* the required field current to produce full-load current of 314.92 A is 23.9 A. Hence, $I_{f2} = 23.9$ A

$\therefore$ $$\bar{I}_{f1} = 37.7\angle(90° - 1.8°) = 1.184 + j37.68$$

and $$\bar{I}_{f2} = 23.9\angle(180° - 36.87°) = -19.12 + j14.34$$

$\therefore$ $$\bar{I}_f = \bar{I}_{f1} + \bar{I}_{f2} = -17.93 + j52.02 = 55.03\angle 109.02°\text{ A}$$

From *OCC*, the open-circuit voltage for the field current of 53.03 A is 1,400.84 V.

$$\therefore \quad \text{Percentage regulation} = \frac{1{,}400.84 - 1{,}270.17}{1{,}270.17} \times 100 = 10.26$$

(ii)

$$\cos\beta = 0.8(\text{leading}) \text{ and } \sin\beta = 0.6$$

$$\beta = \cos^{-1}(0.8) = 36.87°$$

$$E_f = 1326.25 \text{ V and } \delta = 1.8°$$

$$\therefore \quad \bar{I}_{f1} = 37.7\angle(90° + 1.8°)\text{A} = -1.184 + j37.68$$

and

$$\bar{I}_{f2} = 23.9\angle(180° + 36.87°) = -19.11 - j14.34$$

$$\therefore \quad \bar{I}_f = \bar{I}_{f1} + \bar{I}_{f2} = -20.29 + j23.34 = 30.93\angle 131° \text{ A}$$

From *OCC*, the open-circuit voltage for the field current of 30.93 A is 1,300.3 V.

$$\therefore \quad \text{Percentage regulation} = \frac{1{,}300.3 - 1{,}270.17}{1{,}270.17} \times 100 = 2.37\%$$

■

6.8.4 Zero Power Factor Method or Potier Method

Figure 6.20 shows the phasor diagram corresponding to zero power factor lagging load, where terminal voltage ($\bar{V}_t$) is taken as the reference phasor. The armature current ($\bar{I}_a$) lags behind the voltage $\bar{V}_t$ by a 90° angle at zero power factor lagging. $\bar{I}_a R_a$ is drawn parallel to $\bar{I}_a$ and $\bar{I}_a X_{aL}$ is drawn perpendicular to $\bar{I}_a$. Therefore, we have

$$\bar{V}_t + \bar{I}_a R_a + \bar{I}_a X_{aL} = \bar{E}_g \tag{6.18}$$

In Equation (6.18), $\bar{E}_g$ is the generated voltage per phase, shown in Figure 6.20, $\bar{E}_r$ is the armature reaction mmf which is in phase with $\bar{I}_a$, $\bar{E}_f$ is the mmf of the main field winding, that is, field mmf, and $\bar{F}_r$ is the resultant mmf.

We have the following relation from Figure 6.20:

$$\bar{F}_r = \bar{F}_f + \bar{F}_{ar} \tag{6.19}$$

Figure 6.21 shows the phasor diagram when R_a is neglected.

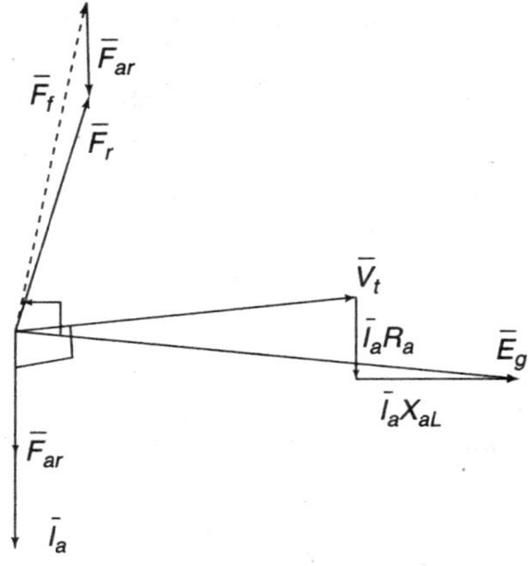

Figure 6.20 Phasor Diagram for Zero pf Load

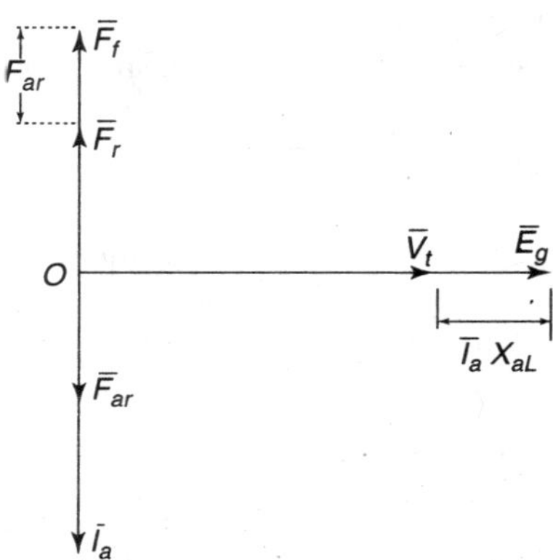

Figure 6.21 Phasor Diagram with R_a neglected

In Figure 6.21, $\bar{V}_t$ is the terminal voltage, $\bar{I}_a X_{aL}$ is the reactance voltage drop and $\bar{E}_g$ is the generated emf. Since all these are in phase, we have

$$\bar{V}_t = \bar{E}_g - \bar{I}_a X_{aL} \tag{6.20}$$

Since $\bar{F}_f$, $\bar{F}_r$ and $\bar{F}_{ar}$ are all in phase, their magnitudes are related by

$$\bar{F}_f = \bar{F}_r + \bar{F}_{ar} \tag{6.21}$$

Equations (6.20) and (6.21) are the basis for Potier triangle.

From Equation (6.21), we have

$$\frac{\bar{F}_f}{T_f} = \frac{\bar{F}_r}{T_f} + \frac{\bar{F}_{ar}}{T_f}$$

i.e.,

$$\bar{I}_f = \bar{I}_r + \bar{I}_{ar} \tag{6.22}$$

where T_f is the effective number of turns per pole on the rotor field.

This method gives more accurate results because it is based on the separation of armature leakage reactance effects. The plot of armature terminal voltage per phase against the field current when the machine is operated with constant rated lagging power factor gives the zero power factor characteristic (*ZPFC*). This *ZPFC* is also called Potier characteristic.

In this method, the reduction in armature terminal voltage due to armature reaction and armature leakage reactance drop can be obtained. With armature terminal voltage ($\bar{V}_t$) if we add $\bar{I}_a R_a$ and $\bar{I}_a X_L$, we get $\bar{E}_g$. Further, if we add $\bar{I}_a X_{ar}$ (drop due to armature reaction) to it, we get $\bar{E}_f$. To obtain *ZPFC*, any of the following methods are followed:

(i) loading alternator with pure reactors,

(ii) driving a similar machine as a synchronous motor at no-load at practically zero power factor or

(iii) connecting an alternator with a three-phase line with ammeters and wattmeters to measure current and power adjusting the field current so that full-load armature current and zero wattmeter reading are obtained.

The shape of *ZPFC* is very much similar to *OCC*, but it is displaced downwards and also to the right.

6.8.4.1 Potier Triangle

Figure 6.22 shows the plot of *OCC* and *ZPFC*. A point *x* is considered on the *ZPFC* corresponding to rated terminal voltage (V_t) and a field current of OA = $I_f = \dfrac{F_f}{T_f}$. Let the value of the armature reaction mmf be expressed in equivalent field current of $BA = I_{ar} = \dfrac{F_{ar}}{T_f}$. The equivalent field current of the resultant mmf becomes $OB = I_r = \dfrac{F_r}{T_f}$, which results a generated voltage $E_g = Bz$ shown in the no-load saturation curve.

For zero lagging power factor operation, we have

$$\bar{E}_g = \bar{V}_t + \bar{I}_a X_{aL}$$

The vertical distance *yz* represents the leakage reactance voltage drop $I_a X_{aL}$, where I_a represents the rated armature current:

$$X_{aL} = \frac{\text{Voltage } yz \text{ per phase}}{\text{Rated armature current}}$$

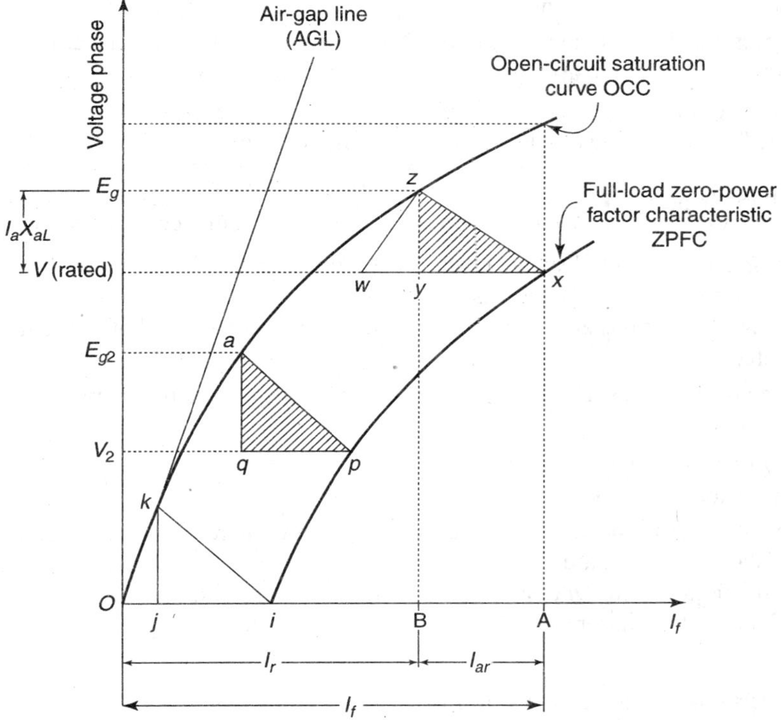

Figure 6.22 Potier Triangle

The *xyz* represents Potier triangle. Let us take a terminal voltage V_t' ($V_t' < V_t$) when alternator operates with rated armature current for zero power factor operation. Here, I_aX_{aL} and the armature mmf remain unaltered because I_a is the rated value. The same Potier triangle will be located between terminal voltage (V_t') point on *ZPFC* and the corresponding E_g' point on the *OCC* for all the conditions of operation with rated armature current at zero lagging power factor. The Potier triangle is moved downwards and new Potier triangle becomes *pqr*. The point *x* is moved to *p* and the point *z* is moved to *r*, where *xy* is kept horizontal and it becomes *pq*. The point *r* determines new generated voltage E_g which is E_g'. When the point *x* reaches the point *i*, the new Potier triangle becomes *ijk* which is the limiting position and it corresponds to short-circuit conditions. This because the terminal voltage V_t at *i* becomes zero. The initial part of *OCC* is almost linear and hence another triangle *Oki* is formed by the base, hypotenuse of the Potier triangle and the *OCC*. A similar triangle *zwx* where *zw* is parallel to *Ok* through the vertex of the Potier triangle lying on the *OCC* can be constructed from the Potier triangle *xyz*.

It is possible to find the armature reaction mmf and the approximate leakage reactance voltage of the machine by using *ZPFC* in conjunction with the *OCC*. The following points are to be followed for the construction of Potier triangle:

(i) A point *x* is taken on the *ZPFC* corresponding to the rated terminal voltage of the alternator which is preferably at the knee of the *OCC*.

(ii) A line *xw* is drawn which is equal to *iO* where *i* is the point for zero and full-load current. *Oi* represents the short-circuit excitation.

(iii) A line *wz* parallel to *Ok* is drawn through *w* which meets *OCC* at *w*.

(iv) A perpendicular *zy* is drawn to *zw*.

(v) Here, *zy* represents the leakage reactance on the scale and *xy* represents the armature reaction mmf or field current equivalent to armature reaction mmf at the rated current.

An equivalent leakage reactance (X_p) is obtained by combining the effect of field leakage flux and armature leakage flux. X_p is known as Potier reactance which is greater than armature leakage reactance.

$$\therefore \quad \text{Potier reacatnace} = \frac{zy \text{ (voltage dropper phase)}}{I_a \text{(zero power factor rated armature current per phase)}}$$

X_p is approximately equal to leakage reactance X_{aL} for cylindrical rotor machines, whereas X_p may be three times X_{aL} for salient-pole machines.

The following assumptions are considered for the construction of the Potier triangle:

(i) R_a is neglected.

(ii) *OCC* is taken on no-load very accurately, and it represents the relation between mmf and voltage on no-load.

(iii) $I_a X_{aL}$ does not depend on excitation.

(iv) Armature reaction mmf has constant value.

The two points *x* and *I* on *ZPFC* are sufficient enough where *x* point corresponds to field current, which circulates rated armature current, and point *i* corresponds to short-circuit condition of the machine. Here, the field current *Oi* circulates the short-circuit current.

Figure 6.23 shows the phasor diagram for a lagging power factor cosβ.

In Figure 6.23,

$\overrightarrow{OP} = \overline{V}_t$ [terminal phase voltage at full-load],

$\overrightarrow{OQ} = \overline{I}_a$ [rated armature current lagging behind $\overline{V}_t$ by an angle β],

$\overrightarrow{PR} = \overline{I}_a R_a$ [drawn parallel to $\overrightarrow{OQ}$],

$\overrightarrow{RS} = \overline{I}_a X_{aL}$ [drawn perpendicular to RS] and

$\overrightarrow{OS} = \overline{E}_g$ [generated emf E_g].

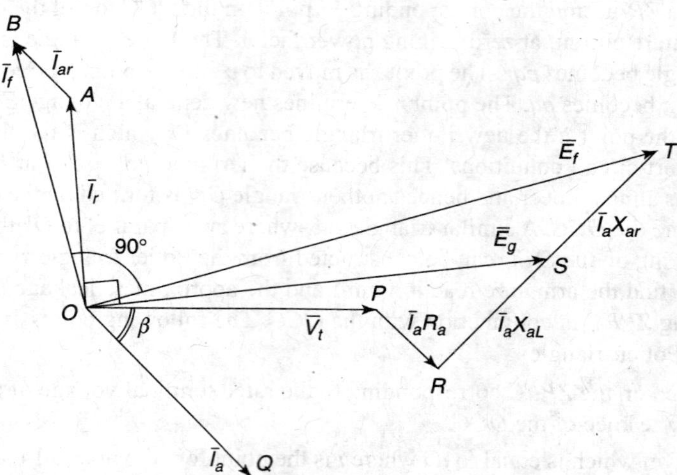

Figure 6.23 Phasor diagram of an alternator at lagging power factor

The field excitation current ($\bar{I}_r$) corresponding to $\bar{E}_g$ is determined from the *OCC*. $\overrightarrow{OA}(=\bar{I}_r)$ is drawn perpendicular to $\overrightarrow{OS}$. $\overrightarrow{AB}$ is drawn parallel to $\overrightarrow{OQ}$, which represents the field current equivalent to full-load armature reaction $\bar{I}_{ar}$. $\overrightarrow{OB}$ represents the total field current ($\bar{I}_f$). When load is thrown off, terminal voltage becomes equal to the generated emf and it corresponds to the field current $\overrightarrow{OB}$. The emf $\bar{E}_f(= OT)$ is determined from *OCC* because it corresponds to the field current $\overrightarrow{OB}$. The phasor $\overrightarrow{OT}$ lags behind the phasor $\overrightarrow{OB}$ by a 90° angle. The voltage regulation is determined by the following relation:

$$\text{Voltage regulation} = \frac{E_f - V_t}{V_t}\ \text{p.u.}$$

Example 6.4 A 200 kVA, 1.1 kV, three-phase, star-connected alternator having effective alternator resistance/phase of 0.62 Ω gave the following test results:

Field current (A)	20	35	50	80	100	120
OC terminal voltage (V)	692.82	1,120	1,450	1,750	1,953	2,180
Zero power factor characteristic (V)	0	450	1,000	1,398	1,600	1,850

Using the zero power factor method, calculate voltage regulation for a load current of 100 A at (i) 0.8 power factor lagging and (ii) 0.8 power factor leading.

Solution

Figure E6.2 shows the plot of *OCC* and *ZPFC*. The Potier triangle is drawn as usual. From Figure E6.2, $db = 300$ V (line to line) and $bc = 10$ A.

Therefore,
$$db\ (\text{per phase}) = \frac{300}{\sqrt{3}} = 173.2\ \text{V}$$

$$I_a X_L = 173.2$$

i.e.,
$$X_L = \frac{173.2}{I_a} = \frac{173.2}{100} = 1.73\Omega$$

$$V_t = \frac{1{,}100}{\sqrt{3}} = 635.08\ \text{V}$$

(i) At 0.8 power factor lagging:

$$E_f = \sqrt{\left(V_t \cos\beta + I_a R_a\right)^2 + \left(V_t \sin\beta + I_a X_L\right)^2}$$

$$= \sqrt{\left(635.08 \times 0.8 + 100 \times 0.62\right)^2 + \left(635.08 \times 0.6 + 100 \times 1.73\right)^2}$$

$$= 794.94\ \text{V}$$

$$\tan\left(\delta + \beta\right) = \frac{V_t \sin\beta + I_a X_L}{V_t \cos\beta + I_a R_a} = \frac{635.08 \times 0.6 + 100 \times 1.73}{635.08 \times 0.8 + 100 \times 0.62} = \frac{554.084}{570.064}$$

$$= 0.972 = \tan 44.18°$$

i.e.,
$$\delta + \beta = 44.18°$$

$$(E_f)_{\text{line}} = \sqrt{3} \times 794.94 = 1{,}376.87\ \text{V}$$

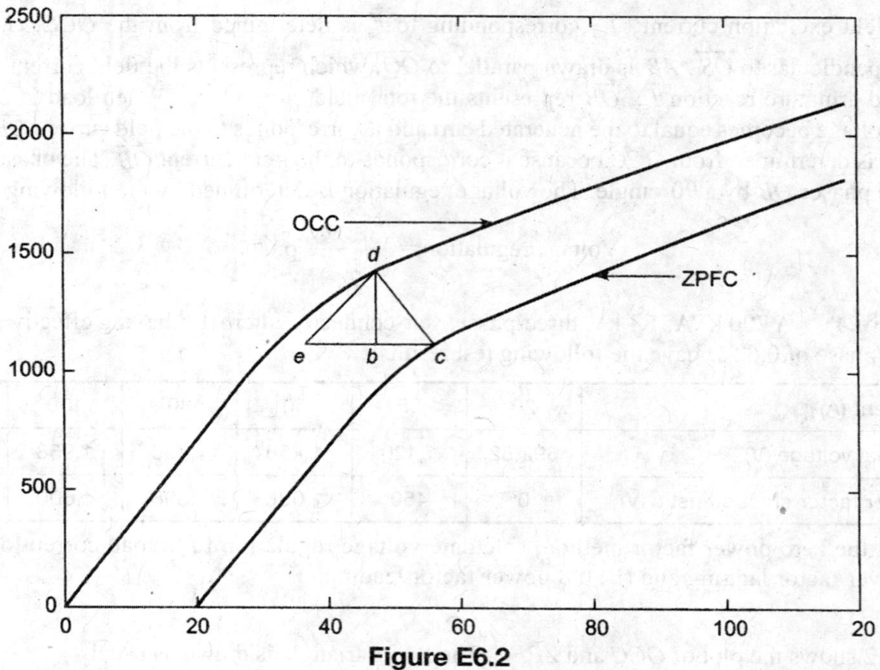

Figure E6.2

From *OCC*, the field current corresponding to line voltage 1,376.87 V is 40 A which leads E_f by a 90° angle.

$$\bar{I}_r = 40\angle(90° + 44.18°) = 40\angle 134.18° \text{A} \quad \text{and} \quad \bar{I}_{ar} = 9.2\angle 0° \text{ A}$$

$$\bar{I}_f = \bar{I}_r - \bar{I}_{ar} = 40\angle 134.18° - 9.2\angle 0° = -27.87 + j28.68 - 9.2$$

$$= -37.07 + j28.68 = 46.86\angle 142.27°$$

The *OCC* voltage corresponding to field current of 46.86 A is 1,390V (line to line) and its phase value is 802.51 V.

$$\therefore \qquad \text{Voltage regulation} = \frac{E_f - V_t}{V_t} \times 100 = \frac{802.51 - 635.08}{635.08} \times 100 = 26.36\%$$

(ii) At 0.8 power factor leading:

$$E_f = \sqrt{(V_t \cos\beta + I_a R_a)^2 + (V_t \sin\beta - I_a X_L)^2}$$

$$= \sqrt{(635.08 \times 0.8 + 100 \times 0.62)^2 + (635.08 \times 0.6 - 100 \times 1.73)^2}$$

$$= 606.84 \text{ V}$$

$$\tan(\delta + \beta) = \frac{V_t \sin\beta + I_a X_L}{V_t \cos\beta + I_a R_a} = \frac{635.08 \times 0.6 - 100 \times 1.73}{635.08 \times 0.8 + 100 \times 0.62} = \frac{208.048}{570.064}$$

$$= 0.365 = \tan 20.05°$$

i.e.,
$$\delta + \beta = 20°$$
$$(E_f)_{\text{line}} = \sqrt{3} \times 606.84 = 1{,}051\,\text{V}$$

From OCC, the field current corresponding to line voltage 1,051 V is 32 A which leads E_f by a 90° angle.

$$\bar{I}_r = 32\angle\left(90° - 20.05°\right) = 32\angle 70°\,\text{A} \quad \text{and} \quad \bar{I}_{ar} = 9.2\angle 0°\,\text{A}$$

$$\bar{I}_f = \bar{I}_r - \bar{I}_{ar} = 32\angle 70° - 9.2\angle 0° = 10.94 + j30.07 - 9.2$$
$$= 1.74 + j30.07 = 31\angle 86.78°$$

The OCC voltage corresponding to field current of 31 A is 995 V (line to line) and its phase value is 574 V.

$$\therefore \quad \text{Voltage regulation} = \frac{E_f - V_t}{V_t} \times 100 = \frac{574 - 635.08}{635.08} \times 100 = -9.61\%$$ ■

6.8.5 ASA Method

The emf and mmf methods are not capable of giving reliable results due to the unrealistic assumption that the magnetic circuit is unsaturated. Therefore, ASA method was proposed as the modification of mmf method, which gave more accurate results as saturation of magnetic circuit is taken into account in this method.

For computation of regulation by ASA, two quantities are required which are F_{sc} and X_L. F_{sc} is the field mmf required to circulate full-load armature current under short-circuit condition and X_L is the leakage reactance. Now F_{sc} can be obtained by *mmf* or *emf* method and X_L can be computed by Potier method. The voltage phasor diagram with armature current taken as reference axis is shown in Figure 6.24. Figure 6.24 shows the phasor diagram of the *mmf*s.

Generated voltage $\bar{E}_g$ is obtained by adding $\bar{I}_a R_a$ and $\bar{I}_c X_L$ drop to the voltage phasor ($\bar{V}_t$):

$$\bar{E}_g = \bar{V}_t + \bar{I}_a R_a + j\bar{I}_a X_L$$

Figure 6.25 shows the voltage phasor diagram. Now with O as centre and E_g as radius, an arc is drawn to meet vertical axis at e. Now, a horizontal line is drawn through e, which meets air-gap line and OCC characteristics at c and d, respectively. F_r is the field excitation to generate rated voltage on open circuit.

$\overrightarrow{O'A}$ represents field excitation to generate rated voltage/phase under open-circuit condition. $\bar{F}_{sc}$ is vectorially added to $\bar{F}_r$ to get net *mmf* ($\bar{F}$). The angle between the vertical line and $\bar{F}_{sc}$ is approximately equal to power factor angle.

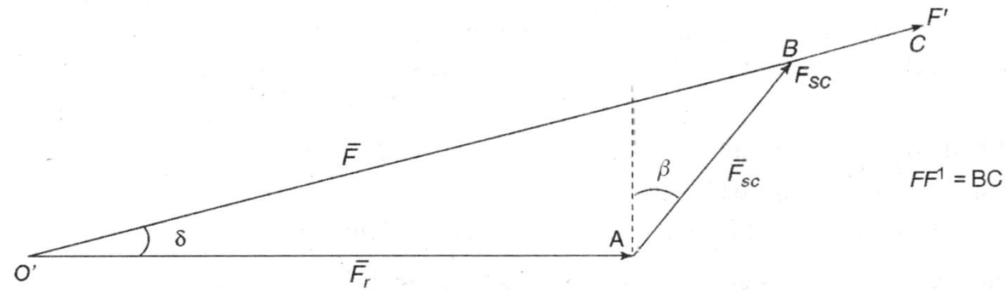

Figure 6.24 Phasor Diagram of the mmf

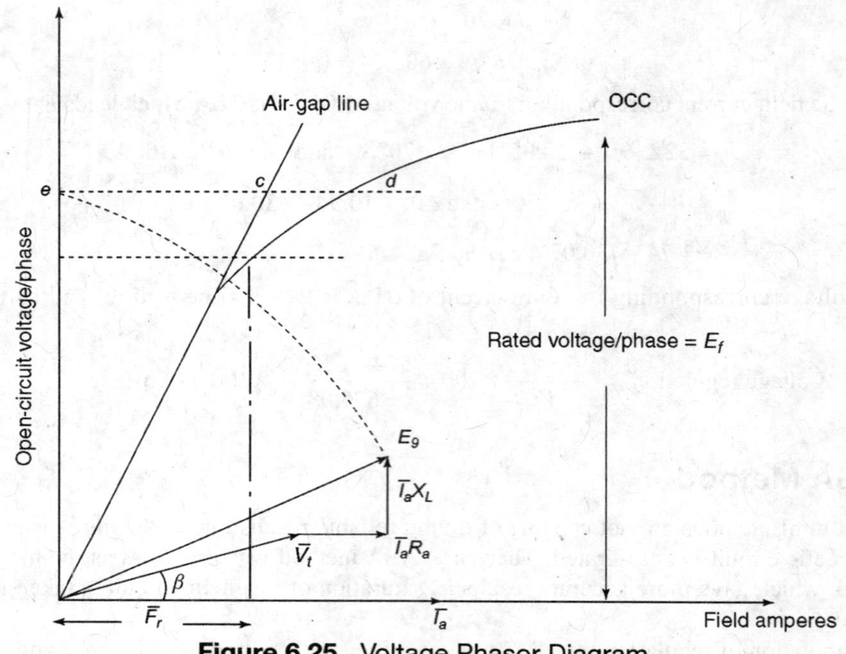

Figure 6.25 Voltage Phasor Diagram

Now, $\overrightarrow{O'B}$ is extended such that $\overrightarrow{BC}$ represents the additional excitation *mmf* required for the compensation of saturation in magnetic circuit.

Now from open-circuit voltage $\bar{E}_f$ computed from *OCC*, regulation can be computed as follows:

$$\frac{E_f - V_t}{V_t} \text{ p.u.}$$

This method is also called *AIEE* method and it gives accurate results for both cylindrical and salient-pole machines. The angle between $\bar{E}_f$ and $\bar{V}_t$ is approximately equal to $\angle BO'A$ which is known as power angle.

6.9 LOAD CHARACTERISTICS OF ALTERNATORS

In DC generators, terminal voltage varies with respect to the magnitude of load current or the armature current. In alternator, the variation of terminal voltage depends not only on the load current but also on the power factor (leading or lagging). In case of alternators, the following load characteristics are important:

(i) terminal voltage versus field current,
(ii) terminal voltage versus load current and
(iii) alternator excitation against load current for a given terminal voltage.

The above three characteristics have been shown in Figures 6.26(a), 6.26(b) and 6.26(c), respectively.

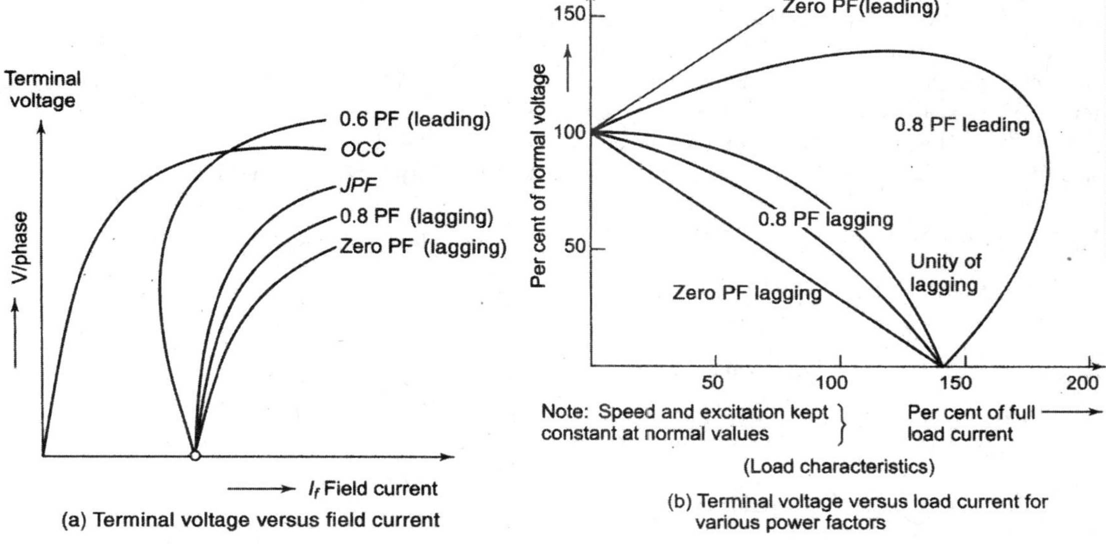

(a) Terminal voltage versus field current

(b) Terminal voltage versus load current for various power factors

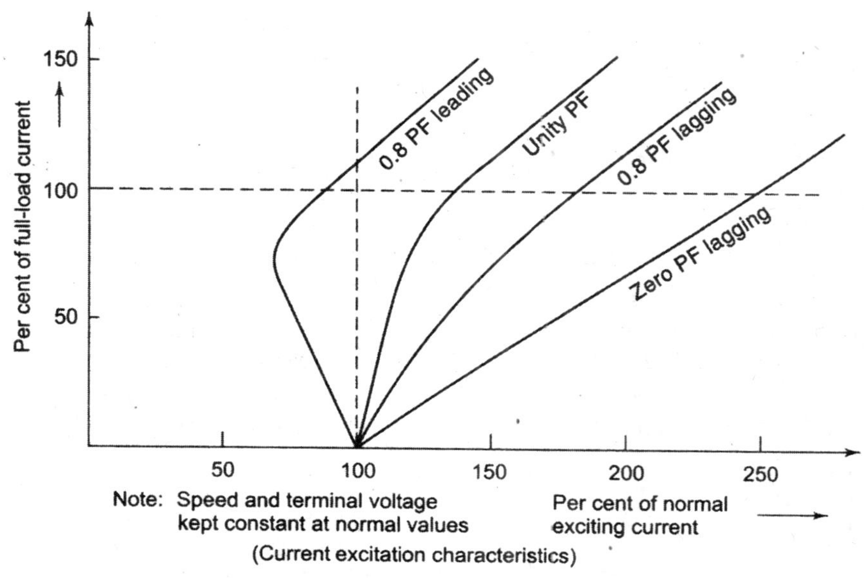

(c) Field current versus load current for various power factors

Figure 6.26 Load Characteristics of Alternator

6.10 OUTPUT POWER EQUATION OF AN ALTERNATOR

Figure 6.27 shows the single-line diagram of an alternator connected to an infinite bus.

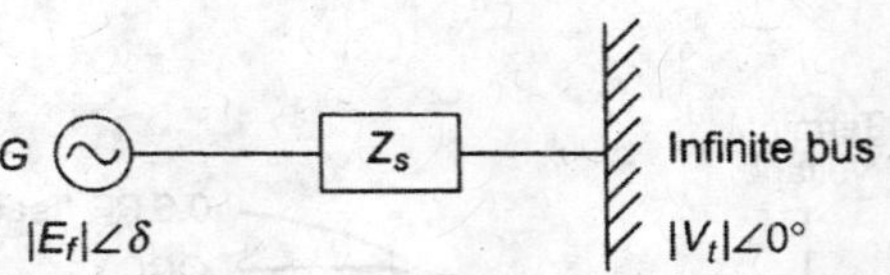

Figure 6.27 Generator Connected to Infinite Bus

Let E_f be the magnitude of excitation voltage of the generator, δ be the power angle, V_t be the magnitude of infinite bus voltage, I_a be the armature current, Z_s be the magnitude of the synchronous impedance (= $R_a + jX_s$) of the alternator, R_a be the effective armature resistance and X_s be the synchronous reactance.

Here, $$\theta = \tan^{-1}\left(\frac{X_s}{R_a}\right)$$

$$\bar{Z}_s = Z_s\angle\theta \quad \text{and} \quad \alpha + \theta = 90°$$

∴ $$\bar{Z}_s = R_a + jX_s = \sqrt{R_a^2 + X_s^2}\angle\tan^{-1}\left(\frac{X_s}{R_a}\right) = Z_s\angle\theta = Z_s\angle(90° - \alpha)$$

∴ $$\bar{I}_a = \frac{E_f\angle\delta - V_t\angle 0°}{Z_s\angle\theta}$$

The complex input power/phase is expressed by

$$\bar{S}_{og} = \bar{P}_{og} + j\bar{Q}_{og} = \bar{V}_t\bar{I}_a^* = V_t\angle 0°\left(\frac{E_f\angle\delta - V_t\angle 0°}{Z_s\angle\theta}\right)^* = V_t\angle 0°\left(\frac{E_f\angle -\delta - V_t\angle 0°}{Z_s\angle -\theta}\right)$$

$$= \frac{V_tE_f}{Z_s}\angle(\theta - \delta) - \frac{V_t^2}{Z_s}\angle\theta$$

$$= \left\{\frac{V_tE_f}{Z_s}\cos\,(\theta - \delta) - \frac{V_t^2}{Z_s}\cos\theta\right\} + j\left\{\frac{V_tE_f}{Z_s}\sin(\theta - \delta) - \frac{V_t^2}{Z_s}\sin\theta\right\} \tag{6.23}$$

Equation (6.23) shows the complex power output per phase of the alternator. Equating real and imaginary parts of Equation (6.23), the real and reactive power can be obtained.

6.10.1 Real Power Output per Phase of Alternator (P_{og})

Equating real parts of Equation (6.23), we have

$$P_{og} = \frac{V_tE_f}{Z_s}\cos(\theta - \delta) - \frac{V_t^2}{Z_s}\cos\theta$$

Again, $$\cos\theta = \frac{R_a}{Z_s}$$

∴ $$P_{og} = \frac{V_tE_f}{Z_s}\cos(\theta - \delta) - \frac{V_t^2}{Z_s}\cos\theta = \frac{V_tE_f}{Z_s}\cos(\theta - \delta) - \frac{V_t^2R_a}{Z_s^2} \tag{6.24}$$

And also $$P_{og} = \frac{V_t E_f}{Z_s}\cos(90° - \alpha - \delta) - \frac{V_t^2 R_a}{Z_s^2} = \frac{V_t E_f}{Z_s}\sin(\alpha + \delta) - \frac{V_t^2 R_a}{Z_s^2} \quad (6.25)$$

P_{og} is also called the electrical power developed by the alternator.

6.10.2 Reactive Power Output per Phase of Alternator (Q_{og})

Equating imaginary parts of Equation (6.23), we have

$$Q_{og} = \frac{V_t E_f}{Z_s}\sin(\theta - \delta) - \frac{V_t^2}{Z_s}\sin\theta$$

Again, $$\sin\theta = \frac{X_s}{Z_s}$$

∴ $$Q_{og} = \frac{V_t E_f}{Z_s}\sin(\theta - \delta) - \frac{V_t^2}{Z_s}\sin\theta = \frac{V_t E_f}{Z_s}\sin(\theta - \delta) - \frac{V_t^2 X_s}{Z_s^2} \quad (6.26)$$

and also $$Q_{og} = \frac{V_t E_f}{Z_s}\sin(90\Upsilon - \alpha - \delta) - \frac{V_t^2 X_s}{Z_s^2} = \frac{V_t E_f}{Z_s}\cos(\alpha + \delta) - \frac{V_t^2 X_s}{Z_s^2} \quad (6.27)$$

6.10.3 Maximum Power Output per Phase of Alternator

For maximum power output of the generator per phase, we have

$$\frac{dP_{og}}{d\delta} = 0 \quad \text{and} \quad \frac{d^2 P_{og}}{d\delta^2} < 0$$

Differentiating Equation (6.25) with respect to δ, we have

$$\frac{d}{d\delta}\left[\frac{V_t E_f}{Z_s}\sin\ (\alpha + \delta) - \frac{V_t^2 R_a}{Z_s^2}\right] = 0$$

i.e., $$\frac{V_t E_f}{Z_s}\cos(\alpha + \delta) = 0$$

i.e., $$\cos(\delta + \alpha) = 0$$

i.e., $$\delta + \alpha = 90°$$

i.e., $$\delta = 90° - \alpha = \theta \quad (6.28)$$

∴ $$(P_{og})_{max} = \frac{V_t E_f}{Z_s} - \frac{V_t^2 R_a}{Z_s^2} = \frac{V_t}{Z_s}\left[E_f - \frac{V_t}{Z_s}R_a\right] \quad (6.29)$$

6.10.4 Power Output Equation of Alternator Neglecting Armature Resistance

Practically R_a is very small and it can be neglected.

∴ $$\theta = \tan^{-1}\left(\frac{X_s}{R_a}\right) = 90° \quad (6.30)$$

and
$$Z_s = X_s \tag{6.31}$$

Using Equations (6.30) and (6.31), Equation (6.25) can be expressed as follows:

$$P_{og} = \frac{E_f V_t}{X_s}\cos(90° - \delta)$$

i.e.,
$$P_{og} = \frac{E_f V_t}{X_s}\sin\delta$$

i.e.,
$$P = P_m \sin\delta \tag{6.32}$$

where
$$P_m = \frac{E_f V_t}{X_s} = \text{maximum power limit} \tag{6.33}$$

Using Equations (6.30) and (6.31), Equation (6.26) can be written as follows:

$$Q_{og} = \frac{E_f V_t}{X_s}\sin(90° - \delta) - \frac{V_t^2}{X_s}$$

i.e.,
$$Q_{og} = \frac{V_t}{X_s}[E_f \cos\delta - V_t] \tag{6.34}$$

(i) If $E_f \cos\delta > V_t$, the alternator is overexcited and it delivers reactive power to the infinite bus. Under this condition, the alternator is operating at lagging power factor.

(ii) If $E_f \cos\delta = V_t$, the alternator is normally excited and it neither delivers reactive power to infinite bus bar nor absorbs reactive power from the infinite bus bar. Under this condition, the alternator is operating at unity power factor.

(iii) If $E_f \cos\delta < V_t$, the alternator is underexcited and it absorbs reactive power from the infinite bus. Under this condition, the alternator is operating at leading power factor.

6.11 INPUT POWER EQUATION OF ALTERNATOR

The input complex power per phase to an alternator is expressed by

$$\bar{S}_{ig} = \bar{P}_{ig} + j\bar{Q}_{ig} = \bar{E}_f \bar{I}_a^* = E_f\angle\delta\left(\frac{E_f\angle\delta - V_t\angle 0}{Z_s\angle\theta}\right)^*$$

$$= E_f\angle\delta\left(\frac{E_f}{Z_s}\angle\theta - \delta - \frac{V_t}{Z_s}\angle\theta\right)$$

$$= \frac{E_f^2}{Z_s}\angle\theta - \frac{E_f V_t}{Z_s}\angle(\theta + \delta)$$

$$\bar{P}_{ig} + j\bar{Q}_{ig} = \frac{E_f^2}{Z_s}(\cos\theta + j\sin\theta) - \frac{E_f V_t}{Z_s}\left[\cos(\theta + \delta) + j\sin(\theta + \delta)\right]$$

$$= \frac{E_f^2}{Z_s}\cos\theta - \frac{E_f V_t}{Z_s}\cos(\theta+\delta) + j\left\{\frac{E_f^2}{Z_s}\sin\theta - \frac{E_f V_t}{Z_s}\sin(\theta+\delta)\right\} \tag{6.35}$$

Equation (6.35) shows the complex input power to an alternator. Equating real and imaginary parts of Equation (6.35), we can get real and reactive power input.

6.11.1 Real Power Input per Phase to Alternator

Equating real parts of Equation (6.35), we have

$$P_{ig} = \frac{E_f^2}{Z_s}\cos\theta - \frac{E_f V_t}{Z_s}\cos(\theta+\delta) \tag{6.36}$$

Again $\cos\theta = \frac{R_a}{Z_s}$, and we have from Equation (6.36)

$$P_{ig} = \frac{E_f^2 R_a}{Z_s^2} - \frac{E_f V_t}{Z_s}\cos(\theta+\delta) \tag{6.37}$$

Again $\theta = 90° - \alpha$, and we have from Equation (6.37)

$$P_{ig} = \frac{E_f^2 R_a}{Z_s^2} - \frac{E_f V_t}{Z_s}\cos(90° - \alpha + \delta)$$

$$P_{ig} = \frac{E_f^2 R_a}{Z_s^2} - \frac{E_f V_t}{Z_s}\sin(\alpha - \delta) \tag{6.38}$$

6.11.2 Reactive Power Input per Phase to Alternator

$$Q_{ig} = \frac{E_f^2}{Z_s}\sin\theta - \frac{E_f V_t}{Z_s}\sin(\theta+\delta) \tag{6.39}$$

Again $\sin\theta = \frac{X_s}{Z_s}$, and we have from Equation (6.39)

$$Q_{ig} = \frac{E_f^2 X_s}{Z_s^2} - \frac{E_f V_t}{Z_s}\sin(\theta+\delta) \tag{6.40}$$

Again $\theta = 90° - \alpha$, and we have from Equation (6.40)

$$Q_{ig} = \frac{E_f^2 X_s}{Z_s^2} - \frac{E_f V_t}{Z_s}\sin(90° - \alpha + \delta)$$

i.e.,

$$Q_{ig} = \frac{E_f^2 X_s}{Z_s^2} - \frac{E_f V_t}{Z_s}\cos(\alpha - \delta) \tag{6.41}$$

6.12 TWO-REACTION THEORY

Andrew Blondel had proposed the two-reaction theory, which resolves the given armature mmfs into two mutually perpendicular components as follows:

(i) One component is located along the axis of salient-pole rotor known as direct-axis (or *d*-axis) component.

(ii) The other component is located perpendicular to the axis of salient-pole rotor known as quadrature-axis (or *q*-axis) component.

The *d*-axis and *q*-axis components of armature mmf are denoted by F_d and F_q, respectively, where the component F_d may be magnetizing or demagnetizing and the component F_q results in cross-magnetizing effect. If the angle between armature current (I_a) and the excitation voltage (E_f) is γ, we have

$$F_d = F_a \sin\gamma \qquad \text{and} \qquad F_q = F_a \cos\gamma$$

where F_a is the amplitude of armature mmf.

6.13 TWO-REACTION THEORY OF SALIENT-POLE ALTERNATOR

Air is uniform in the cylindrical rotor synchronous machine, whereas it is non-uniform for a salient-pole synchronous machine. Figure 6.28 shows a two-pole salient-pole rotor synchronous machine where the rotor rotates in the anticlockwise direction. The direct-axis and quadrature-axis have been shown in Figure 6.28. The direct-axis and quadrature-axis flux paths involve two small air gaps and two large air gaps, respectively. The first one is the path of minimum reluctance, whereas the second one is the path of maximum reluctance.

Figure 6.29(a) shows that the rotor flux is vertically upwards and it induces a voltage E_f in the stator. For a lagging power factor load, the stator current ($\bar{I}_a$) lags behind the generated voltage ($\bar{E}_f$) by an angle γ. The armature current produces stator magnetomotive force ($\bar{F}_s$), which lags $\bar{I}_a$ by a 90° angle. The stator magnetomotive force ($\bar{F}_s$) produces stator magnetic field ($\bar{B}_s$) along its own direction, as shown in Figure 6.29(b). The stator magnetomotive force can be divided into two components: direct-axis component ($\bar{F}_d$) and quadrature-axis component ($\bar{F}_q$).

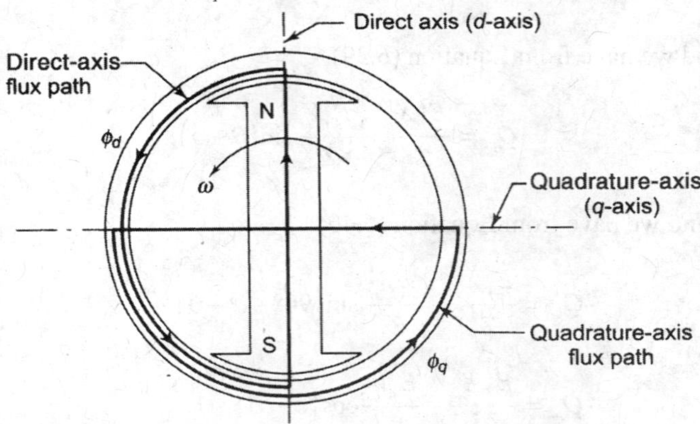

Figure 6.28 Direct Axis and Quadrature Axis

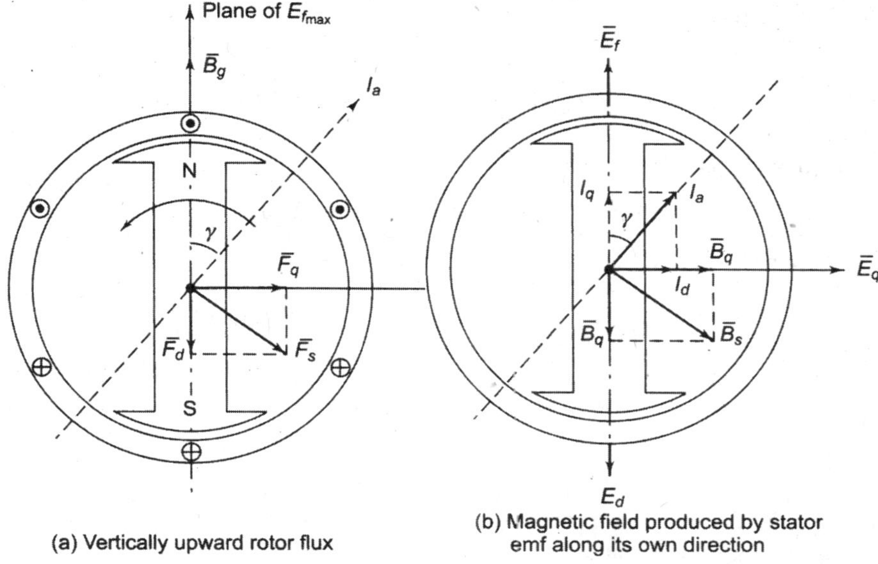

(a) Vertically upward rotor flux

(b) Magnetic field produced by stator emf along its own direction

Figure 6.29 Stator mmf

Let ϕ_d and ϕ_q be the magnitude of direct-axis flux and quadrature-axis flux, respectively. If R_d and R_q are the reluctance along the direct-axis flux path and quadrature-axis flux path, respectively, we can write the following:

$$\bar{\phi}_d = \frac{\bar{F}_d}{R_d} \text{ and } \bar{\phi}_q = \frac{\bar{F}_q}{R_q}$$

Again, R_d is less than R_q. Therefore, $\bar{F}_d$ produces more flux than $\bar{F}_q$. The fluxes $\bar{\phi}_d$ and $\bar{\phi}_q$ produce voltages in the stator winding by armature reaction.

If $\bar{E}_{ad}$ and $\bar{E}_{aq}$ are the direct-axis and quadrature-axis components of armature reaction voltage, we have

$$\bar{E}_{ad} = -jX_{ad}\bar{I}_d \tag{6.42}$$

and

$$\bar{E}_{aq} = -jX_{aq}\bar{I}_q \tag{6.43}$$

where X_{ad} and X_{aq} are the armature reaction reactance in the direct axis and quadrature axis per phase, respectively, and $X_{ad} < X_{aq}$.

The sum of emf induced by the field excitation and these two emfs give the total voltage induced in the stator:

$$\bar{E}' = \bar{E}_f + \bar{E}_{ad} + \bar{E}_{aq} \tag{6.44}$$

i.e.,

$$\bar{E}' = \bar{E}_f - jX_{ad}\bar{I}_d - jX_{aq}\bar{I}_q \tag{6.45}$$

Again,

$$\bar{E}' = \bar{V}_t + \bar{I}_a R_a + j\bar{I}_a X_l \tag{6.46}$$

where X_l is the leakage reactance of the armature.

Now, $$\bar{I}_a = \bar{I}_d + \bar{I}_q \tag{6.47}$$

where I_d is the d-axis component of I_a lagging behind E_f by 90° and I_q is the q-axis component of I_a in phase with E_f.

From Equations (6.45) and (6.46), we have

$$\bar{E}_f = \bar{V}_t + \bar{I}_a R_a + jX_l \bar{I}_a + jX_{ad} \bar{I}_d + jX_{aq} \bar{I}_q \tag{6.48}$$

Substituting Equation (6.47) in Equation (6.48), we have

$$\bar{E}_f = \bar{V}_t + \left(\bar{I}_d + \bar{I}_q\right) R_a + j\left(X_l + X_{ad}\right)\bar{I}_d + +j\left(X_l + X_{aq}\right)\bar{I}_q$$

$$= \bar{V}_t + \left(\bar{I}_d + \bar{I}_q\right) R_a + jX_d \bar{I}_d + +jX_q \bar{I}_q \tag{6.49}$$

where $$X_d = X_l + X_{ad} \text{ and } X_q = X_l + X_{aq}.$$

The reactances X_d and X_q are known as direct-axis and quadrature-axis synchronous reactance.

Finally, from Equation (6.49), we have

$$\bar{E}_f = \bar{V}_t + \bar{I}_a R_a + jX_d \bar{I}_d + jX_q \bar{I}_q \tag{6.50}$$

Equation (6.50) represents the final form of the voltage equation for a salient-pole synchronous generator.

Figure 6.30 shows the phasor diagram of a salient-pole synchronous generator on the basis of two-reaction theory.

Figure 6.31 shows the simplified phasor diagram.

In Figure 6.31, the angle $\delta + \beta$ is unknown and the values of $\bar{I}_d$ and $\bar{I}_q$ are generally not given. Hence, calculation of $\bar{I}_d$ and $\bar{I}_q$ is very difficult. Without prior knowledge of $\delta + \beta$, let us construct the phasor diagram.

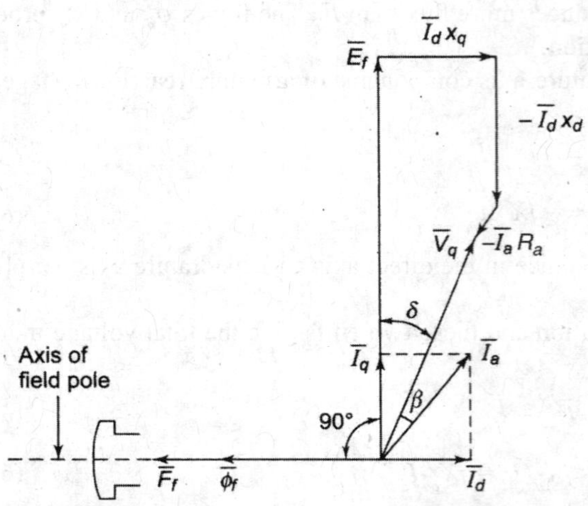

Figure 6.30 Phasor Diagram of Salient-pole Synchronous Generator

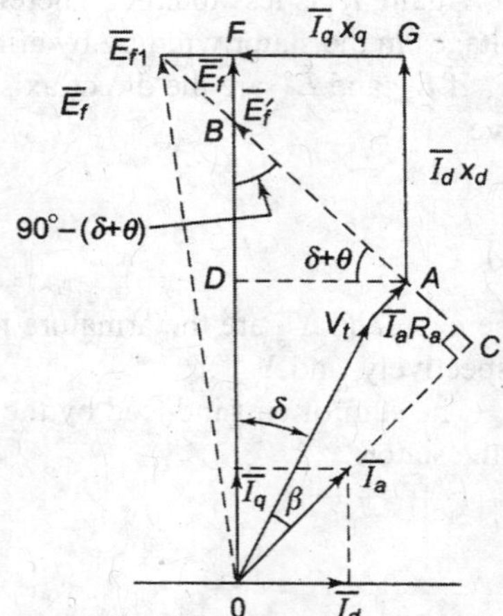

Figure 6.31 Simplified Phasor Diagram

Let us draw Figure 6.32 for the same machine having simplified phasor diagram shown in Figure 6.31 and for the same operating condition. *AB* is drawn normal to $\bar{I}_a$ and hence it is a reactive drop. *AD* is drawn perpendicular to *OB*. From Figure 6.32, the triangles *ADB* and *OCB* are right-angled triangles and hence $\angle BAD = \delta + \beta$.

$$\therefore \qquad AD = AB\cos(\delta+\beta) = XI_a\cos(\delta+\beta) = FG = X_q I_q$$

i.e., $$X = X_q$$

$$\therefore \qquad BF = DF - DB = X_d I_d - X_q I_a \sin(\delta+\beta) = X_d I_d - X_q I_d = (X_d - X_q)I_d$$

In Figure 6.32, $\bar{E}_f$ and $\bar{E}'_f$ are in the same direction.
We can rewrite Equation (6.50) as follows:

$$\bar{E}_f = \bar{V}_t + \bar{I}_a R_a + jX_q \bar{I}_a + j\left(X_d - X_q\right)\bar{I}_d \tag{6.51}$$

Figure 6.32 shows the phasor diagram at lagging power factor load. From Figure 6.32, we have

$$\tan\gamma = \frac{CB}{OC} = \frac{CA+AB}{OP+PC} = \frac{V_t \sin\beta + I_a X_q}{V_t \cos\beta + I_a R_a} \tag{6.52}$$

where $$\gamma = \beta + \delta \tag{6.53}$$

γ is called the internal power factor angle by which I_a lags behind E_f.
From Figure 6.32, we have

$$E_f = V_t \cos\delta + I_q R_a + I_d X_d \tag{6.54}$$

Figure 6.33 shows the phasor diagram at leading power factor load.

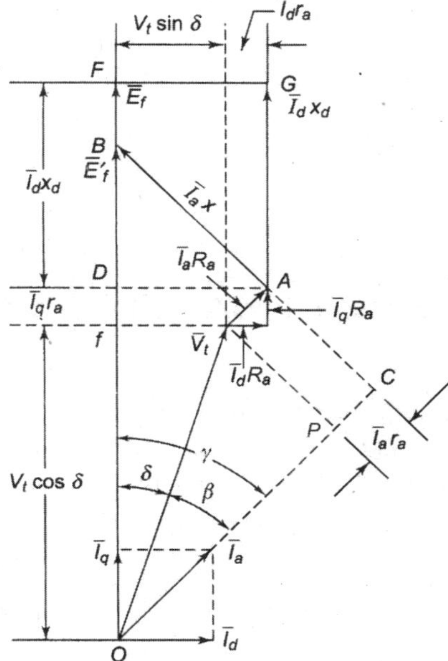

Figure 6.32 Phasor Diagram at Lagging Power Factor for Salient-pole Alternator

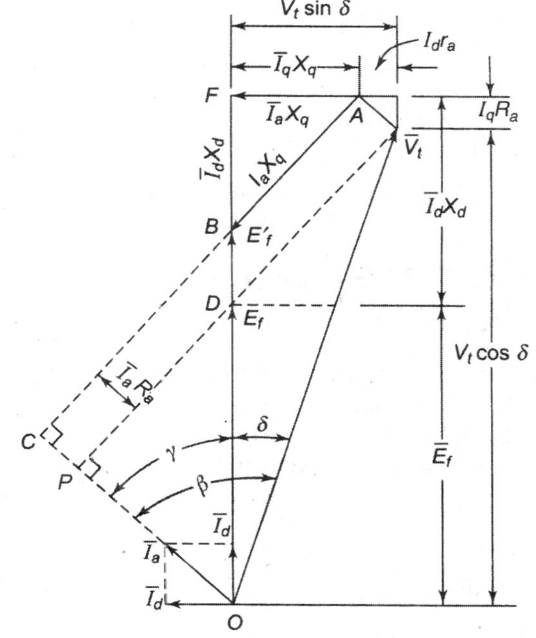

Figure 6.33 Phasor Diagram at Leading Power Factor for Salient-pole Alternator

From Figure 6.33, we have

$$\tan\gamma = \frac{V_t \sin\beta - I_a X_q}{V_t \cos\beta + I_a R_a} \tag{6.55}$$

where
$$\gamma = \beta - \delta \tag{6.56}$$

γ is called the internal power factor angle by which I_a leads E_f.
From Figure 6.33, we have

$$E_f = V_t \cos\delta - I_q R_a - I_d X_d \tag{6.57}$$

Example 6.5 A three-phase star-connected, 50 Hz, alternator has direct-axis synchronous reactance of 0.7 pu and quadrature-axis synchronous reactance of 0.5 p.u. If the generator delivers kVA at rated voltage at full-load and 0.8 power factor lagging, find the percentage regulation. Armature resistance 0.017 p.u.

Solution

$I_a = 1$ p.u., $V_t = 1$ p.u., $X_d = 0.7$ p.u., $X_q = 0.5$ p.u., $R_a = 0.017$ p.u., $\cos\beta = 0.8$ (lagging)

$$\tan\gamma = \frac{V_t \sin\beta + I_a X_q}{V_t \cos\beta + I_a R_a} = \frac{1\times 0.6 + 1\times 0.5}{1\times 0.8 + 1\times 0.017} = 1.346 = \tan 53.39°$$

$\therefore$
$$\gamma = 53.39°$$

Again,
$$\gamma = \beta + \delta$$

$\therefore$
$$\delta = 53.39° - \beta = 53.39° - \cos^{-1}(0.8) = 53.39° - 36.87° = 16.52°$$

$$I_d = I_a \sin\gamma = 1\times\sin 53.39° = 0.8027 \quad \text{and} \quad I_q = I_a \cos\gamma = 1\times\cos 53.39° = 0.5963$$

$\therefore$
$$E_f = V_t \cos\delta + I_q R_a + I_d X_d = 1\times 0.958 + 0.5963\times 0.017 + 0.8027\times 0.7 = 1.53 \text{ p.u.}$$

$\therefore$
$$\text{Percentage regulation} = \frac{1.53 - 1}{1}\times 100 = 53.00\%$$

■

6.14 TORQUE-ANGLE CHARACTERISTIC OF SALIENT-POLE ALTERNATOR

The armature resistance (R_a) is very small and it has negligible impact on the power output and the angle δ of a synchronous machine. Figure 6.34 shows the phasor diagram of a salient-pole alternator after neglecting R_a. We can derive the torque-angle characteristic of a salient-pole alternator.

The complex output power per phase is expressed by

$$S_{1-phase} = V_t I_a^* \tag{6.58}$$

Let us take E_f as reference vector.

$\therefore$
$$\bar{V} = V_t \angle(-\delta) = V_t \cos\delta - jV_t \sin\delta \tag{6.59}$$

Again,
$$\bar{I}_a = I_q - jI_d$$

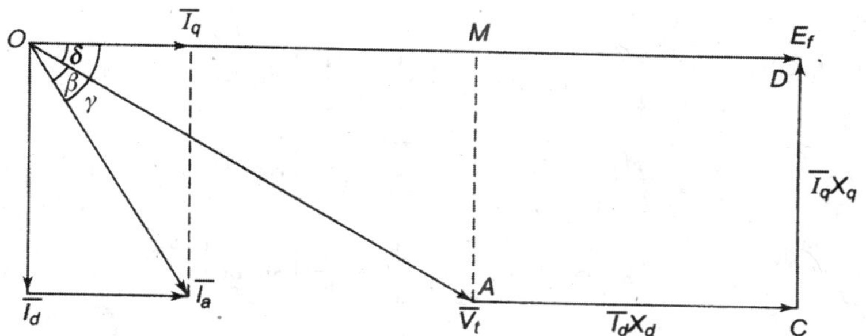

Figure 6.34 Phasor Diagram of Salient-pole Alternator, Neglecting R_a

$$\overline{I}_a^* = I_q + jI_d \tag{6.60}$$

Using Equations (6.59) and (6.60), Equation (6.58) can be expressed as follows:

$$S_{1-phase} = \left(V_t \cos\delta - jV_t \sin\delta\right)\left(I_q + jI_d\right) \tag{6.61}$$

From Figure 6.34, we have

$$CD = X_q I_q \quad \text{and} \quad AM = V_t \sin\delta$$

Again, $$AM = CD$$

Therefore, $$X_q I_q = V_t \sin\delta$$

$\therefore$ $$I_q = \frac{V_t \sin\delta}{X_q} \tag{6.62}$$

Again, $$AC = MD = OD - OM = E_f - V_t \cos\delta$$

$\therefore$ $$I_d = \frac{E_f - V_t \cos\delta}{X_d} \tag{6.63}$$

Using Equations (6.62) and (6.63), Equation (6.61) can be expressed as follows:

$$S_{1-phase} = \left(V_t \cos\delta - jV_t \sin\delta\right)\left(\frac{V_t \sin\delta}{X_q} + j\frac{E_f - V_t \cos\delta}{X_d}\right)$$

$$= \left(\frac{V_t^2}{X_q}\sin\delta\cos\delta + \frac{V_t E_f}{X_d}\sin\delta - \frac{V_t^2}{X_d}\sin\delta\cos\delta\right)$$

$$+j\left(\frac{V_t E_f}{X_d}\cos\delta - \frac{V_t^2}{X_d}\cos^2\delta - \frac{V_t^2}{X_q}\sin^2\delta\right)$$

$$= \left[\frac{V_t E_f}{X_d}\sin\delta + \frac{V_t^2}{2}\left(\frac{1}{X_q} - \frac{1}{X_d}\right)\sin 2\delta\right]$$
$$+j\left[\frac{V_t E_f}{X_d}\cos\delta - \frac{V_t^2}{2X_d}(1+\cos 2\delta) - \frac{V_t^2}{2X_q}(1-\cos 2\delta)\right]$$
$$= \left[\frac{V_t E_f}{X_d}\sin\delta + \frac{V_t^2}{2}\left(\frac{1}{X_q} - \frac{1}{X_d}\right)\sin 2\delta\right]$$
$$+j\left[\frac{V_t E_f}{X_d}\cos\delta - \frac{V_t^2}{2X_d X_q}\left\{(X_d + X_q) - (X_d - X_q)\cos 2\delta\right\}\right] \quad (6.64)$$

Again,

$$S_{1-phase} = P_{1-phase} + jQ_{1-phase} \quad (6.65)$$

Comparing Equations (6.64) and (6.65), the real power per phase in watt is as follows:

$$P_{1-phase} = \frac{V_t E_f}{X_d}\sin\delta + \frac{V_t^2}{2}\left(\frac{1}{X_q} - \frac{1}{X_d}\right)\sin 2\delta$$

The total real power in watts is given by

$$P_{3-phase} = 3P_{1-phase} = \frac{3V_t E_f}{X_d}\sin\delta + \frac{3V_t^2}{2}\left(\frac{1}{X_q} - \frac{1}{X_d}\right)\sin 2\delta \quad (6.66)$$

Comparing Equations (6.64) and (6.65), the reactive power per phase is the following:

$$Q_{1-phase} = \frac{V_t E_f}{X_d}\cos\delta - \frac{V_t^2}{2X_d X_q}\left\{(X_d + X_q) - (X_d - X_q)\cos 2\delta\right\}$$

The total reactive power in VAr is given by

$$Q_{3-phase} = 3Q_{1-phase} = \frac{3V_t E_f}{X_d}\cos\delta - \frac{3V_t^2}{2X_d X_q}\left\{(X_d + X_q) - (X_d - X_q)\cos 2\delta\right\} \quad (6.67)$$

In Equation (6.66), the first term is similar to cylindrical rotor machine. Only difference is in the second term which depends on the saliency that is defined by the quantity $\left(\frac{1}{X_q} - \frac{1}{X_d}\right)$. If $X_q = X_d$, saliency will disappear. In spite of no field current, this saliency exists. Equations (6.66) and (6.67) can be used for both the salient-pole synchronous motor and the synchronous generator. The torque angle (δ) is positive and negative for the generator and the motor, respectively. Figure 6.35 shows the plot of P and Q versus δ curves for a salient-pole machine.

For a three-phase synchronous machine, the developed electromagnetic torque is given by

$$T_{em} = \frac{P_{3-phase}}{\omega_s} = \frac{3}{\omega_s}\left[\frac{V_t E_f}{X_d}\sin\delta + \frac{V_t^2}{2}\left(\frac{1}{X_q} - \frac{1}{X_d}\right)\sin 2\delta\right] \quad (6.68)$$

In Equation (6.68), there are two terms. The first term represents the torque (T_{exc}) due to field excitation, and the second term represents the reluctance torque (T_{rel}):

$$T_{exc} = \frac{3V_t E_f}{\omega_s X_d}\sin\delta \quad (6.69)$$

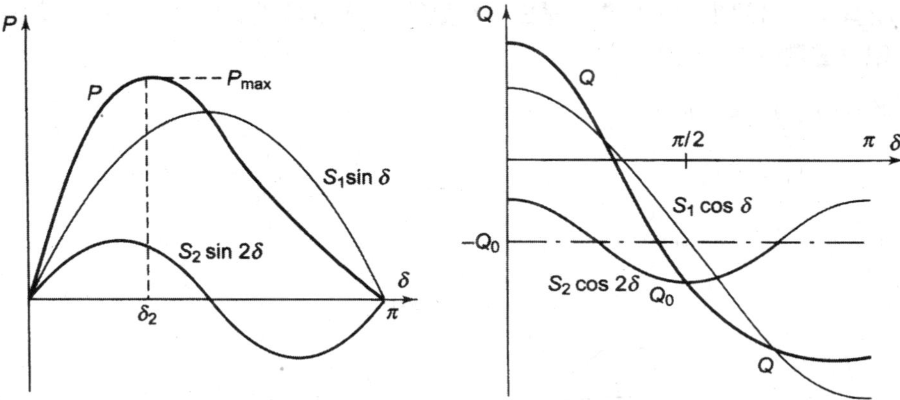

Figure 6.35 *P-Q* Versus δ Plot for Salient-pole Machine

and $$T_{rel} = \frac{3V_t^2}{2\omega_s}\left(\frac{1}{X_q} - \frac{1}{X_d}\right)\sin 2\delta \tag{6.70}$$

where $$\omega_s = 2\pi n_s.$$

The reluctance torque occurs due to saliency of the field poles, and it tends to align the direct axis with that of armature mmf. The reluctance torque does not depend on the excitation. It exists only when the machine is connected to a system and receives reactive power from the synchronous machines operating in parallel with the terminal voltage.

If $E_f = 0$, the first term of Equation (6.68) becomes zero. The machine still has the capability to generate real power (P) and it supplies only about 25 per cent of its rated real power. But it absorbs an excessive amount of reactive power. In this condition, the poles shift relative to stator field, which increase the reluctance of the flux path due to increase of torque. Figure 6.30 shows the P-δ curve for a salient-pole machine. From Figure 6.36, it is obvious that the peak power occurs when δ <90° and δ_{max} depends on the relative magnitudes of V_t, E_f and saliency.

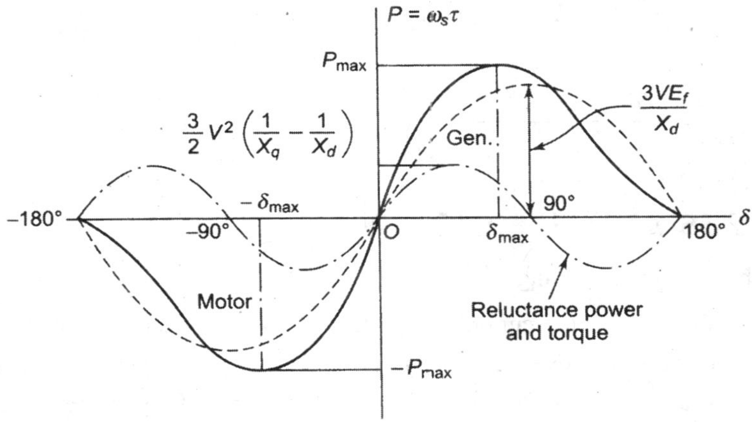

Figure 6.36 *P*-δ Curve for a Salient-pole Machine

6.15 MAXIMUM REACTIVE POWER FOR SALIENT-POLE ALTERNATOR

From Equation (6.67), the reactive power is given by

$$Q_{3-phase} = 3Q_{1-phase} = \frac{3V_t E_f}{X_d}\cos\delta - \frac{3V_t^2}{2X_d X_q}\left\{\left(X_d + X_q\right) - \left(X_d - X_q\right)\cos 2\delta\right\}$$

To get maximum reactive power,

$$\frac{dQ_{3-phase}}{d\delta} = 0$$

i.e.,

$$\frac{3V_t E_f}{X_d}\sin\delta + \frac{6V_t^2}{2X_d X_q}\left(X_d - X_q\right)\sin 2\delta = 0$$

i.e.,

$$E_f \sin\delta + \frac{V_t}{X_q}\left(X_d - X_q\right)2\sin\delta\cos\delta = 0$$

i.e.,

$$\cos\delta = -\frac{X_q E_f}{2V_t\left(X_d - X_q\right)} \tag{6.71}$$

Using Equation (6.71), Equation (6.67) can be written as follows:

$$\left(Q_{3-phase}\right)_{max} = \frac{3V_t E_f}{X_d}\left[-\frac{X_q E_f}{2V_t\left(X_d - X_q\right)}\right] - \frac{3V_t^2}{2X_d X_q}\left(X_d + X_q\right) + \frac{3V_t^2}{2X_d X_q}\left(X_d - X_q\right)\cos 2\delta$$

$$= -\frac{3E_f^2 X_q}{2X_d\left(X_d - X_q\right)} - \frac{3V_t^2}{2X_d X_q}\left(X_d + X_q\right) + \frac{3V_t^2}{2X_d X_q}\left(X_d - X_q\right)\left[\frac{X_q^2 E_f^2}{2V_t^2\left(X_d - X_q\right)^2} - 1\right]$$

$$= -\frac{3V_t^2}{2X_d X_q}\left\{\left(X_d + X_q\right) + \left(X_d - X_q\right)\right\} - \frac{3E_f^2 X_q}{2X_d\left(X_d - X_q\right)} + \frac{3X_q E_f^2}{4X_d\left(X_d - X_q\right)}$$

i.e.,

$$\left(Q_{3-phase}\right)_{max} = -\frac{3V_t^2}{X_q} - \frac{3E_f^2 X_q}{4X_d\left(X_d - X_q\right)} \tag{6.72}$$

6.16 LOSSES AND EFFICIENCY

In an alternator there are the following losses:

- *Copper losses:* Copper losses occur only in the armature winding and field coils.
- *Core loss:* The loss due to eddy current and hysteresis losses is known as core loss.
- *Friction and windage loss:* The loss due to the bearing and brush friction and to the power required to circulate the cooling air is known as friction and windage loss.

- *Load loss:* The armature leakage flux causes eddy current and hysteresis loss in the iron surrounding the armature conductors. This loss is known as load loss. During calculation of copper loss if effective armature resistance is taken into account, the load loss is included in the calculation and further calculation of load loss is not required.

  Efficiency of the alternator is calculated by

$$\text{Alternator efficiency} = \frac{\text{output}}{\text{output} + \text{losses}} = \frac{\text{kVA} \times \text{pf}}{\text{kVA} \times \text{pf} + \text{losses}}$$

where pf represent power factor

During the maximum efficiency, iron loss is equal to copper loss or variable loss.

6.16.1 Determination of Losses

The following two methods are used to determine the losses:

(i) measurement of losses method and

(ii) use of a calibrated DC motor to drive the alternator.

6.16.1.1 Measurement of Losses Method

It is impractical to measure the efficiency of alternators by direct loading and physically it is difficult to get the proper load. Measurement of mechanical power is somewhat difficult. If there is any inaccuracy in measurement, efficiency calculation of the machine will be erroneous.

The measurement of losses method has the following advantages:

- The required source of power is 5 per cent of the rating of the alternator and hence only a part of the losses is required to be provided for.
- It is not required to put an electric load on the alternator.
- Electrical instruments can be used for all measurements and hence the greater accuracy can be obtained.
- The error occurs only for a small portion of the name plate rating and hence there is no reflection of any inaccuracy during calculation of final frequency.

6.16.1.2 Use of a Calibrated DC Motor to Drive the Alternator

This method is quite accurate and simple to determine losses. In this method, a calibrated DC motor is used to drive the alternator. The meaning of calibration of the motor is that all its losses have been determined for varying conditions of operation. Therefore, its output is readily obtained. The following steps are required:

- The alternator is driven at rated synchronous speed without field excitation. The output of the DC motor provides the input of the alternator and windage losses.
- The previous step is repeated with field excitation so that the alternator operates normally. The difference in the motor outputs of the above two steps gives core loss.
- After short-circuiting the armature, the field current is adjusted to get the rated line current. The difference in output of this step to that of the first step gives the armature copper loss as well as load loss.
- Using DC measurement, the field copper loss is measured.

6.17 DETERMINATION OF X_d AND X_q

From no-load test or slip test, X_d and X_q can be determined very easily. Figure 6.37 shows the slip test of an alternator. In this test, near about 25 per cent of rated voltage is applied to the three-phase stator winding and the field winding is kept open circuited.

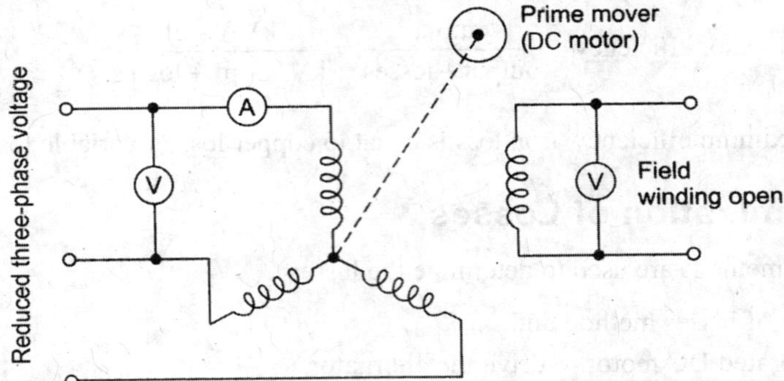

Figure 6.37 Slip Test Connection Diagram to Determine X_d and X_q

With the help of a DC motor, the alternator is driven at a speed slightly more than or less than synchronous speed in the same direction of rotating magnetic field. If the voltmeter connected across the open-circuited field winding terminals shows a small reading, the rotation of the alternator is proper. A small slip between the rotating magnetic field is produced by the armature, and the actual salient field pole is produced due to the rotation of the rotor at a speed (n_r) near about the synchronous speed (n_s), where $n_s - n_r$ is the relative speed between armature mmf and field poles. Figure 6.38 shows the instant when the peak of the armature mmf wave is in line with the axis of the actual salient field poles. This case occurs because the stator mmf slowly passes the actual field poles.

The ammeter at this stage measures the minimum magnetizing current because the reluctance offered by the air gap is minimum. At this position, the field voltmeter reading will be zero because the flux linked with the field winding is maximum but the rate of change of flux is minimum. X_d is determined by the armature terminal voltage per phase to armature current per phase at the same instant.

Figure 6.39 shows the position of the armature wave after one-quarter of the slip cycle, which is in line with the quadrature axis and the reluctance offered by the long air gap is maximum. Hence, the magnetizing current is also large to maintain the same air-gap flux. At this stage, the flux linked with the field winding is minimum and the rate of change of flux is maximum. Hence, the reading of the field voltmeter is maximum. X_q is determined by the ratio of armature terminal voltage per phase to armature current per phase. Figure 6.40 shows the typical oscillogram from slip test.

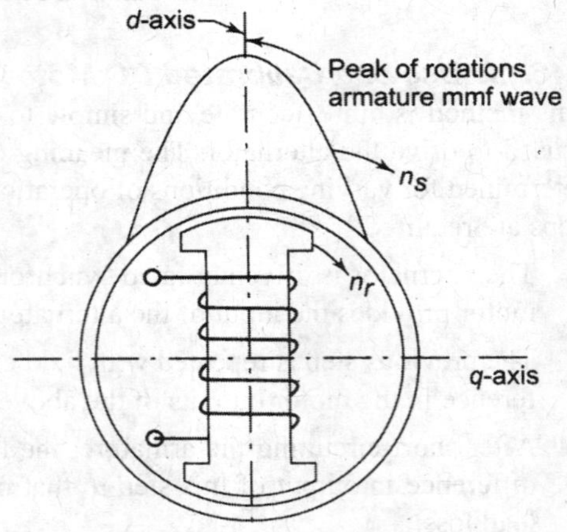

Figure 6.38 Peak of Armature mmf Wave

d-axis
n_r
Peak of rotations armature mmf wave
q-axis
n_s

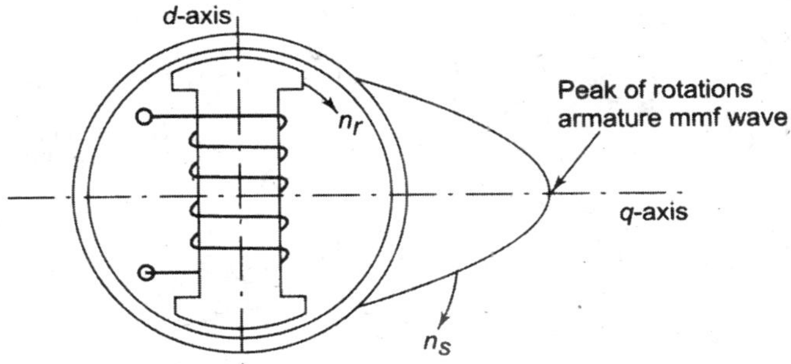

Figure 6.39 Position of Armature Wave After One-quarter of a Slip Cycle

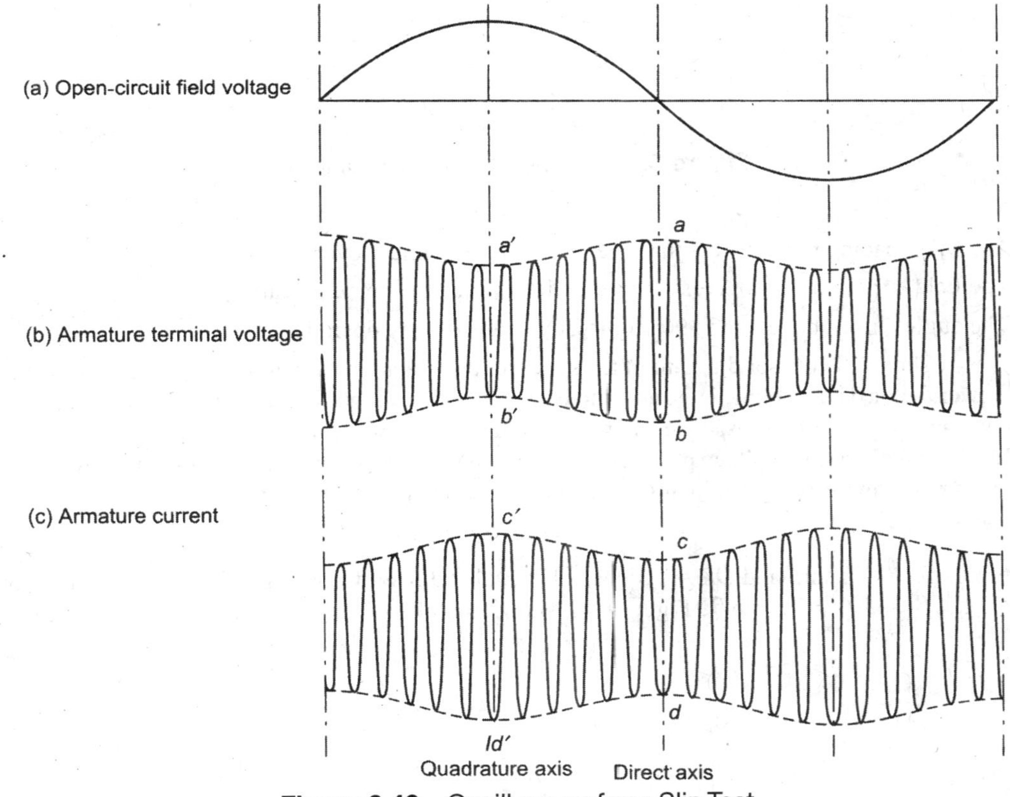

(a) Open-circuit field voltage
(b) Armature terminal voltage
a'
a
b'
b
(c) Armature current
c'
c
d
Id'
Quadrature axis
Direct axis

Figure 6.40 Oscillogram from Slip Test

6.18 CAPABILITY CURVES

The limit within which the synchronous machines operate safely is called *capability curves*, which are also known as *operating charts* or *capability charts*. The following considerations are required:

- The MVA loading must be less than the rating of the alternator.

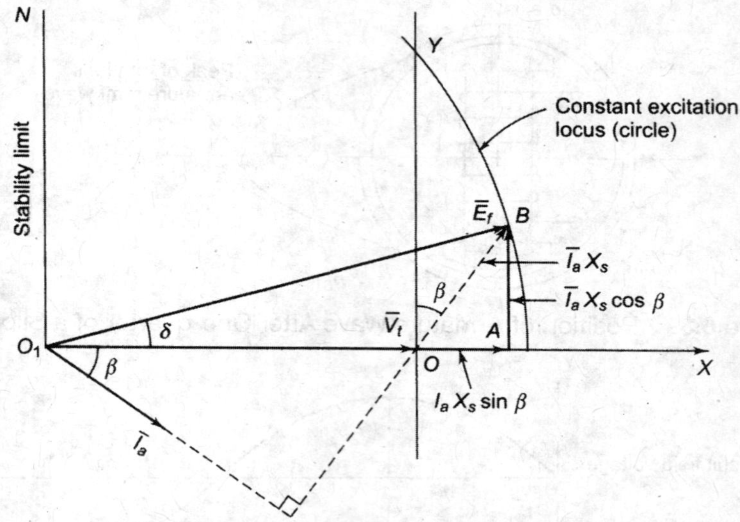

Figure 6.41 Generator Phasor Diagram

- The MW loading must be less than the rating of the prime mover.
- The field current must not cross the value determined by the field heating.
- The angle δ must be less than 90° for steady-state stability operation.

Figure 6.41 shows the phasor diagram of the synchronous machine having the cylindrical rotor at a lagging power factor. The machine is assumed to be connected at an infinite bus bar so that terminal voltage remains constant. Armature resistance and saturation are taken to be negligible order.

Let V_t be the terminal voltage per phase. The length O_1O shown in Figure 6.41 is fixed. A set of axes OX and OY is drawn having origin at O, which is at the tip of V_p. From Figure 6.35, we have

$$OB = I_a X_s$$

Therefore, $OA = I_a X_s \sin\beta$ and $OA = I_a X_s \cos\beta$. The real and reactive power outputs of the generator are $P = V_t I_a \cos\beta$ and $Q = V_t I_a \sin\beta$. Figure 6.42 shows the phasor diagram where the multiplication of

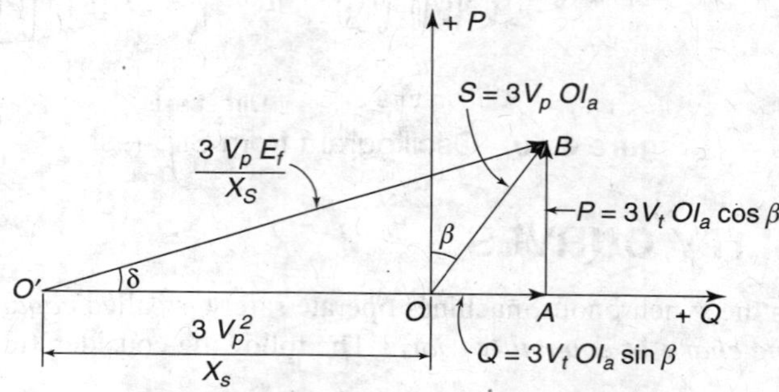

Figure 6.42 Phasor Diagram

each phasor is shown by the quantity $\frac{3V_t}{X_s}$. Therefore, OAB becomes a complex power triangle in Figure 6.42 in terms of three phase values. The following are to be noted:

$$AB = 3V_t I_a \cos\beta = P = 3-\text{phase active power in W}$$

$$OA = 3V_t I_a \sin\beta = Q = 3-\text{phase reactive power in VAr}$$

$$OB = 3V_t I_a = S = 3-\text{phase voltamperes (VA)}$$

Q is taken as positive for lagging current.

Figure 6.43 shows the typical capability curve for a cylindrical rotor generator which is plotted in s-plane. The vertical axis is P and horizontal axis is Q. The locus will be a circle with the centre O for constant current I_a and radius $OA = 3V_t I_a$. For constant excitation, the locus is a circle with centre O_1 having radius $O_1 A = \frac{3V_t E_f}{X_s}$. The radial straight lines are constant power factor lines. OO_1 represents short-circuit current at rated voltage for $E_f = 0$. The theoretical stability limit is the straight line $O_1 N$ which is at right angles to OO_1 and $\delta = 90°$. In Figure 6.36, the operation of the alternator is limited by the maximum field current between p and q and radius is $\frac{3V_t E_f}{X_s}$ with centre O_1. The operation of the

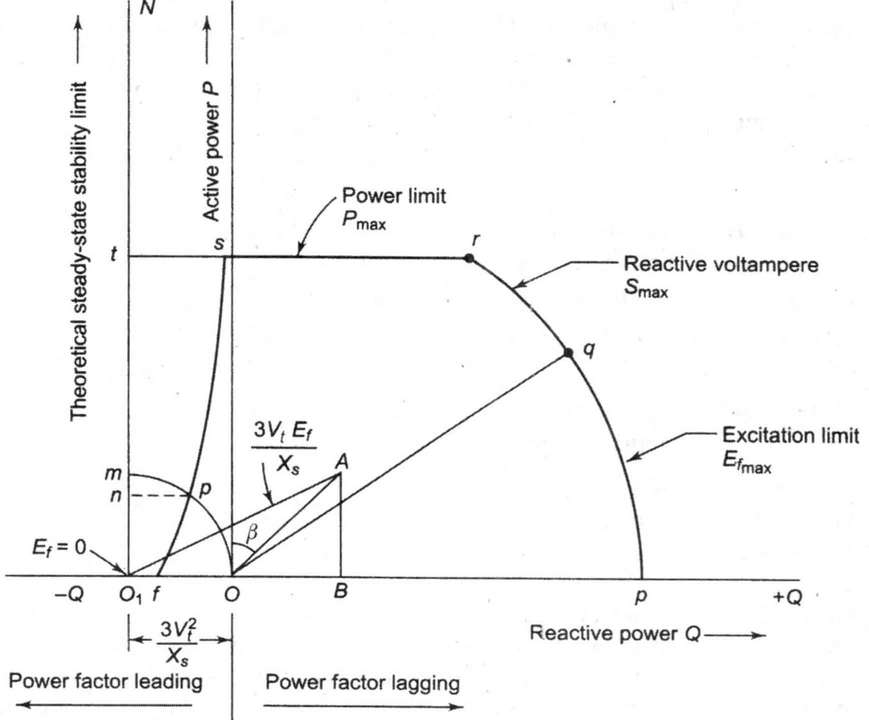

Figure 6.43 Capability Curve for a Cylindrical Rotor

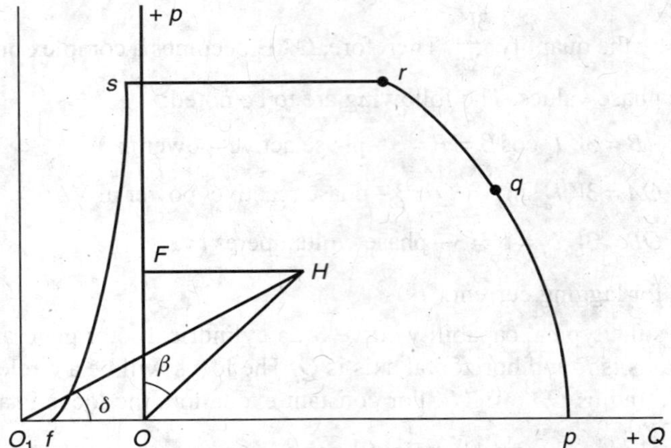

Figure 6.44 Capability Curve with an Operating Point

alternator is limited by the MVA limit between q and r and radius is $3V_fI_a$ with centre O, where I_a is the maximum permissible current. The operation of the alternator is limited by the power factor of the prime mover between r and s. The operation of the alternator is limited by the practical stability limit between s and t. The limit of stability occurs where $\delta = 90°$ for theoretical purpose. Since there is a safety margin between the theoretical and practical limits and practically used, the practical stability limit is usually taken as 10 per cent less than the theoretical stability limit for a safety margin. Therefore, the complete zone of the alternator is *pqrsfOp*. From the point of view of heating and stability, the operation of the alternator within the area *pqrsfOp* is safe. The desired values of P, Q and S can be found out only after location of the operating point.

The following information is available after obtaining the operating point H, as shown in Figure 6.44:

(i) Since the point H is inside the capability curve, there will not be any possibility of the machine to be overheated and it will not fall out of synchronism.

(ii) H and O_1 are connected. The line O_1H makes an angle δ with the I_f axis.

(iii) A line HF is drawn parallel to OO_1. This gives a power equal to OF which is the active power.

(iv) A line OH is drawn to F and this makes an angle β with the P-axis; that is, $\angle HOF = \beta$.

(v) The armature current (I_a) is equal to OH.

(vi) VA output = $OH \times$ operating voltage.

(vii) W output = $OF \times$ operating voltage.

(viii) VAr output = $HF \times$ operating voltage.

(ix) O_1H denotes the E_f.

Example 6.6 A 6.6 kV, 1,200 kVA, three-phase, 50 Hz, star-connected alternator has synchronous reactance of 9.5 Ω per phase. It delivers full load at 0.8 power factor lagging. It is connected to infinite bus bars. If the steam supply is gradually increased, calculate (i) the output at which power factor becomes unity and (ii) the maximum load which it can supply without losing synchronism and the corresponding power factor. Armature resistance is negligible.

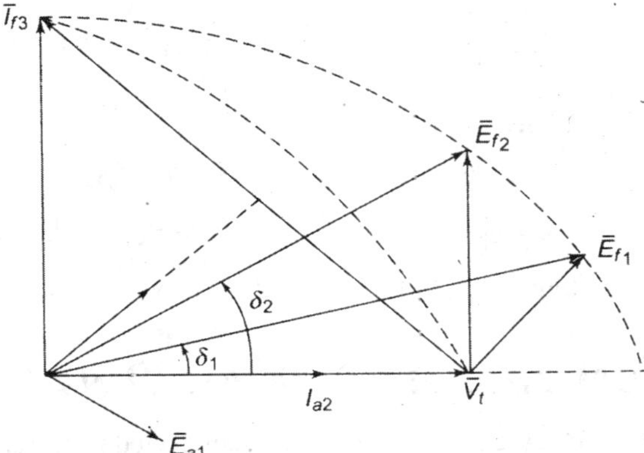

Figure E6.3

Solution

$$V_t = \frac{6.6\times 10^3}{\sqrt{3}} = 3{,}810.5\,\text{V}, \quad I_{a1} = \frac{1{,}500\times 10^3}{\sqrt{3}\times 6.6\times 10^3} = 131.21\,\text{A},\ R_a = 0\,\Omega,\ X_s = 9.5\,\Omega$$

$$P = \frac{E_f V_t}{X_s}\sin\delta = V_t I_a \cos\beta$$

$$E_f = \sqrt{\left(V_t\cos\beta + I_{a1}R_a\right)^2 + \left(V_t\sin\beta + I_{a1}X_s\right)^2}$$

$$= \sqrt{\left(3{,}810.5\times 0.8 + 0\right)^2 + \left(3{,}810.5\times 0.6 + 131.21\times 9.5\right)^2} = 4{,}666.19\,\text{V}$$

Since V_t remains constant and X_s remains unaltered, $E_f \sin\delta$ and $I_a \cos\beta$ increase with increase in P. Figure E6.3 shows this fact.

(i) At unity power factor, let armature current per phase be I_{a2}.

$$I_{a2}X_s = \sqrt{E_f^2 - V_t^2} = \sqrt{4{,}666.19^2 - 3{,}810.5^2} = 2{,}693.21\,\text{V}$$

[Excitation remains unaltered]

i.e.,
$$I_{a2} = \frac{2{,}693.21}{X_s} = \frac{2{,}693.21}{9.5} = 283.5\ \text{A}$$

i.e.,
$$\text{Power output} = 3\times 3{,}810.5\times 283.5\times 1 = 3.24\,\text{MW}$$

(ii) For maximum power output, $\delta = 90°$

From Figure E6.3, we have

$$E_f^2 + V_t^2 = \left(I_{a4}X_s\right)^2$$

i.e.,
$$I_{a4} = \frac{\sqrt{E_f^2 + V_t^2}}{X_s} = \frac{\sqrt{4{,}666.19^2 + 3{,}810.5^2}}{9.5} = 634.15\,\text{A}$$

$$\therefore \qquad \tan\xi = \frac{E_f}{V_t} = \frac{4,666.19}{3,810.5} = 1.224 = \tan 50.75°$$

i.e.,
$$\xi = 50.75°$$

$$\beta_4 = 90° - \xi = 90° - 50.75° = 39.25°$$

$$\text{Power factor} = \cos\beta_4 = \cos 39.25° = 0.774 \text{ (leading)}$$

$$\therefore \qquad \text{Power output} = 3 \times 3,810.5 \times 634.15 \times 0.774 = 5.611 \text{ MW}$$

■

6.19 EXCITATION CIRCLE OF AN ALTERNATOR

Figure 6.45(a) shows the phasor diagram of an alternator connected to an infinite bus bar. The terminal voltage and generated emf are related by

$$\overline{E}_f = \overline{V}_t + \overline{I}_a\overline{Z}_s \qquad (6.73)$$

Also from excitation circle

$$\overrightarrow{OO'} = \overline{V}_t,\ \overrightarrow{O'Q} = \overline{I}_a\overline{Z}_s \text{ and } \overrightarrow{OQ} = \overline{E}_f$$

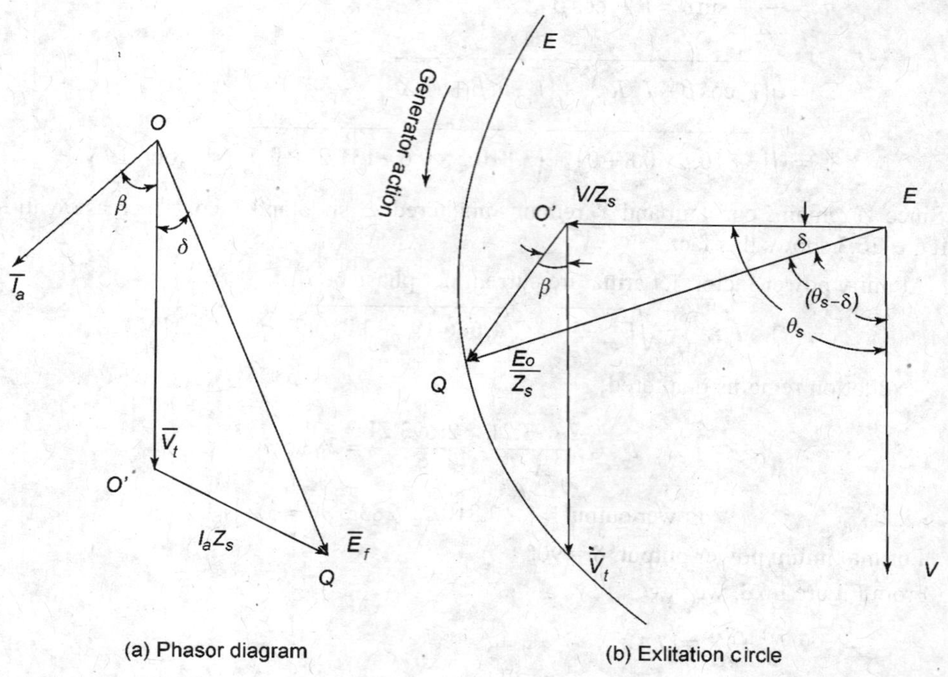

Figure 6.45 Phasor Diagram and Excitation Circle of an Alternator Connected to Infinite Bus Bar

The angle δ by which E_0 leads V_t is called torque angle. Taking V_t as reference phasor, we have

$$E_f\angle\delta = V\angle 0° + (I_a\angle -\beta)(Z_s\angle\theta_s) \tag{6.74}$$

Dividing Equation (6.74) throughout by $Z_s\angle\theta_s$, we have

$$I_a\angle -\beta = \frac{E_f}{Z_s}\angle(\delta - \theta_s) - \frac{V}{Z_s}\angle -\theta_s$$

i.e.,
$$\frac{E_f}{Z_s}\angle(\delta - \theta_s) = I_a\angle -\beta + \frac{V}{Z_s}\angle -\theta_s$$

i.e.,
$$\frac{E_f}{Z_s}\angle -(\theta_s - \delta) = \frac{V}{Z_s}\angle -\theta_s + I_a\angle -\beta \tag{6.75}$$

Now, the division of the phasor by complex quantity Z_s leads to rotation of phasor by θ_s in the clockwise direction. Therefore, in the excitation circle, we get triangle $OO'Q$ of Figure 6.45(a) rotated through an angle θ_s in the clockwise direction as shown in Figure 6.45(b).

Now,
$$OO' = \frac{V_t}{Z_s}, OQ = \frac{E_f}{Z_s}, O'O = I_a$$

and thus,
$$\overrightarrow{OQ} = \overrightarrow{OO'} + \overrightarrow{O'Q}$$

If the excitation of an alternator on infinite bus is held constant, then $\frac{E_f}{Z_s}$ remains constant. $\frac{V_t}{Z_s}$ is on infinite bus, and it also remains constant.

Therefore, the tip of the current phasor $O'Q$ lies on a circle of radius $\frac{E_f}{Z_s}$ with centre at O. The phasor representing $\frac{V_t}{Z_s}$ is drawn lagging V_t by an angle θ_s, where θ_s is an impedance angle given by a $\tan\theta_s = \frac{X_s}{R_a}$. The locus of current is known as excitation circle when synchronous machine is working as a generator at lagging power factor, that is, $\frac{E_f}{Z_s} > \frac{V_t}{Z_s}$. For different values of E_0, excitation circles will be concentric with centre at point O.

6.20 PRIME MOVER CHARACTERISTIC

The load-speed characteristic of prime movers must be drooping for successful operation of alternators. *Drooping* means the speed of the prime mover should fall slightly with increase in load. The speed droop is defined by

$$\text{Speed drop} = \frac{N_{nl} - N_{fl}}{N_{nl}} \text{ p.u.} \tag{6.76}$$

where N_{nl} and N_{fl} represent no-load and full-load speed.

The speed drop is also known as *governor droop* or *inherent speed regulation curve.* Usually, speed droop varies from 2 per cent to 4 per cent from no-load to full-load speed, and speed load characteristics are generally linear. The prime mover determines the quantity of power generation by a machine.

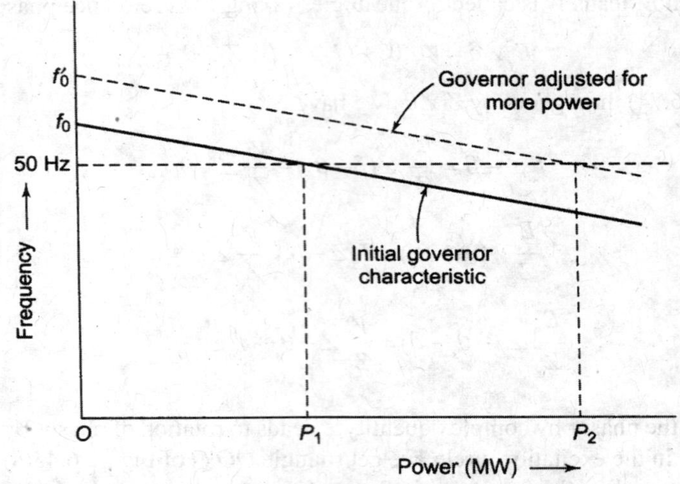

Figure 6.46 Shift of Frequency Load Characteristic

Adjusting spring tension of the governor mechanism, the torque of the machine can be varied because the speed of the prime mover is fixed. Figure 6.46 shows the effect of changing the governor mechanism where the speed load characteristic is shifted to a new position parallel to the initial position.

During parallel operation of two alternators, an increase in governor set causes the following effects on them:

- The system frequency increases.
- The power supplied by one alternator increases, whereas the power supplied by the other alternator decreases.

During parallel operation of two alternators, increasing the field current of the second alternator causes the following:

- The system terminal voltage increases.
- The reactive power supplied by that alternator increases, whereas the reactive power supplied by the other alternator decreases.

6.20.1 Sharing of Powers by Two Alternators

Figure 6.47 shows the frequency load characteristic of two alternators when they are operating in parallel. Let W_A be the full-load power rating of Alternator A, W_B be the full-load power rating of Alternator B, P_A be the power sharing by Alternator A, P_B be the power sharing by Alternator B, P be the power supplied by two alternators, f_{A0} be the no-load frequency of Alternator A, f_{B0} be the no-load frequency of Alternator B, $f_{A(fl)}$ be the full-load

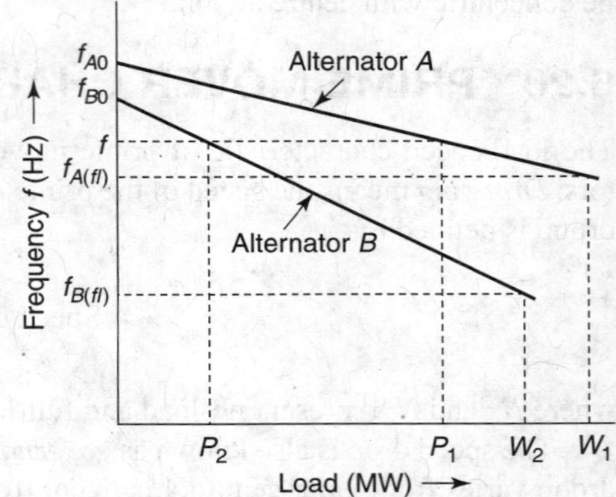

Figure 6.47 Frequency Load Characteristics

frequency of Alternator A, $f_{B(fl)}$ be the full-load frequency of Alternator B and f_c be the common operating frequency of the Alternator A and Alternator B when they are running parallel.

Alternator A

Drop in frequency from no-load to full-load $= f_{A0} - f_{A(fl)}$

Drop in frequency per unit rating $= \dfrac{f_{A0} - f_{A(fl)}}{W_A}$

Drop in frequency for a load $P_A = \dfrac{f_{A0} - f_{A(fl)}}{W_A} \times P_A$

Therefore, operating frequency of Alternator A $= f_{A0} - \dfrac{f_{A0} - f_{A(fl)}}{W_A} \times P_A$ (6.77)

Alternator B

Drop in frequency from no-load to full-load $= f_{B0} - f_{B(fl)}$

Drop in frequency per unit rating $= \dfrac{f_{B0} - f_{B(fl)}}{W_B}$

Drop in frequency for a load $P_A = \dfrac{f_{B0} - f_{B(fl)}}{W_B} \times P_B$

Therefore, operating frequency of Alternator B $= f_{B0} - \dfrac{f_{B0} - f_{B(fl)}}{W_B} \times P_B$ (6.78)

From Equations (6.77) and (6.78), we have

$$f_c = f_{A0} - \frac{f_{A0} - f_{A(fl)}}{W_A} \times P_A = f_{B0} - \frac{f_{B0} - f_{B(fl)}}{W_B} \times P_B \quad (6.79)$$

The following relation is also valid:

$$P = P_A + P_B \quad (6.80)$$

With the help of Equations (6.79) and (6.80), P_A, P_B and f_c can be calculated.

Example 6.7 Two synchronous generators A and B operate in parallel for a station capacity of A is 55 MW and that of B is 30 MW. Full-load regulation of A and B are 4 per cent and 4.5 per cent respectively. Find the load sharing when connected load is 55 MW. Assume no-load frequency of 50 Hz.

Solution

Let P_A and P_B be the load taken by generator A and B respectively in MW.

$$P_A + P_B = 55 \quad (1)$$

Frequency at no-load is $f_0 = 50$ Hz

For load of 55 MW, the drop in frequency = 4% of $f_0 = \dfrac{4}{100} \times 50 = 2$

For a load of 1 MW, the drop in frequency $= \dfrac{2}{55}$

For a load of P_A MW, the drop in frequency $= \frac{2P_A}{55}$

Therefore, operating frequency of generator A is

$$f_A = 50 - \frac{2P_A}{55} \tag{2}$$

Similarly, operating frequency of generator B is $f_B = 50 - \left(\frac{4.5 \times 50}{100}\right)\frac{P_B}{30}$

i.e.,
$$f_B = 50 - \frac{2.25}{30} P_B \tag{3}$$

The generators must operate at the same frequency because they are operating in parallel. From Equations (2) and (3), we have

$$f_A = f_B$$

i.e.,
$$50 - \frac{2P_A}{55} = 50 - \frac{2.25}{30} P_B$$

i.e.,
$$\frac{2P_A}{55} = \frac{2.25}{30} P_B$$

i.e.,
$$P_A = \left(\frac{55}{2} \times \frac{2.25}{30}\right) P_B = 2.0625 P_B \tag{4}$$

From Equations (1) and (4), we have

$$2.0625 P_B + P_B = 55$$

i.e.,
$$P_B = \frac{55}{3.0625} = 17.96$$

From Equation (1), we have

$$P_A = 55 - P_B = 55 - 17.96 = 37.04$$

Therefore, the load sharing by Alternator A and Alternator B are 37.04 MW and 17.96 MW respectively ■.

Example 6.8 Two Alternators A and B operate in parallel for a station capacity of A is 60 MW and that of B is 120 MW. Alternators A and B are fitted with governors having a droop of 3.5 per cent. Calculate the sharing a load of 120 MW by Alternator A and Alternator B.

Solution

Let f_0 be the original frequency in Hz. Since drooping frequency is 3.5 per cent of f_0, drooping frequency becomes $0.035f_0$. Let P_A and P_B be the load taken by generators A and B, respectively, in MW.

$$P_A + P_B = 120 \tag{1}$$

Frequency at load is $f_0 = 50$ Hz

For no-load of 60 MW, the drop in frequency = 3 per cent of $f_0 = 0.035f_0$

For a load of 1 MW, the drop in frequency $= \frac{0.035 f_0}{60}$

For a load of P_A MW, the drop in frequency $= \dfrac{0.035}{60} f_0 P_A$

Therefore, operating frequency of generator A is

$$f_A = f_0 - \frac{0.035}{60} f_0 P_A \quad (2)$$

Similarly, operating frequency of generator B is

$$f_B = f_0 - \frac{0.035}{120} f_0 P_B \quad (3)$$

The generators must operate at the same frequency because they are operating in parallel. From Equations (2) and (3), we have

$$f_A = f_B$$

i.e.,

$$f_0 - \frac{0.035}{60} f_0 P_A = f_0 - \frac{0.035}{120} f_0 P_B$$

i.e.,

$$P_A = \left(\frac{60}{120}\right) P_B$$

i.e.,

$$P_A = \frac{1}{2} P_B \quad (4)$$

From Equations (1) and (4), we have

$$0.5P_B + P_B = 120$$

i.e.,

$$P_B = \frac{120}{1.5} = 80$$

From Equation (1), we have

$$P_A = 55 - P_B = 120 - 80 = 40$$

Therefore, the load sharing by Alternator A and Alternator B is 40 MW and 80 MW, respectively. ■

Example 6.9 Two identical, 3,000 kVA, three-phase alternators operate in parallel. The governors of the first and second machines are such that their frequencies drop uniformly from 50 Hz to 48.5 Hz and 50 Hz to 48 Hz, respectively, on full load. (i) How will the machines share a load of 3,500 kW? (ii) What is the maximum load at unity power factor that can be delivered without overloading either machine?

Solution

Full-load rating of each machine with unity power factor = 3,000 × 1 = 3,000 kW = 3 MW

Let x be the load taken by the first machine in MW and y be the load taken by the second machine in MW.

∴

$$x + y = 3.5 \quad (1)$$

For machine 1:

Drop in frequency per unit rating $= \dfrac{50-48.5}{3}$ and hence the drop in frequency for load of x MW $= \dfrac{50-48.5}{3} x$.

Operating frequency of machine 1 is $f_1 = 50 - \frac{50-48.5}{3}x$

For machine 2:

Drop in frequency per unit rating $= \frac{50-48}{3}$ and hence the drop in frequency for load of y MW $= \frac{50-48}{3}y$.

Operating frequency of machine 2 is $f_2 = 50 - \frac{50-48}{3}y$

(i) For parallel operation of machine 1 and machine 2,

$$f_1 = f_2$$

i.e.,
$$50 - \frac{50-48.5}{3}x = 50 - \frac{50-48}{3}y$$

i.e.,
$$\frac{1.5}{3}x = \frac{2}{3}y$$

i.e.,
$$y = 0.75x$$

Solving Equations (1) and (2), we have

$$x = 2 \text{ MW} \quad \text{and} \quad y = 1.5 \text{ MW}$$

■

(ii) For parallel operation, the minimum value of frequency at which the system can operate without overloading either machine is 48.5 Hz. Since the full-load frequency of the first machine is 48.5 Hz, the total load taken by it is 3 MW or 3,000 kW.

Therefore, for a drop in frequency equal to 50 − 48.5, the total load taken by the machine

$$= \frac{3{,}000}{2.0} \times 1.5 = 2{,}250 \text{ kW}$$

Total load taken by the two machines without overloading either machine
= 3,000 + 2,250 = 4,250 kW

■

6.21 INFINITE BUS

Many alternators operate in parallel in power systems. The alternators are located in different places. The group of alternators at one place should be treated as an equivalent big alternator. This group of alternators is connected to the bus through a transmission line. Similarly, other groups of alternators are also connected to the bus but separated by the transmission line. Now, the bus becomes the source of power. If loads are put on the bus, power will be supplied in such a way that the voltage and frequency of the bus remain constant regardless of the load. This type of system behaves like a large generator and has zero impedance as well as infinite rotational inertia. Such a system is known as infinite bus bar system or simply infinite bus. Therefore, the voltage and frequency of an infinite bus do not depend on the quantity of real and reactive power drawn from this bus. An infinite bus has the following characteristics:

- The terminal voltage always remains constant.
- The frequency always remains constant.
- The impedance is very small.

6.22 NEED FOR PARALLEL OPERATION OF ALTERNATORS

The parts of the north-eastern United States and south-eastern Canada faced widespread power blackouts on 14 August 2003. American states of New York, New Jersey, Vermont, Michigan, Ohio, Pennsylvania, Connecticut, Massachusetts and others were affected.

Among the major urban agglomerations touched by the electrical power outage in the United States were the cities of New York, Albany, Buffalo in New York, Cleveland and Columbus in Ohio, and Detroit. Ottawa and Toronto in Canada were also affected.

Power was suddenly lost around 4 pm Eastern Standard Time. The blackout resulted in the shutting down of nuclear power plants in New York state and Ohio. Air traffic slowed down as flights in the affected airports were halted. Terrorism was quickly ruled out as a cause of the incident by federal authorities. *Approximately, 50 million people were affected by the outage.*

Previous incidents include 5 November 1965 outage caused by a faulty relay at a power plant in Ontario which affected a large area of land stretching from Toronto to New York. Another one followed on 14 July 1977 as a result of a lightning strike, which affected New York city. The power supply in nine western US states was also affected in August 1996 as a result of a high demand of electricity, a heat wave and sagging electrical power lines.

The demand of power is increasing day by day. It is a challenge for power engineers to meet the power demand of customers. A single alternator cannot meet the power demand. To meet the excess power demand, additional alternators are connected in parallel. If a single alternator can meet the power demand, an outage of the alternator will cause interruption of power supply. On the other hand, paralleling of alternators ensures the supply of a part of the total demand when one alternator is out of order. A number of alternators are paralleled to export the output over long cross-country transmission lines. A public utility power plant always tries its best to produce economical power for its customers throughout 24 hours of the day, 365 days a year, year in and year out. During paralleling of alternators, it is not mandatory that the actual machine voltages must be the same because paralleling point voltages can be made the same with the help of transformers. Therefore, different kinds of alternators can be made parallel.

6.22.1 Reasons for Paralleling of Alternators

The reasons for paralleling are shown follows:

- A single alternator may not meet the local or regional power requirements.
- It is possible to shut down one or more alternators for scheduled or emergency maintenance when the alternators are operating in parallel. The load can be supplied.
- At part load, one or more alternators are shut down and the remaining load is carried out with few machines efficiently because alternators are inefficient at part load.
- It is possible to handle load growth by adding alternators without disturbing the original installation.
- The available machine prime movers and alternators can be matched to obtain economical cost and reliable use.

6.22.2 Conditions for Paralleling of Alternators

The following conditions must be satisfied for paralleling of alternators:

- At the paralleling point, voltages must be the same, whereas voltages may not be same at the alternators.
- At the paralleling point, phase sequence must be same.

- During paralleling, the incoming machine should be in phase and it should stay in phase after paralleling under normal conditions.
- The line frequencies must be same at the paralleling point.
- The prime movers should have relatively similar and drooping speed–load characteristics.

6.22.3 Identical Voltage Requirement

The voltages at the paralleling points must be identical. During paralleling of two DC generators, the voltages at the paralleling point must be the same. Any mismatch of the paralleling point voltage will cause the circulating current to flow between them. This will cause sufficient voltage drops through the armature resistances. This circulating current causes the I^2R_a loss through the armature winding resistances, which results in heating of the generators and hence additional load on the prime movers. Therefore, this mismatch should be avoided.

During paralleling of different generated voltages of synchronous generators or alternators, the voltages at the paralleling point must identical. This is possible by using transformers. After completing paralleling, the values of E_{gp} must match. The incoming E_{gp} and V_p must be the same just before closing the paralleling switch.

6.22.4 Identical Phase Sequence

The phase of the incoming machine must be identical to that of the bus. If the bus has phase sequence A–B–C, the incoming machine must have phase sequence A–B–C. On the other hand, if the incoming machine has the phase sequence A–C–B, it will cause a disastrous short circuit. If the phase sequence A of the incoming machine and the bus is fully synchronized and phase sequences B and C of the incoming machine mate with the phase sequences C and B of the bus respectively, this mismatch between two phase sequences out of three phase sequences will result in short-circuit current, which is ultimately limited by the combined synchronous impedances of the windings involved. This short-circuit current will be even more at the moment of switch closure. This is because the armature reaction cross-magnetization needs time to be established. This out of phase sequence should be avoided very carefully.

6.22.5 In-phase Requirement

After getting the correct phase sequence, the incoming machine must be in phase with the bus. Figure 6.48(a) shows the in-phase condition, whereas Figure 6.48(b) shows the out of phase condition. The alternator will pull into synchronism after a few cycles of hunting if the out of phase condition is a small angle. Therefore, the alternator must be in-phase or nearly in-phase condition.

6.22.6 Identical Frequency Requirement

After satisfying all the above conditions, the frequency of the incoming machine must be identical to the frequency of the bus. If the frequency difference is very small and the alternator is in phase with the bus, the paralleling condition will be successful.

6.22.7 Relation of Prime Mover Torque Speed

The prime movers of the alternators must have drooping and similar speed torque characteristic for perfect load sharing. Rising speed torque characteristic must be avoided.

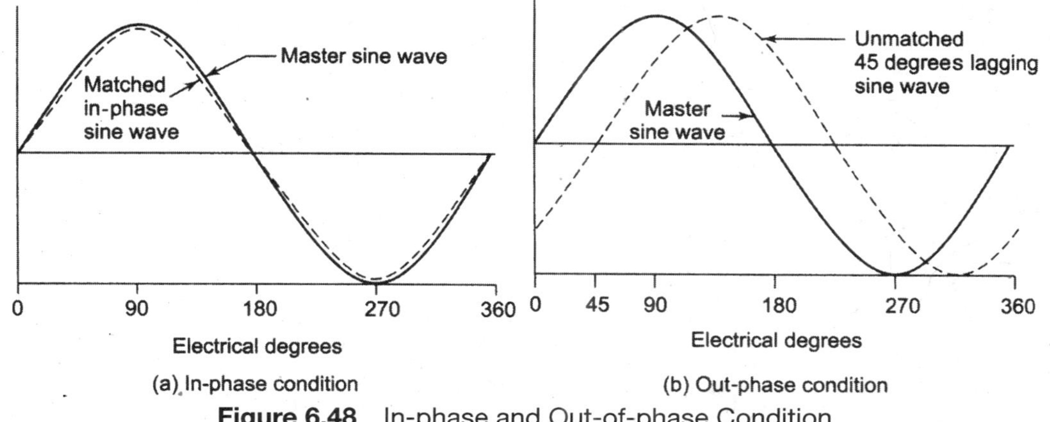

(a) In-phase condition (b) Out-phase condition

Figure 6.48 In-phase and Out-of-phase Condition

6.23 SYNCHRONIZING PROCEDURES OF ALTERNATOR

For successful paralleling of alternators, there are some procedures. These procedures are discussed as follows.

6.23.1 Voltage Matching

To match the voltage, one voltmeter is switched from one to the other or the voltmeters are matched.

6.23.2 Phase Sequence Matching

We cannot use the voltmeters for matching phase sequence. For successful parallel operation of the alternator with the bus, phase sequence of the incoming alternator must be equal to that of the bus. To get the correct phase sequence, dark lamp, bright lamp and two bright, one dark lamp methods are followed. These methods are discussed below.

6.23.2.1 Dark Lamp Phase Sequence Indication

A typical dark lamp method is shown in Figure 6.49. With the synchronizing switch kept open, there is no opposing voltage for the light because the incoming alternator is yet to be turned on. The lamps will glow dimly because $1/\sqrt{3}$ of line-to-line voltage across two phases of the operating alternator is seen by the lamps. If the incoming alternator is brought to the speed, it will have its own voltage and frequency and the lamps will see the difference in voltage between the alternator and the bus. If the lamps go bright and dark together, the desired phase sequence is said to be obtained. If they twinkle on and off, the phase sequence is not correct and any two out of three phase leads between the alternator and the switch must be interchanged.

6.23.2.2 Bright Lamp Phase Sequence Indication

Figure 6.50 shows the bright lamp synchronizing method. If all the lamps go bright and then dark, the phase sequence is correct.

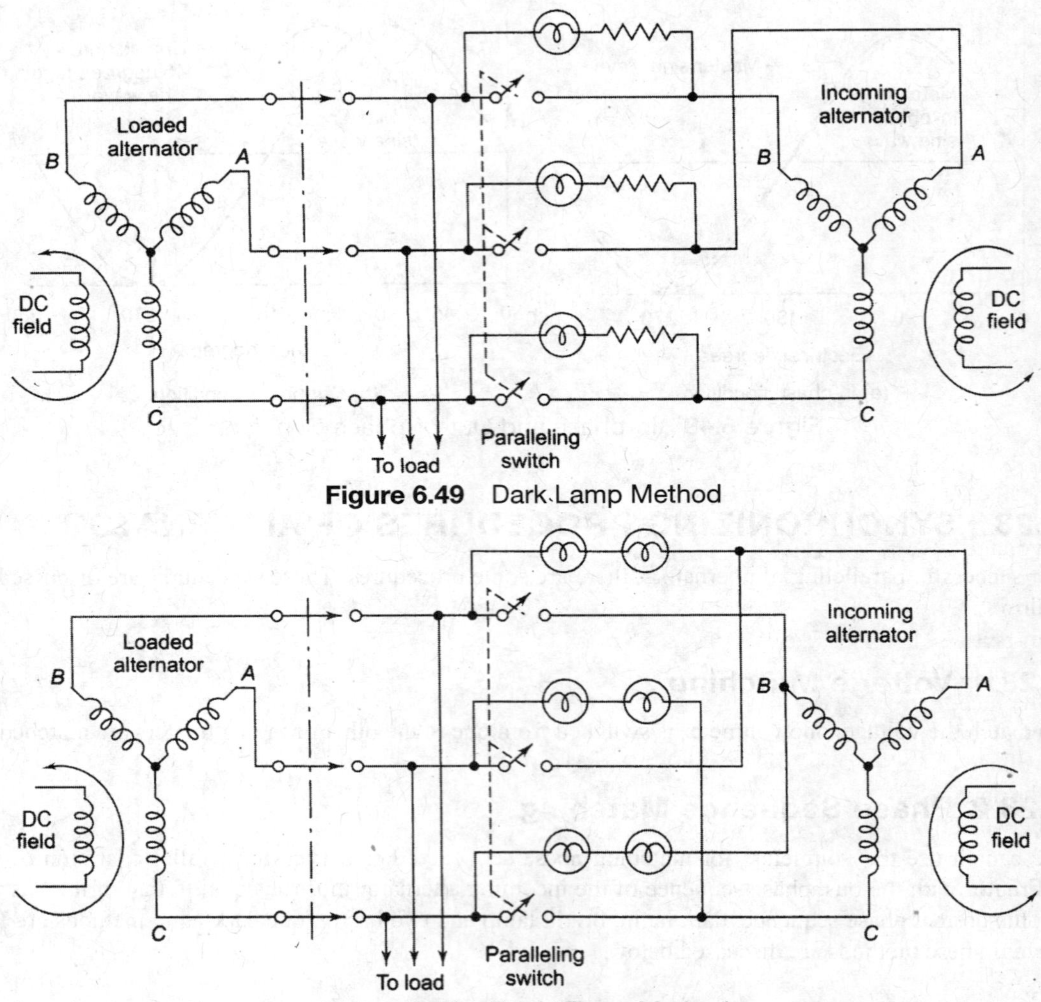

Figure 6.49 Dark Lamp Method

Figure 6.50 Bright Tamp Method

6.23.2.3 Two Bright and One Dark Phase Sequence Indication

Figure 6.51 shows two bright and one dark lamp synchronizing method. If the lamps twinkle back and forth, the phase sequence is correct. If they come and go together, the phase sequence is wrong.

6.23.3 Determination of In-phase

The various lamp methods that can be applied to find the indication of in-phase and out of phase are explained below.

6.23.3.1 Dark Lamp In-phase Indication

To get the in-phase condition, the speed of incoming machine will approach the frequency of the bus. The lamp will be dark. Initially, lamps glow faintly because the incoming machine will build up the

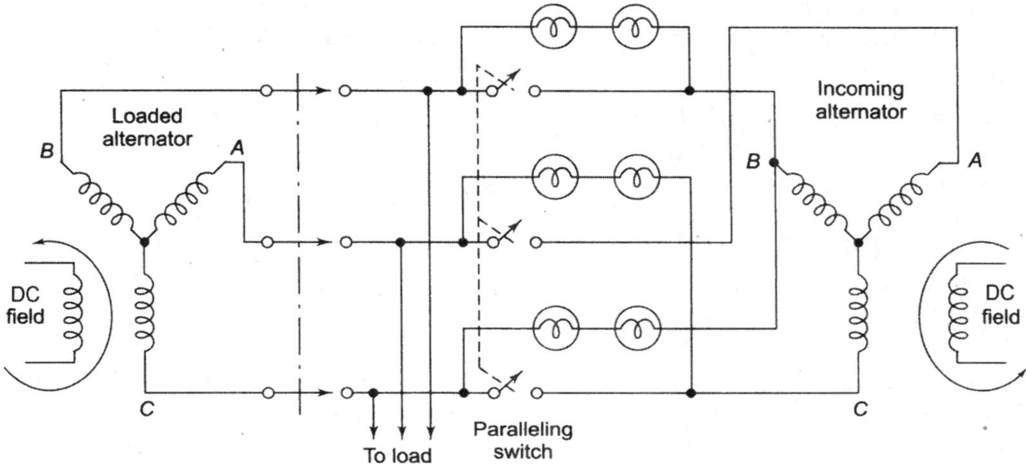

Figure 6.51 Two Bright and One Dark Lamp Method

voltage and the frequency of the incoming machine will approach the frequency of the bus. Since the frequency of the incoming machine is not equal to the frequency of the bus, the lamps will go dark and bright. Finally, the lamps will go dark and will stay so for a few seconds. This dark period is the actual in-phase condition and the paralleling switch may be closed.

Figure 6.52(a) shows the phasor diagram when there is perceptible difference in voltage between the incoming alternator and the bus. The lamps will glow perceptibly. If the two phasors are 180° out of phase, the lamps will glow with full brilliance. Figure 6.52(b) shows the phasor diagram when the voltage of the incoming alternator is nearly in phase with the bus but exactly not in phase. This is near the dark phase situation. Figure 6.52(c) shows the situation when there is no phase difference in voltages between the incoming machine and the bus. The lamps will go out.

6.23.3.2 Bright Lamp In-phase Indication

This method is used for single-phase alternators, which human eyes can perceive more sensitively. In single-phase circuits, the maximum voltage is produced at exactly the correct phase angle. If the lamp-

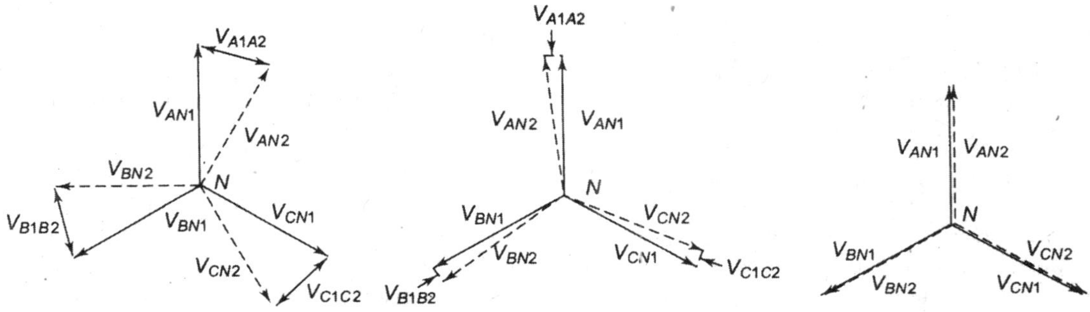

(a) Alternators 2 lagging alternator 1 by large-phase angle, lamps dim (b) Alternator 2 leading alternator 1 by small-phase angle, lamps out (c) Alternator in phase, lamps out, no difference in voltage

Figure 6.52 Voltage Diagrams of the Incoming Machine and the Bus

connected lines are 180° out of phase in three-phase circuits, maximum brightness is produced. There is a 60° error because *A* to *B*, *B* to *C* and *C* to *A* connections are 120° out of phase at synchronism. Therefore, bright lamp synchronization locates wrong phase relation.

6.23.3.3 Two Bright and One Dark In-phase Indication

In this method, two lamps will glow bright and one lamp will glow dark. When one is dark and other two are at the same brightness, in-phase condition is said to be achieved.

6.23.4 Frequency Synchronization

In any of the lamp methods, it is possible to compare the frequency or speed of the incoming machine and the bus. At the synchronizing frequency, there is no change in the light aspect. If the lamp blink cycle becomes longer, the speed of the incoming machine is less than that of the bus. If the lamp cycle becomes shorter due to any increase of speed of the incoming machine, the speed is higher than that of the synchronous speed. It should be noted that the flashing rate becomes apparent if the frequency of the incoming machine becomes nearer to the frequency of the bus. This flashing rate becomes slow when the frequency of the incoming machine is very close to the frequency of the bus. When the frequency of the incoming machine is equal to the frequency of the bus, this rate of change of flashing vanishes.

6.23.5 Synchroscope Synchronization

During installation of alternators, its phase sequence should be checked very carefully. Synchroscope compares the voltage from one phase of the incoming machine with the corresponding phase of the three-phase system. The phase difference between the voltages of the incoming machine and the infinite bus is indicated by the pointer. The pointer is stationary if the frequencies of the incoming machine and the alternator are identical. If the frequency is higher or lower than that of the incoming machine, the rotation of the pointer will be too fast or too slow, respectively, as shown in Figure 6.53. The difference between the frequency of the incoming machine and the infinite bus gives the speed of rotation of the pointer. Adjusting the prime mover input to the incoming machine, it is possible to control the frequency and phase positions.

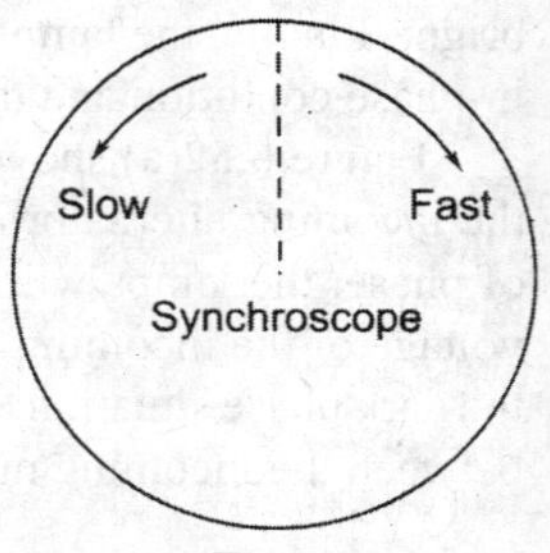

Figure 6.53 Synchronization by Synchroscope

6.24 DISTRIBUTION OF LOAD

When an alternator is running in parallel with other alternators, the load taken up by it is totally determined by their driving torque or the power input of the prime mover. If any change in excitation is carried out, it does not change its kW output but merely changes kVA, or merely changes the power factor at which the load is delivered.

6.24.1 Effect of Change in Excitation

Let us assume that the two alternators operating in parallel are identical; that is, they are supplying half of the active load and reactive load, or equal to the power factor of the load. If the excitation of the alternator 1 is increased, $E_{f1}>E_{f2}$ and it causes circulating current (I_C), which flows through the armature and round the bus bars. From Figure 6.54, I_C is added vectorially to the load current of alternator 1 and subtracted from the load current of alternator 2, which causes a change in load current. Therefore, alternator 1 and alternator 2 will deliver the load current at power factors $\cos\beta_1$ and $\cos\beta_2$, respectively,

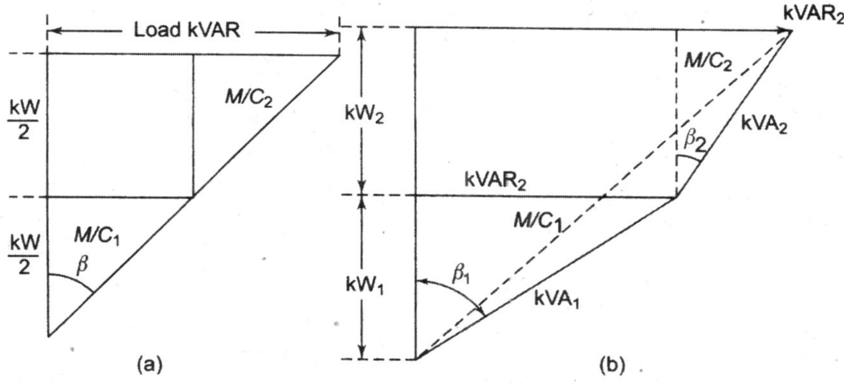

Figure 6.54 Effect of Change in Excitation

where $\cos\beta_1 > \cos\beta_2$. Although the two machines deliver the load currents at different power factors, it has no effect on kW output, but kVAr supplied by alternator 1 is increased, whereas kVAr supplied by alternator 2 is decreased as shown in Figure 6.54.

6.24.2 Effect of Change of Steam Supply

Let us assume that the excitation of the two alternators remains unaltered during their operation in parallel. Let the steam supply to alternator 1 be increased; that is, input to its prime mover is increased. Alternator 1 cannot overrun Alternator 2 because the speeds of the two alternators are tied by their synchronous bond. E_{f1} advances E_{f2} by a small angle δ. Hence, resultant voltage (E_r) is produced and it acts on the local circuit, resulting in a current I_r which lags E_r by a 90° angle. Therefore, power per phase of alternator 1 is increased whereas power per phase of alternator 2 is decreased. Since the increase in steam input has no effect on the division of reactive power, the active power output of alternator 1 is increased whereas active power output of alternator 2 is decreased, as shown in Figure 6.55.

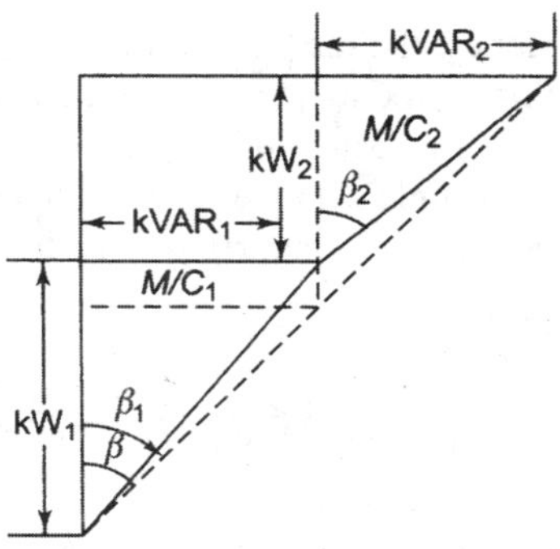

Figure 6.55 Effect of Change of Steam Supply

6.25 SYNCHRONIZING POWER AND SYNCHRONIZING TORQUE COEFFICIENT

After properly synchronizing a synchronous machine to an infinite bus, the synchronous machine has a tendency to stay in *synchronism.* If an alternator transfers power to an infinite bus at a steady-state power angle δ_0 and a sudden transient disturbance due to increase in load will occurs the rotor of the alternator will accelerate and at a load angle $\delta_0 + d\delta$, the alternator supplies new power $P + dP$. Therefore, the operating point shifts to a new line. The steady-state power input remains unchanged. Therefore, the additional load decreases the speed of the machine and hence the alternator comes back to the steady-state position. Similarly, due to sudden transient if the rotor retards, the load angle will decrease. The operating point will shift to a new line and the load on the machine becomes $P - dP$. Once again the input to the machine remains unaltered; therefore, the rotor is accelerated due to reduction in load.

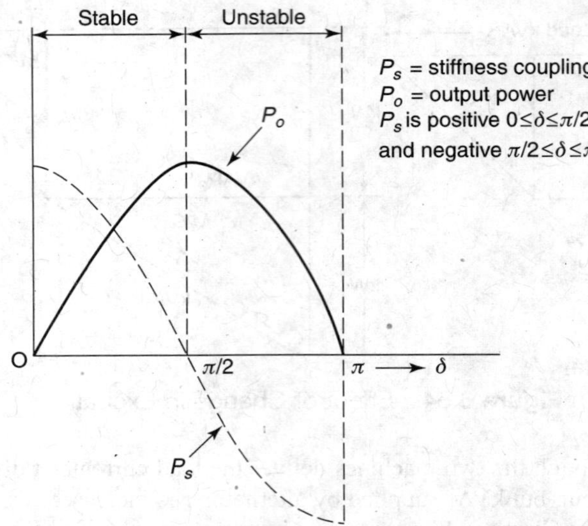

Figure 6.56 Variation of Synchronizing Power with δ

Hence, the machine comes back to synchronism. Therefore, we can conclude that the effectiveness of this correction action depends upon the change in power transfer for a given change in load angle. The synchronizing power coefficient, which is defined by the rate at which synchronous power varies with load angle (δ) gives a measure of effectiveness. The synchronizing power coefficient is also called stiffness of coupling, rigidity factor or stability factor. It is denoted by P_{syn}.

$$P_{syn} = \frac{dP}{d\delta}$$

For a cylindrical rotor, the power output per phase is given by

$$P = \frac{V_t}{Z_s}\left[E_f \cos(\theta - \delta) - V_t \cos\theta\right]$$

$$\therefore \qquad P_{syn} = \frac{dP}{d\delta} = \frac{V_t E_f}{Z_s}\sin(\theta - \delta) \qquad (6.81)$$

Figure 6.56 shows the variation of the synchronizing power with δ for cylindrical machine.

The synchronizing torque coefficient is defined by

$$T_{syn} = \frac{dT}{d\delta} = \left(\frac{1}{2\pi\, n_s}\right)\frac{dP}{d\delta} = \frac{V_t E_f}{2\pi\, n_s Z_s}\sin(\theta - \delta) \qquad (6.82)$$

Since $R_a \ll X_s$ for many synchronous machines, the synchronous power and torque coefficient per phase are given by

$$P_{syn} = \frac{V_t E_f}{X_s}\cos\delta \qquad \text{and} \qquad T_{syn} = \frac{V_t E_f}{2\pi\, n_s X_s}\cos\delta$$

For a salient-pole machine,

$$P = \frac{V_t E_f}{X_s}\sin\delta + \frac{1}{2}V_t^2\left(\frac{1}{X_q} - \frac{1}{X_d}\right)\sin 2\delta \qquad (6.83)$$

$$P_{syn} = \frac{dP}{d\delta} = \frac{V_t E_f}{X_s}\cos\delta + V_t^2\left(\frac{1}{X_q} - \frac{1}{X_d}\right)\cos 2\delta \tag{6.84}$$

Example 6.10 A 3 MVA, three-phase, 10-pole, star-connected alternator is connected to 6.6 kV, 50 Hz bus bars and it has synchronous reactance of 5 Ω per phase. Determine the synchronizing power and synchronizing torque per mechanical degree of rotor displacement at no-load. Assume normal excitation.

Solution

$$P = 10,\ p = \frac{P}{2} = 5,\ f = 50 \text{ Hz},\ \delta = 0° \text{ at no-load}$$

$$E_f = V_t = \frac{6,600}{\sqrt{3}} \text{ V}$$

$$N_s = \frac{120f}{P} = \frac{120 \times 50}{10} = 600 \text{ rpm}$$

$$P_{syn} = \left(\frac{3V_t E_f}{X_s} \times \cos\delta\right) \times \frac{p\pi}{180} = \frac{3}{5} \times \frac{6,600}{\sqrt{3}} \times \frac{6,600}{\sqrt{3}} \times \cos 0° \times \frac{5 \times 3.14}{180}$$

$$= 7,59,880 \text{ W/mecahnical degree}$$

$$T_{syn} = \frac{P_{syn}}{2\pi\, n_s} = \frac{7,59,880}{2 \times 3.14 \times \frac{600}{60}} = 12,100 \text{ N-m}$$

■

6.26 UNITS OF SYNCHRONIZING POWER COEFFICIENT

The unit of synchronizing power coefficient is watts per electrical radian.

$$P_{syn1} = \frac{V_t E_f}{X_s}\cos\delta \text{ W/elec. radian}$$

Again,

$$\pi \text{ radian} = 180°$$

$$1 \text{ radian} = \frac{180}{\pi} \text{degrees}$$

∴

$$P_{syn2} = \frac{dP}{d\delta}\ W/\left(\frac{180}{\pi}\text{degrees}\right)$$

i.e.,

$$P_{syn2} = \left(\frac{dP}{d\delta}\right) \times \frac{\pi}{180}\ W/\text{degrees}$$

If p is the total number of pairs of poles of the machine, we have

$$\theta_{\text{electrical}} = p\theta_{\text{mechanical}}$$

Therefore, synchronizing power coefficient per mechanical radian is given by

$$P_{syn3} = p\frac{dP}{d\delta} W \tag{6.85}$$

and synchronizing power coefficient per mechanical degree is given by

$$P_{syn4} = \left(\frac{p\pi}{180}\right)\frac{dP}{d\delta}\ W \tag{6.86}$$

6.27 SIGNIFICANCE OF SYNCHRONIZING POWER COEFFICIENT

The measurement of stiffness of the electromagnetic coupling between the rotor and the stator is carried out by the synchronizing power coefficient. If the value of P_{syn} is large, the coupling is stiff or rigid. For a too rigid coupling, with change of load or supply, the machine will be under shock which may damage the rotor of the winding.

For three phases, the synchronizing power and torque are given by

$$P_{syn} = \frac{3V_t E_f}{X_s} \cos\delta \tag{6.87}$$

and

$$T_{syn} = \frac{3V_t E_f}{2\pi\, n_s X_s} \cos\delta \tag{6.88}$$

Equation (6.87) suggests that P_{syn} is inversely proportional to the synchronous reactance. If a machine has large air gap, it has relatively small reactance and coupling is too stiff. Once again, P_{syn} is proportional to E_f; therefore, an overexcited machine is more stiff compared to an underexcited machine. Equation (6.87) also suggests that a restoring action is generated when $\delta = 0$, that is, at no-load. For $\delta = \pm 90°$, the restoring action is zero.

The machine will operate at the steady-state limit of stability and in the condition of unstable equilibrium at these values of δ. If the machine is provided with special fast-acting excitation system, then it is possible to run the machine at the steady-state limit of stability. At the steady-state limit of stability, its ability to resist small changes of load is zero.

6.28 HUNTING

For satisfactory operation of a synchronous machine, the mechanical speed of the rotor should be equal to the stator field. Any departure from these conditions will immediately cause synchronizing forces that tend to maintain the equality. The oscillation of the rotor about its final equilibrium position is called *hunting.* When synchronous machine is coupled with a system where torque variation contains harmonics, hunting falls out of step when the frequency of the torque component is equal to that of the frequency of free oscillations.

6.28.1 Causes of Hunting

- Due to occurrence of fault in the system that is supplied by the generator.
- Due to sudden changes of load.
- Due to sudden changes of field current.
- Due to cycle variations of the load torque.

6.28.2 Effects of Hunting

- Hunting increases the chance of loosing synchronism.
- Hunting causes variations of the supply voltage.

- Hunting increases the possibility of resonance.
- Due to hunting, huge mechanical stresses may be developed in the rotor shaft.
- Machine loss increases during hunting.
- Temperature of the machine rises during hunting.

6.28.3 Reduction of Hunting

To reduce hunting, the following techniques are commonly used:

- Damper windings.
- Use of fly wheels.
- Synchronous machine can be designed with suitable synchronizing power coefficient to reduce hunting.

6.29 OSCILLATIONS OF SYNCHRONOUS MACHINES

For every synchronous machine, a natural period of oscillation comes into picture due to variation in load or variation in mechanical speed, which creates phase swinging of the machine. When the period of oscillations becomes equal to the natural period, the developed period of oscillations becomes capable enough to swing the machine out of synchronism.

Let J be the moment of inertia, T_{sy} be the torque per mechanical radian, E_f be the generated voltage per radian, P_{sy} be the synchronous power, T_{sy} be the synchronous torque, Z_s be the synchronous impedance, n_s be the synchronous speed, I_{sc} be the short-circuit current and P be the number of poles.

The total synchronizing power developed by all three phases is $\dfrac{3\delta E_f^2}{Z_s}$ when the rotor is displaced by an angle δ. Synchronous power per electrical radian displacement is $\dfrac{3E_f^2}{Z_s}$.

The displacement in electrical radian $= \dfrac{P}{2} \times \dfrac{3E_f^2}{Z_s} = \dfrac{3PE_f^2}{2Z_s}$.

$$T_{sy} = \frac{\dfrac{3PE_f^2}{2Z_s}}{2\pi\, n_s} = \frac{3PE_f^2}{2Z_s \times (2\pi\, n_s)} = \frac{3PE_f^2}{4\pi\, n_s Z_s} = \frac{3PE_f}{4\pi\, n_s} \times \frac{E_f}{Z_s} = \frac{3PE_f I_{sc}}{4\pi\, n_s} \left[\because I_{sc} = \frac{E_f}{Z_s}\right]$$

$$T_{sy} = \frac{3E_f I_{sc}}{4\pi\, n_s} \times \frac{2f}{n_s} \quad \left[\because P = \frac{2f}{n_s}\right]$$

Natural time (T) is given by

$$T = 2\pi \times \sqrt{\frac{J}{T_{sy}}} = 2\pi \times \sqrt{\frac{Jn_s^2}{0.478 E_f I_{sc} f}}$$

$$T = 9.11 n_s \times \sqrt{\frac{J}{E_f I_{sc} f}} \tag{6.89}$$

Example 6.11 A 24 MVA, 11 kV, three-phase, 50 Hz, 1,500 rpm, alternator is paralleled with others of much greater capacity. The moment of inertia of the rotor is 3×10^5 kg-m^2. The alternator has synchronous reactance of 35 per cent. Determine the natural period of oscillations of the rotor. Short-circuit current is four times of full-load current.

Solution

$$N_s = 1{,}500/60 = 25 \text{ rps}, \; J = 3\times 10^5 \text{ kg-m}^2, \; f = 50 \text{ Hz}, \; I_{sc} = 4\times \text{full-load current}$$

$$E_f = \frac{11{,}000}{\sqrt{3}} = 6{,}350.8\,\text{V}$$

$$I_{f1} = \frac{24{,}000}{\sqrt{3}\times 11} = 1{,}259.67\text{A} \quad \text{and} \quad I_{sc} = 4\times 1{,}259.67 = 5{,}038.68 \text{ A}$$

$$T = 9.11 n_s \times \sqrt{\frac{J}{E_f I_{sc} f}} = 9.11\times 25\times \sqrt{\frac{3\times 10^5}{6{,}350.8\times 5{,}038.68\times 50}} = 3.12\square$$

■

6.30 SUDDEN SHORT CIRCUIT OF SYNCHRONOUS GENERATOR

A sudden short circuit on the terminals of synchronous generator is connected assuming the followings:

(i) Armature reactance is negligible.

(ii) All three phases are short circuited simultaneously.

(iii) The machine is under no-load condition before short circuit.

When the machine is suddenly short circuited on terminals, the following effects occur:

(i) The armature current increases in a very short time.

(ii) Power factor is nearly zero lag.

Due to these two effects, mmf in direct and quadrature axis changes. Due to change in armature mmf in the direct axis, flux linkage with the main winding changes. This change is opposed by an induced current in the field winding according to Lenz's law. So until the armature current attains steady-state value, the field winding acts like secondary of transformer for the armature winding. If the damper w nding is also present, shown in Figure 6.57(a), transient current is also induced in it.

A large demagnetizing mmf generated by initial armature current tends to decrease the pole flux from its original value. But this flux cannot decrease suddenly. To maintain the flux constant, the field current must increase to a large value to neutralize the demagnetizing effect of armature current. Eddy current induced in the damper bars and adjacent metal parts also helps to maintain the field flux constant. These conditions are similar to three winding transformers with armature acting as primary and field winding of leakage reactance x_f and damper winding having leakage reactance x_{cd} acting as two secondaries.

Figure 6.57(b) shows the field current after short circuit. Figure 6.57(c) shows the approximate equivalent circuit of an alternator immediately after short circuit.

The mutual or magnetizing reactance is assumed to be same for the three windings and is denoted by x_{ad}. So the equivalent subtransient reactance of the circuit is

$$X''_d = x_l + \frac{1}{\dfrac{1}{x_{ad}} + \dfrac{1}{x_f} + \dfrac{1}{x_{cd}}} \tag{6.90}$$

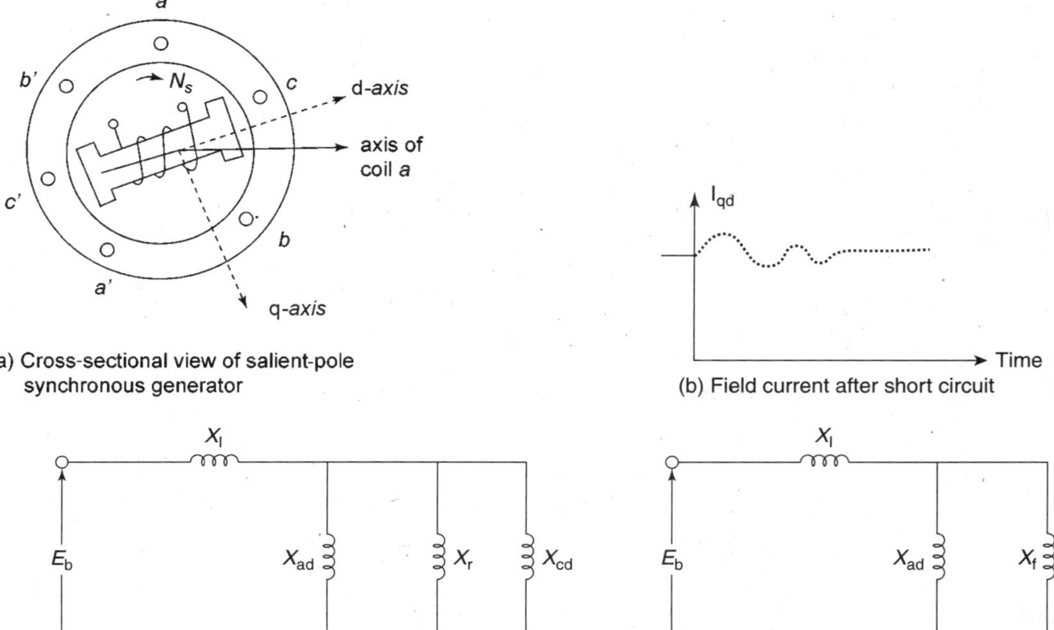

(a) Cross-sectional view of salient-pole synchronous generator

(b) Field current after short circuit

(c) Equivalent circuit of an alternator immediately after short circut

(d) Appropriate equivalent circuit

Figure 6.57 Sudden Shore Circuit of Synchronous Generator

The effect of damper winding and eddy currents in other metallic parts disappears after a few cycles of short-circuit current due to large resistance associated with them. So x_f can also be neglected. The steady-state equivalent circuit is shown in Figure 6.57(d), and the direct-axis transient reactance is given by

$$X'_d = x_l + \frac{1}{\dfrac{1}{x_{ad}} + \dfrac{1}{x_f}} \tag{6.91}$$

where d-axis synchronous reactance is $x_d = x_l + x_{ad}$.

6.31 SHORT-CIRCUIT RATIO

The ratio of the field current required to generate rated voltage on open circuit required to circulate rated armature current during short circuit is termed as short-circuit ratio (SCR).

The *OCC* at rated speed and *SCC* of a three-phase generator are to be drawn at first for calculation of SCR. Figure 6.58 shows the *OCC* and *SCC* of an alternator.

From Figure 6.58, we have

$$\text{SCR} = \frac{\text{field current for rated open-circuit voltage}}{\text{field current for rated short-circuit current}} = \frac{Op}{Os} \tag{6.92}$$

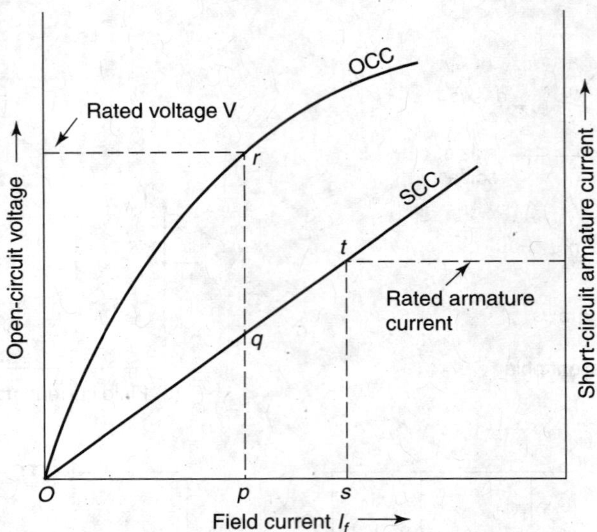

Figure 6.58 OCC and SCC of an alternator

The triangles *Opq* and *Ost* are similar. Therefore, we have

$$\frac{Op}{Os} = \frac{pq}{st} \tag{6.93}$$

From Equations (6.92) and (6.93), we have

$$\text{SCR} = \frac{Op}{Os} = \frac{pq}{st} \tag{6.94}$$

The ratio of open-circuit voltage for a given field current to armature short-circuit current for the same field current is termed as direct-axis synchronous reactance (X_d).

From Figure 6.58, the given field current is taken as *Op*. The direct-axis synchronous reactance in ohm is given by

$$X_d = \frac{pr}{pq} \tag{6.95}$$

The base impedance is defined by

$$\text{Base impedance} = \frac{\text{per phase rated voltage}}{\text{per phase rated current}} = \frac{V_{rated}}{I_{rated}} = \frac{pr}{st} \tag{6.96}$$

The pu value of X_d can be obtained from Equations (6.95) and (6.96):

$$X_{d(p.u.)} = \frac{\dfrac{pr}{pq}}{\dfrac{pr}{st}} = \frac{st}{pq} \tag{6.97}$$

From Equations (6.94) and (6.97), we have

$$\text{SCR} = \frac{pq}{st} = \frac{1}{\dfrac{st}{pq}} = \frac{1}{X_{d(p.u.)}} \tag{6.98}$$

From Equation (6.98), we can conclude that reciprocal of the per unit value of direct-axis synchronous reactance gives the SCR. The degree of saturation in a saturated magnetic circuit decides the SCR. Since SCR depends on the rated voltage of *OCC* and rated armature current on *SCC*, SCR becomes a single value.

The operating characteristics, physical size and cost of the synchronous machine are also decided by the SCR and hence it is an important factor for synchronous machine. Due to change in load, there will be large variation of terminal voltage for an alternator for a low value of SCR. Hence, the machine becomes very load sensitive. Field current is used to be varied over a wide range for keeping the terminal voltage constant. For low value of SCR, synchronizing power becomes small, which keeps the machine in synchronism. This gives a low stability limit and hence the machine becomes less stable for low value of SCR while running in parallel with other alternators. The armature current at short-circuit condition becomes small for a low SCR.

For high value of SCR, the synchronous machine not only has a better voltage regulation but also has an improved steady-state stability limit. The short-circuit fault current in the armature becomes high. For a synchronous machine, the excitation voltage (E_f) is related by

$$E_f \propto \text{field} - \text{flux per pole}$$

i.e.,

$$E_f \propto \frac{\text{field flux per pole}}{\text{reluctance of air gap}}$$

Also, we know that

$$\text{Synchronous iductance } (L_s) \propto \frac{1}{\text{reluctance of air gap}}$$

$$\therefore \qquad \text{SCR} \propto \frac{1}{L_s}$$

i.e., SCR $\propto$ reluctance of air gap

If the length of air gap is increased, SCR may be increased. Hence, the field mmf is to be increased for the same air-gap length to get the same E_f. Either I_f or the number of turns must be increased to increase the field mmf. Therefore, a greater height of field poles is required. Hence, the overall diameter of the machine also increases. Therefore, the size, weight and cost of the machine are increased for a large SCR. The following are the typical values of SCR:

(i) Cylindrical rotor machines: 0.5 to 0.9
(ii) Salient-pole machines: 1.0 to 1.5
(iii) Synchronous compensators: 0.4

6.32 PROTECTION OF GENERATORS

The alternators used in power systems produce very high voltage, and their protection is very expensive due to the following reasons:

(i) The generators are producing very high voltages and are connected to the bus bars.
(ii) Other different equipments are also associated with the generators.
(iii) The generators are very costly, expensive and very important factor in power systems.

The generator should not be solidly grounded. The single line to ground fault would be dangerously high if it is solidly grounded. Besides this, high value of fault current, the resulting asymmetry in the rotating magnetic field inside the generator, causes the large vibrations leading to mechanical damage. Therefore, the neutral of the generator is always grounded through a practical value of resistor which is connected through a step-down transformer known as grounding transformer.

The various electrical faults in a transformer are as follows:

(i) *Stator faults:*

(a) Phase fault

(b) Inter-turn fault on the same phase

(c) Ground fault

(ii) *Rotor faults:*

Short circuit to ground fault

Stator phase and ground faults are protected by longitudinal transverse protection. Transverse differential protection is used to protect the inter-turn fault on the same phase. A single ground fault on the rotor winding does not affect the working of the alternator. However, a second ground fault of the rotor winding causes distortion in the distribution of the main flux. A suitable arrangement with overcurrent relay, bridge rectifier and isolation transformer is used to protect against rotor faults.

The various abnormal operating conditions are the following:

(i) *Electrical:*

(a) *Unbalanced loading:*

For the unbalanced loading of the generator, the stator current will have the negative sequence component of current which rotates at a synchronous speed ($-N_s$) opposite to the direction of the field flux (N_s). The relative velocity between the two is $2N_s$ and this causes the double frequency. Therefore, the hysteresis and eddy current losses will increase. The negative sequence component must be extracted at first, and the protection against unbalanced loading is implemented by the inverse-time overcurrent relay.

(b) *Loss of excitation:*

The loss of excitation occurs due to any of the following reasons:

1. Loss of field to the main exciter
2. Short circuit in the field winding
3. Accidental tripping of the field breaker
4. Field circuit breaker lath failure
5. Loss of AC supply to excitation system

The generator delivers active power and reactive power to the grid. The active power is due to the prime mover and the reactive power is due to the field.

If the field excitation is lost, the alternator will operate as induction generator drawing reactive power to the grid. It will remain synchronized to the grid. The machine will operate at speed slightly above the synchronous speed. This operation needs the flow of slip frequency current in the rotor, damper winding, and also in the slot wedges and the surface of the solid rotor body. This will cause abnormal heating of the rotor and the overloading of the stator. Any mho-type distance relay with offset characteristic is used for protection against loss of excitation.

(ii) *Mechanical:*

(a) *Loss of prime mover:*

Due to the loss of prime move, the machine will draw active power from the bus bar and it will supply the reactive power to the bus bar and will remain synchronized with the grid. The alternator will work as a synchronous motor. Running in this mode is harmful to the prime mover because there will be churning steam in the turbine which causes the objectionable temperature rise and damage to the blades. Directional is used for protection against loss of prime mover.

(b) *Overspeeding:*

The overspeeding occurs due to the fault of the generator. In this case there is output to the grid but the input to the alternator remains the same. If the mechanical input is not quickly reduced by the speed-governing mechanism, this causes the alternator to accelerate dangerously at a high speed. The speed of the machine should be displayed continuously.

ADDITIONAL SOLVED PROBLEMS

Example 6.12 A star-connected three-phase alternator delivers a three-phase star-connected load at a power factor of 0.8 lagging. A wire connects the load and the alternator. The terminal voltage at no-load is 2,500 V and at full-load of 1,460 kW is 2,200 V. Determine the terminal voltage when it delivers a three-phase star-connected load having a resistance of 6 Ω and a reactance of 8 Ω per phase, respectively. Assume constant current and field excitation.

Solution

$$\cos\beta = 0.8 \text{ (lagging)}, \sin\beta = 0.6$$

$$I_L = \frac{1,460\times10^3}{\sqrt{3}\times2,200\times0.8} = 478.94 \text{ A} = I_a$$

No-load phase voltage $= \dfrac{2,500}{\sqrt{3}}$ V and full-load phase voltage $= \dfrac{2,200}{\sqrt{3}}$ V

Therefore, voltage drop/phase for a current of 478.94 A is $= \dfrac{2,500}{\sqrt{3}} - \dfrac{2,200}{\sqrt{3}} = 173.20 \text{ V}$

Voltage drop/phase for a current of 1 A $= \dfrac{173.20}{478.94} = 0.3616 \text{ V}$

If the armature supplies a current of I_a A per phase at 0.8 power factor lagging, the voltage drop/phase = 0.36161 V

$$\text{Terminal voltage/phase} = \frac{2,500}{\sqrt{3}} - 0.3616\, I$$

∴ $$Z_L = \sqrt{6^2+8^2} = 10\,\Omega$$

Load terminal voltage = 10I

∴ $$10I = 1,443.37 - 0.3616\, I_a$$

i.e., $$10.3616\, I_a = 1,443.37$$

i.e., $$I_a = \frac{1,443.37}{10.3616} = 139.3 \text{ A}$$

Therefore, terminal voltage per phase = 139.3 × 10 = 1,393 V

∴ Line values of terminal voltage $= \sqrt{3}\times1,393 = 2,412.75$ V ■

Example 6.13 A 10 kVA, 400 V, three-phase star-connected synchronous generator has an armature resistance of 0.8 Ω per phase and synchronous reactance of 1.4 Ω per phase. (i) Find the voltage regulation of the alternator at full-load at power factors of (a) 0.8 lagging and (b) 0.8 leading. (ii) Calculate the power factor at which the voltage regulation becomes zero.

Solution

$R_a = 0.8\ \Omega$/phase and $X_s = 1.4\ \Omega$/phase

(i) $I_L = \dfrac{10\times 10^3}{\sqrt{3}\times 400} = 14.43\text{ A} = I_a$

$V_t = \dfrac{400}{\sqrt{3}} = 231\text{ V}$

(a) $\cos\beta = 0.8$ (lagging), $\sin\beta = 0.6$

$$E_f = \sqrt{(V_t\cos\beta + I_aR_a)^2 + (V_t\sin\beta + I_aX_s)^2}$$

$$= \sqrt{(231\times 0.8 + 14.43\times 0.8)^2 + (231\times 0.6 + 14.43\times 1.4)^2} = 252.52\text{ V}$$

$\therefore$ Percentage regulation $= \dfrac{E_f - V_t}{V_t}\times 100 = \dfrac{252.52 - 231}{231}\times 100 = 9.316\%$

(b) $\cos\beta = 0.8$ (leading), $\sin\beta = 0.6$

$$E_f = \sqrt{(V_t\cos\beta + I_aR_a)^2 + (V_t\sin\beta - I_aX_s)^2}$$

$$= \sqrt{(231\times 0.8 + 14.43\times 0.8)^2 + (231\times 0.6 - 14.43\times 1.4)^2} = 229.28\text{ V}$$

$\therefore$ Percentage regulation $= \dfrac{E_f - V_t}{V_t}\times 100 = \dfrac{229.28 - 231}{231}\times 100 = -0.7446\%$

(ii) Let $\cos\beta$ be the power factor at which voltage regulation becomes zero. The power factor must be leadi ıg.

$$E_f^2 = (V_t\cos\beta + I_aR_a)^2 + (V_t\sin\beta - I_aX_s)^2$$

$$= V_t^2 + 2V_tI_a(R_a\cos\beta - X_s\sin\beta) + I_a^2(R_a^2 + X_s^2)$$

$$= 231^2 + 2\times 231\times 14.43(0.8\times\cos\beta - 1.4\sin\beta) + 14.43^2(0.8^2 + 1.4^2)$$

For zero regulation, $E_f = V_t = 231\text{V}$

$$231^2 = 231^2 + 2\times 231\times 14.43(0.8\cos\beta - 1.4\sin\beta) + 14.43^2(0.8^2 + 1.4^2)$$

i.e., $0.8\cos\beta - 1.4\sin\beta = \dfrac{-541.38}{2\times 231\times 14.43} = -0.08$

i.e., $1.4\sin\beta - 0.8\cos\beta = 0.08$

i.e., $a\sin(\beta - \theta) = 0.08$

where $a = 1.612$ and $\theta = 29.75°$.

$$\therefore \quad 1.612\sin(\beta - 29.75°) = 0.08$$

$$\text{i.e.,} \quad \sin(\beta - 29.75°) = 0.05 = \sin 2.86°$$

$$\text{i.e.,} \quad \beta - 29.75° = 2.86°$$

$$\text{i.e.,} \quad \beta = 29.75° + 2.86° = 32.61°$$

$$\therefore \quad \cos\beta = 0.842 \text{ [leading]}$$

Therefore, the required power factor is 0.842 (leading). ■

Example 6.14 A 40 kVA, 440 V, three-phase star-connected, 50 Hz, alternator has an effective armature resistance of 0.3 Ω/phase and leakage reactance of 0.6Ω/phase. Determine (i) internal emf, (ii) no-load emf, (iii) percentage voltage regulation at full load and (iv) value of armature reactance due to armature reaction at rated load and unity power factor. The synchronous reactance of the alternator is 3.3 Ω/phase.

Solution

$$I_L = \frac{40 \times 10^3}{\sqrt{3} \times 440} = 52.48\text{A} = I_a/\text{phase}$$

$$V_t = \frac{440}{\sqrt{3}} = 254\text{ V}$$

$\cos\beta = 1$, $\sin\beta = 0$, $R_a = 0.3\Omega$/phase and $X_L = 0.6\Omega$/phase

(i) $$E_f = \sqrt{(V_t\cos\beta + I_aR_a)^2 + (V_t\sin\beta + I_aX_L)^2}$$

$$= \sqrt{(254 \times 1 + 52.48 \times 0.3)^2 + (254 \times 0 + 52.48 \times 0.6)^2} = 271.57\text{ V}$$

$$(E_f)_{\text{Line}} = \sqrt{3} \times 271.57 = 470.37\text{ V}$$ ■

(ii) $X_s = 3.3$ Ω/phase

$$E_f = \sqrt{(V_t\cos\beta + I_aR_a)^2 + (V_t\sin\beta + I_aX_s)^2}$$

$$= \sqrt{(254 \times 1 + 52.48 \times 0.3)^2 + (254 \times 0 + 52.48 \times 3.3)^2} = 320.55\text{ V}$$

$$(E_f)_{\text{Line}} = \sqrt{3} \times 320.55 = 555.208\text{ V}$$ ■

(iii) Voltage regulation $= \dfrac{320.55 - 254}{254} = 0.2620\text{ p.u.} = 26.20\%$ ■

(iv) $X_{ar} = X_s - X_L = 3.3 - 0.8 = 2.5\ \Omega$ ■

Example 6.15 A three-phase, 10 kVA, 440 V, 50 Hz, star-connected alternator supplies the rated load at 0.8 power factor lagging. Determine the power angle and voltage regulation if the armature resistance and synchronous reactance per phase are 0.6 Ω and 9 Ω respectively.

Solution

Full-load current $\left(I_L\right) = \dfrac{11,000}{\sqrt{3} \times 440} = 14.43$ A and $V_t = \dfrac{440}{\sqrt{3}} = 254$ V

$\cos\beta = 0.8$ (lagging), $\sin\beta = 0.6$, R_a/phase = 0.6 Ω and X_s/phase = 9 Ω

Now, $$\beta = \cos^{-1}(0.8) = 36.87°$$

$$E_f = \sqrt{\left(V_t \cos\beta + I_a R_a\right)^2 + \left(V_t \sin\beta + I_a X_s\right)^2}$$

$$= \sqrt{\left(254 \times 0.8 + 14.43 \times 0.6\right)^2 + \left(254 \times 0.6 + 14.43 \times 9\right)^2} = 352.93 \text{ V}$$

Percentage regulation $= \dfrac{352.93 - 254}{254} \times 100 = 38.94\%$

From phasor diagram (Figure E6.4),

$$\tan\left(\beta + \delta\right) = \frac{BD}{OB} = \frac{BC + CD}{OA + AB} = \frac{V_t \sin\beta + I_a X_s}{V_t \cos\beta + I_a R_a} = \frac{254 \times 0.6 + 14.43 \times 9}{254 \times 0.8 + 14.43 \times 0.6} = 1.332$$

i.e., $$\beta + \delta = 53.10°$$

i.e., $$36.87° + \delta = 53.10°$$

i.e., $$\delta = 16.23°$$ ■

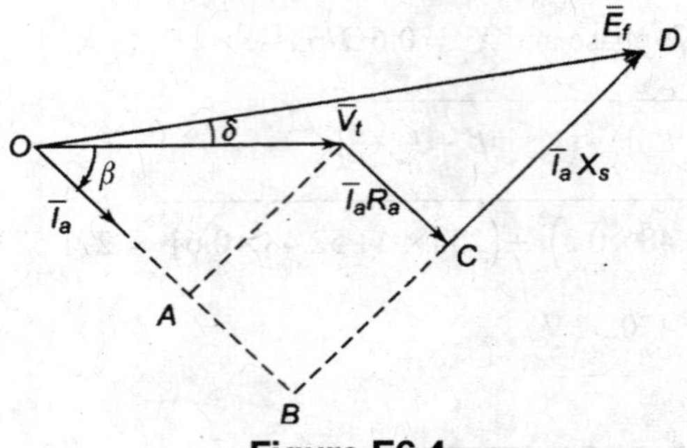

Figure E6.4

Example 6.16

I_f	2	4	6	7	8	10	1	14	16	18
E_0	150	280	390	440	470	520	565	592	–	–
SC line current (A)	10	20	32	41	47	58	70	78	–	–
Zero power factor terminal voltage	–	–	–	0	75	200	310	390	455	500

The table above gives the data for open circuit and load zero power factor tests on an eight-pole, 440 V, 50 Hz, three-phase star-connected alternator having effective armature resistance between any two terminals of 0.4 Ω. Calculate the regulation at full-load current of 41 A at 0.8 power factor lagging using each of the following methods:

(i) synchronous impedance method,
(ii) mmf method and
(iii) Potier triangle method.

Solution

$$V_t = \frac{440}{\sqrt{3}} = 254\,\text{V}$$

$$R_a \text{ per phase} = \frac{0.4}{2} = 0.2\,\Omega$$

The *OCC*, *SCC* and *ZPFC* have been plotted and shown in Figure E6.5. The voltages along the *Y*-axis are the phase voltages and the currents along the *X*-axis are the field currents.

(i) 254 V occurs for a field current of 7 A and short-circuit current of 41 A.

$$\therefore \quad Z_s = \frac{\frac{440}{\sqrt{3}}}{41} = 6.19 \quad \text{and} \quad X_s = \sqrt{Z_s^2 - R_a^2} = \sqrt{6.19^2 - 0.2^2} = 6.186\,\Omega$$

$$\cos\beta = 0.8 \text{ (lagging)}, \ \beta = 36.87°$$

$$E_f = \sqrt{(V_t \cos\beta + I_a R_a)^2 + (V_t \sin\beta + I_a X_s)^2}$$

$$= \sqrt{(254 \times 0.8 + 41 \times 0.2)^2 + (254 \times 0.6 + 41 \times 6.19)^2} = 457.9\,\text{V}$$

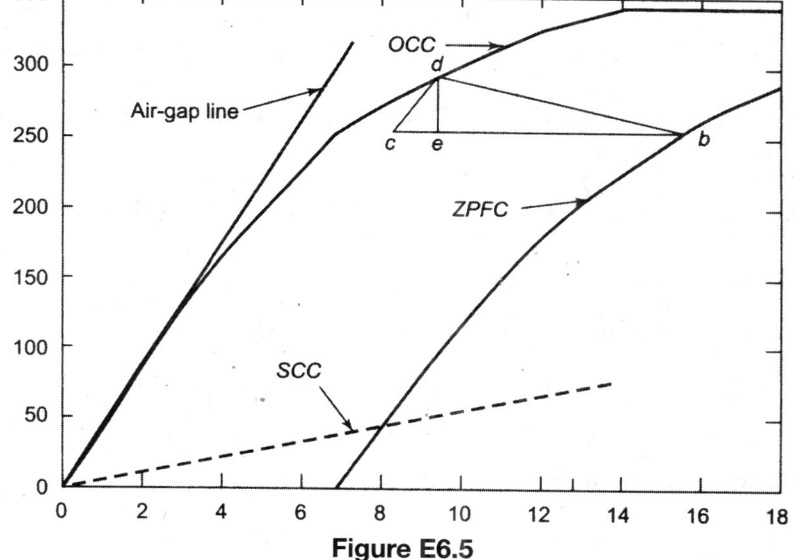

Figure E6.5

$$\tan(\delta+\beta)=\frac{V_t\sin\beta+I_aX_s}{V_t\cos\beta+I_aR_a}=\frac{254\times0.6+41\times6.19}{254\times0.8+41\times0.2}=1.9214=\tan62.5^\circ$$

i.e., $\delta+\beta=62.5^\circ$

i.e. $\delta=62.5^\circ-\beta=62.5^\circ-\cos^{-1}(0.8)=62.5^\circ-36.87^\circ=25.63^\circ$

$$\text{Voltage regulation}=\frac{E_f-V_t}{V_t}\times100=\frac{457.9-254}{254}\times100=80.27\%$$

(ii) Field current to obtain rated voltage of 254 V is 7 A.

$$I_{f1}=7\angle(90^\circ+\beta)=7\angle(90^\circ+36.87^\circ)=7\angle126.87^\circ=-4.2+j5.56$$

$$I_{f2}=7\angle\ 180^\circ-\beta\ =7\angle(180^\circ-36\ 87^\circ)=7\angle143\ 13^\circ=-5.59+j4.20$$

$$I_f=I_{f1}+I_{f2}=-4.2+j5.56-5.59+j4.20=-9.79+j9.76$$

$$=13.82\angle135^\circ$$

The phase voltage corresponding to a field current of 13.82 A is 340 V.

$$\therefore\ \text{Voltage regulation}=\frac{E_f-V_t}{V_t}\times100=\frac{340-254}{254}\times100=33.86\%$$

(iii) $bc=7$ A, $ce=1$ A, $de=42$ V

$\therefore$ $I_aX_L=42$; i.e., $X_L=\frac{42}{41}=1.024$ A

$$E_f=\sqrt{(V_t\cos\beta+I_aR_a)^2+(V_t\sin\beta+I_aX_s)^2}$$

$$=\sqrt{(254\times0.8+41\times0.2)^2+(254\times0.6+41\times1.024)^2}=287.18\text{ V}$$

From *OCC*, the field current corresponding to 287.18V is 9.48 A which leads E_f by a $90^\circ+36.87^\circ$ angle or 126.87°.

$$\tan(\delta+\beta)=\frac{V_t\sin\phi+I_aX_s}{V_t\cos\phi+I_aR_a}=\frac{254\times0.6+41\times1.024}{254\times0.8+41\times0.2}=0.9195=\tan42.6^\circ$$

i.e., $\delta+\beta=42.6^\circ$

$$I_r=9.48\angle(90^\circ+42.6^\circ)=9.48\angle132.6^\circ\text{A}\quad\text{and}\quad I_{ar}=6.0\angle0^\circ\text{ A}$$

$$I_f=I_r-I_{ar}=9.48\angle132.6^\circ-6\angle0^\circ=-6.417+j6.978-6$$

$$=-12.417+j6.978=14\ 24\angle150\ 66^\circ$$

The *OCC* voltage corresponding to field current of 14.24 A is 341 V.

$$\therefore\ \text{Voltage regulation}=\frac{E_f-V_t}{V_t}\times100=\frac{341-254}{254}\times100=34.25\%$$

■

Example 6.17 A three-phase star-connected synchronous motor supplies a load current of 9 A at 0.9 power factor lagging at 440 V per phase. Find the load angles and the components of armature current I_d and I_q if $X_d = 9\ \Omega$ and $X_q = 5\ \Omega$. Neglect armature resistance. Calculate the voltage regulation of the alternator.

Solution

$V_t = 440$ V, $\cos\beta = 0.9$ (lagging), $\sin\beta = 0.436$ I_a per phase = 9A, $X_d = 9\ \Omega$, $X_q = 5\ \Omega$, $R_a = 0$

$$\beta = \cos^{-1}(0.9) = 25.84°$$

$$\tan\gamma = \frac{V_t \sin\beta + I_a X_q}{V_t \cos\beta + I_a R_a} = \frac{440 \times 0.436 - 9 \times 9}{440 \times 0.9 + 9 \times 0} = 0.689 = \tan 34.56°$$

$\therefore$ $$\gamma = 34.56°$$

Again, $$\gamma = \beta + \delta$$

$\therefore$ $$\delta = 34.56° - \beta = 34.56° - \cos(0.9) = 34.56° - 25.84° = 8.72°$$

$$I_d = I_a \sin\gamma = 9 \times \sin 34.56° = 5.10\ \text{A}$$

$$I_q = I_a \cos\gamma = 9 \times \cos 34.56° = 7.41\ \text{A}$$

$\therefore$ $$E_f = V_t \cos\delta + I_q R_a + I_d X_d = 440 \times \cos 8.72° + 7.41 \times 0 + 5.10 \times 9 = 480.81\ \text{V}$$

$\therefore$ Percentage regulation $= \dfrac{480.81 - 440}{440} \times 100 = 9.275$ ■

Example 6.18 A three-phase, 440 V, 50 Hz, delta-connected alternator has direct-axis reactance of 0.12 Ω and 0.09 Ω per phase, respectively. If the alternator supplies 900 A at 0.8 power factor lagging, calculate the following:

(i) the excitation emf neglecting saliency and

(ii) the excitation emf taking into account the saliency.

In the first case, assume $X_s = X_d$.

Neglect the armature resistance.

Solution

$$X_d = 0.12\Omega,\ X_q = 0.09\Omega,\ V_L = 400\ \text{V} = V_{ph} = V_t,\ I_L = 900\text{A}$$

$\therefore$ $$I_a = \frac{I_L}{\sqrt{3}} = \frac{900}{\sqrt{3}} = 519.61\text{A}$$

$$\cos\beta = 0.8(\text{lagging}),\ R_a = 0$$

(i) $$X_s = X_d = 0.12\Omega$$

$$E_f = \sqrt{(V_t \cos\beta + I_a R_a)^2 + (V_t \sin\beta + I_a X_s)^2}$$

$$= \sqrt{(400 \times 0.8 + 519.6 \times 0)^2 + (400 \times 0.6 + 519.6 \times 0.12)^2} = 440.24\ \text{V}$$

$\therefore$ $$(E_f)_{\text{Line}} = 440.24\ \text{V}$$

(ii) $$\tan\gamma = \frac{V_t \sin\beta + I_a X_q}{V_t \cos\beta + I_a R_a} = \frac{400\times0.6 + 519.61\times0.09}{400\times0.8 + 519.61\times0} = 0.896 = \tan 41.86^\circ$$

$\therefore$ $$\gamma = 41.86^\circ$$

Again, $$\gamma = \beta + \delta$$

$\therefore$ $$\delta = 41.86^\circ - \beta = 41.86^\circ - 36.86^\circ = 5^\circ$$

$$I_d = I_a \sin\gamma = 519.61\times\sin 41.86^\circ = 346.74\text{ A}$$

$$I_q = I_a \cos\gamma = 519.61\times\cos 41.86^\circ$$

$$= 387\text{ A}$$

$\therefore$ $$E_f = V_t \cos\delta + I_q R_a + I_d X_d$$

$$= 400\times\cos 5^\circ + 387\times0 + 346.74\times0.12 = 440\text{ V}$$

Example 6.19 A three-phase, 50 Hz, 11 kV, star-connected alternator has synchronous reactance of 5 Ω per phase, and negligible armature resistance has armature current of 220 V at unity power factor during its operation on constant frequency and constant voltage bus bars. If the steam input remains constant and emf is increased by 30 per cent, calculate the new values of current and power factor.

Solution

$$V_t = \frac{11\times10^3}{\sqrt{3}} = 6350.8\text{ V}, I_{a1} = 220\text{ A}, X_s = 5\Omega/\text{phase}, R_a = 0, \cos\beta_1 = 1$$

At unity power factor,

$$E_{f1} = \sqrt{V_t^2 + (I_a X_s)^2} = \sqrt{6,350.8^2 + (220\times5)^2} = 6,445.36\text{ V}$$

Since emf is increased by 30 per cent, the new emf is $E_{f2} = 1.3\ E_{f1} = 1.3 \times 6,445.36 = 8,378.97$ V, which is greater than V_t. Therefore, the power factor is lagging in nature.

Again, at infinite bus $V_t = 6,350.8$ V

The power output remains unaltered because steam input does not change.

$$V_t I_{a2} \cos\beta_2 = V_t I_{a1} \cos\beta_1$$

i.e., $$I_{a2} \cos\beta_2 = I_{a1} \cos\beta_1 = 220\times1 = 220$$

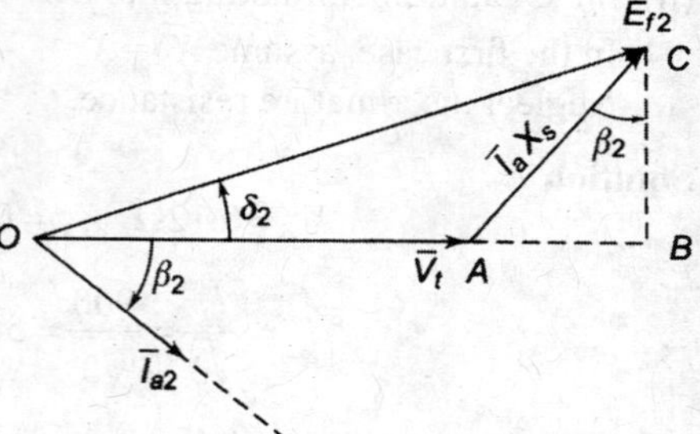

Figure E6.6

From Figure E6.6, we have

$$OC^2 = OB^2 + BC^2$$

i.e., $$E_{f2}^2 = (OA + AB)^2 + BC^2 = (V_t + AC\sin\beta_2)^2 + (AC\cos\beta_2)^2$$

$$= (V_t + I_{a2} X_s \sin\beta_2)^2 + (I_{a2} X_s \cos\beta_2)^2$$

i.e., $$8,378.97^2 = (6,350.8 + 5I_{a2}\sin\beta_2)^2 + (220\times5)^2$$

i.e., $$6,350.8+5I_{a2}\sin\beta_2=\sqrt{8,378.97^2-(220\times 5)^2}=8,306.45$$

i.e., $$I_{a2}\sin\beta_2=\frac{8,306.45-6,350.8}{5}=391.13$$

$$I_{a2}^2=(I_{a2}\cos\beta_2)^2+(I_{a2}\sin\beta_2)^2=220^2+391.13^2$$

i.e., $$I_{a2}=448.76\text{ A}$$

∴ $$\cos\beta_2=\frac{220}{I_{a2}}=\frac{220}{448.76}=0.4902\text{ (lagging)}$$ ■

Example 6.20 A three-phase, 11 kV, 1,500 kVA, 50 Hz, star-connected alternator delivers full-load at 0.8 power factor lagging. It has synchronous reactance of 6 Ω per phase and negligible armature resistance. The emf is increased by 25 per cent at this load with increase in excitation. The machine is connected to an infinite bus bar. Calculate the new values of current and power factor.

Solution

$$V_t=\frac{11\times 10^3}{\sqrt{3}}=6,350.8\text{ V},\quad X_s=6\ \Omega/\text{phase}, R_a=0, \cos\beta_1=0.8\ [\text{lagging}]$$

$$I_L=\frac{1,500\times 10^3}{\sqrt{3}\times 11\times 10^3}=78.73\text{ A}$$

∴ $$I_{a1}\text{ (per phase)}=I_L=78.73\text{ A}$$

Figure E6.7 shows the phasor diagram.

$$E_{f1}=\sqrt{(V_t\cos\beta_1)^2+(V_t\sin\beta_1+I_aX_s)^2}$$

$$=\sqrt{(6,350.8\times 0.8)^2+(6,350.8\times 0.6+78.73\times 6)^2}$$

$$=6,644.98\text{ V}$$

Since emf is increased by 25 per cent, the new emf is $E_{f2}=1.25\ E_{f1}=8,306.22$ V, which is greater than V_t. Therefore, the power factor is lagging in nature.

Again, at infinite bus $V_t=6,350.8$ V.

The power output remains unaltered because steam input does not change.

$$V_tI_{a2}\cos\beta_2=V_tI_{a1}\cos\beta_1$$

i.e., $$I_a\cos\beta_2=I_{a1}\cos\beta_1=78.73\times 0.8=62.98\text{ A}$$

$$OC^2=OB^2+BC^2$$

i.e., $$E_{f2}^2=(OA+AB)^2+BC^2=(V_t+AC\sin\beta_2)^2+(AC\cos\beta_2)^2$$

$$=(V_t+I_{a2}X_s\sin\beta_2)^2+(I_{a2}X_s\cos\beta_2)^2$$

i.e., $$8,306.22^2=(6,350.8+6I_{a2}\sin\beta_2)^2+(62.98\times 6)^2$$

i.e., $$6,350.8+6I_{a2}\sin\beta_2=\sqrt{8,306.22^2-(62.98\times 6)^2}=8,297.62$$

i.e., $I_{a2}\sin\beta_2 = \dfrac{8,297.62-6,350.8}{6} = 324.47$

i.e., $I_{a2}^2 = \left(I_{a2}\cos\beta_2\right)^2 + \left(I_{a2}\sin\beta_2\right)^2 = 62.98^2 + 324.47^2$

i.e., $I_{a2} = 330.53\text{ A}$

i.e., $\cos\beta_2 = \dfrac{62.98}{I_{a2}} = \dfrac{62.98}{330.52} = 0.190\text{ (lagging)}$

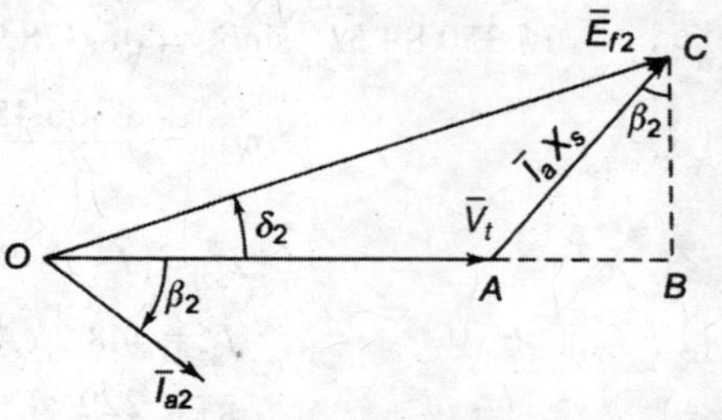

Figure E6.7 ■

Example 6.21 A three-phase, 4,000 kVA, 1,500 rpm, 50 Hz, star-connected alternator runs in parallel with other machines. Its synchronous reactance is 25 per cent. Calculate synchronizing power per unit mechanical angle of phase displacement for (i) no-load and (ii) full-load at 0.8 power factor lagging. Also, calculate the synchronizing torque if mechanical displacement is 0.6°. Neglect the armature resistance.

Solution

$$V_t\left(\text{per phase}\right) = \frac{V_L}{\sqrt{3}} = \frac{11,000}{\sqrt{3}} = 6,350.8\text{ V}$$

$$I_L = \frac{4,000\times10^3}{\sqrt{3}\times11\times10^3} = 209.94\text{ V}$$

$$I_a\text{(per phase)} = I_L = 209.94\text{ A}$$

$$X_s = X_{\text{sp.u.}} \times \frac{V_t}{I_a} = \frac{25}{100}\times\frac{6,350.8}{209.94} = 7.56\Omega,\quad R_a = 0\,\Omega$$

$$P = \frac{120f}{N_s} = \frac{120\times50}{1500} = 4 \text{ and } p = \frac{P}{2} = 2$$

(i) At no-load, $E_f = V_t = 6,350.8\text{ V}$ and $\delta = 0°$

$$P_{\text{syn}} = \left(\frac{3V_tE_f}{X_s}\cos\delta\right)\times\frac{p\pi}{180} = \frac{3\times6,350.8^2}{7.56}\times\cos0°\times\frac{2\times3.14}{180}$$

$$= 5,58,397.5\text{ W/mechanical degree}$$

$$T_{\text{syn}} = \frac{P_{\text{syn}}}{2\pi\,n_s} = \frac{5,58,397.5}{2\times3.14\times\dfrac{1,500}{60}} = 3,556.67\text{ N-m}$$

For mechanical displacement 0.6°,

$$T'_{\text{syn}} = 3,556.67\times0.6 = 2,134.6\text{ N-m}$$

(ii) Full-load, 0.8 power factor lagging:

$$E_f = \sqrt{\left(V_t\cos\beta + I_aR_a\right)^2 + \left(V_t\sin\beta + I_aX_s\right)^2}$$

$$= \sqrt{\left(6,350.8\times0.8+0\right)^2 + \left(6,350.8\times0.6+209.94\times7.56\right)^2} = 7,412.64\text{ V}$$

$$\tan(\delta+\beta)=\frac{V_t\sin\beta+I_aX_s}{V_t\cos\beta}=\frac{6{,}350.8\times0.6+209.94\times7.56}{6{,}350.8\times0.8}=1.0624=\tan46.73°$$

i.e., $\delta+\beta=46.73°$

i.e., $\delta=46.73°-\beta=46.73°-\cos^{-1}(0.8)=46.73°-36.87°=9.86°$

$$P_{\text{syn}}=\left(\frac{3V_tE_f}{X_s}\cos\delta\right)\times\frac{p\pi}{180}=\frac{3\times6{,}350.8\times7{,}412.64}{7.56}\times\cos9.86°\times\frac{2\times3.14}{180}$$

$$=6{,}42{,}133.28\ \text{W/mechanical degree}$$

$$T_{\text{syn}}=\frac{P_{\text{syn}}}{2\pi n_s}=\frac{6{,}42{,}133.28}{2\times3.14\times\frac{1{,}500}{60}}=4{,}090\ \text{N-m}$$

For mechanical displacement 0.6°,

$$T'_{\text{syn}}=4{,}090\times0.6=2{,}454\ \text{N-m}$$

■

Example 6.22 Two 3-phase Alternators A and B are operating in parallel. The respective ratings of A and B are 150 MW and 300 MW having drooping characteristic of 3 per cent and 4 per cent, respectively, from no-load to full-load. If the governors are operating at 50 Hz at no-load, find the sharing of the load of 450 MW by A and B. Also, calculate the system frequency at this load. Repeat this problem if the drop of both governors is 3 per cent.

Solution

Let P_A and P_B be the load taken by generators A and B, respectively, in MW.

$$P_A+P_B=450 \tag{1}$$

Frequency at no-load is f_0 = 50 Hz

For Alternator A

The drop in frequency from no-load to full-load = 3% of $f_0=\frac{3}{100}\times50=1.5$

For a load of 150 MW, the drop in frequency = 1.5 Hz

For a load of 1 MW, the drop in frequency $=\frac{1.5}{150}$

For a load of P_A MW, the drop in frequency $=\frac{1.5P_A}{150}$

Therefore, operating frequency of generator A is

$$f_A=50-\frac{1.5P_A}{150} \tag{2}$$

For Alternator B

Similarly, operating frequency of generator B is $f_B=50-\left(\frac{4\times50}{100}\right)\frac{P_B}{300}$

i.e.,

$$f_B=50-\frac{P_B}{150} \tag{3}$$

The generators must operate at the same frequency because they are operating in parallel. From Equations (2) and (3), we have

$$f_A = f_B$$

i.e.,

$$50 - \frac{1.5P_A}{150} = 50 - \frac{P_B}{150}$$

i.e.,

$$1.5P_A = P_B \quad (4)$$

From Equations (1) and (4), we have

$$1.5P_A + P_A = 450$$

i.e.,

$$P_A = \frac{450}{2.5} = 180$$

From Equation (1), we have

$$P_B = 55 - P_A = 450 - 180 = 270$$

Therefore, the load sharings by Alternator A and Alternator B are 180 MW and 270 MW, respectively.

From the above result, it is clear that Alternator A is overloaded whereas Alternator B is underloaded due to the difference in drop of the governors.

When the Alternators A and B have droop of 3 per cent, load shared by Alternators A and B will be 150 MW and 300 MW, respectively. Therefore, in this case, Alternators A and B will share the load as per their ratings. ■

Example 6.23 Two identical 3,000 kVA Alternators (A and B) are operating in parallel. The governor of the first machine is set such that the frequency drop for the first alternator is from 50 Hz to 48 Hz and that of Alternator B is from 50 Hz to 47.5 Hz. Determine the sharing of load of 5,000 kW by them. Also, calculate the maximum load at unity power factor which can be delivered without overloading any alternator.

Solution

(i) Full-load capacity of each machine at unity power factor = 3,000 × 1 = 3,000 kW = 3 MW

Let P_A and P_B be the load taken by generators A and B, respectively, in MW.

$$P_A + P_B = 5 \quad (1)$$

Frequency at no-load is $f_0 = 50$ Hz

For Alternator A

$$\text{The frequency drop in per unit rating} = \frac{50-48}{3} = \frac{2}{3}$$

$$\text{For a load of } P_A \text{ MW, the drop in frequency} = \frac{2P_A}{3}$$

Therefore, operating frequency of generator A is

$$f_A = 50 - \frac{2P_A}{3} \quad (2)$$

For Alternator B

The frequency drop in per unit rating $= \dfrac{50 - 47.5}{3} = \dfrac{2.5}{3}$

For a load of P_A MW, the drop in frequency $= \dfrac{2.5P_B}{3}$

Therefore, operating frequency of Generator A is

$$f_A = 50 - \frac{2.5P_B}{3} \tag{3}$$

The generators must operate at the same frequency because they are operating in parallel. From Equations (2) and (3), we have

$$f_A = f_B$$

i.e.,
$$50 - \frac{2P_A}{3} = 50 - \frac{2.5P_B}{3}$$

i.e.,
$$P_A = \frac{2.5}{2}P_B = \frac{5}{4}P_B \tag{4}$$

From Equations (1) and (4), we have

$$P_B + \frac{5}{4}P_B = 5$$

i.e.,
$$\frac{9}{4}P_B = 5$$

i.e.,
$$P_B = \frac{20}{9}\text{ MW}$$

From Equation (1), we have

$$P_A = 5 - P_B = 5 - \frac{20}{9} = \frac{25}{9}\text{ MW}$$

Loads shared by Alternator A and Alternator B are $\frac{25}{9}$ MW and $\frac{20}{9}$ MW, respectively. ■

(ii) The minimum value of frequency at which the system can operate with overloading either alternator is 48 Hz.

The full-load frequency of Alternator A is 48 Hz. The total load taken by it is 3,000 kW. For a drop in frequency of 50–48 Hz, i.e. 2 Hz, the load carried by Alternator B will be

$$= \frac{3{,}000}{2.5} \times 2 = 2{,}400\text{ kW}$$

Therefore, total load carried by Alternator A and Alternator B = 3,000 + 2,400 = 5,400 kW. ■

Example 6.24 Two three-phase Alternators A and B are operating in parallel. The ratings of Alternator A and Alternator B are 1,200 kW and 1,600 kW, respectively. The frequency load characteristic of Alternator A varies from 51.2 Hz to 49.6 Hz and that of Alternator B varies from 51.6 Hz to 49.4 Hz at full load. (i) Find the sharing of common load of 2,500 kW by the Alternator A and Alternator B.

(ii) Due to alteration of one machine, if the alternators share the load in the ratio of their ratings, find the common frequency at this condition.

(i) Let P_A and P_B be the load taken by Generators A and B, respectively, in MW.

$$P_A + P_B = 2{,}500 \tag{1}$$

For Alternator A

Here, $f_{0A} = 51.2$ Hz

Drop in frequency for a load of 1,200 kW = (51.2 − 49.6) Hz = 1.6 Hz

Drop in frequency for a load of P_A kW $= \dfrac{1.6P_A}{1{,}200}$ Hz

Actual frequency of operation for Alternator A is

$$f_A = 51.2 - \frac{1.6P_A}{1{,}200} \tag{2}$$

For Alternator B

Here, $f_{0B} = 51.6$ Hz

Drop in frequency for a load of 1,600 kW = (51.6 − 49.4) Hz = 2.2 Hz

Drop in frequency for a load of P_B kW $= \dfrac{2.2P_B}{1{,}600}$ Hz

Actual frequency of operation for Alternator A is

$$f_B = 51.6 - \frac{2.2P_B}{1{,}600} \tag{3}$$

The generators must operate at the same frequency because they are operating in parallel. From Equations (2) and (3), we have

$$f_A = f_B$$

i.e.,
$$51.2 - \frac{1.6P_A}{1{,}200} = 51.6 - \frac{2.2P_B}{1{,}600}$$

i.e.,
$$\frac{2.2P_B}{1{,}600} - \frac{1.6P_A}{1{,}200} = 0.4$$

i.e.,
$$\frac{2.2P_B}{16} - \frac{1.6P_A}{12} = 40$$

i.e.,
$$\frac{12}{16} \times 2.2P_B - 1.6P_A = 480$$

i.e.,
$$1.65P_B - 1.6P_A = 480 \tag{4}$$

From Equation (1), we have

$$P_A = 2{,}500 - P_B \tag{5}$$

Using Equation (5), from Equation (4) we have

$$1.65P_B - 1.6(2{,}500 - P_B) = 480$$

i.e.,
$$1.65P_B + 1.6P_B = 480 + 1.6 \times 2{,}500$$

i.e., $$3.25P_B = 4{,}480$$

i.e., $$P_B = \frac{4{,}480}{3.25} = 1{,}378.46$$

From Equation (5), we have

$$P_A = 2{,}500 - P_B = 2{,}500 - 1{,}378.46 = 1{,}121.54$$

Since $f = f_A = f_B$,

i.e., $$f = 51.2 - \frac{1.6P_A}{1{,}200} = 51.2 - \frac{1.6 \times 1{,}121.54}{1{,}200} = 51.2 - 1.495 = 49.705 \text{ Hz}$$ ■

(ii) Here, $\frac{P_A}{P_B} = \frac{1{,}200}{1{,}600} = \frac{3}{4}$

i.e., $$P_A = 0.75P_B \quad (6)$$

Using Equation (6), from Equation (1) we have

$$0.75P_B + P_B = 2{,}500$$

i.e., $$P_B = \frac{2{,}500}{1.75} = 1{,}428.57 \text{ kW}$$

∴ $$P_A = 2{,}500 - 1{,}428.57 = 1{,}071.43 \text{ kW}$$

Let f be the new common frequency of two alternators.

∴ $$\frac{1{,}071.43}{1{,}200} = \frac{51.2 - f}{51.2 - 49.6}$$

i.e., $$51.2 - f = 1.428$$

i.e., $$f = 51.2 - 1.428 = 49.772 \text{ Hz}$$

Let f_{20} be the new value of no-load frequency of Alternator B.

∴ $$\frac{1{,}378.46}{1{,}600} = \frac{f_{20} - 49.772}{51.6 - 49.4}$$

i.e., $$f_{20} - 49.772 = 1.895$$

i.e., $$f_{20} = 51.667 \text{ Hz}$$ ■

Example 6.25 The respective speed regulations of two 300 kW Alternators A and B which are running in parallel are 100 to 103 per cent and 100 to 104 per cent from full load to no load. Find the sharing of the load of 500 kW by Alternator A and Alternator B. Also, determine the load at which one Alternator ceases to supply any portion of the load.

Solution

Figure E6.8 shows the speed load characteristics of Alternator A and Alternator B. The point M represents the operating point of Alternator A and Alternator B when they are fully loaded. The total loads on Alternator A and Alternator B are 300 + 300 = 600 kW.

During sharing of a load of 700 kW, let the rise of speed of each machine be to some value y per cent of full-load speed. The operating point of Alternator A and Alternator B becomes N and P, respectively, shown in Figure E6.8.

Load on Alternator A = $P_A = OI = KN$, and load on Alternator B = $P_B = OJ = KP$

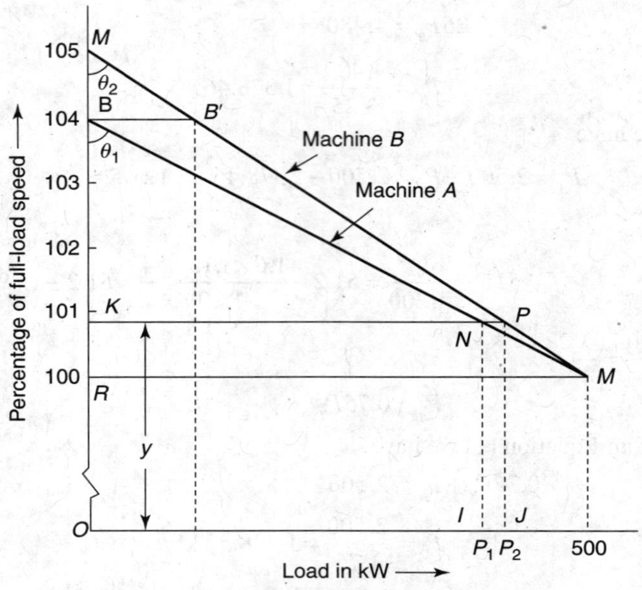

Figure E6.8

From ΔBKD, we have

$$\tan\theta_1 = \frac{KN}{BK} = \frac{OI}{OB-OK} = \frac{P_A}{103-y} \tag{1}$$

From ΔAKD, we have

$$\tan\theta_2 = \frac{KP}{AK} = \frac{OJ}{OA-OK} = \frac{P_B}{104-y} \tag{2}$$

From ΔBRM, we have

$$\tan\theta_1 = \frac{RM}{BR} = \frac{300}{3} \tag{3}$$

From ΔARM, we have

$$\tan\theta_2 = \frac{RM}{AR} = \frac{300}{4} \tag{4}$$

From Equations (1) and (3), we have

$$\frac{P_A}{103-y} = \frac{300}{3}$$

i.e.,

$$P_A = \frac{300}{3}(103-y) \tag{5}$$

From Equations (2) and (4), we have

$$\frac{P_B}{104-y} = \frac{300}{4}$$

i.e., $$P_B = \frac{300}{4}(104 - y) \tag{6}$$

Since Alternator A and Alternator B jointly share a load of 500 kW, we have

$$P_A + P_B = 500 \tag{7}$$

Adding Equations (5) and (6), we have

$$P_A + P_B = \frac{300}{3}(103 - y) + \frac{300}{4}(104 - y)$$

i.e., $$500 = 100(103 - y) + 75(104 - y)$$

i.e., $$175y = 17,600$$

i.e., $$y = 100.57$$

From Equation (5), we have

$$P_A = \frac{300}{3}(103 - 100.57) = 243 \text{ kW}$$

From Equation (6), we have

$$P_B = \frac{400}{3}(104 - 100.57) = 257 \text{ kW}$$

The alternator will cease to supply any load; that is, Alternator B will supply the load when the load is such that the line KNP moves up so that the point K comes to point B and this is represented by the line BB'. Alternator B in this case will supply BB'.

From $\Delta ABB'$, we have

$$\tan\theta_2 = \frac{BB'}{AB} = \frac{BB'}{104 - 103} = BB'$$

From ΔARM, we have

$$\tan\theta_2 = \frac{RM}{AR} = \frac{300}{104 - 100} = \frac{300}{4} = 75$$

When the load drops to 75 kW from 500 kW, the alternator will not supply any portion of the load.

Example 6.26 Two 40 MVA, 11kV, three-phase alternators operate in parallel. The settings of the governors are such that the rise in speed from full-load to no-load is 2.5 per cent in one alternator and 3.5 per cent in other alternator. The characteristics are straight lines in both cases. If each machine is fully loaded, then total load is 90 MW. What would be the load on each machine when the total load is 50 MW?

Solution

From Figure E6.9, we have

$$CE = 45 - AP = 45 - \frac{45}{3.5}x \quad \text{and} \quad ED = 45 - QB = 45 - \frac{45}{2.5}x$$

$\therefore$ $$CD = CE + ED = 45 - \frac{45}{3.5}x + 45 - \frac{45}{2.5}x$$

i.e., $$50 = 90 - \frac{45 \times 6}{3.5 \times 2.5}x$$

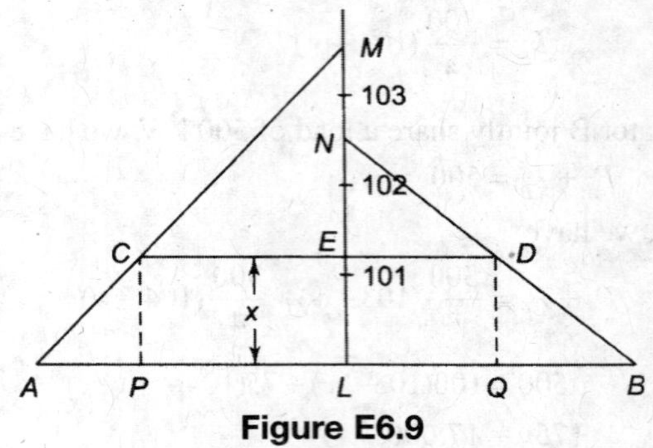

Figure E6.9

i.e., $$x = (90-50)\times\frac{3.5\times 2.5}{45\times 6} = 1.296$$

Load supplied by the first machine $= ED = 45 - \frac{45}{3.5}\times 1.296 = 28.337$ MW

Load supplied by the second machine $= CE = 45 - \frac{45}{2.5}\times 1.296 = 21.672$ MW ■

SIGNIFICANT POINTS

For Lagging pf: $$E_f = [(V_t\cos\beta + I_aR_a)^2 + (V_t\sin\beta + I_aX_s)^2]^{1/2}$$

For Unity pf: $$E_f = \left[\left(V_t\cos\beta + I_aR_a\right)^2 + \left(I_aX_s\right)^2\right]^{\frac{1}{2}}$$

For Leading pf: $$E_f = [(V_t\cos\beta + I_aR_a)^2 + (V_t\sin\beta - I_aX_s)^2]^{1/2}$$

$$\text{Percentage regulation} = \frac{E_f - V_t}{E_f}\times 100 = \frac{I_a\left(R_a\cos\beta - X_s\sin\beta\right)}{V_t}\times 100$$

Power Output:

$$P_{og} = \frac{V_tE_f}{Z_s}\cos(\theta-\delta) - \frac{V_t^2}{Z_s}\cos\theta = \frac{V_tE_f}{Z_s}\cos(\theta-\delta) - \frac{V_t^2R_a}{Z_s^2}$$

$$Q_{og} = \frac{V_tE_f}{Z_s}\sin(\theta-\delta) - \frac{V_t^2}{Z_s}\sin\theta = \frac{V_tE_f}{Z_s}\sin(\theta-\delta) - \frac{V_t^2X_s}{Z_s^2}$$

Maximum Power Output per Phase of the Generator

$$\delta = 90^\circ - \alpha = \theta$$

R_a Is Negligible Order:

$$Q = \frac{V_t}{X_s}[E_f\cos\delta - V_t]$$

(i) If $E_f \cos\delta > V_t$, the alternator is overexcited and it delivers reactive power to the infinite bus. Under this condition, the alternator is operating at lagging power factor.

(ii) If $E_f \cos\delta = V_t$, the alternator is normally excited and it neither delivers reactive power to infinite bus bar nor absorbs reactive power from the infinite bus bar. Under this condition, the alternator is operating at unity power factor.

(iii) If $E_f \cos\delta < V_t$, the alternator is underexcited and it absorbs reactive power from the infinite bus. Under this condition, the alternator is operating at leading power factor.

Input Power Equation

$$P_{ig} = \frac{E_f^2 R_a}{Z_s^2} - \frac{E_f V_t}{Z_s}\cos(\theta+\delta)$$

$$Q_{ig} = \frac{E_f^2 X_s}{Z_s^2} - \frac{E_f V_t}{Z_s}\sin(\theta+\delta)$$

Salient Pole:

For Lagging pf

$$\bar{E}_f = \bar{V}_t + \bar{I}_a R_a + jX_q \bar{I}_a + j(X_d - X_q)\bar{I}_d$$

$$\tan\gamma = \frac{V_t \sin\beta + I_a X_q}{V_t \cos\beta + I_a R_a}$$

For Leading pf

$$\tan\gamma = \frac{V_t \sin\beta - I_a X_q}{V_t \cos\beta + I_a R_a}$$

$$E_f = V_t \cos\delta - I_q R_a - I_d X_d$$

The total reactive power in VAR is given by

$$Q_{3-phase} = 3Q_{1-phase} = \frac{3V_t E_f}{X_d}\cos\delta - \frac{3V_t^2}{2X_d X_q}\left\{(X_d + X_q) - (X_d - X_q)\cos 2\delta\right\}$$

$$T_{exc} = \frac{3V_t E_f}{\omega_m X_d}\sin\delta \quad \text{and} \quad T_{rel} = \frac{3V_t^2}{2\omega_m}\left(\frac{1}{X_q} - \frac{1}{X_d}\right)\sin 2\delta$$

$$\left(Q_{3-phase}\right)_{max} = -\frac{3V_t^2}{X_q} - \frac{3E_f^2 X_q}{4X_d(X_d - X_q)}$$

$$P_{syn} = \frac{dP}{d\delta} = \frac{V_t E_f}{X_s}\cos\delta + V_t^2\left(\frac{1}{X_q} - \frac{1}{X_d}\right)\cos 2\delta$$

$$P_{syn} = \frac{3V_t E_f}{X_s}\cos\delta \text{ (cylindrical rotor)}$$

$$T_{\text{syn}} = \frac{3V_t E_f}{2\pi\, n_s X_s} \cos\delta$$

$$T = 9.11 n_s \times \sqrt{\frac{J}{E_f I_{sc} f}}$$

SHORT QUESTIONS AND ANSWERS

Q.1 What is an alternator?

Ans. An alternator is a synchronous machine, which converts mechanical energy into electrical energy and produces alternating emf.

Q.2 On what principle does it work?

Ans. The alternator works on the principle of Faraday's laws of electromagnetic induction, same as in the DC generator. Whenever a conductor links with a magnetic field, either the conductor is moving or the field is moving and emf is induced in the conductor.

Q.3 What are the different types of alternators? Which is in common use?

Ans. Alternators are of two types:

(i) rotating armature type and

(ii) rotating field and stationary armature type.

Rotating field-type is commonly used.

Q.4 What are the main parts of an alternator?

Ans. The main parts of an alternator are (i) rotor (salient-pole type and cylindrical type), (ii) stator (frame, core and armature conductors) and (iii) exciter.

Q.5 What do you mean by the salient-pole type rotor?

Ans. Salient-pole type rotor means a low and moderate speed rotor having large diameter and small axial length with projected (salient) poles coming out of the rotor frame. The outer surface of which almost follows the inner cylindrical surface of the stator core.

Q.6 Which type of alternator is suitable for coupling with steam turbines and why?

Ans. Smooth cylindrical non-salient-pole type alternator is suitable for coupling with steam turbine because of its high peripheral speed.

Q.7 Which type of alternator is suitable for hydroelectric plants and why?

Ans. Salient-pole type alternator is suitable for hydroelectric plants because of its low and moderate speed.

Q.8 In which alternator is the salient-type rotor used?

Ans. High-speed alternator uses the salient-type rotor.

Q.9 What type of alternator is used for the high-voltage generator?

Ans. Stationary armature-type alternator is used for the high-voltage generator.

Q.10 Why are the stator cores laminated?

Ans. The cores are laminated to minimise eddy current losses.

Q.11 What types of slots are provided in stator cores of alternators?

Ans. (i) Open-type slots and (ii) semi-closed type slots.

Q.12 What are the types of windings used in three-phase alternators?
Ans. (i) Single-layer windings and (ii) double-layer windings.

Q.13 What is meant by breadth factor?
Ans. The breadth factor or the distribution factor is the ratio of the vector sum to the arithmetic sum of the emf generated by the conductors. It is denoted by K_d and is equal to $K_d = \dfrac{\sin \frac{q\alpha}{2}}{q \sin \frac{\alpha}{2}}$, where q is numbers of slots/pole/phase and α is the angle between two adjacent slots.

Q.14 What is the equation of induced emf in an alternator?
Ans. Induced emf per phase, $E_{rms} = 4.44\, K_p\, K_d\, \phi f N$ V
where K_p is pitch factor, K_d is distribution factor, ϕ is flux per pole in Wb, ϕ is frequency in Hz and N is the number of turns per phase.

Q.15 What will be the effect on increasing the field excitation of an alternator?
Ans. The magnetic flux will be increased and accordingly the build-up voltage will be increased further.

Q.16 What is the relation between speed and frequency of an alternator?
Ans. $N_s = \dfrac{120f}{P}$

Q.17 Does changing the number of conductors have any effect on the frequency?
Ans. No, the frequency does not depend upon the number of conductors.

Q.18 How will you adjust the frequency of an alternator?
Ans. Frequency of an alternator can be adjusted by varying the speed of the prime mover (as f is proportional to N_s).

Q.19 What do you mean by hunting alternators?
Ans. Due to the fluctuation of power input to the prime mover, if the driving torque applied to the alternator becomes pulsating, the rotor during rotation will oscillate from its normal speed. This oscillation action is called hunting.

Q.20 What will be the effect of increasing the load on alternator?
Ans. The terminal voltage will drop at unity and lagging power factor. The drop will be comparatively more with lagging load power factor. The terminal voltage will rise with the leading load power factor.

Q.21 When an alternator is supplying a load alone and its excitation is increased, its driving torque remains constant. What will be the effect on power factor?
Ans. Its power factor will remain the same.

Q.22 On what factor does the power factor of an alternator depend?
Ans. Power factor depends on the load of an alternator.

Q.23 What will be the direction of armature reaction flux when a three-phase alternator supplies (i) purely resistive load, (ii) reactive load, (iii) purely inductive load and (iv) purely capacitive load?
Ans. (i) At 90° with the field axis
(ii) Along the load axis
(iii) Opposing the main magnetic field
(iv) Helping the main magnetic field

Q.24 What is called the armature reaction of an alternator?

Ans. The effect of the armature flux on the flux produced by the field ampere-turns is called the armature reaction. The armature flux will distort, oppose or help the field flux causing reduction or increase in the air-gap flux depending upon the power factor of the load.

Q.25 What is synchronous reactance and how does it vary?

Ans. The net reactance of the armature circuit due to the combined effect of leakage reactance and the fictitious reactance due to armature reaction effect is known as synchronous reactance, since it is caused in synchronous speed. It is denoted by the symbol $X_s = \sqrt{Z_s^2 - R_a^2}$, where Z_s is the synchronous impedance. Its unit is ohm.

The synchronous reactance varies inversely with the degree of saturation of the magnetic circuit.

Q.26 What do you mean by synchronous impedance and impedance drop?

Ans. Synchronous impedance is the total resultant opposition of the effective armature resistance and the synchronous reactance of the armature winding and is denoted by $Z_s = \sqrt{R_a^2 + X_s^2}$. Voltage drop due to synchronous impedance is called impedance drop.

Q.27 What is the formula for calculating the generated emf on load condition?

Ans. Generated emf $E_f = \sqrt{(V_t \cos\beta + I_a R_a)^2 + (V_t \sin\beta \pm I_a X_s)^2}$, where V_t is full-load terminal voltage, β is angular displacement between V_t and I_a, $I_a R_a$ is effective armature resistance drop and $I_a X_s$ is synchronous reactance drop.

Q.28 Why is '±' sign used for synchronous reactance drop?

Ans. The '±' sign is used since synchronous reactance drop may be additive or subtractive; '±' sign is used for unity and lagging power factor, whereas '−' is used for leading power factor.

Q.29 What is alternator voltage regulation? Give the equation of percentage regulation of an alternator?

Ans. The change of voltage from full load to no load when the load is thrown off under the same excitation and speed expressed in terms of full-load voltage is called the alternator voltage regulation. Suppose V_t is full load voltage and E_f is no-load voltage.

So, the percentage regulation $= \dfrac{E_f - V_t}{V_t} \times 100$. Being a ratio, it has no unit.

Q.30 What are the losses in alternators?

Ans. The losses in alternators are the following:

(i) Copper losses:

(a) Armature copper losses $(I_a^2 R_a)$

(b) Field copper losses $(I_f^2 R_f)$

(ii) Iron losses:

(a) Hysteresis loss in armature core

(b) Eddy current loss in armature core and pole shoes

(iii) Friction and windage losses:

(a) Friction in bearing and slip ring

(b) Windage and ventilation

SUPPLEMENTARY PROBLEMS

1. A 1,000 kVA, 66 kV, three-phase star-connected alternator has a resistance of 1 Ω and a synchronous reactance of 10 Ω/phase. The machine delivers a load of 600 kW at 0.6 power factor lagging. What is the emf generated in the armature?

   [*Ans.* 38,180 V]

2. A three-phase, star-connected alternator works at 1,600 kVA, 13,500 V. The armature effective resistance and synchronous reactance are 1.5 Ω and 30 Ω/phase, respectively. Calculate the percentage regulation for a load of 1,280 kW at power factors of (i) 0.8 lead, (ii) unity and (iii) 0.8 lag.

   [*Ans.* (i) −11.98%, (ii) 3.22%, (iii) 18.59%]

3. A three-phase, star-connected, 600 kVA, alternator has rated terminal voltage of 3,300 V (line) having stator resistance and reactance per phase of 0.37 Ω and 4.3 Ω respectively. Determine the regulation for a load having power factor of 0.1 lagging.

   [*Ans.* 23.67%]

4. A three-phase star-connected alternator is rated at 1,600 kVA, 13,500 V. The armature effective resistance and synchronous reactance per phase are 1.5 Ω and 30 Ω respectively. Determine the percentage regulation for a load of 1,280 kW at power factors of (i) 0.8 leading, (ii) unity and (iii) at 0.8 lagging.

   [*Ans.* (i) −11.9%, (ii) 3.2%, (iii) 18.6%]

5. A three-phase, star-connected synchronous generator is rated at 1.5 MVA, 11 kV. The armature effective resistance and synchronous reactance are 1.2 Ω and 25 Ω respectively, per phase. Calculate the percentage voltage regulation for a load of 1.4375 MVA at (i) 0.8 power factor lagging and (ii) 0.8 power factor leading. Also, find out the power factor at which the regulation becomes zero.

   [*Ans.* 21.6%, −13.1%, 0.981 (leading)]

6. A three-phase, star-connected, 1,000 kVA, alternator has rated current of 52.5 A. The AC resistance of the winding per phase is 0.45 Ω. The test results are shown below:

   OC test: field current = 12.5 A; voltage between lines = 422 V

   SC test: field current = 12.5 A; line current = 52.5 A

   Determine the full-load voltage regulation of the alternator at (i) 0.8 power factor lagging and (ii) 0.8 power factor leading.

   [*Ans.* (i) 2.628%, (ii) −1.94%]

7. (i) A three-phase, star-connected, 5 kVA, 50 Hz, four-pole alternator has the following test data at rated speed:

Exciting current (in ampere)	1.0	1.5	2.0	2.5	3.0	3.5	4.0	4.5	5.0	6.08
Per phase OC voltage	75	140	173	202	224	240	250	257	260	263

(ii)

Exciting current (A)	1.0	2.0	3.0
SC line current (A)	3.6	7.2	10.8

(iii) Armature resistance per phase is 2.0 Ω.

(a) Draw the OC and SC characteristics on a graph paper and determine the unsaturated value of synchronous reactance per phase in ohm and in per unit.

(b) Also, determine the percentage voltage regulation at rated load at 0.8 power factor leading using synchronous impedance method under unsaturated condition.

[*Ans.* (a) 38.84 Ω, 1.213 p.u., (b) 5.844%]

8. The following test results are obtained on a three-phase 6.6 kV alternator:

OC voltage (kV)	3.1	4.9	6.6	7.5	8.3
Field current (A)	16	25	37.5	50	70

A field current of 20 A is found necessary to circulate full-load current on short circuit of the armature. Calculate the full-load regulation at 0.8 power factor lagging by (i) ampere-turn method and (ii) synchronous impedance method. Neglect the resistance and leakage reactance.

[*Ans.* (i) 15.15%, (ii) 38.7%]

9. Determine the regulation by the zero-power factor method of a 5,000 kVA, 6.6 kV, three-phase, 50 Hz, star-connected alternator at full load, unity power factor having the following test data:

Field current (A)	32	50	75	100	140
DC voltage (kV)	3.1	4.9	6.6	7.5	8.3
Full-load current zero pf test (line voltage in V)	0	1,850	4,250	5,800	7,000

[*Ans.* 3.33%]

10. A 3.5 MVA, slow-speed, three-phase synchronous generator rated at 6.6 kV has 32 poles. Its direct- and quadrature-axis synchronous reactances as measured by the slip test are 9.6 Ω and 6 Ω respectively. Neglecting the armature resistance, determine the regulation and excitation emf needed to maintain 6.6 kV at the terminals when supplying a load of 2.5 MW at 0.8 power factor lagging. What maximum power can generate supply at the rated terminal voltage if the field becomes open-circuited?

[*Ans.* 9,956 V (line to line), 50.85%, 1.361 MW]

11. A 10 kVA, 380 V, 50 Hz, three-phase, star-connected salient-pole alternator has direct-axis and quadrature-axis reactances of 12 Ω and 8 Ω respectively. The armature has resistance of 1 Ω per phase. The generator delivers rated load at 0.8 power factor lagging with the terminal voltage being maintained at rated value. If the load angle is 16.15°, determine (i) direct-axis and quadrature-axis components of armature current and (ii) excitation voltage of the generator.

[*Ans.* (i) 12.14 A, 9.14 A (ii) 633 V]

12. Find the synchronous torque for 2° mechanical displacement in a 4,000 kVA, 50 Hz, three-phase synchronous generator running at 1,500 rpm with a generated emf of 10,000 V. The alternator has synchronous impedance of 20 per cent.

[*Ans.* 8,893.5 N-m]

13. A 20,000 kVA, 10 kV, three-phase, four-pole, 50 Hz, synchronous generator is connected to an infinite bus. The short-circuit current is seven times the normal full-load current. The moment of inertia of the rotating system is 42,000 kg m^2. Find the natural period of oscillation.

[*Ans.* 0.964 second]

14. Two exactly similar 3,000 kVA three-phase synchronous generators work in parallel. The governor of the first machine is such that the frequency drops uniformly from 50 Hz on no-load to 48 Hz on full load. The corresponding uniform speed drop of the second machine is from 50 Hz to 47.5 Hz. Determine the following:

    (a) How will the two machines share a load of 4,500 kW?

    (b) What is the maximum load at unity power factor that can be delivered without overloading either of the two machines?

[*Ans.* (a) 2,500 kW, 2,000 kW (b) 5,400 kW]

MULTIPLE-CHOICE QUESTIONS AND ANSWERS

1. In an alternator, the frequency per revolution is equal to
   (a) number of poles (b) number of pairs of poles
   (c) number of armature conductors (d) forty
2. An alternator on open circuit generates 320 V at 50 Hz when the field current is 3.2 A. Neglecting the salutation, when the frequency is 25 Hz and the field current is 2 A, the open-circuit emf is
   (a) 50 V (b) 100 V (c) 200 V (d) none of these
3. The synchronous impedance method to find voltage regulation of an alternator is not an accurate method because the value of synchronous impedance found is
   (a) always more than its value under normal conditions
   (b) always less than its value under normal conditions
   (c) only the synchronous reactance
   (d) only the armature resistance
4. The experimental data required for Potier method for finding regulation of alternators are
   (a) no-load curve, short-circuit test values
   (b) zero power factor curve, short-circuit test value
   (c) no-load curve, zero power factor curve
   (d) none of the above
5. In a three-phase star-connected alternator, a field current of 40 A produces full-load current of 200 A on short circuit and 1,160 V on open circuit. If the resistance of the alternator is 0.5 Ω the value of synchronous reactance is
   (a) 5–7 Ω (b) 8.73 Ω (c) 3.31 Ω (d) 9 Ω
6. If the prime mover of an alternator supplying load to an infinite bus is suddenly shut down, it will
   (a) stop
   (b) continue to run as an alternator
   (c) continue to run as a synchronous motor running in the reverse direction
   (d) continue to run as a synchronous motor running in the same direction
7. Armature reactance in synchronous generator at rated voltage and zero power factor (lead) is
   (a) magnetizing (b) demagnetizing
   (c) both magnetizing and cross-magnetizing (d) cross-magnetizing
8. Drop in terminal voltage of an alternator due to armature reaction is countered by
   (a) damper winding (b) effect of saliency
   (c) increased prime mover input (d) automatic voltage regulation

9. In a synchronous generator, delivering lagging power factor load,
   (a) The excitation emf leads terminal voltage by the power angle.
   (b) Excitation emf lags terminal voltage by the power angle.
   (c) Excitation emf leads the terminal voltage by the power factor angle.
   (d) None of these.
10. A synchronous motor is operating at constant load while its excitation is adjusted to give UPF current. If the excitation is now increased, the power factor will
   (a) lead (b) lag (c) remain at unity (d) become zero
11. In a synchronous generator,
   (a) The open-circuit voltage leads the terminal voltage by an angle known as power angle.
   (b) The open-circuit voltage leads the terminal voltage by an angle known as overlap angle.
   (c) The open-circuit voltage lags the terminal voltage by an angle known as power angle.
   (d) The open-circuit voltage lags the terminal voltage by an angle known as overlap angle.
12. Find the method which gives the voltage regulation more than the actual value:
   (a) synchronous reactance method (b) emf method
   (c) zero power factor method (d) all of these
13. The phase of an alternator is RYB. If the direction of its field current is reversed, the phase sequence will be
   (a) RYB (b) RBY (c) YRB (d) YBR
14. The armature reaction of an alternator is safely determined by
   (a) amount of current drawn from the alternator
   (b) power factor of the load
   (c) speed of the prime mover driving the alternator
   (d) none of these
15. The stator frame in an alternator serves
   (a) as a return path for the flux (b) to hold the armature stampings
   (c) to protect the whole machine (d) none of these
16. The armature reaction of an alternator will be cross-magnetizing if the power factor of the load is
   (a) less than unity (b) zero leading (c) more than unity (d) unity
17. Power factor of an alternator driven by constant prime mover input can be changed by changing its
   (a) field excitation (b) phase sequence (c) load (d) speed
18. Alternators are rated in terms of
   (a) kW (b) kVA (c) kWh (d) kVAr
19. In case of two alternators running in parallel and perfectly synchronized, the synchronizing power is
   (a) zero (b) positive (c) negative (d) infinite
20. If an alternator is operating at leading power, it voltage regulation is
   (a) more than zero (b) equal to zero (c) negative (d) none of these
21. For the same power rating, the higher voltage alternator is
   (a) larger in size (b) cheaper (c) costlier (d) smaller in size
22. Modern turbo alternators have maximum speed of
   (a) 1,000 rpm (b) 3,000 rpm (c) 4,000 rpm (d) 2,000 rpm

23. In parallel operation of alternators, the synchronizing power is maximum if armatures have
    (a) reactances less than resistances
    (b) reactances equal to resistances
    (c) reactances greater than resistances
    (d) none of these
24. To run two alternators in parallel, the black lamp test is performed to ensure proper
    (a) frequency matching
    (b) phase difference matching
    (c) voltage matching
    (d) phase sequence matching
25. An alternator is said to be underexcited if it operates at
    (a) unity power factor
    (b) leading power factor
    (c) lagging power factor
    (d) none of these

Answers

1 (b)	2 (b)	3 (a)	4 (c)	5 (c)	6 (d)
7 (c)	8 (d)	9 (a)	10 (b)	11 (a)	12 (a)
13 (a)	14 (b)	15 (a)	16 (d)	17 (a)	18 (b)
19 (a)	20 (b), (c)	21 (d)	22 (b)	23 (d)	24 (a)
25 (b)					

7 Synchronous Motors

The details of a synchronous generator or alternator have already been introduced in Chapter 6. In synchronous generators, the field is excited by direct current and no supply is applied to armature. The rotor is rotated by the prime mover at a constant speed known as synchronous speed. Hence emf is induced in the armature. In contrast, in synchronous motor the field is excited by a direct current and stator is connected to a three-phase alternating voltage. Since the motor is rotated at a synchronous speed, it is called synchronous motor. Its speed is constant irrespective of the load. This chapter introduces the various aspects of synchronous motors.

7.1 PRINCIPLES OF OPERATION

In Figure 7.1, the coil *ab* carries current where conductor *a* carries current towards the observer and conductor *b* away from the observer. Therefore, a torque is developed by motor action that drives the conductor from left to right. It will reverse its direction for the next half cycle and the torque then acts from right to left for alternating current that causes the net torque to be zero over any given number of complete cycles. Hence, no continuous motion can result, and thus a synchronous motor is standstill. Therefore, the synchronous motor does not develop any starting torque.

If the conductor *a* is taken under south (*S*) pole, the resulting torque will still be from left to right and continuous motion will result. For continuous operation of a synchronous motor, a given conductor must move from one pole to the other in each half cycle. With the help of a rotating magnetic field, it is very easy to explain the operation of the synchronous motor.

If a three-phase balanced voltage is applied to a three-phase balanced winding, a rotating magnetic field is produced, which rotates at a synchronous speed with respect to the winding. In Figure 7.2, the south pole and the north pole of the rotor (produced by the DC field) lock with the north pole and the south pole of the rotating magnetic field, which causes the rotor to rotate at a synchronous speed with the stator rotating magnetic field. For constant frequency, the synchronous motor must operate at a constant (synchronous) speed.

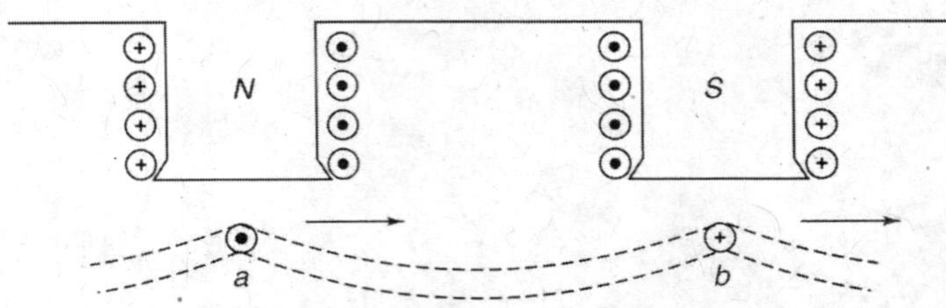

Figure 7.1 Conductor Current in a Synchronous Motor

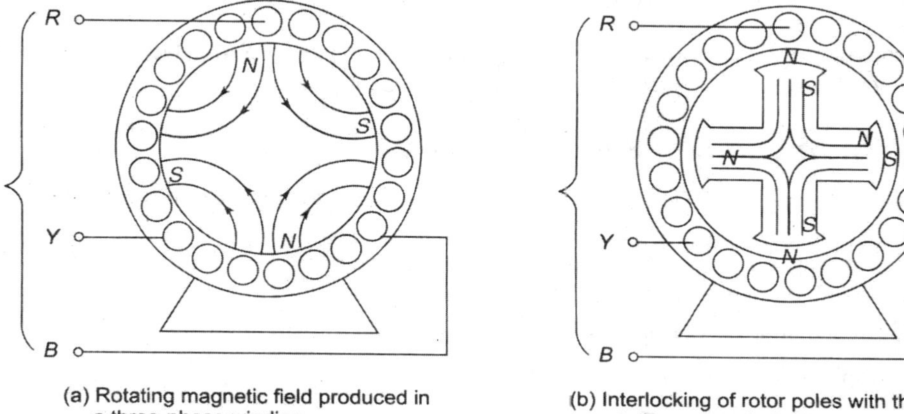

(a) Rotating magnetic field produced in a three-phase winding

(b) Interlocking of rotor poles with the rotating magnetic field

Figure 7.2 Rotating Magnetic Field and Rotor Field Poles Interlocking with Each Other

7.2 ARMATURE REACTION IN SYNCHRONOUS MOTORS

The effect of armature reaction for a leading current is shown in Figure 7.3 where the armature coil has been depicted as moving with the armature from left to right. When the coil axis is in position *Y*, shown by a dotted line, the coil sides are under the centres of the poles and the induced emf is maximum.

The induced emf will also be maximum at the instant when the terminal voltage is substantially 180° from it. Its direction is indicated in the dotted coil. If the current leads this terminal voltage by 90°, it will reach its maximum value one-fourth cycle ahead of the voltage, or at a time when the axis of the coils is in position *X*. For this position of the coil axis, the ampere-turns of the coil act in a direct opposite to those of the north pole. Therefore, the effect of armature reaction due to the leading current in the synchronous motor weakens the field.

Similarly, the effect of armature reaction due to lagging current strengthens the field and a unity power factor current cross-magnetizes the main field. The effect of armature reaction in synchronous motors is just the opposite to the effect in a synchronous generator or an alternator.

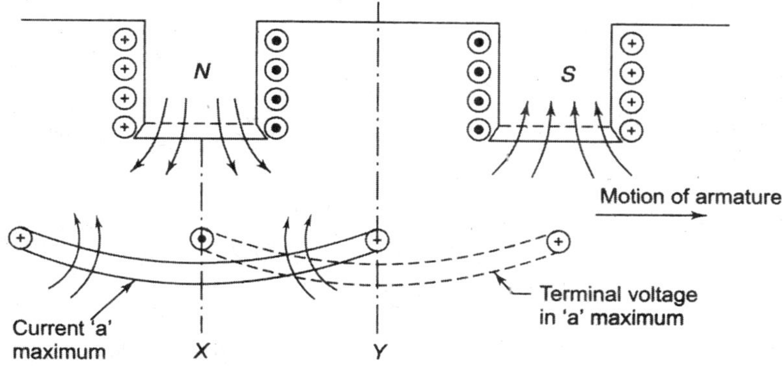

Figure 7.3 Demagnetizing Effect of Leading Current on the Field of Synchronous Motors

7.3 PHASOR DIAGRAM OF SYNCHRONOUS MOTOR

Due to different natures of the load, that is, lagging, leading and unity power factor of the load, three different phasor diagrams of a synchronous motor are obtained.

7.3.1 Effect of Loading Synchronous Motor

Figure 7.4 shows the circuit model of a synchronous motor having cylindrical rotor. In Figure 7.4

$\bar{V}_t$ is the applied voltage per phase,

$\bar{E}_f$ is the excitation voltage per phase,

$\bar{I}_a$ is the armature current per phase,

$\bar{R}_a$ is the armature resistance per phase, and

X_s is the synchronous reactance per phase.

The emf E_f is similar to the back emf of a DC motor. The voltage equation of a synchronous motor is given by

$$\bar{V}_t = \bar{E}_f + \bar{I}_a \bar{Z}_s \tag{7.1}$$

7.3.2 Lagging Power Factor Load

The phasor diagram of a synchronous motor under lagging power factor load is shown in Figure 7.5, where I_a lags behind V_t by an angle β.

In Figure 7.5,

$$\bar{Z}_s = R_a + jX_s = |\dot{Z}_s| \angle\theta\ \Omega$$

where
$$\theta = \tan^{-1}\left(\frac{X_s}{R_a}\right)$$

$E_R = I_a Z_s$ is the voltage drop across synchronous impedance per phase, and I_a lags behind E_R by an angle θ.

Let V_t be the phase voltage applied and E_f be the back emf induced per phase.

The angle between V_t and E_R is $\theta - \beta$.

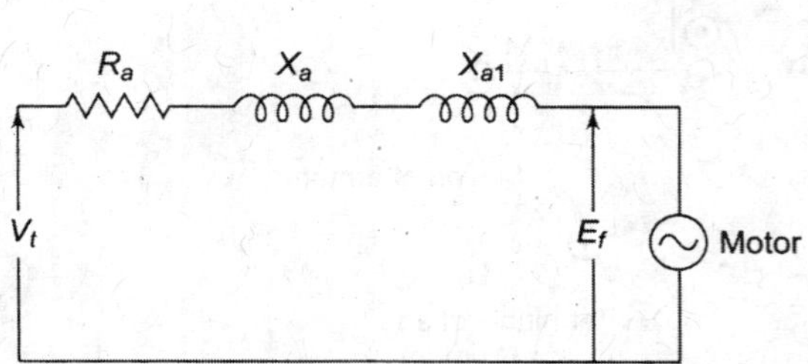

Figure 7.4 Equivalent Circuit Model of a Synchronous Motor

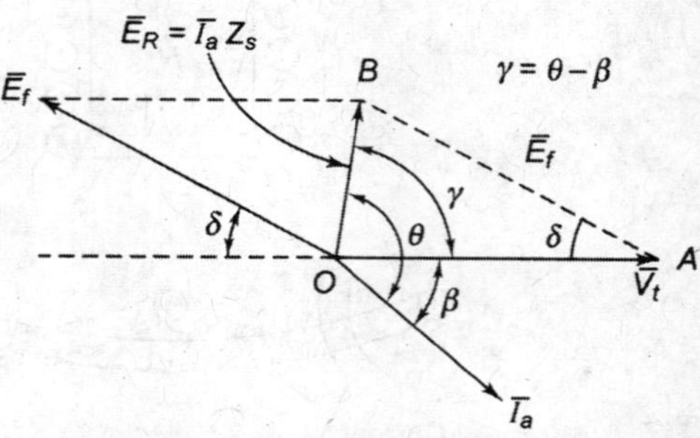

Figure 7.5 Phasor Diagram at Lagging Power Factor Load

From Figure 7.5, using cosine rule in ΔOAB, we have

$$E_f^2 = V_t^2 + E_R^2 - 2V_t E_R \cos(\theta - \beta)$$

Using sine rule in ΔOAB, we have

$$\frac{E_f}{\sin(\theta - \beta)} = \frac{E_R}{\sin \delta} \tag{7.2}$$

i.e.,

$$\sin \delta = \frac{E_R \sin(\theta - \beta)}{E_f} \tag{7.3}$$

E_f has already been calculated. Therefore, sin δ can be calculated using Equation (7.3).

7.3.3 Leading Power Factor Load

The phasor diagram of a synchronous motor under leading power factor load is shown in Figure 7.6, where I_a leads V_t by an angle β.
In Figure 7.6, I_a lags behind E_R by an angle θ.
Let V_t be the phase voltage applied and E_f be the back emf induced per phase.
The angle between V_t and E_R is $\theta + \beta$.
From Figure 7.6, using cosine rule in ΔOAB, we have

$$E_f^2 = V_t^2 + E_R^2 - 2V_t E_R \cos(\theta + \beta) \tag{7.4}$$

Using sine rule in ΔOAB, we have

$$\frac{E_f}{\sin(\theta + \beta)} = \frac{E_R}{\sin \delta}$$

i.e.,

$$\sin \delta = \frac{E_R \sin(\theta + \beta)}{E_f} \tag{7.5}$$

E_f has already been calculated. Therefore, sin δ can be calculated using Equation (7.5).

7.3.4 Unity Power Factor Load

The phasor diagram of a synchronous motor for unity power factor load is shown in Figure 7.7, where I_a is in phase with V_t.

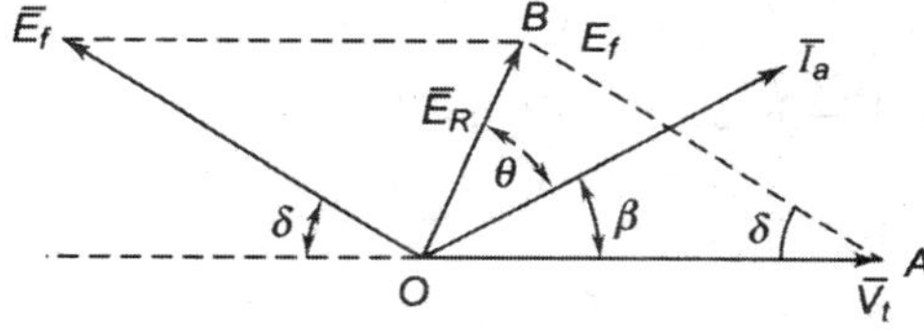

Figure 7.6 Phasor Diagram at Leading Power Factor Load

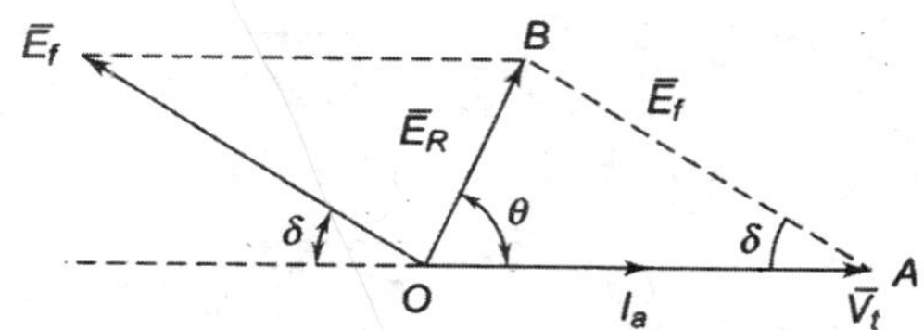

Figure 7.7 Phasor Diagram at Unity Power Factor

In Figure 7.7,

$E_R = I_a Z_s$ is the voltage drop across synchronous impedance per phase.

Let V_t be the phase voltage applied and E_f be the back emf induced per phase.

The angle between V_t and E_R is θ.

From Figure 7.7, using cosine rule in ΔOAB, we have

$$E_f^2 = V_t^2 + E_R^2 - 2V_t E_R \cos\theta \tag{7.6}$$

Using sine rule in ΔOAB, we have

$$\frac{E_f}{\sin\theta} = \frac{E_R}{\sin\delta}$$

i.e.,

$$\sin\delta = \frac{E_R \sin\theta}{E_f} \tag{7.7}$$

E_f has already been calculated. Therefore, sin δ can be calculated using Equation (7.7).

With an increase in load, the phase difference between the applied voltage V_t and the excitation E_f increases. As the angle δ increases with increasing load on the synchronous motor, it is called the *load angle.* The angle δ is also known as the *power angle, coupling angle, torque angle or angle of retardation.*

Example 7.1 A three-phase, 400 V, star-connected synchronous motor has an effective armature resistance and a synchronous reactance per phase of 0.2 Ω and 2 Ω, respectively. It takes 20 A to drive a certain load. Calculate the excitation emf induced in the motor if it works with (i) 0.8 power factor lag, (ii) 0.8 power factor lead and (iii) unity power factor conditions.

Solution

$$V_t = \frac{400}{\sqrt{3}} = 231 \text{ V}$$

R_a/phase = 0.2 Ω and X_s/phase = 2 Ω.

Therefore, Z_s / phase = $\sqrt{R_a^2 + X_s^2} = 2.01\,\Omega$ and $\theta = \tan^{-1}\left(\frac{X_s}{R_a}\right) = 84.3°$

$$E_R = I_a Z_s = 20 \times 2.01 = 40.2 \text{ V}$$

(i) For 0.8 power factor lag:

$\cos\beta = 0.8, \therefore \beta = 36.87°$

$$E_f = \sqrt{V_t^2 + E_R^2 - 2V_t E_R \cos(\theta - \beta)}$$

$$= \sqrt{231^2 + 40.2^2 - 2 \times 231 \times 40.2 \times \cos(84.3° - 36.67°)} = 206.06 \text{ V}$$

(ii) For 0.8 power factor lead

$$E_f = \sqrt{V_t^2 + E_R^2 - 2V_t E_R \cos(\theta + \beta)}$$

$$= \sqrt{231^2 + 40.2^2 - 2 \times 231 \times 40.2 \times \cos(84.3° + 36.67°)} = 254.04 \text{ V}$$

(iii) For unity power factors

$$\cos\beta = 1 \therefore \beta = 0°$$

$$E_f = \sqrt{V_t^2 + E_R^2 - 2V_tE_R\cos\theta}$$

$$= \sqrt{231^2 + 40.2^2 - 2\times 231\times 40.2\times \cos 84.3°} = 230.5\text{ V}$$

■

Example 7.2 A three-phase, 600 V, star-connected synchronous motor has an effective armature resistance and a synchronous reactance per phase of 0.4 Ω and 3.6 Ω, respectively. Calculate the induced emf per phase if the motor works on full load delivering 36 kW. The full-load efficiency is 87 per cent having a power factor of 0.8 leading. Also calculate the load angle.

Solution

$$V_t = \frac{600}{\sqrt{3}} = 346.41\text{ V}$$

$P_{out} = 36$ kW = 36,000 W on full load.
Efficiency (η) = 0.87, R_a/phase = 0.4 Ω and X_s/phase = 3.6 Ω.

$$Z_s = \sqrt{R_a^2 + X_s^2} = \sqrt{0.4^2 + 3.6^2} = 3.622\,\Omega \quad \text{and} \quad \theta = \tan^{-1}\left(\frac{3.6}{0.4}\right) = 83.66°$$

$$P_{in} = \frac{P_{out}}{\eta} = \frac{36,000}{0.87} = 41,379.31\text{ W}$$

$$I_L = \frac{P_{in}}{\sqrt{3}V_L\cos\phi} = \frac{41,379.31}{\sqrt{3}\times 600\times 0.8} = 49.77\text{ A}$$

$I_a = I_L = 49.77$ A because motor is star-connected.

Here, $\cos\beta = 0.8$ (leading), $\therefore \beta = 36.87°$

$$E_R = I_aZ_s = 49.77\times 3.622 = 180.27\text{ V}$$

For leading power factor, we have

$$E_f = \sqrt{V_t^2 + E_R^2 - 2V_tE_R\cos(\theta+\beta)}$$

$$= \sqrt{346.41^2 + 180.27^2 - 2\times 346.41\times 180.27\times \cos(83.66° + 36.87°)}$$

$$= 464.7\text{ V}$$

The generated emf per phase is 464.7 V.

Again, $$\frac{E_f}{\sin(\theta+\beta)} = \frac{E_R}{\sin\delta}$$

i.e., $$\sin\delta = \frac{E_R}{E_f}\times\sin(\theta+\beta) = \frac{180.27}{464.7}\times\sin(83.66° + 36.87°) = 0.334$$

$$\delta = 19.52°$$

The load angle is 19.52°.

■

7.4 OPERATION AT A CONSTANT LOAD WITH VARIABLE EXCITATION

The phasor diagram of a synchronous motor at lagging power factor, unity power factor and leading power factor has already been shown. If the excitation is kept constant, the load current changes with the variation of load. In contrast, if the load is kept constant and the field excitation is varied, the synchronous motor reacts by changing its power factor of operation, which is the most interesting feature of the synchronous motor. Let us consider a synchronous motor operating under a constant load having the corresponding load angle δ.

Let the field excitation be adjusted so that $E_f = V_t$. This excitation is called normal excitation of the motor. The motor is drawing current I_a from the supply and the power input is P_{in}. The power factor of the motor is lagging in nature, as shown in Figure 7.8(a). With change in excitation there is hardly any change in losses. Therefore, power input remains the same for constant load demanding the same power output.

The input power(P_{in}) is expressed by

$$P_{in} = \sqrt{3}V_L I_L \cos\beta = 3V_{ph} I_{ph} \cos\beta$$

The applied voltage of the motor remains constant most of the time. Thus, for constant output, $I_a \cos\beta$ will remain constant whatever be the value of cos β. Therefore, the motor adjusts the value of cosβ, to keep $I_a \cos\beta$ constant. During variable excitation, the synchronous motor reacts by changing its power factor to give a constant output power.

Figures 7.8 (b), 7.8(c) and 7.8 (d) show cases for underexcitation ($E_f < V_t$), overexcitation ($E_f > V_t$) and critical excitation ($E_f = V_t$), respectively. From Figure 7.8, it is clear that I_a varies when excitation is varied from a low to a high value. For critical excitation, the current drawn by the motor is minimum compared to other excitations. Table 7.1 shows the details.

Table 7.1 Comparison of Various Excitations

Type of Excitation	Nature of Power Factors	Comparison of E_f and V_t
Normal excitation	Lagging	$E_f = V_t$
Critical excitation	Unity power factor	$E_f = V_t$
Underexcitation	Lagging power factor	$E_f < V_t$
Overexcitation	Leading power factor	$E_f > V_t$

Example 7.3 A three-phase, 400 V, star-connected synchronous motor drives a constant load and draws a line current of 30 A, having an effective armature resistance and a synchronous reactance per phase of 0.2 Ω and 2 Ω, respectively. Find the power factor at which the motor will operate when the field current is adjusted to give the line values of generated emf from (i) 500 V to (ii) 300 V.

Solution

$$V_t = \frac{400}{\sqrt{3}} = 231 \text{ V}$$

R_a/phase $= 0.2\ \Omega$ and X_s/phase $= 2\ \Omega$.

Therefore, Z_s / phase $= \sqrt{R_a^2 + X_s^2} = 2.01\Omega$ and $\theta = \tan^{-1}\left(\frac{X_s}{R_a}\right) = 84.3°$

$$E_R = I_a Z_s = 30 \times 2.01 = 60.3 \text{ V}$$

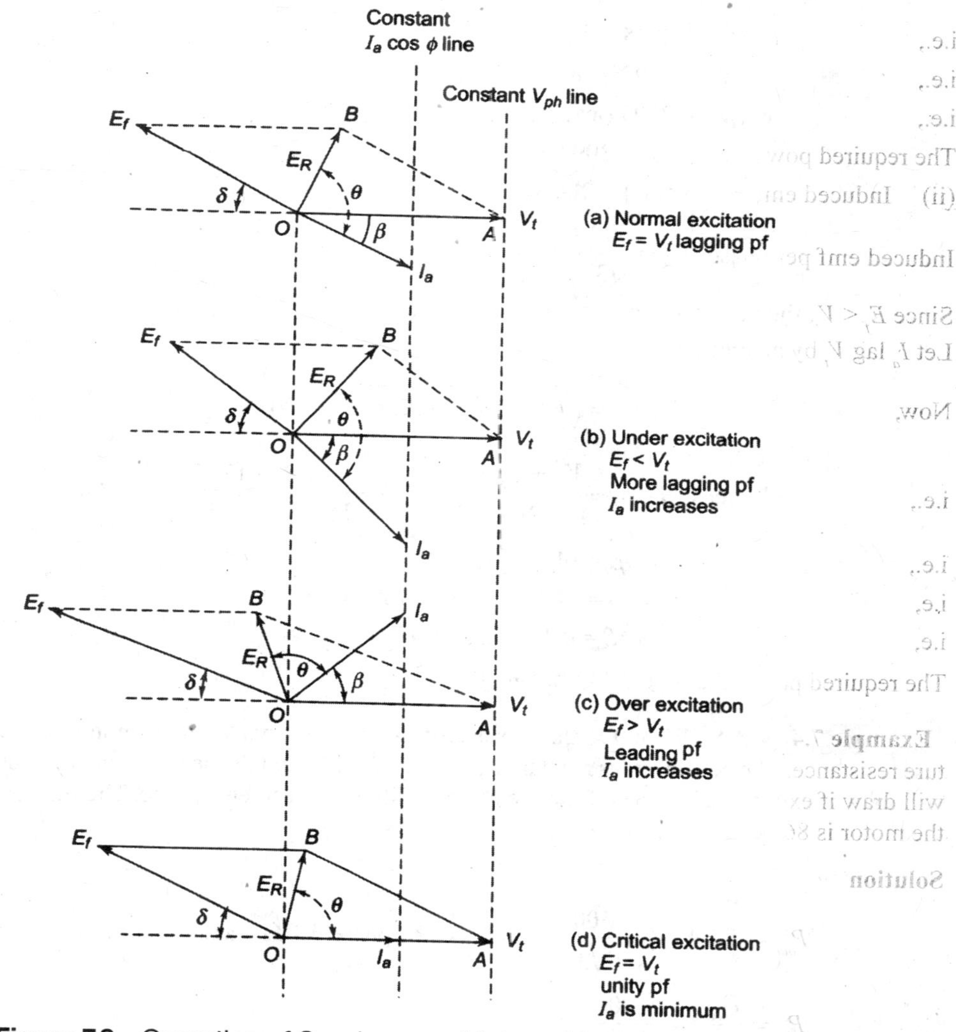

Figure 7.8 Operation of Synchronous Motor at Variable Excitation

(i) Induced emf (line value) = 500 V

Induced emf per phase $\left(E_f\right) = \dfrac{500}{\sqrt{3}} = 288.67$ V

Since $E_f > V_t$, the motor operates at leading power factor.
Let I_a lead V_t by an angle β.

Now,
$$E_f = \sqrt{V_t^2 + E_R^2 - 2V_t E_R \cos\left(\theta + \beta\right)}$$

i.e.,
$$\cos\ (\theta + \beta) = \frac{V_t^2 + E_R^2 - E_f^2}{2V_t E_R} = \frac{231^2 + 60.3^2 - 288.67^2}{2 \times 231 \times 60.3} = -0.9452$$

i.e., $\theta + \beta = 160.95°$

i.e., $\beta = 160.95° - 84.3° = 76.65°$

i.e., $\cos\beta = 0.2309$ (leading)

The required power factor is 0.2309 (leading)

(ii) Induced emf (line value) = 300 V

Induced emf per phase $(E_f) = \dfrac{300}{\sqrt{3}} = 173.20$ V

Since $E_f < V_t$, the motor operates at lagging power factor.
Let I_a lag V_t by an angle β.

Now, $$E_f = \sqrt{V_t^2 + E_R^2 - 2V_tE_R\cos(\theta - \beta)}$$

i.e., $$\cos(\theta - b) = \frac{V_t^2 + E_R^2 - E_f^2}{2V_tE_R} = \frac{231^2 + 60.3^2 - 173.20^2}{2 \times 231 \times 60.3} = 0.969$$

i.e., $\theta - \beta = 14.3°$

i.e, $\beta = 84.3° - 14.3° = 70°$

i.e, $\cos\beta = 0.342$ (lagging)

The required power factor is 0.342 (lagging) ■

Example 7.4 A 8 kW, 500 V, three-phase, star-connected synchronous motor has negligible armature resistance. The synchronous reactance/phase is 12 Ω. Calculate the minimum current that the motor will draw if excitation is varied. Find the corresponding induced emf/phase. The full-load efficiency of the motor is 86 per cent.

Solution

$$P_{out} = 8\text{ kW}, V_t = \frac{500}{\sqrt{3}} = 288.67\text{ V}, R_a/\text{phase} = 0\,\Omega \quad \text{and} \quad X_s/\text{phase} = 12\,\Omega.$$

$$P_{in} = \frac{P_{out}}{\eta} = \frac{8{,}000}{0.86} = 9{,}302.32\text{ W}$$

At unity power factor, armature current becomes minimum.

$\therefore$ $\cos\beta = 1$ or, $\beta = 0°$

$$I_L = I_a = \frac{P_{in}}{\sqrt{3}V_L\cos\beta} = \frac{9{,}302.32}{\sqrt{3} \times 500 \times 1} = 10.74\text{ A}$$

$$E_R = I_aZ_s = 10.74 \times 12 = 128.88\text{ A}$$

$$E_R = \sqrt{V_t^2 + E_R^2 - 2V_tE_R\cos(\theta - \beta)} = \sqrt{V_t^2 + E_R^2} \quad [\text{since } \beta = 0° \quad \text{and} \quad \theta = 90°]$$

$$E_f = \sqrt{288.67^2 + 128.88^2} = 316.13\text{ V}$$

The corresponding induced emf is 316.13 V. ■

7.5 *V* CURVES AND INVERTED *V* CURVES

From Figure 7.8, we have concluded that there is variation of I_a with change of excitation. I_a becomes minimum for unity power factor as depicted in Figure 7.9. Excitation of the field can be changed by changing the rheostat in the field circuit.

The plot of armature current (I_a) with the field current (I_f) takes the shape of English alphabet *V*. That is why the family of curves for various loads are called *V* curves, shown in Figure 7.10(a). In contrast, if the power factor (cosβ) is plotted against the field current, the graph looks like an inverted *V*. The family of curves obtained by plotting power factor against field current for various load conditions is known as inverted *V* curves of synchronous motor, shown in Figure 7.10(b).

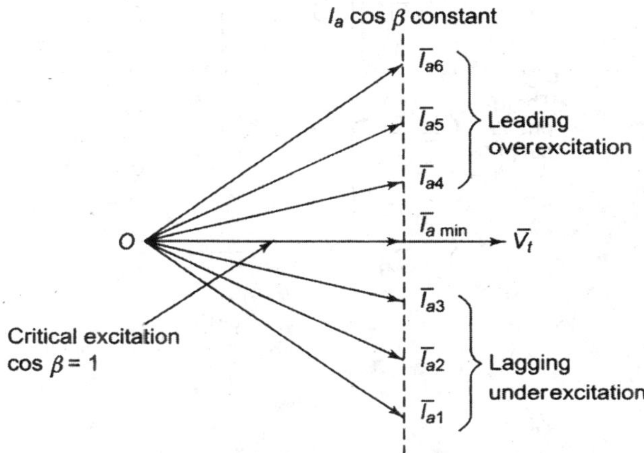

Figure 7.9 Variation of Armature Current with Field Excitation

Figure 7.11 shows the experimental set-up to obtain *V* curves. A three-phase supply is given to stator, as shown in Figure 7.11. The two-wattmeter method is used to measure the input power. Ammeter reading gives the line current, whereas voltmeter reading gives the line voltage. A rheostat is used in the field circuit to adjust the excitation to operate the motor under variable excitation.

7.6 COMPLEX POWER INPUT OF SYNCHRONOUS MOTOR

The phasor diagram of a synchronous motor at lagging power factor is shown in Figure 7.5, whereas Figure 7.4 shows its circuit diagram. At lagging power factor, E_f lags the applied voltage by an angle δ. Let us take V_t as the reference phasor.

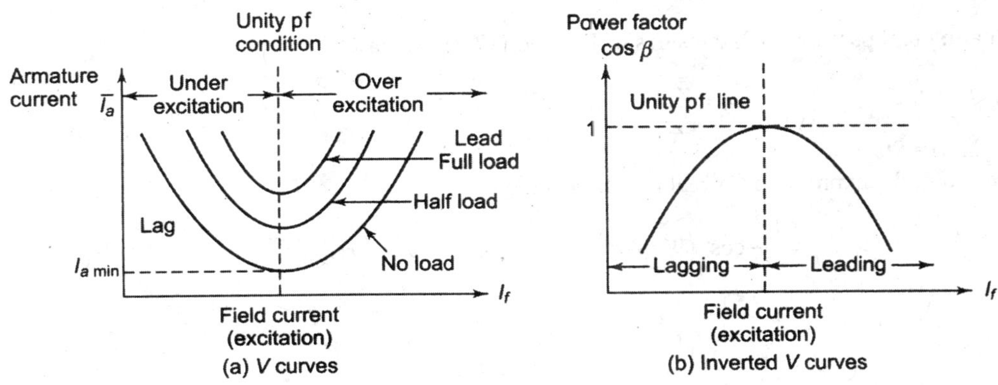

Figure 7.10 *V* and Inverted *V* Curves

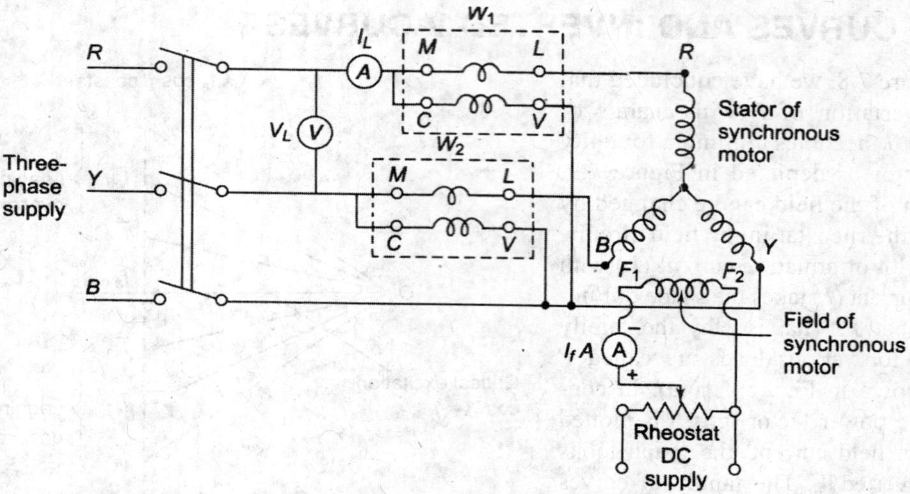

Figure 7.11 Experimental Set-up for *V* Curves

$$\bar{V}_t = V_t\angle 0°,\ \bar{E}_f = E_f\angle -\delta \text{ and } \bar{Z}_s = Z_s\angle\theta$$

From Figure 7.4,

$$\bar{I}_a^* = \left(\frac{\bar{V}_t - \bar{E}_f}{\bar{Z}_s}\right)^* = \frac{V_t\angle 0° - E_f\angle\delta}{Z_s\angle -\theta} = \frac{V_t}{Z_s}\angle\theta - \frac{E_f}{Z_s}\angle(\delta+\theta) \tag{7.8}$$

The complex input power per phase to the synchronous motor is given by

$$S_{in} = P_{in} + jQ_{in} = \bar{V}_t\bar{I}_a^* = V_t\angle 0°\left[\frac{V_t}{Z_s}\angle\theta - \frac{E_f}{Z_s}\angle(\delta+\theta)\right]$$

$$= \frac{V_t^2}{Z_s}\cos\,\theta + j\frac{V_t^2}{Z_s}\sin\,\theta - \left[\frac{V_tE_f}{Z_s}\cos(\delta+\theta) + j\frac{V_tE_f}{Z_s}\sin(\delta+\theta)\right] \tag{7.9}$$

Equating real parts on both the sides of Equation (7.9), we have

$$P_{in} = \frac{V_t^2}{Z_s}\cos\,\theta - \frac{V_tE_f}{Z_s}\cos(\delta+\theta) \tag{7.10}$$

Again, $\theta = 90° - \alpha$

Therefore, the expression for real power becomes

$$P_{in} = \frac{V_t^2}{Z_s}\cos\,(90° - \alpha) - \frac{V_tE_f}{Z_s}\cos(\delta + 90° - \alpha)$$

$$= \frac{V_t^2}{Z_s}\sin\alpha + \frac{V_tE_f}{Z_s}\sin(\delta - \alpha) \tag{7.11}$$

Equating imaginary parts on both the sides of Equation (7.9), we have

$$Q_{in} = \frac{V_t^2}{Z_s}\sin\,\theta - \frac{V_t E_f}{Z_s}\sin\left(\delta+\theta\right) \tag{7.12}$$

Again, $\theta = 90° - \alpha$

Therefore, the expression for reactive power becomes

$$Q_{in} = \frac{V_t^2}{Z_s}\sin\,(90° - \alpha) - \frac{V_t E_f}{Z_s}\sin(\delta + 90° - \alpha)$$

$$= \frac{V_t^2}{Z_s}\cos\alpha - \frac{V_t E_f}{Z_s}\cos\left(\delta - \alpha\right) \tag{7.13}$$

Example 7.5 A three-phase, 3.3 kV, 50 Hz, star-connected synchronous motor has a synchronous impedance of (0.3 + *j*4.9) Ω/phase. Calculate the line current and the power factor for an induced emf of 4 kV and an input power of 900 kW at rated voltage.

Solution

Here, $V_t = \dfrac{3.3\times10^3}{\sqrt{3}} = 1{,}905.25\text{ V},\ E_f = \dfrac{4\times10^3}{\sqrt{3}} = 2{,}309.4\text{ V},\ Z_s = \sqrt{0.3^2+4.9^2} = 4.91$

and $\theta = \tan^{-1}\left(\dfrac{4.9}{0.3}\right) = 86.49°$

$$P_{in}/\text{phase} = 900/3 = 300\text{ kW.}$$

$$300\times10^3 = \frac{V_t^2}{Z_s}\cos\theta - \frac{E_f V_t}{Z_s}\cos\left(\theta+\delta\right)$$

$$= \frac{1{,}905.25^2}{4.91}\times\cos 86.49° - \frac{2{,}309.4\times1{,}905.25}{4.91}\times\cos(86.49°+\delta)$$

$$\cos(86.49°+\delta) = \frac{\left(\dfrac{1{,}905.25^2}{4.91}\times\cos 86.49° - 300\times10^3\right)\times4.91}{2{,}309.4\times1{,}905.25} = -0.2842$$

$$86.49° + \delta = 106.51° \qquad \therefore \delta = 106.51° - 86.49° = 20.02°$$

$$E_R = \sqrt{V_t^2 + E_f^2 - 2V_t E_f\cos\delta}$$

$$= \sqrt{1{,}905.25^2 + 2{,}309.4^2 - 2\times1{,}905.25\times2{,}309.4\times\cos 20.02°} = 833.72\text{ V}$$

$$I_a = \frac{E_R}{Z_s} = \frac{833.72}{4.91} = 169.8\text{ A}$$

$$\therefore \qquad \cos\beta = \frac{900\times10^3}{\sqrt{3}\times3.3\times10^3\times169.8} = 0.927$$

Since, $E_f > V_t$, power factor is 0.927 (leading). ■

7.7 COMPLEX POWER OUTPUT OF A SYNCHRONOUS MOTOR

The complex output power per phase to the synchronous motor is given by

$$S_{om} = P_{om} + jQ_{om} = \bar{E}_f \bar{I}_a^* = \left(E_f \angle -\delta\right)\left[\frac{V_t}{Z_s}\angle\theta - \frac{E_f}{Z_s}\angle(\delta+\theta)\right]$$

$$= \frac{V_t E_f}{Z_s}\cos(\theta-\delta) + j\frac{V_t E_f}{Z_s}\sin(\theta-\delta) - \frac{E_f^2}{Z_s}\cos\theta - j\frac{E_f^2}{Z_s}\sin\theta \quad (7.14)$$

Equating real parts on both the sides of Equation (7.14), we have

$$P_{om} = \frac{V_t E_f}{Z_s}\cos(\delta-\theta) - \frac{E_f^2}{Z_s}\cos\theta \quad (7.15)$$

Again, $\theta = 90° - \alpha$
Therefore, the expression for real power becomes

$$P_{om} = \frac{V_t E_f}{Z_s}\cos\{90° - (\alpha+\delta)\} - \frac{E_f^2}{Z_s}\cos(90° - \alpha) \quad (7.16)$$

$$= \frac{V_t E_f}{Z_s}\sin(\delta+\alpha) - \frac{E_f^2}{Z_s}\sin\alpha$$

Equating imaginary parts on both the sides of Equation (7.14), we have

$$Q_{om} = \frac{V_t E_f}{Z_s}\sin(\theta-\delta) - \frac{E_f^2}{Z_s}\sin\theta \quad (7.17)$$

Again, $\theta = 90° - \alpha$
Therefore, the expression for reactive power becomes

$$Q_{om} = \frac{V_t E_f}{Z_s}\sin\{90° - (\alpha+\delta)\} - \frac{E_f^2}{Z_s}\sin(90° - \alpha)$$

$$= \frac{V_t E_f}{Z_s}\cos\left(\delta+\alpha\right) - \frac{E_f^2}{Z_s}\cos\alpha \quad (7.18)$$

The shaft power (or the net output power) $P_{sh} = P_{om}$ − rotational losses

P_{om} is called the mechanical power or the gross power developed. Rotational losses include friction, windage and core losses.

7.8 MAXIMUM OUTPUT POWER

For maximum output power, the following relations must hold good:

$$\frac{dP_{om}}{d\delta} = 0 \text{ and } \frac{d^2 P_{om}}{d\delta^2} < 0$$

Now,
$$\frac{dP_{om}}{d\delta} = \frac{d}{d\delta}\left\{\frac{V_t E_f}{Z_s}\sin(\delta+\alpha) - \frac{E_f^2}{Z_s}\sin\alpha\right\} = \frac{V_t E_f}{Z_s}\cos(\delta+\alpha)$$

$\therefore$
$$\frac{V_t E_f}{Z_s}\cos(\delta+\alpha) = 0$$

i.e.,
$$\delta + \alpha = 90°$$

i.e.,
$$\delta = 90° - \alpha = \theta \tag{7.19}$$

Therefore, maximum power output occurs at $\delta = \theta$, which is the limit of steady-state stability. The maximum output power of the motor is given by

$$(P_{om})_{max} = \frac{V_t E_f}{Z_s} - \frac{E_f^2}{Z_s^2} R_a \tag{7.20}$$

Example 7.6 A 50 Hz, four-pole, three-phase, star-connected synchronous motor has a synchronous reactance of 12 Ω/phase and negligible armature resistance. The excitation is such as to give an open-circuit voltage of 13.2 kV. The motor is connected to a 11.5 kV, 50 Hz supply. What maximum load can the motor supply before losing synchronism? What is the corresponding motor torque, line current and power factor?

Solution

Supply voltage per phase $V_t = \dfrac{11.5 \times 1{,}000}{\sqrt{3}} \approx 6{,}639.53$ V

Induced emf/phase $E_f = \dfrac{13.2 \times 1{,}000}{\sqrt{3}} = 7{,}621$ V

Since armature resistance is negligible, impedance angle is $\theta = 90°$.
For development of maximum power, $\delta = \theta = 90°$
Impedance drop

$$E_R = \sqrt{V_t^2 + E_f^2 - 2V_t E_f \cos\delta}$$
$$= \sqrt{6{,}639.53^2 + 7{,}621^2 - 2 \times 6{,}639.53 \times 7{,}621 \times \cos 90°}$$
$$= 10{,}107.57 \text{ V}$$

Since motor is star connected, line current = phase current.

$$I_a = \frac{E_R}{Z_s} = \frac{10{,}107.57}{12} = 842.30 \text{ A}$$

The maximum load delivered by the motor is as follows:

$$(P_{om})_{max} = 3 \times \left[\frac{E_f V_t}{Z_s} - \frac{E_f^2}{Z_s}\cos\theta\right] = 3 \times \left[\frac{6{,}639.53 \times 7{,}621}{12} - \frac{7{,}621^2}{12} \times \cos 90°\right] = 12.65 \text{ MW}$$

Since armature resistance is negligible, power input to motor = 12.65 MW

$\therefore$
$$\cos\beta = \frac{P_{in}}{\sqrt{3} V_L I_a} = \frac{12.65 \times 10^6}{\sqrt{3} \times 11{,}500 \times 842.30} = 0.754$$

$$\text{Synchronous speed } (N_s) = \frac{120f}{P} = \frac{120 \times 50}{4} = 1{,}500 \text{ rpm}$$

Developed torque corresponding to the maximum power is as follows:

$$(T_{mech})_{max} = \frac{(P_{om})_{max}}{2\pi \dfrac{N_s}{60}} = \frac{12.65 \times 10^6}{2 \times 3.14 \times \dfrac{1{,}500}{60}} = 80{,}573.25 \text{ N-m}$$ ■

7.9 POWER OUTPUT WHEN ARMATURE RESISTANCE IS NEGLIGIBLE

Neglecting armature resistance, Equation (7.15) can be expressed as follows:

$$P_{om} = \frac{V_t E_f}{X_s} \sin\delta \tag{7.21}$$

$$\therefore \quad P_{om} = P_m \sin\delta \tag{7.22}$$

where $P_m = \dfrac{V_t E_f}{X_s}$ is the maximum power developed by the motor.

∴ Gross torque developed by the motor is as follows:

$$T_g = \frac{P_{om}}{\omega_s} = \frac{P_{om}}{2\pi\, n_s} \tag{7.23}$$

where n_s is the synchronous speed of the motor in rps.

∴ Shaft torque of the motor is given by

$$T_{sh} = \frac{P_{sh}}{\omega_s} = \frac{P_{sh}}{2\pi\, n_s} \tag{7.24}$$

where $P_{sh} = P_{om}$ − windage and friction losses

7.10 INPUT REACTIVE POWER WHEN ARMATURE RESISTANCE IS NEGLIGIBLE

$$Q_{in} = \frac{V_t^2}{X_s} - \frac{V_t E_f}{X_s} \cos\delta = \frac{V_t}{X_s}\left(V_t - E_f \cos\delta\right) \tag{7.25}$$

(i) If $|E_f| \cos \delta > |V_t|$, motor is overexcited. It delivers reactive power to the infinite bus. Under this condition motor is operating at leading power factor.

(ii) If $|E_f| \cos \delta = |V_t|$, motor is normally excited. It neither delivers to nor absorbs reactive power from the infinite bus. Under this condition motor is operating at unity power factor.

(iii) If $|E_f| \cos \delta < |V_t|$, motor is underexcited. It absorbs reactive power from the infinite bus. Under this condition motor is operating at lagging power factor.

7.11 MOTOR CHARACTERISTICS, PERFORMANCE AND CIRCLE DIAGRAM OF A SYNCHRONOUS MOTOR

The study of characteristics, performance and circle diagram of a synchronous motor is of utmost importance. These are discussed below.

7.11.1 Current Locus for Constant Power Input

The current locus for constant power input (P_{in}) is shown in Figure 7.12.

For a given power input (P_{in}) of the synchronous motor, the input power per phase $= \frac{P_{in}}{3}$ W

If V_t be the applied voltage/phase, the power component of the current is as follows:

$$I_a \cos \beta = \frac{P_{in}}{3} \times \frac{1}{V_t} \text{ A} \tag{7.26}$$

A vertical line is at first erected, representing the axis for the voltage V_t. A horizontal line is drawn at a point P such that $OP = P_{in}/3V_t$. The current locus varies on the straight line for various powers. For less excitation, it takes a lagging current (the current vector lying on the right side of the voltage vector V_t), whereas it takes a leading current (the current vector lying on the left side of voltage vector V_t) for overexcitation. Similarly, the current locus is another horizontal line drawn at a distance of $I_a \cos\beta$ from the origin, corresponding to the new value of input power P_{in}.

7.11.2 Current Locus for Constant Power Developed (P_{om})

$$P_{om} = V_t I_a \cos \beta - I_a^2 R_a$$

i.e.,

$$I_a^2 - \left(\frac{V_t}{R_a}\right) I_a \cos \beta = -\frac{P_{om}}{R_a}$$

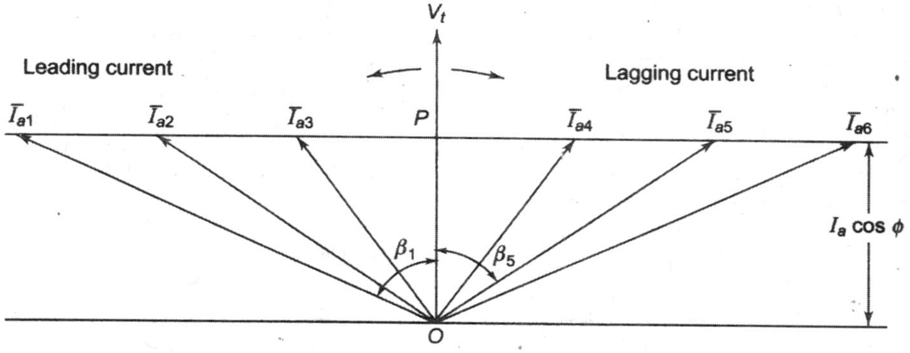

Figure 7.12 Current Locus for Constant Power Input

i.e.,
$$I_a^2 + \left(\frac{V_t}{2R_a}\right)^2 - 2\left(\frac{V_t}{2R_a}\right) I_a \cos\beta = \left(\frac{V_t}{2R_a}\right)^2 - \frac{P_{om}}{R_a} \tag{7.27}$$

In Figure 7.13, $\frac{V_t}{2R_a}$ and I_a are taken as two sides of a triangle, shown by OC and OK, respectively.

The third side is $\sqrt{\left(\frac{V_t}{2R_a}\right)^2 - \frac{P_{om}}{R_a}}$. When P_{om} remains constant, both $\frac{V_t}{2R_a}$ and $\sqrt{\left(\frac{V_t}{2R_a}\right)^2 - \frac{P_{om}}{R_a}}$ are constant. The locus of the current phasor I_a (shown by OK in Figure 7.13), as β changes, is a circle. The radius of the circle is $\sqrt{\left(\frac{V_t}{2R_a}\right)^2 - \frac{P_{om}}{R_a}}$ and the coordinate of the centre is $\left\{\frac{V_t}{2R_a}, 0\right\}$.

From the current locus, it is possible to infer that the power developed increases, the radius CK goes on decreasing, and hence the current locus becomes a circle of smaller diameter. For the development of maximum P_{om}, the radius of the circle must be zero.

$$\sqrt{\left(\frac{V_t}{2R_a}\right)^2 - \frac{(P_{om})_{\max}}{R_a}} = 0$$

i.e.,
$$\frac{(P_{om})_{\max}}{R_a} = \frac{V_t^2}{4R_a^2}$$

i.e.,
$$(P_{om})_{\max} = \frac{V_t^2}{4R_a^2} \tag{7.28}$$

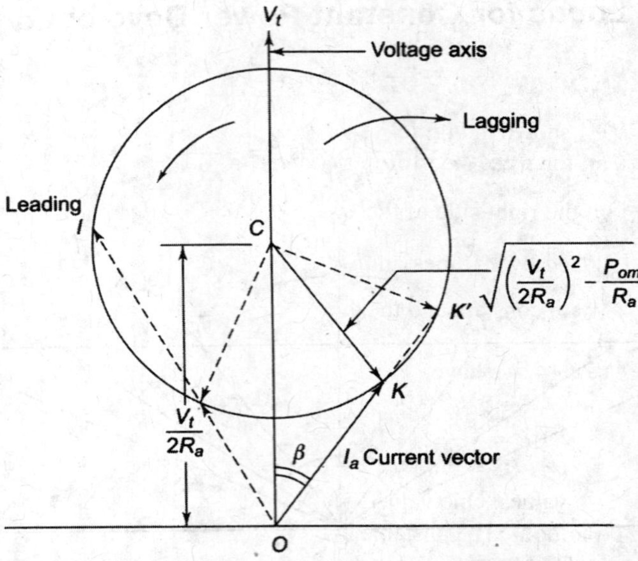

Figure 7.13 Current Locus for Constant Power Developed (P_{om})

When the mechanical power developed is maximum, the current vector $\overrightarrow{OK}$ lies on the voltage vector ($\overrightarrow{OC}$) having a magnitude of $\frac{V_t}{2R_a}$.

Hence, the power input/phase $= V_t \times \frac{V_t}{2R_a} = \frac{V_t^2}{2R_a}$

The armature Cu loss/phase $= \left(\frac{V_t}{2R_a}\right)^2 \times R_a = \frac{V_t^2}{4R_a}$

Power developed $P_{om} = P_i -$ armature copper loss $= \frac{V_t^2}{2R_a} - \frac{V_t^2}{4R_a} = \frac{V_t^2}{4R_a}$

The above equation checks with Equation (7.28). The efficiency at maximum output becomes 50 per cent. Losses become half of the input and rise of temperature would be above the maximum permissible temperature limit. Therefore, $\left(P_{om}\right)_{\max} = \frac{V_t^2}{4R_a}$ is not possible to achieve in practice.

The dotted lines, which are to the left of the voltage vector V_t, show the details for a leading current. Rearranging Equation (7.27), we have

$$I_a^2 - 2\left(\frac{V_t}{2R_a}\right) I_a \cos\beta + \frac{P_{om}}{R_a} = 0$$

i.e.,

$$I_a = \frac{V_t}{2R_a}\cos\beta \pm \sqrt{\left(\frac{V_t}{2R_a}\cos\beta\right)^2 - \frac{P_{om}}{R_a}} \tag{7.29}$$

Armature current has two values. One gives stable operating conditions, whereas the other gives unstable operating conditions.

7.11.3 Current Locus for Constant Excitation (E_f)

The sides of the triangle 1 shown in Figure 7.14 represent the voltage vector V_t, impedance drop I_aZ_s and the induced emf E_f. If all the three sides of this triangle are divided by Z_s, we get triangle 2 with sides $\frac{V_t}{Z_s}$, I_a and $\frac{E_f}{Z_s}$ shown on the right side of the vector V_t in Figure 7.14, which is nothing but triangle 1, with the magnitude reduced by Z_s times and rotated back by an angle of θ. If the excitation remains constant, $\frac{V_t}{Z_s}$ and $\frac{E_f}{Z_s}$ remain constant and the current locus (I_a) is a circle drawn for a radius E_f/Z_s from the point P. From Figure 7.14, we have

$$I_a^2 = \left(\frac{V_t}{Z_s}\right)^2 + \left(\frac{E_f}{Z_s}\right)^2 - 2\left(\frac{V_t}{Z_s}\right)\left(\frac{E_f}{Z_s}\right)\cos\delta \tag{7.30}$$

Therefore, I_a has two values. One value gives stable operation, whereas the other one corresponds to unstable operating conditions. The unstable operating point is represented by current I_a in Figure 7.14 and the angle δ between E_f/Z_s and V_t/Z_s for the unstable point is more than β, whereas the maximum value for angle δ for stable operation is the impedance angle β.

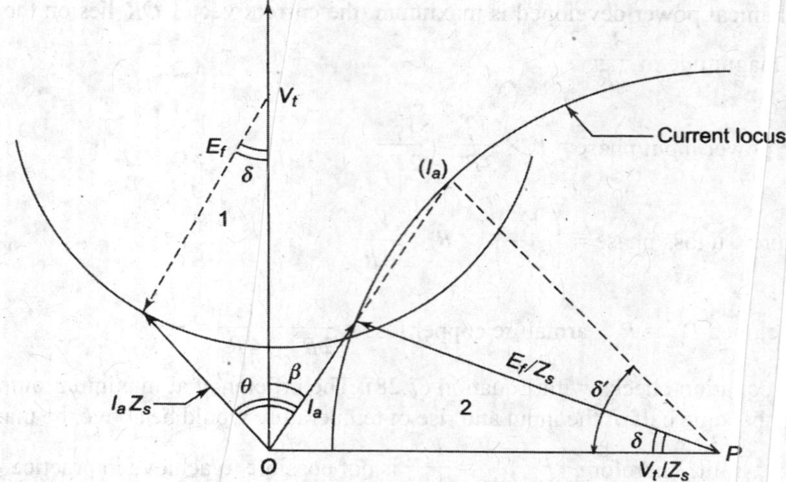

Figure 7.14 Current Locus for Constant Excitation

When the synchronous motor is loaded, the rotor falls back from the stator rotating magnetic field by an angle δ and continues to rotate at synchronous speed. If the load further increases, the angle δ also increases. The motor becomes unstable and it may not be able to rotate along with the rotating field if δ goes beyond β.

7.12 TORQUE OF A SYNCHRONOUS MOTOR

- *Starting torque:* The ability of the motor to accelerate the load is called the starting torque. Sometimes it is also called the breakaway torque.

  In centrifugal pumps it is 10 per cent of the full-load torque, whereas in reciprocating pumps it is 200 to 250 per cent of the full-load torque.
- *Running torque:* The motor under running condition develops this torque, which is determined by the output power and the speed of the driven machine.
- *Pull torque:* The ability of the motor to pull into synchronism during change from induction to synchronous motor operation is called the pull torque.
- *Pull-out torque:* It is the maximum torque that the motor will develop without pulling out of synchronism. Its value varies from 1.25 to 3.5 times the full-load torque.

7.13 SALIENT-POLE SYNCHRONOUS MOTOR—TWO-REACTION MODEL

Salient-pole synchronous generator has already been discussed in Chapter 6. Let us study the phasor diagram of the salient-pole synchronous motor. The voltage equation of synchronous motor is expressed by

$$V_t = E_f + I_aR_a + jI_dX_d + jI_qX_q$$

where X_d and X_q are the direct and quadrature axis reactance of the armature.

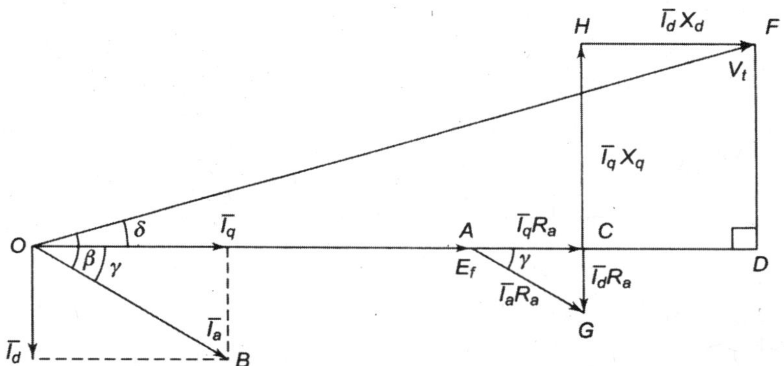

Figure 7.15 Lagging Power Factor

7.13.1 Lagging Power Factor

The phasor diagram for a synchronous motor for lagging power factor is shown in Figure 7.15. In Figure 7.15,

$$OA = E_f,\ AG = I_a R_a,\ GH = I_q X_q,\ HF = I_d X_d,\ OF = V_t,$$

$$OD = OA + AC + CD$$

$$GH = GC + CH$$

i.e.,
$$I_q X_q = I_d R_a + V_t \sin\delta \quad (7.31)$$

Again,
$$\gamma = \beta - \delta \quad (7.32)$$

Therefore,
$$\bar{I}_d = \bar{I}_a \sin\gamma = \bar{I}_a \sin(\beta - \delta) \quad (7.33)$$

and
$$\bar{I}_q = \bar{I}_a \cos\gamma = \bar{I}_a \cos(\beta - \delta) \quad (7.34)$$

From Equations (7.31), (7.33) and (7.34), we have

$$I_a X_q \cos(\beta - \delta) = I_a R_a \sin(\beta - \delta) + V_t \sin\delta$$

i.e.,
$$(V_t - I_a R_a \cos\beta - I_a X_q \sin\beta)\sin\delta = (I_a X_q \cos\beta - I_a R_a \sin\beta)\cos\delta$$

i.e.,
$$\tan\delta = \frac{I_a X_q \cos\beta - I_a R_a \sin\beta}{V_t - I_a R_a \cos\beta - I_a X_q \sin\beta} \quad (7.35)$$

Also,
$$\tan\gamma = \frac{V_t \sin\beta - I_a X_q}{V_t \cos\beta - I_a R_a} \quad (7.36)$$

Again,
$$E_f = V_t \cos\delta - I_q R_a - I_d X_d \quad (7.37)$$

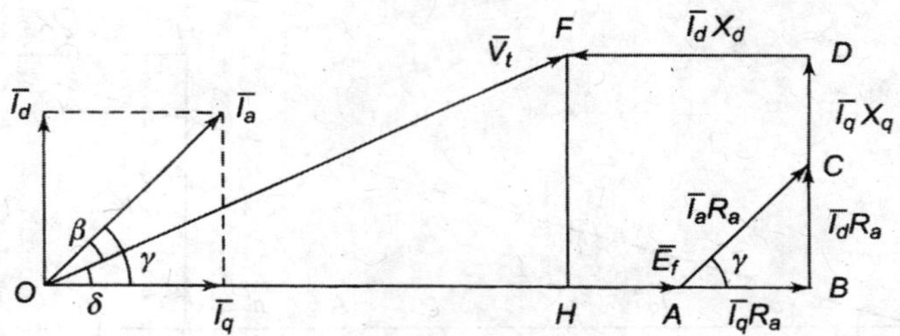

Figure 7.16 Leading Power Factor

7.13.2 Leading Power Factor

The phasor diagram for a synchronous motor for leading power factor is shown in Figure 7.16. In Figure 7.16,

$$OA = E_f,\ AC = I_aR_a,\ CD = I_qX_q,\ DF = I_dX_d,\ OF = V_t,$$

$$AB = AC\cos\gamma = I_aR_a\cos\gamma = I_qR_a$$

$$BC = AC\sin\gamma = I_aR_a\sin\gamma = I_dR_a$$

$$OA = OH + HA = OH + HB - AB = OH + FD - AB$$

i.e., $E_f = V_t\cos\delta + I_dX_d - I_qR_a$

Again, $HF = BD = BC + CD$

i.e., $$V_t\sin\delta = I_dR_a + I_qX_q \tag{7.38}$$

Here, $\gamma = \beta + \delta$

Since, $\overline{I}_d = \overline{I}_a\sin(\beta + \delta)$ and $\overline{I}_q = \overline{I}_a\cos(\beta + \delta)$, we have from Equation (7.38),

$$V_t\sin\delta = I_aR_a\sin(\beta + \delta) + I_aX_q\cos(\beta + \delta)$$

i.e., $(V_t + I_aX_q\sin\beta - I_aR_a\cos\beta)\sin\delta = (I_aX_q\cos\beta + I_aR_a\sin\beta)\cos\delta$

$$\tan\delta = \frac{I_aX_q\cos\beta + I_aR_a\sin\beta}{V_t + I_aX_q\sin\beta - I_aR_a\cos\beta} \tag{7.39}$$

Also, $$\tan\gamma = \frac{V_t\sin\beta + I_aX_q}{V_t\cos\beta - I_aR_a} \tag{7.40}$$

Again, $$E_f = V_t\cos\delta - I_qR_a + I_dX_d \tag{7.41}$$

7.13.3 Unity Power Factor

For unity power factor, $\cos\beta = 1$ and $\sin\beta = 0$.

∴ $$\tan\delta = \frac{I_aX_q}{V_t - I_aR_a} \tag{7.42}$$

Figure 7.17 shows the phasor diagram of the synchronous motor at unity power factor.

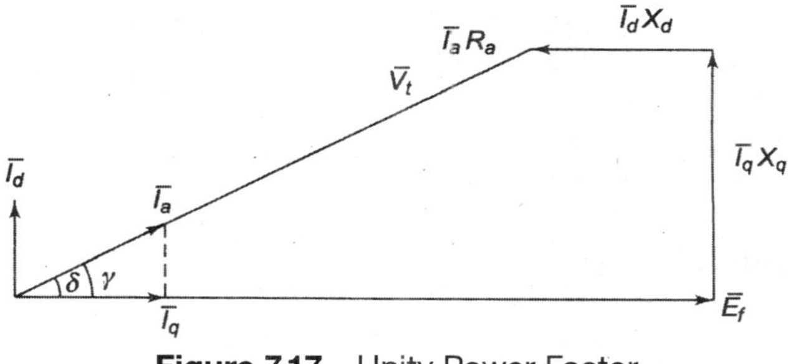

Figure 7.17 Unity Power Factor

If the resistance of the armature is of negligible order, the phasor diagrams of the synchronous motor at lagging, leading and unity power factor are shown in Figures 7.18, 7.19 and 7.20, respectively.

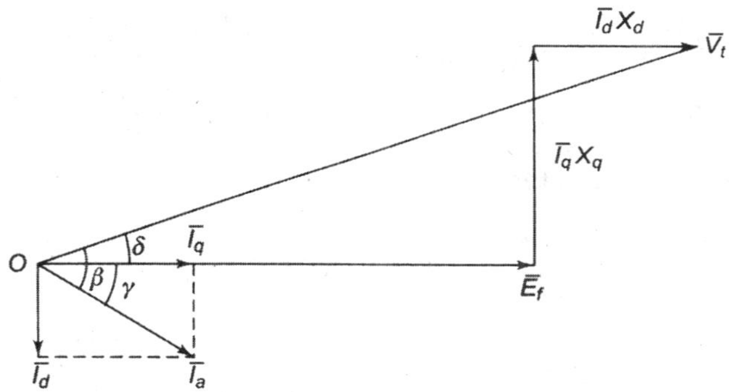

Figure 7.18 Lagging Power Factor

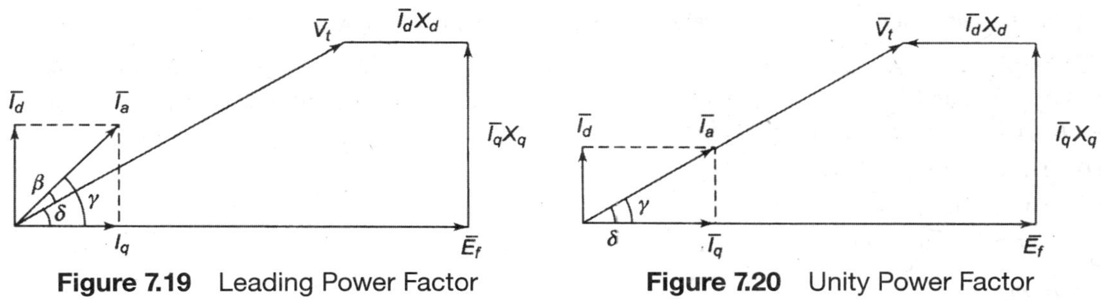

Figure 7.19 Leading Power Factor

Figure 7.20 Unity Power Factor

7.14 POWER DEVELOPED BY A SALIENT-POLE SYNCHRONOUS MOTOR

The input power is the sum of the product of the in-phase components of the current and voltage and the product of quadrature components.

$$P_{in} = I_q V_t \cos\delta + I_d V_t \sin\delta = \frac{V_t \sin\delta}{X_q} \times V_t \cos\delta + \left(\frac{E_f - V_t \cos\delta}{X_d}\right) \times V_t \sin\delta$$

$$= \frac{E_f V_t \sin\delta}{X_d} + \frac{V_t^2}{2}\left(\frac{1}{X_q} - \frac{1}{X_d}\right) \sin 2\delta \tag{7.43}$$

Therefore, the total power input is given by

$$P_{in-3\phi} = \frac{3E_f V_t \sin\delta}{X_d} + \frac{3V_t^2}{2}\left(\frac{1}{X_q} - \frac{1}{X_d}\right) \sin 2\delta \tag{7.44}$$

The power developed is equal to the total power input minus the losses.

7.14.1 Stability and Maximum Load Angle

Differentiating Equation (7.43) with respect to load angle, we have

$$\frac{dP_{in}}{d\delta} = \frac{E_f V_t \cos\delta}{X_d} + V_t^2\left(\frac{1}{X_q} - \frac{1}{X_d}\right) \cos 2\delta \tag{7.45}$$

Equation (7.45) represents the rate of change of P_{in} with respect to δ. This is termed as the stability factor, rigidity factor or the stiffness of coupling.

To obtain the maximum value of P_{in}

$$\frac{E_f V_t \cos\delta}{X_d} + V_t^2\left(\frac{1}{X_q} - \frac{1}{X_d}\right) \cos 2\delta = 0$$

i.e.,

$$\cos\delta = \frac{-E_f X_q}{4V_t\left(X_d - X_q\right)} + \sqrt{\frac{1}{2} + \left[\frac{E_f X_q}{4V_t\left(X_d - X_q\right)}\right]^2} \tag{7.46}$$

Equation (7.46) shows the required value of cos δ to get maximum power input.

Example 7.7 A three-phase, 18 MVA, 10 pole, 50 Hz, 11 kV, star-connected synchronous motor has $X_d = 5\ \Omega$ and $X_q = 3\ \Omega$, respectively. It has negligible armature resistance. Calculate the following on full load at 0.8 power factor lagging:

(i) The excitation voltage.
(ii) The power.
(iii) The maximum value of power angle and the corresponding power.

Solution

(i) $V_t = \dfrac{11\times10^3}{\sqrt{3}} = 6,350.85$ V and $I_a = \dfrac{18\times10^3}{\sqrt{3}\times11} = 944.75$ A

$$\tan\delta = \frac{I_a X_q \cos\beta - I_a R_a \sin\beta}{V_t - I_a X_q \sin\beta + I_a R_a \cos\beta} = \frac{944.75\times3\times0.8 - 0}{6,350.85 - 944.75\times3\times0.6 - 0} = 0.4876$$

i.e., $\delta = 26°$

$$I_d = I_a \sin(\beta - \delta) = I_a \sin(36.87° - 26°) = 178.16 \text{ A}$$

$$I_q = I_a \cos(\beta - \delta) = 944.75 \cos(36.87° - 26°) = 927.8 \text{ A}$$

$$E_f = V_t \cos\delta - I_q R_a - I_d X_d = 6{,}350.8 \times \cos 26° - 927.8 \times 0 - 178.16 \times 5 = 4{,}817.30 \text{ V}$$

(ii) $$P_{3\phi} = 3 \times \left[\frac{E_f V_t}{X_d} \sin\delta + \frac{V_t^2}{2}\left(\frac{1}{X_q} - \frac{1}{X_d}\right) \sin 2\delta\right]$$

$$= 3\left[\frac{4{,}817.30 \times 6{,}350.85}{5} \sin 26° + \frac{6{,}350.85^2}{2}\left(\frac{1}{3} - \frac{1}{5}\right) \sin 52°\right] = 14.40 \text{ MW}$$

$$P_{in} = 18 \times 0.8 = 14.40 \text{ MW}$$

$$P_{3\phi} = P_{in}$$

(iii) $$\cos\delta = -\frac{E_f X_q}{4V_t(X_d - X_q)} + \sqrt{\frac{1}{2} + \left[\frac{E_f X_q}{4V_t(X_d - X_q)}\right]^2}$$

$$= -\frac{4{,}817.3 \times 3}{4 \times 6{,}350.85 \times 2} + \sqrt{\frac{1}{2} + \left(\frac{4{,}817.3 \times 3}{4 \times 6{,}350.85 \times 2}\right)^2} = 0.4777$$

i.e., $\delta = 61.46°$

$$P_{max} = 3\left[\frac{E_f V_t}{X_d} \sin\delta + \frac{V_t^2}{2}\left(\frac{1}{X_q} - \frac{1}{X_d}\right) \sin 2\delta\right]$$

$$= 3\left[\frac{4{,}817.30 \times 6{,}350.85}{5} \sin 61.46° + \frac{6{,}350.85^2}{2}\left(\frac{1}{3} - \frac{1}{5}\right) \sin 122.92°\right] = 22.89 \text{ MW}$$ ■

7.15 DAMPER WINDINGS

Damper windings are provided in different machines such as alternators and induction-start synchronous motors.

The addition of copper damper windings to machines effectively simplifies the characteristics of the machines by largely eliminating the harmonics. Damper windings are of the following types:

(i) *Connected damper winding:* This type of winding is similar to squirrel cage of an induction motor, which is continuous between the poles for a slow-speed machine, whereas additional end rings are brazed to the copper rods for higher-speed machines. In this type of dampers, $X_q \approx X_d$.

(ii) *Non-connected damper winding:* This type of winding in each pole face is independent from those in the adjacent poles, and $X_q \neq X_d$.

(iii) *Special damper winding:* Dampers with this type of winding use double windings, one of which has high resistance and low reactance and the has low resistance and high reactance. This type of damper winding is used in motors to provide better starting characteristics. At low speeds, the high reactance of the low-resistance windings forces the current to flow through the high-resistance winding, which produces a high torque. At higher speeds, the low-resistance winding becomes effective.

7.16 DAMPING EFFECT

Adding damper windings in generators provides a low-resistance path for the flow of current. It also prevents both wave distortion and excessive heating. The best criteria of a polyphase machine to carry unbalanced load are its negative sequence reactance and resistance. The former reflects its ability to prevent unbalancing of the voltage and the latter, its ability to carry the negative sequence current without undue heating of the rotor. These properties are particularly important for such fluctuating loads as electric furnaces. Not only do the dampers reduce the voltage unbalance but they also reduce waveform distortion.

Synchronous generator feeds load through transmission lines having a low ratio of resistance to reactance, and it tends to set up spontaneous hunting that is greater at light loads than at heavy loads.

The rotor oscillates during a system disturbance. The damper bars have a relative movement with respect to the air-gap flux that causes induction of the emf and the flow of current in these bars/windings. The bar current creates the emf according to Lenz's law, which always opposes the relative motion. Therefore, a positive damping term comes to play to reduce the amplitude of the oscillatory motions of the rotor about the operating point. The rotor quickly returns to the steady position. These short-circuited bars are known as 'damper winding or amortisseur winding'. These act like the squirrel cage of an induction motor and provide a starting torque for the motor which otherwise, being of synchronous kind, is non self-starting. Therefore, the damper windings serve a dual purpose. During start-up, synchronous motor achieves a speed close to synchronous. The rotor and stator fields lock with each other as soon as the field excitation is switched on. The damper bars in synchronous machines prevent hunting of the machine.

7.17 HUNTING/SURGING OF SYNCHRONOUS MOTORS

During operation within the region of stability with a fixed excitation operating, an increase in load results in a change in the rotor position from its original value or, the rotor slips backwards relative to the armature field through an angle of $\Delta\delta$ in electrical degrees. An instantaneous adjustment to the changed load conditions is not possible due to the inertia of the rotor and its connected load. Therefore, the rotor swings back somewhat beyond the position of equilibrium and develops torque in excess of the load requirements. This not only checks the initial reduction of speed but also increases the speed slightly above the synchronous speed and there is a periodic swinging of rotor set-up first to one side, then to the other of the new position of equilibrium. This oscillation is superimposed on the uniform synchronous speed of rotation.

Figure 7.21 shows the oscillation of rotor with respect to time. The synchronous motor is operating with a load angle δ before the application of a sudden load. Immediately after the application of load, the new load demands an increase of angle δ_1 to a value δ_2, but the rotor will settle down at δ_2 only after a few oscillations.

The relative to-and-fro motions of the rotor field with respect to the armature field results in production of eddy currents in the pole faces, which are always directed to oppose their cause as per Lenz's law, and hence tend to damp out these oscillations. Changes in δ are accompanied by changes in the

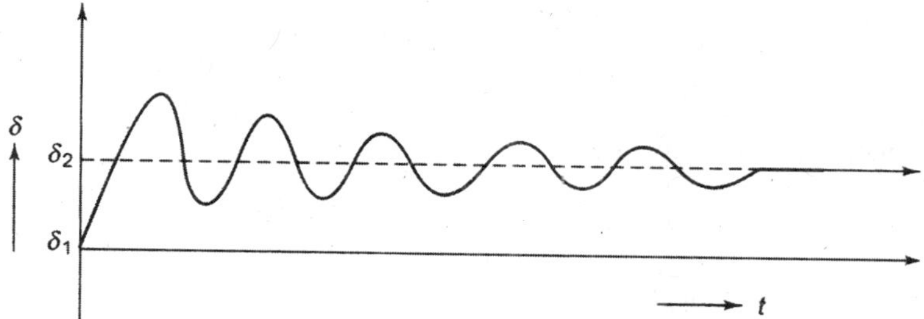

Figure 7.21 Oscillation of Rotor with Respect to Time

armature current. To suppress the objectionable hunting more effectively than by eddy currents in the pole faces, synchronous motors are always provided with damper windings. These windings in the rotor are short-circuited. There is a current produced in the winding due to rotor swing with respect to synchronous speed, which results in a torque opposing the change in speed.

7.18 PERIODICITY OF HUNTING

After sudden application of an additional load equivalent to a torque T_L, the rotor will be displaced from its initial ξ electrical degrees, which is equivalent to an actual space of $\left(\frac{2}{P}\right)\xi$, where P is the number of poles at any given time t. If J is the polar moment of inertia including its connected load, the accelerating torque is $\left(\frac{2J}{P}\right)\frac{d^2\xi}{dt^2}$.

The externally applied torque T_L must be equal to the sum of internally developed torques. In spite of the initial torque, torques due to the elastic pull of the magnetic field upon the field poles and the frictional or braking torque due to the currents in the damping winding come into action.

$$\text{Inertial torque} = \left(\frac{2J}{P}\right)\frac{d^2\xi}{dt^2}$$

Torque due to the elastic pull of magnetic field $= K_1\ \delta P_{om}$ and frictional or braking torque $= K_2\frac{d\xi}{dt}$

The complete equation of motion may be expressed as

$$\left(\frac{2J}{P}\right)\frac{d^2\xi}{dt^2} + K_2\frac{d\xi}{dt} + K_1\ dP_{om} = T_L \tag{7.47}$$

$$dP_{om} = \frac{V_t E_f}{Z_s}\sin\left(\theta - \delta\right) \times d\delta = \frac{V_t E_f}{Z_s}\sin\left(\theta - \delta\right)\ \left[\text{Since } \beta = d\delta\right]$$

Substituting for δP_{om} in Equation (7.47), we have

$$\left(\frac{2J}{P}\right)\frac{d^2\xi}{dt^2} + K_2\frac{d\xi}{dt} + K_1\xi\frac{V_t E_f}{Z_s}\sin\left(\theta - \delta\right) = T_L$$

which may be written as

$$\left(D^2 + 2aD + b\right)\xi = T_L\left(\frac{P}{2J}\right) \tag{7.48}$$

where $a = \frac{K_2 P}{4J}$ and $b = \left(\frac{K_1 P}{2J}\right)\frac{V_t E_f}{Z_s}\sin(\theta - \delta)$

The solution for Equation (7.48) is given by

$$\xi = Ae^{\left(-a+\sqrt{a^2-b}\right)t} + Be^{\left(-a-\sqrt{a^2-b}\right)t} + \frac{T_L}{K}$$

Three possibilities occur, defined by the condition.

(i) $a^2 > b \rightarrow$ overdamped

(ii) $a^2 = b \rightarrow$ critically damped

(iii) $a^2 < b \rightarrow$ underdamped

When $a^2 < b$, it leads to imaginary exponents

$$\xi = e^{-at}\left\{Ae^{j\left(\sqrt{b-a^2}\right)t} + Be^{-j\left(\sqrt{b-a^2}\right)t}\right\} + \frac{T_L}{K}$$

$$= \frac{T_L}{K} + e^{-at}\left\{(A+B)\cos\sqrt{b-a^2}\,t + j(A-B)\sin\sqrt{b-a^2}\,t\right\} \tag{7.49}$$

At $t = 0$, $\xi = 0$ and $\frac{d\xi}{dt} = 0$

Hence,

$$A + B = -\frac{T_L}{K} \tag{7.50}$$

$$j(A-B) = \left(\frac{a}{\sqrt{b-a^2}}\right)\frac{T_L}{K} \tag{7.51}$$

Substituting the values of $A + B$ and $j(A - B)$ from Equations (7.50) and (7.51) in Equation (7.40), we have

$$\xi = \frac{T_L}{K} - \frac{T_L}{K}\left(\frac{b}{b-a^2}\right)^{\frac{1}{2}} e^{-at}\cos\left(\sqrt{b-a^2}\,t - \theta\right) \tag{7.52}$$

where

$$\theta = \tan^{-1}\left[\frac{-a}{\sqrt{b-a^2}}\right] \tag{7.53}$$

Therefore, ξ approaches its ultimate or steady-state value T_L/K by a transient series of oscillations of decreasing amplitude and the frequency of the oscillation is

$$f = \frac{\sqrt{b-a^2}}{2\pi} \tag{7.54}$$

The period of oscillation is given by $= \frac{1}{f} = \frac{2\pi}{\sqrt{b-a^2}}$ (7.55)

7.19 METHODS OF STARTING OF SYNCHRONOUS MOTORS

Since synchronous motors are not self-starting inherently, we should have some means for starting synchronous motors. The methods are discussed here.

(i) *Induction motor start:* Damper windings consisting of copper bronze bars embedded in slots in the pole faces and short-circuited at both ends are usually provided in synchronous motors, which serve not only to damp out oscillations during hunting but are also suitably designed as a squirrel-cage winding for starting purposes. If the rotor reaches near synchronous speed, the field circuit is suddenly energized and the rotor and the stator fields interlock each other. The voltage induced in the field winding at standstill may reach very high values, which is dangerous to life and to the insulation. The field winding is sometimes interrupted at several points by some form to field breaking switch to limit this voltage to a safe value. A better way to limit the induced voltage in the field windings is to reduce the voltage impressed upon the armature terminals to a fraction of normal voltage (30 to 50 per cent) using an auto-transformer.

(ii) *Auxiliary motor start:* Using an auxiliary motor, the synchronous motor is first run as an alternator and is synchronized with the three-phase bus bars. After synchronization, supply to auxiliary motor is cut and the alternator is made to run as a synchronous motor drawing power from three-phase AC mains.

(iii) *Using resistor in the field circuit:* Instead of leaving the field circuit open during the starting stage of a synchronous motor, it can be short-circuited and can be used as a damper winding. When an additional resistance of a value 100 times that of the field resistance is inserted in series with the field the starting torque will be large. The field resistance is reduced as the motor accelerates until it becomes three to four times that of the field itself. This method can be used in machines in which large starting torques are required.

(iv) *Phase connected damper windings:* When the starting torque required is of the order of 200 per cent of the full-load torque, as in the case of tube mills for making cement, machines have been built that employ the principle of the slip ring induction motor. The bars of damper windings embedded in the pole faces are connected to form a balanced three-phase winding, with the terminals connected through slip rings to an adjustable rheostat. The starting resistance is then cut out in successive steps, until the windings are completely short-circuited.

(v) *Super synchronous start:* The armature is not rigidly bolted to the bedplates as in an ordinary motor. It is capable of rotating about the rotor shaft. But under normal conditions, it is held stationary by a brake band surrounding the frame of the armature core. The rotor is mechanically coupled to load. The brake band on stator is released, and when line voltage is impressed on armature terminals, the stator will try to rotate. When the armature reaches near synchronous speed, the rotor field is excited and the machine pulls into synchronism. Thereafter, the brake band on armature is gradually tightened and as the armature slows down, the rotor tries to pick up speed. When armature is held stationary, the rotor attains synchronous speed. This is a misnomer.

7.20 APPLICATIONS OF SYNCHRONOUS MOTOR

Synchronous motors are hardly used below 50 HP in the medium range due to their higher initial cost compared to induction motors. A DC excitation source is required and the starting and control devices are usually more expensive, especially in which automatic operation is required. However, there are certain advantages offered by synchronous motors:

(i) Constant speed operation.
(ii) Power factor control.
(iii) High operating efficiency.

Table 7.2 shows some important characteristics of the synchronous motor along with some applications.

Table 7.2 Synchronous Motor Characteristics and Applications

Type Designation	Synchronous, High Speed above 500 rpm	Synchronous, Low Speed below 500 rpm
1. HP range	25 to several thousands	Usually above 25 to several thousand
2. Starting torque (% of normal)	Up to 120	Low 40
3. Pull-out torque (%)	Up to 200	Up to 180
4. Starting torque (%)	500 to 700	200 to 359
5. Slip	Zero	Zero
6. Power factor	High, but varies with load and with excitation	High, but varies with excitation.
7. Efficiency (%)	Highest of all the motors, 92 to 96 per cent.	Highest of all the motors, 92 to 96 per cent.
8. Typical applications	Fans, blowers, DC generators line-shafts, centrifugal pumps and compressors, reciprocating pumps and compressors, useful for power factor correction, constant speed frequency changes	Lower-speed direct connected loads such as reciprocating compressors when started unloaded, DC generators, rolling mills band mills, bail mills, pumps, useful for power factor control, constant speed. Fly wheel used for pulsating loads.

7.21 SYNCHRONOUS CONDENSERS

Synchronous condensers are synchronous motors used for power factor correction alone without any mechanical output. It draws leading current, as a static condenser does, which is useful for installations of low power factor. The synchronous motors when overexcited act like capacitors. To take leading kVA, it can be connected to a system to improve its power factor. The synchronous condenser is specially designed so that practically all its rated kVA are available for power factor correction. Because of this feature, they are applied particularly to transmission line control.

At load (losses assumed negligible), a synchronous motor operates at a $\delta = 0$, which means that E_f and V_t are in phase (E_f = excitation emf and V_t = voltage at the motor terminals)

When the motor draws zero power factor leading current is shown in Figure 7.22(a).

$$I_a = \frac{E_f - V_t}{X_s}\left(\text{where } E_f > V_t\text{, overexcited}\right) \tag{7.56}$$

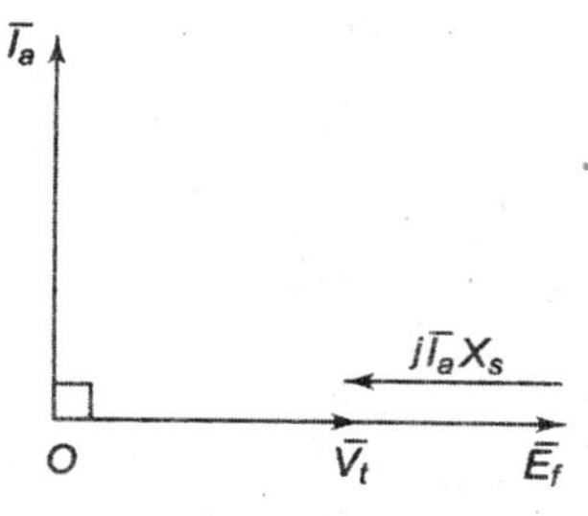

(a) As variable capacitor (overexcited)

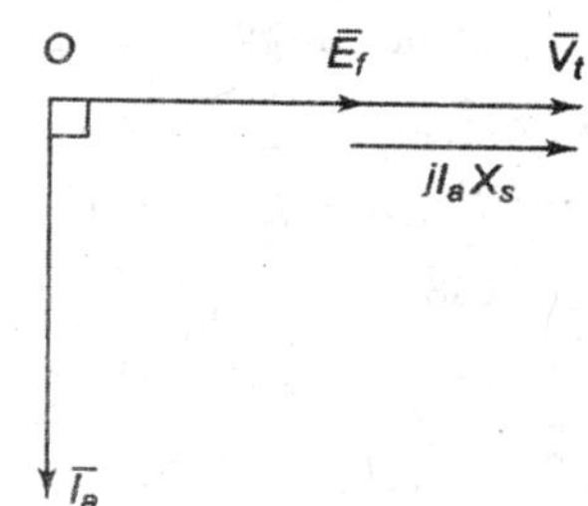

(b) As variable inductor (underexcited)

Figure 7.22 Synchronous Condenser

When the motor draws zero power factor lagging current is shown in Figure 7.22(b)

$$I_a = \frac{V_t - E_f}{X_s} \left(\text{where } E_f < V_t, \text{ underexcited}\right) \tag{7.57}$$

ADDITIONAL SOLVED PROBLEMS

Example 7.8 A three-phase, star-connected, 6,600 V synchronous motor has a synchronous reactance per phase of 15 Ω. For a certain load, the input is 900 kW at normal voltage and the induced line emf is 8,900 V. Neglecting armature resistance, determine the following:

(i) Line current.

(ii) Power factor.

Solution

Synchronous reactance/phase $X_s = 15\ \Omega$

Input to motor = 900 kW

Supply phase voltage $V_t = \dfrac{6{,}600}{\sqrt{3}} = 3{,}810.5$ V

Induced emf/phase $E_f = \dfrac{8{,}900}{\sqrt{3}} = 5{,}138.4$ V

Resistance $R_a = 0$

Since the induced emf is greater than the supply voltage, the motor will operate at leading power factor.
Armature current (I_a) = line current (I_L) because the motor is star connected.
Since power input $= \sqrt{3} V_L I_L \cos\beta = \sqrt{3} V_L I_a \cos\beta$

$$\therefore \qquad I_a \cos\beta = \frac{\text{Power input}}{\sqrt{3} V_L} = \frac{900 \times 1{,}000}{\sqrt{3} \times 6{,}600} = 78.73 \text{ A}$$

$$\text{Internal angle } \theta = \tan^{-1}\frac{X_s}{R_s} = \tan^{-1}\frac{15}{0} = \tan\infty = 90°$$

$$\text{Impedance drop } E_r = I_a Z_s = I_a\sqrt{R_a^2 + X_s^2} = I_a\sqrt{0^2 + 15^2} = 15I_a$$

From right-angled ΔABC (Figure E7.1), we have

$$AB^2 = BC^2 + CA^2$$

$$(5{,}138.4)^2 = CA^2 + (E_R \cos\beta)^2 = CA^2 + (15\, I_a \cos\beta)$$

$$= CA^2 + (15 \times 78.73)^2$$

$$\therefore \quad CA^2 = (5{,}138.4)^2 - (1{,}180.95)^2$$

$$\therefore \quad CA = 5{,}000.85 \text{ V}$$

But $\quad OC = CA - OA = 5{,}000.85 - 3{,}810.5 = 1{,}190.35$ V

Now, $E_R = OB = \sqrt{OC^2 + CB^2} = \sqrt{(1{,}190.35)^2 + (1{,}180.95)^2}$

$$= 1{,}676.77 \text{ V}$$

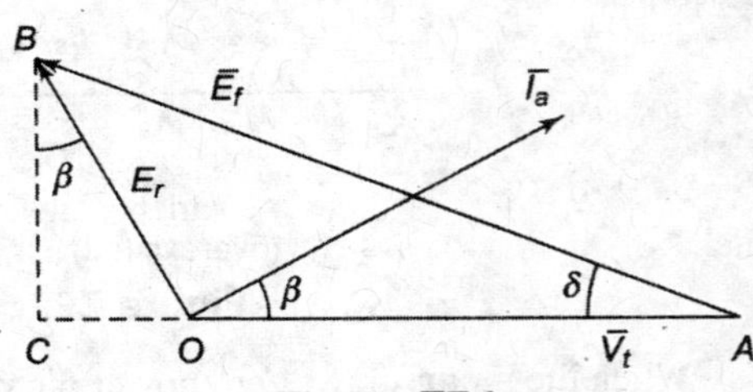

Figure E7.1

(i) Line current = phase current (I_a)

$$= \frac{E_r}{Z_s} = \frac{1{,}676.77}{15} = 111.78 \text{ A}$$

(ii) Power factor $(\cos\beta) = \dfrac{I_a \cos\beta}{I_a} = \dfrac{78.73}{111.78} = 0.7043$ (leading) ■

Example 7.9 A 6.6-kV star-connected, three-phase synchronous motor works at constant voltage and constant excitation. Its synchronous reactance is 18 Ω per phase; neglect resistance. When the input is 1,111.145 kW, the power factor is 0.8 leading. Find the power factor when the input is changed to 1,500 kW.

Solution

Supply voltage per phase $V_t = \dfrac{6.6 \times 1{,}000}{\sqrt{3}} = 3{,}810.5$ V

$$X_s = 18\ \Omega$$

$$\text{Internal angle } \theta = \tan^{-1}\left(\frac{18}{0}\right) = \tan^{-1}(\infty) = 90°$$

When the input is 1,111.145 kW at 0.8 power factor leading:

Armature current/phase $\quad (I_a) = \dfrac{1{,}111.145 \times 1{,}000}{\sqrt{3} \times 6{,}600 \times 0.8} = 121.5$ A

$$\therefore \quad E_R = I_a Z_s = 121.5 \times 18 = 2{,}187 \text{ V.}$$

Here, $\quad \cos\beta = 0.8$ (leading), $\beta = 36.87°$

$$E_f = \sqrt{V_t^2 + E_R^2 - 2V_t E_R \cos(\theta + \beta)}$$

$$= \sqrt{3{,}810.5^2 + 2{,}187^2 - 2 \times 3{,}810.5 \times 2{,}187 \times \cos(90° + 36.87°)}$$

$$= 5{,}413.24 \text{ V}$$

When power is changed to 1,500 W, the excitation (5,413.24 V per phase) and supply voltage (per phase) remain constant.

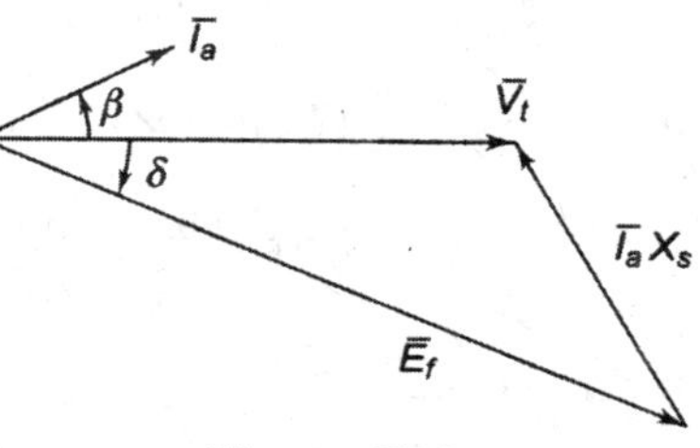

Figure E7.2

$$I_a \cos\beta = \frac{1{,}500 \times 1{,}000}{\sqrt{3} \times 6{,}600} = 131.2 \text{ A}$$

Power/phase $= 1{,}500/3 = 500$ kW

Again, $$P = \frac{E_f V_t}{X_s} \sin\delta$$

i.e., $$\sin\delta = \frac{PX_s}{E_f V_t} = \frac{500 \times 1{,}000 \times 18}{5{,}413.24 \times 3{,}810.5} = 0.4363$$

i.e., $\delta = 25.87°$

From phasor diagram (Figure E7.2) of synchronous motor (neglecting $I_a R_a$), we have

$$(I_a X_s)^2 = E_f^2 + V_t^2 - 2E_f V_t \cos\delta$$

$$I_a = \frac{1}{18}\left[(5{,}413.24)^2 + (3{,}810.5)^2 - 2 \times 3{,}810.5 \times 5{,}413.24 \times \cos 25.87°\right]^{\frac{1}{2}}$$

$$= 143.83 \text{ A}$$

$\therefore$ $$\cos\beta = \frac{I_a \cos\beta}{I_a} = \frac{131.2}{143.83} = 0.9122 \text{ (leading)}$$ ■

Example 7.10 A three-phase, 400 V, star-connected synchronous motor has full load efficiency of 95 per cent and it takes 24 A on full load at unity power factor. What will be the back emf generated and total mechanical power developed in kW on full load and 0.9 power factor leading? The impedances/ phase is (0.2 + *j*2) Ω.

Solution

Efficiency of synchronous motor = 95%

Full-load current I_a = 24 A

Power factor = 1

Impedance/phase = (0.2 + *j*2)

The current at 0.9 power factor $= \frac{24}{0.9} = 26.67$ A

Power input at 0.9 power factor $= \sqrt{3} \times 400 \times 26.67 \times 0.9 = 16{,}629.76$ W

Terminal voltage (V_t) per phase $= \frac{400}{\sqrt{3}} = 231$ V

$\therefore$ $$\tan\theta = \frac{X_s}{R_a} = \frac{2}{0.2} = 10$$

$\therefore \quad \theta = 84.3°$

Impedance drop $(E_R) = I_a Z_s = 26.67 \times \sqrt{0.2^2 + 2^2} = 53.60$ V

$\therefore \quad \cos\beta = 0.9$

$\therefore \quad \beta = \cos^{-1}(0.9) = 25.8°$

Refer to Figure E7.3

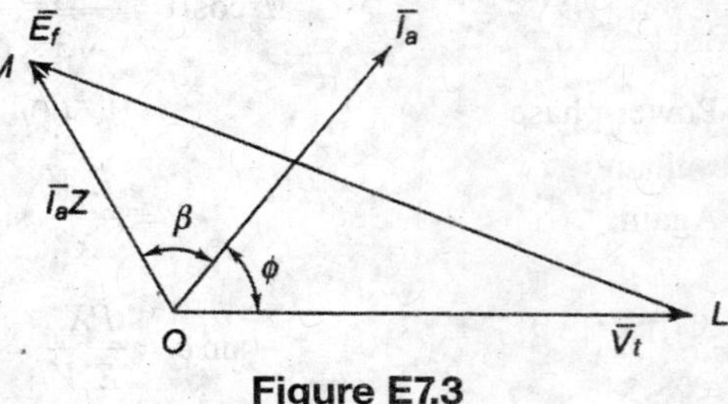

Figure E7.3

$$\angle MOL = \theta + \beta = 84.3° + 25.8° = 110.1°$$

In ΔLOM, we have

$$E_f = \sqrt{V_t^2 + E_R^2 - 2V_t E_R \cos(\theta + \beta)}$$

$$= \sqrt{(231)^2 + (53.6)^2 - 2 \times 231 \times 53.6 \times \cos 110.1°} = 254.45 \text{ V}$$

Induced line emf $= \sqrt{3} \times 254.45 = 440.72$ V

In ΔLOM, $\dfrac{OM}{\sin\delta} = \dfrac{LM}{\sin(\theta + \phi)}$

i.e., $\dfrac{53.6}{\sin\delta} = \dfrac{254.44}{\sin 110.1°}$

i.e., $\sin\delta = \dfrac{53.6}{254.44} \times \sin 110.1° = 0.1978$

i.e., $\delta = \sin^{-1}(0.1978) = 11.4°$

Mechanical power developed = 3 × (emf induced per phase × armature current × cosine of the angle between the armature current and emf induced)

Let γ = angle between the armature current and the emf induced

$= \beta + \delta = 25.8° + 11.4° = 37.2°$

∴ Mechanical power developed

$$= 3 \times 254.44 \times 26.67 \times \cos 37.2° \times 10^{-3} \text{ kW} = 16.215 \text{ kW}$$

■

Example 7.11 A 75 kW 400 V, four-pole, three-phase, star-connected synchronous motor has an effective armature resistance and a synchronous reactance per phase of 0.04 Ω and 0.4 Ω, respectively. Compute the open-circuit emf phase and the gross mechanical power developed for full load at 0.8 power factor leading. Assume an efficiency of 92.5 per cent.

Solution

Given: Motor output = 75 kW; $V_L = 400$ V

R_a/phase = 0.04 Ω, X_s/phase = 0.4 Ω, $\cos\beta = 0.8$ (lead); $\eta = 92.5\%$

Open-circuit emf $E_f = 400$ V

Phase voltage $V_t = \dfrac{400}{\sqrt{3}} = 231$ V

$\cos\beta = 0.8$, $\beta = \cos^{-1}(0.8) = 36.87°$ and $\sin\beta = \sin(36.87°) = 0.6$.

$\therefore$ Motor input $= \dfrac{\text{Motor output}}{\eta} = \dfrac{75}{0.925} = 81.081\text{ kW} = 8{,}1081\text{ W}$

Armature current $(I_a) = \dfrac{8{,}1081}{\sqrt{3} \times 400 \times 0.8} = 146.28\text{ A}$

Resultant voltage $E_R = I_a Z_s = 146.28 \times \sqrt{R_a^2 + X_s^2} = 146.28 \times \sqrt{(0.04)^2 + (0.4)^2} = 58.80\text{ V}$

$$\tan\theta = \frac{X_s}{R_a} = \frac{0.4}{0.04} = 10$$

$\therefore$ Internal angle $\theta = \tan^{-1}(10) = 84.3°$

For leading power factor, we have

$$E_f = \sqrt{V_t^2 + E_R^2 - 2V_t E_R \cos(\theta + \beta)}$$

$$= \sqrt{(231)^2 + (58.80)^2 - 2 \times 231 \times 58.80 \times \cos(84.3° + 36.87°)} = 266.23\text{ V}$$

$\therefore$ Gross mechanical power developed for all the three phases is as follows:

$$(P_{mech})_{gross} = P_{in} - 3I_a^2 R_a = 81{,}081 - 3 \times (146.3)^2 \times 0.04$$

$$= 78{,}512.56\text{ W}$$

■

Example 7.12 A 20-pole, 30 kW, 660 V, 50 HZ, three-phase star-connected synchronous motor is operating with its generated voltage per phase exactly equal to the phase voltage applied to its armature. At loaded condition the motor is retarded by 5 mechanical degrees from its synchronous position. The synchronous reactance and the effective armature resistance are 10 Ω and 1 Ω per phase, respectively. Calculate the following:

(i) The armature current per phase.
(ii) The power phase and the total power drawn by the motor from the bus.
(iii) The developed power.

Solution

Number of poles $(P) = 20$; power output = 30 kW; $V_L = 660$ V

R_a/phase = 1 Ω and X_s/phase = 10 Ω

(i) Armature current/phase I_a

Induced emf per phase $E_f = V_t = \dfrac{660}{\sqrt{3}} = 381\text{ V}$

Load angle $\delta = \dfrac{P}{2} \times$ angle of retardation in mechanical degrees

$= \dfrac{20}{2} \times 5 = 50$ electrical degrees

Impedance/phase $Z_s = \sqrt{R_a^2 + X_s^2} = \sqrt{1^2 + 10^2} = 10.05\Omega$

$$\theta = \tan^{-1}\left(\frac{X_s}{R_a}\right) = \tan^{-1}\left(\frac{10}{1}\right) = 84.3°$$

$$E_R = \sqrt{V_t^2 + E_f^2 - 2V_tE_f\cos\delta}$$

$$= \sqrt{381^2 + 381^2 - 2\times381\times381\times\cos 50°} = 322 \text{ V}$$

$$I_a = \frac{E_r}{Z_s} = \frac{322}{10.05} = 32 \text{ A}$$

(ii) Since $E_f = V_t$, we have

$$\frac{E_R}{\sin\delta} = \frac{E_f}{\sin(\theta - \beta)}$$

i.e.,

$$\frac{322}{\sin 50°} = \frac{381}{\sin(\theta - \beta)}$$

i.e.,

$$\sin(\theta - \beta) = \frac{381\times\sin 50°}{322} = 0.9064$$

i.e.,

$$\theta - \beta = \sin^{-1}(0.9064) = 65°$$

∴

$$\beta = \theta - 65° = 84.3° - 65° = 19.3° \text{ (lagging)}$$

$$\text{Power per phase} = V_tI_a\cos\beta = 381\times32\times\cos 19.3° = 11{,}507 \text{ W} = 11.5 \text{ kW}$$

$$\text{Total power drawn} = 3\times11.5 = 34.5 \text{ kW}$$

(iii) Power developed:

$$\text{Power developed} = \text{Total power drawn} - 3\,I_a^2R_a$$

$$= 34.5 - \frac{3\times32^2\times1}{1{,}000} = 31.428 \text{ kW}$$ ■

Example 7.13 A three-phase, 6.6 kV, star-connected synchronous motor delivers 346.375 kW at full load having full-load efficiency of 85 per cent. The effective armature resistance and the synchronous reactance per phase of the motor are 0.4 Ω and 4.2 Ω, respectively. If the motor works at 0.8 power factor lead on full load, calculate the following:

(i) Generated emf on full load.
(ii) Load angle.

Solution

$$V_t = \frac{6{,}600}{\sqrt{3}} = 3{,}810.5 \text{ V}$$

$$P_{out} = 346.375 \text{ kW}$$

$$P_{in} = \frac{P_{out}}{\eta} = \frac{346.375}{0.85} = 407.5 \text{ kW}$$

∴

$$I_L = \frac{P_{in}}{\sqrt{3}V_L\cos\beta} = \frac{407.5\times1{,}000}{\sqrt{3}\times3{,}810.5\times0.8} = 77.18 \text{ A}$$

Since the motor is star connected, $I_a = I_L = 77.18$ A

$$Z_s \,/\, \text{phase} = \sqrt{0.4^2 + 4.2^2} = 4.22\,\Omega \quad \text{and} \quad \theta = \tan^{-1}\left(\frac{X_s}{R_a}\right) = \tan^{-1}\left(\frac{4.2}{0.4}\right) = 84.56^\circ$$

$$E_R = I_a Z_s = 77.18 \times 4.22 = 325.7 \text{ V}$$

Here, $\cos\beta = 0.8$ (leading) $\therefore\ \beta = 36.87^\circ$

$$E_f = \sqrt{V_t^2 + E_R^2 - 2V_t E_R \cos(\theta + \beta)}$$

$$= \sqrt{3{,}810.5^2 + 325.7^2 - 2 \times 3{,}810.5 \times 325.7 \times \cos(84.56^\circ + 36.87^\circ)}$$

$$= 3990.03 \text{ V}$$

The generated emf per phase is 3,990.03 V.

Again, $$\frac{E_f}{\sin(\theta + \beta)} = \frac{E_R}{\sin\delta}$$

i.e., $$\sin\delta = \frac{E_R}{E_f} \times \sin(\theta + \beta) = \frac{325.7}{3{,}990.03} \times \sin(84.56^\circ + 36.87^\circ) = 0.06965$$

i.e., $$\delta = 3.99^\circ$$

Therefore, the load angle is 3.99°. ■

Example 7.14 A 22 kW, 400 V, three-phase, star-connected synchronous motor has an effective armature resistance and a synchronous reactance per phase of 0 Ω and 4 Ω, respectively. The line value of the developed back emf is 520 V. Calculate the motor power factor and current drawn by the motor.

Solution

$$P_{in} = 22 \text{ kW}, \quad V_t = \frac{400}{\sqrt{3}} = 231 \text{ V}, \quad E_f = \frac{520}{\sqrt{3}} = 300.22 \text{ V}$$

R_a/phase = 0 Ω and X_s/phase = 4 Ω

$$\therefore \quad Z_s \,/\, \text{phase} = 4\,\Omega \quad \text{and} \quad \theta = \tan^{-1}\left(\frac{X_s}{R_a}\right) = 90^\circ$$

$$I_a \cos\beta = \frac{P_{in}}{\sqrt{3} \times V_L} = \frac{22{,}000}{\sqrt{3} \times 400} = 31.75 \text{ A}$$

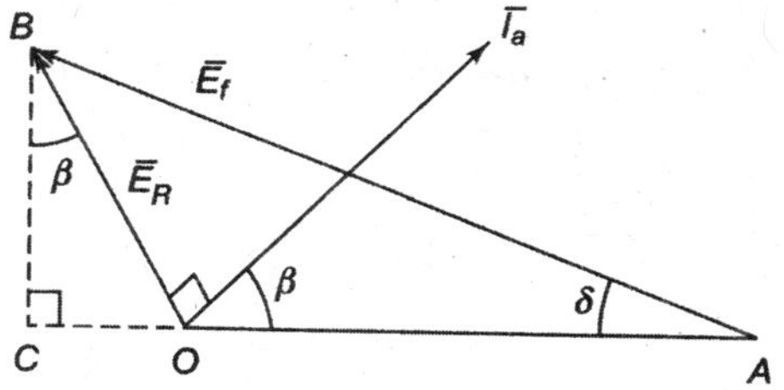

Figure E7.4

Since $E_f > V_t$, power factor is leading.
From phasor diagram (Figure E7.4), we have

$$BC = E_R \cos\beta = I_a Z_s \cos\beta \text{ and } OC = E_R \sin\beta = I_a Z_s \sin\beta.$$

From Δ*CAB*, we have

$$AB^2 = AC^2 + BC^2 = (OA + OC)^2 + BC^2$$

i.e., $$E_f^2 = (V_t + I_a Z_s \sin\beta)^2 + (I_a Z_s \cos\beta)^2$$

i.e., $$300.22^2 = (231 + 4I_a \sin\beta)^2 + (31.75 \times 4)^2$$

i.e., $$231 + 4I_a \sin\beta = \sqrt{300.22^2 - 127^2} = 272.03$$

i.e., $$I_a \sin\beta = 41.03/4 = 10.26\text{ A}$$

$$\tan\beta = \frac{I_a \sin\beta}{I_a \cos\beta} = \frac{10.26}{31.75} = 0.323$$

$$\beta = 17.9°$$

∴ $$\cos\beta = 0.951\text{ (leading)}$$

$$I_a = \frac{31.75}{\cos\beta} = \frac{31.75}{0.951} = 33.38\text{ A}$$

The current drawn by the motor is 33.38 A and the power factor is 0.951 (leading) ■

Example 7.15 A three-phase, 415 V, six-pole, 50 Hz star-connected synchronous motor has emf of 520 V (L-L). The stator winding has a synchronous reactance of 2 Ω per phase and the motor develops a torque of 220 N-m. The motor is operating at 415 V, 50 Hz bus. Calculate the current drawn from the supply and its power factor. Assume R_a/phase = 0 Ω.

Solution

Per phase supply voltage $(V_t) = \frac{415}{\sqrt{3}} = 239.6\text{ V}$

Induced emf per phase $(E_f) = \frac{520}{\sqrt{3}} = 300.22\text{ V}$

Synchronous speed of the motor $(N_s) = \frac{120f}{P} = \frac{120 \times 50}{6} = 1{,}000\text{ rpm}$

Torque developed (T) = 220 N-m

Total power developed $= \frac{T \times 2\pi N_s}{60} = \frac{220 \times 2 \times 3.14 \times 1{,}000}{60} = 23{,}026.67\text{ W}$

Power developed per phase $= \frac{23{,}026.67}{3} = 7{,}675.56\text{ W}$

$$Z_s/\text{phase} = 2\ \Omega \text{ and } \theta = 90°$$

Power developed per phase is given by

$$\frac{E_f V_t}{Z_s}\cos(\theta - \delta) - \frac{E_f^2}{Z_s}\cos\theta = \frac{E_f V_t}{Z_s}\sin\delta \text{ [Since } q = 90°]$$

Therefore, $$\frac{E_f V_t}{Z_s}\sin\delta = 7{,}675.56$$

i.e., $\sin\delta = \dfrac{7{,}575.56 Z_s}{E_f V_t} = \dfrac{7{,}675.56 \times 2}{300.22 \times 239.6} = 0.2134$

∴ $\delta = 12.32°$

Since $E_f > V_t$, we have

$$E_R^2 = \sqrt{V_t^2 + E_f^2 - 2V_t E_f \cos\delta}$$

$$= \sqrt{239.6^2 + 300.22^2 - 2 \times 239.6 \times 300.22 \times \cos 12.32°} = 83.27 \text{ V}$$

Current drawn per phase $(I_a) = \dfrac{E_R}{Z_s} = \dfrac{83.27}{2} = 41.63 \text{ A}$

Again, $\dfrac{E_f}{\sin(\theta+\beta)} = \dfrac{E_R}{\sin\delta}$

i.e., $\sin(\theta+\beta) = \dfrac{E_f}{E_R} \times \sin\delta = \dfrac{300.22}{83.27} \times \sin 12.32° = 0.7692$

i.e., $\theta + \beta = 140.28°$

∴ $\beta = 140.28° - \theta = 140.28° - 90° = 50.28°$

∴ Power factor = cos 50.28° = 0.6390 (leading) ■

Example 7.16 A 500 V, three-phase, star-connected synchronous motor has a resistance and a synchronous reactance of 0.4 Ω and 3.6 Ω, respectively. The open-circuit voltage is 600 V. If friction and core losses are 1 kW, calculate the line current and the power factor when the motor output is 62 kW.

Solution

Supply voltage per phase is $V_t = \dfrac{500}{\sqrt{3}} = 288.7$

Induced emf per phase is $E_f = \dfrac{600}{\sqrt{3}} = 346.41 \text{ V}$

Here, $E_f > V_t$, the power factor is leading.

Total mechanical power developed = 62 + 1 = 63 kW

Power developed per phase $= \dfrac{63}{3} = 21 \text{ kW} = P_{om}$

$$Z_s/\text{phase} = \sqrt{R_a^2 + X_s^2} = \sqrt{0.4^2 + 3.6^2} = 3.622\,\Omega$$

$$\theta = \tan^{-1}\left(\frac{X_s}{R_a}\right) = \tan^{-1}\left(\frac{3.6}{0.4}\right) = 83.66°$$

Power developed per phase is given by

$$\frac{E_f V_t}{Z_s}\cos(\theta-\delta) - \frac{E_f^2}{Z_s}\cos\theta = P_{om}$$

i.e., $\dfrac{288.7 \times 346.41}{3.622} \times \cos(83.66° - \delta) - \dfrac{346.41^2}{3.622} \times \cos 83.66° = 21{,}000$

i.e., $$\cos(83.66° - \delta) = \left(\frac{3.622}{288.7 \times 346.41}\right) \times \left(\frac{346.41^2}{3.622}\cos 83.66° + 21{,}000\right) = 0.8930$$

i.e., $$\delta = 83.66° - 26.74° = 56.92°$$

We can write,

$$E_R = \sqrt{V_t^2 + E_f^2 - 2V_t E_f \cos\delta}$$

$$= \sqrt{288.7^2 + 346.41^2 - 2 \times 288.7 \times 346.41 \times \cos 56.92°} = 306.88\text{ V}$$

Since the motor is star connected, line current = phase current.

$$I_a = \frac{306.88}{3.622} = 84.73\text{ A}$$

Again, $$\frac{E_f}{\sin(\theta + \beta)} = \frac{E_R}{\sin\delta}$$

i.e., $$\sin(\theta + \beta) = \frac{E_f}{E_R} \times \sin\delta = \frac{346.41}{306.88} \times \sin 56.92° = 0.9458$$

i.e., $$\theta + \beta = 108.95°$$

i.e., $$\beta = 108.95° - 83.66° = 25.29°$$

∴ Power factor = cos25.29° = 0.904 [leading]

Example 7.17 The excitation of a 415 V, three-phase, mesh-connected synchronous motor is such that the induced emf is 520 V. The impedance per phase is (0.5 + *j*4.0) Ω. If the friction and iron losses are constant at 1,000 W, calculate the power output, line current, power factor and efficiency for maximum power output.

Solution

Supply voltage per phase V_t = 415 V, induced emf per phase E_f = 520 V

Synchronous impedance/phase $(Z_s) = \sqrt{0.5^2 + 4^2} = 4.03\,\Omega$

Impedance angle $(\theta) = \tan^{-1}\left(\frac{X_s}{R_a}\right) = \tan^{-1}\left(\frac{4}{0.5}\right) = 82.9°$

Maximum power will be developed for the given motor,

$$\delta = \theta = 82.9°$$

Here, $E_f > V_t$, therefore, the power factor is leading.

$$E_R = \sqrt{V_t^2 + E_f^2 - 2V_t E_f \cos\delta} = \sqrt{415^2 + 520^2 - 2 \times 415 \times 520 \times \cos 82.9°}$$

$$= 623.92\text{ V}$$

$$I_a = \frac{E_R}{Z_s} = \frac{623.92}{4.03} = 154.82\text{ A}$$

Line current (I_L) $= \sqrt{3} \times 154.82 = 268.15\text{ A}$

The maximum power developed by the motor is,

$$\left(P_{om}\right)_{\max} = 3\times\left[\frac{E_f V_t}{Z_s} - \frac{E_f^2}{Z_s}\cos\theta\right] = 3\times\left[\frac{415\times 520}{4.03} - \frac{520^2}{4.03}\times\cos 82.9°\right] = 1,35,765.38\text{ W}$$

Output power = power developed − iron and friction losses

$$= 1,35,765.38 - 1,000 = 1,34,765.38\text{ W}$$

$$\text{Total Cu-loss} = 3I_a^2 R_a = 3\times 154.82^2\times 0.5 = 35953.85\text{ W}$$

Input power = power developed + Cu losses

$$= 1,34,765.38 + 35,953.85 = 1,70,719.23\text{ W}$$

$$\text{Power factor }(\cos\beta) = \frac{P_{in}}{\sqrt{3}V_L I_L} = \frac{170719.23}{\sqrt{3}\times 415\times 268.15} = 0.8857$$

$$\text{Efficiency} = \frac{\text{Output}}{\text{Input}} = \frac{134765.38}{170719.23} = 0.7894\text{ p.u.} = 78.94\%$$

■

Example 7.18 A three-phase, 400 V, 40 kVA, 50 Hz, star-connected synchronous motor has a synchronous reactance of 4 Ω. Its friction and windage losses are 1.8 kW and core losses are 1.2 kW. The shaft load is 12 kW initially and the power factor of the motor is 0.8 leading.

(i) Calculate the line current, armature current and excitation emf.

(ii) Calculate the new values of armature current and the motor power factor if the shaft load is increased to 22 kW. Excitation voltage is kept constant.

(iii) Synchronizing power/electrical degree.

Solution

$$P_{in} = 12 + 1.8 + 1.2 = 15\text{ kW}$$

$$V_t = \frac{400}{\sqrt{3}} = 231\text{ V}$$

$$I_L = \frac{P_{in}}{\sqrt{3}\times V_L\cos\beta} = \frac{15\times 10^3}{\sqrt{3}\times 400\times 0.8} = 27.06\text{ A}$$

Since the motor is star connected, armature current = line current.

∴ Armature current = 27.06 A

Here,

$$Z_s = 4\ \Omega \text{ and } \theta = 90°$$

$$E_R = 27.06\times 4 = 108.24\text{ V}$$

$$\beta = \cos^{-1}(0.8) = 36.87°$$

$$E_f = \sqrt{V_t^2 + E_R^2 - 2V_t E_R\cos(\theta+\beta)}$$

$$= \sqrt{231^2 + 108.24^2 - 2\times 231\times 108.24\times\cos(90° + 36.87°)} = 308.35\text{ V}$$

The line current, armature current and excitation voltage are 27.06 A, 27.06 A and 308.35 V, respectively.

(ii) E_f is kept constant and shaft load is interconnected to 22 kW.

$$E_f = 308.35 \text{ V}$$

Since $E_f > V_t$, the power factor is leading in nature.

$$P_{in} = 22 + 1.8 + 1.2 = 25 \text{ kW}$$

Again,
$$P_{in} = \frac{3E_f V_t}{X_s} \sin\delta$$

$$\sin\delta = \frac{P_{in} X_s}{3E_f V_t} = \frac{25\times10^3\times4}{3\times308.35\times231} = 0.4679$$

∴ $$\delta = 27.9°$$

∴ $$E_R = \sqrt{V_t^2 + E_f^2 - 2V_t E_f \cos\delta}$$

$$= \sqrt{231^2 + 308.35^2 - 2\times231\times308.35\times\cos 27.9°} = 150.14 \text{ V}$$

∴ $$I_a = \frac{E_R}{X_s} = \frac{150.14}{4} = 37.53 \text{ A}$$

We know that
$$\sin(\theta+\beta) = \frac{E_f}{E_R}\times\sin\delta = \frac{308.35}{150.14}\times\sin 27.9° = 0.9611$$

i.e.,
$$\theta+\beta = 106.04°$$

i.e.,
$$\beta = 106.04° - \theta = 106.05° - 90° = 16.04°$$

∴ $$\cos\beta = 0.961 \text{ (leading)}.$$

Therefore, the new armature current and power factor are 37.53 A and 0.961 (leading).

(iii)
$$P_{syn}/\text{electrical degree} = \left(\frac{dP}{d\delta}\right)\times\frac{\pi}{180} = \frac{3E_f V_t}{X_s}\times\cos\delta\times\frac{\pi}{180}$$

$$= \frac{3\times308.35\times231}{4}\times\cos 27.9°\times\frac{3.14}{180} = 823.59 \text{ W}$$ ■

Example 7.19 A three-phase, 400 V, 40 kVA star-connected synchronous motor is supplying 15 kW load with 0.8 power factor lagging. The windage and friction losses are 1.5 kW and the core losses are 1.0 kW. Calculate the following:

(i) Armature current and excitation voltage.

(ii) Armature current and power factor if the excitation is increased by 40 per cent and power supplied to the load remains constant.

Solution

(i)
$$P_{in} = 15 + 1.5 + 1.0 = 17.5 \text{ kW}$$

$$V_t = \frac{400}{\sqrt{3}} = 231 \text{ V}$$

$$I_L = \frac{P_{in}}{\sqrt{3}\times V_L\cos\beta} = \frac{17.5\times10^3}{\sqrt{3}\times400\times0.8} = 31.57 \text{ A}$$

Since the motor is star connected, armature current = line current.

$$I_a = 31.57 \text{ A.}$$

Here, $Z_s = 4\ \Omega$ and $\theta = 90°$

$$E_R = 31.57 \times 4 = 126.28 \text{ V}$$

$$\beta = \cos^{-1}(0.8) = 36.87°$$

$$E_f = \sqrt{V_t^2 + E_R^2 - 2V_tE_R\cos(\theta - \beta)}$$

$$= \sqrt{231^2 + 126.28^2 - 2 \times 231 \times 126.28 \times \cos(90° - 36.87°)} = 185.21 \text{ V}$$

Again, $\sin\delta = \dfrac{E_R}{E_f} \times \sin(\theta - \beta) = \dfrac{126.28}{231} \times \sin(90° - 36.87°) = 0.4373$

So, $\delta = 25.93°$

(ii) $E_{f1} = 1.4\ E_f = 1.4 \times 185.2 = 259.28$ V

Since the power supplied to the load remains constant, we have

$$E_{f1} \sin\delta_1 = E_f \sin\delta$$

i.e., $\sin\delta_1 = \dfrac{E_f}{E_{f1}} \times \sin\delta = \dfrac{1}{1.4} \times 0.4373 = 0.3124$

i.e., $\delta_1 = 18.20°$

Since $E_{f1} > V_t$, power factor is leading.

$$E_R = \sqrt{V_t^2 + E_f^2 - 2V_tE_f\cos\delta_1}$$

$$= \sqrt{231^2 + 259.28^2 - 2 \times 231 \times 259.28 \times \cos 18.20°} = 82.416 \text{ V}$$

$$I_a = \frac{E_R}{Z_s} = \frac{82.416}{4} = 20.6 \text{ A}$$

Again, $\dfrac{E_f}{\sin(\theta + \beta_1)} = \dfrac{E_R}{\sin\delta_1}$

i.e., $\sin(\theta + \beta_1) = \dfrac{E_f}{E_R} \times \sin\delta_1 = \dfrac{259.28}{82.416} \times \sin 18.20° = 0.9826$

i.e., $\theta + \beta_1 = 100.7°$

i.e., $\beta_1 = 100.7° - \theta = 100.7° - 90° = 10.7°$

Therefore, power factor = $\cos\beta_1 = \cos 10.7° = 0.9826$ (leading) ■

Example 7.20 A three-phase, 30 kW, 400 V, star-connected synchronous motor operates on full load at 0.8 power factor lagging. The machine has a synchronous reactance of 4 Ω and negligible armature resistance. Calculate the new value of current and the power factor if excitation is increased by 50 per cent.

Solution

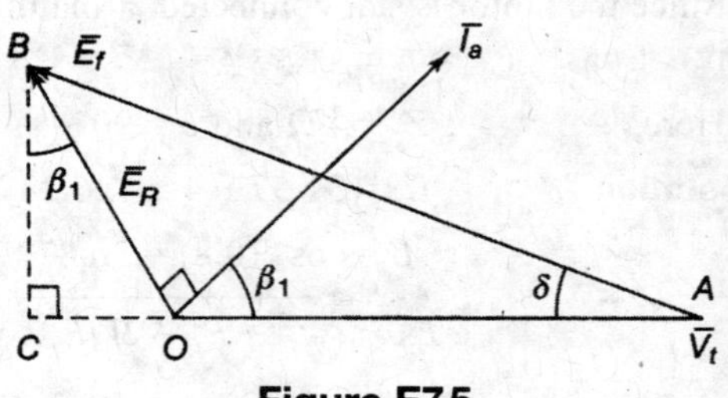

Figure E7.5

$$I_a = \frac{P_{in}}{\sqrt{3}V_L \cos\beta} = \frac{30\times 10^3}{\sqrt{3}\times 400\times 0.8} = 54.13\,\text{A}$$

$\therefore$ $$V_t = \frac{400}{\sqrt{3}} = 231\,\text{V}$$

Here, $Z_s = 4\ \Omega$ and $\theta = 90°$

Since $\cos\beta = 0.8$ (lagging), $\beta = 36.87°$

$$E_R = I_a Z_s = 54.13\times 4 = 216.52\ \text{V}$$

$$E_f = \sqrt{V_t^2 + E_R^2 - 2V_tE_R\cos(\theta-\beta)}$$

$$= \sqrt{231^2 + 216.52^2 - 2\times 231\times 216.52\times\cos(90° - 36.87°)} = 200.55\ \text{V}$$

$$E_{f1} = 1.5\times E_f = 1.5\times 200.55 = 300.825\ \text{V}$$

Since $E_{f1} > V_t$, power factor is leading.

Let the armature current be I_{a1} and the power factor be $\cos\beta_1$ (leading).

$\therefore$ $$I_{a1}\cos\beta_1 = \frac{30\times 10^3}{\sqrt{3}\times 400} = 43.30\,\text{A}$$

From Figure E7.5, we have

$$BC = E_R\cos\beta_1 \text{ and } OC = E_R\sin\beta_1$$

$$E_f^2 = CA^2 + CB^2 = \left(V_t + I_{a1}Z_s\sin\beta_1\right)^2 + \left(I_{a1}Z_s\cos\beta_1\right)^2$$

i.e., $$300.825^2 = \left(231 + 4I_{a1}\sin\beta_1\right)^2 + \left(4\times 43.30\right)^2$$

i.e., $$231 + I_{a1}\sin\beta_1 = 245.96$$

i.e., $$I_{a1}\sin\beta_1 = 14.96$$

i.e., $$\tan\beta_1 = \frac{I_{a1}\sin\beta_1}{I_{a1}\cos\beta_1} = \frac{14.96}{43.30} = 0.3455$$

i.e., $$\beta_1 = 19°$$

i.e., $$\cos\beta_1 = 0.9455 \text{ (leading)}$$

$$I_{a1} = \frac{43.30}{\cos\beta_1} = \frac{43.30}{0.9455} = 45.8\,\text{A}$$

Therefore, the armature current and the power factor of the motor are 45.8 A and 0.9455 (leading) ■

Example 7.21 A three-phase, 18 MVA, 10-pole, 50 Hz, 11 kV, star-connected synchronous motor has $X_d = 5\ \Omega$ and $X_q = 3\ \Omega$, respectively. It has negligible armature resistance. Calculate the following on full load at 0.8 power factor leading:

(i) Excitation voltage.

(ii) Power.

(iii) Synchronizing power/electrical degree
(iv) Synchronizing power/mechanical degree
(v) Maximum value of power angle and the corresponding power

Solution

(i) $V_t = \dfrac{11\times10^3}{\sqrt{3}} = 6{,}350.85\text{ V}$ and $I_a = \dfrac{18\times10^6}{\sqrt{3}\times11\times10^3} = 944.75\text{ A}$

$$\tan\delta = \frac{I_aX_q\cos + I_aR_a\sin\beta}{V_t + I_aX_q\sin\beta - I_aR_a\cos\beta} = \frac{944.75\times3\times0.8-0}{6{,}350.85+944.75\times3\times0.6-0} = 0.2816$$

So, $\delta = 15.73°$

$$I_d = I_a\sin(\beta+\delta) = 944.75\times\sin(36.87°+15.73°) = 750.52\text{ A}$$

$$I_q = I_a\cos(\beta+\delta) = 944.75\times\cos(36.87°+15.72°) = 573{:}95\text{ A}$$

$$E_f = V_t\cos\delta - I_qR_a + I_dX_d = 6{,}350.85\times\cos15.72° - 573.95\times0 + 750.24\times5 = 9{,}864.51\text{ V}$$

(ii) $$P_{3-phase} = 3\times\left[\frac{E_fV_t}{X_d}\times\sin\delta + \frac{V_t^2}{2}\left(\frac{1}{X_q}-\frac{1}{X_d}\right)\sin2\delta\right]$$

$$= 3\times\left[\frac{9{,}864.51\times6{,}350.85}{5}\times\sin15.72° + \frac{6{,}350.85^2}{2}\times\left(\frac{1}{3}-\frac{1}{5}\right)\times\sin31.44°\right]$$

$$= 14.3918\text{ MW}$$

$$P_{in} = 18\times0.8 = 14.40\text{ MW}$$

$$P_{3-phase} = P_{in}$$

(iii) $$P_{syn} = \left(\frac{dP}{d\delta}\right)\times\frac{\pi}{180}\text{ W/electrical degree}$$

$$= 3\left[\frac{E_fV_t}{X_d}\cos\delta + V_t^2\left(\frac{1}{X_q}-\frac{1}{X_d}\right)\cos2\delta\right]$$

$$= 3\times\left[\frac{9{,}864.51\times6{,}350.85}{5}\times\cos15.72° + 6{,}350.83^2\times\left(\frac{1}{3}-\frac{1}{5}\right)\times\cos31.44°\times\frac{3.14}{180}\right]$$

$$= 8{,}71{,}308.54\text{ W/electrical degree}$$

(iv) Synchronizing power/mechanical degree

$$= \left(\frac{dP}{d\delta}\right)\times\frac{\pi}{180}\times p = 8{,}71{,}308.54\times10 = 87{,}13{,}085.4\text{ W}$$

(v) $\cos\delta = -\dfrac{E_f X_q}{4V_t(X_d - X_q)} + \sqrt{\dfrac{1}{2} + \left[\dfrac{E_f X_q}{4V_t(X_d - X_q)}\right]^2}$

$$= -\frac{9,864.51 \times 3}{4 \times 6,350.85 \times 2} + \sqrt{\frac{1}{2} + \left(\frac{9,864.51 \times 3}{4 \times 6,350.85 \times 2}\right)^2} = 0.3336$$

i.e., $\delta = 70.51°$

$$\therefore \quad P_{max} = 3 \times \left[\frac{E_f V_t}{X_d}\sin\delta + \frac{V_t^2}{2}\left(\frac{1}{X_q} - \frac{1}{X_d}\right)\sin 2\delta\right]$$

$$= 3 \times \left[\frac{9,864.51 \times 6,350.85}{5} \times \sin 70.51° + \frac{6,350.85^2}{2} \times \left(\frac{1}{3} - \frac{1}{5}\right) \times \sin 141.02°\right] \text{ W}$$

$= 40.51$ MW ■

SIGNIFICANT POINTS

- *For lagging power factor:* $\sin\delta = \dfrac{E_R \sin(\theta - \beta)}{E_f}$
- *For leading power factor:* $\sin\delta = \dfrac{E_R \sin(\theta + \beta)}{E_f}$
- *For unity power factor:* $\sin\delta = \dfrac{E_R \sin\theta}{E_f}$

where pf = $\cos\beta$ and δ is the power angle.

- *Comparison of various excitations:*

Type of excitation	Nature of powers factor	Comparison of E_f and V_t
Normal excitation	Lagging	$E_f = V_t$
Critical excitation	Unity power factor	$E_f = V_t$
Underexcitation	Lagging power factor	$E_f < V_t$
Overexcitation	Leading power factor	$E_f > V_t$

- *Expression for real power:*

$$P_{in} = \frac{V_t^2}{Z_s}\cos\theta - \frac{V_t E_f}{Z_s}\cos(\delta + \theta) = \frac{V_t^2}{Z_s}\sin\alpha + \frac{V_t E_f}{Z_s}\sin(\delta - \alpha)$$

- *Expression for reactive power:*

$$Q_{in} = \frac{V_t^2}{Z_s}\sin\theta - \frac{V_t E_f}{Z_s}\sin(\delta + \theta) = \frac{V_t^2}{Z_s}\cos\alpha - \frac{V_t E_f}{Z_s}\cos(\delta - \alpha)$$

- *Maximum output power:*

$$(P_{om})_{max} = \frac{V_t E_f}{Z_s} - \frac{E_f^2}{Z_s^2} R_a$$

- *When armature resistance is negligible:*

$$P_{om} = \frac{V_t E_f}{X_s} \sin\delta$$

$$\therefore \quad P_{om} = P_m \sin\delta$$

where $P_m = \dfrac{V_t E_f}{X_s}$ is the maximum power developed by the motor.

Gross torque developed by the motor is,

$$T_g = \frac{P_{om}}{\omega_s} = \frac{P_{om}}{2\pi n_s}$$

where n_s is the synchronous speed of the motor in rps.

Shaft torque of the motor is given by

$$T_{sh} = \frac{P_{sh}}{\omega_s} = \frac{P_{sh}}{2\pi n_s}$$

where $P_{sh} = P_{om}$ − windage and friction losses

Various cases:

(i) If $|E_f| \cos\delta > |V_t|$, motor is overexcited. It delivers reactive power to the infinite bus. Under this condition motor is operating at leading power factor.

(ii) If $|E_f| \cos\delta = |V_t|$, motor is normally excited. It neither delivers to nor absorbs reactive power from the infinite bus. Under this condition motor is operating at unity power factor.

(iii) If $|E_f| \cos\delta < |V_t|$, motor is underexcited. It absorbs reactive power from the infinite bus. Under this condition, motor is operating at lagging power factor.

- *Salient-pole type:*

For lagging power factor:

$$\tan\delta = \frac{I_a X_q \cos\beta - I_a R_a \sin\beta}{V_t - I_a R_a \cos\beta - I_a X_q \sin\beta}$$

$$\tan\gamma = \frac{V_t \sin\theta - I_a X_q}{V_t \cos\theta - I_a R_a}$$

$$E_f = V_t \cos\delta - I_q R_a - I_d X_d$$

For leading power factor:

$$\tan\delta = \frac{I_a X_q \cos\beta + I_a R_a \sin\beta}{V_t + I_a X_q \sin\beta - I_a R_a \cos\beta}$$

$$\tan\gamma = \frac{V_t \sin\beta + I_a X_q}{V_t \cos\beta - I_a R_a}$$

$$E_f = V_t \cos\delta - I_q R_a + I_d X_d$$

For unity power factor:

$$\tan\delta = \frac{I_a X_q}{V_t - I_a R_a}$$

Power developed:

$$P_{in-3\phi} = \frac{3E_f V_t \sin\delta}{X_d} + \frac{3V_t^2}{2}\left(\frac{1}{X_q} - \frac{1}{X_q}\right)\sin 2\delta$$

Condition for maximum power:

$$\cos\delta = \frac{-E_f V_t}{4V_t\left(X_d - X_q\right)} + \sqrt{\frac{1}{2} + \left[\frac{E_f V_t}{4V_t\left(X_d - X_q\right)}\right]^2}$$

SHORT QUESTIONS AND ANSWERS

Q.1 What is a synchronous motor and where is it used?

Ans. A synchronous motor is a machine in which a DC supply is given to the rotor and a three-phase AC power supply is given to the stator. It is used to improve the power factor of the line as well as for the load at constant speed.

Q.2 Why is it called synchronous motor?

Ans. Since the motor runs at synchronous speed in synchronism with the supply frequency, the motor is called synchronous motor.

Q.3 What is synchronous speed?

Ans. Yes, it runs at constant synchronous speed that is given by the equation $N_s = \frac{120f}{p}$ rpm where f= supply frequency, P = number of poles and N_s = synchronous speed of the motor in rpm.

Q.4 Can we vary the speed of a synchronous motor in running condition?

Ans. No.

Q.5 How can the speed of a synchronous motor be changed?

Ans. The speed of a synchronous motor can be changed by varying the frequency with special arrangement.

Q.6 In which motor do the stator field and the rotor field rotate simultaneously?

Ans. Stator field and rotor field rotate simultaneously in a synchronous motor.

Q.7 In which motor is the air gap more than that of the other motors?

Ans. The air gap is more in salient-pole synchronous motor compared to other motors.

Q.8 From where is DC excitation given in the field of synchronous motor?

Ans. The DC excitation is provided by an exciter (DC generator) driven either by mounting on the same shaft or by a separate motor.

Q.9 What do you mean by exciter? How much voltage is induced by this exciter?

Ans. Exciter is a DC shunt or compound generator by which the voltage is induced upto 250 volts.

Q.10 Is the synchronous motor self-started?

Ans. No

Q.11 Why is it necessary to employ special starting equipment for starting a synchronous motor?

Ans. Special starting equipment is necessary because starting torque of synchronous motor is practically zero.

Q.12 What are the methods of starting a synchronous motor?

Ans. Various methods of starting a synchronous motor are as follows:

(i) By means of a DC compound motor directly coupled or belted to the main motor.

(ii) By means of a pony motor (a small induction motor directly coupled or bolted to the main motor).

(iii) The motor is made self-starting by providing damper winding on motor poles.

Q.13 How does a synchronous motor improve power factor?

Ans. By keeping load constant and running the motor overexcited, leading current is drawn from the supply mains thus improving the power factor.

Q.14 What will happen to the load current if the field excitation is changed under constant load?

Ans. The load current will be increased under both over- and underexcitation.

Q.15 What will happen if a synchronous motor is underexcited?

Ans. The motor will draw lagging current thus decreasing the power factor.

Q.16 When will be the magnitude of the load current be minimum?

Ans. The magnitude of the load current will be minimum when the power factor is unity.

Q.17 What are *V* curves?

Ans. The plot of armature current (I_a) with the field current (I_f) takes the shape of English alphabet *V*. This family of curves for different field excitations is called *V* curves.

SUPPLEMENTARY PROBLEMS

1. A three-phase, 400 V, star-connected synchronous motor takes 52.5 A at a power factor of 0.8 leading. Calculate the power supplied and the induced emf. The motor impedance per phase is $(0.25 + j\,3.2)\ \Omega$.

   [*Ans.* 29.1 kW, 612 V (line)]

2. The synchronous reactance per phase of a three-phase star-connected 6.6 kV synchronous motor is 20 Ω. For a certain load the input is 915 kW at normal voltage and the induced emf is 8,942 V. Neglecting armature resistance, evaluate the line current and the power factor.

   [*Ans.* 97 A, 0.8256 (leading)]

3. A three-phase synchronous motor of 8 kW at 1.1 kV has a synchronous reactance of 8 Ω/phase. Calculate the minimum current and the corresponding induced emf for full-load condition. The efficiency of the machine is 0.8. Neglect the armature resistance.

   [*Ans.* 5.25 A, 637 V]

4. A 400 V, three-phase, 50 Hz, star-connected synchronous motor has a rated armature current of 25 A. The armature resistance is negligible. X_d = 5 Ω/phase and X_q = 3.2 Ω/phase. For an input of 10 kW, calculate the power angles δ for excitation voltages (i) 420 V and (ii) 400 V, respectively.

[*Ans.* (i) 45.52° (ii) 46.55°]

5. A three-phase synchronous motor absorbing 60 kW is connected in parallel with a load of 240 kW having a lagging power factor of 0.8. If the combined load has a power factor of 0.9, calculate the value of leading kVA supplied by the motor and its working power factor.

[*Ans.* 69.284 kVA, 0.866 leading]

6. A three-phase star-connected synchronous motor has a synchronous reactance 3.8 Ω/phase and it works at 1.1 kV bus bar. If the excitation is such that the generated emf is 1.2 kV, find its working power factor. Neglect its armature resistance. The motor takes 100 kW from supply mains.

[*Ans.* 0.99]

7. An 18.75 kW (input), 220 V, 50 Hz, three-phase, four-pole star-connected synchronous motor runs at a load angle of 4° and a back emf of 110 V/phase. If the motor impedance is $0.1 + j\,1.5$ Ω/phase, calculate (i) armature current/phase, (ii) motor power factor and (iii) total Cu loss.

[*Ans.* (i) 661.398 A, (ii) 0.074 lagging, (iii) 131.234 kW]

8. A three-phase, 6.6 kV, star-connected synchronous motor takes a current of 72 A at 0.8 power factor leading having effective resistance and reactance per phase of 0.1 Ω and 0.9 Ω, respectively. Calculate the emf induced and the total power input. Also find the motor efficiency if the stray losses are 25.6 kW. The field Cu losses may be neglected.

[*Ans.* 6,658.04 V (line), 95.87%]

MULTIPLE-CHOICE QUESTIONS AND ANSWERS

1. At constant load, the magnitude of armature current of a synchronous motor has large values for
   (a) low values of field excitation only (b) high values of field excitation only
   (c) both low values and high values of field excitation (d) none of these
2. In a synchronous motor, minimum armature current corresponds to
   (a) lagging power factor (b) leading power factor
   (c) zero power factor (d) unity power factor
3. Damper winding is provided in synchronous motors
   (a) to provide starting torque (b) to prevent hunting
   (c) to reduce speed (d) to increase speed
4. Ina synchronous motor, the inverted *V* curve represents the relation between
   (a) field current and power factor (b) field current and armature current
   (c) armature current and power factor (d) none of these
5. The power input to the DC field of a 15 kVA (input) synchronous motor is 6 per cent of rated AC input. The direct current that excites the field at 120 V is
   (a) 12.5 A (b) 7.5 A
   (c) 15 A (d) 10 A

6. A synchronous motor, connected to an infinite bus, is working at a leading power factor. Its excitation voltage is
   (a) $> V_t$
   (b) $= V_t$
   (c) $< V_t$
   (d) $\leq V_t$
7. The motor with the highest efficiency is
   (a) induction motor with low slip
   (b) induction motor with medium slip
   (c) synchronous motor
   (d) DC compound motor
8. The motor that is not self-starting is
   (a) synchronous motor
   (b) DC series motor
   (c) induction motor with medium slip
   (d) induction motor with high slip
9. A synchronous capacitor is
   (a) an ordinary static capacitor bank
   (b) an overexcited synchronous motor
   (c) an overexcited synchronous motor running without load
   (d) none of these
10. A synchronous motor is supplying a load at unity power factor. If the load on the motor is increased keeping its excitation and the terminal voltage constant, the power factor
    (a) will remain the same
    (b) will become leading
    (c) will become lagging
    (d) none of the above
11. Armature reaction in a synchronous motor at rated voltage and zero power factor (lead) is
    (a) magnetizing
    (b) cross-magnetizing
    (c) both (a) and (b)
    (d) demagnetizing
12. A synchronous motor will deliver maximum power when
    (a) load angle is equal to input power angle
    (b) input power factor is unity
    (c) load angle is 45°
    (d) load angle is 0°
13. A synchronous motor can operate over a wide range of power factors, that is, from lagging to leading power factors. This is achieved by
    (a) changing the field excitation
    (b) varying the speed
    (c) varying the applied voltage
    (d) changing the load
14. A salient-pole synchronous motor is fed from an infinite bus and is running at no load. Its field current is reduced to zero. The motor would
    (a) stop
    (b) run at reduced speed
    (c) run at synchronous speed
    (d) run above synchronous speed
15. In power factor improvement, the synchronous motor is to be
    (a) lightly loaded and overexcited
    (b) lightly loaded and overexcited
    (c) heavily loaded and underexcited
    (d) lightly loaded and underexcited
16. A three-phase synchronous motor having constant excitation drives a load taking power from an infinite bus at leading power factor. If the shaft load increases, then
    (a) the power angle decreases while the power factor increases
    (b) both power angle and power factor decrease
    (c) both power angle and power factor increase
    (d) the power angle increases and the power factor decreases

17. A synchronous motor and an alternator are delivering about 5 per cent of their rated power at unity power factor. If these machines are loaded to their rated capacity, their armature currents
    (a) decrease, motor operates at lagging power factor and alternator at leading power factor
    (b) increase, motor operates at leading power factor and alternator at lagging power factor
    (c) increase, both operate at lagging power factors
    (d) increase, motor operates at lagging power factor and alternator at leading power factor
18. A synchronous motor is running from an infinite bus of voltage V_t in steady state, at about 50 per cent of its rated load with a power angle δ_1 between V_t and I_f (induced voltage). The load is suddenly decreased to 25 per cent. E_f attains its new steady state power angle δ_2 with V_t by initially
    (a) falling behind and making a complete rotation
    (b) advancing and making a complete rotation
    (c) falling behind, followed by oscillation about δ_2
    (d) advancing, followed by oscillation about δ_2
19. An ideal synchronous motor has no starting torque because
    (a) rotor is made up of salient-poles
    (b) relative velocity between the stator and the rotor mmfs is zero
    (c) relative velocity between stator and rotor mmfs is not zero
    (d) rotor winding is not highly reactive
20. A synchronous motor with negligible armature resistance runs at a load angle of 20° at the rated frequency. If supply frequency is increased by 10 per cent, keeping other parameters constant, the new load angle will be
    (a) 16°
    (b) 18°
    (c) 20°
    (d) 22°

Answers

1 (a)	2 (d)	3 (a), (b)	4 (a)	5 (b)	6 (a)
7 (c)	8 (a)	9 (c)	10 (c)	11 (c)	12 (a)
13 (a)	14 (b)	15 (b)	16 (a)	17 (b)	18 (c)
19 (b)	20 (d)				

Polyphase Induction Motors

8

The basics as well as the details of the structures and the types of rotors of polyphase induction motors have been introduced in Chapter 3. The concept of rotating magnetic field in the air gap has also been introduced in the chapter. In induction motors, we need to apply a three-phase supply at the stator, whereas no supply is required at the rotor. That is why the three-phase induction motors are known as singly excited motors. Since these types of motors rotate at a speed below the synchronous speed (known as asynchronous speed), they are also called asynchronous motors. This chapter introduces the details of the theoretical aspects of the induction motors.

8.1 ROTOR CURRENT

If the electromagnetic force (emf) generated in the rotor at standstill is E_2 per phase. The frequency of the generated emf at standstill is also equal to the supply frequency because the slip at standstill is 1. Let the slip be s when the speed of the rotor is N_r. The emf induced in the rotor per phase is sE_2. The frequency of the induced emf of the rotor is sf.

Rotor leakage reactance at standstill $= 2\pi fl_2\ \Omega = X_2\ \Omega$

Rotor leakage reactance at any slip s is $= 2\pi\, sfl_2 = sX_2\ \Omega$

If R_2 be the resistance of the rotor, the rotor leakage impedance at standstill is $= \sqrt{R_2^2 + X_2^2}\ \Omega$

At any slip s, rotor leakage impedance given by $\sqrt{R_2^2 + s^2X_2^2}\ \Omega$

Per phase rotor current at standstill is given by $\dfrac{E_2}{\sqrt{R_2^2 + X_2^2}}$ A

Per phase rotor current at any slip 's' is given by $(I_2) = \dfrac{sE_2}{\sqrt{R_2^2 + (sX_2)^2}}$ A

Example 8.1 A three-phase slip-ring induction motor gives a reading of 70 V across the slip rings on open circuit when it is at rest with normal stator voltage applied. The rotor has impedance of $(0.7 + j8)$ Ω/phase when it is star connected. Calculate the rotor current and the rotor power factor for the following cases:

(i) At standstill with slip rings joined to a star-connected starter with a phase impedance of $(6 + j5)$ Ω/phase.

(ii) Running normally with a slip of 3.5 per cent.

Solution

Impedance of the rotor per phase is $0.7 + j8\ \Omega$ and phase impedance of the star-connected starter is $6 + j5\ \Omega$.

(i) *At standstill:*

Induced emf in rotor winding per phase is $E_2 = \dfrac{70}{\sqrt{3}} = 40.41$ V

Resistance per phase in rotor circuit is $R_2 = 0.7 + 6 = 6.7\ \Omega$

Reactance per phase in the rotor circuit is $X_2 = 8 + 5 = 13\ \Omega$

Impedance per phase in the rotor circuit is $Z_2 = \sqrt{R_2^2 + X_2^2} = \sqrt{6.7^2 + 13^2} = 14.625\ \Omega$

Rotor current per phase $= \dfrac{E_2}{Z_2} = \dfrac{40.41}{14.625} = 2.763$ A

Rotor power factor $(\cos\theta_2) = \dfrac{R_2}{Z_2} = \dfrac{6.7}{14.625} = 0.458$ (lag)

(ii) *Running at 3.5 per cent slip:*

Induced emf in rotor winding per phase $sE_2 = 0.035 \times 40.41 = 1.414$V

Rotor impedance per phase $= \sqrt{R_2^2 + s^2X_2^2} = \sqrt{0.7^2 + (0.035\times 8)^2} = 0.754$ V

Rotor current per phase $= \dfrac{sE_2}{Z_2} = \dfrac{1.414}{0.754} = 1.875$ A

Rotor power factor $(\cos\theta_2) = \dfrac{R_2}{Z_2} = \dfrac{0.7}{0.754} = 0.928$ [lag] ■

Example 8.2 The rotor emf (between slip rings) of a three-phase induction motor having star-connected rotor at standstill is 60 V. The rotor resistance and the standstill reactance are 0.6 Ω and 4 Ω, respectively. Determine the following:

(i) Rotor current/phase at starting when the slip rings are short-circuited.

(ii) Rotor current/phase if a star-connected rheostat of resistance 7 Ω/phase is connected across the slip rings.

(iii) Full-load rotor current when rotor power factor of the full-load slip is 4 per cent.

(iv) Rotor power factor at full load.

Solution

Rotor emf at standstill (between slip rings) = 60 V.

Therefore, rotor emf per phase at standstill $(E_2) = \dfrac{60}{\sqrt{3}} = 36.64$ V

Rotor resistance per phase $(R_2) = 0.6\ \Omega$

Rotor reactance per phase (X_2) at standstill = 4 Ω

(i) At starting:

Rotor current per phase (I_2) at starting $= \dfrac{E_2}{\sqrt{R_2^2 + X_2^2}} = \dfrac{36.64}{\sqrt{0.6^2 + 4^2}} = 9.06$ A

(ii) Total rotor resistance $\left(R_2'\right) = R_2 + 7 = 0.6 + 7 = 7.6\ \Omega$

Rotor current per phase (I_2) with rheostat resistance is given by

$$I_2 = \frac{36.64}{\sqrt{7.6^2 + 4^2}} = 4.266 \text{ A}$$

(iii) Full-load rotor current at slip s of 4 per cent

$$= \frac{sE_2}{\sqrt{R_2^2 + (s_{fl} X_2)^2}} = \frac{0.04 \times 36.64}{\sqrt{0.6^2 + (0.04 \times 4)^2}} = \frac{1.4656}{0.6209} = 2.360 \text{ A}$$

(iv) Rotor power factor ($\cos\theta_2$) at full load

$$= \frac{R_2}{\sqrt{R_2^2 + (s_{fl} X_2)^2}} = \frac{0.6}{\sqrt{0.6^2 + (0.04 \times 4)^2}} = 0.966 \text{ [lag]}$$

■

8.2 ROTOR POWER

Let the rotor current I_2 lags the rotor voltage E_2 by an angle θ_2. The rotor power factor is given by

$$\cos\theta_2 = \cos\left\{\tan^{-1}\left(\frac{sX_2}{R_2}\right)\right\}$$

The per phase power input to the rotor is given by

$$P_g = E_2 I_2 \cos\theta_2 \tag{8.1}$$

But

$$\cos\theta_2 = \frac{R_2}{\sqrt{R_2^2 + (sX_2)^2}} = \frac{R_2/s}{\sqrt{(R_2/s)^2 + X_2^2}} \tag{8.2}$$

Using Equation (8.2), Equation (8.1) can be expressed as

$$P_g = E_2 I_2 \cos\theta_2 = E_2 I_2 \left[\frac{R_2/s}{\sqrt{\left(R_2/s\right)^2 + X_2^2}}\right] = \frac{E_2}{\sqrt{\left(R_2/s\right)^2 + X_2^2}} \times I_2 \times \left(\frac{R_2}{s}\right)$$

i.e.,

$$P_g = I_2^2\left(\frac{R_2}{s}\right) \tag{8.3}$$

P_g is called the air gap power that is transferred from the stator to the rotor across the air gap. The expression of P_g from Equation (8.3) can be expressed as

$$P_g = I_2^2\left(\frac{R_2}{s}\right) = I_2^2 R_2 + I_2^2\left(\frac{R_2}{s}\right) - I_2^2 R_2 = I_2^2 R_2 + I_2^2 R_2\left(\frac{1-s}{s}\right)$$

= (rotor ohmic loss) + (internal mechanical power developed in rotor)

$$P_g = sP_g + (1-s)P_g \tag{8.4}$$

The rotor ohmic loss $= I_2^2 R_2 = sP_g = s$ (power input to rotor) (8.5)

Internal mechanical power (P_m) developed in rotor

$$= I_2^2 R_2 \left(\frac{1-s}{s}\right) = (1-s)P_g$$

i.e., $$P_m = (1-s)(\text{power input to rotor}) \tag{8.6}$$

The internal (or gross) torque developed per phase is given by

$$T_e = \frac{P_m}{\omega_r} \tag{8.7}$$

where ω_r is the rotor speed in mechanical rad/sec.

The rotor speed (ω_r) can be expressed in terms of synchronous speed ω_s as follows:

$$\omega_r = (1-s)\omega_s \tag{8.8}$$

Using Equations (8.6) and (8.8), Equation (8.7) can be expressed as

$$T_e = \frac{P_m}{\omega_r} = \frac{(1-s)P_g}{(1-s)\omega_s} = \frac{P_g}{\omega_s} \tag{8.9}$$

where ω_s is synchronous speed in mechanical rad/sec.

Equation (8.9) can also be expressed in the following form:

$$T_e = \frac{P_g}{\omega_s} = \frac{\frac{I_2^2 R_2}{s}}{2\pi\, n_s} = \frac{I_2^2}{2\pi\, n_s}\left(\frac{R_2}{s}\right) \tag{8.10}$$

The shaft power (P_{sh}) can be obtained from P_g as follows:

$P_{sh} = P_m$ – mechanical losses (friction and windage losses)

and also $P_{sh} = P_g$ – rotor ohmic losses – friction and windage losses

Therefore, output or shaft torque can be expressed by

$$T_{sh} = \frac{P_{sh}}{\text{Rotor Speed}} = \frac{P_{sh}}{(1-s)\,\omega_s} \tag{8.11}$$

If the stator input is known, the air gap power (P_g) is expressed by

P_g = stator power input – stator $I^2 R$ loss – stator core loss

Example 8.3 A three-phase, six-pole, 50 Hz star-connected induction motor delivers useful power 25 kW while running at a speed of 950 rpm. It is connected to a supply of 400 V and takes a current of 60 A. Its stator resistance per phase is 0.14 Ω. Mechanical losses are 900 W. Calculate (i) shaft torque, (ii) gross torque developed, (iii) rotor Cu loss, (iv) stator Cu loss, (v) stator iron losses and (vi) overall efficiency. The power factor of the motor is 0.75 (lagging).

Solution

$$N_s = \frac{120f}{P} = \frac{120 \times 50}{6} = 1{,}000 \text{ rpm}$$

$$s = \frac{N_s - N_r}{N_s} = \frac{1{,}000 - 950}{1{,}000} = 0.05$$

$$P_m = P_{\text{out}} + \text{mechanical losses} = 25 \times 10^3 + 900 = 25{,}900 \text{ W}$$

(i) Shaft torque $= \dfrac{P_{\text{out}}}{\omega_r} = \dfrac{P_{\text{out}}}{2 \times \pi \times \dfrac{N_r}{60}} = \dfrac{25 \times 10^3}{2 \times \pi \times \dfrac{950}{60}} = 251.42$ N-m

(ii) Gross torque $= \dfrac{P_m}{\omega_r} = \dfrac{P_m}{2 \times \pi \times \dfrac{N_r}{60}} = \dfrac{25{,}900}{2 \times \pi \times \dfrac{950}{60}} = 260.47$ N-m

(iii) Rotor Cu loss $P_r = \dfrac{sP_m}{1-s} = \dfrac{0.05 \times 25{,}900}{1 - 0.05} = 1{,}363.15$ W

(iv) Stator Cu loss $= 3I_1^2R_1 = 3 \times 60^2 \times 0.14 = 1{,}512$ W

(v) $P_g = \dfrac{P_r}{s} = \dfrac{1{,}363.15}{0.05} = 27{,}263$ W

Total input $(P_{\text{in}}) = \sqrt{3} \times 400 \times 60 \times 0.75 = 31{,}176.91$ W

Stator iron losses $= P_{\text{in}} - P_g - 3I_1^2R_1 = 2{,}401.91$ W

(vi) Overall efficiency $= \dfrac{P_{\text{out}}}{P_{\text{in}}} = \dfrac{25 \times 10^3}{31{,}176.91} = 0.8018$ p.u. $= 80.18\%$

8.3 EXPRESSION FOR TORQUE

Due to the interaction of the stator and rotor fields, torque is produced. The expression for torque is calculated as follows at any slip s.

Let ϕ be the flux of rotating magnetic field,

I_2 be the rotor current/phase and

$\cos\theta_2$ be the rotor power factor.

The torque of an induction motor is given by

$$T \propto \phi\, I_2 \cos\theta_2 \tag{8.12}$$

Again, rotor emf/phase at standstill is $E_2 \propto \phi$. Therefore, the torque expression becomes

$$T \propto E_2\, I_2 \cos\theta_2$$

i.e., $$T = kE_2\, I_2 \cos\theta_2 = kE_2 \times \frac{sE_2}{\sqrt{R_2^2 + s^2X_2^2}} \times \frac{R_2}{\sqrt{R_2^2 + s^2X_2^2}}$$

i.e., $$T = \frac{ksR_2E_2^2}{R_2^2 + s^2X_2^2} \tag{8.13}$$

where k is a constant.

The above expression gives the torque at any slip s.

8.4 STARTING TORQUE

To get the expression for starting torque, let us put $s = 1$ in Equation (8.13)

$$T_{est} = \frac{kR_2E_2^2}{R_2^2 + X_2^2} \tag{8.14}$$

8.4.1 Starting Torque of a Squirrel-cage Motor

The resistance of a squirrel-cage induction motor is fixed and it is less compared to its reactance. There is no provision to add external resistance to squirrel-cage rotor. Although the starting current of the induction motor is very high, it lags by a very large angle behind E_2. Therefore, the starting torque per ampere is very poor. It is approximately 1.5 times its full-load torque. These motors are not suitable for high starting load.

8.4.2 Starting Torque of Slip-ring Motor

In a slip-ring-type motor, there is a provision to add resistance in the rotor from the external circuit of a star-connected rheostat. By adding additional resistance, the power factor of the motor can be improved. As the motor gains speed, the additional resistances can be cut out. This additional resistance increases the rotor resistance and reduces the current. Since the effect of improvement of the power factor during starting predominates the increase in impedance or decrease in current, the torque of the motor is improved. After some point, the increase in impedance predominates the power factor, and hence torque decreases.

8.5 EFFECT OF CHANGE IN SUPPLY VOLTAGE ON STARTING TORQUE

From Equation (8.14), the expression for starting torque is rewritten as

$$T_{est} = \frac{kR_2E_2^2}{R_2^2 + X_2^2} \tag{8.15}$$

Again, emf at standstill is $E_2 \propto \phi \propto V$, we have the following from Equation (8.15)

$$T_{est} = \frac{kR_2V^2}{R_2^2 + X_2^2} \tag{8.16}$$

Equation (8.16) suggests that the starting torque is proportional to square of the supply voltage.

8.6 EFFECT OF CHANGE IN SUPPLY VOLTAGE ON TORQUE AND SLIP

The expression for torque at any value of s is given by

$$T = \frac{KsR_2E_2^2}{R_2^2 + s^2X_2^2} \quad (8.17)$$

Again at standstill, emf induced in a rotor is given by

$$E_2 \propto \phi \propto V$$

$$\therefore \quad T = \frac{K'sR_2V^2}{R_2^2 + s^2X_2^2} \quad (8.18)$$

where K' is another constant.

Since slip s is low at full load, the term $s^2X_2^2$ can be neglected in comparison to R_2^2.

Therefore, from Equation (8.18), we have

$$T = \frac{K'sR_2V^2}{R_2^2} = \frac{K's\,V^2}{R_2}, \quad \text{i.e.,} \quad T \propto sV^2$$

Therefore, any change in supply voltage also changes the value of torque. If supply voltage is decreased or increased, torque decreases or increases. Slip will increase or decrease simultaneously to maintain a constant torque.

8.7 CONDITION FOR MAXIMUM TORQUE

From Equation (8.13), we have

$$T = \frac{KsR_2E_2^2}{R_2^2 + s^2X_2^2} = \frac{KR_2E_2^2}{\frac{R_2^2}{s} + sX_2^2} = KE_2^2\left[\frac{R_2}{\left(\frac{R_2}{\sqrt{s}} - \sqrt{s}X_2\right)^2 + 2R_2X_2}\right]$$

To get maximum torque, the denominator of $\frac{R_2}{\left(\frac{R_2}{\sqrt{s}} - \sqrt{s}X_2\right)^2 + 2R_2X_2}$ must be minimum.

Let the denominator of the above expression be minimum at $s = s_{mT}$. Therefore, we have

$$\frac{R_2}{\sqrt{s_{mT}}} - \sqrt{s_{mT}}X_2 = 0$$

i.e.,

$$s_{mT} = \frac{R_2}{X_2} \quad (8.19)$$

Equation (8.19) gives the required condition for maximum efficiency.

The expression for maximum torque (T_{emax}) becomes

$$T_{e\,max} = \frac{KE_2^2}{2X_2} \quad (8.20)$$

The maximum torque is also known as pull-out or breakdown torque.

From Equation (8.20) we can conclude the following points:

(i) The maximum torque is independent of R_2.

(ii) The maximum torque is inversely proportional to X_2.

(iii) The slip corresponding to maximum torque is dependent on R_2, Equation (8.19).

We can also conclude that the starting torque is maximum when $R_2 = X_2$.

8.8 TORQUE-SLIP AND TORQUE-SPEED CHARACTERISTICS

The general expression for torque at any slip s is expressed by

$$T = \frac{KsR_2E_2^2}{R_2^2 + s^2X_2^2} \quad (8.21)$$

The following conclusions can be drawn from the above equation:

- At $s = 0$, the value of T is zero.
- For low value of s, the term sX_2 is very small, and hence this term can be neglected compared to R_2. For constant value of R_2, T is proportional to s.
- For $s = \frac{R_2}{X_2} = s_{mT}$, the torque becomes maximum.
- For further increase of s, the motor slows down. It may also stop. The operating zone is $0 < s \le s_{mT}$.
- For high value of s, the term R_2 can be neglected compared to sX_2. For constant value of X_2, $T \propto \frac{1}{s}$. Therefore, torque-slip characteristic becomes a rectangular hyperbola.

From the above discussion, it can be concluded that for small value of slip s, $T \propto s$ and for high value of slip s, $T \propto \frac{1}{s}$. Figure 8.1 shows the torque-slip characteristic of an induction motor at different rotor resistances. For $R_2 = 2\ \Omega$, the torque is directly proportional to the slip over the region OA_1. In the region A_1B_1, torque is inversely proportional to slip s, and hence the value of torque decreases with increasing value of slip s. The value of torque for $s = 1$ is known as starting torque (T_{est}), which corresponds to zero rotor speed. From Figure 8.1 it can be visualized that the starting torque (T_{est}) is less than the maximum torque (T_{emax}). In practice, poor starting torque is not desirable. The motor takes more time to attain full-load speed when torque is small, and hence it draws excessively large current during the period of acceleration. The torque-slip characteristics for other values of rotor resistances have been shown in Figure 8.1. The stable zone for curve 1 is OA_1 and the unstable zone for this curve is A_1B_1. Since T_{emax} is dependent on R_2; hence, only the position of T_{emax} is changed due to variation of R_2, whereas its magnitude remains unaltered.

Figure 8.2 shows the plot of input current, power factor, percentage efficiency and torque versus slip s. The maximum torque and the starting torque have also been shown.

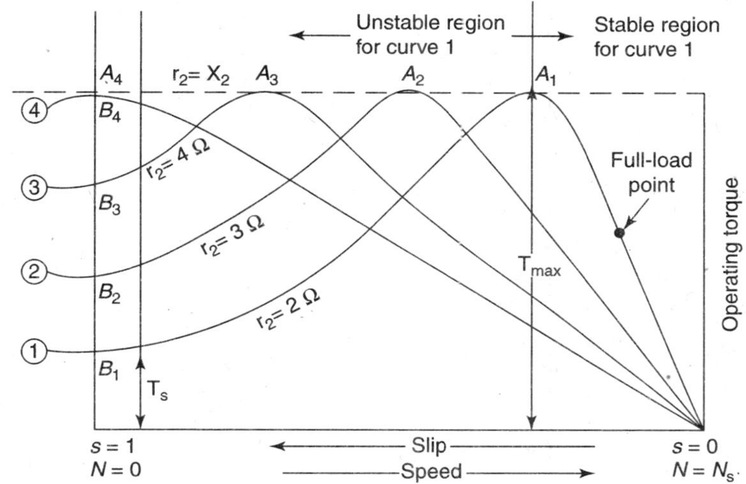

Figure 8.1 Torque-slip Characteristic

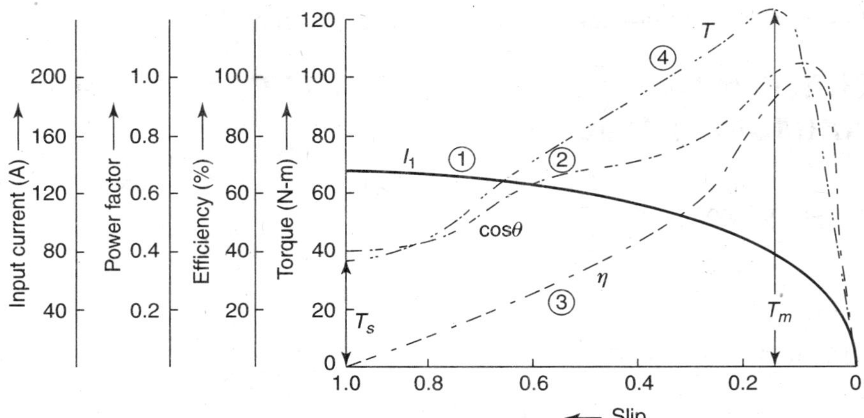

Figure 8.2 Plot of Input Current, Power Factor, Percentage Efficiency and Torque Versus Slip *s*

Figure 8.3 shows the three different modes of operations of the induction motor.

(i) *Motoring mode* ($0 \leq s \leq 1$): In this mode of operation, the rotor rotates in the same direction of operating magnetic field. The speed of the rotor is less than the synchronous speed.

(ii) *Generating mode* ($s < 0$): In this mode of operation, the stator terminals of an induction motor are connected to a constant frequency voltage source. The rotor is being rotated above the synchronous speed with the help of a prime mover.
It may be noted that the stator should not be disconnected from the supply.

(iii) *Braking mode* ($s > 1$): In this mode, slip s is greater than 1. The rotor is being driven by the prime mover in the direction opposite to the rotating magnetic field.

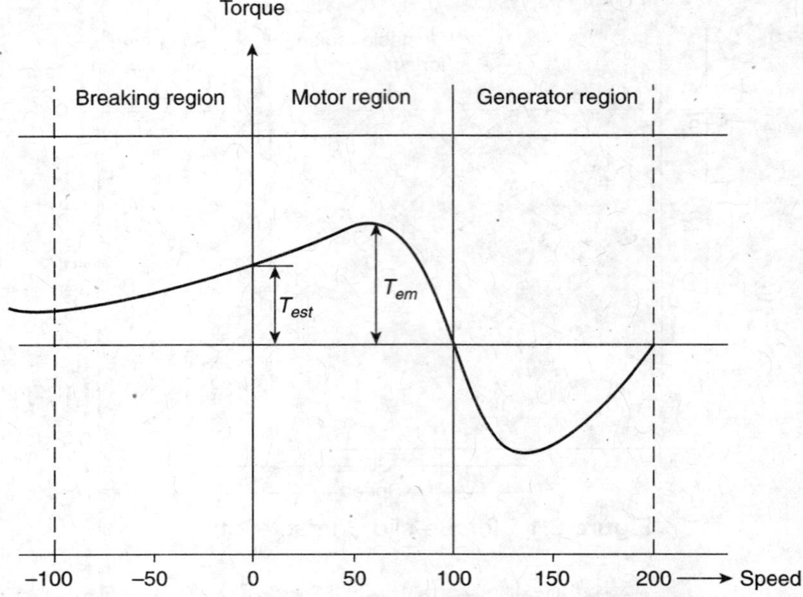

Figure 8.3 Torque-speed Characteristic Curve of an Induction Motor

8.9 RELATION BETWEEN FULL-LOAD TORQUE AND MAXIMUM TORQUE

Let s_{fl} and s_{mT} be the values of s at which full-load torque (T_{efl}) and maximum torque ($T_{e\,max}$) occur. The general expression for a torque at any slip s is given by

$$T = \frac{KsR_2E_2^2}{R_2^2 + s^2X_2^2}$$

At full load, $s = s_{fl}$ and the expression for the torque (T_{efl}) becomes

$$T_{efl} = \frac{Ks_{fl}R_2E_2^2}{R_2^2 + s_{fl}^2X_2^2} \tag{8.22}$$

The expression for the maximum torque ($T_{e\,max}$) is rewritten as

$$T_{e\max} = \frac{KE_2^2}{2X_2} \tag{8.23}$$

From Equations (8.22) and (8.23), we have

$$\frac{T_{efl}}{T_{e\max}} = \frac{\dfrac{Ks_{fl}R_2E_2^2}{R_2^2 + s_{fl}^2X_2^2}}{\dfrac{KE_2^2}{2X_2}} = \frac{2s_{fl}R_2X_2}{R_2^2 + s_{fl}^2X_2^2} = \frac{2s_{fl}R_2\left(\dfrac{X_2}{X_2^2}\right)}{\dfrac{R_2^2 + s_{fl}^2X_2^2}{X_2^2}} = \frac{2s_{fl}\left(\dfrac{R_2}{X_2}\right)}{\left(\dfrac{R_2}{X_2}\right)^2 + s_{fl}^2} = \frac{2s_{fl}s_{mT}}{s_{fl}^2 + s_{mT}^2}$$

i.e.,
$$\frac{T_{efl}}{T_{e\max}} = \frac{2}{\frac{s_{fl}}{s_{mT}} + \frac{s_{mT}}{s_{fl}}} \tag{8.24}$$

8.10 RELATION BETWEEN STARTING LOAD TORQUE AND MAXIMUM TORQUE

From Equations (8.14) and (8.23), we have

$$\frac{T_{est}}{T_{e\max}} = \frac{\frac{KR_2E_2^2}{R_2^2+X_2^2}}{\frac{KE_2^2}{2X_2}} = \frac{2R_2X_2}{R_2^2+X_2^2} = \frac{2\left(\frac{R_2}{X_2}\right)}{\left(\frac{R_2}{X_2}\right)^2+1}$$

i.e.,
$$\frac{T_{est}}{T_{e\max}} = \frac{2s_{mT}}{1+s_{mT}^{\;2}} \tag{8.25}$$

Example 8.4 A three-phase, 3.3 kV, 24-pole, 50 Hz star-connected induction motor has a slip-ring rotor. The rotor resistance and reactance per phase are 0.03 Ω and 0.5 Ω, respectively. Determine the following:

(i) The speed at maximum torque.

(ii) The ratio of full-load torque to maximum torque if full-load torque is obtained at 244 rpm.

Solution

Rotor resistance per phase $(R_2) = 0.03\ \Omega$
Rotor reactance per phase $(X_2) = 0.5\ \Omega$

(i) Let the maximum torque occur at $s = s_{mT}$

Again, $s_{mT} = \frac{R_2}{X_2} = \frac{0.03}{0.5} = 0.060 = 6.0\%$

Synchronous speed $(N_s) = \frac{120f}{P} = \frac{120\times 50}{24} = 250\,\text{rpm}$

Rotor speed at $s = s_{mT}$ is given by

$N = N_s(1-s_{mT}) = 250\times(1-0.06) = 235\,\text{rpm}$

(ii) Full-load slip $(s_{fl}) = \frac{250-244}{250} = 0.024$

Again, $\frac{T_{efl}}{T_{e\max}} = \frac{2s_{fl}s_{mT}}{s_{fl}^2+s_{mT}^2} = \frac{2\times 0.024\times 0.06}{0.024^2+0.06^2} = 0.6896$ ■

Example 8.5 A 12-pole, three-phase, 50 Hz induction motor runs on full load with a slip of 4.5 per cent having rotor resistance and standstill reactance per phase 0.016 Ω and 0.078 Ω, respectively. Find the following:

(i) The available maximum torque in terms of full-load torque.
(ii) The speed at which the maximum torque occurs.

Solution

Rotor resistance per phase (R_2) = 0.016 Ω

Rotor reactance per phase at standstill (X_2) = 0.078 Ω

Full-load slip (s_{fl}) = 4.5 per cent = 0.045

(i)
$$s_{mT} = \frac{R_2}{X_2} = \frac{0.016}{0.078} = 0.205$$

$$\frac{T_{efl}}{T_{e\max}} = \frac{2s_{fl}s_{mT}}{s_{fl}^2 + s_{mT}^2} = \frac{2\times 0.045\times 0.205}{0.045^2 + 0.205^2} = 0.4188$$

i.e.,
$$T_{efl} = 0.4188\ T_{emax}$$

i.e.,
$$T_{emax} = 2.387\ T_{efl}$$

(ii) Synchronous speed $(N_s) = \frac{120f}{P} = \frac{120\times 50}{12} = 500\text{rpm}$

Rotor speed at $s = s_{mT}$ is expressed by

$$N_s = N_s\left(1 - s_{mT}\right) = 500\times\left(1 - 0.205\right) = 397.5\,\text{rpm}$$ ■

8.11 CONDITION FOR THE STARTING TORQUE TO BE EQUAL TO THE MAXIMUM TORQUE

To get the maximum torque at the time of starting, that is, $T_{est} = T_{max}$, the value of slip s must be equal to 1. The condition to be satisfied becomes

$$R_2 = X_2 \tag{8.26}$$

Since R_2 is always less than X_2, the extra resistance must be added in the rotor circuit of the induction motor to achieve this condition shown by Equation (8.26). This is only possible in slip-ring-type induction motors. The operating point corresponding T_{max} will be shifted towards $s = 1$ after addition of resistance in rotor circuit. The point A in Figure 8.1 gets shifted to B when the rotor resistance is increased from 2 Ω to 3 Ω. For $R_2 = X_2$, the operating point for T_{emax} occurs at $s = 1$ in curve 4, shown in Figure 8.1, and $T_{est} = T_{emax}$.

8.12 EFFECT OF VARIATION OF ROTOR RESISTANCE AND ROTOR REACTANCE ON MAXIMUM TORQUE, EFFICIENCY AND POWER FACTOR OF AN INDUCTION MOTOR

The effect on maximum torque, efficiency and power factor of an induction motor due to variation of rotor resistance and rotor reactance is discussed in this article.

8.12.1 Effect of Rotor Resistance on Maximum Torque

From Equation (8.20), the maximum torque is expressed by

$$T_{e\max} = \frac{KE_2^2}{2X_2}$$

and the condition for maximum torque is expressed by Equation (8.19), that is,

$$s_{mT} = \frac{R_2}{X_2}$$

From the expressions, it is clear that any variation of R_2 will not alter the value of T_{emax}. But it will change the value of s_{mT} at which T_{emax} occurs. In contrast, an increase in the value of X_2 will decrease the value of T_{emax}.

8.12.2 Effect of R_2 and X_2 on the Power Factor of an Induction Motor

The power factor of the rotor circuit of an induction motor is expressed by

$$\cos\theta_2 = \frac{R_2}{\sqrt{R_2^2 + (sX_2)^2}} \tag{8.27}$$

The rotor power factor $\cos\theta_2$ increases with an increase in the rotor resistance (R_2), and hence θ_2 decreases. The reflected rotor current ($\bar{I}_2'$) moves towards $\bar{E}_1$, as shown in Figure 8.4. Therefore, the power factor $\cos\theta_1$ of the induction motor improves with an increase in rotor resistance.

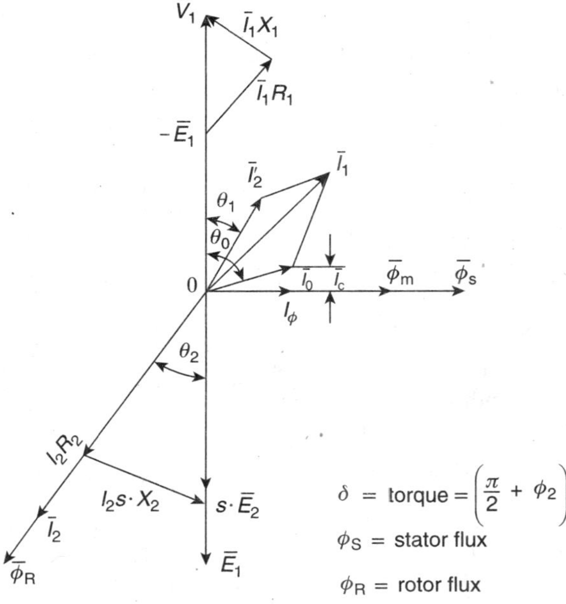

Figure 8.4 Phasor Diagram

The rotor power factor $\cos\theta_2$ decreases with an increase in rotor resistance (X_2), and hence θ_2 increases. The reflected rotor current ($\bar{I}_2'$) moves away from $\bar{E}_2$, as shown in Figure 8.4, which is the phasor diagram of the induction motor. This has been drawn on similar lines as that for a transformer with secondary short circuited. Therefore, the power factor $\cos\theta_1$ of the induction motor decreases with an increase in rotor reactance. This effect of X_2 is exactly opposite to the effect of R_2.

ϕ_m is the stator flux in the air gap (mutual flux).
R_1, X_1 are the stator resistance and the reactance per phase, respectively.
R_2, X_2 are the rotor resistance and the reactance per phase, respectively.
I_1, I_2 are the stator and rotor currents per phase, respectively.
V_1 is the supply voltage per phase on the stator phase.
E_1 is the induced emf in the stator winding per phase.
E_2 is the rotor standstill voltage per phase.
I_0 is the induced motor no-load current per phase.
I_2' is the rotor current reflected on the stator side per phase.
$\cos\theta_1$ and $\cos\theta_2$ are the stator and rotor power factors.
$\cos\theta_0$ is the no-load power factor.
$I_\phi = I_0\sin\theta_0$ is the magnetizing component of no-load current.
$I_c = I_0\cos\theta_0$ is the core-loss component of no-load current.

8.12.3 Effect of R_2 and X_2 on the Efficiency of an Induction Motor

The rotor current decreases slightly with an increase in rotor resistance (R_2) for a given supply voltage, and hence rotor Cu loss increases with an increase in R_2. The slip (rotor Cu loss/rotor input) increases with an increase in rotor resistance. This decreases gross mechanical power developed $\frac{I_2^2 R_2(1-s)}{s}$ with an increase in rotor resistance. Hence, the efficiency of induction motor decreases with an increase in rotor resistance.

Due to the increase in rotor reactance (X_2), the rotor current decreases, and hence there is a reduction in the gross mechanical power developed by the motor. This leads to the reduction of efficiency of induction motor due to the increase in rotor reactance (X_2). The effect of X_2 on the slip of induction motor is not significant.

The variation of stator current per unit (p.u.), power factor (pf) and efficiency of induction motor with speed are shown in Figure 8.2.

Curve 1 represents the stator current, which is five times the full-load current. The current in the rotor decreases when the speed is close to the synchronous speed, and the motor takes only the no-load current. Therefore, the stator current decreases with an increase in speed.

Curve 2 shows the variation of power factor with speed. Slip decreases with an increase in speed, and hence rotor reactance decreases causing an improvement in the rotor power factor. This improves the overall power factor of the motor.

The output is zero at the starting, and hence efficiency is zero. Slip decreases with an increase in the speed. This results in an increase in the power factor developed $\left[I_2^2 R_2 \frac{(1-s)}{s}\right]$, and hence the output increases. Efficiency increases with an increase in the speed to a certain extent. The slip becomes very small when the speed is very close to the synchronous speed. The equivalent resistance representing the power developed increases. This causes a reduction in the motor current. Therefore, the product $\left[I_2^2 R_2\left(\frac{1}{s}-1\right)\right]$ decreases when the rotor speed approaches the synchronous speed.

8.13 EFFECT OF CHANGE IN SUPPLY VOLTAGE ON TORQUE AND SPEED

From Equation (8.17), we have

$$T = \frac{ksR_2E_2^2}{R_2^2 + s^2X_2^2}$$

We know that $E_2 \propto \phi \propto V$, where V is the supply voltage.

$\therefore\ T \propto sV^2$

We can say that the torque at any speed is proportional to the square of the applied voltage. Torque decreases by 20 per cent for 10 per cent decrease of the supply voltage. Changes in supply voltage affect the torque during starting as well as running. T decreases when V decreases. To maintain the torque constant, s must be increased, that is, the speed falls.

An increase in the supply voltage decreases the power factor, secondary copper (Cu) loss and slip, but increases torque, magnetic density, magnetizing current and iron loss.

A decrease in the supply voltage increases the power factor, secondary Cu loss and slip, but decreases torque, magnetic density, magnetizing current and iron loss.

8.14 EFFECT OF CHANGE IN SUPPLY FREQUENCY ON TORQUE AND SPEED

If the change of supply frequency occurs, the change in speed of the induction motor comes into picture. For a frequency drop of 10 per cent, the speed of the motor decreases by 10 per cent. If an induction motor designed for 50 Hz is connected to 60 Hz system, then it runs $(60 - 50) \times 100/50 = 20$ per cent faster than the normal speed. Therefore, either gears will be used to increase the speed or an expensive 50 Hz supply system can be used. The terminal voltage of the machine must be increased by 6/5 times the rated voltage. The new breakdown torque will be equal to the original breakdown torque and the starting torque will be reduced slightly.

A 60 Hz motor will operate satisfactory on 50 Hz supply if its terminal voltage is reduced to 5/6 times of its rated voltage.

An increase in the supply frequency increases the power factor, but decreases the torque. On the other hand, a decrease in the supply frequency decreases the power factor, but increases the torque.

8.15 LOSSES

Figure 8.5 shows the different losses of an induction motor.

Fixed loss = power input at no-load – (stator I^2r loss at no load).

This loss can be obtained by performing no-load test of the induction motor.

Total ohmic losses under variable losses can be obtained using blocked-rotor test of induction motor. It should be noted that the brush contact loss for wound rotor induction motor (WRIM) = slip-ring current × 1 V.

Stray load loss occurs in iron and conductors. It is very difficult to measure stray load loss. To account for this, efficiency is taken as 0.5 per cent less than the calculated value on full load and pro-rata for other loads.

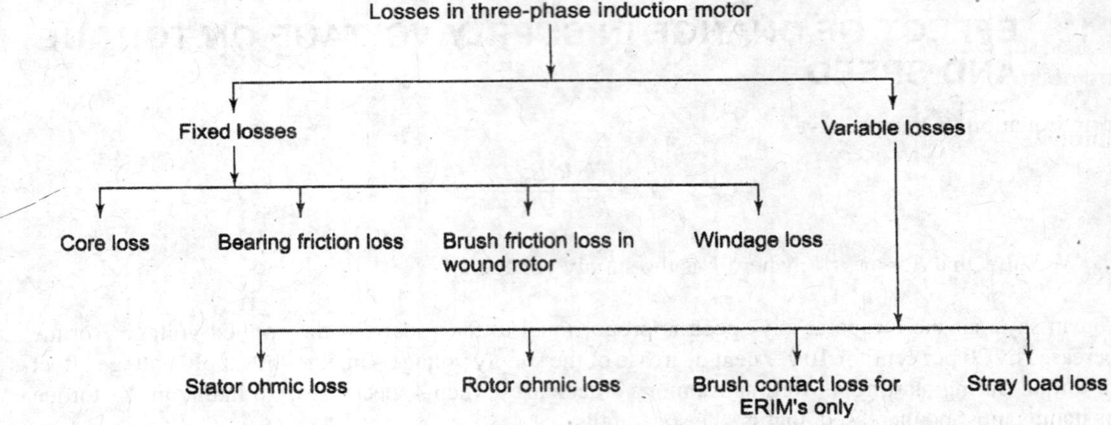

Figure 8.5 Different Losses of an Induction Motor

8.16 POWER STAGES

Figure 8.6 shows the power stages of an induction motor.

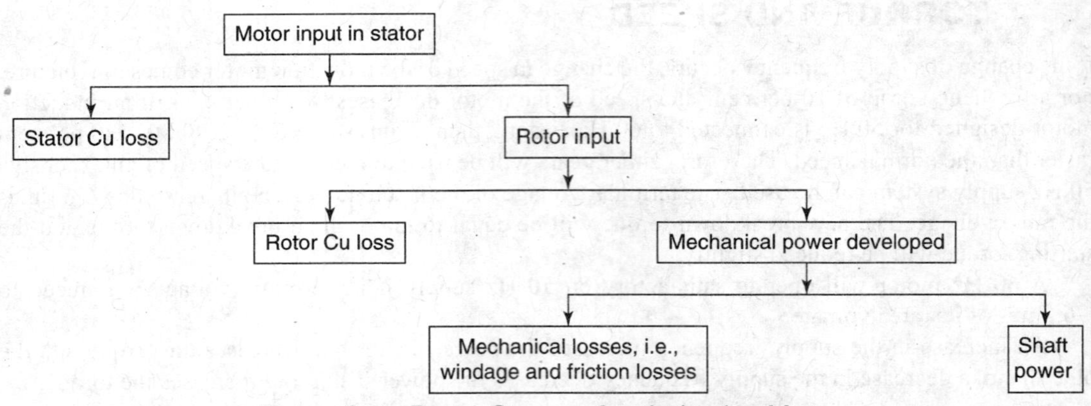

Figure 8.6 Power Stages of an Induction Motor

8.17 EFFICIENCY

The efficiency of a three-phase induction motor is given by

$$\eta = \frac{P_{sh}}{P_{sh} + P_f + P_{Cu}} \times 100 \tag{8.28}$$

where P_{sh} = shaft power

P_f = fixed losses = core losses + friction and windage losses

P_{Cu} = stator and rotor ohmic losses + brush contact loss

Example 8.6 An induction motor has an efficiency of 0.9 when delivering an output of 37 kW. At this load, the stator Cu loss and rotor Cu loss each equals the iron loss. The mechanical losses are one-third of the no-load loss. Calculate the slip.

Solution

$\eta = 0.9$; $P_{out} = 37$ kW; stator Cu loss = rotor Cu loss = iron loss

$$\text{Mechanical losses} = \frac{1}{3} \times \text{no-load loss}$$

$$\text{Motor input } (P_{in}) = \frac{P_{out}}{\eta} = 41.111 \text{ kW}$$

Total losses [including stator Cu and iron losses, rotor Cu loss and rotor mechanical loss (rotor iron loss being negligible)].

$$= P_{in} - P_{out} = 41.111 - 37 = 4.111 \text{ kW} = 4{,}111 \text{ W}$$

No-load loss of an induction motor consists of stator iron loss and mechanical loss (stator Cu loss at no-load being negligible). These two losses are independent of the load.

Let P_i = each of the stator Cu loss, rotor Cu loss and stator iron loss, and P_{m1} = mechanical loss

$$\text{No-load loss} = P_i + P_{m1}$$

Since mechanical losses are one-third of the no-load losses, we have

$$P_{m1} = \frac{1}{3}(P_i + P_{m1})$$

i.e.,

$$P_{m1} = \frac{P_i}{2}$$

Total loss = stator Cu loss + stator iron loss + rotor Cu loss + mechanical loss

i.e.,

$$4{,}111 = 3P_i + \frac{P_i}{2}$$

i.e.,

$$P_i = 1{,}174.6 \text{ W} = 1.1746 \text{ kW}$$

Now, rotor output = motor output + mechanical losses $= 37 + \frac{P_i}{2}$

$$= 37 + \frac{1.1746}{2} = 37.5873 \text{ kW}$$

Rotor input = rotor output + rotor Cu losses

$$= 37.5873 + 1.1746 = 38.762 \text{ kW}$$

$$\text{Slip } (s) = \frac{\text{Rotor Cu losses}}{\text{Rotor input}} = \frac{1.1746}{38.762} = 0.03 \text{ p.u. or } 3\%$$

■

8.18 SYNCHRONOUS WATT

The torque that develops a power of 1 W at the synchronous speed of the motor is known as synchronous watt. If a motor develops 400 synchronous watts, it means that the rotor input is 400 W and rotor will run at a synchronous speed. Synchronous wattage of an induction motor is equal to the power transfer from stator to rotor through the air gap.

$$\text{Rotor input in Watt} = T_{sw} \times 2\pi N_s$$

$$\therefore \quad T_{sw} = \frac{\text{Rotor input in Watt}}{2\pi N_s} \tag{8.29}$$

8.19 MEASUREMENT OF SLIP

For the measurement of slip of an induction motor, any of the following three methods can be used:

(i) Electromechanical counter.

(ii) Mechanical differential counter.

(iii) Stroboscopic method.

8.19.1 Electromechanical Counter

Figure 8.7 shows an electromechanical counter that is used to compare the speed of an induction motor with that of a small synchronous motor. This is a very simple and direct method. The number of poles of the synchronous motor is equal to that of the induction motor.

Figure 8.7 also shows that at the end of each shaft there is a cylinder of phenolic or other insulating materials. A small contactor is connected to a circular slip ring, and this combination is fitted to the insulator. A voltage source in series with an electric pulse counter, which may be either of electromechanical relay type or an electronic digital pulse counter, supplies power to the slip rings. The synchronous motor runs at a synchronous speed, which is the speed of the rotating field of the induction motor. When the induction motor slips, the contractors of the synchronous and induction motors close the circuit and register a pulse. The slip in rpm is equal to the number of pulses registered per minute.

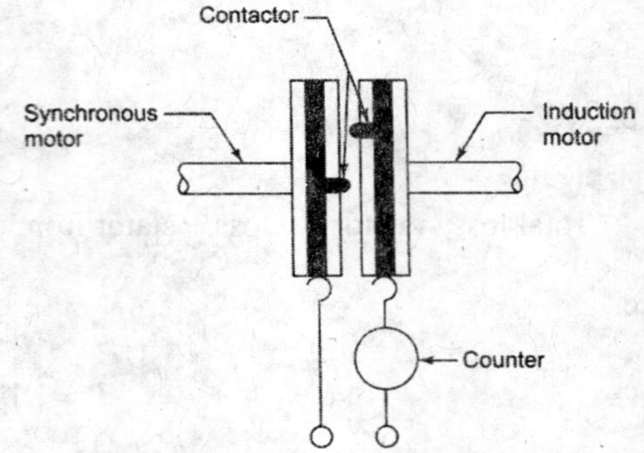

Figure 8.7 Electromechanical Counter

8.19.2 Mechanical Differential Counter

Figure 8.8 shows the mechanical differential counter. The output gear of the mechanical differential counter rotates at a speed that is equal to the difference of two input gears to which the synchronous and induction motor shafts are connected.

To record the slip speed directly, a mechanical counter or a low-speed electric or mechanical tachometer is used. The friction and the drag in the gearing of the differential counter occur due to the slight loading the induction motor. This method should not be implemented to small induction motors

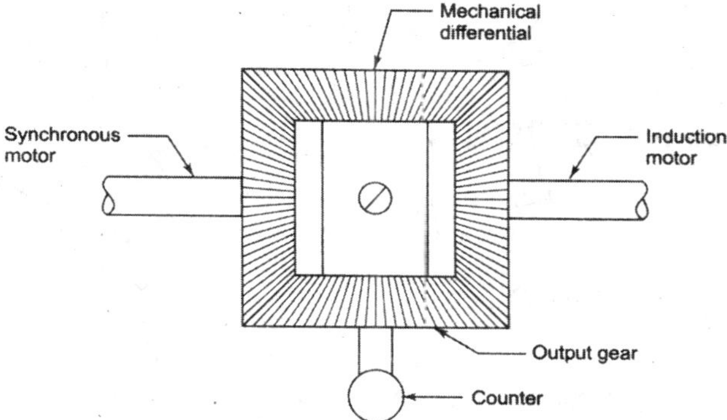

Figure 8.8 Mechanical Differential Counter

for measuring the slip. For small induction motors, generally an electric differential counter with less effective drag is used.

8.19.3 Stroboscopic Method

Figure 8.9 shows the stroboscopic method. A black disc with white sectors is attached to the shaft of the induction motor. The number of white sectors of the black disc is equal to the number of poles of the induction motor. Another disc of equal number of sector-shaped slits is attached to the shaft of a self-starting synchronous motor. It is in turn fitted with a revolution counter that can be thrown in and out of gear.

If the number of passages of sectors be n, the slip in per cent is given by

$$\text{Slip in per cent} = \frac{100n}{N_s N_r} \tag{8.30}$$

where N_s is the number of sectors and N_r be the number of revolutions recorded by the counter during the interval of observation.

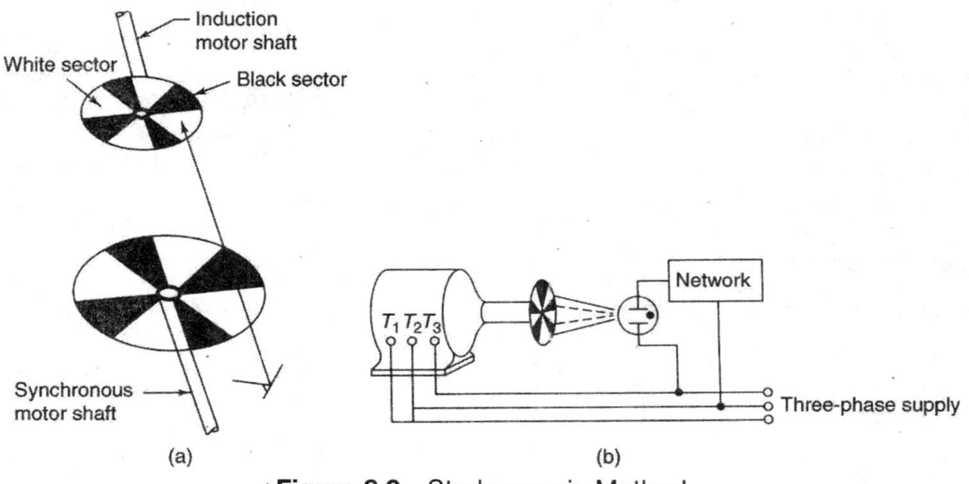

Figure 8.9 Stroboscopic Method

The observation can be simplified for large values of slip using only one sector, and n represents the slip in revolution.

8.20 EQUIVALENT CIRCUIT

The equivalent circuit of an induction motor is similar to that of a transformer. The main difference is that the rotor of an induction motor rotates and mechanical power is developed.

8.20.1 Stator Equivalent Circuit

Counter emfs are generated in all the three phases of the stator due to the rotating air gap flux wave. The stator terminal voltage (V_1) must overcome the counter emf. The counter emf is $-E_1$ and the stator leakage impedance drop is I_1 $(R_1 + jX_1)$. Therefore, the following relation gives the stator primary voltage:

$$\bar{V}_1 = -\bar{E}_1 + \bar{I}_1(R_1 + jX_1)$$

$$\therefore \quad \bar{V}_1 = \bar{V}_1' + \bar{I}_1(R_1 + jX_1) \tag{8.31}$$

where $\bar{V}_1' = -\bar{E}_1$

I_1 is the stator current and $\bar{I}_1 = \bar{I}_1' + \bar{I}_e$

where $\bar{I}_e$ = exciting current and

$\bar{I}_1'$ = is the load component and counteracts the rotor mmf I_2N_2'

Again, $$\bar{I}_0 = \bar{I}_c + \bar{I}_\phi \tag{8.32}$$

In Equation (8.32), $\bar{I}_c$ is the core loss component and $\bar{I}_\phi$ is the magnetizing component lagging $\bar{V}_1'$ by an angle of 90°.

In the induction motor, $\bar{I}_0$ supplies the resultant air gap flux (ϕ) and no-load losses (= core loss + friction, windage loss + small stator and rotor $I^2 R$ loss). But in the transformer, $\bar{I}_0$ supplies core loss only. Therefore, $\bar{I}_0$ should not be called no-load current in the induction motor as in the transformer.

The product $\bar{V}_1'\bar{I}_c$ in a three-phase induction motor gives the core loss per phase. Figure 8.10 shows the equivalent circuit of an induction motor in which I_c and I_ϕ are taken for shunt branch consisting of core loss branch (R_c) in parallel with magnetizing reactance (X_ϕ).

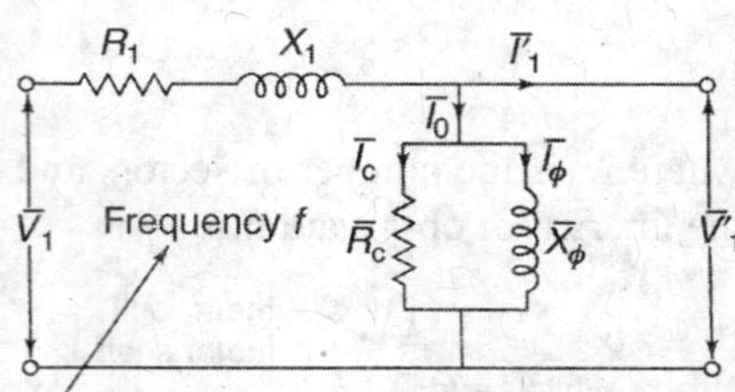

Figure 8.10 Stator Equivalent Circuit for a Polyphase Induction Motor

8.20.2 Rotor Equivalent Circuit

The per phase rotor current is given by

$$I_2 = \frac{sE_2}{R_2 + jsX_2} \tag{8.33}$$

$$= \frac{E_2}{\frac{R_2}{s} + jX_2} \tag{8.34}$$

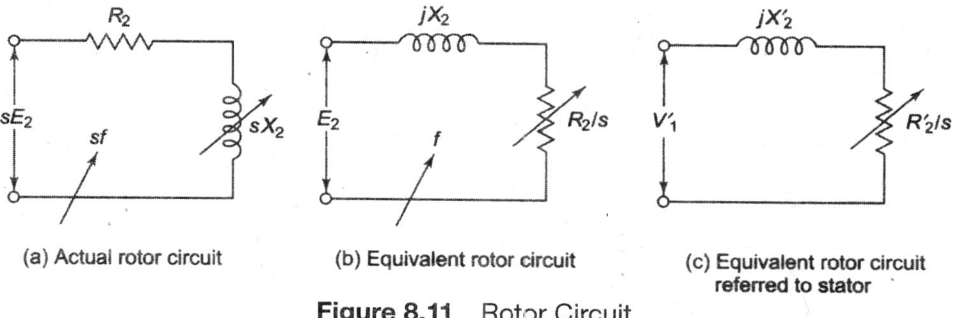

Figure 8.11 Rotor Circuit

Figure 8.11 shows the rotor circuit.

The complete equivalent circuit of an induction motor referred to the stator is shown in Figure 8.12.

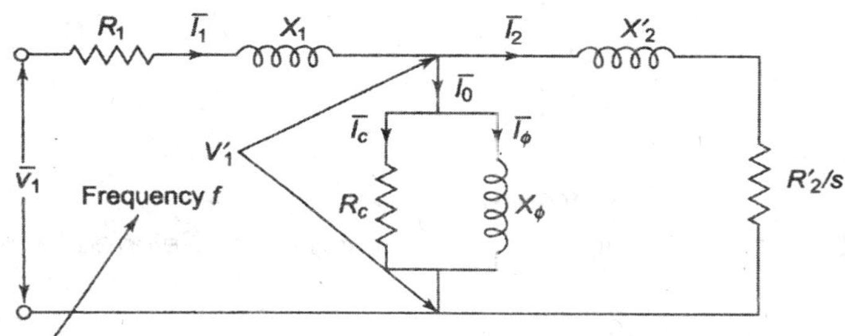

Figure 8.12 Complete Equivalent Circuit of a Three-phase Induction Motor Referred to the Stator

X'_2 and R'_2 are respective values of the rotor leakage reactance and resistance referred to the stator. If the prime notation used for the referred quantities is dropped, Figure 8.13(a) shows the equivalent circuit of a three-phase induction motor and Figure 8.13(b) shows the modified form of the induction motor equivalent circuit.

In Figure 8.13 all the motor quantities have been referred to the stator side.

In Figure 8.13(b),

$$\frac{R_2}{s} = R_2 + R_2\left(\frac{1-s}{s}\right) \tag{8.35}$$

For $s = 1$, Figure 8.13(b) becomes the equivalent circuit of a short-circuited two-winding transformer, and for $s = 0$, Figure 8.13(b) becomes the equivalent circuit of an open-circuited transformer.

In Figure 8.13(b), R_2 is the actual rotor resistance (referred to the stator), whereas $R_2\left(\frac{1-s}{s}\right)$ is the electrical analogue of the variable mechanical load.

In transformers, the approximate equivalent circuit is obtained by moving the shunt branch across the primary terminals because the no-load current is 2 to 5 per cent of full-load current and the per unit leakage reactance is low. The no-load current is 30 to 50 per cent of full-load current in an induction motor, and hence the per unit leakage reactance is high. If the shunt branch is neglected or moved towards stator terminals, considerable error will be introduced.

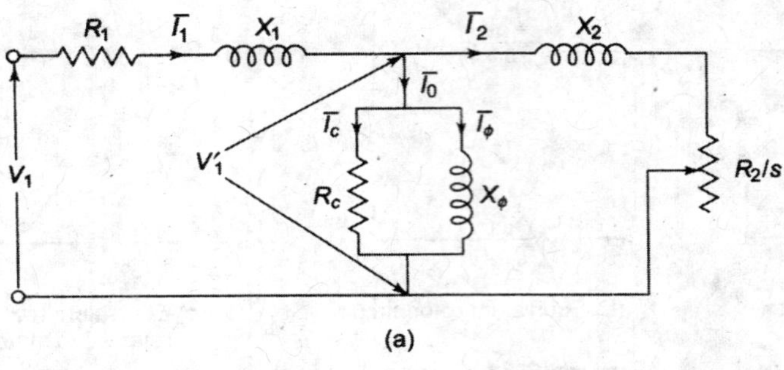

(a)

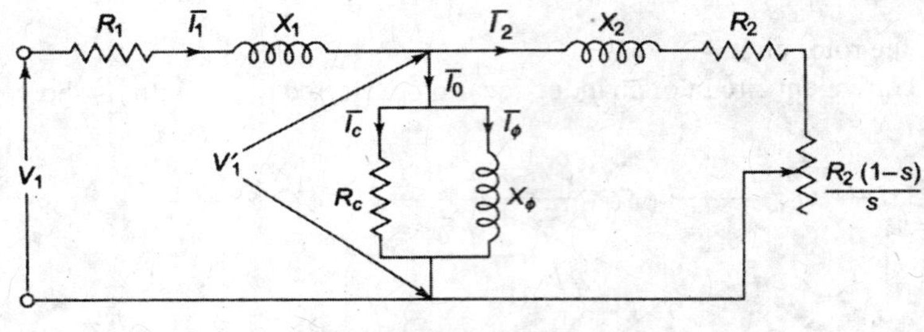

(b) Modified form of (a)

Figure 8.13 Equivalent Circuit of an Induction Motor Referred to the Stator

During normal operating conditions of a constant voltage and frequency, the core loss in induction motor is usually constant. So, the core loss resistance R_c, representing the motor core loss, can be removed from the equivalent circuit. The equivalent circuit shown in Figure 8.13(a) is redrawn and shown in Figure 8.14.

In equivalent circuit the quantities are quantities per phase. In Figure 8.13, Z is the per phase impedance offered to the stator by the rotating the air gap field.

$$\therefore \quad Z = R + jX = \frac{\left(\frac{R_2}{s} + jX_2\right)\left(jX_\phi\right)}{\frac{R_2}{s} + j\left(X_2 + X_\phi\right)} \tag{8.36}$$

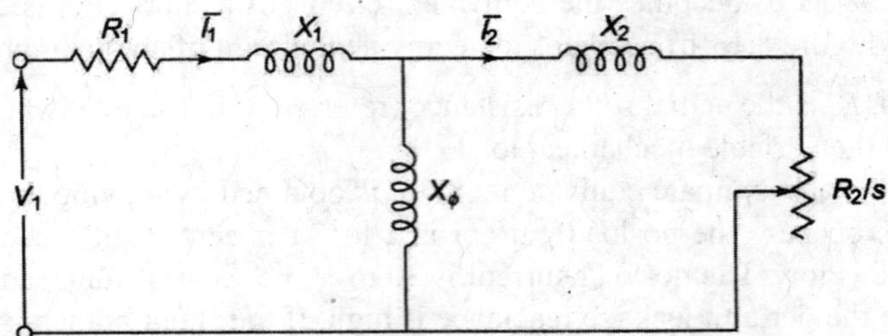

Figure 8.14 Equivalent Circuit of a Three-phase Induction Motor Without Core Loss

Impedance seen by the stator applied voltage,

$$Z_T = R_1 + jX_1 + Z \tag{8.37}$$

$$\therefore \qquad \text{Stator current } (I_1) = \frac{V_1}{Z_T}$$

From Figure 8.14, the total power loss in parallel combination of $\frac{R_2}{s} + jX_2$ and jX_ϕ is $\frac{I_2^2 R_2}{s}$.

Since $Z = R + jX$, the power lost in Z is equal to $I_1^2 R$.
Per phase air gap power

$$P_g = \frac{I_2^2 R_2}{s} = I_1^2 R$$

So, shaft output power, torque and efficiency can be easily calculated.

Example 8.7 The per phase constants in ohms referred to the stator of a three-phase star-connected, 400 V, 50 Hz, four-pole induction motor are shown in Figure E8.1.

$$R_1 = 0.16,\ X_1 = 0.48\ ,\ R_2 = 0.14,\ X_2 = 0.45\ \text{ and } X_m = 30$$

Fixed losses (core, friction and windage losses) = 450 W

Determine the following:

(i) Stator current.
(ii) Rotor speed.
(iii) Output torque.
(iv) Efficiency when the motor is operated at rated voltage and frequency at a slip of 3.5 per cent.

Solution

$$Z = R + jX = \frac{\left(\frac{0.14}{s} + j0.45\right)(j30)}{\frac{0.14}{s} + j0.45 + j30}$$

$$= 3.816 + j0.945$$

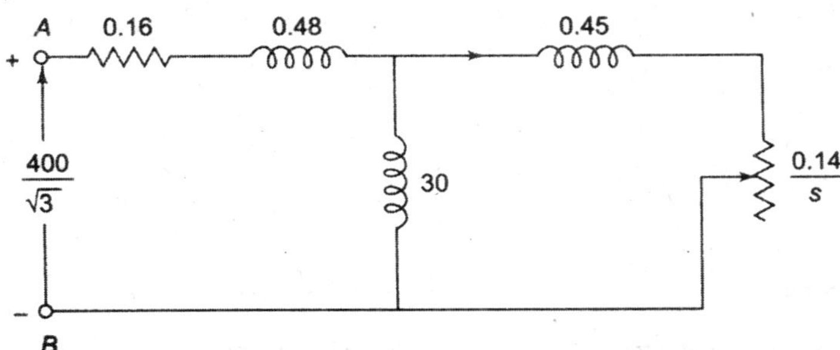

Figure E8.1

Here, $R = 3.816\ \Omega$

$$Z_{AB} = 0.16 + j0.48 + R + jX = 0.16 + j0.48 + 3.816 + j0.945$$

$$Z_{AB} = 3.976 + j1.425 = 4.223\angle 19.72^\circ$$

(i) $$\text{Stator current} = \frac{400\angle 0^\circ}{\sqrt{3} \times 4.223\angle 19.72^\circ} = 54.68\angle -19.72^\circ \text{ A}$$

(ii) $$\text{Synchronous speed } (N_s) = \frac{2f}{P} = \frac{2 \times 50}{4} = 25 \text{ rps}$$

∴ $$\text{Rotor speed } (N_r) = (1 - s)N_s = (1 - 0.035) \times 25 = 24.125 \text{ rps}$$

(iii) $$P_g = mI_1^2 R = 3 \times 54.68^2 \times 3.816 = 34{,}228.40 \text{ W}$$

$$P_m = (1 - s)P_g = (1 - 0.035) \times 34{,}228.40 = 33{,}030.40 \text{ W}$$

$$P_{sh} = P_m - \text{fixed losses} = 33{,}030.40 - 450 = 32{,}580.4 \text{ W}$$

∴ $$\text{Output torque} = \frac{P_{sh}}{\omega_r} = \frac{P_{sh}}{2\pi\ n_r} = \frac{32{,}580.4}{2 \times 3.14 \times 24.125} = 215 \text{ N-m}$$

(iv) $$\text{Stator ohmic losses} = 3I_1^2 R_1 = 3 \times 54.68^2 \times 0.16 = 1{,}435.15 \text{ W}$$

$$\text{Rotor ohmic losses} = sP_g = 0.035 \times 34{,}228.4 = 1{,}198 \text{ W}$$

$$\text{Total losses} = \text{stator Cu loss} + \text{rotor Cu loss} + \text{fixed losses}$$

$$= 1{,}435.15 + 1{,}198 + 450 = 3{,}083.15 \text{ W}$$

$$\text{Power input } (P_{in}) = P_{sh} + \text{total losses} = 32{,}580.4 + 3{,}083.15 = 35{,}663.55 \text{ W}$$

∴ $$\text{Efficiency} = \frac{P_{sh}}{P_{in}} = \frac{32{,}580.4}{35{,}663.55} = 0.9135 \text{ p.u.} = 91.35\%$$ ■

8.21 THEVENIN'S EQUIVALENT CIRCUIT OF AN INDUCTION MOTOR

Figure 8.15 shows the Thevenin's equivalent circuit of Figure 8.14.

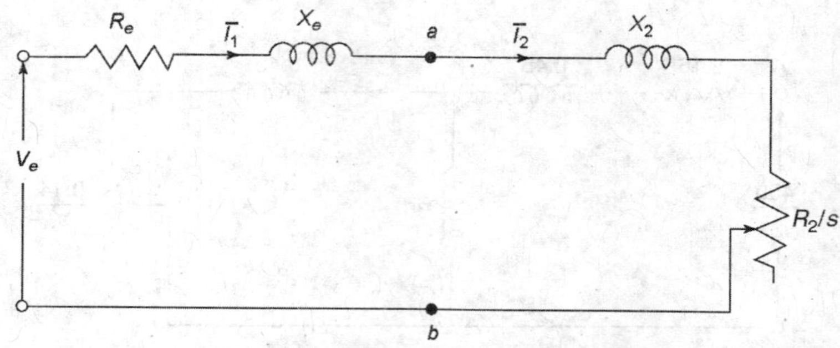

Figure 8.15 Thevenin's Equivalent Circuit

$\therefore$
$$V_e = \frac{V_1(jX_\phi)}{R_1 + j(X_1 + X_\phi)} \tag{8.38}$$

and
$$Z_e = \frac{(R_1 + jX_1)(jX_\phi)}{R_1 + j(X_1 + X_\phi)} \tag{8.39}$$

Since $X_1 + X_\phi$ is much greater than R_1, for most of the induction motors, R_1 can be neglected.
Equations (8.38) and (8.39) become

$$V_e = \frac{V_1 X_\phi}{X_1 + X_\phi} = \frac{V_1 X_\phi}{X'_1} \tag{8.40}$$

and
$$Z_e = R_e + jX_e = \frac{R_1 X_\phi}{X_1 + X_\phi} + \frac{jX_1 X_\phi}{X_1 + X_\phi}$$

$$= \frac{R_1 X_\phi}{X'_1} + j\frac{X_1 X_\phi}{X'_1} \tag{8.41}$$

where $X_1' = X_1 + X_\phi$ is the stator self-reactance per phase.
From Figure 8.15 the rotor current is given by

$$I_2 = \frac{V_e}{\left(R_e + \dfrac{R_2}{s}\right) + j(X_2 + X_e)} \tag{8.42}$$

Using Equation (8.42), the torque is given by

$$T_{e\,max} = \frac{m}{\omega_s}\left[\frac{V_e^2}{\left(R_e + \dfrac{R_2}{s}\right)^2 + (X_2 + X_e)^2}\right]\frac{R_2}{s}$$

i.e.,
$$T_{e\,max} = \frac{\dfrac{KR_2}{s}}{\left(R_e + \dfrac{R_2}{s}\right)^2 + (X_2 + X_e)^2} \text{ N-m} \tag{8.43}$$

where $K = \dfrac{mV_e^2}{\omega_s} = \dfrac{mV_e^2}{2\pi n_s}$, and $T_{e\,max}$ denotes maximum internal torque.

$$T_{e\,max} = \frac{mP_g}{2\pi n_s} = \frac{mI_2^2 R_2}{2\pi n_s s} \tag{8.44}$$

So, P_g, that is, $\dfrac{I_2^2 R_2}{s}$ will be maximum when $\dfrac{R_2}{s}$ will be equal to the impedance seen by $\dfrac{R_2}{s}$,
i.e., $\dfrac{R_2}{s_{mT}} = \sqrt{R_e^2 + (X_e + X_2)^2}$ (From Figure 8.15)

$$\frac{R_2}{s_{mT}} = \sqrt{R_e^2 + X^2} \tag{8.45}$$

where $X = X_2 + X_e$ and s_{mT} is the slip at which maximum torque occurs.

$$\therefore \qquad s_{mT} = \frac{R_2}{\sqrt{R_e^2 + X^2}}$$

Using Equation (8.45), we have from Equation (8.43)

$$T_{em} = \frac{K\sqrt{R_e^2 + X^2}}{\left[R_e + \sqrt{R_e^2 + X^2}\right]^2 + X^2}$$

$$= \frac{K\sqrt{R_e^2 + X^2}}{R_e^2 + R_e^2 + X^2 + 2R_e\sqrt{R_e^2 + X^2} + X^2}$$

$$= \frac{K\sqrt{R_e^2 + X^2}}{2R_e(R_e^2 + X^2) + 2R_e\sqrt{R_e^2 + X^2}}$$

$$\therefore \qquad T_{em} = \frac{K}{2\left[R_e + \sqrt{R_e^2 + X^2}\right]} \tag{8.46}$$

where $K \propto V_e^2$, that is, proportional the square of stator voltage.

The stator load component of current (I_{2st}) is given by

$$I_{2st} = \frac{V_e}{\sqrt{(R_e + R_2)^2 + (X_e + X_2)^2}} \tag{8.47}$$

The stator load current at any slip s is given by

$$I_2 = \frac{V_e}{\sqrt{\left(R_e + \dfrac{R_2}{s}\right)^2 + (X_e + X_2)^2}} \tag{8.48}$$

From Equations (8.47) and (8.48), we have

$$\frac{I_{2st}}{I_2} = \frac{\sqrt{\left(R_e + \dfrac{R_2}{s}\right)^2 + (X_e + X_2)^2}}{\sqrt{(R_e + R_2)^2 (X_e + X_2)^2}} \tag{8.49}$$

For negligible stator resistance R_1, $R_e = 0$. From Equation (8.49), we have

$$\frac{I_{2st}}{I_2} = \sqrt{\frac{(R_2 / s)^2 + X^2}{R_2{}^2 + X^2}} \tag{8.50}$$

where $X = X_e + X_2$

With $R_e = 0$, from Equation (8.45), we have

$$R_2 = s_{mT}\, X \tag{8.51}$$

Substituting the value of R_2 from Equation (8.51) into Equation (8.50), we have

$$\frac{I_{2st}}{I_2} = \sqrt{\frac{(s_{mT}X/s)^2 + X^2}{(s_{mT}X)^2 + X^2}} = \sqrt{\frac{(s_{mT}/s)^2 + 1}{(s_{mT})^2 + 1}} = \sqrt{\frac{s_{mT}^2 + s^2}{s^2(s_{mT}^2 + 1)}} \tag{8.52}$$

For $R_e = 0$, we have from Equation (8.48)

$$I_2 = \frac{V_e}{\sqrt{(R_2/s)^2 + X^2}} \tag{8.53}$$

The current (I_{2mT}) corresponding to maximum torque is given by

$$I_{2mT} = \frac{V_e}{\sqrt{(R_2/s_{mT})^2 + X^2}} \tag{8.54}$$

From Equations (8.53) and (8.54), we have

$$\frac{I_{2mT}}{I_2} = \sqrt{\frac{(R_2/s)^2 + X^2}{(R_2/s_{mT})^2 + X^2}} = \sqrt{\frac{(s_{mT}X/s)^2 + X^2}{(s_{mT}X/s_{mT})^2 + X^2}} = \sqrt{\frac{1}{2}\left[1 + \left(\frac{s_{mT}}{s}\right)^2\right]} \tag{8.55}$$

Example 8.8 A three-phase, 50 Hz, four-pole, 12 kW, 400 V slip-ring induction motor, with its slip rings short-circuited, develops rated output at rated voltage and frequency. At a slip of 5 per cent, the maximum torque occurs with zero external resistance and it is 1.5 times the full-load torque. Neglecting stator resistance and rotational losses, calculate the following:

(i) Slip and rotor speed at full-load torque.
(ii) Rotor ohmic loss at full-load torque.
(iii) Starting torque at rated voltage and frequency.
(iv) Starting current in terms of full-load current.
(v) Stator current in terms of full-load current at maximum torque.
(vi) Efficiency at full load.

Solution

(i)
$$\frac{T_{efl}}{T_{e\max}} = \frac{2s_{mT}s_{fl}}{s_{fl}^2 + s_{mT}^2}$$

i.e.,
$$\frac{1}{1.5} = \frac{2\times 0.05 s_{fl}}{0.05^2 + s_{fl}^2}$$

i.e., $s_{fl}^2 - 0.15 s_{fl} + 0.0025 = 0$

i.e.,
$$s_{fl} = \frac{0.15 \pm \sqrt{0.15^2 - 4\times 0.0025}}{2} = \frac{0.15 \pm \sqrt{0.0125}}{2} = \frac{0.15 \pm 0.1118}{2}$$

i.e., $s_{fl} = 0.1309$ or 0.0191

Neglecting higher value, $s_{fl} = 0.0191$

$$\text{Rotor speed } (N_r) = (1 - 0.0191)\times\left(\frac{120\times 50}{4}\right) = 1{,}471.35 \text{ rpm}$$

(ii) Here, P_{sh} = 12 kW and rotational losses are neglected.

$$\therefore \quad P_m = P_{sh}$$

$$\therefore \quad P_g = \frac{P_m}{1-s} = \frac{12\times 10^3}{1-0.0191} = 12{,}233.66 \text{ W}$$

$\therefore$ Rotor ohmic loss = sP_g = 0.0191 × 12,233.66 = 233.66 W

$$\text{(iii) Full-load torque } (T_{efl}) = \frac{P_g}{2\pi\left(\frac{N_s}{60}\right)} = \frac{12{,}233.66}{2\times 3.14\times \frac{1{,}500}{60}} = 77.92 \text{ N-m}$$

$$\therefore \quad \frac{T_{est}}{T_{e\max}} = \frac{2}{s_{mT} + \frac{1}{s_{mT}}} = \frac{2}{0.05 + \frac{1}{0.05}} \qquad [s_{mT} = 5\%]$$

$$\therefore \quad T_{est} = \frac{2T_{e\max}}{20.05} = \frac{2\times 1.5\times T_{efl}}{20.05} = \frac{3T_{efl}}{20.05} = \frac{3\times 77.92}{20.05} = 11.659 \text{ N-m}$$

(iv)
$$\frac{I_{2st}}{I_{2fl}} = \sqrt{\frac{s_{fl}^2 + s_{mT}^2}{s_{fl}^2\left(1+s_{mT}^2\right)}} = \sqrt{\frac{0.0191^2 + 0.05^2}{0.0191^2\left(1+0.05^2\right)}} = 2.7988$$

i.e.,
$$I_{2st} = 2.7988 I_{2fl}$$

(v)
$$\frac{I_{2mT}}{I_{2fl}} = \sqrt{\frac{1}{2}\left[1+\left(\frac{s_{mT}}{s_{fl}}\right)^2\right]} = \sqrt{\frac{1}{2}\left[1+\left(\frac{0.05}{0.0191}\right)^2\right]} = 1.981$$

$$I_{2mT} = 1.981\, I_{2fl}$$

(vi)
$$\text{Efficiency} = \frac{P_{sh}}{P_i} = \frac{P_{sh}}{P_g} \quad \text{(since stator losses are negligible)}$$

$$= \frac{12\times 10^3}{12{,}233.66} = 0.9809 \text{ p.u.} = 98.09\%$$

■

8.22 STARTING OF INDUCTION MOTORS

Figure 8.16 shows the approximate equivalent circuit of the induction motor at the time of starting. The slip of the induction motor becomes unity at this time. If the motor is directly connected to the supply, the starting current of the motor becomes five to seven times its full-load current. The current (I_0) in the shunt branch of the equivalent circuit becomes small compared to the starting current (I_{st}), and hence shunt

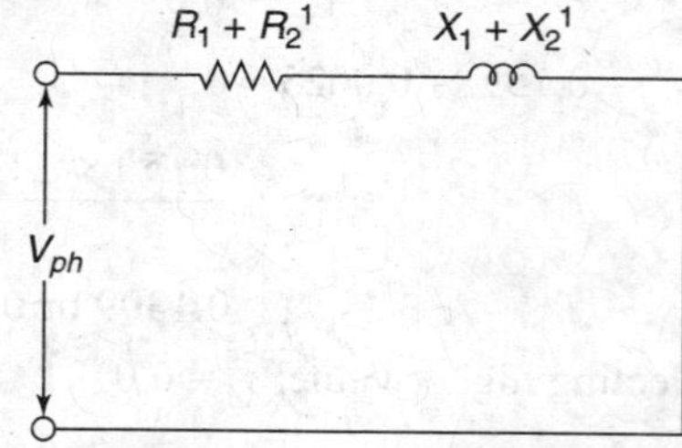

Figure 8.16 Approximate Equivalent Circuit of an Induction Motor During Starting

branch portion has been omitted in the approximate equivalent circuit, as shown in Figure 8.16. The starting current is expressed by

$$I_{st} = \frac{V_{ph}}{Z_{sc}} \tag{8.56}$$

where

$$Z_{sc} = \sqrt{(R_1 + R_2')^2 + (X_1 - X_2')^2} \tag{8.57}$$

Neglecting magnetizing current, the starting torque becomes

$$T_{est} = \frac{3I_{st}^2 R_2'}{\omega_s} \tag{8.58}$$

Neglecting magnetizing current, the full-load torque (T_{efl}) is expressed by

$$T_{efl} = \frac{3I_{fl}^2 R_2'}{\omega_s s_{fl}} \tag{8.59}$$

Small induction motors up to 2 kW capacity may be connected directly to the supply mains, whereas the higher-capacity machines are strictly prohibited to be connect to the supply mains directly because of their high starting current, which is five to seven times the full-load current. This excessive current results in a large voltage drop in the distribution network and stops other machines, which are already running on the supply mains.

8.23 STARTING OF SQUIRREL-CAGE MOTORS

The methods for starting of squirrel-cage induction motor are discussed in the following sub-sections.

8.23.1 Direct-on-line Starting of Induction Motors

The motor is directly switched on to the supply mains. This method is restricted to small motors above 2 kW. Actually, for these motors, the starting torque is about five to six times of the full-load torque.

From Equations (8.58) and (8.59), we have

$$\frac{T_{est}}{T_{efl}} = \frac{I_{2st}^2\left(\frac{R_2}{1}\right)}{I_{2fl}^2\left(\frac{R_2}{s_{fl}}\right)} = \left(\frac{I_{2st}}{I_{2fl}}\right)^2 s_{fl} \tag{8.60}$$

Equation (8.60) is valid when rotor resistance is constant.

Let I_{2st} and I_{2fl} be the per phase rotor currents at starting and full-load conditions, respectively.

$\therefore$ $I_{st} \times$ (effective stator turns) $= I_{2st} \times$ (effective rotor turns)

$\therefore$ $I_{st} = I_{2st}$ (effective rotor to stator turns ratio)

$\therefore$ $I_{fl} = I_{2fl}$ (effective rotor to stator turns ratio)

Equation (8.60) can be expressed as

$$\frac{T_{est}}{T_{efl}} = \left(\frac{I_{st}}{I_{fl}}\right)^2 s_{fl} \tag{8.61}$$

Let V_1 be the per phase stator voltage,

Z_1 be the per phase leakage impedance referred to the stator at standstill and

I_{sc} be the per phase short-circuit current at standstill.

$$\therefore \qquad I_{sc} = \frac{V_1}{Z_1}$$

$\therefore$ for direct switching, $I_{st} = I_{sc} = \dfrac{V_1}{Z_1}$

Equation (8.61) can be written as

$$\frac{T_{est}}{T_{efl}} = \left(\frac{I_{sc}}{I_{fl}}\right)^2 s_{fl} \tag{8.62}$$

Assuming $I_{sc} = 5I_{fl}$ and $s_{fl} = 2.8$ per cent, from Equation (8.62), we have

$$\frac{T_{est}}{T_{efl}} = (6)^2 \times 0.028 \simeq 1$$

Therefore, the starting current must be six times its full-load current for $T_{est} = T_{efl}$. The motor must accelerate quickly and attain the full-load speed under this condition. Otherwise, the motor will be overheated due to excessive current taken from the supply. Hence, the motor preferably should be started at no load so that further increase in the starting current must be avoided. This method is useful for small-size motors.

8.23.2 Stator Resistor (or Reactor) Starting

Figure 8.17 shows the reactor starting of squirrel-cage induction motor. The per phase voltage to the stator is reduced to xV_1, where $x < 1$.

$$I_{st} = \frac{xV_1}{Z_1} = xI_{sc}$$

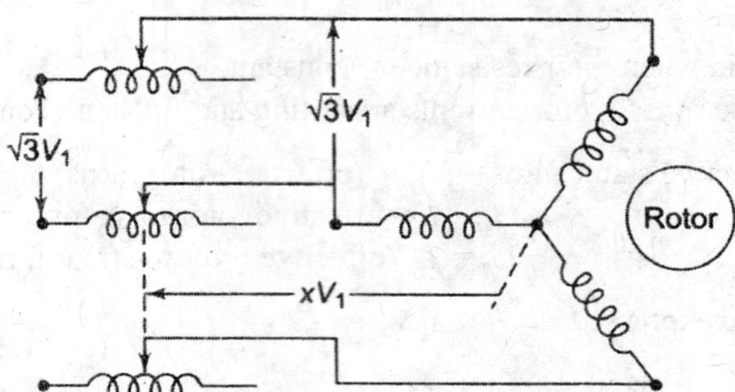

Figure 8.17 Reactor Starting of a Squirrel-cage Induction Motor

We can have

$$\frac{T_{est}}{T_{efl}} = \left(\frac{I_{st}}{I_{fl}}\right)^2 s_{fl} = \left(\frac{xI_{sc}}{I_{fl}}\right)^2 s_{fl}$$

$$\frac{T_{est}}{T_{efl}} = x^2 \left(\frac{I_{sc}}{I_{fl}}\right)^2 s_{fl} \tag{8.63}$$

Since in an induction motor, torque $\propto$ (voltage)^2

$$\therefore \quad \frac{\text{Starting torque with reactor starting}}{\text{Starting torque with direct switching}} = \left(\frac{xV_1}{V_1}\right)^2 = x^2 \tag{8.64}$$

Series resistor is cheaper than series reactor. Series reactor has lower energy loss, and it is more effective in reducing voltage because the power factor of induction motor at starting is quite low.

8.23.3 Autotransformer Starting

Figure 8.18 shows autotransformer starting of squirrel-cage inductor motors, where the fraction of the supply voltage (V_1) ie., xV_1 is applied to the stator terminals at the time of starting, which reduces the motor current drawn from the supply. When the motor has got its acceleration near to its operating point, full-line voltage is applied to the induction motor.

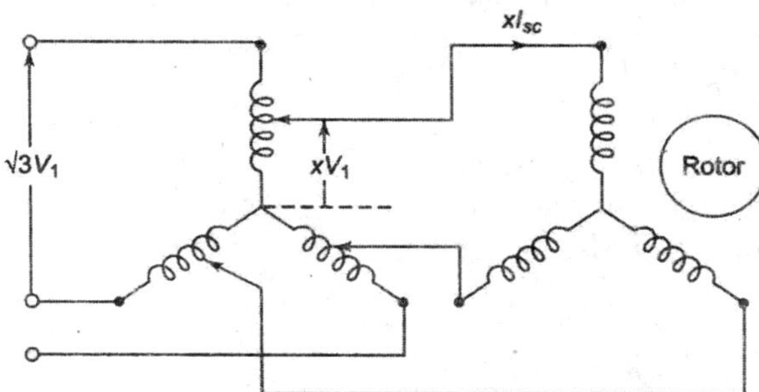

Figure 8.18 Autotransformer Starter

So, per phase starting current in motor winding when autotransformer is present $= \dfrac{xV_1}{Z} = xI_{sc}$.

$\therefore$ No-load current of autotransformer,

Input VA/phase = output VA/phase

$$\therefore \quad I_{st}V_1 = xV_1 (xI_{sc})$$

$\therefore$ Per phase starting current taken from the supply mains is given by

$$I_{st} = x^2 I_{sc} \tag{8.65}$$

$$\therefore \quad \frac{T_{est}}{T_{efl}} = \left[\frac{(xI_{sc})^2}{I_{fl}^2}\right] s_{fl} = x^2 \left(\frac{I_{sc}}{I_{fl}}\right)^2 s_{fl} = x^2 \left(\frac{1}{x^2} \cdot \frac{I_{st}}{I_{fl}}\right)^2 s_{fl}$$

$$\therefore \quad \frac{T_{est}}{T_{efl}} = \frac{1}{x^2} \left(\frac{I_{st}}{I_{fl}}\right)^2 s_{fl} \tag{8.66}$$

From Equation (8.65) $I_{st} = x^2 I_{sc}$

$$\therefore \quad xI_{sc} = \frac{1}{x} I_{st} \tag{8.67}$$

Equation (8.67) gives the per phase motor starting current in terms of I_{st}.

For an induction motor, torque $\propto$ (voltage)^2

$$\therefore \quad \frac{\text{Starting torque with an autotransformer}}{\text{Starting torque with direct switching}} = \left(\frac{xV_1}{V_1}\right)^2 = x^2 \tag{8.68}$$

8.23.4 Star-delta Starter

Figure 8.19 shows a star-delta starter for starting the induction motor. In position 1, A', B', C' are short-circuited. In position 2, A' is connected to B, B' is connected to C and C' is connected to A to form delta connection.

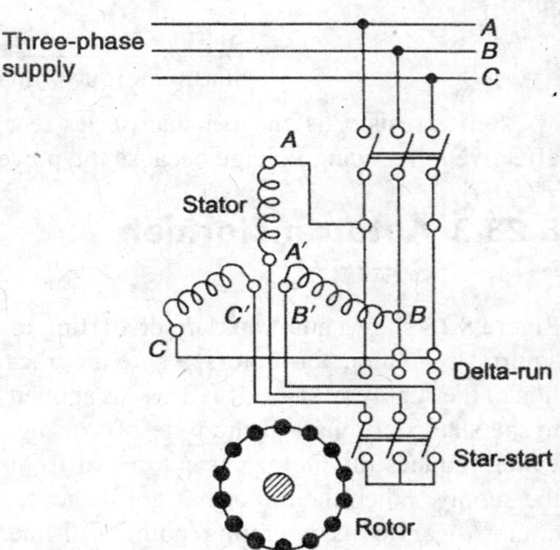

Figure 8.19 Star-delta Starter

Let V_2 be the line voltage. The per phase motor starting current I_{sty} with stator winding in star is given by $I_{sty} = \frac{V_L}{\sqrt{3}Z}$.

If stator windings are delta connected, the per phase motor starting current (I_{std}) with direct switching is given by

$$I_{std} = \frac{V_L}{Z_1} = I_{scd}$$

So, the starting line current with direct switching would be equal to $\sqrt{3}\frac{V_L}{Z_1} = \sqrt{3}I_{scd} = \sqrt{3}I_{std}$

$$\frac{\text{Starting line current with star-delta starter}}{\text{Starting line current with direct switching in delta}} = \frac{I_{sty}}{\sqrt{3}I_{std}} = \frac{\frac{1}{\sqrt{3}}\left(\frac{V_L}{Z_1}\right)}{\sqrt{3}\left(\frac{V_L}{Z_1}\right)} = \frac{1}{3}$$

$$\frac{\text{Starting torque with star-delta starter}}{\text{Starting torque with direct switching in delta}} = \frac{\left(\frac{V_L}{\sqrt{3}}\right)^2}{V_L^2} = \frac{1}{3}$$

$$\frac{\text{Starting torque with star-delta starter}}{\text{Starting torque with direct switching in delta}} = \frac{\frac{1}{\omega_s}(I_{sty})^2\frac{R^2}{1}}{\frac{1}{\omega_s}(I_{fld})^2\frac{R^2}{s_{fl}}} = \frac{\left(\frac{1}{\sqrt{3}}I_{std}\right)^2}{(I_{fld})^2}\cdot s_{fl}$$

$$= \frac{1}{3}\cdot\left(\frac{I_{scd}}{I_{fld}}\right)^2\cdot s_{fl} \tag{8.69}$$

Star-delta starters are widely used because they cheaper than autotransformers.

Example 8.9 The ratio of maximum torque to full-load torque in a three-phase squirrel-cage induction motor is 2.5:1. Determine the ratio of actual starting torque to full-load torque for the following cases:

(i) Direct starting.
(ii) Star-delta starting.
(iii) Autotransformer starting having voltage per phase at starting as 65 per cent of supply.

The rotor resistance and reactance per phase are 0.4 Ω and 4 Ω, respectively.

Solution

$$\frac{T_{e\,max}}{T_{efl}} = \frac{2.5}{1}$$

$$R_2 = 0.4\ \Omega \text{ and } X_2 \text{ (at standstill)} = 4\ \Omega$$

Again, $$T_{e\,max} = \frac{KE_2^2}{2X_2}$$

$$\therefore \quad T_{efl} = \frac{T_{e\,max}}{2.5} = \frac{KE_2^2}{5X_2}$$

(i) $$T_{est} = \frac{KE_2^2 R_2}{R_2^2 + X_2^2}$$

$$\frac{T_{est}}{T_{efl}} = \frac{KE_2^2 R_2}{R_2^2 + X_2^2} \times \frac{5X_2}{KE_2^2} = \frac{5R_2X_2}{R_2^2 + X_2^2} = \frac{5 \times 0.4 \times 4}{0.4^2 + 4^2} = 0.495$$

$$T_{est} = 49.5 \text{ per cent of } T_{efl}$$ ■

(ii) For star-delta starting, rotor induced voltage/phase at starting $= \dfrac{E_2}{\sqrt{3}}$

$$\frac{T_{est}}{T_{efl}} = \frac{K\left(\frac{E_2}{\sqrt{3}}\right)^2 R_2}{R_2^2 + X_2^2} \times \frac{5X_2}{KE_2^2} = \frac{1}{3}\left(\frac{5R_2X_2}{R_2^2 + X_2^2}\right) = \frac{1}{3} \times \frac{5 \times 0.4 \times 4}{0.4^2 + 4^2} = 0.165$$

$$\therefore \quad T_{est} = 16.5 \text{ per cent of } T_{efl}$$ ■

(iii) The voltage per phase at starting is 0.65 E_2

$$\frac{T_{est}}{T_{efl}} = \frac{K(0.65E_2)^2 R_2}{R_2^2 + X_2^2} \times \frac{5X_2}{KE_2^2} = \frac{0.65^2 \times 5R_2X_2}{R_2^2 + X_2^2} = \frac{0.65^2 \times 5 \times 0.4 \times 4}{0.4^2 + 4^2} = 0.2091$$

$$\therefore \quad T_{est} = 20.91 \text{ per cent of } T_{efl}$$ ■

Example 8.10 A three-phase squirrel-cage induction motor has a short-circuit current equal to five times its full-load current. Calculate the starting torque as a percentage of full-load torque if the motor is started by the following:

(i) Direct switching to the supply.
(ii) A star-delta starter.
(iii) An autotransformer (the applied voltage is $\frac{1}{\sqrt{2}}$ times the supply voltage).
(iv) A resistance in the stator circuit.

The stator current in (iii) and (iv) is limited to 2.5 times the full-load current. The full-load slip is 2.5 per cent.

Solution

Short-circuit current $(I_{sc}) = 5I_{fl}$

Full-load slip (s_{fl}) = 2.5 per cent = 0.025

(i) $$\frac{T_{est}}{T_{efl}} = \left(\frac{I_{sc}}{I_{fl}}\right)^2 s_{fl} = \left(\frac{5I_{fl}}{I_{fl}}\right)^2 s_{fl} = 25 \times 0.025 = 0.625$$

$\therefore$ T_{est} = 62.5 per cent of full-load torque.

(ii) $$\frac{T_{est}}{T_{efl}} = \frac{1}{3} \times \left(\frac{I_{sc}}{I_{fl}}\right)^2 s_{fl} = \frac{1}{3}\left(\frac{5I_{fl}}{I_{fl}}\right)^2 s_{fl} = \frac{1}{3} \times 5^2 \times 0.025 = 0.208$$

$\therefore$ T_{est} = 20.8 per cent of full-load torque.

(iii) Here, $x = \dfrac{1}{\sqrt{2}}$

$$\frac{T_{est}}{T_{efl}} = x^2 \left(\frac{I_{sc}}{I_{fl}}\right)^2 s_{fl} = \left(\frac{1}{\sqrt{2}}\right)^2 \left(\frac{2.5I_{fl}}{I_{fl}}\right)^2 s_{fl} = 0.0781$$

$\therefore$ T_{est} = 7.81 per cent of full-load torque.

(iv) $$\frac{T_{est}}{T_{efl}} = \frac{1}{3}\left(\frac{I_{st}}{I_{fl}}\right)^2 s_{fl} = \frac{1}{3}\left(\frac{2.5I_{fl}}{I_{fl}}\right)^2 s_{fl} = 0.052$$

$\therefore$ T_{est} = 5.2 per cent of full-load torque. ■

8.24 STARTING OF SLIP-RING INDUCTION MOTORS

Figure 8.20 shows the resistance steps for a three-phase slip-ring induction motor.

Let r_m be the motor resistance,

V be the motor line voltage,

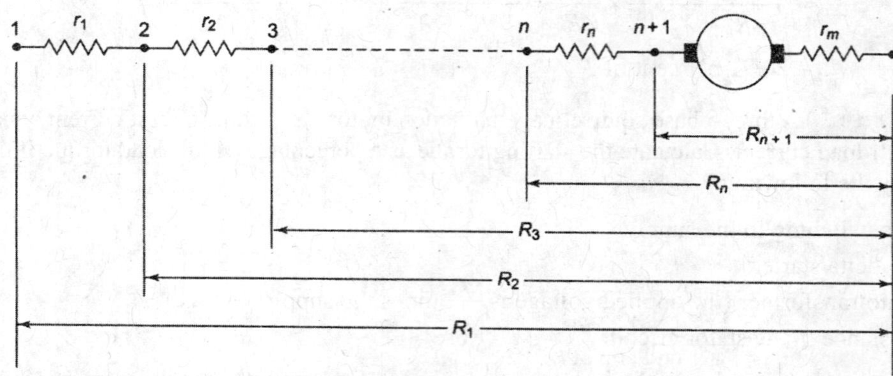

Figure 8.20 Resistance Steps for a Three-phase Slip-ring Induction Motor

E be the induced emf of the rotor standstill emf,

s be the slip,

X be the rotor standstill reactance,

n be the number of sections or studs/elements,

I_1 be the upper current limit during starting and

I_2 be the lower current limit during starting.

Stator resistance and leakage reactance are neglected here.

Stud 1: The machine draws a line current I_1 after switching the motor to supply. Let the slip be s_1.

$$I_1 = \frac{s_1 E}{\sqrt{R_1^2 + s_1^2 X^2}} \tag{8.70}$$

The slip and current fall because the machine picks up speed. Let s_2 be the slip when the value of current just before notching is I_2.

$$I_2 = \frac{s_2 E}{\sqrt{R_1^2 + s_2^2 X^2}} \tag{8.71}$$

Stud 2: Let the resistance r_1 be the cut-out, and the resistance of the circuit becomes R_2. The current rises to I_1. Assuming no change in speed during notching operation, the slip, just before notching, remains s_2.

$$I_1 = \frac{s_2 E}{\sqrt{R_2^2 + s_2^2 X^2}}$$

The current falls to I_2 just before notching at stud 2. Let the slip be s_3.

$$I_2 = \frac{s_3 E}{\sqrt{R_2^2 + s_3^2 X^2}}$$

Therefore, we have

$$I_1 = \frac{s_1 E}{\sqrt{R_1^2 + s_1^2 X^2}} = \frac{s_2 E}{\sqrt{R_2^2 + s_2^2 X^2}} = \cdots\cdots\cdots = \frac{s_{n+1} E}{\sqrt{R_m^2 + s_{n+1}^2 X^2}} \tag{8.72}$$

From Equation (8.72), we have

$$\frac{R_1}{s_1} = \frac{R_2}{s_2} = \cdots\cdots\cdots = \frac{R_m}{s_{n+1}} \tag{8.73}$$

and $$I_2 = \frac{s_2 E}{\sqrt{R_1^2 + s_2^2 X^2}} = \frac{s_3 E}{\sqrt{R_2^2 + s_2^2 X^2}} = \cdots\cdots\cdots = \frac{s_{n+1} E}{\sqrt{R_n^2 + s_{n+1}^2 X^2}} \tag{8.74}$$

From Equation (8.73), we have

$$\frac{s_2}{s_1} = \frac{s_3}{s_2} = \cdots\cdots = \frac{s_m}{s_n} = \frac{R_2}{R_1} = \frac{R_3}{R_2} = \cdots\cdots = \frac{R_m}{R_n} \tag{8.75}$$

The slip at starting (s_1) is 1 and slip at $(n + 1)$th stud (s_{n+1}) = slip at full load (s_m). This will be valid if and only if the upper limit of current is the full-load current.

$$\therefore \quad \frac{s_2}{s_1} = \frac{s_3}{s_2} = \cdots\cdots = \frac{s_m}{s_n} = \frac{R_2}{R_1} = \frac{R_3}{R_2} = \cdots\cdots = \frac{R_m}{R_n} = \alpha$$

$$\therefore \quad s_m = \alpha^n$$

i.e.,
$$\alpha = s^{\frac{1}{n}} \tag{8.76}$$

Again,
$$\frac{R_m}{R_1} = \alpha^n$$

$$\therefore \quad R_m = R_1 \alpha^n \tag{8.77}$$

Further, we have

$$R_2 = \alpha R_1$$
$$R_3 = \alpha R_2 = \alpha^2 R_1$$
$$R_4 = \alpha R_3 = \alpha^3 R_1$$
$$\cdots\cdots\cdots\cdots\cdots\cdots$$
$$R_n = \alpha R_{n-1} = \alpha^{n-1} R_1$$

The sections are

$$r_1 = R_1 - R_2 = R_1(1-\alpha)$$
$$r_2 = R_2 - R_3 = R_1(\alpha - \alpha^2) = \alpha r_1$$
$$r_3 = R_3 - R_4 = R_1(\alpha^2 - \alpha^3) = \alpha^2 r_1$$
$$\cdots\cdots\cdots\cdots\cdots\cdots$$
$$r_n = R_n - R_{n+1} = R_1(\alpha^{n-1} - \alpha^n) = \alpha^{n-1} r_1$$

From Equations (8.72) and (8.74), we have

$$\left(\frac{I_2}{I_1}\right)^2 = \frac{s_2^2\left(R_1^2 + s_1^2 X^2\right)}{s_1^2\left(R_1^2 + s_2^2 X^2\right)} = \frac{\frac{R_1^2}{s_1^2} + X_2}{\frac{R_1^2}{s_2^2} + X^2}$$

$$= \alpha^2 \text{ (for normal motors)} \tag{8.78}$$

Equation (8.78) suggests that the ratio of the rotor current limits is approximately α, which is the approximate ratio of the stator currents.

Example 8.11 The rotor resistance per phase of a three-phase, 60 kW induction motor is 0.020 Ω. Design a starter for this induction motor having six notches, where the upper current limit has to be the full-load current so that the slip is 2.5 per cent.

Solution

Number of studs = 6, number of elements (n) = 6 − 1 = 5, s_{fl} = 2.5 per cent = 0.025

$$\alpha = (s_{fl})^{\frac{1}{n}} = \sqrt[5]{0.025} = 0.478$$

$$R_1 = \frac{\text{Rotor resistance/phase}}{s_{fl}} = \frac{0.020}{0.025} = 0.8$$

$$r_1 = R_1(1-\alpha) = 0.8 \times (1-0.478) = 0.4176\ \Omega$$

$$r_2 = \alpha r_1 = 0.478 \times 0.4176 = 0.19961\ \Omega$$

$$r_3 = \alpha^2 r_1 = 0.478^2 \times 0.4176 = 0.09541\ \Omega$$

$$r_4 = \alpha^3 r_1 = 0.478^3 \times 0.4176 = 0.04560\ \Omega$$

$$r_5 = \alpha^4 r_1 = 0.478^4 \times 0.4176 = 0.02180\ \Omega$$

Rotor resistance/phase = 0.020 Ω

Total = 0.80 Ω

8.25 NO-LOAD TEST OR OPEN-CIRCUIT TEST

This test is similar to the open-circuit test of a transformer. In this test, the motor is uncoupled from its load and the rated voltage at rated frequency is applied to the stator. The motor runs without load. To measure the input power, two-wattmeter method is used. An ammeter and a voltmeter are connected to measure the no-load current and voltage. Since the motor runs at no load, the total input power is equal to constant iron-loss, friction and windage losses of the motor.

During the no-load condition the power factor of the induction motor is less than 0.5. So, one wattmeter will show negative reading. The direction of the current coil terminals should be reversed to take the reading. From this test, R_c and X_ϕ can be determined. Figure 8.21 shows the schematic diagram of the no-load test of an induction motor.

Let V_L = input line voltage (V)
P = three-phase power (W)
I_0 = input line current (A)
V_p = input phase voltage (V)

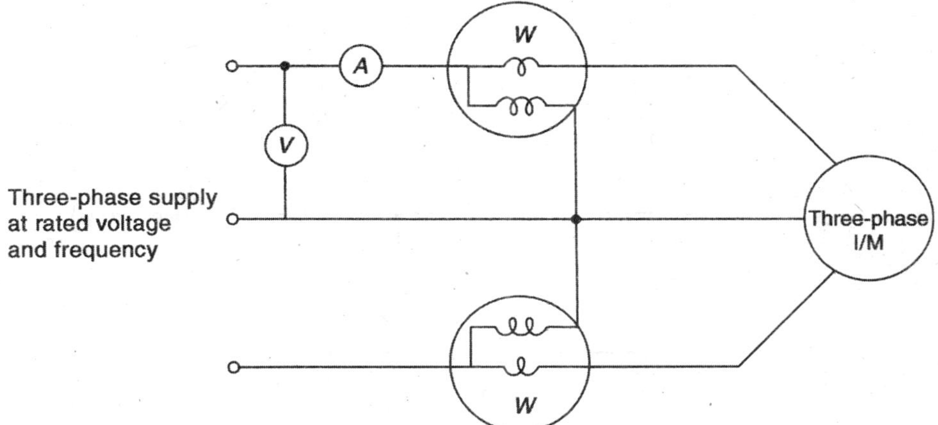

Figure 8.21 No-load Test of a Three-phase Induction Motor

$\therefore \quad P = \sqrt{3} V_L I_0 \cos\theta_0$

$\therefore \quad \cos\theta_0 = \dfrac{P}{\sqrt{3} V_L I_0}$

$\therefore \quad I_c = I_0 \cos\theta_0 \text{ A}$

and $\quad I_\phi = I_0 \sin\theta_0 \text{ A}$

$\therefore \quad R_c = \dfrac{V}{I_c} \ \Omega$ and $X_\phi = \dfrac{V_p}{I_\phi} \ \Omega$

Please see Appendix E for separation of no-load losses.

8.26 BLOCKED-ROTOR OR SHORT-CIRCUIT TEST

Figure 8.22 shows the schematic diagram of blocked-rotor test of an induction motor. The induction motor is locked so that it cannot move. The rotor winding is short-circuited. In squirrel-cage motors, the rotor bars are permanently short-circuited. In a slip-ring motor, the rotor windings are short-circuited by the slip rings.

Let $\quad P_{sc}$ = total power input during this test (W)

I_{sc} = line current on short circuit (A)

V_{sc} = line voltage on short circuit (V)

We know that $P_{sc} = \sqrt{3} V_{sc} I_{sc} \cos\theta_{sc}$

where $\cos\theta_{sc}$ = short-circuit power factor

Equivalent resistance of the rotor referred to the stator $(R_{01}) = \dfrac{P_{sc}}{I_{sc}^2} \ \Omega$

and the equivalent impedance of the rotor referred to the stator $(Z_{01}) = \dfrac{V_{sc}}{I_{sc}} \ \Omega$

Equivalent reactance of the rotor referred to the stator $(X_{01}) = \sqrt{Z_{01}^2 - R_{01}^2} \ \Omega$

Figure 8.23 shows the equivalent circuit of induction motor during the blocked-rotor test.

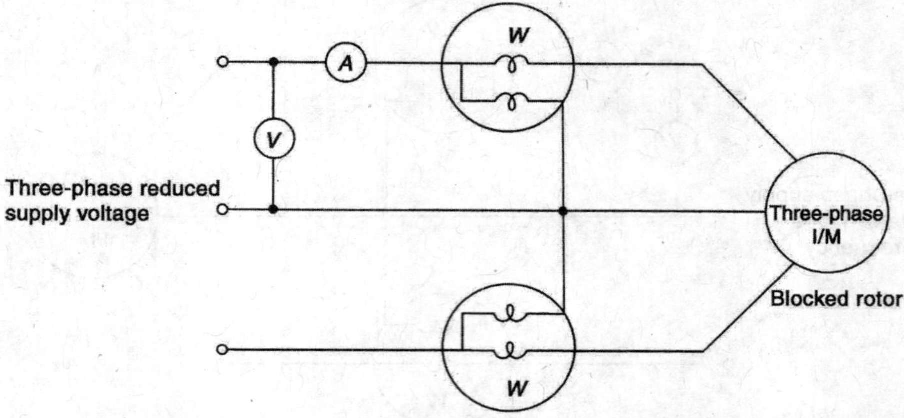

Figure 8.22 Blocked-rotor Test of a Three-phase Induction Motor

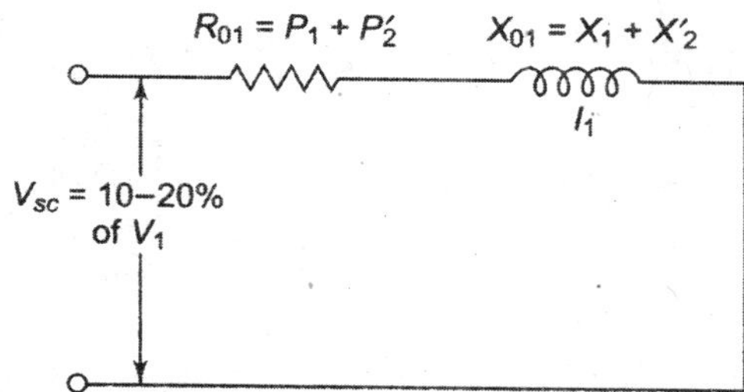

Figure 8.23 Equivalent Circuit of Induction Motor During Blocked-rotor Test

Example 8.12 A 3 kW, 400 V/200 V, delta-star, 50 Hz, three-phase, six-pole induction motor is found to draw a line current of 25 A at a power factor of 0.4, when a blocked-rotor test is conducted at the rated voltage. Determine the stator and the rotor winding resistances in ohms per phase, if the torque developed by the motor under the above conditions is 25 N-m.

Solution

Rating: 3 kW, 400 V/200 V; delta-star. 50 Hz; $P = 6$; $I_L = 25$ A; power factor = 0.4; torque developed (T) = 25 N-m

Stator and rotor winding resistances are R_1 and R_2, respectively.

Under rotor condition, equivalent circuit is shown in Figure E8.2

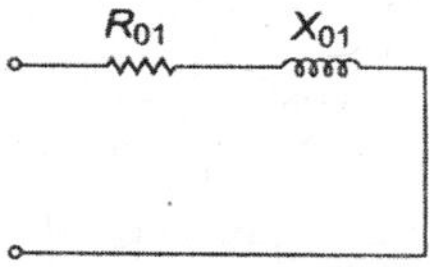

R_{01} X_{01}

Figure E8.2

$$Z_{01} = \frac{400}{(25/\sqrt{3})} = 27.7\ \Omega/\text{phase}$$

$$R_{01} = Z_{01} \times \text{power factor} = 27.7 \times 0.4 = 11.0\ \Omega/\text{phase}$$

where $R_{01} = R_1 + R_2'$

Turns ratio, $$a = \frac{400}{\left(\frac{200}{\sqrt{3}}\right)} = 3.464$$

Rotor current, $$I_2 = \frac{I_1}{\sqrt{3}} \times a = 50\text{ A}$$

Synchronous speed, $$N_s = \frac{120f}{P} = \frac{120 \times 50}{6} = 1{,}000\text{ rpm}$$

Since $$\left(\frac{2\pi N_s}{60}\right) \times T = 3I_2^{\,2}R_2$$

$$R_2 = \frac{2\pi N_s T}{60 \times 3 \times I_2^2} = \frac{2\pi \times 1{,}000 \times 25}{60 \times 3 \times 50^2} = 0.349\ \Omega/\text{phase}$$

$$R_2' = a^2 R_2 = (3.464)^2 \times 0.349 = 4.187\ \Omega$$

Also, $$R_{01} = R_1 + R'_2$$

∴ $$R_1 = R_{01} - R'_2 = 11.08 - 4.187 = 6.893\ \Omega/\text{phase}$$

■

8.27 DIRECT TESTING OF INDUCTION MOTORS

This test is useful to obtain the actual performance characteristics of an induction motor, where load is applied by means of a brake or by coupling it to an AC generator. The efficiency of the generator must be known so that the efficiency and other characteristics of the induction motor can be computed. This method facilitates the measurement of the temperature rise under actual load conditions, which is the prime advantage of this method. Figure 8.24 shows the circuit diagram of an induction motor for load test. The ammeters and wattmeters in Figure 8.24 must have the capacity to withstand the currents of about 25 per cent more than the full-load current and the wattmeters must have the unity power factor. The supply voltage is connected to an autotransformer and its output is connected to the stator terminals of the induction motor. The autotransformer is initially set to zero volts and it is gradually increased from zero to its rated value. The readings of the meters are noted when the motor is running at no load. The load on the motor is usually increased with the help of a brake. The current, voltage, power, torque and the speed of the motor are noted for every setting of the current. The input voltage at the stator is maintained at the rated value throughout the test. Loading of the motor is continued until it reaches a current equal to the full-load current or 25 per cent higher than the full-load current. Let r be the radius of the drum, in meters.

Total input to the motor $= W = W_1 + W_2 = \sqrt{3}V_L I_L \cos\theta$

$$\therefore \quad \cos\theta = \frac{W}{\sqrt{3}V_L I_L} \tag{8.79}$$

$$\text{The output torque } (T) = (T_1 - T_2)rg \text{ N-m} \tag{8.80}$$

where T_1 and T_2 are the tensions in the belt (in kg) used for blocking the rotor.

$$\text{The output power } (P_0) = T\omega = T \times \frac{2\pi N}{60} \text{W} \tag{8.81}$$

$$\text{The efficiency } (\eta) = \frac{P_0}{\omega} \tag{8.82}$$

$$\text{and the operating slip} = \frac{N_s - N}{N_s} \tag{8.83}$$

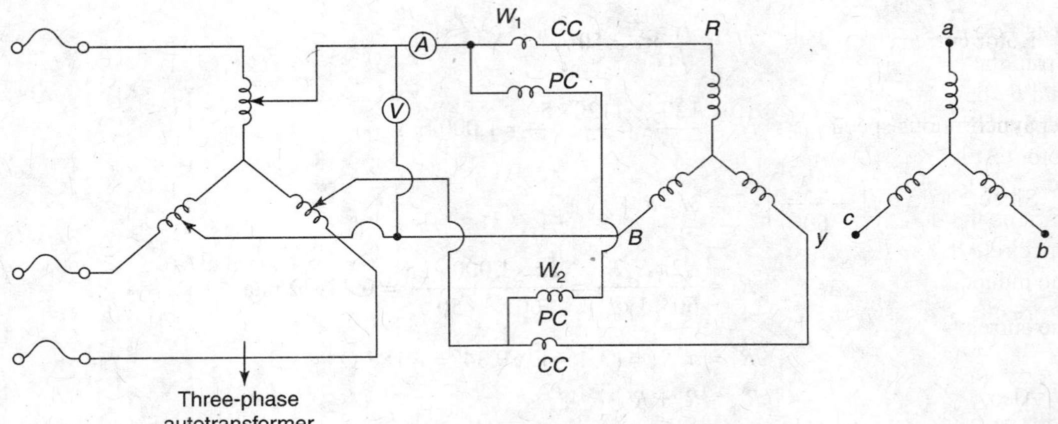

Figure 8.24 Circuit Diagram for Direct Testing of Induction Motor

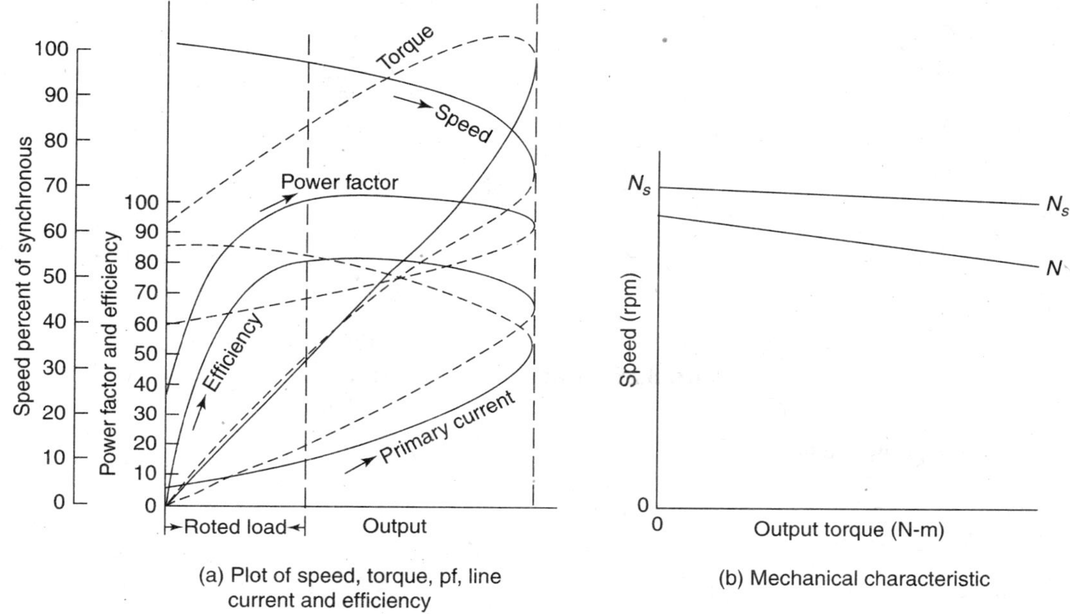

(a) Plot of speed, torque, pf, line current and efficiency

(b) Mechanical characteristic

Figure 8.25 Performance Characteristics of an Induction Motor

The performance characteristics showing the relation between the output power in kW and the efficiency, slip, torque, power factor and the line current of the induction motor shown in Figure 8.25. Figure 8.25 shows the mechanical characteristics of the motor where the speed is a function of torque.

8.28 CIRCLE DIAGRAM

It is possible to derive the performance characteristics of the induction motor from a circular locus. From the no-load test, I_0 and θ_0 have been calculated. From blocked-rotor test, short-circuit current (I_{sc}) and θ_{sc} have been calculated. These are the necessary data to draw the circle diagram. It is possible to separate the stator and the rotor Cu losses by drawing a torque line. The no-load test and the blocked-rotor test have already been discussed in Sections 8.25 and 8.26, respectively.

The locus of the tip of the stator current phasor of an induction motor lies on a semicircle and moves on it as the load on the motor is increased from the no-load to full-load value. This locus is known as the circle diagram of the induction motor. Figure 8.26(a) is considered in which R_2 is the variable and the inductance L is fixed. If a constant AC voltage is applied to this circuit and of R_2 is varied, the tip of the current phasor describes a circle having a diameter of $\frac{V_{ph}}{X_2}$, as shown in Figure 8.26(b). When $R_2 = 0$, $I = \frac{V_{ph}}{X_2}$ lags behind V by an angle of 90° and it is represented by AB. When $R_2 = \infty$, $I = 0$ and coincides with point A. For other values of R_2 lying between zero and infinity, the current phasor lags behind the voltage phasor by an angle θ, shown by the line AC.

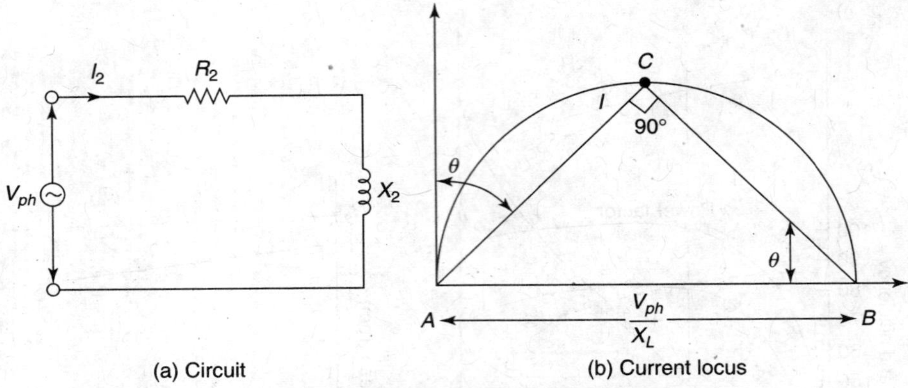

(a) Circuit (b) Current locus

Figure 8.26 Basic Circuit Diagram

From Figure 8.26(b), we have

$$\frac{AC}{AB} = \sin\theta$$

i.e.,
$$AC = AB \sin\theta$$

But
$$\sin\theta = \frac{X_L}{Z}$$

where
$$Z = \sqrt{R_2^2 + X_2^2}$$

$\therefore$
$$AC = \left(\frac{V}{X_2}\right) \times \left(\frac{X_2}{Z}\right) = \frac{V_{ph}}{Z} = I$$

Therefore, the tip of the current phasor lies on the semicircle having a diameter of $\frac{V_{ph}}{X_2}$.

Figure 8.27 shows the approximate equivalent circuit of the induction motor.

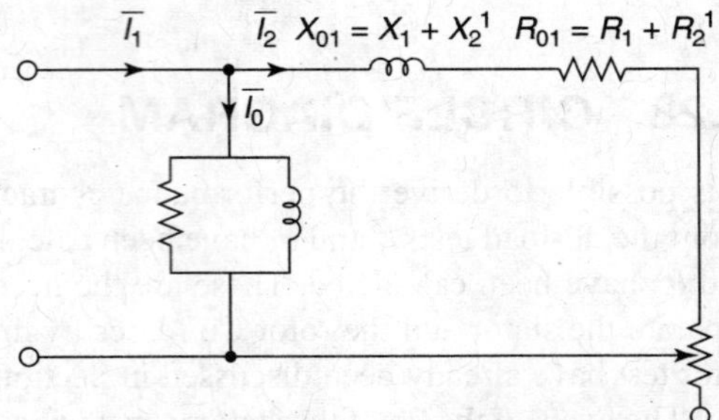

Figure 8.27 Approximate Equivalent Circuit

The end of I_2' lies on a circle with a diameter of $\frac{V_{ph}}{X_{01}}$, where I_2' is the rotor current referred to the stator, I_0 is the exciting current and I_1 is the stator current. The stator current is the vector sum of I_2' and I_0. When I_2' is lagging and $\theta_2 = 90°$, the position of I_2' will be along OA, as shown in Figure 8.28, which is at a right angle to the voltage vector OE. The point C will move along the circle, as shown by dotted line for any other value of θ_2. I_0 lags behind V_t by an angle θ_0. If the excitation circuit parameters remain constant, I_0 and θ_0 also become constant. The tip of the stator current (I_1) lies on the other semi-circle shown by solid lines displaced from the dotted semicircle by an amount of I_0. The diameter of this circle is also $\frac{V_{ph}}{X_{01}}$ and it is parallel to the line OA.

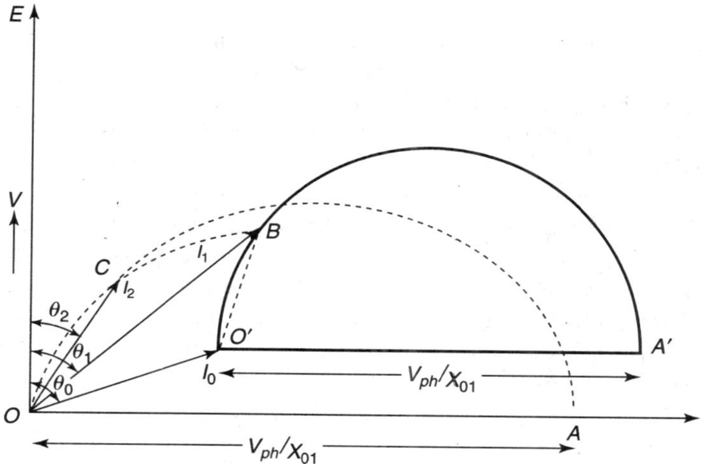

Figure 8.28 Basic Circle Diagram

The data required for the construction of circle can be obtained from no-load test, short-circuit test and stator-resistance test. The following steps are followed:

Step 1: I_0 and θ_0 are computed from no-load test. The vector for I_0 is drawn, which is lagging behind the applied voltage by an angle θ_0, as shown in Figure 8.29.

Step 2: I_{sc} and θ_{sc} can be obtained from the blocked-rotor test. To determine the short-circuit point (S) corresponding to the rated voltage, the length of vector OS is calculated as follows:

$$OS = \frac{I_{sc}}{V_{sc}} \times V_{ph}$$

where V_{ph} is the rated voltage per phase and V_{sc} is also in per phase.

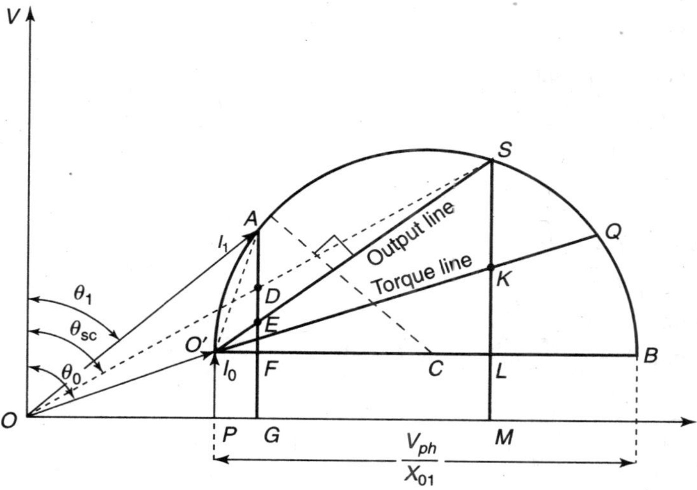

Figure 8.29 Circle Diagram

Here, the scale 1 cm = 100 A is taken and *OS* in amperes is converted into length *OS* using this scale. *OS* is drawn at angle θ_{sc} from the vertical. The two points *O'* and *S* lie on the required circle. To obtain the centre of the circle, *O'S* is bisected at right angles. Its bisector gives the point *C*, which is the centre of the circle. *O'B* is drawn, which is the diameter of the circle and it is at a right angle to the voltage vector *V*. Here, *O'P* is the vertical component of the no-load current *OO'*, which represents the no-load input supplying the core loss, friction and windage loss and a negligibly small amount of stator I^2R loss. The vertical component *SM* of the short-circuit current *OS* is proportional to the motor input on short circuit.

Step 3: The line that separates the stator and the rotor Cu losses in known as torque line. During blocked-rotor test, all power supplied to the motor gives the core loss and Cu loss in the stator and rotor windings. *LM* out of *SM* represents the fixed loss, that is, stator core loss and friction and windage loss. *SL* is the sum of *SK* and *KL*, where *SK* and *KL* represent the rotor Cu loss and the stator Cu loss, respectively. The point *K* is such that the following relation is maintained.

$$\frac{SK}{KL} = \frac{\text{rotor Cu loss}}{\text{stator Cu loss}} = \frac{W_s 3I_2^2 R_1}{3I_s^2 R_1}$$

where R_1 is the stator resistance/phase and it is obtained from stator-resistance test. W_s is the short-circuit motor input, which is approximately equal to motor Cu losses.

The point *K* is joined to *O'* and this line *O'K* is called the torque line, which meets the circle at the point *Q* when it is extended. At the point *Q*, rotor input is zero corresponding to slip s = infinity.

It is assumed that the motor is running taking a current *OL*. The perpendicular *FG* represents fixed losses, *FE* represents stator Cu loss, *EA* represents the rotor input, *ED* is the rotor Cu loss, *DA* is the rotor output, and *AG* represents the motor input. The following expression can be written:

$\sqrt{3}V_L \cdot AG =$ motor input

$\sqrt{3}V_L \cdot FG =$ fixed losses

$\sqrt{3}V_L \cdot FE =$ stator Cu loss

$\sqrt{3}V_L \cdot ED =$ rotor Cu loss

$\sqrt{3}V_L \cdot DG =$ total loss

$\sqrt{3}V_L \cdot DA =$ mechanical output

$\sqrt{3}V_L \cdot EA =$ rotor input $\propto$ torque

$$\text{Efficiency} = \frac{DA}{AG}$$

$$\text{Slip }(s) = \frac{\text{rotor Cu loss}}{\text{rotor input}} = \frac{DE}{EA}$$

$$1 - s = \frac{DA}{EA} = \frac{\text{rotor output}}{\text{rotor input}} = \frac{N}{N_s} = \frac{\text{actual speed}}{\text{synchronous speed}}$$

$$\text{Power factor} = \frac{AG}{OA}$$

Maximum Quantities: To get the maximum quantities, let us consider Figure 8.30. At point *F*, maximum output occurs because at this point the tangent is parallel to the output line *AB*. To locate *F*, a line is drawn from *C* to *F* such that it is perpendicular to *AB*. From point *F* a vertical line is drawn such that it cuts the output line at the point G. GF gives the maximum output. Maximum torque occurs at the point H, where the tangent is parallel to the torque line. To locate H, a line is drawn from C to H such that it is perpendicular to AE. From H a vertical line is drawn that cuts the torque line at I. HI gives the maximum torque. At point J, tangent is horizontal and JK gives the maximum input power. The induction motor will be unstable at J because the point J is beyond H.

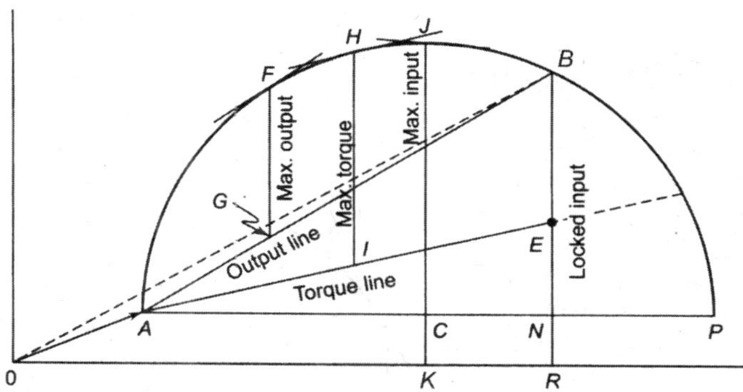

Figure 8.30 Maximum Quantities

Example 8.13 A three-phase, 400 V, induction motor gave the following results:

No-load test: 400 V, 1,200 W, 8A
Short-circuit test: 150 V, 3,900 W, 36 A

The rotor copper loss at standstill is half the total copper loss. Draw the circle diagram. Determine the full-load value of the current, pf and slip when the normal rating is 14 kW. Also, calculate motor input at full load, stator copper loss, rotor copper loss, rotor input, mechanical power output, efficiency of the motor, ratio of maximum torque to full-load torque and maximum power.

Solution

From no-load test data

$$\cos\theta_0 = \frac{1,200}{\sqrt{3}\times 400\times 8} = 0.2165$$

i.e., $\theta_0 = 77.5^\circ$

From short-circuit test data

$$\cos\theta_{sc} = \frac{3,900}{\sqrt{3}\times 150\times 36} = 0.4169$$

i.e., $\theta_{sc} = 65.35^\circ$

The short-circuit current with normal voltage

$$I_{SN} = 36\times\frac{400}{150} = 96 \text{ A}$$

The power taken at this normal voltage $= 3,900\times\left(\frac{400}{150}\right)^2 = 27,733.34$ W

The circle diagram is shown in Figure E8.3. In Figure E8.3, 1 cm = 5 A. OY axis represents the voltage vector, which is perpendicular to OX axis.

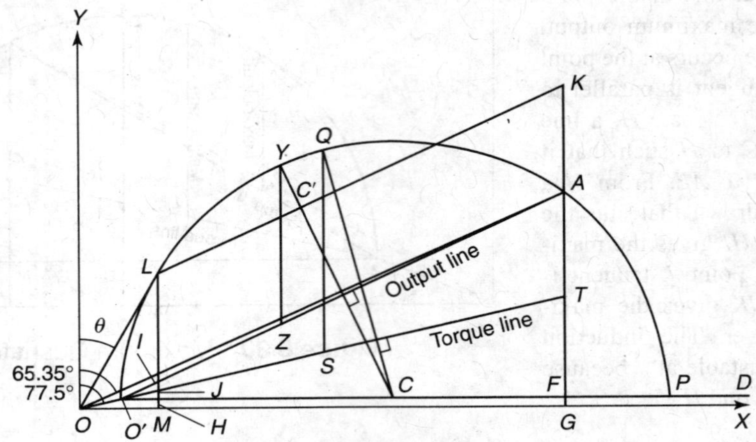

Figure E8.3

In Figure E8.3,

(i) The vector OO' represents the current at no load; that is, $OO' = 8\text{ A} = \frac{8}{5} = 1.6$ cm, which is drawn at 77.5° angle with respect to the vertical line.

(ii) The short-circuit current is represented by the vector OA; that is, $OA = I_{SN} = 96$ A. It measures $\frac{96}{5} = 19.2$ cm. It is drawn at 65.35° angle with respect to the vertical line.

(iii) $O'A$ is added which represents the output line. $O'D$ is drawn which is parallel to OX axis.

(iv) $O'A$ is bisected by the line CC'. CC' intersects $O'D$ at the point C. With C as a centre and CO' as a radius, the semicircle $O'AP$ is drawn. From A, the vertical line AG is drawn, which measures the input power at the blocked rotor condition. The length of AG is 7.6 cm. The power scale becomes 1 cm $= \frac{27,733.34}{7.6} = 3,649.12$ W. GF represents the fixed losses and AF is the total copper loss during the blocked rotor test.

(v) Motor output at full load = 14 kW = 14,000 W $= \frac{14,000}{3,649.12} = 3.84$ cm. Now GA is extended up to K, where $AK = 3.84$ cm. From K, the line KL is drawn, which is parallel to $O'A$. The point L lies on the semicircle, which is at the full-load condition.

(vi) Since the rotor copper loss is half the total loss, AF is bisected at the point T. $O'T$ is added, which represents the torque line.

(vii) LM is drawn perpendicular to OX, where $LM = 4.75$ cm, which intersects the output line and the torque line at the points I and J, respectively.

(viii) From C, a perpendicular CQ is drawn on the torque line, where Q lies on the semicircle. From Q, a vertical line QS is drawn, which intersects the torque line at the point S. The length of $QS = 7.2$ cm.

(ix) CC' is extended up to the point Y, and from Y, a vertical line is drawn, which intersects the output line at a point Z. The length of $YZ = 5.6$ cm.

(x) The vector OL is drawn, which represents the line current at full-load condition. The length of $OL = 5.5$ cm.

Line current at full load = 5.5 × 5 = 27.5 A

By measurement, $\phi = 30°$

Therefore, pf at full load = cos 30° = 0.866

Motor input at full load = LM = 4.75 × 3,649.12 W = 17.34 kW

Stator copper loss at full load = JH = 0.30 × 3,649.12 W = 1.094 kW

Rotor copper loss at full load = IJ = 0.30 × 3,649.12 W = 1.094 kW

Rotor input at full load = LJ = 4.15 × 3,649.12 W = 15.143 kW

Mechanical output at full load = IL = 3.8 × 3,649.12 W = 13.867 kW

$$\text{Slip} = \frac{\text{rotor copper loss}}{\text{rotor input}} = \frac{IJ}{LJ} = \frac{0.3}{4.15} = 0.0723 \text{ or, } 7.23 \text{ per cent}$$

$$\text{Efficiency of the motor at full load} = \frac{\text{motor output at full load}}{\text{motor intput at full load}}$$

$$= \frac{LI}{LM} = \frac{3.8}{4.76} = 0.7983 \text{ p.u.} = 79.83\%$$

$$\frac{\text{Maximum torque}}{\text{Full-load torque}} = \frac{QS}{LJ} = \frac{7.2}{4.15} = 1.735$$

Maximum power = YZ = 5.6 × 3,649.12 W = 20.43 kW

8.29 SPEED CONTROL OF INDUCTION MOTOR

Three-phase induction motors run at a constant speed like DC shunt motors. The variation in speed from no load to full load is called the speed regulation of induction motor, which is usually less than 3 per cent for large-size motors. The methods used for speed control of DC shunt motors and induction motors vary widely. It is possible to obtain a smooth control of speed for DC shunt motors with the help of field control method without sacrificing efficiency, which is inexpensive. But smooth control of speed for induction motors cannot be obtained as efficiently and inexpensively as in DC shunt motors. The different methods for obtaining speed control of induction motors are mainly classified into two categories:

(i) Speed control from rotor side.

(ii) Speed control from stator side.

8.29.1 Speed Control from Rotor Side

The following are the three methods of controlling speed of induction motor from the rotor side:

(i) Rotor rheostat control.

(ii) Operating the motors in cascade.

(iii) Injecting voltage into the rotor circuit.

8.29.1.1 Rotor Rheostat Control

This method is very similar to the speed control of DC shunt motors by armature control. The speed of the motor in this method is varied by inserting external resistances in the rotor circuit of the motor.

This method is suitable only for slip-ring induction motors. At running condition we have the following relation:

$$sE_2 = I_2\sqrt{R_2^2 + s^2 X_2^2}$$

and

$$E_2(1-s) = E_2 - I_2\sqrt{R_2^2 + s^2 X_2^2} \tag{8.84}$$

Also, we know that

$$E_2 = K\phi_m N_s \tag{8.85}$$

From Equations (8.84) and (8.85), we have

$$K\phi_m N_s(1-s) = E_2 - I_2\sqrt{R_2^2 + s^2 E_2^2}$$

i.e.,

$$K\phi_m N_r = E_2 - I_2\sqrt{R_2^2 + s^2 X_2^2}$$

i.e.,

$$N_r = \frac{E_2 - I_2\sqrt{R_2^2 + s^2 X_2^2}}{K\phi_m} \tag{8.86}$$

Equation (8.86) is very similar to the equation for the speed control of DC motors, that is,

$$N = \frac{V_t - I_a(R_a + R)}{K\phi_m} \tag{8.87}$$

where R is the external resistance connected in series with the armature. Equation (8.86) is very similar to Equation (8.87). Equation (8.86) suggests that the speed control of induction motor can be done varying the rotor resistance (R_2). By adding resistance in the rotor circuit, the speed can be continuously reduced to slip-ring induction motor when the supply voltage and frequency are kept constant. This is due to the fact that the energy available in the rotating magnetic field is wasted in the external resistance. Hence, the energy available in the shaft gets reduced. This causes reduction in operating efficiency of the motor, which is a serious disadvantage. This method is used only for speed changes for short durations because this method involves wastage of energy.

8.29.*.2 Cascade Connection of Induction Motors

Figure 8.31 shows the cascade connection of two wound-rotor-type induction motors denoted by M_1 and M_2. These are mechanically coupled directly or through gears and are electrically connected in cascade. The slip rings of M_1 are connected to the stator of motor M_2, whereas the slip rings of M_2 are connected to a variable three-phase resistor. The turns ratio of the stator to the rotor is kept unity to reduce the current in slip rings. The cascade set is started using the variable resistance. This can also be used for the variation of speed. Let P_1 and P_2 be the number of poles on M_1 and M_2, respectively, and the corresponding slips of the two motors be s_1 and s_2, respectively. Let frequencies of the stator supply be f_1 and f_2, respectively.

The synchronous speed of $M_1 = N_{s1} = \dfrac{120 f_1}{P_1}$

The synchronous speed of $M_2 = N_{s2} = \dfrac{120 f_2}{P_2}$

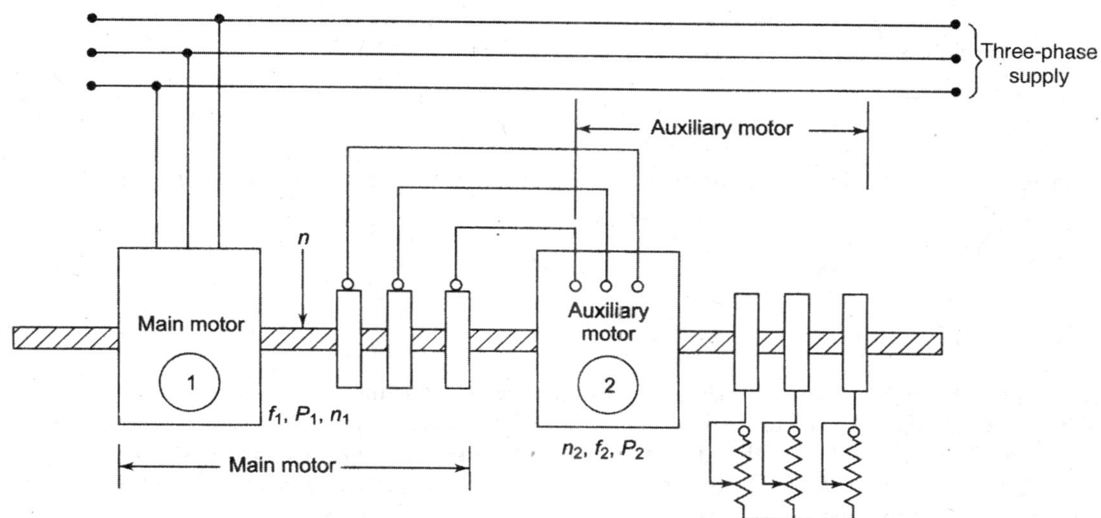

Figure 8.31 Cascade Connections of Induction Motors

$$\text{The actual speed of } M_1 = N = \left(\frac{120 f_1}{P_1}\right)(1 - s_1) \tag{8.88}$$

$$\text{The actual speed of } M_2 = N = \left(\frac{120 f_2}{P_2}\right)(1 - s_2) \tag{8.89}$$

The followings cases are discussed.

Case I:

Let the torques exerted by the two motors be in the same direction. The actual speed of the two motors must be the same due to their mechanical coupling, and hence we have

$$N = \left(\frac{120 f_1}{P_1}\right)(1 - s_1) = \left(\frac{120 f_2}{P_2}\right)(1 - s_2) \tag{8.90}$$

Since $f_2 = s_1 f_1$, we have from Equation (8.90)

$$\left(\frac{120 f_1}{P_1}\right)(1 - s_1) = \left(\frac{120 s_1 f_1}{P_2}\right)(1 - s_2)$$

i.e.,
$$\frac{1 - s_1}{P_1} = \frac{s_1(1 - s_2)}{P_2}$$

i.e.,
$$\frac{1 - s_1}{P_1} = \frac{s_1(1 - s_2)}{P_2}$$

i.e.,
$$P_2(1 - s_1) = P_1 s_1 (1 - s_2)$$

i.e.,
$$s_1(P_1 - P_1 s_2 + P_2) = P_2$$

i.e.,
$$s_1 = \frac{P_2}{P_1 - P_1 s_2 + P_2} \tag{8.91}$$

The motor M_2 operates with a very small slip (s_2) with starting resistance short-circuited. This gives stable operation, and hence $s_2 \approx 0$. Therefore, we have from Equation (8.91)

$$s_1 = \frac{P_2}{P_1 + P_2} \tag{8.92}$$

Hence, the actual speed of the motor M_1 and of the set as a whole approaches the following value:

$$N = \left(\frac{120 f_1}{P_1}\right)\left[1 - \frac{P_2}{P_1 + P_2}\right]$$

i.e.,

$$N = \frac{120 f_1}{P_1 + P_2} \tag{8.93}$$

Equation (8.93) suggests that the speed of the cascade set is the same as the single motor, where the number of poles is equal to the sum of the number of poles of motors M_1 and M_2 connected in cascade. For $P_1 = P_2$, the speed of the set will be half of that of either motor while running on its own. For $P_1 \neq P_2$, the cascade set can be operated stably at three different speeds. These are the nominal speeds of motor M_1 when operating alone, the nominal speed of motor M_2 when operating alone and the speed of cascade connection shown by Equation (8.93).

Case II:
Let the torques exerted by the two motors be in the opposite directions. We have the following expressions:

$$\frac{1 - s_1}{P_1} = \frac{s_1[1 - (2 - s_2)]}{P_2}$$

i.e.,

$$\frac{1 - s_1}{P_1} = \frac{-s_1}{P_2} [\because s_2 \approx 0]$$

i.e.,

$$P_2(1 - s_1) + P_1 s_1 = 0$$

i.e.,

$$(P_1 - P_2)s_1 = -P_2$$

i.e.,

$$s_1 = -\frac{P_2}{P_1 - P_2}$$

i.e.,

$$s_1 = \frac{P_2}{P_2 - P_1} \tag{8.94}$$

Therefore, the speed of the cascade set becomes

$$N = \frac{120 f_1}{P_1}(1 - s_1) = \frac{120 f_1}{P_1}\left(1 - \frac{P_2}{P_2 - P_1}\right) = -\frac{120 f_1}{P_2 - P_1}$$

i.e.,

$$N = \frac{120 f_1}{P_1 - P_2} \tag{8.95}$$

The actual speed of the cascade set can be expressed by

$$N = \frac{120 f_1}{P_1 \pm P_2} \tag{8.96}$$

8.29.1.3 Speed Control of Induction Motors by Injecting Voltage in the Rotor Circuit

From Equation (8.86) we can conclude that it is possible to control the speed of an induction motor by varying the rotor voltage (E_2), which is possible only by injecting voltage into the rotor at the slip frequency of the rotor. The phase of the injected voltage can be anything, that is, it can be in phase or phase opposition to the rotor voltage. If the injected voltage is in phase with the rotor voltage, the net voltage and consequently the current in the rotor circuit get increased; this amounts to decrease the net rotor resistance; hence, the speed of the induction motor increases. On the other hand, if the injected voltage is in phase opposition with the rotor voltage, the net voltage and consequently the current in the rotor circuit get decreased, resulting in an increase in the net rotor resistance. This leads to a decrease in the speed of the induction motor. Therefore, the speed of the induction motor can be increased or decreased with this method. (For a detailed discussion on Schrage motors, please see Section 8.43.)

8.29.2 Speed Control of Induction Motors from Stator Side

The speed of the induction motor can be controlled from the stator side as follows:

(i) By changing the supply frequency.
(ii) By changing the number of stator poles.

8.29.2.1 Speed Control by Changing the Supply Frequency

This method is applicable to the induction motors that are run from an independent voltage source. Hence, this is rarely used. The synchronous speed of an induction motor is expressed by $N_s = \frac{120f}{P}$. If the supply frequency is changed, N_s will be changed, and hence the running speed of the induction motor can be controlled. Since the generated voltage is proportional to the frequency, V/f ratio is maintained constant so that the flux can be kept constant. This method of speed control is used in the case of induction motors used for propulsion of ships because the motors are driven by independent voltage sources in this case.

8.29.2.2 Speed Control by Changing the Number of Poles

Since the squirrel-cage-type rotor adopts itself to any reasonable number of stator poles, this method is applicable to squirrel-cage-type induction motors. The change in number of poles gives the change in synchronous speed because synchronous speed is a function of the number of poles of induction motors. It is possible to achieve the change in number of stator poles by having two or more number of independent stator windings in the same slot. Since each winding is wound for different number of poles, this results in different synchronous speeds. A stator with 36 slots (say) can be wound either for four poles or six poles. If the supply frequency is 50 Hz, a winding with four poles results in the synchronous speed of 1,500 rpm and that with six poles results in the synchronous speed of 1,000 rpm. This method has usage in the case of elevator motors, traction motors and small motors used for driving machine tools. Figure 8.32 shows the most common arrangement for multispeed operation that uses a single winding (stator) in which the coil groups belonging to each phase are connected to give a

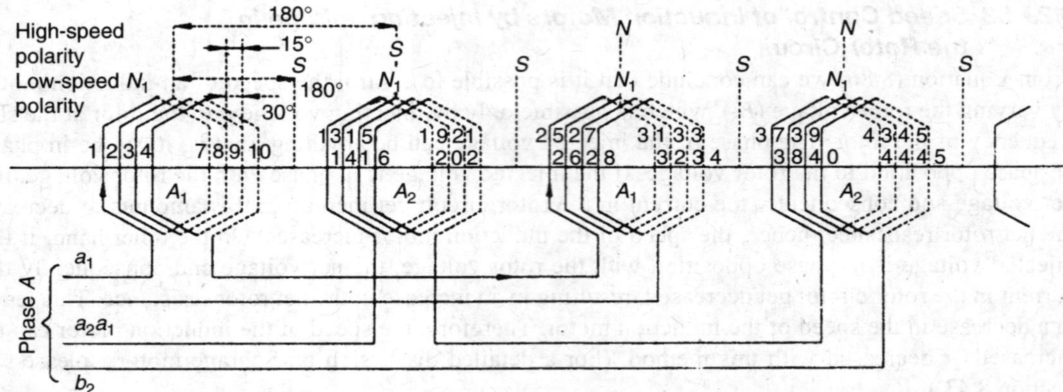

Figure 8.32 Pole Changing Connection Diagram (48-slot, Four-pole, Eight-pole Machine)

speed ratio 2:1. In Figure 8.32, the developed form of the coils of one phase of a three-phase winding is distributed in 48 slots.

The coils are arranged into two different groups so that the two different circuits each having terminals a_1a_2 and b_1b_2 in series, as shown in Figure 8.33. In Figure 8.33(a), all the four coils will be magnetized in the same sense, and hence this results in the production of poles of opposite polarity midway between them, in a direction which is opposite to the arrow shown in Figure 8.32.

Figure 8.33(b) shows the series connections of coils in the order a_1a_2, b_1b_2 such that the current in the coils designated as A_1 and A_2 will be reversed that produces a polarity of the south pole. This results only in four poles resulting in the doubling of the synchronous speed.

Figure 8.34 shows the parallel connection of the coils. Figure 8.34(a) shows the parallel connection of a_1a_2 and b_1b_2. This results in eight poles. On the other hand, Figure 8.34(b) shows parallel connection of a_1a_2 and b_1b_2, which forms four poles only. The slot pitch angles are 15 electrical degrees and 30 electrical degrees for a four-pole connection and eight-pole connection, respectively, whereas respective phase spreads are 60 and 120 electrical degrees, respectively. The pitch factor for a four-pole connection is only half of that for an eight-pole connection.

The change in operating flux density with the change in speed is the major drawback of this method. Therefore, the design of multispeed motor is the compromise between the advantage of variation of the speed at one end and sacrificing the operating characteristics of the motor at the other.

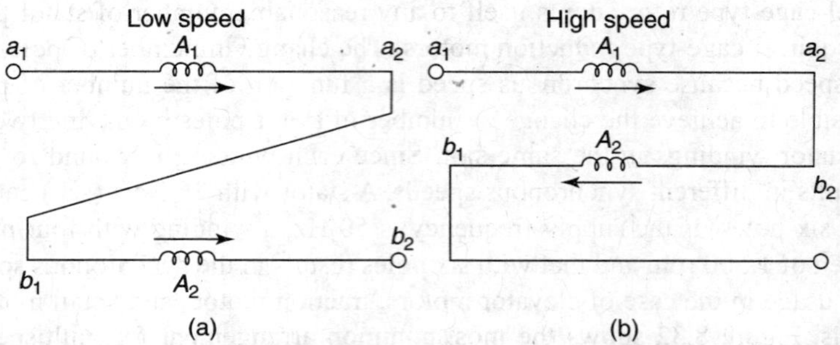

Figure 8.33 Series Connection of Three-phase Circuits for Pole Changing

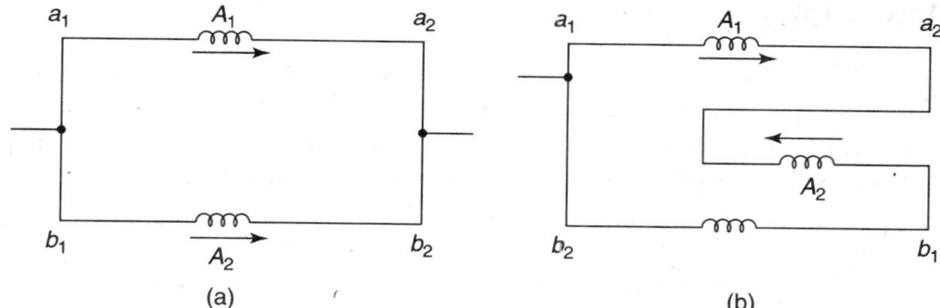

Figure 8.34 Parallel Connection of Three-phase Circuits for Pole Changing

8.30 COMPARISON BETWEEN WOUND-ROTOR AND CAGE-ROTOR INDUCTION MOTORS

Table 8.1 shows the comparison of the advantages and the disadvantages of the slip-ring and squirrel-cage induction motors.

Table 8.1 Comparison of the Advantages and Disadvantages of Slip-ring and Squirrel-cage Induction Motors

Types of Motors	Advantages	Disadvantages
Wound-rotor (slip-ring) motor	• It is possible to get high starting torque by introduction of resistances in the rotor circuit. • Starting current is comparatively less and it is 2 to 2.5 times the full-load current or even less, if required. • It is capable of starting with load, demanding a high torque as in fans, pumps, compressors, hoists, conveyors, cranes, lifts and so on. • Speed control is possible by varying the resistances in the external circuit of the rotor. • The motor can be started direct-on-line (DOL) without the necessity of starters, because of the low starting current of the motor.	• Higher cost. • Lower efficiency. • Losses by way of heat dissipation in external resistors, especially when run at reduced speeds below rated speed. • Requires higher degree of maintenance.
Squirrel-cage motor	• Cheaper cost. • Ruggedness of construction. • No moving contacts. • Slightly higher efficiency than wound motor. • By using a double-cage rotor, it is possible to obtain higher starting torque. • Trouble-free performance, minimum maintenance. • Losses are relatively less.	• Low starting torque. • It cannot be used in cases in which considerable load torque is required. • Starting current is about six to eight times the full-load current. • Speed control is not possible. • Because of the high starting current, it requires a starting device such as an autotransformer or a star–delta starter.

8.31 CRAWLING

During the starting of the induction motor, peculiar behaviour may be observed because there are certain relationships between the number of poles of the stator and rotor slots in the cage motors. The motor may exhibit a tendency to run stably at low speed (one-seventh of the normal); this is called *crawling*. The winding harmonics introduced into the gap by the magnetomotive force (mmf) waveform of the three-phase winding carrying currents are of amplitude $\frac{1}{n^2}$ of the order n in terms of the fundamental amplitude. The movement of the harmonics is with the direction of the fundamental, but $\frac{1}{n}$th of the synchronous speed if $n = 6m + 1$, where m is an integer and against that of the fundamental if $n = 6m - 1$.

So, the fifth harmonic of 4 per cent amplitude rotates against the fundamental at one-fifth of the synchronous speed and the seventh harmonic of slightly more than 2 per cent amplitude rotates at one-seventh of the synchronous speed in the same direction as the fundamental. These harmonic mmfs produce harmonic torques having the same general shape as that of fundamental, but with synchronous speeds one-fifth backwards and one-seventh forwards, respectively as that of the fundamental. The resultant torque-speed curve will be a combination of the fundamental and fifth and seventh harmonic currents, and it will be seen that a marked *saddle effect* is produced, as shown in Figure 8.35.

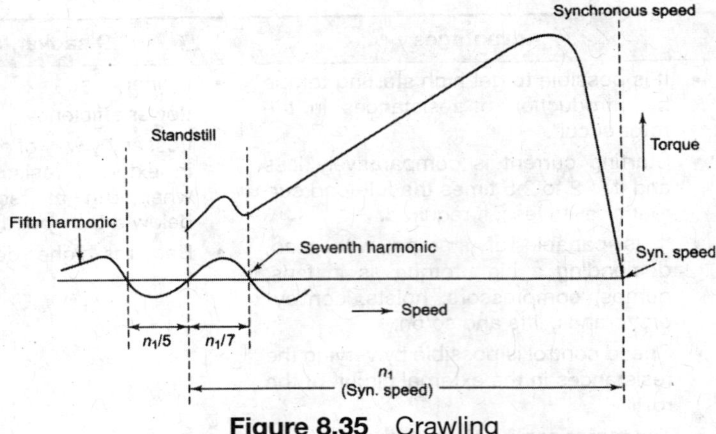

Figure 8.35 Crawling

If the harmonics are sufficiently pronounced, the seventh harmonic may prevent the motor speed exceeding about one-seventh of the normal, since the downward slope of the resultant torque at this speed is a stable running condition over the torque range between the maximum and minimum points. Therefore, the motor crawls.

8.32 MAGNETIC LOCKING (COGGING)

The rotor in the squirrel-cage motor sometimes exhibits the tendency to remain in one place, and does not rotate at all, when energized from mains. This phenomenon happens when the number of rotor and stator slots have a common factor. In such case, there will be certain rotor positions that make the air gap reluctance a minimum, and under such positions, the rotor tends to remain stationary. Particularly, if the input voltage is low, this will have a serious effect on the starting of the motor. The teeth of the rotor and the stator appear to be locked magnetically, and this exerts a reverse torque when the rotor teeth are moving away from those on the stator; this is known as cogging. To overcome this, the number of rotor slots is made prime to the number of slots on the stator to overcome this effect. Many induction motors are built with skewed or twisted rotor slots. Skewing reduces cogging and other torque defects and tends to eliminate motor noise.

8.33 DEEP-CAGE ROTORS

A cage rotor with deep and narrow bars has been shown in Figure 8.36, where the bar is assumed to be made up of a number of narrow layers connected in parallel. In Figure 8.36, there are three layers A, B and C. The layer A is linked with minimum flux; hence, it has minimum leakage inductance. The layer C is linked with maximum flux; hence, it has maximum leakage inductance. The bottom layer offers more impedance compared to the topmost layer during the starting of the induction motor, and hence maximum current flows through the topmost layer and minimum current flows through the bottom layer. Due to unequal distribution of current, the effective rotor resistance increases and the leakage reactance decreases. The starting torque is higher and the starting current is lower due to high rotor resistance during starting. During normal operating conditions, the reactances of the different layers are small and can be neglected compared with the resistance of the rotor. Therefore, the impedances of all the layers are equal and the flow of current through all the bars is equal. Since the resulting large cross-sectional area makes the rotor resistance quite small, it results in good efficiency at low slip.

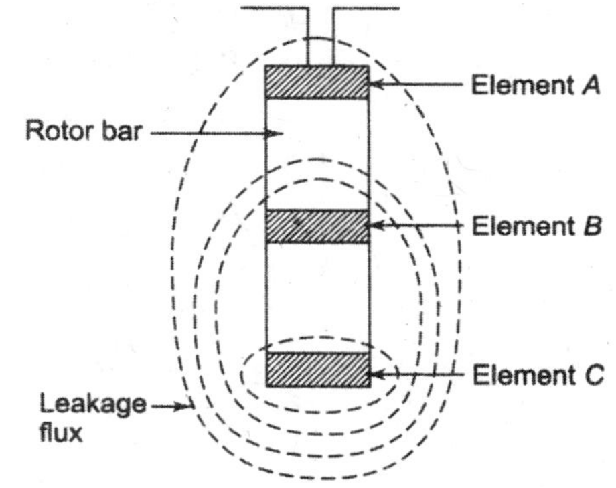

Figure 8.36 Deep-bar Cage Rotor Bar

8.34 DOUBLE-CAGE ROTORS

Double-cage rotors are used to fulfill the requirement of high starting torque. Figure 8.37 shows the arrangement of two cages for the production of large torque.

In Figure 8.37, cage K_1 is close to the periphery of the rotor (outer cage) and has a high resistance. The winch cage K_2 has a low resistance, but is set deeply in slots having a considerable leakage flux on account of long narrowed slit. Thus, the outer cage has a high resistance together with a low reactance normal to an ordinary cage winding, while the inner cage has a low resistance and large reactance. At starting, the leakage reactance of the inner cage is large enough to cause the rotor current to flow chiefly in the outer cage, the high resistance of which produces considerable I^2R losses and consequently good starting torque. When the speed is normal, the reactance of both cages is almost negligible, so that the rotor current is carried by the two cages in parallel giving a low effective resistance. The total torque is the summation of two individual torques and by varying the several resistances and reactances, a wide range of operational characteristics is possible. Figures 8.38 and 8.39 show the equivalent circuit and the torque-slip characteristics, respectively for a double-cage motor.

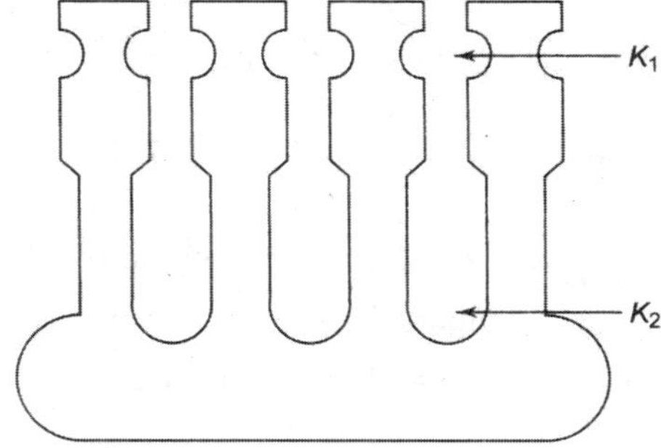

Figure 8.37 Double-cage Motor

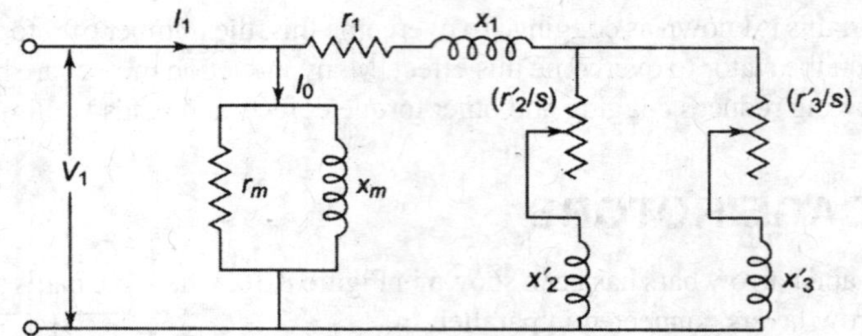

Figure 8.38 Equivalent Circuit of a Double-cage Rotor

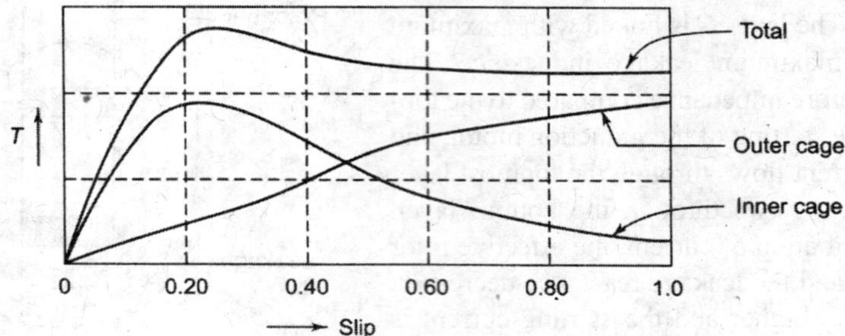

Figure 8.39 Torque-slip Characteristic of a Double-cage Rotor

Example 8.14 The equivalent standstill impedance per phase of the first cage and of the second cage of a squirrel-cage induction motor are $(5 + j1)\ \Omega$ and $(1 + j5)\ \Omega$, respectively. Compare the torques of the two cages at standstill and at a slip of 5 per cent.

Solution

Impedance of the outer cage is $Z_0 = (5 + j1)\ \Omega$ and that of the inner cage is $Z_i = (1 + j5)\ \Omega$

(i) At standstill:

Slip = 1

$$Z_o = \frac{5}{1} + j1 = 5 + j \qquad \therefore |Z_o| = \sqrt{5^2 + 1^2} = 5.099\ \Omega$$

and

$$Z_i = \frac{1}{1} + j5 = 1 + j5 \qquad \therefore |Z_i| = \sqrt{1^2 + 5^2} = 5.099\ \Omega$$

Since $|Z_o| = |Z_i|$, currents will be divided equally in both inner and outer cage.

Power input (P_o) to the outer cage is $= I_o^2 R_o = 5I_o^2$ W

Power input (P_i) to the inner cage is $= I_i^2 R_i = I_i^2$ W

$$\therefore \quad \frac{\text{Torque of the outer cage}}{\text{Torque of the inner cage}} = \frac{T_o}{T_i} = \frac{P_o}{P_i} = \frac{5I_o^2}{I_i^2} = \frac{5}{1} \quad [\text{Here } I_o = I_i]$$

(ii) At 5 per cent slip:

Slip = 1

$$Z_o = \frac{5}{0.05} + j1 = 100 + j \qquad \therefore |Z_o| = \sqrt{100^2 + 1^2} \approx 100\ \Omega$$

and $$Z_i = \frac{1}{0.05} + j5 = 20 + j5 \qquad \therefore |Z_i| = \sqrt{20^2 + 5^2} = 20.61\ \Omega$$

Currents will be divided the two cages in inverse ratio of impedances.

$$\frac{I_o}{I_i} = \frac{20.61}{100}$$

Power input (P_o) to the outer cage is $= I_o^2 R_o = 5I_o^2$ W

Power input (P_i) to the inner cage is $= I_i^2 R_i = I_i^2$ W

$$\therefore \quad \frac{\text{Torque of the outer cage}}{\text{Torque of the inner cage}} = \frac{T_o}{T_i} = \frac{P_o}{P_i} = 5 \times \left(\frac{I_o}{I_i}\right)^2 = 5 \times \left(\frac{20.61}{100}\right)^2 = 0.2124$$

■

8.35 APPLICATIONS

Squirrel-cage induction motors have low starting torque and constant speed. This kind of motor is preferred for driving fans, blowers, water pumps, grinders, lathe machines, drilling machines and printing machines.

Since slip-ring induction motors have high starting torques, they are preferred for lifts, cranes, hoists and compressors.

8.36 COMPARISON BETWEEN SYNCHRONOUS AND INDUCTION MOTORS

Table 8.2 shows the comparison of synchronous and induction motors.

Table 8.2 Comparison of Synchronous and Induction Motors

Synchronous Motor	Induction Motor
• Complicated construction.	• Simple construction particularly in case of cage motor.
• It is not self-starting.	• It is self-starting.
• An additional DC source is required for field excitation.	• No additional DC source is required.
• It operates at synchronous speed and never at any other speed.	• It cannot operate at speed less than synchronous speed.
• Irrespective of the load, the speed is always synchronous.	• With increase in load, speed decreases.
• Speed control is not possible.	• Speed control is possible.
• It is possible to operate at different power factors by changing its excitation.	• It always operates at lagging power factor.
• It is possible to use it as a synchronous condenser.	• It is not possible to use it as a synchronous condenser.
• Hunting starts with changing of load because the motor is more sensitive to load changes.	• Since it is less sensitive to load changes, the hunting is absent.
• The synchronous motor is costlier and needs frequent maintenance.	• Motor is cheaper and almost maintenance free, particularly in squirrel-cage induction motor.

8.37 FACTORS GOVERNING THE PERFORMANCE OF INDUCTION MOTORS

The following factors govern the operating or performance characteristics of induction motors:

- *Rotor resistance:* The rotor resistance is determined basically by the material used for the construction and also the area of cross section of the rings in the case of squirrel-cage induction motor. Copper and aluminum are the commonly used materials.
- *Air-gap length:* The air-gap length is one of the sources of low power factor at which the induction motor operates. If the air gap is reduced too much, the following phenomena occur.
  (i) Increase in noise.
  (ii) Pulsation losses.
  (iii) Magnetic pull.
  (iv) Reduction in overload capacity.
- *Shapes of both stator and rotor slots and teeth:* The reactance is affected by the shapes of rotor teeth and slots. The reactance affects the starting current and maximum torque. If any induction motor has deeper slots, it will have more leakage reactance, and hence the value of starting current, maximum torque and slip will be reduced.

  The slots in the rotor are given a slight skew, which reduces noise, vibration, cogging and synchronous cups. The methods used for reduction of harmonic torques are chording, integral slot winding, increasing the air gap length and skewing of rotor slots.

8.38 EFFECTS OF OPERATING CONDITIONS

The effects of operating conditions are discussed below.

- *Unbalanced supply voltage:* An unbalanced supply voltage produces a rotating magnetic field moving at a non-uniform rate and strength. This will cause unbalanced heating in the stator winding, which will lead to unequal heating.
- *Loading:* Due to the increase in load, the speed falls that leads to an increase in the torque until a point corresponding to maximum torque is reached. If the load is further increased, the speed will fall further decreasing the driving torque, and the rotor comes to a standstill.
- *Variation in line frequency and line voltage:* The variation in line frequency and supply voltage affects the speed, torque of the motor, starting current, full-load current, starting torque, maximum torque as well as operating temperature of various parts of the machine. These lead to unsatisfactory operation of the machine.
- *Effects of break in one phase:* If one phase of the polyphase induction motor breaks due to its normal operation, the motor will continue to run as single phase provided the load does not exceed 57.7 per cent of its normal rating and the rise of the temperature remains the same as that when carrying the rated load as a three-phase induction motor.

8.39 RATINGS OF INDUCTION MOTOR

The following terms decide the rating of induction motors:

(i) Power output (kW).
(ii) Speed.

(iii) Voltage.
(iv) Frequency.
(v) Phase (single/three).
(vi) Line current.
(vii) Temperature rise in specified time.
(viii) Class of motors, that is, A, B, C, D, E or F to be mentioned on the name plate.

Class A motors: Low-impedance, squirrel-cage motors.
Class B motors: High-reactance, squirrel-cage motors.
Class C motors: High-starting torque, low-starting current, double-squirrel-cage motors.
Class D motors: High-resistance, squirrel-cage motors.
Class E motors.
Class F motors.

8.40 COMMON FAULTS IN THREE-PHASE INDUCTION MOTORS

The following are the common faults in three-phase induction motors:

(i) Hot bearing,
(ii) Motor connected, but does not start,
(iii) Motor runs, but dies down,
(iv) Motor does not come up to speed,
(v) Motor takes long time to accelerate,
(vi) Wrong rotation,
(vii) Motor overheats while running under loads,
(viii) Motor vibrates,
(ix) Unbalanced line currents,
(x) Scrapping noise,
(xi) Magnetic noise and
(xii) Motor sparking at slip rings.

8.41 MOST PROBABALE REASONS FOR WHICH THREE-PHASE INDUCTION MOTORS FAIL TO START

The following are the most probable reasons for which three-phase induction motors fail to start:

(i) No supply voltage,
(ii) Voltage too low,
(iii) One phase open,
(iv) Motor may be overloaded,
(v) Defective control gear,
(vi) Starting torque of load too high and
(vii) Mechanical locking in the bearings or at the air gap due to worn-out bearings.

8.42 MOST PROBABALE REASONS FOR WHICH THREE-PHASE INDUCTION MOTORS FAIL TO CARRY LOAD

The following are the causes for which the induction motor fails to carry the load:

(i) Mechanical locking in the hot bearings or at the air gap due to excessive friction,
(ii) Mechanical overloading of motors,
(iii) Low motor voltage and
(iv) Single phasing on running position of the starter.

8.43 SCHRAGE MOTOR

Schrage motors were manufactured by the general electric company of USA, having a capacity rating of up to 100 kW and for voltage rating up to 440 V. Figure 8.40 shows the winding details of a Schrage motor for a two-pole model. This type of motor is a rotor-fed induction motor having the phase-wound secondary on the stator. This is the main difference between this motor and the three-phase induction motor. There is additional winding carried by the top of the rotor slots that is known as auxiliary winding connected to the commutator like in the case of DC winding. This is another difference between this motor and the three-phase induction motor. Figure 8.41 shows that the commutator provided with the two sets of brushes A_1, B_1, C_1 and A_2, B_2, C_2 that are supported on the yoke rings. These are mounted in parallel planes in such a way that it is possible to shift either set of brushes relative to each other in either direction. There are displacements of 120 electrical degrees among the sets of brushes connected A_1A_2, B_1B_2 and C_1C_2, respectively and each set of brushes is connected to the two ends of the one phase of the stator winding.

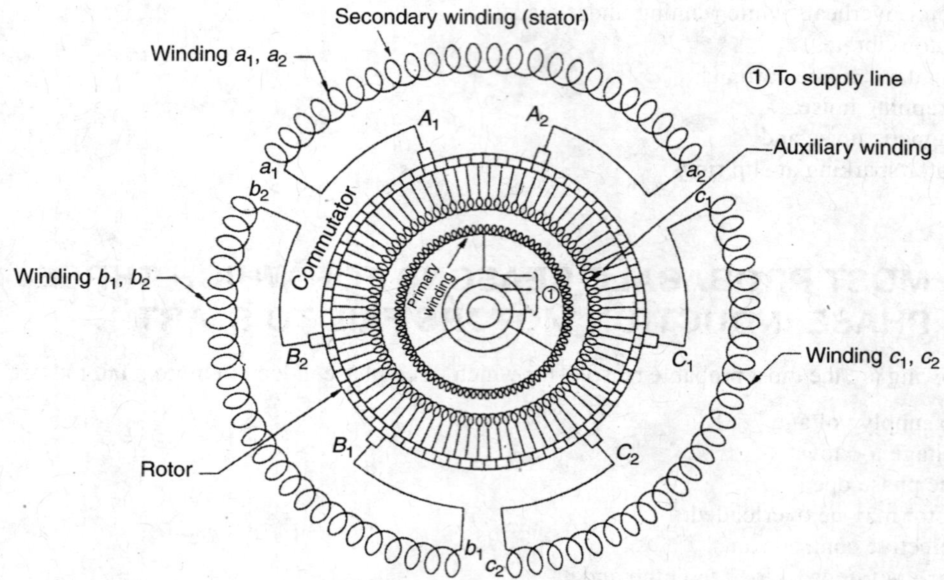

Figure 8.40 Schrage Motor

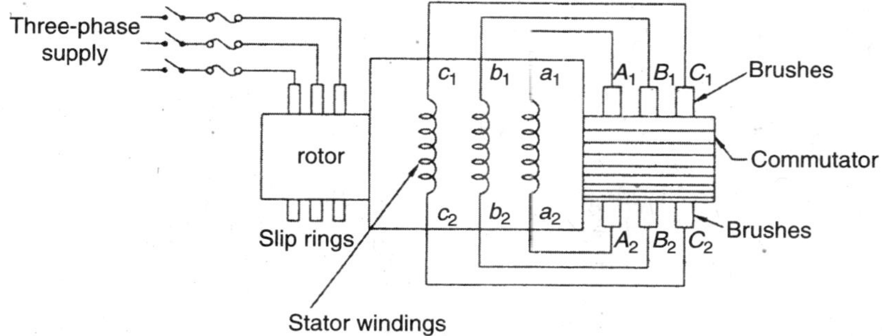

Figure 8.41 Block Diagram of Schrage Motor

Initially the angle of separation between the brush sets A_1A_2, B_1B_2 and C_1C_2 are taken in such a way that the windings a_1a_2, b_1b_2 and c_1c_2 touch each other, as shown in Figure 8.41. Therefore, each phase of the secondary winding is short-circuited. Due to the application of reduced voltage at the time of starting, rotating field will be set up revolving in space at a speed equal to synchronous speed if the rotor is initially at rest. The motor is fed by a three-phase supply and the secondary, that is, the stator is stationary. There will be reaction of the stator on the rotor. This makes the latter to rotate in a direction opposite to the direction of rotation of the revolving field where the speed of rotation of the revolving field is measured with respect to the speed of the rotor conductors. The rotor moves with a speed (N_r) in the opposite direction to the revolving fields having speed N_s. Therefore, the revolving field is dragged back at the speed of the rotor, and hence the speed of revolving field in the space becomes ($N_s - N_r$). This is known as the slip speed.

Before getting the idea of injection of voltage into the secondary windings, that is, the stator, let us discuss the behaviour of the commutator in a rotating magnetic field. Figure 8.42 shows a simple two-pole DC generator in which the direction of the induced emf in the conductors under the influence of the north pole has been shown by crosses and in other conductors has been shown by dots during the rotation of the armature in the clockwise direction. When the brush B_1B_2 coincides with the magnetic neutral axis, the emf measured across the brushes becomes maximum. The emf across the brushes will be zero when the brush axis is placed in the magnetic neutral axis M_1M_2. This is due to the fact that the emf induced in all the conductors that are on one side of the brush axis will not be in the direction shown in Figure 8.42. This is due to the fact that in half of the conductors the induced emf is represented by crosses, while in the other half it is represented by dots. Due to rotation of the brush axis, the voltage across the brushes will vary from positive maximum to zero and then to negative maximum, and finally comes to zero for occupation of position 1, M_1, *2* and M_1 successively by the brush B_1. We can conclude that an alternating emf will be generated across the brushes having the frequency same as that of the rotation of the brush axis with respect to the field system. The similar result can be obtained by the rotating field system with respect to the field system. If the rotating magnetic field is taken instead of rotating field system, the frequency of the voltage across the brush will be the same as the frequency of the rotating field with respect to the brush axis. It can be concluded that the commutator performs as a frequency converter during its placement in the rotating magnetic field. The rotating magnetic field in Schrage motor is set up by the rotor, which carries an additional DC winding. The induced voltage across the brush A_1A_2, B_1B_2 and C_1C_2 corresponds to the slip frequency, which is equal to the frequency of the rotor. Since the speed of rotating magnetic field in the space is equal to the slip speed, the emf induced to the secondary (stator) also corresponds to the slip frequency. The voltage injected into the stator winding

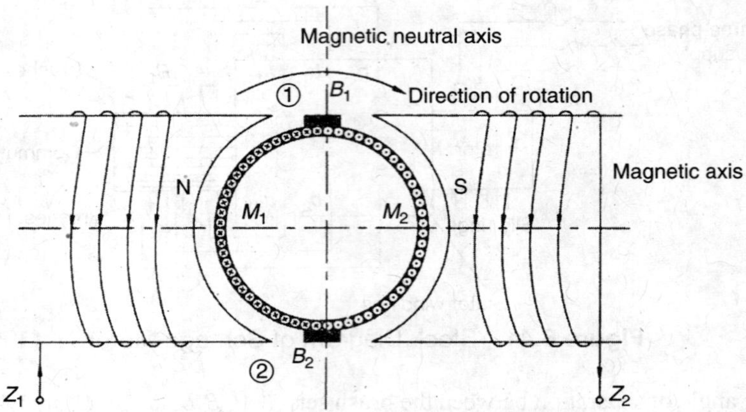

Figure 8.42 Two-pole DC Generator

will always be at the slip frequency when the brushes are connected to the stator windings. It is possible to adjust the phase of the voltage injected into the secondary winding, that is, the stator either in phase or antiphase (out of phase) with the stator-induced voltage.

Figure 8.43 shows the phasor diagram of the emf injected into the secondary.

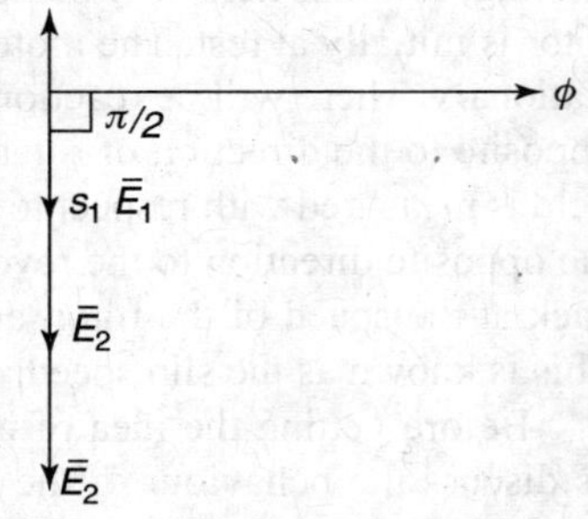

Figure 8.43 Phasor Diagram

Let us consider the following:

E_2 be the magnitude of standstill secondary voltage,
s be the no-load slip at the motor,
E be the magnitude of the voltage injected into the motor secondary and
s_1 be the new operating slip after injecting the voltage.

If the injected voltage is in phase opposition to E_2, the resultant voltage in the rotor circuit becomes $sE_2 + E$, as shown in Figure 8.43. We have the following expression for the new slip s_1.

$$s_1E_2 = (sE_2 + E)$$

i.e.,

$$s_1 = s + \frac{E}{E_2} \tag{8.97}$$

Equation (8.97) suggests that the slip increases and the speed decreases. The following expression is valid when the injected voltage is in phase with the secondary voltage.

$$s_1 = s - \frac{E}{E_2} \tag{8.98}$$

Therefore, the slip decreases and it may be negative at supersynchronous speed. Combining Equation (8.97) and (8.98), we have

$$s_1 = s \pm \frac{E}{E_2} \tag{8.99}$$

Since the no-load slip s is small, Equation (8.99) reduces to

$$s_1 = \pm \frac{E}{E_2} \tag{8.100}$$

The operating speed of the motor can be expressed as

$$N_r = N_s(1 - s_1)$$

i.e.,

$$N_r = N_s\left(1 \mp \frac{E}{E_2}\right) \tag{8.101}$$

A reduced applied voltage has been applied to the primary, that is, rotor at the time of starting. If the motor is switched on directly adjusting the brush axis, the maximum voltage is injected to phase opposition to the secondary voltage, and hence the motor operates at the lowest speed. This reduces the starting current. The starting current becomes about 125 to 175 per cent of the normal high-speed current and the starting torque corresponds to 200 per cent of the normal torque.

The following are the advantages of this machine:

(i) High average efficiency.
(ii) High power factors.
(iii) Continuous speed regulation within the required range.
(iv) The driven machinery can be run at the best possible speed.

These machines are used in printing industry and are also suitable for small presses.

8.44 POWER FACTOR COMPENSATION

The phase of the injected emf can be in phase or in phase opposition with the secondary voltage when yoke is in such a way that the centre of the winding belt a_1a_2 coincides with the centre of the corresponding phase belt of the secondary. There should be an existence of the similar condition for the phase belts b_1b_2 and c_1c_2. This causes the resultant voltage in the secondary to be in phase with the secondary no-load voltage. The power factor of the secondary is solely governed by the resistance and the reactance of the secondary.

If there is non-symmetrical placement of the brushes (A_1A_2) with respect to the centre line of the corresponding phase belt, the injected voltage will not be either in phase or out of phase with the secondary voltage and will differ from it also. The phase of the secondary current is also shifted. Therefore, it can be concluded that the improvement of the power factor can be effected by suitable adjustment of the yoke of the brushes.

8.45 LINEAR INDUCTION MOTOR

Whenever a relative motion exists between the field and short-circuited conductors, currents are induced in them resulting in electromagnetic forces, and under the influence of these forces the conductors try to move in such a way so that the induced currents are eliminated.

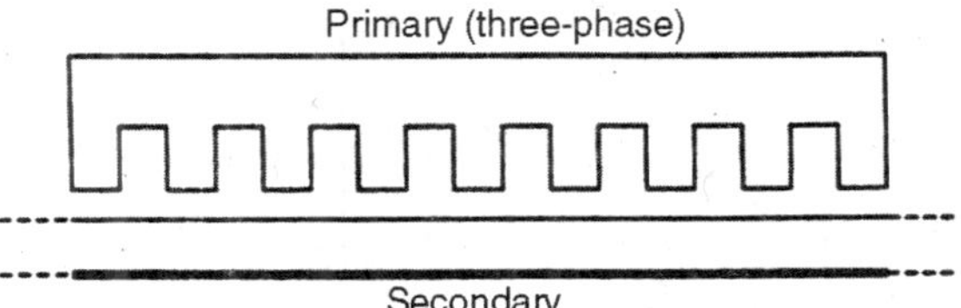

Figure 8.44 Short-single Primary

Figure 8.44 shows the linear induction motor consisting of a field system having three-phase distributed winding, which are placed in slots. The field system may be single or double primary system. The secondary of linear induction motor is usually a conducting plate made of either copper or aluminum.

Due to application of three-phase supply voltage to the three-phase primary winding from a balanced three-phase source, a magnetic field moving in a straight line from one end to the other at a linear synchronous speed V_s is produced that can be expressed by

$$V_s = 2\tau f \text{ m/s} \tag{8.102}$$

where τ is the pole pitch in meters and f is the supply frequency in hertz.

The flux moves linearly and it drags the rotor plate along with it in the same direction with a speed V m/s. The relative speed of travel of the flux with respect to the rotor plate decreases. Slip s of the motor is expressed by

$$s = \frac{V_s - V}{V_s} \tag{8.103}$$

where V_s is the linear synchronous speed and V is the actual speed of the rotor poles. Thrust for force is expressed by

$$F = \frac{P_2}{V_s} \tag{8.104}$$

where P_2 represents the actual power supplied to the rotor.

Figure 8.45 shows the thrust-speed characteristic of linear induction motors.

Linear induction motors have the following advantages:

(i) Simpler in construction.
(ii) Power to weight ratio is better.
(iii) Initial cost is low.
(iv) No over heating of rotor.
(v) Maintenance cost is low due to absence of rotating parts.
(vi) No limitation of maximum speed due to centrifugal forces.
(vii) Due to adhesion between the wheel and the rail, there is no limitation of thrust.

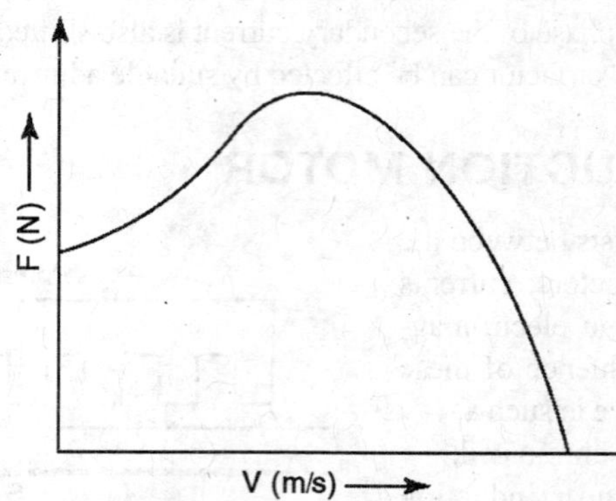

Figure 8.45 Thrust-speed Characteristics

The following are the disadvantages of linear induction motors:

(i) Due to transverse edge and end effects, utilization of motor is poor.
(ii) Capital cost of reaction rail fixed along the centre line of the track is very high.
(iii) Complications arise due to the provision of three-phase collector system along the track.
(iv) High cost.
(v) A lot of difficulties are faced for maintaining adequate clearance at points and crossings.
(vi) The poor efficiency and power factor is due to the more magnetizing current because the large air gap and non-magnetic reaction rail are required.

Linear induction motors have the following applications:

(i) Electromagnetic pumps,
(ii) Conveyors,
(iii) Trolley cars,
(iv) High-speed rail traction,
(v) As booster accelerator and
(vi) Metallic belt conveyors and others.

8.46 INDUCTION GENERATOR

Whenever the induction motor operates at a speed higher than the synchronous speed, the slip becomes negative and it operates as an induction generator. The induced emf and rotor current become zero when the induction motor is driven by another machine at a speed equal to the synchronous speed. The torque developed in this case becomes zero. The slip becomes negative if the speed is increased above the synchronous speed. The rotor emf ($\bar{E}_2$) and rotor current ($\bar{I}_2$) become negative in this case. The reflected current also changes its direction and the resultant current ($\bar{I}_1$) = $\bar{I}_0 + \bar{I}_2'$ has a component out of phase with the supply voltage. The gross mechanical power developed and the corresponding torque become negative in this case, which means that the rotor does not supply mechanical power to the shaft, but absorbs it from the shaft. Therefore, the machine operates as induction generator absorbing excitation current (I_0) from the supply mains. Figure 8.46 shows the circle diagram of the induction generator. For induction generator, the slip must be replaced by a negative sign of equivalent circuit and other characteristic of induction motor. The machine absorbs exciting current ($\bar{I}_0$) from the supply mains when it operates as an induction generator. During the operation of the induction motor as an induction generator at supersynchronous speed, the equivalent circuit and other characteristic equations of induction motor are equally applicable to the induction generator with a difference that the slip will be replaced by a negative sign. The rotor current is expressed by

$$\bar{I}_2 = \frac{-sE_2}{R_2 - jsX_2} \tag{8.105}$$

i.e.,

$$I_2 = \frac{sE_2}{\sqrt{R_2^2 + (sE_2)^2}} \tag{8.106}$$

For an approximate equivalent circuit

$$I_2' = \frac{V_1}{\left[\left(R_1 - \frac{R_2'}{s}\right)^2 + (X_1 + X_2')^2\right]^{\frac{1}{2}}}$$

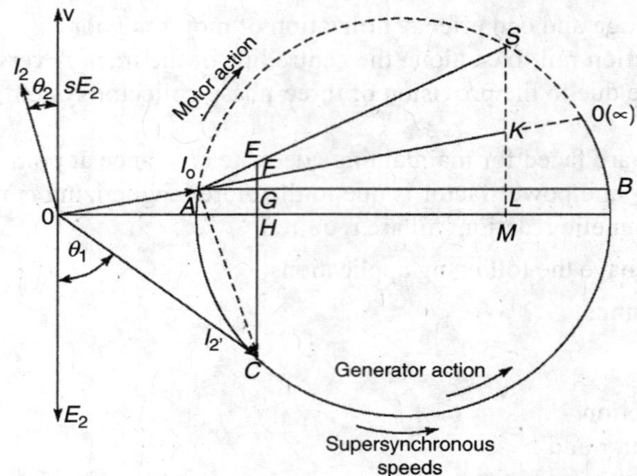

Figure 8.46 Circle Diagram of an Induction Generator

The current locus at sub and supersynchronous speeds are two semicircles. These two meet to form a complete circle, as shown in Figure 8.46.

From Figure 8.46,

CH = electrical output,

EF = rotor Cu loss,

FG = stator Cu loss,

GH = mechanical losses + iron losses in the stator and

CE = input to the machine

The advantages, disadvantages and applications of induction generators are discussed below.

Advantages:

- Simple and rugged construction.
- Cost is cheaper.
- Easy maintenance.
- No hunt or drop out of synchronism.
- Depending upon the loading of the lines, induction regulators adjust voltage to a constant value.

Disadvantages:

- Deliver only leading current.
- Unable to operate independently.
- Do not help in stability.
- Due to disconnection of synchronous machines at the far end and excitation of the induction machines by line capacitance, dangerously high voltages may occur.

Applications:

- For breaking purposes of railway work.
- For installation in small power stations.
- For non uniform speed of the prime mover.

8.47 ELECTRICAL BRAKING OF POLYPHASE INDUCTION MOTORS

Electrical braking is the most important requirement of quick-speed control of a running motor in modern electric drives. The quickness and accuracy of the operations in controlling the speed determines the quality and productivity of the product. Electrical braking is very smooth compared to mechanical braking. The direction of rotation and direction of the developed torque are exactly reverse to each other in electrical braking. The speed-torque characteristics are totally dependent on the method of braking. The three methods of braking are as follows:

(i) Regenerative braking,
(ii) Plugging or counter current braking and
(iii) Dynamic braking.

8.47.1 Regenerative Braking

Regenerative braking is a cheap method of braking in which the machine is operated as an induction generator and electrical energy is pumped back to the supply. The generator action only occurs when the machine runs at supersynchronous speed. It is possible to develop regenerative braking by pole changing on the stator of a squirrel-cage motor.

8.47.2 Plugging or Counter-current Braking

Plugging or counter-current braking is a more commonly used method. In this method, the direction of rotation of the machine can be reversed only after reducing the speed to zero. Interchange of any two supply leads to the stator results in plugging. This reverses the direction of the rotation of the stator mmf, with rotor still running in the initial condition. The developed electromagnetic torque works in a direction opposite to the direction of rotation of the rotor. This helps to bring the motor at rest quickly. If the supply is not removed at zero speed, the motor runs in the opposite direction.

Figure 8.47 shows the operation of slip-ring induction motor during plugging.

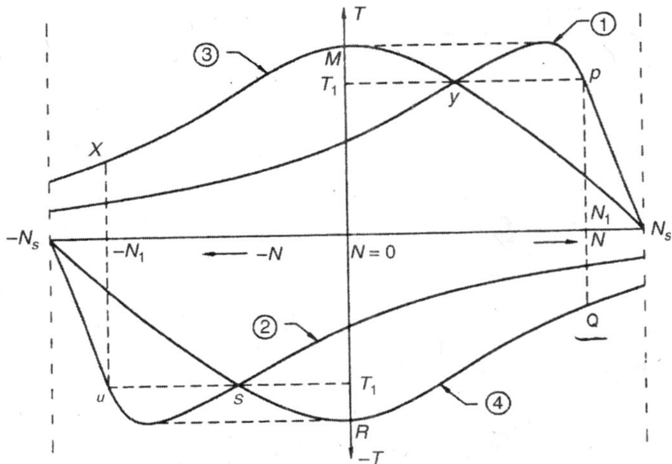

Figure 8.47 Operation of Slip-ring Induction Motor During Plugging

The slip becomes 2*s* during the process of plugging. The induced emf and the corresponding induced current in the rotor become high. Additional resistance in the rotor circuit is inserted to limit the current drawn by the stator during plugging.

Curve 1 in Figure 8.47 is the natural speed-torque characteristic of the motor, whereas curve 2 in Figure 8.47 is the natural characteristic of the motor when running in the opposite direction. Curves 3 and 4 are the speed-torque characteristics with additional resistances inserted in the rotor circuit corresponding to curve 1 and curve 2, respectively. It is assumed that the plugging is done at the load torque T_1, corresponding to speed N_1 for the operating point *P*. It is not possible to change the speed of the motor immediately at the instant of plugging. This is due to the inertia of the rotating system. The operating point *P* shifts from curve 1 to the point *Q* on curve 4 when additional resistances are inserted in the rotor circuit at the time of plugging, so that the stator current can be limited. Due to development of braking torque by the motor, the speed drops, and hence the operating point shifts along curve 4 until the point *R* is reached corresponding to zero speed. The motor should be de-energized if the reverse direction of rotation is not required. If it is not done, the motor starts and gains speed in the opposite direction of rotation along the characteristic *RS* on curve 4. If the additional resistance is cut down at the point *S*, the operating point moves slowly along curve 2 from *S* and steady state is reached at the point *U*. If plugging is again carried out, the braking and acceleration will follow along the track *UXMYP*.

8.47.3 Dynamic Braking

The main disadvantages of plugging are that there is powerless in the extra resistance used for limiting the stator current and the requirement of special precaution to de-energize the motor at zero speed to prevent reversal of rotation.

Figure 8.48 shows the schematic diagram for the dynamic braking of induction motor. In dynamic braking, the stator terminals are disconnected from an AC source and then switched on to a DC source itself through a transformer rectifier set, as shown in Figure 8.48.

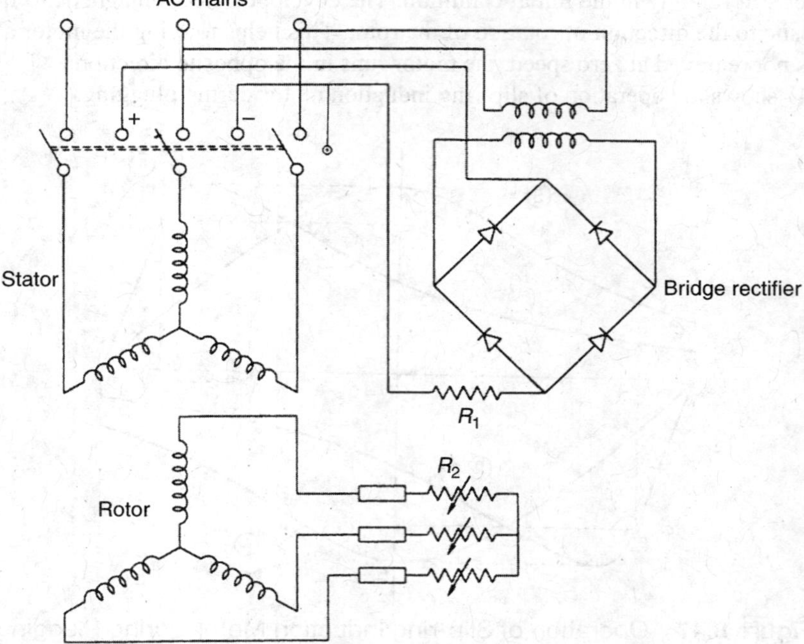

Figure 8.48 Dynamic Braking of an Induction Motor

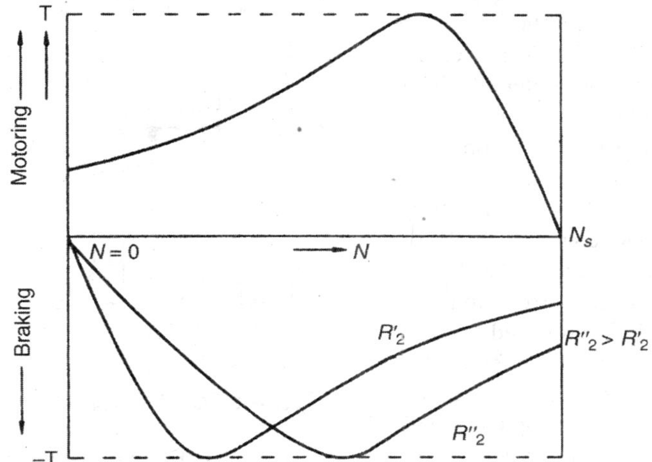

Figure 8.49 Speed-torque Characteristic During Dynamic Braking

To limit the stator current, additional resistance (R_1) is inserted in series with the stator. For slip-ring induction motor, additional resistance (R_2) is inserted in the rotor circuit to limit the current as well as to achieve a series of different braking characteristics, as shown in Figure 8.49.

Stationary magnetic field is produced due to the flow of DC current in stator winding. Induced current in the rotor is produced due to rotation of short-circuited rotor in this magnetic field. The rotor magnetic field and the rotor will rotate at the same speed, but in opposite directions. The speed of the rotor magnetic field in space becomes zero. Therefore, the rotor field becomes stationary with respect to the stator. Interaction of the rotor and stator fields develops the braking torque, which depends on the rotor speed, rotor resistance and direct current through the stator. Therefore, the speed of the motor in dynamic braking can be controlled by varying DC excitation.

8.48 SYNCHRONOUS-INDUCTION MOTOR

If the rotor of a slip-ring induction motor is fed from a DC source, it is possible to run this motor at a synchronous speed.

The rotor of a slip-ring induction motor becomes an electromagnet when it is fed from a DC source. The axis of this electromagnet becomes fixed.

Figure 8.50 shows the schematic diagram of a synchronous induction motor for starting and running conditions. The additional resistance is put on the rotor circuit and it is started as an ordinary slip-ring induction motor and the switch S is at the start position. The additional resistances are slowly cut out. The motor runs with a small slip. The switch is then thrown to run position and the rotor is fed by a DC source by the excitation that is mounted on the same shaft of induction motor. This creates fixed poles on the rotor due to DC excitation, which get magnetically locked with the poles produced by the rotating magnetic field. This rotating magnetic field is produced by the stator winding fed from a three-phase AC source. This helps to run the motor at a synchronous speed.

The synchronous-induction motor can be started with load when it starts as a slip-ring induction motor. The main criterion is that it should have good starting characteristics in addition to other synchronous motor features such as constant speed and power factor control. It is necessary to design the rotor of synchronous-induction motor in such a way that it will be possible to get the maximum mmf with moderate DC excitation power and sinusoidal mmf distribution preferably. This helps to distribute the loss evenly around the rotor so that sufficient damping is obtained against hunting and better characteristics.

If the rotor of a well-designed induction motor is fed from a DC source, it makes a poor synchronous motor due to large synchronous impedance so that excitation current can be limited. It is preferred to have a large synchronous impedance for a synchronous motor operation. This will produce a large variation in power factor with load. The overall load capacity of a synchronous motor becomes greater when the synchronous impedance becomes smaller. To obtain this, a large air gap with consequent stronger excitation is used. It is not possible to make good the starting characteristics of induction motor with more excitation power.

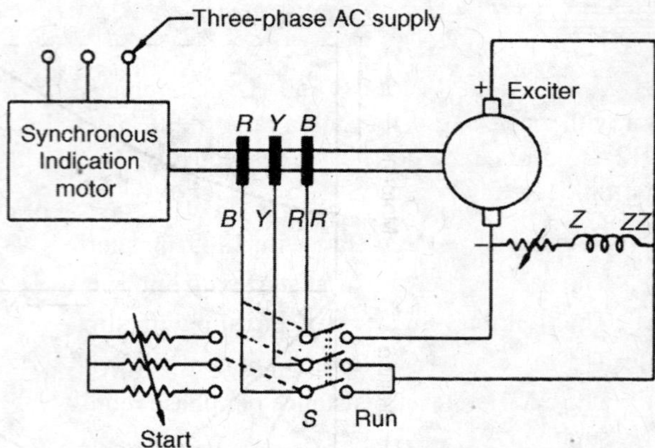

Figure 8.50 Synchronous Induction Motor

Figure 8.51 shows a number of methods for DC excitation to rotor winding. The uniform distribution of rotor heating is dependent on the type of connection. The connections shown in Figures 8.51(a), 8.51(b), 8.51(c) and 8.51(d) produce unbalanced heating. It is possible to get even heating to a large extent by using two-phase or single-phase rotor windings with normal three-phase rotor windings. These have been shown in Figures 8.51(e), 8.51(f) and 8.51(g), where the two-phase connections, Figures 8.51(e) and 8.51(f), produce uniform at the harmonics. The usual usage of the primary chording in synchronous-induction machines reduces the effect of prominent harmonics.

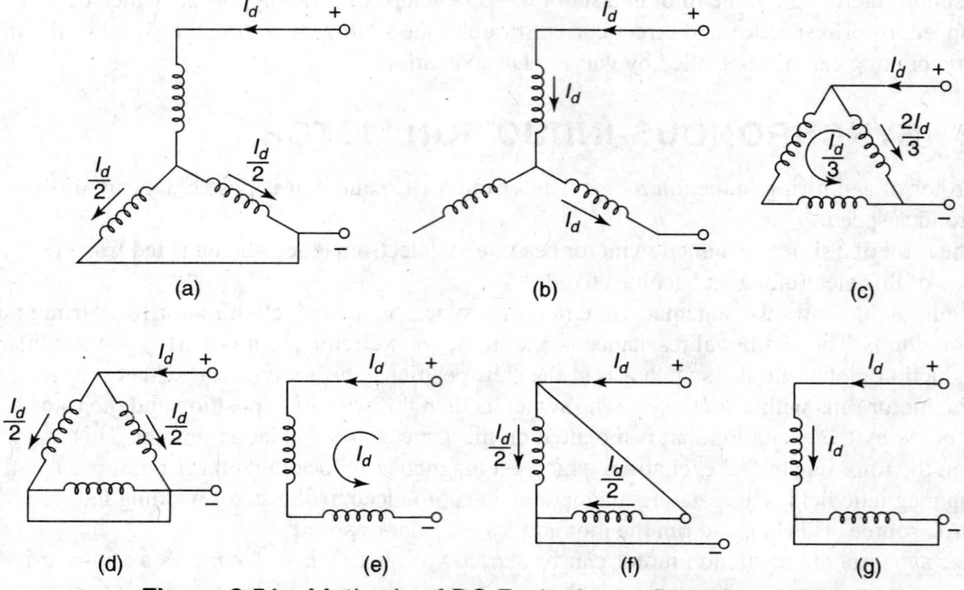

Figure 8.51 Methods of DC Excitation to Rotor Winding

ADDITIONAL SOLVED PROBLEMS

Example 8.15 A 1,100 V, 50 Hz delta connected induction motor has a star-connected slip-ring rotor with a phase transformation ratio of 3.8. The rotor resistance and standstill leakage reactance are 0.012 Ω and 0.25 Ω per phase, respectively. Neglecting stator impedance and magnetizing current, determine the following:

(i) The rotor current at start with slip ring shorted.
(ii) The rotor power factor at start with slip ring shorted.
(iii) The rotor current at 4 per cent slip with slip ring shorted.
(iv) The rotor power factor at 4 per cent slip with slip ring shorted.
(v) The external rotor resistance per phase required to obtain a starting current of 100 A in the stator supply lines.

Solution

$$\text{Turns ratio} = 3.8$$

∴ Phase voltage induced in the rotor winding at standstill

$$E_2 = \frac{1,100}{3.8} = 289.5 \text{ V}$$

Rotor resistance per phase $(R_2) = 0.012\ \Omega$

Rotor reactance (at standstill) per phase $(X_2) = 0.25\ \Omega$

∴ Rotor impedance at standstill $= \sqrt{0.012^2 + 0.25^2} \approx 0.25\ \Omega$

(i) Rotor starting current per phase at normal voltage with slip ring shorted

$$= \frac{289.5}{0.25} = 1,158 \text{ A}$$

(ii) Rotor power factor at start with slip ring shorted

$$= \cos\left(\tan^{-1}\frac{0.25}{0.012}\right) = 0.048 \text{ (lagging)}$$

(iii) At 4 per cent slip, rotor emf $= 0.04 \times E_2 = 0.04 \times 289.5 = 11.58$ V

Rotor impedance at 4 per cent slip $= sX_2 = 0.04 \times 0.25 = 0.01\ \Omega$

Rotor current at 4 per cent slip $= \sqrt{0.012^2 + 0.01^2} = 0.0156\Omega$

$$\therefore \text{ Rotor current at 4 per cent slip} = \frac{11.58}{0.0156} = 742.3\text{A}$$

(iv) Rotor pf at 4 per cent slip $= \dfrac{R_2}{Z_2} = \dfrac{0.012}{0.0156} = 0.769.$

(v) Rotor current corresponding to stator line current of 100 A

$$= \frac{100}{\sqrt{3}} \times 3.8 = 219.4 \text{ A}$$

$$\text{Rotor impedance} = \frac{289.5}{219.4} = 1.32\,\Omega$$

$$\text{Rotor resistance required} = \sqrt{1.32^2 - 0.25^2} = 1.296\,\Omega$$

∴ Required external rotor resistance per phase

$$= 1.296 - 0.012 = 1.284\ \Omega$$

■

Example 8.16 The rotor resistance and rotor reactance at standstill per phase of a six-pole, three-phase, 50 Hz induction motor are 0.026 Ω and 0.12 Ω, respectively. Calculate the following:

(i) The value of speed at which maximum torque occurs.

(ii) The required external resistance to get 75 per cent of maximum torque at starting.

Solution

$$\text{Number of poles } (P) = 6$$

$$\text{Rotor resistance per phase } (R_2) = 0.026\ \Omega$$

$$\text{Rotor reactance at standstill per phase } (X_2) = 0.12\ \Omega$$

$$\text{Synchronous speed } (N_s) = \frac{120f}{P} = \frac{120 \times 50}{6} = 1{,}000\,\text{rpm}$$

$$\text{Slip at maximum torque } (s_{mT}) = \frac{R_2}{X_2} = \frac{0.026}{0.12} = 0.2167$$

(i) The speed of rotor at maximum torque is

$$N = N_s(1-s) = 1{,}000 \times (1 - 0.2167) = 783.3\text{ rpm}$$

(ii) Let r be the external resistance inserted in the rotor circuit to get the starting torque 75 per cent of full-load torque.

$$s_{mT} = \frac{0.026 + r}{0.12}$$

Again,

$$\frac{T_{est}}{T_{e\max}} = \frac{2s_{mT}}{s_{mT}^2 + 1}$$

i.e.,

$$\frac{75}{100} = \frac{2s_{mT}}{s_{mT}^2 + 1}$$

i.e.,

$$s_{mT}^2 - 2.67s_{mT} + 1 = 0$$

$$s_{mT} = \frac{2.67 \pm \sqrt{(-2.67)^2 - 4}}{2} = \frac{2.67 \pm 1.768}{2}$$

∴ s_{mT} = either 2.219 or 0.451

Neglecting higher value, $s_{mT} = 0.451$ (feasible)

$$\therefore \qquad 0.451 = \frac{0.026 + r}{0.12}$$

$$\therefore \qquad r = 0.451 \times 0.12 - 0.026 = 0.02812\ \Omega$$

The required external resistance per phase to be inserted in rotor is 0.02812 Ω. ■

Example 8.17 The rotor resistance and standstill reactance per phase of a three-phase induction motor are 0.018 Ω and 0.12 Ω, respectively. The full-load slip is 2.5 per cent at normal voltage. Find the following:

(i) Percentage reduction in supply voltage so that full-load torque is obtained at half of full-load speed.

(ii) Power factor.

Solution

$$\text{Full-load slip } (s_{fl}) = 2.5 \text{ per cent} = 0.025$$

$$\text{Rotor resistance per phase } (R_2) = 0.018\ \Omega$$

$$\text{Rotor reactance at standstill per phase} = 0.12\ \Omega$$

If N_s be the synchronous speed, the full-load speed is given by

$$N_{fl} = N_s(1 - s_{fl}) = (1 - 0.025)N_s = 0.975N_s$$

$$\text{Speed at half of full-load speed} = \frac{0.975N_s}{2} = 0.4875N_s$$

$$\text{Slip at half of full-load speed } \left(s_{\frac{1}{2}fl}\right) = \frac{N_s - 0.4875N_s}{N_s} = 0.5125$$

We know that $T_{efl} = \dfrac{ks_{fl}R_2E_2^2}{R_2^2 + s_{fl}^2X_2^2}$ and $T_{e\left(\frac{1}{2}\right)fl} = \dfrac{ks_{\frac{1}{2}fl}R_2E_2'^2}{R_2^2 + s_{\frac{1}{2}fl}^2X_2^2}$

where E'_2 is the rotor induced emf at half of full load.

Since $$T_{efl} = T_{e\left(\frac{1}{2}\right)fl}$$

i.e., $$\frac{ks_{fl}R_2E_2^2}{R_2^2 + s_{fl}^2X_2^2} = \frac{ks_{\frac{1}{2}fl}R_2E_2'^2}{R_2^2 + s_{\frac{1}{2}fl}^2X_2^2}$$

i.e., $$\left(\frac{V_{fl}}{V_{\frac{1}{2}fl}}\right)^2 = \frac{s_{\frac{1}{2}fl}}{s_{fl}}\left(\frac{R_2^2 + s_{fl}^2X_2^2}{R_2^2 + s_{\frac{1}{2}fl}^2X_2^2}\right) \quad \text{[since, } E_2 \propto V,\ V \text{ is the supply voltage]}$$

$$= \left(\frac{0.5125}{0.025}\right) \times \left\{\frac{0.018^2 + (0.025 \times 0.12)^2}{0.018^2 + (0.5125 \times 0.12)^2}\right\} = 1.662$$

i.e., $$\frac{V_{fl}}{V_{\frac{1}{2}fl}} = 1.662$$

i.e., $$\frac{V_{\frac{1}{2}fl}}{V_{fl}} = \frac{1}{1.662} = 0.6017$$

(i) Percentage reduction in supply voltage

$$= \frac{V_{fl} - V_{\frac{1}{2}fl}}{V_{fl}} \times 100 = \left(1 - \frac{V_{\frac{1}{2}fl}}{V_{fl}}\right) \times 100 = 39.83\%$$

(ii) Rotor power factor ($\cos\theta_2$) at half of full load

$$= \frac{R_2}{\sqrt{R_2^2 + \left(s_{\frac{1}{2}fl}\right)^2 X_2^2}} = \frac{0.018}{\sqrt{(0.018)^2 + (0.5125 \times 0.12)^2}} = 0.2809\,(\text{lag})$$ ■

Example 8.18 A 70 kW, 400 V, three-phase, four-pole, 50 Hz wound-rotor induction motor operates at rated voltage and frequency with wound rotor short-circuited, and gives full-load slip of 0.035. At maximum torque, it has the slip of 0.22. Neglecting stator resistance and rotational losses, calculate (i) the maximum torque and (ii) full-load rotor ohmic losses.

For rated power output, calculate (iii) slip at maximum torque (iv) full-load slip and (v) full-load torque if its rotor resistance becomes 1.5 times its value when it is short-circuited.

Solution

$$\text{Synchronous speed } (N_s) = \frac{120f}{P} = \frac{120 \times 50}{4} = 1{,}500 \text{ rpm}$$

$$\text{Rotor speed } (N_r) = (1 - s_{fl})N_s = (1 - 0.035) \times 1{,}500 = 1{,}447.5 \text{ rpm}$$

Since the rotational losses are neglected, the mechanical power developed is also 70 kW.

∴ $$P_m = 70 \text{ kW} = 70{,}000 \text{ W}$$

Again, $$P_m = \omega_r T_{efl}$$

∴ $$T_{efl} = \frac{P_m}{\omega_r} = \frac{P_m}{2\pi\left(\frac{N_r}{60}\right)} = \frac{70{,}000}{2 \times 3.14 \times \frac{1{,}447.5}{60}} = 462.03 \text{ N-m}$$

(i) If $T_{e\max}$ be the maximum torque at slip s_{mT}, we have

$$\frac{T_{efl}}{T_{e\max}} = \frac{2s_{mT}s_{fl}}{s_{fl}^2 + s_{mT}^2}$$

$$\therefore \qquad T_{e\max} = T_{efl}\left(\frac{s_{fl}^2 + s_{mT}^2}{2s_{mT}s_{fl}}\right) = 462.03 \times \left[\frac{0.22^2 + 0.035^2}{2 \times 0.22 \times 0.035}\right] = 1,488.85 \text{ N-m}$$

(ii) Full-load rotor ohmic loss $= \dfrac{s_{fl}}{1-s_{fl}} \times P_m = \dfrac{0.035}{1-0.035} \times 70,000 = 2,538.86$ W

(iii) Rotor resistance $(R_1') = 1.5R_2$

The new slip at maximum torque is given by

$$s'_{mT} = \frac{R_2'}{X_2} = \frac{1.5R_2}{X_2} = 1.5\left(\frac{R_2}{X_2}\right) = 1.5s_{mT} = 1.5 \times 0.22 = 0.33$$

(iv) Let the full-load torque be T'_{efl} when $R_2' = 1.5R_2$.

$$\frac{T'_{efl}}{T'_{e\max}} = \frac{2s'_{fl}s'_{mT}}{\left(s'_{fl}\right)^2 + \left(s'_{mT}\right)^2} = \frac{2 \times 0.33 \times s'_{fl}}{0.33^2 + \left(s'_{fl}\right)^2} = \frac{0.66s'_{fl}}{\left(s'_{fl}\right)^2 + 0.1089}$$

i.e.,

$$T'_{efl} = \frac{0.66s'_{fl}T'_{em}}{\left(s'_{fl}\right)^2 + 0.1089} = \frac{0.66s'_{fl} \times 1,488.85}{\left(s'_{fl}\right)^2 + 0.1089} = \frac{982.64s'_{fl}}{\left(s'_{fl}\right)^2 + 0.1089}$$

Again,

$$T'_{efl} = \frac{p_m}{\omega_r} = \frac{70,000}{\left(1-s'_{fl}\right)\omega_s} = \frac{70,000}{\left(1-s'_{fl}\right) \times 2 \times 3.14 \times \frac{1,500}{60}} = \frac{70,000}{157\left(1-s'_{fl}\right)}$$

$$\therefore \qquad \frac{70,000}{157\left(1-s'_{fl}\right)} = \frac{982.64s'_{fl}}{\left(s'_{fl}\right)^2 + 0.1089}$$

i.e.,

$$\left(s'_{fl}\right)^2 + 0.1089 = \frac{982.64 \times 157s'_{fl}\left(1-s'_{fl}\right)}{70,000} = 2.204\left(s'_{fl} - s'^2_{fl}\right)$$

i.e.,

$$3.204s'^2_{fl} - 2.204s'_{fl} + 0.1089 = 0$$

$$\therefore \qquad s'_{fl} = \frac{2.204 \pm \sqrt{2.204^2 - 4 \times 3.204 \times 0.1089}}{2 \times 3.204} = \frac{2.204 \pm 1.860}{2 \times 3.204}$$

Therefore, s'_{fl} is either 0.6342 or 0.05368

The feasible value of slip at full-load torque is 0.05368.

(v) $T'_{efl} = \dfrac{70,000}{157\left(1-s'_{fl}\right)} = \dfrac{70,000}{157(1-0.05368)} = 471.15$ N-m

The full-load torque is 471.15 N-m. ■

Example 8.19 The maximum and starting torques of a three-phase induction motor are 2.2 times and 1.7 times of its full-load torque; respectively. Calculate the percentage reduction in the rotor circuit resistance to get the maximum torque twice the full-load torque. Take the full-load slip of 6 per cent. Neglect stator impedance.

Solution

Here, $$T_{est} = 1.7 T_{efl} \text{ and } T_{e\max} = 2.2 T_{efl}$$

Again, $$\frac{T_{est}}{T_{e\max}} = \frac{2 s_{mT} s_{st}}{s_{st}^2 + s_{mT}^2}$$

i.e., $$\frac{1.7\, T_{efl}}{2.2\, T_{efl}} = \frac{2 s_{mT}}{1 + s_{mT}^2} \quad \text{[since slip at starting is 1]}$$

i.e., $$\frac{1.7}{2.2} = \frac{2 s_{mT}}{1 + s_{mT}^2}$$

i.e., $$s_{mT}^2 - 2.588 s_{mT} + 1 = 0$$

i.e., $$s_{mT} = \frac{2.588 \pm \sqrt{2.588^2 - 4 \times 1 \times 1}}{2} = \frac{2.588 \pm 1.642}{2}$$

i.e., $$s_{mT} = \text{either } 2.115 \text{ or } 0.473$$

Neglecting higher value of s_{mT}, we have

∴ $$s_{mT} = 0.473$$

Again, $$s_{mT} = \frac{R_2}{X_2} = 0.473$$

∴ $$R_2 = 0.473 X_2$$

Let the rotor resistance per phase be $R_2' \Omega$ so that the slip at maximum torque be s_{mT}'.

∴ $$\frac{T_{efl}}{T_{e\max}} = \frac{2 s_{mT}' s_{fl}}{s_{fl}^2 + (s_{mT}')^2} = \frac{2 s_{mT}' \times 0.06}{0.06^2 + (s_{mT}')^2}$$

i.e., $$\frac{1}{2} = \frac{0.12 s_{mT}'}{3.6 \times 10^{-3} + (s_{mT}')^2}$$

i.e., $$(s_{mT}')^2 - 0.24 s_{mT}' + 3.6 \times 10^{-3} = 0$$

i.e., $$s_{mT}' = \frac{0.24 \pm \sqrt{0.24^2 - 4 \times 3.6 \times 10^{-3}}}{2} = \frac{0.24 \pm 0.2078}{2}$$

∴ $$s_{mT}' = \text{either } 0.2239 \text{ or } 0.0161$$

Neglecting lower value of s'_{mT}, we have

$$s'_{mT} = 0.2239$$

Since $$s'_{mT} = \frac{R'_2}{X_2}$$

$\therefore$ $$\frac{R'_2}{X_2} = 0.2239$$

i.e., $$R'_2 = 0.2239X_2$$

$\therefore$ Percentage reduction in rotor circuit resistance

$$= \frac{0.473X_2 - 0.2239X_2}{0.473X_2} \times 100 = 52.66\%$$

■

Example 8.20 The starting and maximum torques of a three-phase induction motor are 1.6 times and 2.4 times of the full-load torque, respectively. Calculate the following:

(i) The slip at maximum torque.
(ii) The full-load slip.
(iii) The rotor current at starting in terms of full-load rotor current assuming constant rotor resistance. Neglect stator impedance.

Solution

Here, $$T_{est} = 1.6T_{efl} \quad \text{and} \quad T_{e\max} = 2.4T_{efl}$$

$$\frac{T_{est}}{T_{efl}} = \frac{1.6}{2.4}$$

(i) $$\frac{T_{est}}{T_{e\max}} = \frac{2s_{mT}}{s_{mT}^2 + 1} = \frac{1.6}{2.4} = \frac{2}{3}$$

i.e., $$s_{mT}^2 - 3s_{mT} + 1 = 0$$

i.e., $$s_{mT} = \frac{3 \pm \sqrt{3^2 - 4 \times 1 \times 1}}{2} = \frac{3 \pm 2.236}{2}$$

Therefore, $$s_{mT} = \text{either } 2.618 \text{ or } 0.382$$

Neglecting higher value of s_{mT}, we have $s_{mT} = 0.382$

(ii) $$\frac{T_{efl}}{T_{e\max}} = \frac{2s_{mT}s_{fl}}{s_{fl}^2 + s_{mT}^2}$$

i.e., $$\frac{1}{2.4} = \frac{2 \times 0.382s_{fl}}{s_{fl}^2 + 0.382^2}$$

i.e., $s_{fl}^2 - 1.8336 s_{fl} + 0.1459 = 0$

i.e., $$s_{fl} = \frac{1.8336 \pm \sqrt{1.8336^2 - 4 \times 1 \times 0.1459}}{2} = \frac{1.8336 \pm 1.667}{2}$$

i.e., $$s_{fl} = \text{either } 1.7503 \text{ or } 0.0833$$

Neglecting higher value of s_{fl}, we have

$$s_{fl} = 0.0833$$

(iii) $$T_{est} = \frac{1}{\omega_s} I_{2st}^2 R_2 \quad \text{and} \quad T_{efl} = \frac{1}{\omega_s} I_{2fl}^2 \frac{R_2}{s_{fl}}$$

$\therefore$ $$\frac{T_{est}}{T_{efl}} = \left(\frac{I_{2st}}{I_{2fl}}\right)^2 s_{fl}$$

i.e., $$\left(\frac{I_{2st}}{I_{2fl}}\right)^2 = \frac{T_{est}}{T_{efl}} \times \frac{1}{s_{fl}} = \frac{1.6}{0.0833}$$

i.e., $$I_{2st} = 4.3826 I_{2fl}$$ ■

Example 8.21 The shaft output of a three-phase induction motor is 20 kW at 1,440 rpm. Total stator loss is 650 W. Friction and windage losses amount to 1.2 per cent of shaft output. Determine the following:

(i) The rotor input and stator input.

(ii) The starting torque with rated voltage starting if developed maximum torque occurs at 1,350 rpm.

Solution

(i) Full-load slip $(s_{fl}) = \frac{1,500 - 1,440}{1,500} = 0.04$

Stator losses = 650 W

Friction and windage loss = 1.2 per cent of shaft output = $\frac{1.2}{100} \times 20 \times 10^3 = 1.2 \times 200 = 240$ W

Mechanical power developed (P_m) = 20,000 + 240 = 20,240 W

$\therefore$ $$\text{Rotor input } (P_g) = \frac{P_m}{1-s} = \frac{20,240}{1-0.04} = 21,083.34 \text{ W}$$

$$\text{Stator input} = P_g + \text{stator losses} = 21,083.34 + 650 = 21,733.34 \text{ W}$$

(ii) $$s_{mT} = \frac{1,500 - 1,350}{1,500} = 0.1$$

$$T_{efl} = \frac{P_g}{\omega_s} = \frac{21,083.34}{2 \times 3.14 \times \frac{1,500}{60}} = 134.29 \text{ N-m}$$

$$\frac{T_{efl}}{T_{e\max}} = \frac{2s_{mT}s_{fl}}{s_{fl}^2 + s_{mT}^2} = \frac{2\times 0.1\times 0.04}{0.1^2 + 0.04^2}$$

∴ $$T_{e\max} = 194.72 \text{ N-m}$$

Again, $$\frac{T_{est}}{T_{e\max}} = \frac{2s_{mT}}{1+s_{mT}^2}$$

$$T_{est} = \frac{2\times 0.1}{0.1^2+1}\times 194.72 = 38.56 \text{ N-m}$$

■

Example 8.22 A 120 kW, three-phase, four-pole, 400 V, 50 Hz induction motor (wound-rotor-type) with rotor winding short-circuited has a full-load slip of 0.05, and the slip at which maximum torque occurs is 0.23 during operation at rated voltage and frequency. Neglecting stator resistance and rotational losses, find (i) maximum torque, (ii) starting torque and (iii) full-load rotor ohmic loss.

If rotor resistance is made 2.5 times of its previous value by adding external resistance, calculate (iv) slip at full-load output, (v) full-load torque and (vi) full-load rotor power factor. Full-load output in both cases remains the same.

Solution

Here, $$P_{sh} = 120 \text{ kW}$$

Since rotational losses are neglected, we have

$$P_m = P_{sh} = 120 \text{ kW}$$

Here, $$s_{fl} = 0.05 \quad \text{and} \quad s_{mT} = 0.23$$

∴ $$P_g = \frac{P_m}{1-s_{fl}} = \frac{120}{1-0.05} = 126.316 \text{ kW}$$

$$\text{Synchronous speed } (N_s) = \frac{120\times 50}{4} = 1{,}500 \text{ rpm}$$

∴ $$T_{efl} = \frac{P_g}{\omega_s} = \frac{P_g}{2\pi\times\dfrac{N_s}{60}} = \frac{126.316\times 10^3}{2\pi\times\dfrac{1{,}500}{60}} = 804.56 \text{ N-m}$$

(i) Again, $$\frac{T_{efl}}{T_{e\max}} = \frac{2s_{mT}s_{fl}}{s_{fl}^2+s_{mT}^2} = \frac{2\times 0.23\times 0.05}{0.23^2+0.05^2}$$

∴ $$T_{e\max} = 2.408T_{efl} = 2.408\times 804.56 = 1{,}937.38 \text{ N-m}$$

(ii) $$\frac{T_{est}}{T_{e\max}} = \frac{2s_{mT}}{s_{mT}^2+1}$$

∴ $$T_{est} = \frac{2s_{mT}}{s_{mT}^2+1}\times T_{em} = \frac{2\times 0.23}{0.23^2+1}\times 1{,}937.38 = 846.42 \text{ N-m}$$

■

(iii) Rotor ohmic loss $= sP_g = 0.05 \times 126.316 = 6.3158$ kW

(iv) Full-load output is $T_e\omega_r = \dfrac{3V^2 s_{fl}(1-s_{fl})}{R_2}$

With rotor resistance 2.5 times of R_2, the full load output becomes $\dfrac{3V^2 s'_{fl}(1-s'_{fl})}{2.5R_2}$

Since full-load output in both cases remains the same, we have

$$\frac{3V^2 s_{fl}(1-s_{fl})}{R_2} = \frac{3V^2 s'_{fl}(1-s'_{fl})}{2.5R_2}$$

i.e., $$0.05\times(1-0.05) = \frac{s'_{fl}(1-s'_{fl})}{2.5}$$

i.e., $$(s'_{fl})^2 - s'_{fl} + 0.11875 = 0$$

i.e., $$s'_{fl} = \frac{1\pm\sqrt{1-4\times 0.11875}}{2} = \frac{1\pm 0.7245}{2}$$

i.e., $$s'_{fl} = \text{either } 0.86225 \text{ or } 0.13775$$

Neglecting higher value of s'_{fl}, we have

$$s'_{fl} = 0.13775$$

(v) $$T_{efl} = \frac{1,20,000}{(1-0.13775)\times 2\times 3.14\times \dfrac{1,500}{60}} = 886.44 \text{ N-m}$$

(vi) $$\text{Previously } s_{mT} = \frac{R_2}{X_2} = 0.23$$

Since $R'_2 = 2.5R_2$, the slip at which maximum torque occurs is given by

$$s'_{mT} = \frac{R'_2}{X_2} = 2.5\left(\frac{R_2}{X_2}\right) = 2.5s_{mT} = 2.5\times 0.23 = 0.575$$ ■

Example 8.23 The rated output of a three-phase, 50 Hz, four-pole induction motor is 12 kW at 1,440 rpm. Neglecting stator resistance and rotational losses, determine the starting current. The maximum torque is developed at 1,300 rpm.

Solution

$$\text{Synchronous speed } (N_s) = 1,500 \text{ rpm}$$

$$\text{Slip}(s_{fl}) = \frac{1,500-1,440}{1,500} = 0.04$$

$$P_{sh} = 12 \text{ kW} = 12{,}000 \text{ W}$$

$$\therefore \quad T_{efl} = \frac{P_m}{(1-s)\omega_s} = \frac{12{,}000}{(1-0.04)\times 2\times 3.14\times \frac{1{,}500}{60}} = 79.62 \text{ N-m}$$

The slip (s_{mT}) at which maximum torque occurs is $= \dfrac{1{,}500 - 1{,}300}{1{,}500} = 0.134$

Now,

$$\frac{T_{efl}}{T_{em}} = \frac{2s_{mT}s_{fl}}{s_{fl}^2 + s_{mT}^2}$$

i.e.,

$$\frac{79.62}{T_{e\max}} = \frac{2\times 0.134\times 0.04}{0.134^2 + 0.04^2}$$

i.e.,

$$T_{e\max} = 145.247 \text{ N-m}$$

Again,

$$\frac{T_{est}}{T_{e\max}} = \frac{2s_{mT}}{1+s_{mT}^2}$$

i.e.,

$$T_{est} = \frac{2\times 0.134}{1+0.134^2}\times 145.247 = 38.24 \text{ N-m}$$

■

Example 8.24 A three-phase induction motor drives a load running at a constant torque and at a speed of 1,450 rpm. If supply voltage drops to 90 per cent, find the motor speed and losses. Take linear torque-slip characteristic over the working range.

Solution

$$\text{Slip } (s_1) = \frac{1{,}500 - 1{,}450}{1{,}500} = 0.034$$

For linear working range

$$T_e = \frac{3V^2 s}{\omega_s R_2}$$

Since the load requires a constant torque

$$\frac{3V^2 s_1}{\omega_s R_2} = \frac{3(0.9V)^2 s_2}{\omega_s R_2}$$

i.e.,

$$s_1 = (0.9)^2 s_2$$

i.e.,

$$s_2 = \frac{s_1}{(0.9)^2} = \frac{0.034}{(0.9)^2} = 0.042$$

$$\therefore \quad \text{Motor speed} = 1{,}500\times(1-0.042) = 1{,}437 \text{ rpm}$$

For $s_1 = 0.034$, the rotor current is given by

$$I_{21} = \frac{s_1 V}{R_2} = \frac{0.034V}{R_2}$$

For $s_2 = 0.042$, the rotor current is given by

$$I_{22} = \frac{s_2(0.9V)}{R_2} = \frac{0.042 \times 0.9V}{R_2}$$

$$\therefore \quad \frac{I_{21}}{I_{22}} = \frac{0.034}{0.042 \times 0.9} = 0.899$$

Now, $$\frac{\text{Rotor Cu loss with reduced voltage}}{\text{Rotor Cu loss with normal voltage}} = \left(\frac{I_{22}}{I_{21}}\right)^2 = 1.237$$

$$\%\text{Reduction in speed} = \frac{1{,}450 - 1{,}437}{1{,}450} \times 100 = 0.896\%$$

$$\%\text{Reduction in ohmic loss} = \frac{1.237 - 1}{1.237} \times 100 = 19.16\%$$

■

Example 8.25 A three-phase, four-pole, 50 Hz, 200 V induction motor has a rotor resistance and a standstill reactance of 0.11 Ω and 0.99 Ω, respectively. Calculate the following:

(i) Total torque at 4 per cent slip.
(ii) The maximum torque.
(iii) The speed at maximum torque if the ratio of the rotor to stator turns is 0.70. Neglect stator impedance.

Solution

(i) $$\text{Stator voltage / phase} = \frac{200}{\sqrt{3}} = 115.47 \text{ V}$$

Standstill rotor emf per phase = 115.47 × 0.70 = 80.83 V

Full-load slip (s_{fl}) = 0.04.

$$\therefore \quad I_{2fl} = \frac{s_{fl}E_2}{\sqrt{R_2^2 + (s_{fl}X_2)^2}} = \frac{0.04 \times 80.83}{\sqrt{0.11^2 + (0.04 \times 0.99)^2}} = 27.66 \text{ A}$$

$$\text{The total Cu loss} = 3I_{2fl}^2 R_2 = 3 \times 27.66^2 \times 0.11 = 252.47 \text{ W}$$

$$P_g = \frac{252.47}{s_{fl}} = \frac{252.47}{0.04} = 6{,}311.75 \text{ W}$$

$$\text{Synchronous speed}\,(N_s) = \frac{120f}{P} = \frac{12\ldots}{4} = 1,\ldots$$

$$\therefore \quad T_e = \frac{P_g}{\omega_s} = \frac{P_g}{2\pi\frac{N_s}{60}} = \frac{6,311.75}{2\times 3.14\times\frac{1,500}{60}} = 40.20 \text{ N-m}$$

(ii)
$$s_{mT} = \frac{R_2}{X_2} = \frac{0.11}{0.99} = \frac{1}{9}$$

$$\therefore \quad I_{2m} = \frac{s_{mT}E_2}{\sqrt{2R_2^2}} = \frac{\frac{1}{9}\times 80.83}{\sqrt{2\times 0.11^2}} = 57.73\,\text{A}$$

$$\text{Total rotor Cu loss} = 3I_{2m}^2R_2 = 3\times 57.73^2\times 0.11 = 1,099.808 \text{ W}$$

$$P_g = \frac{1,099.808}{s_{mT}} = \frac{1,099.808}{\frac{1}{9}} = 9,898.27 \text{ W}$$

$$\therefore \quad T_e = \frac{P_g}{\omega_s} = \frac{9,898.27}{2\times 3.14\times\frac{1,500}{60}} = 63.04 \text{ N-m}$$

(iii)
$$\text{Speed at maximum torque} = (1-s_{mT})N_s = \left(1-\frac{1}{9}\right)\times 1,500 = 1,333.34 \text{ rpm}$$ ■

Example 8.26 The useful full-load torque of three-phase, four-pole, 50 Hz induction motor is 180 N-m. The rotor emf is observed to make 90 cycles per minute. Calculate (i) motor output, (ii) Cu loss in rotor, (iii) motor input and (iv) efficiency, if mechanical torque lost in windage and friction is 22 N-m and stator losses are 800 W.

Solution

$$\text{Synchronous speed}\,(N_s) = \frac{120f}{P} = \frac{120\times 50}{4} = 1,500 \text{ rpm}$$

$$\text{Frequency of rotor emf}\,(f_r) = \frac{90}{50} = 1.5 \text{ Hz}$$

$$\therefore \quad s_{fl} = \frac{f_r}{f} = \frac{1.5}{50} = 0.03$$

$$\text{Rotor speed}\,(N_r) = (1-s_{fl})\,N_s = (1-0.03)\times 1,500 = 1,455 \text{ rpm}$$

(i)
$$\text{Motor output} = T_{sh}\times\frac{2\pi\,N_r}{60} = 180\times\frac{2\times 3.14\times 1,455}{60} = 27,412.2 \text{ W}$$

...rque $(T_e) = 180 + 22 = 202$ N-m

(ii) Again,

$$T_e = \frac{P_g}{\omega_s} = \frac{P_g}{2\times 3.14\times \frac{N_s}{60}}$$

i.e.,

$$202 = \frac{P_g}{2\times 3.14\times \frac{1,500}{60}}$$

i.e.,

$$P_g = 202\times 2\times 3.14\times \frac{1,500}{60} = 31,714 \text{ W}$$

(iii) Motor input $(P_i) = 31,714 + 800 = 32,514$ W.

(iv) $$\text{Efficiency}(\eta) = \frac{27,412.2}{32,514} = 0.8430 \text{ p.u.} = 84.30\%$$ ■

Example 8.27 A 50 Hz, 440 V, three-phase, four-pole induction motor develops half the rated torque at 1,490 rpm with the applied voltage magnitude remaining at the rated value. What should be its frequency if the motor has to develop the same torque at 1,600 rpm? Neglect stator and rotor winding resistances, leakage reactances and iron losses.

Solution

Here, $f = 50$ Hz; $P = 4$; $N_r = 1{,}490$ rpm; New speed $(N_n) = 1{,}600$ rpm

$$\text{Synchronous speed, } N_s = \frac{120f}{P} = \frac{120\times 50}{4} = 1,500 \text{ rpm}$$

$$\text{Slip at a speed of } 1,490 \text{ rpm } (s) = \frac{N_s - N_r}{N_s} = \frac{1,500-1,490}{1,500}\times 100\%$$

$$= 0.667\%$$

The torque developed by an induction motor is given by

$$T \propto sV^2$$

Slip s for constant torque and constant applied voltage remains unchanged.

$$\text{New synchronous speed } (N_{sn}) = \frac{N_n}{1-s} = \frac{1,600}{1-0.00667} = 1,610.7 \text{ rpm}$$

$$\therefore \quad \text{New frequency } (f_n) = \frac{P\times N_{sn}}{120} = \frac{4\times 1,610.7}{120} = 53.7 \text{ Hz}$$ ■

Example 8.28 A 25 HP, 400 V, 50 Hz, four-pole, star-connected three-phase induction motor has the following impedances per phase in ohms referred to the stator side:

$$R_1 = 0.641,\ R_2 = 0.332,\ X_1 = 1.106,\ X_2 = 0.464 \text{ and } X_{mag} = 26.30$$

Rotational losses are assumed constant and are 1.1 kW and core losses are assumed to be of negligible order. If the slip is 2.2 per cent at rated voltage and frequency, find (i) speed, (ii) stator current, (iii) power factor, (iv) output and input-power and (v) efficiency of the motor.

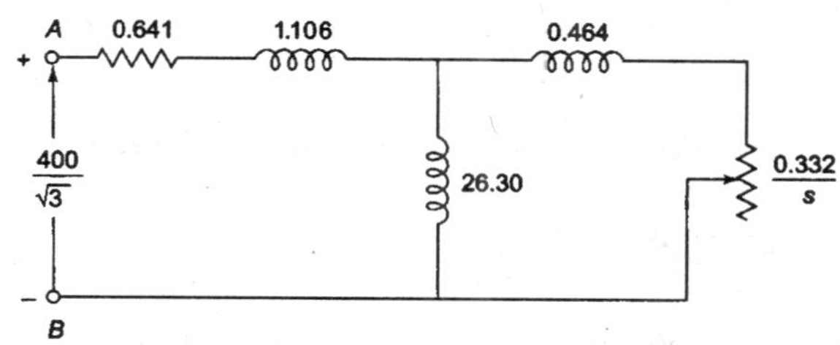

Figure E8.4

Solution

$$Z_{AB} = 0.641 + j1.106 + \frac{\left(\frac{0.332}{s} + j0.464\right)(j26.30)}{\frac{0.332}{s} + j0.464 + j26.30}$$

$$= 0.641 + j1.106 + \frac{(15.09 + j0.464)(j26.30)}{(15.09 + j26.764)}$$

$$= 0.641 + j1.106 + 11.05 + j6.69$$

Here, $R = 11.05\Omega$

$\therefore$ $Z_{AB} = 11.691 + j7.796 = 14.05\angle 33.7°$

(i) $N_r = N_s(1-s) = \frac{120f}{P}(1-s) = \frac{120 \times 50}{4} \times (1 - 0.022) = 1{,}467 \text{ rpm}$

(ii) Stator current $(I_1) = \frac{400\angle 0°}{\sqrt{3} \times 14.05\angle 33.7°} = 16.44\angle -33.7°$

(iii) Power factor $= \cos(-33.7°) = 0.832$ (lagging)

(iv) $P_g = 3I_1^2 R = 3 \times 16.44^2 \times 11.05 = 8{,}959.57$ W

$$P_{in} = (1-s)P_g = (1-0.022)\times 8{,}959.57 = 8{,}762.46 \text{ W}$$

$$P_{sh} = P_{in} - 1.1\times 10^3 = 8{,}762.46 - 1{,}100 = 7{,}662.46 \text{ W}$$

$\therefore$ Output power (P_o) = P_{sh} = 7,662.46 W

Total losses = stator ohmic loss + rotor ohmic loss + fixed losses

$$= 3I_1^2 R_1 + sP_g + 1{,}100$$

$$= 3\times 16.44^2 \times 0.641 + 0.022\times 8{,}959.57 + 1{,}100$$

$$= 1{,}816.85 \text{ W}$$

$\therefore$ Input power = (7,662.46 + 1,816.85) W = 9,479.31 W

(v) $$\text{Efficiency} = \frac{\text{Output power}}{\text{Input power}} = \frac{7{,}662.46}{9{,}479.31} = 0.8083 \text{ p.u.} = 80.83\%$$ ■

Example 8.29 An three-phase induction motor runs at a slip frequency of 2 Hz when supplied from a three-phase 400 V, 50 Hz supply. For the same developed torque, find the slip frequency at which motor will run when supplied from a three-phase 340 V, 40 Hz system. Slip at which the machine develops maximum torque using 50 Hz supply is 0.1. Neglect the stator impedance. Do not neglect rotor resistance in the expression of torque.

Solution

Let Rating of induction motor = 400 V, 50 Hz

Slip frequency = 2 Hz

Slip at maximum torque, $s_{mT} = 0.1$

When the slip frequency is 2 Hz, the slip at this frequency $s_1 = \frac{2}{50} = 0.04$

Let the slip at 340 V, 40 Hz be s_2

$$\text{Torque } T \propto \frac{sE_2^2}{R_2^2 + (sX_2)^2} \quad (1)$$

Here, the stator impedance is neglected and as such $V = E_2$

Also, the slip at maximum torque, $s_{mT} = \frac{R_2}{X_2} = 0.1$

$\therefore$ $$R_2 = 0.1\,X_2$$

Substituting this value in Equation (1), we have

$$T \propto \frac{sV^2}{(0.01+s^2)X_2^2}$$

Since the developed torque for both the cases is same,

$$\frac{s_1 V_1^2}{(0.01+s_1^2)X_2^2} = \frac{s_2 V_2^2}{(0.01+s_2^2)X_2^2}$$

i.e.,
$$\frac{0.04\times(400)^2}{0.01+(0.04)^2} = \frac{s_2\times(340)^2}{0.01+s_2^2}$$

i.e.,
$$\frac{6400}{0.0116} = \frac{s_2\times(340)^2}{0.01+s_2^2}$$

i.e.,
$$6{,}400(0.01 + s_2^2) = 0.0116\times s_2 \times 1{,}15{,}600$$

i.e.,
$$64 + 6{,}400\, s_2^2 = 1{,}340.96\, s_2$$

i.e.,
$$6{,}400 s_2^2 - 1{,}340.96 s_2 + 64 = 0$$

i.e.,
$$s_2 = \frac{1{,}340.96 \pm \sqrt{(1{,}340.96)^2 - 4\times 6{,}400\times 64}}{2\times 6{,}400}$$

$$= \frac{1{,}340.96 \pm 399.7}{2\times 6{,}400} = 0.136 \quad \text{or} \quad 0.0735$$

Since the slip cannot be high, let $s_2 = 0.0735$ at 40 Hz.
Hence, slip frequency = 0.0735 × 40 = 2.94 Hz ■

Example 8.30 An three-phase induction motor runs at a slip frequency of 2 Hz when supplied from a three-phase, 400 V, 50 Hz supply. For the same developed torque, find the slip frequency at which it will run when supplied from a three-phase 340 V, 40 Hz system. Slip at which the machine develops maximum torque using 50 Hz supply is 0.1. Neglect the stator impedance and assume linear torque-slip characteristics between zero torque and maximum torque in the working region and neglect rotor resistance in the expression of torque.

Solution The maximum torque in a three-phase induction motor ignoring stator impedance is given by

$$T_{em} = \frac{3}{\omega_s}\cdot\frac{V^2}{2X_2}$$

For 400 V supply, $s_{mT} = 0.1$ (1)

When
$$f = 50 \text{ Hz}, \quad \frac{R_2}{X_2} = s_{mT}$$

i.e.,
$$0.1 \propto \frac{R_2}{50}$$

For $$f = 40 \text{ Hz}, \ s'_{mT} \propto \frac{R_2}{40} \qquad (2)$$

From Equations (1) and (2), it can be written as

$$s'_{mT} = 0.1 \times \frac{50}{40} = 0.125$$

$$\text{Maximum torque, } T_{em} = \frac{3}{\omega_s} \cdot \frac{V^2}{2X_2}$$

i.e., $$T_{em} \propto \frac{V^2}{f^2} (\because \omega_s \propto f \text{ and } X_2 \propto f)$$

Since torque-slip characteristics is linear between zero torque and maximum torque, the developed torque (T_d) at a slip of 0.04

$$T_d \propto \frac{T_{em}}{s_{mT}} \times 0.04 \propto \left(\frac{400}{50}\right)^2 \times \frac{0.04}{0.1} \qquad (3)$$

For 340 V, 50 Hz

$$T_d \propto \frac{T_{em}}{s'_{mT}} \times 0.04 \times \left(\frac{340}{40}\right)^2 \times \frac{s}{0.125} \qquad (4)$$

For the same developed torque, from Equations (3) and (4) it can be written as

$$\left(\frac{400}{50}\right)^2 \times \frac{0.04}{0.1} = \left(\frac{340}{40}\right)^2 \times \frac{s}{0.125}, \text{ where } s \text{ is the new slip}$$

So, $$s = \left(\frac{400}{50}\right)^2 \times \frac{0.04}{0.1} \times 0.125 \times \left(\frac{40}{340}\right)^2 = 0.0443$$

$$\text{Slip frequency} = sf = 0.0443 \times 40 = 1.772 \text{ Hz}$$

■

Example 8.31 A 50 kVA, 400 V three-phase, 50 Hz squirrel-cage induction motor has full-load slip of 5 per cent. Its standstill impedance is 0.866 Ω/phase. It is started using a tapped autotransformer. If the maximum allowable supply current at the time of starting is 100 A, calculate the tap position and the ratio of starting torque to full-load torque.

Solution

s_{fl} = 5 per cent or 0.05; standstill impedance = 0.866 Ω per phase, I_{st} = 100 A

$$\text{Full-load current } (I_{fl}) = \frac{\text{Output in kVA} \times 1{,}000}{\sqrt{3} \times \text{line voltage}}$$

$$= \frac{50 \times 1{,}000}{\sqrt{3} \times 400} = 72.17 \text{ A}$$

$$\text{Short-circuit current, } I_{sc} = \frac{(400/\sqrt{3})}{0.866} = 266.7 \text{ A}$$

Tap position of the transformer

$$K = \sqrt{\frac{I_{st}}{I_{sc}}} = \sqrt{\frac{100}{266.7}} = 0.6123 \text{ p.u. or } 61.23\%$$

Now,

$$\frac{T_{est}}{T_{efl}} = K^2 \left(\frac{I_{sc}}{I_{fl}}\right)^2 s_{fl} = (0.6123)^2 \times \left(\frac{266.7}{72.2}\right)^2 \times 0.05 = 0.256$$ ■

Example 8.32 Estimate approximately the starting torque of a three-phase induction motor in terms of full-load torque when started by (i) star-delta starter and (ii) an autotransformer starter with 50 per cent tapping. The short-circuit current of motor is five times the full-load current and the full-load slip is 5 per cent.

Solution

Given $I_{sc} = 5I_{fl}$; $s_{fl} = 5$ per cent or 0.05

(i) Starting torque with star-delta starter

$$T_{est} = \frac{1}{3}\left(\frac{I_{sc}}{I_f}\right)^2 s_f \times T_{efl} = \frac{1}{3} \times 5^2 \times 0.05 \times T_{efl} = 0.4167 T_{efl}$$

i.e., 41.67 per cent of full-load torque.

(ii) Starting torque with an autotransformer starter with 50 per cent tapping ($K = 0.5$)

$$T_{est} = K^2 \left(\frac{I_{sc}}{I_f}\right)^2 s_f T_{efl} = 0.5^2 \times 5^2 \times 0.05 \times T_{efl} = 0.3125 T_{efl} = 31.25 \text{ per cent of full-load torque}$$ ■

Example 8.33 The starting torque and maximum torque of a three-phase induction motor are 140 per cent and 220 per cent of the full-load torque. Neglecting stator resistance and assuming constant rotor resistance, calculate the following:

(i) Slip at maximum torque. (ii) Full-load slip.

(iii) Rotor current at starting in terms of the full-load current.

Solution

$$T_{est} = 1.4 T_{efl}; \; T_{em} = 2.2 T_{efl}$$

Since

$$\frac{T_{est}}{T_{em}} = \frac{2 s_{mT}}{s_{mT}^2 + 1}$$

i.e.,

$$\frac{1.4}{2.2} = \frac{2 s_{mT}}{s_{mT}^2 + 1}$$

i.e., $$1.4\,(s_{mT}^2+1)=4.4\,s_{mT}$$

i.e., $$s_{mT}^2-3.143s_{mT}+1=0$$

i.e., $$s_{mT}=\frac{3.143\pm\sqrt{(3.143)^2-4}}{2}=\frac{3.143\pm2.424}{2}=0.359 \text{ or } 2.78$$

i.e., $s_{mT}=0.359$ (neglecting higher value).

(ii) Since $$\frac{T_{efl}}{T_{est}}=\frac{2s_{fl}s_{mT}}{s_{fl}^2+s_{mT}^2}$$

i.e., $$\frac{1}{2.2}=\frac{2s_{fl}\times0.359}{s_{fl}^2+(0.359)^2}=\frac{0.718s_{fl}}{s_{fl}^2+0.1289}$$

i.e., $$s_{fl}^2-1.579s_{fl}+0.1289=0$$

i.e., $$s_{fl}=\frac{1.579\pm\sqrt{(1.579)^2-4\times0.1289}}{2}=\frac{1.579\pm1.406}{2}=0.0865 \text{ or } 1.49$$

Hence, $s_{fl}=0.0865$ (neglecting higher value)

(iii) Since $$T_{efl}=\frac{1}{\omega_s}I_{2fl}^2\frac{R_2}{s_{fl}}$$

Thus, $$T_{est}=\frac{1}{\omega_s}I_{2st}^2R_2$$

$$\frac{T_{est}}{T_{efl}}=\left[\frac{I_{2st}}{I_{2fl}}\right]^2 s_{fl}$$

i.e., $$\left[\frac{I_{2st}}{I_{2fl}}\right]^2=\frac{1.4}{0.0865}=16.185$$

i.e., $$I_{2st}=4.023\,I_{2fl}$$ ■

Example 8.34 A three-phase squirrel-cage induction motor has a slip of 4 per cent at full load. Its starting current is five times the full-load current. The stator impedance and magnetizing current may be neglected. The rotor resistance is assumed to be constant.

(i) Calculate the maximum torque and the slip at which it would occur.

(ii) Calculate the starting torque.

Express torque in per unit (p.u.) of the full-load torque.

Solution

It is given that:

Slip at full load (s_{fl}) = 4 per cent = 0.04

$$I_{st}=5I_{fl}$$

(i) Since
$$\frac{T_{efl}}{T_{est}} = \frac{2s_{fl}s_{mT}}{s_{fl}^2 + s_{mT}^2} \tag{1}$$

where
$$s_{mT} = \frac{R_2}{X_2}$$

Stator impedance and no-load current are neglected. We have

$$I_{fl} = I_2' = \frac{s_{fl}E_1}{\sqrt{R_2^2 + s_{fl}^2 X_2^2}} \tag{2}$$

But
$$I_{st} = \frac{E_1}{\sqrt{R_2^2 + X_2^2}} \tag{3}$$

Dividing Equation (3) by Equation (2), we have

$$\frac{I_{st}}{I_{fl}} = \frac{1}{s_{fl}}\left[\frac{R_2^2 + s_{fl}^2 X_2^2}{R_2^2 + X_2^2}\right]^{1/2} = \frac{1}{s_{fl}}\left[\frac{s_{mT}^2 + s_{fl}^2}{s_{mT}^2 + 1}\right]^{1/2}$$

i.e.
$$5 = \frac{1}{0.04}\left(\frac{s_{mT}^2 + 0.04^2}{s_{mT}^2 + 1}\right)^{\frac{1}{2}}$$

i.e.,
$$25 = \frac{1}{(0.04)^2} \times \left(\frac{s_{mT}^2 + 0.04^2}{s_{mT}^2 + 1}\right)$$

i.e.,
$$25 \times (0.04)^2 \times (s_{mT}^2 + 1) = s_{mT}^2 + 0.04^2$$

i.e.,
$$0.04(s_{mT}^2 + 1) = s_{mT}^2 + 0.0016$$

i.e.,
$$0.04s_{mT}^2 + 0.04 = s_{mT}^2 + 0.0016$$

i.e.,
$$s_{mT}^2(1 - 0.04) = 0.04 - 0.0016$$

i.e.,
$$s_{mT} = 0.2$$

Hence, maximum torque occurs when $s = s_{mT} = 20\%$

But
$$T_{e\max} = \frac{1}{2}k_t$$

and
$$T_{efl} = k_t\left(\frac{0.04 \times 0.2}{0.04^2 + 0.2^2}\right) = \frac{k_t}{5.2}$$

Hence,
$$k_t = 5.2\ T_{efl}$$

$$T_{e\,\max} = \frac{1}{2} \times 5.2\ T_{fl} = 2.6 T_{efl}$$

(ii)

$$T_{est} = k_t \left(\frac{s_{mt}}{1 + s_{mT}^2} \right) \quad [\text{putting} s_f = 1 \text{ in Equation } (1)]$$

$$= k_t \left(\frac{0.2}{1 + 0.2^2} \right) = 5.2 T_{efl} \times \frac{0.2}{1.04} = T_{efl}.$$

Therefore, T_{em} is 2.6 p.u. and it occurs at a slip of 0.2 and T_{est} is 1 p.u. ■

Example 8.35 A three-phase, four-pole, 60 kW, 50 Hz induction motor connected to rated supply voltage and running without load consumes 3 kW. When prevented from rotating it draws rated current at 30 per cent rated supply and takes a power input of 4 kW. Assuming that under rated load conditions, the stator and rotor Cu losses are equal and that the mechanical losses are 30 per cent of the no-load losses. Determine (a) slip at rated load and (b) starting torque of the motor with rated applied voltage.

Solution

$$\text{Mechanical losses} = \frac{30}{100} \times 3{,}000 = 900 \text{ W}$$

Stator core loss = power input at no load − mechanical loss − stator I^2R loss at no load

Neglecting stator I^2R loss at no load,

$$\text{Stator core loss} = 3{,}000 - 900 = 2{,}100 \text{ W}$$

Power input during blocked-rotor test = stator I^2R loss + rotor I^2R loss = 4,000 W

$$\text{Stator } I^2R \text{ loss} = \text{rotor } I^2 \text{ loss} = \frac{4{,}000}{2} = 2{,}000 \text{ W}$$

(a) At rated load, the air gap power

P_g = output power + mechanical loss + rotor $I^2 R$ loss

$\therefore$ $P_g = 60{,}000 + 900 + 2{,}000 = 62{,}900$ W

But rotor I^2R loss = 2,000 W = sP_g

$$\therefore \quad \text{Slip at rated load, } (s) = \frac{2{,}000}{62{,}900} = 0.0318$$ ■

(b) At rated voltage, power input to motor during blocked-rotor test,

$$= 4 \times \left(\frac{100}{30} \right)^2 = 44.444 \text{ kW}$$

Air gap power,

$$P_g = \text{power input} - \text{stator core loss} - \text{stator } I^2R \text{ loss}$$
$$= 44{,}444 - 2{,}100 - 2{,}000 = 40{,}344 \text{ W}$$

Synchronous speed,

$$\omega_s = \frac{4\pi f}{P} = \frac{4\pi \times 50}{4} = 50\pi \text{ rad/s}$$

$$\therefore \qquad \text{Starting torque } (T_{st}) = \frac{P_g}{\omega_s} = \frac{40,344}{50\pi} = 256.96 \text{ N-m}$$ ■

Example 8.36 A 3.8 kW, 200 V, 50 Hz, four-pole, three-phase star-connected induction motor gave the following results:

No-load test: 200 V (line), 5 A (line), 360 W

Short-circuit test: 100 V (line), 27 A (line), 1,750 W

The rotor copper loss at standstill is half the total copper loss. Draw the circle diagram. Determine the full-load value of the current, pf and slip maximum torque in terms of full-load torque.

Solution

From no-load test data,

$$\cos\theta_0 = \frac{360}{\sqrt{3}\times 200\times 5} = 0.208$$

i.e.,

$$\theta_0 = 78^\circ$$

From short-circuit test data,

$$\cos\theta_{sc} = \frac{1,750}{\sqrt{3}\times 100\times 27} = 0.374$$

i.e.,

$$\theta_{sc} = 68^\circ$$

Short-circuit current with normal voltage $I_{SN} = 27\times\dfrac{200}{100} = 54$ A

Power taken becomes at this normal voltage $= 1,750\times\left(\dfrac{200}{100}\right)^2 = 7,000$ W

The circle diagram is shown in Figure E8.5. In Figure E8.5, 1 cm = 4 A. OY axis represents the voltage vector which is perpendicular to OX axis.

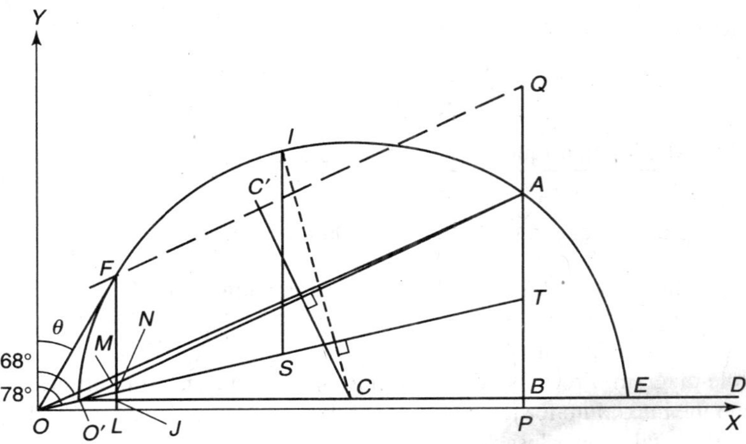

Figure E8.5

In Figure E8.5,

(i) The vector OO' represents the current at no load; that is, $OO' = 5\text{ A} = \frac{5}{4} = 1.25$ cm which is drawn at 78° angle with respect to the OY axis.

(ii) The short-circuit current is represented by the vector OA; that is, $OA = I_{SN} = 54$ A. It measures $\frac{54}{4} = 13.5$ cm. It is drawn at 68° angle with respect to OY axis.

(iii) $O'A$ is added which represents the output line. $O'D$ is drawn which is parallel to OX axis.

(iv) $O'A$ is bisected by the line CC'. CC' intersects $O'D$ at the point C. With C as a centre and CO' as a radius, the semicircle $O'AE$ is drawn. From A, the vertical line AP is drawn which measures the input power at the blocked rotor condition. The length of AP is 5.0 cm. The power scale becomes 1 cm $= \frac{7{,}000}{5} = 1{,}400$ W. BP represents the fixed losses and AB is the total copper loss during the blocked rotor test.

(v) Motor output at full load = 3.8 kW = 3,800 W $= \frac{3{,}800}{1{,}400} = 2.714$ cm. Now BA is extended up to Q where $AQ = 2.714$ cm. From Q, the line QF is drawn which is parallel to $O'A$. The point F lies on the semicircle, which is at the full-load condition.

(vi) Since the rotor copper loss is half the total loss, AB is bisected at the point T. $O¢T$ is added which represents the torque line.

(vii) FL is drawn perpendicular to OX where $FL = 3.5$ cm, which intersects the output line and the torque line at the points M and N, respectively.

(viii) From C a perpendicular CI is drawn on the torque line, where I lies on the semicircle. From I, a vertical line IS is drawn, which intersects the torque line at the point S. The length of $IS = 5.3$ cm.

(ix) The vector OF is drawn which represents the line current at full-load condition. The length of $OF = 4$ cm.

Line current at full load = 4 × 4 = 16 A

By measurement, $\theta = 30°$

∴ power factor at full load = cos 30° = 0.866

$$\text{Slip} = \frac{\text{rotor copper loss}}{\text{rotor input}} = \frac{MN}{FN} = \frac{0.2}{3.05} = 0.0656 \quad \text{or,} \quad 6.56 \text{ per cent}$$

$$\frac{\text{Maximum torque}}{\text{Full-load torque}} = \frac{IS}{FN} = \frac{5.3}{3.05} = 1.737$$

Therefore, maximum torque = 173.7 per cent of full-load torque.

Example 8.37 A three-phase, 15 kW, 400 V, six-pole induction motor gave the following results:

No-load test: 400 V, 11.2 A, pf = 0.19

Short-circuit test: 100 V, 25.5 A, pf = 0.39

Draw the circle diagram. From this diagram, find the line current, slip, efficiency, power factor at full load. Also, find the maximum torque in terms of full-load torque. Assume that the rotor copper loss at standstill is half the total copper loss.

Solution

No-load power factor = 0.19, $\theta_0 = \cos^{-1}(0.19) = 79°$

Short-circuit power factor = 0.39, $\theta_{sc} = \cos^{-1}(0.39) = 67°$

Short-circuit current (*ISN*) due to normal voltage $= 25.5 \times \dfrac{400}{100} = 102$ A

Short-circuit power input with this current $= \sqrt{3} \times 400 \times 102 \times 0.39 = 27{,}560.4$W

Current scale is taken as 1 cm = 5 A. Figure E8.6 shows the circle diagram. The vertical axis *OY* represents the voltage vector which is perpendicular to *OX* axis.

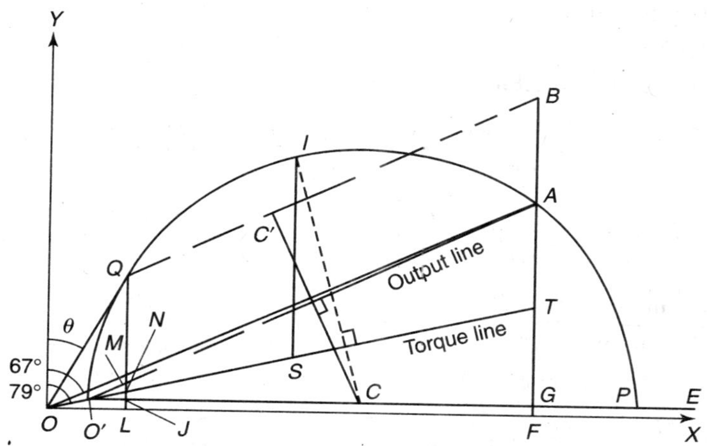

Figure E8.6

In Figure E8.6,

(i) The vector OO' represents the current at no load; that is, $OO' = 11.2 \text{ A} = \dfrac{11.2}{5} = 2.24$ cm which is drawn at 79° angle with respect to the vertical line.

(ii) The short-circuit current is represented by the vector *OA*; that is, $OA = I_{SN} = 102$ A. It measures $\dfrac{102}{5} = 20.4$ cm. It is drawn at 67° angle with respect to the vertical line.

(iii) $O'A$ is added which represents the output line. $O'D$ is drawn which is parallel to *OX* axis.

(iv) $O'A$ is bisected by the line CC'. CC' intersects $O'D$ at the point *C*. With *C* as a centre and CO' as a radius, the semicircle $O'AP$ is drawn. From *A*, the vertical line *AF* is drawn which measures the input power at the blocked rotor condition. The length of *AF* = 7.9 cm. The power scale becomes 1 cm $= \dfrac{27{,}560.4}{7.2} = 3{,}488.66$ W. *GF* represents the fixed losses and *AG* is the total copper loss during the blocked rotor test.

(v) Motor output at full load = 15 kW = 15,000 W $= \dfrac{15{,}000}{3{,}488.66} = 4.3$ cm. Now *GA* is extended up to *B*, where *AB* = 4.3 cm. From *B* the line *BQ* is drawn which is parallel to $O'A$. The point *Q* lies on the semicircle which is at the full-load condition.

(vi) Since the rotor copper loss is half the total loss, AG is bisected at the point T. $O'T$ is added which represents the torque line.

(vii) QL is drawn perpendicular to OX, where $QL = 6.2$ cm, which intersects the output line and the torque line at the points M and N, respectively.

(viii) From C a perpendicular CI is drawn on the torque line where I lies on the semicircle. From I a vertical line IS is drawn, which intersects the torque line at the point S. The length of $IS = 7.9$ cm.

(ix) The vector OQ is drawn, which represents the line current at full-load condition. The length of $OQ = 7.4$ cm.

Line current at full load = 7.4 × 5 = 37 A

By measurement, $\theta = 30°$.

Therefore, power factor at full load = cos 30° = 0.866

$$\text{Slip} = \frac{\text{rotor copper loss}}{\text{rotor input}} = \frac{MN}{QN} = \frac{0.45}{5.35} = 0.0740 \quad \text{or,} \quad 7.40 \text{ per cent}$$

$$\text{Efficiency} = \frac{QM}{QL} = \frac{4.9}{6.2} = 0.79 \text{ pu} = 79\%$$

$$\frac{\text{Maximum torque}}{\text{Full-load torque}} = \frac{IS}{QN} = \frac{7.9}{5.4} = 1.43$$

∴ Maximum torque = 143 per cent of full-load torque

Example 8.38 A three-phase, 5.8 kW, 400 V, four-pole, 50 Hz, slip-ring induction motor gave the following results:

No-load readings: 400 V, 6.2 A, pf = 0.088

Short-circuit test: 100 V, 12.4 A, 725 W

Draw the circle diagram. If the ratio of primary to secondary turns is 2.64, stator resistance per phase is 0.68 Ω and rotor resistance per phase is 0.186 Ω, calculate the following:

(a) full-load current,
(b) full-load slip,
(c) full-load power factor,
(d) full-load efficiency,
(e) maximum torque in terms of full-load torque and
(f) maximum power.

Solution

$$\cos\theta_0 = 0.088 \quad \text{i.e., } \theta_0 = 84.95°$$

$$\cos\theta_{sc} = \frac{725}{\sqrt{3}\times 100\times 12.44} = 0.336 \quad \text{i.e., } \phi_{sc} = 70.33°$$

$$\text{Short-circuit current with normal voltage, } I_{SN} = 12.44\times\frac{400}{100} = 49.76 \text{ A}$$

$$\text{Total input with normal voltage} = 725\times\left(\frac{400}{100}\right)^2 = 12{,}000 \text{ W}$$

Let us take the current scale 1 cm = 2.5 A. The circle diagram is shown in Figure E8.7. Axis *OY* is the voltage axis which is perpendicular to *OX*.

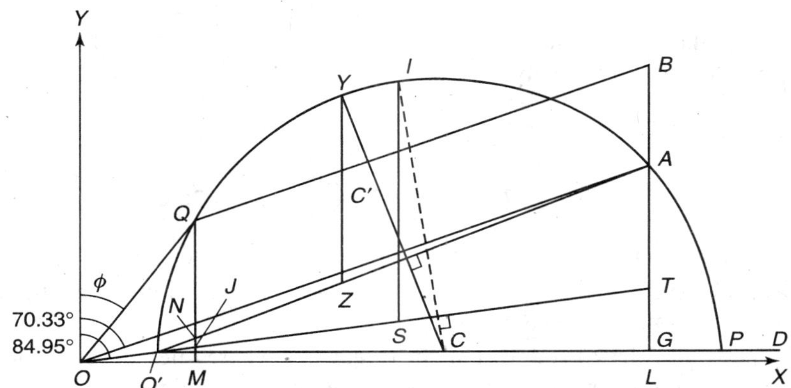

Figure E8.7

In Figure E8.7,

(i) The vector OO' represents the current at no load; that is, $OO' = 6.2 \text{ A} = \frac{6.2}{2.5} = 2.48$ cm which is drawn at 84.95° angle with respect to the vertical line.

(ii) The short-circuit current is represented by the vector OA; that is, $OA = I_{SN} = 49.76$ A. It measures $\frac{49.76}{2.5} = 19.9$ cm. It is drawn at 70.33° angle with respect to the vertical line.

(iii) $O'A$ is added which represents the output line. $O'D$ is drawn which is parallel to OX axis.

(iv) $O'A$ is bisected by the line CC'. CC' intersects $O'D$ at the point C. With C as a centre and CO' as a radius, the semicircle $O'AP$ is drawn. From A, the vertical line AL is drawn which measures the input power at the blocked rotor condition. The length of AG is 6.7 cm. The power scale becomes $1 \text{ cm} = \frac{12,000}{6.7} = 1791.05$ W.

(v) Motor output at full load = 5.8 kW = 5,800 W $= \frac{5,800}{1,791.05} = 3.24$ cm. Now LA is extended up to B where $AB = 3.24$ cm. From B, the line BQ is drawn which is parallel to $O'A$. The point Q lies on the semicircle which is at the full-load condition.

(vi) For drawing the torque line, the ratio of primary to secondary turns is 2.64, stator resistance per phase (R_1) is 0.68 Ω and rotor resistance per phase (R_2) is 0.186 Ω.

$$\frac{\text{Rotor copper loss}}{\text{Stator copper loss}} = 2.64^2 \times \frac{0.186}{0.68} = 1.9$$

$$\therefore \quad \frac{\text{Rotor copper loss}}{\text{Total copper loss}} = \frac{1.9}{2.9}$$

$$\therefore \quad \text{Rotor copper loss} = \frac{1.9}{2.9} \times \text{total copper loss}$$

Let T be the point which divides AG such that $\frac{AT}{TG} = \frac{1.9}{1}$, where AT and TG represent rotor and stator copper losses, respectively.

$$\therefore \quad \frac{AT}{AG} = \frac{1.9}{2.9}$$

i.e.,

$$AT = \frac{1.9}{2.9} \times AG = \frac{1.9}{2.9} \times 6.5 = 4.26 \text{ cm}$$

$O'T$ is drawn, which represents the torque line.

(vii) QM is drawn perpendicular to OX, where $QM = 4.8$ cm, which intersects the output line and torque line at the points N and J, respectively.

(viii) From C, a perpendicular CI is drawn on the torque line where I lies on the semicircle. A vertical line IS is drawn from I, which intersects the torque line at the point S. The length of $IS = 8.3$ cm.

(ix) CC' is extended up to the point Y, and from Y, a vertical line is drawn which intersects the output line at the point Z. The length of $YZ = 6.4$ cm.

(x) The vector OQ is drawn which represents the line current at full-load condition. The length of $OQ = 6.1$ cm.

(a) Line current at full load = 6.1 × 2.5 = 15.25 A

(b) Full-load slip $= \frac{\text{rotor copper loss}}{\text{rotor input}} = \frac{NJ}{QJ} = \frac{0.4}{4.4} = 0.09$ or, 9 per cent

By measurement, $\theta = 37°$

(c) Therefore, pf at full load = cos 37° = 0.798

(d) Efficiency of the motor at full load $= \frac{\text{motor output at full load}}{\text{motor intput at full load}}$

$$= \frac{QN}{QM} = \frac{4}{4.8} = 0.8334 \text{ pu} = 83.34\%$$

(e) $\frac{\text{Maximum torque}}{\text{Full-load torque}} = \frac{IS}{QJ} = \frac{8.3}{4.4} = 1.88$

(f) Maximum power = YZ = 6.4 × 1,791.05 W = 11.46 kW

SIGNIFICANT POINTS

At any slip s, rotor impedance is $= \sqrt{R_2^2 + s^2 X_2^2}\ \Omega$

Per phase rotor current at standstill is given by $= \frac{E_2}{\sqrt{R_2^2 + X_2^2}}$ A

Per phase rotor current at any slip (s) is given by $(I_2) = \frac{sE_2}{\sqrt{R_2^2 + (sX_2)^2}}$ A

The per phase power input to the rotor is given by

$$P_g = E_2 I_2 \cos\theta_2 \quad \text{i.e.,} \quad P_g = I_2^2 \frac{R_2}{s}$$

Internal mechanical power (P_m) developed in rotor is $I_2^2 R_2 \left(\frac{1-s}{s}\right) = (1-s)P_g$

$$P_m = (1-s) \text{ (power input to rotor)}$$

Internal mechanical torque (T_e) developed in rotor,

$$T_e = \frac{P_m}{\omega_r} = \frac{(1-s)P_g}{(1-s)\omega_s} = \frac{P_g}{\omega_s}$$

ω_s is synchronous speed in mechanical radians per second.

$$T_e = \frac{P_g}{\omega_s} = \frac{\frac{I_2^2 R_2}{s}}{2\pi\ N_s} = \frac{1}{2\pi\ N_s}\frac{I_2^2 R_2}{s}$$

The shaft power (P_{sh}) can be obtained from P_g as follows:

$$P_{sh} = P_m - \text{mechanical losses (friction and windage losses)}$$

$\therefore$ $$P_{sh} = P_g - \text{Rotor ohmic losses} - \text{friction and windage loses.}$$

Therefore, output or shaft torque is given by

$$T_{sh} = \frac{P_{sh}}{\text{Rotor speed}} = \frac{P_{sh}}{(1-s)\omega_s}$$

The torque at any slip s

$$T = \frac{ksR_2E_2^2}{R_2^2 + s^2X_2^2}$$

Starting torque $$T_{est} = \frac{ksR_2E_2^2}{R_2^2 + X_2^2}$$

The expression for maximum torque is $T_{em} = \dfrac{kE_2^2}{2X_2}$

The ratio of full-load torque to maximum torque is

$$\frac{T_{efl}}{T_{e\max}} = \frac{2}{\dfrac{s_{fl}}{s_{mT}} + \dfrac{s_{mT}}{s_{fl}}}$$

The ratio of starting torque to maximum torque is

$$\frac{T_{est}}{T_{e\max}} = \frac{2s_{mT}}{1+s_{mT}^2}$$

SHORT QUESTIONS AND ANSWERS

Q.1 What do you mean by AC three-phase induction motor?

Ans. If the motor has three-phase windings and runs on AC three-phase supply on induction principle, it is said to be an AC three-phase induction motor.

Q.2 On what principle does the induction motor work?

Ans. If a short-circuited conductor is placed within a rotating magnetic field, a current is set up in the conductor due to electromagnetic induction and it tends to move.

Q.3 What are the types of induction motors?

Ans. There are two types of induction motors:

(a) Squirrel-cage induction motor
  (i) Single-cage
  (ii) Double cage

(b) Wound-rotor induction motor

Q.4 What are the main parts of AC three-phase induction motor?

Ans. There are two main parts:

(a) *Stator:* A static part made of steel stampings carrying a three-phase winding in its slot.

(b) *Rotor:* A rotating part consisting of shaft, rotor stampings, rotor windings, slip rings and so on.

Q.5 What is the difference between a squirrel-cage-type rotor and phase-wound rotor?

Ans. In a squirrel-cage-type rotor, there are no windings, but a large number of copper bars are arranged on end rings at each end, revetted and soldered, forming the whole system. Therefore, the short-circuited body appears to be the cage of a squirrel.

Phase-wound rotor has a second winding like stator accommodated in the slots of rotor core, the terminals of which are brought out and connected to slip rings mounted on the same shaft.

Q.6 What are the types of slots used in squirrel-cage rotor?

Ans. There are the two types of slots used in a squirrel-cage rotor:

(i) Semiclosed-type slots.

(ii) Closed-type slots.

Q.7 Why are semiclosed-type slots preferred in wound rotors?

Ans. The semiclosed-type slots are preferred in wound rotors because it reduces the magnetic reluctance.

Q.8 Why are rotor slots made skewed by a small angle to the shaft axis?

Ans. The rotor slots are made skewed by a small angle to the shaft axis because

(i) it prevents magnetic locking of rotor teeth with stator teeth,

(ii) it reduces the humming noise when running,

(iii) it gives more uniform torque when running and

(iv) it avoids crawling of an induction motor.

Q.9 State the advantages and the disadvantages of using a large number of slots in an induction motor.

Ans. *Advantages:* Variable air gap reluctance is reduced thus resulting in reduction of pulsating exciting current, irregular torque, tooth losses and noise.

Disadvantages: Tooth thickness is reduced which increases the cost of manufacture.

Q.10 State the advantages and the disadvantages of decreasing the air gap length of a three-phase induction motor?

Ans. *Advantages:* It reduces the requirement of magnetizing current to set up the air gap flux and also improves power factor.

Disadvantages: It increases the possibility of mechanical difficulties, motor noise and tooth losses.

Q.11 How does an induction motor work?

Ans. A rotating magnetic field revolving at a synchronous speed is produced when a three-phase supply is applied to the stator of an induction motor. Since this rotating magnetic field cuts the short-circuited copper bars of the squirrel-cage rotor winding, a heavy current flows in the rotor bars and creates another field. The effect of the rotor field is to oppose the main rotating field. As the rotor is free to move, it tends to follow the rotating field in the same direction. As a result, the rotor starts and picks up speed until it approaches the synchronous speed. But the rotor actually never reaches the synchronous speed and always runs just below the synchronous speed.

Q.12 What will be the rotor frequency at the time of starting and at running condition?

Ans. During starting, the rotor frequency is the same as the supply frequency. During running it is less than that of the supply frequency.

Q.13 Why is it not possible to run an induction motor on synchronous speed?

Ans. If the rotor achieves the synchronous speed, the relative speed between the rotating magnetic field and the rotor will be zero and there will be no induced emf on rotor conductors. Therefore, torque developed by rotor will be zero.

Q.14 What are the factors on which the torque of an induction motor depends?

Ans. The torque of an induction motor depends on the flux produced by the stator, the rotor current per phase and the rotor power factor.

Q.15 What are the types of torques taken into consideration for an electric motor?

Ans. Starting torque and running torque are the two types of torques taken into consideration for an electric motor.

Q.16 What is the relation between torque and voltage?

Ans. Torque is directly proportional to the square of the voltage.

Q.17 Is shunt-torque characteristic applicable to induction motor?

Ans. The speed of an induction motor slows down slightly when torque is increased. Therefore, the drooping nature of speed-torque curve, known as shunt characteristic, is applicable to the induction motor.

Q.18 What is the condition for maximum starting torque for an induction motor?

Ans. The condition for maximum starting torque is that the rotor resistance equals its standstill reactance.

Q.19 What do you mean by breakdown torque of an induction motor?

Ans. The maximum torque developed by an induction motor at which the motor becomes unstable with increase in slip is known as breakdown torque, which is nearly twice the full-load torque.

Q.20 What are the conditions for getting good starting and running torques in an induction motor?
Ans. Usually, the rotor winding resistance per phase is about 1 per cent or 2 per cent of the rotor winding reactance per phase at standstill. To obtain good starting torque, high rotor resistance with minimum standstill reactance and reduced supply voltage is preferable. To obtain good running torque, a low rotor resistance and constant supply voltage is preferable.

Q.21 Where is the slip-ring (wound-rotor) induction motor used?
Ans. A slip-ring induction motor is used when high starting torque is required.

Q.22 What is the advantage of double-cage induction motor over single-cage induction motor?
Ans. The advantage of double-cage induction motor is to obtain high starting torque without sacrificing electrical efficiency under normal running conditions.

Q.23 What is the effect of variation of supply frequency in an induction motor?
Ans. If supply frequency increases, it increases power factor, but decreases the torque. If supply frequency decreases, it decreases power factor, but increases the torque. But variation of supply frequency has no effect on the per cent slip.

Q.24 What is the effect of variation of supply voltage in an induction motor?
Ans. If supply voltage increases, it decreases power factor, secondary Cu loss and slip but increases torque, magnetic density, magnetizing current and iron loss. If supply voltage decreases, it increases power factor, secondary Cu loss and slip but decreases torque, magnetic density, magnetizing current and iron loss.

Q.25 What factors determine the direction of rotation of an induction motor?
Ans. The phase sequence of the supply mains and the order in which the supply lines are connected to the stator winding determine the direction of rotation of an induction motor.

Q.26 How can you reverse the direction of rotation of AC three-phase induction motors?
Ans. For reversing the direction of rotation of AC three-phase induction motors, it is required to transpose or change over any two leads of supply lines.

Q.27 State the most common cause of blow outs in an induction motor and explain it?
Ans. Single-phasing is the most common cause of blow outs in an induction motor, which means the opening of one wire either by blown out fuse or any disconnection during the running of three-phase induction motor when the other two wires act as single-phase operation.

Q.28 What are the different methods of speed control of the three-phase induction motors?
Ans. The different methods of speed control of three-phase induction motors are as follows:

(i) By varying the frequency.
(ii) By changing the number of poles.
(iii) By inserting the resistance in the rotor circuit.
(iv) By cascade connection.
(v) By injecting emfs in the rotor circuit.

Q.29 What will be the effect on speed if the number of poles increases?
Ans. Speed will decrease because speed is inversely proportional to the number of poles.

Q.30 What will be the effect on speed if frequency increases?
Ans. Speed will increase because speed is directly proportional to frequency.

Q.31 What is the disadvantage of rheostatic control?
Ans. The disadvantage of rheostatic control is that it causes more Cu loss.

Q.32 What are the different types of starters used for AC induction motors?

Ans. The different types of starters are given below:

(i) Direct-on-line starter,
(ii) Rotor-resistance starter,
(iii) Autotransformer starter and
(iv) Direct-on-line starter.

Q.33 Why are three-phase induction motors widely used?

Ans. The three-phase induction motors are widely used because they are less expensive, simple, reliable, robust and require less maintenance.

Q.34 What are the standard power factors of an induction motor at different loads?

Ans. The standard power factors of an induction motor are: 0.85 at full load, 0.8 at 75 per cent of full load, 0.7 at half of full load, 0.5 at 25 per cent of full load and 0.1 at no load.

Q.35 What are the applications of three-phase squirrel-cage and slip-ring induction motors?

Ans. Following are the applications of three-phase squirrel-cage induction motors:

(i) Blowers,
(ii) Most of machine tools,
(iii) Lathe,
(iv) Printing press and
(v) Crushers and others.

Following are the applications of three-phase slip-ring induction motors:

(i) Conveyers,
(ii) Hoists,
(iii) Compressors,
(iv) Rolling mills and
(v) Mines and others.

Q.36 What will happen if single-phase phasing occurs of a working induction motor?

Ans. During working of a three-phase induction motor, if single phase occurs, the motor will continue to run as a single-phase machine. There will be a peculiar noise, which results from the large current increase in the two running lines and in all the three windings till it is isolated by the overload device or the single-phase preventor. Single-phase working of the motor reduces the capacity of the motor and also places extra duty on the windings if this condition persists for a long time, and ultimately this will burn out the coil.

Q.37 What will happen if single-phase phasing occurs of a non-working induction motor?

Ans. If the motor is non-working and single phase occurs, the motor will fail to start. It results in a humming sound if it is connected to three-phase supply lines and current drawn under this condition will be 87 per cent of the starting current under normal conditions. If the supply voltage is applied for a long time, the motor will burn out.

Q.38 What phase is likely to burn out in single phasing of (i) delta-connected and (ii) star-connected motors?

Ans. The phase connected across the live or operative lines is likely to burn out carrying nearly 300 per cent of its normal current in the first case, whereas in the second case, the two phases falling in series combination connected across the live or operative lines are likely to burn out carrying nearly 250 per cent of its normal full-load current.

Q.39 What will happen when a three-phase motor continues to run even after the fuse on one phase is blown?

Ans. Current will be induced in the short-circuited rotor winding continuously. The stator winding will be heavily overloaded for the same loading condition and it gets unduly heated up causing a probable burn out of the phase winding.

Q.40 How can you protect the motor against single phasing?

Ans. To protect the motor against single phasing, it is preferred to incorporate a combined overload and single-phasing relay or a phase-failure relay in the control gear and starter.

SUPPLEMENTARY PROBLEMS

1. A three-phase induction motor having a star-connected rotor has an induced emf of 80 V between slip rings at standstill on open circuit. The rotor has a resistance and reactance per phase of 1 Ω and 4 Ω, respectively. Determine the current/phase and power factor when (a) slip rings are short-circuited, (b) slip rings are connected to a star-connected rheostat of 3 Ω/phase.

   [*Ans.* (a) 11.2 A, 0.243 (lagging) (b) 81.6 A, 0.707 (lagging)]

2. A 3.3 kV, 24-pole, 50 Hz, three-phase star-connected induction motor has a slip-ring rotor resistance of 0.016 Ω and a standstill reactance of 0.265 Ω per phase. Find (a) the speed at maximum torque and (b) ratio of full load to maximum torque if full-load torque is obtained at 247 rpm.

   [*Ans.* (a) 235 rpm (b) 0.382]

3. A 18.4 kW, six-pole, 50 Hz, three-phase slip-ring induction motor runs at 960 rpm on full load with a rotor current per phase of 35 A. If the Cu loss in the short-circuit gear is 250 W and mechanical losses are 1,000 W, calculate the resistance per phase of the three-phase winding.

   [*Ans.* 0.152 Ω/phase]

4. A three-phase, 50 Hz induction motor designed for voltage V_1 is switched on to 40 Hz supply of voltage V_2. Find the following:
   (i) The ratio of starting torques, starting currents and maximum torque when $V_2 = 1.5\ V_1$.
   (ii) The ratio of V_1 to V_2 to give equal values of starting torques both at 40 Hz and 50 Hz. Ignore stator impedance and rotor resistance.

   $\left[\textit{Ans.}\ \text{(i)}\ \frac{5}{6}, 0.23, 0.28\ \text{(ii)} \left(\frac{50}{40}\right)^{\frac{3}{2}}\right]$

5. A 3 kW, 400 V/200 V, delta/star, 50 Hz, three-phase, six-pole induction motor is found to draw a line current of 25 A at a power factor of 0.4, when a blocked-rotor test is conducted at the rated voltage. Calculate the stator and rotor winding resistances in ohm per phase, if the torque developed by the motor under the above conditions is 25 N-m.

   [*Ans.* 6.88 Ω/phase, 0.35 Ω/phase]

6. A 230 V, 20 HP, 60 Hz, six-pole, three-phase induction motor driving a constant torque load at rated frequency, rated voltage and rated horse power, has a speed of 1,175 rpm and an efficiency of 92.1 per cent. Determine the new operating speed if a system disturbance causes 10 per cent drop in voltage and 6 per cent drop in frequency. Assume that the friction, windage and stray power losses remain constant.

   [*Ans.* 1,101 rpm]

7. A three-phase, four-pole, 50 Hz induction motor at standstill has 120 V across its star-connected rotor terminal. The rotor resistance and reactance per phase are 0.2 Ω and 10 Ω, respectively. Calculate the speed when the rotor is drawing a current of 16 A at a particular load. Also, calculate the speed at which the torque is maximum and the corresponding value of rotor input.

[*Ans.* 1,428.75 rpm, 1,200 rpm, 73657.28 W]

8. The full-load efficiency and power factor of a 16 HP, 440 V, three-phase induction motor are 85 per cent and 0.8 lag, respectively. The blocked rotor line current is 45 A at 220 V. Calculate the ratio of starting to full-load current if the motor is provided with a star-delta starter. Neglect magnetizing current.

[*Ans.* 2.256]

9. A 440 V, three-phase, six-pole, 50 Hz, delta-connected induction motor has the following equivalent circuit parameters at normal frequency

$$R_1 = 0.2\,\Omega,\ R_2' = 0.18\ \Omega, X_1 = X_2' = 0.58\ \Omega$$

The machine is subjected to an occasional fall of 40 per cent simultaneously in both voltage and frequency. At what total mechanical load is it safe to drive so that the machine just does not stall under the conditions? When operating at normal voltage and frequency, calculate the slip when delivering the same torque.

[*Ans.* 822.3 W, 0.066]

10. A three-phase induction motor has the following particulars:

1,500 kW, 2,300 V, star-connected, four-pole, 50 Hz, $R_1 = 0.02\ \Omega$, $R'_2 = 0.12\ \Omega$, $R_C = 451.2\ \Omega$, $X_m = 50\ \Omega$, $X_1 = X'_2 = 0.32\ \Omega$, all parameters are referred to the stator. Calculate the following:

(i) Stator current, (ii) Mechanical power output and (iii) Efficiency at a slip of 0.03746.

[*Ans.* (i) 408.6 A, (ii) 1,576.488 kW, (iii) 91.03%]

11. The ratio of maximum torque to full-load torque in a three-phase squirrel-cage induction motor is 2.2:1. Determine the ratio of actual starting torque to full-load torque for the following cases:

(i) Direct starting, (ii) Star-delta starting and (iii) Autotransformer starting with a tapping of 70 per cent. The rotor resistance and standstill reactance per phase are 0.5 Ω and 5 Ω, respectively.

[*Ans.* (i) 0.4356, (ii) 0.1452, (iii) 0.2134]

MULTIPLE-CHOICE QUESTIONS AND ANSWERS

1. If R_2 be the standstill rotor resistance of an induction motor running at a slip s, the mechanical load equivalent to an electrical resistance R_L in a stationary equivalent circuit is equal to

(a) $\frac{R_2}{s}$ (b) $\frac{R_2(1-s)}{s}$

(c) $\frac{R_2}{s^2}$ (d) $\frac{R_2(1-s)}{s^2}$

2. The starting torque of a three-phase induction motor can be increased by increasing

(a) the rotor resistance (b) the rotor reactance

(c) the stator resistance (d) the stator reactance

3. In a three-phase induction motor, torque is related with supply voltage as
   (a) $T \propto V^{\frac{1}{2}}$ (b) $T \propto \frac{1}{V}$
   (c) $T \propto V$ (d) $T \propto V^2$
4. In a polyphase squirrel-cage induction motor, increased starting torque can be obtained by
   (a) increasing the frequency of operation (b) using deep-bar rotors
   (c) increasing the number of poles (d) using a double-cage rotor
5. The ratio among rotor input, rotor output and rotor Cu losses is
   (a) $1{:}1 - s{:}s$ (b) $1{:}s{:}1 - s$
   (c) $s{:}1{:}1 - s$ (d) $1 - s{:}s{:}1$
6. Skewing of the rotor slots in an induction motor is to
   (a) reduce the magnetic hum of the motor (b) increase the breadth factor
   (c) reduce the locking tendency of the rotor (d) increase the distribution factor
7. A four-pole, 50 Hz, three-phase induction motor has blocked rotor reactance per phase, which is four times the rotor resistance per phase. The speed at which maximum torque develops is
   (a) 1,125 rpm (b) 1,500 rpm
   (c) 1,050 rpm (d) 1,210 rpm
8. If an induction motor has a rotor resistance of 0.02 Ω and a standstill reactance of 0.1 Ω, the value of total resistance of the starter for the rotor circuit for maximum torque to be exerted at starting is
   (a) 0.08 Ω (b) 0.02 Ω
   (c) 6 Ω (d) 4 Ω
9. In an induction motor, if the ratio of the rotor output to rotor input is 0.96, the per cent slip is
   (a) 0.04% (b) 4%
   (c) 5.16% (d) 5.87%
10. In an induction motor, if the rotor resistance is equal to the standstill reactance, the maximum torque is
    (a) equal to starting torque (b) more than starting torque
    (c) less then starting torque (d) none of these
11. In an induction motor the ratio of starting current to full-load current is 12.65 and the ratio of starting torque to full-load torque is 1.6. The per cent slope at full load is
    (a) 0.01% (b) 12.64%
    (c) 1% (d) 2%
12. The ratio of starting torque to maximum torque is 80 per cent. The speed at which maximum torque is 12 per cent will be
    (a) $\frac{5}{3}$ (b) $\frac{3}{5}$
    (c) $\frac{1}{3}$ (d) 2
13. The short-circuit current at normal voltage is five times the full-load current. The load slip is 4 per cent. The starting torque when star-delta switch is used is equal to

(a) 4 times full-load torque
(b) 0.33 times full-load torque
(c) 0.43 times full-load torque
(d) 0.21 times full-load torque

14. The power factor of a squirrel-cage large induction motor is
(a) low at light loads only
(b) low at heavy loads only
(c) low at both light and heavy loads
(d) low at rated load only

15. While breaking three-phase induction motor, its stator terminals were switched over to a DC source from AC supply. The method of breaking is
(a) regeneration breaking
(b) counter current breaking
(c) eddy current breaking
(d) none of these

16. A three-phase induction motor runs at a supersynchronous speed. For self-excitation the machine
(a) draws real power from the mains
(b) draws reactive power from the mains
(c) feeds reactive power to the mains
(d) generates emf at the expense of residual magnetism

17. For a high starting torque, the most suited induction motor is
(a) squirrel-cage type
(b) slip-ring type
(c) deep-bar squirrel-cage type
(d) double-cage induction motor

18. A star-delta starter is equivalent to an autotransformer starter with a tapping of
(a) 86.6%
(b) 57.7%
(c) 57%
(d) 58%

19. In a slip-ring induction motion, if the rotor resistance is increased then
(a) starting torque and efficiency increase
(b) starting torque and efficiency decrease
(c) starting torque increases, but efficiency decreases
(d) starting torque decreases, but efficiency increases

20. In a three-phase induction motor, the slip for maximum torque in terms of rotor resistance R_2 is
(a) independent of R_2
(b) inversely proportional to R_2
(c) proportional to R_2^2
(d) directly proportional to R_2

21. The starting current of an induction motor is five times the full-load current, while its full-load slip is 4 per cent. The ratio of starting torque to full-load torque is
(a) 2.00
(b) 0.50
(c) 1.00
(d) 0.25

22. The stator and the rotor of a three-phase induction motor behave like a/an
(a) ordinary two-winding transformer
(b) variable voltage, constant frequency transformer
(c) constant voltage variable frequency transformer
(d) variable voltage, variable frequency transformer, V/f remaining constant.

23. For satisfactory performance of a three-phase, 480 V, 60 Hz induction motor, the supply voltage at 50 Hz should be equal to
(a) 480 V
(b) 420 V
(c) 400 V
(d) 350 V

24. In star-delta starting of a squirrel-cage large induction motor compared with DOL needs to be expanded, starting current and torque are reduced by factors of

(a) $\frac{1}{\sqrt{3}}, \frac{1}{3}$ (b) $\frac{1}{3}, \frac{1}{3}$
(c) $\frac{1}{\sqrt{3}}, \frac{1}{\sqrt{3}}$ (d) $\frac{1}{3}, \frac{1}{\sqrt{3}}$

25. In a three-phase induction motor, the mechanical power developed, in terms of the air gap power (P_g), is
(a) $(s-1)P_g$ (b) $\frac{P_g}{1-s}$
(c) $\frac{1}{s}$ (d) $(1-s)P_g$

26. A starting torque of 100 N-m is developed by an autotransformer starter with a tapping of 40 per cent. If the tapping of the autotransformer is at 80 per cent, the starting torque will be
(a) 200 N-m (b) 50 N-m
(c) 400 N-m (d) 25 N-m

27. A three-phase induction motor when started picks up speed, but runs stably at about half the normal speed. This is because of
(a) unbalance in the supply voltages (b) non-sinusoidal nature of the supply voltage
(c) stator circuit asymmetry (d) rotor circuit asymmetry

28. The phenomena of crawling in a three-phase cage rotor induction motor may be due to
(a) stator circuit third-time harmonic (b) stator circuit third-space harmonic
(c) rotor circuit third-time harmonic (d) rotor circuit third-space harmonic

29. A three-phase induction motor is driving a constant torque load of 1 p.u. at 5 per cent slip. It has a maximum torque of 2 p.u. at 10 per cent slip. Torque speed variation in stable zone is assumed to be linear for the stable operation of the motor. The minimum per supply voltage is approximately
(a) 0.25 (b) 0.50
(c) 0.707 (d) 0.87

30. Slip test is performed to determine
(a) slip (b) direct axis reactance and quadrature axis
(c) positive sequence reactance and negative sequence reactance (d) subtransient reactance

31. The power input to a 415 V, 50 Hz, six-pole, three-*phase* induction motor running at 975 rpm is 40 kW. The stator losses are 1 kW and friction and windage losses 2 kW. The efficiency of the motor is
(a) 92.5% (b) 91.67%
(c) 90% (d) 88%

32. Rotor of an induction motor is analogue of a
(a) mechanical clutch (b) gear trains
(c) dashpot (d) spring

33. The efficiency of an induction motor *A* is 89 per cent and that of *B* is 80 per cent. The size of motor *A* is
    (a) equal to *B*
    (b) smaller than *B*
    (c) bigger than
    (d) none of these
34. The inner cage of a double-cage induction motor has
    (a) high inductance and resistance
    (b) high inductance and low resistance
    (c) low inductance and resistance
    (d) low inductance and high resistance
35. The equivalent circuit of a double-cage rotor induction motor is equivalent to two rotor circuits connected in
    (a) parallel
    (b) series
    (c) series parallel
    (d) either (a) or (b)

Answers

1 (b)	2 (a)	3 (d)	4 (d)	5 (a)	6 (a), (c)
7 (a)	8 (a)	9 (b)	10 (a)	11 (c)	12 (b)
13 (b)	14 (a)	15 (c)	16 (a)	17 (b)	18 (c)
19 (b)	20 (d)	21 (c)	22 (c)	23 (c)	24 (b)
25 (d)	26 (c)	27 (b)	28 (b)	29 (a)	30 (c)
31 (b)	32 (a)	33 (c)	34 (b)	35 (a)	

9 Single-phase Motors and Special Machines

Single-phase supply is mainly used for lighting purposes in offices, shops, schools, houses. Besides lighting purposes, it has tremendous uses in our day-to-day life. Domestic applications include mixers, automatic washing machines, electric iron, hair dryers and others. The power rating of these types of machines is very small. These motors are also called fractional kilowatt motors because some of them have rating less than a kilowatt. The aim of this chapter is to introduce the construction, working principle and applications of these motors.

9.1 CLASSIFICATION OF SINGLE-PHASE INDUCTION MOTORS

Fractional kilowatt motors have the following classifications:

(i) *Single-phase induction motors:*
 (a) Resistance split type, (b) capacitor split type and (c) shaded pole type

(ii) *Single-phase synchronous motors:*
 (a) Reluctance motors and (b) hysteresis motors

(iii) *Single-phase induction motors:*
 (a) AC series motors and (b) universal motors

(iv) *Special purpose motors:*
 (a) Stepper motors and (b) servomotors

9.2 PRODUCTION OF ROTATING FIELD

Figure 9.1 shows two windings *A* and *B*. These windings are displaced in such a way that the magnetic fields produced in space are 90° apart. Let the magnetic field produced by the two windings be equal in magnitude and 90° apart in time. These are represented by:

$$\phi_A = \phi_m \sin \omega t$$
$$\phi_B = \phi_m \sin(\omega t - 90°)$$

These two magnetic fields produce a resultant magnetic field having a constant magnitude ϕ_m, which is rotating, as shown in Figure 9.2.

If the windings *A* and *B* are displaced in space by 90° and produce fields such that they are not equal in magnitude and not 90° apart in time, the resultant magnetic field is a rotating magnetic field having variable magnitude throughout the revolution. The non-uniform magnetic field produces a non-uniform torque causing the operation to be noisy. A motor producing non-uniform rotating field has less starting torque compared with that of a motor of the same rating producing a uniform rotating field.

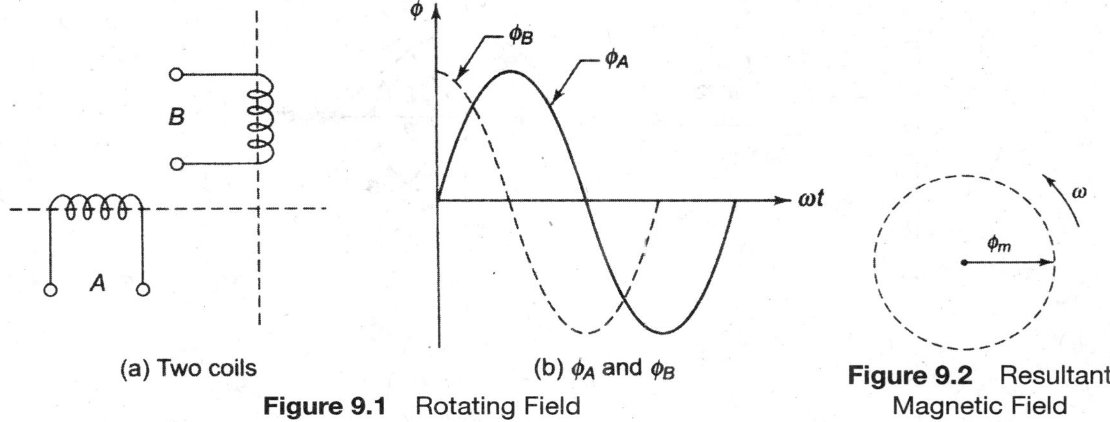

(a) Two coils (b) ϕ_A and ϕ_B

Figure 9.1 Rotating Field

Figure 9.2 Resultant Magnetic Field

9.3 WORKING PRINCIPLE OF SINGLE-PHASE INDUCTION MOTOR

For operation as a motor, there must be two fluxes that will interact with each other and produce the torque. The field winding in DC motor produces the main flux. The DC supply is given to the armature and armature flux is produced. The electromagnetic torque is produced due to the interaction of the above two fluxes. However, in single-phase induction motors, no DC supply is applied to the rotor and single-phase AC supply is applied to the stator only. Therefore, stator produces alternating flux which is known as main flux. This flux links with the rotor conductors and emf is induced in it due to transformer action. Since the rotor is a closed circuit, this emf drives the current through it resulting in rotor flux. Since the second flux is produced according to induction principle due to induced emf, the motor is termed as *induction motor*. The main differences of DC and AC motors are:

- No separate DC supply is required to rotor in single-phase induction motor, whereas separate DC supply is required to armature in a DC motor.
- DC motors are self-starting, whereas single-phase induction motors are not self-starting.

With the help of double revolving field theory and cross-magnetic field theory, we can explain why single-phase induction motors are not self-starting.

9.4 DOUBLE REVOLVING FIELD THEORY

The pulsating magnetic fields produced by the single-phase induction motors can be resolved into two rotating magnetic fields with the help of double revolving magnetic field theory. All the magnetic fields have equal magnitude and they rotate in opposite directions. The resultant torque is the sum of the two torques.

The equation of alternating magnetic field having axis fixed in space is represented by

$$\alpha(\delta) = \alpha_{max} \sin \omega t \cos \delta \tag{9.1}$$

where α_{max} is the amplitude of the sinusoidally distributed air gap flux.

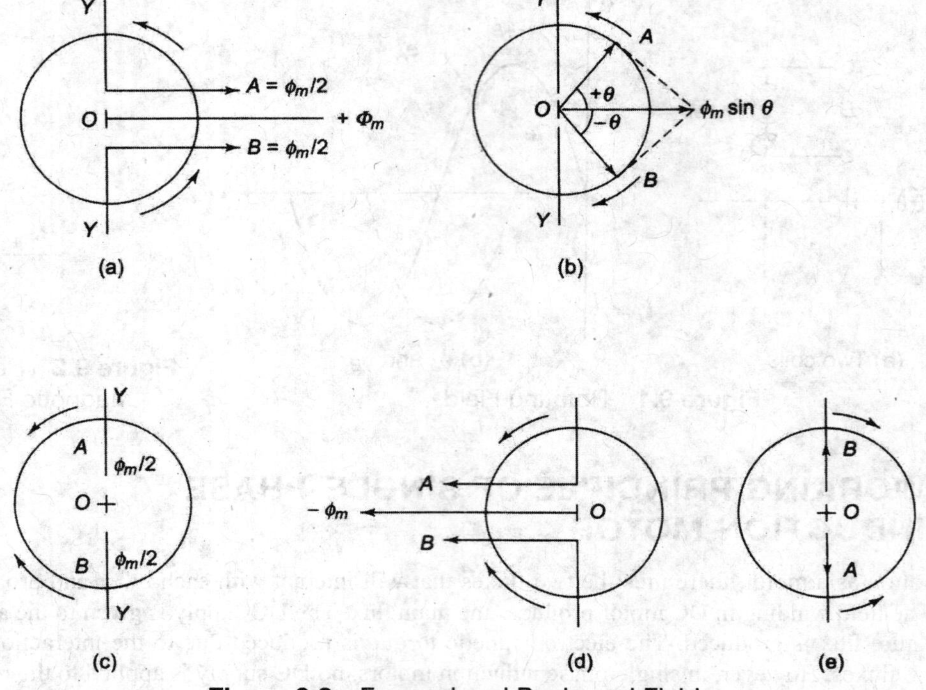

Figure 9.3 Forward and Backward Fields

Equation (9.1) can be written as:

$$\alpha(\delta)=\frac{1}{2}\alpha_{max}\sin(\omega t+\delta)+\frac{1}{2}\alpha_{max}\sin(\omega t-\delta) \tag{9.2}$$

Equation (9.2) has two terms. The first term and second term represent the revolving field moving in the negative and positive direction, respectively, having amplitude equal to 1/2 α_{max}. The fields in the positive and negative α-direction are called forward and backward rotating fields both rotating at a synchronous speed ω_s ($2\pi n_s$).

Therefore, the pulsating magnetic field is resolved into forward and backward rotating magnetic fields having equal magnitude and rotating at synchronous speed. Therefore, the rotor is stationary.

The above facts are represented by Figure 9.3.

In Figure 9.3(a), both fluxes A and B are in phase with the reference. Therefore, the net flux is maximum. In Figure 9.3(b), both fluxes are at an angle θ with the reference. In Figure 9.3(c), both fluxes are at an angle of 180º. Hence, the resultant flux is zero. In Figure 9.3(d), both fluxes are at an angle of 0º but in the direction opposite to Figure 9.3(a). Hence, resultant flux is once again maximum but in the opposite direction of Figure 9.3(a). In Figure 9.3(e), both fluxes are at an angle of 180º but the position of fluxes A and B are opposite to Figure 9.3(a). Figure 9.4 represents the resultant flux.

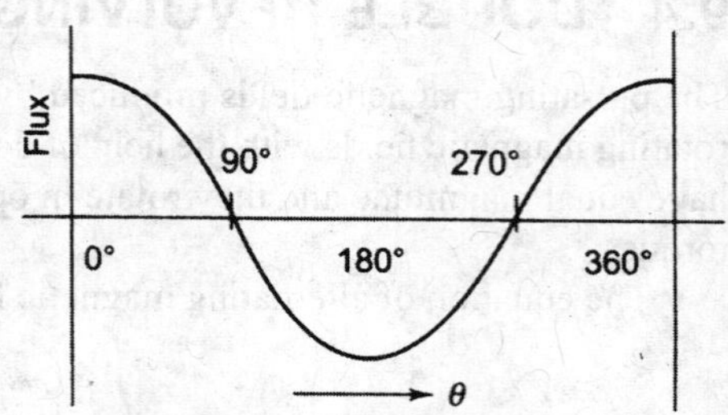

Figure 9.4 Resultant Flux

Therefore, we can conclude that the single-phase induction motor with single stator winding cannot produce self-starting torque. Although the rotor is stationary, let us give it an initial rotation in either direction. The torque due to rotating magnetic field in the initial direction will be greater than the torque due to rotating magnetic field in the other direction. Therefore, a net positive torque is produced in the direction of initial rotation, and hence the motor will continue to run in the same direction of initial rotation.

9.5 ROTOR SLIP WITH RESPECT TO TWO ROTATING FIELDS

After starting the induction motor by means of auxiliary means, a resultant torque is produced and it will keep running in the same direction of one of the fields. The direction in which the rotor is started initially is known as forward field by definition.

The slip (s_f) of the rotor with respect to the forward rotating field is given by

$$s_f = \frac{n_s - n}{n_s} = 1 - \frac{n}{n_s} \tag{9.3}$$

where n_s and n are the synchronous speed and rotor speed, respectively.

The slip (s_b) of the rotor with respect to the backward rotating field is given by

$$s_b = \frac{n_s - (-n)}{n_s} = \frac{n_s + n}{n_s} = 1 + \frac{n}{n_s} \tag{9.4}$$

From Equation (9.4), we have:

$$s_b = 1 + \frac{n}{n_s} = 2 - \left(1 - \frac{n}{n_s}\right) = 2 - s \tag{9.5}$$

In single-phase induction motor, the speed of the rotor is less than the synchronous speed. Equations (9.3) and (9.5) represent the motor operation and the braking region. Figures 9.5(a) and 9.5(b) shows the rotor equivalent circuits corresponding to the forward and backward rotating fluxes, respectively.

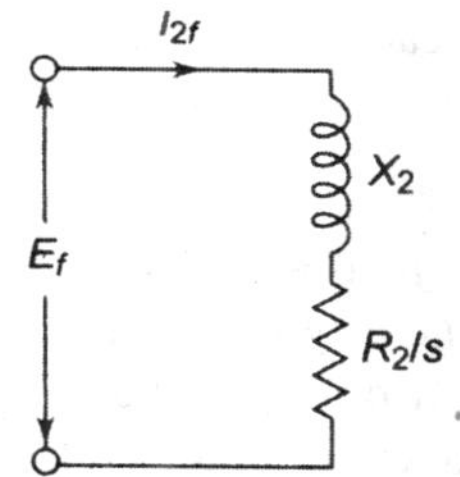

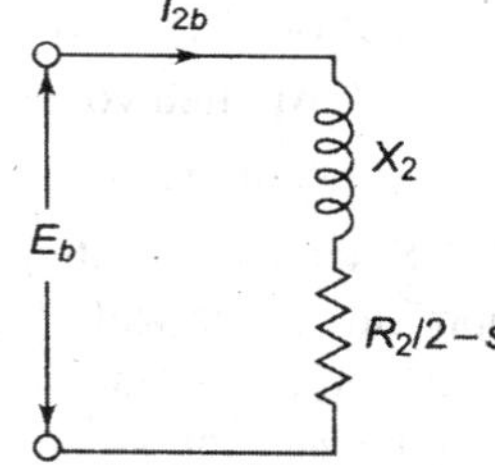

Figure 9.5 Rotor Equivalent Circuit

Figure 9.6 shows the torque-speed characteristics of single-phase induction motor.

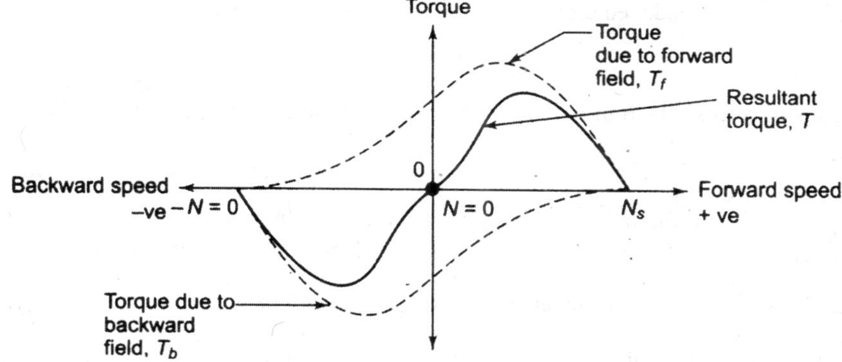

Figure 9.6 Torque-speed Characteristics of Single-phase Induction Motor

9.6 EQUIVALENT CIRCUIT OF SINGLE-PHASE, SINGLE-WINDING INDUCTION MOTOR

Auxiliary windings are connected to induction motors to make them self-starting. Therefore, single-phase induction motors, in fact, are two-phase motors. Figure 9.7 shows the equivalent circuit of a single-phase induction motor having stationary rotor. In Figure 9.7, core loss branch has not been shown because core loss branch is assumed to be lumped with the mechanical and stray losses as a part of rotational losses of the motor.

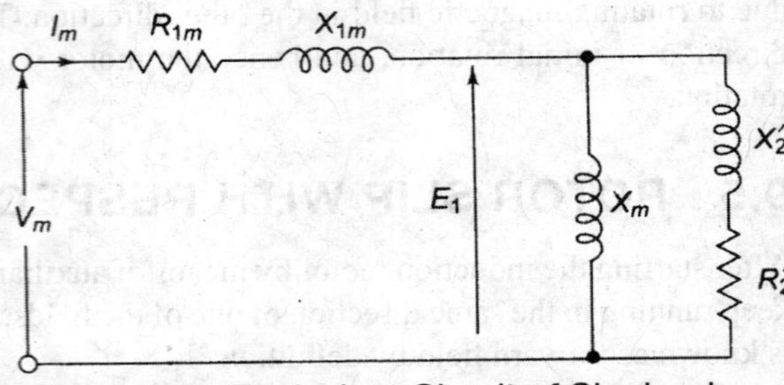

Figure 9.7 Equivalent Circuit of Single-phase Induction Motor at Standstill

In Figure 9.7:

R_{1m}: Resistance of main rotor winding

X_{1m}: Leakage reactance of the main stator winding

X_M: Magnetising reactance

R'_2: Standstill rotor resistance to main stator winding

X'_2: Standstill rotor leakage reactance to main stator winding

V: Applied voltage

I_m: Main winding current

Since the pulsating magnetic field is resolved into backward and forward fields having equal and opposite fluxes with the motor, the magnitude of each rotating flux is one-half of the alternating flux. Therefore, it is assumed that the two rotating fluxes are acting on two separate rotors. Hence, we can assume that the single-phase induction motor consists of two rotors having a common stator winding and two imaginary rotors. At standstill, the impedance of each rotor referred to stator winding is $\left(\frac{R_2'}{2}+j\frac{X_2'}{2}\right)$. Figure 9.8 shows the equivalent circuit of single-phase induction motor at standstill with the effect of separating forward and backward rotating fluxes.

The emfs induced in the main stator winding by the forward and backward fluxes are E_{mf} and E_{mb}, respectively. The resultant-induced emf (E_m) in the stator winding is given by

$$E_m = E_{mf} + E_{mb} \tag{9.6}$$

Since the circuits of rotors due to the forward and backward fields are identical at standstill, we have

$$E_{mf} = E_{mb} \tag{9.7}$$

If the motor is started with the help of auxiliary windings and the auxiliary windings are switched out after reaching certain speed, the effective rotor resistance depends on the slip. Let s be the slip due to the forward field. The effective rotor resistances due to forward and backward fields are $\frac{R_2'}{2s}$ and $\frac{R_2'}{2(2-s)}$, respectively. The equivalent circuit is shown in Figure 9.9, taking into account forward and backward fields.

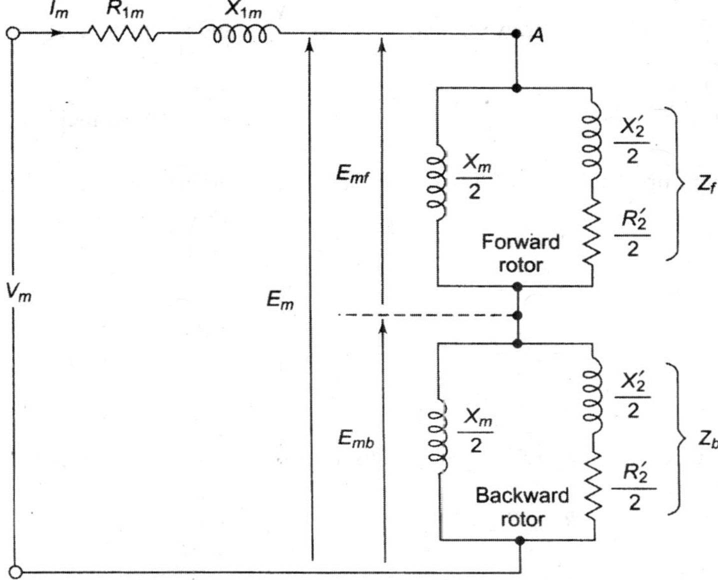

Figure 9.8 Effect of Separating Forward and Backward Rotating Fluxes at Standstill

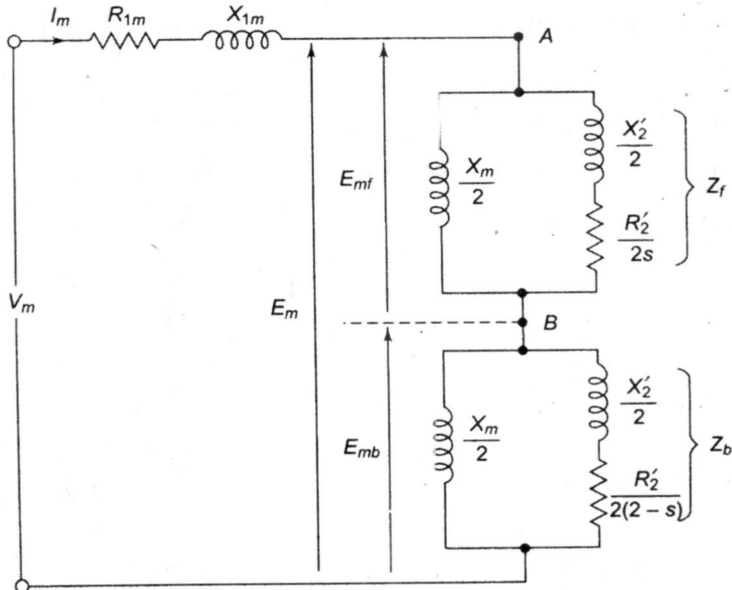

Figure 9.9 Equivalent Circuit Due to Forward and Backward Fields During Running

The rotor impedance due to the forward field is $\left(\frac{R_2'}{2s}+j\frac{X_2'}{2}\right)$ and it is in parallel with $j\frac{X_M}{2}$. Again, the rotor impedance due to the backward field is $\left[\frac{R_2'}{2(2-s)}+j\frac{X_2'}{2}\right]$ and it is in parallel with $j\frac{X_M}{2}$.

Now, equivalent impedance (Z_f) of imaginary rotor due to forward rotating field is

$$Z_f = R_f + jX_f = \left(\frac{R_2'}{2s}+j\frac{X_2'}{2}\right) \| \left(j\frac{X_M}{2}\right)$$

$$= \frac{\left(\frac{R_2'}{2s}+j\frac{X_2'}{2}\right)\times\left(j\frac{X_M}{2}\right)}{\left(\frac{R_2'}{2s}+j\frac{X_2'}{2}\right)+\left(j\frac{X_M}{2}\right)} \tag{9.8}$$

and the equivalent impedance (Z_b) of imaginary rotor due to backward rotating field is

$$Z_b = R_b + jX_b = \left[\frac{R_2'}{2(2-s)}+j\frac{X_2'}{2}\right] \| \left(j\frac{X_M}{2}\right)$$

$$= \frac{\left[\frac{R_2'}{2(2-s)}+j\frac{X_2'}{2}\right]\times\left(j\frac{X_M}{2}\right)}{\left[\frac{R_2'}{2(2-s)}+j\frac{X_2'}{2}\right]+\left(j\frac{X_M}{2}\right)} \tag{9.9}$$

Figure 9.10 shows the simplified circuit.

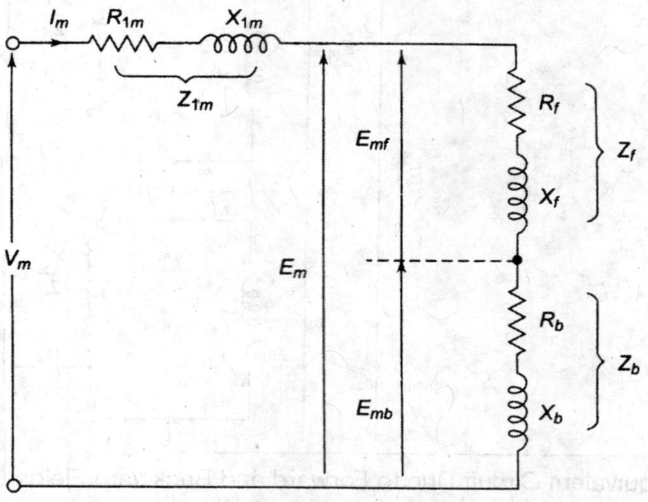

Figure 9.10 Simplified Circuit

From Figure 9.10, the current in stator winding is given by

$$I_m = \frac{V_m}{Z_{lm} + Z_f + Z_b} \tag{9.10}$$

9.7 POWER DEVELOPED AND LOSSES OF SINGLE-PHASE, SINGLE-WINDING INDUCTION MOTOR

To calculate the losses as well as the power of single-phase, single-winding induction motor, let us consider Figure 9.10. The torque due to backward field opposes the torque due to forward field. The total air gap power of single-phase induction motor is given by

$$P_g = P_{gf} - P_{gb} \tag{9.11}$$

where P_{gf} and P_{gb} are the air gap power due to forward and backward fields, respectively.

Now, $$P_{gf} = I_m^2 R_f \text{ and } P_{gb} = I_m^2 R_b$$

$\therefore$ $$P_g = P_{gf} - P_{gb} = I_m^2 R_f - I_m^2 R_b = I_m^2 (R_f - R_b) \tag{9.12}$$

The torque due to forward field is given by

$$\tau_f = \frac{1}{\omega_s} P_{gf} = \frac{P_{gf}}{2\pi\, n_s} \tag{9.13}$$

Again, the torque due to backward field is given by

$$\tau_f = \frac{1}{\omega_s} P_{gb} = \frac{P_{gb}}{2\pi\, n_s} \tag{9.14}$$

where ω_s represents the synchronous speed.

The resultant torque is given by

$$\tau_{in} = \tau_f - \tau_b = \frac{1}{2\pi\, n_s}(P_{gf} - P_{gb}) = \frac{I_m^2 (R_f - R_b)}{2\pi\, n_s} \tag{9.15}$$

The Cu loss of the single-phase motor is given by

$$P_r = P_{rf} + P_{rb} = sP_{gf} + (2-s)P_{gb} \tag{9.16}$$

where P_{rf} and P_{rb} are the Cu loss of the rotor due to forward and backward fields, respectively. The mechanical power developed in induction motor is given by

$$P_{mech} = \omega\tau_{in} = (1-s)\omega_s\tau_{in} = (1-s)(P_{gf} - P_{gb}) = (1-s)I_m^2 (R_f - R_b) \tag{9.17}$$

Shaft output power is given by

$$P_{out} = P_{mech} - \text{Core loss} - \text{Mechanical loss} - \text{Stray loss}$$
$$= P_{mech} - P_{rots}$$

where $$P_{rot} = \text{Friction loss} + \text{Windage loss} + \text{Core loss} = \text{Rotational losses}$$

Example 9.1 The equivalent circuit parameters of a 230 V, 50 Hz, single-phase induction motor having friction, windage loss and core loss of 50 W are given below:

$$R_{1m} = 2.4\,\Omega,\ R_2' = 4.7\,\Omega,\ X_{1m} = 3.2\,\Omega,\ X_2' = 2.8\,\Omega \quad \text{and} \quad X_M = 90\,\Omega$$

Calculate (i) input current, (ii) power factor, (iii) developed power, (iv) output power and (v) efficiency for a slip of 0.04.

Solution

$$s = 0.04,\ R_{1m} = 2.4\ \Omega,\ R_2' = 4.7\ \Omega,\ X_{1m} = 3.2\Omega,\ X_2' = 2.8\Omega,\ X_M = 90\Omega$$

$$\frac{R_2'}{2s} = \frac{4.7}{2\times 0.04} = 58.75\Omega \quad \text{and} \quad \frac{R_2'}{2(2-s)} = \frac{4.7}{2\times(2-0.04)} = 1.199\Omega$$

Again, $\frac{1}{2}X_2' = \frac{1}{2}\times 2.8 = 1.4\Omega$ and $\frac{1}{2}X_M = \frac{1}{2}\times 90 = 45\Omega$

For forward field,

$$Z_f = R_f + jX_f = \frac{\left(\frac{R_2'}{2s} + j\frac{X_2'}{2}\right)\times\left(j\frac{X_M}{2}\right)}{\left(\frac{R_2'}{2s} + j\frac{X_2'}{2}\right)+\left(j\frac{X_M}{2}\right)} = \frac{(58.75 + j1.4)(j45)}{58.75 + j1.4 + j45}$$

$$= 21.23 + j28.235$$

and the equivalent impedance (Z_b) of imaginary rotor due to backward rotating field is

$$Z_b = R_b + jX_b = \frac{\left[\frac{R_2'}{2(2-s)} + j\frac{X_2'}{2}\right]\times\left(j\frac{X_M}{2}\right)}{\left[\frac{R_2'}{2(2-s)} + j\frac{X_2'}{2}\right]+\left(j\frac{X_M}{2}\right)} = \frac{(1.99 + j1.4)(j45)}{1.99 + j1.4 + j45}$$

$$= 1.87 + j1.44$$

$$Z_{1m} = R_{1m} + jX_{1m} = 2.4 + j3.2\Omega$$

$$Z = Z_{1m} + Z_f + Z_b = 2.4 + j3.2 + 21.23 + j28.235 + 1.87 + j1.44$$

$$= 26.627 + j32.875$$

$$= 42.30\angle 51°$$

(i) Input current $(I_m) = \frac{V_m}{Z} = \frac{230\angle 0°}{42.3\angle 51°} = 5.437\angle -51°$

(ii) Power factor $= \cos(-51°) = 0.629$ [lagging]

(iii) Developed power $= I_m^2\left(R_f - R_b\right)(1-s) = 5.59^2(21.23 - 1.87)(1-0.04) = 580.76$ W

(iv) Output power $= 580.76 - 50 = 530.76$ W

(v) Input power $= VI_m\cos\theta = 230\times 5.437\times 0.629 = 786.57$ W

$$\text{Efficiency} = \frac{\text{Output power}}{\text{Input power}} = \frac{530.76}{786.57} = 0.6747\ \text{p.u.} = 67.47\%$$

■

9.8 DETERMINATION OF EQUIVALENT CIRCUIT PARAMETERS

To determine the parameters of the equivalent circuit of induction motor, the blocked rotor test and no-load test are carried out. These tests are similar to the tests for induction motor except for capacitor start motor because in capacitor start motor, these tests are conducted with auxiliary circuit open.

9.8.1 Blocked Rotor Test

During this test, rotor is kept blocked and a low voltage supply is given to the stator to maintain the rated current in main winding. The voltage (V_{sc}), current (I_{sc}) and power (P_{sc}) are measured. During the blocked rotor test, $s = 1$ and $\left(\frac{R_2'}{2} + j\frac{X_2'}{2}\right)$ can be neglected compared with $\frac{X_m}{2}$. Figure 9.11 shows the equivalent circuit of the single-phase induction motor during blocked rotor test.

From Figure 9.11, the equivalent series resistance (R_0) of the motor is given by

$$R_0 = R_{1m} + \frac{R_2'}{2} + \frac{R_2'}{2} = R_{1m} + R_2'$$

$$\therefore \quad R_2' = R_{1m} - R_0 \tag{9.18}$$

The effective rotor resistance at the line frequency can be calculated because R_{1m} is already measured.

The equivalent series reactance (X_0) of the motor is given by

$$X_0 = X_{1m} + \frac{X_2'}{2} + \frac{X_2'}{2} = X_{1m} + X_2' \tag{9.19}$$

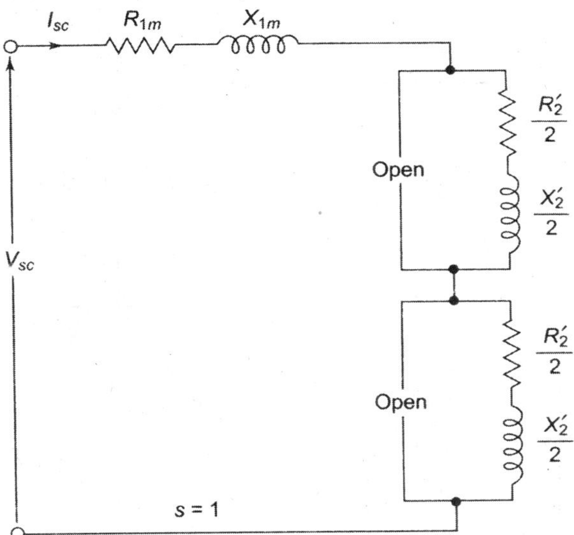

Figure 9.11 Equivalent Circuit of the Single-phase Induction Motor During Blocked Rotor Test

As we cannot separate X_{1m} and X'_2, let us make the following simplified assumption:

$$X_{1m} = X_2' = \frac{1}{2}X_0 = \frac{\sqrt{Z_0^2 - R_0^2}}{2} \tag{9.20}$$

Therefore, the parameters R_2, X_{1m} and X'_2 can be calculated from blocked rotor test.

9.8.2 No-load Test

In this test, motor is run without load at rated voltage and frequency. The rated voltage is applied and the voltage (V_0), current (I_0) and power (P_0) are measured. During no-load test $s \approx 0$ and hence $\frac{R_2'}{2s}$ is large compared with $\frac{X_M}{2}$. Figure 9.12 shows the equivalent circuit of the induction motor. From Figure 9.12, the equivalent reactance at no-load is given by

$$X_0 = X_{1m} + \frac{X_M}{2} + \frac{X_2'}{2} \tag{9.21}$$

X_M can be calculated because X_{1m} and X'_2 have already been calculated from blocked rotor test. The no-load power factor is given by

$$\cos\theta_0 = \frac{P_0}{V_0 I_0} \tag{9.22}$$

The no-load equivalent impedance is given by

$$Z_0 = \frac{V_0}{I_0} \tag{9.23}$$

and the no-load equivalent reactance is given by

$$X_0 = Z_0\sqrt{1-\cos^2\theta_0} \tag{9.24}$$

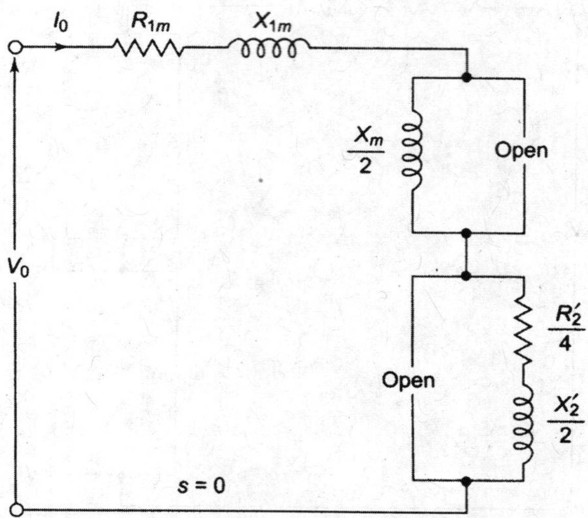

Figure 9.12 Equivalent Circuit of Induction Motor

Example 9.2 The test results of a 230 V, single-phase induction motor are given below:
Blocked rotor test: 110 V, 9.5 A, 450 W
No-load test: 230 V, 4.4 A, 120 W
The starting winding is kept open during blocked rotor test and stator winding resistance is 1.4 Ω. Find the equivalent circuit parameters and the core, friction and windage losses.

Solution

Blocked rotor test:

$$V_{sc} = 110\text{ V},\ I_{sc} = 9.5\text{ A},\ P_{sc} = 450\text{ W}$$

$$Z_0 = \frac{V_{sc}}{I_{sc}} = \frac{110}{9.5} = 11.58\ \Omega,\ R_0 = \frac{P_{sc}}{I_{sc}^2} = \frac{450}{9.5^2} = 4.986\ \Omega$$

$$X_0 = \sqrt{Z_0^2 - R_0^2} = \sqrt{11.58^2 - 4.986^2} = 10.45\ \Omega$$

$$X_{1m} = X_2' = \frac{1}{2}X_0 = 5.225\ \Omega,\ R_{1m} = 1.4\ \Omega$$

$$R_2' = R_0 - R_{1m} = 4.986 - 1.4 = 3.586\ \Omega$$

No-load test:

$$V_0 = 230\text{ V},\ I_0 = 4.4\text{ A},\ P_0 = 120\text{ W}$$

$$\cos\theta_0 = \frac{P_0}{V_0 I_0} = \frac{120}{230\times 4.4} = 0.1186\ \therefore \sin\theta_0 = 0.993$$

$$\therefore \quad Z_0 = \frac{V_0}{I_0} = \frac{230}{4.4} = 52.27\ \Omega \text{ and } X_0 = Z_0 \sin\theta_0 = 52.27\times 0.993 = 51.9\ \Omega$$

∴ Core, friction and windage losses

$$= P_0 - I_0^2\left(R_{1m} + \frac{R_2'}{4}\right) = 120 - 4.4^2\left(1.4 + \frac{3.586}{4}\right) = 74.23\text{ W}$$

■

9.9 SPLIT-PHASE INDUCTION MOTORS

Figure 9.13 shows a split-phase induction motor which has a cage rotor and two stator windings (main winding and auxiliary winding). The main winding has very low resistance and high reactance, whereas the auxiliary winding has very high resistance and low reactance. The impedance of the auxiliary winding is very highly resistive in nature. The main winding is reactive in nature. The current (I_m) through the main winding lags behind the voltage by an angle θ_m, whereas the current through the starting winding (I_{st}) is almost in phase with the voltage, as shown in Figure 9.14. Therefore, a time phase difference between I_m and I_{st} exists which is of the order of 30° and not 90°. This phase difference is sufficient to produce a rotating magnetic field. The rotating magnetic is not uniform because the currents in the two windings are not the same. The starting torque is small and it is of the order of 1.5 to 2 times the rated running torque. During starting, both the windings are connected in parallel and supply is given. When the speed of the motor is about 70 to 80 per cent of synchronous speed, the starting winding is automatically disconnected from the supply. That is why auxiliary winding is also called starting winding. A centrifugal switch is used

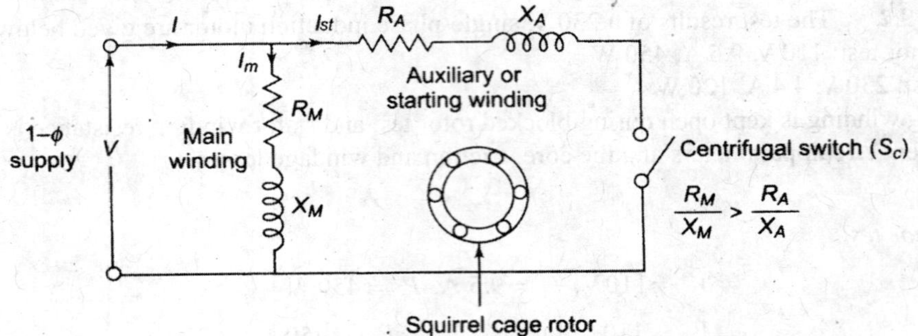

Figure 9.13 Split-phase Induction Motor

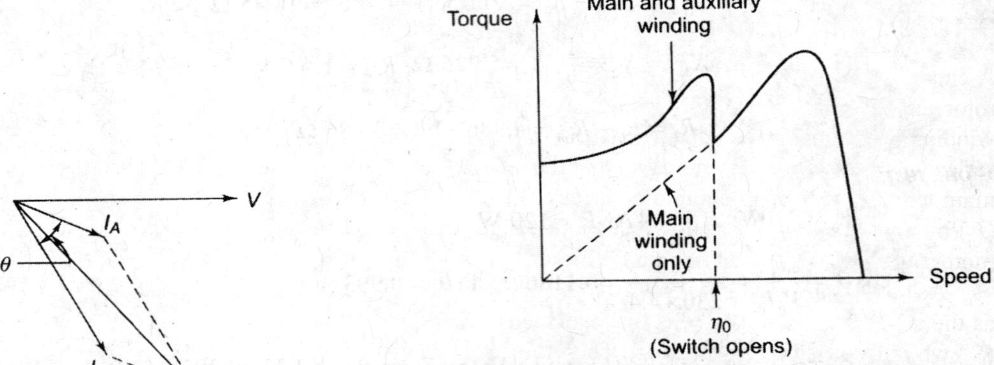

Figure 9.14 Phasor Diagram

Figure 9.15 Torque-speed Characteristic of Split-phase Induction Motor

to disconnect the starting winding for motors having rating about 100 W or more. For smaller motors, a relay is connected in series with the main winding. During starting a heavy current flows through the coil which makes its contacts close and the starting winding is brought in circuit. When the motor achieves 70 to 80 per cent of synchronous speed, the current through the relay decreases and the relay opens resulting in disconnection of auxiliary windings from the supply. Now the motor runs on the main winding. Figure 9.15 shows the torque-speed characteristic of split-phase induction motor.

The starting torque and the maximum or pull-out torque of this type of motors are of the order of 1.5 to 2.5 times of the full-load torque at about 75 per cent of synchronous speed. The staring current of the split-phase motors is generally 7 to 8 times of full-load current. To get the reversal of rotation, the motor at first is brought to rest and the line connections of either the main or auxiliary windings are reversed.

9.10 CAPACITOR MOTORS

A capacitor is used in series of auxiliary windings of single-phase induction motor so that the phase difference between the currents in main and auxiliary windings gets an appreciable value. The three different types of capacitors are discussed as follows.

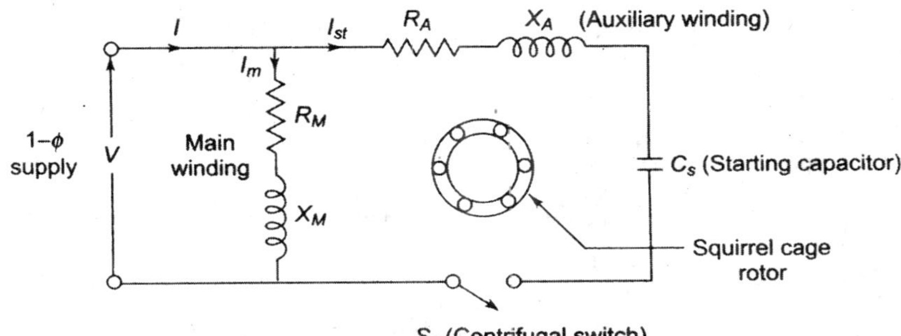

Figure 9.16 Capacitor-start Motor

9.10.1 Capacitor-start Motors

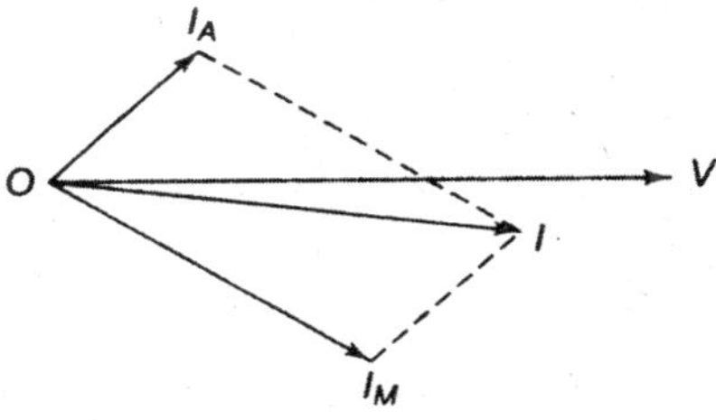

Figure 9.17 Phasor Diagram

A capacitor-start motor is shown in Figure 9.16 having cage rotor and two stator windings – main winding and auxiliary winding. A properly selected capacitor (C_s) is connected in series with the auxiliary winding so that the current through the main winding (I_m) lags behind the current of auxiliary winding (I_{st}) by an angle of 90°, as shown in Figure 9.17 and hence the motor acts like a balanced two-phase motor.

As soon as the motor reaches 70 to 80 per cent of synchronous speed, the auxiliary winding, as well as the starting capacitor (C_s), is disconnected from the supply automatically by the centrifugal switch (S_c) which is mounted on the shaft. Therefore, the capacitor (C_s) is used only for starting purposes. The starting torque developed by the capacitor-start motor is higher than the split-phase induction motor which is of the order of 3 to 4.5 times of the full-load torque. To obtain the high starting torque, the value of starting capacitor must be large. The starting capacitor is of electrolytic type and is of the order of 250 μF. Figure 9.18 shows the torque-speed characteristic of capacitor-start motor. To get reversal of rotation, the motor must be brought to rest at first and the line connection must be interchanged. This type of motor is suitable for loads of high inertia and hence is suitable for compressors, pumps, refrigerators, conveyors and others.

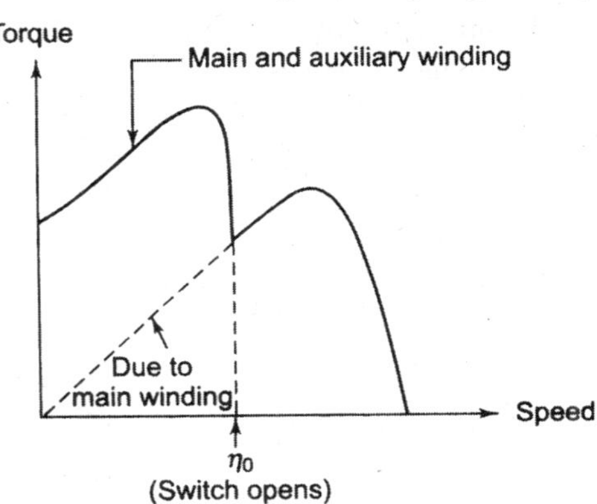

Figure 9.18 Torque-speed Characteristic of Capacitor-start Motor

9.10.2 Capacitor-start Capacitor-run Motors

Figure 9.19(a) shows the capacitor-start capacitor-run induction motor having cage rotor and two stator windings – main winding and auxiliary windings. The motor has two capacitors C_s and C_r and these are connected in parallel at starting. The angle between the currents through the main winding and auxiliary winding is 90°. The value of C_s should be large to get high starting torque. It is short time rated and electrolytic in nature. Since the rated line current is smaller than the normal current during normal operation, the

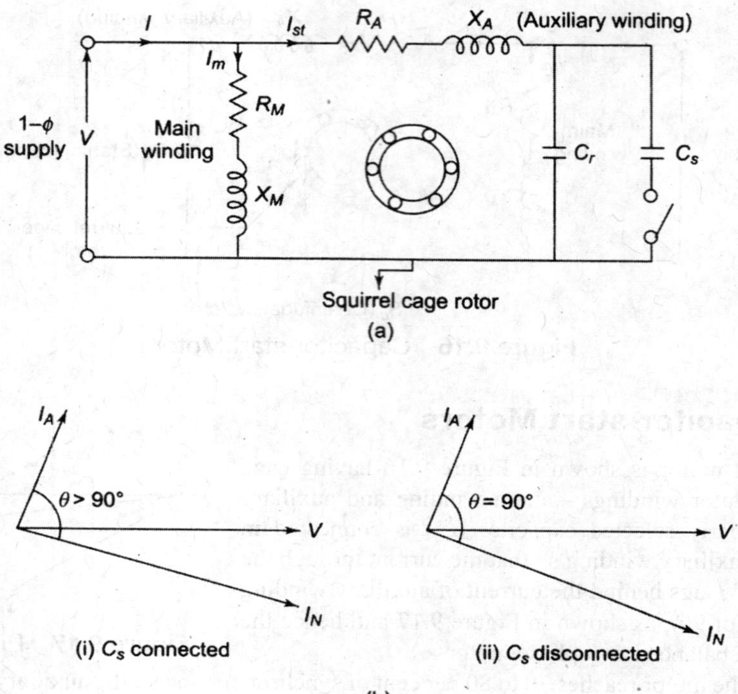

Figure 9.19 (a) Capacitor-start Capacitor-run Motor (b) Its Phasor Diagram

value C_r must be small. The capacitor stays in the circuit permanently and it has oil-filled paper construction. The capacitor is disconnected from the supply as the motor achieves 70 to 80 per cent of synchronous speed. This type of motor is quiet and smooth running and has higher efficiency compared with the motors running only on main windings. Figure 9.19(b) shows the phasor diagrams of capacitor-start capacitor-run induction motor.

Figure 9.20 shows the torque-speed characteristic of capacitor-start capacitor-run induction motor. These motors are used for refrigeration, compressors and pumping equipment.

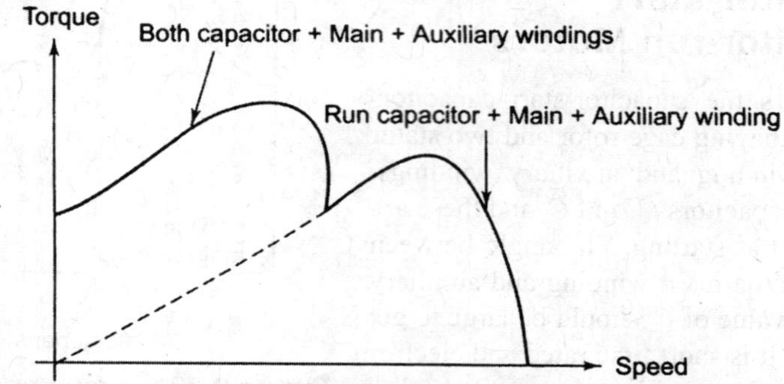

Figure 9.20 Torque-speed Characteristics of Capacitor-start Capacitor-run Induction Motor

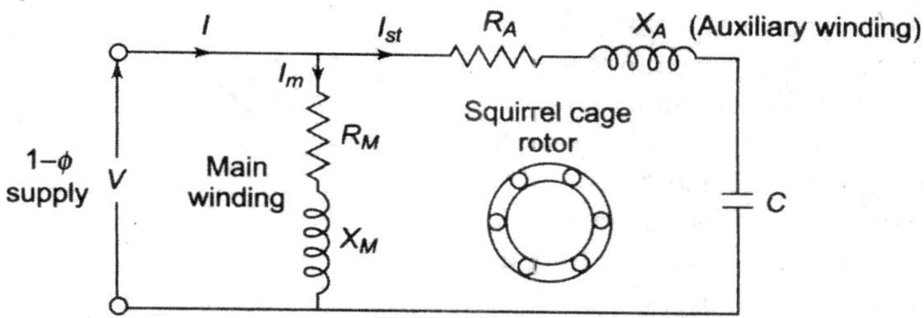

Figure 9.21 Permanent Split Capacitor Induction Motor

9.11 PERMANENT SPLIT CAPACITOR MOTORS

Figure 9.21 shows the permanent split capacitor induction motor having cage rotor and two stator windings – main winding and auxiliary windings. This type of motor has one capacitor (C) connected in series with the auxiliary windings permanently, and hence this type of motor is called single value capacitor motor. This type of motor has no starting switch and operates in the same way as capacitor-start capacitor-run induction motor. It produces a uniform torque and is less noisy during operation.

This type of motor is used for air conditioners to drive refrigerator compressors and fans and blowers of heaters. This type of motor has the following advantages:

- Higher efficiency
- Higher pull-out torque
- Higher power factor
- No centrifugal switch

The main limitations are:

- Low starting torque
- For continuous running, electrolytic capacitor cannot be used and paper capacitor of equivalent rating has larger size and costs more.

9.12 SHADED POLE MOTOR

The effect of shaded coil is to retard a portion of flux in the poles in time phase, so that there is a sweeping of the flux across the pole face in the direction of the shading of the coil, as shown in Figure 9.22.

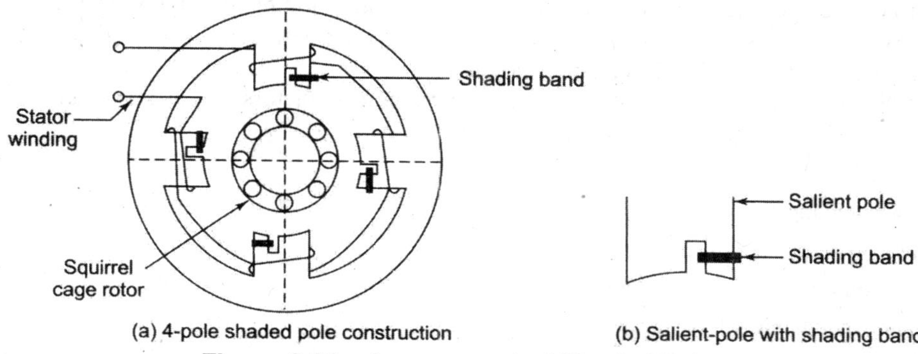

Figure 9.22 Arrangement of Shaded Poles

The fluxes in the unshaded part of the pole and in the shaded part of the pole cut the rotor conductors and induce currents, which in turn produce a torque sufficient to start the motor. The shaded pole produces a weak starting torque and is used with fractional horse power motors, such as fan motors and shaving motors. Figure 9.23 shows the torque-speed characteristics of a shaded pole motor.

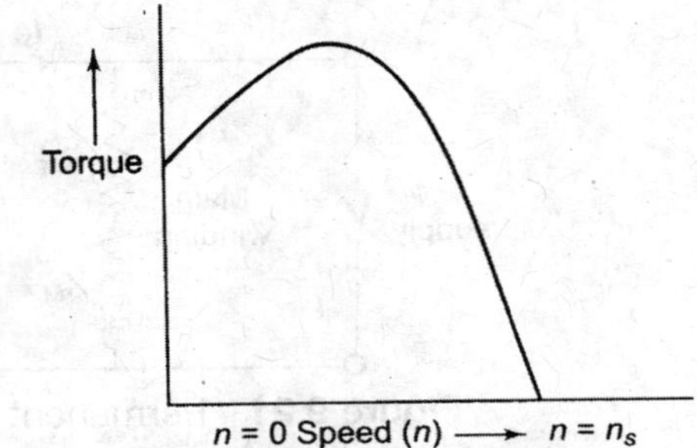

Figure 9.23 Torque-speed Characteristics of a Shaded Pole Motor

This type of motor is cheap, having low starting torque, low power factor and low efficiency. This type of motor is used for small fans, advertising displays, toy motors, film projectors, record players, hair dryers, photocopying machines, and so on.

9.13 SINGLE-PHASE SYNCHRONOUS MOTORS

We know that synchronous motors have constant speed characteristic. Small-capacity synchronous motors are smaller in size and do not require DC excitation to rotor. There are two types of single-phase synchronous motors, namely, reluctance motors and hysteresis motors.

9.13.1 Reluctance Motors

Figure 9.24 shows a reluctance motor which is similar to a single-phase induction motor. The stator has both the windings – main winding and auxiliary windings. This type of motors has variable air gap and no DC supply is required to rotor. Salient-pole type of structure is given to the rotor, and it carries short-circuited copper bars acting like a squirrel cage rotor of induction motor. The starting torque is produced by any of the phase splitting methods. Air gap decides the reluctance of the magnetic path. Hence, more the air gap, more is the reluctance and less the air gap, less is the reluctance. The reluctance is greatest when the magnetic field of the stator magnetic field aligns with the cut out portion of the rotor, whereas it is least when it aligns with the centre portion of the rotor teeth. Therefore, the reluctance of the magnetic circuit is a function of the rotor position with respect to the axis of the stator field, that is, a function of the air gap. The rotor starts to rotate like a normal induction motor after giving the supply. The starting torque of this type of motor is very high and is solely dependent on the position of the rotor. Figure 9.25 shows the torque-speed characteristic of reluctance motor.

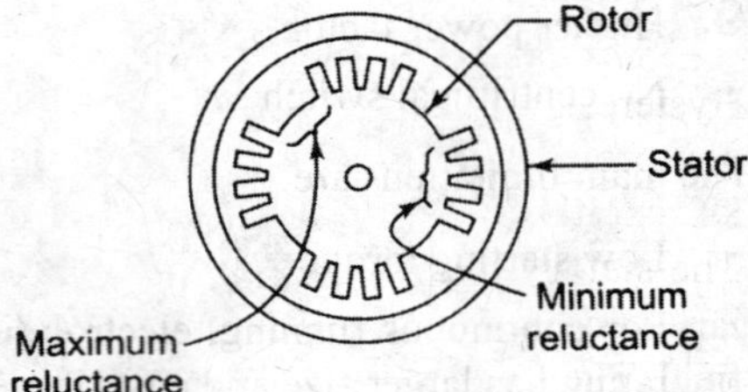

Figure 9.24 Reluctance Motor

This type of motor is used in control apparatus, automatic regulators, clocks and all kinds of timing devices, recording instruments, teleprinters and gramophones.

This type of motor has the following advantages:

- Less maintenance
- No requirement of DC supply
- Robust construction
- Constant speed characteristic

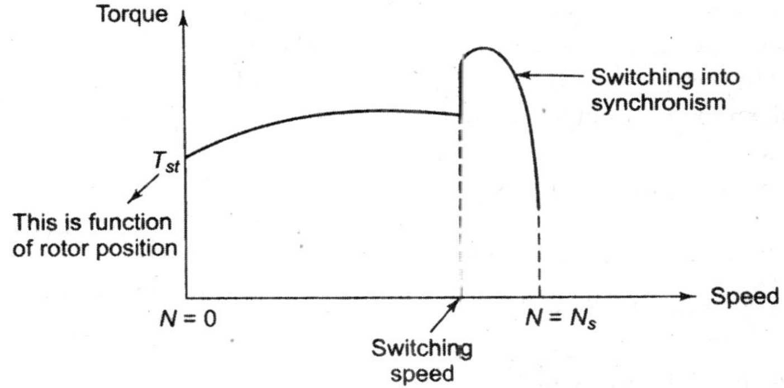

Figure 9.25 Torque-speed Characteristic of Reluctance Motor

This type of motor has the following limitations:

- Less efficiency
- Poor power factor
- Less capacity to drive loads compared with 3-phase synchronous motors

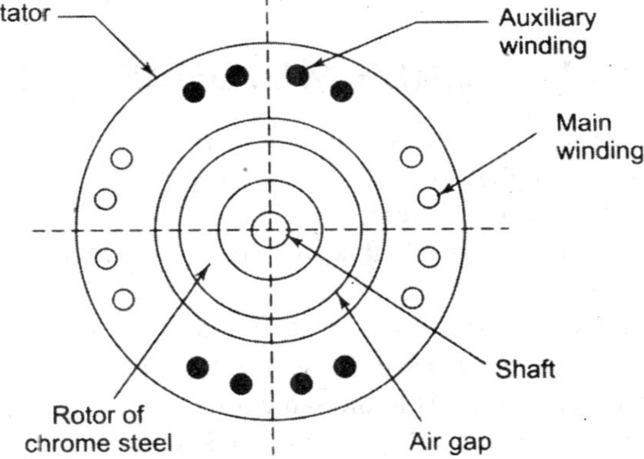

Figure 9.26 Hysteresis Motor (Cross-section View)

9.13.2 Hysteresis Motor

Hysteresis motor is a synchronous motor having non-projected poles which does not need any DC excitation to rotor, as shown in Figure 9.26. The stator carries the main winding and auxiliary windings to produce the rotating magnetic field. The stator can also be of shaded pole type. The rotor is made of hard magnetic material such as chrome steel or alnico for high retentivity and is of smooth cylindrical structure. The rotor has no winding and the material of the rotor must have high hysteresis loop area, as shown in Figure 9.27.

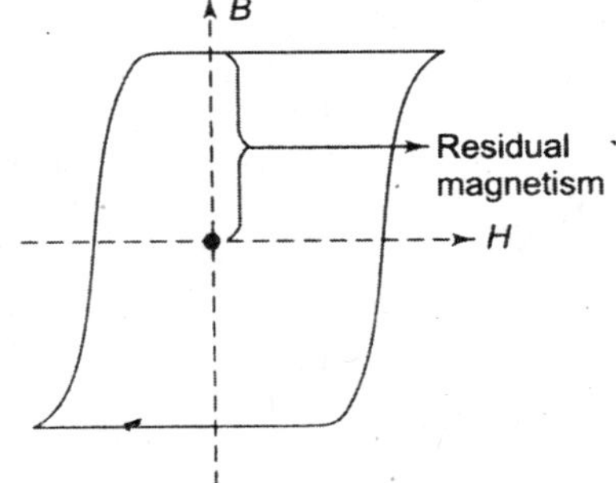

Figure 9.27 Hysteresis Loop for Rotor Material

After giving supply to the stator, the rotating magnetic field is produced that induces poles in the rotor. Since hysteresis phenomena is dominant in rotor material, rotor pole axis lags behind the rotating magnetic field. Therefore, the rotor poles get attracted towards the moving stator field poles and hence rotor gets subjected to torque known as *hysteresis torque* which is constant at all speeds. The rotor pole strength remains constant when the stator field axis moves forward. Therefore, higher is the hysteresis torque if higher is the retentivity. At the beginning, the rotor starts rotating due to the combined effect of *hysteresis torque* and torque due to eddy current induced in the rotor. The rotor pulls into synchronism when the speed of the rotor is near the synchronous speed. Magnetic locking between the rotor and stator is ensured by high retentivity. The motor either rotates at synchronous speed or not at all due to the principle of magnetic locking. The starting torque and running torque are almost equal in this motor.

Figure 9.28 shows the torque-speed characteristic of hysteresis motor.

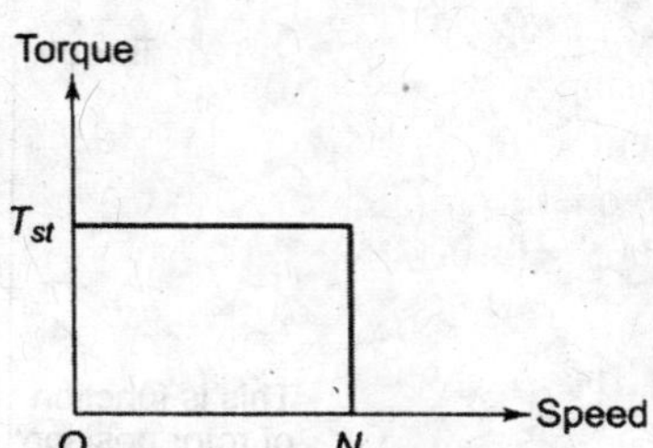

Figure 9.28 Torque-speed Characteristic of Hysteresis Motor

This type of motor has the following advantages:

- There are no mechanical vibrations because the rotor has no teeth and no winding.
- It can accelerate the high inertia load.
- There is possibility of multi-speed operation using gear train.
- The operation is quiet and noiseless due to absence of vibrations.

This type of motor is used for sound-recording instruments, sound-producing equipments, tape recorder, high-quality record players, teleprinters, timing devices and so on, because this type of motor has noiseless operation.

9.14 SERIES MOTOR OR UNIVERSAL MOTOR

A series motor can be run from either DC or AC (single-phase) supply, as shown in Figure 9.29 provided both stator and rotor cores are laminated to limit iron loss.

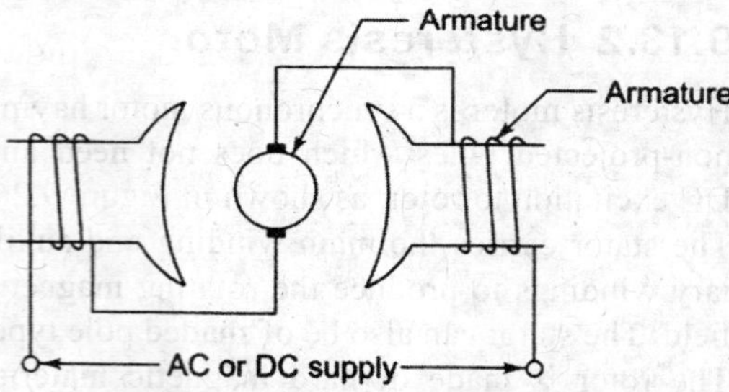

Figure 9.29 Universal Motors

Since both field flux and armature current change their magnitude and sign simultaneously, their product is always positive, even if both parameters individually are negative. Therefore, the torque produced by series motor is in the same direction for each half cycle of AC and is pulsating. We know:

$$T_{inst} \propto i\phi$$

$$\propto I_m \phi_m \sin \omega t \sin \omega t$$

$$\propto I_m \phi_m \sin^2 \omega t$$

$$\propto I_m \phi_m \left(\frac{1-\cos 2\omega t}{2}\right)$$

i.e.,
$$T_{av} \propto \left[\frac{1}{2} I_m \phi_m - \frac{1}{2} I_m \phi_m \cos 2\omega t\right] \quad (9.25)$$

i.e.,
$$T_{av} \propto \left[\left(\frac{I_m}{\sqrt{2}}\right)\left(\frac{\phi_m}{\sqrt{2}}\right)\right] \quad [\because \text{ average of cosine is zero}]$$

Since $\phi \propto I_a$, we have

$$T_{av} \propto \phi I_a$$

i.e.,
$$T_{av} \propto I_a^2$$

From Equation (9.25), it is seen that the instantaneous torque has an alternating component, whose frequency is double that of supply frequency. Figure 9.30 shows the field polarity is alternating with armature current variation.

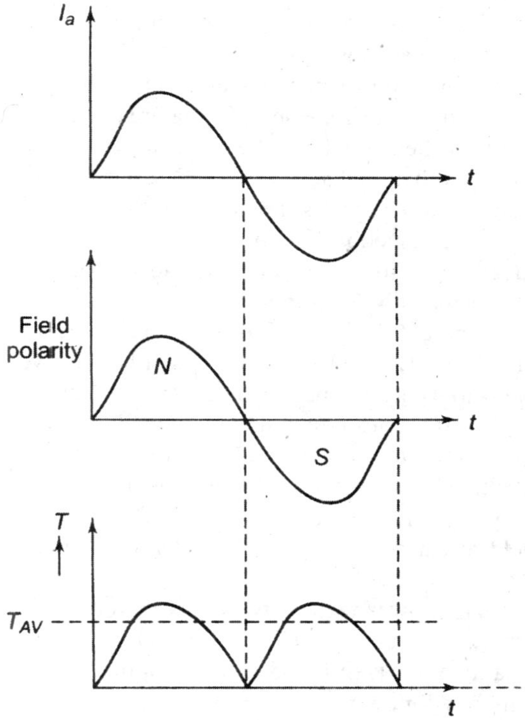

Figure 9.30 The Field Polarity Alternating with Armature Current Variation

9.14.1 Circuit Model and Phasor Diagram

The net emf induced in the armature is the sum of the transformer emf and that due to leakage inductance (x_a) and is given by

$$jI_a x_a + jE_t = jI_a(x_a + jE_t/I_a)$$

Since E_t is caused by cross-flux whose origin is an armature current, the two must be in phase.

$$\frac{E_t}{I_a} = x_{at}$$

The net inductive emf in the armature circuit is

$$jI_a(x_{at} + x_a) = jI_a X_a \quad \text{where} \quad X_a = x_a + x_{at}$$

The circuit model of the AC series motor is shown in Figure 9.31.

9.14.2 Torque

As stated earlier, the torque developed is pulsating in nature with an average value and a predominant second harmonic component. The average torque is given by

$$T_{av} = E_a I_a/\omega_m.$$

Being a series motor, $E_b \propto I_a$ and the torque is therefore proportional to square of armature current, which becomes directly proportional at high armature current, because of saturation.

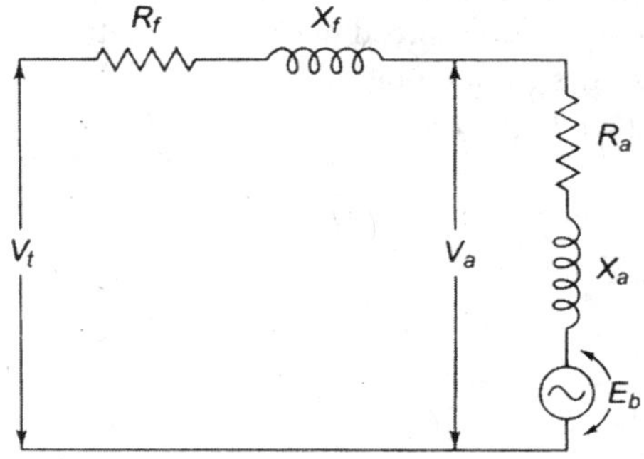

Figure 9.31 Circuit Model of AC Series Motor for AC Operation

9.14.3 Phasor Diagram and Performance Characteristics

In AC series motor, there will be a series drop in armature and field, along with reactance drop due to both in addition to rotational emf. Vector diagram of AC series motor is shown in Figure 9.32.

R_a, R_f = resistances of armature and field winding, respectively

X_a, X_f = reactances of armature and field winding, respectively

and E_b = rotational back emf produced in the armature

It will be seen that the field and armature reactances increase the angle of lag and, therefore, cause reduction in power factor. For given value of torque and applied voltage, value of armature current, as well as of the field flux, will be constant. It may also be observed that the back emf in AC series motor is less than that in DC series motor. Since field flux remains constant, reduction in emf in AC series motor would suggest that its speed for given developed torque is less than that of DC series motor.

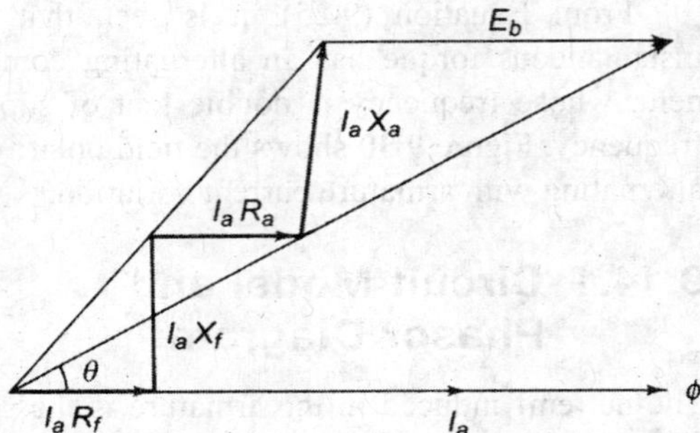

Figure 9.32 Phasor Diagram of AC Series Motor

Figure 9.33 shows speed-torque characteristics of AC and DC series motors. It will be observed that the whole performance revolves on the point that back emf in AC series motor should be made, as far as possible, equal to the back emf in DC series motor. This would mean reducing the armature and field reactances. Field reactance can be reduced by using less number of turns on the field, which will in turn reduce field flux. However, to develop the same torque, the number of turns on the armature has to be increased in the same ratio. This would reduce field reactance and increase armature reactance. To cancel the effect of armature reactance, compensating winding is used.

The power factor of a series motor is rather poor because of the large reactances of field and armature (x_f and x_a). The power factor at low speed is very poor which is due to the small value of rotational back emf, and as such applied voltage has to be balanced by field and armature reactances which will now be high. However, power factor of AC series motor is higher at higher speeds and it is about 0.95.

The arrangement of main field winding connected in series with the compensating winding for an AC series motor is shown in Figure 9.34. The axis of this winding is along the brush axis and the winding must be spread over the full pitch for effective neutralisation of cross-flux. The arrangement also shows an interpole winding connected in series with the other two windings, whose effect is only over the interpolar region.

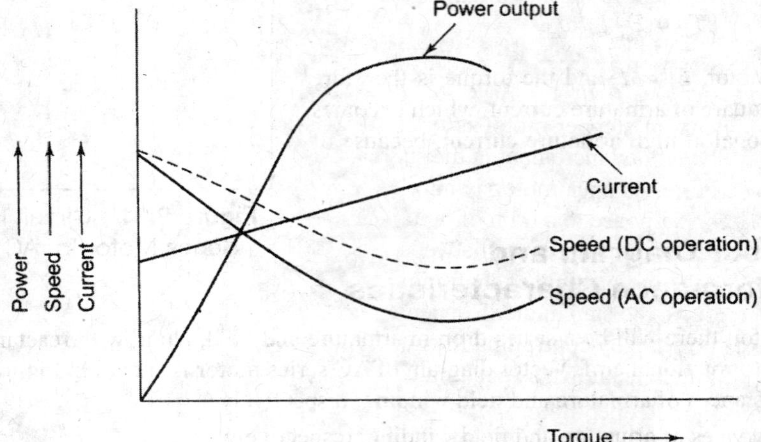

Figure 9.33 Performance Characteristic of Series Motor AC Operation

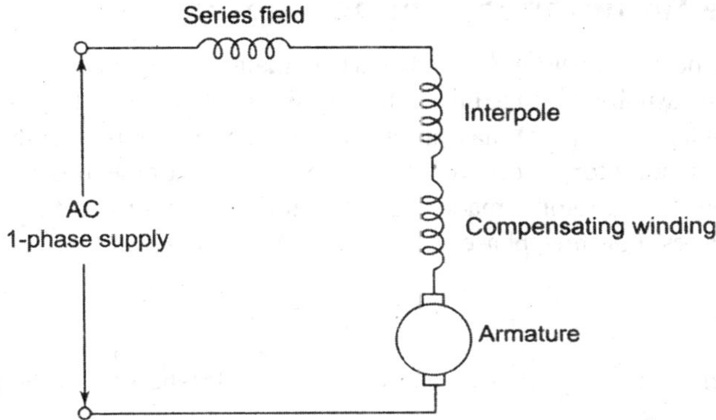

Figure 9.34 Series Motor with Compensating Winding and Interpole Winding Connected to AC Mains

The no-load speed of a universal motor is very high (about 20,000 rpm). Hence, the motor is smaller in size than other types for a given load. Universal motors are used where light weight is important as in vacuum cleaners, portable tools and so on.

9.14.3.1 Constructional features

(i) It has less number of turns on the field and more number of turns on the armature. The number of turns on the armature is regulated by adapting lap winding and using large number of poles, especially for traction motors.

(ii) It has compensating winding, which cancels the magnetic effect of armature current.

(iii) Since field is weak, it has small air gap.

(iv) To reduce transformer emf which gives commutating problem, these motors are run preferably from $16\frac{2}{3}$ cycles for 25 cycles supply.

(v) These motors are suitable for maximum 400-V supply voltage.

9.15 STEPPER MOTOR

The important property of stepper motor is to convert a train of input pulses, that is, a square wave pulse, into a precisely defined increment in the shaft position where each pulse moves the shaft through a fixed angle. Therefore, stepper motor is an electromechanical device that actuates a train of step movements of shaft in response to a train of input pulses where the step movement may be angular and linear and one-to-one relationship exists between the input pulse and step movement of the shaft. The angle through which the motor turns or shaft moves for each pulse is known as the *step angle* which is expressed in degrees. As the step angle is dependent on the number of input pulses, the motor is suitable for controlling position by controlling the number of input pulses. Therefore, this system is used to control the position which is known as position control system. There are three types of stepper motors:

- Variable reluctance stepper motor
- Permanent magnet stepper motor
- Hybrid stepper motor

9.15.1 Variable Reluctance Stepper Motor

This is the most basic type of stepper motor and helps to explain the principle of operation of stepper motor. The stator of the motor is usually for three phases having six salient poles with concentrated exciting windings around each one of them. The stator has laminated construction and is assembled in a single stack. The number of poles of stator and rotor are different to secure the ability of the motor for bi-directional rotation and self-starting capability. The rotor is made of slotted steel laminations. Let N_s and N_r be the number of stator poles and rotor poles. For three phase, the number of rotor poles in terms of stator poles is given by

$$N_r = N_s \pm \left(\frac{N_s}{q}\right)$$

where q is the number of stator phases or stacks. Figure 9.35 shows the stepper motor having four salient-pole rotor.

The basic diagram has been shown in Figure 9.36 where the coils are wound in such a way that diametrically opposite poles are connected in series. The three phases are energised from a DC supply with the help of switches.

9.15.1.1 Operation

The operation depends on various reluctance positions of rotor with respect to stator. If one phase of the stator is excited, it produces a magnetic field. Its axis lies along the poles, the phase around which it is excited. The rotor moves in the direction of minimum reluctance and such a position indicates the position where the axis of magnetic field of stator tallies with the axis passing through any two poles of the rotor. The phases *A, B* and *C* are energised in sequence one after the other with the help of switches *SW*1, *SW*2 and *SW*3, respectively.

When the phase *AA′* is excited with the switch *SW*1 closed, the magnetic axis lies along the vertical direction or along the poles formed by *AA′*. The rotor adjusts itself in the direction of minimum reluctance position, as shown in Figure 9.37.

Next the phase *BB′* is excited with switch *SW*2 closed and the switch *SW*1 open to de-energise *AA′*. The stator magnetic axis shifts along the poles formed due to *BB′*. Now the rotor will try to align in the direction of minimum reluctance position and hence turns through an angle of 30° in the anti-clockwise direction.

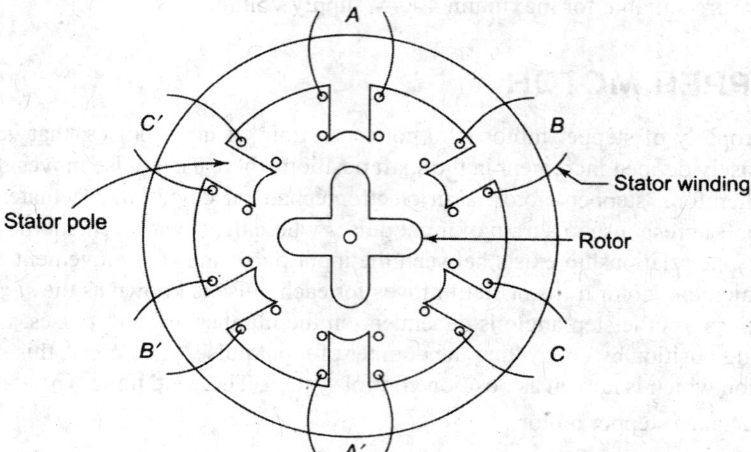

Figure 9.35 Schematic Arrangement of Variable Reluctance Stepper Motor

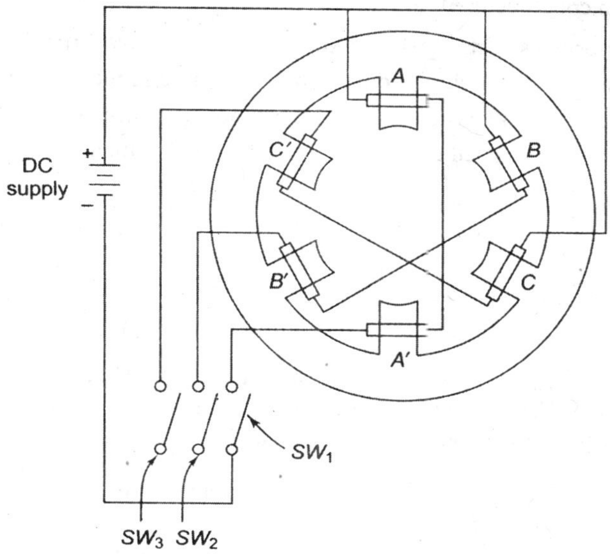

Figure 9.36 Basic Driving Circuit

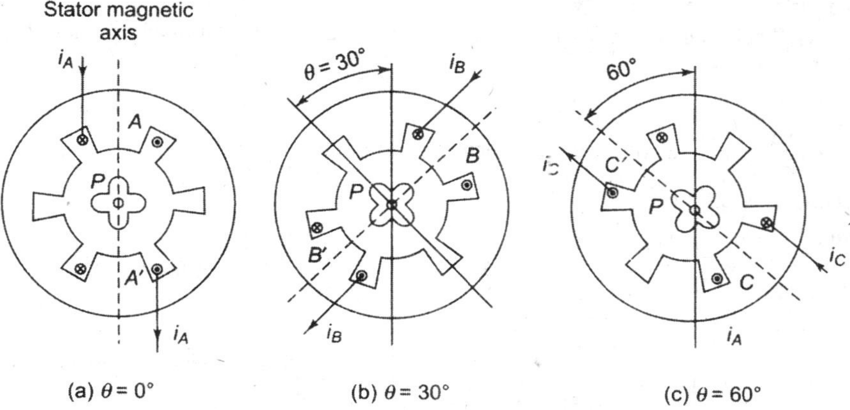

Figure 9.37 Operation of Variable Reluctance Stepper Motor

Next the phase CC' is excited with the switch $SW3$ closed and the switch $SW2$ open. The stator magnetic axis shifts along the poles formed due to CC'. Now the rotor will try to align in the direction of minimum reluctance position and hence turns through an angle of 30° in the anti-clockwise direction.

If we successively excite the three phases in a specified sequence, the motor takes 12 steps to complete one revolution. Let i be the current through the phase excited and L be the inductance of the relevant phase at an angle θ. The torque developed by the motor is expressed by:

$$T_m = \frac{1}{2} i^2 \frac{dL}{d\theta}$$

From the above, we can conclude the following:

- By exciting the stator phases, it is possible to rotate the rotor in a particular direction.
- The rotor rotates in the anti-clockwise direction, if the phase sequence is *A-B-C-A*....
- The rotor rotates in the clockwise direction, if the phase sequence is *A-C-B-A*....
- If all the three phases are excited once, the distance through which the rotor is moved is known as rotor tooth pitch.
- Rotor tooth pitch $= \dfrac{360°}{N_r}$
- The step angle $(\alpha_s) = \dfrac{360°}{qN_r}$

In the above discussion, we have seen that the rotor rotates through an angle of 30° when each phase is excited simultaneously. Now if two phases are excited simultaneously, that is, *AA′* and *BB′*, the stator magnetic field axis shifts to a mid-position rather than along *BB′* resulting in the rotor being aligned along it, that is, through an angle of 15°. Usually, the step angle is reduced by $\frac{1}{2}, \frac{1}{5}, \frac{1}{10}, \frac{1}{16}$ and $\frac{1}{32}$. This technique is known as *micro-stepping*.

9.15.1.2 Multi-stack stepper motor

Figure 9.38 shows the cross-sectional view of multi-stack stepper motor. For a *m* stack motor, the motor is divided into a number of magnetically isolated sections known as stacks along its axial length. If *m* is the number of stacks, that is, the number of phases, and N_r is the number of rotor poles, the step angle is given by

$$\alpha_s = \frac{360°}{mN_r}$$

Following are the advantages of variable reluctance induction motor:

- High torque-to-inertia ratio
- Fast dynamic response

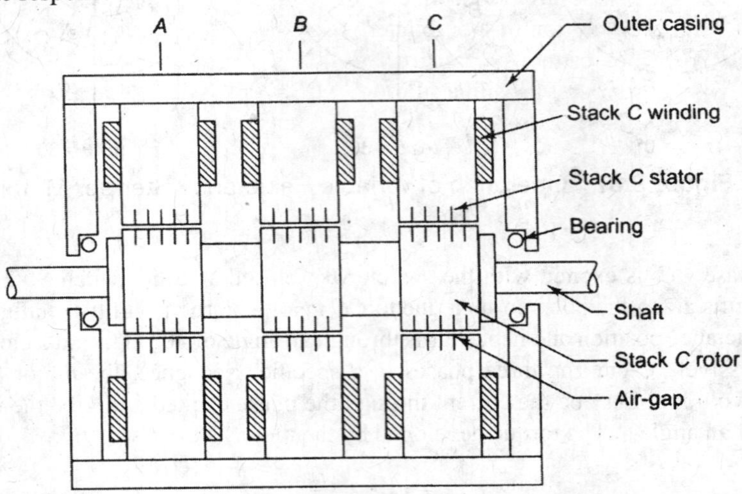

Figure 9.38 Cross-sectional View of Multi-stack Stepper Motor

- High rates of acceleration
- Simple and low cost
- Efficient cooling arrangement
- Robust rotor construction due to absence of brushes

9.15.2 Permanent Magnet Stepper Motor

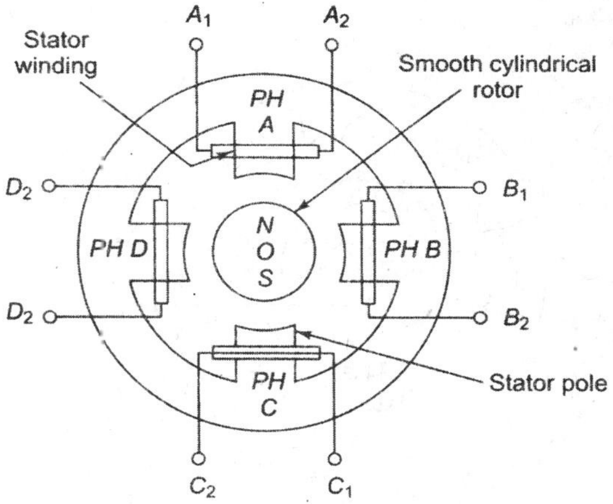

Figure 9.39 Four-phase Permanent Magnet Stepper Motor

This type of motor has multi-polar stator, as shown in Figure 9.39, having four stator poles. The exciting coils are wound around the poles. The number of slots per pole per phase is usually taken as one in such multi-polar machines.

The rotor is generally taken as cylindrical pole type, but it may be salient or smooth cylindrical type. The rotor is made of ferrite material which is permanently magnetized, and hence this motor is called permanent magnet stepper motor. Figure 9.40 shows the basic driving circuit for four-phase permanent magnet stepper motor to provide the voltage pulses to the stator winding of stepper motor, as shown in Figure 9.39.

Just after the application of the voltage pulses to various phases with the help of driving circuit, the rotor starts rotating through a step for each input voltage pulse. After closing the switch *SW*1 to excite phase *A*, we get *N* pole in phase *A* and the rotor starts to rotate in such a way that the magnetic axis of the permanent magnet rotor adjusts with the magnetic axis of stator as depicted in Figure 9.37(a). Next phase *B* is excited after closing the switch *SW*2 and disconnecting phase *A*. The rotor further adjusts its own magnetic axis with *N* poles of phase *B*. Therefore, it rotates through an angle of 90°, as shown in Figure 9.37(b). Similarly, if the phases *C* and *D* are excited, the rotor rotates through an angle of 90° in clockwise direction for excitation of each phase. Therefore, repetition of such sequence results in a step motion of a permanent magnet stepper motor. The major disadvantage of stepper motors is that they cannot be constructed with permanent magnet rotors with large number of poles and small steps are not possible. In order to overcome this, variable reluctance type stepper motor is used.

9.15.3 Hybrid Stepper Motor

The hybrid stepper motor is based on the principles of the permanent magnet and variable reluctance stepper motors. The rotor flux in hybrid stepper motor is generated by the permanent magnet and is directed by the rotor teeth to the appropriate parts of the air gap. The permanent magnet that is placed in the middle of the rotor is magnetised in the axial direction.

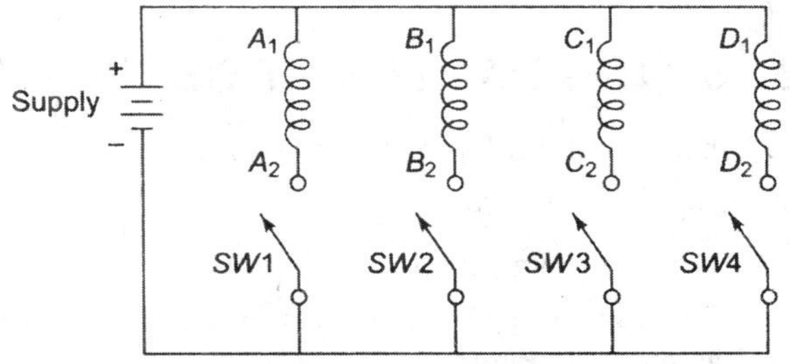

Figure 9.40 Basic Driving Circuit for Permanent Magnet Four-phase Stepper Motor

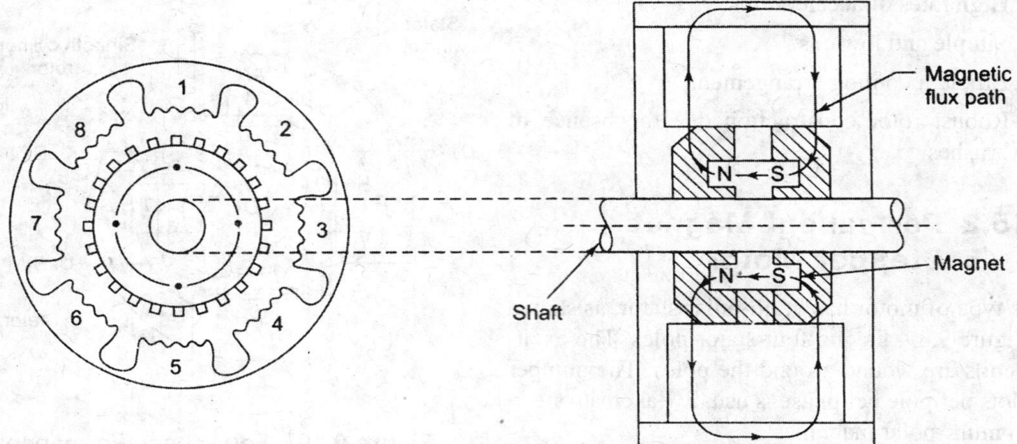

Figure 9.41 Hybrid Stepper Motor

Each pole of the magnet is surrounded by soft-toothed laminations. Figure 9.41 shows the construction of hybrid stepper motor.

There are generally eight poles on the stator and each pole has between two and six teeth and there is two-phase winding. The coils on poles 1, 3, 5 and 7 are connected in series to form phase *A*, whereas coils on poles 2, 4, 6 and 8 are connected in series to form phase *B*. If phase *A* carries positive current, stator poles 1 and 5 become South while 3 and 7 become North. Therefore, the rotor teeth with South and North polarity align with the teeth of stator poles 3 and 7 and 1 and 5, respectively. If phase *B* is excited and phase *A* is de-energised, the rotor will move by one quarter of tooth pitch. Torque is produced in a hybrid motor due to interaction of stator- and rotor-produced fluxes. Since the rotor field is produced by permanent magnet, it remains constant. Hence, the motor torque is proportional to the phase current. Following are advantages of the hybrid stepper motor:

- Step angle is very small. It is 1.8°.
- Higher torque per unit volume. It is greater than that of reluctance motor.
- There is some detent torque due to the permanent magnet.

9.16 CHARACTERISTICS OF STEPPER MOTORS

The characteristics of stepper motor are classified as follows:

- Static characteristic
- Dynamic characteristic

Torque displacement characteristic and torque current characteristic are known as static characteristic, as shown in Figure 9.42.

The torque stepping rate characteristic is known as dynamic characteristic, as shown in Figure 9.43.

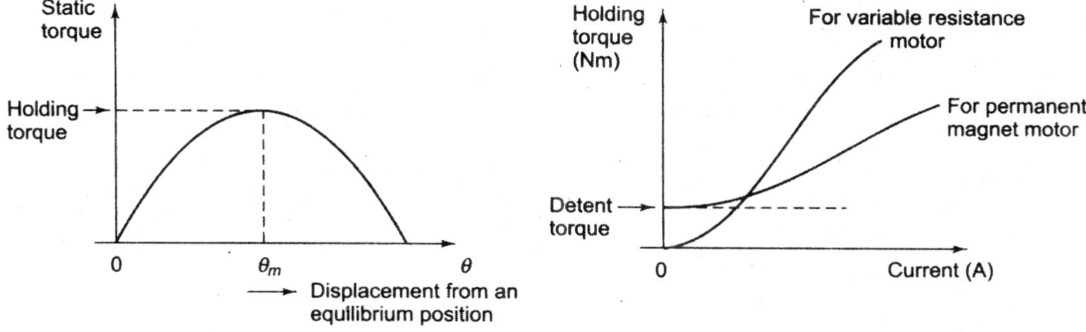

(a) Torque-displacement characteristic (b) Torque-current characteristic

Figure 9.42 Static Characteristic

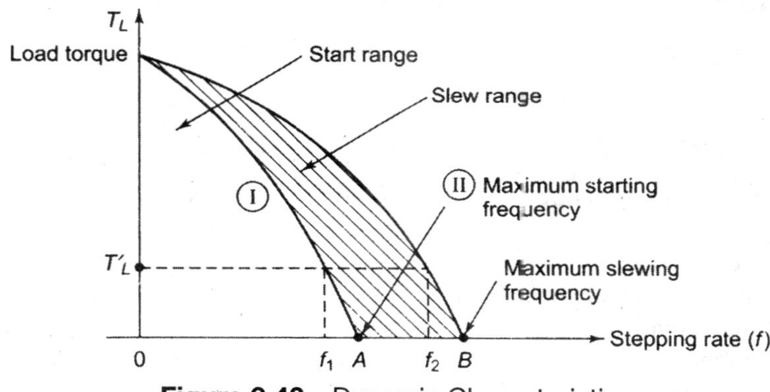

Figure 9.43 Dynamic Characteristic

9.17 DC SERVOMOTORS

DC motor converts electrical energy to mechanical energy. The motor is the opposite of a generator though a DC motor can be used both as a generator and as a motor.

AC motors have the following disadvantages:

(i) Non-linear characteristic.

(ii) Difficult to control its position through continuous speed or torque changes.

These disadvantages are absent in DC motors. The disadvantages of DC motors are as follows:

(i) Maximum torque is limited by commutation and not by heating as with other motors.

(ii) During collection of armature current by brush through commutator, arcing occurs. In course of time, this sparking causes pitting. Modern science has advanced a lot to reduce wear and tear and maintenance costs.

(iii) Brush and commutator requirement makes the DC motor expensive.

(iv) A basic requirement for motor is high torque and low inertia (mass) or a high torque-to-inertia ratio. This is achieved by reducing diameter and increasing length. Again modern technology has produced low-weight rotors, ironless rotors and electronic rotors.

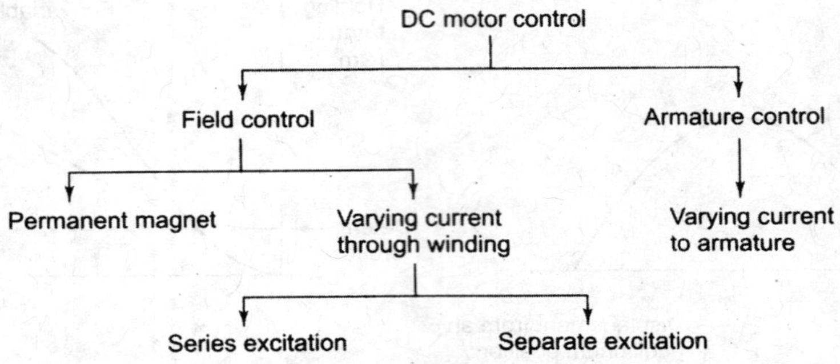

Figure 9.44 Excitation of DC Servomotors

9.17.1 Excitation of DC Servomotors

Figure 9.44 shows armature and field control of DC motor. Table 9.1 compares armature and field control of DC motor.

9.17.2 Armature-controlled DC Servomotor

Figure 9.45 shows an armature-controlled DC motor. Control signal is applied to the armature terminals. Let

R_a be the armature resistance in ohm,
L_a be the inductance of armature in henry,
i_a be the armature current,
i_f be the field current,
e_a be the applied voltage,
e_b be the motor torque and
J, f be the inertia and friction parameters of the motor and load, respectively.

Table 9.1 Comparison of Armature and Field Control of DC Motor

Armature Control	Field Control
• Large amount of current is required as the source has to meet the full power requirements of motor	• Requirement of current is small
• Back emf provides damping	• Damping has to be provided by motor and load
• Time constant on account of inductance of winding is less	• Time constant will be more because large winding is required to produce the necessary flux
• Speed of response of motor to changing current is fast	• Speed of response to changing current is slow
• Easy to provide constant field current	• Difficult to provide constant current to armature because of back emf
• Efficiency is high	• Efficiency is low

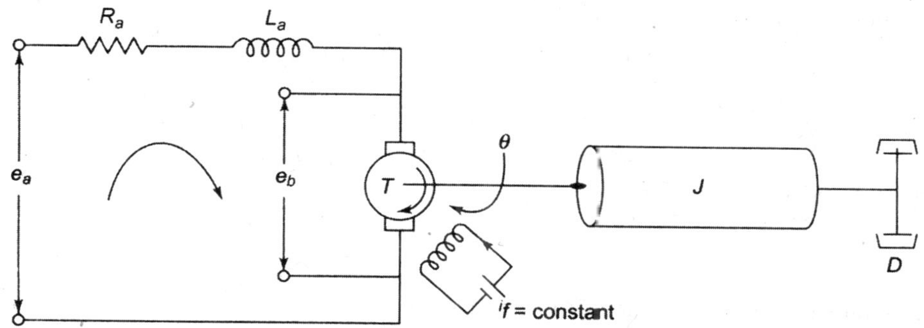

Figure 9.45 Equivalent Diagrams of Armature-controlled DC Servomotor

Now,

(i) Air gap flux (ϕ) is proportional to i_f

$$\therefore \qquad \phi \propto i_f$$

$$\text{i.e.,} \qquad \phi = k_f i_f \tag{9.26}$$

(ii) Torque is proportional to flux (ϕ) and armature current (i_a).

$$\therefore \qquad T \propto \phi i_a$$

$$\text{i.e.,} \qquad T \propto K_f i_f i_a$$

$$\text{i.e.,} \qquad T = K_1 K_f i_f i_a$$

$$\text{i.e.,} \qquad T = K_T i_a \tag{9.27}$$

$$\text{where} \qquad K_T = K_1 K_f i_f i_a \tag{9.28}$$

(iii) Back end (e_b) is proportional to speed.

$$\therefore \qquad e_b \propto \frac{d\theta}{dt}, \qquad \text{i.e.,} \qquad e_b = K_b \frac{d\theta}{dt} \tag{9.29}$$

Applying KVL to armature circuit, we have:

$$e_a = i_a R_a + L_a \frac{di_a}{dt} + e_b \tag{9.30}$$

The load torque equation is given by

$$J \frac{d^2\theta}{dt^2} + f \frac{d\theta}{dt} = T_m = K_T\, i_a \tag{9.31}$$

To find $\frac{\theta(s)}{E_a(s)}$, let us take Laplace transform of Equations (9.29), (9.30) and (9.31).

From Equation (9.29), we have:

$$E_b(s) = K_b s\theta(s) \tag{9.32}$$

From Equation (9.30), we have:

$$(L_a s + R_a) I_a(s) = E_a(s) - E_b(s) \tag{9.33}$$

From Equation (9.31), we have:

$$(Js^2 + fs)\theta(s) = T_m(s) = K_T I_a(s) \tag{9.34}$$

From Equation (9.33), we have:

$$I_a(s) = \frac{E_a(s) - E_b(s)}{sL_a + R_a} \tag{9.35}$$

Putting the value of E_b (s) in Equation (9.35), we have:

$$I_a(s) = \frac{E_a(s) - K_b s\theta(s)}{sL_a + R_a} \tag{9.36}$$

Substituting I_a (s) into Equation (9.34), we have:

$$(Js^2 + fs)\theta(s) = \left[\frac{E_a(s) - K_b s\theta(s)}{sL_a + R_a}\right] K_T$$

i.e.,

$$\left[\frac{(sL_a + R_a)(Js^2 + fs)}{K_T} + K_b s\right]\theta(s) = E_a(s)$$

$$\left[\frac{(sL_a + R_a)\ (Js^2 + fs) + K_T K_b s}{K_T}\right]\theta(s) = E_a(s)$$

$$\therefore \quad \frac{\theta(s)}{E_a(s)} = \frac{K_T}{\left[(sL_a + R_a)\ (Js + f) + K_T K_b\right]s} \tag{9.37}$$

From Equation (9.37), we get:

$$\frac{\theta(s)}{E_a(s)} = \left[\frac{\dfrac{K_T}{(sL_a + R_a)\ (Js + f)}}{1 + \dfrac{K_T K_b}{(sL_a + R_a)\ (Js + f)}}\right] \cdot \frac{1}{s} \tag{9.38}$$

The block diagram of Equation (9.38) is shown in Figure 9.46.

In Equation (9.37) if L_a is small, neglecting L_a, we have:

$$\frac{\theta(s)}{E_a(s)} = \frac{K_T}{R_a\left(Js^2 + sf\right) + K_T\ K_b s} = \frac{K_T / R_a}{Js^2 + s\left(f + \dfrac{K_T K_b}{R_a}\right)} = \frac{K_T / R_a}{Js^2 + sf'}$$

i.e.,

$$\frac{\theta(s)}{E_a(s)} = \frac{\dfrac{K_T}{R_a}}{s(Js + f')} \tag{9.39}$$

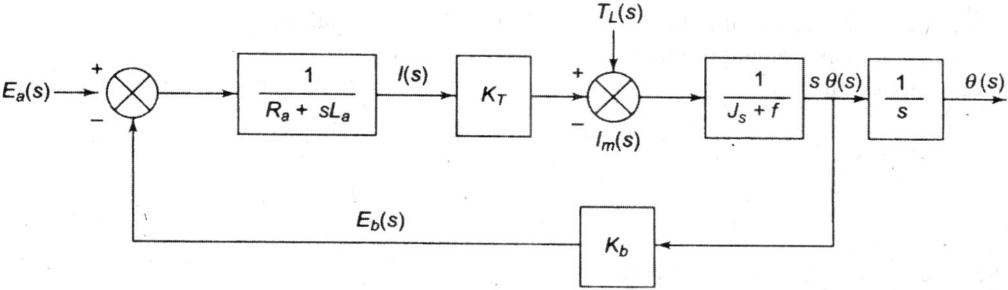

Figure 9.46 Block Diagram of Armature-controlled DC Motor

where
$$f' = f + \frac{K_T K_b}{R_a} \tag{9.40}$$

Equation (9.39) can also be written as:

$$\frac{\theta(s)}{E(s)} = \frac{K_m}{s(sT_m + 1)} \tag{9.41}$$

where
$$K_m = \frac{K_T}{R_a f'} \tag{9.42}$$

and
$$T_m = \frac{J}{f'} \tag{9.43}$$

9.17.3 Field-controlled DC Servomotors

Figure 9.47 shows field-controlled DC servomotor. In this figure,

R_f is the field winding resistance in ohm,
L_f is the field winding inductance in henry,
e_f is the field-controlled voltages in volt,
i_f is the field current in ampere,
T_M is the torque developed by the motor N-m,
J is the equivalent moment of inertia of motor and load referred to motor shaft in kg-m^2,

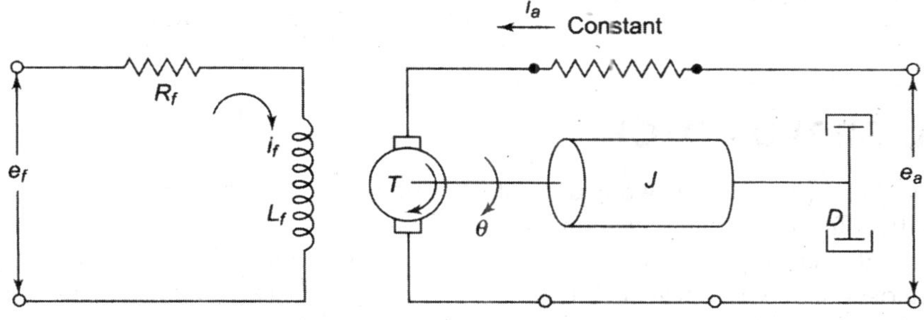

Figure 9.47 Field-controlled DC Servomotors

f is the viscous coefficient of motor and load referred to motor shaft in $\frac{\text{N-m}}{\text{rad/sec}}$ and

θ is the angular displacement of motor shaft in radiation.

Since armature current is fed from a constant current source in the field-controlled motor, we have

$$T_M = K_1 K_f i_f i_a = K_T i_f \tag{9.44}$$

where
$$K_T' = K_1 K_f i_a. \tag{9.45}$$

is another constant.

The field circuit equation is expressed by:

$$L_f \frac{di_f}{dt} + R_f i_f = e_f \tag{9.46}$$

The torque equation is expressed by:

$$J\frac{d^2\theta}{dt^2} + f\frac{d\theta}{dt} = T_M = K_f' i_f \tag{9.47}$$

Taking Laplace transform of Equations (9.46) and (9.47) and assuming zero initial conditions, we have

$$(sL_f + R_f)I_f(s) = E_f(s) \tag{9.48}$$

and
$$(Js^2 + fs)\theta(s) = T_M(s) = T_M(s) = K_T' I_f(s) \tag{9.49}$$

From Equations (9.48) and (9.49), we have

$$\frac{\theta(s)}{E_f(s)} = \frac{K_T}{(L_f s + R_f)(Js + f)} \tag{9.50}$$

$$\frac{\theta(s)}{E_f(s)} = \frac{K_M}{s(T_f s + 1)(T_M s + 1)} \tag{9.51}$$

where $K_M = \frac{K_T'}{R_f f}$ is the motor gain constant, $T_f = \frac{L_f}{R_f} =$ time constant of field circuit and

$T_m = \frac{J}{f} =$ mechanical time constant.

Figure 9.48 shows block diagram of field-controlled DC motor, as shown in Figure 9.47.

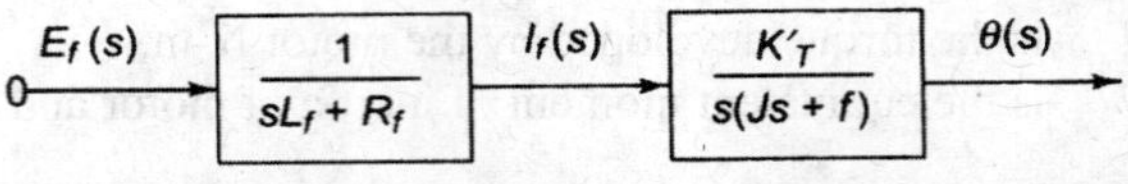

$$E_f(s) \rightarrow \boxed{\frac{1}{sL_f + R_f}} \xrightarrow{I_f(s)} \boxed{\frac{K'_T}{s(Js + f)}} \rightarrow \theta(s)$$

Figure 9.48 Block Diagram of Field-controlled DC Motor

9.18 AC SERVOMOTORS

An AC servomotor is basically a two-phase induction motor, as shown in Figure 9.49. A two-phase servomotor differs from a normal induction motor as follows.

The rotor of the servomotor is built with high resistance so that its *X/R* ratio is small and the torque-speed characteristic is nearly linear in contrast to the highly non-linear characteristic with large *X/R* ratio. Conventional induction motor with high *X/R* ratio is used for servo applications, but because of

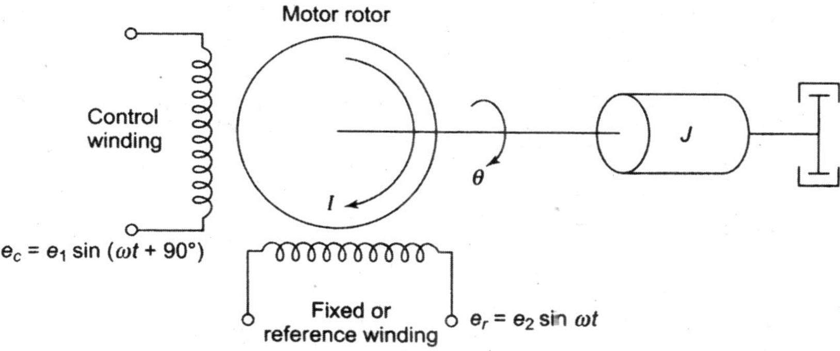

Figure 9.49 Two-phase AC Motor

the positive slope for part of the characteristic, the system using such a motor becomes unstable. The rotor construction is usually squirrel cage or drag-cup type having small diameter to reduce inertia.

The two-phase AC motor has two windings at right angles to each other and the two windings are supplied with voltages with 90° phase difference through a capacitor. The diameter of the rotor is small but the length of the rotor is long to reduce inertia for faster motor response. The frequency is normally 50 or 400 Hz. Voltage whose amplitude can be varied but of the same frequency is applied to the other winding. The flux will be rotating at the frequency of supply at synchronous speed $(\omega_s) = \frac{120f}{P}$. This rotating flux cuts to produce a varying flux and reacts with the stator flux causing the rotor to rotate in a smooth manner. From the speed-torque curve, the stall torque $T_s = K_m e_c$

$$\text{No load speed } (\omega_{nl}) = \frac{K_m e_c}{D_m}$$

K_m is different for different e_c

$$\text{The ratio } \frac{\text{Stalled rotor torque}}{\text{No load speed}} = \frac{T_s}{\omega_{nl}} \text{ is called } m$$

As in the case of DC motors,

$$T_D = K_m e_c - D_m \omega$$

i.e.,

$$T_D = J\frac{d^2\theta}{dt^2} + D\frac{d\theta}{dt} = K_m e_c$$

Figures 9.50 and 9.51 show flux by control and reference winding and torque-speed characteristic, respectively.

Transfer function,

$$G(s) = \frac{\theta(s)}{E_C(s)} = \frac{K_m}{Js^2 + Ds} = \frac{K}{s(T_m s + 1)} \tag{9.52}$$

$K = K_m / D$ = motor gain constant
$T_m = J/D$ = motor time constant

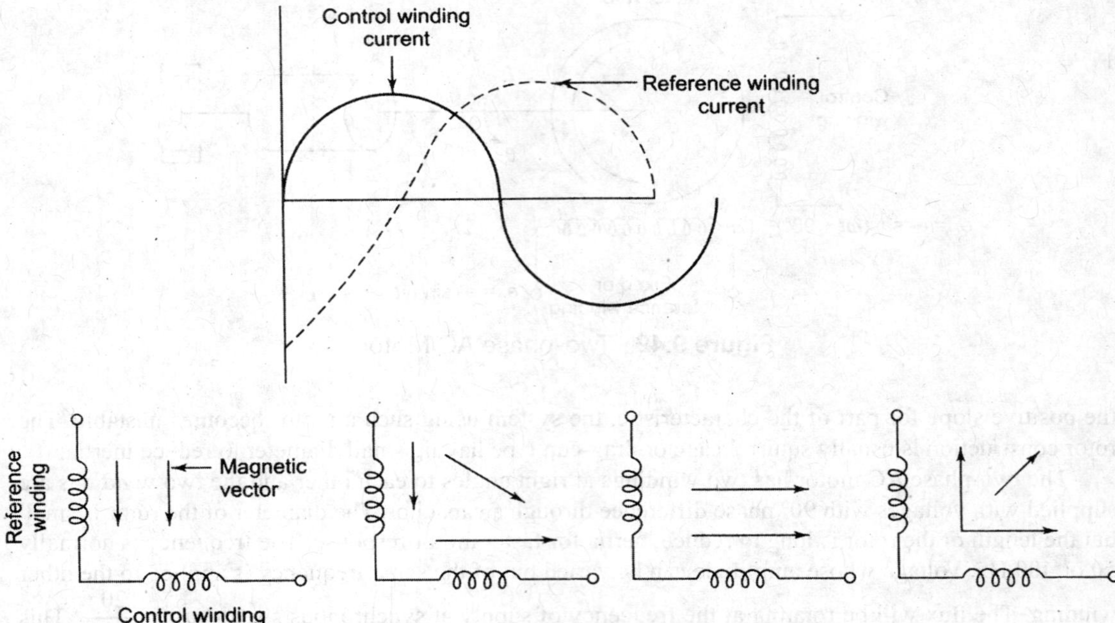

Figure 9.50 Flux by Control and Reference Winding

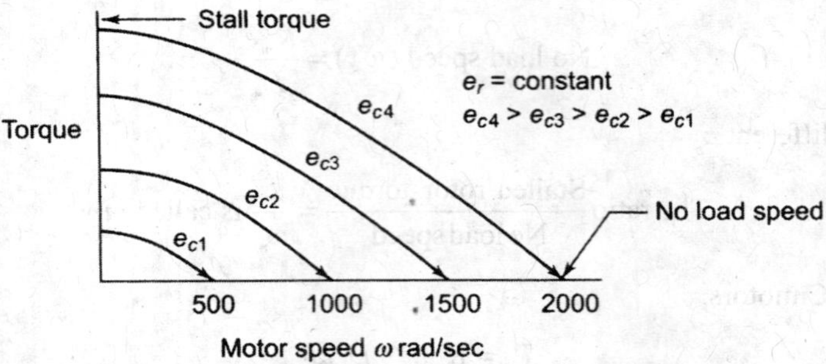

Figure 9.51 Torque-speed Characteristic

The block diagram is shown in Figure 9.52.

D is the viscous damping coefficient due to motor and load. The damping is high if the slope of the *T*–*ω* curve is steep. For $|T_m s| << 1$, the servomotor acts as an integrator $\left(\frac{1}{s}\right)$.

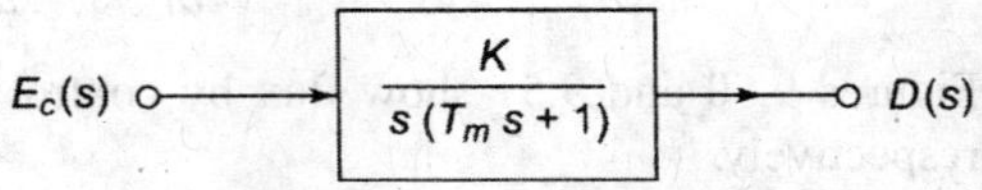

$$E_c(s) \longrightarrow \boxed{\frac{K}{s(T_m s + 1)}} \longrightarrow D(s)$$

Figure 9.52 Block Diagram

Example 9.3 An AC servomotor has both windings excited with 115 V AC. It has a stall torque of 2 lb ft and its coefficient of viscous friction is 0.2 lb ft sec.

(i) Determine its no-load speed.

(ii) It is connected to a constant load of 0.8 lb-ft and the coefficient of viscous friction of 0.05 lb-ft sec. through a gear pass with a ratio of 4. Calculate the speed at which the motor will run.

Solution

$$\text{No-load speed } \omega_{nl} = \frac{K_m V_m}{F_m} = \frac{2}{0.2} = 10 \text{ rad/sec}$$

(ii)
$$\text{Total viscous friction } (F_{\omega t}) = F_m + N^2 F_l = 0.2 + 4^2 \times 0.05 = 1 \text{ lb-ft sec}$$

$$T_l = K_m V_m - F\omega_t \omega$$

or,
$$0.8 = 2 - \omega$$

or,
$$\omega = 1.2 \text{ rad/ sec}$$

■

9.19 SERVOMECHANISM

Servomechanism is a power amplifying feedback control system. The controlled variable in it is a mechanical position or time derivative of position such as velocity and acceleration.

The essential parts of servomechanism are as follows:

(i) A transducer to convert angular position to an electrical signal.

(ii) An amplifier to improve the signal strength.

(iii) Device to develop torque such as motor.

(iv) Devices to transmit torque to load such as gears.

9.19.1 Positional Servomechanism

In a positional servomechanism, the position of the load must be maintained both under steady-state and transient conditions.

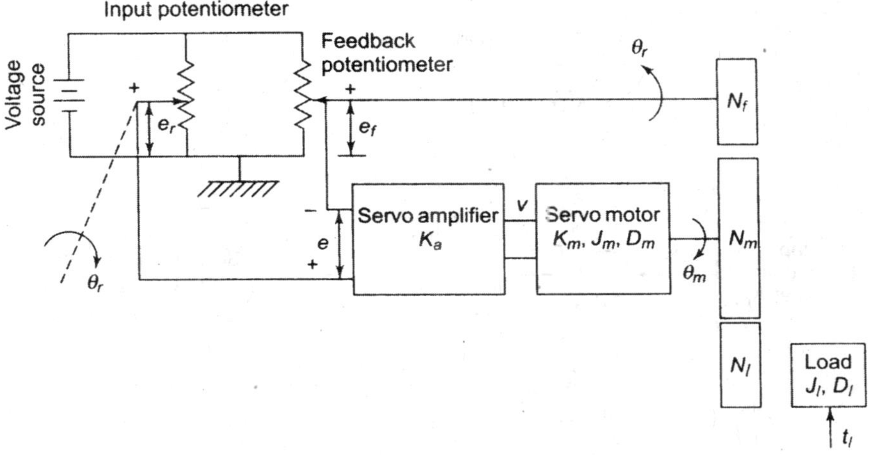

Figure 9.53 Positional Servomechanism

In Figure 9.53,

$\theta_r(t)$ is reference angular position, rad

e_r is input reference voltage at which the marked position of load is placed

e_f is feedback voltage due to feedback gear angular position

e is error signal $= e_R - e_f$

K_a is gain of servoamplifier

K_m is motor constant lb-ft or N-m/V or A

J_m is motor inertia, kg-m^2

D_m is damping friction of motor in N-m/rad/s

N_m is gear teeth of motor

θ_m is motor shaft angular position in rad

N_t is gear teeth of load

J_l is moment of inertia of load kg-m^2

D_l is damping viscous friction of load N-m/rad/s

r_l is additional torque on load

N_f is gear teeth of feedback

θ_f is angular position of feedback gear in rad

The two potentiometers must be identical in their electrical characteristics and connected in parallel with their positions set at the initial marked position of load. Figure 9.54 shows the block diagram of a servomotor.

$$e_r(t) = K_p\theta_r(t) \quad \text{and} \quad e_f(t) = K_p\theta_m(t)$$

$$E_r(s) = K_p\theta_r(s) \quad \text{and} \quad E_f(s) = K_p\theta_m(s)$$

$$e_r(t) = e_f(t) = e \quad \text{and} \quad v = K_a e$$

$$E_r(s) - E_f(s) = E(s) \quad \text{and} \quad V = K_a E(s)$$

i.e.,

$$V(s) = K_a K_p[\theta(s) - \theta_m(s)]$$

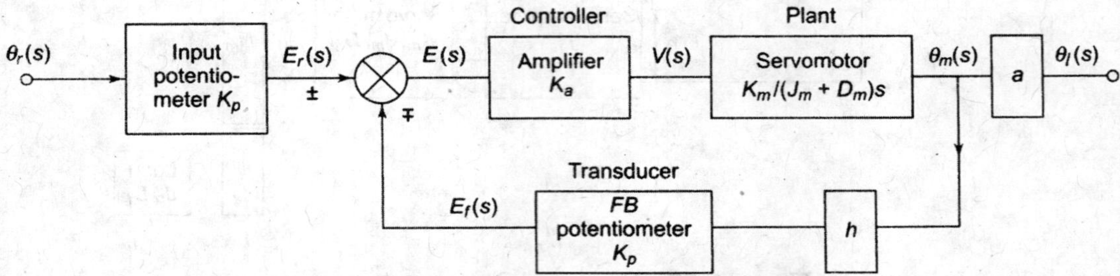

Figure 9.54 Block Diagram of a Servomotor

Gear train to load $(a) = \dfrac{N_m}{N_l}$

Gear train to feedback $(h) = \dfrac{N_m}{N_f}$

It is possible to transfer the load due to feedback gear to the motor as follows:

$$J = J_m + h^2 J_f \quad \text{and} \quad D = D_m + h^2 D_f$$

If we include the load,

$$J = J_m + h^2 J_f + a^2 J_1 \quad \text{and} \quad D = D_m - h^2 D_f + h^2 D_f$$

Since we are interested to feedback loop only, we can write:

$$J = J_m + h^2 J_f \quad \text{and} \quad D = D_m + h^2 D_f$$

$$\theta_m(s) = V(s)\left[\frac{K_m}{s(sJ+D)}\right] = \frac{K_p K_a K_m[\theta_r(s) - \theta_m(s)]}{s(Js+D)}$$

i.e., $$\theta_m(s) = \frac{K\theta_r(s)}{s(Js+D)} - \frac{K\theta_m(s)}{s(Js+D)} \text{ where } K = K_p K_a K_m$$

i.e., $$\left[1 + \frac{K}{s(Js+D)}\right]\theta_m(s) = \frac{K\theta_r(s)}{s(Js+D)}$$

i.e., $$\text{Transfer function } G(s) = \frac{\theta_m(s)}{\theta_r(s)} = \frac{\dfrac{K}{s(Js+D)}}{1 + \dfrac{K}{s(Js+D)}} = \frac{K}{s(Js+D)+K} = \frac{K}{Js^2 + Ds + K}$$

i.e., $$G(s) = \frac{\dfrac{K}{J}}{s^2 + \dfrac{D}{J}s + \dfrac{K}{J}} \quad (9.53)$$

The block diagram is reduced, as shown in Figure 9.55.

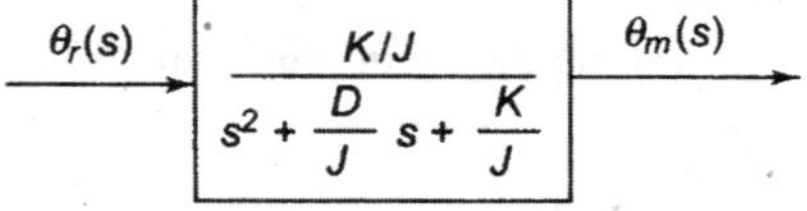

Figure 9.55 Reduced Block Diagram of a Servomotor

The servomotor may be either DC or two-phase AC servomotor. In a DC motor, the armature voltage is controlled by a DC amplifier and the field is provided by a permanent magnet.

Figure 9.56 shows the waveform diagram.

In AC system, the reference voltage is AC and the amplifier is an AC amplifier. The motor demodulates the signal so that the feedback signal is DC. The final waveform θ_l (s) will be same as for DC.

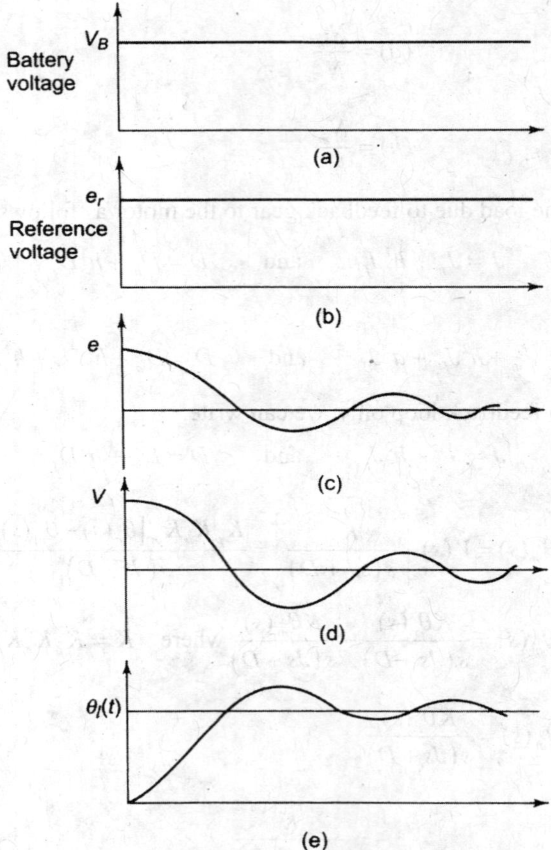

Figure 9.56 Waveform of a Servomotor

9.19.2 Rate Servomechanism

The derivative of the position, that is, velocity, is monitored and kept constant here, having velocity as the controlled variable. Figure 9.57 shows the schematic diagram of a rate servomechanism.

The system consists of a motor whose shaft speed $\theta(\omega)$ is to be held constant, the load attached to the shaft, a tachometer coupled to the motor shaft whose generated voltage is fed along with reference voltage and the difference error voltage amplified and applied to the DC motor to correct it to the desired speed.

Taking the input–output relationship of each block, for the amplifier block, output voltage is given by

$$V = K_a e$$

For the tachometer,

$$E_t = k_g \dot{\theta} = e_f$$

$$E_t = K_g \dot{\theta} = e_f$$

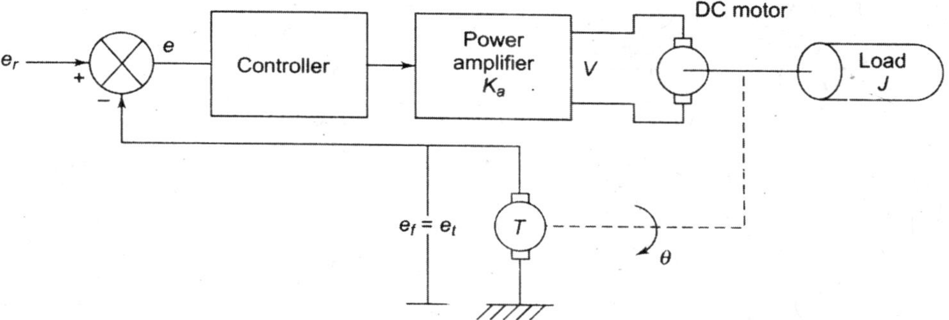

Figure 9.57 Rate Servomechanism

where K_a is the amplifier gain and K_g is tachometer constant.

$$e = e_r - e_f = e_r - K_g\dot{\theta}$$

The torque developed by the motor must be equal to the torque due to inertia and damping.

$$\left(\frac{J}{K_T}\right)\frac{d^2\theta}{dt^2} + \left(\frac{D}{K_T}\right)\frac{d\theta}{dt} = V = K_a\,(e_r - K_g\dot{\theta})$$

Solving for e_r, we have:

$$e_r = \left(\frac{D}{K_aK_T} + K_g\right)\frac{d\theta}{dt} + \left(\frac{J}{K_aK_T}\right)\frac{d^2\theta}{dt^2}$$

In the steady state $\frac{d^2\theta}{dt^2} = 0$, therefore, we have,

$$\dot{\theta}_{ss} = \frac{d\theta}{dt} = e_r\left(\frac{K_aK_T}{D + K_aK_TK_g}\right) = e_r\left(\frac{\frac{K_aK_T}{D}}{1 + \frac{K_aK_TK_g}{D}}\right) = \frac{e_rK_aK_m}{1 + K_aK_gK_m} \tag{9.54}$$

where $\frac{K_T}{D} = K_m$ is called the velocity gain constant which relates steady-state motor speed to armature voltage and hence is a velocity gain. Its unit is (rad/s)V. The time constant (τ_c) of the closed-loop system is defined by:

$$\tau_c = \frac{\dot{\theta}}{\ddot{\theta}} = \frac{\frac{J}{D}}{1 + \frac{K_aK_TK_g}{D}} = \frac{\frac{J}{D}}{1 + K_aK_mK_g} \tag{9.55}$$

where $K_aK_mK_g$ is the open-loop gain.

If $K_aK_mK_g$ is very large, or, $K_aK_mK_g >> 1$, we get, from Equation (9.54):

$$\dot{\theta}_{ss} = \frac{e_r}{K_g} \tag{9.56}$$

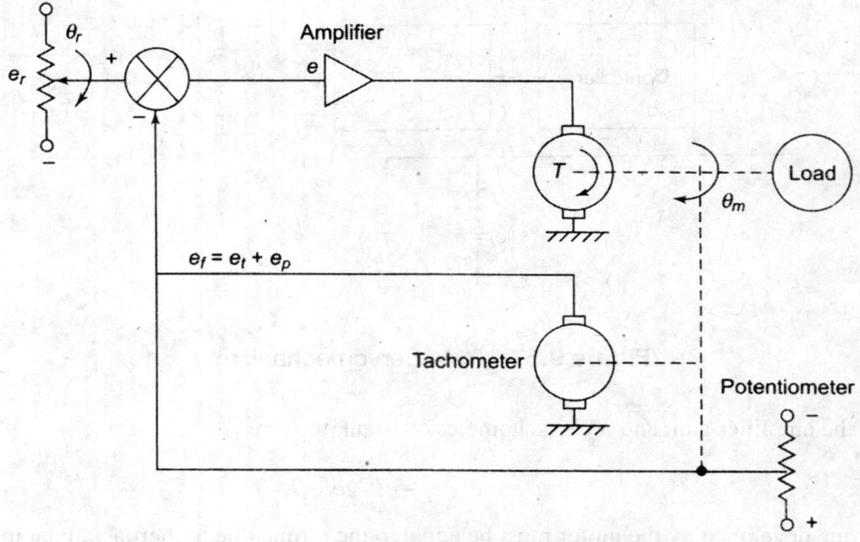

Figure 9.58 Damping by Derivative Feedback

The steady-state speed depends only on input voltage and the characteristic of the tachometer.

9.19.3 Damping by Derivative Feedback

The slope of the speed-torque curve gives the viscous damping or damping coefficient. The feedback introduced by the voltage generated by a tachometer is a derivative feedback as $V_g = K_g w$. Figure 9.58 shows damping by derivative feedback.

The error voltage is given by

$$e = e_r - (e_t + e_p) = K_p(\theta_r - \theta_m) - K_g\theta_m$$

Equating the torques, we get:

$$J\ddot{\theta}_m + D\dot{\theta}_m = K_p K_a K_m(\theta_r - \theta_m) - K_g K_d K_m \dot{\theta}_m \quad (9.57)$$

K_d is the amplifier gain of the voltage from tachometer.

Solving θ_r,

$$\theta_r = \frac{JK_m}{K}\ddot{\theta}_m + \left(\frac{K + K_g K_d K_m}{K}\right)\dot{\theta}_m + \theta_m \quad (9.58)$$

where $K = K_p K_a K_m$ is the open-loop gain.

Damping ratio from the second-order Equation (9.58) after dividing throughout by J/K is

$$2\xi\omega_n = \frac{D + K_g K_d K_m}{K} \quad \text{and} \quad \omega_n = \sqrt{\frac{K}{J}}$$

$$\therefore \quad \xi = \frac{D + K_g K_d K_m}{2\sqrt{KJ}} \quad (9.59)$$

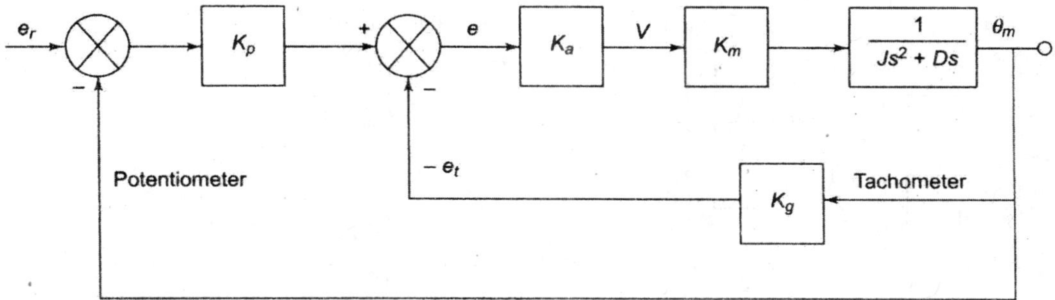

Figure 9.59 Block Diagram

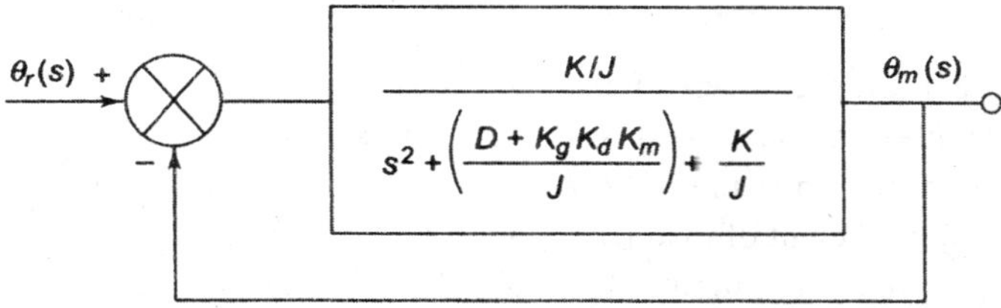

Figure 9.60 Simplified Block Diagram

For potentiometer alone, we can write:

$$2\xi\omega_n = \frac{D}{J} \quad \text{and} \quad \omega_n = \sqrt{\frac{K}{J}}$$

$$\therefore \quad \xi = \frac{D}{2\sqrt{KJ}} \tag{9.60}$$

Therefore, damping ratio is increased by $K_g K_d K_m$ by feeding back negatively the tachometer voltage. Figure 9.59 shows the block diagram.

If we take Laplace transform of Equation (9.58), we get

$$\frac{\theta_m(s)}{\theta_r(s)} = \frac{\frac{K}{J}}{s^2 + \left(\frac{D + K_g K_d K_m}{J}\right)s + \frac{K}{J}} \tag{9.61}$$

Figure 9.60 shows simplified block diagram of Equation (9.61).

Example 9.4 Figure E9.1 shows a schematic diagram of an automatic control system where the angular position of an antenna is controlled.

The system has the following data:

The gear ratio $(N_1/N_2) = 0.1$, voltage gain of the amplifier $(A) = 1{,}000$
Motor torque constant = 0.25 N-m/A of field current
Field winding resistance $(R_f) = 20$ ohm
Self-inductance of winding $(L_f) = 1$ H

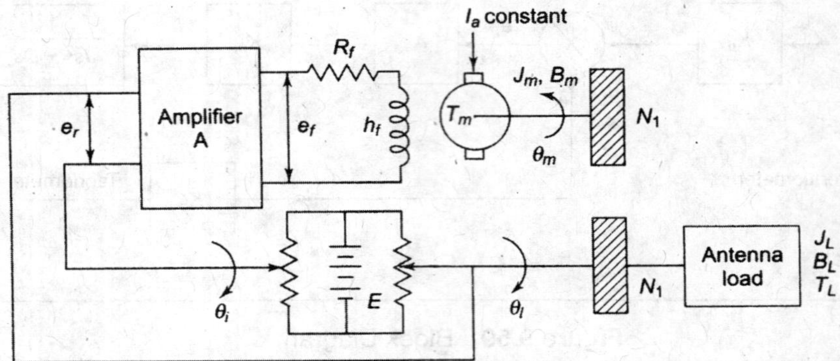

Figure E9.1

Error detector sensitivity = 0.02 V/1° error
Moment of inertia of load (J_L) = 50 × 10^{-6} kg-m²
Moment of inertia of motor (J_M) = 0.5 × 10^{-6} kg-m²
Viscous friction coefficient of load (B_L) = 0.9 × 10^{-5} N-m/rad/s
Viscous friction coefficient of motor (B_m) = 0.1 × 10^{-7} N-m/rad/s

Determine the transfer function relating input displacement θ_i and output displacement θ_0.

Solution

$$\frac{\theta_L}{\theta_m} = \frac{\dot{\theta}_L}{\dot{\theta}_m} = \frac{\ddot{\theta}_L}{\ddot{\theta}_m} = \frac{N_1}{N_2}$$

Equivalent moment of inertia referred to motor shaft is

$$J = J_m + \left(\frac{N_1}{N_2}\right)^2 J_L = 1 \times 10^{-6} \text{ kg-m}^2$$

Similarly, equivalent viscous friction referred to motor shaft is

$$D = D_m + \left(\frac{N_1}{N_2}\right) B_L = 1 \times 10^{-7} \text{ Nm/rad/sec}$$

Figure E9.2 shows the block diagram of the system.

The open-loop transfer function is given by

$$G(s) = \frac{K_s A K_T \left(\frac{N_1}{N_2}\right)}{(R_f + sL_f)(Js^2 + Bs)} = \frac{0.2 \times 1{,}000 \times 0.25 \times 0.1 \times \frac{180}{\pi}}{(20+s)(s^2+0.1s)}$$

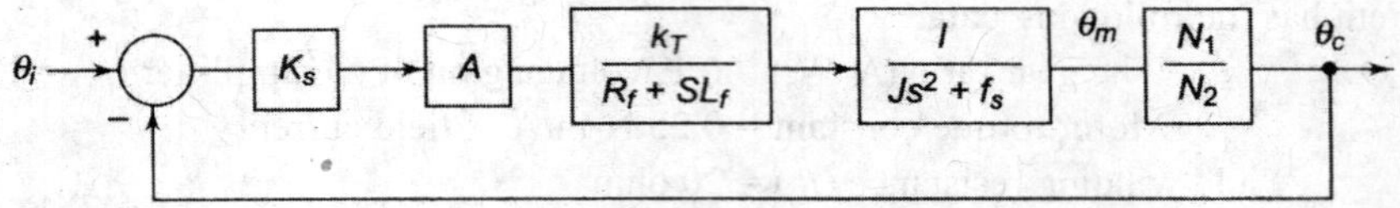

Figure E9.2

The closed-loop transfer function is given by

$$\frac{\theta_c(s)}{\theta_i(s)} = \frac{G(s)}{1+G(s)} = \frac{286.5\times 10^6}{(s+20)(s^2+0.1s)+286.5\times 10^6}$$

SIGNIFICANT POINTS

The slip (s_f) of the rotor with respect to the forward rotating field is given by

$$s_f = \frac{n_s - n}{n_s} = 1 - \frac{n}{n_s}$$

where n_s and n are the synchronous speed and rotor speed, respectively.

The slip (s_b) of the rotor with respect to the backward rotating field is given by

$$s_b = \frac{n_s - (-n)}{n_s} = \frac{n_s + n}{n_s} = 1 + \frac{n}{n_s}$$

$$s_b = 1 + \frac{n}{n_s} = 2 - \left(1 - \frac{n}{n_s}\right) = 2 - s$$

Equivalent impedance(Z_f) of imaginary rotor due to forward rotating field is

$$Z_f = R_f + jX_f = \left(\frac{R_2'}{2s} + j\frac{X_2'}{2}\right) \Bigg\| \left(j\frac{X_M}{2}\right) = \frac{\left(\frac{R_2'}{2s} + j\frac{X_2'}{2}\right)\times\left(j\frac{X_M}{2}\right)}{\left(\frac{R_2'}{2s} + j\frac{X_2'}{2}\right)\times\left(j\frac{X_M}{2}\right)}$$

and the equivalent impedance (Z_b) of imaginary rotor due to backward rotating field is

$$Z_b = R_b + jX_b = \left[\frac{R_2'}{2(2-s)} + j\frac{X_2'}{2}\right] \Bigg\| \left(j\frac{X_M}{2}\right)$$

$$= \frac{\left[\frac{R_2'}{2(2-s)} + j\frac{X_2'}{2}\right]\times\left(j\frac{X_M}{2}\right)}{\left[\frac{R_2'}{2(2-s)} + j\frac{X_2'}{2}\right]+\left(j\frac{X_M}{2}\right)}$$

If P_{gf} and P_{gb} are the air gap power due to forward and backward field, respectively, we can write:

$$P_{gf} = I_m^2 R_f \text{ and } P_{gb} = I_m^2 R_b$$

$$\therefore \quad P_g = P_{fg} - P_{gb} = I_m^2 R_f - I_m^2 R_b = I_m^2 (R_f - R_b)$$

The torque due to forward field is given by

$$\tau_f = \frac{1}{\omega_s} P_{gf} = \frac{P_{gf}}{2\pi n_s}$$

Again, the torque due to backward field is given by

$$\tau_f = \frac{1}{\omega_s} P_{gb} = \frac{P_{gb}}{2\pi n_s}$$

where ω_s represents the synchronous speed.

The resultant torque is given by

$$\tau_{in} = \tau_f - \tau_b = \frac{1}{2\pi n_s}(P_{gf} - P_{gb}) = \frac{I_m^2 (R_f - R_b)}{2\pi n_s}$$

The Cu loss of the single-phase motor is given by

$$P_r = P_{rf} + P_{rb} = sP_{gf} + (2-s)P_{gb}$$

where P_{rf} and P_{rb} are the Cu loss of the rotor due to forward and backward field, respectively.

The mechanical power developed in induction motor is given by

$$P_{mech} = \omega\tau_{in} = (1-s)\omega_s\tau_{in} = (1-s)(P_{gf} - P_{gb}) = (1-s)I_m^2(R_f - R_b)$$

Shaft output power is given by

$$P_{out} = P_{mech} - \text{Core loss} - \text{Mechanical loss} - \text{Stray loss}$$

$$= P_{mech} - P_{rot}$$

where P_{rot} = Friction loss + Windage loss + Core loss = Rotational losses

SHORT QUESTIONS AND ANSWERS

Q.1 How are the 1-phase AC motors made self-starting?

Ans. 1-phase AC motors are made self-starting by providing auxiliary winding in parallel with main winding.

Q.2 Where and why is a centrifugal switch used in a split-phase motor?

Ans. It is placed on the shaft of the rotor and electrically connected with the auxiliary winding. When the motor achieves 70 to 80 per cent of its synchronous speed, it disconnects the auxiliary winding.

Q.3 What types of windings are used in split-phase motor?

Ans. Double-layer chain or concentric type winding is used in split-phase motor.

Q.4 Why and where is a capacitor used in the one-phase AC motor?

Ans. It is used in series with the auxiliary winding to give a phase displacement between starting and main winding so that starting torque is improved.

Q.5 What type of rotor is used in a capacitor-start motor?

Ans. Squirrel cage rotor is used in a capacitor-start motor.

Q.6 How can the direction of rotation of capacitor-start motor be reversed?

Ans. By reversing either the running or starting winding that leads to the supply lines, the direction of rotation of a capacitor-start motor can be reversed.

Q.7 What is a permanent capacitor motor?

Ans. It is a one-phase AC motor having two windings and a permanent capacitor. It has no centrifugal switch.

Q.8 State the uses of permanent capacitor motors?

Ans. Permanent capacitor motors are used for ceiling fans and table fans where starting torque is not high.

Q.9 What is a shaded-pole motor?

Ans. It is a single-phase AC motor. A portion of the stator core is encircled by a copper ring.

Q.10 What type of row is used in capacitor-start capacitor-run shaded-pole motor?

Ans. Squirrel cage rotor is used in capacitor-start capacitor-run shaded pole motor.

Q.11 What is a universal motor? Why is it called so?

Ans. The 1–phase fractional horse power motors ranging from $\frac{1}{100}$ to $\frac{1}{2}$ HP in which field winding and armature windings are connected in series are called universal motors. Since these motors can be used with both DC and AC supply, they are called universal motors.

SUPPLEMENTARY PROBLEMS

1. The constants of a 1/4-HP, 230-V, 4-pole, 50-Hz, single-phase induction motors are as follows:

$$R_1 = 10.0\ \Omega,\ X_1 = 12.8\ \Omega,\ R_2' = 11.65\ \Omega$$

$$X_2' = 12.8\ \Omega,\ X_m = 258.0\ \Omega$$

The total load is such that the machine runs at 3 per cent slip, when the applied voltage is at 210 V. The iron losses is 35.5 W at 210 V. Find the following:

(a) input current,
(b) power developed,
(c) shaft power (if mechanical losses are 7 W) and
(d) efficiency

[*Ans.* (a) 1.8586 A, (b) 166.2 W, (c) 159.2 W, (d) 62.85%]

2. A 230-V, 380-W, 4-pole, single phase induction motor gave the following test results:

No-load test:	230 V	84 W	2.8 A
Blocked rotor test:	110 V	460 W	6.2 A

The stator winding resistance is 4.6 Ω and during the blocked rotor test, two auxiliary windings are kept open. Determine the equivalent circuit parameters?

[*Ans.* $R_1 = 4.6\ \Omega$, $R'_2 = 7.067\ \Omega$, $X_1 = X'_2 = 6.68\ \Omega$, $X_m = 143.75\ \Omega$]

MULTIPLE-CHOICE QUESTIONS AND ANSWERS

1. In a double revolving field theory of single-phase induction motor, if the slip of the forward motor is s, then the slip of backward motor is

(a) $2s$ (b) s (c) $2 - s$ (d) $s - 2$

2. In case of a split-phase motor, the phase shift between current in two windings is around

(a) 30° (b) 70° (c) 90° (d) 120°

3. Which one of the following capacitor-start split-phase motor will have the largest value of capacitance?

(a) $\frac{1}{2}$ HP; 3,450 rpm
(b) $\frac{1}{4}$ HP; 1,725 rpm
(c) $\frac{1}{2}$ HP; 1,140 rpm
(d) $\frac{3}{4}$ HP; 1,140 rpm

4. Compared with a resistor split-phase motor, a capacitor-start motor has
   (a) higher starting torque (b) lower starting torque
   (c) higher running torque (d) lower running torque
5. If a capacitor-start induction motor is switched on to supply with its capacitor replaced by an inductor of equivalent reactance, the motor will
   (a) not start at all (b) start and run at rated speed
   (c) start and run slowly (d) start with humming noise
6. Capacitor-start capacitor-run induction motor is basically a
   (a) AC series motor (b) two-phase motor
   (c) synchronous motor (d) commutator
7. The main drawback of two capacitor-type induction motor is
   (a) low power factor (b) poor starting torque
   (c) pulsating torque (d) high cost
8. In comparison with resistance-start induction motor, the permanent capacitor motor,
   (a) has better power factor (b) is more expensive
   (c) has better starting torque (d) all of these
9. In comparison with capacitor-start capacitor-run motor, the permanent capacitor motor has
   (a) poor power factor (b) poor starting torque
   (c) simple construction and low cost (d) all of these
10. The rotating magnetic field in a shaded pole motor is developed by using
   (a) shading coils (b) salient poles
   (c) a capacitor (d) damper windings
11. The phase splitting in a shaded pole motor is obtained by placing a shading coil at the slot cut around the
   (a) smaller part of the hole (b) larger part of the hole
   (c) both the parts of the hole (d) any of the above
12. In a shaded pole motor, shading coils are used to
   (a) reduce winding losses (b) reduce friction losses
   (c) produce rotating magnetic field (d) protect against sparking
13. In a shaded pole squirrel cage induction motor, the flux in the shaded part always
   (a) leads the flux in the unshaded pole segment
   (b) is in phase with the flux in the unshaded pole segment
   (c) lags the flux in the unshaded pole segment
   (d) none of the above
14. Consider the following statements regarding fractional horse power shaded pole motor:
   1. Its direction of rotation is from unshaded to shaded portion of the poles.
   2. Its direction is from shaded to unshaded portion of the poles.
   3. It can remain stalled for short periods without any harm.
   4. It has a very poor power factor.

   Of these statements
   (a) 1, 3 and 4 are correct (b) 2, 3 and 4 are correct
   (c) 2 and 4 are correct (d) 1 and 4 are correct
15. The direction of shaded pole motor can be reversed by
   (a) reversing supply terminals
   (b) shifting any shading coil to the other half of the pole

(c) shifting two shading coils to the other half of the pole
(d) shifting all the shading coils to the other half of the pole

16. The wattage rating for a ceiling motor will be in the range
(a) 50 to 250 W (b) 250 to 500 W (c) 50 to 150 W (d) 10 to 20 W

17. What will happen if a single-phase supply is connected to an ordinary DC series motor?
(a) It will not run at all (b) It will run, but not satisfactorily
(c) The armature will burn out (d) The field winding will burn out

18. AC series motors are built with as few field turns as possible to reduce
(a) flux (b) eddy current losses (c) reactance (d) speed

19. A universal motor runs at
(a) higher speed with DC supply and with less sparking
(b) higher speed with AC supply and with less sparking
(c) same speed with both AC and DC supplies
(d) higher speed with AC supply but with increased sparking at the brushes

20. Speed of a universal motor is
(a) dependent on frequency of supply (b) proportional to frequency of supply
(c) independent of frequency of supply (d) none of the above

21. The speed of a universal motor can be controlled by
(a) introducing a variable resistance in series with the motor
(b) tapping the fields at various points
(c) centrifugal mechanism
(d) all of the above

22. In repulsion motors, brush angle with respect to magnetic axis is kept at
(a) 0° to 15° (b) 15° to 45° (c) 90° (d) 180°

23. The torque-speed characteristic of a repulsion motor resembles which of the following DC motor characteristic?
(a) separately excited (b) shunt (c) series (d) compound

24. In a single-phase repulsion motor power factor is
(a) always leading (b) high at low speed (c) high at high speed (d) always unity

25. In a single-phase repulsion motor by shifting the brush position
(a) speed and direction of rotation can be controlled
(b) only speed can be controlled
(c) speed, torque and direction of rotation can be controlled
(d) only torque can be controlled by varying the speed at constant output power

Answers

1 (c)	2 (a)	3 (d)	4 (a)	5 (a)	6 (b)
7 (d)	8 (d)	9 (d)	10 (a)	11 (a)	12 (c)
13 (c)	14 (a)	15 (d)	16 (c)	17 (b)	18 (c)
19 (a)	20 (c)	21 (b)	22 (b)	23 (c)	24 (c)
25 (c)					

Basic Definition, Hysteresis and Eddy Current Losses

Appendix

A

A.1 RESISTANCE AND RESISTIVITY

Resistance is the property of a substance due to which it opposes the flow of current through it. The unit of resistance is ohm (Ω).

A conductor shown in Figure A.1 is taken length = l m and area of cross-section = a m^2.

If R is its resistance, we have

$R \propto l$ when a is constant.

$R \propto \frac{1}{a}$ when l is constant.

$R \propto \frac{1}{a}$ when both l and a are not constant.

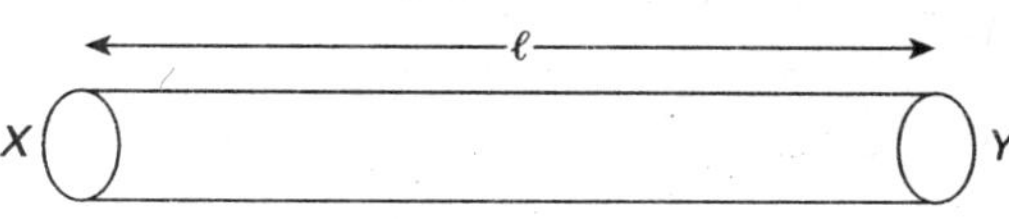

Figure A.1

$$\therefore \qquad R = \rho \frac{l}{a} \tag{A.1}$$

In Equation (A.1) ρ is a constant of proportionality, which is known as resistivity of the material.

If $l = 1$ m and $a = 1$ m^2, from Equation (A.1), we have $R = \rho$.

From Equation (A.1), we have

$$\rho = \frac{Ra}{l} = \frac{\Omega \text{m}^2}{\text{m}} = \Omega - \text{m} \tag{A.2}$$

The unit of resistivity (ρ) is Ω-m.

A.2 INDUCTANCE

Usually, inductors are made of many turns of fine wire often wounded on a magnetic material that is capable of storing more energy, and it is property of one of the circuit elements by which energy is capable of being stored in magnetic flux field.

A.2.1 Self-inductance

The flux linked with the coil changes when the current in a coil is changing. In this case, an emf is produced in the coil. Let the permeability of the coil be taken as constant. The induced emf of the coil is proportional to the rate of change of current:

$$e \propto \frac{di}{dt}$$

i.e.,

$$e = L\frac{di}{dt} \tag{A.3}$$

In Equation (A.3), L is the constant of proportionality, which is known as self-inductance of the coil. As per Faraday's law of electromagnetic induction, the induced emf in a coil is expressed by

$$e = N\frac{d\phi}{dt} \tag{A.4}$$

where N is the number of turns of the coil.
From Equations (A.3) and (A.4), we get

$$L\frac{di}{dt} = N\frac{d\phi}{dt}$$

i.e.,

$$L = N\frac{d\phi}{di} \tag{A.5}$$

If ϕ versus i graph is taken as linear, we have from Equation (A.5)

$$L = \frac{N\phi}{i} \tag{A.6}$$

Since

$$\phi = \frac{Ni}{S} = \frac{Ni}{\frac{l}{\mu a}} = \left(\frac{N\mu a}{l}\right)i \tag{A.7}$$

'S' is the reluctance of the coil that is equal to $\frac{l}{\mu a}$.
From Equation (A.7), we have

$$\frac{\phi}{i} = \frac{N\mu a}{l} \tag{A.8}$$

Using Equation (A.8), we have from Equation (A.6)

$$L = \frac{N\phi}{i} = \frac{N^2\mu a}{l} = \frac{\mu N^2 a}{l} \tag{A.9}$$

From Equation (A.9), we can conclude that inductance of any coil is dependent on its length and area of cross-section.

From Equation (A.6), self-inductance can also be defined as the flux linkage of the coil per ampere current flow through it. The unit of self-inductance is henry (H).

A.2.2 Mutual Coupling

Circuits become mutually coupled when the interchanging of energy takes place from one circuit to other. The coupling between the two circuits may be conductive, electromagnetic, or electrostatic.

A.2.3 Magnetic Coupling

The magnetic coupling has been shown in Figure A.2 in which a portion of magnetic flux established by one coil interlinks with the other. Hence, energy may be transferred from one circuit to the other through the medium of magnetic flux, which is common to both.

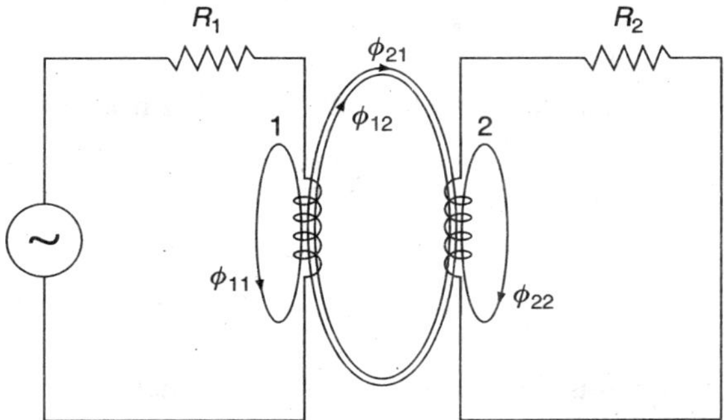

Figure A.2 Magnetic Coupling

In Figure A.2, let ϕ_1 and ϕ_2 be the total flux established by V_1 and V_2, respectively.

ϕ_1 and ϕ_2 can be expressed as follows:

$$\phi_1 = \phi_{11} + \phi_{12} \tag{A.10}$$

$$\phi_2 = \phi_{21} + \phi_{22} \tag{A.11}$$

In Equations (A.10) and (A.11), ϕ_{11} and ϕ_{22} are the only fluxes that are linked to coil 1 and coil 2, respectively, whereas ϕ_{12} and ϕ_{21} in these two equations are the mutual fluxes that are linked to the turns of coil 2 and coil 1, respectively.

A.2.4 Mutual Inductance

The induced voltage in coil 2 due to a change in flux ϕ_{12} can be expressed by

$$e_2 = N_2 \frac{d\phi_{12}}{dt} \tag{A.12}$$

Since ϕ_{12} is related to i_1, e_2 is proportional to the rate of change of i_1:

$$e_2 \propto \frac{di_1}{dt}$$

i.e.,

$$e_2 = M_{12} \frac{di_1}{dt} \tag{A.13}$$

In Equation (A.13), M_{12} is the constant of proportionality. It is known as mutual inductance between two coils, which is dimensionally equivalent to self-inductance L. The unit of mutual inductance is henry (H).

From Equations (A.12) and (A.13), we have

$$M_{12} \frac{di_1}{dt} = N_2 \frac{d\phi_{12}}{dt}$$

i.e.,

$$M_{12} = N_2 \frac{d\phi_{12}}{di_1} \tag{A.14}$$

Also,
$$M_{21} = N_1 \frac{d\phi_{21}}{di_2} \tag{A.15}$$

For linear variation of flux and current, we have the following from Equations (A.14) and (A.15):

$$M_{12} = N_2 \frac{\phi_{12}}{i_1} \tag{A.16}$$

and
$$M_{21} = N_1 \frac{\phi_{21}}{i_2} \tag{A.17}$$

Let the permeability of the mutual flux path be constant. We have $M_{12} = M_{21} = M$.

Therefore, the mutual inductance between two coils is defined as the weber-turns in one coil due to current through the other coil.

A.3 COEFFICIENT OF COUPLING

Coefficient of coupling gives an idea of how much of the flux produced by one coil is linking with the other; i.e., it indicates the extent to which the two inductors are coupled independently of the size of the inductor concerned.

Let the two inductive coupled coils 1 and 2 shown in Figure A.2 be the number of turns N_1 and N_2, respectively. The expression for inductance of coil 1 and coil 2 becomes

$$L_1 = \frac{\mu_1 N_1^2 a}{l} \tag{A.18}$$

and
$$L_2 = \frac{\mu_2 N_2^2 a}{l} \tag{A.19}$$

In Figure A.2, ϕ_1 is the total flux established in coil 1 due to flow of current i_1 through it. If a part of it, i.e., $\phi_{12} = k_1\phi_1$, is linked with coil 2, we have

$$\phi_1 = \frac{N_1 i_1 \mu a}{l} \tag{A.20}$$

$$M = \frac{k_1 \phi_1 N_2}{i_1} \tag{A.21}$$

where $k_1 \leq 1$.

From Equations (A.20) and (A.21), we have

$$M = \frac{k_1 N_1 N_2 \mu a}{l} \tag{A.22}$$

In Figure A.2, ϕ_2 is the total flux established in coil 2 due to flow of current i_2 through it. If a part of it, i.e., $\phi_{21} = k_2\phi_2$, is linked with coil 1, we have

$$\phi_2 = \frac{N_2 i_2 \mu a}{l} \tag{A.23}$$

If a fraction, $k_2 \leq 1$, of this flux is linked with coil 1, then

$$M = \frac{k_2\phi_2 N_1}{i_2} = k_2 \frac{N_2 N_1 \mu a}{l} \tag{A.24}$$

From Equations (A.22) and (A.24), we have

$$M^2 = k_1 k_2 \left(\frac{N_1^2 \mu a}{l}\right)\left(\frac{N_2^2 \mu a}{l}\right) = k_1 k_2\ L_1 L_2$$

$$\therefore \quad M = \sqrt{k_2 k_2}\ \sqrt{L_1 L_2} = k\sqrt{L_1 L_2} \tag{A.25}$$

where $k = \sqrt{k_1 k_2}$

From Equation (A.25),

$$k = \frac{M}{\sqrt{L_2 L_2}} \tag{A.26}$$

The constant k is known as the coefficient of coupling which has been expressed in Equation (A.26). The ratio of mutual inductance M to the square root of the product of inductance of coil 1 and coil 2 determines the value of k.

The coils are called tightly coupled or closely coupled, when the flux due to one coil is fully linked with the other. The value of k is unity in this case. If the flux in one coil does not link with the other coil, $k = 0$. The coils in this case are said to be magnetically isolated from each other or loosely coupled.

A.4 DOT CONVENTION

The emf induced due to mutual inductance may be aiding or opposing the emf induced due to self-inductance in a circuit, which totally depends on the relative directions of currents, i.e., the relative modes of windings of the coils as well as physical placement of one winding with respect to other. Using dot convention, the nature of mutually induced emf can be determined very easily. The actual mode of the winding is not required to show. Figure A.3 shows the sign of mutually induced emf.

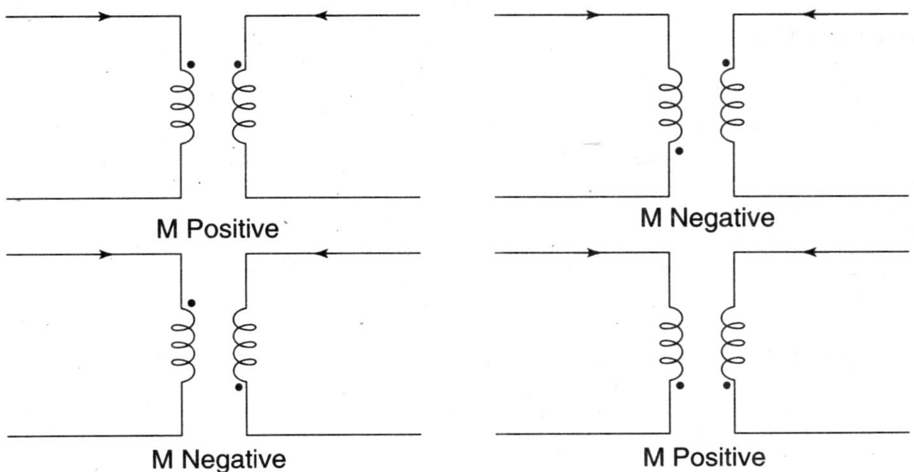

Figure A.3 Dot Convention

A.5 INDUCTIVE COUPLING IN SERIES

If two inductors are coupled in series and mutual inductance exists between them, the equivalent inductance of this series coupling can be determined for series aiding and series opposing. This is explained below.

(i) Series Aiding

The connection of two coils in series aiding is shown in Figure A.4 when the flux produced by the two coils is additive in nature as per dot convention.

Let

L_1 be the self-inductance of coil 1.
L_2 be the self-inductance of coil 2.
M is the mutual inductance between coil 1 and coil 2.

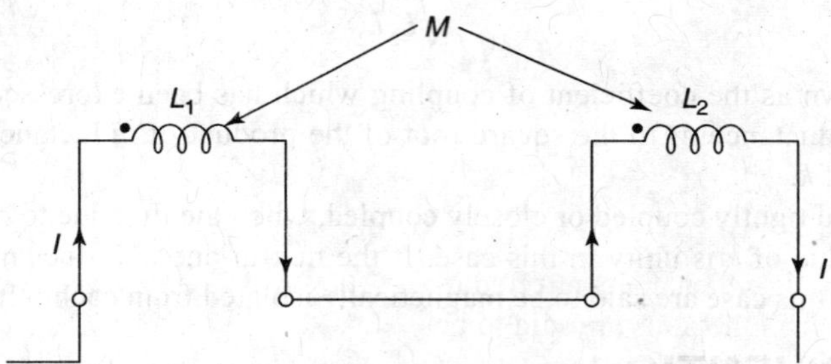

Figure A.4 Inductive Coupling in Series (Flux Aiding)

For Coil 1:

Self-induced emf $$e_1 = -L_1 \frac{di}{dt} \tag{A.27}$$

Mutually induced emf

$$e_1' = -M \frac{di}{dt} \tag{A.28}$$

[This is due to change of current in coil 2]

For Coil 2:

Self-induced emf $$e_2 = -L_2 \frac{di}{dt} \tag{A.29}$$

Mutually induced emf

$$e_2' = -M \frac{di}{dt} \tag{A.30}$$

[This is due to change of current in coil 1]

Therefore, the total induced emf of the above combination can be written as

$$e = -L_1 \frac{di}{dt} - L_2 \frac{di}{dt} - 2M \frac{di}{dt}$$

$$= -\left(L_1 + L_2 + 2M\right) \frac{di}{dt} \tag{A.31}$$

If L is the equivalent inductance of the coil, it can be written as

$$e = -L \frac{di}{dt} \tag{A.32}$$

From Equations (7.31) and (7.32), we have

$$-L \frac{di}{dt} = -\left(L_1 + L_2 + 2M\right) \frac{di}{dt}$$

$$\therefore \quad L = L_1 + L_2 + 2M \tag{A.33}$$

(ii) Series Opposing

Figure A.5 shows the coils which are connected in series. The fluxes or emfs are in opposite direction as per dot convention. Hence, the coils are said to be in series opposing.

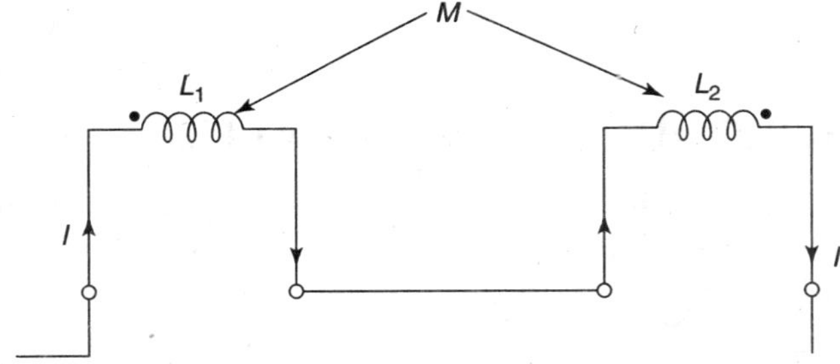

Figure A.5 Inductive Coupling in Series (Flux Opposing)

For Coil 1:

Self-induced emf
$$e_1 = -L_1 \frac{di}{dt} \tag{A.34}$$

Mutually induced emf in coil 1 due to change of current in coil 2 can be written as

$$e_1' = M \frac{di}{dt} \tag{A.35}$$

For Coil 2:

Self-induced emf
$$e_2 = -L_2 \frac{di}{dt} \tag{A.36}$$

Mutually induced emf in coil 2 due to change of current in coil 1 can be written as

$$e_2' = M\frac{di}{dt} \tag{A.37}$$

The total emf induced in the combination can be expressed as

$$e = -L_1\frac{di}{dt} - L_2\frac{di}{dt} + 2M\frac{di}{dt}$$

$$\therefore \quad e = -(L_1 + L_2 - 2M)\frac{di}{dt} \tag{A.38}$$

If L is the equivalent inductance of the combination, it can be written as

$$e = -L\frac{di}{dt} \tag{A.39}$$

From Equations (A.38) and (A.39), we have

$$-L\frac{di}{dt} = -(L_1 + L_2 - 2M)\frac{di}{dt}$$

Therefore,
$$L = L_1 + L_2 - 2M \tag{A.40}$$

A.6 INDUCTIVE COUPLING IN PARALLEL

When two coils are coupled in parallel and mutual inductance exists, the equivalent inductance of the parallel combination can be determined for parallel aiding and parallel opposing as discussed below.

(i) Parallel Aiding

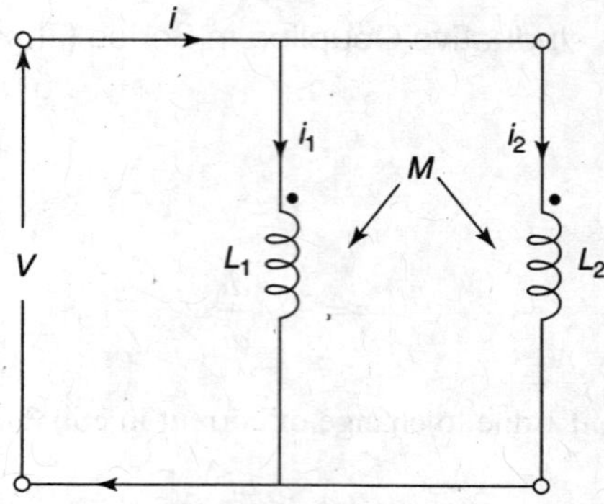

Figure A.6 Inductive Coupling in Parallel (Flux Aiding)

The parallel connection of two coils is shown in Figure A.6. The flux is additive as per dot convention. Using KVL, we have

$$V = L_1 \frac{di_1}{dt} + M \frac{di_2}{dt} \tag{A.41}$$

and

$$V = L_2 \frac{di_2}{dt} + M \frac{di_1}{dt} \tag{A.42}$$

From Equations (A.41) and (A.42), we have

$$L_1 \frac{di_1}{dt} + M \frac{di_2}{dt} = L_2 \frac{di_2}{dt} + M \frac{di_1}{dt} \tag{A.43}$$

Here, $i = i_1 + i_2$

$\therefore \quad i_2 = i - i_1 \qquad \text{(A.44)}$

Using Equation (A.44), we have from Equation (A.43)

$$L_1 \frac{di_1}{dt} + M \frac{d}{dt}(i - i_1) = L_2 \frac{d}{dt}(i - i_1) + M \frac{di_1}{dt}$$

i.e.,

$$L_1 \frac{di_1}{dt} + M \frac{di}{dt} - M \frac{di_1}{dt} = L_2 \frac{di}{dt} - L_2 \frac{di}{dt} + M \frac{di_1}{dt}$$

i.e.,

$$\left(L_1 + L_2 - 2M\right) \frac{di_1}{dt} = \left(L_2 - M\right) \frac{di}{dt}$$

$\therefore$

$$\frac{di_1}{dt} = \left(\frac{L_2 - M}{L_1 + L_2 - 2M} \right) \frac{di}{dt} \tag{A.45}$$

Similarly,

$$\frac{di_2}{dt} = \left(\frac{L_1 - M}{L_1 + L_2 - 2M} \right) \frac{di_1}{dt} \tag{A.46}$$

Using Equations (A.45) and (A.46), we have from Equation (A.41)

$$V = L_1 \left(\frac{L_2 - M}{L_1 + L_2 - 2M} \right) \frac{di}{dt} + M \left(\frac{L_1 - M}{L_1 + L_2 - 2M} \right) \frac{di}{dt}$$

i.e.,

$$V = \left(\frac{L_1 L_2 - L_1 M + L_1 M - M^2}{L_1 + L_2 - 2M} \right) \frac{di}{dt}$$

i.e.,

$$V = \left(\frac{L_1 L_2 - M^2}{L_1 + L_2 - 2M} \right) \frac{di}{dt} \tag{A.47}$$

Let L be the equivalent inductance of the parallel combination. We have

$$V = L \frac{di}{dt} \tag{A.48}$$

From Equations (A.47) and (A.48), we have

$$L\left(\frac{di}{dt}\right) = \frac{L_1L_2 - M^2}{L_1 + L_2 - 2M}\left(\frac{di}{dt}\right)$$

$$\therefore \quad L = \frac{L_1L_2 - M^2}{L_1 + L_2 - 2M} \tag{A.49}$$

Equation (A.49) shows the equivalent inductance of Figure A.6.

(ii) Parallel Opposing

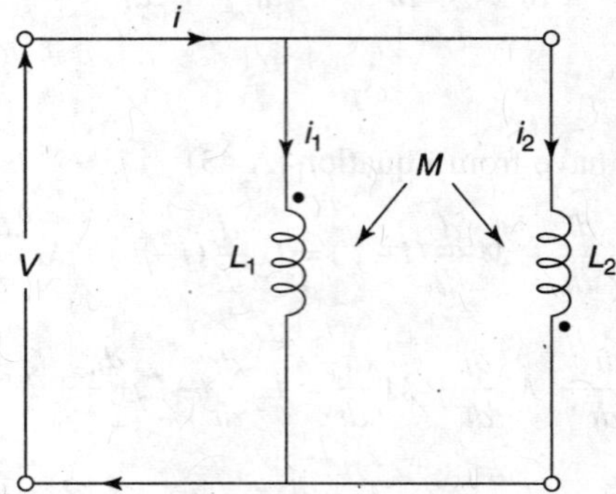

Figure A.7 Inductive Coupling in Parallel (Flux Opposing)

Figure A.7 shows the parallel connections of two coils when the mutually induced emf in one coil due to variation of current in other coil is negative.

Using KVL, from Figure A.7, we have

$$V = L_1\frac{di_1}{dt} - M\frac{di_2}{dt} \tag{A.50}$$

and

$$V = L_2\frac{di_2}{dt} - M\frac{di_1}{dt} \tag{A.51}$$

From Equations (A.50) and (A.51), we have

$$L_1\frac{di_1}{dt} - M\frac{di_2}{dt} = L_2\frac{di_2}{dt} - M\frac{di_1}{dt} \tag{A.52}$$

Since $i = i_1 + i_2$

$$\therefore \quad \frac{di}{dt} = \frac{di_1}{dt} + \frac{di_2}{dt}$$

i.e.,

$$\frac{di_2}{dt} = \frac{di}{dt} - \frac{di_1}{dt} \tag{A.53}$$

Using Equation (A.53), we have from Equation (A.52):

$$L_1\frac{di_1}{dt} - M\left(\frac{di}{dt}-\frac{di_1}{dt}\right) = L_2\left(\frac{di}{dt}-\frac{di}{dt}\right) - M\frac{di_1}{dt}$$

i.e.,

$$\left(L_1 + L_2 + 2M\right)\frac{di_1}{dt} = \left(L_2 + M\right)\frac{di}{dt}$$

$\therefore$

$$\frac{di_1}{dt} = \left(\frac{L_2 + M}{L_1 + L_2 + 2M}\right)\frac{di}{dt} \tag{A.54}$$

Similarly, we have

$$\frac{di_2}{dt} = \left(\frac{L_1 + M}{L_1 + L_2 + 2M}\right)\frac{di}{dt} \tag{A.55}$$

Using Equations (A.54) and (A.55), we have from Equation (A.50):

$$V = L_1\left(\frac{L_2 + M}{L_1 + L_2 + 2M}\right) - M\left(\frac{L_1 + M}{L_1 + L_2 + 2M}\right)\frac{di}{dt}$$

$\therefore$

$$V = \left(\frac{L_1L_2 - M^2}{L_1 + L_2 + 2M}\right)\frac{di}{dt} \tag{A.56}$$

If L is the equivalent inductance of Figure A.7, we have

$$V = L\frac{di}{dt} \tag{A.57}$$

From Equations (A.56) and (A.57), we hav

$$L\frac{di}{dt} = \left(\frac{L_1L_2 - M^2}{L_1 + L_2 + 2M}\right)\frac{di}{dt}$$

$\therefore$

$$L = \frac{L_1L_2 - M^2}{L_1 + L_2 + 2M} \tag{A.58}$$

A.7 AC OPERATION OF MAGNETIC CIRCUIT

DC sources are not used for excitation of magnetic circuit of transformers and other AC machines. AC sources are used. The steady-state current is calculated by the applied voltage and resistance of the circuit when DC excitation is applied. The inductance in this case plays the role only for the transient part. The adjustment of the magnetic flux takes place as per the value of current to satisfy the relationship of B–H curve or magnetization curve. For the case of AC excitation, inductance comes into picture for steady-state performance. The flux is determined by the impressed voltage and frequency. The adjustment of magnetization current takes place as per the value of this flux to maintain the relationship imposed by the magnetization curve.

Sometimes it is required to preserve the linear relationship of magnetization curve. Usually, the normal flux density in the magnetic circuit is out beyond the magnetization curve for the partial saturating circuit. Hence, it is difficult to get the accurate analysis on the basis of constant self-inductance. The equivalent parameters which remain substantially constant are used. From Faraday's law, we have

$$e = N\frac{d\phi}{dt} \tag{A.59}$$

An iron core is considered which is excited by a winding having N turns carrying a current i. Let the magnetic flux (ϕ) vary sinusoidally with time t as shown by the following relation:

$$\phi = \phi_{\max} \sin 2\pi ft \tag{A.60}$$

In Equation (A.60), $\phi_{\max}$ and f are the maximum value of flux in the cycle and the supply frequency. The induced emf is expressed by

$$e = N\frac{d\phi}{dt} = 2\pi fN\phi_{\max} \cos 2\pi ft \tag{A.61}$$

The rms or effective value of induced voltage is expressed by

$$E_{\text{rms}} = \frac{2\pi fN\phi_{\max}}{\sqrt{2}} = 4.44 fN\phi_{\max} \tag{A.62}$$

The polarity of this induced emf is as per Lenz's law which opposes the change in flux. The induced emf leads the flux by an angle 90°. The applied voltage is opposed by the induced emf and the resistance drop. Usually, the resistance drop does not exceed a few per cent of the applied voltage in case of AC machines. Hence, resistance drop may be neglected. Therefore, the induced emf and the applied voltage are considered equal in magnitude. The value of $\phi_{\max}$ is determined from Equation (A.62).

A.8 HYSTERESIS AND EDDY CURRENT LOSSES

There are types of power losses in the form of heat in iron core when the magnetic circuits are subjected to time-varying flux densities. Theses losses cannot be neglected and have significance for determining heating, rating and efficiency of rotating machines, transformers and other AC-operated devices. The first loss is termed as hysteresis loss, which is associated with the phenomenon of hysteresis. Due to removal of the mmf, all the energy of the magnetic field is not returned to the circuit when the ferromagnetic materials are involved. The hysteresis loss is proportional to the area of the hysteresis loop and to the number of loops traversed per second when the variation of flux density is from $+B_{\max}$ to $-B_{\max}$ at the frequency f.

A.8.1 Determination of Hysteresis Loss

A ring of circumference l m and area of cross-section a m^2 having number of turns N of an insulated wire is considered here. The current through the wire is taken as I amperes. The magnetizing force is expressed by

$$H = \frac{NI}{l} \tag{A.63}$$

i.e.,

$$I = \frac{Hl}{N} \tag{A.64}$$

If the magnetic flux density at this instant is B, the total flux through the ring is expressed by

$$\phi = Ba \tag{A.65}$$

If the current through the solenoid is alternating in nature, the emf induced is expressed by

$$e_1 = -N\frac{d\phi}{dt} = -N\frac{d(Ba)}{dt} = -Na\frac{dB}{dt} \tag{A.66}$$

The emf expressed by Equation (A.66) opposes the flow of current as per Lenz's law. If the source supply has an equal and opposite emf, it is possible to maintain the constant flow of current (I). Therefore, the applied voltage becomes

$$e = -e_1 = Na\frac{dB}{dt} \tag{A.67}$$

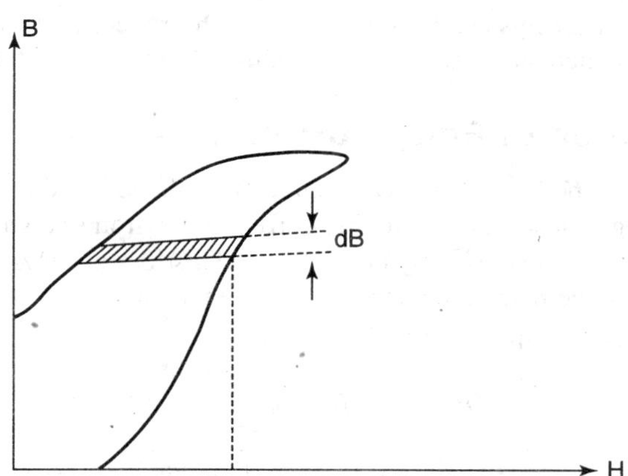

Figure A.8 Plot of B vs H

Figure A.8 shows the plot of flux density (B) and magnetizing force (H). A short interval dt is considered here, and the change of flux density at this time is also shown. Therefore, the energy consumed in this short interval is expressed by

$$dE = eI\,dt = Na\frac{dB}{dt} \times I \times dt = Na\frac{dB}{dt} \times \frac{Hl}{N} \times dt = alH\,dB \quad \text{J}$$

Therefore, the total energy consumed during one complete cycle of magnetization is expressed by

$$W = al\int_0^{B_{max}} H\,dB \tag{A.68}$$

In Equation (A.68), al represents the volume of the ring and $H\,dB$ represents the area of elementary strip of B–H curve. $\int H\,dB$ is the total area of the hysteresis loop. Energy consumed per cycle is the product of volume of the ring and the area of the hysteresis loop.

The area of the loop is measured in units of H and B, i.e., AT/m and Wb/m^2, to determine the hysteresis loss. The area of the loop is measured in square metre which is multiplied by the scales of H and B during measurement the area of the loop in units of H and B.

If 1 m represents p AT/m on the H-axis and q Wb/m² on the B-axis, the hysteresis loss is expressed by

$$W_h = \text{area of hysteresis loop in m}^2 \times p \times q \text{ /m}^2\text{/cycle} \quad \text{(A.69)}$$

Using Steinmetz method, it is possible to determine the hysteresis loss. As per this method, the hysteresis loss per cubic metre per cycle of magnetization of a magnetic material is dependent on (i) the maximum value of flux density and (ii) magnetic quality of the material. Therefore,

Hysteresis loss $\propto B_{max}^{1.6}$ J/m³ / cycle

i.e., hysteresis loss $= \eta B_{max}^{1.6}$ J/m³ / cycle (A.70)

$= \eta B_{max}^{1.6} fv$ J/s or W (A.71)

where η is known as Steinmetz hysteresis coefficient which is a constant for each specimen, f is the frequency of reversal of magnetization and v is the volume of magnetic material in m³.

A.8.2 Determination of Eddy Current Loss

Because the core itself is composed of conducting material, voltage is induced in it due to the variation of flux. This induced voltage causes the circulating current in the iron known as eddy current. This current creates i^2r loss which is known as eddy current loss. This loss is dependent on the rate of change of flux and the resistance of the path. Therefore, it can be taken as this loss varies as the square of the maximum flux density and the frequency and it is expressed by

$$P_e = K_e f^2 B_{max}^2 t^2 v \text{ W or J/s} \quad \text{(A.72)}$$

where the constant K_e is known as eddy current coefficient and it is dependent on the type of core material, f is the number of complete magnetization cycles per second, B_{max} is the maximum flux density (Wb/m²), t is the thickness of laminations in metres, and v is the volume of core material in m³.

The core resistance must be increased to reduce the eddy currents. The magnetic cores are made of laminated thin sheets with an insulating layer (surface oxide or varnish) between successive laminations. Because eddy currents have magnetic effect, this makes the flux density at the centre lower compared to the surface which is known as screening effect. If the cores are properly laminated, this effect can be neglected at power frequencies. At high power frequencies, this effect may be significant.

These two losses altogether are known as iron loss or core loss. Core loss is present both in DC and AC machines.

Appendix

Reluctance Motor B

Figure B.1 shows a reluctance motor because there is no winding in the rotor which is main difference between the doubly excited system and the reluctance motor.

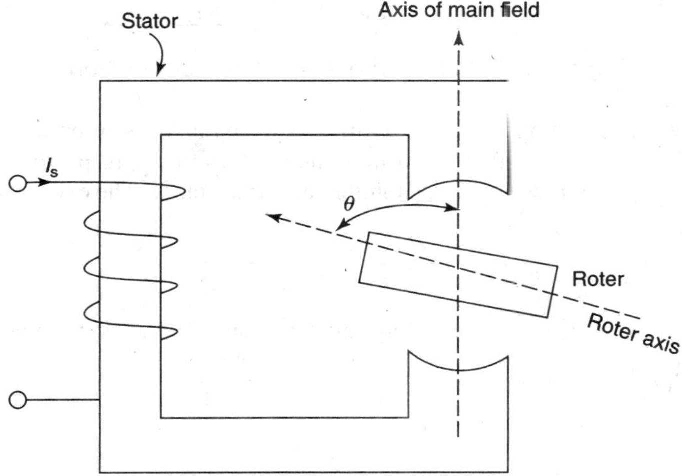

Figure B.1 Reluctance Motcr

The stator inductance is expressed by

$$L_{ss} = L_0 + 2L \cos 2\theta \tag{B.1}$$

This is a function of rotor position θ.

The stator current is

$$i_s = I_{sm} \sin \omega t \tag{B.2}$$

The position of the rotor at any time t is expressed by

$$\theta = \omega_m t + \delta \tag{B.3}$$

where δ is the initial angle of the rotor wrt stator axis and ω_m is the angular velocity of the rotor.

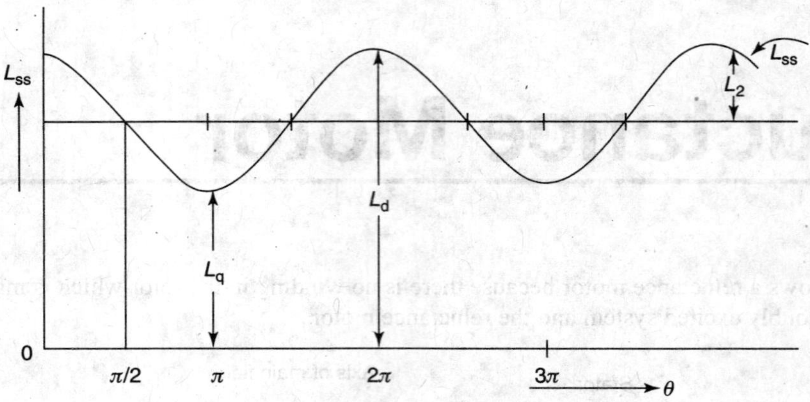

Figure B.2 Variation of L_{ss} with Rotor Position

Figure B.2 shows the variation of stator self-inductance with the position of the rotor. For $\theta = 0°$ and $\theta = 90°$, L_d and L_q are the maximum and minimum values of L_{ss}, respectively. L_d represents the direct axis inductance and L_q represents the quadrature axis inductance. The expression for L is given by

$$L = \frac{L_d - L_q}{2} \tag{B.4}$$

Here $i_r = 0$ because there is no rotor winding and the torque expression becomes

$$T = \frac{1}{2} i_s^2 \frac{\delta(L_{ss})}{\delta\theta}$$

$$= \frac{1}{2} I_{sm}^2 \sin^2 \omega t \ \frac{\delta}{\delta\theta}[L_0 + L\cos 2\theta]$$

$$= -LI_{sm}^2 \sin^2 \omega t \ \sin 2\theta$$

$$= -LI_{sm}^2 \left(\frac{1-\cos 2\omega t}{2}\right) \sin 2(\omega_m t + \delta)$$

$$= -\frac{1}{2} LI_{sm}^2 \sin 2(\omega t + \delta) + \frac{1}{2} LI_{sm}^2 \cos 2\omega t \ \sin 2(\omega_m t + \delta)$$

$$= -\frac{1}{2} LI_{sm}^2 \sin 2(\omega t + \delta) + \frac{1}{4} LI_{sm}^2 [\sin 2\{(\omega_m + \omega)t + \delta\} + \sin 2\{(\omega_m - \omega)t + \delta\}] \tag{B.5}$$

In Equation (B.5), there are three sinusoidal terms. The average value of each of these three sinusoidal terms is zero over a cycle except in certain conditions.

Case 1:

If $\omega_m = 0$, we have

$$T = \frac{1}{2} L I_{sm}^2 \sin 2\delta = \frac{1}{4}(L_d - L_q) I_{sm}^2 \sin 2\delta$$

The average torque is

$$T = -\frac{1}{4} I_{sm}^2 (L_d - L_q) \sin 2\delta \tag{B.6}$$

Case 2:

If $\omega_m = \pm\omega$, we have

$$T_{av} = \frac{1}{4} L I_{sm}^2 \sin 2\delta$$

$$= \frac{(L_d - L_q)}{8} I_{sm}^2 \sin 2\delta \tag{B.7}$$

Equation (B.7) shows that the reluctance motor develops average torque when it rotates at a synchronous speed (at the supply frequency) in either direction. The average torque is proportional to sin2δ.

MMF of Distributed Winding

Appendix **C**

The distributed winding in an armature is used in most of the rotating machines. Due to the flow of current in different coils which forms the armature winding, a magnetic field is produced having the same number of poles as the field winding.

The magnetic field's pattern along the air-gap periphery, which is set up by the distributed winding, is dependent on the following:

(i) winding arrangement,
(ii) nature of the current flowing through the winding and
(iii) the configuration of the magnetic circuit.

The distribution of air-gap flux in the machine helps to determine the following:

(i) the wave and magnitude of the generated emf and
(ii) the electrical torque.

The production of mmf by distributed winding in the slots along the air-gap periphery is studied at first.

Figure C.1(a) shows that each slot has six conductors where each conductor carries a current 'i'. Due to the traversing path 1, the ampere–conductor enclosed becomes $2i$ and the change of mmf has been shown in Figure C.1(b). If the paths 2 and 3 are traversed, the ampere–conductor becomes $4i$ and $6i$, respectively. Figure C.1(b) shows the mmf variation which changes linearly from 0 to $6i$ over the slot width. Figure C.1(c) shows the simplified form of Figure C.1(b) where the change of mmf is considered stepped at the middle of the slot.

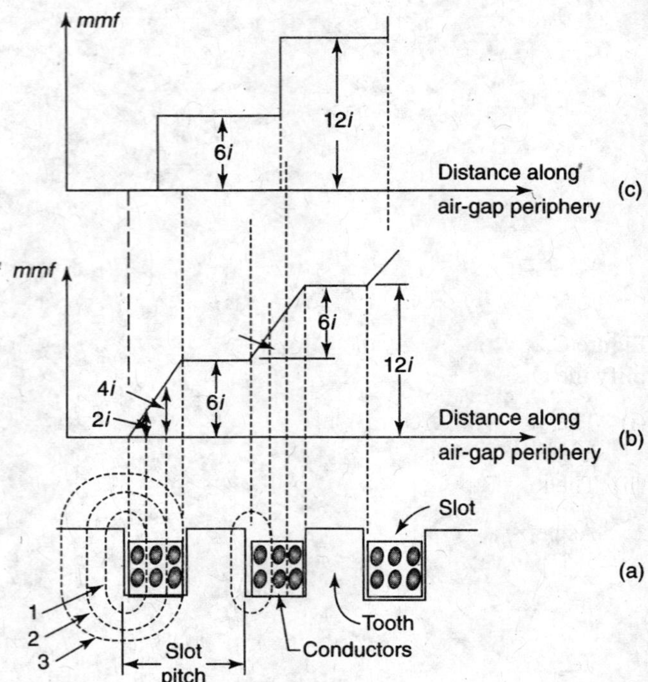

Figure C.1 Mmf by Distributed Winding in the Slots along the Air-Gap Periphery

Figure C.2(a) shows a full-pitched coil on the stator for a two-pole machine having uniform air gap. Let the coil consist of N turns where each turn carries a current 'i'. The direction of currents has been shown in Figure C.2(a).

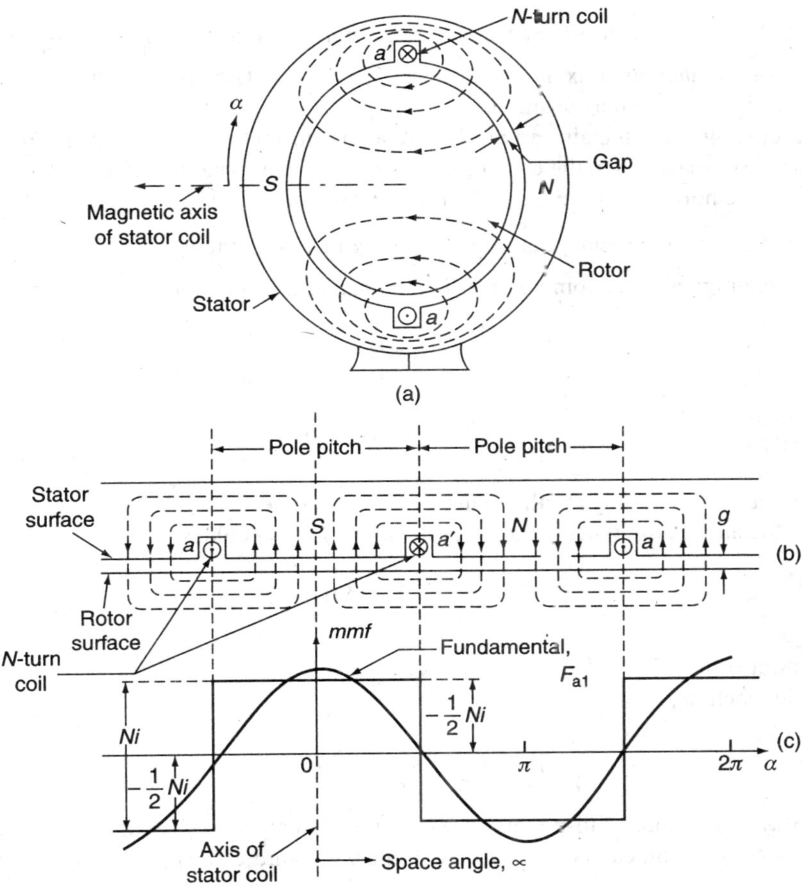

Figure C.2 Full-Pitched Coil and Air-Gap mmf

Figure C.2(b) shows the developed view of Figure C.2(a). The following assumptions are made to simplify the analysis:

(i) The stator and rotor iron have much higher permeability compared to air so that the magnetic flux is offered by the air gap only.

(ii) The magnetic lines of force cross the air gap radially.

As per Ampere's circuital law, we have

$$\oint \vec{H} \cdot d\vec{l} = \text{total current enclosed}$$

i.e., $2xH = Ni$ where 'x' is the air-gap length.

$$\text{i.e.,} \quad xH = \frac{Ni}{2} \tag{C.1}$$

Equation (C.1) shows that the magnetic potential difference across the air gap is $\frac{1}{2}Ni$. Figure C.2(c) shows the variation of magnetic flux along the air-gap periphery. The magnetic flux leaving the rotor or entering the stator is considered as positive.

Figure C.2(c) shows that the air-gap mmf wave at any instant is rectangular. There is no variation of mmf with time and space when the current in the coil is direct in nature. If it is alternating in nature, mmf varies with time not with space; i.e., it is time variant only.

Using Fourier series, the rectangular mmf waveform having magnitude $\frac{1}{2}Ni$ can be resolved into its fundamental and higher order components. The fundamental component part becomes

$$F_{a1} = \frac{4}{\pi}\cdot\frac{Ni}{2}\cos\alpha = F_{1p}\cos\alpha \tag{C.2}$$

i.e.,

$$F_{1p} = \frac{4}{\pi}\cdot\frac{Ni}{2}\ \text{ATs/pole} \tag{C.3}$$

The alignment of F_{1p} is always with the magnetic axis of the coil.

When 'i' is the alternating type, the maximum value of F_{1p} becomes

$$F_{1pm} = \frac{4}{\pi}\cdot\frac{N\sqrt{2}I}{2}\ \text{ATs/pole} \tag{C.4}$$

The maximum value of 'i' is $I_{max} = \sqrt{2}I_{rms} = \sqrt{2}I$.

For a P-pole machine,

$$F_{1pm} = \frac{4}{\pi}\cdot\frac{N\sqrt{2}I}{P}\ \text{ATs/pole} \tag{C.5}$$

Figure C.3(a) shows three full-pitched coils on the stator, where windings are distributed on it. Let each coil has N turns and each turn carry a current 'i'. Figure C.3(b) shows the developed view of Figure C.3(a).

Figure C.3((c)(i)) shows the mmf variation for the coil a–a' only which is rectangular wave of magnitude $\pm\frac{1}{2}Ni$. Figures C.3((c)(ii)) and C.3((c)(iii)) show the mmf variation of coils b–b' and c–c', respectively. When the individual mmfs are summed up, the combined mmf can be obtained. From Figure C.3((c)(iv)), the combined mmf to the left of the coil a is $-\frac{3}{2}Ni$. Its value is $-\frac{1}{2}Ni$ between coil sides a and b, while it is $\frac{1}{2}Ni$ between coil sides b and c. The combined value of mmf between coil sides c and a' is $\frac{3}{2}Ni$. Figure C.3((c)(iv)) shows the complete combined value of mmf.

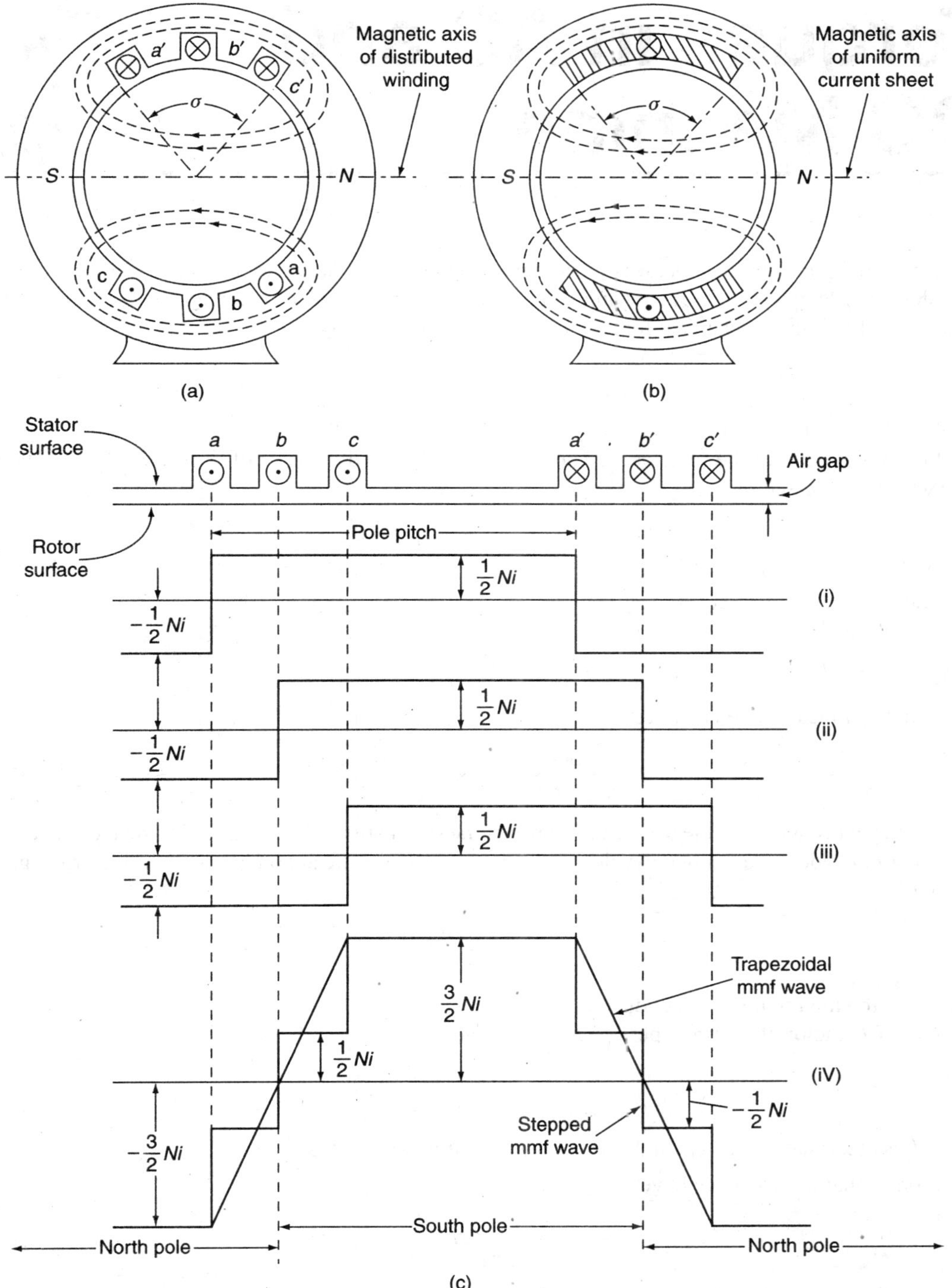

Figure C.3 Full-Pitched Coil, Its Developed View and mmf Variation

Torques in AC and DC Machine

Appendix

D

When a current-carrying conductor is placed in a magnetic field, it experiences a mechanical force. If a conductor of length 'l' metres carrying a current 'i' ampere is placed in a magnetic field having flux density B Wb/m^2, the mechanical force (F) exerted on it is given by

$$F = Bil \quad \text{N} \tag{D.1}$$

Let the diameter of the rotor be D. The force acts at a radial distance of $\frac{D}{2}$ metres. The torque per conductor is expressed by

$$T = Bil\left(\frac{D}{2}\right) \tag{D.2}$$

The average flux density B_{av} is given by

$$B_{av} = \frac{\phi P}{\pi Dl} \tag{D.3}$$

The developed torque becomes

$$T_e = \frac{\phi P}{\pi Dl} \times l \times i \times \frac{D}{2} = \frac{1}{2\pi}\phi Pi \tag{D.4}$$

For the entire winding, the torque developed is the summation of the torques for the individual conductors or coil sides. This torque developed due to electromagnetic action is known as electromagnetic torque (T_e).

$$T_e = \frac{1}{2\pi}\phi ZPi \tag{D.5}$$

where Z is the total number of conductors.

For a DC motor, the current per conductor is given by

$$i = \frac{I_a}{A}$$

where I_a is the armature current and A is the number of parallel paths.

From Equation (D.5), we have

$$T_e = \frac{1}{2\pi}\phi ZP\left(\frac{I_a}{A}\right) \tag{D.6}$$

Equation (D.6) represents the torque developed by the armature of a DC motor.

Separation of No-load Losses of an Induction Motor

Appendix

E

To separate the no-load losses (iron losses and mechanical losses) of a three-phase squirrel cage induction motor, the circuit connection is made as shown in Figure E.1.

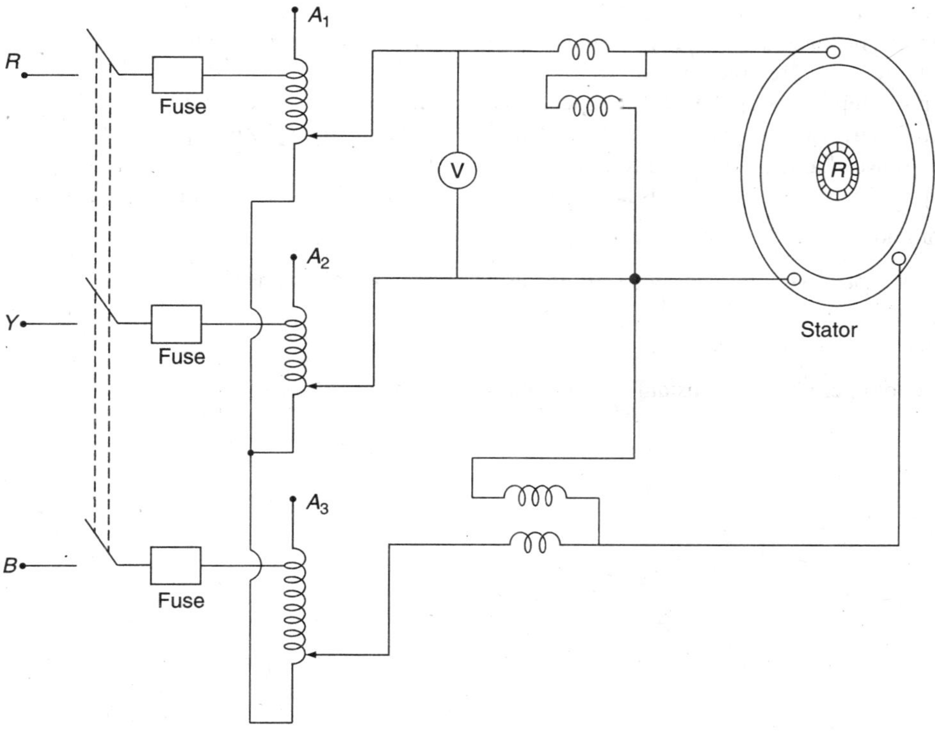

Figure E.1 Circuit Diagram

The name plate details of the motor and the autotransformer should be noted at first. The required apparatuses are ammeter, voltmeter, wattmeter, three-phase autotransformer and rheostat. The autotransformer's initial position must be in the minimum voltage position and the motor need not to be loaded throughout the experiment. The required table (Table E.1) is shown in Figure E.1.

Table E.1 Data for Separation of No-Load Loss of Three-Phase Squirrel Cage Induction Motor

Sl. No.	No-load voltage (V_0) (V)	No-load current (I_0) (A)	No-load input power W_1 (W)	W_2 (W)	$W = W_1 + W_2$ (W)	Stator copper loss $3I_0^2R_s$ (W)	Constant loss/phase $W_c = \frac{W - I_0^2 R_s}{3}$ (W)	Core loss/phase $W_i = W_c - W_m$ (W)

The following steps are followed:

Step 1: Make the circuit diagram connection as shown in Figure E.1.
Step 2: Give the three-phase supply and start the motor.
Step 3: The autotransformer is varied from its initial position, and the input power, voltage and current are noted down.
Step 4: Step 1 to Step 3 are to be repeated and readings are to be tabulated.
Step 5: The stator copper loss and constant loss are calculated by the respective formula.
Step 6: Mechanical losses are found out by the suitable graph (constant losses and input voltage).
Step 7: The mechanical loss is subtracted from the constant loss to obtain the core loss.

Input power $W = W_1 + W_2$

Stator copper loss $= 3I^2R_s$, where R_s is the stator resistance per phase

Constant loss per phase $(W_c) = \dfrac{W - 3I^2R_s}{3}$

Core loss per phase = constant loss per phase – mechanical loss

Separation of Losses of an Induction Motor

Appendix **F**

The following losses occur in an induction motor in addition to the I^2R loss.

(i) Brush friction loss = P_b
(ii) Eddy current loss = P_e
(iii) Hysteresis loss = P_h
(iv) Pulsation loss = P_p
(v) Mechanical losses in windage and bearing friction = P_f

The rotor core loss at standstill is $P_{e2} + P_{h2}$ when the rotor is open circuited. This loss at any slip 's' becomes $s^2 P_{e2} + sP_{h2}$. The pulsation loss occurs at high frequency in the stator and rotor which is produced due to variations of gap reluctance when the tooth tips pass each other. Using no-load loss test, the brush friction loss can be separated.

The following procedures are adopted to separate all losses for the slip ring motor:

No-Load Test:

1. The power supplied to the stator at normal voltage and at rest is measured, and it is corrected for I^2R loss. This gives the iron losses

$$P_1 = P_{e1} + P_{h2} + P_{e2} + P_{h2} \tag{F.1}$$

2. The power supplied to the stator at normal voltage and frequency is measured. The rotor is short circuited and kept running on no load. The corrected stator input becomes

$$P_2 = P_{e1} + P_{h1} + P_p + P_f + P_b \tag{F.2}$$

The rotor eddy current is small and P_{h2} is supplied by the stator. But in practice, the whole of it is returned as a driving torque and partly goes for $P_f + P_b$.

3. The rotor circuit is opened suddenly as the machine is running in Step 2. The stator input is measured. The stator input becomes

$$P_3 = P_{e1} + P_{h1} + P_{h2} \tag{F.3}$$

Transformation Test Ratio:

4. Let V_1 be the normal voltage applied to stator and v_1 be the measured stator voltage when the normal voltage V_2 is applied to the rotor. Now, a voltage $(V_1 + v_1)/2$ is applied to the stator on open circuit and the mutual flux is kept constant.

The rotor input is measured when it is running on no load and the stator is shot circuited. The function of the stator and rotor is reversed as compared to Step 2. The corrected power input becomes

$$P_4 = P_{e2} + P_{h2} + P_p + P_f + P_b \tag{F.4}$$

5. The stator is made short circuited suddenly during the machine running in Step 4. The rotor input becomes $P_5 = P_{e2} + P_{h2} + P_{h1}$. The mechanical losses $(P_6) = P_f + P_b$ are determined from no-load test.

Blocked Rotor Test:

6. The supplied power to the stator at reduced voltage and full-load stator current is measured. The rotor is kept short circuited and locked. The I^2R loss of the machine is determined.

From the transformation ratio test:

Rotor voltage and stator voltage can be obtained and hence turns ratio can be determined.

From no-load test:

Wattmeter constant = ________________

C.T. ratio = ________________

Serial number	L–L voltage (V)	W_1 (W)	W_2 (W)	$W_1 + W_2$ (W)	Corrected power (W)

Blocked rotor test:

Wattmeter constant = ________________

C.T. ratio = ________________

Serial number	L–L voltage (V)	I (A)	W_1 (W)	W_2 (W)	$W_1 + W_2$ (W)	Corrected power (W)

From no-load test:

$P_6 = P_f + P_b$ W

$P_{e1} = P_1 - P_5$ W

$P_{e2} = P_1 - P_3$ W

$$P_{h1} = \frac{P_2 - P_5 + P_1}{2} - (P_1 - P_5) \quad \text{W}$$

$$P_{h2} = \frac{P_1 - P_2 + P_4}{2} - (P_1 - P_3) \quad \text{W}$$

$$P_b = P_4 - P_7 - \frac{P_1 - P_2 + P_4}{2} \quad \text{W}$$

The corresponding values at normal voltage can be obtained.

It is difficult to measure the brush friction loss when there is no internal short-circuiting device. $P_b = 0$ is assumed in this case. The rotor eddy current is small because it is directly proportional to s^2. The voltage is maintained constant to obtain the accurate result.

Appendix G

Tertiary Windings

There may be the third winding in the transformer which is known as tertiary winding in addition to the primary and secondary windings. These transformers are known as triple-wound transformers. The following are the purposes of tertiary windings:

(a) Substation auxiliaries can be supplied at a voltage which are different from those of the primary and secondary windings.

(b) Compensating devices can also be supplied.

(c) Three supply systems operating at different voltages can be interconnected.

(d) It can be used to measure the HV testing transformer.

Figure G.1 shows the equivalent circuit of a three-winding transformer, where Z_p, Z_s and Z_t are the primary, secondary and tertiary windings of the transformer.

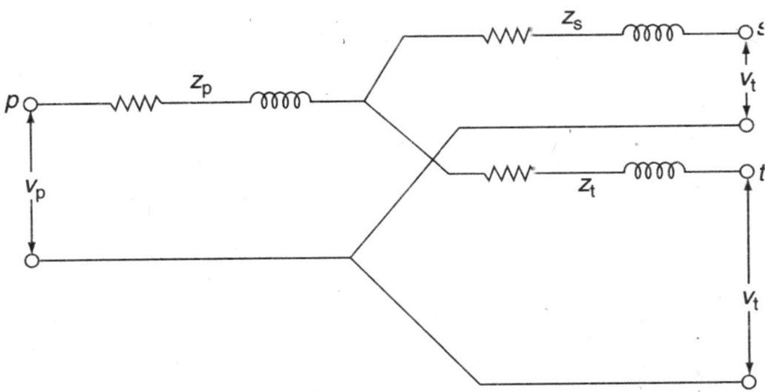

Figure G.1 Equivalent Circuit

The external circuits are connected to the terminals 1, 2 and 3 and the common level terminal O. If exciting current is neglected, $I_1 + I_2 + I_3 = 0$.

SC impedance of windings primary and secondary with tertiary open is given by

$$Z_{ps} = Z_p + Z_s \tag{G.1}$$

SC impedance of windings primary and tertiary with secondary open is given by

$$Z_{pt} = Z_p + Z_t \tag{G.2}$$

SC impedance of windings secondary and tertiary with primary open is given by

$$Z_{st} = Z_s + Z_t \tag{G.3}$$

From Equations (G.1), (G.2) and (G.3), we have

$$Z_p = \frac{1}{2}(Z_{ps} + Z_{pt} - Z_{st}) \tag{G.4}$$

$$Z_s = \frac{1}{2}(Z_{st} + Z_{ps} - Z_{pt}) \tag{G.5}$$

$$Z_t = \frac{1}{2}(Z_{pt} + Z_{st} - Z_{ps}) \tag{G.6}$$

Solid State Control of Drives

Appendix **H**

SPEED CONTROL OF DC AND AC DRIVE MOTORS

Motors have become an integral part of our day-to-day lives, with rating ranging from fractional horsepower to hundreds of horsepower, depending on use. We have read about the characteristics of the motors. Now, we will study their speed control, both manual and automatic. Manual control too has been dealt with, and electronic control – manual and automatic – has been introduced. The control operation of any motor includes start, stop, reverse speed and change speed.

SPEED CONTROL OF DC MACHINES

The fundamentals of control of speed of a DC motor have already been discussed. We can use diodes and semiconductor controlled rectifiers (SCRs) to control the applied voltage of a DC motor. A general block diagram for the use of diodes and SCRs is shown in Figures H.1 (a) and (b), respectively.

The control of applied DC voltage from the diode rectifier is normally done manually. This has now generally been replaced by SCRs because the angle of conduction can be controlled accurately by push-button operation.

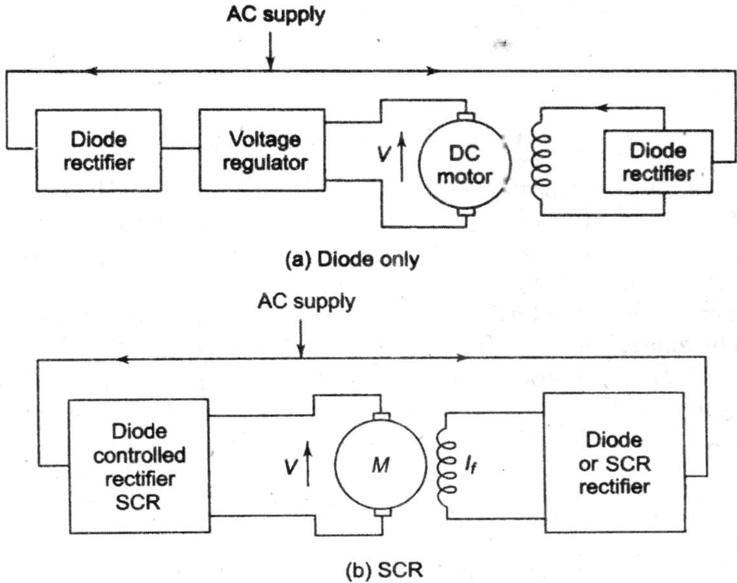

Figure H.1 Block Diagram for the Use of Diode and SCRs

A single-phase half-wave controlled rectifier circuit is shown in Figure H.2.

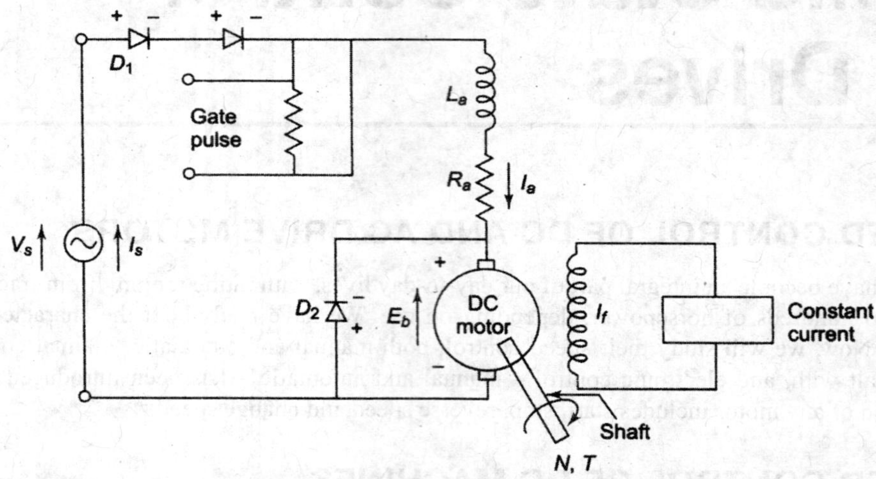

Figure H.2 Single-phase Half-wave Controlled Rectifier

The angle of conduction is controlled by the gate pulse. The mean value of DC voltage is given by

$$V_{mean} = \frac{V_m}{2\pi}(1+\cos\alpha) \tag{H.1}$$

When $\alpha = 0$, the output voltage is $\frac{V_m}{\pi}$ – the same as diode, which is half of the maximum at $\alpha = \frac{\pi}{2}$, that is, $\frac{V_m}{2\pi}$ and 0 at π.

$$\therefore \quad \text{Conduction angle } \theta_C = \alpha - \theta_\pi \tag{H.2}$$

The back emf is zero at standstill. Therefore, full rectified voltage should not be applied. The conduction of SCR must be controlled by starting the gate pulse from π radians of supply voltage. As the DC voltage to the motor increases from zero, the motor picks up speed. Hence, the back emf E_b develops and gets applied to the cathode of the SCR. The SCR will conduct as long as the applied voltage V_s is greater than E_b. Therefore, the peak repetitive inverse voltage can approach a value equal to twice the supply voltage. Hence, a diode is connected in series with the SCR to increase the peak inverse voltage (PIV) of SCR (shown dotted). A freewheeling diode D_2 is connected across the motor to conduct during the negative half-cycle to nullify the armature inductive current stored during the positive half-cycle.

$$v_s = L\frac{di_s}{dt} + Ri_s + E_b \tag{H.3}$$

where v_s and i_s are instantaneous values and E_b is the back emf.

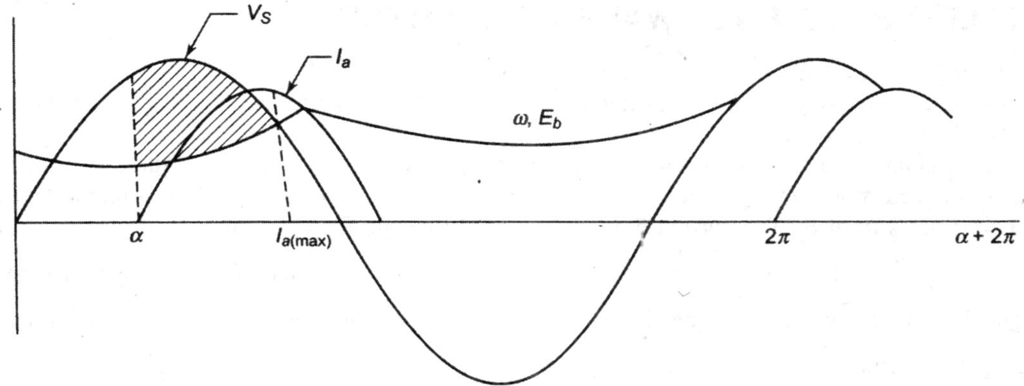

Figure H.3 Instantaneous Values of v_s, i_s and the Back emf E_b

From the curves in Figure H.3, it can be seen that the armature current, I_a, increases until applied voltage from the SCR is greater than the back emf E_b, and reaches a maximum when the two are equal. If $V_s - E_b$ is negative, then jI_a stops flowing. The commutating ciode takes over when the applied pulse turns negative.

The mechanical power developed is E_bI_a. Armature current flows only for a very short period; hence, the mean armature current is low. After the armature current stops flowing, the motor coasts along until the next firing angle, $2\pi + \alpha$, on account of the stored energy in the inertia J, comprising the rotor of the motor and load.

Single-phase half-wave converters are used with fractional horsepower motors, 1 HP and less.

DISADVANTAGES

(i) Large pulses of armature current are required to meet the load torque requirements. Hence, losses in the armature are high resulting in beating.

(ii) As with the single-phase AC rectifier, the DC current flows through the transformer causing core saturation.

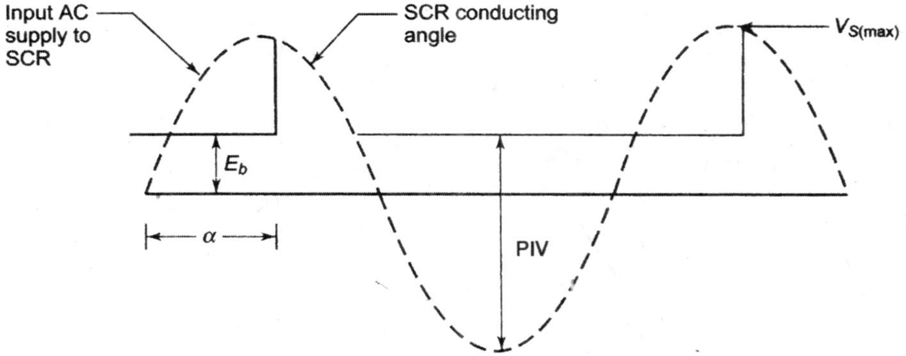

Figure H.4 SCR Conducting Angle

SINGLE-PHASE FULL-WAVE CONVERTER

The circuit given in Figure H.5 is used for motors from 1 kW to 20 kW.

The same circuit may also be redrawn with D_1 and SCR 2 interchanged. In this case, the free wheeling diode D_3 is not required.

The armature current flows when the applied voltage from SCR is more than the back emf E_b – possible between α_1 and α_2 of Figure H.6. If one half of the bridge is fired at α_1, the armature current flow commences from θ_2. The current flows from peak at α_2 and falls to zero at θ_2. The current flow from α_2 to θ_2 is due to the stored energy $\frac{1}{2}LI^2$ in the armature inductance L_f. This energy has to be dissipated by the freewheeling diode. The same condition is repeated when the second section of the bridge conducts during the other half.

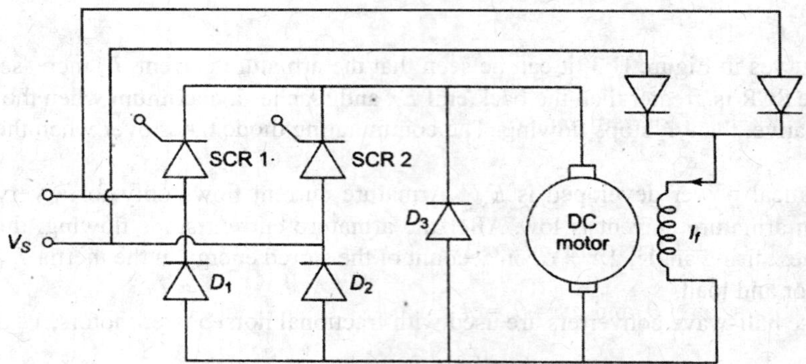

Figure H.5 Half-controlled Bridge

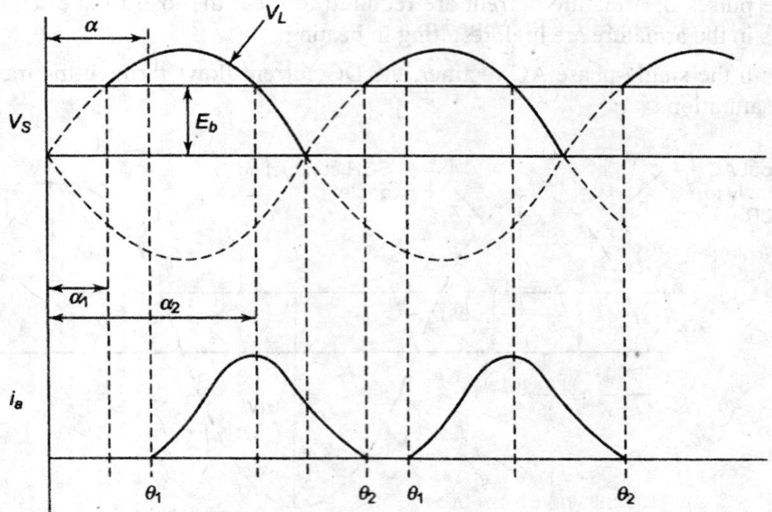

Figure H.6 Instantaneous Values of v_s and i_a

From Equation (H.3), the differential equation for the equivalent motor circuit is given by

$$L\frac{di}{dt} + Ei + E_b = V_m \sin \omega t$$

where i is i_a, known as armature current.

The solution of this differential equation is as follows:

$$I_a(t) = A_e^{-R_a t/L} + \frac{V_m}{\sqrt{R_a^2 + \omega^2 L^2}} \sin(\omega t - \alpha) - \frac{E_b}{R_a} \quad \text{(H.4)}$$

where A is constant and $\alpha = \tan^{-1}\frac{\omega L}{R_a}$. To find A at θ_1, $i(t) = 0$.

$$\therefore \quad A = \frac{E_b}{R_a} - \frac{V_m}{Z}\sin(\theta_1 - \phi) \quad \text{(H.5)}$$

(i) Current discontinuous $\theta_2 - \theta_1 < \pi$

$$i_a(t) = \frac{V_m}{R}\{\cos\alpha \sin(\omega t - \alpha\} - \frac{E_b}{V_m} + \left[\frac{E_b}{V_m}\cos\alpha \sin(\omega t - \alpha)\right] \times e^{\left(R_a/\omega L\right)(\omega t - \theta_1)} \quad \text{(H.6)}$$

(ii) $\theta_2 - \theta_1 < \pi$

$$i_a(t) = \frac{V_m}{R}\{\cos\alpha \sin(\omega t - \alpha)\} - \frac{E_b}{V_m} + \frac{2\cos\alpha \sin(\theta_2 - \phi) e^{\left(R_a/\omega L\right)(\omega t - \theta_1)}}{[1 - e^{-(\pi R_a / \omega L)}]} \quad \text{(H.7)}$$

between the range $\theta\ \omega t = \theta_1$ and $\omega t = \theta_2 = \pi$.

$$\text{Torque } T = \frac{E_b I_{av}}{2\pi n} = \frac{(V_L - I_{av} R_a)}{2\pi n} I_{av} \quad \text{(H.8)}$$

V_L is the rectified voltage of DCR applied to a motor. For the period of conduction $\frac{\theta_2 + \theta_1}{2\pi}$

$$\text{Torque } T = \frac{\theta_2 + \theta_1}{2\pi} \cdot \frac{(V_L - I_{av} R_a}{2\pi\pi} I_{av} \quad \text{(H.9)}$$

$$I_{av} = \frac{1}{\pi}\int_{\theta_1}^{\theta_2} \frac{V_m \sin\omega_t - E_b}{R_a} dt \quad \text{(H.10)}$$

For limited conduction

$$I_{av} = \frac{V_m}{\pi R_a}\left[\cos\theta_1 - \cos\theta_2 - \frac{E_b}{V_m} + (\theta_1 - \theta_1)\right] \quad \text{(H.11)}$$

For continuous conduction

$$I_{av} = \frac{V_m}{\pi R_a}\left(\cos\theta_1 - \frac{E_b}{V_m} + \pi\right) \quad \text{(H.12)}$$

The voltage and the current waveform are the same in Figure H.6. The main advantage of a bridge circuit over a biphase circuit is that PIV is V_m instead of $2V_m$. V_{mean} for both the circuits is almost same

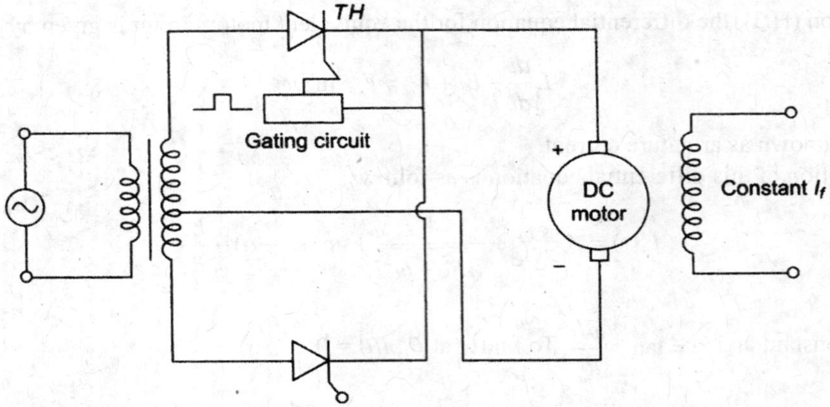

Figure H.7 Biphase Full-wave Motor Control Circuit

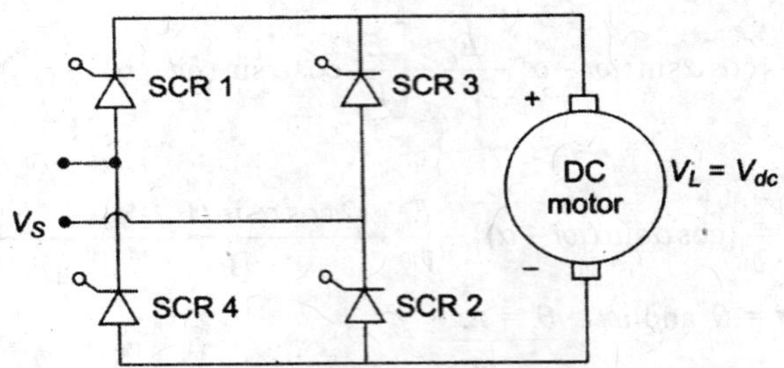

Figure H.8 Fully Controlled Bridge Circuit

0.9 V_{rms}. Thyristors 1 and 2 are fired from the same pulse source and so are 3 and 4. Finally, a bridge circuit has the advantage of lower ripple percentage as well. Figures H.7 and H.8 show a biphase full-wave motor control circuit and a fully controlled bridge circuit, respectively.

THREE-PHASE HALF-WAVE RECTIFIER CONTROL CIRCUIT

Three-phase controlled rectification is used for controlling motors from a few kW to hundreds of kW, and up to 2,000 kW for a double bridge connection. The double bridge connection scores over the Ward Leonard set, eliminating the need for a DC generator and AC driving motor; hence, reduction of the moving parts. Figure H.9 shows a three-phase half-wave control circuit.

Without gate pulse, each SCR will conduct for 120° or $\frac{2\pi}{3}$. Line commutation between V_3 and V_1 occurs at 30° or $\frac{\pi}{6}$ and continues up to $\frac{2\pi}{3}$, that is, 150° or $\frac{5\pi}{6}$. So the thyristor firing angle has to be between $\frac{\pi}{6}$ and $\frac{5\pi}{6}$. We have already derived, the mean voltage in Equation (H.2).

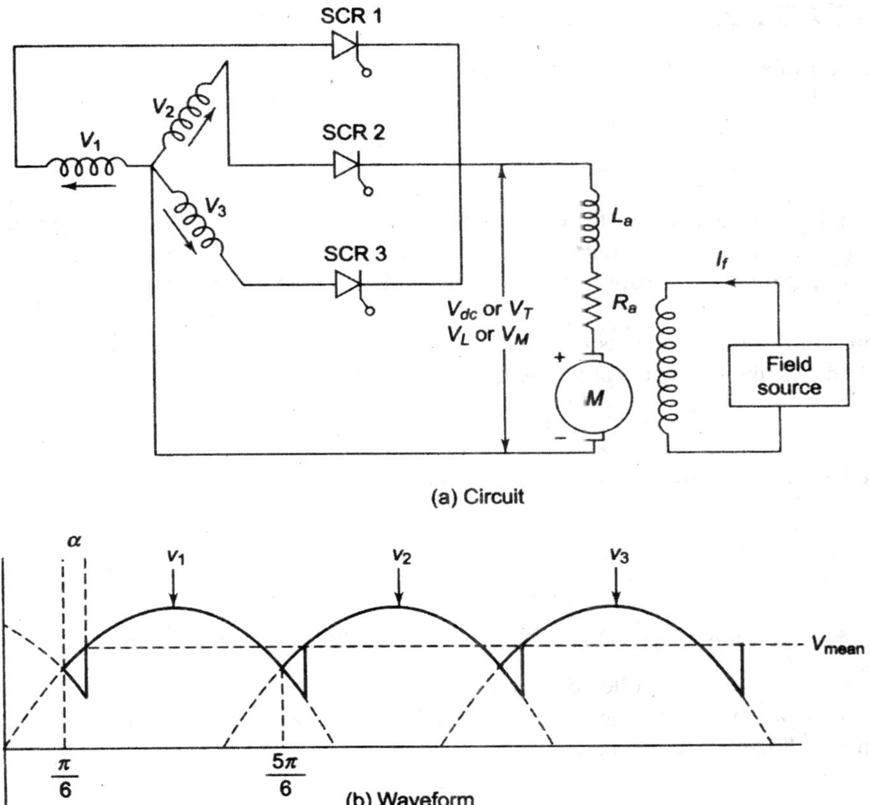

Figure H.9 Three-phase Half-wave Control Circuit

$$V_{\text{mean}} = V_{DC} = \frac{3\sqrt{3}}{2\pi} V_{\text{max (line)}} \cos\alpha$$

$$= 0.827\, V_{\text{max (line)}} \cos\alpha \tag{H.13}$$

So, α can be adjusted between $\frac{\pi}{6}$ and $\frac{\pi}{2}$.

The ripple frequency is 150 Hz.

From Figure H.9, it can be seen that there is a period of overlap between $V_1 - V_2 - V_3 - V_1$, defined as the commutation angle or the angle of overlap. During the overlap period, both V_1 and V_2 are positive and diodes D_1 and D_2 will conduct such that the current will be an addition of the two diode currents. This overlap continues until the current of one diode becomes zero and the incoming diode reaches the mean value. Firing of SCR is reckoned from the point of commutation, so the angle of overlap or commutation is reduced. This makes for a linear variation of load current during the period of commutation. It may also be noted that when there is line commutation from one phase to another, there is a momentary line-to-line short because the two line voltages are equal and opposite at the point of commutation.

POWER FACTOR

The power factor of a load fed from an AC supply is given by

$$\text{PF} = \frac{\text{Mean power}}{V_{\text{rms}} I_{\text{rms}}} \qquad \text{(H.14)}$$

In a normal AC system, power factor is easily defined as the cosine of the angle between the voltage and the current. It cannot be defined that simply in a rectifier system because there are mean and harmonics. In controlled rectification, the load current is delayed by the firing angle with respect to supply phase voltage.

The input supply voltage is sinusoidal and may be assumed to have no harmonics. Hence, the power output is only due to the fundamental frequency component.

$$\text{Power} = V_{(\text{rms})1} I_{(\text{rms})1} \cos \alpha_1 \qquad \text{(H.15)}$$

where suffix 1 indicates fundamental. Substituting in Equation (H.14)

$$\text{PF} = \frac{V_{(\text{rms})1} I_{(\text{rms})1} \cos \alpha_1}{V_{\text{rms}} I_{\text{rms}}} \qquad \text{(H.16)}$$

where $\frac{I_{(\text{rms})1}}{I_{\text{rms}}}$ is called input distortion caused by harmonics, and ϕ_1 is equal to input displacement factor and α_1 will equal α in fully controlled connections.

Hence, the power factor is reduced by the harmonic content of the output and is always less than 1 even for a simple diode where voltage and current are in phase.

THREE-PHASE HALF-CONTROLLED AND FULLY CONTROLLED BRIDGE CIRCUIT

If the diodes D_1 to D_3 are replaced by three SCRs, it becomes a fully controlled bridge circuit. There is no need for a freewheeling diode also.

Note the six notches in Figure H.11 compared to the three in Figure H.10.

Fully controlled bridge circuits have a problem of starting. Normally, gate firing pulses are delivered in sequence. So, if SCR 1 is triggered first and SCR 5 later, the circuit will not be complete and no output pulse will be delivered. So, two thyristors of one circuit should be fired simultaneously.

HARMONICS AND HEATING

Rectified output pulses are harmonics of the fundamental supply frequency. For the output having *p* pulses, the output has up to *p*th harmonic of the fundamental. Controlled rectifiers have a higher harmonic content than the uncontrolled rectifiers.

The flow of harmonic current in load is caused by the harmonic voltage. For controlled rectifiers, harmonics shift the precise angle of firing between phases, which is known as 'jitter'. Hence, commutation difficulties occur due to harmonics.

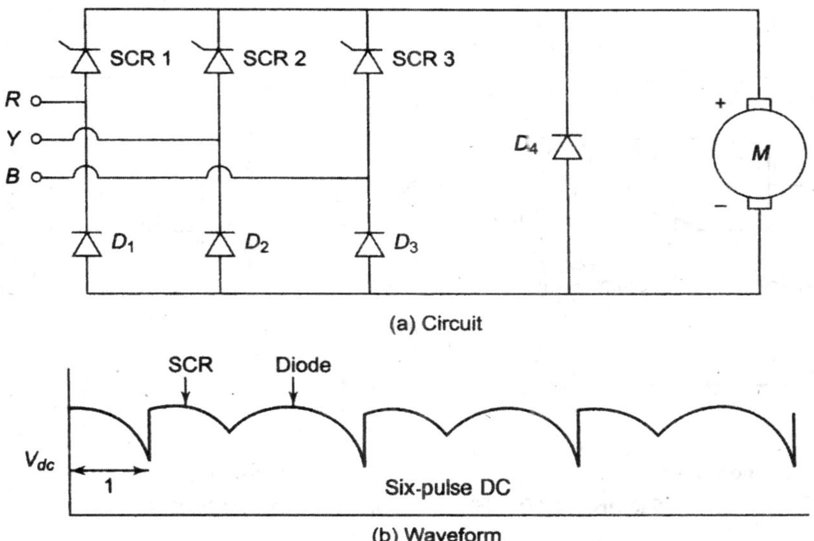

Figure H.10 Half-controlled Three-phase Bridge Circuit

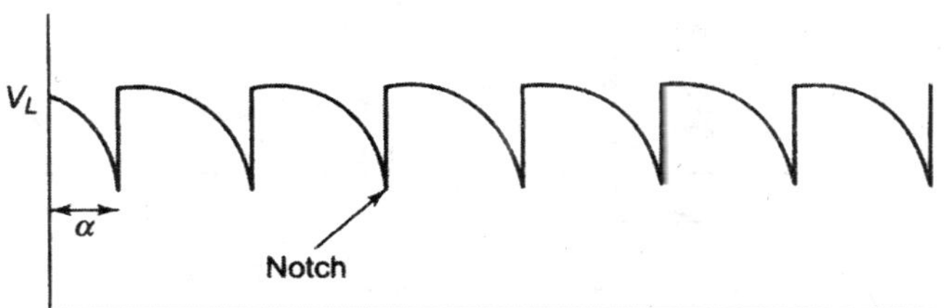

Figure H.11 Pulse Fully Controlled Waveform

The armature or copper loss proportional to the square of the armature current occurs due to flow of harmonic currents in armature. So, harmonics increase copper loss and heat up the motor. Harmonics also cause an increase in the iron loss of the armature core and the nonlaminated parts of any interpole due to the cross magnetization component of armature reaction. Therefore, filters are essential in controlled rectifiers.

AC MOTOR CONTROL

Thyristors can also be used for speed control of AC motors by means of phase control, which is obtained by applying a variable voltage to the motor. It is not possible to get speed control of a synchronous motor by changing the input voltage control. Its speed is controlled using full-wave inverter circuits. Figures H.12(a) and (b) show the schematic arrangement for the speed control of single-phase and three-phase inductions. The root mean square (RMS) input voltage can be changed by varying the firing angle of the thyristor.

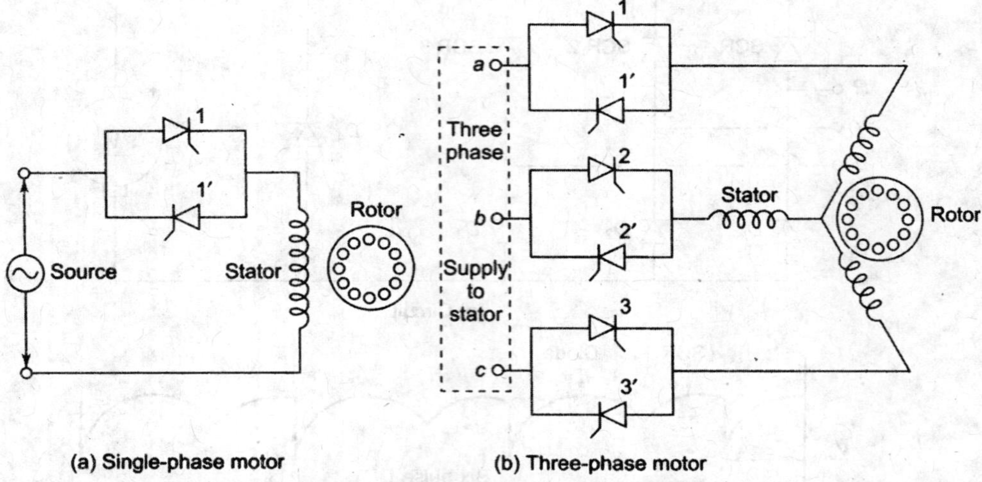

Figure H.12 Phase Control of an Actuator

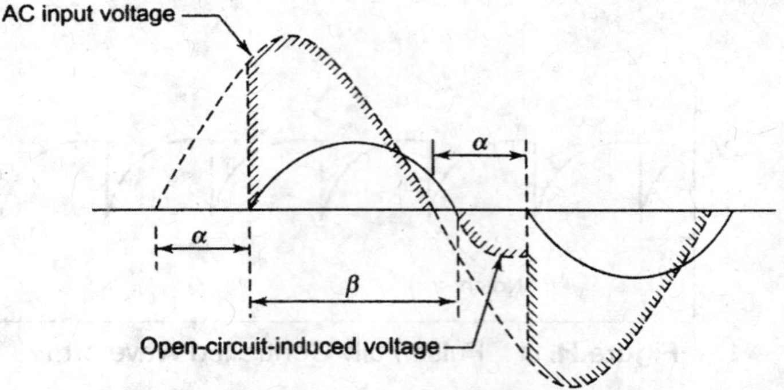

Figure H.13 Current and Voltage Waveform

Two SCRs are connected antiparallel instead of using triac because the motor is an inductive-load motor. For single-phase circuits, a unijunction transistor (UJT) relaxation oscillator can be used. The input current and the voltage waveform are shown in Figure H.13. The applied voltage will, for every half-cycle, have an angle of conduction β less than π. During this period, the rotor currents will induce a voltage in the stator phase winding.

CHARACTERISTIC OF PHASE-CONTROLLED AC MOTOR

Figure H.14 shows the characteristics of an induction motor with variable applied voltage. This method of control is very simple and economical, providing a wide range of speed control if the load torque increases with speed, as shown by the curves in Figure H.14. If the load torque is constant, the speed variation is limited, as shown by curve 2. The disadvantage of this method of speed control is that efficiency decreases with a decrease in speed.

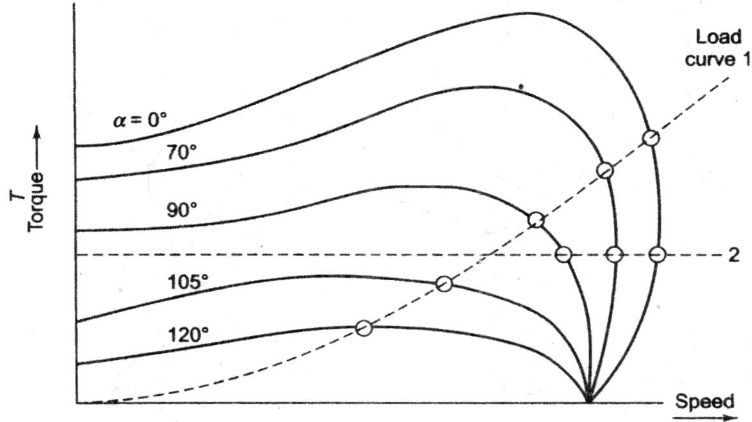

Figure H.14 Speed-torque Characteristic

The speed control of a slip ring induction motor can be achieved by varying the rotor resistance where the variation can be obtained either by phase control or by an ON-OFF control. The method of varying the rotor resistance using phase control is shown in Figure H.15. The corresponding speed-torque characteristics are shown in Figure H.16.

The power consumed by resistance R_o is reduced with an increase in the firing angle. Low input power factor is the main drawback of this method. The lagging power factor of the load reduces the pull-out torque. To improve rotor power factor, a capacitor C is connected across the line. This method of control can be applied for both constant and variable torque loads and a wide range of speed control is achieved with good starting torque. The frequency of the input voltage to the bridge varies with the speed of the motor. Thus, the firing frequency and the firing angle must be synchronized with the rotor frequency.

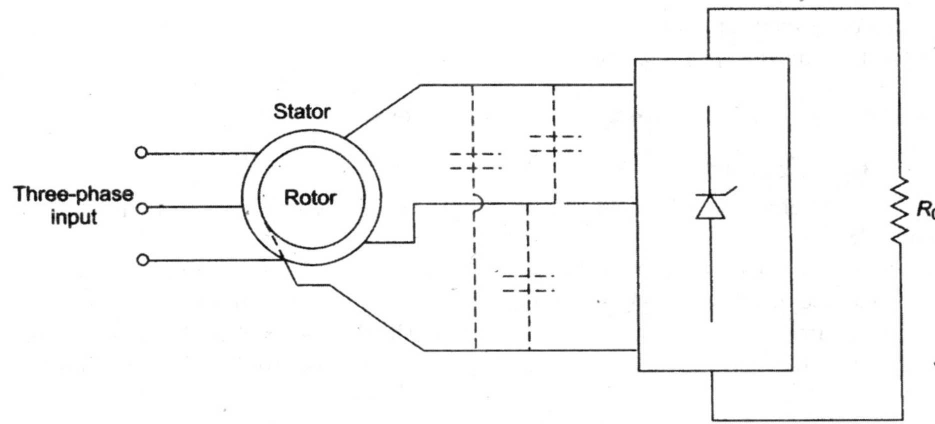

Figure H.15 Controlled Bridge

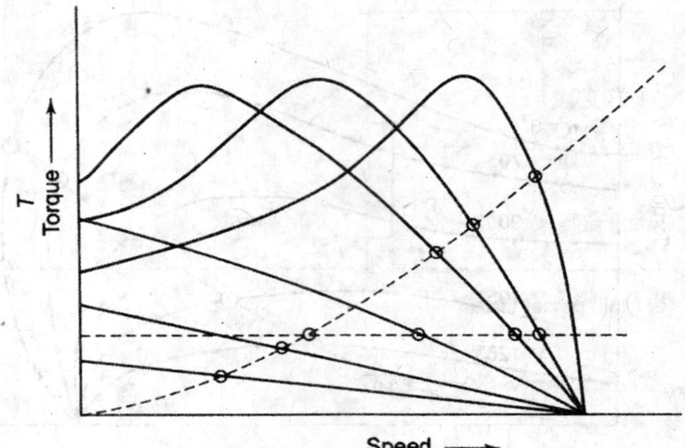

Figure H.16 Speed-torque Characteristics

The second method of speed control is by an ON-OFF control. In this method, power is applied to the load for a specific period known as time (T_{on}) and removed for a period T_{off} where $T_{on} + T_{off}$ represents the duty cycle. A mechanism to introduce a ($T_{on} + T_{off}$) cycle is introduced at the input of the stator of induction motor. For a given ON-OFF time, the RMS voltage applied to the stator is given by

$$V_{rms} = \frac{V_m}{\sqrt{2}}\sqrt{\frac{T_{on}}{T_{on}+T_{off}}} \quad \text{(H.17)}$$

where V_m is the peak amplitude of AC voltage. By changing the duty cycle, the value of V_{rms} can be altered. Using the steady-state equivalent circuit of the induction motor as in Figure H.17, the resultant torque can be estimated for any applied voltage at any given slip. The power dissipated in $\frac{R_r^t}{s}$ gives the torque.

R_s = Stator resistance per phase
X_s = Stator reactance per phase
X'_r = Rotor reactance per phase referred to the stator slip s
$\frac{R_r^t}{x}$ = Rotor resistance per phase referred to the stator slip s
X_m = Magnetizing reactance

Torque-speed characteristics are shown in Figure H.18. These are similar to those obtained for the phase control.

Another type of ON-OFF control method for speed control is shown in Figure H.19.

In this method, normal voltage is applied to the stator and the control is applied to rotor windings. Rotor windings are subjected to open and short circuits by the switch S_1 of the chopper circuit. If T_{01} is the developed torque under normal operation, T_{02}, the torque under rotor ON-OFF control, will be

$$T_{02} = T_{01}\left(\frac{T_{on}}{T_{on}+T_{off}}\right) \quad \text{(H.18)}$$

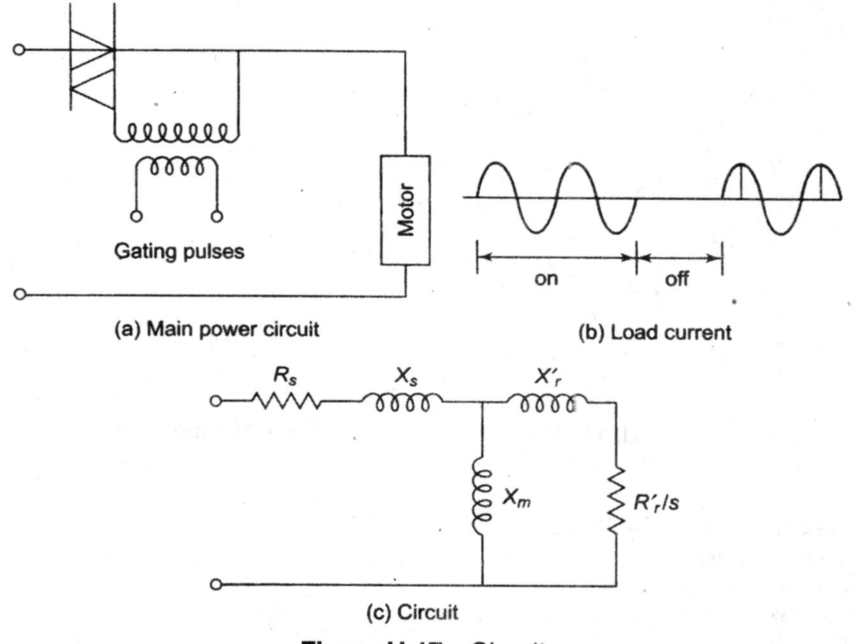

Figure H.17 Circuit

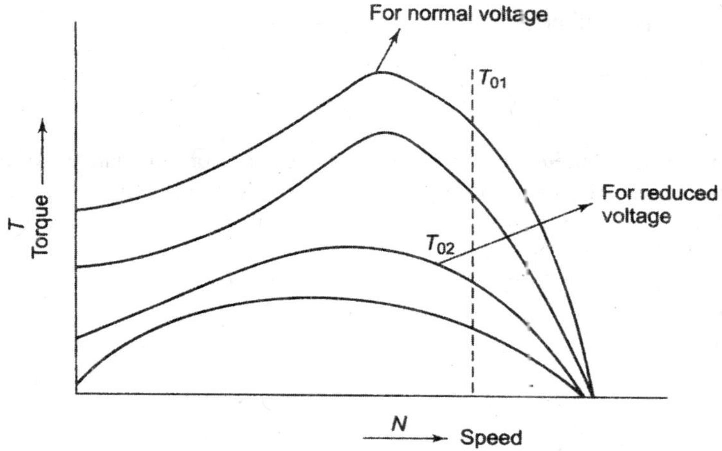

Figure H.18 Torque-speed Characteristics

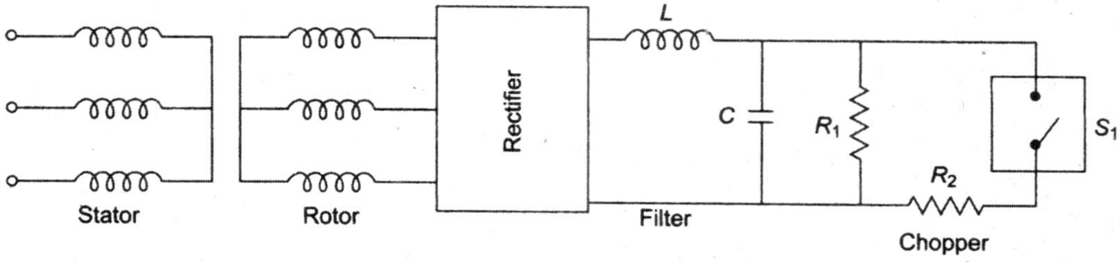

Figure H.19 ON-OFF Control

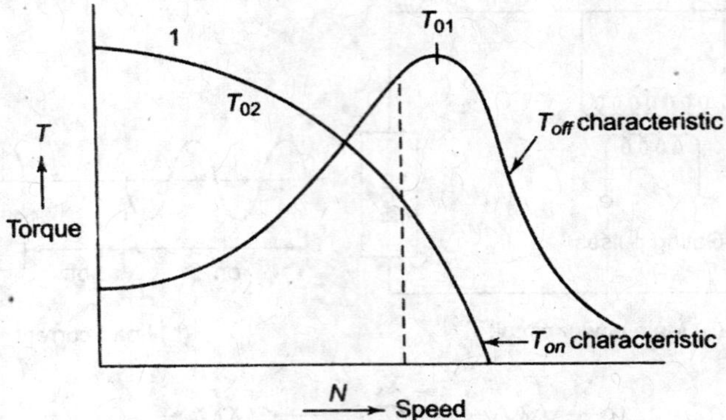

Figure H.20 Torque-speed Characteristic

A chopper circuit is used for this purpose. When the chopper switch is open, a large resistance R_1 is in circuit, thereby increasing the rotor resistance. When the chopper is on, r_2, which is a small resistance, gets connected across R_1, thereby reducing the rotor resistance. By this method, the torque of the motor is varied during ON and OFF periods. Figure H.20 shows the torque-speed characteristic.

When the switch S_1 is ON, rotor resistance is low and the characteristic is as per curve 1, representing T_{02}. When S_1 is open, the characteristic is per curve 2, representing T_{01}

The net average torque will be

$$T_{net} = \frac{T_{01}T_{off} + T_{02}T_{03}}{T_{on} + T_{off}} \tag{H.19}$$

The ON-OFF methods applied to control stator voltages or rotor impedance are less efficient at high values of slip, as the power dissipation in the order is given by power input multiplied by slip. Hence, the efficiency will be

$$\frac{\text{Mechanical power output}}{\text{Power input}} = 1 - s \tag{H.20}$$

Multiple Choice Questions

1. If the secondary winding of the ideal transformer shown in Figure 1 has 40 turns, the number of turns in the primary winding for maximum power transfer to the 2 Ω resistor will be
   (a) 20 (b) 40 (c) 80 (d) 160

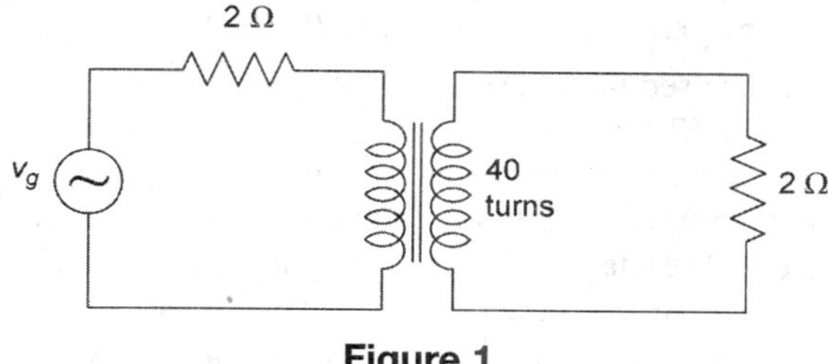

Figure 1

2. If the rated voltage from the power lines is applied to the primary of a single-phase transformer which is operated on no-load, then
   (a) both input voltage and current are sinusoidal
   (b) both input voltage and current are non-sinusoidal
   (c) input voltage is non-sinusoidal and current are sinusoidal
   (d) input voltage is sinusoidal and current are non-sinusoidal
3. For the purpose of analysis, an exact equivalent circuit of a transformer is usually replaced by an approximate equivalent circuit. In doing so errors loss is of differential nature. Due to this, the analysis by approximate equivalent circuit gives fairly satisfactory results. Under the circumstances, which one of the following is an equivalent circuit as compared to the exact equivalent circuit?
   (a) This accounts for somewhat greater primary winding copper loss and lesser core loss.
   (b) This accounts for somewhat lesser primary winding copper loss and greater core loss.
   (c) This accounts for somewhat greater secondary winding copper loss and lesser core loss.
   (d) This accounts for somewhat lesser secondary winding copper loss and greater core loss.
4. At 50 Hz operation, a single-phase transformer has hysteresis loss of 200 W and eddy current loss of 100 W. Its core loss at 60 Hz operation will be
   (a) 432 W (b) 406 W (c) 384 W (d) 360 W
5. A 2 kVA transformer has iron loss of 150 W and full-load cupper loss of 250 W. The maximum efficiency of the transformer would occur when the total loss is
   (a) 500 W (b) 400 W (c) 300 W (d) 275 W

6. If the frequency of input voltage is increased keeping the magnitudes of voltage unchanged, then
   (a) both hysteresis and eddy current losses in the core will increase
   (b) hysteresis loss will increase and eddy current loss will decrease
   (c) hysteresis loss will decrease and eddy current loss will increase
   (d) hysteresis loss will decrease and eddy current loss will remain unchanged
7. The voltage regulation of transformer at full load and 0.8 power factor lagging is 2.5 per cent. The voltage regulation at full-load and 0.8 power factor leading will be
   (a) –2.5% (b) zero (c) 1% (d) 2.5%
8. For successful operation of two single-phase transformers, the most essential condition is that their
   (a) percentage impedances are equal (b) polarities are properly connected
   (c) turns ratios are exactly equal (d) kVA ratings are equal
9. In a transformer, if the iron losses and copper losses are 40.5 kW and 50 kW, respectively, then at what fraction of load will the efficiency be maximum?
   (a) 0.60 (b) 0.57 (c) 0.70 (d) 0.90
10. Can a 50 Hz transformer be used for 25 Hz, if the input voltage is maintained constant at the rated value corresponding to 50 Hz?
    (a) Yes. As the voltage is constant, current levels will not change.
    (b) No. Flux will be doubled, which will drive the core to excessive saturation.
    (c) No. Owing to decreased reactance of transformer, input current will be doubled at load.
    (d) Yes. At constant voltage, insulation will not be overstressed.
11. The short circuit test is performed on a transformer with a certain impressed voltage at rated frequency. If the short circuit test is now performed with the same magnitude of impressed voltage, but at a frequency higher than the rated frequency, then
    (a) both the magnitude of the current and the power factor will increase
    (b) the magnitude of the current will decrease but the power factor will increase
    (c) the magnitude of the current will increase but the power factor will decrease
    (d) both the magnitude of the current and the power factor will decrease
12. Which of the following regarding a transformer is correct?
    (a) In transformers, a laminated core is used to reduce copper losses.
    (b) Core losses in a transformer can be determined by an open circuit test.
    (c) Short circuit test is conducted on a transformer to determine constant losses.
    (d) In a shell-type transformer, the primary and secondary windings are wound on separate limbs.
13. The main purpose of the conservator in a transformer is to
    (a) store extra oil to compensate for the loss of oil due to leakage
    (b) achieve better cooling of the transformer
    (c) take up the expansion of oil due to heating
    (d) have the Buchhloz relay fitted
14. Which of the following statements are incorrect?
    i. The maximum voltage regulation of transformer occurs at leading power factor.
    ii. The voltage regulation of a transformer is maximum when the load power factor (lagging) angle has the same value as the angle of equivalent impedance.
    iii. The voltage regulation of a transformer at zero power factor is always zero.
    iv. The voltage regulation of a transformer can be negative at leading power factor.

Select the correct answer using the codes given below:
(a) i and iii (b) ii and iii (c) ii and iv (d) i and iv

15. When a short circuit test is conducted on a single-phase transformer, 30 per cent of the rated voltage is required to allow full-load current. The short circuit power factor is found to be 0.2. The percentage regulation at unity power factor (UPF) is
(a) 30 (b) 29.5 (c) 15 (d) 6

16. Capacitance voltage transformer (CVT) is used to
(a) improve the power factor of the transmission
(b) reduce losses in a transmission line
(c) connect instruments on the low tension side
(d) reduce the incidence of overvoltage surges on transmission lines

17. A transformer designed for operation on 60 Hz supply is worked on 50 Hz supply system without changing its voltages and current ratings. When compared with full load, efficiency at 60 Hz, the transformer efficiency on full load at 50 Hz will
(a) increase marginally (b) increase by a factor of 1.2
(c) remain unaltered (d) decrease marginally

18. A transformer has a core loss of 200 W and full-load copper loss of 800 W. The maximum efficiency of the transformer will occur at
(a) 0.5 times full-load current (b) 0.6 times full-load current
(c) 0.7 times full-load current (d) 0.8 times full-load current

19. If the primary winding of a 100 kVA transformer is connected to the direct supply,
(a) the voltage induced in the secondary will be direct voltage
(b) back emf will be induced in the primary winding
(c) the primary winding may burn
(d) low voltage will be induced in the secondary

20. A 480/120 V, 5 kVA, two-winding transformer is to be used as an autotransformer to supply power at 480 V from 1,600 V supply. The kVA rating of the autotransformer will be
(a) 5 (b) 15 (c) 25 (d) 50

21. A 100 kVA transformer has copper loss of 100 W at full load and iron loss of 100 W. Under which of the following conditions, the efficiency of the transformer will be
(a) half load, unity power factor
(b) half load, 0.8 power factor (lagging)
(c) $\frac{5}{4}$th of full load, unity power factor
(d) $\frac{5}{4}$th of full load, 0.8 power factor (lagging)

22. For the transformer cores, cold-rolled silicon steel is preferred to hot-rolled silicon steel because of
(a) lower cost (b) higher tensile strength
(c) higher flux density (d) affinity for oil

23. In transformers, interleaved disc connection is used for
(a) welding transformers
(b) step-down transformers
(c) low-voltage low-current transformers
(d) high-voltage transformers

24. The function of breather in a transformer is to
    (a) provide oxygen to the oil
    (b) provide cooling air
    (c) arrest moisture entering the transformer
    (d) filter the transformer oil
25. The transformer has the turns ratio of 2:1 (primary to secondary). If peak-to-peak value of secondary voltage is 220 V, the primary voltage will be
    (a) 220 V rms (b) 440 V peak-to-peak
    (c) 300 V rms (d) 220 V rms
26. Which relay is used to protect the transformer from inter turn fault?
    (a) Air blast relay (b) Buchholz relay (c) Mho relay (d) None of these
27. High-frequency transformers use
    (a) aluminium core (b) iron core (c) copper core (d) air core
28. The efficiency of a well-designed transformer may be expected in the range
    (a) 75–80% (b) 65–70% (c) 85–90% (d) 95–99%
29. Two similar, 250 kVA, single-phase transformers gave the following results when tested by back-to-back method:
    Mains wattmeter, $W_1 = 5.0$ kW
    Primary series circuit wattmeter, $W_2 = 7.5$ kW (at full-load current)
    The individual transformer efficiency at 75 per cent full load and 0.8 pf load will be nearly
    (a) 97% (b) 95% (c) 93% (d) 90%
30. What is a typical use of an autotransformer?
    (a) toy transformer (b) control transformer
    (c) variable transformer (d) isolating transformer
31. The insulating properties of transformer oil are adversely affected by the presence of
    (a) oxygen (b) copper (c) moisture (d) ozone
32. In a transformer on no-load, the input voltage
    (a) leads magnetising current by 90°
    (b) lags magnetising current by 90°
    (c) is in phase with the magnetising current
    (d) is always at 45° to the magnetising current
33. An autotransformer is preferred in cases where
    (a) load is fluctuating (b) power factor is varying
    (c) ratio of transformation is low (d) ratio of transformation is high
34. In transformers, disc coils are generally preferred for
    (a) leading power factor (b) high-voltage winding
    (c) constant load (d) low-voltage winding
35. Buchholz relay works on the principle of
    (a) production of eddy currents whenever load changes are rapid
    (b) ionization of transformer oil
    (c) magnetic hum due to sudden variation of load
    (d) generation of gas due to breakdown of the insulation

36. The results of open-circuit test to a single-phase transformers are
    Voltage applied: 100 V
    Current drawn: 0.5 A
    Power consumed: 30 W

    The corresponding reactance of the transformer is
    (a) 200 Ω (b) 250 Ω (c) 400 Ω (d) 500 Ω

37. For a 100 per cent efficient transformer if the number of turns in the primary and secondary are 1,000 and 100, respectively and the input to the transformer is 100 W, the power output is
    (a) 100 W (b) 1,000 W (c) 10 kW (d) 100 kW

38. Under operating conditions, the secondary of a current transformer is always short circuited because
    (a) it protects the primary circuit
    (b) it is safe to human beings
    (c) it avoids core saturation and high-voltage induction
    (d) none of the above

39. Which of the following transformer is smallest?
    (a) 1 kVA, 50 Hz
    (b) 1 kVA, 200 Hz
    (c) 1 kVA, 400 Hz
    (d) 1 kVA, 600 Hz

40. What special precaution should be taken when using a booster transformer?
    (a) It should never be left open-circuit.
    (b) There should be no fusing in the high-voltage side.
    (c) It should never be left close circuit.
    (d) None of the above.

41. Helical coils are very well suited for
    (a) high-voltage winding of small-rating transformer
    (b) low-voltage winding of large-rating transformer
    (c) high-voltage winding of large-rating transformer
    (d) none of the above

42. In parallel operation of two single-phase transformers, if the impedance triangles of the transformers are not identical in shape and size
    (a) power factors at which the transformers operate will be same but different from the power factor of common load
    (b) power factors at which the transformers operate and the power factor of common load will be the same
    (c) power factor of one transformer and the power factor of common load will be the same
    (d) power factors at which the transformers operate will be different from one another and again these will be different from the power factor of common load

43. On which factor does the short circuit voltage of a transformer mainly depend?
    (a) on the ohmic resistance of the primary winding
    (b) on the ohmic resistance of the secondary winding
    (c) cross-sectional area of the iron core
    (d) magnitude of the leakage flux

44. A 100 kVA transformer has copper loss of 1,000 W at full load and iron loss of 1,000 W. At half the full load and 0.8 power factor (lagging), the efficiency of the transformer will be
(a) 98.91% (b) 96.97% (c) 95% (d) 91.91%

45. Eddy current losses in a transformer are minimised by laminating the core, interlamination insulation being provided by
(a) thick paper insulation (b) thin sheets of mica
(c) a thin coat of core plate varnish (d) none of the above

46. Use of higher flux density in the transformer design
(a) increases the weight/kVA (b) decreases the weight/kVA
(c) reduces iron loss (d) improves insulation

47. In a transformer, the primary and secondary voltages are
(a) 60° out of phase (b) 90° out of phase
(c) 180° out of phase (d) always in phase

48. The efficiency of a transformer at full load 0.8 pf lagging is 95 per cent. Its efficiency at full load 0.8 pf leading will be
(a) 80% (b) 90% (c) 95% (d) 100%

49. Steel used in the construction of transformer core has
(a) low permeability and low hysteresis loss (b) low permeability and high hysteresis loss
(c) high permeability and high hysteresis loss (d) high permeability and low hysteresis loss

50. Transformer oil should be designed free from
(a) moisture (b) sulphur (c) alkalis (d) all of these

51. The leakage flux of a transformer depends on
(a) supply frequency (b) load current (c) mutual flux (d) none of these

52. The no-load current of a transformer in terms of full-load current is typically
(a) 1–4% (b) 4–12% (c) 12–16% (d) 16–20%

53. Good transformer oil should not have water more than
(a) 4 ppm (b) 8 ppm (c) 12 ppm (d) 20 ppm

54. A transformer is connected to a constant voltage supply. As the supply frequency increases, the magnitude of the flux in the core
(a) decreases (b) increases towards saturation
(c) becomes zero (d) becomes constant

55. The size of the transformer core depends on
(a) frequency (b) flux density of the core
(c) both (a) and (b) (d) cross-sectional area of the core

56. If we impress a DC voltage of 230 V to an unloaded 230 V, 50 Hz transformer, the transformer will
(a) burn out
(b) give very high secondary voltage
(c) give secondary voltage according to turns ratio
(d) none of the above

57. Ferrite cores have less eddy current losses than iron losses because ferrites have
(a) low permeability (b) high hysteresis
(c) high resistance (d) all of these

58. If a two-winding step-down transformer is converted into an autotransformer by using additive polarity, then
    (a) kVA rating gets reduced
    (b) kVA rating increases considerably
    (c) kVA rating remains unaltered
    (d) a large number of secondary taps are needed

59. Incorrect polarity in parallel operation results in
    (a) open circuit
    (b) short circuit
    (c) load sharing proportional to their kVA rating
    (d) regeneration of power

60. In parallel operation of transformers, to reduce copper loss
    (a) they should have equal turns ratio
    (b) their phases should be same
    (c) they should have zero impedance
    (d) none of the above

61. Power transformers are designed to have maximum efficiency at
    (a) a little more than full load
    (b) near full load
    (c) half load
    (d) quarter load

62. Scott connection is used to transform
    (a) single-phase supply into two-phase supply
    (b) two-phase supply into single-phase supply
    (c) three-phase supply into two-phase supply
    (d) two-phase supply into three-phase supply

63. The iron losses of a transformer may be calculated by knowing the weight of
    (a) yokes alone
    (b) cores alone
    (c) cores and yoke
    (d) copper winding alone

64. Other things remaining same, the efficiency of an auto-transformer compared to a two-winding transformer is
    (a) lower
    (b) the same
    (c) higher
    (d) dependent of load

65. Special silicon steel is used for laminations in power transformer to reduce
    (a) eddy current loss
    (b) hysteresis loss
    (c) copper losses
    (d) both (a) and (b)

66. The reactance of a transformer is defined by its
    (a) leakage flux
    (b) common core flux
    (c) permeability of the core material
    (d) size of the cores

67. Concentric windings are used in core type transformer with
    (a) LT winding placed next to core
    (b) HT winding placed next to core
    (c) LT winding on the outer side
    (d) HT winding on the outer side

68. Oil natural cooling with tubes is used for transformers of capacity
    (a) 100 kVA
    (b) 500 kVA
    (c) 1,000 kVA
    (d) 3,000 kVA

69. The magnetic flux in a transformer follows a path of
    (a) high reluctance
    (b) low reluctance
    (c) high conductivity
    (d) low conductivity

70. The magnetic coupling between the primary and secondary of a transformer may be increased by
(a) increasing the number of laminations (b) changing the turns ratio
(c) using the magnetic core of low reluctance (d) none of the above

71. For minimum weight of a transformer, the weight of iron should be
(a) less than the weight of copper (b) greater than the weight of copper
(c) equal to the weight of copper (d) none of these

72. A constant current transformer should not have
(a) a high value of reactance
(b) a movable secondary winding
(c) a high value of resistance
(d) primary and secondary windings surrounding the core

73. Transformers transform
(a) frequency (b) voltage (c) current (d) power

74. An ideal power transformer will have maximum efficiency at a load such that
(a) copper loss is less than iron loss (b) copper loss is equal to iron loss
(c) copper loss is higher than iron loss (d) none of the above

75. Two transformers operating in parallel will share the load depending on their
(a) rating (b) leakage reactance
(c) efficiency (d) per unit impedance

76. If the applied voltage of a certain transformer is increased by 50 per cent and the frequency is reduced to 50 per cent (assuming that the magnetic circuit remains unsaturated), the maximum core flux density will
(a) change to three times the original value
(b) change to 1.5 times the original value
(c) change to 0.5 times the original value
(d) remain the same as original value

77. The low-voltage winding of a core-type transformer is subdivided into two equal halves, each half having the original width of the single winding with the high-voltage winding (instead of having the usual construction of low-voltage winding adjacent to the core and surrounded by the high-voltage winding). Such an interlacing of coils would make the combined primary and secondary leakage reactance (in terms of the primary) nearly
(a) twice (b) equal (c) half (d) one-fourth

78. In a transformer fed from a fundamental frequency of voltage source, the source of harmonics is the
(a) overload (b) poor insulation
(c) iron loss (d) saturation of core

79. A 40 kVA transformer has a core loss of 400 W and full-load copper loss of 800 W. The proportion of full load at maximum efficiency is
(a) 50% (b) 62.3% (c) 70.7% (d) 100%

80. A single-phase induction regulator is a constant input transformer to obtain smooth variation of the output voltage by varying the
(a) ratio of turns between the primary and secondary windings
(b) frequency
(c) flux density in the core
(d) angle between the magnetic axes of the primary and secondary windings

81. An autotransformer having a transformation ratio of 0.8 supplies a load of 10 kW. The power transferred inductively from primary to the secondary is
(a) 10 kW (b) 8 kW (c) 2 kW (d) zero

82. Consider the following statements:
i. skewing the slots
ii. fractional slot winding
iii. short-chorded winding
iv. distributed winding
Of these statements
(a) i and ii are correct
(b) ii, iii and iv are correct
(c) i, iii and iv are correct
(d) i and iii are correct

83. When a transformer winding suffers a short circuit, the adjoining turns of the same winding experience
(a) an attractive force (b) a repulsive force (c) no force

84. The core flux of a practical transformer with a resistive load
(a) is strictly constant with load changes
(b) increases linearly with load
(c) increases as the square root of the load
(d) decreases with increased load

85. A 1 kVA, 230 V/100 V, single-phase 50 Hz transformer having negligible winding resistance and leakage reactance is operating under saturation while 250 V, 50 Hz sinusoidal supply is connected to the high-voltage winding. A resistive load is connected to the low-voltage winding, which draws rated current. Which one of the following quantities will not be sinusoidal?
(a) voltage induced across the low-voltage winding
(b) core flux
(c) load current
(d) current drawn from the source

86. Two three-phase transformers are to be connected for parallel operation. Which one of the following arrangements is impossible?
(a) Transformer A: primary Y; secondary Y
Transformer B: primary Δ; secondary Δ
(b) Transformer A: primary Δ; secondary Y
Transformer B: primary Δ; secondary Δ
(c) Transformer A: primary Y; secondary Δ
Transformer B: primary Δ; secondary Δ
(d) Transformer A: primary Δ; secondary Δ
Transformer B: primary Δ; secondary Δ

87. If the waveform of the voltage impressed on the primary of a Y-Δ bank contains a 5th harmonic, the wave-forms of the resultant voltages of the primary and secondary would be

	Primary	*Secondary*
(a)	peaked	peaked
(b)	peaked	flat-topped
(c)	flat-topped	peaked
(d)	flat-topped	flat-topped

88. A 120 kVA 6,000/400 V star/star three-phase, 50 Hz transformer has an iron loss of 1,800 W. The maximum efficiency occurs at 3/4 full load. The efficiency of the transformer at full load is 0.8 pf will be
(a) 98% (b) 95% (c) 92% (d) 90%

89. i. star–star ii. delta–delta
iii. star–delta iv. delta–star
The core section of a large transformer can work in parallel with another transformer. Identify from the above.
(a) i and ii (b) iii and iv (c) i and iv (d) ii and iii

90. Three-phase transformers may be
i. star–star connection ii. delta–delta connection
iii. star–delta connection iv. delta–star connection
A star–star connected transformer can work in parallel with another transformer. Identify from the above.
(a) i and ii (b) iii and iv (c) i and iv (d) ii and iii

91. The most suitable and economical connection for a small high-voltage transformer is
(a) star–delta connection (b) delta–delta connection
(c) delta–star connection (d) star–star connection

92. In a three-phase Δ/Y transformer shown in Figure 2, the phase displacement of secondary line voltages with corresponding primary line voltages will be

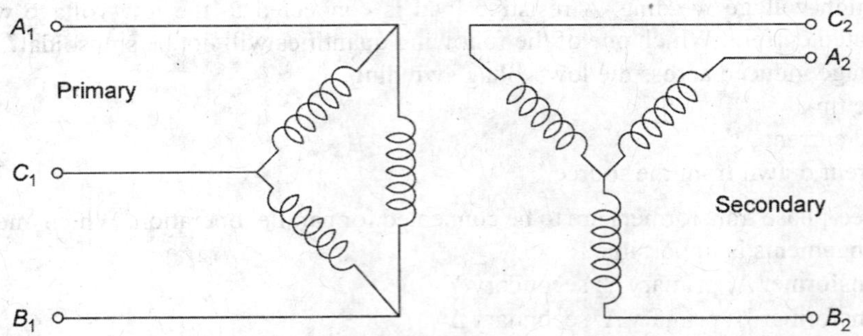

Figure 2

(a) zero (b) 30° lag (c) 30° lead (d) 180°

93. Three three-phase transformers each rated at 75 MVA, 132 kV/11 kV have the following different methods of cooling:
i. self oil cooled
ii. forced oil cooled
iii. forced air cooled
The correct sequence in ascending order in terms of the weights of these transformers is
(a) i, ii, iii (b) ii, iii, i (c) iii, i, ii (d) iii, ii, i

94. A three-phase delta/star transformer is supplied at 6,000 V on the delta-connected side. The terminal voltage on the secondary side when supplying full load at 0.8 lagging power factor is 415 V. The equivalent resistance and reactance drops for the transformer are 1 per cent and 5 per cent, respectively. The turns ratio of the transformer is
(a) 14 (b) 24 (c) 42 (d) 20

95. In a 110 V compound generator, the armature, shunt and series windings are 0.06 Ω, 27.5 Ω and 0.04 Ω, respectively. The load consists of 200 lamps each rated at 55 W, 110 V. Find the total emf when the machine is connected for long shunt operation
(a) 12 V (b) 72 V (c) 100 V (d) 120 V

96. The DC generator used for arc welding purposes is a
(a) series generator
(b) shunt generator
(c) cumulatively compound generator
(d) differentially compound generator

97. A 220 V DC machine has an armature resistance of 1 Ω. If the full-load current is 20 A, the difference in the induced voltages when the machine is running as a motor and as a generator is
(a) 20 V (b) zero (c) 40 V (d) 50 V

98. A separately excited DC generator is feeding a shunt motor. If the load torque on the motor is halved approximately
(a) armature currents of both motor and generator are halved
(b) armature current of motor is halved and that of generator is unaltered
(c) armature current of generator is halved and that of motor is unaltered
(d) armature currents of both motor and generator are unaltered

99. An electric train employing a DC series motor is running at a fixed speed, when a sudden slight drop in the mains voltage occurs. This would result in
(a) drop in speed and rise in current (b) rise in speed and drop in current
(c) rise in speed and rise in current (d) drop in speed and rise in current

100. A DC machine has conductors per pair of poles Z, length of each armature conductor L, linear velocity of armature conductor up and air gap flux density distribution $B = B_{max} \sin \theta$. The instantaneous emfs in successive conductors under one pole should be a series whose first two terms are $e_1 = B_m L_u \sin \theta$ and $e_2 = B_m L_u \sin (\theta + 2\pi/Z)$. The last term in this series will be
(a) $B_m L_u \sin (\theta + \pi/Z)$
(b) $B_m L_u \sin (\theta + \pi/2Z)$
(c) $B_m L_u \sin (\theta + \pi - \pi/Z)$
(d) $B_m L_u \sin (\theta + \pi - 2\pi/Z)$

101. Consider the following statements regarding speed control of DC motors:
i. Ward-Leonard method is suitable for constant torque drive.
ii. Field control method facilities speed control below `base speed'.
iii. 'Armature resistance control' method is more efficient when compared to Ward-Leonard method.
iv. Field control method is suitable for constant horse power drive.

Of these statements:
(a) i, ii and iii are correct
(b) i, iii and iv are correct
(c) ii, iii and iv are correct
(d) i and iv are correct

102. A DC shunt motor runs at 500 rpm of 220 V. A resistance of 4.5 Ω is connected in series with the armature for speed control. The armature resistance is 0.05 Ω. The current to stall the motor will be

(a) 44 A
(b) 50 A
(c) 44.4 A
(d) 60 A

103. Consider the following statements:
If the terminal voltage of a DC shunt motor is halved with the load torque varying as the square of the speed, then

i. speed is halved
ii. speed remains unaltered
iii. armature current is doubled
iv. armature current remains unaltered

Of these statements:

(a) i and iii are correct
(b) i and iv are correct
(c) ii and iii are correct
(d) ii and iv are correct

104. The armature current of a DC motor fed from a thyristor power converter contains ripple. This ripple in the armature current affects

(a) the commutating capability of the motor
(b) the overload capacity of the motor
(c) the torque capability of the motor
(d) the controllability of the speed of the motor

105. The stator of a 2/4 pole changing cage motor is initially wound for 2 poles. The reconnection of the stator winding is done for 4-poles. The reconnection of the stator winding to 4 poles through a change-over switch while the motor is running would be

(a) constant torque drive
(b) constant HP drive
(c) plugging to standstill
(d) regeneration breaking to half the original speed

106. The applied voltage of a DC machine is 230 V. Then the back emf for maximum power developed is

(a) 115 V
(b) 200 V
(c) 230 V
(d) 460 V

107. A lap wound armature winding fitted with a commutator and a pair of brushes on it is rotated at a speed N_t in a rotated magnetic field having P poles and rotating at a speed N_r in space N_t and N_r in both being in the same direction. The frequency of the induced voltage across the brushes on the commutator is

(a) $\frac{N_t P}{120}$
(b) $\frac{(N_t - N_r)P}{120}$
(c) $\frac{(N_t + N_r)P}{120}$
(d) $\frac{N_r P}{120}$

108. If the armature of a DC motor is supplied from a constant current source and its field fed from a constant voltage source, then the torque-speed characteristics of the motor will be

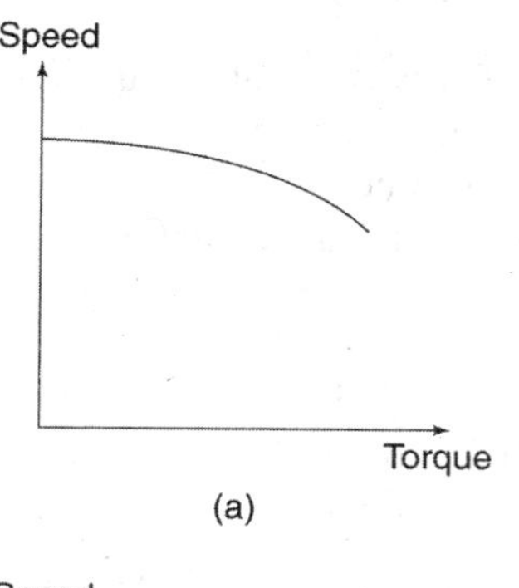

(a)

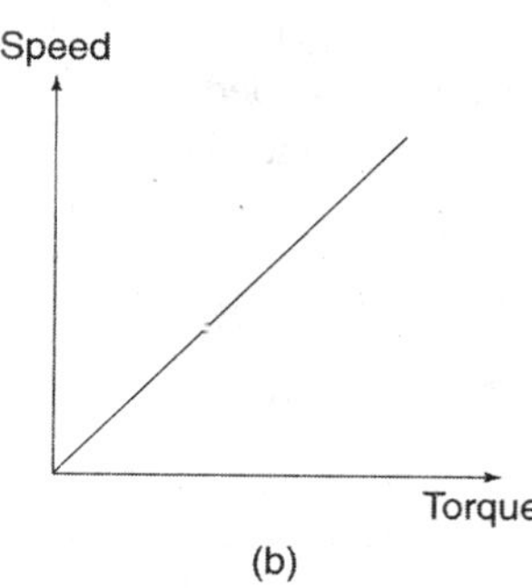

(b)

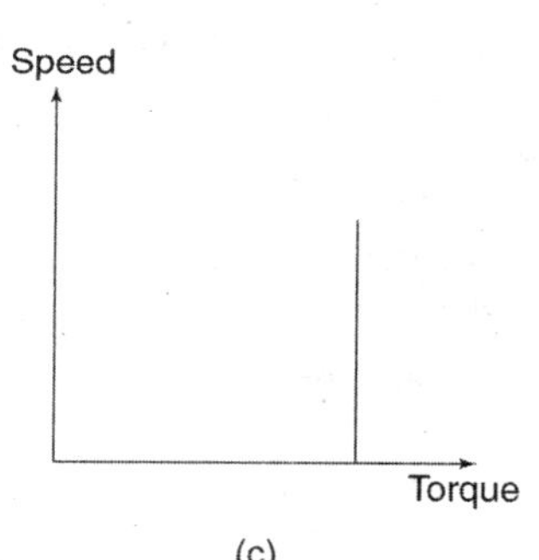

(c)

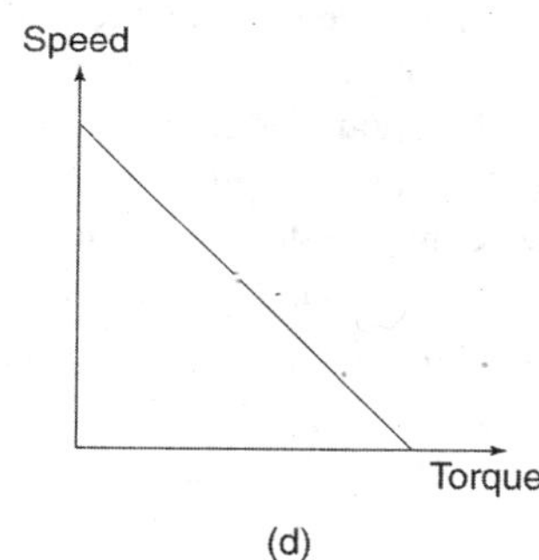

(d)

109. Consider the following statements:
   i. In a DC series generator, the full-load voltage is more than the no-load voltage.
   ii. In a separately excited DC generator, the full-load voltage is more than the no-load voltage.
   iii. In a DC shunt generator, the full-load voltage is less than the no-load voltage.
   Of these statements:
   (a) i, ii and iii are correct
   (b) i and ii are correct
   (c) ii and iii are correct
   (d) i and iii are correct

110. The armature resistance of a six pole lap wound DC machine is 0.05 Ω. If the armature is rewound using a wave winding, the armature resistance will be
   (a) 0.45 Ω
   (b) 0.30 Ω
   (c) 0.15 Ω
   (d) 0.10 Ω

111. To obtain steady arc in arc welding, one should use
   (a) DC series generator
   (b) DC shunt generator
   (c) DC differentially compounded generator
   (d) DC cumulatively compounded generator

112. Match List I (DC motor) with List II (characteristics) and select the correct answer using the codes given below:

List I	*List II*
A. Cumulatively compounded motor	i. Fairly constant speed irrespective of load
B. Differentially compounded motor	ii. It may start in reverse direction
C. Series motor	iii. Definite no-load speed
D. Shunt motor	iv. Never started without load

Codes:

(a) A B C D
ii iii i iv

(b) A B C D
iii ii iv i

(c) A B C D
i iii ii iv

(d) A B C D
iv i iii ii

113. A belt-driven cumulative compounded DC generator is delivering power to the DC mains. If the belt snaps, then the machine will run as
(a) cumulative compounded motor in the same direction
(b) differentially compounded motor in the same direction
(c) cumulative compounded motor in the opposite direction
(d) differentially compounded motor in the opposite direction

114. For a p-pole machine, the relation between electrical (θ_e) and mechanical (θ_m) degrees is given by
(a) $\theta_e = \dfrac{2}{p\theta_m}$ (b) $\theta_e = p\theta_m$ (c) $\theta_e = \theta_m$ (d) $\theta_e = \left(\dfrac{p}{2}\right)\theta_m$

115. If the field winding of an unloaded DC shunt motor gets opened while running, it will
(a) stop (b) run with reduced speed
(c) run with increased speed (d) race

116. List I shows the record of terminal voltage of a 10 kW, 220 V. 1,440 rpm DC shunt generator is being run at rated speed, obtained with field circuit switch open as well as closed. List II gives the possible reasons. Match List I with List II and select the correct answer using the codes given below the list:

List I	*Open*	*Closed*	*List II*
A.	8	8	i. Field circuit resistance more than critical
B.	8	12	ii. Load circuit resistance less than critical
C.	8	0	iii. Field polarity reversed

Codes:

(a) A B C
i ii iii

(b) A B C
ii iii i

(c) A B C
ii i iii

(d) A B C
ii i iii

117. A DC cumulatively compounded motor delivers rated load torque at rated speed. If the series field is short circuited, then the armature current and speed will
(a) both decrease (b) both increase
(c) increase and decrease, respectively. (d) decrease and increase, respectively

118. A separately excited DC machine, having an armature resistance of 2 Ω was working on a 220 V supply and drawing 10 A armature current from the source when the supply voltage suddenly drooped to 200 V. Assuming that the field circuit source voltage remained unaffected, how will the armature current of the machine react to the change?
(a) It will initially rise to 11 A and then settle down to 10 A.
(b) It will fall momentarily to 0.09 A and then slowly attain 10 A.
(c) It will reduce to zero first and then settle back to 10 A.
(d) It will remain unaffected by the change and continue to be 10 A.

119. A DC cumulatively compounded generator was operating supplying power satisfactorily to an infinite bus when the mechanical power supply from the prime mover failed. The machine will then run as a
(a) differentially compounded motor with speed reversed
(b) differentially compounded motor with the direction of speed the same as before
(c) cumulatively compounded motor with the same direction of speed as before
(d) cumulatively compounded motor with speed reversed

120. In DC machines, the field system has to be provided on stator unlike synchronous machines wherein it could be on any member, because
(a) it reduces field structure iron losses
(b) it gives more uniform air-gap flux distribution
(c) commutator action is not possible otherwise
(d) DC machines are comparatively low rating

121. In a DC generator operating on load with its brushes on the geometrical neutral axis (GNA), the magnetic neutral axis (MNA) is shifted in the direction of rotation. Now if the brushes are given a lead of 90° (electrical), then the MNA will
(a) shift forward by 90°
(b) shift forward by more than 90°
(c) shift forward by less than 90°
(d) coincide with the GNA

122. A bipolar DC machine with interpoles has a main-pole flux of ϕ per pole and an interpole flux of ϕ_i per pole. The yoke of the machine is divided into four quadrants by the main-pole axis and the commutation axis

The flux distribution in the quadrants will be
(a) $\frac{1}{2}(\phi+\phi_i)$ in all the four quadrants
(b) $\frac{1}{2}(\phi-\phi_i)$ in all the four quadrants
(c) $\frac{1}{2}(\phi+\phi_i)$ in two diametrically opposite quadrants and $\frac{1}{2}(\phi-\phi_i)$ in remaining two quadrants
(d) $\frac{1}{2}(\phi+\phi_i)$ in two adjacent quadrants and $\frac{1}{2}(\phi-\phi_i)$ in remaining two quadrants

123. A cumulatively compounded DC motor runs at 1,000 rpm at no-load. On full load, the flux increases by 10 per cent, whereas the full-load drop in the combined resistance of the armature and series field is 5 per cent. Neglecting magnetic saturation, the full-load speed will be nearly
(a) 863 rpm (b) 909 rpm (c) 1,000 rpm (d) 1,050 rpm

124. A DC overcompound generator is supplying power to an infinite bus. If the prime mover is accidentally cut-off, the DC machine will
(a) stop running
(b) run as cumulatively compounded motor in the reverse direction
(c) run as differentially compounded motor in the reverse direction
(d) run as differentially compounded motor in the same direction

125. A DC shunt motor is run under the following conditions:
i. Normal voltage with no additional resistance in armature or field circuits.
ii. 50 per cent normal voltage with no additional resistance in armature or filed circuits.
iii. Normal voltage with a small external resistance in the armature circuit of a value equal to the armature resistance.
iv. Normal voltage with a high resistance in series with the field winding.
The speed of the motor would increase in the sequence
(a) i, ii, iii and iv
(b) i, iv, iii and ii
(c) ii, i, iv and iii
(d) ii, iii, i and iv

126. The residual magnetism of a self-excited DC generator is lost. To build up its emf again
(a) the field winding must be replaced
(b) the armature connection must be reversed
(c) the field winding connection must be reversed
(d) field winding must be excited by low-voltage DC supply

127. The magnetic circuit of a DC generator with negligible armature resistance and brush contact voltage drop remains unsaturated when it delivers full-load terminal current at rated terminal voltage. The brushes are kept in the interpolar axis and speed is maintained constant. When the load is switched off, the terminal voltage would
(a) remain the same
(b) increase
(c) decrease
(d) become zero

128. Under which of the following conditions is a DC motor provided with compensating winding used?
i. Wide range of speed control above normal
ii. Wide range of steady load variation with no speed control
iii. Wide range of rapid variation in load
Select the correct answer using the codes given below:
(a) i, ii and iii (b) ii and iii (c) i and ii (d) i and iii

129. Match List I with List II and select the correct answer using the codes given below the lists:

List I	*List II*
A. Interpoles	i. Commutation in DC machines
B. Three point starter	ii. Lap winding
C. Dummy coils	iii. DC shunt motor
D. Equalising rings	iv. Wave winding

Codes:

(a) A B C D
i ii iii iv

(b) A B C D
ii iii iv i

(c) A B C D
i iii iv ii

(d) A B C D
iii iv i ii

130. A four-pole dynamo with wave wound armature has 51 slots containing 20 conductors in each slot. The induced emf is 357 V and the speed is 8,500 rpm. The flux per will be
(a) 3.5 mWb (b) 1.2 mWb (c) 1.4 mWb (d) 21 mWb

131. Out of the following factors for a DC machine
i. interpole
ii. armature resistance
iii. armature reactance
iv. reduction in field current
v. reduction in field current
The factors that are responsible for decrease in the terminal voltage of a shunt generator are
(a) i, ii and iv
(b) ii, iii and iv
(c) iii, iv and v
(d) ii, iv and v

132. In a DC generator, the windage loss is proportional to
(a) supply voltage
(b) square of supply voltage
(c) square of armature current
(d) flux density

133. Out of the following losses in a DC machine, which one has the highest proportion?
(a) armature copper loss
(b) field copper loss
(c) hysteresis loss
(d) eddy current loss

134. A shunt generator running at 1,000 rpm has induced emf at 200 V. If the speed increases to 1,200 rpm, the induced emf will be nearly
(a) 150 V
(b) 175 V
(c) 240 V
(d) 290 V

135. The yoke of a DC generator is usually made of
(a) cast iron
(b) copper
(c) stainless steel
(d) silicon steel

136. What will happen if a 220 V DC series motor is connected to 230 V AC supply?
(a) The armature winding of motor will burn.
(b) The motor will vibrate violently.
(c) The motor will run with less efficiency and more sparking.
(d) The motor will not run.

137. In a DC machine, the dummy coils of wave winding are differentiated from other coils by
(a) their thickness
(b) their length
(c) the tapped ends of conductor
(d) none of the above

138. Two similar shunt machines are changed to compound machines. If machine A is wound with twice as many series field turns per pole as machine B, then
(a) machine A will have lesser speed change with load than that of B
(b) machine A will have greater speed change with load than that of B
(c) machine A will have equal speed change with load than that of B
(d) none of the above

139. In series/parallel control of DC series motors
(a) speed in series arrangement is four times the speed in parallel arrangement
(b) speed in series arrangement is one-fourth times the speed in parallel arrangement
(c) speed in series arrangement is double the speed in parallel arrangement
(d) speed in series arrangement is half of the speed in parallel arrangement

140. Multiplex lap winding is used
(a) to reduce commutation difficulties
(b) to increase voltage of machine
(c) to increase number of parallel paths
(d) to reduce armature reaction effect

141. A series motor is working drawing a load current of 1 A from the lines. If the load is reduced such that the current drawn is halved, the speed of the machine (neglecting saturation and armature resistance) would be
(a) unchanged (b) reduced by 50%
(c) reduced by 100% (d) increased by 100%

142. A cumulatively compound DC generator is supplying 20 A and 200 V. Now if the series field winding is short circuited, the terminal voltage will
(a) remain unaltered at 200 V (b) rise to 220 V
(c) shoot up to a high value (d) will reduce to 150 V

143. A DC armature having 1,152 lap-connected conductors carrying 120 A rotates at 2,500 rpm. If the machine has 12-pole field with flux per pole 0.075 Wb, the power output of the armature will be
(a) 21.5 kW (b) 43 kW (c) 86 kW (d) 172 kW

144. DC series motors are preferred to traction due to
(a) continuous long runs (b) near full load operation
(c) high starting torque (d) high efficiency and low cost

145. A DC generator can be termed as
(a) rotating amplifier (b) prime mover (c) power pump (d) booster

146. If A be the commercial efficiency, B the electrical efficiency and C the mechanical efficiency of a DC generator then which of the following relations is valid?
(a) $A = B \times C$ (b) $B = A \times C$ (c) $C = A \times B$ (d) $C = A + B$

147. The material used for brushes in 220 V DC machine is
(a) aluminium (b) carbon (c) electrographite (d) copper-carbon

148. Hysteresis loss in a DC machine is proportional to
(a) 1/N (b) N (c) N^2 (d) $1/N^2$

149. Brushes in DC machines are always placed along:
(a) geometrical neutral axis (b) magnetic neutral axis
(c) perpendicular to GNA (d) perpendicular to MNA

150. In lap winding of a DC machine, equaliser rings are used to avoid
(a) unequal contribution of current at the brushes resulting in sparkless commutation
(b) noise developed in the machine
(c) harmonics and eddy currents in windings
(d) short circuit conditions of the closed machine

151. Commutation may be improved by
   (a) reducing the number of turns in armature and segments of commutator
   (b) increasing the resistance of brushes
   (c) neutralizing the reactance voltage by producing a reverse emf in the coil undergoing commutator
   (d) all of these

152. The maximum number of equalizer rings needed in an electrical machine equals:
   (a) $\frac{2 \times \text{number of conductors}}{\text{number of polers}}$
   (b) $\frac{\text{number of conductors}}{\text{number of polers}}$
   (c) $\frac{\text{number of conductors}}{2 \times \text{number of polers}}$
   (d) none of these

153. The factor that does not cause decrease in the terminal voltage of a DC shunt generator is
   (a) armature resistance
   (b) armature reactance
   (c) reduction in field current
   (d) none of these

154. If instead of DC voltage, an AC voltage is applied on to the field of a DC generator then the output of DC generator
   (a) will be AC voltage
   (b) will be DC voltage
   (c) will be zero
   (d) machine will burn out

155. Speed of a DC shunt mòtor is proportional to
   (a) $I_a/2$
   (b) $1/I_a$
   (c) I_a
   (d) independent of I_a

156. Two DC series motors connected in series draw current I from the supply and run at a speed N. When the same motors are connected in parallel taking current I from the supply, the speed of each motor will be
   (a) N
   (b) $2N$
   (c) $4N$
   (d) $N/2$

157. A cumulative compound DC motor runs at 1,500 rpm on full load. If the series field is short circuited, its speed
   (a) becomes zero
   (b) remains unchanged
   (c) increases
   (d) decreases

158. Regenerative breaking is used only when the load has
   (a) a small value
   (b) retarding motion
   (c) an overhauling characteristic
   (d) none of the above

159. Neglecting all losses, the developed torque (T) of a DC separately excited motor, operating under constant terminal voltage, is related to its output power (P) as under
   (a) $T \infty \sqrt{P}$
   (b) $T \infty P$
   (c) $T^2 \infty P^3$
   (d) T independent of P

160. A separately excited DC motor has an armature resistance of 0.5 Ω. It runs off a 250 V DC supply drawing an armature current of 20 A at 1,500 rpm. The torque developed for an armature current of 10 A will be ________ for the same field current
   (a) 15 Nm
   (b) 30 Nm
   (c) 45 Nm
   (d) 90 Nm

161. *Assertion A:* A differentially compound DC motor has the possibility of starting in the reverse direction.

   Assertion B: Owing to higher inductance in the series field, the resultant main flux reverses during starting.

Mark the answer
(a) Both A and B are true and B is the correct explanation of A.
(b) Both A and B are true but B is not the correct explanation of A.
(c) A is true but B is false.
(d) A is false but B is true.

162. Match List I (parts of electrical machines) with List II (the approximate nature of the air-gap mmf pattern produced by them) and select the correct answer using the codes given below the list:

List I	*List II*
A. DC machine, stator	i. Sinusoidal
B. DC machine, rotor	ii. Trapezoidal
C. Salient-pole synchronous machine, stator	iii. Triangular
D. Squirrel-cage rotor of Induction motor	

Codes:

	A	B	C	D
(a)	i	ii	i	iii
(b)	iii	i	iii	i
(c)	ii	iii	i	iii
(d)	ii	iii	i	i

163. A DC shunt generator, when driven at its rated speed, is found to be not generating any voltage. Which of the following would account for this?
 i. There is no residual magnetism.
 ii. The connection of the field winding is not proper with respect to the armature.
 iii. The resistance of the field circuit is greater than the critical resistance.
 iv. The load resistance is less than the critical armature resistance.
 Codes:
 (a) iii and iv
 (b) i, ii and iv
 (c) i, ii and iii
 (d) i, ii, iii and iv

164. To have sparkles commutation, the armature reaction effect in a DC machine is neutralized by
 (a) using compensating winding and commutating poles
 (b) shifting the brush axis from the geometrical neutral axis to the magnetic neutral axis
 (c) fixing the brush axis in line with the poles
 (d) increasing the field excitation

165. In a DC shunt generator working on load, the brushes are moved forward in the direction of rotation. As a result of this, commutation will
 (a) improve but terminal voltage will fall
 (b) worsen and terminal voltage will fall
 (c) improve and terminal voltage will rise
 (d) worsen and terminal voltage will rise

166. A DC over-compounded generator was operating satisfactorily and supplying power to an infinite bus when the prime mover failed to supply any mechanical power. The machine would then run as a
 (a) cumulatively compounded motor with speed reversed
 (b) cumulatively compounded motor with direction of rotation as before
 (c) differentially compounded motor with speed reversed
 (d) differentially compounded motor with direction of speed as before

167. If a 230 V DC series motor is connected to a 230 V AC supply
   (a) motor will vibrate violently
   (b) motor will run less with less efficiency and more sparking
   (c) motor will not run
   (d) fuse will be blown

168. The interpoles in DC machines have a tapering shape in order to
   (a) reduce the overall weight
   (b) reduce the saturation in the interpole
   (c) economise on the material required for interpoles and their windings
   (d) increase the acceleration of commutation

169. Consider the following statements:
   i. Sparking at brushes
   ii. Acceleration time
   iii. Starting current
   iv. Voltage dip in supply
   Of these statements:
   (a) i and ii are correct
   (b) ii, iii and iv are correct
   (c) i, iii and iv are correct
   (d) i, ii, iii and iv are correct

170. As compared to the use of a single series DC motor for electric traction for a given starting time t, the series-parallel control using two similar motors with time $t/2$ for each series and parallel operation would give a saving in starting energy of
   (a) 100%
   (b) 50%
   (c) 25%
   (d) zero

171. Consider the following statements:
   Permanent magnet DC motors used in cassette tape recorders have
   i. magnets on stator and armature on the rotor
   ii. magnets on the rotor and armature on the stator
   iii. electronic commutation and no brushes
   iv. mechanical commutation and brushes
   v. automatic speed governors
   Of these statements:
   (a) i, iii and v are correct
   (b) i, iv and v are correct
   (c) ii, iii and v are correct
   (d) i and iv are correct

172. The compensating winding in a DC machine
   (a) is located in armature slots for compensating of the armature
   (b) is located on commutating poles for improving the commutation
   (c) is located on pole shoes for avoiding the flashover at the commutator surface
   (d) is located on pole shoes to avoid the sparking at the brushes

173. A 240 V DC series motor takes 40 A when giving its rated output at 1,500 rpm. Its resistance is 0.3 Ω. The value of resistance that must be added to obtain rated torque at 1,000 rpm is
   (a) 6 Ω
   (b) 5.7 Ω
   (c) 2.2 Ω
   (d) 1.9 Ω

174. A permanent magnet DC commutator motor has a no load speed of 6,000 rpm when connected to a 120 V DC supply. The armature resistance is 2.5 Ω and other losses may be neglected. The speed of the motor with supply voltage of 60 V developing a torque 0.5 Nm is
   (a) 3,000 rpm
   (b) 2,673 rpm
   (c) 2,836 rpm
   (d) 5,346 rpm

175. An electric motor with constant output power will have a torque-speed characteristic in the form of
(a) straight line through the origin
(b) straight line parallel to the speed axis
(c) circle about the origin
(d) rectangular hyperbola

176. A DC series motor fed from rated supply voltage is overloaded and its magnetic circuit is saturated. By which curve of Figure 3 will the torque-speed characteristic of this motor be approximately represented?

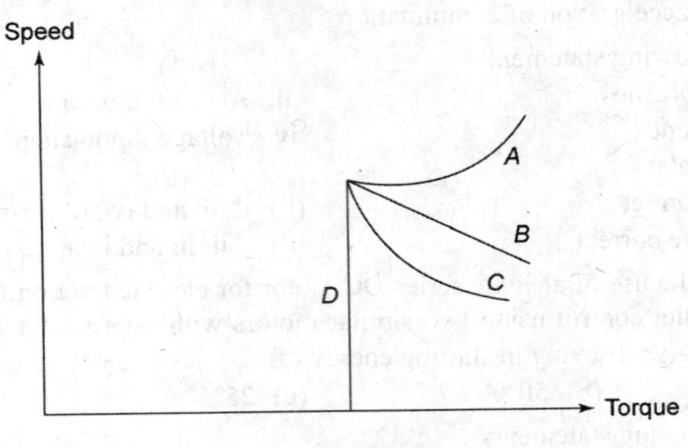

Figure 3

(a) Curve A (b) Curve B (c) Curve C (d) Curve D

177. A 200 V, 2,000 rpm, 10 A, separately excited cDC motor has an armature resistance of 2 Ω. Rated DC voltage is applied to both the armature and field winding of the motor. If the armature draws 5 A from the source, the torque developed by the motor is
(a) 4.30 Nm (b) 4.77 Nm (c) 0.45 Nm (d) 0.50 Nm

178. Two 550 kVA alternators operate in parallel to supply the following loads:
(i) 250 kW at 0.95 pf lagging
(ii) 100 kW at 0.85 pf lagging
One machine is supplying 200 kW at 0.9 pf lagging. The pf of other machine must be
(a) 0.89 leading (b) 0.95 leading (c) 0.95 lagging (d) 0.89 lagging

179. Match List I with List II and select the correct answer using the codes given below the lists:

List I (Parts of a turbogenerator used in thermal power plants)	*List II* (Materials from which these parts are made)
A. Steel core	i. Copper
B. Rotor core	ii. Copper alloy
C. Stator winding	iii. Silicon steel
D. Slip rings	iv. Mild steel
	v. Aluminum

Codes:

	A	B	C	D
(a)	iii	v	ii	i
(b)	iv	v	i	ii
(c)	iv	iii	i	ii
(d)	iii	iv	i	ii

180. The most appropriate operating speed in rpm of generators used in thermal, nuclear and hydro plants would, respectively, be
(a) 3,000, 300 and 1,500
(b) 3,000, 3,000 and 300
(c) 1,500, 1,500 and 3,000
(d) 1,000, 900 and 750

181. A large AC generator supplying power to an infinite bus has a sudden short circuit occurring at its terminals. Assuming the prime mover input and the voltage behind the transient reactance to remain constant immediately after the fault, the acceleration of the generator rotor is
(a) inversely proportional to the moment of inertia of the machine
(b) inversely proportional to the square of the voltage
(c) directly proportional to the square of the short circuit current
(d) directly proportional to the short circuit power

182. The fifth harmonic component of the induced voltage in three-phase AC generator can be entirely eliminated by using a winding pitch of
(a) 2/3
(b) 4/5
(c) 5/6
(d) 6/7

183. Which of the following limit the reactive power output of a synchronous generator?
i. Armature current
ii. Field current
iii. Load angle
iv. Prime mover
Select the correct answer using the codes given below:
Codes:
(a) i and ii
(b) ii and iii
(c) iii and iv
(d) i and iv

184. During the slip-test for determining the direct and the quadrature axis synchronous reactance of an alternator, the voltage across the open field circuit terminals is
(a) DC voltage
(b) AC voltage of supply frequency
(c) AC voltage of slip frequency
(d) a modulated voltage

185. A three-phase fault occurs at the terminals of an unloaded alternator at an instant when the AC component in one phase is at its maximum value. The variation in DC component during fault period in that phase will depend upon
(a) subtransient time constant
(b) transient time constant
(c) open-circuit time constant
(d) armature time constant

186. Which one of the following motors running under the specified conditions will have a leading power factor?
(a) slip ring induction motor at no load
(b) under-excited synchronous motor
(c) over-excited synchronous motor
(d) double-cage induction motor at load

187. A synchronous machine working in the motor mode is fed from an infinite bus and is delivering half of full-load. If an increase in field current causes an increase in the armature current, then the motor will
(a) deliver reactive power and active power to the bus
(b) absorb reactive power and active power from the bus
(c) absorb reactive power from the bus and deliver active power to the bus
(d) absorb reactive power to the bus and deliver active power from the bus

188. The voltage across the open-circuited field terminals of a synchronous machine under slip test is
(a) DC
(b) AC of slip frequency
(c) a modulated supply frequency AC voltage with slip frequency envelope
(d) AC of supply frequency

189. Two three-phase AC generators are such that one has twice the linear dimensions of other. The field windings of each are excited to give identical sinusoidal air-gap flux density waveform. Both have the same number of stator slots and identical winding patterns. The conductor/slot in big generator is K times that of the smaller one. The value of K to get equal no-load voltage at the same frequency is
(a) 8 (b) 4 (c) 1/2 (d) 1/4

190. A three-phase and four-pole alternator has 48 stator slots carrying the three-phase distributed winding. Each coil of the winding is short chorded by one slot pitch. The winding factor is given by
(a) $\dfrac{1}{16 \cot 7.5°}$ (b) $\dfrac{1}{8 \cot 7.5°}$ (c) $\dfrac{1}{16 \sin 7.5°}$ (d) $\cos 7.5°$

191. A three-phase synchronous motor connected to infinite bus is operating at half of full-load with normal excitation. When the load on the synchronous motor is suddenly increased,
(a) its speed will first decrease and then become synchronous
(b) its speed will first increase and then become synchronous
(c) its speed will fluctuate around synchronous speed and then become synchronous
(d) its speed will remain unchanged

192. Two mechanically coupled alternators deliver power at 50 Hz and 60 Hz, respectively. The highest speed of the alternators is
(a) 3,600 rpm (b) 3,000 rpm (c) 600 rpm (d) 500 rpm

193. When a balanced three-phase distributed-type armature winding is carrying three-phase, balanced currents, the strength of the resultant rotating magnetic field is
(a) three times the amplitude of each constituent pulsating magnetic field
(b) equal to the amplitude of each constituent pulsating magnetic field
(c) half the amplitude of each constituent pulsating magnetic field
(d) one and a half times the amplitude of each constituent pulsating magnetic field

194. An alternator is capable of delivering power at a particular frequency. The frequency can be increased by
(a) increasing the current supplied to the electromagnetic field
(b) reversing the armature rotation
(c) increasing armature speed
(d) reversing field polarity

195. The simplest way to eliminate the harmonic induction torque is
   (a) integral slot winding (b) skewing
   (c) chording (d) none of the above

196. For an alternator when the power factor of the load is unity,
   (a) armature flux will have square waveform (b) armature flux will be demagnetising
   (c) armature flux will be cross-magnetising (d) armature flux will reduce to zero

197. The driving power from the prime mover driving the alternator is lost but the alternator remains connected to the supply network and the field supply remains on. The alternator will
   (a) get burnt
   (b) behave as an induction motor but will rotate in the opposite direction
   (c) behave as a synchronous motor but will rotate in the same direction
   (d) behave as a synchronous motor but will rotate in a reverse direction to that corresponding to generator action

198. If the input of the prime mover of an alternator is kept constant but the excitation is changed, then
   (a) active component of the output is changed
   (b) reactive component of the output is changed
   (c) power factor of the load remains constant
   (d) power factor of the load changes from lagging to leading

199. An exciter for a generator is
   (a) series motor (b) shunt motor (c) series generator (d) shunt generator

200. When two alternators are running in parallel, their kVAR load share is changed by changing their __________ while their kW load share is changed by changing their __________
   (a) excitation, driving torque (b) driving torque, excitation
   (c) excitation, excitation (d) driving torque, driving torque

201. The Potier's triangle separates
   (a) iron losses and copper losses
   (b) field mmf and armature mmf
   (c) stator voltage and rotor voltage
   (d) armature leakage reactance and armature reaction mmf

202. Alternators used in aircraft systems usually have frequency of
   (a) 25 Hz (b) 50 Hz (c) 100 Hz (d) 400 Hz

203. Two alternators A and B are sharing a resistive load (pf is unity) equally. Now if the excitation of alternator A is increased,
   (a) alternator A will become lagging and alternator B will become leading
   (b) alternator A will become leading and alternator B will become lagging
   (c) both alternators will continue to operate on unity power factor
   (d) both alternators will operate on lagging power factor

204. The advantage of providing damper winding in alternators is
   (a) elimination of harmonic effects
   (b) provide low resistance path for the currents to unbalancing of voltage
   (c) oscillations are provided when two alternators operate in parallel
   (d) all of the above

205. In a synchronous machine, if the field flux is ahead of the armature field axis, in the direction of rotation, the machine working as
(a) asynchronous motor (b) asynchronous generator
(c) synchronous motor (d) synchronous generator

206. In a synchronous generator, which of the following will have emf close to sine waveform?
(a) concentrated winding in full pitch coils (b) concentrated winding in short pitch coils
(c) distributed winding in full pitch coils (d) distributed winding in short pitch coils

207. The effect of cross magnetising in an alternator is to make the output
(a) true sinusoidal (b) non sinusoidal (c) harmonic free (d) none of these

208. Synchronising torque comes into operation in all the cases except
(a) phase difference between two voltages
(b) frequency difference between two voltages
(c) voltage difference between two voltages
(d) reduction in excitation current in one of the alternators

209. The advantage of using short-pitched windings in an alternator is that it
(a) suppresses the harmonics in generated emf
(b) reduces the total voltage around the armature coils
(c) saves copper used in windings
(d) improves cooling by better circulation of air

210. For the same power rating, an alternator operating at lower voltage will be
(a) less noisy (b) costlier (c) larger in size (d) more efficient

211. Salient pole-type rotors as compared to cylindrical type rotors are
(a) smaller in diameter and larger in axial length
(b) larger in diameter and smaller in axial length
(c) smaller in diameter and larger in axial length
(d) larger in diameter and larger in axial length

212. When a generator designed for operation at 60 Hz is operated at 50 Hz, then
(a) the operating voltage must be derated to (50/60) of the original voltage
(b) the operating voltage must be derated to $(50/60)^2$ of the original voltage
(c) the kVA rating can be upgraded to (60/50) of the rated value
(d) generator will not take any load

213. The maximum power developed in the synchronous motor will depend on
(a) rotor excitation only
(b) maximum value of coupling angle
(c) supply voltage only
(d) rotor excitation supply voltage and maximum value of coupling angle

214. An overexcited synchronous motor draws current at
(a) lagging power factor (b) leading power factor
(c) unity power factor (d) depends on the nature of the load

215. If a synchronous motor drops too far behind, the power it takes from the supply also increases too much, and the armature tries to get accelerated, until it is in correct position. Sometimes, some motor overshoots the marks and then the process of acceleration-retardation continues. This phenomenon is known as
(a) synchronization (b) hunting (c) pulling out (d) swinging

216. In case of a synchronous motor we have:
    i. load
    ii. speed
    iii. DC excitation

    The magnitude of stator back emf depends on
    (a) i only
    (b) i and ii only
    (c) iii only
    (d) i, ii and iii only

217. The breakdown torque of a synchronous motor varied as
    (a) 1/(applied voltage)
    (b) 1/(applied voltage)^2
    (c) applied voltage
    (d) (applied voltage)^2

218. Hunting in a synchronous motor cannot be due to
    (a) variable frequency
    (b) variable load
    (c) variable supply voltage
    (d) windage friction

219. A high starting torque synchronous motor has
    (a) simplex rotor
    (b) phase wound damper
    (c) five slip rings
    (d) all of the above

220. The parameters connected with the operation of a synchronous motor are
    i. speed
    ii. power factor
    iii. armature current

    When the excitation of the motor is varied, which parameters vary along with it
    (a) i only
    (b) ii only
    (c) ii and iii only
    (d) i, ii and iii only

221. The negative-phase sequence in a three-phase synchronous motor exist when
    (a) motor is overloaded
    (b) motor is underloaded
    (c) unbalanced voltage is supplied
    (d) motor armature is hot

222. If a synchronous motor fails to pull into synchronism after supplying DC field current, the probable cause may be
    (a) high core losses
    (b) low field current
    (c) high field current
    (d) low short circuit ratio

223. In a synchronous machine in case the axis of field flux is in line with the armature flux, then
    (a) machine is working as synchronous motor
    (b) machine is working as synchronous generator
    (c) machine is said to be floating
    (d) machine will vibrate violently

224. If other factors remain constant, the speed of a synchronous motor in its operating (and load) range is correctly described by which of the following?
    (a) The speed varies directly only in proportion to the number of poles.
    (b) The speed is independent of the frequency of the voltage supply.
    (c) The speed depends on the frequency of the voltage supply and the number of poles.
    (d) The speed depends on the magnitude of the voltage supply and the number of its pole

225. While starting the synchronous motor by induction motor action, very high emf is induced in the field. This induced emf may damage the insulation of the field winding and the slip rings. The insulation damage can be prevented by
(a) short circuiting the field winding by field discharge resistance
(b) splitting the field winding into several sections
(c) either of (a) or (b)
(d) none of the above

226. The construction of a synchronous motor resembles which of the following machines?
(a) induction motor (b) rotor converter
(c) alternator (d) series motor

227. The rotor of a synchronous motor can only run at synchronous speed of the stator magnetic field due to
(a) Faraday's law of electromagnetic induction
(b) Lenz's law
(c) magnetization of rotor poles by stator magnetic field
(d) interlocking action between stator and rotor fields

228. An inverted V-curve of a synchronous motor is the variation of
(a) field current and DC excitation of constant load
(b) supply voltage and field current at constant excitation
(c) power factor and supply voltage during hunting
(d) supply voltage and excitation current at constant load

229. Damper windings are provided on
(a) pole faces (b) separate armature
(c) rotor shaft (d) stator frame

230. When a synchronous motor is running at synchronous speed, the damper winding produces
(a) damping torque (b) eddy current torque
(c) torques aiding the developed torque (d) no torque

231. In a synchronous generator, a divided winding rotor is preferable to a conventional winding rotor because of
(a) higher efficiency (b) increased steady state stability limit
(c) higher short circuit current (d) better damping

232. The per-unit impedance of a synchronous machine is 0.242. If the base voltage is increased by 1.1 times, the per-unit value will be
(a) 0.266 (b) 0.242 (c) 0.220 (d) 0.200

233. If the dimensions of all the parts of a synchronous generator, and the number of field and armature turns are doubled, then the generated voltage will change by a factor of
(a) 1 (b) 2 (c) 4 (d) 8

234. Consider the following statements:
The effects of slot harmonics are effectively reduced by
i. skewing the slots
ii. fractional slot winding
iii. short-chorded winding
iv. distributed winding
Of these statements
(a) i and ii are correct (b) ii, iii and iv are correct
(c) i, iii and iv are correct (d) i and iii are correct

235. Which of the following methods would give a higher than actual value of regulation of an alternator?
(a) ZPF method (b) MMF method (c) EMF method (d) ASA method

236. A turbo alternator set feeds power to a three-phase constant-voltage, constant-frequency bus. If the steam supply to the set is cut off, then the set will
(a) continue to run at rated speed in the same direction
(b) continue to run at reduced speed in the same direction
(c) run at rated speed in reverse direction
(d) come to stop

237. An ideal synchronous motor has no starting torque because the
(a) rotor is made up of salient poles
(b) relative velocity between the stator and rotor mmfs is zero
(c) relative velocity between the stator and rotor mmfs is not zero
(d) rotor winding is highly reactive

238. Electrical machines are generally designed to have maximum efficiency
(a) no load (b) 50% load (c) near about full load (d) full load

239. Which of the following statements are true?
i. High-speed alternators usually have salient pole rotors.
ii. Cylindrical rotor alternators have large length-to-diameter ratio.
iii. Salient pole machines have large number of poles.
iv. To ensure proper cooling, cylindrical rotor alternators use radial ducts only.
(a) i and ii only (b) ii and iii only (c) iii and iv only (d) i and iv only

240. Which of the following statements are true?
i. A synchronous motor supplying its rated load has its excitation increased resulting in power factor becoming more lagging.
ii. In a synchronous motor, the rotor mmf and stator mmf are stationary with respect to each other.
iii. The short circuit characteristic of an alternator is always nonlinear.
(a) i and ii only (b) ii and iii only (c) i and iii only (d) ii only

241. Consider the following statements:
i. Asynchronous motor has no starting torque but when started it always runs at a fixed speed.
ii. Neglecting residual magnetism, both cylindrical rotor and salient pole synchronous motor produce some mechanical power even through the field is unexcited.
iii. A single-phase reluctance motor is not self-starting even if paths for eddy current are provided in the rotor.
iv. A single-phase hysteresis motor is self-starting.
Of the above statements:
(a) Only i and ii are correct (b) Only i and iii are correct
(c) Only ii, iii and iv are correct (d) Only i, ii and iv are correct

242. The essential conditions for running two alternators in parallel are the following:
i. Terminal voltage should be same.
ii. Frequency should be same.
iii. Phase sequence should be same.

Of the above statements, which of the following options hold
(a) i only (b) ii only (c) i and ii only (d) i, ii and iii

243. The magnitude of emf induced in the stator due to revolving flux will depend on the
(a) speed of the motor
(b) DC excitation current
(c) load of the motor
(d) speed and rotor flux

244. The magnitude of the induced emf in the stator of a synchronous motor
(a) is equal to the supply voltage
(b) is less than the supply voltage
(c) can be increased than the supply voltage
(d) can be increased or decreased than the supply voltage

245. In all cases of electromagnetic induction, an induced voltage will cause a current to flow in a closed circuit in such a direction that the magnetic field caused by that current will oppose the change that produces the current is the original statement of
(a) Lenz's law
(b) Faraday's law of magnetic induction
(c) Fleming's law of induction
(d) Ampere's law

246. A centre-zero ammeter connected in the rotor circuit of a six-pole 50 Hz induction motor makes 30 oscillations in one minute. The rotor speed is
(a) 970 rpm (b) 990 rpm (c) 1,010 rpm (d) 1,030 rpm

247. Consider the following statements:
In a three-phase induction motor connected to a three-phase supply, if one of the lines suddenly gets disconnected, then the
i. motor will come to standstill
ii. motor will continue to run at the same speed with line current unchanged
iii. motor will continue to run at a slightly reduced speed with increased line current
iv. rotor currents will have both sf and $(2-s)f$ component frequencies where s is slip and f is supply frequency
Of these statements:
(a) i and iv are correct
(b) i and ii are correct
(c) iii and iv are correct
(d) ii and iii are correct

248. In a self-excited induction generator, to keep the frequency of generated voltage constant with the increase in load the speed of the induction machine should be
(a) increased
(b) decreased
(c) maintained less than the rated synchronous speed
(d) maintained more than the rated synchronous speed

249. In a three-phase induction machine, motoring generating and braking operations take place in the range of slip 's' given by
(a) motoring: $1 > s > 0$
generating: $0 > s > -1$
braking: $s > 1$
(b) motoring: $1 > s > 0$
generating: $0 > s > -1$
braking: $0 > s > -1$
(c) motoring: $0 > s > -1$
generating: $0 > s > -1$
braking: $0 > s > -1$
(d) motoring: $0 > s > -1$
generating: $s > 1$
braking: $1 > s > 0$

250. A voltmeter gives 120 oscillations per minute when connected to the rotor of an induction motor. The stator frequency is 50 Hz. The slip of the motor is
   (a) 2% (b) 2.5% (c) 4% (d) 5%

251. Semi-closed slots or totally closed slots are used in induction motors, essentially to
   (a) improve starting torque
   (b) increase pull-out torque
   (c) increase efficiency
   (d) reduce magnetising current and improve power factor

252. A starting torque of 80 Nm is developed in an induction motor by an auto-transformer starter with a tapping of 30 per cent. If the tapping of autotransformer starter is 60 per cent, then the starting torque will be
   (a) 40 Nm (b) 160 Nm (c) 240 Nm (d) 320 Nm

253. The torque-slip characteristic of a polyphase induction motor becomes almost linear at small values of slip, because in this range of slips
   (a) effective rotor circuit resistance is very large compared to the rotor reactance
   (b) rotor resistance is equal to stator resistance
   (c) rotor resistance is equal to the rotor reactance
   (d) rotor reactance is equal to the stator reactance

254. Breakdown torque of a three-phase induction motor of negligible stator impedance is
   (a) directly proportional to the rotor resistance
   (b) inversely proportional to the rotor resistance
   (c) directly proportional to the reactance
   (d) inversely proportional to the rotor leakage reactance

255. If a three-phase induction motor, fed from a rated-voltage and rated-frequency supply, is driven by a prime-mover in the same direction when operating as a motor, at super synchronous speed, the machine will operate as
   (a) a reactive power generator (b) a frequency converter
   (c) an active power generator (d) a rotary converter

256. A reduction of the voltage applied to a three-phase induction motor
   (a) increases the breakdown torque and the starting torque
   (b) increases the breakdown torque but decreases the starting torque
   (c) decreases the breakdown torque
   (d) decreases both breakdown and starting torque

257. Consider the following statements regarding the equivalent circuit parameters of an induction motor:
   i. Leakage reactance is dependent on supply current.
   ii. Magnetizing reactance is dependent on the air-gap flux.
   iii. Core loss is dependent on the input voltage and frequency.
   iv. Rotor resistance is dependent on the speed.
   Of these statements:
   (a) ii and iii are correct
   (b) i, iii and iv are correct
   (c) i, ii, iii and iv are correct
   (d) i, ii and iv are correct

258. Crawling of an induction motor is due to
(a) space harmonics in the stator flux
(b) the number of teeth on stator being equal to the number of teeth on the rotor
(c) the fluctuation in load torque
(d) unequal distribution of winding

259. A three-phase induction machine operates on three-phase fixed frequency AC mains at a slip of 1.5. Consider the following statements:
i. It draws electrical power from the mains.
ii. It draws the mechanical power from the mains.
iii. It delivers electrical power to the mains.
iv. It delivers mechanical power to the mains.
Of these statements:
(a) i and ii are correct (b) i and iv are correct
(c) ii and iii are correct (d) iii and iv are correct

260. The magnetising current component of the no-load current of an induction motor is much larger than a corresponding transformer because of
(a) additional friction and windage loss in motor
(b) different winding configuration on stator and rotor
(c) increased flux requirement
(d) an air-gap in the magnetic circuit

261. The rotor of a four-pole, three-phase cage induction motor is replaced by a three-phase, four-pole wound rotor. When fed with normal supply, the machine will
(a) not run (b) run at very low speed
(c) run at slightly lower than normal speed (d) run at slightly higher than normal speed

262. A three-phase induction motor operates on a variable frequency, variable voltage supply such that V/f is constant. The breakdown torque of the drive will
(a) decrease with decrease with frequency
(b) remain constant irrespective of the frequency
(c) be slightly lower at very low frequencies
(d) increase with decrease in frequency

263. The stator of a 415 V, 50 Hz, six-pole, 970 rpm three-phase slip ring induction motor is fed with normal voltage. The rotor with slip rings open-circulated is rotated at 500 rpm along the normal direction of its rotation. The frequency of the voltage across slip rings will be
(a) 15 Hz (b) 25 Hz (c) 26.5 Hz (d) 75 Hz

264. Which class of insulating materials can withstand maximum temperature?
(a) A class (b) B class (c) C class (d) E class

265. Which of the following induction motors will have the least shaft diameter?
(a) 20 HP, 2,880 rpm (b) 20 HP, 1,440 rpm (c) 20 HP, 960 rpm (d) 20 HP, 730 rpm

266. An induction motor is
(a) self-starting with zero torque
(b) self-starting with very high torque
(c) self-starting with smaller torque
(d) none of the above

267. An induction motor will take least starting current when
   (a) directly switched on
   (b) started by reactance starter
   (c) started by autotransformer with minimum of 65 per cent tapping
   (d) started by star–delta starter

268. A 20 kW, four-pole, 400 V, 60 Hz, three-phase, 1,728 rpm induction motor is connected to 400 V, 50 Hz supply. The motor will
   (a) not run
   (b) burn out
   (c) run at more than 1,720 rpm
   (d) run at less than 1,720 rpm

269. A four-pole, three-phase induction motor is running at 4 per cent slip at full load. If the speed of the motor is 720 rpm, the supply frequency is
   (a) $16\frac{2}{3}$ HZ
   (b) 25 Hz
   (c) 50 Hz
   (d) 100 Hz

270. When the supply voltage of an induction motor is increased by 10 per cent, the maximum running torque will
   (a) remain unaltered
   (b) increase by about 10 per cent
   (c) increase by 20 per cent
   (d) decrease by about 10 per cent

271. When a three-phase induction motor is loaded,
   (a) induced emf in the rotor decreases and its frequency increases
   (b) induced emf on the rotor increases and its frequency falls
   (c) induced emf in the rotor as well as frequency falls
   (d) induced emf in the rotor as well as frequency increases

272. Squirrel cage induction motor has
   (a) zero starting torque
   (b) very small starting torque
   (c) medium starting torque
   (d) very high starting torque

273. In an induction motor, rotor slots are usually not quite parallel to the shaft but are given a slight skew for the following reasons:
   (a) to reduce the magnetic hum
   (b) to reduce the locking tendency of the rotor
   (c) to reduce the locking tendency and the magnetic hum
   (d) to increase the speed of the rotor

274. The purpose of blades in squirrel cage induction motor is
   (a) to reduce the magnetic resistance of the rotor
   (b) to cool the rotor
   (c) to reduce the electrical resistance of the rotor cage
   (d) none of the above

275. In the following motor, external resistance can be added to start the motor
   (a) slip ring induction motor
   (b) squirrel cage induction motor
   (c) salient pole synchronous motor
   (d) wound rotor synchronous motor

276. Torque developed by three-phase induction motor equals
   (a) copper losses in the rotor
   (b) gross mechanical power developed
   (c) air-gap power transferred to rotor
   (d) power input to rotor minus rotor losses

277. When an induction motor runs at rated load and rated speed, the iron losses are
   (a) negligible
   (b) very high
   (c) independent of supply frequency
   (d) independent of supply voltage

278. The starting torque of an induction motor is maximum when
(a) rotor resistance equals to rotor reactance
(b) rotor resistance is twice the rotor reactance
(c) rotor resistance is half the rotor reactance
(d) rotor resistance is $\sqrt{2}$ times rotor reactance

279. Improvement of power factor in an induction motor results in
(a) decreased torque
(b) increased torque
(c) increased current
(d) increased torque and decreased current due to increased impedance

280. For smooth starting of three-phase squirrel-cage induction motor, the following starting method is preferred:
(a) rotor resistance (b) star–delta (c) autotransformer (d) stator resistance

281. The low power factor of an induction motor at low load results in
(a) high speed (b) low magnetising component of I_0
(c) high magnetising component of I_0 (d) high working component of I_0

282. The three-phase induction motor is mainly suitable for following applications:
(a) for running different machine tools where several speeds are required
(b) for running paper machine requiring exact speed control
(c) for running electric vehicles
(d) for running rolling mills needing exact speed control

283. Any odd harmonics in the current of an induction motor will result in magnetic field, which
(a) oscillates at harmonics frequency
(b) rotates in backward direction
(c) rotates in forward direction at the harmonic speed
(d) is stationary relative to the field of the fundamental

284. The number of different speeds that can be obtained from two induction motors in cascade is
(a) 2 (b) 3 (c) 4 (d) 6

285. The efficiency of induction motor *X* is 90 per cent and that of *Y* is 80 per cent. The size of motor *X* is
(a) larger than Y (b) equal to Y
(c) smaller than Y (d) may be either larger or smaller

286. Imbalance in the shaft of an induction motor occurs due to
(a) slip rings (b) overheating of winding
(c) non-uniform air gap (d) rigid construction

287. Negative sequence currents are set up in an induction motor when
(a) bearing friction is more (b) windage losses are excessive
(c) air gap is more (d) unbalanced three-phase supply is given

288. For a 60-HP motor, which type of rotor will be preferred?
(a) wound rotor
(b) squirrel cage rotor using round copper bars
(c) squirrel cage rotor using rectangular copper bars
(d) die-casted aluminium rotor

289. Two slip-ring induction motors having number of poles as P_1 and P_2, respectively, are connected in cascade. If the supply frequency is f, the speed of the main motor will be

(a) $\frac{f}{P_1+P_2}$ (b) $\frac{2f}{P_1+P_2}$ (c) $\frac{120f}{P_1+P_2}$ (d) $\frac{60f}{P_1+P_2}$

290. A three-phase induction motor takes A amperes on full load at a power factor cos ϕ. The main component responsible for producing magnetism is

(a) A sin ϕ (b) A cos ϕ (c) A tan ϕ (d) A/cos ϕ

291. The stator of a three-phase, four-pole wound rotor induction motor is connected to 50 Hz source but its rotor is energised from 20 Hz source. The two no-load speeds of the motor will be

(a) 1,440 rpm and 960 rpm (b) 2,100 rpm and 900 rpm
(c) 2,400 rpm and 600 rpm (d) 1,560 rpm and 1,440 rpm

292. Following are the trouble shooting operations:

i. Check the motor.
ii. Check the thermal overload relay.
iii. Check the fuses.
iv. Check the motor control panel.

If the motor fails to start, the correct sequence in which these checks should be carried out is

(a) i, ii, iii and iv (b) i, iii, iv and ii (c) iii, ii, iv and i (d) ii, iv, i and iii

293. Match List I (type of winding) with List II (feature) and select the correct answer using the codes given below the list:

List I	*List II*
A. Fractional slot windings	i. Only two-thirds of pole pitch are occupied by the winding
B. Chorded and distributed winding	ii. Effects of slot harmonics are reduced or even eliminated
C. Damper windings	iii. The effects of phase-belt harmonics are reduced
D. Single-phase windings	iv. Additional asynchronous torque capability in projected pole motors

Choose the correct code:

(a) A B C D
ii iii iv i

(b) A B C D
ii iii i iv

(c) A B C D
iii ii i iv

(d) A B C D
iii ii iv i

294. While conducting a 'slip' test for the determination of direct and quadrature-axis synchronous machine, the rotor of the machine is run with a slip 's' and stator supply frequency 'f'. The frequency of

i. voltage induced across open field terminals
ii. envelope of the armature terminal voltage
iii. envelope of armature current
iv. armature

will be

(a) sf, sf, sf and f, respectively
(b) sf, f, sf and sf, respectively
(c) f, sf, f and f, respectively
(d) f, $(1-s)f$, $(2-s)f$ and sf, respectively

295. If two induction motors are identical except that the air-gap of motor 'A' is 50 per cent greater than that of motor 'B', then
(a) no-load power factor of A will be better than that of B
(b) no-load power factor of A will be poorer than that of B
(c) core losses of A will be more than those of B
(d) flux of A will be smaller than that of B

296. Which of the following statements regarding skewing of motor bars in a squirrel-cage induction motor are correct?
i. It prevents cogging.
ii. It produces more uniform torque.
iii. It increases starting torque.
iv. It reduces motor 'hum' during its operation.
Select the correct answer using the codes given below:
(a) ii, iii and iv (b) i, ii and iii (c) i, ii and iv (d) i, iii and iv

297. The rotor power output of a three-phase induction motor is 15 kW and the corresponding slip is 4 per cent. The rotor copper loss will be
(a) 600 W (b) 625 W (c) 650 W (d) 700 W

298. A three-phase induction machine draws active power 'P' and reactive power 'Q' from the grid. If it is operated as a generator, P and Q will, respectively, be
(a) positive and negative (b) negative and negative
(c) positive and positive (d) negative and positive

299. A three-phase induction motor fed from a three-phase voltage controller is suitable for driving loads whose torque is
(a) constant irrespective of speed
(b) inversely proportional to the square of the speed
(c) directly proportional to the square of speed
(d) inversely proportional to speed

300. A three-phase induction motor when started picks up speed but runs stably at about half the normal speed. This is because of
(a) unbalance in the supply voltage (b) non-sinusoidal nature of supply voltage
(c) stator circuit asymmetry (d) rotor circuit asymmetry

Answers

1 (c)	2 (d)	3 (b)	4 (c)	5 (c)	6 (d)
7 (c)	8 (b)	9 (d)	10 (b)	11 (d)	12 (b)
13 (c)	14 (a)	15 (d)	16 (c)	17 (d)	18 (a)
19 (c)	20 (c)	21 (b)	22 (c)	23 (d)	24 (c)
25 (b)	26 (b)	27 (d)	28 (d)	29 (a)	30 (d)
31 (c)	32 (a)	33 (d)	34 (b)	35 (d)	36 (b)
37 (b)	38 (c)	39 (d)	40 (b)	41 (b)	42 (d)
43 (d)	44 (b)	45 (c)	46 (b)	47 (c)	48 (c)

49 (d)	50 (d)	51 (b)	52 (a)	53 (b)	54 (b)
55 (c)	56 (a)	57 (c)	58 (b)	59 (b)	60 (a)
61 (b)	62 (c)	63 (c)	64 (c)	65 (c)	66 (b)
67 (a)	68 (d)	69 (b)	70 (c)	71 (c)	72 (c)
73 (d)	74 (d)	75 (a)	76 (a)	77 (d)	78 (d)
79 (c)	80 (d)	81 (d)	82 (d)	83 (a)	84 (a)
85 (d)	86 (b)	87 (b)	88 (b)	89 (b)	90 (a)
91 (d)	92 (b)	93 (b)	94 (b)	95 (c)	96 (d)
97 (c)	98 (a)	99 (d)	100 (d)	101 (d)	102 (a)
103 (c)	104 (a)	105 (d)	106 (a)	107 (a)	108 (b)
109 (d)	110 (a)	111 (c)	112 (b)	113 (b)	114 (d)
115 (d)	116 (c)	117 (b)	118 (c)	119 (b)	120 (c)
121 (d)	122 (c)	123 (a)	124 (d)	125 (d)	126 (d)
127 (a)	128 (d)	129 (c)	130 (b)	131 (d)	132 (c)
133 (a)	134 (c)	135 (a)	136 (c)	137 (c)	138 (b)
139 (b)	140 (c)	141 (d)	142 (c)	143 (c)	144 (c)
145 (a)	146 (a)	147 (c)	148 (b)	149 (b)	150 (a)
151 (b)	152 (a)	153 (a)	154 (a)	155 (d)	156 (c)
157 (c)	158 (b)	159 (b)	160 (a)	161 (c)	162 (d)
163 (d)	164 (a, b)	165 (a)	166 (d)	167 (b)	168 (c)
169 (d)	170 (b)	171 (d)	172 (d)	173 (d)	174 (b)
175 (d)	176 (b)	177 (b)	178 (a)	179 (d)	180 (b)
181 (a)	182 (b)	183 (a)	184 (c)	185 (d)	186 (c)
187 (d)	188 (b)	189 (c)	190 (b)	191 (c)	192 (c)
193 (d)	194 (c)	195 (c)	196 (c)	197 (c)	198 (b)
199 (d)	200 (a)	201 (d)	202 (d)	203 (a)	204 (d)
205 (d)	206 (d)	207 (b)	208 (d)	209 (b)	210 (c)
211 (b)	212 (a)	213 (d)	214 (b)	215 (b)	216 (c)
217 (c)	218 (d)	219 (d)	220 (c)	221 (c)	222 (b)
223 (c)	224 (c)	225 (c)	226 (c)	227 (d)	228 (a)

229 (a)	230 (d)	231 (b)	232 (d)	233 (d)	234 (a)
235 (c)	236 (a)	237 (c)	238 (c)	239 (b)	240 (d)
241 (b)	242 (d)	243 (b)	244 (d)	245 (a)	246 (b)
247 (c)	248 (a)	249 (a)	250 (c)	251 (d)	252 (d)
253 (a)	254 (d)	255 (c)	256 (d)	257 (a)	258 (a)
259 (a)	260 (d)	261 (c)	262 (a)	263 (b)	264 (c)
265 (a)	266 (c)	267 (d)	268 (d)	269 (b)	270 (c)
271 (c)	272 (b)	273 (c)	274 (b)	275 (a)	276 (d)
277 (a)	278 (a)	279 (d)	280 (c)	281 (c)	282 (a)
283 (b)	284 (c)	285 (a)	286 (c)	287 (d)	288 (a)
289 (a)	290 (a)	291 (b)	292 (c)	293 (a)	294 (a)
295 (b)	296 (c)	297 (b)	298 (d)	299 (c)	300 (a)

References

Chapman, S. J. 1999. *Electric Machinery Fundamentals*. New York: McGraw-Hill.

Clayton, A. E. and N. H. Hancock 1962. *Performance and Design of DC Machines*. London: ELBS Pitman Edn.

Dorf, R. C. 1993. *The Electrical Engineering Handbook*. Boca Raton, Florida, 33431: CRC Press.

Draper, A. 1967. *Electric Machines*. London: Longman.

Elgerd, O. 1982. *Electric Energy System Theory: An Introduction*. New York: McGraw-Hill.

Fink, D. G. 1993. *Standard Handbook for Electrical Engineers*. New York: McGraw-Hill.

Fitzerald, A. E., C. Kingsley and S. D. Umans 1983. *Electric Machinery*, 4th Edn. New York: McGraw-Hill.

Hamidi, E. S. 1994. *Design of Small Electrical Machines*. New York: Wiley.

Kemp, P. 1963. *Alternating Current Electrical Engineering*. London: Macmillan and Co. Ltd.

Kloss, A. 1984. *A Basic Guide to Power Electronics*. New York: Wiley.

Langsdorf, A. C. 1955. *Theory and Design of Small Induction Motors*. New York: McGraw-Hill.

Leonard, W. 1995. *Control of Electrical Drives*. New York: Springer.

Lord, H. W. 1971. *Pulse Transformers, IEEE Transactions on Magnetics*, Vol. Mag—71. New York, pp. 17–28.

Puchestein, A. F., T. C. Llyod and A. G. Conrad 1964. *Alternating Current Machines*. Bombay: Asia Publishing House.

Rashid, M. H. 1993. *Power Electronics*. Upper Saddle River: N.J.: Prentice Hall.

Richardson, D. V. and A. J. Caisse 1997. *Rotating Electric Machinery and Transformer Technology*. Reston: Reston Publishing Company.

Sarma, M. S. 1979. *Synchronous Machines*. New York: Gordon and Breach Science Publishers.

Say, M. G. 1983. *Alternating Current Machines*, 5th Edn. Sir Isaac Pitman and Sons Ltd.

Taylor, E. O. 1971. *The Performance and Design of AC Computator Motors*. Allahabad India: A. H. Wheeler & Co.P Ltd.

Veinott, C. 1959. *Theory and Design of Small Induction Motors*. New York: McGraw-Hill.

Index

A
Accelerated commutation, 282–84
AC machines, EMF generated in, 242
Air blast cooling
  for high-capacity transformers, 50
  and oil-immersed self-cooled transformer, 49
Air gap, 230
All-day efficiency of transformer, 89
Alternator
  equivalent circuit, 433
  excitation circle, 476–77
  infinite bus, 482
  input power equation, 458–59
    reactive power input, 459
    real power input, 459
  load characteristics, 454–55
  load distribution, 488–89
  losses and efficiency, 468–69
  output power equation
    maximum power output, 457
    power output equation, 457–58
    reactive power output, 457
    real power output, 456–57
  parallel operation
    conditions, 483–84
    identical frequency requirement, 484
    identical phase sequence, 484
    identical voltage requirement, 484
    in-phase requirement, 484–85
    prime mover torque speed, 484
    reasons, 483
  phasor diagram
    lagging power factor, 434
    leading power factor, 435
    unity power factor, 435
  powers sharing, 478–82
  prime mover characteristic, 477–78
  voltage equation, 434
  voltage regulation determination, 337, 447–53
    Ampere-turn method, 443–47
    ASA method, 453–54
    direct loading, 438
    emf/synchronous impedance method, 438–43
Alternator at lagging power factor, 450
Ampere-turns (AT), 9
Armature core, 188–89
Armature mmf wave, 470
Armature reaction
  in DC machines, 270–74
  unity power factor load, 431–32
  zero lagging and leading, 432
Armature resistance measurement, 440–41
Armature torque of motor, 339–40
Armature windings, 189, 219–21
  leakage reactance, 431
  resistance, 430–31
Automatic starters, 360
  back emf, 361–62
  series current-limit, 363–64
  shunt current-limit, 362–63
  time element, 361
Autotransformers
  advantages of, 100–101
  applications, 101
  construction, 97–99
  conversions, 100
  copper saving, 99–100
  disadvantages of, 100–101

B
Booster transformer, 111–12
Brake test of DC motor, 372–73
Breather, 51
Bright lamp method, 485, 486
Brushes
  and bearings, 190
  drop, 262
  shift on terminal voltage, 300
Brushless DC motor
  bipolar/full-wave, 396
  bipolar or full-wave, 396
  features, 396–97
  features of, 396–97
  unipolar/half-wave, 394–96
Buchholz relay, 52
Bushings, 47

C
Capability curves, 470–74
Circle diagram of polyphase induction motors
  approximate equivalent circuit, 618
  exciting current, 618
  maximum quantities, 620–21
  no-load test, 619
  short-circuit current, 620
  short-circuit test, 619
  stator current phasor, 617
  stator-resistance test, 619
Circuit connection and open-circuit characteristic, 439, 440
Co-energy of system, 26–27

Coil arrangement and waveform of mmf, 270, 271
Commutation, 277–79
  accelerated, 282–84
  improving methods
    emf commutation, 286–87
    resistance commutation, 285–86
  linear, 279–81
  retarded, 281–82
  sinusoidal, 284
Commutator, 189
Compensating winding, 277
Complex power, 12–13
Compound generator, 301–3
  parallel operation, 308–10
Compound motors, 343–44
  characteristics of, 346
Compound-wound generator, 261
Concentric windings, 45
Conductors
  current in, 1
  force, in magnetic field, 2–3
  magnetic flux, 1–2
Conservator, 51
Construction of DC machines, 186
  armature core, 188–89
  armature windings, 189
  brushes and bearings, 190
  commutator, 189–90
  essential parts of, 190
  magnetic frame/yoke, 187
  pole coils, 188
  pole cores and pole shoes, 187–88
Continuous disc windings, 46
Continuously variable reactor, 106
Cooling of transformer, 49–50
Copper loss, 266
Copper saving in autotransformer, 99–100
Core-type laminations, 41–43
Corkscrew rule, 1–2
Coupling angle, 530
Coupling field energy, 23–24
Crawling, 630
Cross-field generators, 391–94
Cross-magnetizing Ampere-turns per pole, 275–76
Cross-over windings, 46
Current-carrying conductors
  force between, 2
  isolated, flux distribution of, 2
Current transformer, 107
Cylindrical rotating machine, 32–33
  cases, 33–37
  magnetic actuator, 36
  pulsating torque, 34
  synchronous motor/machine, 33
Cylindrical rotor, capability curve, 473

D

Damper windings, 219, 549–50
Damping effect, 550
Dark lamp method, 485, 486
DC compound motor, characteristics, 346
DC generators, 207–8
  abnormal sound, causes, 312
  armature, heating causes, 312
  armature reaction, 270–74
  characteristics, 287
  classification, 261–62
  condition for maximum efficiency, 268–70
  efficiency, 267–68
  emf generated in, 246
  excessive heating, causes, 311
  external and internal characteristic
    with armature reaction, 299
    neglecting armature reaction, 298–99
  internal/total characteristic, 297–98
  overload, indications and causes, 310–11
  parallel operations, 303–4
  power stages, 267
  rapid brush, reasons, 312
  sparking at brushes, 311
  two-segment commutator, 209–13
  types, 261
  uses, 310
  voltage regulation, 297
DC machines construction, 186
DC motors
  armature torque, 339–40
  automatic starters, 360
    back emf, 361–62
    series current-limit, 363
    shunt current-limit, 362–63
    time element, 361
  back emf, 338
  brake test, 372–73
  compound motor, 343
  Field's test, 386–88
  Hopkinson's test, 377–80
  manual starter
    four-point starter, 359–60
    three-point starter, 358–59
  maximum mechanical power, 339
  principle, 213–14
  retardation/running test, 383–86
  rotational losses, 341–43
  separation of losses, 380–83
  series motor characteristics, 345–46
  shunt/separately excited, characteristics, 345
  speed control
    armature resistance, 347–50
    field resistance, 350–55
  speed regulation, 347
  speed with back emf and flux, 344–45
  starter, 357
  Swinburne's test, 373–75
    generating mode, 375–76
    motoring mode, 375
  torque and speed, 347
  uses, 388–89
  voltage equation, 337–38
  ward-leonard control/voltage control, 355–57
DC series motor, 345–46
DC shunt motor starter design, 364–66
DC windings types
  simple lap, 196–200

wave, 200–205
Deep-cage rotors, 631
Delta connections, 16
currents and power, 18–20
phasor diagram, 19
voltages, 18–20
Delta–delta connection, 158–59
Delta–star connection, 160–61
Delta–zig-zag star connection, 161–62
Diametrical connection, 171–72
Direct testing of polyphase induction motors
autotransformer, 616
drum radius, 616
efficiency, 616
input voltage, 616
operating slip, 616
output power, 616
performance characteristics, 617
Disc coils, 47
Distribution factor, 238–39
Double-cage rotors, 631–32
Double-delta connection, 171
Double-layer spiral coil, 46
Double-layer winding, 223
Double-star connection, 171
Doubly excited systems, 31–32
Drooping, 477
Drum windings, 192–94
armature conductors, 195
coil sides per layer, 194
coil span, 194–95
commutator pitch, 195
winding pitch, 195, 196
Dry-type transformer, 49
Dummy coils, 207
Dynamically induced emf, 5–6
Dynamometer, cross-field machines, 390–91

E
Eddy current loss, 266
Efficiency of transformer, 85
all-day, 89
maximum, conditions for, 86
kVA supplied, 87–88
load current, 87
Electrical and mechanical degrees, 234–36
Electrical energy input, 27–28
Electrical faults, 498
Electric braking, 366
of polyphase induction motors
dynamic braking, 644–45
plugging/counter-current braking, 643–44
regenerative braking, 643
of series motors, 368–69
of shunt motors, 367–68
Electromagnetic torque, 184–85
Electromagnetism, 1
mechanical forces in system, 28–30
Electromechanical energy conversion system, 24
Emf commutation, 286–87
Emf equation
long-shunt compound generator, 263
series generator, 263
short-shunt compound generator, 263–64
shunt generator, 262
Emf generated in AC machines, 242–43
Emf generated in DC generator, 246
Energy
conversion principle, 22–23
in coupling field, 23–24
Equalizer rings, 287–88
Equalizing connections for lap winding, 205–7
Equivalent circuit of polyphase induction motors
rotor, 596–99
stator, 596

F
Faraday's first and second laws, 4–5
Field energy, 24–26
Field flux and flux distribution, 270, 271
Field's test, 386–87
Fleming's left-hand rule, 3
Fleming's right-hand rule, 6
Flux density, 8
Flux per pole, 240–41
Four-point starters, 359–60
Fractional slot winding, 224–27
Full-pitched coil, generated emf in, 241–42

G
Generated emf in full-pitched coil, 241–42
Generator phasor diagram, 472
Generators protection, 497–99

H
Helical windings, 46
Hopkinson's test (back-to-back test), 377–79
Hunting
causes, 492
effects, 492–93
reduction, 493
Hydrogenerators, 218
Hysteresis loss, 266

I
Incoming machine and bus, voltage diagrams, 487
Induced EMF, 5
and current, 3–4
Induction generator
advantages and applications of, 642
circle diagram of, 641–42
disadvantages of, 642
excitation current and rotor current, 641
rotor emf, 641
Induction machines, 243–44
Induction motors
electrical braking of polyphase
dynamic braking, 644–45
plugging/counter-current braking, 643–44
factors governing performance
air-gap length, 634
rotor resistance, 634
slots and teeth shapes of, 634
operating conditions effects

Induction motors (*Contd.*)
break in one phase effects, 634
line frequency and line voltage, variation, 634
loading, 634
unbalanced supply voltage, 634
Infinite bus, 482
Inherent voltage regulation of transformer, 78
In-phase and out-of-phase condition, 484, 485
Inrush phenomenon, 112–13
Integral slot winding, 223–24
Isolated current carrying conductors, 2

K
Kapp's regulation, 84

L
Laminations amalgamation, 40
Lap winding, 196–200
equalizing connections, 205–7
uses, 207
Lenz's law, 5
Linear commutation, 279–81
Linear induction motor
applications of, 641
disadvantages of, 641
induced currents, 639
short-single primary, 639
slip, 640
synchronous speed, 640
thrust-speed characteristics, 640
Linear systems
magnetic force estimation, 30
Load angle, 530
Loaded and unsaturated DC machines, 273–74
Load sharing by transformers
equal voltage ratios, 92–94
unequal voltage ratios, 95–96
Long-shunt compound generator, 263
Losses in DC generator, 266
constant/standing, 267
copper, 266
core loss, 266
iron loss, 266
magnetic loss, 266
mechanical, 266
stray, 266
Losses in transformer, 84
copper loss, 85
core/iron loss, 85

M
Magnetic circuit, 7
Magnetic field intensity, 8
Magnetic flux, 8–9
Corkscrew rule, 1–2
right-hand rule, 1
Magnetic force, in linear systems, 30
Magnetic frame or yoke, 187
Magnetic locking (cogging), 630–31
Magnetomotive force, 7
Magnet withdrawn, 4
Manual starters
four-point, 359–60
three-point, 358–59
Maximum torque of polyphase induction motors
condition, 583–84
rotor resistance effect, efficiency, 589
Maximum voltage regulation, 79
Mechanical faults, 498–99
Mechanical forces, in electromagnetic system, 28
cases, 29–30
Mechanical losses, 266
Moving coil reactor, 105
Moving shunt reactor, 106

N
Negligible armature resistance, output power, 540
No-load curve for self-excited generators, 290–93
No-load release-type series motor starter, 364

O
Off-load tap changing transformers, 108
Oil-immersed forced air-cooled transformers, 50
Oil-immersed forced oil-cooled transformers, 50
Oil-immersed self-cooled transformers, 49–50
Oil-immersed water-cooled transformers, 50
On-load tap
changer with single primary, 109–10
changing transformers, 109
Open-delta connection, 163–65
Operating conditions effects of induction motor
break in one phase effects, 634
line frequency and line voltage, variation, 634
loading, 634
unbalanced supply voltage, 634
Output power equation of alternator
maximum power output, 457
reactive power output, 457
real power output, 456–57
Overloaded generators indications, 310
causes, 311–12

P
Parallel operation of alternator
conditions, 483–84
identical frequency requirement, 484
identical phase sequence, 484
identical voltage requirement, 484
in-phase requirement, 484, 485
prime mover torque speed, 484
reasons, 483
Parallel operation of single-phase transformer, 92
Performance and circle diagram of synchronous motor
constant excitation, 543–44
constant power developed, 541–43
constant power input, 541
Permeability, 9
Phasor diagram of alternator
lagging power factor, 434
leading power factor, 435
unity power factor, 435
Phasor diagram of mmf, 453

[illegible]otor
Phasor diagram, synchro[illegible]528
    equivalent circuit[illegible]d, 528–29
    lagging power [illegible]oad, 529
    leading pow[illegible] load, 529–31
    loading [illegible]57
    unity[illegible] current braking, 368
  Pigtail[illegible] single-phase transformer, 90
  Pitc[illegible]
  [illegible]88
  [illegible] and shoes, 187–88
  [illegible]ase induction machines
  [illegible]r gap, 230
  squirrel-cage rotor, 228–29
  wound rotor, 229–30
Polyphase induction motors
  applications, 633
  blocked-rotor/short-circuit test, 614–15
  circle diagram
    approximate equivalent circuit, 618
    exciting current, 618
    maximum quantities, 620–21
    no-load test, 619
    short-circuit current, 620
    short-circuit test, 619
    stator current phasor, 617
    stator-resistance test, 619
  direct testing
    autotransformer, 616
    drum radius, 616
    efficiency, 616
    input voltage, 616
    operating slip, 616
    output power, 616
    performance characteristics, 617
  efficiency, 592
  electrical braking of, 643–45
  equivalent circuit
    rotor, 596–99
    stator, 596
  factors governing performance of, 634
  full-load and maximum torque, 586–87
  losses, 591–92
  maximum torque
    condition, 583–84
    rotor resistance effect, efficiency, 589
  no-load test/open-circuit test, 613–14
  operating conditions effects, 634
  power factor
    rotor resistance and reactance, effect, 589–90
  power stages, 592
  ratings of, 634–35
  rotor current
    electromagnetic force (emf), 577
  rotor power
    air gap power, 579
    mechanical power, 580
    phase power input, 5[illegible]9
    stator input, 580
  rotor resistance and reactance, effect, 590
  slip measurement
    electromechanical counter, 594
    mechanical differential counter, 594–95
    stroboscopic method, 595–96
  starting load torque and maximum torque, 587
  starting of, 604–5
  starting torque
    change in supply voltage, effect, 582
    condition, 588
    slip-ring motor, 582
    squirrel-cage motor, 582
  synchronous watt, 594
  Thevenin's equivalent circuit of, 600–603
  torque and slip
    change in supply frequency on, 591
    change in supply voltage on, 583, 591
    characteristics, 584–86
  torque expression, 581–82
Potential transformer, 107
Potier method, 447–53
Potier triangle, 448–51
Power angle, 530
Power factor of polyphase induction motors
  rotor resistance and reactance, effect, 589–90
Power triangle, 12
Preventive autotransformer, 110–11
Pull-out torque, 544
Pulse transformer, 102
  response characteristics, 103–4
  usage of, 104

R

R–C series circuit, 10
Reactance voltage, 284–85
Regenerative braking, 369
Reluctance, 9
  torque, 185
Resistance commutation, 285–86
Retardation/running test, 383–85
Retarded commutation, 281–82
Rheostatic/dynamic braking, 367–68
Right-hand rule, 1
Ring windings, 190–92
Rotating electrical machines
  classification, 184
  constructional features, 185–86
  electromagnetic torque, 184–85
  reluctance torque, 185
Rotating magnetic field concept, 246–49
Rotational losses of DC machines, 341
Rotatory electromagnetic system, 31
Rotor
  currents, 31–32
  faults, 498
  frequency, 233–34
  of polyphase induction motors, emf, 577
  salient pole-type, 216–17
  smooth cylindrical, 217
  speed of MMF, 234

Rotor power of polyphase induction motors
  air gap power, 579
  mechanical power, 580
  phase power input, 579
  stator input, 580
Rotor side speed control of induction motor
  cascade connection, 624–26
  injecting voltage in rotor circuit, 627
  rotor rheostat control., 623–24
Running torque, 544

S
Salient-pole alternator
  losses and efficiency, 468–70
  maximum reactive power, 468
  torque-angle characteristic, 464–67
  two-reaction theory, 460–64
Salient-pole synchronous motor, 544
  lagging power factor, 545
  leading power factor, 546
  power developed, 547–48
  stability and maximum load angle, 548–49
  unity power factor, 546–47
Salient pole-type rotor, 216–17
Sandwich windings, 47
Saturable reactor, 106
Schrage motor, 636
  advantages of, 639
  block diagram of, 637
  phasor diagram, 638
  power factor compensation, 639
  two-pole DC generator, 638
Scott connection, 165–67
Self-excited generator, 261
  no-load curve, 290–91
Separately excited DC motor, 345
Separately excited generators, 261
  advantages and disadvantages, 294
  internal and external characteristic, 290
  no-load saturation characteristic, 289
Separation of losses in DC machine, 380–83
Series generator, 263
  circuit diagram, 301
  parallel operation, 307
Series motor, electric braking
  plugging/reverse current, 368
  regenerative, 369–72
  rheostatic, 368
Series R–L–C circuit, 11
Series R-L circuit, 9–10
Series-wound generator, 261
Shell-type transformers, 41–44
Short-circuit ratio, 495–97
Short-circuit test, 67–68
Short-shunt compound generator, 263–64
Shunt generator, 262
  conditions for, 295
  failure reasons, 295
  internal and external characteristic, 295–97
  parallel operation, 304–6
  voltage build-up, 294
Shunt motors
  characteristics, 345
  electric braking
    plugging/reverse, 367–68
    regenerative, 368
    rheostatic/dynamic, 367
Shunt-wound generator, 261
Simple DC generator and two-segment commutator, 209–12
Single-layer windings, 221–23
Single-phase circuits
  series L-C-R circuit, 11
  series R–C circuit, 10
  series R–L–C circuit, 11
  series R-L circuit, 9–10
Single-phase transformer, 39–41
  core-type transformer, 41–43
  parallel operation of, 92
  polarity test of, 90
  shell type, 43–44
  spiral core type, 44
Single-turn generator, 208
Singly excited machine, 229
Sinusoidal commutation, 284
Six-phase star connection, 171
Slip in induction motor, 232–33
Slip measurement of polyphase induction motors
  electromechanical counter, 594
  mechanical differential counter, 594–95
  stroboscopic method, 595–96
Slip-ring induction motors
  starting of
    approximate ratio, 612
    full-load current, 611
    resistance steps for three-phase, 610
    stud, 611
Slip ring rotor, 229–30
Slip test connection diagram, 470
Smooth cylindrical type rotor, 217
Speed control of DC motors
  armature resistance control, 347–49
  field resistance control, 350–51
Speed control of induction motor
  rotor side, 623–27
  stator side, 627–29
Speed of rotor mmf, 234
Speed regulation, 347
Speed–time curve, 384
Spiral core-type transformer, 44
Spiral windings, 45
Squirrel-cage motors
  autotransformer, 607–8
  direct-on-line, 605–6
  star-delta, 608
  stator resistor, 606–7
Squirrel-cage rotor, 228–29
Standing losses, 267
Star connections
  currents and power, 16–18
  domestic and industrial load, 20

phasor diagram, 17
voltages, currents and power, 16–18
Star–delta connection, 159–60
Star–star connection, 156–58
Starter principle, 357
Starters for DC series motors, 364
Starting/breakaway torque, 544
Starting torque of polyphase induction motors
change in supply voltage, effect, 582
condition, 588
slip-ring motor, 582
squirrel-cage motor, 582
Statically induced emf, 6–7
Stator, 215–16
faults, 498
Stator side speed control of induction motor
changing number of poles, 627–29
changing supply frequency, 627
Steam supply, effect of change, 489
Step-up and step-down transformer, 39, 56
Stray losses, 266
Sumpner's test, 91
Swinburne's test, 373–75
generating mode, 375–76
motoring mode, 375
Synchronizing power
coefficient units, 492
and synchronizing torque coefficient, 489–91
Synchronizing procedures, alternator
frequency synchronization, 488
in-phase determination, 486–88
phase sequence matching, 485–86
synchroscope synchronization, 488
voltage matching, 485
Synchronous condensers, 554–55
Synchronous generators, 244–46
armature reaction
unity power factor load, 431–32
zero lagging power factor load, 432
zero leading power factor load, 432
armature winding, parameters of
leakage reactance, 431
resistance, 430–31
capability curves, 471–76
determination of, 470–71
reactance and impedance, 433
short-circuit ratio, 495–97
sudden short circuit, 494–95
Synchronous-induction motor, 633
DC Excitation to rotor winding, 645, 646
slip-ring induction, 645
starting characteristics, 646
Synchronous machines
armature windings, 219–20, 243
double-layer winding, 223
fractional slot winding, 224–27
integral slot winding, 223–24
single-layer, 221–23
classification, based on prime mover
hydrogenerators, 218
turbogenerators, 218
construction of, 214–15
rotor, 216–18
stator, 215–16
damper windings, 219
excitation system, 218–19
field winding, 243
frequency and synchronous speed, 219
Synchronous machines oscillations, 493–94
Synchronous motors, 33
applications of, 553–54
armature reaction, 527
classification, 218
complex power input, 535–37
complex power output, 538
conductor current, 526
constant excitation, 543–44
constant load operation, 532–34
constant power developed, current locus, 541–43
current locus, constant power input, 541
curves, 535–36
hunting/surging, 550–51
input reactive power at negligible armature resistance, 540–41
leading current, demagnetizing effect, 527
maximum output power, 538–40
motor characteristics, 541–44
operation principles, 526
performance and circle diagram, 541–44
periodicity of hunting, 551–52
power output at negligible armature resistance, 540
rotating magnetic field and rotor field poles, 526–27
starting methods, 553
torque, 544
Synchronous speed, in induction motor, 232

T

Tap changing transformers, 108
Tapped reactor, 105
Tapping, 48
Terminal voltage, brush shift effect, 300
Testing of DC machines, 372
Three-phase circuits, 13–15
advantages of, 15
generation of voltage, 14
interconnection of, 15
phase sequence, 15
phasor diagram, 14
power, measurement of, 21–22
waveform diagram, 14
Three-phase induction motors
common faults, 635
principle of operation, 230–32
Three-phase transformers
advantages of, 154
connection
delta–delta, 158–59
delta–star, 160–61
delta–zig-zag star, 161–63
star–delta, 159–60
star–star, 156–58
connections, 173–76

Three-phase transformers (*Contd.*)
construction of, 155
core-type, 156
shell-type, 156
open-delta/V–V connection, 163–65
operation principle, 154–55
parallel operations of, 170
rating, 176
Scott connection/T–T connection, 165–67
six-phase conversion
diametrical connection, 171–72
double-delta connection, 171–72
double-star connection, 171
star connection, 171–72
two-phase conversion, 167–70
Three-point starters, 358–59
Three-winding transformer, 173
Three-wire generator, 389
Torque and slip of polyphase induction motors
change in supply frequency on, 591
change in supply voltage on, 583, 591
Torque and speed of DC series motor, 347
Torque angle/angle of retardation, 530
Transformers
air blast and oil-immersed self-cooled transformer, 49, 50
Buchholz relay, 52
bushings, 48
conservator and breather, 51–52
cooling of, 49–50
definition, 38
efficiency of, 85–86
all-day efficiency, 89–90
maximum efficiency condition, 86–89
emf equation, 54–56
equivalent circuit, 64–66
equivalent reactance, 61–62
equivalent resistance, 60–61
exact voltage drop, 73–74
leakage reactance and impedance voltage drop, 77
on load, 59
load sharing, 92–96
losses in, 84
copper loss, 85
core/iron loss, 85
magnetic leakage, 61
methods of cooling, 50
no load, 57–59
oil, 51
open circuit test/no-load test, 66–67
per unit resistance, 77
principle, 38–39
rating, 176
resistance and leakage reactance, 63–64
separation of core/iron losses, 70–72
short circuit/impedance test, 67–69
step-up and step-down transformer, 56–57
Sumpner's test, 91
tank, 52
tapping, 48
terminals and leads, 48
theory of, 53
total approximate voltage drop, 72–73
types of, 39
unequal voltage ratios, 95–96
voltage regulation
calculation for, 78–84
inherent voltage regulation, 78
Kapp's regulation, 84
voltage regulation down, 78
voltage regulation up, 78
Turbo alternators, 217
Turbogenerators, 218
Two bright and one dark lamp method, 486, 487
Two current-carrying conductors, 2
Two-reaction theory, 460
Two-wattmeter connection, 21
Two-winding transformer, 41

V

Variable excitation, 532–34
V curves experimental set-up, 536
Voltage regulation determination
Ampere-turn method, 443–47
ASA method, 453–54
direct loading, 438
direct load testing circuit diagram, 438
synchronous impedance method, 438–42
zero power factor method/Potier method, 447–53
Voltage regulation down/up, 78

W

Ward-Leonard control, 355
Wave winding, 200–203
uses, 207
Welding transformers, 104
reactors, 105–6
Windings, 44–45
concentric, 45
continuous disc, 46–47
cross-over, 46
factor, 240
helical, 46
sandwich, 46–47
spiral, 45–46
Wound-rotor, 229–30
and cage-rotor induction motors, 629
Wound-rotor induction motor, 229

Z

Zero power factor method, 447–53
Zero voltage regulation, 79